OPERATIONS RESEARCH

OPERATIONS RESEARCH

R.VEERACHAMY

Formerly Professor & Chairman

Department of Economics

Bangalore University

Professor & Head

T. John Business School, Bangalore.

V. RAVI KUMAR

Asst. Professor

Department of Mechanical Engineering

Amrita School of Engineering

Bangalore.

NEW DELHI

Published and Printed by
Techar Pvt. Ltd.
4435-36/7, Ansari Road, Daryaganj
New Delhi - 110 002 (India)
E-mail: tech_info@techsarworld.com
Website: www.techsarworld.com

ISBN: 978-93-80026-40-4

Reprint 2025

Dedicated

To

My Mother

Preface

Continuous decision making is an essential part of any business organization. Modern business management invariably applies operations research techniques to solve real-life problems of their organizations. Though the management can readily use the ready-made quantitative tools, for a wise application one must have a thorough understanding of the involved principles and the nature of the problem described in the ready-made tool. Having these pre-requisites in mind, nearly all the universities and professional institutes throughout the world designed their course curriculum giving prime importance to operations research. In this book, a modest attempt has been made to discuss all the commonly used operations research techniques in a wide spectrum of decision making environment.

The book is completely self-contained, with all the essential quantitative techniques used in operations research. Each chapter is designed to be modular, so that the book can be tailored to the needs of a course. The authors have taken due care to avoid excessive theoretical description in favour of real-life applications. The exposition takes great care by providing several real-life worked out examples in all the chapters, to guide the students step-by-step in solving even the most complicated problems. All the diagrams in this text are scaled so that the student can read the relevant answers from the graph also. Wherever possible excel templates are being provided to enable the student to get the answers for the problems given in the exercises by merely substituting the input data into the template. To provide exposure in the concerned topic lot of problems from various Indian universities question papers are provided for all the chapters. The book specially designed for the use as a textbook for both graduate and post graduate course in management and engineering.

Although enough care has been taken to keep the book error free, we personally take sole responsibility for the leftover errors, if any. The authors gratefully indebted to the student community at large, without whose encouragement and constructive suggestions the book would not have taken this shape. Critical evaluation and valuable suggestions that will help the author to enhance the utility of book in latter edition is most welcome and would be greatly appreciated. We also owe my gratitude to the dedicated team of I.K. International Publisher (P) Limited for their sincere effort in bringing this text.

R. VEERACHAMY
&
V. RAVI KUMAR

Contents

CHAPTER

1

Nature and Scope of Operations Research

- *Introduction*
- *History of operations research (OR)*
- *The meaning and the definition of OR*
- *Characteristic features of OR*
- *Methodology of operations research*
- *Problem formulation*
- *Modelling on operations research*
- *Collection of input data*
- *Solution finding*
- *Validation of the model*
- *Implementation*
- *Points to remember*

1.1 INTRODUCTION

Continuous decision making is an essential part of the managerial process. The manager has an authority to plan, organize and control the activities of the business. For example, as a student he will have to take the correct decision about the course he should take for higher studies. As a businessman, he should take the right decision regarding the type of business venture that he wants to take. As a farmer, he should take the right decision regarding the crop that he is planning to grow. Decision making in big business establishments is extremely difficult because it affects many people directly and indirectly. Traditionally, decision making has been considered as an art, a talent that is being acquired on the basis of their past experience in the concerned field. However, the present-day decision making in a more complex environment needs not only the experience but also a systematic and scientific approach. The mere trial-and-error based commonsense approach may prove to be fatal on several occasions. The complexity of interacting forces in business necessitates the use of scientific methods in decision making.

In this context it is desirable to have a thorough knowledge about the basic tools which are necessary to solve business related problems. Operations research is a specially designed kit to provide the essential scientific tools for decision making in a systematic way. In all business related problems quality decision making is an essential feature. Such quality decisions heavily depend upon the quality of the input information and the quality of tools that are being used.

1.2 HISTORY OF OPERATIONS RESEARCH (OR)

In order to understand the meaning of Operations Research one must know something about its history and evaluation over time. Though it is very difficult to predict the origin of the operations research, the scientific approach to management decision making could be traced during the 18th century industrial revolution. The operations research, as it exists today, gave its birth during the Second World War. British Military requested a group of scientists to examine the optimum strategies and tactics for verity of its military related operations during the said war period. These scientists developed many new quantitative techniques to assist military operations. Partly because of the scientists involved were talented men, partly because of the pressures of war time necessity, and partly because of the synergism generated from the interaction of different disciplines, these teams of scientists were remarkably successful in improving the effectiveness of complex military operations.

The name *operational research* was derived from the context in which it was used, first research activity on operation areas of the armed forces. This great beginning initiated by British armed forces was effectively utilized by American military management to design their military operations under the name *operations research.* The following are some of the areas of military related investigations made by using operations research techniques.

- To determine the optimum convoy size to minimize losses from submarine attacks of enemies.
- To determine the correct locations to display radar units in order to maximize the potential coverage against possible enemy attacks.
- To determine the correct colour of the aircraft in order to minimize the chance of being deducted by the enemy submarines.

After the end the World War II, many of these scientists who had been active in the military OR groups diverted their attention towards the use of OR techniques for solving civilian problems. As a consequence some of the techniques that had been applied to the complex war related problems were successfully transferred and assimilated for the use of the industry. Some of the scientists returned to the universities and concentrated their efforts on providing a sound foundation for many of the techniques that had been hastily developed during the war period. Some others moved to the private sector economy and adopted the techniques to the needs of the industry. Petroleum companies were the first user of OR techniques in the civilian area. Later as researchers started applying these techniques to smaller common problems like inventory management, allocation, replacement, scheduling, etc. Today service organizations like banking, insurance companies, hospitals, libraries, etc., are using these techniques.

In the early 1950s, civilian OR activities had reached a level of development and emerged as a unique discipline by itself. The Operations Research Society of America (ORSA) was founded in 1952 to serve the professional needs of scientists working in OR area. In the year 1953, the Institute of Management Sciences (TIMS) was established to strengthen this development.

The invention of digital computers accelerated the growth of operations research techniques to a greater extent. Such a drastic development in computation techniques gave birth to refinement of some of the old operations research techniques. However, clear-cut well structured OR techniques were operational since 1950s. Typically, they were repeatedly used for optimum solutions in the area of resource allocation, inventory control and construction related problems. In 1960s more and more colleges in UK started introducing the OR as a separate discipline in their curriculum. Depending upon unique development patterns at each institute OR programs sometimes appeared within the departments of industrial engineering, sometimes in business school, and occasionally in mathematics or economics. Since 1960, well-structured planning problems also attracted the use of OR.

1.3 THE MEANING AND THE DEFINITION OF OR

The term operations research has a number of quite distinct meanings. To some OR, is the composition of certain body of problems, techniques, and solutions that has been accumulated under the name OR over the past 30 years. To others it is an activity or process for solving certain problems. The Operational Research Society of Great Britain has adopted the following definition for OR.

"Operational research is the application of the methods of science to complex problems arising in the direction and management of large systems of men, machines, materials and money in industry, business, government, and defence. The distinctive approach is to develop a scientific model of the system, incorporating measurements of factors such as chance and risk, with which to predict and compare the outcomes of alternative decisions, strategies or controls. The purpose is to help management to determine its policy and actions scientifically".

The Operations Research Society of America has offered a shorter, but similar definition as highlighted below.

"Operations research is concerned with scientifically deciding how to best design and operate man-machine systems, usually under conditions requesting the allocation of scarce resources."

1.4 CHARACTERISTIC FEATURES OF OR

Decision Making

Primarily OR deals with the optimum decision making problems. Irrespective of the situation, OR gives a systematic procedure for solving the problem. The step-by-step procedure for the effective use of OR techniques are highlighted below.

- Define the problem and establish a suitable criteria like cost minimization, profit maximization, revenue maximization etc.
- State the alternative courses actions available in archiving the set target goals.
- Evaluate the validity of the alternative courses of action and select the optimum value of the variable outlined in step one.

Scientific Approach

In OR there is no place for whims and fancies to guess the answer. They are all obtained on the basis of well defined universally accepted scientific methods. It is a formalized reasoning process consisting of the following steps.

- The proposed problem must be clearly defined without any room for misrepresentation.
- Observations are made on varying conditions to analyze the behaviour of the system.
- On the basis of observations, testable hypotheses are framed.
- To test the said hypotheses an experiment is designed and executed.
- Finally accept or reject the hypothesis on basis the experimental results.

Objective

OR aims to get the optimum result to the problem under reference. To achieve such a task a measure of effectiveness is defined clearly. On the basis of this effectiveness measure the optimum solution is obtained.

Inter-disciplinary Team Approach

Managerial problems are inter-disciplinarily in nature. They have economic, psychological and engineering characteristics with them. Such a mixture needs a blend of specialized people from the areas of mathematics, statistics, management, computer science, economics and so on.

1.5 METHODOLOGY OF OPERATIONS RESEARCH

The dominant feature of OR is its ability in expressing the problem in a well stretched mathematical format. This distinctive feature enables us to use other scientific methods developed in other fields in solving the problem. It consists of the following steps.

- Formulate the statement of the problem under reference.
- Specify the assumptions and formulate the mathematical model of the problem.
- Acquire the needed input data.
- Solve the problem for optimality.
- Validate the model and implement the solution.

1.6 PROBLEM FORMULATION

The first step in any quantitative approach is to formulate a clear-cut problem statement. In many cases formulation itself proves to be the biggest challenge. It is often said "correctly identified problem is half solved." For example, the observations on excessive cost of production do not necessarily constitute a problem. It is only the indication of something else going wrong somewhere. It may be the improper inventory management or excessive wastage and the things like that. Sometimes, such symptoms may originate from other sources on which we do not have control. Some other occasions the problem under reference may be due to some other problem emerging from some other section. Under such circumstances the correct specification of the problem is essential for finding the correct corrective solution.

1.7 MODELLING ON OPERATIONS RESEARCH

Once the problem is defined, necessarily we go for model formulation. A model by definition is a theoretical abstraction of real life situation. Real life problems are more complex because of the involvement of innumerable inherent factors. The decision maker will have to abstract the situation by using crucial input and output variables. Having selected the crucial input and output variables, he combines them all by using the theoretical relationships among them. Thus a model is simple manifestation of the complex real life situation. Such a simple model does carry with it the implication that a model is always, a representation that is less than perfect. If that is the case then why one should go for modelling. There are many conceivable reasons for use of models rather than the real situation itself. Often the motivation behind modelling is economy. To save time, to save money or some other valuable resource we often prefer modelling rather than the real situation. Sometimes modelling is done to avoid risk associated with tampering of a real object. Sometimes the real environment is so complicated that a representative model is needed just to understand it or to communicate with others. Such models are quite prevalent in life sciences, physical chemistry and physics.

The very first step is the construction of the model itself. This step requires a set of coordinated decisions. What aspects of the real system should be incorporated and what aspects should be ignored is the real problem in formulation. The selection of the essential attributes of the real system and the omission of the irrelevant ones require a kind of selective perception, having the purpose in mind. Depending up on the purpose in mind we construct different verities of model as shown below.

a) Iconic Models

These models are essentially scaled up or scaled down representation of the original object under reference. Mini toy automobile is a scaled down iconic representation of the big original automobile. Similarly, the atomic structure of an atom is the scaled up iconic representation of the very small atom. Remember, in iconic models the models need not have all the characters of the original situation. Only we project the needed characters in the model. For example, if we want to study the parking space problem of the big newly planned bus stand then we use mini iconic models resembling the verities of automobile that could be parked for our study. Though the mini ones are models of the real ones they need not have interiors like engines, seating arrangements etc.

The major merit of economic model is its mini representation of the original that too with only the needed characteristics of the original. It resembles virtually the thing it represents. Hence translating any findings by using iconic model to the original real life situation is easier. However, such models do not permit manipulations like alteration for experimentation.

b) Analog Models

In analog models we map the given characteristics of the original with more comfortable characteristics of another. In such a mapping all the original geographical features are mapped into easily understandable characteristics of a different kind. For example, to represent sea in a geographical mapping we use only blue colour as a rule. Similarly, to represent green forest we always use green colour for such areas. Thus in geographical mapping all the geographical features is mapped correctly by another set of well defined rules. However, in general though analog models are less specific they are easier for manipulation compared to iconic models.

c) Symbolic Models

In symbolic models many real life situations are depicted by mathematical symbols. For example, in economics to describe the market we symbolize the price by p and quantity demanded q and express the demand law by a downward sloping straight line of the type $q = 100 - 2p$. In such modelling we manipulate the price and predict the corresponding demand. However, these models are more abstract than the other two types.

d) Static and Dynamic Models

A static model is one in which the decision variables do not involve sequence of decisions over multiple periods of time. All the variables are related to a particular point of time. For example, the current demand for a certain product dependents up on the current price in the market $\{D_t = f(P_t)\}$. In a static model we optimize the decision variables at one shot in the sense that the obtained solution values are applicable for all points of time.

On the other hand in dynamic models decision variables do involve sequence of decisions over multiple periods. For example, the current supply of sugarcane by farmers as a function of the last year's seasonal price of sugarcane is a dynamics representation of the situation. $\{S_t = f(P_{t-1})\}$

e) Linear and Non-Linear Models

In a model the specification equations are linear $\{D_t = 200 - 2P_t\}$ then such models are called linear models. Whenever the model involves non-linear equations then they are called non-linear models. We use calculus as an optimization technique in non-linear models.

f) Integer and Non-Integer Models

If all the decision variables are free to assume non-integer optimum values then such models are called non-integer models. Clearly variables like volume, temperature, height, weight etc., are free to take both integer and fractional values as well. Such models are discussed under linear programming. If the decision variables are number of houses, number of workers etc., then clearly fractional value has no meaning. In such models we restrict our solutions only to integer values. Such programming problems are discussed under integer programming.

g) Deterministic and Stochastic Models

If things were to happen with certainty then such models are called deterministic. In such models the optimum value of the objective function is certain to occur. On the other hand, models involve uncertainty (otherwise called randomness) they are called stochastic models. In waiting line theory the arrival and service rates are stochastic and hence described by Poisson and exponential distribution respectively. Thus waiting line characteristics are stochastic in nature.

1.8 COLLECTION OF INPUT DATA

Once the model is formulated the researcher plans for collection of the necessary input data. Since the data quality decides the accuracy of the model, the data obtained must be as good as possible. The importance of data quality is often referred as GIGO to mean Gold In Gold Out, Garbage In Garbage Out.

1.9 SOLUTION FINDING

Having formulated the model and collected the data, the next stage of research is to obtain the solution. If the given problem is small in size then the researcher can go for complete enumeration for finding the optimum solution. For larger problems one will have to use systematic search method for optimum solution. At every stage of iteration one will have to test for optimality till the final optimum solution is obtained. On several occasions the optimum solution may be many. In such a case the researcher have the choice to choose the optimum depending up on the situation. To handle such problems for optimum results we have enough algorithms. In the absence of algorithm step-by-step logical rule could be of great help. Once the problem is solved it must be tested for sensitivity by using appropriate sensitivity analysis.

1.10 VALIDATION OF THE MODEL

After obtaining the optimum solution the models validity must be tested for its adequacy in explaining the reality. Structural assumptions of the model are to be examined in assessing the validity. Unless the model's structural stability is established we cannot use it for future.

1.11 IMPLEMENTATION

Implementation of the solution is the last but difficult step in the operations research. It may be beneficial for one and non-beneficial to others. Compromising solutions to all sections of the organization definitely need not be the optimum all the time. On several occasions the high level of mathematical abstraction will ignore the human relation altogether. Therefore, a model that secures moderate theoretical abstraction could be of great help in overcoming such a difficult implementation problem. The following flow chart illustrates the sequential approach to solve the given problem.

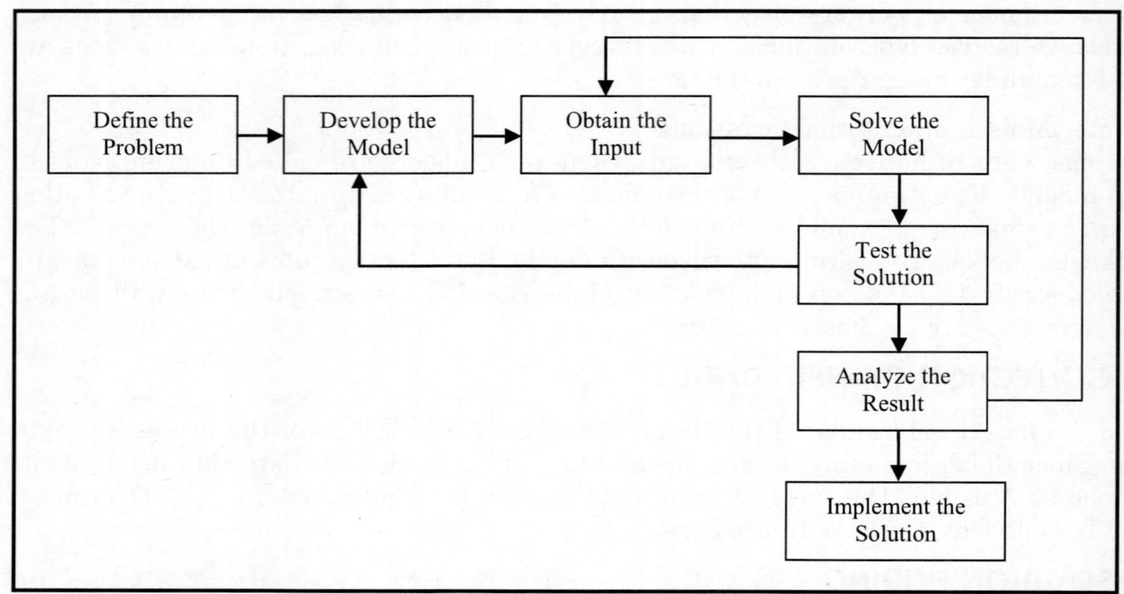

Figure 1.1

1.12 POINTS TO REMEMBER

After developing a model, both the developer and the final user of OR should be consciously aware of the inherent limitations. Hence should exercise constant vigilance against the common traps listed below.

- **Do not go for a complicated model when a simple one will serve the purpose**

  According to this principle, given two or more models each adequately performs the indented function then the simpler model is more useful and therefore be preferred. For example, if the intension is to study the parking problem of different varieties of vehicles in a given space then the needed iconic models are simple miniature outline of all type of vehicles without any inner components like engines.

- **Beware of modelling the problem to suit the technique**

  OR professionals are often criticized for destroying reality to suit the tools that they are indented to use or familiar with. For example, experts in linear programming may tend to view every problem that they encounter as a simple linear programming problem. In reality, not all operations research problems are involving only linear function. Similarly, all operations research problems are not searching for optimum solution. So, one must be careful while using the tool to the given situation.

- **Models must be validated before implementation**

  It would be futile to establish appropriateness of the given problem before actually using it. There are number of commonly employed techniques for doing this, depending upon the nature of the model.

  One method for validating the predictive model is "retrospective testing". Accordingly, the model is pre-tested with the help of the past data. For example, if the model is constructed to forecast the monthly sales of a certain commodity, it could be tested using the past data for its ability in predicting the future.

  If the model cannot be validated before implementation, then perhaps it may be implemented in phases. For example, a new model for inventory control may be implemented for a certain selected group of items while the older system is retained for the majority of items. As the model validates, more items can be included in the list.

  Finally it is worth to remember that real things change over time and hence the model designed for a particular time period may not be useful in future unless the model is structurally stable.

- **A model should never be taken too literally**

  Since modelling is the simplest manifestation of the reality, the deduced conclusions should not be taken literally to reflect the real situation by 100%. The only preventive measure is to be meticulous in expressing the original assumptions at the time of formulation, and again when the conclusions are reached. A model should neither be appreciated to much for its ability, nor criticized for failing to do.

CHAPTER

2

Linear Programming: I Problem Formulation & Graphical Solution

2.1 INTRODUCTION

A large number of decision related problems in business involve allocation of scarce resources to various activities, with the objective of either maximizing the revenue or minimizing the cost. Till recently, calculus has been the traditional tool in solving non-linear optimization problems with and without constraints. However, when the given objective function is linear, the calculus method of optimization fails due to non-fulfilment of both first and second order conditions. In fact, for a linear objective function the first order derivative invariably becomes a constant and hence cannot be equated to zero. Moreover, the second order condition becomes zero and hence cannot take either the positive or negative values as necessitated in calculus technique.

To handle such situations luckily today we have a number of specially designed tools. Linear programming addressed in this section is a special tool wherein we consider the maximization as well as minimization of linear objective functions subject to a number of linear inequalities otherwise called constraints. Linear programming was first introduced and popularized by a Russian mathematician L.V. Kantorovich. Later George B. Dantzig has developed this technique into a foolproof method in the year 1947 for scheduling the complicated procurement activities of the United States Air Force. Since then the technique of linear programming, as a device to solve business related problems have made tremendous progress. In this section we propose to introduce some of the linear programming techniques for solving business related problems.

2.2 LINEAR PRODUCTION FUNCTIONS

Linear production activity is a process in which one or more outputs are produced by using one or more inputs in fixed proportions. Thus, the relevant linear production function exhibits constant returns to scale. In such cases, a rise in all the inputs by a constant proportion necessarily leads to a rise in all the outputs by the same constant proportion as shown in Table 2.1.

Table 2.1

Linear production function			
Inputs		Output	
Labour (L)	Capital (K)	X	Y
10	40	10	5
20	80	20	10

In Table 2.1 a combination of 10 units of labour (L) and 40 units of capital (K) yields 10 units of output X and 5 units of output Y. Similarly, in the second row, the increase in both L and K by two times exhibits twice the output levels.

Linear production functions with two inputs and one output

Consider a linear production activity 1 in which a single product X is produced by using two inputs namely capital (K) and labour (L) in a fixed proportion as shown in Table 2.2.

Table 2.2

Linear production activity 1		
Inputs		Output
L	K	X
1	4	1
2	8	2
3	12	3
4	16	4

In the production activity described in Table 2.2, 1 unit of labour is combined with 4 units of capital to obtain 1 unit output X (1st row).

Similarly, for 2 units of output X, we use 2 units of L and 8 units of K (2nd row).

Therefore, in general for a given X units of output, the input requirements are given by L = 1 × X, K = 4 × X.

The following Figure 2.1 shows the relevant diagrammatic representation for the data given in Table 2.2 by using the familiar iso-quants.

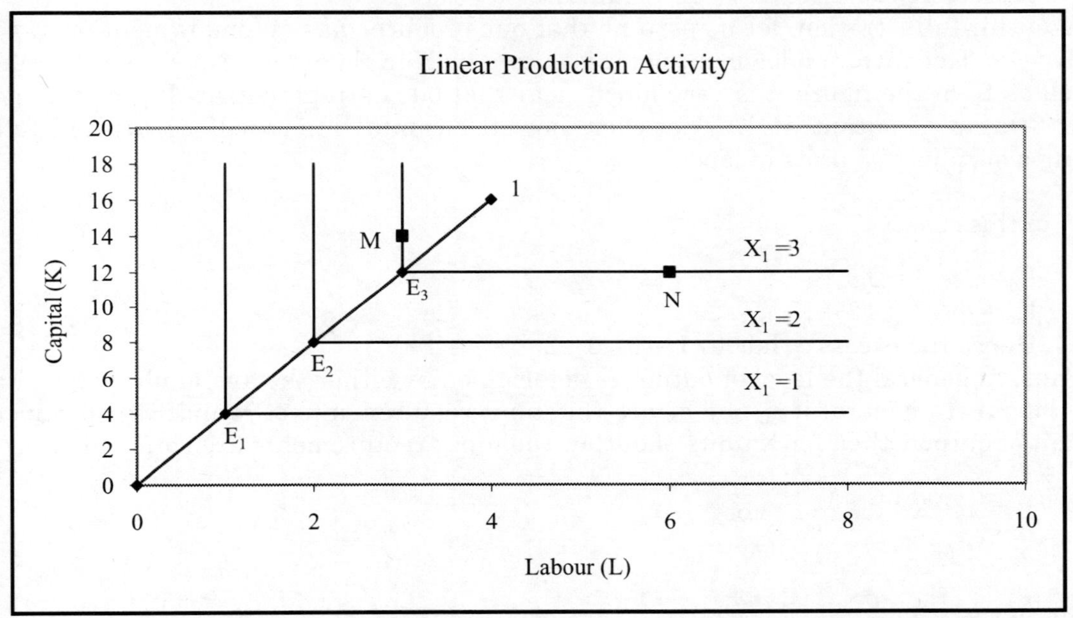

Figure 2.1

In Figure 2.1 we show three iso-quants for output levels 1, 2, and 3 units as described in Table 2.2. All the iso-quants take 'L' shape as shown in the figure. The expansion path of the output X is described by the straight line ray O1 originating from the origin. Efficient production takes place only at the corner points of these respective 'L'-shaped iso-quants lying on the expansion path as shown in the figure.

Given the technology as described in Table 2.2, suppose that our entrepreneur has 3 units of L and 12 units of K at his disposal for the production of the product X. From the production activity table given, it is clear that our entrepreneur can produce only 3 units of output X by using labour and the capital fully. This production point is shown in the figure by E_3 on the expansion path. At this point the demand for labour is equal to the supply of labour ($L_d = 1 \times 3 = 3$). Similarly, the demand for capital is equal to the supply of capital ($K_d = 4 \times 3 = 12$).

As a second illustration, now let us assume that our producer has 3 units of L and 14 units of K at his disposal. This factor endowment is shown in Figure 2.2 by the point M. Here again from the activity data table it follows that our producer will produce only 3 units of output. However, this time there will be an excess supply of 2 units of unutilized capital. Here the labour is used fully and production takes place at E_3 once again.

Remember that the output is solely determined by the scarce input, namely labour in this case. Here the input requirements are:

$L_d = 1 \times 3 = 3 = L_s$
$K_d = 4 \times 3 = 12 < K_s$
Hence the excess K = 14 – 12 = 2 units

As a third illustration, let us assume that our producer has an endowment of L = 6 and K = 12. This factor endowment point is shown by N in the figure. Now from the table as well as from the figure it is very much clear that our entrepreneur will produce once again 3 units of output at E_3. However, now the capital input will be utilized fully leaving a surplus of 3 units of labour.

Thus, in this case

$L_d = 1 \times 3 = 3 < L_s$
$K_d = 4 \times 3 = 12 = K_s$
Hence the excess of labour L = 6 – 3 = 3

Thus, in general the level of output is determined by either the availability of capital or by labour. In general if a_L and a_K are the input requirements of L and K respectively per unit of output, then for X units of output the input requirements are obtained as

$L = a_L \times X$ or

$X = L/a_L$

Similarly,

$K = a_K \times X$ or

$X = K/a_K$

However, since the output X is limited either by capital or by labour, in general, the production function is written as

$X = \min [(L/a_L), (K/a_K)]$

In particular, for the three illustrative cases, the respective production levels are obtained as

(i) X = Minimum of $\{(L/a_L), (K/a_K)\}$ = Minimum of $\{(3/1), (12/4)\} = 3$
(ii) X = Minimum of $\{(L/a_L), (K/a_K)\}$ = Minimum of $\{(3/1), (14/4)\} = 3$
(iii) X = Minimum of $\{(L/a_L), (K/a_K)\}$ = Minimum of $\{(6/1), (12/4)\} = 3$

Rule of thumb to identify the surplus inputs

1. If the endowment point is on the right side (N) of the expansion path as shown in Figure 2.2 then labour will be in surplus. Hence the scarce input capital determines the output level and it will be utilized fully.

2. If the endowment point is on the left side (M) of the expansion path then capital will be in surplus. Hence, the scarce input labour determines the output level and it will be utilized fully.

3. If the endowment point lies exactly on the expansion path (C) then both the inputs will be fully utilized.

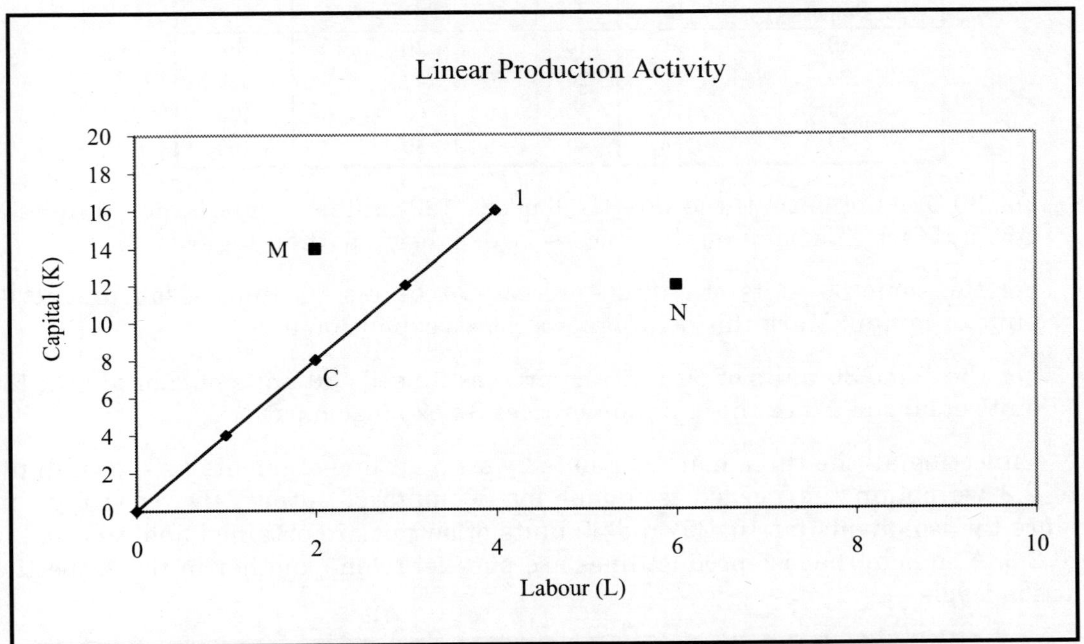

Figure 2.2

2.3 DERIVATION OF ISO-QUANT COUNTERS

If the number of production activities available is limited with fixed input ratios, then the corresponding expansion paths are straight lines as shown in the figure. The slope of the expansion paths is determined by the respective K/L ratios. Table 2.3 and the associated Figure 2.3 show a three-process production varying from the capital intensive one on the top to the labour intensive one at the bottom.

Table 2.3

Input requirements per unit of output			
Process	K	L	X
1	4	1	1
2	2	2	1
3	1	4	1

On the basis of data given in the above Table 2.3, the input requirements for 10, 20, 30, and 40 units of output are obtained and reported for all the three processes in Table 2.4.

Table 2.4

	Input requirements					
	Process 1		Process 2		Process 3	
Output	K	L	K	L	K	L
10	40	10	20	20	10	40
20	80	20	40	40	20	80
30	120	30	60	60	30	120
40	160	40	80	80	40	160

- For 30 units of output, the process 1 needs 120 units of capital and 30 units of labour. Mark this point on the process 1's expansion path in Figure 2.3.
- For the same 30 units of output, process 2 requires 60 units of labour and 60 units of capital. Mark this point on process 2's expansion path.
- For the same 30 units of output, the process 3 needs 30 units of capital and 120 units of labour. Mark this point on process 3's expansion path.

By connecting all the three marked points by straight line segments as shown in the Figure 2.3 we obtain the needed iso-quant for 30 units of output. By repeating this procedure the iso-quants for 10, 20, and 40 units of output are obtained and reported in Figure 2.3. Also note that iso-product lines are parallel to one another in the respective production levels.

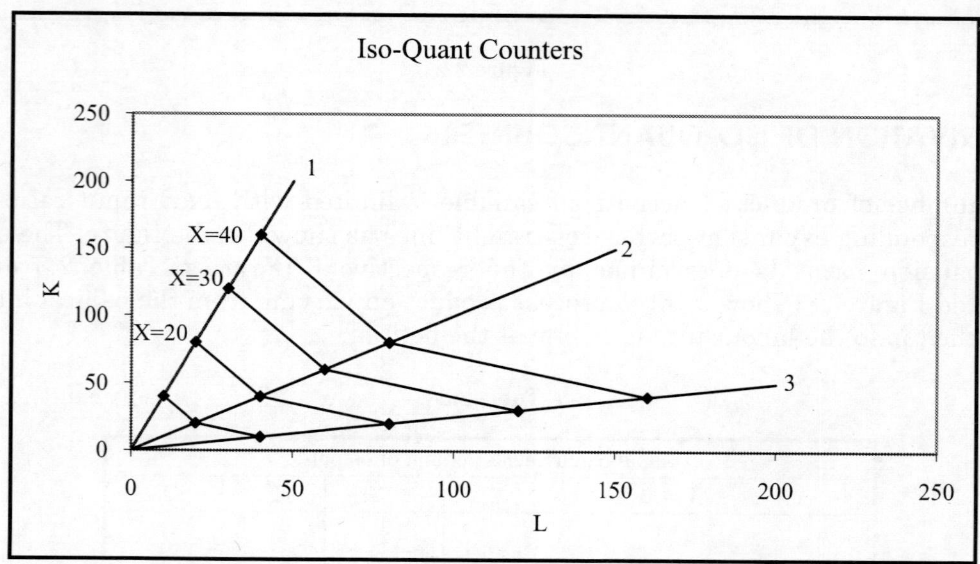

Figure 2.3

Note that instead of 3 if there are 300 processes for the same product, then our iso-quants will have perfectly convex shape as we normally assume in economics.

2.4 PROCESS MIXING AND FULL EMPLOYMENT

With a single production process the fuller utilization of both capital and labour can happen only by accident. Such a coincidence occurs provided the factor endowment point lies exactly on the expansion path O1. If the endowment point were to lies left side of the expansion path O1 then the output is solely determined by the availability L (K will be in excess). Similarly, whenever the endowment point lies on the right side of the expansion path O1 then the output is limited by the availability of K (L will be in excess).

Such excess capacities can always be removed by introducing at least one more production process into our system as shown below in Figure 2.4. In Figure 2.4 the point E represents the factor endowment available. Further, we notice that this endowment point E lies in-between the process 1 and 2. Now for the fuller and efficient utilization of both K and L the entrepreneur must produce the output by using both the production processes in a certain combination.

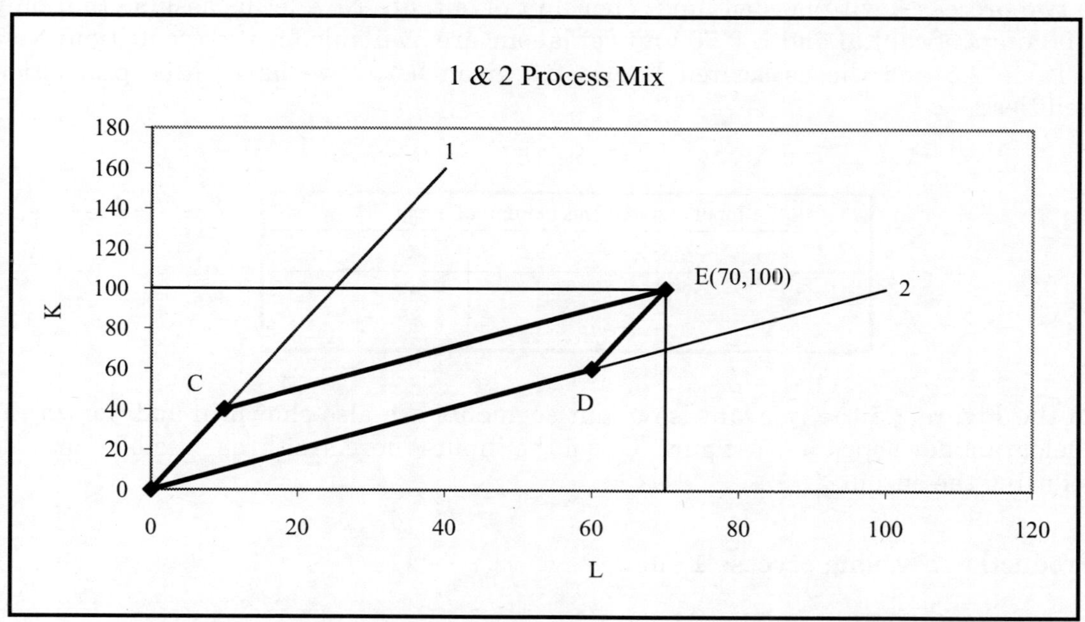

Figure 2.4

The knowledge of vector addition helps us in getting such a process mix result diagrammatically. Let E be the endowment point with 100 units of capital and 70 units of labour. To construct a parallelogram, from E draw a line parallel to O2 towards O1. Let this line meet the O1 line at C. Similarly, from E again draw another line parallel to

O1 but towards O2. Let this line meet the O2 line at D. Now OCED is the needed parallelogram. If OC is considered one vector and OD is considered another vector then the resultant will be the diagonal vector OE. Thus, vector OC represents the amount output X produced in process 1 and the vector OD represents amount of output X produced in process 2 by using all the 70 units of labour and 100 units of capital. Moreover, only this process mix will give the maximum output.

2.5 LIMITATIONS OF INPUTS

The following numerical example illustrates use of parallelogram principle in solving the problem.

■ Example 1

Suppose we have two processes 1 and 2 so that the process 1 requires 4 units of labour and 1 unit of capital to produce one unit of output and process 2 requires 2 units of labour and 2 units of capital to produce one unit of output. In Table 2.5 we represent these two processes with needed inputs per unit of output. Now let us assume that only K = 100 units of capital and L = 70 units of labour are available for the production. Now from Table 2.5 and the associated Figure 2.5 given below we have three production possibilities.

Table 2.5

Input requirements per unit of output		
Process	K	L
1	4	1
2	2	2

In the Figure 2.5 the relevant iso-quant segments are also shown in addition to the parallelogram developed in the figure. Given the inputs there are three possible methods of producing the output X.

1) Production by using process 1 alone

$X_1 = \min \{(100/4), (70/1)\} = \text{Min } (25, 70) = 25$

Table 2.6

Demand and supply positions of inputs using process 1				
Inputs	$X_1 = 25$	Demand	Supply	Excess
L	25 x 1 = 25	25	70	45
K	25 x 4 = 100	100	100	0

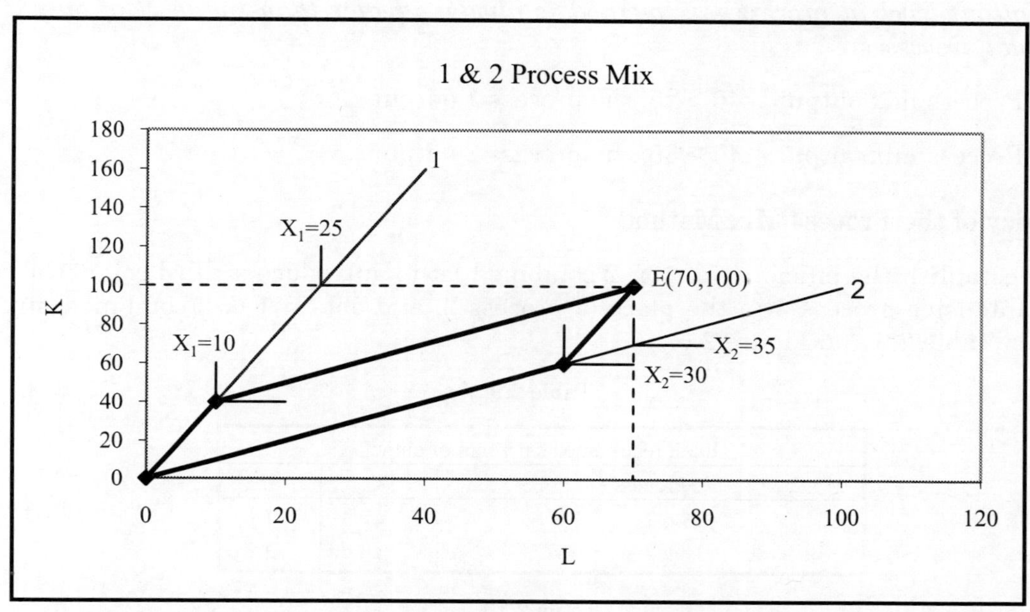

Figure 2.5

2) Production by using process 2 alone

$$X_2 = \min \{(100/2), (70/2)\} = \text{Min } (50, 35) = 35$$

Table 2.7

Demand and supply positions of inputs using process 2				
Inputs	$X_2 = 35$	Demand	Supply	Excess
L	35 x 2 = 70	70	70	0
K	35 x 2 = 70	70	100	30

3) Production by using mixing both processes 1 and 2

Output by process 1 is $X_1 = 10$

Output by process 2 is $X_2 = 30$

Total production $X = X_1 + X_2 = 40$

Table 2.8

Demand and supply positions of inputs using process 1 & 2					
Inputs	X_1=10	X_2 = 30	Demand	Supply	Excess
L	10 x 1 = 10	30 x 2 = 60	70	70	0
K	10 x 4 = 40	30 x 2 = 60	100	100	0

***Note:** Output level in process mix method is always greater than the level of output in individual processes.*

Process mix output = 40 > 25, the process 1 output

Process mix output = 40 > 35, the process 2 output

Efficiency of the Process Mix Method

Now to establish the efficiency of 1 & 2 combined let us introduce a third comparatively labour intensive process 3 in the place of process 2, and obtain 1 & 3 combined mix as shown in Table 2.9 and Figure 2.6.

Table 2.9

Input requirements per unit of output		
Process	K	L
1	4	1
3	1	4

Table 2.10

Demand and supply positions of inputs under the process 1 & 3 mix					
Inputs	$X_1 = 22$	$X_2 = 12$	Demand	Supply	Excess
L	22 x 1 = 22	12 x 4 = 48	70	70	0
K	22 x 4 = 88	12 x 1 = 12	100	100	0

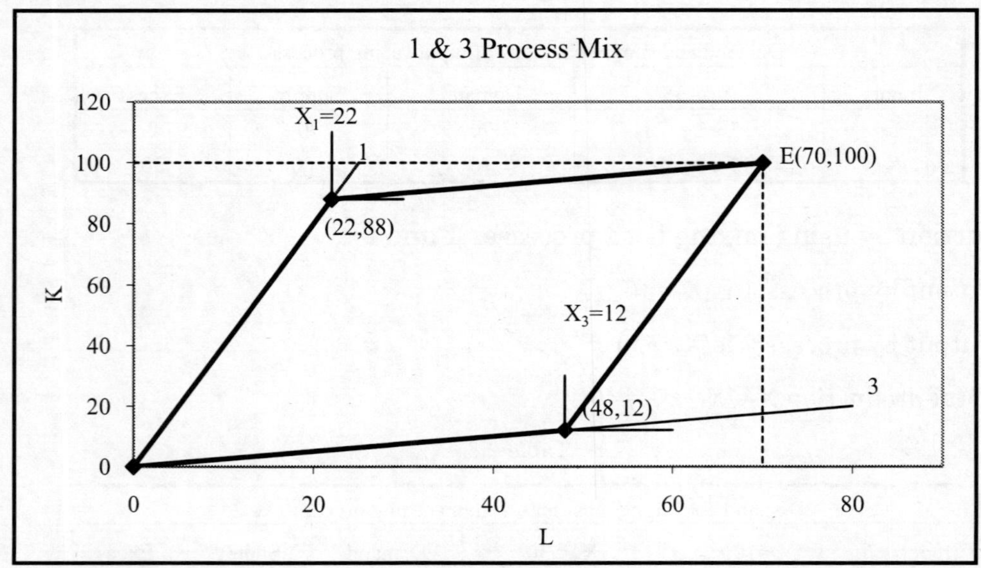

Figure 2.6

Here clearly the newly introduced process 3 is comparatively more labour intensive than the process 2. Figure 2.6 illustrates the result of this process mix. For full utilization of both capital and labour as per the figure the producer will have to produce 22 units of output X by using the process 1 and 12 units output X by using process 3. Thus the total output of this product mix is = 22 +12 = 34. Clearly this output is less than the output corresponding to the product mix using process 1 and 2 i.e., 40 < 34. However, note that on both the occasions the inputs are fully utilized.

Note 1: For efficient process mix one must choose the closest two production processes on either side of the given endowment point.

Note 2: The process mix is purely technical in the sense that there is no influence of output and input prices in this analysis.

2.6 LINEAR PROGRAMMING PROBLEM FORMULATION

To formulate the needed linear programming problem let us reconsider the problem that we have discussed in the previous section but in a different perspective. Let process 1 produces the product 1 and the process 2 produces the product 2. Further let X_1 be the quantity of product 1 that is being produced using process 1 and X_2 be the quantity of product 2 that is being produced using process 2. Now from the production data given in Table 2.5, since each unit of the product produced by process 1 requires 4 units of capital and each unit of product produced by process 2 requires 2 units of capital, the total capital requirement for X_1 units of product 1 and X_2 units of product 2 by using processes 1 and 2 respectively is obtained as $4X_1 + 2X_2$. However, since the availability of capital is restricted to 100 units only, this requirement cannot exceed 100 units of capital. So the capital constraint of the problem may be written as

$$4X_1 + 2X_2 \leq 100$$

Similarly, since each unit of the product produced by process 1 requires 1 unit of labour and each unit of product produced by process 2 requires 2 units of labour, the labour requirement for X_1 and X_2 units of output is obtained as $1X_1 + 2X_2$. However, since the availability of labour is only 70 units this requirement cannot exceed 70 units. So the labour constraint may be written as

$$1X_1 + 2X_2 \leq 70$$

Graphical Solution for Determining the Optimum Output

To draw these constraints in graph let us ignore the less than inequalities and take only the equalities in both the constraints for further discussion. So the respective linear equality constraints are:

$$4X_1 + 2X_2 = 100 \text{ (Capital)}$$

$$X_1 + 2X_2 = 70 \text{ (Labour)}$$

Since these equations are straight lines, to draw them on a graph sheet we need a minimum of two points for each one of them. In Table 2.11, two such points are obtained and reported for both the straight lines.

Table 2.11

Constraints			
$1X_1 + 2X_2 = 70$		$4X_1 + 2X_2 = 100$	
X_1	X_2	X_1	X_2
0	35	0	50
70	0	25	0

After plotting these points on a graph sheet we connect the corresponding points and get the needed straight lines as shown in the Figure 2.7.

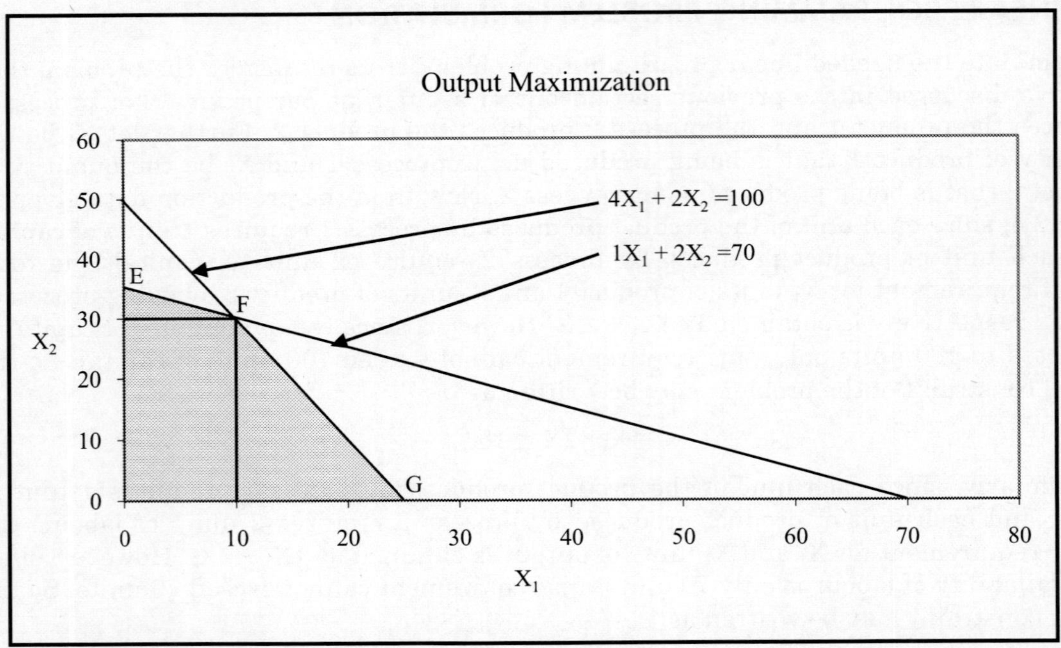

Figure 2.7

Since our original constraints are less than or equal to (≤) type, the shaded area that lies on and below these two straight lines satisfy the *less than or equal to* inequalities in the original constraints. To reorganize this condition fulfilment we shade this area as shown in the figure. This shaded region is often called the feasible region for our linear programming. The corner points O(0,0), E(0,35), F(10,30), G(25,0) are called the basic feasible solution points.

1) At O, the origin, nothing is produced and hence both the resources are fully unutilized.

2) At E only 35 units of X_2 is produced by using the process 2. Since this point is on the labour constraint, labour gets fully utilized. However, capital is in surplus in this case.

3) At point F, 10 units of X_1 and 30 units of X_2 are being produced by using process 1 and 2 respectively. Since this point is on the intersection of both the constraints both the resources are fully utilized this time.

4) At point G, 25 units of X_1 is produced using process 1. Further since the point F is on the capital line it confirms the fuller utilization of capital. At the same time since this point is below the labour line labour is underutilized.

At this stage remember that the efficiency of production necessitates the full utilization of both the resources. Clearly, at F both the resources are fully utilized. Thus, the efficient production yields 10 units of X_1 and 30 units of X_2. Other than this point all other corner points of the feasible region will leave some resource or the other underutilized.

Note: *For a moment consider that product 1 and 2 are one and the same but produced in two different processes as it was assumed in our activity analysis. From the graph it follows that the producer will produce 10 units of X_1 by using process 1 and 30 units of X_2 by using process 2. Total output from both the processes $X_1 + X_2 = 10 + 30 = 40$. This is exactly the same total output that the producer got in the process mix activity analysis discussed in the previous section. When this assumption is valid, Table 2.12 gives the summary results of the total output levels at all the four corners of the feasible region.*

Table 2.12

Corner points	X_1	X_2	Total output
O	0	0	0
E	0	35	35
F	10	30	40
G	25	0	25

Clearly, we once again confirm that F is the more efficient production point yielding a maximum of 40 units of output.

2.7 INFLUENCE OF PRODUCT PRICE

Till now we were discussing the optimum output strategies, purely on the technical ground having the fuller utilization of both the resources in mind. The role of the product prices is being totally ignored while deciding the optimum product mix. In this section we propose to introduce products prices into our model and recalculate the appropriate product mix that maximizes the total revenue to the producer rather than the full utilization of the resources.

Suppose each unit of X_1 is priced at Rs.10 and X_2 is priced at Rs.15 in the market then the corresponding total revenue function is obtained as: $R = 10 X_1 + 15 X_2$. Luckily, this newly introduced revenue function is also a straight line and hence amenable for inclusion into the graph. So to obtain the maximum revenue yielding product mix, we introduce the so obtained revenue function also into our graph for an arbitrarily chosen R-value, say 150. As usual we determine two points, one on the X-axis and another on Y-axis and draw the needed revenue straight line.

Table 2.13

$150 = 10 X_1 + 15 X_2$	
X_1	X_2
0	10
15	0

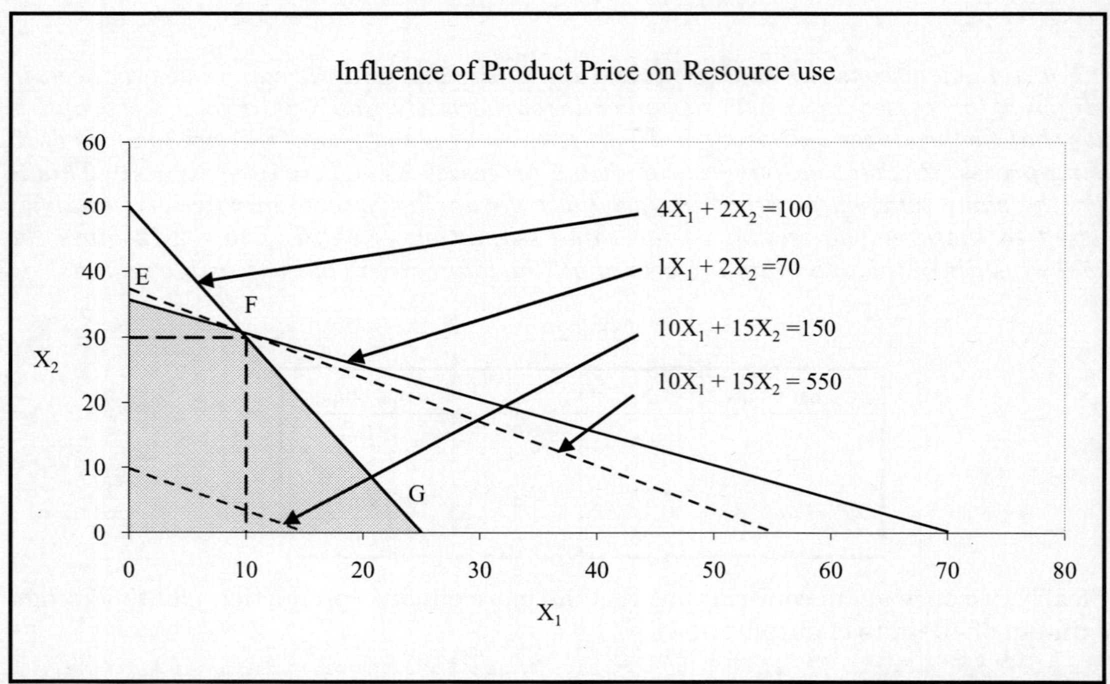

Figure 2.8

In Figure 2.8 the dotted line at the bottom represents this revenue line. Once this line is drawn we move this line up by keeping the slope intact as long as it touches any one of corners of the feasible reign OEFG. Such an optimum solution is reached once again corresponding to the point of intersection F as shown in Figure 2.8. Remember, that if we move the revenue line further up definitely we will be moving away from the feasible region. On completion, the whole of the linear programming problem may be summarized as shown below.

Maximize: $R = 10X_1 + 15X_2$

Subject to:

$$4X_1 + 2X_2 \leq 100 \text{ (Constraint K)}$$
$$1X_1 + 2X_2 \leq 70 \text{ (Constraint L)}$$
$$X_1 \; ; \; X_2 \leq 0$$

Now from the Figure 2.8 we read $X_1 = 10$ and $X_2 = 30$ as the equilibrium levels of output at F. So the corresponding maximum value of revenue is obtained by substituting these values in the formulated revenue function.

$$R = 10 X_1 + 15 X_2$$
$$= 10 \times 10 + 15 \times 30$$
$$= \text{Rs. } 550.$$

Alternative Method

In this alternative method we do not insert the revenue function into the graph to get the optimum solution. Instead we use an important linear programming theorem.

The optimum solution to a given linear programming problem will always be one or more corners of the convex feasible region.

So according to this theorem we first read all the corners of the feasible region from the graph and substitute these values into the revenue function. From the graph, the corner points are O (0, 0), E (0, 35), F (10, 30), G (25, 0). Table 2.14 gives the summary revenue results for all the four corner points after substituting the coordinates.

Table 2.14

Corner points	X_1	X_2	$R = 10X_1 + 15X_2$
O	0	0	0
E	0	35	525
F	10	30	550
G	25	0	250

Now from the table the highlighted maximum revenue yielding outputs are $X_1 = 10$ and $X_2 = 30$. The associated maximum revenue is Rs. 550.

Since this final optimum solution is at the point of intersection of both the labour and capital constraints clearly both the resources are fully utilized at this equilibrium. Table 2.15 shows the necessary calculations and confirms this result.

Table 2.15

Inputs	Process 1 Demand	Process 2 Demand	Total	Supply	Excess
Labour	10 x 1 =10	30 x 2 = 60	70	70	0
Capital	10 x 4 = 40	30 x 2 = 60	10	100	0

Influence of Price

A revisit

In the previous set of prices luckily the technical optimum solution that uses both the resources fully and the economic solution that maximizes the revenue are one and the same. This need not be the case all the time. The product mix and the associated input usage heavily depend upon the product pricing. To illustrate this dependency of price let us increase the price of X_1 to Rs. 40 and decrease the price of X_2 to Rs. 10 and restate our problem and obtain the solution once again as shown below.

Maximize: $R = 40X_1 + 10X_2$

Subject to: $4X_1 + 2X_2 \leq 100$ (Constraint K)

$1X_1 + 2X_2 \leq 70$ (Constraint L)

$X_1 \; ; \; X_2 \geq 0$

i) Graphical Solution

Table 2.16 and the associated Figure 2.9 illustrate the suitably modified graphical solution.

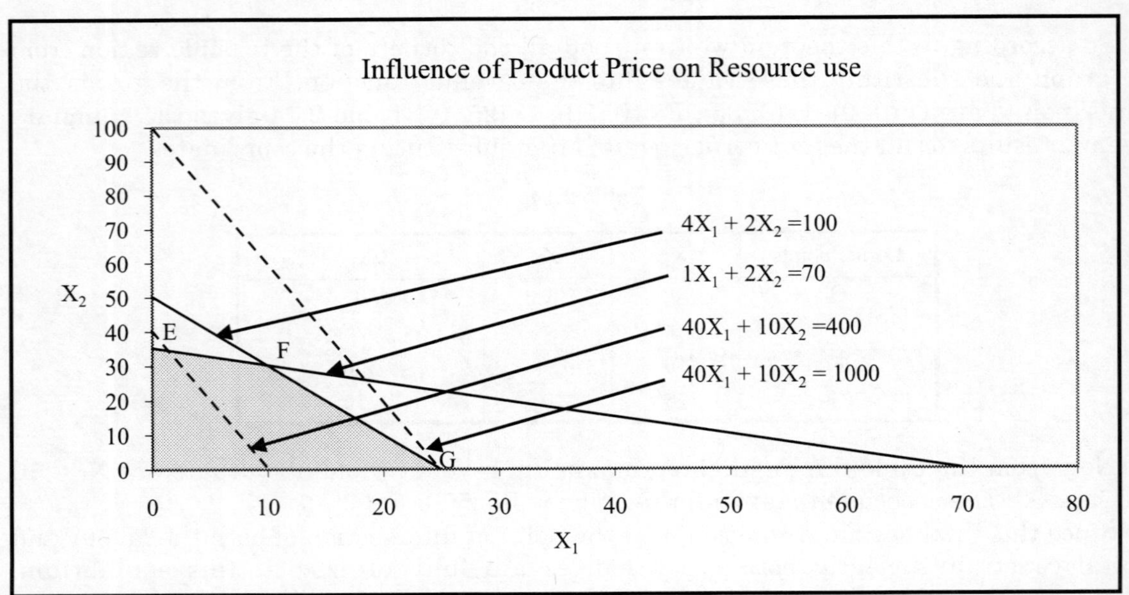

Figure 2.9

Table 2.16

$400 = 40X_1+10X_2$	
X_1	X_2
0	40
10	0

From the graph the optimum outs at the point G are: $X_1 = 25$ and $X_2 = 0$

So the maximum revenue = $R = 40X_1 + 10X_2 = 40 \times 25 + 10 \times 0 =$ Rs. 1000.

ii) Alternative Method

The relevant corner points are obtained from the graph and substituted in the revenue function in Table 2.17. The needed resource requirements are obtained and reported in Table 2.18.

Table 2.17

Corner points	X_1	X_2	$R = 40X_1 + 10X_2$
O	0	0	0
E	0	35	350
F	10	30	700
G	25	0	1000

Table 2.18

Process in use: 1			
Inputs	Demand	Supply	Excess
Labour	25 x 1 = 25	70	45
Capital	25 x 4 = 100	100	0

Clearly, this time economic revenue maximization decision differs from that of the (physical) engineering decision of full employment. While engineering solution is at F our economic solution has moved to point G yielding the highest revenue of Rs.1000. The physical efficiency norm of full employment thus is no longer valued now. Because of the very attractive price of the product 1 this time the producer decides to produce only product 1. Tables 2.17 and 2.18 give the summary result of the associated resource use. The labour resource is in excess supply this time though the capital is fully utilized. Thus, the producer uses the more capital-intensive process 1 and utilizes the capital fully. The labour is in excess supply to the extent of 45 units.

Influence of Price

A second revisit

This time let us increase the price of item X_2 from Rs.15 to Rs. 40 and leave the price of X_1 at its original level of Rs.10 and restate and resolve the problem once again as follows.

$$\begin{aligned}
\text{Maximize: } & R = 10X_1 + 40X_2 \\
\text{Subject to: } & 4X_1 + 2X_2 \leq 100 \text{ (Constraint K)} \\
& 1X_1 + 2X_2 \leq 70 \text{ (Constraint L)} \\
& X_1 ; \quad X_2 \geq 0
\end{aligned}$$

i) Graphical Solution

Table 2.19 and the associated Figure 2.10 illustrate the corresponding graphical solution.

Table 2.19

$400 = 10X_1+40X_2$	
X_1	X_2
0	10
40	0

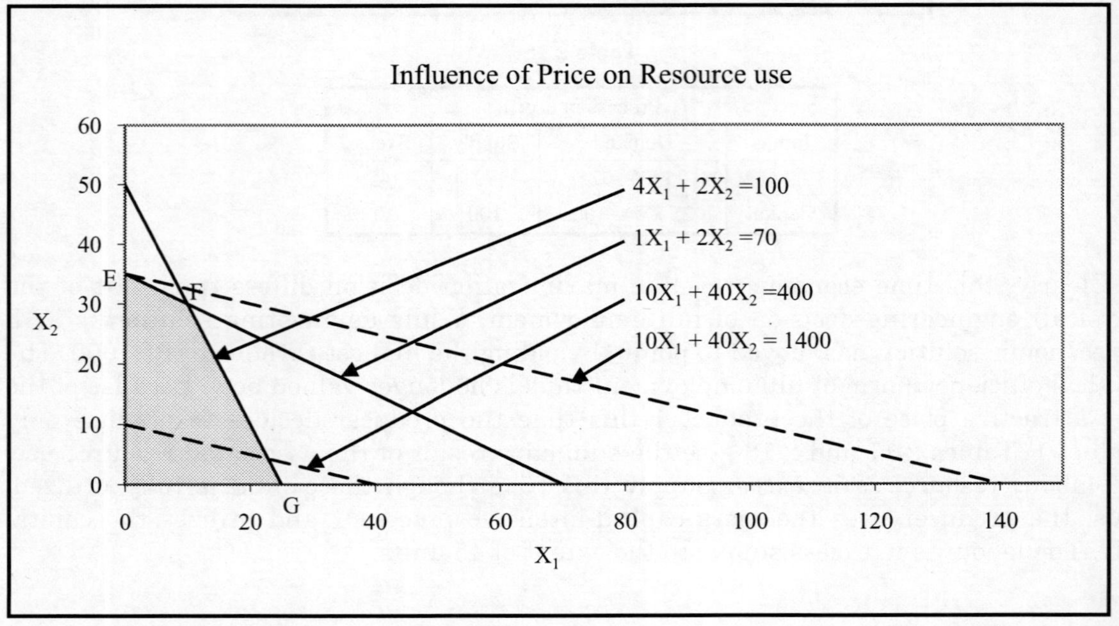

Figure 2.10

From the graph the optimum outs are: $X_1 = 0$ and $X_2 = 35$

So the maximum revenue = $R = 10X_1 + 40X_2 = 10\times0+40\times35$ = Rs. 1400.

ii) Alternative Method

The relevant corner points are obtained from the graph and substituted in the revenue function in Table 2.20. The needed resource requirements are obtained and reported in Table 2.21.

Table 2.20

Corner points	X_1	X_2	$R = 10X_1 + 40X_2$
O	0	0	0
E	0	35	1400
F	10	30	1300
G	25	0	250

Clearly, this time also economic revenue maximization decision differs from that of the (physical) engineering decision of full employment. While engineering solution is again at F, our economic solution has moved to point E yielding highest revenue of Rs.1400. Because of the very attractive price of the product 2, this time the producer decides to produce only product 2. Table 2.21 gives the summary result of the associated resource use. The capital resource is in excess supply this time though the labour is fully utilized.

Table 2.21

Process in use: 2			
Inputs	Demand	Supply	Excess
Labour	35 x 2 =70	70	0
Capital	35 x 2 = 70	100	30

Thus, the producer uses the more labour-intensive process 2 and utilizes the labour fully this time. The capital is in excess supply to the extent of 30 units.

2.8 DUAL PROBLEM

In linear programming for every maximization problem there exists a corresponding minimization problem and vice versa. The given original problem is often called the primal and the new problem that originates from the primal is often referred as the dual problem. Further, if the original primal problem is a maximization problem then its dual version will be a minimization type and vice versa. In this context one must remember the following three important properties of this dual problem.

1. If the prime problem has the solution, so as the dual.

2. The maximum value of the prime problem always equals to the minimum value of the dual problem and vice versa.

3. While solving the primal problem, the dual problem is generally facilitated in the sense that the solution for the dual is obtainable from the last iteration table of the primal.

■ Example 2

Formulation of the Dual Problem

Now in this section let us formulate the dual problem for the prime problem discussed in the previous section. While doing so we consider the following two facts. To produce one unit of the product 1, we need 1 unit of labour and 4 units of capital. Now let P_K and P_L are the imputed prices for capital K and labour L respectively. The total cost of production per unit product 1 is computed as the sum of the labour cost and the capital cost. As per the production table discussed in the previous section the cost of production of one unit of product 1 will be equal to $4P_K + P_L$. Since the product 1 is sold in the market at Rs. 10, the total cost at the imputed price rate must be *greater than or equal to* the quoted product price Rs. 10.

i.e. $4 P_K + P_L \geq 10$

Similarly, since the product 2 is sold in the market at Rs.15, the total cost at the imputed price rate must be *greater than or equal to* the quoted product price Rs. 15.

i.e. $2 P_K + 2P_L \geq 15$

Now since we have 100 units of capital and 70 units of labour at our disposal the total cost pertaining to the full utilization of both the resources is written as:

$C = 100 P_K + 70 P_L$

Clearly, this time our objective is to minimize C, the cost of production, by optimally imputing values to P_K and P_L. Thus, the whole of the dual problem is summarized as:

$$
\begin{aligned}
\text{Minimize: } & C = 100P_K + 200P_L \\
\text{Subject to: } & 4P_K + P_L \geq 10 \\
& 2P_K + 2P_L \geq 15 \\
& P_K \;;\; P_L \geq 0
\end{aligned}
$$

Now we use the usual graphical and tabular methods to get the final optimum solution.

i) Graphical Solution

As usual let us first convert all the two inequalities in our model to equalities and draw them on the graph sheet. To do this we identify two points on each one of them as shown in the Table 2.22.

Table 2.22

$2P_K + 2P_L = 15$		$4P_K + P_L = 10$	
P_L	P_K	P_L	P_K
0	7.5	10	0
7.5	0	0	2.5

Now using the data in Table 2.22 we draw the needed constraints on a graph sheet as shown in Figure 2.11.

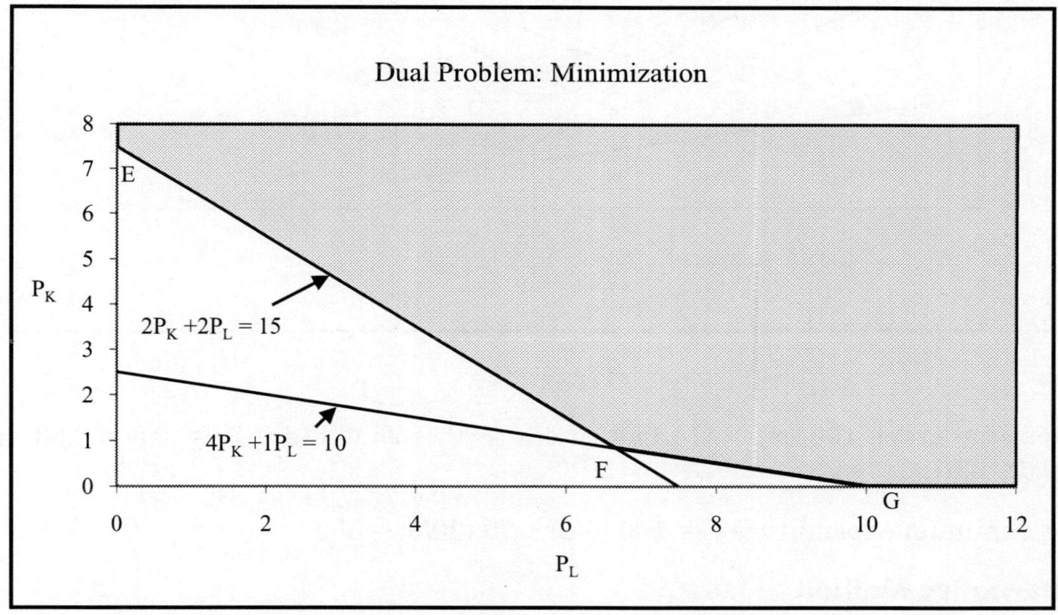

Figure 2.11

In our dual problem since the constraints are greater than or equal to type the shaded area on and above the constraints constitute the necessary feasible region as shown in Figure 2.11. The next task is introduce the objective function into this figure for an arbitrarily chosen C = 700 as usual. Once this is done we move this objective function up or down till this line become tangent to the outer most corner of the feasible region as shown in Figure 2.12.

Table 2.23

$700 = 100\ P_K + 70\ P_L$	
P_L	P_K
0	7
10	0

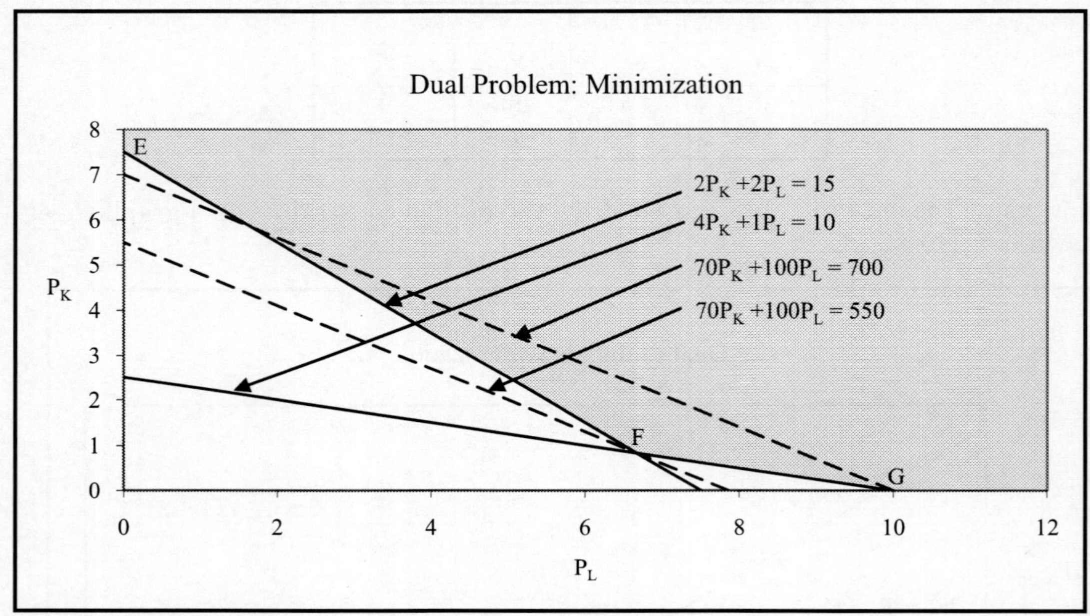

Figure 2.12

From the graph the imputed input prices at the point F are $P_K = 5/6 = 0.83$ and $P_L = 20/3 = 6.67$.

So the minimum expenditure C = 100 (5/6) + 70 (20/3) = 550.

ii) Alternative Method

From the graph the corner points are: E(0,7.5), F(20/3,5/6), G(10,0). The objective function values at these corner points are obtained and reported in Table 2.24.

Table 2.24

Corner points	P_L	P_K	$C = 70P_L + 100P_K$
E	0	7.5	750
F	6.66667	0.8333	550
G	10	0	700

Thus, at F the minimum cost = Rs. 550 the corresponding imputed prices are P_L = 6.67 and P_K = 0.83.

Note: *The value of the objective function in the primal is equal to the value of the objective function in the dual: R = C = 550. Now to highlight some more important issues of the influence of price on the equilibrium output let us consider a linear programming problem with two decision variables and three constraints.*

Example 3

A firm produces two products 1 and 2 by combining three inputs in certain fixed proportion as shown in Table 2.25.

Table 2.25

Output	Input X	Input Y	Input Z
1	1	2	2
2	3	2	0

Now let X_s = 90, Y_s = 80 and Z_s = 60 be the quantities of these inputs available to the producer. Further let the product 1 be sold at Rs.3 per unit and product 2 is sold at Re.1 per unit in the market. Now let X_1 and X_2 are the quantities of the products 1 and 2 that maximize the total revenue R. The statement of the corresponding linear programming in its complete form may be stated as

Maximize: $R = 3X_1 + 1X_2$

Subject to:

$1X_1 + 3X_2 \leq 90$ (Constraint X)

$2X_1 + 2X_2 \leq 80$ (Constraint Y)

$2X_1 + 0X_2 \leq 60$ (Constraint Z)

$X_1 ; X_2 \geq 0$

i) Graphical Solution

As usual let us convert the inequality constraints into equality constraints and draw them all on the graph sheet as shown in Figure 2.13 by plotting two points for each one of them as shown in Table 2.26.

Table 2.26

$1X_1+3X_2=90$		$2X_1+2X_2=80$		$2X_1-0X_2=60$	
X_1	X_2	X_1	X_2	X_1	X_2
0	30	0	40	30	0
90	0	40	0	0	-

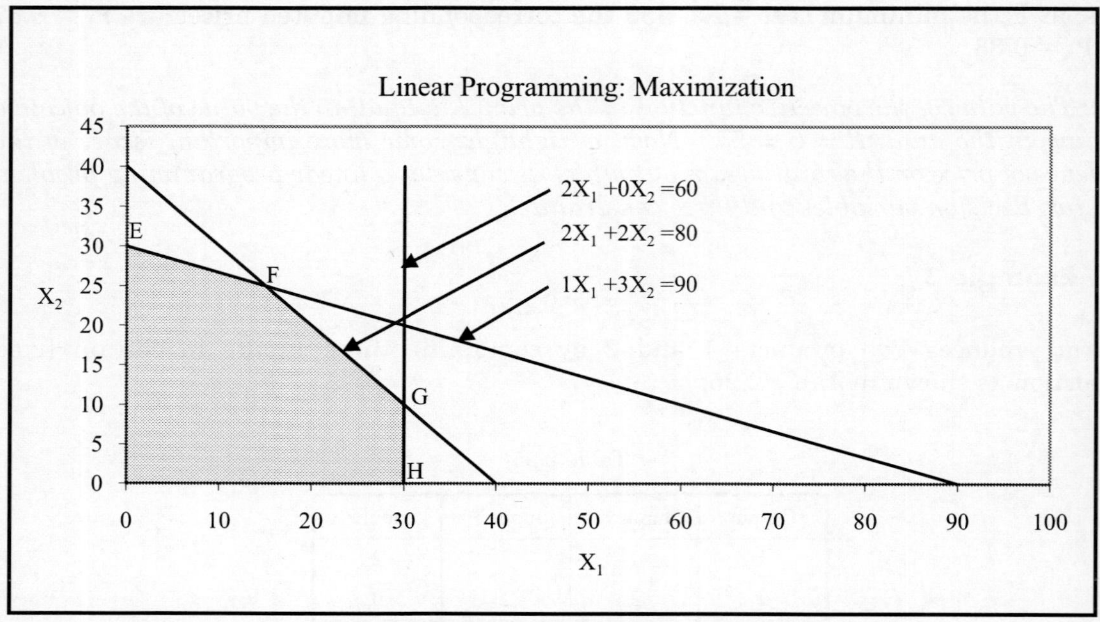

Figure 2.13

Now let us insert the needed objective function for an arbitrarily chosen value of R = 30 into the graph as shown in Figure 2.14 and shift it up till it becomes tangent to the corner at G as shown.

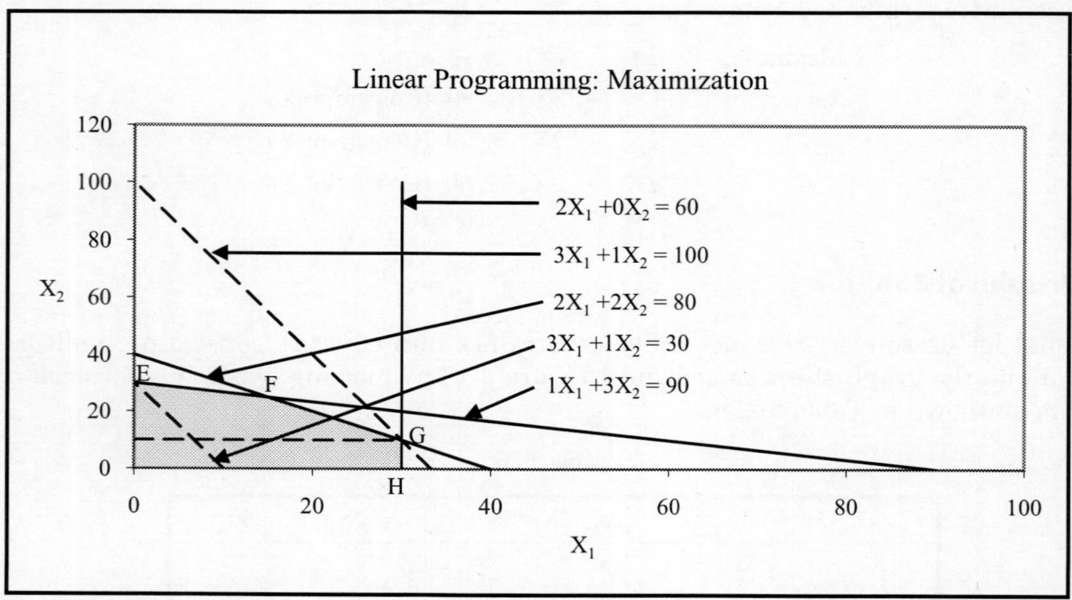

Figure 2.14

Table 2.27

$3X_1+1X_2=30$	
X_1	X_2
0	30
10	0

From the graph the equilibrium values are $X_1 = 30$ and $X_2 = 10$. Thus the maximum revenue is obtained as

$$R = 3X_1 + 1X_2$$

$$3\times30 + 1\times10 = \mathbf{100}$$

ii) Alternative Method

As usual by the alternative method the solution is obtained by substituting all the corner points into the objective function. The results are shown in Table 2.28.

Table 2.28

Corner points	X_1	X_2	$R = 3X_1 + 1X_2$
O	0	0	0
E	0	30	30
F	15	25	70
G	30	10	100
H	30	0	90

The highest revenue yielding corner solution is at G and the associated maximum revenue Rs. 100.

Binding and Non-binding Inputs

In the above illustration at the equilibrium point G, the Y and Z constraints intersect. Since this point lies on both these constraints these resources are fully utilized at this point. Such fully utilized resources are often called binding inputs. In economic sense such inputs command positive shadow or imputed prices in the corresponding dual problem. However, at G we are well below the X constraint exhibiting its under-utilization. Such resources are called non-binding inputs. They do not command any positive shadow pricing in the corresponding dual problem. Such non-binding inputs command only zero imputed pricing in the corresponding dual problem.

2.9 DUAL PROBLEM ONCE AGAIN

■ Example 4

For the data given in example 3 above, formulate dual and solve it for imputed prices and the associated minimum cost.

Table 2.29

Output	Input X	Input Y	Input Z
1	1	2	2
2	3	2	0
Total endowment	90	80	60

Solution

Let P_X, P_Y, and P_Z are the imputed input prices of X, Y, and Z respectively. From the data given above the per unit cost of product 1 is obtained as

$$1P_X + 2P_Y + 2P_Z$$

As the product 1 is sold at Rs. 3 per unit, the resource cost of producing one unit of product 1 must be at least greater than or equal to Rs. 3. So the relevant constraint for the product 1 is written as

$$1P_X + 2P_Y + 2P_Z \geq 3$$

Similarly, since the product 2 is sold at Re. 1 per unit its production cost at imputed rate must be at least equal to or greater than 1. So the relevant constraint pertains to product 2 is written as

$$3P_X + 2P_Y + 0P_Z \geq 1$$

Now the total cost of all the three resources is obtained as $C = 90P_X + 80P_Y + 60P_Z$. While making payments this cost is to be minimized. So the complete dual problem may be summarized as follows.

$$\begin{aligned}
\text{Minimize: } & C = 90P_X + 80P_Y + 60P_Z \\
\text{Subject to: } & 1P_X + 2P_Y + 2P_Z \geq 3 \\
& 3P_X + 2P_Y + 0P_Z \geq 1 \\
& P_X \; ; \; P_Y \; ; \; P_Z \geq 0
\end{aligned}$$

Now since our problem involves three decision variables a two-dimensional graphical solution is beyond the scope. In the absence of a graph we will not be in a position to get the needed corners of our feasible region as well. However, in our primal problem corresponding to this dual, since the input X is non-binding, it commands no imputed positive shadow price in the dual. So we can safely assume $P_X = 0$ in the given dual formation. After incorporating this additional useful information our dual problem is restated as follows.

$$\text{Minimize:} \quad C = 80P_Y + 60P_Z$$

$$\text{Subject to:} \quad \begin{aligned} 2P_Y + 2P_Z &\geq 3 \\ 2P_Y + 0P_Z &\geq 1 \\ P_Y \; ; \; P_Z &\geq 0 \end{aligned}$$

Now after converting the inequalities into equalities we can solve the problem by using two dimensional graphical methods. Table 2.30 identifies two points on each one of the straight lines constraints given in the problem.

Table 2.30

$2P_Y + 2P_Z = 3$		$2P_Y + 0P_Z = 1$	
P_Y	P_Z	P_Y	P_Z
0	1.5	0.5	0
1.5	0	0	∞

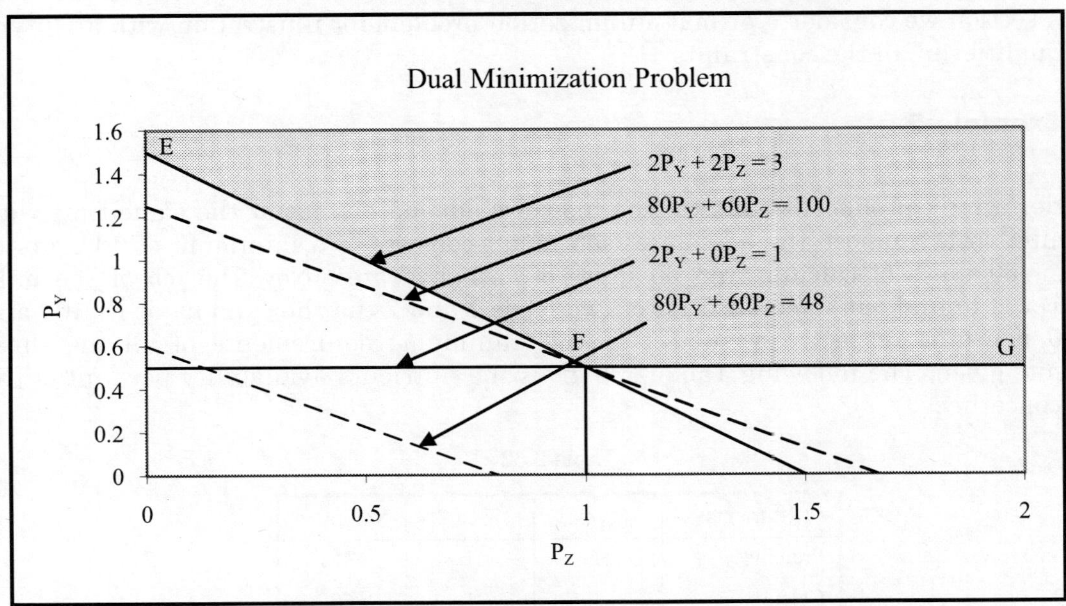

Figure 2.15

Once the constraints are drawn in the graph, we introduce the objective function for an arbitrarily chosen C = 48. For the final solution we move the objective function up or down till we get the tangential equilibrium as shown in Figure 2.15 at F.

From the graph the imputed prices are $P_X = 0$, $P_Z = 1$, $P_Y = 0.5$ and $C = 80\times0.5+60\times1= 100$. Here again it is important to note that R in primal and C in dual are equal to one another (R = C = 100).

Alternative Method

Table 2.31

Corner points	P_Y	P_Z	$C = 80\ P_Y + 60\ P_Z$
E	1.5	0	120
F	0.5	1	100
G	0.5	2	160

From the Table 2.31, the imputed prices are $P_X = 0$, $P_Z = 1$, $P_Y = 0.5$ and $C = 80\times0.5+60\times1= 100$.

2.10 CONSTRAINT COST MINIMIZATION

In the previous section both the primal problems had ≤ inequalities in all the constraints. In the corresponding dual problems the constraints were formed with ≥ inequalities. This need not be the case in all the problems. Sometimes the primal problem can have ≥ inequalities and their associated dual can have ≤ inequalities. Further, the primal can be a minimization and the dual can be a maximization problem. In this section we consider a primal minimization problem for illustration with all ≥ type of inequalities in all the constraints.

■ Example 5

Consider a milk producer wants to feed his cows adequately but at the same time with minimum cost. Suppose the adequate feed menu consists of a minimum of 40 units of protein, 60 units of calcium and 60 units of carbohydrates. Now the job of the milk producer is to find out the quantities of two feeds 1 and 2 that he can buy at Rs.100 and Rs.200 per unit respectively subject to the minimum requirements of all the three nutrients given. The following Table 2.32 gives the nutrients availability per unit of the feeds concerned.

Table 2.32

Nutrients	Feed 1	Feed 2	Minimum
Protein	1	1	40
Calcium	3	1	60
Carbohydrate	1	6	60

Solution

Now let X_1 and X_2 be the amounts of two feeds 1 and 2 that the producer buys from the market to feed his cows. So the objective function of the cost of production and the nutrient constraints is summarized as follows.

$$\text{Minimize: } C = 100X_1 + 200X_2$$

$$\begin{aligned} \text{Subject to: } \quad 1X_1 + 1X_2 &\geq 40 \quad \text{(Protein X)} \\ 3X_1 + 1X_2 &\geq 60 \quad \text{(Calcium Y)} \\ 1X_1 + 6X_2 &\geq 60 \quad \text{(Carbohydrate Z)} \\ X_1 \; , \; X_2 &\geq 0 \end{aligned}$$

The constraints and the objective functions are graphed as has been done in all the preceding examples, by identifying two points for each one of them. Since the constraints are ≥ type, in the Figure 2.16 the shaded area on and above the constraints shows the feasible region.

Table 2.33

$1X_1 + 1X_2 = 40$		$3X_1 + 1X_2 = 60$		$1X_1 + 6X_2 = 60$	
X_1	X_2	X_1	X_2	X_1	X_2
0	40	0	60	0	10
40	0	30	0	60	0

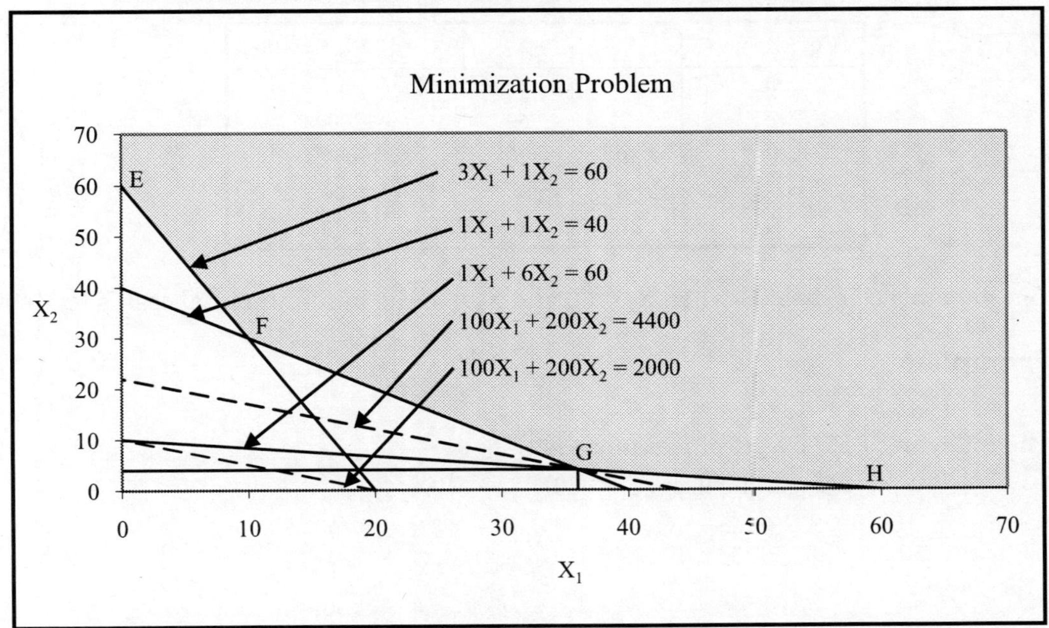

Figure 2.16

Now to find the amount of X_1 and X_2 that meets all the nutrition requirements at minimum cost, we assign an arbitrary C = 2000 and insert the cost constraint as a straight line in the graph drawn already.

Table 2.34

$100X_1+200X_2 = 2000$	
X_1	X_2
0	10
20	0

Now we move this objective line up without changing its slope till it touches the lowest corner point of the feasible region. The coordinates of the so obtained corner point give us the needed final solution. In the Figure 2.16 the solution occurs at the point G and yields $X_1 = 36$ and $X_2 = 4$. Thus the minimum cost for the feeds is obtained as: $C = 100 \times 36 + 200 \times 4 = 4400$.

Alternative Method

Once the corner points are obtained from the graph, the needed minimum value of C is obtained by substituting these values in our objective function and then choosing the least. The following Table 2.35 illustrates all such corner point results.

Table 2.35

Corner points	X_1	X_2	$C = 100X_1 + 200X_2$
O	0	0	0
E	0	60	12000
F	10	30	7000
G	36	4	4400
H	60	0	6000

From the table clearly at G when $X_1 = 36$ and $X_2 = 4$, the minimum total cost is Rs.4400.

■ Example 6

Solve the following LPP by graphical method

$$\text{Minimize: } C = 20X_1 + 10X_2$$

$$\begin{aligned} \text{Subject to: } \quad 1X_1 + 2X_2 &\leq 40 \\ 3X_1 + 1X_2 &\geq 30 \\ 4X_1 + 3X_2 &\geq 60 \\ X_1 ; \quad X_2 &\geq 0 \end{aligned}$$

Solution

The constraints and the objective functions are graphed as has been done in all the preceding examples by identifying two points for each one of them as shown in Table 2.36.

Table 2.36

$1X_1 + 2X_2 = 40$		$3X_1 + 1X_2 = 30$		$4X_1 + 3X_2 = 60$	
X_1	X_2	X_1	X_2	X_1	X_2
0	20	0	30	0	20
40	0	10	0	15	0

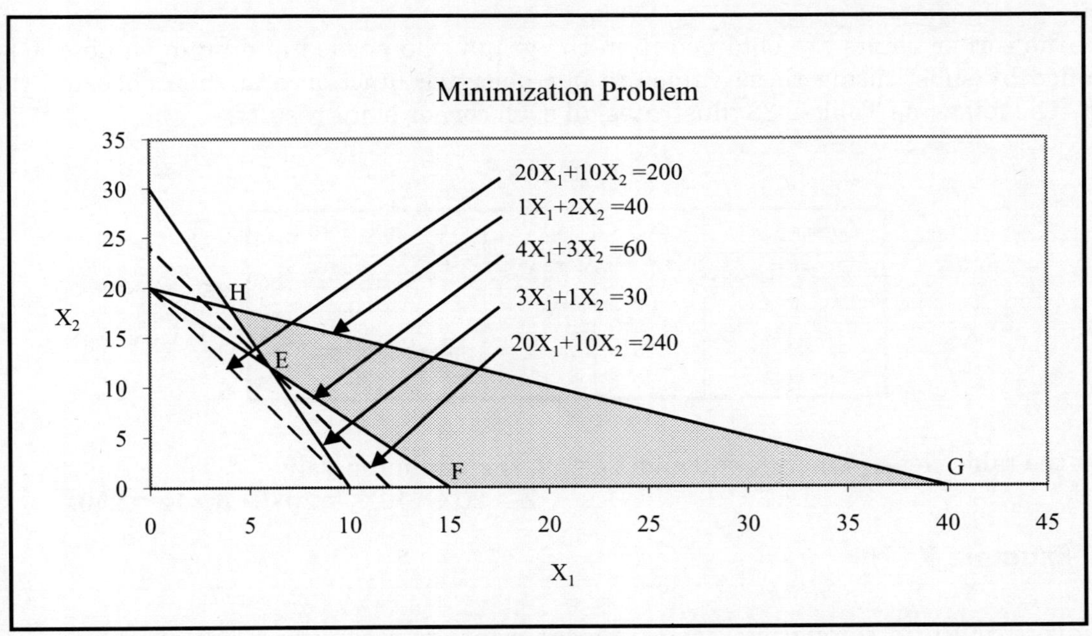

Figure 2.17

Note that the first constraint is *less than or equal to* type. Thus relevant points on and below this line satisfy this constraint. The other two constraints are ≥ type, so the points on and above these lines satisfy these two constraints. The feasible region satisfying all the three constraints is shown by the shaded area EFGH in Figure 2.17.

Now to find the amount of X_1 and X_2 that meets our objective function, as we did in all the previous cases, we assign an arbitrary C = 200 and insert the cost constraint as a straight line into the graph as shown in the Figure 2.17.

Table 2.37

$20X_1+10X_2=200$	
X_1	X_2
0	20
10	0

This line is shown by the dotted line at the bottom in Figure 2.17. The upward shifting of this line touches the corner having the coordinate (6, 12) at E. The needed minimum is obtained by merely substituting these values in the objective function given. So the minimum value of the objective function is obtained as: $Z = 20X_1+10X_2 = 20\times6+10\times12 = 240$.

Alternative Method

Once the corner points are obtained from the graph, the needed minimum value of C is obtained by substituting these values in our objective function and then choosing the least. The following Table 2.38 illustrates all such corner point results.

Table 2.38

Corner points	X_1	X_2	$C = 20X_1 + 10X_2$
E	6	12	240
F	15	0	300
G	40	0	800
H	4	17	250

From the table clearly when $X_1 = 6$ and $X_2 = 12$, the minimum value

$$Z = 20X_1+10X_2 = 20\times6+10\times12 = 240.$$

■ Example 7

An advertising firm through its advertisement wants to reach two types of customer. It considers customers with more than Rs. 40,000 as group A and customers with less than Rs. 40,000 as group B. One TV advertisement costs Rs. 50,000 and one radio advertisement costs Rs. 20,000. It has a total budget of Rs. 2,00,000 for advertising. Pre-committed contractual obligations require a minimum of three TV programs. Similarly, the radio advertisement cannot go beyond 5. Survey reveal that one TV program can reach 7,50,000 of A type and 1,50,000 of B type audience. Similarly, one radio program can reach 40,000 type A and 2,60,000 type B audience. Formulate a linear program problem and obtain the graphical solution that maximizes the total number of audience.

Solution

Linear programming formulation

Let X_1 and X_2 be the respective TV and radio programs. The relevant linear programming problem could be formulated as

Maximize: $Z = (7{,}50{,}000+1{,}50{,}000)\,X_1+(40{,}000+2{,}60{,}000)X_2$

$$\begin{aligned} \text{Subject to:} \quad & 50{,}000X_1 + 20{,}000X_2 \leq 2{,}00{,}000 \text{ or} \\ & 5X_1 + 2X_2 \leq 20 \\ & X_1 + 0X_2 \geq 3 \\ & 0X_1 + 1X_2 \leq 5 \\ & X_1 \;;\; X_2 \geq 0 \end{aligned}$$

Graphical Solution

The relevant constraints are drawn as usual in the following Figure 2.18, the feasible region and the corner points are identified as shown in Figure 2.18.

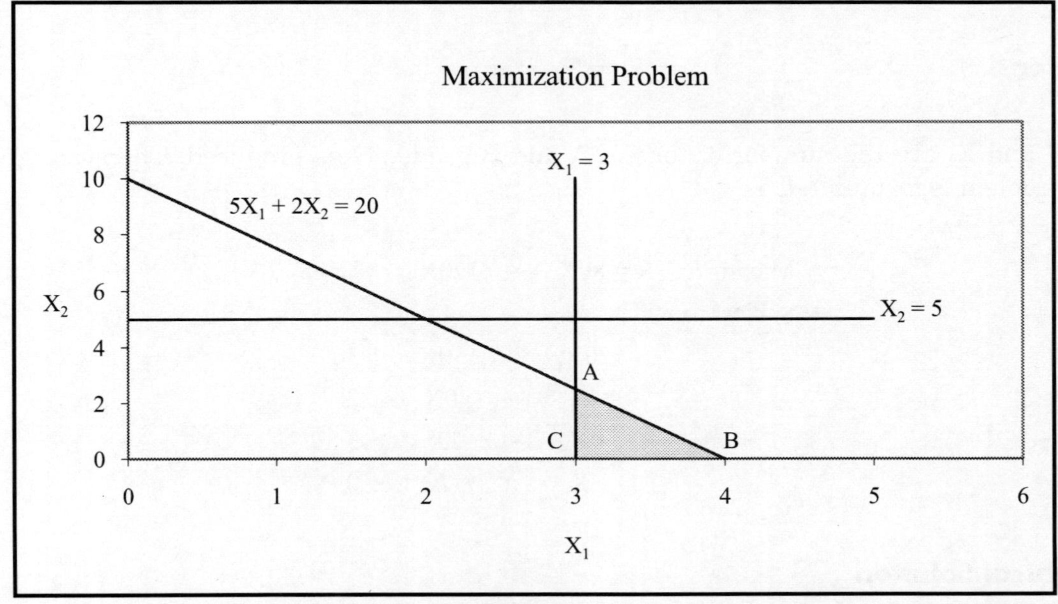

Figure 2.18

In Table 2.39 the objective function values are calculated for all the three corner points and reported.

Table 2.39

Corner points	Co-ordinates	Value of the objective function
A	(3,2.5)	$Z = 9,00,000 \times 3 + 3,00,000 \times 2.5 = 34,50,000$
B	(4,0)	$Z = 9,00,000 \times 4 + 3,00,000 \times 0 = 36,00,000$
C	(3,0)	$Z = 9,00,000 \times 3 + 3,00,000 \times 0 = 27,00,000$

Since 36,00,000 is the maximum among the three values, the solutions are $X_1 = 4$; $X_2 = 0$ and $Z = 36,00,000$.

■ Example 8

A firm manufactures two types of steel A and B. The productive capacity of both the items putting together is 9 tons. The pre-contact obligation is to supply at least 2 tons of A and at the least 3 tons of B to a nearby firm. Each ton of A and B needs 20 and 50 machine hours respectively. All together only 360 machine hours are available. If A is sold at Rs. 80 per ton and B is sold at Rs.120 per ton, determine the product mix that maximizes the total sales revenue.

Solution

Let X_1 and X_2 are the number of tons of A and B grade steels produced. The associated LPP problem is formulated as

$$
\begin{aligned}
\text{Maximize: } R &= 80X_1 + 120X_2 \\
\text{Subject to: } & 1X_1 + 1X_2 \le 9 \\
& 20X_1 + 50X_2 \le 360 \\
& 1X_1 + 0X_2 \ge 2 \\
& 0X_1 + 1X_2 \ge 3 \\
& X_1 \; ; \; X_2 \ge 0
\end{aligned}
$$

Graphical Solution

The relevant constraints are drawn as usual in the following Figure 2.19. The feasible region and the corner points are identified as shown in the figure.

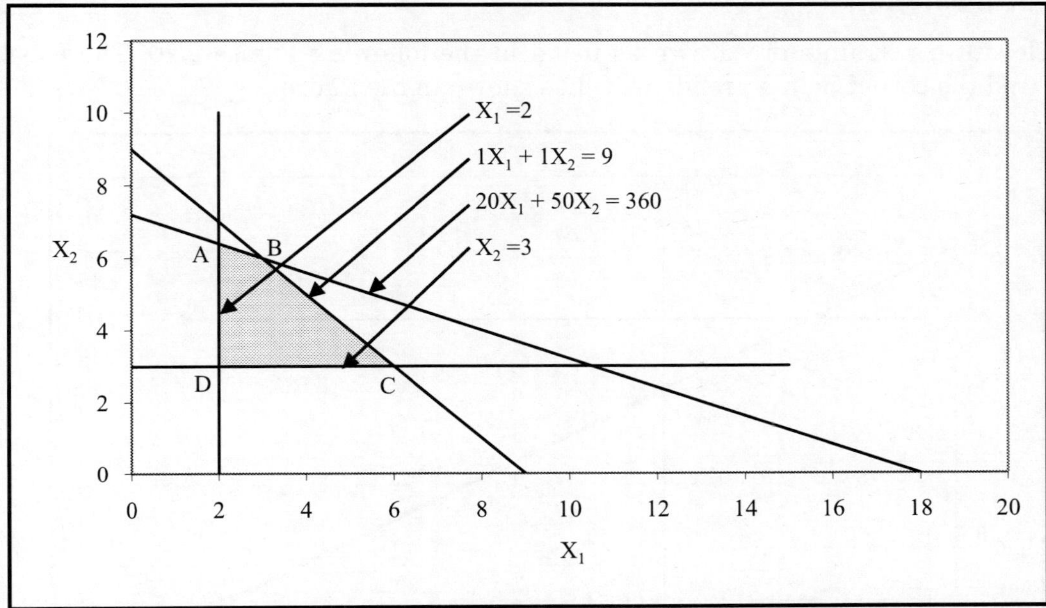

Figure 2.19

In Table 2.40 the objective function values are calculated and reported for all the four corner points.

Table 2.40

Corner points	Co-ordinates	Value of the objective function
A	(2,6.4)	$Z = 80 \times 2 + 120 \times 6.4 = 928$
B	(3,6)	$Z = 80 \times 3 + 120 \times 6 = 960$
C	(6,3)	$Z = 80 \times 6 + 120 \times 3 = 840$
D	(2,3)	$Z = 80 \times 2 + 120 \times 3 = 520$

The maximum revenue yielding solution values are: $X_1 = 3$ and $X_2 = 6$; $R = 960$.

■ Example 9

Solve the following LPP problem graphically.

$$
\begin{aligned}
\text{Maximize: } & R = 4X_1 + 6X_2 \\
\text{Subject to: } & 1X_1 + 1X_2 = 5 \\
& 1X_1 + 0X_2 \geq 2 \\
& 0X_1 + 1X_2 \leq 4 \\
& X_1 \; ; \; X_2 \geq 0
\end{aligned}
$$

Graphical Solution

The relevant constraints are drawn as usual in the following Figure 2.20. The feasible region and the corner points are identified as shown in the figure.

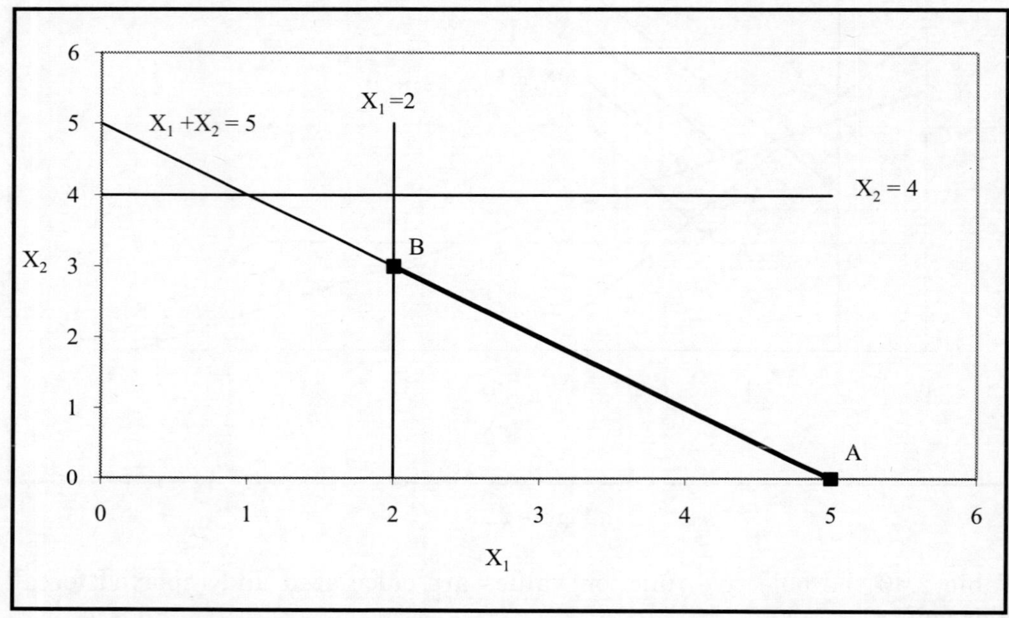

Figure 2.20

This time it is interesting to note that the feasible region is the highlighted line segment AB and not the area. In Table 2.41 we calculate the objective function values at point A and B.

Table 2.41

Corner points	Co-ordinates	Value of the objective function
A	(5,0)	$Z = 4 \times 5 + 6 \times 0 = 20$
B	(2,3)	$Z = 4 \times 2 + 6 \times 3 = 26$

The maximum result yielding solutions values are $X_1 = 2$, $X_2 = 3$, $Z = 26$.

■ Example 10

A cow feed production company wants to have 200 kg of a certain mixture of two ingredients A and B costing Rs. 3 and Rs. 5 per kg respectively. The legal requirement necessitates at least 60 kg of A and not more than 80 kg of B in the mixture produced. Obtain the amount of ingredients A and B so that the cost of manufacturing of the said mixture is minimized.

Solution

Let X_1 and X_2 be the amount of ingredient A and B bought for preparation of the mixed feed. Now from the data we formulate the LPP as follows.

$$\begin{aligned} \text{Minimize: } C = 3X_1 &+ 5X_2 \\ \text{Subject to: } 1X_1 &+ 1X_2 = 200 \\ 1X_1 &+ 0X_2 \leq 80 \\ 0X_1 &+ 1X_2 \geq 60 \\ X_1 &; X_2 \geq 0 \end{aligned}$$

Graphical Solution

The relevant constraints are drawn as usual in the Figure 2.21. The feasible region and the corner points are identified as shown in the figure.

This time it is interesting to note that the feasible region is simply a highlighted point A in the graph. The relevant objective function value is obtained by merely substituting the coordinates of this point as shown in Table 2.42.

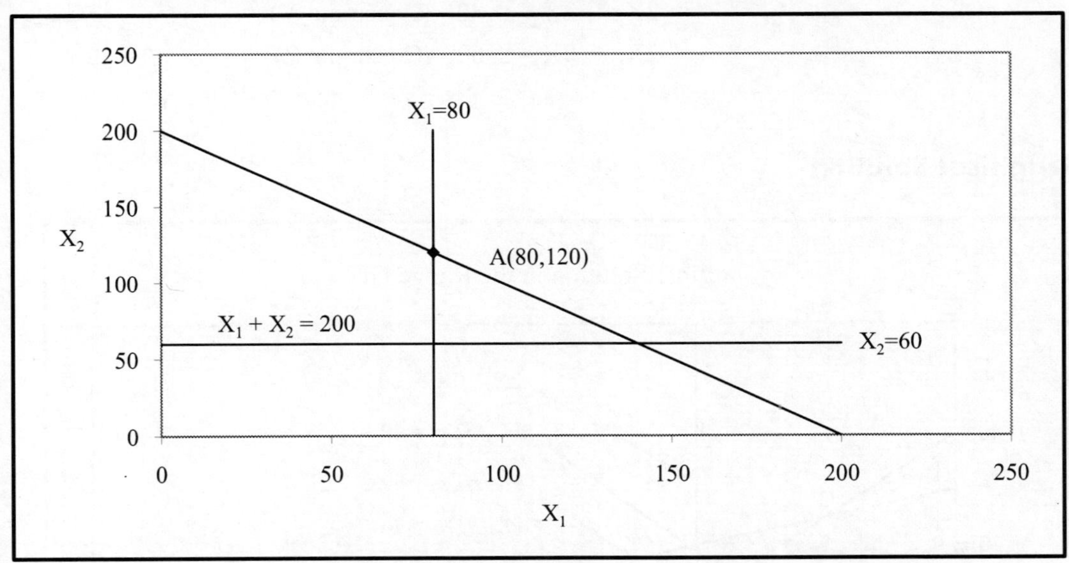

Figure 2.21

Table 2.42

Corner point	Co-ordinates	Value of the objective function
A	(80,120)	$Z = 3 \times 80 + 5 \times 120 = 840$

So the associated solutions values are $X_1 = 80$; $X_2 = 120$; $Z = 3 \times 80 + 5 \times 120 = 840$.

2.11 SPECIAL PROBLEMS IN LINEAR PROGRAMMING

In linear programming we come across several special situations inviting special attention. Some of them really do not pose any practical problem to the programmer but generate curious results. In other cases, optimum solutions do not exist at all. Some such problems are highlighted with suitable illustrations in this section.

1) The Problem of Multiple Optimum

Whenever the objective function has the same slope as that of any one of the constraints then we will have infinite number of optimal solutions over the entire stretch of a line segment. The following example illustrates one such case of multiple solutions.

■ Example 11

Solve the following linear programming problem.

Maximize: $R = 1X_1 + 3X_2$

Subject to: $1X_1 + 3X_2 \leq 90$ (Constraint X)

$2X_1 + 2X_2 \leq 80$ (Constraint Y)

$2X_1 + 0X_2 \leq 60$ (Constraint Z)

$X_1 ; \quad X_2 \geq 0$

i) Graphical Solution

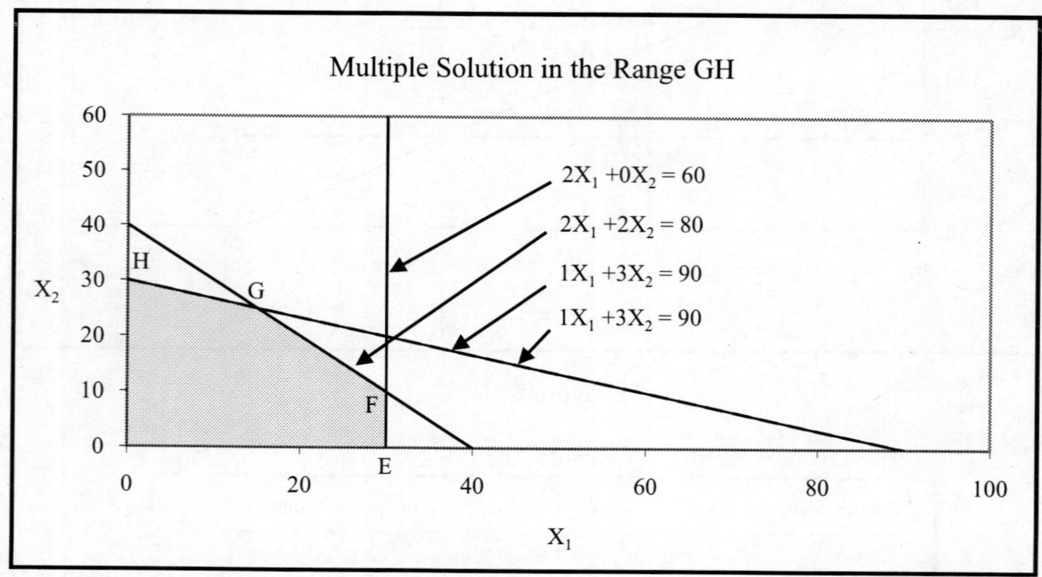

Figure 2.22

The relevant constraints are drawn as usual in Figure 2.22. The feasible region and the corner points are identified in it.

In this problem the slopes of the objective function and that of the slope of the constraint for the input X are one and the same. Thus at the optimum stage both of them coincide one over the other as shown in the Figure 2.22. Under such circumstances all the points on the line segment 'GH' yields the optimal solution. However, such a situation does not pose any additional problem. In fact, the corner points G and H also yield the same level of revenue as that of any other point in the infinite solution set from G to H. In the graph all the points in the line segment GH are equilibrium points yielding the same revenue of Rs. 90.

ii) Alternative Solution

As usual the solution is obtained by substituting the corner points into the objective function. The results are shown in Table 2.43.

Table 2.43

Corner points	X_1	X_2	$R = 1X_1 + 3X_2$
O	0	0	0
E	30	0	30
F	30	10	60
G	15	25	90
H	0	30	90

***Note:** All the points on the line segment GH are revenue-maximizing solution points.*

2) The Problem of Redundant Constraints

Often one or more constraints of our model become unnecessary and thus do not affect the feasible region and hence the optimum solution even in the event of their removal. Such constraints are called redundant constraints. The following example illustrates one such case.

■ Example 12

For the LPP model given below obtain the optimum solution.

Maximize: $R = 3X_1 + 1X_2$

Subject to:

$1X_1 + 3X_2 \leq 126$ (Constraint X)

$2X_1 + 2X_2 \leq 80$ (Constraint Y)

$2X_1 + 0X_2 \leq 60$ (Constraint Z)

$X_1 ;\ X_2 \geq 0$

1. Graphical Solution

The relevant constraints are drawn as usual in Figure 2.23. The feasible region and the corner points are identified as shown in figure.

Here the constraint for X is unnecessary and hence becomes redundant. In fact our feasible region is bounded by Y and Z constraints alone. Also note that in this case both the inputs Z and Y are binding and the input X is non-binding.

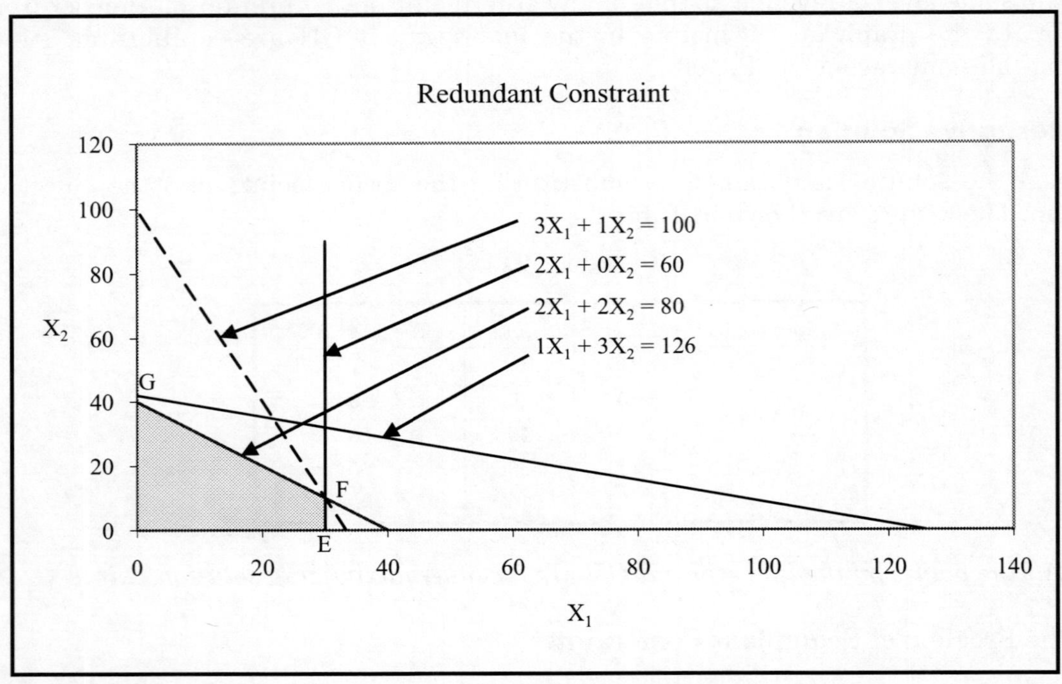

Figure 2.23

2. Alternative Solution

As usual the alternative solution is obtained by substituting the corner points into the objective function as shown in Table 2.44.

Table 2.44

Corner points	X_1	X_2	$R = 3X_1 + 1X_2$
O	0	0	0
E	30	0	30
F	30	10	100
G	0	40	40

Thus the solutions are: $X_1 = 30$, $X_2 = 10$ and $R = 100$.

3. The Problem of No Feasible Solution

Serious problems arise whenever the set of constraints does not form a unique feasible region. When there is no feasible region then there cannot be any optimum solution as well. The following example illustrates one such case.

Example 13

For the following LPP, obtain the optimum solution.

$$\text{Maximize: } R = 3X_1 + 2X_2$$

$$\text{Subject to: } 3X_1 + 1X_2 \geq 120 \text{ (Constraint X)}$$

$$2X_1 + 2X_2 \leq 60 \text{ (Constraint Y)}$$

$$X_1 ;\quad X_2 \geq 0$$

Solution

In this illustration not even a single point could be found satisfying both the constraints. Since there is no basic feasible region, there cannot be a final optimum solution.

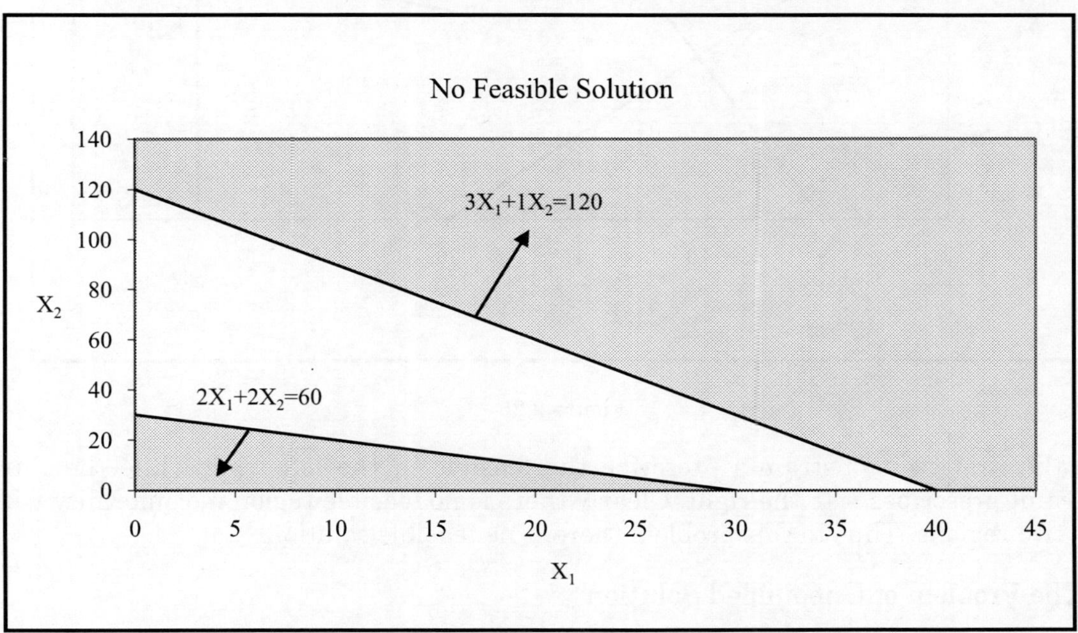

Figure 2.24

■ Example 14

For the following LPP, obtain the optimum solution.

$$
\begin{array}{llcl}
\text{Maximize: } Z = 20X_1 & + & 30X_2 & \\
\text{Subject to: } \quad 2X_1 & + & 2X_2 & \leq 40 \\
4X_1 & - & X_2 & \leq 20 \\
X_1 & & & \geq 30 \\
X_1 & ; & X_2 & \geq 0
\end{array}
$$

Solution

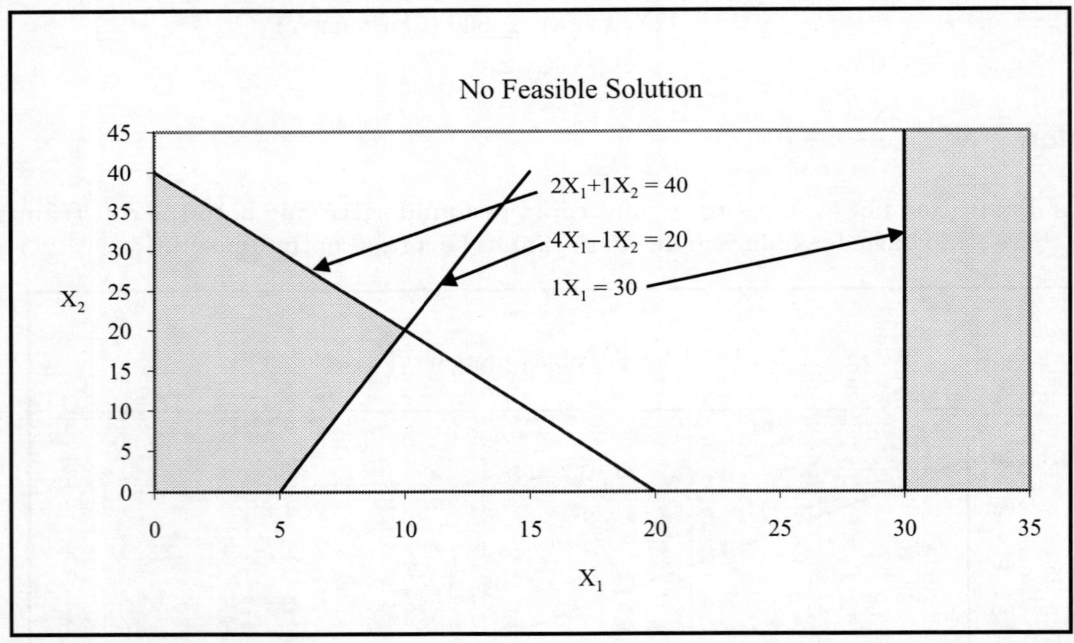

Figure 2.25

In the first two constraints prescribe the feasible in the left wing while the third constraint prescribes it to the right. Clearly there is no feasible region in conformity with these two regions. Thus to this problem there is no feasible solution.

4) The Problem of Unbounded Solution

Sometimes, interestingly, our LPP problem will have unbounded value for the objective function in the sense its value can be increased infinitely without any limit. This is really a serious problem in LPP models. Two such problems are illustrated in the following examples.

Example 15

Solve the following problem graphically.

$$\text{Maximize: } R = 3X_1 + 2X_2$$

$$\text{Subject to: } 1X_1 \geq 5$$

$$+ 1X_2 \leq 5$$

$$X_1 \; ; \; X_2 \geq 0$$

Solution

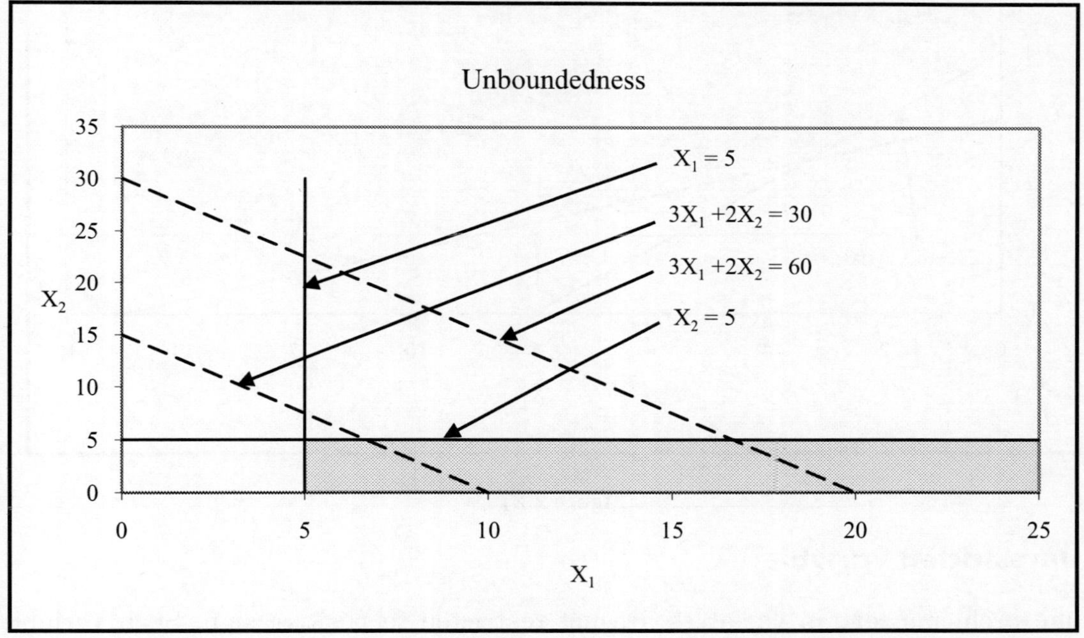

Figure 2.26

In the above graph the highlighted feasible region clearly increases without any bound in the right side and hence there is no optimum solution.

Example 16

Solve the following LPP problem graphically.

$$\text{Maximize: } R = 10X_1 + 20X_2$$

$$\text{Subject to: } 2X_1 + 4X_2 \geq 16$$

$$X_1 + 5X_2 \geq 15$$

$$X_1 \; ; \; X_2 \geq 0$$

Solution

In the graph shown below the highlighted feasible region clearly increases without any bound up and right side and hence there is no optimum solution.

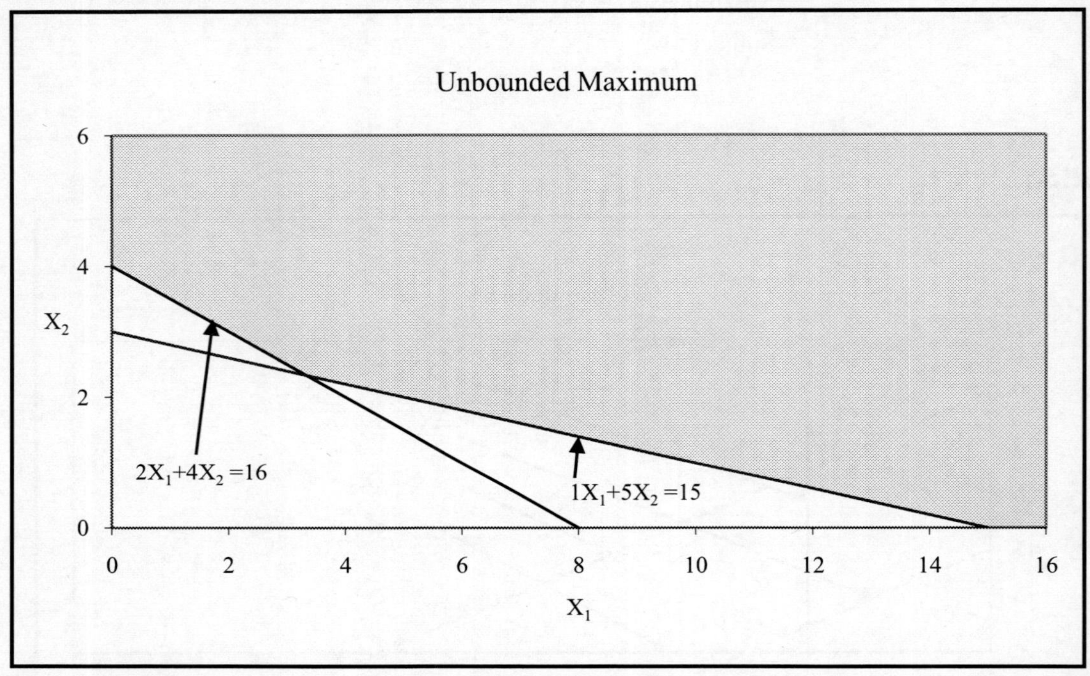

Figure 2.27

5) Unrestricted Variable

In some problems certain variables are not restricted to positive sign. Such variables thus can have negative values as well in the final solution. For example, consider the following illustration.

$$\text{Maximize: } R = 3X_1 + 5X_2$$

$$\begin{aligned} \text{Subject to: } \quad & 2X_1 + 5X_2 \leq 132 \\ & 3X_1 + 2X_2 \leq 100 \\ & X_1 \geq 0 \qquad X_2 \text{ unrestricted} \end{aligned}$$

In the above problem the variable X_2 is unrestricted in the sense that it could be either positive or negative. So we cannot continue our linear programming in such cases. In fact all the included variables must be *greater than or equal to zero*. To achieve this task we propose to introduce two new non-negative variables X_3 and X_4 so that $X_3 - X_4 = X_2$. So our modified problem with non-negative variables may be written as

$$\text{Maximize: } R = 3X_1 + 5X_3 - 5X_4$$
$$\text{Subject to: } 2X_1 + 5X_3 - 5X_4 \leq 132$$
$$3X_1 + 2X_3 - 2X_4 \leq 100$$
$$X_1 \; ; \; X_3 \; ; \; X_4 \geq 0$$

For optimum solution we proceed further with this new formation as usual.

2.12 PRIMAL AND DUAL RELATIONSHIP

The following schematic chart illustrates the relationship between primal and dual.

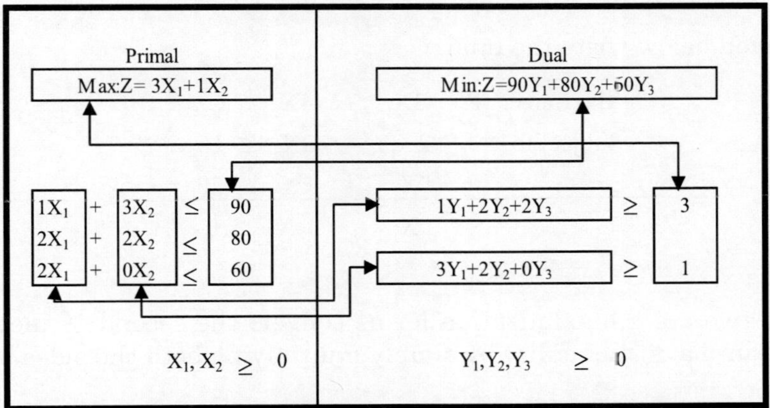

Figure 2.28

Points to remembered

1. The number of variables in the primal is equal to the number of constraints in the dual.
2. The number of constraints in the primal is equal to the number of variables in the dual.
3. The per unit contribution coefficients in the primal becomes the right hand side constraints of the dual.
4. The primal is a maximization problem then the dual will be a minimization problem and vice versa.
5. The coefficients of the first variable of the primal become the coefficients of the first constraint in the dual.
6. Similarly, the coefficients of the second variable of the primal become the coefficients for the dual's second constraint and so on.

7. If the constraints are less than or equal type then constraints of the dual will have greater than or equal sign.

8. The *i*th constraint in the dual will have equality sign if and only if the *i*th variable in the primal is unconstraint in sign.

9. The *i*th variable in the dual is unconstraint in sign if and only if the *i*th constraint in the primal has equality sign.

10. The dual of the dual is the original problem itself.

■ Example 17

For the primal given below write the dual.

$$\begin{aligned}
\text{Maximize: } R &= 3X_1 + 4X_2 \\
\text{Subject to: } & 2X_1 + 3X_2 \leq 16 \\
& 5X_1 + 2X_2 \geq 20 \\
& X_1 \; ; \; X_2 \geq 0
\end{aligned}$$

Solution

Since the primal given is a maximization let us convert the second $\geq$ inequality of the second constraint into a $\leq$ inequality by simply multiplying both the sides by -1.

$$\begin{aligned}
\text{Maximize: } Z &= 3X_1 + 4X_2 \\
\text{Subject to: } & 2X_1 + 3X_2 \leq 16 \\
& -5X_1 - 2X_2 \leq -20 \\
& X_1 \; ; \; X_2 \geq 0
\end{aligned}$$

If Y_1 and Y_2 are the corresponding dual variables then the associated dual is written as

$$\begin{aligned}
\text{Minimize: } C &= 16Y_1 - 20Y_2 \\
\text{Subject to: } & 2Y_1 - 5Y_2 \geq 3 \\
& 3Y_1 - 2Y_2 \geq 4 \\
& Y_1 \; ; \; Y_2 \geq 0
\end{aligned}$$

■ Example 18

For the primal given below write the dual.

$$
\begin{aligned}
\text{Minimize: } & C = 5X_1 - 6X_2 + 4X_3 \\
\text{Subject to: } & 3X_1 + 4X_2 + 6X_3 \geq 9 \\
& X_1 + 3X_2 + 2X_3 \geq 5 \\
& 7X_1 - 2X_2 - 4X_3 \leq 10 \\
& X_1 - 2X_2 + 4X_3 \geq 4 \\
& 2X_1 + 5X_2 - 3X_3 \geq 3 \\
& X_1 ; X_2 ; X_3 \geq 0
\end{aligned}
$$

Solution

Since the primal given is a minimization let us convert all the $\leq$ inequalities into a $\geq$ inequalities by simply multiplying both the sides by –1. The resulting primal is written as

$$
\begin{aligned}
\text{Minimize: } & C = 5X_1 - 6X_2 + 4X_3 \\
\text{Subject to: } & 3X_1 + 4X_2 + 6X_3 \geq 9 \\
& X_1 + 3X_2 + 2X_3 \geq 5 \\
& -7X_1 + 2X_2 + 4X_3 \geq -10 \\
& X_1 - 2X_2 + 4X_3 \geq 4 \\
& 2X_1 + 5X_2 - 3X_3 \geq 3 \\
& X_1 ; X_2 ; X_3 \geq 0
\end{aligned}
$$

Now the needed dual is written as

$$
\begin{aligned}
\text{Maximize: } & R = 9Y_1 + 5Y_2 - 10Y_3 + 4Y_4 + 3Y_5 \\
\text{Subject to: } & 3Y_1 + Y_2 - 7Y_3 + Y_4 + 2Y_5 \leq 5 \\
& 4Y_1 + 3Y_2 + 2Y_3 - 2Y_4 + 5Y_5 \leq -6 \\
& 6Y_1 + 2Y_2 + Y_3 + 4Y_4 - 3Y_5 \leq 4 \\
& Y_1 ; Y_2 ; Y_3 ; Y_4 ; Y_5 \geq 0
\end{aligned}
$$

■ Example 19

For the primal given below write the dual.

$$
\begin{aligned}
\text{Minimize: } & C = 40X_1 + 120X_2 \\
\text{Subject to: } & X_1 - 2X_2 \leq 8 \\
& 3X_1 + 5X_2 = 90 \\
& 15X_1 + 44X_2 \leq 660 \\
& X_1 ; X_2 \geq 0
\end{aligned}
$$

Solution

Since there is an equality constraint we replace this constraint with one ≤ and another ≥ type first as shown below.

$$\begin{aligned} \text{Minimize: } C = 40X_1 &+ 120X_2 \\ \text{Subject to: } X_1 &- 2X_2 \leq 8 \\ 3X_1 &+ 5X_2 \leq 90 \\ 3X_1 &+ 5X_2 \geq 90 \\ 15X_1 &+ 44X_2 \leq 660 \\ X_1 &;\ X_2 \geq 0 \end{aligned}$$

Further, since the primal given is a minimization let us convert all the ≤ constraints into ≥ constraints by simply multiplying both the sides by –1 as usual.

$$\begin{aligned} \text{Minimize: } C = 40X_1 &+ 120X_2 \\ \text{Subject to: } -X_1 &+ 2X_2 \geq -8 \\ 3X_1 &+ 5X_2 \geq 90 \\ -3X_1 &- 5X_2 \geq -90 \\ -15X_1 &- 44X_2 \geq -660 \\ X_1 &;\ X_2 \geq 0 \end{aligned}$$

Now the needed dual is written as

$$\begin{aligned} \text{Maximize: } R = -8Y_1 &+ 90Y_2 - 90Y_3 - 660Y_4 \\ \text{Subject to: } -Y_1 &+ 3Y_2 - 3Y_3 - 15Y_4 \leq 40 \\ 2Y_1 &+ 5Y_2 - 5Y_3 - 44Y_4 \leq 120 \\ Y_1 &;\ Y_2;\ Y_3;\ Y_4 \geq 0 \end{aligned}$$

If $Y_5 = (Y_2 - Y_3)$ then the above problem reduces to

$$\begin{aligned} \text{Maximize: } R = -8Y_1 &+ 90Y_5 - 660Y_4 \\ \text{Subject to: } -Y_1 &+ 3Y_5 - 15Y_4 \leq 40 \\ 2Y_1 &+ 5Y_5 - 44Y_4 \leq 120 \\ Y_1 &;\ ;\ Y_4 \geq 0 \\ Y_5 &\quad \text{Unrestricted in sign} \end{aligned}$$

Example 20

For the primal given below write the dual.

$$\begin{array}{llllll} \text{Maximize:} & R = 3X_1 + & 5X_2 + & 7X_3 & \\ \text{Subject to:} & X_1 + & X_2 + & 3X_3 & \leq 10 \\ & 4X_1 - & X_2 + & 2X_3 & \geq 15 \\ & X_1 ; & X_2 & & \geq 0 \\ & & & X_3 & \text{Unrestricted in sign} \end{array}$$

Solution

Since X_3 is unrestricted we replace it by the difference between two non-negative variables $X_4 - X_5$. So after replacement the primal problem is rewritten as

$$\begin{array}{lllll} \text{Maximize:} & R = 3X_1 + & 5X_2 + & 7(X_4 - X_5) & \\ \text{Subject to:} & X_1 + & X_2 + & 3(X_4 - X_5) & \leq 10 \\ & 4X_1 - & X_2 + & 2(X_4 - X_5) & \geq 15 \\ & X_1 ; & X_2 ; & X_4 ; X_5 & \geq 0 \end{array}$$

Since the problem is maximization, the second constraint is to be multiplied by –1.

$$\begin{array}{llllll} \text{Maximize:} & R = 3X_1 + & 5X_2 + & 7X_4 + & 7X_5 & \\ \text{Subject to:} & X_1 + & X_2 + & 3X_4 + & X_5 & \leq 10 \\ & -4X_1 + & X_2 - & 2X_4 + & X_5 & \leq 15 \\ & X_1 ; & X_2 ; & X_4 ; & X_5 & \geq 0 \end{array}$$

Now the needed dual is written as

$$\begin{array}{llll} \text{Minimize:} & C = 10Y_1 - & 15Y_2 & \\ \text{Subject to:} & Y_1 - & 4Y_2 & \geq 3 \\ & Y_1 + & Y_2 & \geq 5 \\ & 3Y_1 - & 2Y_2 & \geq 7 \\ & -3Y_1 + & 2Y_2 & \geq -7 \\ & Y_1 ; & Y_2 & \geq 0 \end{array}$$

2.13 APPLICATIONS OF LINEAR PROGRAMMING

The linear programming is widely used in the following areas.

1. **Defence related problems**
    a) Transportation problems
    b) Optimum weaponry system
    c) Optimum level of force development
2. **Finance related problems**
    a) Profit planning
    b) Investment policy for maximum return
    c) Investment risk analysis
    d) Auditing
3. **Marketing related problem**
    a) Travelling salesman problem
    b) Plant locations
    c) Media selection
4. **Personal problems**
    a) Determination of optimum organizational level
    b) Job evaluation and allocations
    c) Salary criteria
5. **Production related problems**
    a) Production mix
    b) Production planning
    c) Assembly line balancing
    d) Crude oil refinery
    e) Paper trimming
    f) Agriculture
    g) Hospital scheduling

2.14 SOME USEFUL CONCEPTS IN LINEAR PROGRAMMING

In order to make the concepts more illustrative let us bring a known problem once again. Let us reconsider example discussed earlier in this section.

■ Example 21

$$\text{Maximize: } R = 3X_1 + 1X_2$$

$$\begin{aligned} \text{Subject to:} \quad & 1X_1 + 3X_2 \le 90 \quad \text{(Constraint X)} \\ & 2X_1 + 2X_2 \le 80 \quad \text{(Constraint Y)} \\ & 2X_1 + 0X_2 \le 60 \quad \text{(Constraint Z)} \\ & X_1;\ X_2 \ge 0 \end{aligned}$$

Graphical Solution

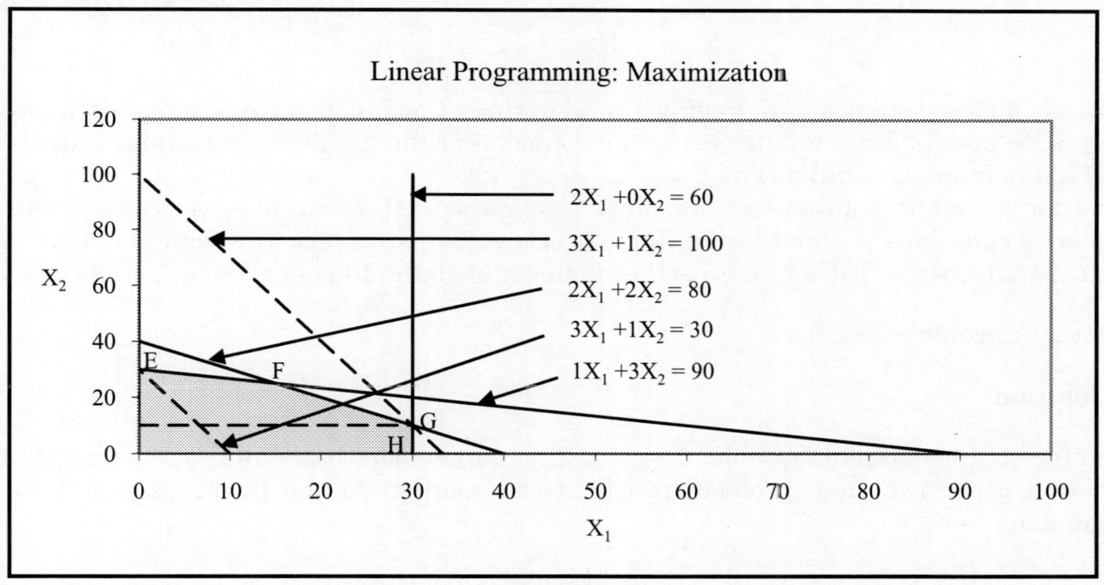

Figure 2.29

Table 2.45

Corner points	X_1	X_2	$R = 3X_1 + 1X_2$
O	0	0	0
E	0	30	30
F	15	25	70
G	30	10	100
H	30	0	90

The highest revenue yielding corner solution is at G and the associated maximum revenue is Rs. 100.

Inclusion of Dummy Variables

Before we proceed for further discussion let us put this problem in the standard form by adding three slacks one each and convert the inequalities into equalities as shown hereunder. These dummies are really the unutilized component of the respective resource. Since these are the leftover inputs (not converted into outputs) they do not add anything to the revenue. Thus in our objective function these variables are included with zero coefficients. Also remember these dummies are either equal to zero (inputs are used fully) or positive (under utilization) and will never be negative.

$$
\begin{array}{llllllll}
\text{Maximize:} & R = 3X_1 & + 1X_2 & + 0S_1 & + 0S_2 & + 0S_3 & & \\
\text{Subject to:} & 1X_1 & + 3X_2 & + S_1 & & & = 90 \\
 & 2X_1 & + 2X_2 & + & S_2 & & = 80 \\
 & 2X_1 & + 0X_2 & + & & S_3 & = 60 \\
 & X_1\ ; & X_2\ ; & S_1\ ; & S_2\ ; & S_3 & \geq 0
\end{array}
$$

In the above system we are having a total of three linear constraints in five variables X_1, X_2, S_1, S_2 and S_3. For a unique solution we know that the number of equations must be equal to the number of unknowns.

So for a unique solution set we must assume at last 2 variables as zero so that number of equations is equal to the number of unknowns. In fact, this could be done in ${}^5C_2 = 10$ possible ways. Table 2.46 gives the summary of all the 10 possible solution sets.

Various Concepts

1. Solution

Values of the decision variables X_j ($j = 1, 2, \ldots, n$) which satisfy all the constraints of a linear programming problem is called the solution to the linear programming problem.

2. Feasible solution

A feasible solution to a given linear programming is an 'n+m' dimensional vector satisfying the constraints and non-negativity conditions given in the problem. In our two dimensional graphical illustration, all points well within the shaded region including the line segments constitute the feasible solution set.

3. Basic solution

A basic solution is a solution obtained by solving equality constraints after setting (n+m) – m variables into zero. In our illustrative example there are n structural (2) and m slack (3) variables. All together there are m+n = 3+2 = 5 variables in our illustration. The basic solutions are obtained by solving all the three constraints after putting any two of variables to zero. Table 2.46 summarizes all possible $10 = {}^5C_2$ basic solutions.

Table 2.46

Sl. no	Corner points	Equations	Basic solutions
1	O	$+ S_1 = 90$ $+ S_2 = 80$ $+ S_3 = 60$	$X_1 = 0$ $S_1 = 90$ $S_2 = 80$ $S_3 = 60$ $X_2 = 0$
2	E	$+ 3X_2 = 90$ $+ 2X_2 + S_2 = 80$ $+ 0X_2 + S_3 = 60$	$X_1 = 0$ $X_2 = 30$ $S_2 = 20$ $S_3 = 60$ $S_1 = 0$
3	-	$+ 3X_2 + S_1 = 90$ $+ 2X_2 = 80$ $+ 0X_2 + S_3 = 60$	$X_1 = 0$ $S_1 = -30$ $X_2 = 40$ $S_3 = 60$ $S_2 = 0$
4	-	$+ 3X_2 + S_1 = 90$ $+ 2X_2 + S_2 = 80$ $+ 0X_2 = 60$	$X_1 = 0$ $S_1 = -\infty$ $S_2 = -\infty$ $X_2 = \infty$ $S_3 = 0$
5	-	$1X_1 = 90$ $2X_1 + S_2 = 80$ $2X_1 + S_3 = 60$	$X_2 = 0$ $X_1 = 90$ $S_2 = -100$ $S_3 = -120$ $S_1 = 0$

(Contd.)

Sl. no	Corner points	Equations	Basic solutions
6	-	$1X_1 + S_1 = 90$ $2X_1 = 80$ $2X_1 + S_3 = 60$	$X_2 = 0$ $S_1 = 50$ $X_1 = 40$ $S_3 = -20$ $S_2 = 0$
7	H	$1X_1 + S_1 = 90$ $2X_1 + S_2 = 80$ $2X_1 = 60$	$X_2 = 0$ $S_1 = 60$ $S_2 = 20$ $X_1 = 30$ $S_3 = 0$
8	F	$1X_1 + 3X_2 = 90$ $2X_1 + 2X_2 = 80$ $2X_1 + 0X_2 + S_3 = 60$	$S_1 = 0$ $X_1 = 15$ $X_2 = 25$ $S_3 = 30$ $S_2 = 0$
9	-	$1X_1 + 3X_2 = 90$ $2X_1 + 2X_2 + S_2 = 80$ $2X_1 + 0X_2 = 60$	$S_1 = 0$ $X_2 = 20$ $S_2 = -20$ $X_1 = 30$ $S_3 = 0$
10	G	$1X_1 + 3X_2 + S_3 = 90$ $2X_1 + 2X_2 = 80$ $2X_1 + 0X_2 = 60$	$S_2 = 0$ $S_3 = 30$ $X_2 = 10$ $X_1 = 30$ $S_3 = 0$

4. Basic and non-basic variables

The variables which are not set to zero while solving the constraints are called basic variables. The variables which are set to zero while solving are called non-basic variables. For example, in solution 1, S_1, S_2 and S_3 are the basic variables and the zero set X_1 and X_2 are non-basic variables.

5. Basic feasible solution

Among the basic solutions (10) the sub-set of (5) solutions that satisfy the non-negativity constraints as well in addition are called basic feasible solutions. In the Table 2.46 only five (shaded) of them satisfy the non-negativity conditions and qualify for the basic feasible solution set. Interestingly, these five basic feasible solutions are nothing but the five corner points O, E, F, G, H of our convex feasible region in our diagrammatic representation shown.

6. Optimum feasible solution

The optimum feasible solution is one among the basic feasible solutions that optimizes the given objective function. From the table given above the highest revenue yielding corner (F) gives the final optimum solution. i.e., $X_1 = 30$, $X_2 = 10$, $S_1 = 30$ and $R=100$.

7. Degenerate solution

Degeneracy in a linear problem occurs, when one or more of the basic variables are zero in the final optimum solution. Normally the number of non-zero basic variables in the final optimum solution is the same as the number of constraints. But when the number of positive variables in the solution set is less than the number of constraints, the degeneracy occurs. In our illustrative example 21 the solution does not degenerate because the number of constraints namely 3 equal to the number of positive basic variables namely 3 in all highlighted basic feasible solution sets. So to illustrate the degeneracy we consider another example 22.

■ Example 22

Solve the following linear programming problem

$$\begin{aligned} \text{Maximize: } & R = 28X_1 + 30X_2 \\ \text{Subject to: } & 6X_1 + 3X_2 \le 18 \\ & 3X_1 + X_2 \le 8 \\ & 4X_1 + 5X_2 \le 30 \\ & X_1 \; ; \; X_2 \ge 0 \end{aligned}$$

Solution

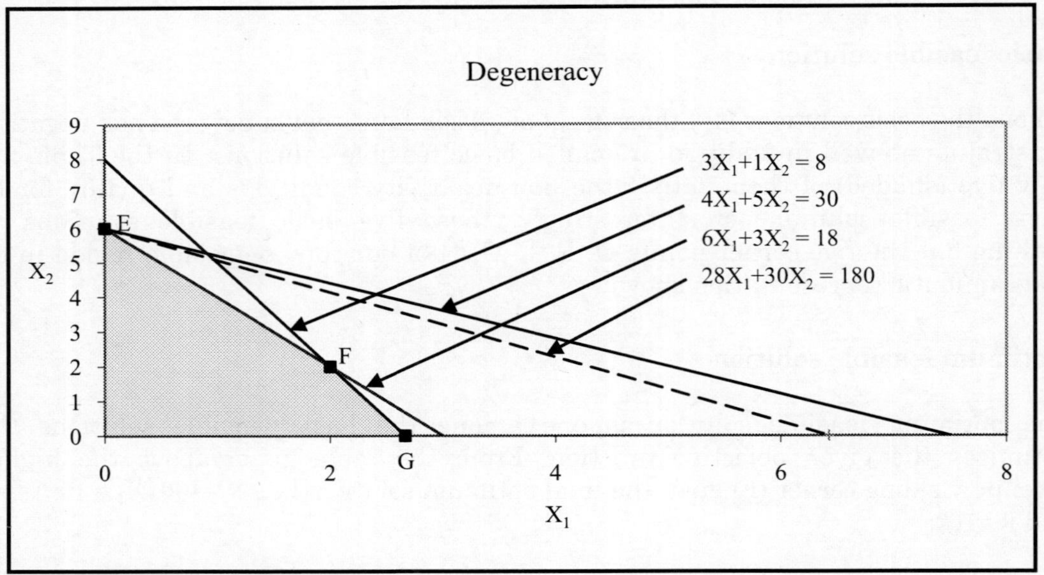

Figure 2.30

The values at the corner points are evaluated and given in Table 2.47.

Table 2.47

Point	X_1	X_2	$R = 28X_1 + 30X_2$
0	0	0	0
E	0	6	180
F	2	2	116
G	2.6667	0	74.6667

At the final solution point E, constraints 1 and 3 intersect. So both these resources are fully utilized. So the associated slacks S_1 and S_3 invariably become zero. In addition since this point is on X_2 axis X_1 is also zero. Further E is just below the constraint 2 the associated slack S_2 is positive. Since it is a three constraint two decision variable model altogether there are 5 variables in the standard form. At the solution point E, the basic variables are $X_1 = 0$, $X_2 = 6$, $S_2 = 2$. Since one of the basis variables is zero, the solution degenerate in the sense that simplex table (refer next section) is found to be repetitive and hence becomes cyclic thereafter during further iterations.

EXERCISES

1. The following are the input requirements per unit of a certain product by three processes A, B, and C

Process	Input requirements per unit of output		
	K	L	Q
A	3	1	1
B	2	2	1
C	1	3	1

   a) Draw a production diagram for all the three processes.

   b) If 100 units of labour and 50 units of capital are available find out the level of output when process A alone is used.

   c) Process B alone is used.

   d) Process C alone is used.

   e) Which is more efficient among the three?

   f) Also state the more efficient product mix process and calculate the optimum level of outputs.

   g) Formulate a suitable linear programming problem and obtain the optimum values by using diagrammatic method.

2. Using a single production process a company produces two varieties of products A and B. The said production process has 45,000 hours of processing time for the use. It takes 5 hours to produce one unit of A and 3 hours to produce the product B. The market survey reveals that 7000 units of A and 10,000 units of B can be sold. If the price per unit of A is Rs. 600 and that of B is Rs. 400. Formulate a linear programming problem and obtain its solution to maximize the total revenue.

3. A steel rolling mill produces two types of steel A and B by using its full capacity of 80 hours of processing time per week. It takes 20 minutes to produce one unit of A and 15 minutes to produce one unit of B. In addition to machine time, per unit of A consumes 3 kg of raw material and per unit B consumes 3 kg. The raw material availability is 12,000 kg per week. On both the accounts the product A costs Rs. 10 and sold at Rs. 15 per unit. Similarly, the product B costs Rs. 15 and sold at Rs. 20 per unit. Formulate this into a linear programming problem and obtain the corresponding solution that accounts for the maximum profit.

4. A company manufactures two products A and B by using two production processes. Given the following data, formulate the linear programming problem and obtain the profit maximizing output of levels of both the products.

Description	Product A	Product B
Machine time 1 used per unit	3	2
Machine time 2 used per unit	4	6
Selling price per unit	200	300
Machine cost per unit	100	150
Raw material cost per unit	50	50

5. A bakery produces two types of biscuit A and B and sells them at Rs. 10 and Rs. 20 per pack respectively. The manufacturing involves three processing units. Given the following data formulate a linear programming problem.

Description	Blending	Baking	Packing
Biscuit A	3	2	150
Biscuit B	4	6	200
Full capacity per day	10 hours	20 hours	8 hours

6. The sky line company agrees two charter planes for a tourist group. The group needs at least 160 first class and 300 economy class seats. The company has two types of model. Model 234 has 20 first class and 30 economy class seats. The model 456 has 30 first class and 60 economy class seats. Each flight of 234 costs Rs. 1,00,000 and 456 costs Rs. 1,50,000. How many each of the type should be used to minimize the operational cost? Draft a linear programming problem.

7. The manager of a milk diary in Bangalore wants to find the correct blend of two types of feed. Both contain various percentages of four essential nutrients as follows. Formulate a correct linear programming. Also formulate its dual.

Ingredient	% per kg		Minimum requirement
	Feed 1	Feed 2	
1	40	20	4
2	10	30	2
3	20	40	3
4	30	10	6
Cost per kg	5	3.5	

8. Solve the following linear programming problem by Graphical Method. Also formulate its dual.

$$\text{Maximize: } R = 45X_1 + 80X_2$$

$$\text{Subject to: } 5X_1 + 20X_2 \le 400$$

$$10X_1 + 15X_2 \le 450$$

$$X_1 \; ; \; X_2 \ge 0$$

9. Solve the following linear programming problem by Graphical Method. Also formulate its dual.

$$\begin{aligned} \text{Minimize: } & C = 100X_1 + 200X_2 \\ \text{Subject to: } & 1X_1 + 1X_2 \leq 40 \\ & 3X_1 + 1X_2 \leq 60 \\ & 1X_1 + 6X_2 \leq 60 \\ & X_1 ; X_2 \geq 0 \end{aligned}$$

10. Solve the following linear programming problem by Graphical Method. Also formulate its dual.

$$\begin{aligned} \text{Minimize: } & C = 1X_1 + 3X_2 \\ \text{Subject to: } & 1X_1 + 3X_2 \leq 90 \\ & 2X_1 + 2X_2 \leq 80 \\ & 2X_1 + 0X_2 \leq 60 \\ & X_1 ; X_2 \geq 0 \end{aligned}$$

11. Solve the following linear programming problem by Graphical Method. Also formulate its dual.

$$\begin{aligned} \text{Minimize: } & C = 100X_1 + 200X_2 \\ \text{Subject to: } & 1X_1 + 1X_2 \leq 40 \\ & 2X_1 + 4X_2 \leq 70 \\ & 1X_1 + 6X_2 \leq 60 \\ & X_1 ; X_2 \geq 0 \end{aligned}$$

12. Solve the following linear programming problem by Graphical Method. Also formulate its dual.

$$\begin{aligned} \text{Minimize: } & C = 2X_1 + 1X_2 \\ \text{Subject to: } & 3X_1 + 1X_2 \geq 3 \\ & 4X_1 + 3X_2 \geq 6 \\ & 1X_1 + 2X_2 \geq 2 \\ & X_1 ; X_2 \geq 0 \end{aligned}$$

CHAPTER

3

Linear Programming: II Simplex Method

- *Simplex algorithm*
- *Maximization and Minimization problems with all constraints ≤ type*
- *Use of excel spreadsheets for simplex calculations*
- *Use of artificial variables in linear programming with ≥ constraints*
- *Minimization problems with all constraints ≥ type*
- *Maximization and Minimization problems with all types of inequality*
- *Some special types of Problem*
- *Integrating primal and dual optimum solutions*
- *Built-in excel spreadsheet solution to the linear programming problem*
- *Some important linear programming theorems*
- *Formation of general linear programming primal and Dual problem in matrix form*
- *Standard form of a linear programming problem in matrix form*
- *Exercises*

3.1 SIMPLEX ALGORITHM

The two-dimensional graphical method outlined in the previous section is useful only in simple cases with two decision variables, measured one on the X-axis and another on the Y-axis. Whenever we come across more than two decision variables, we use a specialized tool often called simplex algorithm. The computational procedure of the simplex method is based on the following important property already highlighted in the graphical solution.

The optimum solution to a linear programming problem, if it exists, occurs at one or more corner points of the convex feasible region.

In the light of this important property simplex procedure provides us a step-by-step search process towards the final optimum solution. The search process conveniently starts from the very origin with a known initial solution involving only the slack variables. For iterative convenience the original data is arranged in a tabular form commonly known as simplex tableau. At every stage of iteration the so obtained feasible solution is checked for its optimality.

If the optimality test is satisfied then iteration gets terminated and final solution is read from the resulting table. On the other hand if the optimality test fails we proceed further and further till we get the optimum solution by repeating the iterative procedure. Convergent iterative computational process is the secret in getting the needed solution. The following example 1 illustrates this method step-by-step.

Slack Variables

Before we proceed to the simplex procedure it is important to formulate the problem in its standard form by introducing appropriate slack variables to all the constraints to bring in equalities in action. For example, if it is a *less than or equal to* type of inequality then we add a suitable positive value to the left side and convert the inequality into an equality. The added variable is often called slack variable. In economic sense it represents the amount of unutilized resource of the concerned constraint.

Since unused slacks (inputs) add no value to the revenue, we include the slacks in the revenue function as well with 0 coefficients. In fact wherever the input variables are binding, the added slack will have zero values in their respective constraints. Only the non-binding input variables will have the non-zero slack values in the formation. However, before solving the problem, we will not know the constraints which deserve really slack addition, it is a common practice to use slacks for all the given constraints and convert them all into equalities first. The following example 1 illustrates the method of converting the given LPP into its standard form by incorporating slacks.

Normally we use the simplex method for problems involving more than two decision variables. However, in this illustrative section we prefer to use a known two decision variable problem for which the graphical solution is already available. Such a choice will facilitate us for step-by-step comparison of simplex method with the corresponding diagrammatic representation.

3.2 MAXIMIZATION PROBLEMS WITH ALL CONSTRAINTS ≤ TYPE

■ Example 1

A manufacturing company is producing two products 1 and 2 by using two machines A and B. The product 1 needs 5 hours of machine time A and 10 hours of machine time B per unit of output. Similarly, product 2 requires 20 hours of machine time A and 15 hours of machine time B per unit of output. The total machine times available with the company are 400 and 450 hours respectively. The company makes the revenue of Rs. 45 per unit of product 1 and Rs. 80 per unit of product 2 in the market. Formulate a suitable linear programming problem and obtain the suitable product mix that maximizes the total revenue by using simplex method.

Solution

Let us first summarize the given data in a tabular form to facilitate easy accessibility.

Table 3.1

Machines	Machine hours per unit of output		Machine time available
	1	2	
A	5	20	400
B	10	15	450
Revenue per unit	Rs. 45	Rs. 80	

If X_1 and X_2 are the amounts of product 1 and 2 produced then, from the data given we formulate the linear programming problem in its usual form as shown below. With the information given, the linear programming problem is formulated and reported below.

$$\begin{aligned} \text{Maximize: } & 45X_1 + 80X_2 \\ \text{Subject to: } & 5X_1 + 20X_2 \leq 400 \text{ (Constraint A)} \\ & 10X_1 + 15X_2 \leq 450 \text{ (Constraint B)} \\ & X_1 \; ; \; X_2 \geq 0 \end{aligned}$$

Graphical Solution

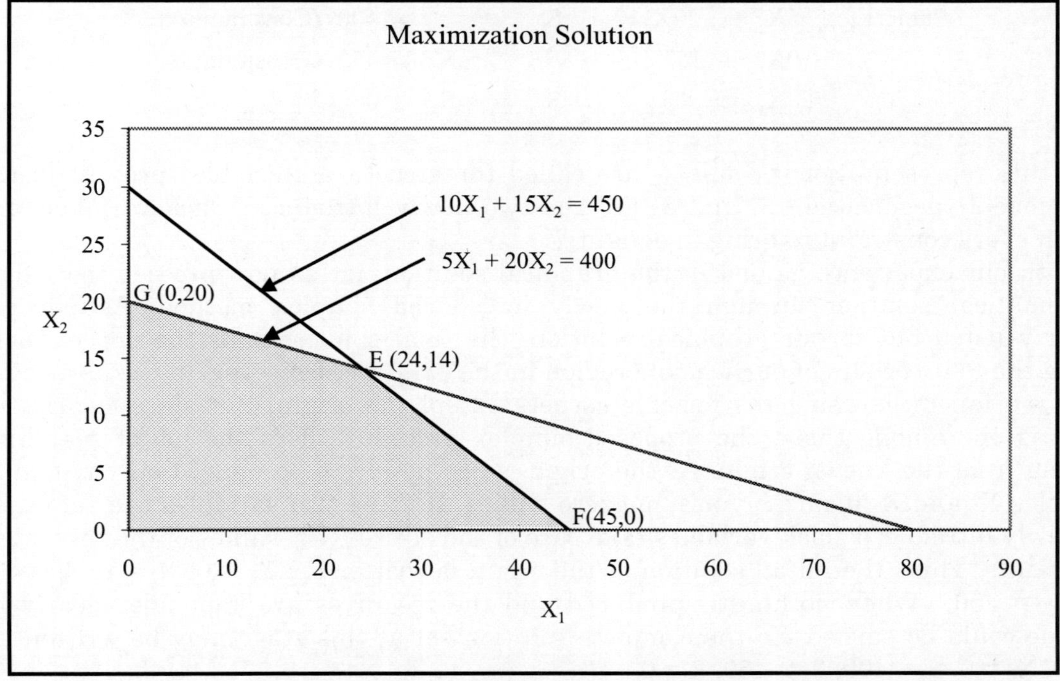

Figure 3.1

Now the needed maximum revenue yielding outputs are obtained by substituting the coordinates of the four corners of our feasible region as shown in Table 3.2.

Table 3.2

Corner points	X_1	X_2	$R = 45X_1 + 80X_2$
O	0	0	0
G	0	20	1600
E	24	14	2200
F	45	0	2025

So the solution values are: $X_1 = 24$; $X_2 = 14$; R = Rs. 2200.

Simplex Solution

Now in the given problem we are having two ≤ constraints. As a first step let us add two slack variables S_1 and S_2, one each, into the left sides of the respective constraints and bring in equalities. Since these slacks take care of the unutilized resource, constraint wise, it adds no revenue to the producer. So in the objective function we include these two slacks with zero coefficients as shown below.

$$\begin{aligned}
\text{Maximize: } & 45X_1 + 80X_2 + 0S_1 + 0S_2 \\
\text{Subject to: } & 5X_1 + 20X_2 + S_1 = 400 \quad \text{(Constraint A)} \\
& 10X_1 + 15X_2 + S_2 = 450 \quad \text{(Constraint B)} \\
& X_1 ;\ X_2 ;\ S_1 ;\ S_2 \geq 0
\end{aligned}$$

In this representation X_1 and X_2 are called the structural variables present in the given model specification. S_1 and S_2 are the two newly introduced slack variables one each in every constraint to bring in equality.

With our experience gained in the graphical solution, let us now proceed towards a mathematical solution through the newly introduced simplex method. As per the property listed out in our graphical solution, the final solution must be one or more among the four corners of our feasible region in the diagram. However, in the absence of graph we fail in guessing the feasible corners except the origin. But since clearly the origin is one among them, the proposed simplex procedure does the job of searching starting from the known origin. At the origin, since nothing is produced our structural variables X_1 and X_2 invariably assume zero values. If these two variables are zero then the newly introduced slack variables take care of the entire RHS values of the respective constraints. Thus, the initial solution at the origin is written as: $X_1 = 0$, $X_2 = 0$, $S_1 = 400$ and $S_2 = 450$. When nothing is produced and the resources are kept idle, clearly no revenue could be earned. So the complete solution set at this stage may be written as: $X_1 = 0$, $X_2 = 0$, $S_1 = 400$, $S_2 = 450$, $R = 0$.

Now it is time to start our first iteration to improve our revenue from zero to a positive number. In order to facilitate simultaneous handling of all the relevant data set, let us summarize all the relevant data in a compact tabular form often called simplex tableau as shown below in tableau 1.

The highlighted first column is numbered from 1 onwards in order to refer the row entries. Similarly, the highlighted first row is named A, B, etc., to help us in referring the correct column. Remember, this is exactly the same way the computer excel work sheet is named for correct reference.

To begin with, let us name the variables in row 2 of the tableau 1. Since our iteration starts from the origin our choice for basis column is none other than the two slacks S_1 and S_2. So list these variables under the head "Basis" as shown in tableau 1. Now under the column head "C_i" we enter the per unit contribution (revenue) coefficients of these two slack variables entered under the head "Basis". Since slacks command no revenue, below the head C_i we make two zero entries. Now let us move to the right side and fill the heads "X_1 to S_2" in row 2 as shown in the tableau 1. Simply copy the coefficients of the two redefined constraints in the same order in row number 3 and row number 4 as shown in the tableau 1. Also fill the RHS, resource availability data of both the constraints under the head X_b. Here for this head we use "X_b" to represent that these are the solution values of the variables listed in the Basis column. In cell B5 enter the row head as "C_j". In this row we represent coefficients of our redefined objective function corresponding to the variables listed in the top-heading row. Now after having completed all the entries we calculate Z_jth row by using the formula as follows

$$Z_j = \sum_{i=1}^{m} c_i x_{ij} \qquad j = 1, 2, \ldots\ldots\ldots n$$

In particular,

$$Z_1 = \text{Cell C6} = \text{A3} \times \text{C3} + \text{A4} \times \text{C4}$$
$$= 0(5) + 0(10) = 0$$

$$Z_2 = \text{Cell D6} = \text{A3} \times \text{D3} + \text{A4} \times \text{D4}$$
$$= 0(20) + 0(15) = 0$$

$$Z_3 = \text{Cell E6} = \text{A3} \times \text{E3} + \text{A4} \times \text{E4}$$
$$= 0(1) + 0(0) = 0$$

$$Z_4 = \text{Cell F6} = \text{A3} \times \text{F3} + \text{A4} \times \text{F4}$$
$$= 0(0) + 0(1) = 0$$

$$Z_5 = \text{Cell G6} = \text{A3} \times \text{G3} + \text{A4} \times \text{G4}$$
$$= 0(400) + 0(450) = 0$$

The yielded 0 values are entered from cell C6 to G6. These entries in the Z_j row from cell C6 to F6 show the amounts by which the revenue contribution would be reduced per unit addition of the variables listed in the top row 1. The last $C_j - Z_j$ row is simply the subtracted result of Z_j's from the respective C_j's. Thus it represents the net revenue contributions per unit of the respective variables when they are brought into the basis.

The above iterative calculation automatically reports the associated revenue in cell G6. Since we begin our first iteration at the origin, the associated zero revenue calculation is brought out in the cell G6. The "Mini Ratio" heading in cell H2 will be used in the subsequent iteration.

Tables 3.3, 3.4, 3.5

	A	B	C	D	E	F	G	H
1					Tableau 1			
2	C_i	Basis	X_1	X_2	S_1	S_2	X_b	Mini ratio
3	0	S_1	5	20	1	0	400	400/20=20
4	0	S_2	10	15	0	1	450	450/15=30
5		C_j	45	80	0	0		
6		Z_j	0	0	0	0	0	
7		$C_j - Z_j$	45	80	0	0		
8					Tableau 2			
9	C_i	Basis	X_1	X_2	S_1	S_2	X_b	Ratio
10	80	X_2	0.25	1	0.05	0	20	80
11	0	S_2	6.25	0	−0.75	1	150	24
12		C_j	45	80	0	0		
13		Z_j	20	80	4	0	1600	
14		$C_j - Z_j$	25	0	−4	0		
15					Tableau 3			
16	C_i	Basis	X_1	X_2	S_1	S_2	X_b	
17	80	X_2	0	1	0.08	−0.04	14	
18	45	X_1	1	0	−0.12	0.16	24	
19		C_j	45	80	0	0		
20		Z_j	45	80	1	4	2200	
21		$C_j - Z_j$	0	0	−1	−4		

Step 1: In the tableau 1 above the row $C_j - Z_j$ shows the net revenue contribution per unit inclusion of the variable under reference. A look at the row shows that when one unit of X_2 is included in the production line (Basis) it will increase the revenue by Rs. 80. However, the inclusion of one unit of X_1 into the Basis adds only Rs. 45. So the rational producer will prefer to introduce more net revenue yielding output namely X_2 into the Basis first. To reveal this rational choice we shade this box as shown in tableau 1. The associated column is often called the Key column.

Step 2: Since we have already decided to go for X_2 inclusion in step 1, the corresponding input requirements per unit of X_2 are 20 hours of machine time A and 15 hours of machine time B as revealed by the coefficients in the body matrix corresponding to X_2. Having decided for X_2 the level of X_2 that one could produce is obtained as X_2 = Mini (400/20, 450/15) = Mini (20, 30) = 20. This calculation is made available in the last column under the head "Mini Ratio". This choice of going for X_2 is constrained by the availability of the machine time A. For 20 units of output, at the rate of 20 per unit, all the 400 hours of machine time A gets exhausted. At this point it is important to note that if we go for the maximum of 30 units of X_2, though machine time B is available machine time A is not available. The following table shows the current overall position. The levels of slacks at this point of time are also shown in the Table 3.6.

Table 3.6

Demand and supply positions of inputs at F							
	Demand						
Machine	$X_1 = 0$	$X_2 = 20$	Total demand	Supply	Excess	S_1	S_2
A	0 x 5 = 0	20 x 20 = 400	400	400	0	0	–
B	0 x 10 = 0	20 x 15 = 300	300	450	150	–	150
Profit = $P_1X_1 + P_2X_2$ = 45 x 0 + 80 x 20 = 1600							

To show this rational choice we shade minimum value box as shown in the tableau 1. This row is often referred as the Key row. The entry corresponding to the meeting point of these shaded column and shaded row, namely 20 is often called the pivot element. We shade this box again to highlight the pivot element. After having decided to produce X_2, the slack that is to be replaced is none other than the one in the pivotal row, namely S_1. Now it is time to prepare an empty tableau 2 with column and row headings intact for our next iteration.

In this new tableau 2, replace the slack S_1 by structural variable X_2 to bring the new situation in action. Also replace the net contribution coefficient 0 corresponding to slack S_1 by 80, the incoming contribution coefficient of X_2 in column C_j. However, the contribution coefficient for S_2 should be restored at the same old zero level in tableau 2.

To incorporate this new situation in tableau 2 we divide the entire row 2 in tableau 1 by the pivot element, namely 20, except the last "Mini ratio" column and report them all in row 10 in tableau 2. The very purpose of doing this is to make the pivot element as unity in tableau 2. The following are the calculated results for the row 10 in tableau 2.

Cell C10 = C3/D3 = 5/20 = 0.25

Cell D10 = D3/D3 = 20/20 =1

Cell E10 = E3/D3 = 1/20 = 0.05

Cell F10 = F3/D3 = 0/20 = 0

Cell G10 = G3/D3 = 400/20 = 20

Next let us make the cell just below the pivot element, namely, D11 in tableau 2 to zero by appropriate elementary row operation. To do this we multiply row 10 in tableau 2 by 15 (cell value D4 in tableau 1) and subtract it from row 4 in tableau 1. The results are entered in row 11 in tableau 2.

Conveniently one could use the following ready-made formula as well to get all the elements of such rows.

$$\begin{bmatrix}\text{Former element in the} \\ \text{remaining row} \\ \text{in tableau 1}\end{bmatrix} - \begin{bmatrix}\text{Former intersectional} \\ \text{element of the} \\ \text{remaining row in} \\ \text{tableau 1}\end{bmatrix} \times \begin{bmatrix}\text{New corresponding} \\ \text{element in the} \\ \text{replaced row in} \\ \text{tableau 2}\end{bmatrix}$$

The resulting calculations are shown below

Cell C11 = C4 – D4 × C10 = $10 - 15 \times 0.25 = 6.25$

Cell D11 = D4 – D4 × D10 = $15 - 15 \times 1 = 0$

Cell E11 = E4 – D4 × E10 = $0 - 15 \times 0.05 = -0.75$

Cell F11 = F4 – D4 × F10 = $1 - 15 \times 0 = 1$

Cell G11 = G4 – D4 × G10 = $450 - 15 \times 20 = 150$

Step 3: After having completed row 11 in tableau 2 as described above complete the row 12 entries by copying the corresponding entries from tableau 1.

Step 4: After having completed step 3 we recalculate Z_jth row once again by using the formula.

$$Z_j = \sum_{i=1}^{m} c_i x_{ij} \qquad j = 1, 2, \ldots\ldots\ldots n$$

In particular,

Z_1 = Cell C13 = A10 × C10 + A11 × C11

$= 80(0.25) + 0(6.25) = 20$

Z_2 = Cell D13 = A10 × D10 + A11 × D11

$= 80(1) + 0(0) = 80$

Z_3 = Cell E13 = A10 × E10 + A11 × E11

$= 80(0.05) + 0(-0.75) = 4$

Z_4 = Cell F13 = A10 × F10 + A11 × F11

$= 80(0) + 0(0) = 0$

Z_5 = Cell G13 = A10 × G10 + A11 × G11

$= 80(20) + 0(150) = 1600$

Note that at this stage the optimum revenue is nothing but the G13 cell entry namely 1600.

Step 5: After having completed row 13 in tableau 2 we calculate $Z_j - C_j$ and report it in row 14 in tableau 2.

The calculations are listed below

Cell C14 $= C12 - C13 = 45 - 20 = 25$

Cell D14 $= D12 - D13 = 80 - 80 = 0$

Cell E14 $= E12 - E13 = 0 - 4 = -4$

Cell F14 $= F12 - F13 = 0 - 0 = 0$

At this stage the output levels are reported in column below X_b in tableau 2. The solution values are: $X_2 = 20$, $S_2 = 150$, $Z = 1600$. Now at this stage we moved from origin O to G in our corresponding diagrammatic exposition.

The left over B machine time given by the slack $S_2 = 150$ in tableau 2 at this stage naturally induce the producer to go for further production and improve the revenue further. But this time in the second tableau 2, the index row $C_j - Z_j$ reveals that there is a potential for going for X_1 with a net reward of Rs. 25 per unit inclusion. Now it is time to prepare an empty tableau 3 with row and column headings intact as usual.

From here onwards, our next iteration starts. Having the tableau 2 as reference we will repeat all the steps that we have underwent till this time once again. In tableau 2 the pivot element is 6.25. So our new incoming variable is X_1 and the associated outgoing variable is S_2.

Accordingly, we replace S_2 by X_1 in the basis column in tableau 3. In C_i column also we change the corresponding per unit contribution as 45 in the place of zero. Now to make the pivot element unity in tableau 3 the relevant row operation is to divide the entire row 11 in tableau 2 by 6.25. The cell-wise calculations are shown below and reported in row number 18 in tableau 3.

Row 18 calculations

Cell C18 $= C11 / C11 = 6.25 / 6.25 = 1$

Cell D18 $= D11 / C11 = 0 / 6.25 = 0$

Cell E18 $= E11 / C11 = -0.75 / 6.25 = -0.12$

Cell F18 $= F11 / C11 = 1 / 6.25 = 0.16$

Cell G18 $= G11 / C11 = 150 / 6.25 = 24$

Once the entries for row 18 are entered, as usual to make the element just above unity in column 17 in tableau 3 to zero, the relevant row operation is to multiply the unity element in row 18 in tableau 3 by 0.25 (cell C10 value in tableau 2) and subtract the result from the respective elements in row 10 in tableau 2. The calculation details are shown below and reported in row 17 in tableau 3.

Row 17 calculations

$$\text{Cell C17} = \text{C10} - \text{C10} \times \text{C18} = 0.25 - 0.25 \times 1 = 0$$

$$\text{Cell D17} = \text{D10} - \text{C10} \times \text{D18} = 1 - 0.25 \times 0 = 1$$

$$\text{Cell E17} = \text{E10} - \text{C10} \times \text{E18} = 0.05 - (0.25 \times -0.12) = 0.08$$

$$\text{Cell F17} = \text{F10} - \text{C10} \times \text{F18} = 0 - (0.25 \times 0.16) = -0.04$$

$$\text{Cell G17} = \text{G10} - \text{C10} \times \text{G18} = 20 - (0.25 \times 24) = 14$$

Once this is done we copy the C_j as usual from the previous table. Now let us calculate the Z_j for row 20 as usual.

Row 20 calculations

$$\text{Cell C20} = \text{A17} \times \text{C17} + \text{A18} + \text{C18} = 80 \times 0 + 45 \times 1 = 45$$

$$\text{Cell D20} = \text{A17} \times \text{D17} + \text{A18} + \text{D18} = 80 \times 1 + 45 \times 0 = 80$$

$$\text{Cell E20} = \text{A17} \times \text{E17} + \text{A18} + \text{E18} = 80 \times 0.08 + 45 \times (-0.12) = 1$$

$$\text{Cell F20} = \text{A17} \times \text{F17} + \text{A18} + \text{F18} = 80 \times (-0.04) + 45 \times 0.16 = 4$$

$$\text{Cell G20} = \text{A17} \times \text{G17} + \text{A18} + \text{G18} = 80 \times 14 + 45 \times 24 = 2200$$

After completing this step let us calculate $C_j - Z_j$ and report the results in row 21 in tableau 3. In the tableau 3, the $C_j - Z_j$ row, all the entries are either zero or negative. Hence no further improvement in revenue is possible.

The final iterative solution values are reported in cell G17 and G18. So $X_1 = 24$; $X_2 = 14$ and as per cell reference G20 the optimum revenue = Rs. 2200. Now we successfully moved from G to E in the diagram. Remember, that our diagrammatic optimum solution is the same as that of the simplex optimum solution.

■ Example 2

a) For the data given below formulate a linear programming problem and obtain the maximum revenue by simplex method.

b) Also solve the above problem graphically. From the objective function corresponding to each iteration, draw suitable inferences.

Table 3.7

Inputs	Inputs per unit of output		Inputs available
	1	2	
L	1	2	70
K	4	2	100
Profit per unit	Rs. 10	Rs. 15	

Solution a)

As usual let X_1 and X_2 be the amounts of product 1 and 2 produced by the producer to obtain the optimum revenue. From the data given we formulate the linear programming problem first as shown below.

$$\begin{aligned} \text{Maximize: } & 10X_1 + 15X_2 \\ \text{Subject to: } & 1X_1 + 2X_2 \leq 70 \text{ (Labour)} \\ & 4X_1 + 2X_2 \leq 100 \text{ (Capital)} \\ & X_1 , \quad X_2 \geq 0 \end{aligned}$$

Now since we are having two *less than or equal to* type constraints, we introduce two slacks S_1 and S_2 to the respective constants and bringing in equalities and formulate the linear programming problem as shown below.

$$\begin{aligned} \text{Maximize: } & 10X_1 + 15X_2 + 0S_1 + 0S_2 \\ \text{Subject to: } & 1X_1 + 2X_2 + S_1 = 70 \quad \text{(Labour)} \\ & 4X_1 + 2X_2 + S_2 = 100 \quad \text{(Capital)} \\ & X_1 \; ; \; X_2 \; ; \; S_1 \; ; \; S_2 \geq 0 \end{aligned}$$

As a first step let us transfer the data into the simplex tableau 1 as shown below and mark the key column, key row and pivot element by shading.

Tables 3.8, 3.9, 3.10

	A	B	C	D	E	F	G	H
1					Tableau 1			
2	C_i	Basis	X_1	X_2	S_1	S_2	X_b	Mini ratio
3	0	S_1	1	2	1	0	70	35
4	0	S_2	4	2	0	1	100	50
5		C_j	10	15	0	0		
6		Z_j	0	0	0	0	0	
7		$C_j - Z_j$	10	15	0	0		
8					Tableau 2			
9	C_j	Basis	X_1	X_2	S_3	S_4	X_b	Mini ratio
10	15	X_2	0.5	1	0.5	0	35	70
11	0	S_2	3	0	−1	1	30	10
12		C_j	10	15	0	0		
13		Z_j	7.5	15	7.5	0	525	
14		$C_j - Z_j$	2.5	0	−7.5	0		
15					Tableau 3			
16	C_j	Basis	X_1	X_2	S_3	S_4	X_b	Mini ratio
17	15	X_2	0	1	0.667	−0.16667	30	
18	10	X_1	1	0	−0.33	0.333333	10	
19		C_j	10	15	0	0		
20		Z_j	10	15	6.667	0.833333	550	
21		$C_j - Z_j$	0	0	−6.67	−0.83333		

Since 35 is the minimum ratio in the minimum ratio column and 15 is the highest value in the $C_j - Z_j^{th}$ row the intersection element 2 is our pivot element in tableau 1.

Now it is time to prepare an empty tableau 2 with row and column headings intact. To make this pivot entry 15 as unity in tableau 2 we divide the entire row 3 in tableau 1 by 3, except the last entry. The so obtained results are reported in row 10 in tableau 2.

The next step is to make the left over pivot column element cell (C11 value) just below the pivot element in tableau 2 as zero. In our illustration this element is 2 in the row 4 of tableau 1. To do this, the needed row operation is to multiply row 10 in tableau 2 by 2 and subtract it from row 4 of tableau 1 completely except the last column entry. The resulting result after this row operation is reported in tableau 2 in row 11 with the needed zero entry in the right place.

At this stage we bring in X_2 into the basis and drop the slack S_1. This change is effected in tableau 2. In the C_i row the appropriate per unit contribution value of zero of S_1 is replaced by the per unit contribution coefficient of X_2, namely, 15.

The per unit contribution coefficients in the C_j^{th} row in tableau 2 are simply the corresponding row values in tableau 1. So copy and complete this row 12 in tableau 2.

The Z_j values are calculated as shown below and reported in row 13 in tableau 2.

$$Z_1 = A10 \times C10 + A11 \times C11 = 15 \times 0.5 + 0 \times 3 = 7.5$$
$$Z_2 = A10 \times D10 + A11 \times D11 = 15 \times 1 + 0 \times 0 = 15$$
$$Z_3 = A10 \times E10 + A11 \times E11 = 15 \times 0.5 + 0 \times (-1) = 7.5$$
$$Z_4 = A10 \times F10 + A11 \times F11 = 15 \times 0 + 0 \times 1 = 0$$
$$Z_5 = A10 \times G10 + A11 \times G11 = 15 \times 35 + 0 \times 30 = 52$$

At this stage the optimum value of the objective function is readable from cell G13 as Rs. 525.

Now to test the optimality we once again calculate $C_j - Z_j$. This has been done and reported in the row 14 in tableau 2. Since 2.5 is the row's highest positive value it becomes the key column for our next iteration. Since 10 is the associated minimum ratio the highlighted element 3 is our new pivot element in tableau 2.

Our next iteration starts now with the introduction of an empty tableau 3. Now to make the pivot element 3 in tableau 3 as unity the relevant row operation is to divide the entire row 11 by 3 in tableau 2. The resulting row results are reported in row 18 in tableau 3. Now we must make the element just above this unity element to zero in tableau 3. Since the corresponding element is 0.5 in tableau 2 the relevant row operation is to multiply row 18 in tableau 3 by 0.5 and subtract them from the corresponding elements of row 10 in tableau 2. The results are reported in row 17 in tableau 3 with the needed zero in the correct place.

At this stage clearly X_1 is incoming and S_2 are outgoing variables. So S_2 in tableau 2 is being replaced by X_1 in tableau 3 in the basis column. Also note that the 0 contribution of S_2 is being replaced by 10, the per unit contribution of X_1 in tableau 3. After making these changes we copy once again C_j row from the previous tableau. After copying we calculate the Z_j in the usual manner and report it in row 20. Lastly we calculate the $C_j - Z_j$ as usual and report it row 21.

In the above tableau 3, the index row $C_j - Z_j^{th}$ all the entries are either zero or negative. Thus our iterative search for improvement ends here. The optimum solutions are obtained from the table as $X_1 = 10$, $X_2 = 30$ and $R = 550$.

Graphical Solution

For the graphical solution as usual we convert the constraints into equalities and draw suitable diagram as shown below. After highlighting the feasible region the coordinates of the corner points are obtained and the associated revenues are being obtained and reported in Table 3.11. In it, corresponding to the corner point F, the revenue reaches its maximum.

Table 3.11

Corner points	X_1	X_2	$R = 10X_1 + 15X_2$
O	0	0	0
E	0	35	525
F	10	30	550
G	25	0	250

$X_1 = 30$; $X_2 = 10$; $R = \text{Rs. } 550$

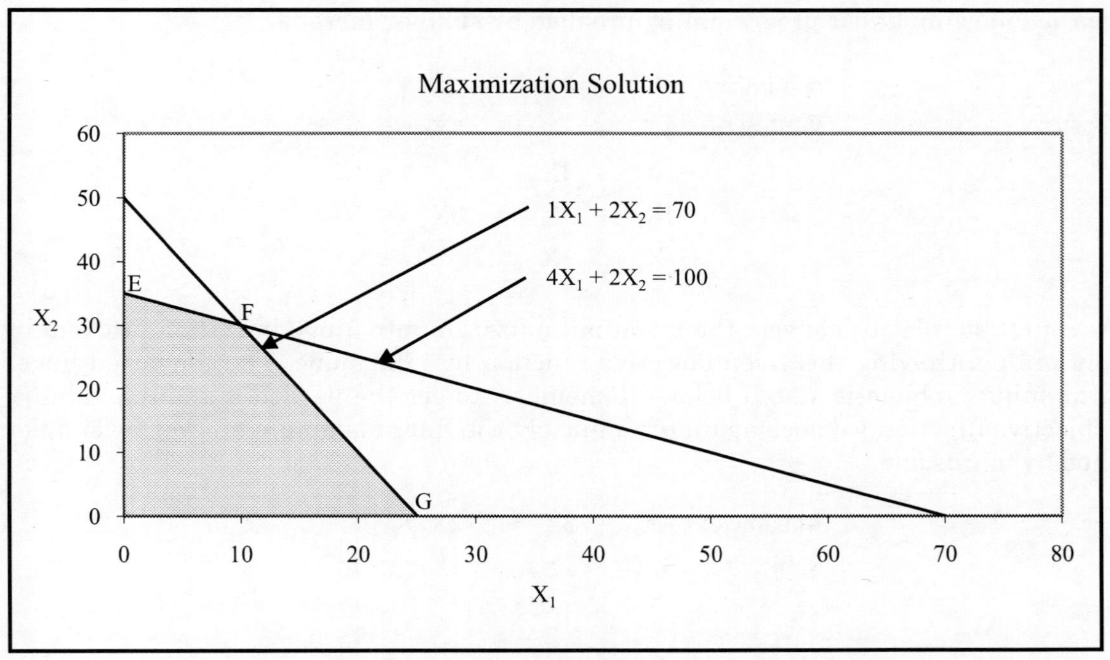

Figure 3.2

Note: *The revenue of the initial table is simply origin. The second iteration result is the corner point E (0,35). The third and final optimum result is the corner point F (10, 30). Thus the simplex method is simply a search procedure through the corner points of the feasible region starting from the very origin.*

3.3 MINIMIZATION PROBLEMS WITH ALL CONSTRAINTS ≤ TYPE

Minimization Problems

Minimization problems are solved by simplex method by applying any one of the following methods.

1. Pre-multiply the given objective function by minus and then maximize it in the usual manner by using the simplex procedure.
2. Alternatively, calculate $Z_j - C_j$ instead of $C_j - Z_j$ in the last index row and proceed further in the usual manner.

However, finally to get the needed minimum value of the objective function we once again pre-multiply the final maximum answer obtained by simplex method by minus one.

■ Example 3

Solve the following linear programming problem by simplex method.

$$\begin{aligned} \text{Minimize:}\quad & X_1 - 3X_2 + 2X_3 \\ \text{Subject to:}\quad & 3X_1 - X_2 + 3X_3 \le 7 \\ & -2X_1 + 4X_2 \le 12 \\ & -4X_1 + 3X_2 + 8X_3 \le 10 \\ & X_1 \; ; \; X_2 \ge 0 \end{aligned}$$

As a first step let us convert the given minimization into a maximization problem by simply pre-multiplying the given objective function by minus one. The converted linear programming problem is stated below. Remember, to get the needed minimum value of the objective function we once again pre-multiply the final maximum answer by simplex method by minus one.

$$\begin{aligned} \text{Maximize:}\quad & -X_1 + 3X_2 - 2X_3 \\ \text{Subject to:}\quad & 3X_1 - X_2 + 3X_3 \le 7 \\ & -2X_1 + 4X_2 \le 12 \\ & -4X_1 + 3X_2 + 8X_3 \le 10 \\ & X_1 \; ; \; X_2 \ge 0 \end{aligned}$$

Once this is done we convert the inequalities into equalities in the usual manner. Since all the three constraints are less than type let us first introduce three slack variables one each and convert the problem into its standard form amenable to start our first iteration by simplex method. As usual these slacks are introduced in the objective function with zero coefficients to insist that they do not add anything to the total revenue. Note that since the objective function is already converted into a maximization problem we calculate $C_j - Z_j$ in the last row and not $Z_j - C_j$.

$$\begin{aligned}
\text{Maximize: } & -X_1 + 3X_2 - 2X_3 + 0S_1 + 0S_2 + 0S_3 \\
\text{Subject to: } & 3X_1 - X_2 + 3X_3 + S_1 \leq 7 \\
& -2X_1 + 4X_2 + S_2 \leq 12 \\
& -4X_1 + 3X_2 + 8X_3 + S_3 \leq 10 \\
& X_1 \; ; \; X_2 \quad X_3 \; ; \; S_1 \; ; \; S_2 \; ; \; S_3 \geq 0
\end{aligned}$$

3.4 USE OF EXCEL SPREADSHEETS FOR SIMPLEX CALCULATIONS

Since we will have to do lots of calculations at every stage of iteration it is advantageous to use excel spreadsheet which obey all sorts of functional formulas. The needed steps are illustrated below.

1. After entering all the information in tableau 1 on an excel worksheet as usual look at the $C_j - Z_j$ row and fix the largest positive value. In our illustration the highest value is 3 in cell D8. Shade this cell using formatting cell command to highlight this as the key column.

2. Once this is done enter the formula in cell J3 =I3//D3. Press the enter key. This will calculate the needed ratio corresponding to this cell. In our problem this value is –7. Copy this formula to J4 and J5 cells to get the left over ratios. Among the three entries choose the least after ignoring the minus values in this ratio column. In our illustration such a minimum value is 3 in the cell J4. Shade this cell to highlight the key row as shown in tableau 1.

3. At the point of intersection corresponding to the key row and key column there is 4 in cell D4 Shade this cell and cell it as pivotal element.

4. Now it is time to go for tableau 2 to continue our search process.

5. So just below tableau 1 prepare another empty tableau 2 with column and row heading intact.

6. In tableau 2 to make our pivot element unity, enter the formula in cell D12 =D4/$D$4. Press the enter key. This will make the cell D12 as unity as desired.

7. Copy this formula on either side of row 12 leaving the last ratio cell. Automatically the needed calculations will appear in this row.

8. In cell B12 put the incoming X_2 in the place of the outgoing S_2. In cell A12, in the place of zero contribution co-efficient, put X_2's per unit contribution coefficient, namely, 3.

Table 3.12

	A	B	C	D	E	F	G	H	I	J
1					Tableau 1					
2	C_i	Basis	X_1	X_2	X_3	S_1	S_2	S_3	X_b	M.R
3	0	S_1	3	−1	3	1	0	0	7	−7
4	0	S_2	−2	4	0	0	1	0	12	3
5	0	S_3	−4	3	8	0	0	1	10	3.333
6		C_j	−1	3	−2	0	0	0		
7		Z_j	0	0	0	0	0	0	0	
8		$C_j–Z_j$	−1	3	−2	0	0	0		
9					Tableau 2					
10	C_i	Basis	X_1	X_2	X_3	S_1	S_2	S_3	X_b	M.R
11	0	S_1	2.5	0	3	1	0.25	0	10	4
12	3	X_2	−0.5	1	0	0	0.25	0	3	−6
13	0	S_3	−2.5	0	8	0	−0.75	1	1	−0.4
14		C_j	−1	3	−2	0	0	0		
15		Z_j	−1.5	3	0	0	0.75	0	9	
16		$C_j–Z_j$	0.5	0	−2	0	−0.75	0		
17					Tableau 3					
18	C_i	Basis	X_1	X_2	X_3	S_1	S_2	S_3	X_b	M.R
19	−1	X_1	1	0	1.2	0.4	0.1	0	4	
20	3	X_2	0	1	0.6	0.2	0.3	0	5	
21	0	S_3	0	0	11	1	−0.5	1	11	
22		C_j	−1	3	−2	0	0	0		
23		Z_j	−1	3	0.6	0.2	0.8	0	11	
24		$C_j–Z_j$	0	0	−2.6	−0.2	−0.8	0		

9. Our next task is to make entries just above and below the unit entry cell D12 to zero by appropriate row operation. To make the entries below unity to zero in cell D13 enter the formula = D5−$D$5*D12 and press the enter key. This will make the cell D13 zero as desired. Copy this formula on either side except the last cell in the right to complete the other entries of this row.

10. Now to make the other zero in cell D11, enter the formula D11 =D3−$D$3*D12 and press the enter key. This will make cell D11 as zero. Copy this formula on either side except the last cell in the right to make the other entries.

11. In row 14 copy the C_j entries from row 6 as it is.

12. Now to calculate the new Z_j's, in the cell C15 enter the formula, Sumproduct($A$11:$A$13,C11:C13). Press the enter key. This will calculate the necessary Z_j in C15.

13. Copy this formula for the entire row and complete the other entries in this row.

14. In cell C16 enter C16 = C14−C15. Enter and copy this formula for the whole row to obtain the index row.

After completing this step tableau 2 will look like the one shown in tableau 2 with all its entries intact. Till this time the value of the objective function is displayed in cell I15 = 9.

In index row 16 since we have 0.5 as positive entry we will have to repeat the whole exercise as described earlier and get a tableau 3. In this tableau 3 since no positive entries are found in the index row our iteration stops at this stage. The objective function value in cell I23 = 11. The X_1 and X_2 values are reported in cells I19 and I20.

Thus the maximum solutions are $X_1 = 4$; $X_2 = 5$; $X_3 = 0$; $S_3 = 11$; $Z = 11$. So the needed minimum to the original problem is obtained by merely multiplying the obtained maximum by minus one. Thus, the needed minimum is equal to −11.

3.5 USE OF ARTIFICIAL VARIABLES IN LINEAR PROGRAMMING WITH ≥ CONSTRAINTS

So far we have considered the problems with ≤ inequalities only in our constraints for both maximization and minimization problems. Such inequalities are converted into equalities by introducing merely slack variables with unit positive coefficients in the left side of the inequalities concerned. Such an inclusion gave us the needed basic solution which facilitates to begin our simplex search from the very origin of the feasible region. Other than this type, very often we come across "=" and '≥' inequalities in several occasions. In the case of greater than type of inequalities, we are supposed to subtract surplus variables from the left side to bring in equalities in the constraints. However, this alone does not provide the needed positive unit matrix to facilitate the origin start as usual. So we propose to introduce another set of variables called artificial (or dummy) in addition to surplus variables in the left side of the relevant constraints with the badly needed positive unit coefficients. Remember, that these new variables have no physical meaning in the problem and get themselves eliminated in the final solution. However, occasionally we find one or more artificial variables in the final solution. Under such circumstances our problem solving effort will fail.

In the case of equality, there is no need to add any more variable to the left side to bring in equality. However, if it does not contain unit coefficient variable in the left side we cannot start the problem. Again we have to subtract a surplus variable and add an artificial variable with hefty penalty as in the case of greater than inequality. The following example illustrates the methodology of such conversions.

3.6 MINIMIZATION PROBLEMS WITH ALL CONSTRAINTS ≥ TYPE

■ Example 4

Solve the following linear programming problem.

$$\begin{aligned} \text{Minimize:} \quad & 2X_1 + 1X_2 \\ \text{Subject to:} \quad & 3X_1 + 1X_2 \geq 3 \\ & 4X_1 + 3X_2 \geq 6 \\ & 1X_1 + 2X_2 \geq 2 \\ & X_1 \ ; \ X_2 \geq 0 \end{aligned}$$

Graphical solution

As usual the linear constraints are drawn and the feasible region is marked in the Figure 3.3. Once this is done the corner points are noted down from the graph and the objective function values are calculated in Table 3.13.

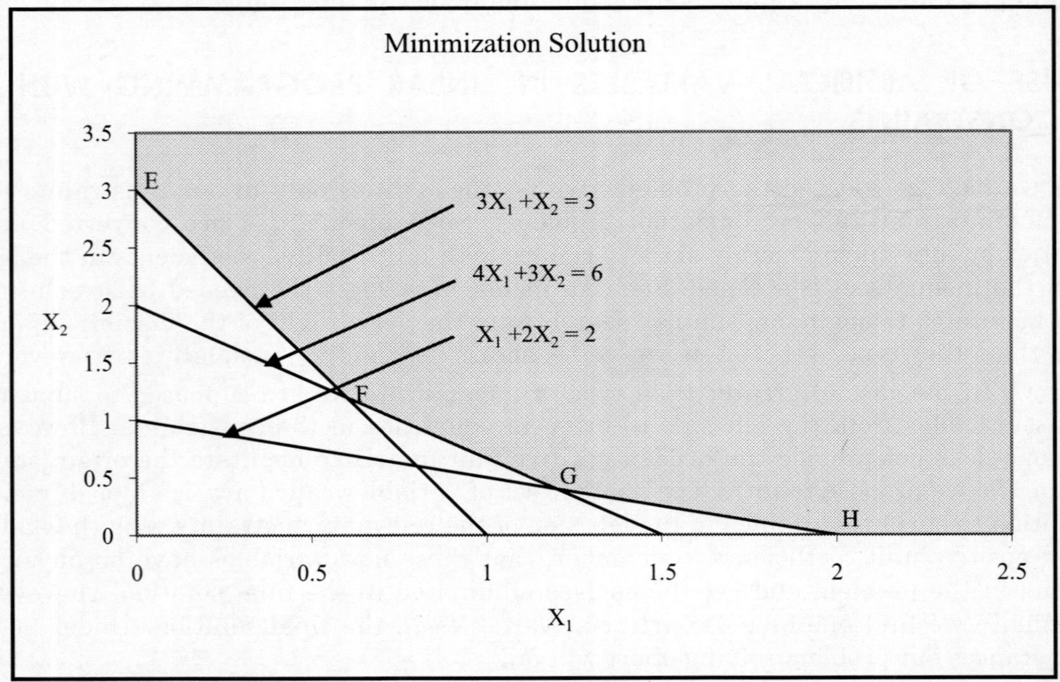

Figure 3.3

Table 3.13

Corner points	X_1	X_2	$C = 2X_1 + 1X_2$
E	0	3	3
F	0.6	1.2	2.4
G	1.2	0.4	2.8
H	2	0	4

Clearly the origin is not in the faceable region. Hence, we cannot start our simplex search by giving full credit to slacks. Alternatively, to convert these three *greater than or equal to* inequalities in the constraints into equalities, let us subtract three surplus variables S_1, and S_2 and S_3 one each from the left hand side as shown below.

$$\begin{aligned}
\text{Minimize: } & 2X_1 + 1X_2 + 0S_1 + 0S_2 + 0S_3 \\
\text{Subject to: } & 3X_1 + 1X_2 - S_1 = 3 \\
& 4X_1 + 3X_2 - S_2 = 6 \\
& 1X_1 + 2X_2 - S_3 = 2 \\
& X_1\ ;\ X_2\ ;\ S_1\ ;\ S_2\ ;\ S_3 \geq 0
\end{aligned}$$

The inclusion of surplus variables does not solve the problem. The basis matrix defined is not a positive unit matrix due to the negative coefficients of the surplus variables. In diagrammatic terms this is to say that our origin is not there in our purview of the feasible region. To overcome this problem let us add three artificial variables A_1 and A_2 and A_3 with needed plus unit coefficients in the left-hand side of constraints in addition to the surplus variables introduced already. Note that the effect of this inclusion is absorbed by the three surplus variables already introduced without affecting the respective equality signs.

$$\begin{aligned}
\text{Minimize: } & 2X_1 + 1X_2 + 0S_1 + 0S_2 + 0S_3 + A_1 + A_2 + A_3 \\
\text{Subject to: } & 3X_1 + 1X_2 - S_1 + A_1 = 3 \\
& 4X_1 + 3X_2 - S_2 + A_2 = 6 \\
& 1X_1 + 2X_2 - S_3 + A_3 = 2 \\
& X_1\ ;\ X_2\ ;\ S_1\ ;\ S_2\ ;\ S_3\ ;\ A_1\ ;\ A_2\ ;\ A_3 \geq 0
\end{aligned}$$

In this modified model it is to be noted that $A_1 < S_1$, $A_2 < S_2$ and $A_3 < S_3$.

Solution: Two-Phase Method

In the Two-Phase Method, the optimization is done in two phases. In the first phase we consider the following modified linear programming problem with objective function containing zero coefficients for all other variables and (plus) unit coefficients for artificial variables. However, for maximization problem these coefficients for artificial variables must be minus unity. Since it is a minimization problem, we propose to convert this into a maximization problem by calculating $Z_j - C_j$ in the simplex tableau instead of $C_j - Z_j$.

Phase I: Problem Formulation

$$
\begin{array}{llllllllll}
\text{Minimize:} & 0X_1 + 0X_2 + 0S_1 + 0S_2 + 0S_3 + A_1 + A_2 + A_3 \\
\text{Subject to:} & 3X_1 + 1X_2 - S_1 + A_1 = 3 \\
& 4X_1 + 3X_2 - S_2 + A_2 = 6 \\
& 1X_1 + 2X_2 - S_3 + A_3 = 2 \\
& X_1 ;\ X_2 ;\ S_1 ;\ S_2 ;\ S_3 ;\ A_1 ;\ A_2 ;\ A_3 \geq 0
\end{array}
$$

As a first step let us enter the data in tableau 1 in the respective places as usual to begin our first iteration.

Table 3.14

Tableau 1											
C_i	Basis	X_1	X_2	S_1	S_2	S_3	A_1	A_2	A_3	X_b	M.R
1	A_1	3	1	−1	0	0	1	0	0	3	1
1	A_2	4	3	0	−1	0	0	1	0	6	1.5
1	A_3	1	2	0	0	−1	0	0	1	2	2
	C_j	0	0	0	0	0	1	1	1		
	Z_j	8	6	−1	−1	−1	1	1	1	11	
	$Z_j–C_j$	8	6	−1	−1	−1	0	0	0		

In the tableau 1 as per marking of key row and key column X_1 enters the basis in the place of the artificial variable A_1. From this starting tableau 1 we obtain tableau 2 as usual by using Simplex procedure. However, while doing so we completely drop A_1 in tableau 2 for not giving any more change for its inclusion again.

Table 3.15

Tableau 2										
C_i	Basis	X_1	X_2	S_1	S_2	S_3	A_2	A_3	X_b	Min
0	X_1	1	0.333	−0.33	0	0	0	0	1	3
1	A_2	0	1.667	1.333	−1	0	1	0	2	1.2
1	A_3	0	1.667	0.333	0	−1	0	1	1	0.6
	C_j	0	0	0	0	0	1	1		
	Z_j	0	3.333	1.667	−1	−1	1	1	3	
	$Z_j–C_j$	0	3.333	1.667	−1	−1	0	0		

In the above tableau 2 as per marking of key row and key column X_2 enters the basis in the place of the artificial variable A_3. Form this tableau 2 we obtain tableau 3 as usual by using Simplex procedure. However, while doing so we completely drop A_3 in tableau 3 for not giving any more change for its inclusion again.

In the tableau 3 since S_1 enters the basis in the place of the artificial A_2, the A_2 column in tableau 3 is completely drooped in tableau 4 for not giving any more change for its inclusion again while doing the Simplex calculations.

Table 3.16

Tableau 3									
C_i	Basis	X_1	X_2	S_1	S_2	S_3	A_2	X_b	M.R
0	X_1	1	0	−0.4	0	0.2	0	0.8	−2
1	A_2	0	0	1	−1	1	1	1	1
0	X_2	0	1	0.2	0	−0.6	0	0.6	3
	C_j	0	0	0	0	0	1		
	Z_j	0	0	1	−1	1	1	1	
	Z_j–C_j	0	0	1	−1	1	0		

Table 3.17

Tableau 4								
C_i	Basis	X_1	X_2	S_1	S_2	S_3	X_b	M.R
0	X_1	1	0	0	−0.4	0.6	1.2	
0	S_1	0	0	1	−1	1	1	
0	X_2	0	1	0	0.2	−0.8	0.4	
	C_j	0	0	0	0	0		
	Z_j	0	0	0	0	0	0	
	Z_j–C_j	0	0	0	0	0		

Since we have successfully eliminated all the three artificial variables in tableau 4 we go back to our original objective function and redo the problem starting from this simplex tableau 4.

Phase II

To incorporate this new situation we modify C_i column and C_j row in tableau 4 with new coefficients in the objective function corresponding to the structural variables X_1 and X_2 alone. For surplus variables, Tableau 4 entries are taken as it is. After carrying out this minor modification we apply once again Simplex method to get our final solution starting from tableau 5 afresh. The new Z_j and $Z_j - C_j$ calculations are shown in tableau 5. Since the index now in tableau 5 is having 0.4. We repeat the Simplex calculations and obtain the final tableau 6.

Table 3.18

Tableau 5								
C_i	Basis	X_1	X_2	S_1	S_2	S_3	X_b	M.R
2	X_1	1	0	0	−0.4	0.6	1.2	2
0	S_1	0	0	1	−1	1	1	1
1	X_2	0	1	0	0.2	−0.8	0.4	−0.5
	C_j	2	1	0	0	0		
	Z_j	2	1	0	−0.6	0.4	2.8	
	$Z_j - C_j$	0	0	0	−0.6	0.4		

Table 3.19

Tableau 6							
C_i	Basis	X_1	X_2	S_1	S_2	S_3	X_b
2	X_1	1	0	−0.6	0.2	0	0.6
0	S_3	0	0	1	−1	1	1
1	X_2	0	1	0.8	−0.6	0	1.2
	C_j	2	1	0	0	0	
	Z_j	2	1	−0.4	−0.2	0	2.4
	$Z_j - C_j$	0	0	−0.4	−0.2	0	

In tableau 6 since $Z_j - C_j$ row all the values are zero or negative our iteration stops here. **From the last tableau 6, the optimum solutions are:** $X_1 = 0.6$, $X_2 = 1.2$ and $Z = 2.4$.

Charne's Big M method

In this method we start the problem straightaway by incorporating a hefty positive penalty M (M is positive) for minimization and negative (−M) for maximization for all the artificial variables in the objective function so that they all get eliminated automatically in the final solution. So the coefficient of structural variables will be intact in this method from the very beginning. So our problem may be restated as

Problem Formulation

$$\text{Minimize: } 2X_1 + 1X_2 + 0S_1 + 0S_2 + 0S_3 + MA_1 + MA_2 + MA_3$$

$$\text{Subject to: } 3X_1 + 1X_2 - S_1 + A_1 = 3$$

$$4X_1 + 3X_2 - S_2 + A_2 = 6$$

$$1X_1 + 2X_2 - S_3 + A_3 = 2$$

$$X_1 ; X_2 ; S_1 ; S_2 ; S_3 ; A_1 ; A_2 ; A_3 \geq 0$$

Since unit vectors are made available artificially by inclusion of artificial variables in the constraints, we start our search by giving full credit to these artificial variables in the basis as shown in tableau 1. However, this time the heavy per unit penalty occupy C_i column in the very beginning as shown in the tableau 1.

Table 3.20

Tableau 1											
C_i	Basis	X_1	X_2	S_1	S_2	S_3	A_1	A_2	A_3	X_b	Min
M	A_1	3	1	−1	0	0	1	0	0	3	1
M	A_2	4	3	0	−1	0	0	1	0	6	1.5
M	A_3	1	2	0	0	−1	0	0	1	2	2
	C_j	2	1	0	0	0	M	M	M		
	Z_j	8M	6M	−M	−M	−M	M	M	M	11M	
	Z_j–C_j	8M−2	6M−1	−M	−M	−M	0	0	0		

In this tableau 1, clearly 8M−2 is not only positive but also bigger than other entries in the last row (M is a very big positive number). The first row is the key row and first column is the key column. Accordingly, since X_1 enters the basis and the artificial variable A_1 leaves basis, we completely drop it in the next iteration tableau 2 for not giving any more change for its re-entry.

Table 3.21

Tableau 2										
C_i	Basis	X_1	X_2	S_1	S_2	S_3	A_2	A_3	X_b	Min
2	X_1	1	1/3	−1/3	0	0	0	0	1	3
M	A_2	0	5/3	4/3	−1	0	1	0	2	6/5
M	A_3	0	5/3	1/3	0	−1	0	1	1	3/5
	C_j	2	1	0	0	0	M	M		
	Z_j	2	10/3M+2/3	5/3M−2/3	−M	−M	M	M	2+3M	
	Z_j–C_j	0	10/3M−1/3	5/3M−2/3	−M	−M	0	0		

In this tableau 2 above clearly (10/3M)−1/3 is the largest positive value than any other entry. The third row is the key row and third column is key column. Thus, X_2 enters the basis and the artificial variable A_3 leaves basis. We completely drop it in the next tableau 3.

In this tableau 3 clearly M−1/5 is the largest positive value than any other entry. The third row is the key row and sixth column is key column. Thus, S_3 enters the basis and the artificial variable A_2 leaves basis. We completely drop it in the next tableau 4.

Table 3.22

Tableau 3									
C_i	Basis	X_1	X_2	S_1	S_2	S_3	A_2	X_b	Min
2	X_1	1	0	−2/3	0	1/5	0	4−5	4
M	A_2	0	0	1	−1	1	1	1	1
1	X_2	0	1	1/5	0	−3/5	0	3/5	−ve
	C_j	2	1	0	0	0	M		
	Z_j	2	1	M−3/5	−M	M−1/5	M	M+11/5	
	Z_j–C_j	0	0	M−3/5	−M	M−1/5	0		

Table 3.23

Tableau 4							
C_i	Basis	X_1	X_2	S_1	S_2	S_3	X_b
2	X_1	1	0	−3/5	1/5	0	3/5
0	S_3	0	0	1	−1	1	1
1	X_2	0	1	4/5	−3/5	0	6/5
	C_j	2	1	0	0	0	
	Z_j	2	1	−2/5	−1/5	0	12/5
	Z_j–C_j	0	0	−2/5	−1/5	0	

Now we have successfully eliminated all the three artificial variables. Further no positive value is found in the $Z_j - C_j$ th row. Thus, this is the final iteration table.

So the optimum values are: $X_1 = 3/5 = 0.6$, $X_2 = 6/5 = 1.2$ and $Z = 12/5 = 2.4$.

■ Example 5

Find the optimum solution for the following LPP

$$\begin{aligned} \text{Minimize: } & C = 100X_1 + 200X_2 \\ \text{Subject to: } & 1X_1 + 1X_2 \geq 40 \\ & 3X_1 + 1X_2 \geq 60 \\ & 1X_1 + 6X_2 \geq 60 \\ & X_1 ; X_2 \geq 0 \end{aligned}$$

Solution

Let us first convert the given problem into equalities by introducing three surplus variables as usual. Further, such an inclusion does not solve the unit matrix requirement to start our simplex method, we introduce three artificial variables also with needed unit coefficients.

$$\begin{aligned} \text{Minimize: } & C = 100X_1 + 200X_2 + 0S_1 + 0S_2 + 0S_3 + 1A_1 + 1A_2 + 1A_3 \\ \text{Subject to: } & 1X_1 + 1X_2 - S_1 + A_1 = 40 \\ & 3X_1 + 1X_2 - S_2 + A_2 = 60 \\ & 1X_1 + 6X_2 - S_3 + A_3 = 60 \\ & X_1 ; X_2 ; S_1 ; S_2 ; S_3 ; A_1 ; A_2 ; A_3 \geq 0 \end{aligned}$$

So our Phase I problem could be stated as follows.

$$\begin{aligned} \text{Minimize: } C = \; & 0X_1 + 0X_2 + 0S_1 + 0S_2 + 0S_3 + 1A_1 + 1A_2 + 1A_3 \\ \text{Subject to: } & 1X_1 + 1X_2 - S_1 + A_1 = 40 \\ & 3X_1 + 1X_2 - S_2 + A_2 = 60 \\ & 1X_1 + 6X_2 - S_3 + A_3 = 60 \\ & X_1 \; ; \; X_2 \; ; \; S_1 \; ; \; S_2 \; ; \; S_3 \; ; \; A_1 \; ; \; A_2 \; ; \; A_3 \geq 0 \end{aligned}$$

Phase I

Since this is a minimization we convert it into maximization by calculating $Z_j - C_j$ in the index row instead of $C_j - Z_j$.

Table 3.24

Tableau 1											
C_i	Basis	X_1	X_2	S_1	S_2	S_3	A_1	A_2	A_3	X_b	Min
1	A_1	1	1	−1	0	0	1	0	0	40	40
1	A_2	3	1	0	−1	0	0	1	0	60	60
1	A_3	1	6	0	0	−1	0	0	1	60	10
	C_j	0	0	0	0	0	1	1	1		
	Z_j	5	8	−1	−1	−1	1	1	1	160	
	$Z_j–C_j$	5	8	−1	−1	−1	0	0	0		

Table 3.25

Tableau 2										
C_i	Basis	X_1	X_2	S_1	S_2	S_3	A_1	A_2	X_b	Min
1	A_2	0.833	0	−1	0	0.167	1	0	30	36
1	A_3	2.833	0	0	−1	0.167	0	1	50	17.65
0	X_2	0.167	1	0	0	−0.17	0	0	10	60
	C_j	0	0	0	0	0	1	1		
	Z_j	3.667	0	−1	−1	0.333	1	1	80	
	$Z_j–C_j$	3.667	0	−1	−1	0.333	0	0		

Table 3.26

Tableau 3									
C_i	Basis	X_1	X_2	S_1	S_2	S_3	A_1	X_b	Min
1	A_1	0	0	−1	0.294	0.118	1	15.29	52
0	X_1	1	0	0	−0.353	0.059	0	17.65	−50
0	X_2	0	1	0	0.059	−0.176	0	7.059	120
	C_j	0	0	0	0	0	1		
	Z_j	0	0	−1	0.294	0.118	1	15.29	
	$Z_j–C_j$	0	0	−1	0.294	0.118	0		

Table 3.27

Tableau 4								
C_i	Basis	X_1	X_2	S_1	S_2	S_3	X_b	Min
0	S_2	0	0	−3.4	1	0.4	52	52
0	X_1	1	0	−1.2	0	0.2	36	−50
0	X_2	0	1	0.2	0	−0.2	4	120
	C_j	0	0	0	0	0		
	Z_j	0	0	0	0	0	0	
	$Z_j–C_j$	0	0	0	0	0		

Phase II

Table 3.28

Tableau 5								
C_i	Basis	X_1	X_2	S_1	S_2	S_3	X_b	Min
0	S_2	0	0	−3.4	1	0.4	52	
100	X_1	1	0	−1.2	0	0.2	36	
200	X_2	0	1	0.2	0	−0.2	4	
	C_j	100	200	0	0	0		
	Z_j	100	200	−80	0	−20	4400	
	$Z_j–C_j$	0	0	−80	0	−20		

So the optimum values are: $X_1 = 36$, $X_2 = 4$, $Z = 4400$.

3.7 MAXIMIZATION PROBLEMS WITH ALL TYPES OF INEQUALITY

Sometimes we will have all types of inequality in our constraints in maximization problems. There may be a few equality constraints, few greater than or equality constraints and few less than or equality constraints. The following example illustrates this type of problem.

■ Example 6

A computer science BE students as a hobby buys computer components and assembles them into computer monitor, computer CPU, and UPS. To do this job he gets time only during weekend holidays. In a month he can make 500 units of one type or all the items together. Computer monitor and CPU combined order per month is at least 150 units. Since power failure if very often occurs UPS sales alone exceeds the sales of computer monitor and CPU by 100 units. The revenue from computer monitor, CPU and UPS are Rs. 75, Rs. 125 and Rs. 150 per unit respectively. Formulate this statement into a linear programming problem and obtain its solution.

Solution

Let X_1, X_2, X_3 be the number of units of computer monitor, CPU and UPS assembled per month. The associated linear programming problem is shown below.

$$\begin{aligned}
\text{Maximize:}\quad & 75X_1 + 125X_2 + 150X_3 \\
\text{Subject to:}\quad & X_1 + X_2 + X_3 \le 500 \\
& X_1 + X_2 \ge 150 \\
& X_1 - X_2 - X_3 = 100 \\
& X_1 ;\ X_2 ;\ X_3 \ge 0
\end{aligned}$$

For simplex format, introduce one slack to the first constraint. Subtract one surplus to constraint 2. To have the initial basis we add two artificial variables to second and third constraints respectively. But now note that in the objective function minus unit coefficients are being introduced for artificial variables because of maximization problem.

1) Two-Phase Method

$$\begin{aligned}
\text{Maximize:}\quad & 0X_1 + 0X_2 + 0X_3 + 0S_1 + 0S_2 - 1A_1 - 1A_2 \\
\text{Subject to:}\quad & X_1 + X_2 + X_3 + S_1 = 500 \\
& X_1 + X_2 - S_2 + A_1 = 150 \\
& X_1 - X_2 - X_3 + A_2 = 100 \\
& X_1 ;\ X_2 ;\ X_3 ;\ S_1 ;\ S_2 ;\ A_1 ;\ A_2 \ge 0
\end{aligned}$$

Table 3.29

Tableau 1										
C_i	Basis	X_1	X_2	X_3	S_1	S_2	A_1	A_2	X_b	Ratio
0	S_1	1	1	1	1	0	0	0	500	500
−1	A_1	1	1	0	0	−1	1	0	150	150
−1	A_2	1	−1	−1	0	0	0	1	100	100
	C_j	0	0	0	0	0	−1	−1		
	Z_j	−2	0	1	0	1	−1	−1	−250	
	C_j-Z_j	2	0	−1	0	−1	0	0		

Table 3.30

Tableau 2									
C_i	Basis	X_1	X_2	X_3	S_1	S_2	A_1	X_b	Ratio
0	S_1	0	2	2	1	0	0	400	200
−1	A_1	0	2	1	0	−1	1	50	25
0	X_1	1	−1	−1	0	0	0	100	−100
	C_j	0	0	0	0	0	−1		
	Z_j	0	−2	−1	0	1	−1	−50	
	C_j-Z_j	0	2	1	0	−1	0		

Table 3.31

Tableau 3								
C_i	Basis	X_1	X_2	X_3	S_1	S_2	X_b	Ratio
0	S_1	0	0	1	1	1	350	200
0	X_2	0	1	0.5	0	−0.5	25	25
0	X_1	1	0	−0.5	0	−0.5	125	−100
	C_j	0	0	0	0	0		
	Z_j	0	0	0	0	0	0	
	C_j-Z_j	0	0	0	0	0		

Phase II

Table 3.32

Tableau 4								
C_i	Basis	X_1	X_2	X_3	S_1	S_2	X_b	Ratio
0	S_1	0	0	1	1	1	350	350
125	X_2	0	1	0.5	0	−0.5	25	50
75	X_1	1	0	−0.5	0	−0.5	125	−250
	C_j	75	125	150	0	0		
	Z_j	75	125	25	0	−100	12,500	
	C_j-Z_j	0	0	125	0	100		

Table 3.33

Tableau 5								
C_i	Basis	X_1	X_2	X_3	S_1	S_2	X_b	Ratio
0	S_1	0	−2	0	1	2	300	150
150	X_3	0	2	1	0	−1	50	−50
75	X_1	1	1	0	0	−1	150	−150
	C_j	75	125	150	0	0		
	Z_j	75	375	150	0	−225	18,750	
	C_j-Z_j	0	−250	0	0	225		

Table 3.34

Tableau 6								
C_i	Basis	X_1	X_2	X_3	S_1	S_2	X_b	Ratio
0	S_2	0	−1	0	0.5	1	150	150
150	X_3	0	1	1	0.5	0	200	−50
75	X_1	1	0	0	0.5	0	300	−150
	C_j	75	125	150	0	0		
	Z_j	75	150	150	112.5	0	52,500	
	C_j-Z_j	0	−25	0	−112.5	0		

So the optimum values are: $X_1 = 300$, $X_3 = 200$, $Z = 52{,}500$.

2) Big M method

Since it is a maximization problem big minus (−M) penalties are being introduced in the objective function.

$$\begin{aligned}
\text{Maximize: } & C = 75X_1 + 125X_2 + 150X_3 + 0S_1 + 0S_2 - MA_1 - MA_2 \\
\text{Subject to: } & X_1 + X_2 + X_3 + S_1 = 500 \\
& X_1 + X_2 - S_2 + A_1 = 150 \\
& X_1 - X_2 - X_3 + A_2 = 100 \\
& X_1 ;\ X_2 ;\ X_3 ;\ S_1 ;\ S_2 ;\ A_1 ;\ A_2 \geq 0
\end{aligned}$$

Table 3.35

Tableau 1										
C_i	Basis	X_1	X_2	X_3	S_1	S_2	A_1	A_2	X_b	Ratio
0	S_1	1	1	1	1	0	0	0	500	500
−M	A_1	1	1	0	0	−1	1	0	150	150
−M	A_2	1	−1	−1	0	0	0	1	100	100
	C_j	75	125	150	0	0	−M	−M		
	Z_j	−2M	0	M	0	M	−M	−M	−250M	
	C_j–Z_j	75+2M	125	150+M	0	M	0	0		

Table 3.36

Tableau 2									
C_i	Basis	X_1	X_2	X_3	S_1	S_2	A_1	X_b	Ratio
0	S_1	0	2	2	1	0	0	400	200
−M	A_1	0	2	1	0	−1	1	50	25
75	X_1	1	−1	−1	0	0	0	100	−100
	C_j	75	125	150	0	0	−M		
	Z_j	75	−2M−75	−M−75	0	M	−M	7500−50M	
	C_j–Z_j	0	200+2M	225+M	0	−M	0		

Table 3.37

Tableau 3								
C_i	Basis	X_1	X_2	X_3	S_1	S_2	X_b	Ratio
0	S_1	0	0	1	1	1	350	350
125	X_2	0	1	0.5	0	−0.5	25	50
75	X_1	1	0	−0.5	0	−0.5	125	−250
	C_j	75	125	150	0	0		
	Z_j	75	125	25	0	−100	12,500	
	C_j–Z_j	0	0	125	0	100		

Table 3.38

Tableau 4								
C_i	Basis	X_1	X_2	X_3	S_2	S_2	X_b	Ratio
0	S_1	0	−2	0	1	2	300	150
150	X_3	0	2	1	0	−1	50	−50
75	X_1	1	1	0	0	−1	150	−150
	C_j	75	125	150	0	0		
	Z_j	75	375	150	0	−225	18,750	
	$C_j–Z_j$	0	−250	0	0	225		

Table 3.39

Tableau 5								
C_i	Basis	X_1	X_2	X_3	S_1	S_2	X_b	Ratio
0	S_2	0	−1	0	0.5	1	150	150
150	X_3	0	1	1	0.5	0	200	−50
75	X_1	1	0	0	0.5	0	300	−150
	C_j	75	125	150	0	0		
	Z_j	75	150	150	112.5	0	52,500	
	$C_j–Z_j$	0	−25	0	−112.5	0		

So the optimum values are: $X_1 = 300$, $X_3 = 200$, $S_5 = 150$, $Z = 52{,}500$.

3.8 MINIMIZATION PROBLEMS WITH ALL TYPES OF INEQUALITY

Sometimes we will have all types of inequality in our constraints in minimization problems as well. There may be few equality constraints, few greater than or equality constraints and few less than or equality constraints. The following example illustrates this type of problem.

■ Example 7

The final product of a firm must weigh exactly 150 kg. It uses two raw materials A and B with costs Rs. 2 and Rs. 8 per unit. At least 14 units of B and no more than 20 units of A must be used. Each unit of A weighs 5 kg and each unit of B weigh 10 kg. How much of each raw material be used per unit of the final product if its cost is to be minimized?

Solution

Let us formulate an appropriate linear programming problem first. Let X_1 and X_2 be the quantities of A and B bought to make one unit of the product under consideration. Then the problem is formulated as

$$\begin{aligned} \text{Minimize:} \quad & 2X_1 + 8X_2 \\ \text{Subject to:} \quad & 5X_1 + 10X_2 = 150 \\ & X_1 \le 20 \\ & X_2 \ge 14 \\ & X_1 \; ; \; X_2 \ge 0 \end{aligned}$$

1. In equality constraints we introduce only artificial dummy variable. This variable is included in the objective function with a very large positive coefficient so that it gets eliminated in the final solution by big M method.
2. In less than inequality we introduce slacks with positive unit coefficient. However, it takes 0 coefficients in the objective function as usual.
3. In the case of greater than constraint we introduce surplus variables with negative coefficients to bring in equality on both the sides. Moreover, artificial dummies are introduced with very big coefficients for big M method in the objective function.

So the problem is reformulated as

$$\begin{aligned} \text{Minimize:} \quad & 2X_1 + 8X_2 + 0S_1 + 0S_2 + 1A_1 + 1A_2 \\ \text{Subject to:} \quad & 5X_1 + 10X_2 + A_1 = 150 \\ & X_1 + S_1 = 20 \\ & X_2 - S_2 + A_2 = 14 \\ & X_1 \; ; \; X_2 \; ; \; S_1 \; ; \; S_2 \; ; \; A \; ; \; A_2 \ge 0 \end{aligned}$$

1. Two-Phase Method

$$\begin{aligned} \text{Minimize:} \quad & 0X_1 + 0X_2 + 0S_1 + 0S_2 + 1A_1 + 1A_2 \\ \text{Subject to:} \quad & 5X_1 + 10X_2 + A_1 = 150 \\ & X_1 + S_1 = 20 \\ & X_2 - S_2 + A_2 = 14 \\ & X_1 \; ; \; X_2 \; ; \; S_1 \; ; \; S_2 \; ; \; A_1 \; ; \; A_2 \ge 0 \end{aligned}$$

Phase I

Table 3.40

Tableau 1									
C_i	Basis	X_1	X_2	S_1	S_2	A_1	A_2	X_b	Min
1	A_1	5	10	0	0	1	0	150	30
0	S_1	1	0	1	0	0	0	20	20
1	A_2	0	1	0	−1	0	1	14	–
	C_j	0	0	0	0	1	1		
	Z_j	5	0	1	−1	0	−1		
	Z_j–C_j	5	0	1	−1	−1	−2		

Table 3.41

Tableau 2									
C_i	Basis	X_1	X_2	S_1	S_2	A_1	A_2	X_b	Min
1	A_1	0	10	−5	0	1	0	50	5
0	X_1	1	0	1	0	0	0	20	–
1	A_2	0	1	0	−1	0	1	14	14
	C_j	0	0	0	0	1	1		
	Z_j	0	11	−5	−1	1	1	64	
	Z_j–C_j	0	11	−5	−1	0	0		

Table 3.42

Tableau 3								
C_i	Basis	X_1	X_2	S_1	S_2	A_1	X_b	Min
0	X_2	0	1	−0.5	0	0	5	−10
0	X_1	1	0	1	0	0	20	20
1	A_2	0	0	0.5	−1	1	9	18
	C_j	0	0	0	0	1		
	Z_j	0	0	0.5	−1	1	9	
	Z_j–C_j	0	0	0.5	−1	0		

Table 3.43

Tableau 4							
C_i	Basis	X_1	X_2	S_1	S_2	X_b	Min
0	X_2	0	1	0	−1	14	−10
0	X_1	1	0	0	2	2	20
0	S_1	0	0	1	−2	18	18
	C_j	0	0	0	0		
	Z_j	0	0	0	0	0	
	Z_j–C_j	0	0	0	0		

Phase II

Table 3.44

Tableau 5							
C_i	Basis	X_1	X_2	S_1	S_2	X_b	Min
8	X_2	0	1	0	−1	14	
2	X_1	1	0	0	2	2	
0	S_1	0	0	1	−2	18	
	C_j	2	8	0	0		
	Z_j	2	8	0	−4	116	
	Z_j–C_j	0	0	0	−4		

So the optimum values are: $X_1 = 2$, $X_2 = 1.4$, $Z = 116$.

2. Big M Method

$$\text{Minimize: } 2X_1 + 8X_2 + 0S_1 + 0S_2 + MA_1 + MA_2$$

$$\begin{aligned}\text{Subject to: } 5X_1 + 10X_2 + A_1 &= 150\\ X_1 + S_1 &= 20\\ X_2 - S_2 + A_2 &= 14\\ X_1 ;\ X_2 ;\ S_1 ;\ S_2 ;\ A_1 ;\ A_2 &\geq 0\end{aligned}$$

Table 3.45

Tableau 1									
C_i	Basis	X_1	X_2	S_1	S_2	A_1	A_2	X_b	Min
M	A_1	5	10	0	0	1	0	150	15
0	S_1	1	0	1	0	0	0	20	–
M	A_2	0	1	0	−1	0	1	14	14
	C_j	2	8	0	0	M	M		
	Z_j	5M	11M	0	−M	M	M	164M	
	$Z_j–C_j$	5M−2	11M−8	0	−M	0	0		

Table 3.46

Tableau 2								
C_i	Basis	X_1	X_2	S_1	S_2	A_1	X_b	Min
M	A_1	5	0	0	10	1	10	1
0	S_1	1	0	1	0	0	20	–
8	X_2	0	1	0	−1	0	14	–
	C_j	2	8	0	0	M		
	Z_j	5M	8	0	10M−8	M	10M+112	
	$Z_j–C_j$	5M−2	0	0	10M−8	0		

Table 3.47

Tableau 3							
C_j	Basis	X_1	X_2	S_1	S_2	X_b	Min
0	S_2	0.5	0	0	1	1	2
0	S_1	1	0	1	0	20	20
8	X_2	0.5	1	0	0	15	30
	C_j	2	8	0	0		
	Z_j	4	8	0	0	90	
	$Z_j–C_j$	2	0	0	0		

Table 3.48

Tableau 4						
C_j	Basis	X_1	X_2	S_1	S_2	X_b
2	X_1	1	0	0	2	2
0	S_1	0	0	1	−2	18
8	X_2	0	1	0	−1	14
	C_j	2	8	0	0	
	Z_j	2	8	0	−4	116
	$Z_j – C_j$	0	0	0	0	

All the entries in the last row are either 0 or negative we have reached the final solution. **Thus, $X_1 = 2$; $S_1 = 18$; $X_2 = 14$ and the minimum cost = Rs.116.**

■ Example 8

Find the optimum solution by ***Two-Phase*** method.

$$\begin{aligned} \text{Minimize:} \quad & 6X_1 + 4X_2 + 3X_3 \\ \text{Subject to:} \quad & 4X_1 + 5X_2 + 3X_3 = 40 \\ & 2X_1 + X_2 + 6X_3 = 60 \\ & 3X_1 + 4X_2 + 2X_3 = 60 \\ & X_1 \; ; \; X_2 \; ; \; X_3 \geq 0 \end{aligned}$$

Since the given constraints are equal type we introduce three surplus variables and at the same time three artificial variables with unit coefficients in order to have a unit matrix to initiate our simplex calculations.

Two-Phase Method Problem Formulation

$$\begin{aligned} \text{Minimize:} \quad & 6X_1 + 4X_2 + 3X_3 + 0S_1 + 0S_2 + 0S_3 + 1A_1 + 1A_2 + 1A_3 \\ \text{Subject to:} \quad & 4X_1 + 5X_2 + 3X_3 - S_1 + A_1 = 40 \\ & 2X_1 + X_2 + 6X_3 - S_2 + A_2 = 60 \\ & 3X_1 + 4X_2 + 2X_3 - S_3 + A_3 = 60 \\ & X_1 \; ; \; X_2 \; ; \; X_3 \; ; \; S_1 \; ; \; S_2 \; ; \; S_3 \; ; \; A_1 \; ; \; A_2 \; ; \; A_3 \geq 0 \end{aligned}$$

Phase I: Problem formulation

$$
\begin{aligned}
\text{Minimize: } & 0X_1 + 0X_2 + 0X_3 + 0S_1 + 0S_2 + 0S_3 + 1A_1 + 1A_2 + 1A_3 \\
\text{Subject to: } & 4X_1 + 5X_2 + 3X_3 - S_1 + A_1 = 40 \\
& 2X_1 + X_2 + 6X_3 - S_2 + A_2 = 60 \\
& 3X_1 + 4X_2 + 2X_3 - S_3 + A_3 = 60 \\
& X_1 ; X_2 ; X_3 ; S_1 ; S_2 ; S_3 ; A_1 ; A_2 ; A_3 \geq 0
\end{aligned}
$$

Solution

Table 3.49

Tableau 1												
C_i	Basis	X_1	X_2	X_3	S_1	S_2	S_3	A_1	A_2	A_3	X_b	Min
1	A_1	4	5	3	−1	0	0	1	0	0	40	13.3
1	A_2	2	1	6	0	−1	0	0	1	0	50	8.33
1	A_3	3	4	2	0	0	−1	0	0	1	60	30
	C_j	0	0	0	0	0	0	1	1	1		
	Z_j	9	10	11	−1	−1	−1	1	1	1		
	$Z_j–C_j$	9	10	11	−1	−1	−1	0	0	0		

Table 3.50

Tableau 2											
C_i	Basis	X_1	X_2	X_3	S_1	S_2	S_3	A_1	A_3	X_b	Min
1	A_1	3	4.5	0	−1	0.5	0	1	0	15	3.33
0	X_3	0.33	0.17	1	0	−0.17	0	0	0	8.33	50
1	A_3	2.33	3.67	0	0	0.33	−1	0	1	43.3	11.8
	C_j	0	0	0	0	0	0	1	1		
	Z_j	5.33	8.17	0	−1	0.83	−1	1	1	58.3	
	$Z_j–C_j$	5.33	8.17	0	−1	0.83	−1	0	0		

Table 3.51

Tableau 3										
C_i	Basis	X_1	X_2	X_3	S_1	S_2	S_3	A_3	X_b	Min
0	X_2	0.667	1	0	−0.22	0.111	0	0	3.333	−15
0	X_3	0.222	0	1	0.037	−0.19	0	0	7.778	210
1	A_3	−0.11	0	0	0.815	−0.07	−1	1	31.11	38.18
	C_j	0	0	0	0	0	0	1		
	Z_j	−0.11	0	0	0.815	−0.07	−1	1	31.11	
	$Z_j–C_j$	−0.11	0	0	0.815	−0.07	−1	0		

Table 3.52

Tableau 4									
C_i	Basis	X_1	X_2	X_3	S_1	S_2	S_3	X_b	Min
0	X_2	0.636	1	0	0	0.091	−0.27	11.82	−15
0	X_3	0.227	0	1	0	−0.18	0.045	6.364	210
0	S_1	−0.14	0	0	1	−0.09	−1.23	38.18	38.18
	C_j	0	0	0	0	0	0		
	Z_j	0	0	0	0	0	0	0	
	$Z_j{-}C_j$	0	0	0	0	0	0		

Phase II

Table 3.53

Tableau 5									
C_i	Basis	X_1	X_2	X_3	S_1	S_2	S_3	X_b	Min
4	X_2	0.636	1	0	0	0.091	−0.27	11.82	
3	X_3	0.227	0	1	0	−0.18	0.045	6.364	
0	S_1	−0.14	0	0	1	−0.09	−1.23	38.18	
	C_j	6	4	3	0	0	0		
	Z_j	3.227	4	3	0	−0.18	−0.95	66.36	
	$Z_j{-}C_j$	−2.77	0	0	0	−0.18	−0.95		

So the optimum solutions are: $X_1 = 0$, $X_2 = 11.8$, $X_3 = 6.36$, $Z = 66.4$.

■ Example 9

Reformulate the following problem suitably for a big M method

$$\begin{aligned} &\text{Minimize: } C = 18X_1 + 20X_2 + 21X_3 \\ &\text{Subject to: } X_1 + X_2 + X_3 = 1000 \\ &X_1 \le 300 \\ &X_2 \ge 150 \\ &X_3 \ge 200 \\ &X_1 ;\ X_2 ;\ X_3 \ge 0 \end{aligned}$$

Solution

$$\begin{aligned} &\text{Minimize: } C = 18X_1 + 20X_2 + 21X_3 + 0S_1 + 0S_2 + 0S_3 + MA_1 + MA_2 + MA_3 \\ &\text{Subject to: } 1X_1 + 1X_2 + 1X_3 + A_1 = 1000 \\ &X_1 + S_1 = 300 \\ &X_2 - S_2 + A_2 = 150 \\ &X_3 - S_3 + A_3 = 200 \\ &X_1 ;\ X_2 ;\ X_3 ;\ S_1 ;\ S_2 ;\ S_3 ;\ A_1 ;\ A_2 ;\ A_3 \ge 0 \end{aligned}$$

■ Example 10

Reformulate the following problem suitably for a big M method

$$
\begin{aligned}
\text{Minimize: } C &= X_1 + 2X_2 + 3X_3 - X_4 \\
\text{Subject to: } & X_1 + 2X_2 + 3X_3 = 15 \\
& 2X_1 + X_2 + 5X_3 = 20 \\
& X_1 + 2X_2 + X_3 + X_4 = 10 \\
& X_1 \; ; \; X_2 \; ; \; X_3 \ge 0
\end{aligned}
$$

Solution

In this problem since for X_4 in the third constraint have the needed unit coefficient already; we include only two artificial variables A_1 and A_2 with needed unit coefficients in constraint 1 and 2 only.

$$
\begin{aligned}
\text{Minimize: } C &= X_1 + 2X_2 + 3X_3 - X_4 + MA_1 + MA_2 \\
\text{Subject to: } & X_1 + 2X_2 + 3X_3 + A_1 = 15 \\
& 2X_1 + X_2 + 5X_3 + A_2 = 20 \\
& X_1 + 2X_2 + X_3 + X_4 = 10 \\
& X_1 \; ; \; X_2 \; ; \; X_3 \; ; \; X_4 \; ; \; A_1 \; ; \; A_2 \ge 0
\end{aligned}
$$

3.9 SOME SPECIAL TYPES OF PROBLEM

1. Negative Right Hand Side

If the constraint is of the type with negative number in the right hand side it should be converted into a positive number by multiplying both the sides of the equation by a minus one.

■ Example 11

If the constraint is $-2X_1+3X_2 \ge -15$ then after multiplication by (-1) it becomes $2X_1 - 3X_2 \le 15$.

2. Choice of the Key Column

While using the C_j-Z_j for key column selection if there is a tie among two or more values then follow the following rules for selection.

- If the tie is between decision variables, the choice can be arbitrary.
- If the tie is between slack variables, the choice is arbitrary.
- If the tie is between a decision variable and slack, the choice must be the decision variable.

3. Tie for the Key Row

For the selection of the key row we use the minimum ratio. Sometimes this ratio may not be unique and therefore there will be a tie in the selection. This problem is commonly addressed as degeneracy. To overcome this problem one is advised to observe the following steps.

- Select the rows, which are having the tie.
- Find the coefficient of the slack variables and divide each one of them by the respective coefficients in the key column and break the tie. If the ratios does not break the tie, try for similar ratios for the coefficients of the decision variables.
- Compare the resulting ratios, column by column.
- Select the ratio, which has the smallest ratio.

The following example illustrates one such case having the row choice problem.

■ Example 12

Solve the following problem by Simplex method.

$$\begin{aligned} \text{Maximize: } & R = 25X_1 + 30X_2 + 40X_3 \\ \text{Subject to: } & 2X_1 + 1X_2 + 2X_3 \leq 50 \\ & 0.5X_1 + 3X_2 + 1X_3 \leq 25 \\ & 1X_1 + 2X_2 + 1X_3 \leq 26 \\ & X_1 \; ; \; X_2 \; ; \; X_3 \geq 0 \end{aligned}$$

Solution

$$\begin{aligned} \text{Maximize: } & R = 25X_1 + 30X_2 + 40X_3 + 0S_1 + 0S_2 + 0S_3 \\ \text{Subject to: } & 2X_1 + 1X_2 + 2X_3 + S_1 = 50 \\ & 0.5X_1 + 3X_2 + 1X_3 + S_2 = 25 \\ & 1X_1 + 2X_2 + 1X_3 + S_3 = 26 \\ & X_1 \; ; \; X_2 \; ; \; X_3 + S_1 + S_2 + S_3 \geq 0 \end{aligned}$$

Table 3.54

Tableau 1									
C_i	Basis	X_1	X_2	X_3	S_1	S_2	S_3	X_b	Ratio
0	S_1	2	1	2	1	0	0	50	25
0	S_2	0.5	3	1	0	1	0	25	25
0	S_3	1	2	1	0	0	1	26	26
	C_j	25	30	40	0	0	0		
	Z_j	0	0	0	0	0	0	0	
	C_j-Z_j	25	30	40	0	0	0		

In the table above though there is no problem for column selection, we have a tie in the row selection. To overcome this difficulty we take the coefficients of the slack variable pertaining to these rows as shown below.

Table 3.55

	S_1	S_2	S_3
S_1	1	0	0
S_2	0	1	0

Now let us divide these coefficients by the respective coefficients of the key column already selected as shown below.

Table 3.56

	S_1	S_2	S_3
S_1	1/2	0/2	0/2
S_2	0/1	1/1	0/1

In the above table since 0/1 = 0 is the least in the first column, we select S_2 as the key row. Accordingly, S_2 will leave and X_3 will enter into the basis.

Table 3.57

Tableau 2										
C_i	Basis	X_1	X_2	X_3	S_1	S_2	S_3	X_b	Ratio
0	S_1	1	−5	0	1	−2	0	0	0
40	X_3	0.5	3	1	0	1	0	25	50
0	S_6	0.5	−1	0	0	−1	1	1	2
	Z_j	25	30	40	0	0	0		
	C_j	20	120	40	0	40	0	1000	
	$Z_j–C_j$	5	−90	0	0	−40	0		

Table 3.58

Tableau 3									
C_i	Basis	X_1	X_2	X_3	S_1	S_2	S_3	X_b	Ratio
25	X_1	1	−5	0	1	−2	0	0	
40	X_3	0	5.5	1	−0.5	2	0	25	
0	S_3	0	1.5	0	−0.5	0	1	1	
	Z_j	25	30	40	0	0	0		
	C_j	25	95	40	5	30	0	1000	
	$Z_j–C_j$	0	−65	0	−5	−30	0		

Thus the final solutions are: $X_1 = 0$, $X_3 = 25$, $R = 1000$.

Note: *The revenue has not increased in the last iteration.*

4. Multiple Solutions

Whenever the objective function has the same slope as that of one of the constraints then we will have infinite number of optimal solutions. In the final Simplex table this is indicated by a zero coefficient in the $Z_j - C_j^{th}$ row for a non-basic variable. This will happen irrespective of whether the non-basic is a decision variable or a slack.

The following example illustrates one such case.

■ Example 13

Solve the following problem by Simplex method.

$$\text{Maximize: } R = 1X_1 + 3X_2$$

$$\text{Subject to: } \begin{aligned} 1X_1 + 3X_2 &\le 90 \\ 2X_1 + 2X_2 &\le 80 \\ 2X_1 + 0X_2 &\le 60 \\ X_1;\ X_2 &\ge 0 \end{aligned}$$

Solution

Let us convert the inequalities into equalities by introducing three slack variables first as shown below.

$$\text{Maximize: } R = 1X_1 + 3X_2 + 0S_1 + 0S_2 + 0S_3$$

$$\text{Subject to: } \begin{aligned} 1X_1 + 3X_2 + S_1 &= 90 \\ 2X_1 + 2X_2 + S_2 &= 80 \\ 2X_1 + 0X_2 + S_3 &= 60 \\ X_1;\ X_2;\ S_1;\ S_2;\ S_3 &\ge 0 \end{aligned}$$

Table 3.59

Tableau 1

C_i	Basis	X_1	X_2	S_1	S_2	S_3	X_b	Ratio
0	S_1	1	3	1	0	0	90	30
0	S_2	2	2	0	1	0	80	40
0	S_3	2	0	0	0	1	60	–
	C_j	1	3	0	0	0	0	
	Z_j	0	0	0	0	0	0	
	$C_j - Z_j$	1	3	0	0	0		

Table 3.60

Tableau 2								
C_i	Basis	X_1	X_2	S_1	S_2	S_3	X_b	Ratio
3	X_2	0.3333	1	0.3333	0	0	30	90
0	S_2	1.3333	0	−0.6667	1	0	20	15
0	S_3	2	0	0	0	1	60	30
	C_j	1	3	0	0	0	0	
	Z_j	1	3	1	0	0	90	
	$C_j–Z_j$	0	0	−1	0	0		

So the optimum values are: $X_1 = 0$, $X_2 = 30$, $S_1 = 0$, $S_2 = 20$, $S_3 = 60$ and $R = 90$.

In the last simplex table shown above the non-basic decision variable X_1 is having the zero coefficients in the last row. Thus the inclusion of this variable into the basic will not improve the total revenue any longer. To confirm let us go one step further and enter X_1 into the basis in the place of S_2.

Table 3.61

Tableau 3								
C_i	Basis	X_1	X_2	S_1	S_2	S_3	X_b	Ratio
3	X_2	0	1	0.5	0	0	25	
1	X_1	1	0	−0.5	1	0	15	
0	S_3	0	0	1	−2	1	30	
	C_j	1	3	0	0	0	0	
	Z_j	1	3	1	0	0	90	
	$C_j–Z_j$	0	0	−1	0	0		

So the optimum values are: $X_1 = 15$, $X_2 = 25$, $S_1 = 0$, $S_2 = 0$, $S_3 = 30$ and $R = 90$.

From these two solutions we can generate infinite number of solution as convex combinations as shown below.

$$\lambda\begin{bmatrix}15\\25\\0\\0\\30\end{bmatrix}+(1-\lambda)\begin{bmatrix}0\\30\\0\\20\\60\end{bmatrix}=\begin{bmatrix}15\lambda\\25\lambda+30(1-\lambda)\\0\\20(1-\lambda)\\30\lambda+60(1-\lambda)\end{bmatrix}=\begin{bmatrix}15\lambda\\30-5\lambda\\0\\20(1-\lambda)\\60-30\lambda\end{bmatrix}\quad 0<\lambda<1$$

- If $\lambda = 1$ then from the above equation the solutions are $X_1 = 15$, $X_2 = 25$, $S_1 = 0$, $S_2 = 0$, $S_3 = 30$, $R = 90$.
- If $\lambda = 0$ then the solutions are: $X_1 = 0$, $X_2 = 30$, $S_1 = 0$, $S_2 = 20$, $S_3 = 60$, $R = 90$.

- If $\lambda = 0.5$ then the solutions are: $X_1 = 7.5$, $X_2 = 27.5$, $S_1 = 0$, $S_2 = 10$, $S_3 = 45$, $R = 90$.
- By assigning different values to λ in the range 0 to 1 one can generate infinite number of solutions.

5. Degenerate Solution

In a basic solution, if any one or more of the basic variables takes the zero values, then the solution is called degenerate solution. The following example illustrates one such situation.

■ Example 14

Show that the following linear programming problem is degenerate.

$$\begin{aligned} \text{Maximize: } R &= 28X_1 + 30X_2 \\ \text{Subject to: } & 6X_1 + 3X_2 \leq 18 \\ & 3X_1 + X_2 \leq 8 \\ & 4X_1 + 5X_2 \leq 30 \\ & X_1 \; ; \; X_2 \geq 0 \end{aligned}$$

Graphical Solution

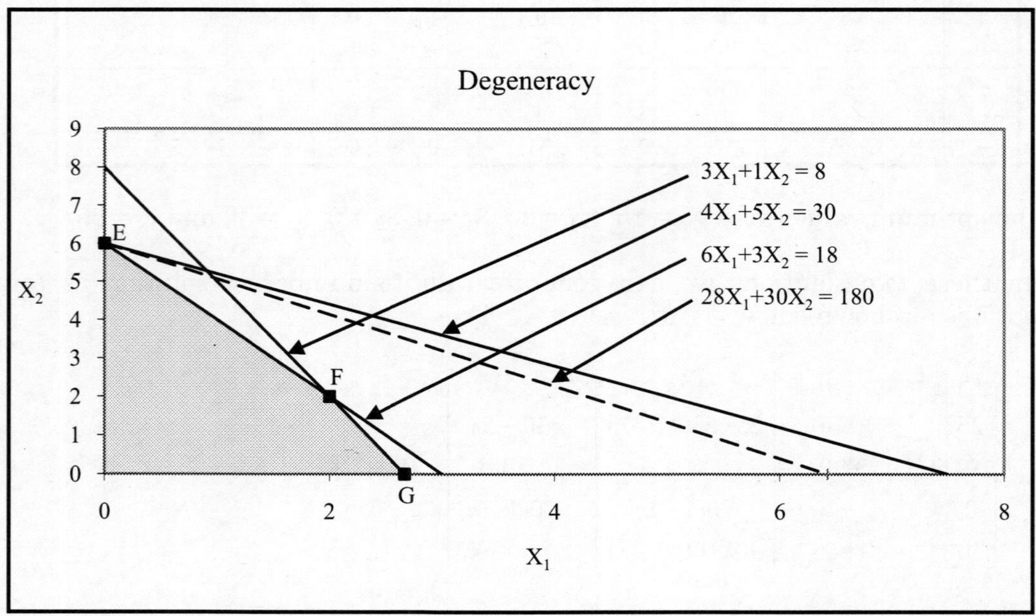

Figure 3.4

A careful look at the solution point E in the above diagram, it is clear that at E, constraints 1 and 3 are satisfied. So the associated slacks S_1 and S_3 are zero. In addition since this point is on X_2 axis X_1 is also zero. Since E is just below the constraint 2 the associated slack S_2 is positive. Since it is a three-constraint two-decision variable model altogether there are 5 variables in the standard form. At the point E the basic variables are $X_1 = 0$, $X_2 = 6$, $S_2 = 2$. Since one of the basis variables is zero the solution degenerate. In such cases the simplex solution is found to be repetitive and hence cyclic as highlighted below.

Now let us consider this problem using simplex method

$$\text{Maximize: } R = 28X_1 + 30X_2 + 0S_1 + 0S_2 + 0S_3$$

$$\text{Subject to: } 6X_1 + 3X_2 + S_1 = 18$$

$$3X_1 + X_2 + S_2 = 8$$

$$4X_1 + 5X_2 + S_3 = 30$$

$$X_1 \;;\; X_2 \;;\; S_1 \;;\; S_2 \;;\; S_3 \geq 0$$

Solution

Table 3.62

Tableau 1								
C_i	Basis	X_1	X_2	S_1	S_1	S_3	X_b	Ratio
0	S_1	6	3	1	0	0	18	6
0	S_2	3	1	0	1	0	8	8
0	S_3	4	5	0	0	1	30	6
	C_j	28	30	0	0	0		
	Z_j	0	0	0	0	0	0	
	$C_j–Z_j$	28	30	0	0	0		

In the above tableau 1 in the index row since 30 is the largest there no problem in selecting the key column. But for the selection of key row there is a tie between two rows having the minimum ratio 6. Fortunately since both of them are slacks we have the option of choosing either S_1 or S_3 as the outgoing variable for bringing in X_2. We prefer S_1 as the outgoing variable in Table 3.62.

Table 3.63

Tableau 2								
C_i	Basis	X_1	X_2	S_1	S_2	S_3	X_b	Ratio
30	X_2	2	1	0.3333	0	0	6	6
0	S_2	1	0	−0.3333	1	0	2	8
0	S_3	−6	0	−1.6667	0	1	0	6
	C_j	28	30	0	0	0		
	Z_j	60	30	10	0	0	180	
	$C_j–Z_j$	−32	0	−10	0	0		

In tableau 2 since the index row entries are negative or zero our iteration ends here. The corresponding solution values are $X_1 = 0$; $X_2 = 6$ $S_2 = 2$ and R = 180.

In tableau 1 instead of choosing S_1 if we choose S_2 as the outgoing variable then we get the following two iterations.

Table 3.64

Tableau 3								
C_i	Basis	X_1	X_2	S_1	S_2	S_3	X_b	Ratio
0	S_1	3.6	0	1	0	−0.6	0	0
0	S_2	2.2	0	0	1	−0.2	2	0.9091
30	X_2	0.8	1	0	0	0.2	6	7.5
	C_j	28	30	0	0	0		
	Z_j	24	30	0	0	6	180	
	C_j-Z_j	4	0	0	0	−6		

Table 3.65

Tableau 4								
C_i	Basis	X_1	X_2	S_1	S_2	S_3	X_b	Ratio
28	X_1	1	0	0.2778	0	−0.167	0	0
0	S_2	0	0	−0.611	1	0.1667	2	0.9091
30	X_2	0	1	−0.222	0	0.3333	6	7.5
	C_j	28	30	0	0	0		
	Z_j	28	30	1.1111	0	5.3333	180	
	C_j-Z_j	0	0	−1.111	0	−5.333		

In tableau 4 since the index row entries are negative or zero our iteration ends here.

The corresponding solution values are $X_1 = 0$; $X_2 = 6$; $S_2 = 2$ and R= 180.

Some important observations

1. In tableau 2 the slack S_3 has zero solution value though it is in the basis. Similarly, in tableau 4 the basis variable X_1 has zero solution value.

2. When there is a tie in row selection, the next table invariably gives rise to degenerate solution.

3. As the outgoing is a degenerate variable the objective function value will not improve in the corresponding iteration.

■ Example 15

Solve the following problem by Simplex method.

$$\text{Maximize: } R = 3X_1 + 4X_2$$
$$\text{Subject to: } 7X_1 + 5X_2 \leq 40$$
$$3X_1 + 4X_2 \leq 20$$
$$X_1 - 0X_2 \leq 0$$
$$X_1 \,;\; X_2 \geq 0$$

Solution

Let us convert the inequalities into equalities by introducing three slack variables as shown below and start out Simplex iteration as usual starting from the very origin.

$$\text{Maximize: } R = 3X_1 + 4X_2 + 0S_1 + 0S_2 + 0S_3$$
$$\text{Subject to: } 7X_1 + 5X_2 + S_1 = 40$$
$$3X_1 + 4X_2 + S_2 = 20$$
$$X_1 - X_2 + S_3 = 0$$
$$X_1 \,;\; X_2 \,;\; S_1 \,;\; S_2 \,;\; S_3 \geq 0$$

Table 3.66

Tableau 1								
C_i	Basis	X_1	X_2	S_1	S_1	S_3	X_b	Ratio
0	S_1	7	5	1	0	0	40	8
0	S_2	3	4	0	1	0	20	5
0	S_3	1	−1	0	0	1	0	0
	C_j	3	4	0	0	0		
	Z_j	0	0	0	0	0	0	
	$C_j–Z_j$	3	4	0	0	0		

Table 3.67

Tableau 2								
C_i	Basis	X_1	X_2	S_1	S_2	S_3	X_b	Ratio
0	S_1	3.25	0	1	−1.25	0	15	8
4	X_2	0.75	1	0	0.25	0	5	5
0	S_3	1.75	0	0	0.25	1	5	0
	C_j	3	4	0	0	0		
	Z_j	3	4	0	1	0	20	
	$C_j–Z_j$	0	0	0	−1	0		

In the very first table itself the solution gets degenerate because the basic variable $S_3 = 0$. However, in the next step of iteration we get a basic non-degenerate solution.

Table 3.68

Tableau 3									
C_i	Basis	X_1	X_2	S_1	S_2	S_3	X_b	Ratio
0	S_1	3.25	0	1	−1.25	0	15	8
4	X_2	0.75	1	0	0.25	0	5	5
0	S_3	1.75	0	0	0.25	1	5	0
	C_j	3	4	0	0	0		
	Z_j	3	4	0	1	0	20	
	$C_j–Z_j$	0	0	0	−1	0		

So the optimum solutions are: $X_1 = 0$, $X_2 = 5$, $S_1 = 15$, $S_2 = 0$, $S_3 = 5$, $R = 20$.

Note that there no improvement in the revenue during this iteration. Also note that this problem will offer multiple solutions because the corresponding coefficients of the objective function and that of the second constraint are equal.

6. Unbounded Solution

It is quite possible that one or more variables will increase the objective function value indefinitely without violating the basic conditions. Such solutions are called unbounded solutions.

The following example highlights one such situation.

■ Example 16

Solve the following problem by both graphical and Simplex method.

$$\begin{aligned} \text{Maximize: } & R = 3X_1 + 2X_2 \\ \text{Subject to: } & 1X_1 \geq 5 \\ & + 1X_2 \leq 5 \\ & X_1 ;\ X_2 \geq 0 \end{aligned}$$

Graphical Solution

In the following graph shift in the objective function up and up could continuously increase the revenue indefinitely. It never reaches the tangential corner of the feasible region. In fact, the feasible region never ends in the right side.

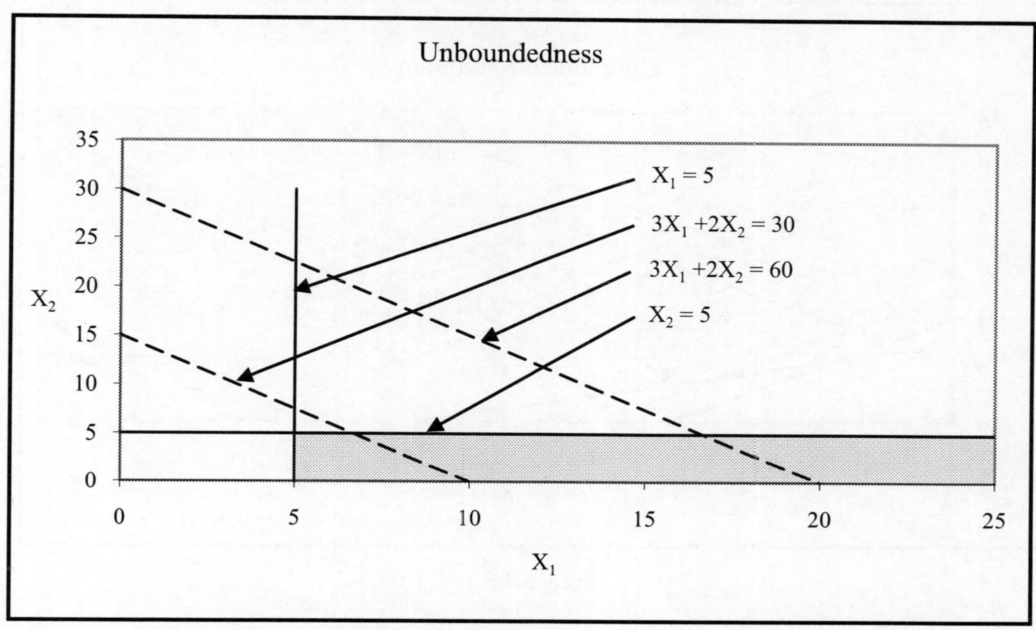

Figure 3.5

■ Example 17

Solve the following LPP problem graphically.

$$\begin{aligned} \text{Maximize: } & R = 10X_1 + 20X_2 \\ \text{Subject to: } & 2X_1 + 4X_2 \geq 16 \\ & X_1 + 5X_2 \geq 15 \\ & X_1 \; ; \; X_2 \geq 0 \end{aligned}$$

Graphical Solution

In the following graph the highlighted feasible region clearly increases without any bound up and right side and hence there is no optimum solution as such.

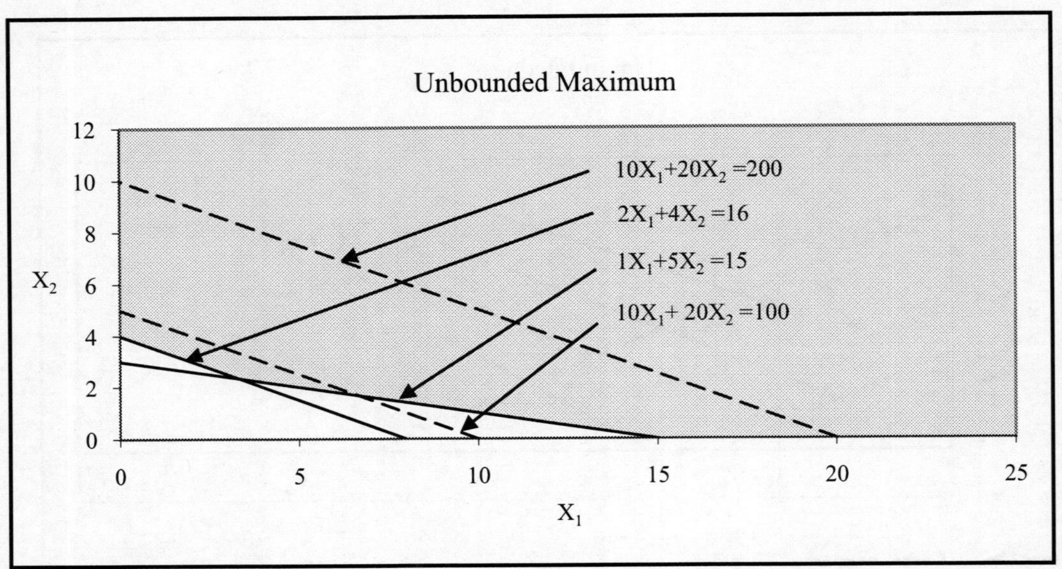

Figure 3.6

■ Example 18

Solve the following problem by Simplex method.

$$\begin{aligned} \text{Maximize:}\quad & R = 3X_1 + 2X_2 \\ \text{Subject to:}\quad & X_1 - X_2 \le 1 \\ & 3X_1 - 2X_2 \le 6 \\ & X_1 \; ; \; X_2 \ge 0 \end{aligned}$$

Solution

$$\begin{aligned} \text{Maximize:}\quad & R = 3X_1 + 2X_2 + 0S_1 + 0S_2 \\ \text{Subject to:}\quad & X_1 - X_2 + S_1 = 1 \\ & 3X_1 - 2X_2 + S_2 = 6 \\ & X_1 \; ; \; X_2 \; ; \; S_1 \; ; \; S_2 \ge 0 \end{aligned}$$

Table 3.69

Tableau 1							
C_i	Basis	X_1	X_2	S_1	S_2	X_b	M.R
0	S_1	1	−1	1	0	1	1
0	S_2	3	−2	0	1	6	2
	C_j	3	2	0	0		
	Z_j	0	0	0	0	0	
	C_j-Z_j	3	2	0	0		

Table 3.70

Tableau 2							
C_i	Basis	X_1	X_2	S_1	S_2	X_b	M.R
3	X_1	1	−1	1	0	1	−1
0	S_2	0	1	−3	1	3	3
	C_j	3	2	0	0		
	Z_j	3	−3	3	0	3	
	$C_j–Z_j$	0	5	−3	0		

Table 3.71

Tableau 3							
C_i	Basis	X_1	X_2	S_1	S_2	X_b	M.R
3	X_1	1	0	−2	1	4	−2
2	X_2	0	1	−3	1	3	−1
	C_j	3	2	0	0		
	Z_j	3	2	−12	5	18	
	$C_j–Z_j$	0	0	12	−5		

In tableau 3 though $C_j–Z_j$ show the possibility of S_1 inclusion, the negative minimum ratio tell us there is nothing to leave (both the ratios are negative). In such cases the value of the objective function will rise indefinitely without any bound.

7. Redundant Constraints

■ Example 19

Solve the following problem by Simplex method.

$$\begin{aligned} \text{Maximize: } R &= 4X_1 + 3X_2 + X_3 \\ \text{Subject to: } & 2X_1 + 5X_2 + 7X_3 = 90 \\ & 4X_1 + 10X_2 + 14X_3 = 180 \\ & X_1\ ;\ X_2\ ;\ X_3 \geq 0 \end{aligned}$$

$$\begin{aligned} \text{Maximize: } R &= 4X_1 + 3X_2 + X_3 + 1A_1 + 1A_2 \\ \text{Subject to: } & 2X_1 + 5X_2 + 7X_3 + A_1 = 90 \\ & 4X_1 + 10X_2 + 14X_3 + A_2 = 180 \\ & X_1\ ;\ X_2\ ;\ X_3\ ;\ A_1\ ;\ A_2 \geq 0 \end{aligned}$$

Phase I

Table 3.72

Tableau 1								
C_i	Basis	X_1	X_2	X_3	A_1	A_2	X_b	Ratio
1	A_1	2	5	7	1	0	90	12.857
1	A_2	4	10	14	0	1	180	12.857
	C_j	0	0	0	1	1		
	Z_j	6	15	21	1	1	270	
	$Z_j–C_j$	6	15	21	0	0		

Table 3.73

Tableau 2							
C_i	Basis	X_1	X_2	X_3	A_1	X_b	Ratio
1	A_1	0	0	0	1	0	12.857
0	X_3	0.2857	0.7143	1	0	12.857	12.857
	C_j	0	0	0	1		
	Z_j	0	0	0	1	0	
	$Z_j–C_j$	0	0	0	0		

At the end of Phase I, the basis still contains one artificial variable A_1 with zero value in the objective function. Since $Z_j–C_j$ is zero we cannot bring in any other variable in and remove A_1 by iteration. At this point it is important to note that the first constraint is redundant in the sense it is simply obtained by dividing the second constant by 2 throughout. So we will have to remove this redundant constraint from the problem itself if at all we are interested in a solution. In phase II shown below the corresponding row is removed and final solution is obtained.

Phase II

Table 3.74

Tableau 3						
C_i	Basis	X_1	X_2	X_3	X_b	Ratio
0	X_3	0.28571	0.71429	1	12.8571	
	C_j	4	3	0		
	Z_j	0	0	0	0	
	$Z_j–C_j$	4	3	0		

Table 3.75

Tableau 4						
C_i	Basis	X_1	X_2	X_3	X_b	Ratio
4	X_1	1	2.5	3.5	45	
	C_j	4	3	0		
	Z_j	4	10	14	180	
	$Z_j–C_j$	0	–7	–14		

So the optimum solutions are: $X_1 = 45$, $X_2 = 0$, $R = 180$.

8. Infeasibility Case

In the case of infeasibility there will not be any feasible region. If it is a simple two-variable problem then one could easily verify feasibility from the very graph. If there are more than two decision variables then test by diagram is not possible. While applying the Simplex method, if an artificial variable is found in the final solution table with positive value then safely one can infer that there is no feasible solution to the problem under reference the following example illustrates one such case.

Example 20

Solve the following linear programming problem

$$\begin{aligned} \text{Maximize: } & R = 20X_1 + 30X_2 \\ \text{Subject to: } & 2X_1 + 1X_2 \le 40 \\ & 4X_1 - 1X_2 \le 20 \\ & X_1 + 0X_2 \ge 30 \\ & X_1 \; ; \; X_2 \ge 0 \end{aligned}$$

Phase I Problem formulation

$$\begin{aligned} \text{Maximize: } & R = 0X_1 + 0X_2 + 0S_1 + 0S_2 + 0S_3 - 1A_1 \\ \text{Subject to: } & 2X_1 + 1X_2 + S_1 = 40 \\ & 4X_1 - X_2 + S_2 = 20 \\ & 1X_1 + 0X_2 - S_3 - A_1 = 30 \\ & X_1 \; ; \; X_2 \; ; \; S_1 \; ; \; S_2 \; ; \; S_3 \; ; \; A_1 \ge 0 \end{aligned}$$

In the following table we workout the problem by applying the two-phase method through phase I first.

Phase I

Table 3.76

Tableau 1									
C_i	Basis	X_1	X_2	S_1	S_2	S_3	A_1	X_b	Ratio
0	S_1	2	1	1	0	0	0	40	20
0	S_2	4	−1	0	1	0	0	20	5
−1	A_1	1	0	0	0	−1	1	30	30
	C_j	0	0	0	0	0	−1		
	Z_j	−1	0	0	0	1	−1		
	$C_j–Z_j$	1	0	0	0	−1	0		

Table 3.77

Tableau 2									
C_i	Basis	X_1	X_2	S_1	S_2	S_3	A_1	X_b	Ratio
0	S_1	0	1.5	1	−0.5	0	0	30	20
0	X_1	1	−0.25	0	0.25	0	0	5	−20
−1	A_1	0	0.25	0	−0.25	−1	1	25	100
	C_j	0	0	0	0	0	−1		
	Z_j	0	−0.25	0	0.25	1	−1		
	$C_j–Z_j$	0	0.25	0	−0.25	−1	0		

Table 3.78

Tableau 3									
C_i	Basis	X_1	X_2	S_1	S_2	S_3	A_1	X_b	Ratio
0	X_2	0	1	0.667	−0.33	0	0	20	
0	X_1	1	0	0.167	0.167	0	0	10	
−1	A_1	0	0	−0.17	−0.17	−1	1	20	
	C_j	0	0	0	0	0	−1		
	Z_j	0	0	0.167	0.167	1	−1		
	$C_j–Z_j$	0	0	−0.17	−0.17	−1	0		

As per the calculation since index row contains negative or zero elements the I phase automatically gets terminated. However, the artificial variable A_1 never got eliminated. So we fail to proceed further for phase II. This is an indication of absence of feasible solution.

3.10 INTEGRATING PRIMAL AND DUAL OPTIMUM SOLUTIONS

■ Example 21

For the problem given below write its dual and solve it by Simplex method.

$$\begin{aligned} \text{Maximize: } & R = 45X_1 + 80X_2 \\ \text{Subject to: } & 5X_1 + 20X_2 \leq 400 \\ & 10X_1 + 15X_2 \leq 450 \\ & X_1 \; ; \; X_2 \geq 0 \end{aligned}$$

Dual formation

$$\begin{aligned} \text{Minimize: } & Z = 400Y_1 + 450Y_2 \\ \text{Subject to: } & 5Y_1 + 10Y_2 \geq 40 \\ & 20Y_1 + 15Y_2 \geq 80 \\ & Y_1 \; ; \; Y_2 \geq 0 \end{aligned}$$

Since we have greater than or equal inequalities in both the constraints let us update the problem with suitable surplus and artificial variables first

$$\text{Minimize: } Z = 400Y_1 + 450Y_2 + 0S_1 + 0S_2 + A_1 + A_2$$
$$\text{Subject to: } 5Y_1 + 10Y_2 - S_1 + A_1 = 40$$
$$20Y_1 + 15Y_2 - S_2 + A_2 = 80$$
$$Y_1 \; ; \; Y_2 \; ; \; S_1 \; ; \; S_2 \; ; \; A_1 \; ; \; A_2 \geq 0$$

Phase I

$$\text{Minimize: } Z = 0Y_1 + 0Y_2 + 0S_1 + 0S_2 + A_1 + A_2$$
$$\text{Subject to: } 5Y_1 + 10Y_2 - S_1 + A_1 = 40$$
$$20Y_1 + 15Y_2 - S_2 + A_2 = 80$$
$$Y_1 \; ; \; Y_2 \; ; \; S_1 \; ; \; S_2 \; ; \; A_1 \; ; \; A_2 \geq 0$$

Table 3.79

Tableau 1

C_i	Basis	Y_1	Y_2	S_1	S_2	A_1	A_2	X_b	Min
1	A_1	5	10	−1	0	1	0	45	9
1	A_2	20	15	0	−1	0	1	80	4
	C_j	0	0	0	0	1	1		
	Z_j	25	25	−1	−1	1	1	125	
	$Z_j–C_j$	25	25	−1	−1	0	0		

Table 3.80

Tableau 2

C_i	Basis	Y_1	Y_2	S_1	S_2	A_1	X_b	Min
1	A_1	0	6.25	−1	0.25	1	25	4
0	Y_1	1	0.75	0	−0.05	0	4	5.333
	C_j	0	0	0	0	1		
	Z_j	0	6.25	−1	0.25	1	25	
	$Z_j–C_j$	0	6.25	−1	0.25	0		

Table 3.81

Tableau 3

C_i	Basis	Y_1	Y_2	S_1	S_2	X_b	Min
0	Y_2	0	1	−0.16	0.04	4	
0	Y_1	1	0	0.12	−0.08	1	
	C_j	0	0	0	0		
	Z_j	0	0	0	0	0	
	$Z_j–C_j$	0	0	0	0		

Table 3.82

Tableau 4							
C_i	Basis	Y_1	Y_2	S_1	S_2	X_b	Min
450	Y_2	0	1	–0.16	0.04	4	
400	Y_1	1	0	0.12	–0.08	1	
	C_j	400	450	0	0		
	Z_j	400	450	–24	–14	2200	
	$Z_j–C_j$	0	0	–24	–14		

Phase II

Table 3.83

Tableau 5							
C_i	Basis	Y_1	Y_2	S_1	S_2	X_b	Min
450	Y_2	0	1	–0.16	0.04	4	
400	Y_1	1	0	0.12	–0.08	1	
	C_j	400	450	0	0		
	Z_j	400	450	–24	–14	2200	
	$Z_j–C_j$	0	0	–24	–14		

So the optimal solutions are : $Y_1 = 1$, $Y_2 = 4$, Minimum $Z = 2200$.

Given the simplex solution for the primal one can read the answer for its dual from the very last iteration table of the primal. Similarly, given the optimum solution for the dual the optimum for the primal is always readable from the last iteration table of the dual. Thus, it is enough to solve either the primal or the dual. To assess the situation let us consider example 23 for discussion.

Solution for the Dual from the Final Primal Solution Table for Example 21

Look the values stated just below slack variables in the primal final solution Table 3.84 corresponding to Z_j^{th} row. These values are the optimum values of the corresponding dual variables. In our illustration these values are highlighted as 1 and 4. Thus the dual solutions from the primal are $Y_1 = 1$ and $Y_2 = 4$. The optimum value of the primal and dual is equal to 2200.

Table 3.84: The primal solution for example 23

C_i	Basis	X_1	X_2	S_1	S_2	X_b
80	X_2	0	1	2/25	–1/25	14
45	X_1	1	0	–3/25	–4/25	24
	C_j	45	80	0	0	
	Z_j	45	80	1	4	2200
	$C_j–Z_j$	0	0	–1	0	

Solution for the Primal from the Final Dual Solution Table for Example 21

Table 3.85: The dual solution

C_i	Basis	Y_1	Y_2	S_1	S_2	X_b
450	Y_2	0	1	−0.16	0.04	4
400	Y_1	1	0	0.12	−0.08	1
	C_j	400	450	0	0	
	Z_j	400	450	−24	−14	2200
	$Z_j–C_j$	0	0	−24	−14	

Look at the values entered just below the surplus variable in the dual final solution Table 3.85 corresponding to Z_jth row. These values, of course, with reversed sign give the optimum values of the corresponding primal variables. In our illustration these values are highlighted as −24 and −14. Thus the primal solutions are $X_1 = 24$ and $X_2 = 14$. The optimum values of the primal and dual are equal to 2200.

■ Example 22

For the primal given obtain the dual. Also solve the dual and read the solution for the primal.

$$\begin{aligned} \text{Minimize: } C &= 200X_1 + 900X_2 + 1500X_3 \\ \text{Subject to: } & X_1 + 2X_2 + X_3 \geq 12 \\ & 3X_1 + 2X_2 + X_3 \geq 10 \\ & X_1 + X_2 + 5X_3 \geq 14 \\ & X_1 \ ; \ X_2 \ ; \ X_3 \geq 0 \end{aligned}$$

Solution

Dual problem

$$\begin{aligned} \text{Maximize: } R &= 12Y_1 + 10Y_2 + 14Y_3 \\ \text{Subject to: } & Y_1 + 2Y_2 + Y_3 \leq 1200 \\ & 3Y_1 + 2Y_2 + Y_3 \leq 900 \\ & Y_1 + Y_2 + 5Y_3 \leq 1500 \\ & Y_1 \ ; \ Y_2 \ ; \ Y_3 \geq 0 \end{aligned}$$

Solution for the dual

Table 3.86

Tableau 1									
C_i	Basis	Y_1	Y_2	Y_3	S_1	S_2	S_3	X_b	Ratio
0	S_1	1	2	1	1	0	0	1200	1200
0	S_2	3	2	1	0	1	0	900	900
0	S_3	1	1	5	0	0	1	1500	300
	C_j	12	10	14	0	0	0		
	Z_j	0	0	0	0	0	0	0	
	C_j-Z_j	12	10	14	0	0	0		

Table 3.87

Tableau 2									
C_i	Basis	Y_1	Y_2	Y_3	S_1	S_2	S_3	X_b	Ratio
0	S_1	0.8	1.8	0	1	0	−0.2	900	1125
0	S_2	2.8	1.8	0	0	1	−0.2	600	214.3
14	Y_3	0.2	0.2	1	0	0	0.2	300	1500
	C_j	12	10	14	0	0	0		
	Z_j	2.8	2.8	14	0	0	2.8	4200	
	C_j-Z_j	9.2	7.2	0	0	0	−2.8		

Table 3.88

Tableau 3									
C_i	Basis	Y_1	Y_2	Y_3	S_1	S_2	S_3	X_b	Ratio
0	S_1	0	1.286	0	1	−0.286	−0.143	728.6	566.7
12	Y_1	1	0.643	0	0	0.357	−0.071	214.3	333.3
14	Y_3	0	0.071	1	0	−0.071	0.214	257.1	3600
	C_j	12	10	14	0	0	0		
	Z_j	12	8.714	14	0	3.286	2.143	6171	
	C_j-Z_j	0	1.286	0	0	−3.286	−2.143		

Table 3.89

Tableau 4									
C_i	Basis	Y_1	Y_2	Y_3	S_1	S_2	S_3	X_b	Ratio
0	S_1	−2	0	0	1	−1	0	300	566.7
10	Y_2	1.556	1	0	0	0.556	−0.111	333.3	333.3
14	Y_3	−0.111	0	1	0	−0.111	0.222	233.3	3600
	C_j	12	10	14	0	0	0		
	Z_j	14	10	14	0	4	2	6600	
	C_j-Z_j	−2	0	0	0	−4	−2		

Solution for the dual: $Y_1 = 0$, $Y_2 = 333.3333$; $Y_3 = 233.3333$; $R = 6600$.

Solution for the primal: $X_1 = 0$; $X_2 = 4$; $X_3 = 2$; $C = 6600$.

Example 23

For the primal obtain the dual. Also obtain its standard form and solve it by Simplex method

$$
\begin{aligned}
\text{Maximize: } & R = 3X_1 + 1X_2 \\
\text{Subject to: } & 1X_1 + 3X_2 \leq 90 \\
& 2X_1 + 2X_2 \leq 80 \\
& 2X_1 + 0X_2 \leq 60 \\
& X_1 \;;\; X_2 \geq 0
\end{aligned}
$$

Solution: Dual

$$
\begin{aligned}
\text{Minimize: } & C = 90Y_1 + 80Y_2 + 60Y_3 \\
\text{Subject to: } & 1Y_1 + 2Y_2 + 2Y_3 \geq 3 \\
& 3Y_1 + 2Y_2 + 0Y_3 \geq 1 \\
& Y_1 \;;\; Y_2 \;;\; Y_3 \geq 0
\end{aligned}
$$

Standard form of the dual problem

$$
\begin{aligned}
\text{Minimize: } & C = 90Y_1 + 80Y_2 + 60Y_3 + 0S_1 + 0S_2 + MA_1 + MA_2 \\
\text{Subject to: } & 1Y_1 + 2Y_2 - 2Y_3 - S_1 + A_1 = 3 \\
& 3Y_1 + 2Y_2 - 0Y_3 - S_2 + A_2 = 1 \\
& Y_1 \;;\; Y_2 \;;\; Y_3 \;;\; S_1 \;;\; S_2 \;;\; A_1 \;;\; A_2 \geq 0
\end{aligned}
$$

Solution by Two-Phase Method: Phase I

Table 3.90

Tableau 1

C_i	Basis	Y_1	Y_2	Y_3	S_1	S_2	A_1	A_2	X_b	Min
1	A_1	1	2	2	−1	0	1	0	3	1.5
1	A_2	3	2	0	0	−1	0	1	1	0.5
	C_j	0	0	0	0	0	1	1		
	Z_j	4	4	2	−1	−1	1	1	4	
	$Z_j–C_j$	4	4	2	−1	−1	0	0		

Table 3.91

Tableau 2

C_i	Basis	Y_1	Y_2	Y_3	S_1	S_2	A_1	X_b	Min
1	A_1	−2	0	2	−1	1	1	2	1
0	Y_2	1.5	1	0	0	−0.5	0	0.5	∞
	C_j	0	0	0	0	0	1		
	Z_j	−2	0	2	−1	1	1	2	
	$Z_j–C_j$	−2	0	2	−1	1	0		

Table 3.92

Tableau 3								
C_i	Basis	Y_1	Y_2	Y_3	S_1	S_2	X_b	Min
0	Y_3	−1	0	1	−0.5	0.5	1	1
0	Y_2	1.5	1	0	0	−0.5	0.5	∞
	C_j	0	0	0	0	0		
	Z_j	0	0	0	0	0	0	
	$Z_j–C_j$	0	0	0	0	0		

Phase II

Table 3.93

Tableau 4								
C_i	Basis	Y_1	Y_2	Y_3	S_1	S_2	X_b	Min
60	Y_3	−1	0	1	−0.5	0.5	1	1
80	Y_2	1.5	1	0	0	−0.5	0.5	∞
	C_j	90	80	60	0	0		
	Z_j	60	80	60	−30	−10	100	
	$Z_j–C_j$	−30	0	0	−30	−10		

So the optimum solutions are: $Y_1 = 0$, $Y_2 = 0.5$, $Y_3 = 1$, $C = 100$.

■ Example 24

For the following primal write its dual:

$$
\begin{aligned}
\text{Maximize: } R = {} & 40X_1 + 30X_2 + 50X_3 \\
\text{Subject to: } & 0.4X_1 + 0.5X_2 + 0.6X_3 \le 20 \\
& 0.0X_1 + 0.2X_2 + 0.1X_3 \le 5 \\
& 0.6X_1 + 0.3X_2 + 0.3X_3 \le 21 \\
& X_1 \; ; \; X_2 \; ; \; X_3 \ge 0
\end{aligned}
$$

Solution

$$
\begin{aligned}
\text{Minimize: } Z = {} & 20Y_1 + 5Y_2 + 21Y_3 \\
\text{Subject to: } & 0.4Y_1 + 0Y_2 + 0.6Y_3 \ge 40 \\
& 0.5Y_1 + 0.2Y_2 + 0.3Y_3 \ge 30 \\
& 0.6Y_1 + 0.1Y_2 + 0.3Y_3 \ge 50 \\
& Y_1 \; ; \; Y_2 \; ; \; Y_3 \ge 0
\end{aligned}
$$

3.11 BUILT-IN EXCEL SPREADSHEET SOLUTION TO THE LINEAR PROGRAMMING PROBLEM

Excel spreadsheet could be used effectively to solve the linear programming problem of any size and any type. By taking an example we will illustrate the method for getting computer excel solution.

■ Example 25

Solve the following linear programming problem using Excel spreadsheet.

$$\text{Maximize: } R = 3X_1 + 1X_2$$

$$\begin{aligned} \text{Subject to: } \quad & 1X_1 + 3X_2 \le 90 \text{ (Constraint X)} \\ & 2X_1 + 2X_2 \le 80 \text{ (Constraint Y)} \\ & 2X_1 + 0X_2 \le 60 \text{ (Constraint Z)} \\ & X_1 \;;\; X_2 \ge 0 \end{aligned}$$

Solution

Step 1: Open an excel worksheet and enter the data on the top of the spreadsheet as shown in the spreadsheet from row 1 to 6. Cells B3 to C5 describe the input requirement per unit of output. B6 and D6 describe the revenue per unit of output as stated in the problem. D3 to D5 we enter the respective input availability.

Step 2: Reserve cells B12 for the output X_1 and C12 for output X_2 by shading.

Step 3: In cell B14 enter the formula for the calculation of the revenue as =B6*B12+C6*C12.

Table 3.94

	A	B	C	D
1		Input requirements per unit of output		
2	Inputs	X_1	X_2	Inputs available
3	X	1	3	90
4	Y	2	2	80
5	Z	2	0	60
6	Profit per unit	3	1	
7				
8		Model description		
9				
10		Decision variables		
11		X_1	X_2	
12				
13				
14	Maximum profit	=B6*B12+C6*C12		
15				
16	Constraints	Inputs used		Inputs available
17	X	=B3*B12+C3*C12	<=	=D3
18	Y	=B4*B12+C4*C12	<=	=D4
19	Z	=B5*B12+C5*C12	<=	=D5
20				

Step 4: In cell B17 enter the left side of the first constraint as =B3*B12+C3*C12
In cell B18 enter the left side of the second constraint as =B4*B12+C4*C12.
In cell B19 enter the left side of the third constraint as =B5*B12+C5*C12.

Step 5: To specify the availability of inputs enter
Cell D17: = D3
Cell D18: = D4
Cell D19: = D5
This will automatically copy the input entries already made in D3, D4 and D5.

Step 6: Make the other descriptive entries as shown in the spreadsheet to make the model self-explanatory.

Spreadsheet solution

Step 7: Select the **Tools** pull down menu on the top of your computer screen.

Step 8: Select the **Solver** option.

Step 9: In the **Solver Parameter** dialog box enter B14 into the **Set target Cell** box. Select **Equal to: Max. Option.**
Enter B12:C12 in to the **By changing Cells** box. Select **Add.**

Table 3.95

	A	B	C	D
1		Input requirements per unit of output		
2	Inputs	X_1	X_2	Inputs available
3	X	1	3	90
4	Y	2	2	80
5	Z	2	0	60
6	Profit per unit	3	1	
7				
8		Model description		
9				
10		Decision variables		
11		X_1	X_2	
12		30	10	
13				
14	Maximum profit	100		
15				
16	Constraints	Inputs used		Inputs available
17	X	60	<=	90
18	Y	80	<=	80
19	Z	60	<=	60
20				

Step 10: When the **Add constraint** dialog box appears enter:

Enter B17:B19 in the **Cell Reference** box.
Select <=
Enter D17:D19 in the **constraint** box
Select **OK**

Step 11: When solver Parameters dialog box appears:

Choose **Options**

Step 12: When **Solver Options** dialog box appears:

Select **Assume Linear Model**
Select **Assume Non-Negative**
Select **OK**

Step 13: When the **Solver Parameters** dialog box appears:

Select **Solve**

Step 14: When the **Solver Results** dialog box appears:

Select Keep **Solver Solution**
Select **OK**

After completing all these steps the final solution sheet will look like the one shown in Table 3.95 with all the needed results.

Excel solution for more than two decision variables

The graphical method is useful only for two-decision variables. However, the spreadsheet method is capable of handling more than two-decision variables as well. The following example illustrates this method.

■ Example 26

Solve the following problem with three-decision variables by spreadsheet method

$$
\begin{array}{llcccccl}
\text{Maximize:} & R = & 40X_1 & + & 30X_2 & + & 50X_3 & \\
\text{Subject to:} & & 0.4X_1 & + & 0.5X_2 & + & 0.6X_3 & \leq 20 \\
 & & 0.0X_1 & + & 0.2X_2 & + & 0.1X_3 & \leq 5 \\
 & & 0.6X_1 & + & 0.3X_2 & + & 0.3X_3 & \leq 21 \\
 & & X_1 & ; & X_2 & ; & X_3 & \geq 0
\end{array}
$$

Table 3.96

	A	B	D	E	E
1		Material requirements			
2	Material	X_1	X_2	X_3	Inputs available
3	Material 1	0.4	0.5	0.6	20
4	Material 2		0.2	0.1	5
5	Material 3	0.6	0.3	0.3	21
6	Profit per unit	40	30	50	
7					
8	Model				
9		Decision variables			
10		X_1	X_2	X_3	
11		27.5	0	15	
12	Maximum profit	1850			
13					
14	Constraints	Inputs used			Inputs available
15	Material 1	20	<=		20
16	Material 2	1.5	<=		5
17	Material 3	21	<=		21

So the optimum solutions are: $X_1 = 27.5$, $X_2 = 0$, $X_3 = 15$ and $Z = 1850$.

3.12 SOME IMPORTANT LINEAR PROGRAMMING THEOREMS

Primal

Maximize: $c'X$

Subject to: $AX \leq b$

$X \geq 0$

Dual

Minimize: $b'Y$

Subject to: $A'Y \geq c$

$Y \geq 0$

Theorem 1

$X^* \geq 0$ is a feasible solution of the primal and $Y^* \geq 0$ is a feasible solution of the corresponding dual then $c'X^* \leq b'Y^*$.

Theorem 2

$X^* \geq 0$ is a feasible solution of the primal and $Y^* \geq 0$ is a feasible solution of the corresponding dual so that $c'X^* = b'Y^*$, then the optimum feasible solution of the primal is the optimum feasible solution of the dual.

Theorem 3

If the primal has a finite optimum solution then the corresponding dual also will have a finite optimum solution so that $c'X^* = b'Y^*$.

This theorem not only establishes the existence of the dual solution but also reveals the method of getting the answer for the dual when the primal is solved. In fact, the values of $C_j - Z_j$ corresponding to the slack variables in the primal are the optimum values of the dual solution.

Theorem 4

If the primal has an unbounded solution, the corresponding dual will not have any feasible solution.

Theorem 5

If the dual has no feasible solution but the primal has one solution then, the solution of said primal is unbounded.

3.13 FORMATION OF GENERAL LINEAR PROGRAMMING PRIMAL PROBLEM IN MATRIX FORM

The general form of a linear programming problem (LPP) with n decision variables and m constraints is normally represented in the following form

$$\begin{aligned}
\text{Maximize: } & c_1X_1 + c_2X_2 + \ldots + c_nX_n \\
\text{Subject to: } & a_{11}X_1 + a_{12}X_2 + \ldots + a_{1n}X_n \leq b_1 \\
& a_{21}X_1 + a_{22}X_2 + \ldots + a_{2n}X_n \leq b_2 \\
& a_{31}X_1 + a_{32}X_2 + \ldots + a_{3n}X_n \leq b_2 \\
& \ldots\ldots\ldots\ldots\ldots\ldots\ldots\ldots\ldots\ldots \\
& \ldots\ldots\ldots\ldots\ldots\ldots\ldots\ldots\ldots\ldots \\
& a_{m1}X_1 + a_{m2}X_2 + \ldots + a_{mn}X_n \leq b_m \\
& X_1 + X_2 + \ldots + X_n \geq 0
\end{aligned}$$

By using summation notation this could also be written as

$$\text{Maximize: } \sum_{j=1}^{n} c_jX_j \qquad \text{(Objective function)}$$

$$\text{Subject to: } \sum_{i=1}^{n} a_{ij}X_j \leq b_i \quad i = 1,2,\ldots n \text{ (Constraints)}$$

$$X_j \geq 0 \qquad \text{(Non-negative conditions)}$$

If X is an n×1, C is an n×1, A is an m×n and b is an m×1 matrix with the respective elements

$$X=\begin{bmatrix} X_1 \\ X_2 \\ \vdots \\ \vdots \\ Xn \end{bmatrix};\ C=\begin{bmatrix} C_1 \\ C_2 \\ \vdots \\ \vdots \\ C_n \end{bmatrix};\ A=\begin{bmatrix} a_{11} & a_{12} & \cdots & \cdots & a_{1n} \\ a_{21} & a_{22} & \cdots & \cdots & a_{2n} \\ \cdots & \cdots & \cdots & \cdots & \cdots \\ \cdots & \cdots & \cdots & \cdots & \cdots \\ a_{m1} & a_{m2} & \cdots & \cdots & a_{mn} \end{bmatrix};\ b=\begin{bmatrix} b_1 \\ b_2 \\ \vdots \\ \vdots \\ b_m \end{bmatrix}$$

Then the linear programming problem in its matrix format is normally written as

$$\text{Maximize:}\quad \begin{bmatrix} c_1 & c_2 & \ldots & \ldots & c_n \end{bmatrix} \begin{bmatrix} X_1 \\ X_2 \\ \vdots \\ \vdots \\ X_n \end{bmatrix}$$

$$\text{Subject to:}\quad \begin{bmatrix} a_{11} & a_{12} & \cdots & \cdots & a_{1n} \\ a_{21} & a_{22} & \cdots & \cdots & a_{2n} \\ \cdots & \cdots & \cdots & \cdots & \cdots \\ \cdots & \cdots & \cdots & \cdots & \cdots \\ a_{m1} & a_{m2} & \cdots & \cdots & a_{mn} \end{bmatrix} \begin{bmatrix} X_1 \\ X_2 \\ \vdots \\ \vdots \\ X_n \end{bmatrix} \le \begin{bmatrix} b_1 \\ b_2 \\ \vdots \\ \vdots \\ b_m \end{bmatrix}$$

$$X_1 \ge 0$$

$$\begin{aligned} \text{Maximize: } & C'X \\ \text{Subject to: } & AX \le b \\ & X \ge 0 \end{aligned}$$

3.14 STANDARD FORM OF A LINEAR PROGRAMMING PROBLEM IN MATRIX FORM

We can always write this in the standard form after introducing m slack variables one each for all the m constraints.

$$\text{Maximize: } c_1X_1 + c_2X_2 + \ldots + c_nX_n + 0S_1 + 0S_2 + \ldots + 0S_m$$

$$\begin{aligned}
\text{Subject to: } & a_{11}X_1 + a_{12}X_2 + \ldots + a_{1n}X_n + S_1 \quad \ldots \quad = b_1 \\
& a_{21}X_1 + a_{22}X_2 + \ldots + a_{2n}X_n + \quad + S_2 \quad \ldots \quad = b_2 \\
& a_{31}X_1 + a_{32}X_2 + \ldots + a_{3n}X_n + \quad + S_3 \quad = b_3 \\
& \ldots\ldots\ldots\ldots\ldots\ldots\ldots\ldots\ldots\ldots\ldots\ldots \\
& \ldots\ldots\ldots\ldots\ldots\ldots\ldots\ldots\ldots\ldots\ldots\ldots \\
& a_{m1}X_1 + a_{m2}X_2 + \ldots + a_{mn}X_n + \quad \ldots + S_m = b_m \\
& X_1 + X_2 + \ldots + X_n \;;\; S_1 \;;\; S_2 \quad \ldots \quad S_m \geq 0
\end{aligned}$$

$$\text{Maximize: } \begin{bmatrix} c_1 & c_2 & \ldots & \ldots & c_c \end{bmatrix} \begin{bmatrix} X_1 \\ X_2 \\ \vdots \\ \vdots \\ X_n \end{bmatrix}$$

$$\text{Subject to: } \begin{bmatrix} a_{11} & a_{12} & \ldots\ldots & \ldots\ldots & a_{1n} \\ a_{21} & a_{22} & \ldots\ldots & \ldots\ldots & a_{2n} \\ \ldots\ldots & \ldots\ldots & \ldots\ldots & \ldots\ldots & \ldots\ldots \\ \ldots\ldots & \ldots\ldots & \ldots\ldots & \ldots\ldots & \ldots\ldots \\ a_{m1} & a_{m2} & \ldots\ldots & \ldots\ldots & a_{mn} \end{bmatrix} \begin{bmatrix} X_1 \\ X_2 \\ \vdots \\ \vdots \\ X_n \end{bmatrix}$$

$$+ \begin{bmatrix} 1 & 0 & 0 & 0 & 0 \\ 0 & 1 & 0 & 0 & 0 \\ \vdots & \vdots & \vdots & \vdots & \vdots \\ 0 & 0 & 0 & 1 & 0 \\ 0 & 0 & 0 & 0 & 1 \end{bmatrix} \begin{bmatrix} S_1 & S_2 & \ldots & \ldots & S_m \end{bmatrix} = \begin{bmatrix} b_1 \\ b_2 \\ \vdots \\ \vdots \\ b_m \end{bmatrix}$$

$$\begin{aligned}
\text{Maximize: } & C'X \\
\text{Subject to: } & AX + IS = b \\
& X \geq 0
\end{aligned}$$

where S is an $1 \times m$ slack variables and I is an $m \times m$ identity matrix.

The elements of C indicate the revenue per unit of the output and hence called the revenue matrix. In the matrix A, typical element a_{ij} refers to the per unit requirement of the ith input for the jth output. This is often referred to as the activity matrix. The matrix b shows the availability of resources so it is called the availability matrix.

Example 27

Express the following linear programming problem in its matrix form

$$\begin{aligned} \text{Maximize:}\quad & R = 3X_1 + 1X_2 \\ \text{Subject to:}\quad & 1X_1 + 3X_2 \leq 90 \\ & 2X_1 + 2X_2 \leq 80 \\ & 2X_1 + 0X_2 \leq 60 \\ & X_1 \,;\; X_2 \geq 0 \end{aligned}$$

$$\text{If } X = \begin{bmatrix} X_1 \\ X_2 \end{bmatrix};\; C = \begin{bmatrix} 3 \\ 1 \end{bmatrix};\; A = \begin{bmatrix} 1 & 3 \\ 2 & 2 \\ 2 & 0 \end{bmatrix};\; b = \begin{bmatrix} 90 \\ 80 \\ 60 \end{bmatrix} \text{ then}$$

$$\text{Maximize:}\quad [3 \;\; 1] \begin{bmatrix} X_1 \\ X_2 \end{bmatrix}$$

$$\text{Subject to:}\quad \begin{bmatrix} 1 & 3 \\ 2 & 2 \\ 2 & 2 \end{bmatrix} [X_1 \;\; X_2] \leq \begin{bmatrix} 90 \\ 80 \\ 60 \end{bmatrix}$$

$$X_i \geq 0$$

$$\begin{aligned} \text{Maximize:}\quad & C'X \\ \text{Subject to:}\quad & AX \leq b \\ & X \geq 0 \end{aligned}$$

Let us convert the inequality constraints into equalities by introducing three slack variables first and put the problem in the standard form.

$$\begin{aligned} \text{Minimize:}\quad & R = 3X_1 + 1X_2 + 0S_1 + 0S_2 + 0S_3 \\ \text{Subject to:}\quad & 1X_1 + 3X_2 + S_1 = 90 \\ & 2X_1 + 2X_2 + S_2 = 80 \\ & 2X_1 + 0X_2 + S_3 = 60 \\ & X_1 \,;\; X_2 \,;\; S_1 \,;\; S_2 \,;\; S_3 \geq 0 \end{aligned}$$

$$\text{Maximize:}\quad [3 \;\; 1] \begin{bmatrix} X_1 \\ X_2 \end{bmatrix} + \begin{bmatrix} 0 \\ 0 \\ 0 \end{bmatrix} [S_1 \;\; S_2 \;\; S_3]$$

$$\text{Subject to:}\quad \begin{bmatrix} 1 & 3 \\ 2 & 2 \\ 2 & 2 \end{bmatrix} [X_1 \;\; X_2] + \begin{bmatrix} 1 & 0 & 0 \\ 0 & 1 & 0 \\ 0 & 0 & 1 \end{bmatrix} [S_1 \;\; S_2 \;\; S_3] \leq \begin{bmatrix} 90 \\ 80 \\ 60 \end{bmatrix}$$

$$X_i, S_i \geq 0$$

This standard form could be stated in matrix form as follows

Maximize: $C'X$
Subject to: $AX + IY = b$
$X, Y \geq 0$

3.15 FORMATION OF GENERAL LINEAR PROGRAMMING DUAL PROBLEM IN MATRIX FORM

The optimum solution of either problem invariably contains the optimum information regarding the other. In fact, if the given original problem in it simplex formulation contains m×n unit matrix then it dual formulation will have n×m unit matrix.

Dual for general primal discussed in the previous section

$$\begin{aligned}
\text{Minimize: } Z = b_1Y_1 &+ b_2Y_2 + \ldots + b_mY_m \\
\text{Subject to: } a_{11}Y_1 &+ a_{21}Y_2 + \ldots + a_{m1}Y_m \leq c_1 \\
a_{12}Y_1 &+ a_{22}Y_2 + \ldots + a_{m2}Y_m \leq c_2 \\
a_{13}Y_1 &+ a_{23}Y_2 + \ldots + a_{m3}Y_m \leq c_2 \\
&\ldots\ldots\ldots\ldots\ldots\ldots\ldots\ldots \\
&\ldots\ldots\ldots\ldots\ldots\ldots\ldots\ldots \\
a_{1n}Y_1 &+ a_{2n}Y_2 + \ldots + a_{mn}Y_n \leq c_n \\
Y_1 &+ Y_2 + \ldots + Y_n \geq 0
\end{aligned}$$

$Y = \begin{bmatrix} Y_1 \\ Y_2 \\ \vdots \\ \vdots \\ Y_m \end{bmatrix}$ is an m×1 matrix then the dual in it s matrix may be written as

Minimize: $b'Y$
Subject to: $A'Y \geq C$
$Y \geq 0$

EXERCISES

1. A company produces two products A and B using a common production process. The production process spans a total of 50,000 man-hours. Product A uses 6 man-hours to produce one unit of the product. Product B uses 4 man-hours per unit of output. The market survey reveals that the company can sell at the most 7000 units of A and 10,000 units of B. If Rs. 500 and Rs. 300 are the respective revenues for A and B formulate a suitable linear programming problem.

2. A company manufactures two products A and B. These products undergo processing through two machines before taking the final shape. Given the following data formulate a suitable linear programming problem and solve it by Simplex method.

	Product A	Product B	Weekly capacity
Machine 1	3	4	220
Machine 2	2	3	400
Selling price	Rs.25	Rs.45	

3. A company manufactures two batteries, namely, Standard and Excel. The revenues are Rs. 40 and Rs. 50 per unit respectively. The Excel consumes twice the time than the Standard model. If Standard is the priority it can make 1000 batteries per day with all its resource at its command. Putting both the varieties together it can produce 800 per day. If the outer casting availability for Standard model is 400 and Excel is 600 formulate a suitable linear programming problem.

4. A food processing company in Bangalore wants to venture for a new food item with low calorie and high portion by blending three basic food items A, B and C. The recommended specifications by medical experts are as follows:

Elements	Basic foods			New food
	A	B	C	
Calories	320	240	180	Less than or equal to 300
Protein	250	230	200	Greater than or equal to 200
Vitamin A	100	150	50	Greater than or equal to 100
Vitamin C	75	120	130	Greater than or equal to 120
Cost	10	20	30	

Formulate a suitable linear programming problem and solve it by simplex method.

5. The police department in a town scheduled its minimum requirements as follows.

Period	Time	Minimum number required
1	0-6	20
2	6-10	45
3	10-14	87
4	14-18	100
5	18-22	45
6	22-24	25

Formulate a suitable linear programming problem. Note that each policeman works 8 consequent hours from the beginning of a given period.

6. The Selwel advertising company in Bangalore wants to venture a new advertising campaign in three different media—TV, radio, and newspaper. Clearly the very purpose of this new venture is to reach as much customers as possible. The results of the market survey conducted in this regard are given below.

Description	TV ordinary	TV prime	Radio	Newspaper
Cost of advertising per unit	Rs.40,000	Rs.75,000	Rs.30,000	Rs.15,000
No. of customers reached	4,00,000	9,00,000	5,00,000	2,00,000
Number of women customers reached	3,00,000	4,00,000	2,00,000	1,00,000

Note that the company does not want to spend more than Rs. 8,00,000 on all the items putting together. Further it requires:

(a) at least 2 million exposures take place among women

(b) advertising on TV is restricted to a maximum of Rs. 5,00,000

(c) at least 3 advertisement units be bought on ordinary and two units on prime time TV

(d) the number of advertising units both on radio and newspaper be 5 and 10 units respectively.

Formulate a linear programming by incorporating all the restrictions addressed.

7. An electric company produces three types of part for automatic washing machine. It purchases castings of the parts from a local foundry and then finishes the parts on drilling, shaping and polishing machines. The selling price of part A, B and C respectively are Rs. 8, Rs. 10, and Rs. 14. All parts made can be sold. Casting for parts A, B, and C cost Rs. 5, Rs. 6 and Rs. 10.

The shop possesses only one of each type of machine. Costs per hour to run each of type of three machines are Rs. 20 for drilling, Rs. 30 for shaping and Rs. 30 for polishing. The capacities for each part (parts per hour) on each machine are shown in the following table.

Machine	Capacity per hour		
	Part A	Part B	Part C
Drilling	25	40	25
Shaping	25	20	20
Polishing	40	30	40

The management of the shop wants to know how many parts of each type it should produce per hour in order to maximize revenue for an hour's run. Formulate this problem as an LP model.

(Delhi University, MBA 1986)

8. A company is making two products A and B. The cost of producing one unit of the product A and B are Rs. 60 and Rs. 80 respectively. As per the agreement, the company has to supply at least 2000 unit of product A to its regular customer. One unit of A requires one machine hour whereas product B has machine hours available abundantly within the company. Total machine hours available for product A are 4000 hours. One unit of each product A and B requires one labour hour each and total of 500 labour hours are available. The company wants to minimize the cost of production by satisfying the given requirements. Formulate the problem as a linear programming problem.

(Allahabad University, MBA, 1989)

9. A firm assembles and sells two different types of outboard motor A and B, using four resources. The production process can be described as follows:

Resources	Capacity per month
Motor unit shop resource	400 type A units or 250 type B units or any linear combination of the two
Type A gear and derive shop resource	175 type A units
Type B gear and derive shop resource	225 type B units
Final assembly resource	200 type A units or 359 type B units or any linear combination of the two

Type a unit bring in revenue of Rs. 90 each, type B units Rs. 60 each. What should be the optimum product mix? Formulate the problem.

(Delhi University, B.A (Econ. Hons) 1989)

10. Four products have to be produced through a plant, the quantity required for the next production period being

Product 1: 2000 units Product 2: 3000 units

Product 3: 3000 units Product 4: 6000 units

There are three production lines on which the products could be produced. The rate of production in units per day and the total available capacity in days are given in the following table. The cost of using the lines is Rs. 600, Rs. 500 and Rs. 4000 per day respectively.

Production Line	Product				Maximum line days
	1	2	3	4	
1	150	100	500	400	20
2	200	100	760	400	20
3	160	80	890	600	18
Total	2000	3000	3000	6000	

Formulate the problem as an LP model to minimize the cost of production.

(Delhi University, MBA, 1994)

11. There are three products X, Y, Z which has three ingredients a, b, c in it. The selling price of X, Y, Z is Rs. 100, Rs. 120 and Rs. 140 each respectively. The cost of the ingredients a, b, and c are Rs. 40, Rs. 50, and Rs. 120 each respectively.

12. Write the dual of:

$$\begin{aligned} \text{Maximize: } & Z = 5X_1 + 3X_2 \\ \text{Subject to: } & 4X_1 + 5X_2 \le 10 \\ & 5X_1 + 2X_2 \le 10 \\ & 3X_1 + 8X_2 \le 12 \\ & X_1 \; ; \; X_2 \ge 0 \end{aligned}$$

(Bangalore University, M.B.A July/August 2005)

13. Solve the following using Simplex method

$$\begin{aligned} \text{Maximize: } & Z = 20X_1 + 30X_2 \\ \text{Subject to: } & 3X_1 + 3X_2 \le 36 \\ & 5X_1 + 2X_2 \le 50 \\ & 2X_1 + 6X_2 \le 60 \\ & X_1 \; ; \; X_2 \ge 0 \end{aligned}$$

(Bangalore University, M.B.A July/August 2005)

14. Write the dual of:

$$\begin{aligned} \text{Maximize: } & Z = 2X_1 + 3X_2 + X_3 \\ \text{Subject to: } & 4X_1 + 3X_2 + X_3 = 5 \\ & 1X_1 + 2X_2 + 5X_3 = 4 \\ & X_1 \; ; \; X_2 \; ; \; X_3 \ge 0 \end{aligned}$$

(Bangalore University, M.B.A October 2004)

15. Obtain the maximum of:

$$\begin{aligned} \text{Maximize: } & Z = 8X_1 + 4X_2 + 6X_3 \\ \text{Subject to: } & 4X_1 + 8X_2 + 0X_3 \ge 10 \\ & 4X_1 + 6X_2 + 2X_3 \ge 4 \\ & X_1 \; ; \; X_2 \; ; \; X_3 \ge 0 \end{aligned}$$

Use either two-phase method or find dual and solve.

(Bangalore University, M.B.A October 2004)

CHAPTER

4

Linear Programming: III Sensitivity Analysis

- *Post-optimality sensitivity analysis*
- *Sensitivity analysis for objective function coefficients*
- *Sensitivity analysis for constraint's RHS values*
- *Sensitivity analysis: a revisit using final Simplex table*
- *Sensitivity analysis for the coefficients in the objective function*
- *Sensitivity of resource change (R.H.S) on the optimum revenue*
- *Sensitivity of new product introduction on the optimal revenue*
- *Exercises*

4.1 POST-OPTIMALITY SENSITIVITY ANALYSIS

After solving a lengthy linear programming problem, one may notice that certain curtail variables either from the objective function or from the constraints has changed over time. Such an observation definitely will pose the question whether the already obtained optimum solution is still valid or not. One way out to this question is simple. We redo the problem for optimality by incorporating the noticed changes into the original problem. The other option is to conduct sensitivity test for the validity of the old result under new situation. Such a test will give us the range in which the variation in the concerned variables leave the already obtained optimum values of the decision variables unaltered. From the final simplex tableau, many such important post optimality questions can be answered. In this section, the sensitivity is addressed under two broad categories.

1. How a given change in any one of the per unit revenue contribution coefficients in the objective function affect the optimal decision variables.
2. How a given change in any one of the input availability given in the RHS in constraints affect the optimal revenue.

Thus, clearly the sensitivity analysis succeeds the optimum solution already obtained to the original linear programming problem. For this reason, sensitivity analysis is often referred as the post optimal test.

4.2 SENSITIVITY ANALYSIS FOR OBJECTIVE FUNCTION COEFFICIENTS

To illustrate the methodology behind the sensitivity analysis let us take the following problem for illustration.

■ Example 1

Solve the following linear programming problem using Excel spreadsheet solver. Also obtain its graphical solution and confirm the so obtained excel result.

$$\text{Maximize: } R = 3X_1 + 1X_2$$

$$\begin{aligned} \text{Subject to: } \quad 1X_1 + 3X_2 &\leq 90 \quad \text{(Constraint X)} \\ 2X_1 + 2X_2 &\leq 80 \quad \text{(Constraint Y)} \\ 2X_1 + 0X_2 &\leq 60 \quad \text{(Constraint Z)} \\ X_1, \quad X_2 &\geq 0 \end{aligned}$$

Excel Solver Solution

Table 4.1

	A	B	C	D
1		Input requirements per unit of output		
2	Inputs	X_1	X_2	Inputs available
3	X	1	3	90
4	Y	2	2	80
5	Z	2	0	60
6	Price per unit	3	1	
7		Model description		
8		Decision variables		
9		X_1	X_2	
10		30	10	
11	Max. revenue	100		
12	Constraints	Inputs used		Inputs available
13	X	60	<=	90
14	Y	80	<=	80
15	Z	60	<=	60
16				

Graphical Solution

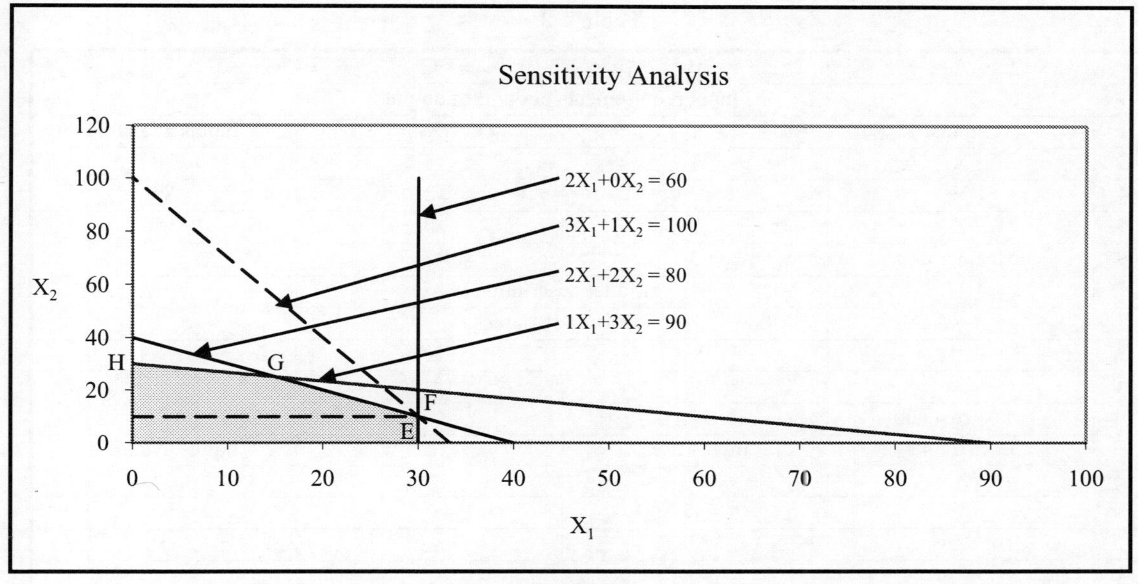

Figure 4.1

Now from both the methods the optimal values of the decision variables are $X_1 = 30$ and $X_2 = 10$. The associated maximum revenue is Rs. 100. Now to see the sensitivity on these optimum output levels we propose to consider 5 different values for the per unit contribution coefficient of the product X_2.

Case 1

Let the price of X_2 be increased to 2 from 1 in the original problem. Then the revised problem is stated as follows.

$$\text{Maximize: } R = 3X_1 + 2X_2$$

$$\begin{aligned} \text{Subject to: } \quad 1X_1 + 3X_2 &\le 90 \quad \text{(Constraint X)} \\ 2X_1 + 2X_2 &\le 80 \quad \text{(Constraint Y)} \\ 2X_1 + 0X_2 &\le 60 \quad \text{(Constraint Z)} \\ X_1, \quad X_2 &\ge 0 \end{aligned}$$

In Table 4.2, though the maximum revenue has gone up to Rs. 110 the values of the decision variables do not change. Thus, we say that our optimum solution addressed in terms of decision variables is insensitive for such a small change in the price of X_2. The firm will produce again $X_1 = 30$ and $X_2 = 10$ and earn revenue of Rs. 110. This need not be the case all the time.

Excel Solution

Table 4.2

	A	B	C	D
1		Input requirements per unit of output		
2	Inputs	X_1	X_2	Inputs available
3	X	1	3	90
4	Y	2	2	80
5	Z	2	0	60
6	Price per unit	3	2	
7		Model description		
8		Decision variables		
9		X_1	X_2	
10		30	10	
11	Max. revenue	110		
12	Constraints	Inputs used		Inputs available
13	X	60	<=	90
14	Y	80	<=	80
15	Z	60	<=	60
16				

Graphical Solution

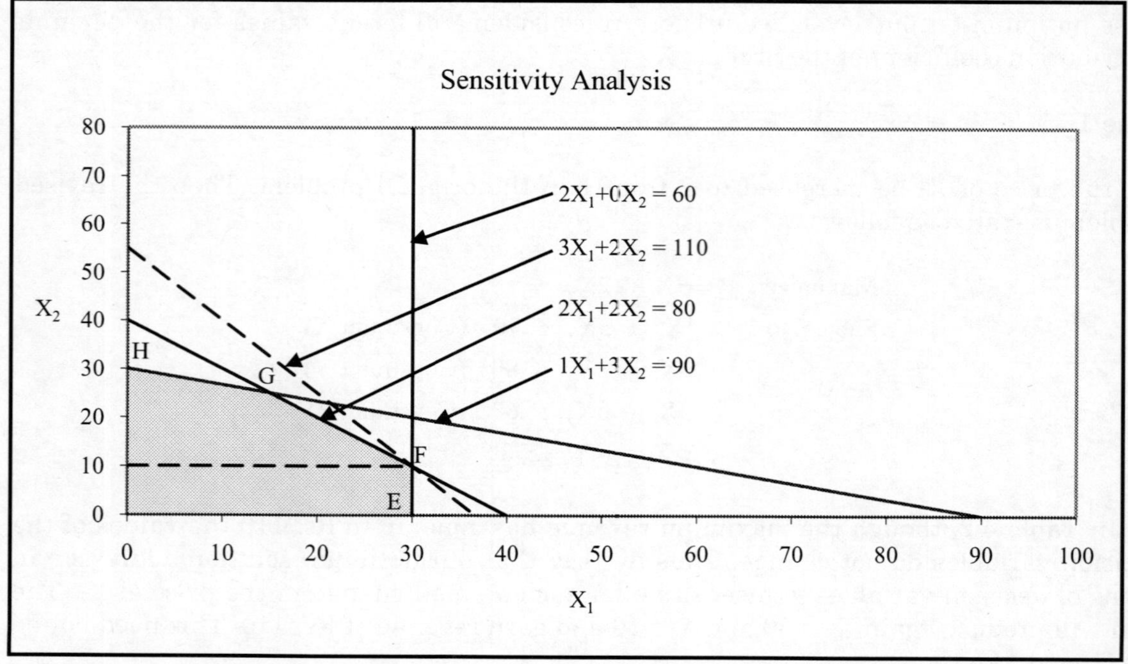

Figure 4.2

Case 2

Let the price of X_2 is increased to 3 from 1 in the original problem. The revised problem is stated as follows.

$$\begin{aligned}
\text{Maximize:}\quad & R = 3X_1 + 3X_2 \\
\text{Subject to:}\quad & 1X_1 + 3X_2 \le 90 \quad \text{(Constraint X)} \\
& 2X_1 + 2X_2 \le 80 \quad \text{(Constraint Y)} \\
& 2X_1 + 0X_2 \le 60 \quad \text{(Constraint Z)} \\
& X_1,\ X_2 \ge 0
\end{aligned}$$

Excel Solution

Table 4.3

	A	B	C	D
1		Input requirements per unit of output		
2	Inputs	X_1	X_2	Inputs available
3	X	1	3	90
4	Y	2	2	80
5	Z	2	0	60
6	Price per unit	3	3	
7		Model description		
8		Decision variables		
9		X_1	X_2	
10		30	10	
11	Max. revenue	120		
12	Constraints	Inputs used		Inputs available
13	X	60	<=	90
14	Y	80	<=	80
15	Z	60	<=	60
16				

In Table 4.3, though the maximum revenue has gone up to Rs. 120 this time, the values of the decision variables do not change now also. However, we have a multiple solution situation for our decision variables inclusive of our original optimum solution at F. Thus, we say that our optimum solution addressed in terms of decision variables is insensitive for change in the price of X_2 to Rs. 3 also.

Graphical Solution

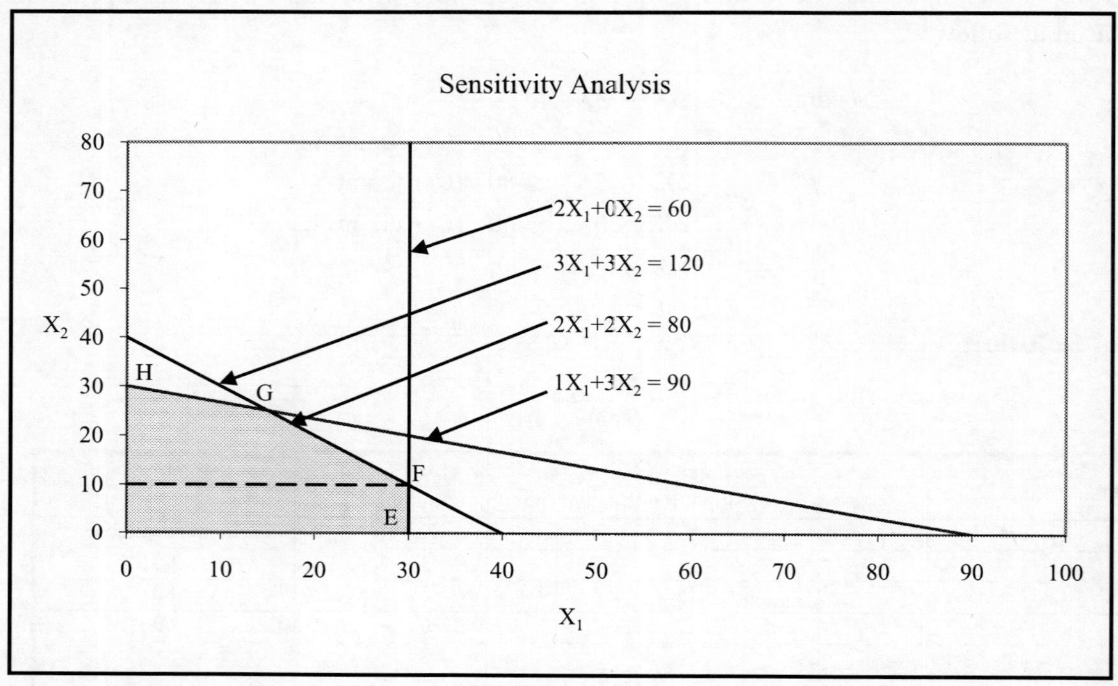

Figure 4.3

Case 3

Let the price of X_2 be increased to 5 from 1 in the original problem. The revised problem is stated as follows.

$$\text{Maximize: } R = 3X_1 + 5X_2$$

$$\text{Subject to: } \quad 1X_1 + 3X_2 \leq 90 \quad \text{(Constraint X)}$$

$$2X_1 + 2X_2 \leq 80 \quad \text{(Constraint Y)}$$

$$2X_1 + 0X_2 \leq 60 \quad \text{(Constraint Z)}$$

$$X_1,\ X_2 \geq 0$$

This time not only the maximum revenue has gone up to Rs. 170 as usual but also the values of the decision variables changed to a new level. The equilibrium corner point moved from F to G this time in the graph. Thus, we say that our optimum solution addressed in terms of decision variables is sensitive for such a change in the price of X_2 from Re. 1 to Rs. 5. So the upper limit for the non-sensitivity of the optimum solution seems to be with price Rs. 3 for X_2.

Excel Solution

Table 4.4

	A	B	C	D
1		Input requirements per unit of output		
2	Inputs	X_1	X_2	Inputs available
3	X	1	3	90
4	Y	2	2	80
5	Z	2	0	60
6	Per unit price	3	5	
7		Model description		
8		Decision variables		
9		X_1	X_2	
10		15	25	
11	Max. revenue	170		
12	Constraints	Inputs used		Inputs available
13	X	90	<=	90
14	Y	80	<=	80
15	Z	30	<=	60
16				

Graphical Solution

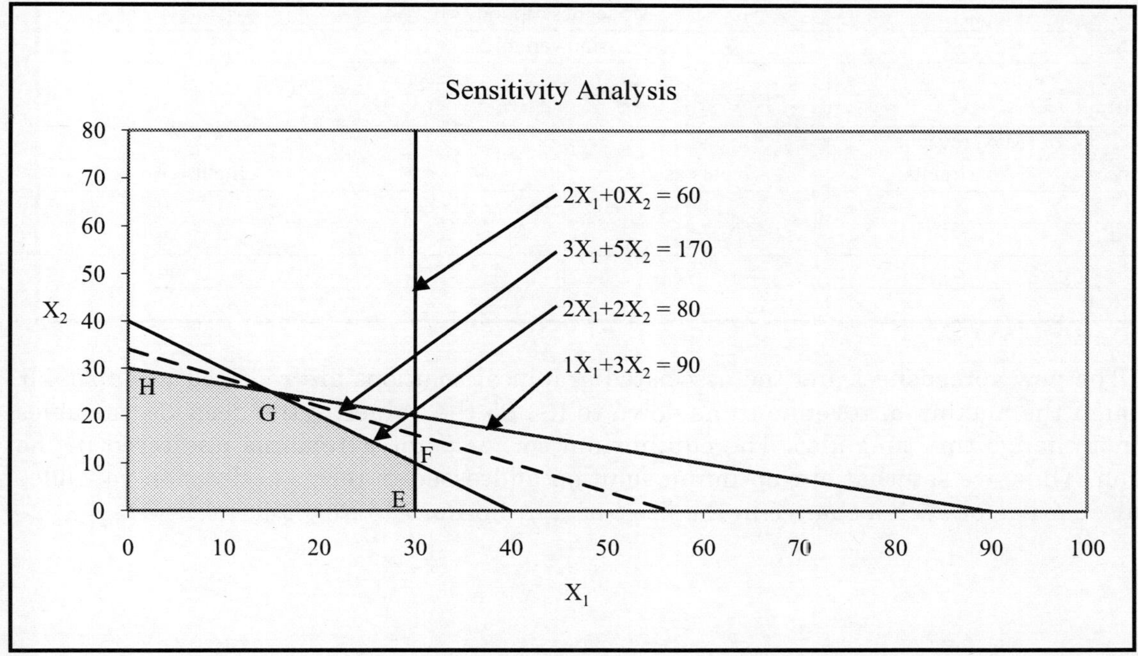

Figure 4.4

Case 4

Now to get the lower limit for non-sensitivity of X_2 let us decrease the price X_2 to 0.5 from the original level of 1 and work out the problem. The new problem is written as

$$\text{Maximize:} \quad R = 3X_1 + 0.5X_2$$

$$\begin{aligned} \text{Subject to:} \quad 1X_1 + 3X_2 &\le 90 \quad \text{(Constraint X)} \\ 2X_1 + 2X_2 &\le 80 \quad \text{(Constraint Y)} \\ 2X_1 + 0X_2 &\le 60 \quad \text{(Constraint Z)} \\ X_1,\ X_2 &\ge 0 \end{aligned}$$

Excel Solution

Table 4.5

	A	B	C	D
1		Input requirements per unit of output		
2	Inputs	X_1	X_2	Inputs available
3	X	1	3	90
4	Y	2	2	80
5	Z	2	0	60
6	Per unit price	3	0.5	
7		Model description		
8		Decision variables		
9		X_1	X_2	
10		30	10	
11	Max. revenue	95		
12	Constraints	Inputs used		Inputs available
13	X	60	<=	90
14	Y	80	<=	80
15	Z	60	<=	60
16				

The new spreadsheet and the associated graphical solutions are reported in Table 4.5. Though the maximum revenue came down to Rs. 95 the values of the decision variables do not change this time also. The equilibrium corner point F remains unaltered in the graph. Thus, we say that our optimum solution addressed in terms of decision variables is insensitive for such a change in the price of X_2 from Re. 1 to Re. 0.50.

Graphical Solution

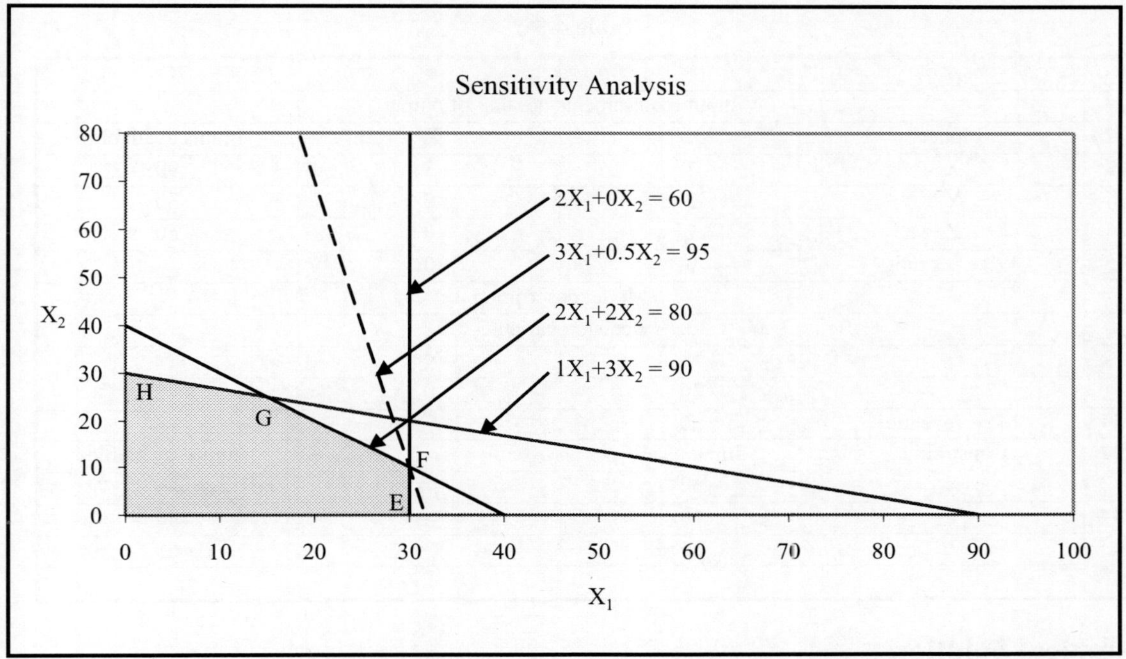

Figure 4.5

Case 5

Let us go further and reduce the price of X_2 to 0 and see the sensitivity.

$$\begin{aligned} &\text{Maximize:} \quad R = 3X_1 + 0X_2 \\ &\text{Subject to:} \quad 1X_1 + 3X_2 \leq 90 \quad \text{(Constraint X)} \\ &\qquad\qquad\quad 2X_1 + 2X_2 \leq 80 \quad \text{(Constraint Y)} \\ &\qquad\qquad\quad 2X_1 + 0X_2 \leq 60 \quad \text{(Constraint Z)} \\ &\qquad\qquad\quad X_1, \; X_2 \geq 0 \end{aligned}$$

Again we come across a multi-equilibrium situation inclusive of our original solution at F. Thus, this value may be taken as the lower limit for the said coefficient. As a final result we would say that our original optimum solution in terms of decision variables is insensitivity in the range from 0 to 3 for the price of X_2. To arrive at this conclusion remember that we have made lot of effort.

Excel Solution

Table 4.6

	A	B	C	D
1		Input requirements per unit of output		
2	Inputs	X_1	X_2	Inputs available
3	X	1	3	90
4	Y	2	2	80
5	Z	2	0	60
6	Price per unit	3	0	
7		Model description		
8		Decision variables		
9		X_1	X_2	
10		30	0	
11	Max. revenue	90		
12	Constraints	Input used		Inputs available
13	X	30	<=	90
14	Y	60	<=	80
15	Z	60	<=	60
16				

Graphical Solution

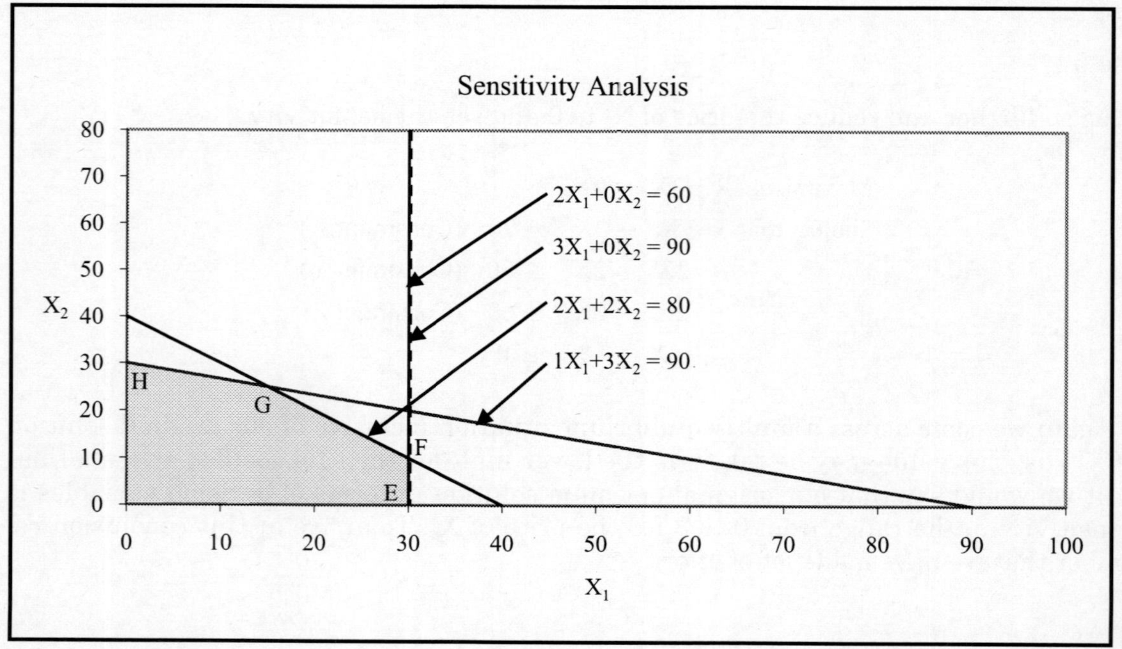

Figure 4.6

Luckily, our spreadsheet solver solution is inclusive of this option also. It gives the lower and upper limit for both the coefficients in the objective function. It also gives sensitivity limits for all the RHS variables as well. It needs only a further step in the excel solver solution iteration. In the final Report box one will have to select **Sensitivity**. After this step as usual click **OK** for the sensitivity table. The solver's sensitivity report for the co-efficients of the objectivity function is reproduced in Table 4.7.

Table 4.7

Adjustable cells						
Cell	Name	Final value	Reduce cost	Objective coefficient	Allowable increase	Allowable decrease
$B$15	X_1	30	0	3	1E+30	2
$C$15	X_2	10	0	1	2	1

From the table one could read both the lower and upper limits for both the coefficients. In the table the entries below the final value column show the optimum solution values for both X_1 and X_2. For this optimum values the sensitivity analysis table shows both allowable increase and allowable decrease from their respective original levels for X_1 and X_2:

Upper Limit = Current Value + Allowable increase

Lower Limit = Current Value – Allowable decrease

In Table 4.8 both lower and upper limits are obtained and reported for both the coefficients.

Table 4.8

Decision variable	Lower limit	Allowable decrease	Current value	Allowable increase	Upper limit
X_1	1	2	3	1E+30	1E+30
X_2	0	1	1	2	3

Note: *In the table for X_2 the allowable increase is stated as 1E+30 to indicate the infinity. Thus for X_2 coefficient the allowable increase is infinity.*

Sensitivity for a Simultaneous Change in both the Coefficients

In the sensitivity analysis so far we have considered only one objective function coefficient at a time by keeping all other variables constant. However, one may have interest in assessing the sensitivity whenever two or more coefficients change simultaneously.

100% rule for objective function coefficients

According to this rule for all objective function coefficients changed simultaneously, the sum of percentage allowable increase and allowable decrease must be equal to or less than 100%. If the sum of the percentages is less than 100%, the optimum solution values remain unaltered.

Suppose the management want to decrease the price of X_1 by Re. 1 from 3 to 2 and increase price of X_2 by Re. 0.5 from 1 to 1.5. From the table the allowable decrease in the price of X_1 is 2 (3 to 1). The actual decrease is Re. 1. Thus for this much of decrease in the price of X_1 the optimum values of the decision variable will not change. The percentage decrease in price of X_1 over the allowable decrease of 2 is obtained as $1/2 \times 100 = 50\%$.

Similarly, since the increase in the price of X_2 by 0.5 is within the limit of the permissible 2, the optimum values of the decision variable will remain unaltered. The percentage increase in price of X_2 over the allowable increase of 2 is obtained as $0.5/2 \times 100 = 25\%$. Therefore the sum of the decrease in the price of X_1 and increase in price of X_2 is equal to 50 + 25 = 75%. Thus these proposed changes in both the prices leave the decision variables unaltered. To get ourselves convinced let us redo the problem once again after incorporating the proposed changes. The modified programming after making necessary changes may be written as

$$\text{Maximize: } R = 2X_1 + 1.5X_2$$

$$\begin{aligned} \text{Subject to: } \quad 1X_1 + 3X_2 &\leq 90 \quad \text{(Constraint X)} \\ 2X_1 + 2X_2 &\leq 80 \quad \text{(Constraint Y)} \\ 2X_1 + 0X_2 &\leq 60 \quad \text{(Constraint Z)} \\ X_1, \quad X_2 &\geq 0 \end{aligned}$$

Excel Solution

Table 4.9

	A	B	C	D
1		Input requirements per unit of output		
2	Inputs	X_1	X_2	Inputs available
3	X	1	3	90
4	Y	2	2	80
5	Z	2	0	60
6	Price per unit	2	1.5	
7		Model description		
8		Decision variables		
9		X_1	X_2	
10		30	10	
11	Max. revenue	75		
12	Constraints	Input used		Inputs available
13	X	60	<=	90
14	Y	80	<=	80
15	Z	60	<=	60
16				

Graphical Solution

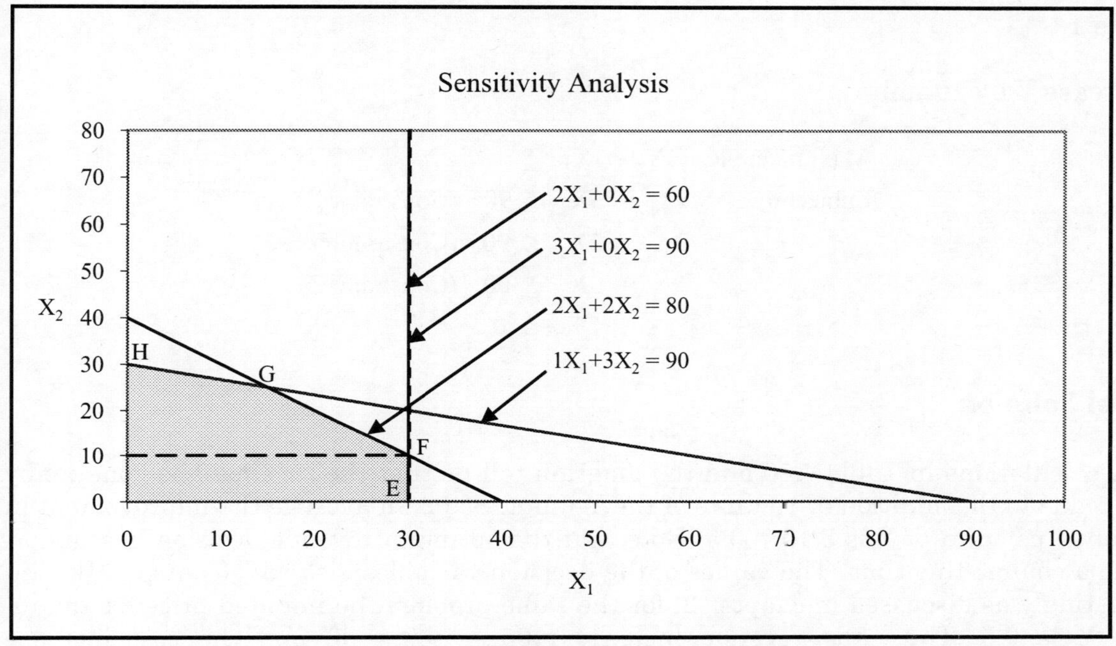

Figure 4.7

In our illustrative case since this sum = 75%, our optimum solution remains unaltered at $X_1 = 30$ and $X_2 = 10$. So the revenue maximum is obtained as: = 30 × 2 + 10 × 1.5 = 75.

Note: *The above rule is silent when the said sum percentage is greater than 100%. So we can only say that if this sum is greater than 100% a different optimum solution may exist. Thus, whenever this sum exceeds 100% the revised problem must be solved to obtain the new optimum solution.*

4.3 SENSITIVITY ANALYSIS FOR CONSTRAINT'S RHS VALUES

In this section the sensitivity is tested not for the sensitivity of the decision variables. Instead we test the rate of change in revenue per unit increase or decrease in the availability of inputs. For example, assume that we increase the labour resource by one unit. Such an increase not only changes the output level but also brings in more revenue. If this increase in revenue is equivalent to the imputed price of the labour input then such an addition adds nothing extra to the producer. Till such increase is ineffective we would say the chance in the availability of inputs is insensitive. The following example illustrates this situation. Let us increase the supply of the input Y by 10 units. For original problem we notice the inputs Y and Z are binding and X is non-binding. So let us consider the sensitivity for binding inputs first.

Sensitivity for Binding Inputs

Case 1

Increase Y by 10 units

$$\begin{aligned}
\text{Maximize: } & R = 3X_1 + 1X_2 \\
\text{Subject to: } & 1X_1 + 3X_2 \leq 90 \quad \text{(Constraint X)} \\
& 2X_1 + 2X_2 \leq 90 \quad \text{(Constraint Y)} \\
& 2X_1 + 0X_2 \leq 60 \quad \text{(Constraint Z)} \\
& X_1, \; X_2 \geq 0
\end{aligned}$$

Excel Solution

The calculations in Table 4.10 and the diagram tell us that the revenue has gone up by Rs. 5 due to the addition of 10 units of the Y input. So on an average the improvement in revenue per unit of Y is 5/10 = 0.5. Note that the optimum revenue decision has undergone a change this time. The values of the decision variables also have changed. But interestingly as discussed in chapter 2, for the same problem the imputed price for the input Y is 0.5. Thus, the revenue increase (105 – 100 = 5) and the cost increase (0.5 × 10 = 5) are one and the same levelling the net revenue unaltered.

Table 4.10

	A	B	C	D
1		Input requirements per unit of output		
2	Inputs	X_1	X_2	Inputs available
3	X	1	3	90
4	Y	2	2	90
5	Z	2	0	60
6	Price per unit	3	1	
7		Model description		
8		Decision variables		
9		X_1	X_2	
10		30	15	
11	Max. revenue	105		
12	Constraints	Input used		Inputs available
13	X	75	<=	90
14	Y	90	<=	90
15	Z	60	<=	60
16				

Graphical Solution

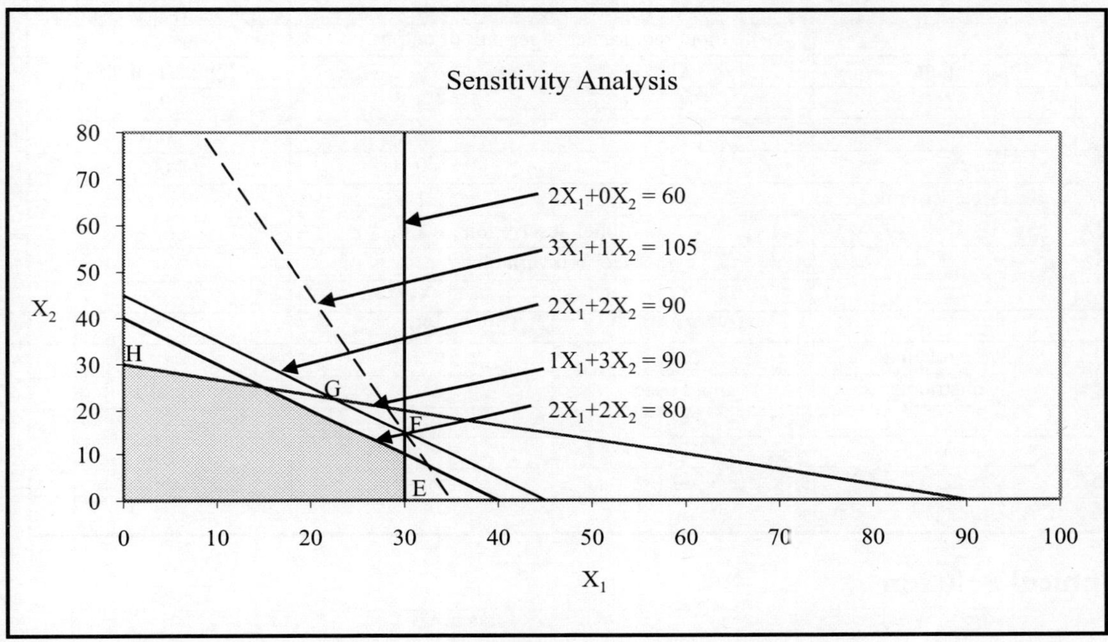

Figure 4.8

Thus really the increase Y by 10 units is insensitive as far as the net revenue is concerned. Note that this time we are not concerned with the values of the decision variables. In fact this time our decision variables value differs considerably.

Case 2

Increase Y by 20 units

Maximize: $R = 3X_1 + 1X_2$

Subject to:

$$1X_1 + 3X_2 \leq 90 \quad \text{(Constraint X)}$$
$$2X_1 + 2X_2 \leq 100 \quad \text{(Constraint Y)}$$
$$2X_1 + 0X_2 \leq 60 \quad \text{(Constraint Z)}$$
$$X_1, \; X_2 \geq 0$$

Excel Solution

The calculation in Table 4.11 tells us that the revenue has gone up by Rs. 10 due to the addition of 20 units of input Y. So on an average the improvement in revenue per unit of Y is 10/20 = 0.5 again. The revenue increase (110 – 100 = 10) and the cost increase (0.5 × 20 = 10) are one and the same. Thus this time also the increase in Y by 20 units leave the net revenue unaltered at Rs. 100. So the decision of going for additional 20 units of resource Y is insensitive once again.

Table 4.11

	A	B	C	D
1		Input requirements per unit of output		
2	Inputs	X_1	X_2	Inputs available
3	X	1	3	90
4	Y	2	2	100
5	Z	2	0	60
6	Price per unit	3	1	
7		Model description		
8		Decision variables		
9		X_1	X_2	
10		30	20	
11	Max. revenue	110		
12	Constraints	Input used		Inputs available
13	X	90	<=	90
14	Y	100	<=	100
15	Z	60	<=	60
16				

Graphical Solution

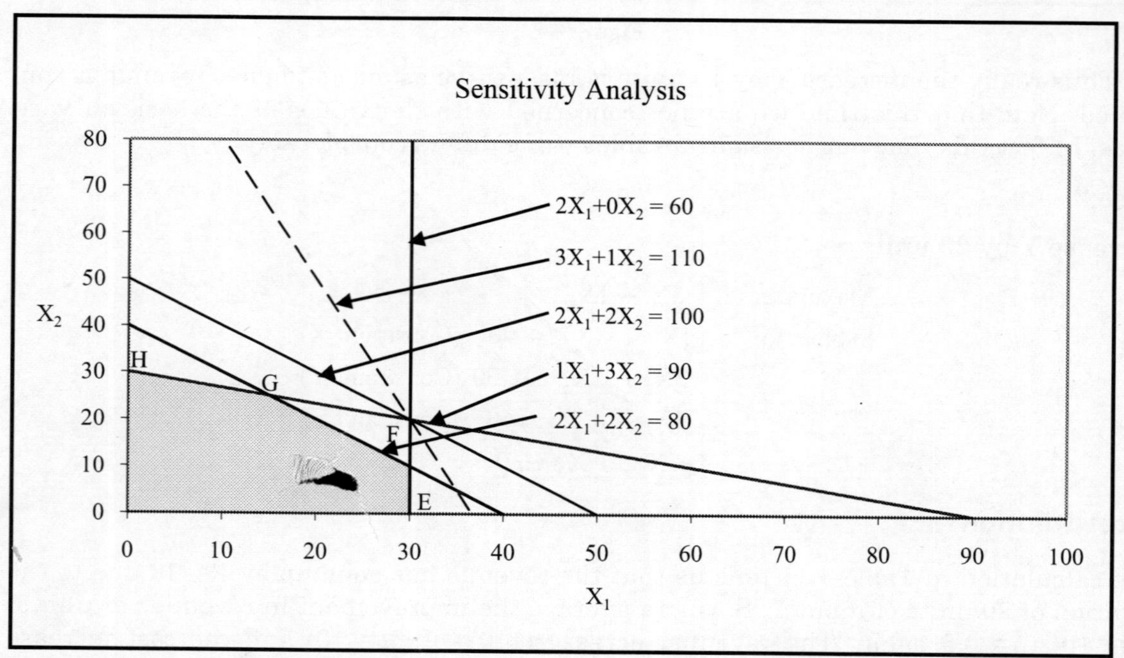

Figure 4.9

Case 3

Increase Y by 21 units

$$\begin{aligned}
\text{Maximize: } & R = 3X_1 + 1X_2 \\
\text{Subject to: } & 1X_1 + 3X_2 \le 90 \quad \text{(Constraint X)} \\
& 2X_1 + 2X_2 \le 101 \quad \text{(Constraint Y)} \\
& 2X_1 + 0X_2 \le 60 \quad \text{(Constraint Z)} \\
& X_1,\ X_2 \ge 0
\end{aligned}$$

Excel Solution

The calculation in Table 4.12 tells us that the revenue has gone up by Rs. 10 due to the addition of 21 units of the Y input. So on an average the improvement in revenue per unit of Y is 10/21 = 0.47619. This value is less than the imputed value for Y, namely, 0.5. The increase in revenue (110 – 100 = 10) is less than the increase in cost (0.5 × 21= 10.5). This time the net revenue will be Rs. 99.5. Thus the decision of adding 21 units of Y is sensitive from net revenue point of view. So the allowable increase is only 20 units of resource Y in this case.

Table 4.12

	A	B	C	D
1		Input requirements per unit of output		
2	Inputs	X_1	X_2	Inputs available
3	X	1	3	90
4	Y	2	2	101
5	Z	2	0	60
6	Price per unit	3	1	
7		Model description		
8		Decision variables		
9		X_1	X_2	
10		30	20	
11	Max. revenue	110		
12	Constraints	Input used		Inputs available
13	X	90	<=	90
14	Y	100	<=	101
15	Z	60	<=	60
16				

Graphical Solution

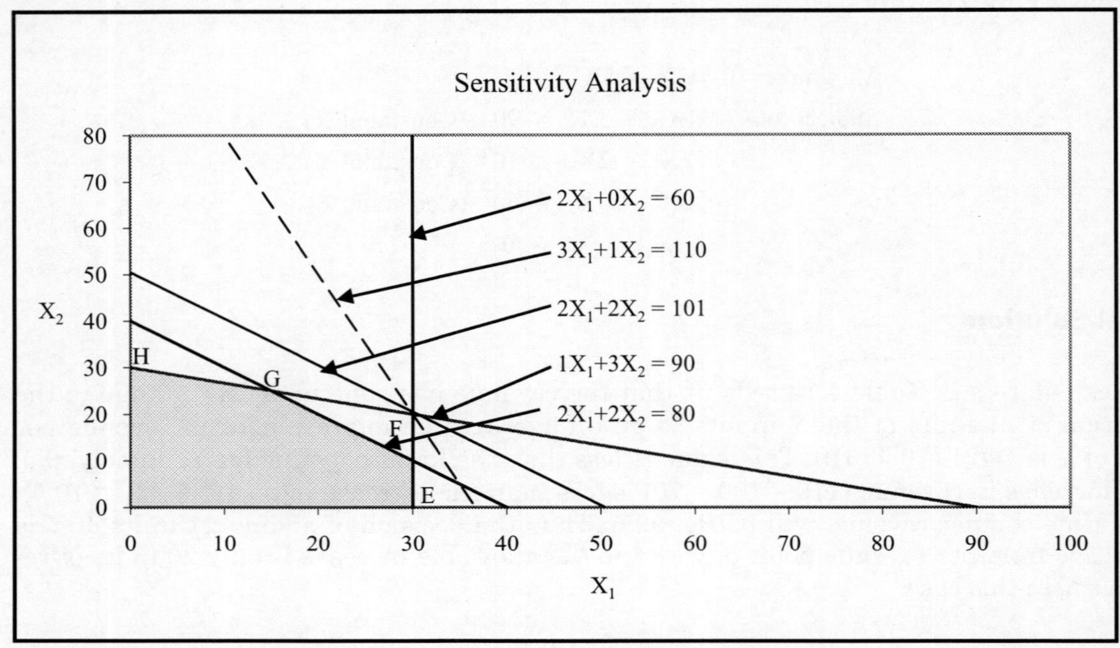

Figure 4.10

Case 4

Decrease Y by 20 units

$$
\begin{aligned}
\text{Maximize:} \quad & R = 3X_1 + 1X_2 \\
\text{Subject to:} \quad & 1X_1 + 3X_2 \leq 90 \quad \text{(Constraint X)} \\
& 2X_1 + 2X_2 \leq 60 \quad \text{(Constraint Y)} \\
& 2X_1 + 0X_2 \leq 60 \quad \text{(Constraint Z)} \\
& X_1, \quad X_2 \geq 0
\end{aligned}
$$

Excel Solution

The calculations in Table 4.13 tell us that the revenue has gone down by Rs. 10 due to the reduction of 20 units of the Y input. So at an average the reduction in revenue per unit of Y is 10/20 = 0.5. This reduction will not affect the net revenue since the post-optimal imputed rate per unit of resource Y is also exactly equal to 0.5. The 20 unit reduction in resource Y will save Rs. 10 (0.5 × 20). Thus the revenue reduction is exactly equal to the cost reduction.

Table 4.13

	A	B	C	D
1		Input requirements per unit of output		
2	Inputs	X_1	X_2	Inputs available
3	X	1	3	90
4	Y	2	2	60
5	Z	2	0	60
6	Price per unit	3	1	
7		Model description		
8		Decision variables		
9		X_1	X_2	
10		30	0	
11	Max. revenue	90		
12	Constraints	Input used		Inputs available
13	X	30	<=	90
14	Y	60	<=	60
15	Z	60	<=	60
16				

Graphical Solution

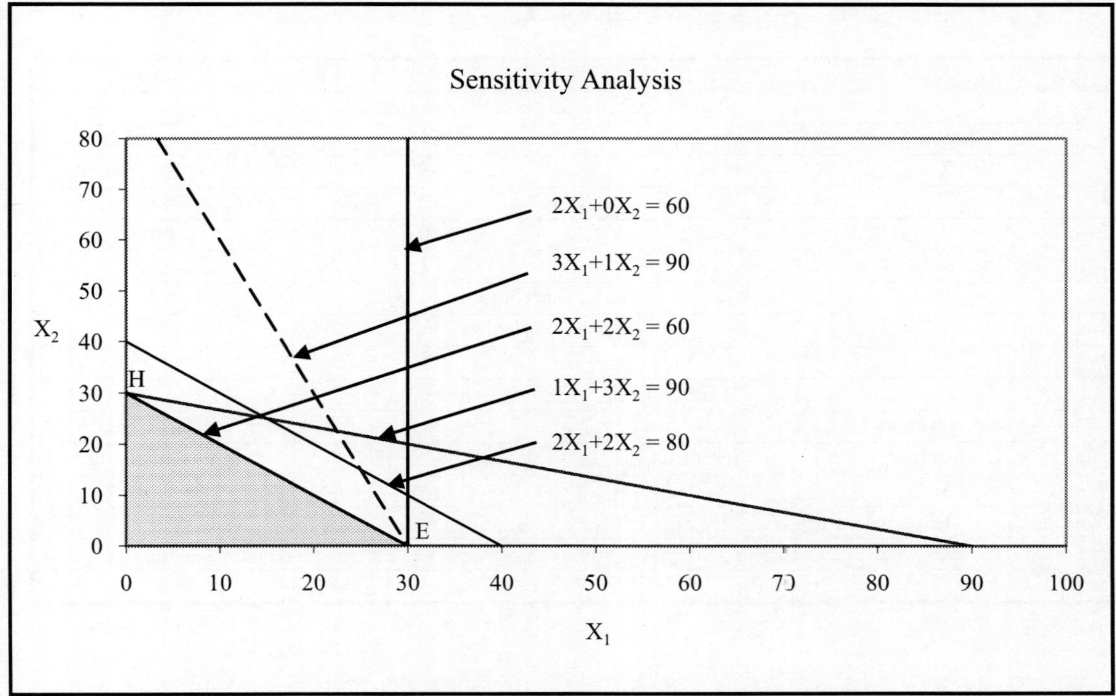

Figure 4.11

Case 5

Decrease in Y by 21 units

$$\text{Maximize: } R = 3X_1 + 1X_2$$

$$\begin{aligned} \text{Subject to: } \quad 1X_1 + 3X_2 &\leq 90 \quad \text{(Constraint X)} \\ 2X_1 + 2X_2 &\leq 51 \quad \text{(Constraint Y)} \\ 2X_1 + 0X_2 &\leq 60 \quad \text{(Constraint Z)} \\ X_1 ,\ X_2 &\geq 0 \end{aligned}$$

Excel Solution

The calculations in Table 4.14 tell us that the revenue has gone down by Rs. 11.5 due to the reduction of 21 units of the Y input. So at an average the reduction in revenue per unit of Y is 11.5/21 = 0.547619. This figure is greater than the imputed price of Y, namely, 0.5. Thus, the allowable decrease is only 20 units in the case of Y. Since the current level of Y is 80, the allowable decrease is 20 and allowable increase is 20, makes the permissible range for Y from 60 to 100.

Table 4.14

	A	B	C	D
1		Input requirements per unit of output		
2	Inputs	X_1	X_2	Inputs available
3	X	1	3	90
4	Y	2	2	59
5	Z	2	0	60
6	Price per unit	3	1	
7		Model description		
8		Decision variables		
9		X_1	X_2	
10		29.5	0	
11	Max. revenue	88.5		
12	Constraints	Inputs used		Inputs available
13	X	29.5	<=	90
14	Y	59	<=	59
15	Z	59	<=	60
16				

Graphical Solution

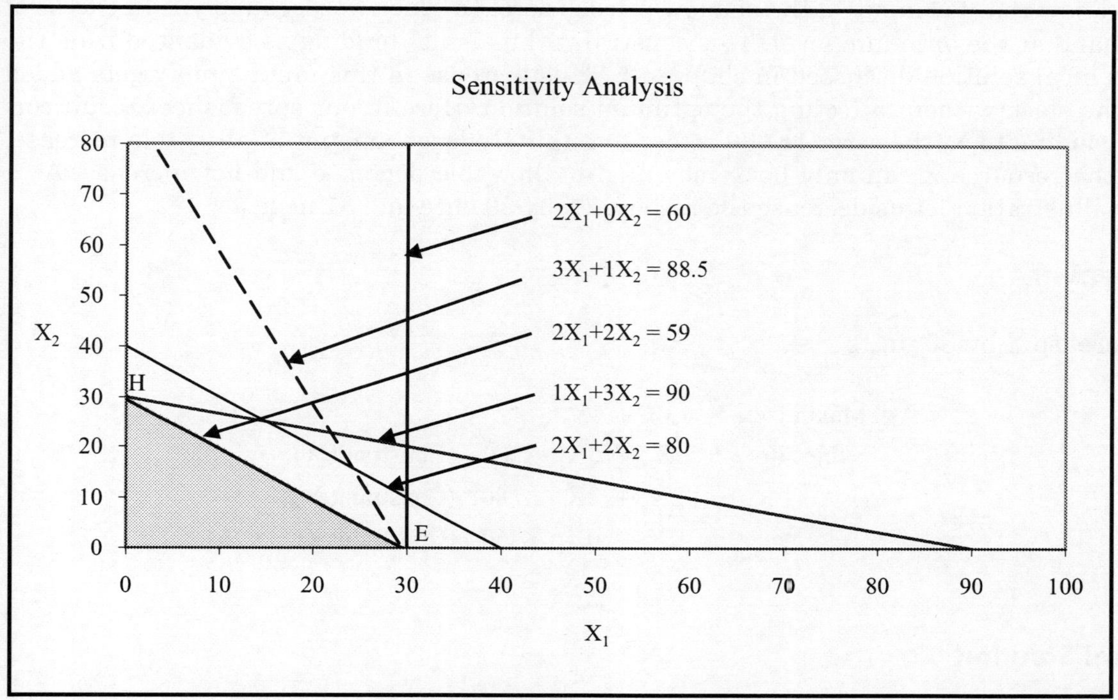

Figure 4.12

Spreadsheet sensitivity data for RHS variables

To arrive at this conclusion we have used 5 illustrations. Such a task is unwarranted. The solver sensitivity result is readily available for all the RHS variables as well. For Y constraint, as we have discussed, the allowable decrease and increase are 20 units each in the following table.

Table 4.15

Constraints						
Cell	Name	Final value	Shadow price	Constraint R.H.S	Allowable increase	Allowable decrease
$B$20	X Used (LHS)	60	0	90	1E+30	30
$B$21	Y Used (LHS)	80	0.5	80	20	20
$B$22	Z Used (LHS)	60	1	60	20	30

In the above table the shadow price per unit for the input Y is 0.5 as stated in our discussion. For Y input this ratio will be constant in the range 60 to 100.

Sensitivity analysis for non-binding inputs

We know that for input X, the shadow price is zero. In other words, supply is in excess of demand at the optimum level. This constraint thus is not binding as evidenced from the graphical solution discussed in chapter 2. Thus increase in this input simply gets added to the slack without affecting the optimum solution. Here in our spreadsheet result corresponding to X, the term 1E+30 refers to infinitely large amount of allowable increase. So the resource X can only have meaningful allowable decrease and not increase. As a last illustration let us decrease the resource X by 30 once and 31 next.

Case 1

Decrease X by 30 units

$$\text{Maximize: } R = 3X_1 + 1X_2$$

$$\begin{aligned} \text{Subject to: } \quad 1X_1 + 3X_2 &\le 60 \quad \text{(Constraint X)} \\ 2X_1 + 2X_2 &\le 80 \quad \text{(Constraint Y)} \\ 2X_1 + 0X_2 &\le 60 \quad \text{(Constraint Z)} \\ X_1, \quad X_2 &\ge 0 \end{aligned}$$

Excel Solution

Table 4.16

	A	B	C	D
1		Input requirements per unit of output		
2	Inputs	X_1	X_2	Inputs available
3	X	1	3	60
4	Y	2	2	80
5	Z	2	0	60
6	Price per unit	3	1	
7		Model description		
8		Decision variables		
9		X_1	X_2	
10		30	10	
11	Max. revenue	100		
12	Constraints	Inputs used		Inputs available
13	X	60	<=	60
14	Y	80	<=	80
15	Z	60	<=	60
16				

Graphical Solution

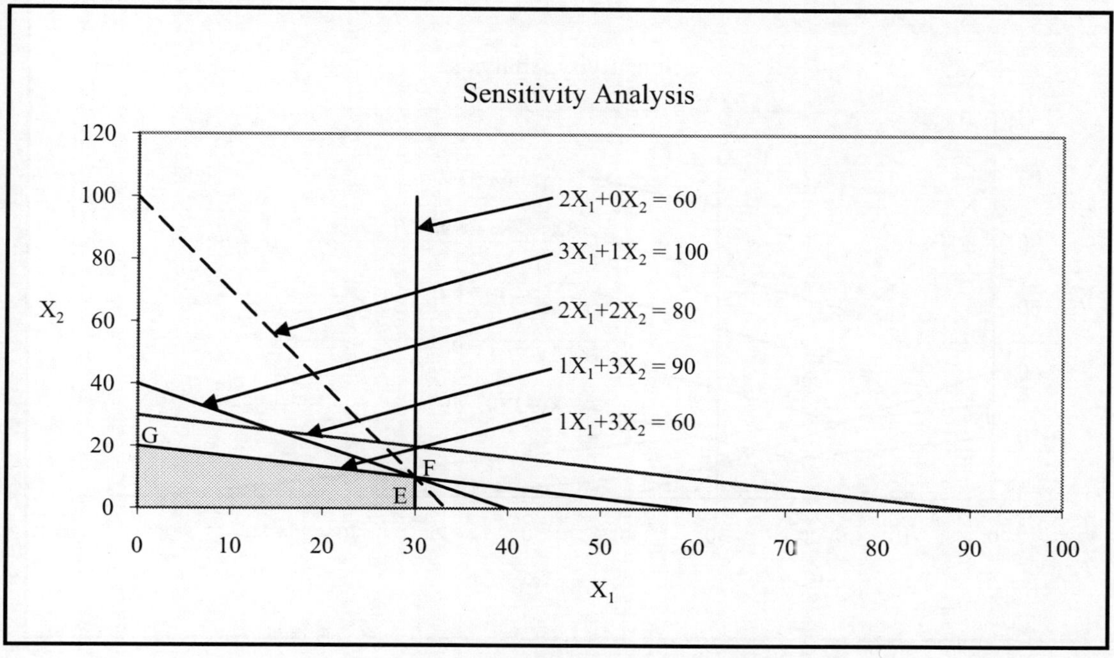

Figure 4.13

Thus though the input has gone down neither revenue has gone down nor optimal decision variables have changed. This is mainly due to the reason that till this time the reduced 30 units were available as slack.

Case 2

Decrease X by 31 units

$$\text{Maximize: } R = 3X_1 + 1X_2$$

$$\begin{aligned} \text{Subject to: } \quad 1X_1 + 3X_2 &\le 59 \quad \text{(Constraint X)} \\ 2X_1 + 2X_2 &\le 80 \quad \text{(Constraint Y)} \\ 2X_1 + 0X_2 &\le 60 \quad \text{(Constraint Z)} \\ X_1, \quad X_2 &\ge 0 \end{aligned}$$

Graphical Solution

From the table, we notice that the revenue has started declining. The input no longer will remain as slack in this range. Now in fact Y has excess supply and commands zero shadow price as shown from the Table 4.17.

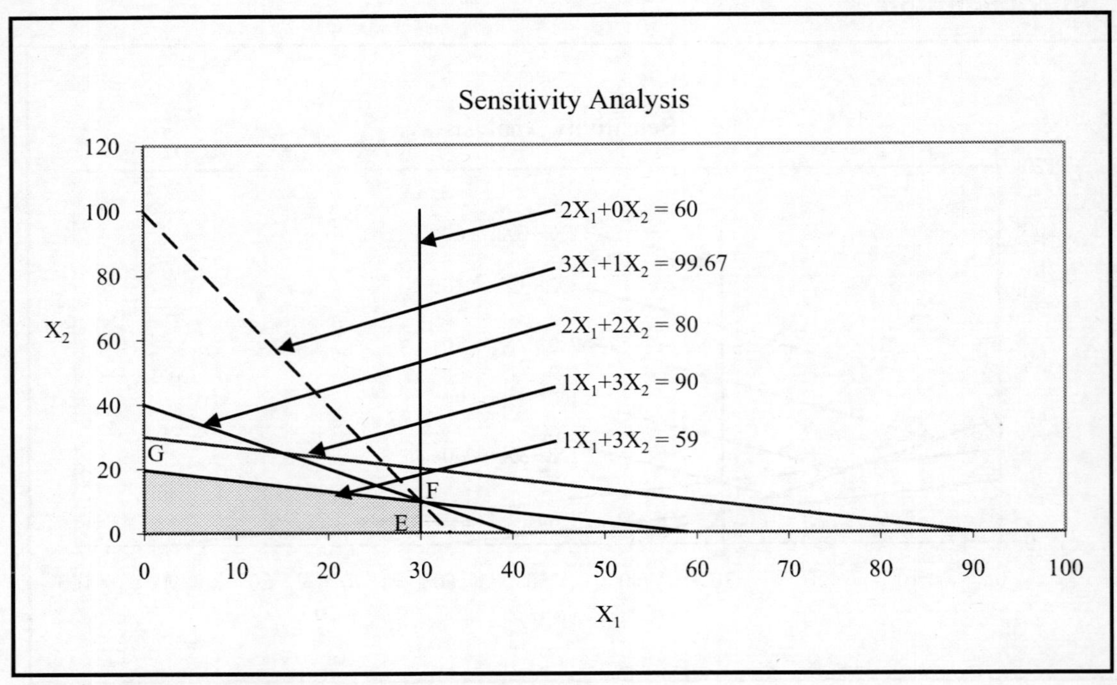

Figure 4.14

Excel Solution

Table 4.17

	A	B	C	D
1		Input requirements per unit of output		
2	Inputs	X_1	X_2	Inputs available
3	X	1	3	59
4	Y	2	2	80
5	Z	2	0	60
6	Price per unit	3	1	
7		Model description		
8		Decision variables		
9		X_1	X_2	
10		30	9.66666667	
11	Max. revenue	99.66666667		
12	Constraints	Inputs used		Inputs available
13	X	59.00000001	<=	59
14	Y	79.33333334	<=	80
15	Z	60	<=	60
16				

Table 4.18

Constraints						
Cell	Name	Final value	Shadow price	Constraint R.H.S	Allowable increase	Allowable decrease
$B$20	X Used (LHS)	59	0.33333	59	1	29
$B$21	Y Used (LHS)	79.333	0	80	1E+30	0.66666667
$B$22	Z Used (LHS)	60	1.33333	60	1	60

4.4 SENSITIVITY ANALYSIS: A REVISIT USING FINAL SIMPLEX TABLE

Till now we took only a two-decision variable problem for illustration. Further, while doing so we invariably used the excel spreadsheet solution for discussion. Now in this section we propose to use the simplex tabular analysis for our discussion.

■ Example 2

Solve the following problem by Simplex method. Also make sensitivity analysis

$$\text{Maximize: } C = 3X_1 + 3X_2 + 4X_3 + 7X_4$$

$$\begin{aligned} \text{Subject to: } X_1 + X_2 + X_3 + X_4 &= 6 \quad \text{(Capital)} \\ 5X_1 + 3X_2 + 2X_3 + X_4 &= 60 \quad \text{(Labour)} \\ X_1 + 3X_2 + 5X_3 + 6X_4 &= 50 \quad \text{(Machine Time)} \\ X_1 ;\ X_2 ;\ X_3 ;\ X_4 &\geq 0 \end{aligned}$$

As a first step we introduce slacks and rewrite the problem as shown below.

$$\text{Maximize: } C = 3X_1 + 3X_2 + 4X_3 + 7X_4 + 0S_1 + 0S_2 + 0S_3$$

$$\begin{aligned} \text{Subject to: } X_1 + X_2 + X_3 + X_4 + S_1 &= 6 \\ 5X_1 + 3X_2 + 2X_3 + X_4 + S_2 &= 60 \\ X_1 + 3X_2 + 5X_3 + 6X_4 + S_3 &= 50 \\ X_1 ;\ X_2 ;\ X_3 ;\ X_4 ;\ S_1 ;\ S_2 ;\ S_3 &\geq 0 \end{aligned}$$

Simplex Solution

Table 4.19

Tableau 1										
C_i	Basis	X_1	X_2	X_3	X_4	S_1	S_2	S_3	X_b	Min
0	S_1	1	1	1	1	1	0	0	9	9
0	S_2	5	3	2	1	0	1	0	60	60
0	S_3	1	3	5	8	0	0	1	50	6.25
	C_j	3	3	4	7	0	0	0		
	Z_j	0	0	0	0	0	0	0	0	
	$C_j - Z_j$	3	3	4	7	0	0	0		

Table 4.20

Tableau 2										
C_i	Basis	X_1	X_2	X_3	X_4	S_1	S_2	S_3	X_b	Min
0	S_1	0.875	0.625	0.375	0	1	0	–0.13	2.75	3.143
0	S_2	4.875	2.625	1.375	0	0	1	–0.13	53.75	11.03
7	X_4	0.125	0.375	0.625	1	0	0	0.125	6.25	50
	C_j	3	3	4	7	0	0	0		
	Z_j	0.875	2.625	4.375	7	0	0	0.875	43.75	
	$C_j - Z_j$	2.125	0.375	–0.38	0	0	0	–0.88		

Table 4.21

Tableau 3										
C_i	Basis	X_1	X_2	X_3	X_4	S_1	S_2	S_3	X_b	Min
3	X_1	1	0.714	0.429	0	1.143	0	–0.14	3.143	
0	S_2	0	–0.86	–0.71	0	–5.57	1	0.571	38.43	
7	X_4	0	0.286	0.571	1	–0.14	0	0.143	5.857	
	C_j	3	3	4	7	0	0	0		
	Z_j	3	4.143	5.286	7	2.429	0	0.571	50.43	
	$C_j - Z_j$	0	–1.14	–1.29	0	–2.43	0	–0.57		

From the above final tableau the optimal solutions are:

$X_1 = 3.143$, $X_2 = 0$, $X_3 = 0$, $X_4 = 5.857$ and $R = 50.43$.

4.5 SENSITIVITY ANALYSIS FOR THE COEFFICIENTS IN THE OBJECTIVE FUNCTION

In our problem at the optimum level, X_1 and X_4 are basic and X_2 and X_3 are non-basic variables. Let us consider each one of them for their sensitivity in this section.

a) Sensitivity analysis for non-basic variable X_2

The last $C_j - Z_j$ row tells us the sensitivity of these two non-basic variables. If the price of X_2 is increased from 3 to anything more than 1.14, $C_2 - Z_2$ entry in the last row will become positive. So this non-basic variable become basic and hence will affect our optimal solution. Thus, the maximum permitted increase without affecting our optimum solution is $3 + 1.14 = 4.14$ for X_2. Note that in this case there is no limit for allowable decrease.

b) Sensitivity analysis for non-basic variable X_3

Similarly, if the price of X_3 is increased from 4 to anything more than 1.29, then $C_3 - Z_3$ entry in the last row will become positive. So this non-basic variable becomes basic and hence will affect our optimal solution. Thus, the maximum permitted increase without affecting our optimum solution is $4 + 1.29 = 5.29$. Note that in this case also there is no limit for allowable decrease.

Sensitivity analysis for basic variables

In this case the changes in the coefficients can be both positive and negative. So it is imperative that for restoration of optimum we must work out a minimum and a maximum for each one of the basic variables under consideration. In the case of the basic variable since every Z_j is going to be affected, we will have to recalculate the entire $C_j - Z_j$ before making sensitivity statement. With new C_j's if none of the $C_j - Z_j$ elements are positive then we conclude that the current solution is optimum. On the other hand, the occurrence of positive values in this row, one will have to continue the iteration till he gets the new optimum.

c) Sensitivity analysis for basic variable X_1

Let Δ_1 stands for the chance in the price of X_1. To find out the range of variation that will not have any effect in the optimum solution let us divide the $C_j - Z_j$ row by the corresponding elements of the X_1th row element as shown below. In the resultant last row the positive coefficient gives us the maximum limit for Δ_1. If there is more than one positive coefficient then select the smallest positive value for this purpose. For the lower limit among the negative coefficients select the one with least absolute value.

Table 4.22

X_1	1	0.714	0.429	0	1.143	0	–0.14
$C_j - Z_j$	0	–1.14	–1.29	0	–2.43	0	–0.57
$(C_j - Z_j)/X_1$	0	–1.6	–3	–	–2.13	–	4

Thus, for X_1 the maximum value is obtained as 3 + 4 = 7. Similarly, the minimum value is obtained as 3 – 1.6 = 2.4.

d) Sensitivity analysis for basic variable X_4

Table 4.23

X_4	0	0.286	0.571	1	–0.14	0	0.143
$C_j - Z_j$	0	–1.14	–1.29	0	–2.43	0	–0.57
$(C_j - Z_j)/X_4$	–	–4	–2.25	0	17	–	–4

After making similar calculations the range of variation permissible for X_4 is (3 – 2.25, 3 + 17) = (0.75, 20). Note that the Simplex calculations and spreadsheet solver results shown in Table 4.27 are identical.

4.6 SENSITIVITY OF RESOURCE CHANGE (R.H.S) ON THE OPTIMUM REVENUE

i) Sensitivity of capital having the slack of 38.43 units at the optimum

In the final table we notice that S_2 is the slack variable for the input capital. Since it is has slack value of 38.43 units, it commands zero shadow prices in the dual. This means that with current optimal solution we are not fully utilizing this resource. Therefore, there is no need to add this resource. Even if we do so the optimum revenue will remain unaltered. However, if we reduce this resource below a certain critical minimum the optimum revenue gets affected. Accordingly, we fix this critical value as minimum for sensitivity. From the final simplex solution table the slack value of capital is 38.43. So the minimum for this input is obtained as 60 – 38.43 = 21.57. Even if we reduce this resource to 21.57 the optimum revenue will not change.

ii) Sensitivity of labour having the slack of 0 units at the optimum

In our illustration both capital and machine time are fully utilized and the corresponding slacks will have zero values at the optimum level. For example, for sensitive of Capital resource we fix a minimum and maximum within which the revenue contribution per unit will remain constant at 2.428 (Shadow price for the input labour = 2.428). These limits are obtained as follows.

Table 4.24

S_1	X_b	X_b/S_1
1.143	3.1429	2.75
–5 .57	38.429	–6.9
–0.14	5.8571	–41

In the above table, 2.75 indicate the limit by which this resource can be brought down without affecting the rate of revenue per unit of output. Similarly, the lowest absolute value among the two negative values in the last column namely 6.9 indicates the extent by which the resource can be increased without affecting the per unit profit of 2.482. So insensitivity limit is obtained as (9 – 2.75 to 9 + 6.9) = (6.25 to 15.9).

iii) Sensitivity of machine time having the slack of 0 units at the optimum

Table 4.25

S_3	X_b	X_b/S_3
–0.14	3.1429	–22
0.571	38.429	67.25
0.143	5.8571	41

By similar reasoning, the insensitivity for Machine Time to keep the per unit revenue at 0.571 is (50-41 to 50+22) = (9 to 72).

4.7 SENSITIVITY OF NEW PRODUCT INTRODUCTION ON THE OPTIMAL REVENUE

Having the available resources and the associated slacks at the optimum level in mind the production manager may be interested in introducing one more new product. One can easily check the feasibility by checking the associated dual constraint satisfied by the current values of the dual variables. If it is fulfilled then it is worth to go for the new venture, otherwise not.

Let the new activity that is to be introduced be X_5. Let the resource requirements per unit of this new product be 7/4, 3, and 7 respectively. Now the new problem may be written as

$$\begin{aligned} \text{Maximize:} \quad & C = 3X_1 + 3X_2 + 4X_3 + 7X_4 + C_5X_5 \\ \text{Subject to:} \quad & X_1 + X_2 + X_3 + X_4 + 7/4X_5 = 6 \\ & 5X_1 + 3X_2 + 2X_3 + X_4 + 3X_5 = 60 \\ & X_1 + 3X_2 + 5X_3 + 6X_4 + 7X_5 = 50 \\ & X_1 \; ; \; X_2 \; ; \; X_3 \; ; \; X_4 \; ; \; X_5 \geq 0 \end{aligned}$$

In the revenue function C_5 is the per unit revenue from the new activity. Now the problem is at what C_5 it is worthwhile to go for the new venture. The corresponding dual relation is

$$1.75Y_1 + 3Y_2 + 7Y_3 \geq C_5$$

Substituting the current values from the Simplex table the above equation may be written as

$$1.75 \times 2.429 + 3 \times 0 + 7 \times 0.571 \geq C_5$$

$$C_5 \leq 8.25$$

Thus the per unit revenue contribution of the new product must be greater than 8.25 for its introduction.

Solution and summary of sensitivity report through solver excel spreadsheet

For the problem discussed above the solution and the sensitivity report by spreadsheet is given below,

Excel Solver Solution Table

Table 4.26

	A	B	C	D	E	F
1			Material requirements			
2	Material	X_1	X_2	X_3	X_4	RHS
3	Labour	1	1	1	1	9
4	Capital	5	3	2	1	60
5	Machine	1	3	5	8	50
6	Profit per unit	3	3	4	7	
7						
8	Model					
9			Decision variables			
10		X_1	X_2	X_3	X_4	
11		3.142857143	0	0	5.857142857	
12						
13	Max. revenue	50.42857143				
14						
15	Constraints	LHS				RHS
16	Labour	9	<=			9
17	Capital	21.57142857	<=			60
18	Machine	50	<=			50
19						

Excel Solver Sensitivity Table

Table 4.27

Adjustable cells

Cell	Name	Final value	Reduced cost	Objective coefficient	Allowable increase	Allowable decrease
$B$15	X_1	3.142857	0	3	4	1.599999999
$C$15	X_2	0	−1.142857	3	1.14285714	1E+30
$D$15	X_3	0	−1.285714	4	1.28571429	1E+30
$E$15	X_4	5.857143	0	7	17	2.25

Constraints

Cell	Name	Final value	Shadow price	Constraint R.H.S	Allowable increase	Allowable decrease
$B$20	Labour	9	2.428571	9	6.8974359	2.75
$B$21	Capital	21.57143	0	60	1E+30	38.42857143
$B$22	Machine	50	0.571429	50	22	41

EXERCISES

1. A firm produces three products A, B, C. The unit contributions of these three products are Rs. 5, Rs. 10, and Rs. 8 respectively. Each unit of A needs 3 kg of raw material, 4 machine hours, and 2 labour hours. Similarly each unit of B requires 5 kg of raw material, 4 machine hours, and 4 labour hours and each unit of C requires 2 kg of raw material, 4 machine hours, and 5 labour hours. Preproduction period 60 kg of raw material, 72 machine hours and 100 labour hours are available. Find out the best production strategy. Also investigate the effect the following decision alternatives on the optimum solution obtained already.
   a. An increase of 12 machine hours
   b. A decrease of 6 kg or raw material
   c. The unit contribution of A is increased to Rs. 6
   d. The unit contribution of B is decreased by Rs. 4
   e. The unit contribution of C is increased by Rs. 3

2. For the linear programming problem given below conduct the sensitivity analysis for the contribution coefficient of the objective function.

$$\begin{aligned} \text{Maximize: } & 10X_1 + 15X_2 \\ \text{Subject to: } & 1X_1 + 2X_2 \le 70 \text{ (Labour)} \\ & 4X_1 + 2X_2 \le 100 \text{ (Capital)} \\ & X_1, \; X_2 \ge 0 \end{aligned}$$

3. For the linear programming problem given below conduct the sensitivity analysis for the right hand side variables.

$$\begin{aligned} \text{Maximize: } & 45X_1 + 80X_2 \\ \text{Subject to: } & 5X_1 + 20X_2 \le 400 \text{ (Constraint A)} \\ & 10X_1 + 15X_2 \le 450 \text{ (Constraint B)} \\ & X_1, \; X_2 \ge 0 \end{aligned}$$

CHAPTER

5

Transportation & Transhipment Problems

5.1 INTRODUCTION

One of the earliest applications of linear programming in the field of transportation was originally developed by Hitchcook in 1941 in a study entitled "the distribution of products from several sources to numerous locations". In the year 1947 T.C. Koopmans made an independent study on "optimum utilization of transportation system". Subsequently, these problems were systematically converted into a linear programming problem and solved by the well-known Simplex method by Prof. George B. Dantzig.

Transportation, assignment, and transhipment are some of the special cases of linear programming problems. Transportation problems are commonly used in planning distribution of goods and services produced in certain centres to various sales designations. Typically, quantity of goods available at each supply point is limited by the production capacity. However, the quantity of goods needed at various sales locations are predicted well in advance. Since transportation invariably involves cost it will have to be minimized to optimize the profit.

Example 1

Let us consider the transportation problem faced by a softdrink company with three plants located at three places in India. The following table provides data pertaining to their respective production capacities.

Table 5.1

Origin	Plant location	Production capacity
1	Mumbai	5000
2	Chennai	6000
3	Kolkata	2500
	Total	13,500

Let the softdrink company have four sales locations as listed below with their respective demands.

Table 5.2

Destination	Distribution centres	Demand forecast
1	Madurai	6000
2	Dharwad	4000
3	Pune	2000
4	Trivandrum	1500
	Total	13,500

The following table gives the per unit transport cost in rupees from the three production centres to four sales locations.

Table 5.3

Origin	Pune	Dharwad	Madurai	Trivandrum
Mumbai	3	2	7	6
Chennai	7	5	2	3
Kolkata	2	5	4	5

Draw a suitable route mapping for this problem.

Solution:

In Figure 5.1 we show all the possible (12) transportation routes from all the three centres with their respective transport costs. Similarly, Figure 5.2 shows the route naming pattern. If the transport takes place from origin i destination j then the corresponding route is known as X_{ij}.

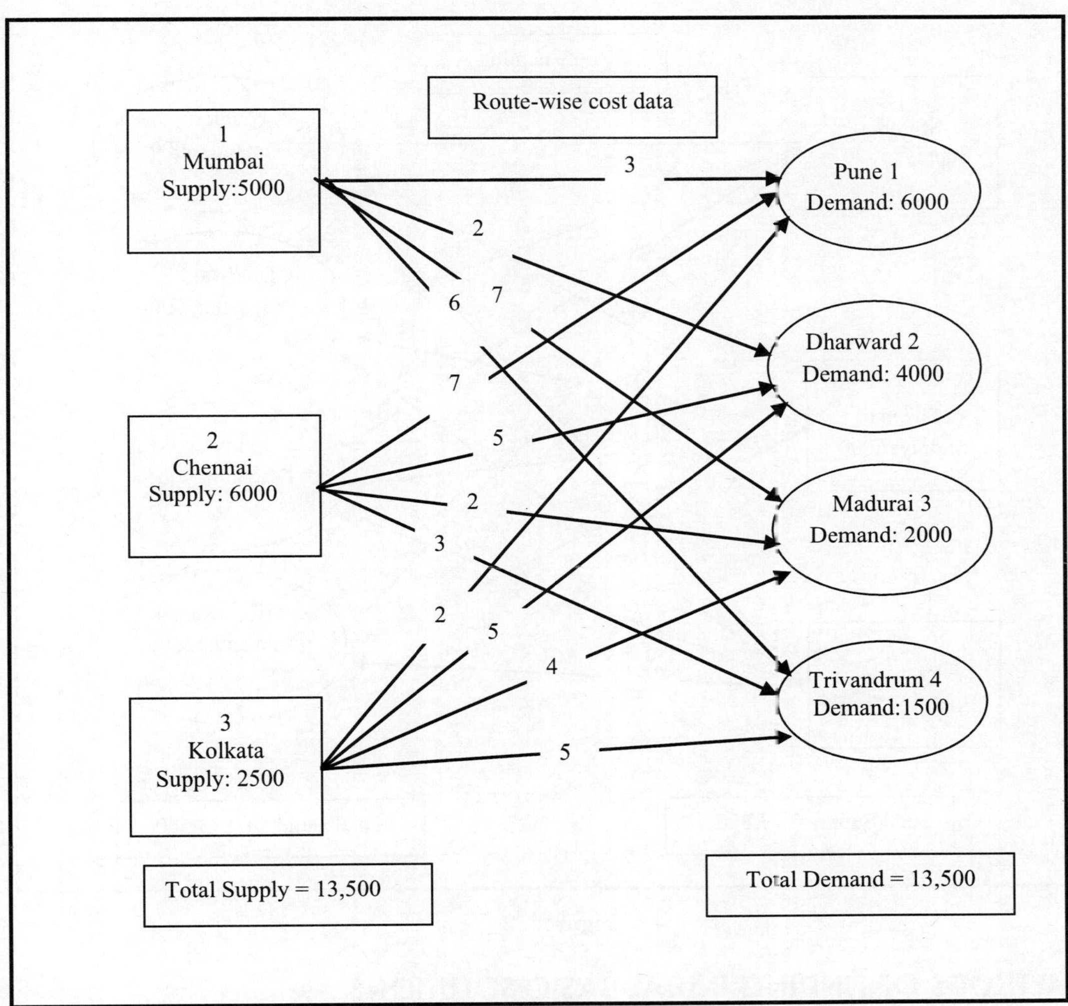

Figure 5.1

Now the company will have to plan for its distribution from these three centres to all the four sales locations having both the demand and supply conditions in mind. While doing so the ultimate aim is to minimize the total transport cost in order to maximize profit.

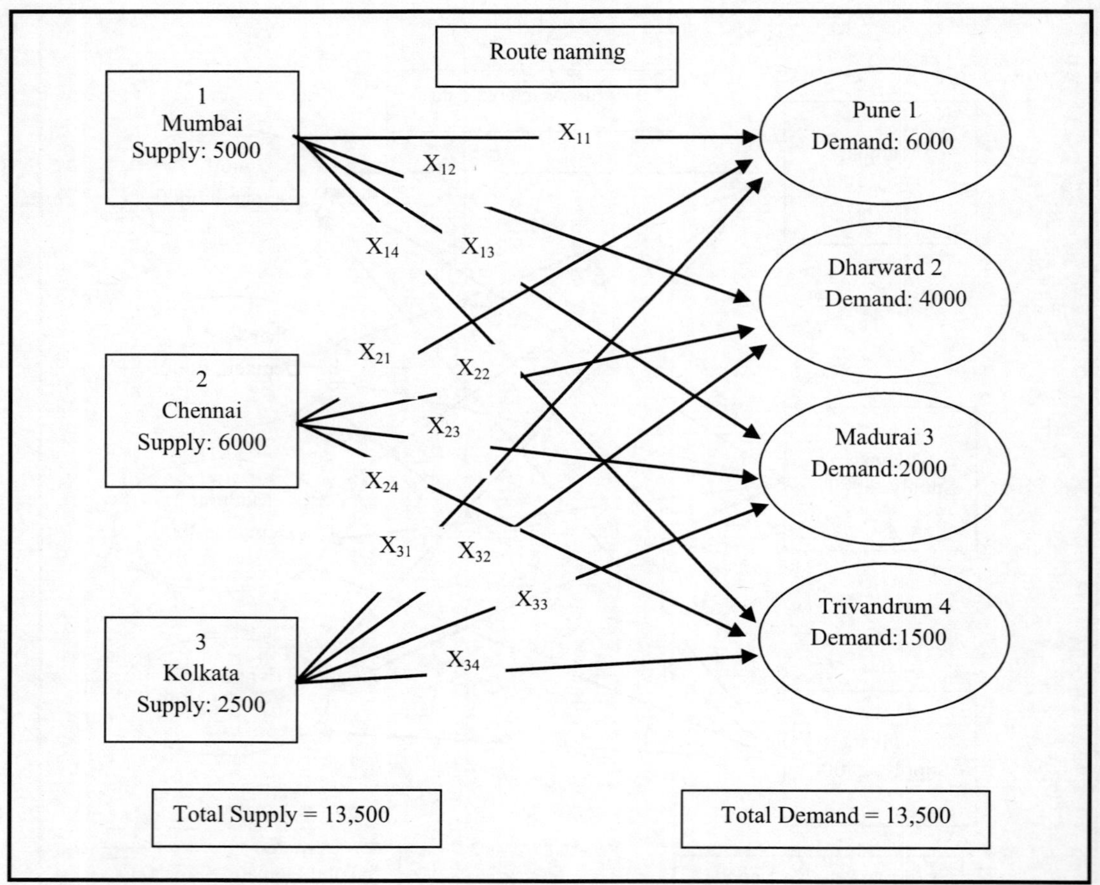

Figure 5.2

5.2 METHODS OF FINDING INITIAL BASIC SOLUTIONS

For a given transportation problem one can find the initial tentative solution by using any one of the following methods :

- Solution by North-West Corner Method
- Solution by Matrix Minimum Method
- Solution by Vogel Approximation Method (VAM Method)

5.3 SOLUTION BY NORTH-WEST CORNER METHOD

In this method the very first allotment starts from the North-West corner cell of the cost data having both the demand (column) and the supply (row) constraints in mind. When this route is fully used, depending upon which number is lower the remainder of either factory capacity or destination requirment is then assigined to the new row(s) or column(s) until it is fully utilized. Let us illustrate this rule and get the initial solution for the transportation problem discussed above.

Example 2

For the given transportation problem obtain an initial solution by North-west Corner Rule.

Table 5.4

Origin	Destinations				
	Pune 1	Dharwad 2	Madurai 3	Trivandrum 4	Total s_i
Mumbai 1	3	2	7	6	5000
Chennai 2	7	5	2	3	6000
Kolkata 3	2	5	4	5	2500
Total d_j	6000	4000	2000	1500	13,500

Solution

The figures shown in the tables are the respective transport cost per unit from respective origins to respective destinations. For example, the cost of transporting 1 unit of the product from origin 1 (Mumbai) to destination 2 (Dharwad) is Rs. 2. The figures at the bottom row refers the demand from the respective destinations (d_j). Similarly, the figures at the right extreme column are the respective supply positions from the respective origins (s_i). In Figure 5.2 the respective algebrical namming pattern is shown.

Table 5.5

Origin	Destinations				
	Pune 1	Dharwad 2	Madurai 3	Trivandrum 4	Total s_i
Mumbai 1	3	2	7	6	5000
Chennai 2	7	5	2	3	6000
Kolkata 3	2	5	4	5	2500
Total d_j	6000	4000	2000	1500	13,500

According to this North-west corner rule we must allot the North-west corner cell first. It involves the transport cost of Rs. 3 per unit from origin Mumbai to destination Pune.

Let this cell allotment be denoted by the variable X_{11}. Now we allot to this cell by the rule Min (d_1,s_1) = Min (6000, 5000) = 5000. So X_{11} = 5000. Once this allocation is made, since Mumbai supply is fully exhausted, we remove the the first row pertaining to Mumbai from the very Table 5.5. However, now there will be 1000 (6000–5000) units of unsatisfied demand for Pune. So after incorporating this correction, destination Pune must be restored in our new table with the left over demand figure of 1000 units. Note that all other remaining entries are kept intact in Table 5.6.

Table 5.6

Origin	Destinations				
	Pune 1	Dharwad 2	Madurai 3	Trivandrum 4	Total s_i
Chennai 2	7	5	2	3	6000
Kolkata 3	2	5	4	5	2500
Total d_j	1000	4000	2000	1500	8500

Now in this new reduced table, the North-west corner cell is to be allotted its full quota as usual. So by the same minimum allotment rule X_{21} = Min (d_1, s_2) = Min(1000, 6000) = 1000. So after allotting 1000 to cell X_{21} we remove this particular Pune column in the next table, since its full quota of 6000 units is completely satisfied (5000 from the first allotment and 1000 from the second allotment). Now note that from origin 2 since already 1000 units are allotted to Pune it will have only 5000 units stock as listed in Table 5.7.

Table 5.7

Origin	Destinations			
	Dharwad 2	Madurai 3	Trivandrum 4	Total s_i
Chennai 2	5	2	3	5000
Kolkata 3	5	4	5	2500
Total d_j	4000	2000	1500	7500

In this table north-west corner cell allotment is given by X_{22} = Min (d_2, s_2) = (4000, 5000) = 4000. After this allotment since Dharwad's full demand is satisfied we remove this column for our next iteration. At same time do not forget to make correction for Chennai's supply. It will have only 1000 (5000–4000) units.

Table 5.8

Origin	Destinations		
	Madurai 3	Trivandrum 4	Total s_i
Chennai 2	2	3	1000
Kolkata 3	4	5	2500
Total d_j	2000	1500	3500

In this table north-west corner cell allotment is given by X_{23} = Min (d_3,s_2) = Min (2000, 1000) = 1000. After this allotment since Chennai's supply is fully exhausted, we remove Chennai row for our next iteration. At same time do not forget to make correction for Madurai's demand. It needs only 1000 units of next allotment.

Table 5.9

Origin	Destinations		
	Madurai 3	Trivandrum 4	Total s_i
Kolkata 3	4	5	2500
Total d_j	1000	1500	2500

In this table north-west corner cell allotment is given by X_{33} = Min (d_3, s_3) = (1000, 2500) = 1000. After this allotment since Madurai's full demand is satisfied we remove this column for our next iteration. At same time do not forget to make correction for Kolkata's supply. It will have only 1500 (2500−1000) units.

Table 5.10

Origin	Destinations	
	Trivandrum 4	Total s_i
Kolkata 3	5	1500
Total d_j	1500	1500

In this table the north-west corner allotment is X_{34} = Min (d_4, s_3) = Min (1500, 1500) = 1500. The leftover supply from Kolkata is enough to meet the demand for Trivandrum. This is our final stage of iteration.

The final consolidated solution is provided in Table 5.11 by the highlighted cells. The figures in the parenthesis show the respective allotments.

Table 5.11

Origin	Destinations				
	Pune 1	Dharwad 2	Mumbai 3	Trivandrum 4	Total s_i
Mumbai 1	3 (5000)	2	7	6	5000
Chennai 2	7 (1000)	5 (4000)	2 (1000)	3	6000
Kolkata 3	2	5	4 (1000)	5 (1500)	2500
Total d_j	6000	4000	2000	1500	13,500
Total Cost =	55,500				

From the table the total transport cost is obtained as

$$3X_{11}+7X_{21}+5X_{22}+2X_{23}+4X_{33}\ 5X_{34} = 3(5000)+7(1000)+5(4000)+2(1000)+4(1000)+5(1500)$$
$$= \text{Rs. } 55{,}500$$

Note: *However, since this method totally disregards the given cost data at the time of allotment, the solution obtained is inefficient.*

5.4 SOLUTION BY MATRIX MINIMUM METHOD

In this method we give due credit to the given per unit transportation cost while allotting the cell values. Further, we do not necessarily start from the north-west corner this time. Instead we begin the allotment with the cell having the least per unit transport cost. Once the allotted column or row is deleted we start again from the least cost cell among the leftover cells. Let us illustrate this method by taking the same example discussed in the previous method.

■ Example 3

For the transportation problem given below obtain an initial solution by Matrix minimum method.

Table 5.12

Origin	Destinations				
	1	2	3	4	Total s_i
1	3	2	7	6	5000
2	7	5	2	3	6000
3	2	5	4	5	2500
Total d_j	6000	4000	2000	1500	13,500

Solution

Table 5.13

Origin	Destinations				
	1	2	3	4	Total s_i
1	3	2	7	6	5000
2	7	5	2	3	6000
3	2	5	4	5	2500
Total d_j	6000	4000	2000	1500	13,500

From the table given above we notice that $C_{12} = C_{31} = C_{23} = 2$ are the least cost cells. So let us allotment to these cells first by using the minimum rule. So $X_{12} = \text{Min}(s_1, d_2) = (5000, 4000) = 4000$. Thus destination 2 gets its full quota of 4000 units from origin1. So destination 2 is dropped for the next iteration. After this allotment origin 1 will have only 1000 units for further allotment as shown below.

Table 5.14

Origin	Destinations			
	1	3	4	Total s_i
1	3	7	6	1000
2	7	2	3	6000
3	2	4	5	2500
Total d_j	6000	2000	1500	9500

X_{31} = Min (6000, 2500) = 2500. Now total supply gets exhausted from origin 3. So origin 3 is removed for next iteration. After receiving 2500 units from origin 3, destination 1 will be left with 3500 units of unfilled demand.

Table 5.15

Origin	Destinations			
	1	3	4	Total s_i
1	3	7	6	1000
2	7	2	3	6000
Total d_j	3500	2000	1500	7000

X_{23} = Min (2000, 6000) = 2000. Now since the demand from destination 3 is fully met with, we remove this column and proceed further. The leftover supply from origin 2 is 4000 units.

Table 5.16

Origin	Destinations		
	1	4	Total s_i
1	3	6	1000
2	7	3	4000
Total d_j	3500	1500	5000

X_{11} = Min (3500, 1000) = 1000. Supply from destination 1 is exhausted so we remove this row. The leftover demand from destination 1 is 2500 units.

Table 5.17

Origin	Destinations		
	1	4	Total s_i
2	7	3	4000
Total d_j	2500	1500	4000

X_{24} = Min (1500, 4000) = 1500. Demand from destination 4 is fully met. So destination 4 is removed in the next iteration. Leftover supply with origin 2 is 2500 units.

Table 5.18

Origin	Destinations	
	1	Total s_i
2	7	2500
Total d_j	2500	2500

X_{21} = Min (2500, 2500) = 2500. Demand from destination 1 is fully met. Supply from origin 2 is fully exhausted. Thus the allotment problem is fully solved and reported in the following table.

Table 5.19

Origin	Destinations				
	1	2	3	4	Total s_i
1	3 (1000)	2 (4000)	7	6	5000
2	7 (2500)	5	2 (2000)	3 (1500)	6000
3	2 (2500)	5	4	5	2500
Total d_j	6000	4000	2000	1500	
	Minimum Total Cost =		42,000		

The total cost of transport = 3 × 1000 + 7 × 2500 + 2 × 2500 + 2 × 4000 + 2 × 2000 + 3 × 1500 = 42,000.

This solution is definitely a better solution than the previous. It has brought down the transportation cost to Rs. 42,000. For northwest corner rule the total transportation cost was obtained as Rs. 55,500.

Note: *Since in this method allotments are made having the least cost in mind the solution is cost efficient to some extent.*

5.5 SOLUTION BY VOGEL (VAM) APPROXIMATION METHOD

Though this method needs a good deal of calculations for finding the initial basic feasible solution, it yields the best possible result closer to the final optimum solution in the very beginning itself. The steps involved in this method are listed below.

- From the given cost matrix we find the penalty for all columns and rows. The penalty is levied as a result of not assigning a non-zero value to the cell with lowest transportation cost. These penalties are obtained by subtracting the lowest cost in a column (or row) from the next lowest cost in the same column (or row).
- In the second stage we identify largest penalty among all rows and columns. If these penalites are equal for two or more rows we select the topmost row for the next stage of computation. If the penalties are equal for two or more columns then we select the extreme left column for the next stage of computation.
- We allocate as much as possible to the cell having the lowest cost in the selected highest penalty row or column in step 2 by applying Min(d_i,s_i) rule as we did it earlier. If the supply is more than the demand with reference to the cell under consideration the column corresponding to this cell is eliminated in the next stage. If the demand equals supply both the column and the row pertaining to cell under reference gets eliminated simultaneously. If the demand exceeds the supply then the corresponding row gets eliminated. In all cases after the allotment the corresponding supply and demand quantities are adjusted for the next stage.
- This procedure is repeated till final result is obtained.

Example 4

For the transportation problem given below obtain an initial solution by VAM Method.

Table 5.20

Origin	Destinations				
	1	2	3	4	Total s_i
1	3	2	7	6	5000
2	7	5	2	3	6000
3	2	5	4	5	2500
Total d_j	6000	4000	2000	1500	13,500

Solution

Table 5.21

Origin	Destinations					
	1	2	3	4	Total s_i	Penalty
1	3	2	7	6	5000	1
2	7	5	2	3	6000	1
3	2	5	4	5	2500	2
Total d_j	6000	4000	2000	1500	13,500	
Penalty	1	3	2	2		

Step 1: In the first column the lowest cost is 2 and the next lowest cost is 3. Thus 3–2 = 1 is the associated penalty corresponding to this column. It is shown at the bottom of column 1 in Table 5.21. Similarly, the penalty values are calculated for the remaining columns and rows and reported in the table.

Step 2: Among the seven penalties reported, the penalty 3 pertaining to destination 2 is the highest and equal to 3 (highlighted). The lowest cost in this column is 2 (highlighted).

Step 3: So our first allotment should go to this cell. X_{12} = Min (4000, 5000) = 4000. So make full allotment to destination 2 from origin 1. Remove the destination 2 from the list. After the allotment the leftover supply with origin 1 is only 1000 units (5000–4000).

Step 4: Get the reduced transportation matrix with due corrections both in the supply and demand conditions. Repeat the penalty calculations again. Also repeat the whole cycle of operation once again.

Table 5.22

Origin	Destinations				
	1	3	4	Total s_i	Penalty
1	3	7	6	1000	3
2	7	2	3	6000	1
3	2	4	5	2500	2
Total d_j	6000	2000	1500	9500	
Penalty	1	2	2		

This time the highest penalty is 3 in the first row. The lowest cost in this row is 3. Accordingly, this cell must be allotted fully by the minimum rule. X_{11} = Min (6000,1000) = 1000. So allot the leftover 1000 unit from origin 1 to destination 1. Delete the origin 1 from the table. Remember, destination 1 needs 5000 more units.

Table 5.23

Origin	Destinations				
	1	3	4	Total s_i	Penalty
2	7	2	3	6000	1
3	2	4	5	2500	2
Total d_j	5000	2000	1500	8500	
Penalty	5	2	2		

X_{31} = Min (5000, 2500) = 2500. So allot the whole supply of 2500 units of origin 3 to destination 1 and delete origin 3. Note still destination 1 needs (5000 – 2500) 2500 units.

Table 5.24

Origin	Destinations				
	1	3	4	Total s_i	Penalty
2	7	2	3	6000	1
Total d_j	2500	2000	1500	6000	
Penalty	7	2	2		

X_{21} = Min (2500, 6000) = 2500. So allot 2500 units to destination 1 from origin 2 and delete destination 1 from the list. Note that the origin 2 is left with only 3500 (6000 – 2500) units at this stage.

Table 5.25

Origin	Destinations			
	3	4	Total s_i	Penalty
2	2	3	3500	1
Total d_j	2000	1500	3500	
Penalty	2	3		

X_{24} = Min (1500,3500) = 1500. So allot 1500 from origin 2 to destination 4 and move destination 4 from the table. At this stage origin 2 is left with only 2000 units.

Table 5.26

Origin	Destinations		
	3	Total s_i	Penalty
2	2	2000	1
Total d_j	2000	2000	
Penalty	2		

X_{23} = Min (2000, 2000) = 2000. So allot the leftover 2000 in origin 2 to destination 3 and complete the allotment.

Table 5.27

Origin	Destinations				
	1	2	3	4	Total s_i
1	3 (1000)	2 (4000)	7	6	5000
2	7 (2500)	5	2 (2000)	3 (1500)	6000
3	2 (2500)	5	4	5	2500
Total d_j	6000	4000	2000	1500	13,500
Total transportation cost =		42,000			

The final solutions values are: X_{11} = 1000; X_{12} = 4000; X_{21} = 2500; X_{23} = 2000; X_{24} = 1500; X_{31} = 2500. The total transport cost is Rs. 42,000.

***Note:** The VAM method total transportation cost is least compared to the other two methods discussed in this section.*

■ Example 5

A company XYZ is currently spending Rs.1000 on transport. Given the per unit transport cost table obtain the optimum transport cost and the associated cost savings by using VAM method.

Table 5.28

Factory	Warehouse				s_i
	W1	W2	W3	W4	
F1	19	30	50	12	7
F2	70	30	40	60	10
F3	40	10	60	20	18
d_j	5	8	7	15	35

Solution by VAM method (non-optimal)

Table 5.29

Factory		Warehouse				s_i	Row Penalty			
		W1	W2	W3	W4		I	II	III	IV
F1		**19** (5)	30	50	**12** (2)	7,2,0,	7	18	38	38
F2		70	30	**40** (7)	**60** (3)	10	10	10	20	20
F3		40	**10** (8)	60	**20** (10)	18,10,0	10	10	**40**	-
d_j		5,0	8,0	7,0	15,5,0	35				
Column Penalty	I	**21**	20	10	8	Total Cost = 859				
	II	-	**20**	10	8					
	III			10	8					
	IV			10	**48**					

First Iteration

In the above table all the needed calculations are shown in a single table. To begin with we calculate the column penalty and report it corresponding to column penalty I in the table. Similarly, the row penalties are calculated and reported row penalty I. Among these columns and row penalties the highest penalty is the highlighted column penalty 21. In this column 19 is the least transportation cost. So we go for allotment to this cell first as per our minimum rule. X_{11} = Min(7,5) = 5. This allotment is shown in the corresponding cell within bracket. Once this is done W_1 is to be removed for the next iteration because of its full demand is met. At the same time the remaining supply from F1 is to adjusted 7−5 = 2. This 2 is entered next to 7 in s_i column.

Second Iteration

Remember, that this time the column penalty is calculated and reported in column penalty II for W2, W3, W4 columns only. However, for row penalty we repeat for all the three origins as usual. Such calculations are reported in row penalty II in the table. Now among row penalty II and column penalty II, the highest is the highlighted as 20. The least cost corresponding to this column is 10. So let us make allotment to this cell as usual. X_{32} = Min(8,18) = 8. This allotment is shown in the cell within the bracket. After making this allotment the left over supply 18−8 = 10 is shown next to 18 in the s_i column. Remember, that this time W2 is also removed for the next iteration.

Third Iteration

This time the column penalty is calculated for the remaining two destinations W3 and W4 only. However, row penalties are to be calculated for all the three origins. This time 40 is the highest row penalty and 20 is the associated least cost cell value deserves allotment. So X_{34} = Min(10,15) = 10. This allotment is shown within bracket in the correct place in the table. The leftover supply from F3 = 10−10 = 0 is shown next to 10 in the s_i column. At this stage the unfulfilled demand from W4 = 15−10 = 5 is written after 15 with a separator.

Fourth Iteration

At this stage we are left with F1 and F2 factories with W3 and W4 warehouses only. The fourth stage column and row penalties reported in the table. Since the column penalty 48 is the highest the associated least cost is 12 and the available supply from F1 is only 2. Thus this allotment is shown in the bracket. Once this is done F1 also gets eliminated. Easily the leftover supply 10 from F2 is distributed between W3 and W4 as 7 and 3 respectively having the column and row total in mind. This completes the entire process of allotment.

The total cost of this allotment = 19 × 5 + 10 × 8 + 40 × 7 + 12 × 2 + 60 × 3 = Rs. 859

Therefore the associated cost saving due to this allotment = 1000 – 859 = Rs. 141.

Note: *Since in this method allotments are made having the penalty of not assigning in mind the obtained result is very much closer to the final optimum solution.*

Basic feasible solution

In a given transportation problem if there are n columns and m rows then we will have n×m cells altogether. In the final solution if there are at least m + n – 1 allotted cells then the solution is said to be a basic feasible solution.

In all our illustrations discussed so far m + n –1 = 3+4–1= 6. In all the three methods, our initial basic solutions were having 6 allotted cells. Thus the obtained results constitute the needed basic feasible solution set.

5.6 LINEAR PROGRAMMING FORMATION OF TRANSPORTATION PRIMAL

In general let there be m origins and n destinations in our transportation problem as described by the per unit cost matrix given below.

Table 5.30

Origin	Destinations						
	1	2	. . .	j	. . .	n	s_i
1	c_{11}	c_{12}	. . .	c_{1j}	. . .	c_{1n}	s_1
2	c_{21}	c_{22}	. . .	c_{2jj}	. . .	c_{2n}	s_2
.	.	.	. . .	.	. . .	.	.
.	.	.	. . .	.	. . .	.	.
i	c_{i1}	c_{i2}	. . .	c_{ij}	. . .	c_{in}	s_n
.	.	.	. . .	.	. . .	.	.
.	.	.	. . .	.	. . .	.	.
m	c_{m1}	c_{m2}	. . .	.	. . .	c_{mn}	s_m
d_j	d_1	d_2	. . .	d_j	. . .	d_n	

j = index for the destination ranging from 1 to n.

i = index for the origin ranging 1 to m.

c_{ij} = per unit cost of shipping from origin i destination j.

s_i = supply available at the origin i.

d_j = demand emerging from the jth destination.

If X_{ij} refers the number of units shipped from the ith origin to j^{th} destination then the objective function of our transportation problem that aims at minimizing the total cost is written as

$$
\begin{aligned}
\text{Minimize:} \quad Z = & c_{11}X_{11} + c_{12}X_{12} + \ldots\ldots + c_{1n}X_{1n} \\
& + c_{21}X_{21} + c_{22}X_{22} + \ldots\ldots + c_{2n}X_{2n} \\
& + c_{31}X_{31} + c_{32}X_{32} + \ldots\ldots + c_{3n}X_{3n} \\
& + \ldots\ldots\ldots\ldots\ldots\ldots\ldots\ldots\ldots\ldots \\
& + \ldots\ldots\ldots\ldots\ldots\ldots\ldots\ldots\ldots\ldots \\
& + c_{m1}X_{m1} + c_{m2}X_{m2} + \ldots\ldots + c_{mn}X_{mn}
\end{aligned}
$$

The supply constraints

$$
\begin{aligned}
X_{11} + X_{12} + \ldots\ldots + X_{1n} &= s_1 \\
X_{21} + X_{22} + \ldots\ldots + X_{2n} &= s_2 \\
X_{31} + X_{32} + \ldots\ldots + X_{3n} &= s_3 \\
\ldots\ldots\ldots\ldots\ldots\ldots\ldots\ldots\ldots \\
X_{i1} + X_{i2} + \ldots\ldots + X_{in} &= s_i \\
\ldots\ldots\ldots\ldots\ldots\ldots\ldots\ldots\ldots \\
X_{m1} + X_{m2} + \ldots\ldots + X_{mn} &= s_m
\end{aligned}
$$

Demand constraints

$$
\begin{aligned}
X_{11} + X_{21} + \ldots\ldots + X_{j1} + \ldots\ldots + X_{m1} &= d_1 \\
X_{12} + X_{22} + \ldots\ldots + X_{j2} + \ldots\ldots + X_{m2} &= d_2 \\
X_{13} + X_{23} + \ldots\ldots + X_{j3} + \ldots\ldots + X_{m3} &= d_3 \\
\ldots\ldots\ldots\ldots\ldots\ldots\ldots\ldots\ldots\ldots\ldots\ldots \\
\ldots\ldots\ldots\ldots\ldots\ldots\ldots\ldots\ldots\ldots\ldots\ldots \\
X_{1n} + X_{2n} + \ldots\ldots + X_{jn} + \ldots\ldots + X_{mn} &= d_n
\end{aligned}
$$

$$\text{Minimize:} \quad \sum_{i=1}^{m}\sum_{j=1}^{n} c_{ij}X_{ij}$$

$$\text{Subject to:} \quad \sum_{j=1}^{n} X_{ij} = s_i \quad i=1, 2, 3 \ldots\ldots\ldots m \text{ (supply constraints)}$$

$$\sum_{i=1}^{m} X_{ij} = d_j \quad j=1,2, 3, 4 \ldots\ldots\ldots n \text{ (demand constraints)}$$

For a feasible solution $\sum_{i=1}^{m} s_i = \sum_{j=1}^{n} d_j$.

5.7 LINEAR PROGRAMMING FORMATION OF TRANSPORTATION DUAL

In general linear programming transportation problem we will be having with (m+n) equality constraints only. For the formulation of the dual we express each one these constraints by a pair of two equivalent constraints, one is *greater than or equal to* type and the other is *less than or equal to* type.

The equivalent primal with modified constraints is written as

$$\text{Minimize: } Z=\sum_{i=1}^{m}\sum_{j=1}^{n} c_{ij}X_{ij}$$

$$\text{Subject to: } \sum_{j=1}^{n} X_{ij} \geq s_i; \quad \sum_{j=1}^{n} X_{ij} \leq s_i \text{ or } \sum_{j=1}^{n}(-X_{ij}) \geq -s_i;\ i=1, 2,\ldots m$$

$$\sum_{i=1}^{m} X_{ij} \geq d_j; \quad \sum_{i=1}^{m} X_{ij} \leq d_j \text{ or } \sum_{i=1}^{m}(-X_{ij}) \geq -d_j;\ j=1, 2,\ldots n$$

$$X_{ij} \geq 0 \text{ for all i and j}$$

We know that in the formation of the dual we will have as many dual variables as that of the number of constraints. Let U_i^+ (i =1,2,3......n) stand for the n dual variables corresponding to n supply constraints having ≥ inequalities as stated above. Similarly, let U_i^- (i = 1, 2 ,3 ... n) stand for the n dual variables corresponding to n supply constraints having ≤ inequalities as stated above.

Also let V_i^+ (i =1,2,3......m) stand for the m dual variables corresponding to m demand constraints having ≥ inequalities as stated above. Similarly, let V_i^- (i = 1, 2, 3...m) stand for the m dual variables corresponding to m supply constraints having ≤ inequalities as stated above. With these notations altogether there are 2(n+m) dual variables in our model.

Now the relevant dual problem may be written as

$$\text{Maximize: } Z=\sum_{i=1}^{m}(U_i^+ - U_i^-)s_i + \sum_{j=1}^{n}(V_j^+ - V_j^-)d_j$$

$$\text{Subject to: } (U_i^+ - U_i^-) + (V_j^+ - V_j^-) \leq c_{ij}$$

$$U_i^+, U_i^-, V_j^+, V_j^- \geq 0 \text{ for all i and j}$$

If we write $(U_i^+ - U_i^-) = U_i$ and $(V_j^+ - V_j^-) = V_j$, the dual problem stated above can also be rewitten as

$$\text{Maximize: } Z=\sum_{i=1}^{m} U_i s_i + \sum_{j=1}^{n} V_j d_j$$

$$\text{Subject to: } U_i + V_j \leq c_{ij}$$

$$U_i, V_j \text{ unrestricted in sign.}$$

***Note:** The dual solution of the transportation problem implicitly measures the comparative location advantage of the cells. Accordingly,* $c_{ij} - U_i - V_j = 0$ *for all occupied cells and* $c_{ij} - U_i - V_j \geq 0$ *for all other empty cells. If the solution does not degenerate then there will be m+n−1 occupied cells enough to formulate m+n−1 equations in m+n variables. By assuming zero for any one of them we solve the entire system for the remaining variables. Once this is done we perform optimality test for all the left over empty cells by calculating* $\Delta_{ij} = c_{ij} - (U_i + V_j)$*. If these cell values are non-negative for all the empty cells then the associated allotment is declared as optimal in the sense it will give the minimum total transport cost. If one or more cell values are negative, reallocation is recommended from the neighbouring costlier cells to the most cost effective cell having the largest negative value. While doing so the amount of transfer is decided having both the column and row total in mind. Accordingly, other cell values in the desired closed loop also gets modified. This method is commonly called MODI method discussed in the next section.*

■ Example 6

A firm has facilities in six places. It has manufacturing facilities at places A, B, and C with productive capacity of 50, 40, and 60 units respectively. The warehouses located at destinations D, E, and F demands 20, 95, and 35 units respectively. Per unit transportation cost are given in the table shown below. Formulate a suitable linear programming problem to minimize the total transportation cost. Also formulate its dual.

Table 5.30a

Plants	Warehouses			s_i
	1	2	3	
1	6	4	1	50
2	3	8	7	40
3	4	4	2	60
d_j	20	95	35	150

Primal formulation

Let X_{ij} stand for the number of units trasported from i^{th} plant to j^{th} warehouse. Now the minimization LPP is formulated as:

$$Z = 6X_{11} + 4X_{12} + 1X_{13} + 3X_{21} + 8X_{22} + 7X_{23} + 4X_{31} + 4X_{32} + 2X_{33}.$$

Subject to: $X_{11} + X_{12} + X_{13} = 50$

$X_{21} + X_{22} + X_{23} = 40$ (Supply constraints)

$X_{31} + X_{32} + X_{33} = 60$

$X_{11} + X_{21} + X_{31} = 20$

$X_{12} + X_{22} + X_{32} = 95$ (Demand constraints)

$X_{13} + X_{23} + X_{33} = 35$

$X_{ij} \geq 0$ for i = 1, 2, 3 and j = 1, 2, 3

Like any other linear programming problem of minimization, to find the dual we will have to convert all the equalities into inequalities. For doing so, the given single equality is replaced by a pair of inequalities of one *greater than or equal to* type and another *less than or equal to* type. Since we have three equality constraints in the supply set after replacement we will have altogether six supply constraints as shown below. Similarly, since we have three demand equalities altogether we will have 6 replaced inequalities as shown below.

$$
\begin{array}{lll}
X_{11} + X_{12} + X_{13} \geq 50 & (1) & \\
X_{11} + X_{12} + X_{13} \leq 50 & (2) & \\
X_{21} + X_{22} + X_{23} \geq 40 & (3) & \\
X_{21} + X_{22} + X_{23} \leq 40 & (4) & \text{Supply Constraints} \\
X_{31} + X_{32} + X_{33} \geq 60 & (5) & \\
X_{31} + X_{32} + X_{33} \leq 60 & (6) & \\
X_{11} + X_{21} + X_{31} \geq 20 & (7) & \\
X_{11} + X_{21} + X_{31} \leq 20 & (8) & \\
X_{12} + X_{22} + X_{32} \geq 95 & (9) & \\
X_{12} + X_{22} + X_{32} \leq 95 & (10) & \text{Demand Constraints} \\
X_{13} + X_{23} + X_{33} \geq 35 & (11) & \\
X_{13} + X_{23} + X_{33} \leq 35 & (12) &
\end{array}
$$

Now to make all the even numbered constraints as greater than or equal to type we multiply them all by −1 and rewrite the primal in the standard form as shown below.

$$Z = 6X_{11} + 4X_{12} + 1X_{13} + 3X_{21} + 8X_{22} + 7X_{23} + 4X_{31} + 4X_{32} + 2X_{33}$$

Subject to :

$$
\begin{array}{rrrrrrrrrcr}
X_{11} & + X_{12} & + X_{13} & & & & & & & \geq & 50 \\
-X_{11} & - X_{12} & - X_{13} & & & & & & & \geq & -50 \\
 & & & X_{21} & + X_{22} & + X_{23} & & & & \geq & 40 \\
 & & & -X_{21} & - X_{22} & - X_{23} & & & & \geq & -40 \\
 & & & & & & X_{31} & + X_{32} & + X_{33} & \geq & 60 \\
 & & & & & & -X_{31} & - X_{32} & - X_{33} & \geq & -60 \\
X_{11} & & & + X_{21} & & & + X_{31} & & & \geq & 20 \\
-X_{11} & & & - X_{21} & & & - X_{31} & & & \geq & -20 \\
 & X_{12} & & & + X_{22} & & & + X_{32} & & \geq & 95 \\
 & -X_{12} & & & - X_{22} & & & - X_{32} & & \geq & -95 \\
 & & X_{13} & & & + X_{23} & & & + X_{33} & \geq & 35 \\
 & & -X_{13} & & & - X_{23} & & & - X_{33} & \geq & -35
\end{array}
$$

Dual formulation

Now from this restated primal the dual in standard notations is formulated as usual as follows

$$\text{Maximize}: Z = 50(U_1^+ - U_1^-) + 40(U_2^+ - U_2^-) + 60(U_3^+ - U_3^-) + 20(V_1^+ - V_1^-) + 95(V_2^+ - V_2^-) + 35(V_3^+ - V_3^-)$$

$$\text{Subject to}: (U_1^+ - U_1^-) + (V_1^+ - V_1^-) \le 6$$

$$(U_1^+ - U_1^-) + (V_2^+ - V_2^-) \le 4$$

$$(U_1^+ - U_1^-) + (V_3^+ - V_3^-) \le 1$$

$$(U_2^+ - U_2^-) + (V_1^+ - V_1^-) \le 3$$

$$(U_2^+ - U_2^-) + (V_2^+ - V_2^-) \le 8$$

$$(U_2^+ - U_2^-) + (V_3^+ - V_3^-) \le 7$$

$$(U_3^+ - U_3^-) + (V_1^+ - V_1^-) \le 4$$

$$(U_3^+ - U_3^-) + (V_2^+ - V_2^-) \le 4$$

$$(U_3^+ - U_3^-) + (V_3^+ - V_3^-) \le 2$$

$$U_1^+, U_1^-, U_2^+, U_2^-, U_3^+, U_3^-, V_1^+, V_1^-, V_2^+, V_2^-, V_3^+, V_3^- \ge 0$$

Or

$$U_1 + V_1 \le 6$$

$$U_1 + V_2 \le 4$$

$$U_1 + V_3 \le 1$$

$$U_2 + V_1 \le 3$$

$$U_2 + V_2 \le 8$$

$$U_2 + V_3 \le 7$$

$$U_3 + V_1 \le 4$$

$$U_3 + V_2 \le 4$$

$$U_3 + V_3 \le 2$$

$U_1, U_2, U_3, V_1, V_2, V_3$ are unrestricted in sign

Or

$$\text{Maximize}: Z = 50U_1 + 40U_2 + 60U_3 + 20V_1 + 95V_2 + 35V_3$$

Subject to :

$$\begin{aligned}
U_1 + V_1 &\le 6 \\
U_1 + V_2 &\le 4 \\
U_1 + V_3 &\le 1 \\
U_2 + V_1 &\le 3 \\
U_2 + V_2 &\le 8 \\
U_2 + V_3 &\le 7 \\
U_3 + V_1 &\le 4 \\
U_3 + V_2 &\le 4 \\
U_3 + V_3 &\le 2
\end{aligned}$$

U_1 ; U_2 ; U_3 ; V_1 ; V_2 ; V_3 all are unrestricted sign

METHODS FOR GETTING FINAL OPTIMUM SOLUTION

From the basic feasible solutions obtained one can get the needed final optimum solution by using any one of the following two methods.

1) The Modified Distribution Method (MODI)
2) Stepping Stone Method

5.8 THE MODIFIED DISTRIBUTION METHOD (OR U-V METHOD)

This method provides the optimum minimum cost solution to the transportation problem by using the constraints of the duality problem to the given primal problem. Let V_1, V_2,............... V_n are the unrestricted dual variables corresponding to the demand constraints and U_1, U_2 ....U_m are the unrestricted dual variables corresponding to the supply constraints. Since the original problem has m rows and n columns, any non-generate basic solution has $X_{ij} > 0$ for atleast m+n−1 variables. Since in the primal all the constants are equalities, the respective slacks (Y_{ij}) are zero. Suppose the dual variables are chosen in such a way that $U_i+V_j = C_{ij}$ in the n+m−1 basic cells and $U_i+V_j \le C_{ij}$ for the remaining cells, form the duality theorem it is possible to prove that the solution to the primal problem is optimal. In MODI method we use these two duality conditions to verify whether the obtained solution is optimal or not. The following 5 steps are used for such a task.

Step 1: Find out the basic feasible solution of the given problem by any one of the methods discussed above.

Step 2: Corresponding to m+n−1 basic alloted cells obtained, frame the simultaneous equations so that $C_{ij} = U_i + V_j$. These m+n−1 equations are to be solved for m + n variables. So equate one among the m + n variables arbitrarily to zero and solve the m+n−1 equations for m+n−1 variables.

Step 3: With the so obtained solution values verify the whether $C_{ij} - U_i - V_j \geq 0$ for all the empty cells. If all these values for empty cells are zero or positive then confirm that the given solution is optimum and stop further iteration.

Step 4: If one or more empty cells are found to have negative calculated values then select the one having the largest negative value for allotment from the neighbouring costlier occupied cells after forming a closed loop for transfer. While doing so allocation to adjacent cells in the closed loop will have to be adjusted so that one of the original basic cell becomes non-basic.

Step 5: After completing the allotment to the deserving cell, compute a fresh set of values for the dual variables using C_{ij} values of the new set of basic cells. After inserting these values into the system test the optimality conditions for all the new sets of empty cells. If all of them are found to have either zero or positive values stop further iteration and confirm that the allotted values are options. On the other hand, if some cells are found to have negative values we repeat step 1 to 5 repeatedly till we get the optimum solution.

LOOPING IN THE TRANSPORTATION SYSTEM

Origins	Destinations			
	1	2	3	4
1				
2				
3				
4				
5				

Figure 5.3

Origins	Destinations			
	1	2	3	4
1				
2				
3				
4				
5				

Figure 5.4

In transportation models to improve the solution towards the optimum from a basic solution we often form a closed loop with independent cells starting from an empty cell that deserves allotment. Here by independent cells we mean non-repetition of the concerned cell along the loop path as shown in Figure 5.3. The cell C_{22} is counted twice along the path specified and hence not independent. However, in Figures 5.4 and 5.5 all the marked cells are independent.

Origins	Destinations			
	1	2	3	4
1				
2				
3				
4				
5				

Figure 5.5

1. Every loop must have only even number of cells in total with a minimum of 4 as in Figure 5.5.
2. All the referred cells are in independent positions as in Figures 5.4 and 5.5.
3. The closed loop may be rectangular as in Figure 5.5 or as shown in Figure 5.4.
4. The starting cell must be an empty independent cell which deserves allotment from the neigbouring alloted cells. Other than this allotment deserving empty cell all other cells must be alloted ones. In all the figures the allotted cells are highlighted except the allotment deserving one.

■ Example 7

For the transportation problem given below obtain the basic solution by matrix minimum method. Also obtain the optimum solution by MODI method.

Table 5.31

Origin	Destinations					
	1	2	3	4	5	Total s_i
1	1	9	13	36	51	50
2	24	12	16	20	1	100
3	14	33	1	23	26	150
Total d_j	100	70	50	40	40	300

Before applying MODI method we must get a basic solution: so let us get the basic solution by Matrix Minimum method.

Basic solution by Matrix minimum method

Table 5.32

Origin	Destinations					
	1	2	3	4	5	Total s_i
1	1	9	13	36	51	50
2	24	12	16	20	1	100
3	14	33	1	23	26	150
Total d_j	100	70	50	40	40	300

Fix X_{11} = 50 and eliminate row 1.

Table 5.33

Origin	Destinations					
	1	2	3	4	5	Total s_i
2	24	12	16	20	1	100
3	14	33	1	23	26	150
Total d_j	50	70	50	40	40	250

Fix X_{33} = 50 and eliminate column 3.

Table 5.34

Origin	Destinations				
	1	2	4	5	Total s_i
2	24	12	20	1	100
3	14	33	23	26	100
Total d_j	50	70	40	40	200

Fix X_{25} = 40 and eliminate column 5.

Table 5.35

Origin	Destinations			
	1	2	4	Total s_i
2	24	12	20	60
3	14	33	23	100
Total d_j	50	70	40	160

Fix X_{22} = 60 and eliminate row 2.

Table 5.36

Origin	Destinations			
	1	2	4	Total s_i
3	14	33	23	100
Total d_j	50	10	40	100

Fix X_{31} = 50 and eliminate column 1.

Table 5.37

Origin	Destinations		
	2	4	Total s_i
3	33	23	50
Total d_j	10	40	50

Fix X_{34} = 40 and eliminate column 4.

Table 5.38

Origin	Destinations	
	2	Total s_i
3	33	10
Total d_j	10	10

Fix X_{32} = 10.

Thus the final basic solution of the matrix minimum method is summarised in the following table. The figures in the parenthesis is the allocation to the respective cells. In our problem n+m−1 = 5+3−1 = 7. Further, we are having in 7 cells. Thus clearly our solution does not degenerate.

Table 5.39

Origin	Destinations					
	1	2	3	4	5	Total s_i
1	1(50)	9	13	36	51	50
2	24	12(60)	16	20	1(40)	100
3	14(50)	33(10)	1(50)	23(40)	26	150
Total d_j	100	70	50	40	40	300
Total cost 2810						

So directly we can proceed for the final optimum solution.

MODI method for optimum solution

As a first step let us introduce three supply constraints related unrestricted dual variables U_1, U_2, U_3 and five demand constraints related unrestricted dual variables V_1, V_2, V_3, V_4, V_5 as shown in the following Table 5.40 and make the necessary cost entries and allotments made in the preliminary solution.

In terms of dual variables the optimality conditions are:

1. $C_{ij} - U_i - V_j = 0$ for all the m+n−1 basic allotted cells
2. $C_{ij} - U_i - V_j \geq 0$ for all the remaing non-basic empty cells.

Table 5.40

Origins		Destinations M1	M2	M3	M4	M5	Supply
		$V_1 = 14$	$V_2 = 33$	$V_3 = 1$	$V_4 = 23$	$V_5 = 22$	
P1	$U_1 = -13$	1 | **50**	9 | (−11)	13 | 25	36 | 26	51 | 42	50
P2	$U_2 = -21$	24 | 31	12 | **60**	16 | 36	20 | 18	1 | **40**	100
P3	$U_3 = 0$	14 | **50**	33 | **10**	1 | **50**	23 | **40**	26 | 4	150
Demand		100	70	50	40	40	300
					Total cost =	2810	

Calculation of solution value for the dual variables using allotted basic cells in Table 5.40

Now let us obtain the dual variable values by using the first condition stated above.

From Table 5.40

$U_1 + V_1 = 1$; $U_2 + V_2 = 12$; $U_2 + V_5 = 1$; $U_3 + V_1 = 14$; $U_3 + V_2 = 33$; $U_3 + V_3 = 1$; $U_3 + V_4 = 23$

Here, we are having 7 equations in eight variables. Since U_3 appears many times to make the calculations simple. Let $U_3 = 0$.

So the computed values are:

$U_1 = -13$; $U_2 = -21$; $U_3 = 0$; $V_1 = 14$; $V_2 = 33$; $V_3 = 1$; $V_4 = 23$; $V_5 = 22$

Now let us enter these computed U values in the first column and V values in the first row as shown in Table 5.40 and calculate the $C_{ij} - U_i - V_j$ for all the empty cells. These values reported at the bottom right hand side corners of the respective cells. For non-basic cells as per the duality theorem, these values must be zero or positive provided the solution is optimum.

The calculations in Table 5.40 reveal that for all the non-basic cells except cell C_{12} satisfy the condition $C_{ij}-U_i-V_j \geq 0$. For the cell C_{12} this value is obtained as $C_{12} - U_1 - V_2 = 9 - (-13) - 33 = -11$. Such an observation tell us that there is a further scope for cost saving to the extend of Rs. 11 per unit by transferring already allotted units from the neighbouring costlier basic cell to this non-basic cell under reference. One such relevant neighbourhood looping for one unit transfer with relevent cost calculations is exhibited in Table 5.41. The associated looping is shown in the Table 5.40 by on row scheme.

Table 5.41

Origin	Destinations	
	1	2
1	1(−1)	9(+1)
3	14(+1)	33(−1)

To avail of this bonus benefit, since the maximum permissible such transfers is only 10 units from the cell C_{32}, we transfer all the 10 units from cell C_{32} to cell C_{12} and make necessary chances in cell C_{11}, cell C_{32} and cell C_{31} having both column and row constraints in mind. The revised solution is shown in Table 5.42. Note that such a reallocation has brought down the total transportation cost from Rs. 2810 by Rs. 2700 as shown in Table 5.42.

Table 5.42

Origins		Destinations					Supply
		M1	M2	M3	M4	M5	
		$V_1 = 14$	$V_2 = 22$	$V_3 = 1$	$V_4 = 23$	$V_5 = 11$	
P1	$U_1 = -13$	1 **40**	9 **10**	13 (25)	36 (26)	51 (53)	50
P2	$U_2 = -10$	24 (20)	12 **60**	16 (25)	20 (7)	1 **40**	100
P3	$U_3 = 0$	14 **60**	33 (11)	1 **50**	23 **40**	26 (15)	150
Demand		100	70	50	40	40	300
					Total cost =	2700	

Form Table 5.42 our second iteration of optimality test follows.

Calculation of solution value for the dual variables using basic cells in Table 5.42

$U_1 + V_1 = 1$; $U_2 + V_2 = 12$; $U_2 + V_5 = 1$; $U_3 + V_1 = 14$; $U_1 + V_2 = 9$; $U_3 + V_3 = 1$; $U_3 + V_4 = 23$

So we are having 7 equations in eight variables. Since U_3 appears many times to make the calculations simple let $U_3 = 0$.

So the computed values are:

$U_1 = -13$; $U_2 = -10$; $U_3 = 0$; $V_1 = 14$; $V_2 = 22$; $V_3 = 1$; $V_4 = 23$; $V_5 = 11$

Now we insert these values into the Table 5.42 and calculate $C_{ij} - U_i - V_j$ once again for all empty cells. For all the non-basic cell values the results are non-negative positive numbers. Thus, the solution obtained in Table 5.42 gets confirmed as the optimum solution.

So from the Table 5.42 the solution values are obtained as: $X_{11} = 40$, $X_{12} = 10$, $X_{22} = 60$, $X_{25} = 40$, $X_{31} = 60$, $X_{33} = 50$, $X_{34} = 40$. Total cost = 2700.

■ Example 8

For the transportation problem given below obtain the basic solution by VAM method. Also obtain optimal solution by MODI method.

Table 5.43

Origin	Destinations				
	1	2	3	4	Total s_i
1	20	22	17	4	120
2	24	37	9	7	70
3	32	37	20	15	50
Total d_j	60	40	30	110	240

Basic Solution: By VAM method

Table 5.44

Origin	Destinations					
	1	2	3	4	Total s_i	Penalty
1	20	22	17	4	120	13
2	24	37	9	7	70	2
3	32	37	20	15	50	15
Total d_j	60	40	30	110	240	
Penalty	4	15	8	3		

Fix $X_{12} = 40$ and eliminate column 2.

Table 5.45

Origin	Destinations				
	1	3	4	Total s_i	Penalty
1	20	17	4	80	13
2	24	9	7	70	2
3	32	20	15	50	15
Total d_j	60	30	110	200	
Penalty	4	8	3		

Fix $X_{14} = 80$ and eliminate origin 1.

Table 5.46

Origin	Destinations				
	1	3	4	Total s_i	Penalty
2	24	9	7	70	2
3	32	20	15	50	5
Total d_j	60	30	30	120	
Penalty	8	11	8		

Fix $X_{23} = 30$ and eliminate destination 3.

Table 5.47

Origin	Destinations			
	1	4	Total s_i	Penalty
2	24	7	40	17
3	32	15	50	17
Total d_j	60	30	90	
Penalty	8	8		

Fix $X_{24} = 30$ and eliminate destination 4.

Table 5.48

Origin	Destinations		
	1	Total s_i	Penalty
2	24	10	24
3	32	50	32
Total d_j	60	60	
Penalty	8		

Fix $X_{31} = 50$ and eliminate origin 3.

Table 5.49

Origin	Destinations		Penalty
	1	Total s_i	
2	24	10	
Total d_j	10	10	
Penalty			

Fix X_{21} = 10 and finish the allotment.

Initial basic feasible solution

Table 5.50

	1	2	3	4	Total s_i
1	20	22(40)	17	4(80)	120
2	24 (10)	37	9(30)	7(30)	70
3	32(50)	37	20	15	50
Total d_j	60	40	30	110	240
	Total cost =	3520			

X_{12} = 40, X_{14} = 80, X_{21} = 10, X_{23} = 30, X_{24} = 30, X_{31} = 50.

In Table 5.50 the number of basic cells = m+n−1 = 6. Thus, the result is a basic feasible solution with a total cost of Rs. 3520.

MODI method for optimum solution

Calculation of solution value for the dual variables using basic cells in Table 5.51

As usual we introduce the dual variables first into the Table 5.51 and obtain the dual constraints as follows.

$U_1 + V_2 = 22$, $U_1 + V_4 = 4$, $U_2 + V_1 = 24$, $U_2 + V_3 = 9$, $U_2 + V_4 = 7$, $U_3 + V_1 = 32$

After putting $U_2 = 0$ the values of the dual variables are: $U_1 = -3$, $U_2 = 0$, $U_3 = 8$, $V_1 = 24$, $V_2 = 25$, $V_3 = 9$, $V_4 = 7$. For tesitng the conditions for non-basic cell, conveniently let us put these values in the first row and first column in our Table 5.51. For optimum solution the non-basic cells must satisfy the condition $C_{ij} - U_i - V_j \geq 0$. The calculated $C_{ij} - V_i - U_j$ are reported in the following Table 5.51.

The mere inspection of Table 5.51 tell us that except C_{11} all other non-basic cells satisfy optimality dual conditions. For the $C_{11} - U_1 - V_1 = 20 - 24 + 3 = -1$. Such an observation tell us that there is further scope for improvement by converting this non-basic cell into a basic cell and some other costilier neighbourhood basic cells into a non-basic cell.

Table 5.51

Origins		Destinations 1	2	3	4	Supply
		$V_1 = 24$	$V_2 = 25$	$V_3 = 9$	$V_4 = 7$	
1	$U_1 = -3$	20 −1	22 **40**	17 11	4 **80**	120
2	$U_2 = 0$	24 **10**	37 12	9 **30**	7 **30**	70
3	$U_3 = 8$	32 **50**	37 4	20 3	15 0	50
Demand		60	40	30	110	240
					Total cost =	3520

The closed loop shown in Figure 5.51 highlights the needed transfer. The inspection of the Table 5.51 tell us that a maximum of 10 units only transferable from C_{21} to C_{11}. The following table shows the result after affecting the needed transfers having the column and row total in mind.

Table 5.52

Origins		Destinations 1	2	3	4	Supply
		$V_1 = 20$	$V_2 = 22$	$V_3 = 6$	$V_4 = 4$	
1	$U_1 = 0$	20 **10**	22 **40**	17 11	4 **70**	120
2	$U_2 = 3$	24 1	37 12	9 **30**	7 **40**	70
3	$U_3 = 12$	32 **50**	37 3	20 2	15 −1	50
Demand		60	40	30	110	240
					Total cost =	3510

To verify the optimality condition at this stage let us redo the exercise once again having Table 5.52 as starting point.

Calculation of solution value for the dual variables using basic cells in Table 5.52

$U_1+V_1 = 20$, $U_1+V_2 = 22$, $U_1+V_4 = 4$, $U_2+V_3 = 9$, $U_2+V_4 = 7$, $U_3+V_1 = 32$

After putting $U_1 = 0$ the values of the dual variables are:

$U_1 = 0$, $U_2 = 3$, $U_3 = 12$, $V_1 = 20$, $V_2 = 22$, $V_3 = 6$, $V_4 = 4$.

Now to test the optimality we put these dual variable values into the Table 5.52 and recalculate the optimality test values specially for all the empty cells. The necessary cell values reveal a possibility of further improvement by transfering the units to the cell C_{34} which is having the value of −1. So in the next stage of iteration this cell must be converted into a basic cell. Table 5.52 shows the loop for the proposed transfer. Mere inspection of Table 5.52 tell us that only a maximum of 50 units form the high cost cell C_{31} could be allotted to the low cost C_{34}. After effecting this transfer having the column and row total in mind is shown in Table 5.53.

Table 5.53

Origins		Destinations 1 ($V_1 = 20$)	2 ($V_2 = 22$)	3 ($V_3 = 6$)	4 ($V_4 = 4$)	Supply
1	$U_1 = 0$	20 / **60**	22 / **40**	17 / 11	4 / **20**	120
2	$U_2 = 3$	24 / 1	37 / 12	9 / **30**	7 / **40**	70
3	$U_3 = 11$	32 / 1	37 / 4	20 / 3	15 / **50**	50
Demand		60	40	30	110	240
					Total cost =	3460

Now let us verify once again this allotment for optimality by using dual variables.

Calculation of solution value for the dual variables using basic cells in Table 5.53

From Table 5.53

$U_1 + V_1 = 20$, $U_1 + V_2 = 22$, $U_1 + V_4 = 4$, $U_2 + V_3 = 9$, $U_2 + V_4 = 7$, $U_3 + V_4 = 15$

After putting $U_1 = 0$ the values of the dual variables are:

$U_1 = 0$, $U_2 = 3$, $U_3 = 11$, $V_1 = 20$, $V_2 = 22$, $V_3 = 6$, $V_4 = 4$.

For testing the conditions for non-basic cells conveniently let us put these values in the first row and first column in our final solution Table 5.53. Now for all non-basic cells $C_{ij} - U_i - V_j \geq 0$ condition is satisfied. Thus, the solution already obtained in the previous step get confirmed as the optimum solution.

$X_{11} = 60$, $X_{12} = 40$, $X_{14} = 20$, $X_{23} = 30$, $X_{24} = 40$, $X_{34} = 50$

Total cost = Rs. 3460

■ Example 9

For the following VAM solution given below find the final solution by MODI method.

Table 5.54

Origin	Destinations				
	1	2	3	4	Total s_i
1	3 (1000)	2 (4000)	7	6	5000
2	7 (2500)	5	2 (2000)	3 (1500)	6000
3	2 (2500)	5	4	5	2500
Total d_j	6000	4000	2000	1500	13,500
	Total cost =		42,000		

As usual let us insert the dual variables into the MODI Table 5.55 first.

MODI Method: First Iteration

Calculation of solution value for the dual variables using basic cells in Table 5.55

Now for the basic cells the duality condition is $C_{ij} = U_i + V_j$

So $U_1 + V_1 = 3$; $U_1 + V_2 = 2$; $U_2 + V_1 = 7$; $U_2 + V_3 = 2$; $U_2 + V_4 = 3$; $U_3 + V_1 = 2$

Solutions are: $U_1 = -4$, $U_2 = 0$, $U_3 = -5$, $V_1 = 7$, $V_2 = 6$, $V_3 = 2$, $V_4 = 3$

Table 5.55

Origins		Destinations 1	2	3	4	Supply
		$V_1 = 7$	$V_2 = 6$	$V_3 = 2$	$V_4 = 3$	
1	$U_1 = -4$	3 **1000**	2 **4000**	7 9	6 7	5000
2	$U_2 = 0$	7 **2500**	5 −1	2 **2000**	3 **1500**	6000
3	$U_3 = -5$	2 **2500**	5 4	4 7	5 7	2500
Demand		6000	4000	2000	1500	13,500
					Total cost =	42,000

Now let us substitute these values in Table 5.55 and test the optimality conditions as usual for all the non-basic cells. Since the cell C_{22} is having a −1 there is a potential for reducing the transport cost by transferring the units from the costlier neighbouring cells. To avail of this cost saving we shift the whole of 2500 units from cell C_{21} to C_{22} as described by the highlighted closed loop. This shift reduces the C_{12} cell value to 1500 units to have the column total intact. To restore the row total the cell C_{11} is changed to 3500 units. The entire scheme of shifting and the resulting Table 5.56 is shown below.

Table 5.56

Origins		Destinations 1	2	3	4	Supply
		$V_1 = 6$	$V_2 = 5$	$V_3 = 2$	$V_4 = 3$	
1	$U_1 = -3$	3 **3500**	2 **1500**	7 8	6 6	5000
2	$U_2 = 0$	7 1	5 **2500**	2 **2000**	3 **1500**	6000
3	$U_3 = -4$	2 **2500**	5 4	4 6	5 6	2500
Demand		6000	4000	2000	1500	13,500
					Total cost =	39,500

Second Iteration

Calculation of solution value for the dual variables using basic cells in Table 5.56

Conditions: $U_1+V_1 = 3$; $U_1+V_2 = 2$; $U_2+V_2 = 5$; $U_2+V_3 = 2$; $U_2+V_4 = 3$; $U_3 + V_1 = 2$

Solutions are: $U_2 = 0$; $V_1 = 6$; $V_3 = 2$; $V_4 = 3$; $U_1 = -3$; $V_2 = 5$; $U_3 = -4$

Since after incorporating these values into Table 5.56 we notice that all the empty non-basic cell values are non-negative. Thus the obtained allotment in the previous step itself yieds the optimum solution.

$X_{11} = 3500$; $X_{12} = 1500$; $X_{22} = 2500$; $X_{23} = 2000$; $X_{24} = 1500$; $X_{31} = 2500$.

Minimum transportation cost = Rs. 39,500.

5.9 DEGENERACY IN TRANSPORTATION MODELS

In the initial solution table of the transportation problem, the number of allotted cells is less than m+n−1 then the solution degenerate. The degeneracy may occur either in the initial solution stage or in the final stage of search for optimum solution. If it occurs in the very initial solution stage then we will end up with too few equations in solving for too many variables while using MODI method. Instead if we apply stepping stone method for final solution then one may not be in a position to find a closed loop with needed independent cells for cost saving transfers. So before we proceed for optimal test in either case degeneracy must be removed.

Theorem

A degenerate basic solution in a transportation problem exists if and only if some partial sum of availibilites (s_i) is equal to a partial sum of requirements (d_j).

In this section we propose to discuss two methods for removing degeneracy.

1. Perturbation Technique for Removal of Degeneracy

In order to avoid degenerate basic solution in a transportation problem we will have to ensure that no partial sum of s_i and d_j are equal. To ensure such a result we modify our problem by making very small additions as shown below and continue the problem in the usual manner.

$$s_i^* = s_i + d, \quad i = 1, 2, \ldots\ldots m$$

$$d_j^* = d_j, \quad j = 1, 2, \ldots\ldots n-1$$

$$d_n^* = d_n + md; \; d > 0$$

The following examples illustrate the usefulness of this method.

■ Example 10

Solve the following transportation problem by north-west corner rule. Degeneracy, if any, remove it by perturbation method.

Table 5.57

Factories	Markets					
	M1	M2	M3	M4	M5	s_i
F1	17	9	14	10	14	50
F2	13	6	11	11	12	50
F3	6	17	9	12	12	30
F4	15	20	11	14	6	50
d_j	30	30	40	60	20	180

Solution

X_{11} = Min (30,50) = 30. So allot 30 units from factory 1 to market 1 and remove market 1 (M1 column) from the table. At this stage remember factory 1 will have only 20 (50 –30) units for the next iteration.

Table 5.58

Factories	Markets				
	M2	M3	M4	M5	s_i
F1	9	14	10	14	20
F2	6	11	11	12	50
F3	17	9	12	12	30
F4	20	11	14	6	50
d_j	30	40	60	20	150

X_{12} = Min (20,30) = 20. So allot the leftover 20 units of factory 1 to market 2 and remove factory 1 from the table. At this stage remember M2 still has unsatisfied demand of 10 units (30–20).

Table 5.59

Factories	Markets				
	M2	M3	M4	M5	s_i
F2	6	11	11	12	50
F3	17	9	12	12	30
F4	20	11	14	6	50
d_j	10	40	60	20	130

X_{22} = Min (50,10) =10. So allot 10 units of factory 2 supply to market 2 and remove market 2 from the table. At this stage note that factory 2 is having the leftover supply of 40 (50–40) units at its disposal.

Table 5.60

Factories	Markets			
	M3	M4	M5	s_i
F2	11	11	12	40
F3	9	12	12	30
F4	11	14	6	50
d_j	40	60	20	120

X_{23} = Min (40,40) = 40. So allot leftover supply of 40 units with factory 2 to market 3. Here since the leftover supply of 40 is exactly equal to the the full demand needs of market 3 both the market 3 and factory 2 are removed simultaneously for the next iteration.

Table 5.61

Factories	Markets		
	M4	M5	s_i
F3	12	12	30
F4	14	6	50
d_j	60	20	80

X_{34} = Min (30,60) = 30. So allot 30 units of supply with factory 3 to market 4 and remove factory 3 from the table. At this stage market 4 is still having an unsatisfied demand of 30 units.

Table 5.62

Factories	Markets		
	M4	M5	s_i
F4	14	6	50
d_j	30	20	50

X_{44} = Min (50,30) = 30. So allot 30 units of factory 4 supply to market 4 and remove market 4 from the table. At this stage factory 4 is left with 20 (50–30) units of supply.

Table 5.63

Factories	Markets	
	M5	s_i
F4	6	20
d_j	20	20

X_{45} = Min (20,20). So allot 20 units of factory 4 to market 5 and complete the assignment.

Initial degeneratred solution

Table 5.64

Factories	Markets					
	M1	M2	M3	M4	M5	s_i
F1	17 (30)	9 (20)	14	10	14	50
F2	13	6 (10)	11 (40)	11	12	50
F3	6	17	9	12 (30)	12	30
F4	15	20	11	14 (30)	6 (20)	50
d_j	30	30	40	60	20	180
Minimum cost =	2090					

Here the number of occupied cell should have been m+n−1= 8. But in our problem we have only seven allotted cells. This is mainly due to the simultaneous removal of both destination and origin at a time during our interaction. Thus, the initial solution degenerates.

***Note:** In the problem the degenerate solution occurs because partial supply sum is equal to partial demand sum i.e.,* $\mathbf{s_1+s_2 = 50+50 = d_1+d_2+d_3 = 30+30+40}$.

Perturbation Treatment and Initial Solution by North-West Corner Rule

Table 5.65

Factories	Markets					
	M1	M2	M3	M4	M5	s_i
F1	17	9	14	10	14	50+d
F2	13	6	11	11	12	50+d
F3	6	17	9	12	12	30+d
F4	15	20	11	14	6	50+d
d_j	30	30	40	60	20+4d	180+4d

X_{11}= Min (30,50+d) = 30.

So remove M1 and adjust F1 supply to 50 +d−30 = 20 +d.

Table 5.66

Factories	Markets				
	M2	M3	M4	M5	s_i
F1	9	14	10	14	20+d
F2	6	11	11	12	50+d
F3	17	9	12	12	30+d
F4	20	11	14	6	50+d
d_j	30	40	60	20+4d	150+4d

X_{12} = Min (30,20+d) = 20+d. So remove F1 and adjust M2 demand to 30−(20+d) =10−d.

Table 5.67

Factories	Markets				
	M2	M3	M4	M5	s_i
F2	6	11	11	12	50+d
F3	17	9	12	12	30+d
F4	20	11	14	6	50+d
d_j	10−d	40	60	20+4d	130+3d

X_{22} = Min (10−d,50+d) = 10−d. So remove M2 and adjust F2 supply to 50+d−(10−d)= 40+2d.

Table 5.68

Factories	Markets			
	M3	M4	M5	s_i
F2	11	11	12	40+2d
F3	9	12	12	30+d
F4	11	14	6	50+d
d_j	40	60	20+4d	120+4d

X_{23} = Min (40,40+2d) = 40. So remove M3 and adjust the supply of F2 to 40 +2d−40 = 2d.

Table 5.69

Factories	Markets		
	M4	M5	s_i
F2	11	12	2d
F3	12	12	30+d
F4	14	6	50+d
d_j	60	20+4d	80+4d

X_{24}= Min (60,2d) = 2d. So remove F2 and adjust the demand at M4 to 60−2d.

Table 5.70

Factories	Markets		
	M4	M5	s_i
F3	12	12	30+d
F4	14	6	50+d
d_j	60−2d	20+4d	80+2d

X_{34} = Min (60−2d,30+d) = 30+d. So remove F3 and adjust the demand for M4 as 60−2d −(30+d) = 30−3d.

Table 5.71

Factories	Markets		
	M4	M5	s_i
F4	14	6	50+d
d_j	30−3d	20+4d	50+d

X_{44} = Min (30−3d,50+d) = 30−3d. So remove M4 and adjust the supply at F4 to 50+d−(30−3d) = 20+4d

Table 5.72

Factories		
	M5	s_i
F4	6	20+4d
d_j	20+4d	20+4d

Table 5.73

Factories	Markets					
	M1	M2	M3	M4	M5	s_i
F1	17(30)	9(20+d)	14	10	14	50+d
F2	13	6(10−d)	11(40)	11(2d)	12	50+d
F3	6	17	9	12(30+d)	12	30+d
F4	15	20	11	14(30−3d)	6(20+4d)	50+d
d_j	30	30	40	60	20+4d	180+4d
Minimum cost =17x30+9x20+6x10+11x40+12x30+14x30+6x20 = 2090						

Notice that in Table 5.73 we are having 8 alloted cells.

■ Example 11

Obtain the initial solution for the following transportation problem by matrix minimum method. Degeneracy if any remove it by perturbation method.

Table 5.74

Factories	Markets					
	M1	M2	M3	M4	M5	s_i
F1	17	9	14	10	14	50
F2	13	6	11	11	12	50
F3	6	17	9	12	12	30
F4	15	20	11	14	6	50
d_j	30	30	40	60	20	180

Solution

Table 5.75

Factories	Markets					
	M1	M2	M3	M4	M5	s_i
F1	17	9	14	10	14	50
F2	13	6	11	11	12	50
F3	6	17	9	12	12	30
F4	15	20	11	14	6	50
d_j	30	30	40	60	20	180

X_{31} = Min (30, 30) = 30

Table 5.76

Factories	Markets				
	M2	M3	M4	M5	s_i
F1	9	14	10	14	50
F2	6	11	11	12	50
F4	20	11	14	6	50
d_j	30	40	60	20	150

X_{22} = Min (30, 50) = 30

Table 5.77

Factories	Markets			
	M3	M4	M5	s_i
F1	14	10	14	50
F2	11	11	12	20
F4	11	14	6	50
d_j	40	60	20	120

X_{45} = Min (20, 50) = 20

Table 5.78

Factories	Markets		
	M3	M4	s_i
F1	14	10	50
F2	11	11	20
F4	11	14	30
d_j	40	60	100

X_{23} = Min (40, 20) = 20

Table 5.79

Factories	Markets		
	M3	M4	s_i
F1	14	10	50
F4	11	14	30
d_j	20	60	80

$X_{43} = \text{Min}\ (20, 30) = 20$

Table 5.80

Factories	Markets	
	M4	s_i
F1	10	50
F4	14	10
d_j	60	60

$X_{14} = \text{Min}\ (60, 50) = 50$

Table 5.81

Factories	Markets	
	M4	s_i
F4	14	10
d_j	10	10

$X_{44} = \text{Min}\ (10, 10) = 10$

Initial solution

Table 5.82

Factories	Markets					
	M1	M2	M3	M4	M5	s_i
F1	17	9	14	10(50)	14	50
F2	13	6(30)	11(20)	11	12	50
F3	6(30)	17	9	12	12	30
F4	15	20	11(20)	6(10)	6(20)	50
d_j	30	30	40	60	20	180
Min. cost = 1560						

In the initial solution we must have m+n−1 = 8 allowed cells. Since we have only 7 cells in the initial allotment in Table 5.82 the solution degenerates.

Removal of degeneracy using Perturbation Treatment by Matrix minimum method

Table 5.83

Factories	Markets					
	M1	M2	M3	M4	M5	s_i
F1	17	9	14	10	14	50+d
F2	13	6	11	11	12	50+d
F3	6	17	9	12	12	30+d
F4	15	20	11	14	6	50+d
d_j	30	30	40	60	20+4d	180+4d

X_{31} = Min (30,30+d) = 30. So remove M1.

Table 5.84

Factories	Markets				
	M2	M3	M4	M5	s_i
F1	9	14	10	14	50+d
F2	6	11	11	12	50+d
F3	17	9	12	12	d
F4	20	11	14	6	50+d
d_j	30	40	60	20+4d	150+4d

X_{22} = Min (50+d, 30) =30. So remove M2.

Table 5.85

Factories	Markets			
	M3	M4	M5	s_i
F1	14	10	14	50−d
F2	11	11	12	20−d
F3	9	12	12	d
F4	11	14	6	50−d
d_j	40	60	20+4d	120−4d

X_{45} = Min (20+4d, 50+d)=20+4d. So remove M5.

Table 5.86

Factories	Markets		
	M3	M4	s_i
F1	14	10	50+d
F2	11	11	20+d
F3	9	12	d
F4	11	14	30−3d
d_j	40	60	100

X_{33} = Min (40, d) = d. So remove F3.

Table 5.87

Factories	Markets		
	M3	M4	s_i
F1	14	10	50+d
F2	11	11	20+d
F4	11	14	30−3d
d_j	40−d	60	100−d

X_{14} = Min (60, 50+d) =50+d. So remove F1.

Table 5.88

Factories	Markets		
	M3	M4	s_i
F2	11	11	20+d
F4	11	14	30−3d
d_j	40−d	10−d	50−2d

X_{23} = Min (40−d, 20+d) = 20+d. So remove F2.

Table 5.89

Factories	Markets		
	M3	M4	s_i
F4	11	14	30−3d
d_j	20−2d	10−d	30−3d

X_{43} =Min (20−2d , 30−3d)=20−2d. So remove M3.

Table 5.90

Factories	Markets	
	M4	s_i
F4	14	10−d
d_j	10−d	10−d

X_{44} = Min (10−d, 10−d) = 10−d. Complete the allotment.

Table 5.91

Factories	Markets					
	M1	M2	M3	M4	M5	s_i
F1	17	9	14	10 (50+d)	14	50+d
F2	13	6 (30)	11(20+d)	11	12	50+d
F3	6 (30)	17	9 (d)	12	12	30+d
F4	15	20	11(20−2d)	14 (10−d)	6 (20+4d)	50+d
d_i	30	30	40	60	20+4d	180+4d
Minimum cost = 1560						

Note: *Now we have eight allotted cells. So the solution does not degenerate.*

2. Epsilon Technique for Removal of Degeneracy

In this method we obtain the initial solution as usual without any treatment. Once the degenerated solution with less number of allotted cell is obtained we add one or more ε (very very small value commonly known as epsilon) to independent cells and make up the cell shortage. Once the shortage is removed we proceed further in the usual manner for optimum solution by using MODI method. Remember, that though ε is theoretically a small positive quantity, its presence in one or more cells will leave the column and row totals unaltered. Such inclusions will not affect the total transportation cost as well. In this context it is important to note the following properties of epsilon.

$$k + \varepsilon = k$$

$$k - \varepsilon = k$$

$$0 + \varepsilon = \varepsilon$$

$$\varepsilon + \varepsilon = \varepsilon$$

$$\varepsilon - \varepsilon = 0$$

$$k \times \varepsilon = \varepsilon$$

In the following two examples we use this epsilon technique for obtaining the final optimum solution.

■ Example 12

Find the initial solution for the transportation problem by VAM method. Also obtain the optimum solution by MODI method.

Table 5.92

Plant	Market				Supply
	M1	M2	M3	M4	
P1	6	4	9	1	40
P2	20	6	11	3	40
P3	7	1	0	14	50
P4	7	1	12	6	30
Demand	90	30	50	30	

Initial solution by VAM method

Since there is excess supply to the extent of 20 units in this problem we first balance it by introducing one dummy destination M5 to take care of this excess supply. In the following Table 5.93 we use VAM method to get the closest initial solution so that we get the optimum solution by using MODI method with less number of iterations.

Table 5.93

Plants		Markets M1	M2	M3	M4	M5	s_i	Row penalty I	II	III	IV	V
P1		6 **30**	4	9	1 **10**	0	40,30,0	1	1	3	5	3
P2		20	6	11	3 **20**	0 **20**	40,20,0	3	**3**	3	**17**	-
P3		7	1	0 **50**	14	0	50,0	0	-	-	-	-
P4		7 **60**	1 **30**	12	6	0	90,60,0	1	1	**5**	1	1
d_j		90	30,0	50,0	30,10,0	20,0	220					
Column penalty	I	1	0	**9**	2	0						
	II	1	3	-	2	0						
	III	1	3	-	2	-						
	IV	1	-	-	5	-						
	V	1	-	-	**5**	-						

In the Table 5.93 initial solution obtained by VAM method degenerate since it contains only seven allotted cells. Remember m+n−1=5+4−1 = 8 in our illustration.

Removal of degeneracy by introducing

So let us first bring in one more additional cell by allotting the negligible amount of the allotment denoted by ε without affecting the column and row totals. For such an inclusion

Table 5.94

Plants		Markets M1	M2	M3	M4	M5	Supply
		$V_1 = 6$	$V_2 = 0$	$V_3 = -2$	$V_4 = 1$	$V_5 = -2$	
P1	$U_1 = 0$	6 **30**	4 / 4	9 / 11	1 **10**	0 / 2	40
P2	$U_2 = 2$	20	6 / 4	11 / 11	3 **20**	0 **20**	40
P3	$U_3 = 2$	7 / −1	1 / −1	0 **50**	14 / 11	0 **ε**	50
P4	$U_4 = 1$	7 **60**	1 **30**	12 / 13	6 / 4	0 / 1	90
Demand		90	30	50	30	20	220
					Total cost =	700	

we identify all the available independent cells and choose the one having the least cost. Here the cells C_{13}, C_{23}, C_{31}, C_{32}, C_{34}, C_{35}, C_{43} are independent. Since only one additional allotted cell is needed to eliminate degeneracy we prefer to allot ε to the cell to C_{35} in Table 5.94 having the zero cost of transport. Such an inclusion removes the shortage of allotted cells. Now from this initial solution we proceed further for the final optimum solution by MODI method.

Test for optimum solution by MODI method

As a first step let us introduce three supply constraints related dual variables U_1, U_2, U_3 and five demand constraints related dual variables V_1, V_2, V_3, V_4, V_5 as shown in Table 5.95.

Now from the Table 5.94 we have the following eight equations in seven unknowns:

$$U_1+V_1 = 6,\ U_1+V_4 = 1,\ U_2+V_4 = 3,\ U_2+V_5 = 0,\ U_3+V_3 = 0,\ U_3+V_5 = 0,\ U_4+V_1 = 7,\ U_4+V_2 = 1$$

By assuming $U_1 = 0$ the solution for the remaining variables are obtained as

$$U_1 = 0,\ U_2 = 2,\ U_3 = 2,\ U_4 = 1,\ V_1 = 6,\ V_2 = 0,\ V_3 = -2,\ V_4 = 1,\ V_5 = -2$$

After obtaining the solution values for the dual variables we insert them all in the Table 5.94 and test the optimality conditions for all the empty cells.

Table 5.95

Plants		Markets					Supply
		M1	M2	M3	M4	M5	
		$V_1 = 6$	$V_2 = 0$	$V_3 = -1$	$V_4 = 1$	$V_5 = -2$	
P1	$U_1 = 0$	6 **30**	4 4	9 10	1 **10**	0 2	40
P2	$U_2 = 2$	20 12	6 4	11 10	3 **20**	0 **20**	40
P3	$U_3 = 1$	7 **(ε)**	1 0	0 **50**	14 12	0 1	50
P4	$U_4 = 1$	7 **60**	1 **30**	12 12	6 4	0 1	90
Demand		90	30	50	30	20	220
					Total cost =	700	

The so obtained worksheet with all needed calculations is reported in Table 5.94. Since there are two highlighted negative unit entries corresponding to cell C_{31} and cell C_{32} we confirm that the solution obtained by VAM method in Table 5.94 is not optimum.

Now among these two equally cost effective cells we choose cell C_{31} for allotment from the neighbouring cells C_{35} as described by the loop. For such transfer the needed closed loop is shown in Table 5.94. Since ε is very small by assumption without affecting the other allotments we can transfer this small unaccountable amount ε from cell C_{35} to cell C_{31} as shown in Table 5.95.

From Table 5.95

$U_1+V_1 = 6$, $U_1+V_4 = 1$, $U_2+V_4 = 3$, $U_2+V_5 = 0$, $U_3+V_1 = 7$, $U_3+V_3 = 0$, $U_4+V_1 = 7$, $U_4+V_2 = 1$

$U_1= 0$, $U_2 = 2$, $U_3 = 1$, $U_4 = 1$, $V_1 = 6$, $V_2 = 0$, $V_3 = -1$, $V_4 = 1$, $V_5 = -2$

Table 5.95 shows the test results. All empty cells have the needed positive values. Thus the obtained solution in Table 5.95 gives the needed least total transportation cost.

The solution values are: $X_{11} = 30$, $X_{14} = 10$, $X_{24} = 20$, $X_{25} = 20$, $X_{33} = 50$, $X_{41} = 60$, $X_{42} = 30$ and the minimum cost = Rs. 700.

■ Example 13

A manufactoring company has three plants 1, 2 and 3 which supply to the distributors at 1, 2, 3, 4 and 5. Monthly plant capacities are 80, 50 and 90 units respectively. Monthly requirement of distributors are 40, 40, 50, 40 and 80 units respectively. The per unit transportation cost in rupees is given below.

Table 5.96

	To					
From	1	2	3	4	5	s_i
1	5	8	6	6	3	80
2	4	7	7	6	6	50
3	8	4	6	6	3	90
d_j	40	40	50	40	80	

Soultion

Since this is an excess demand problem to bring in balance we introduce a dummy origin to absorb this net balance of 30 units. First we obtain the VAM solution in Table 5.97.

Since m+n−1 = 8 and the number of occupied cells only 7 in Table 5.97 the solution degenerate and hence we cannot proceed further to optimality solution. Now let us increase the number of occupied cells by one by placing the ε in the independent cell corresponding to the dummy row and column 2 as shown in Table 5.98.

Table 5.97

<table>
<tr><td rowspan="2" colspan="2">Plants</td><td colspan="5">Markets</td><td rowspan="2">s_i</td><td colspan="7">Row penalty</td></tr>
<tr><td>1</td><td>2</td><td>3</td><td>4</td><td>5</td><td>I</td><td>II</td><td>III</td><td>IV</td><td>V</td><td>VI</td><td>VII</td></tr>
<tr><td colspan="2">1</td><td>5</td><td>8</td><td>6 50</td><td>6</td><td>3 30</td><td>80,50,0</td><td>2</td><td>2</td><td>2</td><td>2</td><td>1</td><td>0</td><td>-</td></tr>
<tr><td colspan="2">2</td><td>4 40</td><td>7</td><td>7</td><td>6 10</td><td>6</td><td>50,10,0</td><td>2</td><td>2</td><td>2</td><td>2</td><td>2</td><td>1</td><td>0</td></tr>
<tr><td colspan="2">3</td><td>8</td><td>4 40</td><td>6</td><td>6</td><td>3 50</td><td>90,50,0</td><td>1</td><td>1</td><td>3</td><td>-</td><td>-</td><td>-</td><td></td></tr>
<tr><td colspan="2">Dummy</td><td>0</td><td>0</td><td>0</td><td>0 30</td><td>0</td><td>30,0</td><td>0</td><td>-</td><td>-</td><td>-</td><td>-</td><td>-</td><td></td></tr>
<tr><td colspan="2">d_j</td><td>40,0</td><td>40,0</td><td>50,0</td><td>40,10,0</td><td>80,30,0</td><td>0</td><td></td><td></td><td></td><td></td><td></td><td></td><td></td></tr>
<tr><td rowspan="7">Column penalty</td><td>I</td><td>4</td><td>4</td><td>6</td><td>6</td><td>3</td><td colspan="8" rowspan="7"></td></tr>
<tr><td>II</td><td>1</td><td>3</td><td>1</td><td>0</td><td>3</td></tr>
<tr><td>III</td><td>1</td><td>-</td><td>1</td><td>0</td><td>0</td></tr>
<tr><td>IV</td><td>1</td><td>-</td><td>1</td><td>0</td><td>3</td></tr>
<tr><td>V</td><td>1</td><td>-</td><td>1</td><td>0</td><td>-</td></tr>
<tr><td>VI</td><td>-</td><td>-</td><td>1</td><td>0</td><td>-</td></tr>
<tr><td>VII</td><td>-</td><td>-</td><td>-</td><td>0</td><td>-</td></tr>
</table>

Now from Table 5.98

$U_1+V_3 = 6$; $U_1+V_5 = 3$; $U_2+V_1 = 4$; $U_2+V_4 = 6$; $U_3+V_2 = 4$; $U_3+V_5 = 3$; $U_4+V_2 = 0$, $U_4+V_4 = 0$

Table 5.98

<table>
<tr><td rowspan="3" colspan="2">Origin</td><td colspan="5">Destination</td><td rowspan="3">Supply</td></tr>
<tr><td>1</td><td>2</td><td>3</td><td>4</td><td>5</td></tr>
<tr><td>$V_1 = 2$</td><td>$V_2 = 4$</td><td>$V_3 = 6$</td><td>$V_4 = 4$</td><td>$V_5 = 3$</td></tr>
<tr><td>1</td><td>$U_1 = 0$</td><td>5
3</td><td>8
4</td><td>6 50</td><td>6
2</td><td>3 30</td><td>80</td></tr>
<tr><td>2</td><td>$U_2 = 2$</td><td>4 40</td><td>7
1</td><td>7
−1</td><td>6 10</td><td>6
1</td><td>50</td></tr>
<tr><td>3</td><td>$U_3 = 0$</td><td>6
4</td><td>4 40</td><td>6
0</td><td>6
2</td><td>3 50</td><td>90</td></tr>
<tr><td>4(Dummy)</td><td>$U_4 = -4$</td><td>0
2</td><td>0 (ε)</td><td>0
−2</td><td>0 30</td><td>0
1</td><td>30</td></tr>
<tr><td colspan="2">Demand</td><td>40</td><td>40</td><td>50</td><td>40</td><td>80</td><td>250</td></tr>
<tr><td colspan="8">Total cost = 920</td></tr>
</table>

The solution values are

$U_1 = 0$, $U_2 = 2$, $U_3 = 0$, $U_4 = -4$, $V_1 = 2$, $V_2 = 4$, $V_3 = 6$, $V_4 = 4$, $V_5 = 3$

Now let us insert these values and test the conditions for optimality in Table 5.98. Clearly the cell $C_{43} = -2$ shows the possibility of taking advantage of cost saving by bringing in some units from the neighbouring cells to this cell. In our next iterative Table 5.99 we move ε from C_{42} to this typical cell C_{43} without affecting the other allotments because of its smallness.

Table 5.99

Origin		Destination 1	2	3	4	5	Supply
		$V_1 = 4$	$V_2 = 4$	$V_3 = 6$	$V_4 = 6$	$V_5 = 3$	
1	$U_1 = 0$	5 1	8 4	**6** **50**	6 0	**3** **30**	80
2	$U_2 = 0$	**4** **40**	7 3	7 1	**6** **10**	6 3	50
3	$U_3 = 0$	6 2	**4** **40**	6 0	6 0	**3** **50**	90
4(Dummy)	$U_4 = -6$	0 2	0 2	**0** ε	**0** **30**	0 3	30
Demand		40	40	50	40	80	250
					Total cost =	920	

After this reallotment we test the allotment for optimality as usual.

$U_1 + V_3 = 6$; $U_1 + V_5 = 3$; $U_2 + V_1 = 4$; $U_2 + V_4 = 6$; $U_3 + V_2 = 4$; $U_3 + V_5 = 3$; $U_4 + V_3 = 0$, $U_4 + V_4 = 0$

The solution values are

$U_1 = 0, U_2 = 2, U_3 = 0, U_4 = -6, V_1 = 4, V_2 = 4, V_3 = 6, V_4 = 6, V_5 = 3$

After inserting these values into Table 5.99 we test for optimality once again. Clearly all the non-empty cells are either zero or positive in this table. So the VAM solution obtained earlier gets confirmed as the optimum solution. The total least cost = Rs. 920. Thus $\alpha_{11} = 50$; $\alpha_{13} = 30$; $\alpha_{21} = 40$; $\alpha_{24} = 10$; $\alpha_{32} = 40$; $\alpha_{35} = 50$; $\alpha_{44} = 30$.

5.10 STEPPING STONE METHOD

As usual we obtain an initial solution by using any one of the methods outlined earlier. Once the initial solution is obtained we test whether the solution degenerates or not. If there is no degenerated solution then we calculate the cost effectiveness for all the non-basic cells for one unit transfer from the neighbourhood as we did it MODI method and choose the one deserving most cost effective for improvement. This process is repeated for many more times till a no-improvement situation is reached. The following example illustrates this method.

Example 14

Let us illustrate this method by taking Example 1 and start from its VAM solution.

VAM Solution

Table 5.100

Origin	Destinations				
	1	2	3	4	Total s_i
1	3 (1000)	2 (4000)	7	6	5000
2	7 (2500)	5	2 (2000)	3 (1500)	6000
3	2 (2500)	5	4	5	2500
Total d_j	6000	4000	2000	1500	13,500
Total transportation cost =			42,000		

Now let us start our stepping stone method from this starting Table 5.100 and calculate the cost that could be saved by assigning one unit for the non-basic empty cell C_{13}. For cell C_{13} add one unit and subtract the same from cell C_{23} and make it 1999 in order to restore the column requirements. Now to satisfy the second row requirements add one unit to the cell C_{21} and make it 2501. Again to fulfil first column requirements subtract one unit from cell C_{11} and make it 999 units. Now note that all column and row requirements are fulfilled. Calculate the one unit cost saving due this shifting as 7−2+7−3 = 9. The relevant operational loop is shown in Table 5.101 by a rectangular cover.

Cost saving calculations for cell C_{13}

Table 5.101

Origin	Destinations				
	1	2	3	4	Total s_i
1	3 (999)	2 (4000)	7(1)	6	5000
2	7 (2501)	5	2 (1999)	3 (1500)	6000
3	2 (2500)	5	4	5	2500
Total d_j	6000	4000	2000	1500	
Cost saving =7−2+7−3 = 9					

Cost saving calculations for cell C_{14}

Table 5.102

Origin	Destinations				
	1	2	3	4	Total s_i
1	3 (999)	2 (4000)	7	6(1)	5000
2	7 (2501)	5	2 (2000)	3 (1499)	6000
3	2 (2500)	5	4	5	2500
Total d_j	6000	4000	2000	1500	
Cost Saving = 6−3+7−3 = 7					

Cost saving calculations for cell C_{22}

Table 5.103

Origin	Destinations				
	1	2	3	4	Total s_i
1	3 (1001)	2 (3999)	7	6	5000
2	7 (2499)	5(1)	2 (2000)	3 (1500)	6000
3	2 (2500)	5	4(1)	5	2500
Total d_j	6000	4000	2000	1500	
Cost saving = 5−2+3−7 = −1					

Cost saving calculations for cell C_{32}

Table 5.104

Origin	Destinations				
	1	2	3	4	Total s_i
1	3 (1001)	2 (3999)	7	6	5000
2	7 (2500)	5	2 (2000)	3 (1500)	6000
3	2 (2499)	5(1)	4	5	2500
Total d_j	6000	4000	2000	1500	
Cost saving = 5−2+3−2 = 4					

Cost saving calculations for cell C_{33}

Table 5.105

Origin	Destinations				
	1	2	3	4	Total s_i
1	3 (1000)	2 (4000)	7	6	5000
2	7(2501)	5	2 (1999)	3 (1500)	6000
3	2 (2499)	5	4(1)	5	2500
Total d_j	6000	4000	2000	1500	
Cost saving = 4−2+7−2 = 7					

Cost saving calculations for cell C_{34}

Table 5.106

Origin	Destinations				
	1	2	3	4	Total s_i
1	3 (1000)	2 (4000)	7	6	5000
2	7 (2501)	5	2 (2000)	3 (1499)	6000
3	2 (2499)	5	4	5(1)	2500
Total d_j	6000	4000	2000	1500	
Cost saving = 5−3+7−2 = 7					

Among the six calculations only for cell C_{22} we get negative value showing the cost saving option. So according to this method this cell must be allotted as much as possible having the column and row constraints in mind. Accordingly, the whole 2500 units of the cell C_{21} is shifted to our empty non-basic cell C_{22}. To satisfy the column requirement these 2500 units are shifted from cell C_{12} to cell C_{11}. The final result is shown in the Table 5.107.

Table 5.107

Origin	Destinations				
	1	2	3	4	Total s_i
1	3 (3500)	2 (1500)	7	6	5000
2	7	5(2500)	2 (2000)	3 (1500)	6000
3	2 (2500)	5	4	5	2500
Total d_j	6000	4000	2000	1500	
Total Cost = 39,500					

The second cycle of calculations starts from Table 5.107

In this initial solution there are 6 basic cells and 6 non-basic cells. Since m+n−1 =3+4−1= 6 the obtained solution does not degenerate. Now to check the optimality of this allotment we calculate per unit cost savings for all the six non-basic cells once again.

Cost saving per unit for non-basic cell C_{13}

Table 5.108

Origin	Destinations				
	1	2	3	4	Total s_i
1	3 (3500)	2 (1499)	7(1)	6	5000
2	7	5(2501)	2 (1999)	3 (1500)	6000
3	2 (2500)	5	4	5	2500
Total d_j	6000	4000	2000	1500	
Cost savings = 7−2+5−2 = 8					

Cost saving per unit for non-basic cell C_{14}

Table 5.109

Origin	Destinations				
	1	2	3	4	Total s_i
1	3 (3500)	2 (1499)	7	6(1)	5000
2	7	5(2501)	2 (2000)	3 (1499)	6000
3	2 (2500)	5	4	5	2500
Total d_j	6000	4000	2000	1500	
Cost saving = 6−3+5−2 =6					

Cost saving per unit for non-basic cell C_{21}

Table 5.110

Origin	Destinations				
	1	2	3	4	Total s_i
1	3 (3499)	2 (1501)	7	6	5000
2	7(1)	5(2499)	2 (2000)	3 (1500)	6000
3	2 (2500)	5	4	5	2500
Total d_j	6000	4000	2000	1500	
Cost saving = 7−5+2−3 =1					

Cost saving per unit for non-basic cell C_{32}

Table 5.111

Origin	Destinations				
	1	2	3	4	Total s_i
1	3 (3501)	2 (1499)	7	6	5000
2	7	5(2500)	2 (2000)	3 (1500)	6000
3	2 (2499)	5(1)	4	5	2500
Total d_j	6000	4000	2000	1500	
Cost saving = 5−2+3−2 = 4					

Cost saving per unit for cell C_{33}

Table 5.112

Origin	Destinations				
	1	2	3	4	Total s_i
1	3 (3501)	2 (1499)	7	6	5000
2	7	5(2501)	2 (1999)	3 (1500)	6000
3	2 (2499)	5	4(1)	5	2500
Total d_j	6000	4000	2000	1500	
Cost saving = 4−2+3−2 =6					

Cost saving per unit for non-basic cell C_{34}

Table 5.113

Origin	Destinations				
	1	2	3	4	Total s_i
1	3 (3501)	2 (1499)	7	6	5000
2	7	5(2501)	2 (2000)	3 (1499)	6000
3	2 (2499)	5	4	5(1)	2500
Total d_j	6000	4000	2000	1500	
Cost saving = 3−2+5−3+5−2 =6					

All the six non-basic cells cost saving calculations are found to be positive. Thus, already obtained result in the previous iteration in Table 5.107 gets confirmed as the final optimum solution. The solutions are: $X_{11} = 3500$, $X_{12} = 1500$, $X_{22} = 2500$, $X_{23} = 2000$, $X_{24} = 1500$, $X_{31} = 2500$, Total transport cost = Rs. 39,500.

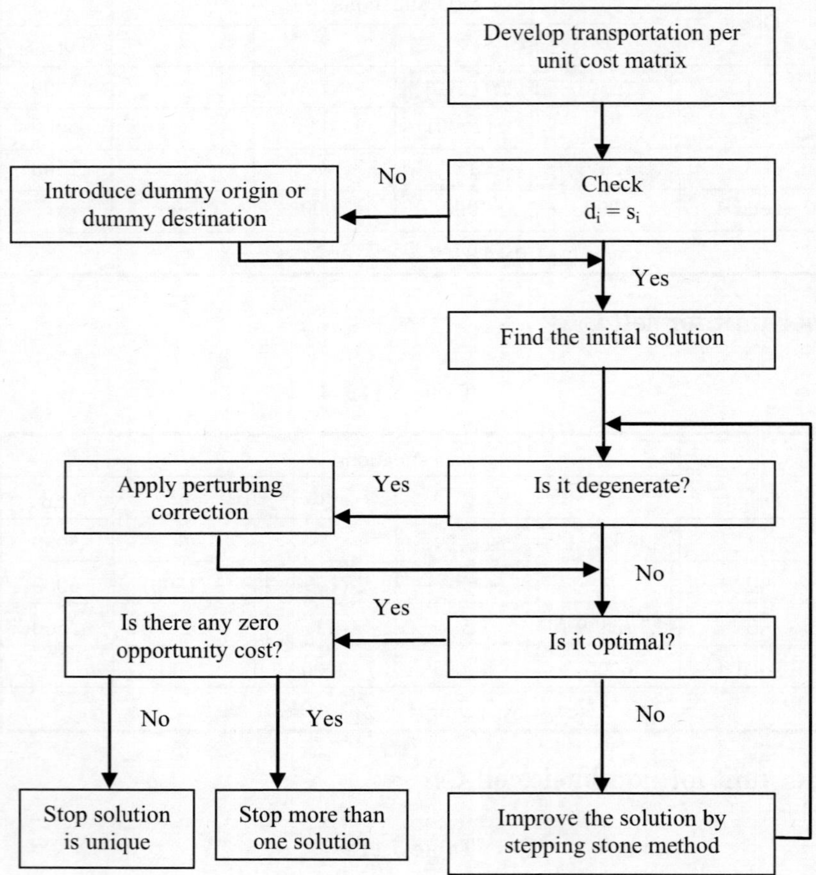

Figure 5.6

Note: *The zero cell saving to any cell implies the existence of multiple solution to the given problem. If any negative numbers appears then the cell having the hightest negative number is to be allotted fully having the constraints in mind. This sequence of adjustment continues till we get positive cost saving figures for all the empty cells. This final step will give us the needed optimum solution. The Figure 5.6 gives the step-by-step procedure for solving transportation problems.*

5.11 LINEAR PROGRAMMING FORMULATION: A NUMERICAL ILLUSTRATION

Example 15

Convert the following transportation problem into a linear programming problem and solve it by using excel spreadsheet.

Table 5.114

	Destinations				
Origin	1	2	3	4	Total s_i
1	3	2	7	6	5000
2	7	5	2	3	6000
3	2	5	4	5	2500
Total d_j	6000	4000	2000	1500	13,500

If X_{ij} = number of units transported from i^{th} origin to j^{th} destination then the total transport cost for all the routes originates from all the three origins listed below

$$\text{Transport cost from origin 1}: \ 3X_{11} + 2X_{12} + 7X_{13} + 6X_{14}$$
$$\text{Transport cost from origin 2}: \ 7X_{21} + 5X_{22} + 2X_{23} + 3X_{24}$$
$$\text{Transport cost from origin 3}: \ 2X_{31} + 5X_{32} + 4X_{33} + 5X_{34}$$

The total supply constraints for all the three centres are listed below. They cannot exceed the total production in the respective centres as stated by the respective less than or equal to constraints.

$$\text{Supply from origin 1}: \ X_{11} + X_{12} + X_{13} + X_{14} \leq 5000$$
$$\text{Supply from origin 2}: \ X_{21} + X_{22} + X_{23} + X_{24} \leq 6000$$
$$\text{Supply from origin 3}: \ X_{31} + X_{32} + X_{33} + X_{34} \leq 2500$$

The demand forecast identities for all the four destinations are reported below.

$$\text{Demand from destination 1}: \ X_{11} + X_{21} + X_{31} = 6000$$
$$\text{Demand from destination 2}: \ X_{12} + X_{22} + X_{32} = 4000$$
$$\text{Demand from destination 3}: \ X_{13} + X_{23} + X_{33} = 2000$$
$$\text{Demand from destination 4}: \ X_{14} + X_{24} + X_{34} = 1500$$

By combining cost, the objective functions and the constraints into one model we can formulate the needed linear programming problems as follows.

$$\text{Min: } 3X_{11} + 2X_{12} + 7X_{13} + 6X_{14} + 7X_{21} + 5X_{22} + 2X_{23} + 3X_{24} + 2X_{31} + 5X_{32} + 4X_{33} + 5X_{34}$$

$$\begin{array}{llllllllllllll}
X_{11} + & X_{12} + & X_{13} + & X_{14} & & & & & & & & & \le 5000 \\
& & & & X_{21} + & X_{22} + & X_{23} + & X_{24} & & & & & \le 6000 \\
& & & & & & & & X_{31} + & X_{32} + & X_{33} + & X_{34} & \le 2500 \\
X_{11} & & & & + X_{21} & & & & + X_{31} & & & & = 6000 \\
& X_{12} & & & & + X_{22} & & & & + X_{32} & & & = 4000 \\
& & X_{13} & & & & + X_{23} & & & & + X_{33} & & = 2000 \\
& & & X_{14} & & & & + X_{24} & & & & + X_{34} & = 1500
\end{array}$$

$X_{ij} \ge$ for i = 1,2,3 and j =1,2,3,4

Spreadsheet solution

As usual on the top portion of the spreadsheet we provide the input data for the excel solver. The transport costs from origins to destinations are entered in cells B3:E5. The supply (s_i) are entered in cells F3:F5. Similarly, the demand from all the four destinations are entered in cells B6:E6. The key elements needed for the excel solver are the decision variables, the objective function, the left and the right hand sides of all the relevant constraints. All of them are described at the bottom portion of the excel sheet.

Decision variables

Cells B12:E14 are reserved for the final solution of the solver. To highlight the importance we shade this area.

Objective function

The formula = SUMPRODUCT(B3:E5,B12:E14) is placed in cell B9 for the automatic calculation of the value of the objective function.

Constraint's LHS

Cells F12:F14 gives the LHS values of the supply constraints and cells B15:E15 gives the LHS values of the demand constraints.

$$F12 = SUM(B12:E12)$$
$$F13 = SUM(B13:E13)$$
$$F14 = SUM(B14:E14)$$
$$B15 = SUM(B12:B14)$$
$$C15 = SUM(C12:C14)$$
$$D15 = SUM(D12:D14)$$
$$E15 = SUM(E12:E14)$$

Constraint's RHS

Cells H12:H14 contain the RHS values of the supply constraints. Cells B17:E17 contains the RHS values of the demand constraints.

H12 = F3
H13 = F4
H15 = F5
B17 = B6
C17 = C6
D17 = D6
E17 = E6

After making the entries as described above we are ready to solve the linear programming problem with the help of solver available in the excel spreadsheet. To solve the problem follow the following steps.

Step 1: Be sure you are in excel worksheet. Select the **Tools** pull down menu

Step 2: Select **Solver** Option

Step 3: In the **Solver parameters** dialog box enter B9 in the set target box

Select Equal to:**Min** option

Enter B12:E14 in the **By Changing cells** box

Choose **Add.**

Table 5.115

	A	B	C	D	E	F	G	H
1			Destination					
2	Origin	1	2	3	4	Supply s_i		
3	1	3	2	7	6	5000		
4	2	7	5	2	3	6000		
5	3	2	5	4	3	2500		
6	Demand d_j	6000	4000	2000	1500	13,500		
7				Excess supply =		0		
8	Model							
9	Min. Cost	39,500						
10			Destination					
11	Origin	1	2	3	4	Total		
12	1	3500	1500	0	0	5000	<=	5000
13	2	0	2500	2000	1500	6000	<=	6000
14	3	2500	0	0	0	2500	<=	2500
15	Total	6000	4000	2000	1500			
16		=	=	=	=			
17		6000	4000	2000	1500			
18								

Step 4: In **Add constraint** dialog box Enter F12:F14 in **Cell Reference** box

Select <=
Enter H12: H14 in the **constraint** box
Choose **Add**
Enter B15: E15 in the **Cell Reference** box
Select =
Enter B17: E17 in the **constraint** box
Select **OK**

Step 5: In **Solver parameter** dialog box choose **Options**

Step 6: In **Solver options** dialog box select **Assume Linear Model**

Select **Assume Non-Negative**
Select **OK**

Step 7: In **Solver Parameters** dialog box choose **Solve**

In **Solver Results** dialog box choose **Keep Solver Solution**

Select **OK**

The final result will look like the one shown in Table 5.115.

In the excel Table 5.115 the shaded portion give us the needed solution.

From the table the solutions are:

$X_{11} = 3500$, $X_{12} = 1500$, $X_{22} = 2500$, $X_{23} = 2000$, $X_{24} = 1500$, $X_{31} = 2500$,
Minimum Transport cost = Rs. 39,500.

Table 5.116

Origin	Destinations	Allotment	Cost per unit	Total cost
1	1	3500	3	10,500
1	2	1500	2	3000
2	2	2500	5	12,500
2	3	2000	2	4000
2	4	1500	3	4500
3	1	2500	2	5000
			Grand Total	39,500

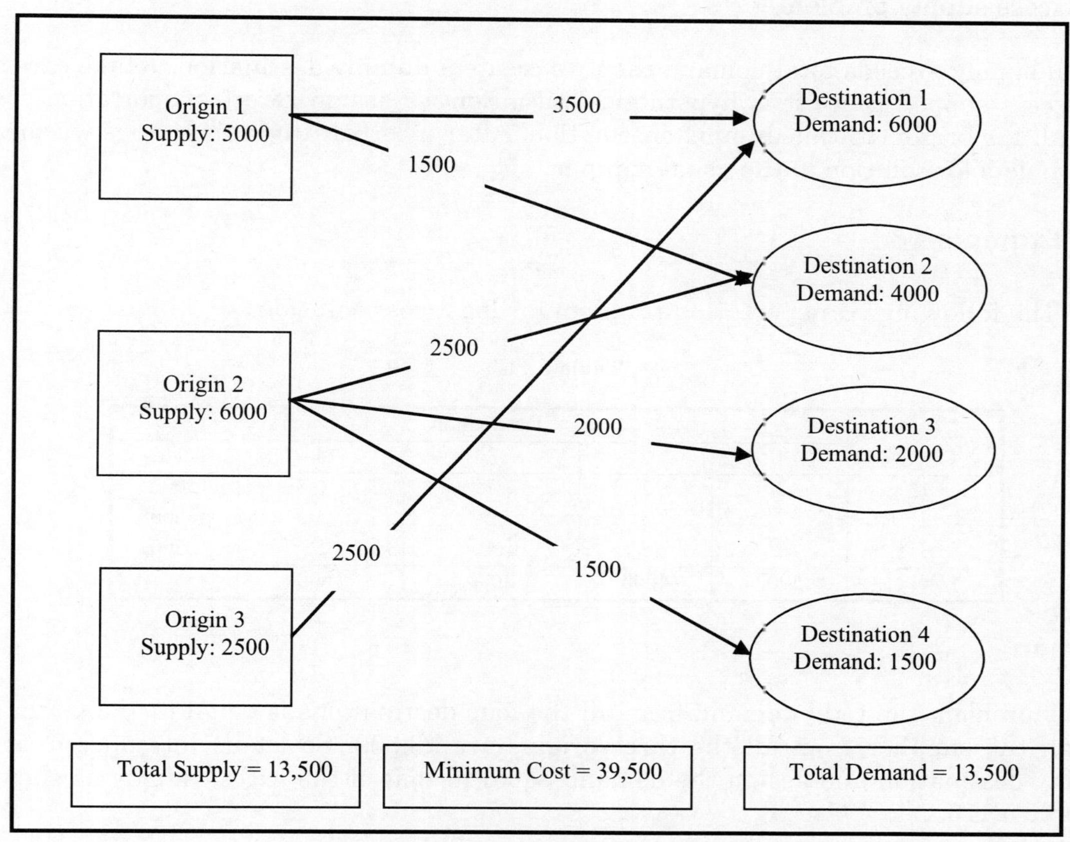

Figure 5.7

The minimum cost of transport is obtained as Rs. 39,500 for all the six efficient routes in Table 5.116. The above network chart illustrates the final solution and the associated route-wise and centre-wise distribution.

5.12 VARIANTS OF TRANSPORTATION PROBLEMS

In all the problems discussed in the previous section the demand and supply were in balance. This need not be the case all the time. We can have variety of unbalanced situations as listed below.

1. Total supply exceeds total demand
2. Total demand exceeds total supply
3. Route maximum allotment
4. Route minimum allotment
5. Route cancellation

With slight modification all the said variants can comfortably be accommodated within the framework discussed.

1. Excess supply problem

If total supply exceeds total demand then we create a dummy destination to take care of the excess supply. Since it is hypothetical situation we assign zero transportation cost from all the origins to this dummy destination. After incorporating this change we solve the problem for solution in the usual manner.

■ Example 16

Solve the following transportation problem for least cost solution.

Table 5.117

	Destinations				
Origin	1	2	3	4	Total s_i
1	3	2	7	6	5000
2	7	5	2	3	6000
3	2	5	4	5	2500
Total d_j	5000	4000	2000	1500	

Solution

In this problem the total demand from all the four destinations is equal to 12,500 units though the supplies from all the three origins are 13,500. So let us introduce a fifth dummy destination and assign the demand equal to that of the excess supply as shown in Table 5.118.

Table 5.118

	Destinations					
Origin	1	2	3	4	5	Total s_i
1	3	2	7	6	0	5000
2	7	5	2	3	0	6000
3	2	5	4	5	0	2500
Total d_j	5000	4000	2000	1500	1000	13,500

Initial solution by matrix minimum method

Table 5.119

	Destinations					
Origin	1	2	3	4	5	Total s_i
1	3	2	7	6	0	5000
2	7	5	2	3	0	6000
3	2	5	4	5	0	2500
Total d_j	5000	4000	2000	1500	1000	13,500

X_{25} = Min(1000,6000) = 1000. So delete destination 5 and reduce the supply from origin 2 to 5000 units.

Table 5.120

	Destinations				
Origin	1	2	3	4	Total s_i
1	3	2	7	6	5000
2	7	5	2	3	5000
3	2	5	4	5	2500
Total d_j	5000	4000	2000	1500	12,500

X_{32} = Min(2500,5000) = 2500. So delete origin 3 and reduce the demand from destination 1 to 2500 units.

Table 5.121

	Destinations				
Origin	1	2	3	4	Total s_i
1	3	2	7	6	5000
2	7	5	2	3	5000
Total d_j	2500	4000	2000	1500	10,000

X_{23} = Min(5000,2000) = 2000. So delete destination 3 and reduce the supply from origin 2 to 3000 units.

Table 5.122

	Destinations			
Origin	1	2	4	Total s_i
1	3	2	6	5000
2	7	5	3	3000
Total d_j	2500	4000	1500	8000

X_{12} = Min(5000,4000) = 4000. So delete destination 2 and reduce the supply from origin 1 to 1000 units.

Table 5.123

	Destinations		
Origin	1	4	Total s_i
1	3	6	1000
2	7	3	3000
Total d_j	2500	1500	4000

X_{24} = Min(3000,1500) = 1500. So delete destination 4 and reduce the supply from origin 1 to 1500 units.

Table 5.124

Origin	Destinations 1	Total s_i
1	3	1000
2	7	1500
Total d_j	2500	2500

X_{11} = Min(1000,2500) = 1000. So delete origin 1 and reduce the demand from destination 1 to 1500 units.

Table 5.125

Origin	Destinations 1	Total s_i
2	7	1500
Total d_j	1500	1500

X_{21} = Min(1500,1500) = 1500. Complete the allotment.

Initial basic solution: The resulting initial solution is reported in Table 5.126.

Table 5.126

Origin	Destinations 1	2	3	4	5	Total s_i
1	3 (1000)	2(4000)	7	6	0	5000
2	7(1500)	5	2(2000)	3(1500)	0 (1000)	6000
3	2(2500)	5	4	5	0	2500
Total d_j	5000	4000	2000	1500	1000	13,500
Total cost = 35,000						

Note that this solution does not degenerate and thus we can proceed further for MODI method for getting the optimum solution.

MODI Method for Optimum Solution

Table 5.127

Origin		Destinations 1	2	3	4	5	s_i
		$V_1 = 7$	$V_2 = 6$	$V_3 = 2$	$V_4 = 3$	$V_5 = 0$	
1	$U_1 = -4$	3, **1000**	2, **4000**	7 (9)	6 (7)	0 (4)	5000
2	$U_2 = 0$	7, **1500**	5 (−1)	2, **2000**	3, **1500**	0, **1000**	6000
3	$U_3 = -5$	2, **2500**	5 (4)	4 (7)	5 (7)	0 (5)	2500
d_j		5000	4000	2000	1500	1000	13,500
Total cost = 35,000							

Conditions: $U_1+V_1 = 3$; $U_2+V_1 = 7$; $U_3+V_1 = 2$; $U_1+V_2 = 2$; $U_2+V_3 = 2$; $U_2+V_4 = 3$; $U_2+V_5 = 0$

Solution: $U_2 = 0$; $V_1 = 7$, $V_3 = 2$; $V_4 = 3$; $V_5 = 0$; $U_3 = -5$; $U_1 = -4$ $V_2 = 6$

Table 5.128

Origin		Destinations 1	2	3	4	5	s_i
		$V_1 = 6$	$V_2 = 5$	$V_3 = 2$	$V_4 = 3$	$V_5 = 0$	
1	$U_1 = -3$	3, **2500**	2, **2500**	7 (8)	6 (6)	0 (3)	5000
2	$U_2 = 0$	7 (1)	5, **1500**	2, **2000**	3, **1500**	0, **1000**	6000
3	$U_3 = -4$	2, **2500**	5 (4)	4 (6)	5 (6)	0 (4)	2500
d_j		5000	4000	2000	1500	1000	13,500
Total cost = 33,500							

After incorporating these solution values into Table 5.127 we test the non-basic empty cells for optimality. In cell C_{22} we notice a negative entry −1 indicating the cost saving potentiality. For efficient transfer the appropriate closed loop is also shown in Table 5.127. The result after incorporating the transfer is shown in Table 5.128.

Let us restart our optimality test for Table 5.128 in the usual manner.

Conditions: $U_1+V_1 = 3$; $U_3+V_1 = 2$; $U_1+V_2 = 2$; $U_2+V_3 = 2$; $U_2+V_4 = 3$; $U_2+V_5 = 0$; $U_2+V_2 = 5$

Solution: $U_2 = 0$; $V_1 = 6$; $V_3 = 2$; $V_4 = 3$; $V_5 = 0$; $U_3 = -4$; $U_1 = -3$; $V_2 = 5$

Let us insert these values into Table 5.128 and test for optimality once again. Since all the non-basic cells are non-negative the previous iteration results are optimum.

$X_{11} = 2500$; $X_{12} = 2500$; $X_{22} = 1500$; $X_{23} = 2000$; $X_{24} = 1500$; $X_{25} = 1000$; $X_{31} = 2500$

Total Cost = 33,500.

Excel Solution to Excess Supply Problem

In this computerised method excess supply poses no additional problem. The excess supply will appear as a slack in some origin.

Table 5.129

	A	B	C	D	E	F	G	H
1				Destination				
2	Origin	1	2	3	4	Supply s_i		
3	1	3	2	7	6	5000		
4	2	7	5	2	3	6000		
5	3	2	5	4	3	2500		
6	Demand d_j	5000	4000	2000	1500	13,500		
7				Excess supply =		1000		
8	Model							
9	Min. cost	33,500						
10				Destination				
11	Origin	1	2	3	4	Total		
12	1	2500	2500	0	0	5000	<=	5000
13	2	0	1500	2000	1500	5000	<=	6000
14	3	2500	0	0	0	2500	<=	2500
15	Total	5000	4000	2000	1500			
16		=	=	=	=			
17		5000	4000	2000	1500			
18								

In this case since the total demand is less by 1000 units at the optimum level 1000 units will not be transported from origin 2.

2. Excess Demand

If the total demand exceeds total supply then we will have to create a dummy origin that takes care of the excess demand. Since it is a hypothetical situation we assign zero transportation cost from all the destinations to this dummy origin. After doing this we attempt the problem for solution in the usual manner.

■ Example 17

Modify the following transportation problem suitably and solve for optimum solution.

Table 5.130

	Destinations				
Origin	1	2	3	4	Total s_i
1	3	2	7	6	5000
2	7	5	2	3	5000
3	2	5	4	5	2500
Total d_j	6000	4000	2000	1500	

Solution

In this problem the total demand from all the four destinations is equal to 13,500 units though the supplies from all the three origins are only 12,500. So let us introduce the fourth dummy origin and assign the supply equal to that of the excess demand. Solve the modified problem as usual.

Table 5.131

	Destinations				
Origin	1	2	3	4	Total s_i
1	3	2	7	6	5000
2	7	5	2	3	5000
3	2	5	4	5	2500
4	0	0	0	0	1000
Total d_j	6000	4000	2000	1500	13,500

Excel Solution to Excess Demand Problem

If the total demand exceeds total supply then our linear programming problem will not have any feasible solution. In this case, we modify the given problem by simply introducing a dummy origin with the supply equal to the excess demand with zero transport cost to all the destinations. In the final solution the dummy origin takes care of the excess demand as shown in Table 5.132.

Table 5.132

	A	B	C	D	E	F	G	H
1			Destination					
2	Origin	1	2	3	4	Supply		
3	1	3	2	7	6	5000		
4	2	7	5	2	3	5000		
5	3	2	5	4	5	2500		
6	Dummy	0	0	0	0	1000		
7	Demand	6000	4000	2000	1500			
8	Minimum cost	33,500		Excess demand =			1000	
9			Destination					
10	Origin	1	2	3	4	Total		
11	1	2500	2500	0	0	5000	<=	5000
12	2	0	1500	2000	1500	5000	<=	5000
13	3	2500	0	0	0	2500	<=	2500
14	Dummy	1000	0	0	0	1000	<=	1000
15	Total	6000	4000	2000	1500			
16		=	=	=	=			
17		6000	4000	2000	1500			

3. Route Cancellation or Non-existence

In many realistic situations, the assumption of the existence of all routes from all origins to all destinations is not possible. Often natural obstructions like hilly region in between the said routes may not be there. Occasionally, due to factors like flood causes route cancellations. The route cancellation can be accommodated by simply dropping the concerned route in the relevant constraint. Alternatively we add an additional constraint with value zero for the concerned route.

■ Example 17 (a)

In the following problem route X_{22} does not exist. Obtain the optimum solution.

Table 5.132(a)

	Destinations				
Origin	1	2	3	4	Total s_i
1	3	2	7	6	5000
2	7	5	2	3	6000
3	2	5	4	5	2500
Total d_j	6000	4000	2000	1500	13,500

Excel Solution to Cancellation of Route X_{22}

1. Solution by dropping the route X_{22}

Table 5.133

	A	B	C	D	E	F	G	H
1				Destination				
2	Origin	1	2	3	4	Supply		
3	1	3	2	7	6	5000		
4	2	7	5	2	3	6000		
5	3	2	5	4	5	2500		
6	Demand	6000	4000	2000	1500			
7				Excess supply =		0		
8	Model							
9	Minimum cost	42,000						
10				Destination				
11	Origin	1	2	3	4	Total		
12	1	1000	4000	0	0	5000	<=	5000
13	2	2500	0	2000	1500	6000	<=	6000
14	3	2500	0	0	0	2500	<=	2500
15	Total	6000	4000	2000	1500			
16		=	=	=	=			
17		6000	4000	2000	1500			
18								

2. Solution by adding additional constraint $X_{22} = 0$

Table 5.134

	A	B	C	D	E	F	G	H
1				Destination				
2	Origin	1	2	3	4	Supply		
3	1	3	2	7	6	5000		
4	2	7	5	2	3	6000		
5	3	2	5	4	5	2500		
6	Demand	6000	4000	2000	1500			
7				Excess supply =		0		
8	Model							
9	Minimum cost	42,000						
10				Destination				
11	Origin	1	2	3	4	Total		
12	1	1000	4000	0	0	5000	<=	5000
13	2	2500	0	2000	1500	6000	<=	6000
14	3	2500	0	0	0	2500	<=	2500
15	Total	6000	4000	2000	1500			
16		=	=	=	=			
17		6000	4000	2000	1500			
18								

3. Solution by putting a hefty penalty transport cost in the cell X_{22}

Table 5.135

	A	B	C	D	E	F	G	H
1			Destination					
2	Origin	1	2	3	4	Supply		
3	1	3	2	7	6	5000		
4	2	7	100	2	3	6000		
5	3	2	5	4	5	2500		
6	Demand	6000	4000	2000	1500			
7				Excess supply =		0		
8	Model							
9	Minimum cost	42,000						
10			Destination					
11	Origin	1	2	3	4	Total		
12	1	1000	4000	0	0	5000	<=	5000
13	2	2500	0	2000	1500	6000	<=	6000
14	3	2500	0	0	0	2500	<=	2500
15	Total	6000	4000	2000	1500			
16		=	=	=	=			
17		6000	4000	2000	1500			
18								

***Note:** Cancellation of non-optimal route will not have any effect on the final solution.*

4. Route Capacity Restriction 1

In certain cases due to prepaid commitments and reservation certain fixed amount (X_{31}=1000) will have to be sent through certain specified routes. Under such circumstances a pre-optimal allotment of 1000 is made to the X_{31} from origin 3 to destination 1 and the route is completely dropped before optimizing. Do not forget to make the necessary correction in both demand supply constraints.

■ Example 18

In the following transportation problem due to heavy traffic the route X_{31} is booked in advance for 1000 units. Reformulate the problem and solve it by excel.

Table 5.136

Origin	Destinations 1	2	3	4	Total s_i
1	3	2	7	6	6000
2	7	5	2	3	5000
3	2	5	4	5	2500
Total d_j	6000	4000	2000	1500	13,500

Solution 1

Let us allot the prepaid quota of 1000 units to route X_{31} and subtract this amount from the total demand for destination 1. Similarly, we subtract this quantity from origin 3 and restate the new modified problem as shown below.

Table 5.137

Origin	Destinations 1	2	3	4	Total s_i
1	3	2	7	6	6000
2	7	5	2	3	3000
3	2	5	4	5	1500
Total d_j	5000	4000	2000	1500	10,500

Now since X_{31} = 1000 is already allotted it deserves no more allotment. So cancel this route and do the problem for other routes. The cancellation can be accomplished by adding an additional constraint X_{31} = 0. After making such an addition we solve the problem in the usual manner using excel worksheet as shown in Table 5.138.

Table 5.138

	A	B	C	D	E	F	G	H
1			Destination					
2	Origin	1	2	3	4	Supply		
3	1	3	2	7	6	5000		
4	2	7	5	2	3	6000		
5	3	2	5	4	5	1500		
6	Demand	5000	4000	2000	1500			
7				Excess supply =		0		
8	Model							
9	Minimum cost	43,500						
10			Destination					
11	Origin	1	2	3	4	Total		
12	1	5000	0	0	0	5000	<=	5000
13	2	0	2500	2000	1500	6000	<=	6000
14	3	0	1500	0	0	1500	<=	1500
15	Total	5000	4000	2000	1500			
16		=	=	=	=			
17		5000	4000	2000	1500			
18								

X_{11} = 5000, X_{31} = 1000 (pre-allotted), X_{22} = 2500, X_{32} = 1500, X_{23} = 2000, X_{24} = 1500, Total cost = 2000 (pre-allotted) + 43,500 = 45,500.

Solution 2

Alternatively, go by the original cost table with the additional constraint X_{31} = 1000 and do the problem by excel.

Table 5.139

	A	B	C	D	E	F	G	H
1			Destination					
2	Origin	1	2	3	4	Supply		
3	1	3	2	7	6	5000		
4	2	7	5	2	3	6000		
5	3	2	5	4	5	2500		
6	Demand	6000	4000	2000	1500			
7				Excess supply =		0		
8	Model							
9	Minimum cost	45,500						
10			Destination					
11	Origin	1	2	3	4	Total		
12	1	5000	0	0	0	5000	<=	5000
13	2	0	2500	2000	1500	6000	<=	6000
14	3	1000	1500	0	0	2500	<=	2500
15	Total	6000	4000	2000	1500			
16		=	=	=	=			
17		6000	4000	2000	1500			
18								

X_{11} = 5000, X_{31} = 1000, X_{22} = 2500, X_{32} = 1500, X_{23} = 2000, X_{24} =1500, Total cost = 45,500.

5. Route Capacity Restriction 2

In a particular route due to heavy traffic the route capacity is restricted as $X_{31} \leq 1000$ then before solving the problem we simply will have to add this additional constraint into our model before solving it by excel. Remember that the unrestricted solution for this cell is X_{31} = 2500. The minimum total cost = Rs. 39,500. Due to the restriction imposed the minimum cost went up to Rs. 45,500.

Table 5.140

	A	B	C	D	E	F	G	H
1			Destination					
2	Origin	1	2	3	4	Supply		
3	1	3	2	7	6	5000		
4	2	7	5	2	3	6000		
5	3	2	5	4	5	2500		
6	Demand	6000	4000	2000	1500			
7				Excess supply =		0		
8	Model							
9	Minimum cost	45,500						
10			Destination					
11	Origin	1	2	3	4	Total		
12	1	5000	0	0	0	5000	<=	5000
13	2	0	2500	2000	1500	6000	<=	6000
14	3	1000	1500	0	0	2500	<=	2500
15	Total	6000	4000	2000	1500			
16		=	=	=	=			
17		6000	4000	2000	1500			
18								

6. Route Capacity Restriction 3

Let in a particular route due transport restriction a minimum of 1000 units are to transported from origin 3 to destination 1, then this additional constraint may be written as $X_{31} \geq 1000$ by the transport department.

Solution

Table 5.141

	A	B	C	D	E	F	G	H
1			Destination					
2	Origin	1	2	3	4	Supply		
3	1	3	2	7	6	5000		
4	2	7	5	2	3	6000		
5	3	2	5	4	5	2500		
6	Demand	6000	4000	2000	1500			
7				Excess supply =		0		
8	Model							
9	Minimum cost	39,500						
10			Destination					
11	Origin	1	2	3	4	Total		
12	1	3500	1500	0	0	5000	<=	5000
13	2	0.00	2500	2000	1500	6000	<=	6000
14	3	2500	0	0	0	2500	<=	2500
15	Total	6000	4000	2000	1500			
16		=	=	=	=			
17		6000	4000	2000	1500			
18								

These are our unrestricted optimum values. Since $X_{31} = 2500 > 1000$ our constraint is readily fulfilled. So in this case the constraint is redundant.

7. Route Capacity Restriction 4

Let $X_{12} \geq 1600$

Table 5.142

	A	B	C	D	E	F	G	H
1			Destination					
2	Origin	1	2	3	4	Supply		
3	1	3	2	7	6	5000		
4	2	7	5	2	3	6000		
5	3	2	5	4	5	2500		
6	Demand	6000	4000	2000	1500			
7				Excess supply =		0		
8	Model							
9	Minimum cost	39,600						
10			Destination					
11	Origin	1	2	3	4	Total		
12	1	3400	1600	0	0	5000	<=	5000
13	2	100	2400	2000	1500	6000	<=	6000
14	3	2500	0	0	0	2500	<=	2500
15	Total	6000	4000	2000	1500			
16		=	=	=	=			
17		6000	4000	2000	1500			
18								

In this case the minimum transport cost has gone up Rs. 39,500 to Rs. 39,600 due to this additional restriction placed on X_{12}. Thus the constraint is not redundant.

5.13 MAXIMIZATION PROBLEM

In certain type of transportation problems we maximize the objective function. In such cases the objective function is framed with net return per unit of output as coefficients in the objective function. The following example illustrates the method of converting the given maximization problem into a minimization problem.

■ Example 19

If per unit returns at the factory outlet including the transport cost are Rs. 8, Rs. 7, Rs. 6 respectively determine the optimum distribution system that maximizes the total return. The transportation cost matrix is given below.

Table 5.143

Origin	Destinations				
	1	2	3	4	Total s_i
1	3	2	7	6	6000
2	7	5	2	3	5000
3	2	5	4	5	2500
Total d_j	6000	4000	2000	1500	13,500

Solution

From the data given we first get the net return per unit at all factory outlets by simply subtracting the respective transportation cost from their respective returns. The so obtained result is shown in Table 5.144.

Table 5.144

Origin	Destinations				
	1	2	3	4	Total s_i
1	1	0	5	4	6000
2	6	4	1	2	5000
3	2	5	4	5	2500
Total d_j	6000	4000	2000	1500	13,500

Now to convert the given maximization into an equivalent minimization problem we obtain the largest cell value, namely, 6 in our Table 5.144 and subtract all elements from this number. The so obtained result is shown in Table 5.145.

Table 5.145

Origin	Destinations				
	1	2	3	4	Total s_i
1	1	0	5	4	6000
2	6	4	1	2	5000
3	2	5	4	5	2500
Total d_j	6000	4000	2000	1500	13,500

This minimization problem is solved for optimum solution by using excel solver Table 5.146.

Excel solution

Table 5.146

	A	B	C	D	E	F	G	H
1			Destination					
2	Origin	1	2	3	4	Supply		
3	1	1	0	5	4	5000		
4	2	6	4	1	2	6000		
5	3	2	5	4	5	2500		
6	Demand	6000	4000	2000	1500			
7				Excess supply =		0		
8	Model							
9	Minimum cost	23,500						
10			Destination					
11	Origin	1	2	3	4	Total		
12	1	3500	1500	0	0	5000	<=	5000
13	2	0	2500	2000	1500	6000	<=	6000
14	3	2500	0	0	0	2500	<=	2500
15	Total	6000	4000	2000	1500			
16		=	=	=	=			
17		6000	4000	2000	1500			
18								

These cell values are multiplied by the respective net return given in the previous table to get the net maximum return.

Maximum net return = 3500×5+2500×4+1500×6+2500×2+2000×5+1500×2 = Rs. 54,500.

■ Example 20

A company has four factories F1, F2, F3 and F4 manufacturing the same product. Production and raw material costs differ from factory to factory as given in the following table in the first two rows. The transportation costs from the factories to sales depots, S1, S2 and S3 are also given. The last two columns in the table give the sales price and the total requirement at each depot. The production capacity of each factory is given in the last row.

Table 5.147

		F1	F2	F3	F4	Sales price	Requirements
Production cost per unit		15	18	14	13		
Raw material cost per unit		10	9	12	9		
	S1	3	9	5	4	34	80
Transport cost per unit	S2	1	7	4	5	32	120
	S3	5	8	3	6	31	150
Supply		10	150	50	100		

Formulate a suitable transportation model.

Solution

By definition profit = Sales price – Production cost – Raw material cost – Transportation cost

Total requirement = 80+ 120+150 = 350

Total supply =10 +150+50+100 = 310

Dummy factory F5 = is to produce 350–310 = 40

Table 5.148

Profit matrix

Factory	Sale depot			Availability
	S1	S2	S3	
F1	6	6	1	10
F2	−2	−2	−4	150
F3	3	2	2	50
F4	8	5	3	100
F5	0	0	0	40
Requirements	80	120	150	350

Table 5.149

Relative loss matrix

Factory	Sale depot			Availability
	S1	S2	S3	
F1	2	2	7	10
F2	10	10	12	150
F3	5	6	6	50
F4	0	3	5	100
F5	8	8	8	40
Requirements	80	120	150	350

5.14 TRANSHIPMENT PROBLEMS

It has been assumed in transportation problems that the items must be shipped only from origin to destinations. In practice it is not possible to distinguish origins and destinations. A destination receiving some consignment may use only a part of it and send the balance to other destinations. So as far as these other destinations are concerned the destination under discussion becomes the origin. Similarly, often from factory consignments are straightaway sent to warehouses for further distribution to retail shops. Thus, in transhipment problem each place is regarded both as origin and destination. It receives items from elsewhere, then it is regarded as destination, if it distributes its consignments to other centres then it is regarded as origin.

Thus in this type of problem one will have three types of node, namely, origin node, transhipment node and final destination node. Moreover, in this general type shipment is made between any pair of nodes, may be origin to warehouse or origin to final destination or one warehouse to another warehouse or from one origin to another origin. If there are m origins and n destinations in our original transportation problem then we will have m+n origin and equal number of destinations in the modified transhipment model. The general linear programming formulation of the transhipment problem is normally written as

$$\text{Minimize}: \sum_{i=1}^{m}\sum_{j=1}^{n} C_{ij}X_{ij}$$

$$\text{Subject to:} \quad \sum_{\text{arrow out}} X_{ij} - \sum_{\text{arrow in}} X_{ij} = s_i \text{ (Origin node)}$$

$$\sum_{\text{arrow out}} X_{ij} - \sum_{\text{arrow in}} X_{ij} = 0 \text{ (Transhipment node)}$$

$$\sum_{\text{arrow out}} X_{ij} - \sum_{\text{arrow in}} X_{ij} = d_j \text{ (Destination node)}$$

$$X_{ij} \geq 0$$

■ Example 21

Consider a transportation problem with two production plants and three warehouses. The transport costs from the plants to warehouses are given in the following table. Obtain the solution for this transportation problem.

Unit transport cost from plant to warehouses

Table 5.150

Plant	Warehouses			Supply
	W1	W2	W3	
P1	1	3	15	150
P2	3	5	25	300
Demand	150	150	150	450

Solution

To this simple transportation problem the optimum solution is obtained as follows:

Table 5.151

	A	B	C	D	E	F	G
1			Destination				
2	Origin	W1	W2	W3	Supply		
3	P1	1	3	15	150		
4	P2	3	5	25	300		
5	Demand	150	150	150			
6	Model		Excess supply =		0		
7	Minimum cost	3450					
8			Destination				
9	Origin	1	2	3	Total		
10	1	0	0	150	150	<=	150
11	2	150	150	0	300	<=	300
12	Total	150	150	150			
13		=	=	=			
14		150	150	150			
15							

Example 22

Consider a transhipment problem with two production plants and three warehouses. The transport costs from the plants to warehouses are given in the following table.

Table 5.152

	Warehouses			
Plant	W1	W2	W3	Supply
P1	1	3	15	150
P2	3	5	25	300
Demand	150	150	150	450

The unit transport cost from plant to plant is given by:

Table 5.153

	Plants	
Plant	P1	P2
P1	0	65
P2	1	0

The unit transport cost from warehouse to warehouse is given by:

Table 5.154

Warehouses	Warehouses W1	W2	W3
W1	0	23	1
W2	1	0	3
W3	65	3	0

The unit transport cost from warehouse to plants is given by

Table 5.155

Warehouse	Plants P1	P2
W1	3	15
W2	25	3
W3	45	55

Obtain the solution for this transhipment problem.

Solution

When each plant is considered a destination and each destination is considered a origin altogether we have 5 destinations and five origins. Now let us convert this into a transhipment problem and obtain the optimum solution. From the above data tables the associated transhipment problem is formulated as follows. The buffer 450 as the total capacity and total requirement given in the original transportation problem is added to each row and column of the transhipment problem. Thus the resulting transhipment problem has m+n = 5 origins and five destinations.

Complete Table

Table 5.156

	P1	P2	W1	W2	W3	Supply
P1	0	65	1	3	15	150+450=600
P2	1	0	3	5	25	300+450=700
W1	3	15	0	23	1	450
W2	25	3	1	0	3	450
W3	45	55	65	3	0	450
Demand	450	450	150+450	150+450	150+450	2700

Table 5.157

	A	B	C	D	E	F	G	H	I
1			Destination						
2	Origin	1	2	3	4	5	Supply		
3	1	0	65	1	3	15	600		
4	2	1	0	3	5	25	750		
5	3	3	15	0	23	1	450		
6	4	25	3	1	0	3	450		
7	5	45	55	65	3	0	450		
8	Demand	450	450	600	600	600	2700		
9	Model			Excess supply =		0			
10	Minimum cost	1200							
11			Destination						
12	Origin	1	2	3	4	5	Total		
13	1	150	0	300	150	0	600	<=	600
14	2	300	450	0	0	0	750	<=	750
15	3	0	0	300	0	150	450	<=	450
16	4	0	0	0	450	0	450	<=	450
17	5	0	0	0	0	450	450	<=	450
18	Total	450	450	600	600	600			
19		=	=	=	=	=			
20		450	450	600	600	600			

Since the optimum transportation cost is Rs. 3450 and the associated transhipment optimum cost is Rs.1200 we conclude that the transhipment always gives better solution.

■ Example 23

The following chart explains a transhipment problem. The respective cost of transportation is also shown in the network diagram. Obtain the optimum solution.

Table 5.158

Plant	Warehouses 3	Warehouses 4
1	2	3
2	3	1

Warehouses	Retail outlets 5	Retail outlets 6	Retail outlets 7	Retail outlets 8
3	2	6	3	6
4	4	4	6	5

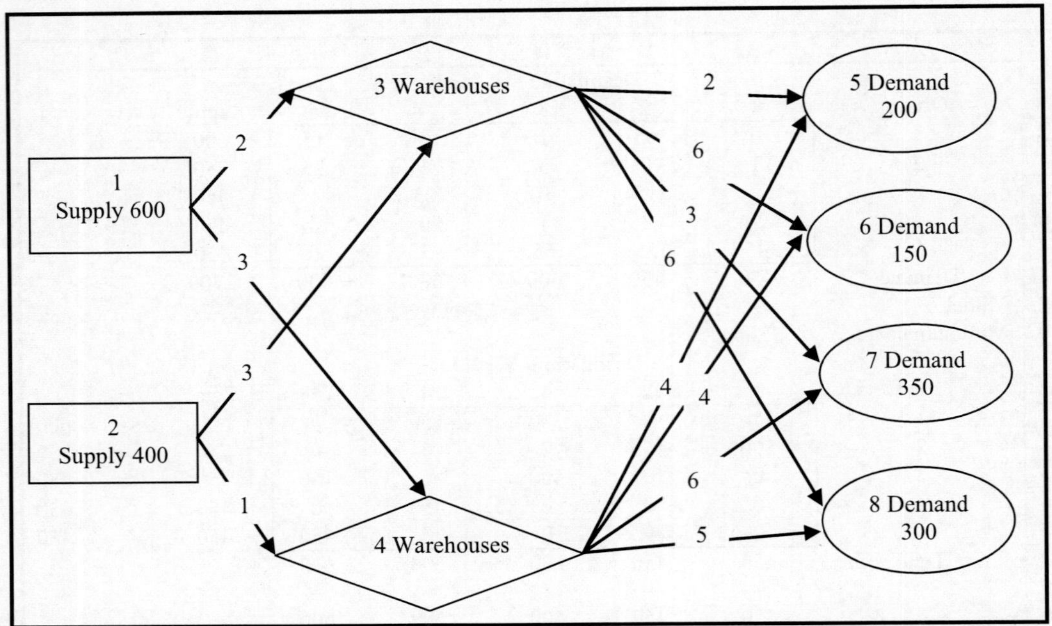

Figure 5.8

Supply constraints

Plant 1: $X_{13} + X_{14} \leq 600$ Plant 2: $X_{23} + X_{24} \leq 400$

Warehouse node 3 shipped in: $X_{13} + X_{23}$

Warehouse node 3 shipped out: $X_{35} + X_{36} + X_{37} + X_{38}$

For optimum transaction the supply and demand must balance.

So $X_{35} + X_{36} + X_{37} + X_{38} = X_{13} + X_{23}$. After transferring all to the LHS it becomes

$- X_{13} - X_{23} + X_{35} + X_{36} + X_{37} + X_{38} = 0$

Similarly, the constraint for node 4 is obtained as

$- X_{14} - X_{24} + X_{45} + X_{46} + X4_{47} + X_{48} = 0$

For destination node again the amount shipped in must be equal to final demand

So

$X_{35} + X_{45} = 200$

$X_{36} + X_{46} = 150$

$X_{37} + X_{47} = 350$

$X_{38} + X_{48} = 300$

Thus our associated linear programming formulation of the problem could be stated as:

$$\text{Min: } 2X_{13} + 3X_{14} + 3X_{23} + 1X_{24} + 2X_{35} + 6X_{36} + 3X_{37} + 6X_{38} + 4X_{45} + 4X_{46} + 6X_{47} + 5X_{48}$$

$$X_{13} + X_{14} \le 600$$

$$+ X_{23} + X_{24} \le 400$$

$$-X_{13} + X_{23} \quad X_{35} + X_{36} + X_{37} + X_{38} = 0$$

$$-X_{14} - X_{24} \quad X_{45} + X_{46} + X_{47} + X_{48} = 0$$

$$X_{11} + X_{35} + X_{45} = 200$$

$$X_{12} + X_{36} + X_{46} = 150$$

$$X_{13} + X_{36} + X_{47} = 350$$

$$X_{14} + X_{36} + X_{48} = 300$$

$X_{ij} \ge 0$ for i,j

Excel Solution

Table 5.159

	A	B	C	D
1		Warehouse node		
2	Production/Destination node	3	4	Supply/Demand
3	1	2	3	600
4	2	3	1	400
5	5	2	4	200
6	6	6	4	150
7	7	3	6	350
8	8	6	5	300
9		Decision variables		
10		3	4	
11	1	600	0	600
12	2	0	400	400
13	5	200	0	200
14	6	0	150	150
15	7	350	0	350
16	8	50	250	300
17	MIN	5200		
18	constraints			
19	1	600	<=	600
20	2	400	<=	400
21	5	200	=	200
22	6	150	=	150
23	7	350	=	350
24	8	300	=	300

5.15 VARIANTS OF TRANSHIPMENT PROBLEMS

Transhipment problem may be formulated with several variants as shown under.

- Excess supply
- Excess demand
- Route capacity restrictions
- Route cancellation
- Maximization of objective functions and so on.

EXERCISES

1. A softdrinks company has three factory outlets for its products. It distributes the product to five zonal centres for sales. The associated transportation costs are given in the following table. The factory capacity is 40, 30 and 50 respectively.

	Zones					
Factory	Z1	Z2	Z3	Z4	Z5	Supply
F1	5	4	6	9	10	40
F2	13	12	6	5	6	30
F3	12	8	8	7	10	50
Demand	20	30	25	25	20	120

(a) Formulate the network chart for this problem.

(b) Formulate a linear programming problem and write its dual.

2. For the following transportation problems obtain the basic feasible solutions by using (a) North-west corner rule (b) Least cost matrix method (c) Vogel's Approximation method.

	Zones					
Factory	M1	M2	M3	M4	M5	Supply
S1	5	4	6	9	10	40
S2	3	4	7	8	12	50
S3	13	12	6	5	6	30
S4	5	8	12	6	9	30
Demand	30	30	30	40	20	150

3. Solve the following transportation problem by least cost matrix method.

Factory	Zones			Supply
	M1	M2	M3	
F1	10	8	8	8
F2	10	7	10	7
F3	11	9	7	9
F4	12	14	10	4
Demand	10	10	8	28

4. A firm manufactures a single product using three of its plants located in three different cities. The respective supply capacities are 60, 35, and 40 respectively. The firm has committed to supply 22, 45, 20, 18 and 30 units to five customers A, B, C, D and E respectively. Given the transport cost matrix obtain the least cost solution by using Vogel's method.

Factory	Individuals				
	A	B	C	D	E
S1	4	1	3	4	4
S2	2	3	2	2	3
S3	3	5	2	4	4

5. A soap manufacturing company in south has three factories one each at Chennai, Coimbatore and Bangalore. Given the transportation cost per unit matrix obtain the least cost solution by both north-west corner and least cost matrix methods and compare the results.

Factory	Cities				Supply
	Cochin	Nellore	Salem	Madurai	
Chennai	5	4	6	9	12
Coimbatore	3	4	7	8	7
Bangalore	13	12	6	5	7
Demand	5	4	4	11	

6. In the above problem supply and demand conditions are revised obtain the new solution by least cost matrix method.

Factory	Cities				Supply
	Cochin	Nellore	Salem	Madurai	
Chennai	5	4	6	9	12
Coimbatore	3	4	7	8	8
Bangalore	13	12	6	5	7
Demand	5	7	5	11	

7. A firm has three production facilities in three cities. The so produced products are supplied to four cities. By using his familiar method a student gets the following solution.

Origin	Destinations				
	Dindigul	Erode	Palani	Kochin	Total s_i
Chennai	7	10(10)	14(20)	8	30
Salem	7	11	12 (5)	16(35)	40
Madurai	5(20)	8(10)	15	9	30
Total d_j	20	20	25	35	100

a) Is the above solution feasible?

b) Is the solution degenerate?

c) Is the solution optimal? If not obtain the optimum solution.

d) Is this solution unique? If not obtain the other solutions.

e) If there is pre-commitment for sending 5 units from Salem to Dindigul obtain the new solution.

f) If there is pre-commitment for sending a minimum of 5 units from Salem to Dindigul obtain the new solution.

g) If Salem to Palani is not viable due to bad weather obtain the new optimum solution.

8. A TV assembling unit has four plants in different city locations with capacity 125, 250, 175 and 100 units respectively. The demand schedules for its four outlets are 100, 400, 90 and 60 respectively. Though the component costs are one and the same in all plant locations due to varying levels of transport cost the profit per set varies considerably. Given the per unit profit table obtain the optimum solution that maximizes the total profit.

Factory	Showrooms			
	S1	S2	S3	S4
F1	90	100	120	110
F2	100	105	130	117
F3	111	109	110	120
F4	130	125	108	113

9. A company producing a component in three different cities sells them at a uniform rate of Rs. 2.5 per unit in all its five sales outlets. Due to varying locations the input costs are Re. 1.00, Re. 0.90, and Re. 0.80 respectively. Given the per unit transportation cost matrix obtain profit maximizing solution.

Factory	Sales outlets				
	S1	S2	S3	S4	S5
C1	0.05	0.07	0.1	0.15	0.15
C2	0.08	0.06	0.09	0.12	0.14
C3	0.1	0.09	0.08	0.1	0.15

10. ABC Tool Company has a sales force of 25 men who operate from three regional offices. The company produces four basic product lines of hand tools. Mr. Jain, sales manager feels that 6 salesmen are needed to distribute product line 1,10 salesmen to distribute product line 2, 4 salesmen to product line 3 and 5 salesmen to product line 4. The cost (in Rs.) per day of assigning salesmen from each of the offices for selling each of the product line is as follows.

Regional office	Product lines			
	1	2	3	4
A	20	21	16	18
B	17	28	14	16
C	29	23	19	20

At present, 10 salesmen are allotted to office A, 9 salesmen to office B and 7 salesmen to office C. How many salesmen should be assigned from each office for selling each product line in order to minimize the cost?

11. Solve the following transportation problem having the cost structure as:

From	To				
	1	2	3	4	Supply
A	10	18	11	7	20
B	9	12	14	6	40
C	8	9	12	10	35
Demand	16	18	31	30	

12. Find out the minimum cost solution for the following transportation problem which has cost structure as:

From	To			
	P	Q	R	Availability
A	16	19	12	14
B	22	13	19	16
C	14	28	8	12
Requirement	10	15	17	

13. A company has factories at F1, F2 and F3 which supply to warehouses at W1, W2 and W3. Weekly factory supplies are 200, 160, and 90 units respectively. Weekly warehouse demands are 180, 120 and 150 units respectively. Unit shipping cost (in rupees) is as follows

	Warehouses			
	W1	W2	W3	Supply
F1	16	20	12	200
F2	14	8	18	160
F3	26	24	16	90
Demand	180	120	150	

14. Solve the following transportation problem to minimize the transportation cost

	Warehouses				
	W1	W2	W3	W4	Production
F1	5	14	9	6	18
F2	15	13	8	8	9
F3	7	12	16	13	11
Requirement	10	7	5	6	

(Bangalore University, July/August, 2006)

15. Determine the most profitable production and distribution schedule and corresponding profit. Surplus production should be taken to yield zero profit.

	Sales depot	Factories				Sale price per unit	Requirements
		F1	F2	F3	F4		
Production cost per unit	-	15	18	14	13		
Raw material cost per unit	-	10	9	12	9		
Transportation cost per unit	S1	3	9	5	4	34	80
	S2	1	7	4	5	32	120
	S3	5	8	3	6	31	150
Production capacity		10	150	50	100		

(Bangalore University, M.B.A July/August 2005)

16. Maximize the following transportation problem. Figures inside relate to profit and not the cost of transportation. Obtain the maximum profit.

	M1	M2	M3	
F1	7	5	5	100
F2	7	11	9	80
F3	10	9	9	60
	110	70	40	

(Bangalore University, M.B.A October 2004)

17. Given the following linear programming problem

Maximize: $Z = 12X_{11} + 8X_{12} + 2X_{13} + 9X_{21} + 10X_{-22} + 9X_{23} + 7X_{31} + 15X_{32} + 6X_{33}$

Subject to: $X_{11} + X_{12} + X_{13} = 40$

$X_{21} + X_{22} + X_{23} = 30$ Demand constraints

$X_{31} + X_{32} + X_{33} = 30$

$X_{11} + X_{21} + X_{31} = 20$

$X_{12} + X_{22} + X_{32} = 50$ Supply constraints

$X_{13} + X_{23} + X_{33} = 30$

$X_{11,}X_{12}, X_{13}, X_{21}, X_{22}, X_{23}, X_{31}, X_{32}, X_{33} \geq 0$

a) Write the dual for the given linear programming.

b) Formulate as a transportation problem and solve it for optimum solution.

CHAPTER

6

Integer Programming

- *Introduction*
- *Simple linear programming solution by graphical method*
- *Mixed integer solution by graphical method*
- *Branch-and-bound method for complete integer solution by graphical method*
- *Branch-and-bound method for mixed integer solution by graphical method*
- *Branch-and-bound method for all integers by Simplex method*
- *Gomory's cutting plane method for complete integer solution by Simplex method*
- *The Gomory's cut for mixed integer problems*
- *Excel solution to integer programming*
- *Use of '0 -1' integer programming*
- *Applications in fixed cost modelling*
- *Application in production and distribution problems*
- *Exercises*

6.1 INTRODUCTION

In the previous chapter, especially in linear programming, the optimum decision variables were not restricted to integers. The optimum decision variables thus can take both integer and non-integer values in simple linear programming problems. The divisibility assumption which permits decimal values to the decision variables normally does not pose any additional problem in many practical situations. However, in certain special cases the divisibility assumption becomes totally unrealistic. For example, if one of the decision variables is the optimum number of houses to be built, then 2.5 houses does not make any sense. Of course under normal circumstances the so obtained decimal values are rounded off to the next highest integer values or to the next lowest integer values depending upon the situation. However, in the case of over rounding off one will have to find additional resources for achieving the rounded-off result. Similarly, in the case of under rounding off, unnecessarily we will be leaving some resource or the other unproductive. To overcome such difficulties, in this section we propose to introduce a specially designed integer programming for solving such problems. Definitely this refined approach is going to be superior to the common practice of rounding off.

Accordingly, as a first step we relax the given integer condition in the IP problem and solve it as a simple LP problem. If we get optimum all integer decision variables solution in the very first step we accept it as the final non-integer solution to the given integer problem. In such a coincidental situation, the optimum value of the LP problem necessarily will be the optimum value for our IP problem as well. But unfortunately, such a nice coincidence need not happen all the time when solving an IP problem as an LP problem. Very often we come across situations in which the decision variables take non-integer optimum values. Under such circumstance we introduce additional constraints and cut the original feasible region into size to have our needed integer solution. Remember, that once the cut constraint is introduced the resulting problem is solved again as an LP problem in the usual manner either by graphical or by Simplex method. Repeatedly we introduce more and more cuts till the needed final all integer optimum solution is obtained. However, remember that there is no separate independent algorithm for solving integer programming.

If we insist all integer solutions for all our decision variables then the programming is often called *integer programming*. On several occasions integer values are insisted only for selective decision variable. Such problems are called *mixed integer programming*. If in the given problem there are only two decision variables then the needed all integer solution is obtained by using simple graphical method.

To illustrate the difference between the unrestricted programming and restricted all integer programming let us consider the following illustrative example.

6.2 SIMPLE LINEAR PROGRAMMING SOLUTION BY GRAPHICAL METHOD

■ Example 1

For the integer programming (IP) problem given below

$$\begin{aligned}
\text{Maximize: } & Z = 8X_1 + 5X_2 \\
\text{Subject to: } & X_1 + X_2 \leq 6 \\
& 9X_1 + 5X_2 \leq 45 \\
& X_1 \;;\; X_2 \geq 0 \quad X_1 \text{ and } X_2 \text{ integers}
\end{aligned}$$

(a) Find the simple linear programming (LP) solution by graphical method.

(b) Obtain all integer solution by rounding off method and illustrate its inefficiency.

(c) Also obtain the corresponding all integer solution by complete enumeration method.

(d) Find the all integer solution by graphical method.

a) Simple LP solution by graphical method

The statement of the LP problem: After removing the integer condition the problem may be written as a simple LP problem as shown below

$$\text{Maximize:} \quad Z = 8X_1 + 5X_2$$
$$\text{Subject to:} \quad X_1 + X_2 \leq 6$$
$$9X_1 + 5X_2 \leq 45$$
$$X_1 \,;\, X_2 \geq 0$$

Graphical Solution as LP

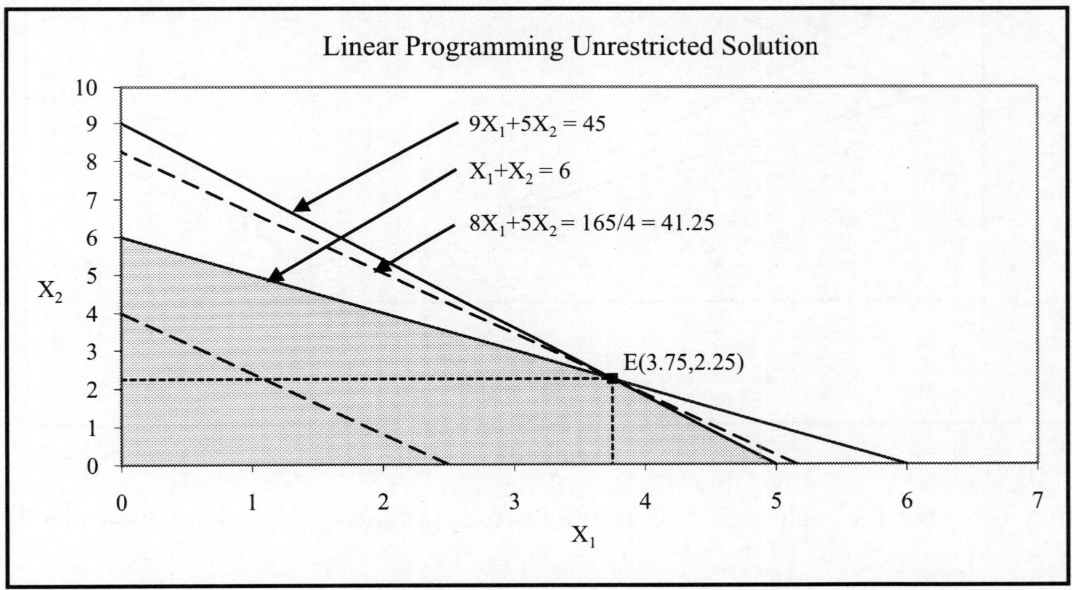

Figure 6.1

As usual we convert both inequality constraints into equality constraints and draw them all on the graph sheet by identifying two points on each. Once this is done insert the given objectivity function with an arbitrarily chosen profit (say 22.5) and move it up with unaltered slope so that it becomes tangent at E as shown in Figure 6.1 the solution values are X_1 =13/4 = 3.25, X_2 = 9/4 = 2.25 and Z = 165/4 = 41.25. Clearly both the decision variables are non-integers.

b) All integer solution by rounding off method

Since X_1 = 3.25 and X_2 = 2.25 the rounding off solution may be written as X_1 = 3 and X_2 = 2. So the value of the objective function is obtained as $Z = 8 \times 3 + 5 \times 2$ = Rs.34. Thus this time the value of the objective function is substantially less (34 < 41.25) than the corresponding L.P. value.

c) All integer solution by complete enumeration

In Figure 6.1 we add all possible integer points within the feasible region as shown in Figure 6.2. Once this is done list out X and Y coordinates for all these feasible integer points. Substitute these values in the objective function given as shown in Table 6.1 and obtain the optimum.

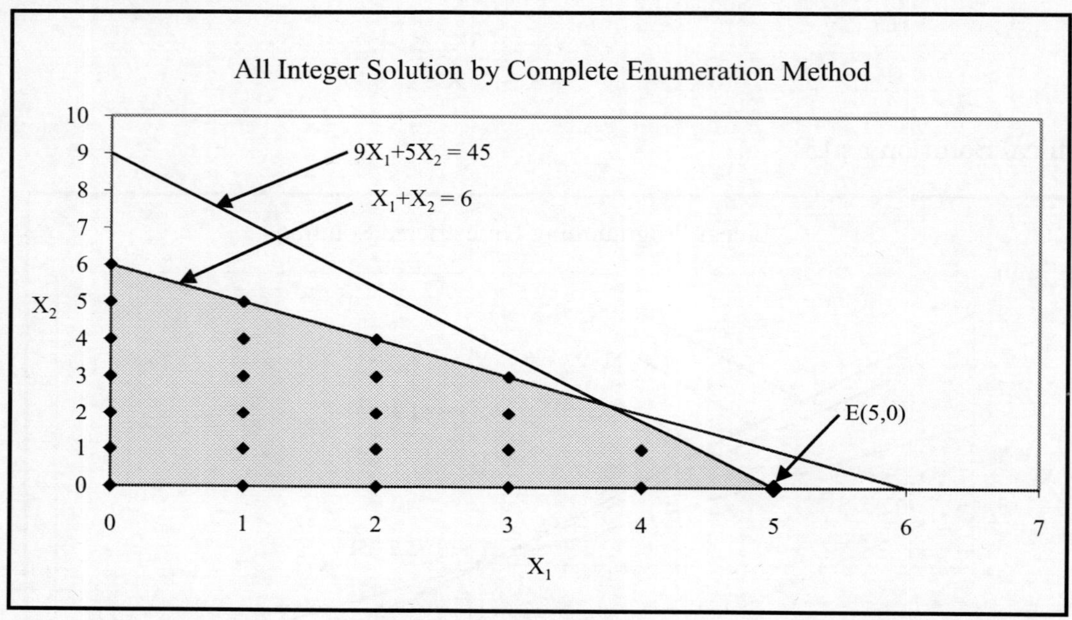

Figure 6.2

From Table 6.1 the highest Z = 40 and the associated integer solution values are $X_1 = 5$ and $X_2 = 0$.

Table 6.1

Sl. No	X_1	X_2	$Z = 8X_1+5X_2$
1	0	0	0
2	1	0	8
3	2	0	16
4	3	0	24
5	4	0	32
6	5	0	40
7	0	1	5
8	1	1	13
9	2	1	21
10	3	1	29
11	4	1	37
12	0	2	10

Sl .No	X_1	X_2	$Z = 8X_1+5X_2$
13	1	2	18
14	2	2	26
15	3	2	34
16	0	3	15
17	1	3	23
18	2	3	31
19	3	3	39
20	0	4	20
21	1	4	28
22	2	4	36
23	0	5	25
24	1	5	33
25	0	6	30

Note: Remember, that this time not only we were able to get all integer solution but also better objective function value compared to rounding off solution result (40 > 34). Also note that this time the under rounding off leaves both the resources under utilized.

d) All integer solution by graphical method

In complete enumeration method outlined searching for all integer points within the feasible region is a tough job. Instead, in Figure 6.3 we bring in the objective function as well for an arbitrary value of 22.5 and shift it up keeping parallel till the outermost integer point is encountered at J as shown in Figure 6.3.

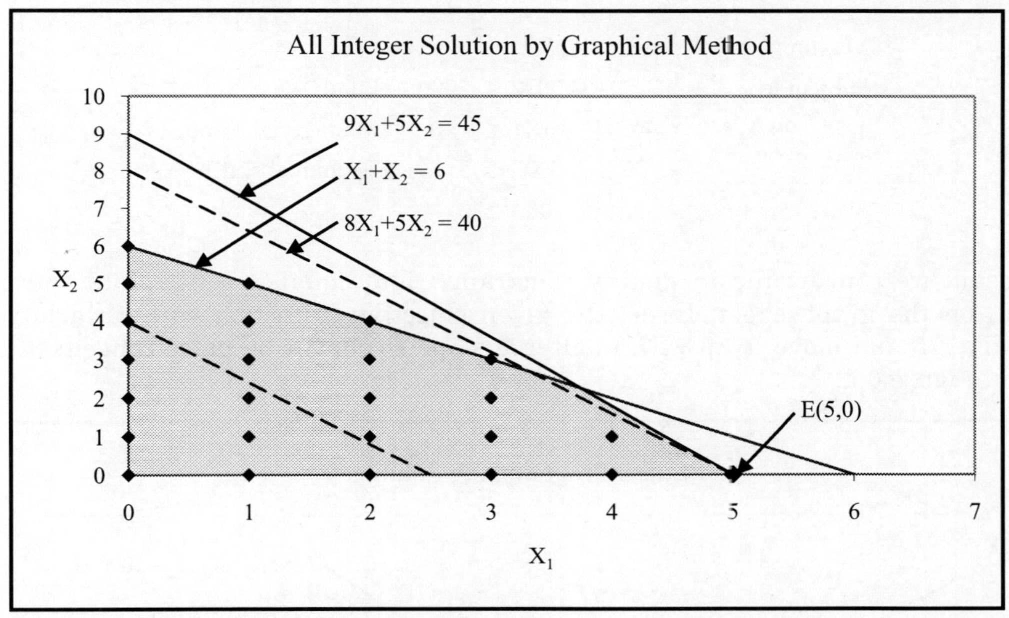

Figure 6.3

Now from Figure 6.3 we obtain Z = 40 and the **associated integer solution values** are $X_1 = 5$ and $X_2 = 0$.

■ Example 2

Solve the following IP problem as LP problem for an unrestricted solution by graphical method. Also obtain the all integer solution by complete enumeration and graphical methods.

$$\begin{aligned}
\text{Maximize: } & R = 10X_1 + 25X_2 \\
\text{Subject to: } & 282X_1 + 400X_2 \leq 2000 \quad \text{(Funds)} \\
& 4X_1 + 40X_2 \leq 140 \quad \text{(Managerial Time)} \\
& X_1 + 0X_2 \leq 5 \quad \text{(Technical skill)} \\
& X_1 \; ; \; X_2 \geq 0 \quad \text{and integer}
\end{aligned}$$

a) Unrestricted (LP) solution by graphical method

First remove the integer restriction in the given problem and restate it as a simple linear programming problem first as shown below.

$$\begin{aligned}
\text{Maximize: } & R = 10X_1 + 25X_2 \\
\text{Subject to: } & 282X_1 + 400X_2 \leq 2000 \quad \text{(Funds)} \\
& 4X_1 + 40X_2 \leq 140 \quad \text{(Managerial Time)} \\
& X_1 + 0X_2 \leq 5 \quad \text{(Technical skill)} \\
& X_1 \; ; \; X_2 \geq 0
\end{aligned}$$

As usual we convert the inequality constraints into equality constraints and draw them all on the graph sheet. Insert the given objectivity function with an arbitrarily chosen R = 75 and move it up with unaltered slope so that it becomes tangent at E as shown in Figure 6.4.

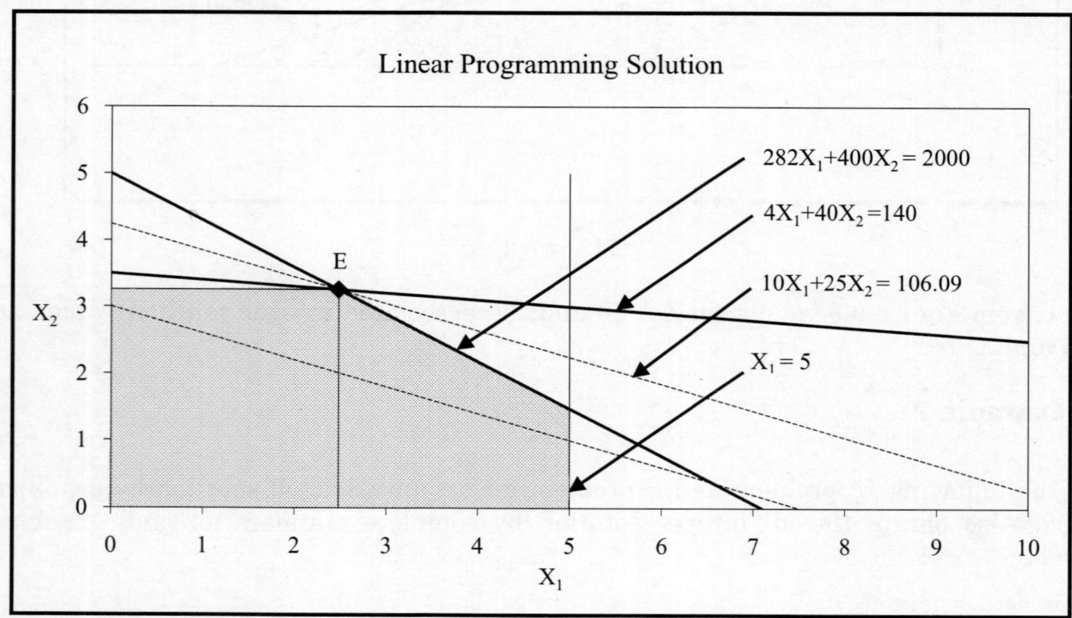

Figure 6.4

Now from the graph the solution values are obtained as: $X_1 = 2.479$, $X_2 = 3.252$ and the value of the objective function is

$$Z = 10 \times 2.479 + 25 \times 3.252 = \text{Rs. } 106.09$$

b) All integer solution by complete enumeration method

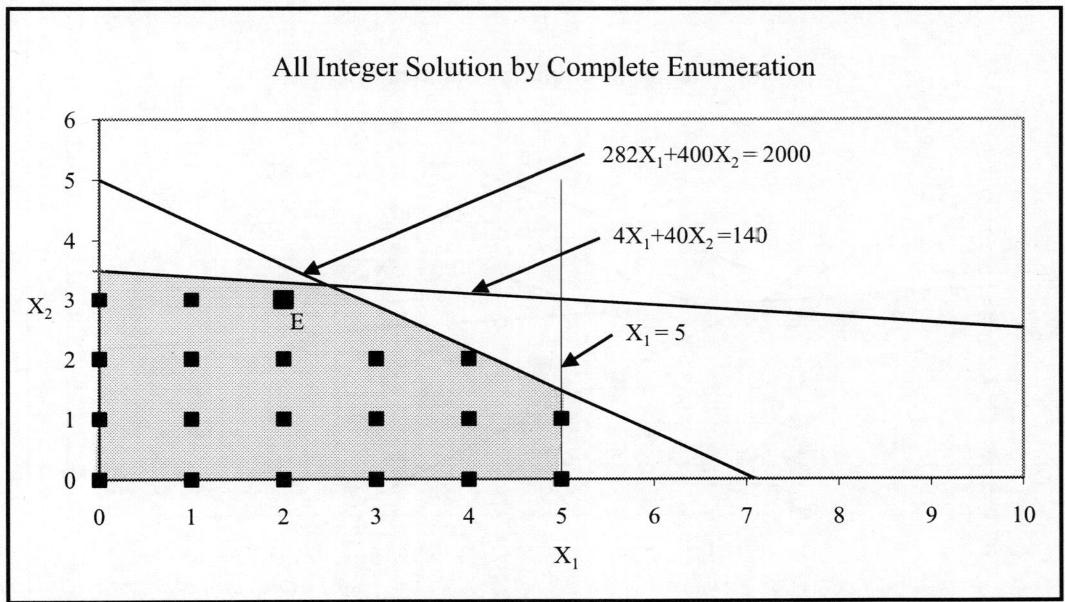

Figure 6.5

Table 6.2

X_1	X_2	Profit	X_1	X_2	Profit
0	0	0	4	1	65
1	0	10	5	1	75
2	0	20	0	2	50
3	0	30	1	2	60
4	0	40	2	2	70
5	0	50	3	2	80
0	1	25	4	2	90
1	1	35	0	3	75
2	1	45	1	3	85
3	1	55	2	3	95

Now from the table the optimum value is: $X_1 = 2$; $X_2 = 3$; $Z = 10\times2+25\times3 = \text{Rs.}95$. Note that this profit is substantially less than our simple linear programming value of Rs.109.09.

c) All integer solution by graphical method

Here within the feasible region all the integer points are marked as shown in the graph. Once this is done we simply move the inserted objective function with unaltered slope up or down as usual till a final outermost integer point is encountered on it as shown in Figure 6.6.

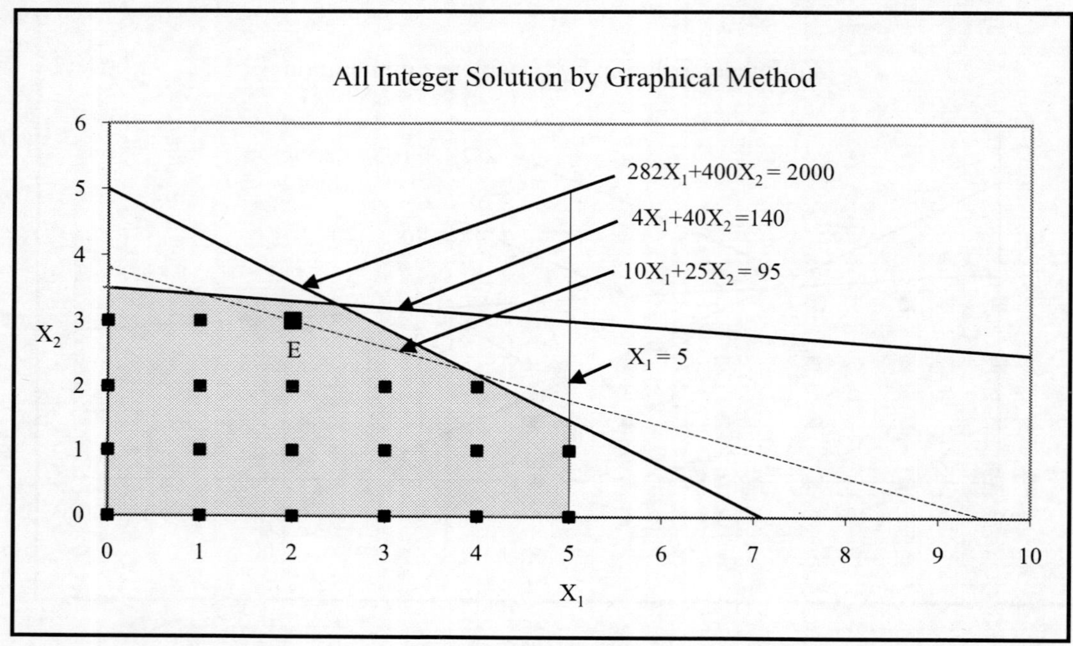

Figure 6.6

Now from the graph the optimum value is: $X_1=2$; $X_2=3$; $Z=10\times2+25\times3=$Rs.95. Note that this profit is substantially less than our simple linear programming value of Rs.109.09.

■ Example 3

For the IP problem given below obtain all integer solution by graphical method.

$$\text{Maximize: } Z = 5X_1 + 3X_2$$

$$\begin{aligned} \text{Subject to:} \quad 4X_1 + 2X_2 &\leq 25 \\ X_1 &\leq 5 \\ X_2 &\leq 8 \\ X_1 ;\ X_2 &\geq 0 \quad X_1 \text{ and } X_2 \text{ integers} \end{aligned}$$

Solution

In Figure 6.7 as usual we introduce all the relevant constraints and mark the feasible region. Once this is done we mark all the integer points within the shaded feasible region. After marking the integer points we introduce the objective function for an arbitrary value of Z (say Z = 30) into the graph as usual. Now we shift this objective function with its slope intact up till it touches the outermost integer point E as shown in the figure. Accordingly, read the solution values as $X_1 = 2$ and $X_2 = 8$ and obtain the maximum value of our objective function by mere substitution as Z = 34.

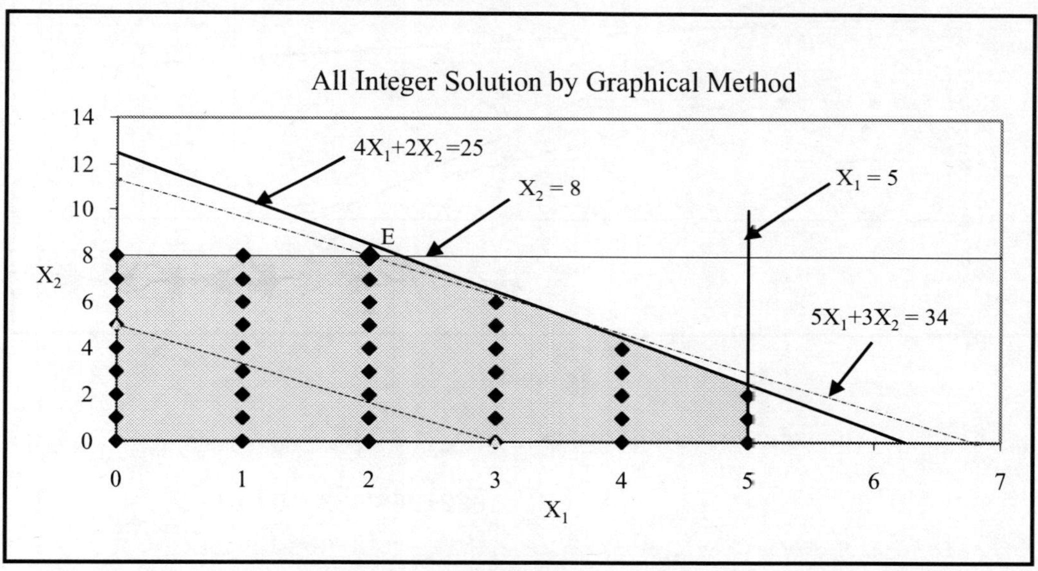

Figure 6.7

6.3 MIXED INTEGER SOLUTION BY GRAPHICAL METHOD

In the previous section we considered the problem with all integer solution only. Sometimes we insist integer solution only for few decision variables. The other decision variables may be allowed to take even the decimal values. The following example illustrates one such case.

■ Example 4

Solve the following mixed integer problem for X_1 integer solution.

Maximize: $R = 10X_1 + 25X_2$

Subject to:

$282X_1 + 400X_2 \leq 2000$ (Funds)

$4X_1 + 40X_2 \leq 140$ (Managerial Time)

$X_1 + 0X_2 \leq 5$ (Technical skill)

$X_1 \; ; \; X_2 \geq 0$ X_1 integer

Solution

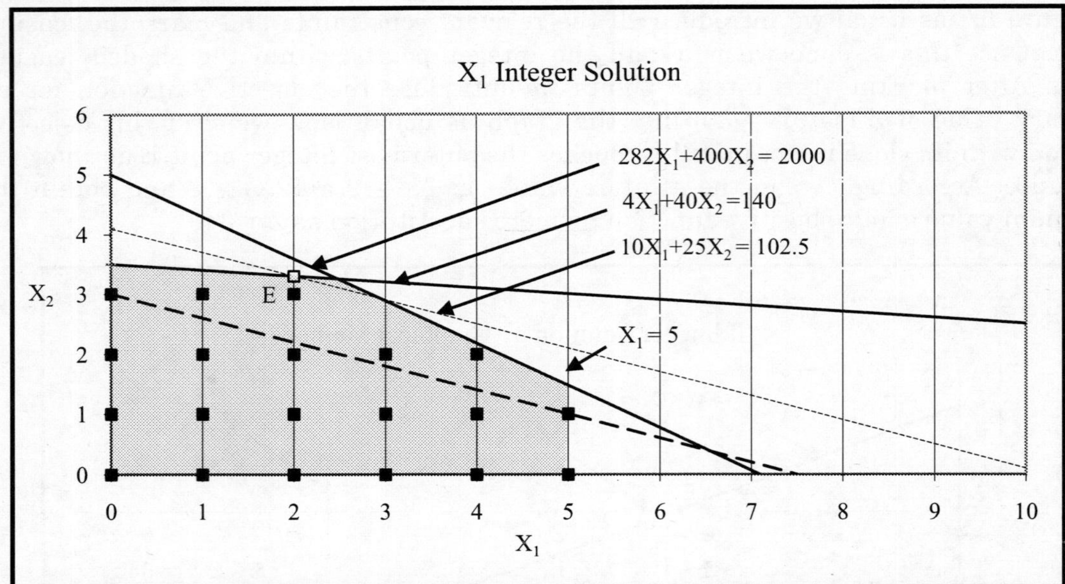

Figure 6.8

From the graph the solution values are: $X_1 = 2$; $X_2 = 3.285$,

$$R = 10\times2+25\times3.3 = 102.125$$

Or

From the graph the solution values are: $X_1 = 3$, $X_2 = 2.655$,

$$R = 10 \times 3 + 25 \times 2.885 = 102.125$$

■ Example 5

Solve the following mixed integer problem for X_2 integer solution

$$\begin{aligned}
\text{Maximize: } & R = 10X_1 + 25X_2 & \\
\text{Subject to: } & 282X_1 + 400X_2 \leq 2000 & \text{(Funds)} \\
& 4X_1 + 40X_2 \leq 140 & \text{(Managerial Time)} \\
& X_1 + 0X_2 \leq 5 & \text{(Technical skill)} \\
& X_1 ; \quad X_2 \geq 0 & X_2 \text{ integer}
\end{aligned}$$

Since we want to have X_1 integer value we insert vertical grid line as shown in Figure 6.8. Now to find the needed integer solution for X_1 and non-integer solution for X_1 we shift the arbitrarily drawn object function up till the needed point E is reached.

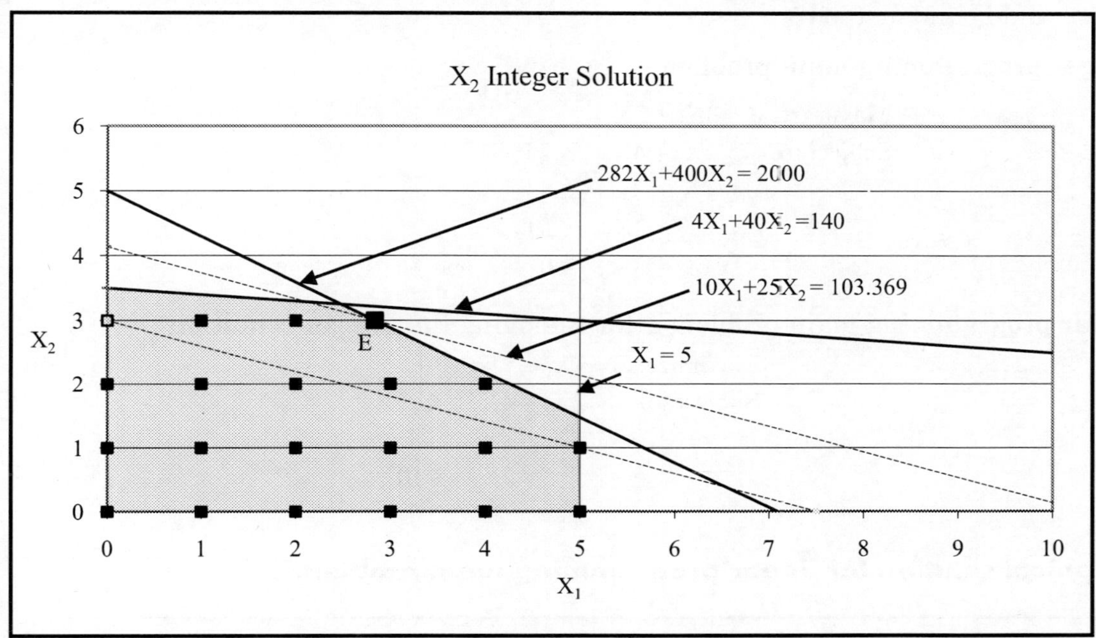

Figure 6.9

Since X_2 is an integer, in the graph we introduce the horizontal grid lines along the integer points as shown in the figure. Now to find the needed integer solution for X_2 and the decimal solution for X_1 we shift the arbitrarily (R = 75) drawn objective function, with its slope intact, up till it passes through the point E on $X_2 = 3$ grid line. Remember that this is the highest achievable point under the given situation. From the graph the associated solution values are: $X_1 = 2.8369$; $X_2 = 3$; $Z = 10\times2.8369+25\times2 = 103.369$.

6.4 BRANCH-AND-BOUND METHOD FOR COMPLETE INTEGER SOLUTION BY GRAPHICAL METHOD

The branch-and-bound method was first developed by A.H. Land and A.G. Doig to solve all integers and mixed integer type of problems. The innovative procedure involves splitting of the given problem into branches. Such splitting continues till an optimum integer solution is obtained. Before we proceed for a branch-and-bound approach it is important to know the following elementary but important observation.

If we solve the LP problem as a relaxation of a pure IP problem and obtain a solution in which all the decision variables are integers, then optimum solution of the so obtained LP problem is also the optimum solution to the corresponding IP problem.

Example 6

For the linear programming problem given in example 1 find all integer solution by branch-and-bound method.

Integer programming main problem

$$\begin{aligned} &\text{Maximize: } Z = 3X_1 + 5X_2 \\ &\text{Subject to: } \quad 2X_1 + 4X_2 \le 25 \\ &\qquad X_1 \le 8 \\ &\qquad 2X_2 \le 10 \\ &\qquad X_1 \ ; \ X_2 \ge 0 \quad X_1 \text{ and } X_2 \text{ are integers} \end{aligned}$$

Linear programming main problem after removing the integer condition

$$\begin{aligned} &\text{Maximize: } Z = 3X_1 + 5X_2 \\ &\text{Subject to: } \quad 2X_1 + 4X_2 \le 25 \\ &\qquad X_1 \le 8 \\ &\qquad 2X_2 \le 10 \\ &\qquad X_1 \ ; \ X_2 \ge 0 \end{aligned}$$

Graphical solution for linear programming main problem

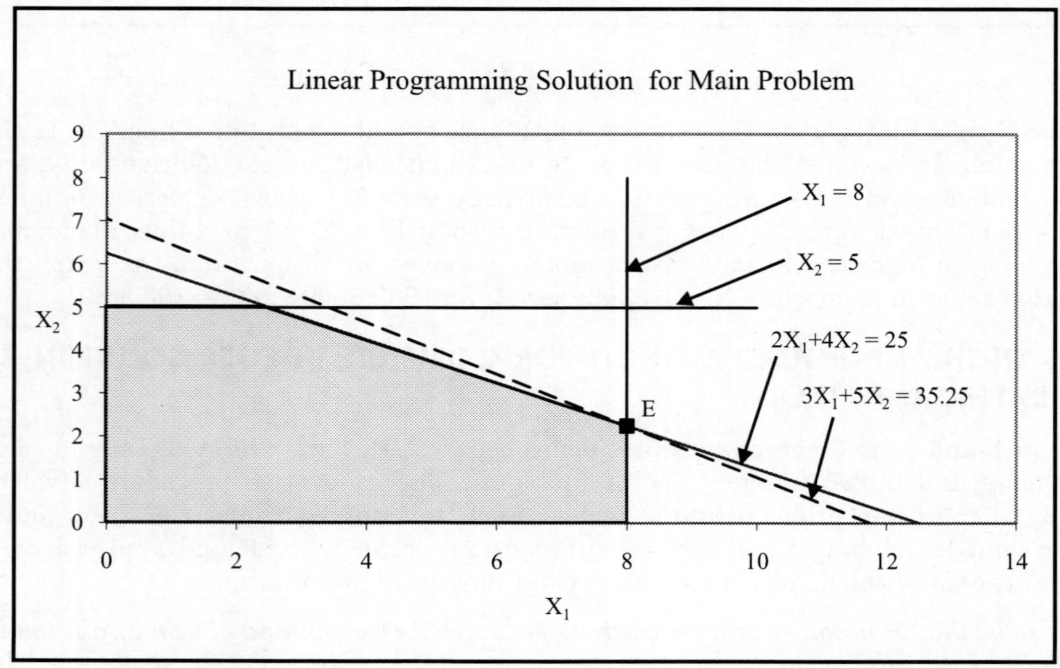

Figure 6.10

From Figure 6.10 the integer relaxed solution is obtained as: $X_1 = 8$, $X_2 = 2.25$ and $Z = 35.25$ at point E. Clearly 35.25 could be the upper bound for our integer restricted objective. At this stage the lower bound could be the $Z_1 = 34$ is obtained by merely substituting the rounded off decision variables namely, $X_1 = 8$ and $X_2 = 2$ into the objective function.

In this unrestricted solution since $X_2 = 2.25$ is the only non-integer decision variable value, we branch out two sum problem from X_2 by incorporating two additional constraints one each so that in both the problems we deliberately avoid the non-integer, namely, $X_2 = 2.25$. The needed two constraints are: $X_2 \leq 2$ and $X_2 \geq 3$. The two sub-problems are:

LP Sub-problem 1: Main Problem added with $X_2 \leq 2$

$$\begin{aligned} \text{Maximize: } & Z = 3X_1 + 5X_2 \\ \text{Subject to: } & 2X_1 + 4X_2 \leq 25 \\ & X_1 \leq 8 \\ & 2X_2 \leq 10 \text{ (redundant)} \\ & X_2 \leq 2 \\ & X_1 \ ; \ X_2 \geq 0 \end{aligned}$$

Graphical solution for linear programming sub-problem 1

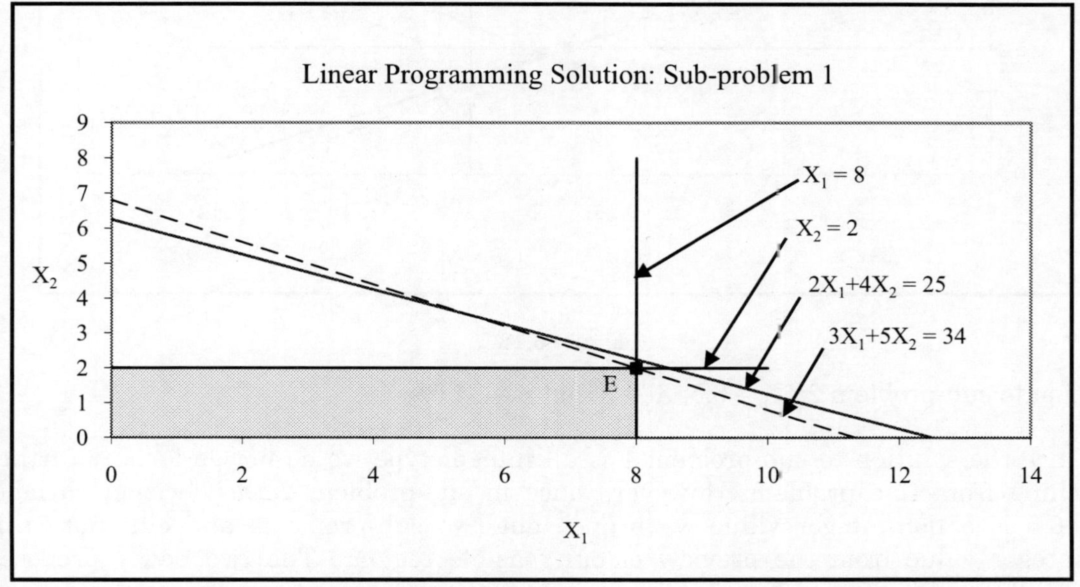

Figure 6.11

Solution to sub-problem 1: $X_1 = 8$, $X_2 = 2$, $Z = 34$

LP Sub-problem 2: Main Problem added with $X_2 \geq 3$

$$\text{Maximize: } Z = 3X_1 + 5X_2$$
$$\text{Subject to: } \begin{aligned} 2X_1 + 4X_2 &\leq 25 \\ X_1 &\leq 8 \\ 2X_2 &\leq 10 \\ X_2 &\geq 3 \\ X_1 ;\ X_2 &\geq 0 \end{aligned}$$

Graphical solution for linear programming sub-problem 2

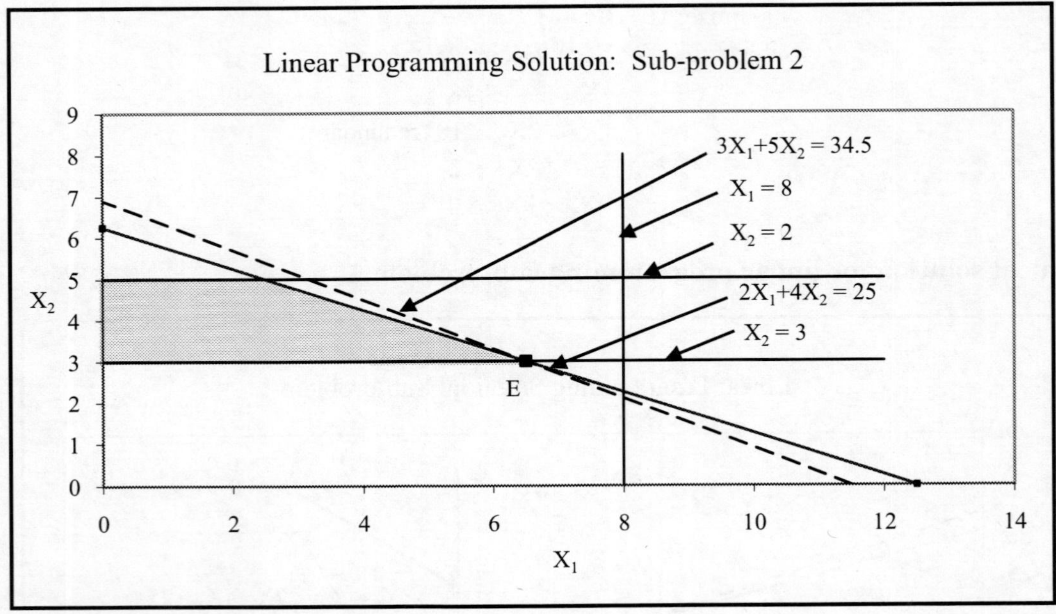

Figure 6.12

Solution to sub-problem 2: $X_1 = 6.5$; $X_2 = 3$ and $Z = 34.5$.

Since the solution to sub-problem 1 is all integer type we do not go for any further branching from this problem. However, since in sub-problem 2 the decision variable $X_1 = 6.5$ is a non-integer value we branch out two sub-problems and eliminate this non-integer value from the preview of our feasible region. The two newly proposed constraints are: $X_1 \leq 6$ and $X_1 \geq 7$.

Sub-problem 3: Sub-problem 2 added with $X_1 \leq 6$

$$\begin{aligned} \text{Maximize: } & Z = 3X_1 + 5X_2 \\ \text{Subject to: } & 2X_1 + 4X_2 \leq 25 \\ & X_1 \leq 8 \quad \text{(redudant)} \\ & 2X_2 \leq 10 \\ & X_2 \geq 3 \\ & X_1 \leq 6 \\ & X_1 \ ; \ X_2 \geq 0 \end{aligned}$$

Graphical solution for linear programming sub-problem 3

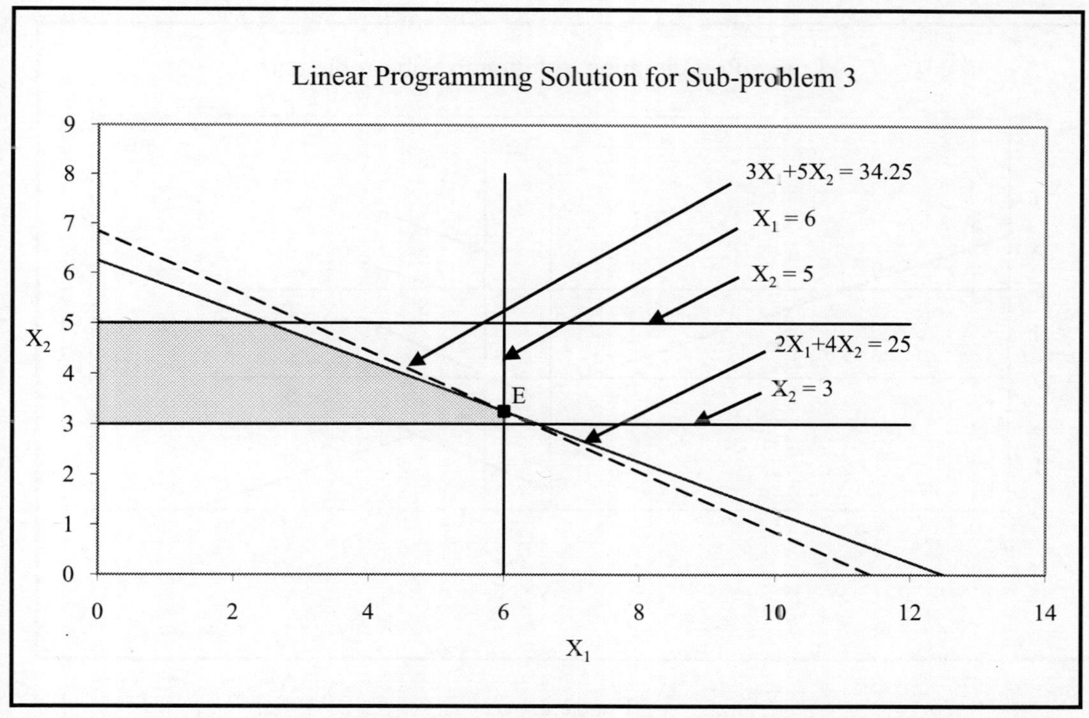

Figure 6.13

Solution to sub-problem 3: $X_1 = 6$; $X_2 = 3.25$ and $Z = 34.25$

Sub-problem 4: Sub-problem 2 added with $X_1 \geq 7$

$$\begin{aligned} \text{Maximize: } & Z = 3X_1 + 5X_2 \\ \text{Subject to: } & 2X_1 + 4X_2 \leq 25 \\ & X_1 \leq 8 \\ & 2X_2 \leq 10 \\ & X_2 \geq 3 \\ & X_1 \geq 7 \\ & X_1 \ ; \ X_2 \geq 0 \end{aligned}$$

Graphical solution for linear programming sub-problem 4

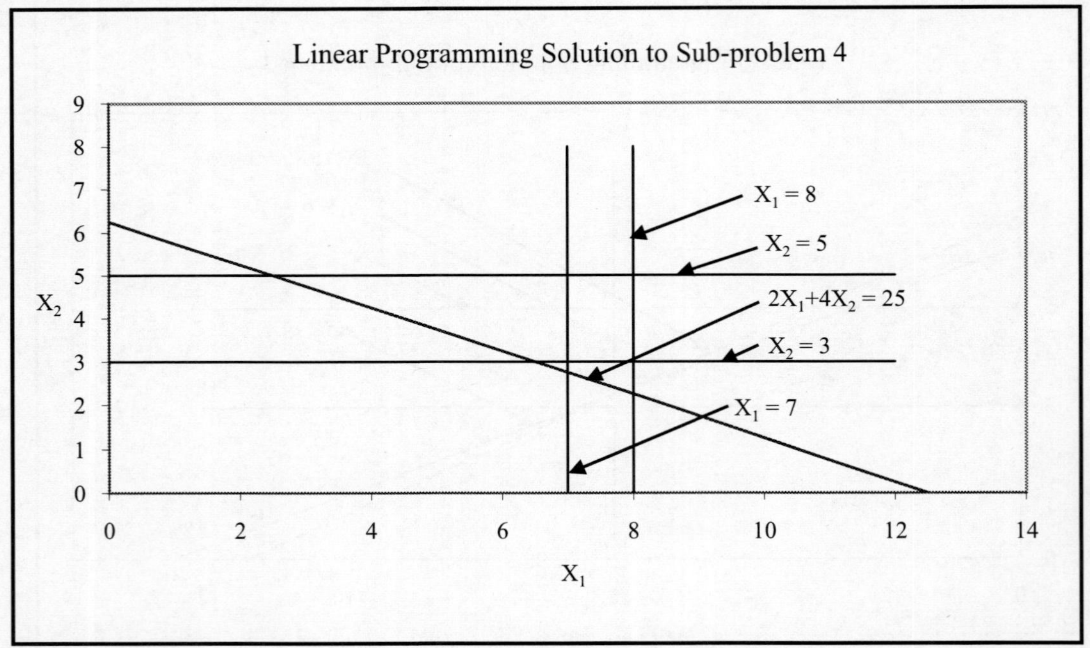

Figure 6.14

Solution to sub-problem 4: Feasible solution does not exist.

Since there is no feasible solution to sub-problem 4 we go back to sub-problem 3 and branch out from the non-integer value $X_2 = 2.5$.

Sub-problem 5: Sub-problem 3 added with $X_1 \le 3$

$$
\begin{aligned}
\text{Maximize: } & Z = 3X_1 + 5X_2 \\
\text{Subject to: } & 2X_1 + 4X_2 \le 25 \\
& X_1 \le 8 \quad \text{(redundant)} \\
& 2X_2 \le 10 \quad \text{(redundant)} \\
& X_2 \ge 3 \\
& X_1 \le 6 \\
& X_2 \le 3 \\
& X_1 \; ; \; X_2 \ge 0
\end{aligned}
$$

Graphical solution for linear programming sub-problem 5

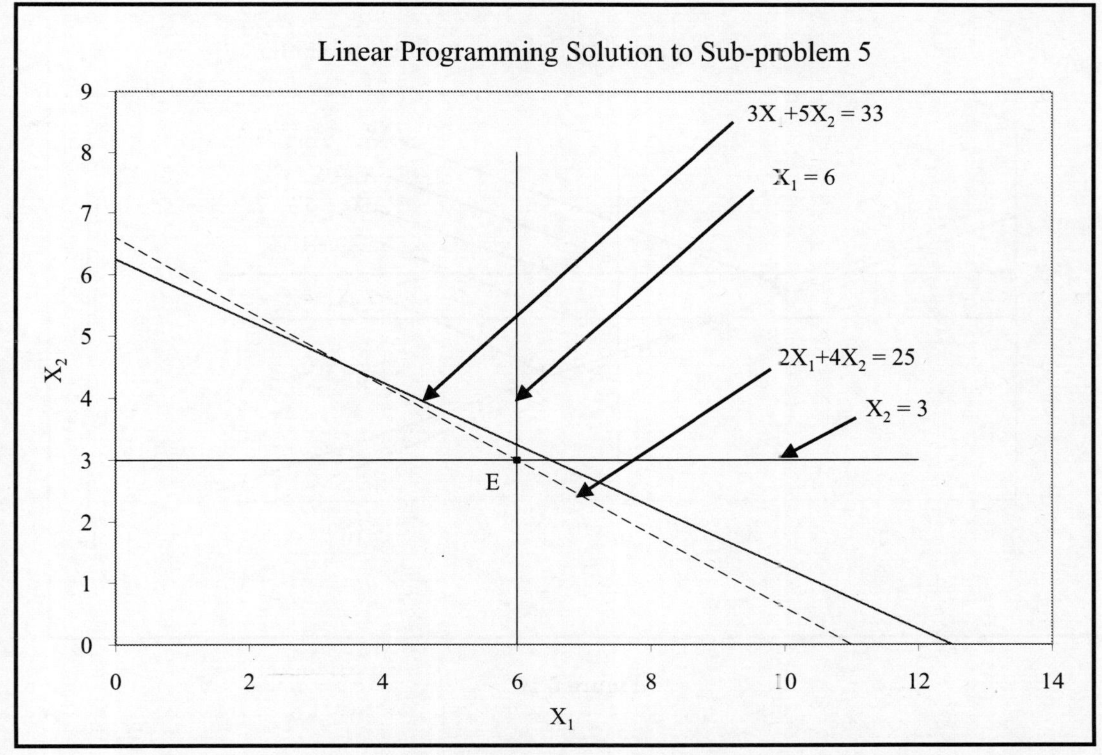

Figure 6.15

Solution to sub-problem 5: $X_1 = 6$; $X_2 = 3$ and $Z = 33$

Sub-problem 6: Sub-problem 3 added with $X_2 \geq 4$

$$\begin{aligned}
\text{Maximize: } & Z = 3X_1 + 5X_2 \\
\text{Subject to: } & 2X_1 + 4X_2 \leq 25 \\
& X_1 \leq 8 \quad \text{(redundant)} \\
& 2X_2 \leq 10 \\
& X_2 \geq 3 \quad \text{(redundant)} \\
& X_1 \leq 6 \\
& X_2 \geq 4 \\
& X_1 \; ; \; X_2 \geq 0
\end{aligned}$$

Graphical solution for linear programming sub-problem 6

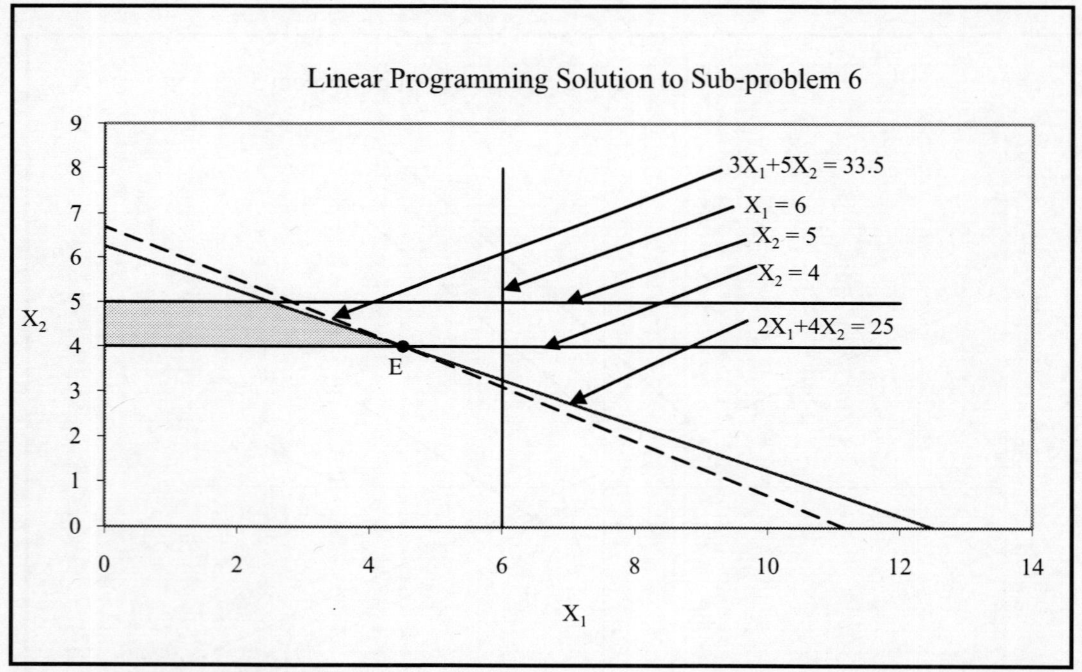

Figure 6.16

Solution to sub-problem 6: X_1 = 4.5; X_2= 4 and Z = 33.5

The following chart illustrates the step-wise branching out with their respective solutions.

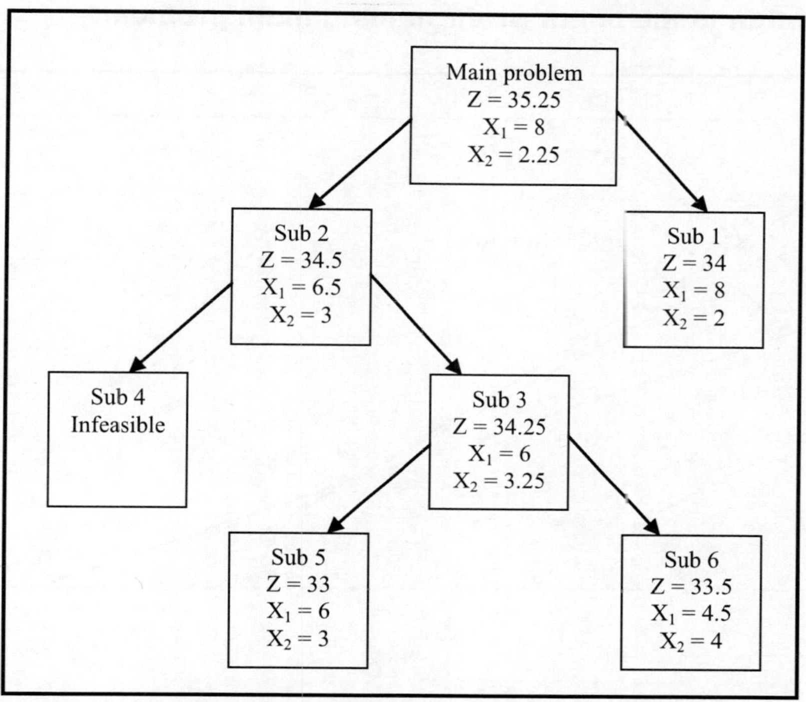

The final all integer solution values given by sub-problem 1 are $X_1 = 8$, $X_2 = 2$, $Z = 34$.

■ Example 7

For the linear programming problem given below find all integer solution by branch-and-bound method.

Original IP problem

$$\text{Maximize: } Z = 8X_1 + 5X_2$$
$$\text{Subject to: } X_1 + X_2 \leq 6$$
$$9X_1 + 5X_2 \leq 45$$
$$X_1 \; ; \; X_2 \geq 0 \quad X_1 \text{ and } X_2 \text{ integers}$$

Converted LP main problem

Convert the given IP problem as simple LP problem as shown below by removing the integer restrictions and obtain its solution by graphical method in the usual manner.

$$\text{Maximize: } Z = 8X_1 + 5X_2$$
$$\text{Subject to: } X_1 + X_2 \leq 6$$
$$9X_1 + 5X_2 \leq 45$$
$$X_1 \; ; \; X_2 \geq 0$$

Graphical solution to the linear programming main problem

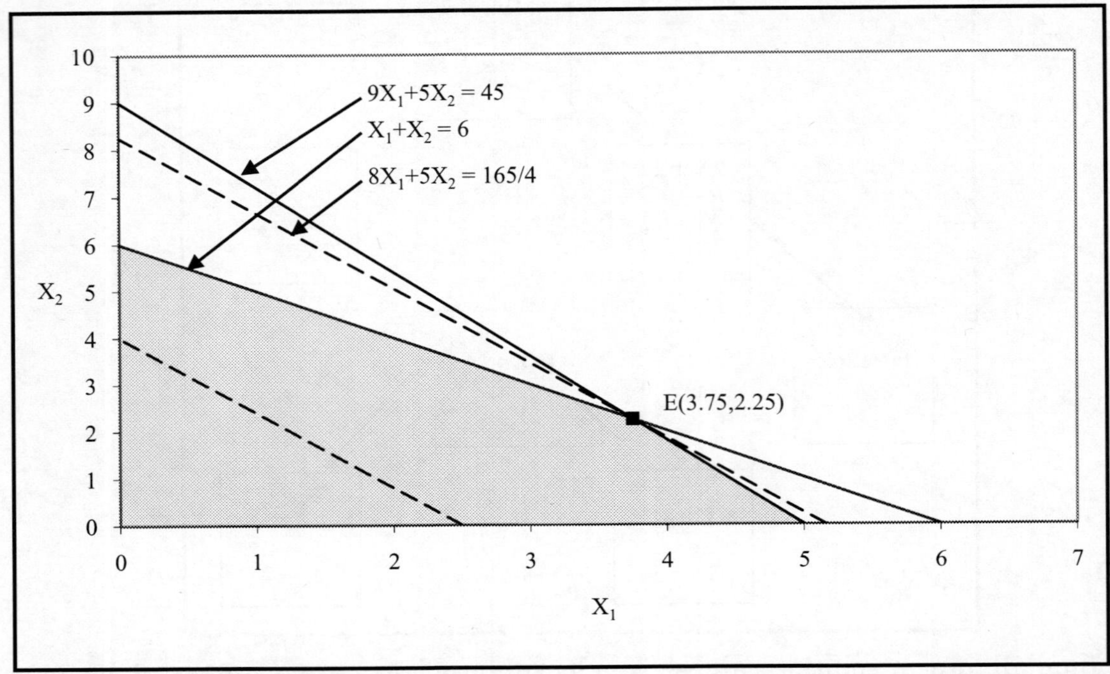

Figure 6.17

Solution values to main problem

$$X_1 = 3.75;\ X_2 = 2.25;\ Z = 165/4 = 41.25.$$

The branch-and-bound method begins by solving the simple linear programming for unrestricted solution. In such a solution if all the decision variables are integers then confirm that solution as the solution for the integer restricted solution. Unfortunately, the optimum solution from Figure 6.17 to the simple linear programming problem is $X = 15/4$, $X_2 = 9/4$ and $Z = 165/4$. Such observations imply that the optimum value for our integer solution cannot exceed this $Z = 165/4$. This value thus sets the upper limit for our proposed integer set problem.

Our next step is to partition the feasible region of the simple linear programming as an attempt to say more about the needed integer solution. Since both the decision variables are non-integers in our LP solution we arbitrarily choose the non-integer solution $X_1 = 15/4 = 3.75$ for branching. Now observe that every point in the feasible region for the IP must have either $X_1 \leq 3$ or $X_1 \geq 4$. Accordingly, by adding one constraint each to our main problem we frame two sub-problems as follow.

Sub-problem 1:

Main LP problem added with $X_1 \leq 3$

$$\begin{aligned} \text{Maximize: } & Z = 8X_1 + 5X_2 \\ \text{Subject to: } & X_1 + X_2 \leq 6 \\ & 9X_1 + 5X_2 \leq 45 \\ & X_1 \leq 3 \\ & X_1 ; X_2 \geq 0 \end{aligned}$$

Graphical solution to sub-problem 1

As usual we add this new constraint to the graph already drawn as shown in Figure 6.18. Clearly the feasible region of our problem is reduced the left side of this new constraint shown in the figure. The highest value of the objective functicn is achieved at the corner point E.

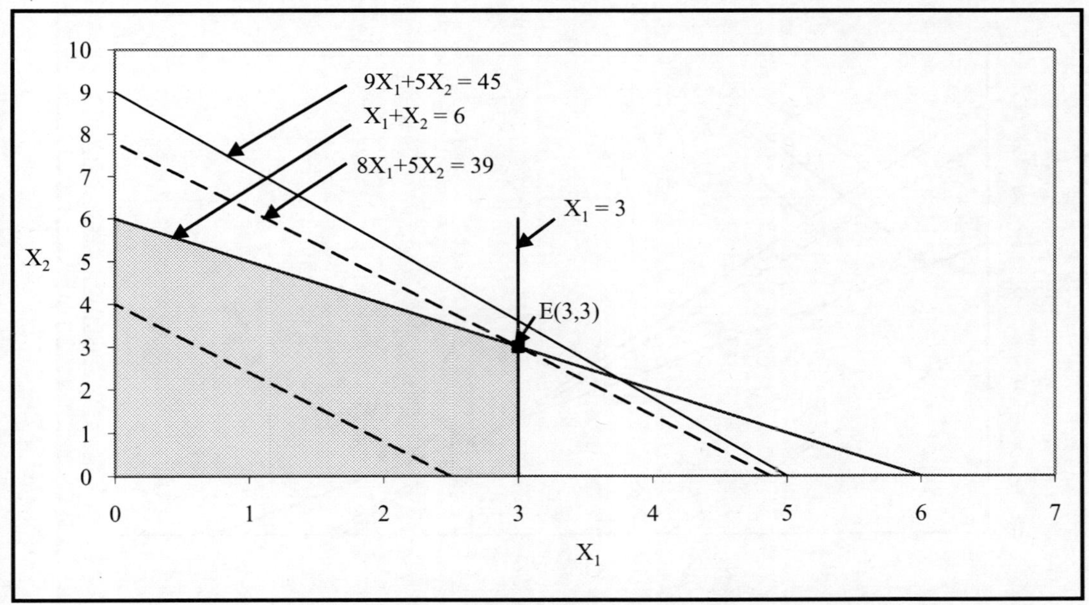

Figure 6.18

The solution values to sub-problem 1

$$X_1 = 3,\ X_2 = 3, \text{ and } Z = 39.$$

We are lucky to have an all integer solution at this stage. However before making any decision we will have to solve the other branch as well.

Sub-problem 2

Main problem LP added with $X_1 \geq 4$

$$\begin{aligned} \text{Maximize: } Z &= 8X_1 + 5X_2 \\ \text{Subject to: } \quad X_1 + X_2 &\leq 6 \\ 9X_1 + 5X_2 &\leq 45 \\ X_1 &\geq 4 \\ X_1, \quad X_2 &\geq 0 \end{aligned}$$

Graphical solution to sub-problem 2

This times inclusion of the greater than or equal to constraint limits our feasible region to the triangular region right side of this constraint as shown in Figure 6.19.

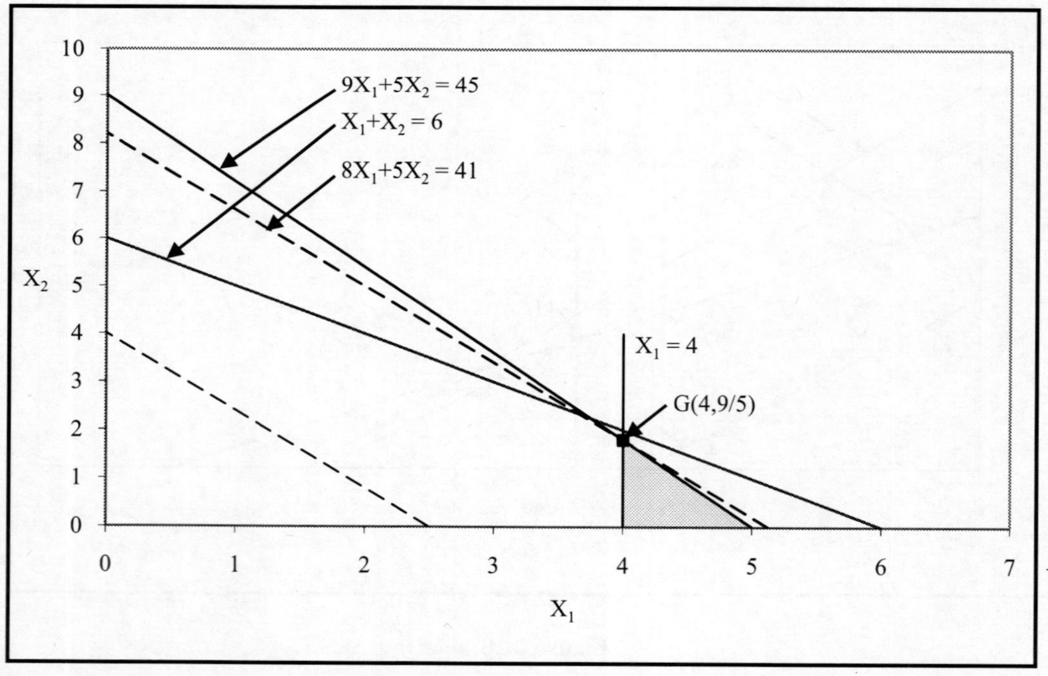

Figure 6.19

The solution values to sub-problem 2 are $X_1 = 4$, $X_2 = 9/5 = 1.8$ and $Z = 41$.

In this solution though the optimum value is more, X_2 is not an integer as wanted. So let us branch out two sub-problems 3 and 4 from sub-problem 2 as shown below.

Sub-problem 3: Sub-problem 2 added with $X_2 \geq 2$

$$\begin{aligned} \text{Maximize: } & Z = 8X_1 + 5X_2 \\ \text{Subject to: } & X_1 + X_2 \leq 6 \\ & 9X_1 + 5X_2 \leq 45 \\ & X_1 \geq 4 \\ & X_2 \geq 2 \\ & X_1 ; X_2 \geq 0 \end{aligned}$$

Graphical solution to sub-problem 3

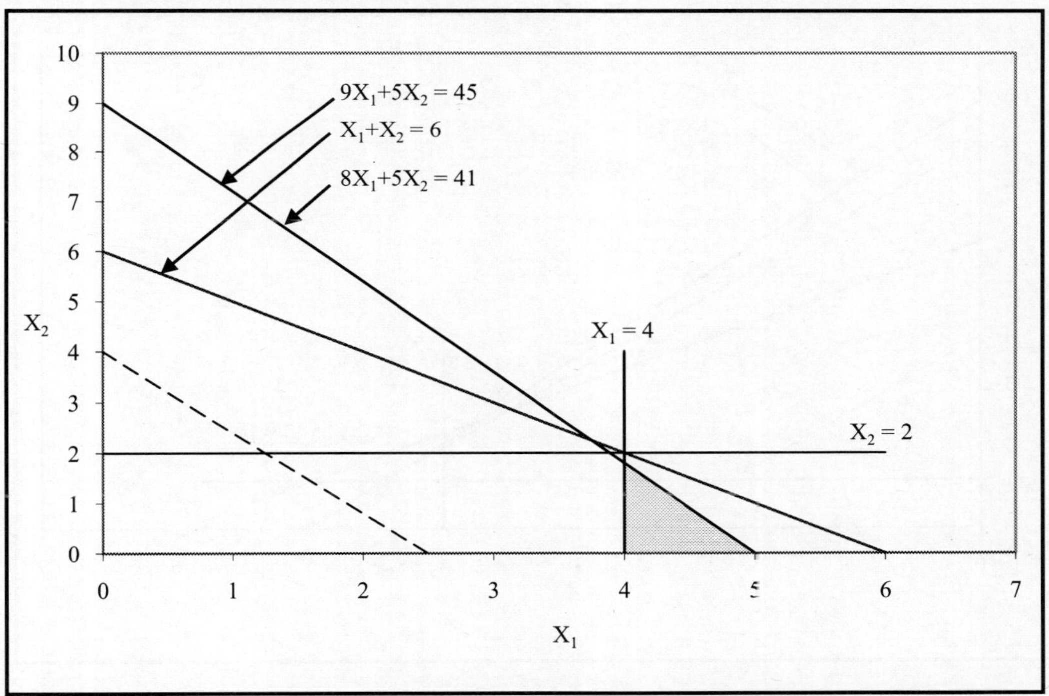

Figure 6.20

The solution values to sub-problem 3 are

No feasible solution

Clearly there is no feasible region in this case as we are ccnfined to the region above $X_2 = 2$ and right side of $X_1 = 4$.

Sub-problem 4: Sub-problem 2 added with $X_2 \leq 1$

$$\begin{aligned}
\text{Maximize: } & Z = 8X_1 + 5X_2 \\
\text{Subject to: } & X_1 + X_2 \leq 6 \\
& 9X_1 + 5X_2 \leq 45 \\
& X_1 \geq 4 \\
& X_2 \leq 1 \\
& X_1 ; X_2 \geq 0
\end{aligned}$$

Graphical solution to sub-problem 4

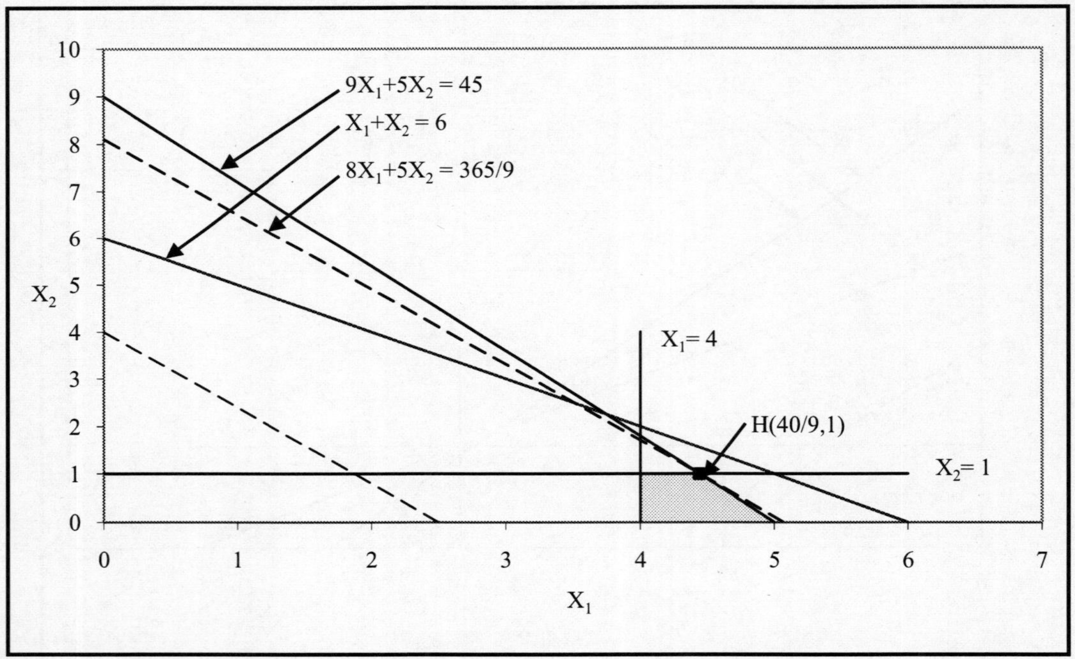

Figure 6.21

The solution values to sub-problem 4 are $X_1 = 40/9 = 4.44$, $X_2 = 1$ and $Z = 365/9$

In this solution since X_1 is a non-integer solution, let us branch out two new problems from X_1 as follows.

Sub-problem 5: Sub-problem 4 added with $X_1 \leq 4$

$$\begin{aligned} \text{Maximize: } & Z = 8X_1 + 5X_2 \\ \text{Subject to: } & X_1 + X_2 \leq 6 \\ & 9X_1 + 5X_2 \leq 45 \\ & X_1 \geq 4 \\ & X_2 \leq 1 \\ & X_1 \leq 4 \\ & X_1 \ ; \ X_2 \geq 0 \end{aligned}$$

Graphical solution to sub-problem 5

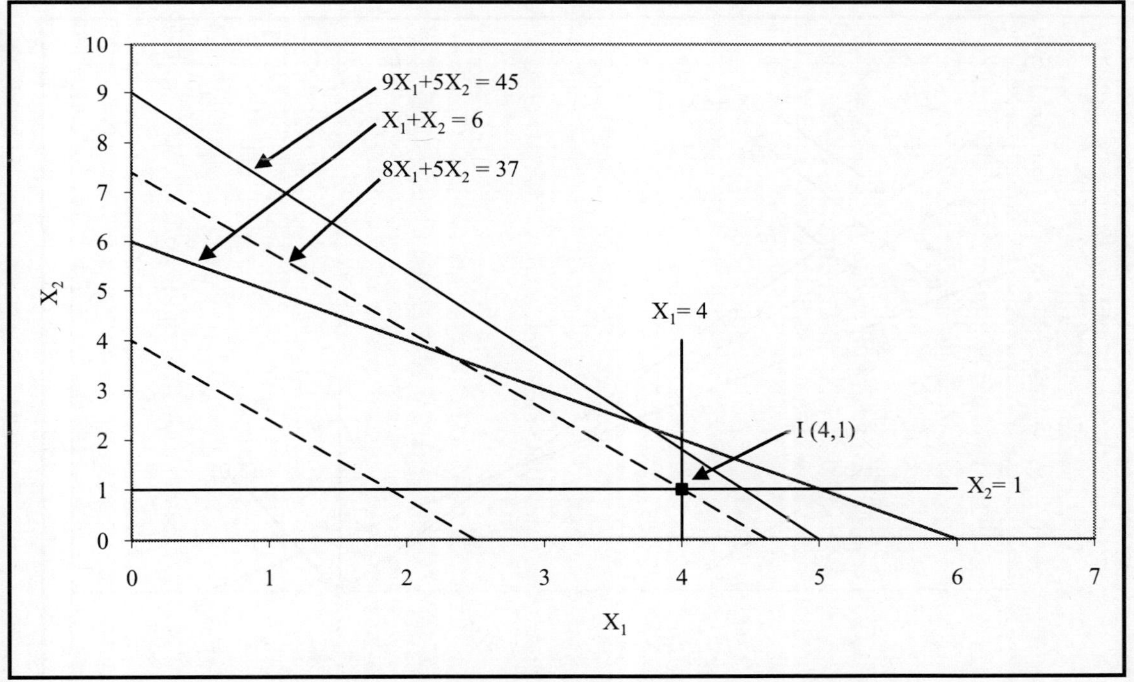

Figure 6.22

Remember that this time $X_1 \leq 4$ and $X_1 \geq 4$ are there in addition to other constraints. So the feasible region is restricted to points on the line $X_1 = 4$ restricted by line $X_1 = 1$.

So from Figure 6.22 the solution values are

$$X_1 = 4, X_2 = 1 \text{ and } Z = 37$$

Sub-problem 6: Sub-problem 4 added with $X_1 \geq 5$

$$\begin{aligned} \text{Maximize: } & Z = 8X_1 + 5X_2 \\ \text{Subject to: } & X_1 + X_2 \leq 6 \\ & 9X_1 + 5X_2 \leq 45 \\ & X_1 \geq 4 \quad \text{(redundant)} \\ & X_2 \leq 1 \\ & X_1 \geq 5 \\ & X_1 ;\ X_2 \geq 0 \end{aligned}$$

Graphical solution to sub-problem 6

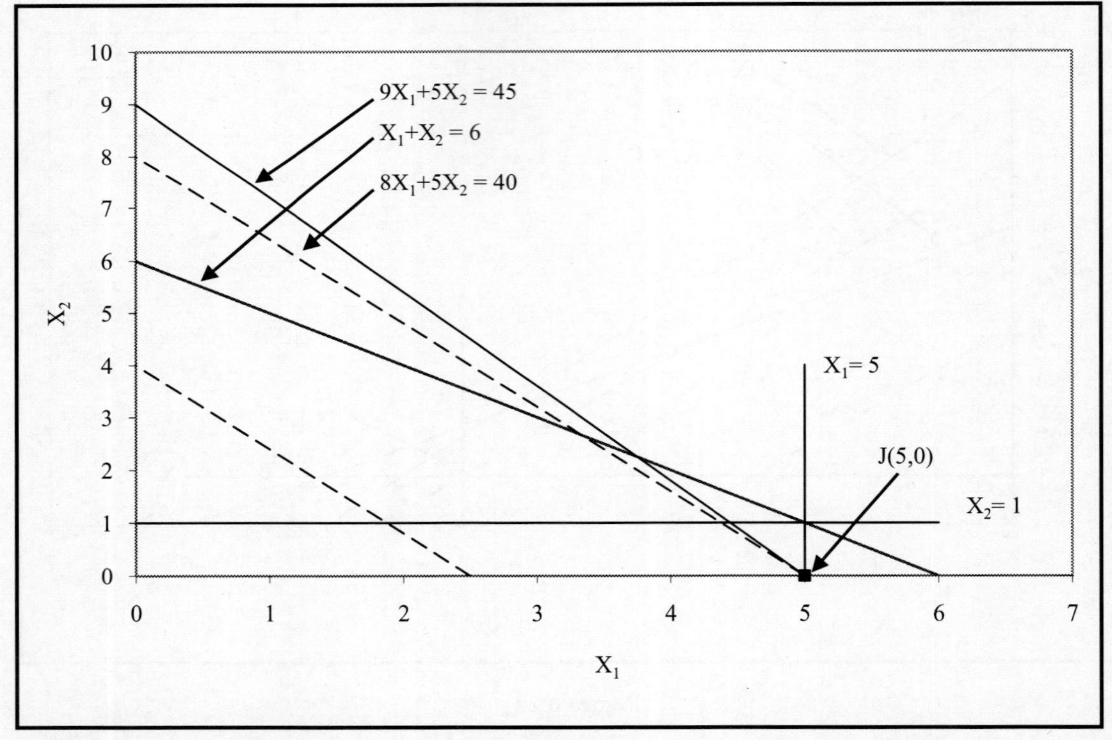

Figure 6.23

So the associated solution values are: $X_1 = 5$, $X_2 = 0$ and $Z = 40$.

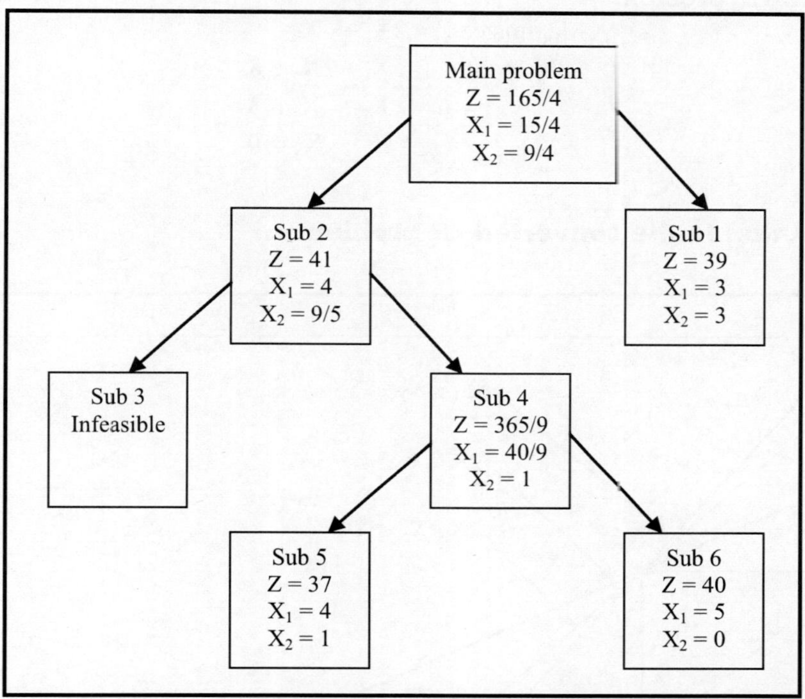

Figure 6.24

This time clearly the only point (5,0) constitute the feasible region as shown in the Figure 6.23. The solution to the last LP problem is the needed integer solution.

The Figure 6.24 illustrates all the branching out steps involved in this problem.

6.5 BRANCH-AND-BOUND METHOD FOR MIXED INTEGER SOLUTION BY GRAPHICAL METHOD

In mixed IP, some values are required to be integers and other variables are allowed to be either integer or non-integer. For solving such problems we branch out only on variable required to be integers. The following example illustrates one such case.

■ Example 8

Solve the following mixed IP.

$$\begin{aligned} \text{Maximize: } & Z = 2X_1 + X_2 \\ \text{Subject to: } & 5X_1 + 2X_2 \leq 8 \\ & X_1 + X_2 \leq 3 \\ & X_1 ;\ X_2 \geq 0 \quad X_1 \text{ integer} \end{aligned}$$

Converted main LP problem

$$\text{Maximize: } Z = 2X_1 + X_2$$
$$\text{Subject to: } 5X_1 + 2X_2 \le 8$$
$$X_1 + X_2 \le 3$$
$$X_1 \; ; \; X_2 \ge 0$$

Graphical solution to the converted LP problem

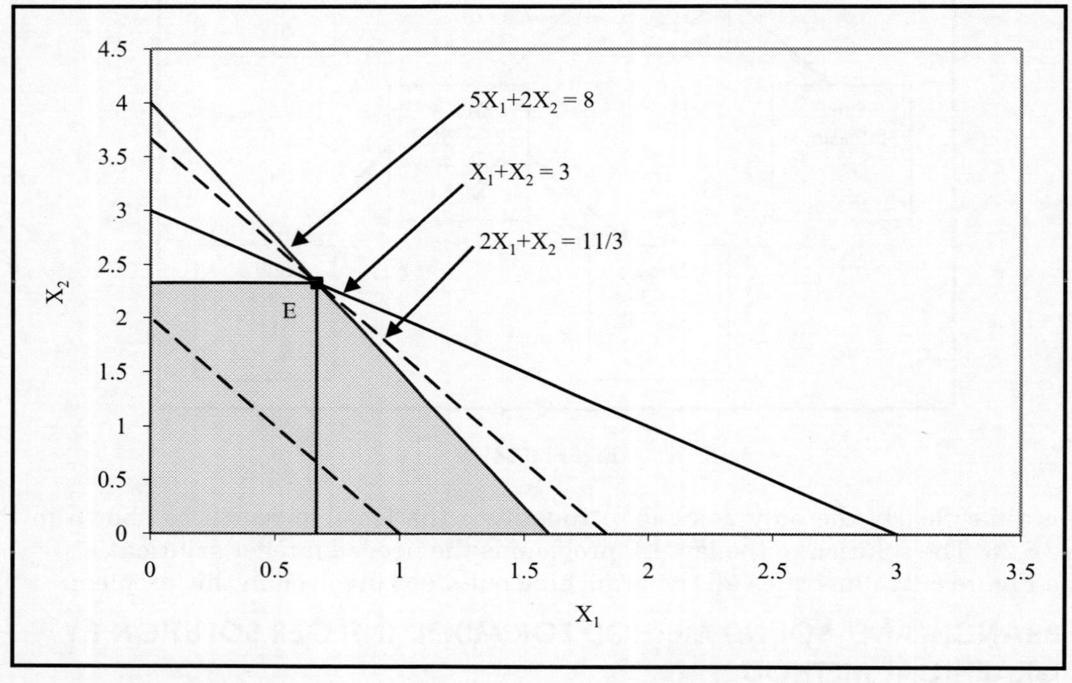

Figure 6.25

The solution values for the main problem are: $X_1 = 2/3$, $X_2 = 7/3$, $Z = 11/3$

Since integer value is insisted only on X_1 this time we propose to branch on X_1 only.

Sub-problem 1: Original LP added with $X_1 \le 0$

$$\text{Maximize: } Z = 2X_1 + X_2$$
$$\text{Subject to: } 5X_1 + 2X_2 \le 8$$
$$X_1 + X_2 \le 3$$
$$X_1 \le 0$$
$$X_1 \; ; \; X_2 \ge 0$$

Solution to the sub-problem 1 by graphical method

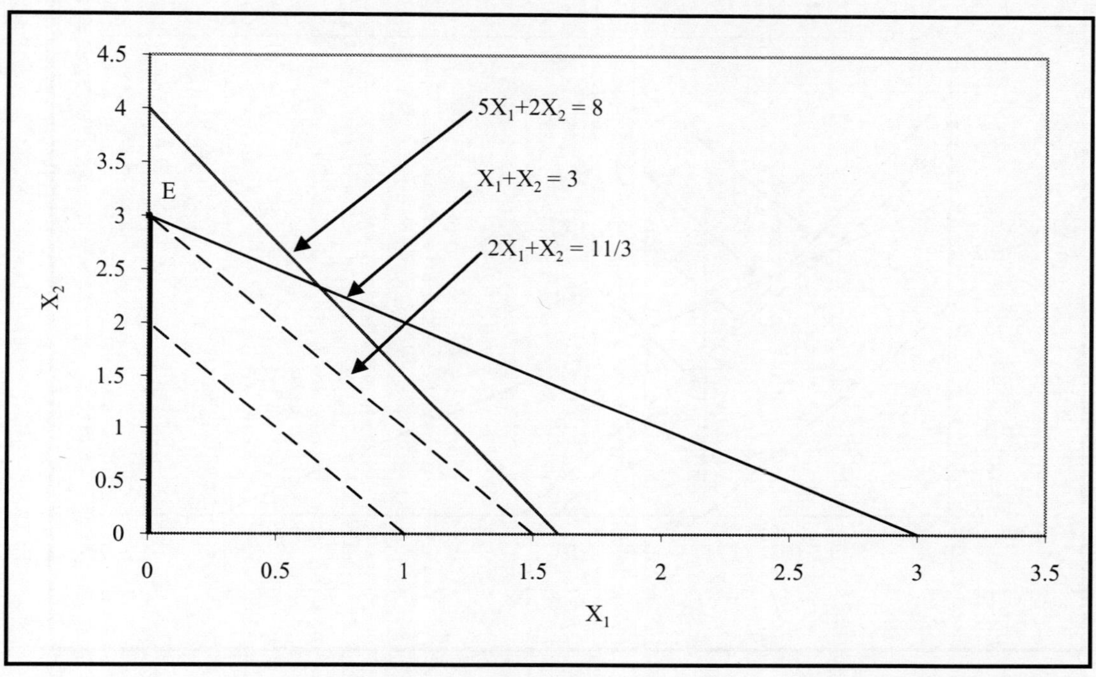

Figure 6.26

In this case since in addition $X_1 \leq 0$ operates all the points on Y-axis less than 3 constitute the feasible region. The associated solution set is obtained as: $X_1 = 0$, $X_2 = 3$ and $Z = 3$.

Sub-problem 2: Original LP added with $X_1 \geq 1$

$$\begin{aligned} \text{Maximize: } & Z = 2X_1 + X_2 \\ \text{Subject to: } & 5X_1 + 2X_2 \leq 8 \\ & X_1 + X_2 \leq 3 \\ & X_1 \geq 1 \\ & X_1\ ;\ X_2 \geq 0 \end{aligned}$$

Solution to the sub-problem 2 by graphical method

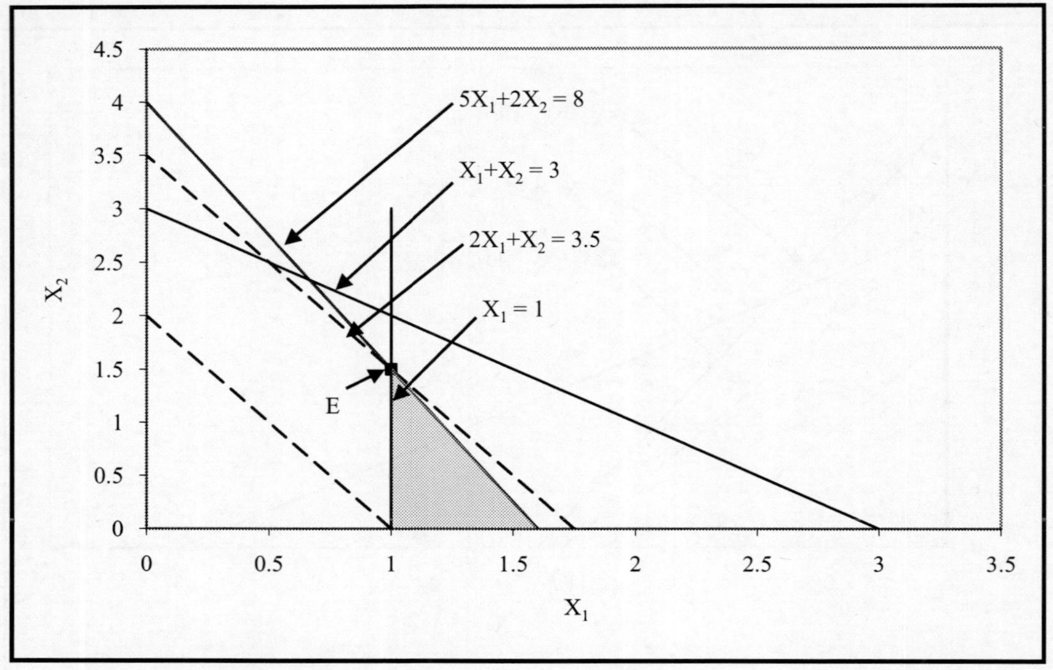

Figure 6.27

In this case the shaded area is the feasible region and the solution set is obtained as: $X_1 = 1$, $X_2 = 1.5$ and $Z = 3.5$. This is indeed the final wanted solution.

6.6 BRANCH-AND-BOUND METHOD FOR ALL INTEGERS BY SIMPLEX METHOD

■ Example 9

For the linear programming problem given below find the all integer solution by branching method.

$$\begin{aligned} \text{Maximize: } & Z = 5X_1 + 3X_2 \\ \text{Subject to: } & 4X_1 + 2X_2 \le 25 \\ & X_1 \le 5 \\ & X_2 \le 8 \\ & X_1 ;\ X_2 \ge 0 \quad X_1 \text{ and } X_2 \text{ integers} \end{aligned}$$

Converted LP Problem

Relax the given integer condition and rewrite the given IP problem as simple LP problem as shown below.

$$\begin{aligned} &\text{Maximize: } Z = 5X_1 + 3X_2 \\ &\text{Subject to: } \quad 4X_1 + 2X_2 \le 25 \\ &\qquad X_1 \le 5 \\ &\qquad X_2 \le 8 \\ &\qquad X_1 ; X_2 \ge 0 \end{aligned}$$

Solution to LPP

$$\begin{aligned} &\text{Maximize: } Z = 5X_1 + 3X_2 + 0S_1 + 0S_2 + 0S_3 + 0S_4 \\ &\text{Subject to: } \quad 4X_1 + 2X_2 + S_1 = 25 \\ &\qquad X_1 + S_2 = 5 \\ &\qquad X_2 + S_3 = 8 \\ &\qquad X_1 ; X_2 ; S_1 ; S_2 ; S_3 ; S_4 \ge 0 \end{aligned}$$

Once this is done we proceed further and solve this LP problem in the usual manner by using Simplex method.

Table 6.3

Tableau 1

C_i	Basis	X_1	X_2	S_1	S_2	S_3	X_b	Ratio
0	S_1	4	2	1	0	0	25	6.25
0	S_2	1	0	0	1	0	5	5
0	S_3	0	1	0	0	1	8	–
	C_j	5	3	0	0	0		
	Z_j	0	0	0	0	0		
	C_j-Z_j	5	3	0	0	0		

Table 6.4

Tableau 2

C_i	Basis	X_1	X_2	S_1	S_2	S_3	X_b	Ratio
0	S_1	0	2	1	–4	0	5	2.5
5	X_1	1	0	0	1	0	5	–
0	S_3	0	1	0	0	1	8	8
	C_j	5	3	0	0	0		
	Z_j	5	0	0	5	0		
	C_j-Z_j	0	3	0	–5	0		

Table 6.5

Tableau 3								
C_i	Basis	X_1	X_2	S_1	S_2	S_3	X_b	Ratio
3	X_2	0	1	0.5	−2	0	2.5	−1.25
5	X_1	1	0	0	1	0	5	5
0	S_3	0	0	−0.5	2	1	5.5	2.75
	C_j	5	3	0	0	0		
	Z_j	5	3	1.5	−1	0	32.5	
	C_j-Z_j	0	0	−1.5	1	0		

Table 6.6

Tableau 4								
C_i	Basis	X_1	X_2	S_1	S_2	S_3	X_b	Ratio
3	X_2	0	1	0	0	1	8	2.5
5	X_1	1	0	0.25	0	−0.5	2.25	–
0	S_2	0	0	−0.25	1	0.5	2.75	8
	C_j	5	3	0	0	0		
	Z_j	5	3	1.25	0	0.5	35.25	
	C_j-Z_j	0	0	−1.25	0	−0.5		

From tableau 4 we obtain the solution for the unrestricted LP problem as:

$$X_1 = 2.25,\ X_2 = 8,\ Z = 35.25$$

Now clearly the solution is not the wanted all integer type. The decision variable $X_1 = 2.25$ is a non-integer solution. So let us split this problem into two sub-problems by introducing two new constraints $X_1 \leq 2$, and $X_1 \geq 3$ so that our non-integer solution $X_1 = 2.25$ is totally avoided in future.

Sub-problem 1: Main problem added with $X_1 \leq 2$

$$\begin{aligned} \text{Maximize: } & Z = 5X_1 + 3X_2 \\ \text{Subject to: } & 4X_1 + 2X_2 \leq 25 \\ & X_1 \leq 5 \\ & X_2 \leq 8 \\ & X_1 \leq 2 \\ & X_1\ ;\ X_2 \geq 0 \end{aligned}$$

Solution

$$\begin{aligned}
\text{Maximize: } & Z = 5X_1 + 3X_2 + 0S_1 + 0S_2 + 0S_3 + 0S_4 \\
\text{Subject to: } & 4X_1 + 2X_2 + S_1 = 25 \\
& X_1 + S_2 = 5 \\
& X_2 + S_3 = 8 \\
& X_1 + S_4 = 3 \\
& X_1 ; X_2 ; S_1 ; S_2 ; S_3 ; S_4 \geq 0
\end{aligned}$$

Table 6.7

Tableau 1								
C_i	Basis	X_1	X_2	S_1	S_2	S_3	X_b	Ratio
0	S_1	4	2	1	0	0	25	6.25
0	S_2	1	0	0	1	0	2	2
0	S_3	0	1	0	0	1	8	–
	C_j	5	3	0	0	0		
	Z_j	0	0	0	0	0		
	C_j-Z_j	5	3	0	0	0		

Table 6.8

Tableau 2								
C_i	Basis	X_1	X_2	S_1	S_2	S_3	X_b	Ratio
0	S_1	0	2	1	−4	0	17	8.5
5	X_1	1	0	0	1	0	2	–
0	S_3	0	1	0	0	1	8	8
	C_j	5	3	0	0	0		
	Z_j	5	0	0	5	0	10	
	C_j-Z_j	0	3	0	−5	0		

Table 6.9

Tableau 3								
C_i	Basis	X_1	X_2	S_1	S_2	S_3	X_b	Ratio
0	S_1	0	0	1	−4	−2	1	8.5
5	X_1	1	0	0	1	0	2	–
3	X_2	0	1	0	0	1	8	8
	C_j	5	3	0	0	0		
	Z_j	5	3	0	5	3	34	
	C_j-Z_j	0	0	0	−5	−3		

The unrestricted solution to the sub-problem 1 are $X_1 = 2$, $X_2 = 8$, $Z = 34$

Sub-problem 2: Main problem added with $X_1 \geq 3$

$$\text{Maximize: } Z = 5X_1 + 3X_2$$
$$\begin{aligned} \text{Subject to: } \quad & 4X_1 + 2X_2 \leq 25 \\ & X_1 \leq 5 \\ & X_2 \leq 8 \\ & X_1 \geq 3 \\ & X_1 ;\ X_2 \geq 0 \end{aligned}$$

Solution

$$\text{Maximize: } Z = 5X_1 + 3X_2 + 0S_1 + 0S_2 + 0S_3 + 0S_4 + MA_1$$
$$\begin{aligned} \text{Subject to: } \quad & 4X_1 + 2X_2 + S_1 = 25 \\ & X_1 + S_2 = 5 \\ & X_2 + S_3 = 8 \\ & X_1 + S_4 + A_1 = 3 \\ & X_1 ;\ X_2 ;\ S_1 ;\ S_2 ;\ S_3 ;\ S_4 ;\ A_1 \geq 0 \end{aligned}$$

Table 6.10

Tableau 1										
C_i	Basis	X_1	X_2	S_1	S_2	S_3	S_4	A_1	X_b	Ratio
0	S_1	4	2	1	0	0	0	0	25	6.25
0	S_2	1	0	0	1	0	0	0	5	5
0	S_3	0	1	0	0	1	0	0	8	INF
−M	A_1	1	0	0	0	0	−1	1	3	3
	C_j	5	3	0	0	0	0	−M		
	Z_j	−M	0	0	0	0	M	−M		
	$C_j - Z_j$	5+M	3	0	0	0	−M	0		

Table 6.11

Tableau 2									
C_i	Basis	X_1	X_2	S_1	S_2	S_3	S_4	X_b	Ratio
0	S_1	0	2	1	0	0	4	13	3.25
0	S_2	0	0	0	1	0	1	2	2
0	S_3	0	1	0	0	1	0	8	INF
5	X_1	1	0	0	0	0	−1	3	−3
	C_j	5	3	0	0	0	0		
	Z_j	5	0	0	0	0	−5	15	
	$C_j - Z_j$	0	3	0	0	0	5		

Table 6.12

Tableau 3									
C_i	Basis	X_1	X_2	S_1	S_2	S_3	S_4	X_b	Ratio
0	S_1	0	2	1	−4	0	0	5	2.5
0	S_4	0	0	0	1	0	1	2	INF
0	S_3	0	1	0	0	1	0	8	8
5	X_1	1	0	0	1	0	0	5	INF
	C_j	5	3	0	0	0	0		
	Z_j	5	0	0	5	0	0	25	
	C_j-Z_j	0	3	0	−5	0	0		

Table 6.13

Tableau 4									
C_i	Basis	X_1	X_2	S_1	S_2	S_3	S_4	X_b	Ratio
3	X_2	0	1	0.5	−2	0	0	2.5	−1.25
0	S_4	0	0	0	1	0	1	2	2
0	S_3	0	0	−0.5	2	1	0	5.5	2.75
5	X_1	1	0	0	1	0	0	5	5
	C_j	5	3	0	0	0	0		
	Z_j	5	3	1.5	−1	0	0	32.5	
	C_j-Z_j	0	0	−1.5	1	0	0		

Table 6.14

Tableau 5									
C_i	Basis	X_1	X_2	S_1	S_2	S_3	S_4	X_b	Ratio
3	X_2	0	1	0.5	0	0	2	6.5	−1.25
0	S_4	0	0	0	1	0	1	2	2
0	S_3	0	0	−0.5	0	1	−2	1.5	2.75
5	X_1	1	0	0	0	0	−1	3	5
	C_j	5	3	0	0	0	0		
	Z_j	5	3	1.5	0	0	1	34.5	
	C_j-Z_j	0	0	−1.5	0	0	−1		

The solutions are $X_1 = 3$, $X_2 = 6.5$ and $Z = 34.5$

In the above two sub-problems the first one gave us completely integer solution. But in the second, only X_1 is an integer. So we take the objective value (34) of the sub-problem which gave complete integer solution as our upper bound. Also remember that the Z value of sub-problem 2 exceeds the upper bound (34.5>34). So we subdivide the sub-problem 2 as sub-problem 3 and 4 and continue our search for integer solution. Since the non-integer solution of sub-problem 2 is $X_2 = 6.5$ we add two additional constraints for $X_2 \leq 6$ and $X_2 \geq 7$ two our sub-problem 2 and form sub-problem 3 and sub-problem 4.

Sub-problem 3: Sub-problem 2 added with $X_2 \le 6$

$$\begin{aligned} \text{Maximize: } & Z = 5X_1 + 3X_2 \\ \text{Subject to: } & 4X_1 + 2X_2 \le 25 \\ & X_1 \le 5 \\ & X_2 \le 8 \quad \text{(redundant)} \\ & X_1 \ge 3 \\ & X_2 \le 6 \\ & X_1 ;\ X_2 \ge 0 \end{aligned}$$

Solution

$$\begin{aligned} \text{Maximize: } & Z = 5X_1 + 3X_2 + 0S_1 + 0S_2 + 0S_3 + 0S_4 + MA_1 \\ \text{Subject to: } & 4X_1 + 2X_2 + S_1 = 25 \\ & X_1 + S_2 = 5 \\ & X_2 + S_3 = 6 \\ & X_1 - S_4 + A_1 = 3 \\ & X_1 ;\ X_2 ;\ S_1 ;\ S_2 ;\ S_3 ;\ S_4 ;\ A_1 \ge 0 \end{aligned}$$

Table 6.15

Tableau 1										
C_i	Basis	X_1	X_2	S_1	S_2	S_3	S_4	A_1	X_b	Ratio
0	S_1	4	2	1	0	0	0	0	25	6.25
0	S_2	1	0	0	1	0	0	0	5	5
0	S_3	0	1	0	0	1	0	0	6	INF
−M	A_1	1	0	0	0	0	−1	1	3	3
	C_j	5	3	0	0	0	0	−M		
	Z_j	−M	0	0	0	0	M	−M		
	C_j-Z_j	5+M	3	0	0	0	−M	0		

Table 6.16

Tableau 2									
C_i	Basis	X_1	X_2	S_1	S_2	S_3	S_4	X_b	Ratio
0	S_1	0	2	1	0	0	4	13	3.25
0	S_2	0	0	0	1	0	1	2	2
0	S_3	0	1	0	0	1	0	6	INF
5	X_1	1	0	0	0	0	−1	3	−3
	C_j	5	3	0	0	0	0		
	Z_j	5	0	0	0	0	−5		
	C_j-Z_j	0	3	0	0	0	5		

Table 6.17

Tableau 3									
C_i	Basis	X_1	X_2	S_1	S_2	S_3	S_4	X_b	Ratio
0	S_1	0	2	1	−4	0	0	5	2.5
0	S_4	0	0	0	1	0	1	2	INF
0	S_3	0	1	0	0	1	0	6	6
5	X_1	1	0	0	1	0	0	5	INF
	C_j	5	3	0	0	0	0		
	Z_j	5	0	0	5	0	0		
	$C_j - Z_j$	0	3	0	−5	0	0		

Table 6.18

Tableau 4									
C_i	Basis	X_1	X_2	S_1	S_2	S_3	S_4	X_b	Ratio
3	X_2	0	1	0.5	−2	0	0	2.5	−1.25
0	S_4	0	0	0	1	0	1	2	2
0	S_3	0	0	−0.5	2	1	0	3.5	1.75
5	X_1	1	0	0	1	0	0	5	5
	C_j	5	3	0	0	0	0		
	Z_j	5	3	1.5	−1	0	0	32.5	
	$C_j - Z_j$	0	0	−1.5	1	0	0		

Table 6.19

Tableau 5									
C_i	Basis	X_1	X_2	S_1	S_2	S_3	S_4	X_b	Ratio
3	X_2	0	1	0	0	1	0	6	−1.25
0	S_4	0	0	0.25	0	−0.5	1	0.25	2
0	S_2	0	0	−0.25	1	0.5	0	1.75	1.75
5	X_1	1	0	0.25	0	−0.5	0	3.25	5
	C_j	5	3	0	0	0	0		
	Z_j	5	3	1.25	0	0.5	0	34.25	
	$C_j - Z_j$	0	0	−1.25	0	−0.5	0		

The solutions are $X_1 = 3.25$, $X_2 = 6$ and $Z = 34.25$

Sub-problem 4: Sub-problem 2 added with $X_2 \geq 7$

$$
\begin{aligned}
&\text{Maximize: } Z = 5X_1 + 3X_2 \\
&\text{Subject to: } 4X_1 + 2X_2 \leq 25 \\
&X_1 \leq 5 \\
&X_2 \leq 8 \\
&X_1 \geq 3 \\
&X_2 \geq 7 \\
&X_1 ;\ X_2 \geq 0
\end{aligned}
$$

This problem does not have feasible region and hence feasible solution. Problem 3 once again contains a non-integer solution. So let us further sub-divide sub-problem 3 into two sub-problems 5 and 6 and solve for further solution.

Sub-problem 5: Sub-problem 3 added with $X_2 \leq 3$

$$
\begin{aligned}
&\text{Maximize: } Z = 5X_1 + 3X_2 \\
&\text{Subject to: } 4X_1 + 2X_2 \leq 25 \\
&X_1 \leq 5 \quad \text{(redundant)} \\
&X_2 \leq 8 \\
&X_1 \geq 3 \\
&X_2 \leq 6 \\
&X_1 \leq 3 \\
&X_1 ;\ X_2 \geq 0
\end{aligned}
$$

The solutions are obtained as $X_1 = 3$, $X_2 = 6$ and $Z = 33$

Sub-problem 6: Sub-problem 3 added with $X_1 \geq 4$

$$
\begin{aligned}
&\text{Maximize: } Z = 5X_1 + 3X_2 \\
&\text{Subject to: } 4X_1 + 2X_2 \leq 25 \\
&X_1 \leq 5 \\
&X_2 \leq 8 \\
&X_1 \geq 3 \quad \text{(redundant)} \\
&X_2 \leq 6 \\
&X_1 \geq 4 \\
&X_1 ;\ X_2 \geq 0
\end{aligned}
$$

The solutions are: $X_1 = 4$, $X_2 = 4.5$ and $Z = 33.5$

The Z values of these two problems are less than the 34 obtained in the beginning. So the all integer solution values are: $X_1 = 2$, $X_2 = 8$, $Z = 34$.

6.7 GOMORY'S CUTTING PLANE METHOD FOR COMPLETE INTEGER SOLUTION

The integer-programming algorithm by Simplex method is simply an extension of the non-integer programming algorithm. To begin with we simply ignore the integer requirements and solve the given problem as usual. If we get all integer solution it is well and good. No additional effort is needed. On the other hand, if the solution is not all integers we add one or more new constraints to the problem one at a time in a sequential order. The new set of feasible solutions includes all the original feasible integers but exclude the original non-integer solution initially found. This new constraint included is often called cut because it cuts out a part of the original feasible region in its diagrammatic representation. The revised problem is solved again by using Simplex method. At this stage if we get all integer solution the procedure gets terminated. If not, one more cut is applied and Simplex procedure is repeated. As many cuts as it requires are added till we get the final integer solution. Though there are several methods of cuts we propose to use R.E. Gomory's fractional cut method in this section. The following example illustrates the cutting by using graphical method.

■ Example 10

Solve the following IP for all integer solution by using Gomory's fractional cut method. Also illustrate the proposed cut geometrically.

$$\begin{aligned} \text{Maximize: } Z &= X_1 - X_2 \\ \text{Subject to: } & X_1 + 2X_2 \le 4 \\ & 6X_1 + 2X_2 \le 9 \\ & X_1 ;\ X_2 \ge 0 \quad \text{and integer} \end{aligned}$$

Solution

Step 1: In integer programming first we convert all the coefficients as integers if they possess fractional values by appropriate multiplication. Since in our problem all the coefficients are integers this step is avoided.

Unrestricted solution by Simplex method

Step 2: Let us introduce two slacks and rewrite the problem in its standard form as a simple linear programming model

$$\begin{aligned} \text{Maximize: } Z &= X_1 - X_2 + 0S_1 + 0S_2 \\ \text{Subject to: } & X_1 + 2X_2 + S_1 = 4 \\ & 6X_1 + 2X_2 + S_2 = 9 \\ & X_1 ;\ X_2 ;\ S_1 ;\ S_2 \ge 0 \end{aligned}$$

Step 3: Solve the problem for unrestricted solution first by using Simplex method in the usual manner.

Table 6.20

Tableau 1							
C_i	Basis	X_1	X_2	S_1	S_2	X_b	Min
0	S_1	1	2	1	0	4	4
0	S_2	6	2	0	1	9	1.5
	C_j	1	−1	0	0		
	Z_j	0	0	0	0		
	C_j−Z_j	1	−1	0	0		

Table 6.21

Tableau 2						
C_i	Basis	X_1	X_2	S_1	S_2	X_b
0	S_1	0	10/6	1	−1/6	5/2
1	X_1	1	2/6	0	1/6	3/2
	C_j	1	−1	0	0	
	Z_j	1	2/6	0	1/6	3/2
	C_j−Z_j	0	−8/6	0	−1/6	

So the optimum unrestricted solutions are $X_1 = 3/2$, $X_2 = 0$, $Z = 1.5$

Gomory's fractional cut

Step 4: Here since $X_1 = 1.5$ is not an integer. It is time to bring in an appropriate cut by introducing a new constraint called cutting plane. For this purpose we choose the non-integer variable X_1 in our final Simplex solution. So from Table 6.21 the relevant constraint is obtained as:

$$X_1 + (2/6)\,X_2 + 0S_1 + 1/6\,S_2 = 3/2$$

Now let us rewrite the coefficients of this constraint as the sum of an integer and non-negative fraction. For example $8/3 = 2 + 2/3$; $-7/5 = -2 + 3/5$. So the above equation is rewritten as

$$(1+0)\,X_1 + (0+2/6)\,X_2 + 0S_1 + (0+1/6)\,S_2 = 1+1/2.$$

This equation may also be rewritten as $2/6X_2 + 1/6S_2 = 1/2 + (1-X_1)$

$2/6X_2 + 1/6S_2 = 1/2$ + some integer. Now since the RHS is equal to 1/2 plus some integer the LHS must be equal or greater than 1/2.

$$2/6X_2+1/6S_2 \geq 1/2$$

To avoid the introduction of artificial variable let us convert *greater than or equal to* into *less than or equal to* by multiplying the above equation by −1.

$$-2/6X_2-1/6S_2 \leq -1/2$$

Now we add a slack S_3 to bring in equality

$$-2/6X_2-1/6S_2 +S_3 = -\ 1/2$$

This is our new equality constraint called cut to be introduced to the final Simplex tableau obtained above.

Simplex method continued in Table 6.22 after introducing the fractional cut

Table 6.22

Tableau 1							
C_i	Basis	X_1	X_2	S_1	S_2	S_3	X_b
1	X_1	1	2/6	0	1/6	0	3/2
0	S_1	0	10/6	1	−1/6	0	5/2
0	S_3	0	−2/6	0	−1/6	1	−1/2
	C_j	1	−1	0	0	0	
	Z_j	1	2/6	0	1/6	0	3/2
	C_j-Z_j	0	−8/6	0	−1/6	0	

In this table, to start Simplex iteration we must have a positive value in the C_j-Z_j row. Since these values are already zero or negative it looks although the optimum has been already reached. Really this is not so because one of the basic variable $S_3 = -1/2$. Therefore one must eliminate this negative basic variable from the final solution. So to obtain the optimum solution let us use duality Simplex method as follows

$$\text{Key column} = \text{Min}\left[(C_j - Z_j)/a_{ij}\right]: a_{ij} \leq 0$$

$$= \text{Min}\left[\left(\frac{-8/6}{-2/6}\right);\left(\frac{-1/6}{-1/6}\right)\right]$$

$$= \text{Min}[4,1] = 1 \ \ (S_2 \text{ column})$$

$$\text{Key row} = \text{Min}[3/2, 5/2, -1/2] = -1/2 \ \ (S_3 \text{ row})$$

Between these two values since S_2 column value is the smallest, S_2 enters and S_3 goes out in Table 6.22. After entering S_2 in the place of S_3 in the usual way we make necessary changes and report the result in Table 6.23.

Table 6.23

Tableau 3							
C_i	Basis	X_1	X_2	S_1	S_2	S_3	X_b
1	X_1	1	0	0	0	1	1
0	S_1	0	2	1	0	−1	3
0	S_2	0	2	0	1	−6	3
	C_j	1	−1	0	0	0	
	Z_j	1	0	0	0	1	1
	C_j-Z_j	0	−1	0	0	−1	

In the above table since there is no positive values in the C_j-S_j and no negative values in X_b column we confirm that the optimum solution has been reached. The optimum solutions are $X_1 = 1$, $X_2 = 0$ and $Z = 1$.

Diagrammatic illustration of the Gomory's cutting plane method for complete integers

Unrestricted solution by graphical method

In Figure 6.28 we show the unrestricted LP solution diagrammatically in the usual way.

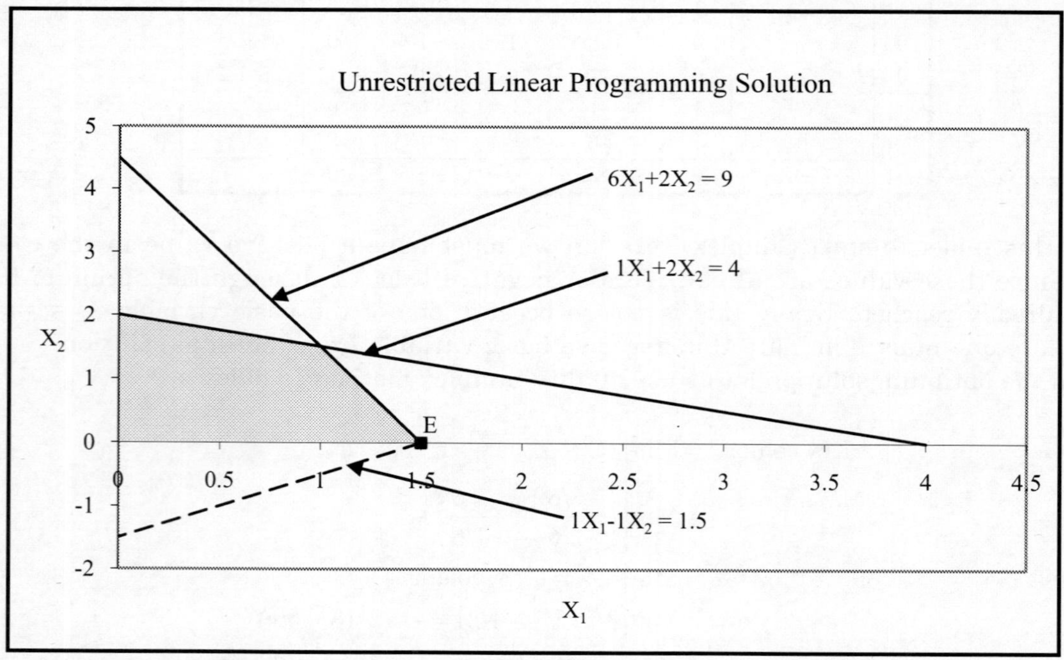

Figure 6.28

From the above given graph we obtain the solution at E as $X_1 = 1.5$, $X_2 = 0$ and the value of the objective function as $Z = 1.5$.

All integer solution after introducing the cut

The cut in terms of decision variables is obtained by eliminating the slack S_2 in the cut obtained by using the constraints used in the problem.

$$-\frac{2}{6}X_2 - \frac{1}{6}S_2 \leq -\frac{1}{2}$$

$$-\frac{2}{6}X_2 - \frac{1}{6}(9 - 6X_1 - 2X_2) \leq -\frac{1}{2}$$

$$-\frac{2}{6}X_2 - \frac{9}{6} + X_1 + \frac{2}{6}X_2 \leq -\frac{1}{2}$$

$$X_1 \leq -\frac{1}{2} + \frac{3}{2}$$

$$X_1 \leq 1$$

In Figure 6.29 a new constraint $X_1 \leq 1$ is added so that the feasible original integer solutions, namely $X_2 = 0$ is intact in the new feasible region but at the same time excludes the unwanted optimum non-integer solution, namely $X_1 = 1.5$. Such an inclusion cuts the feasible region to the shaded area containing all other integer points intact. Thus, here the old solution, namely $X_1 = 1.5$ is excluded in the feasible region.

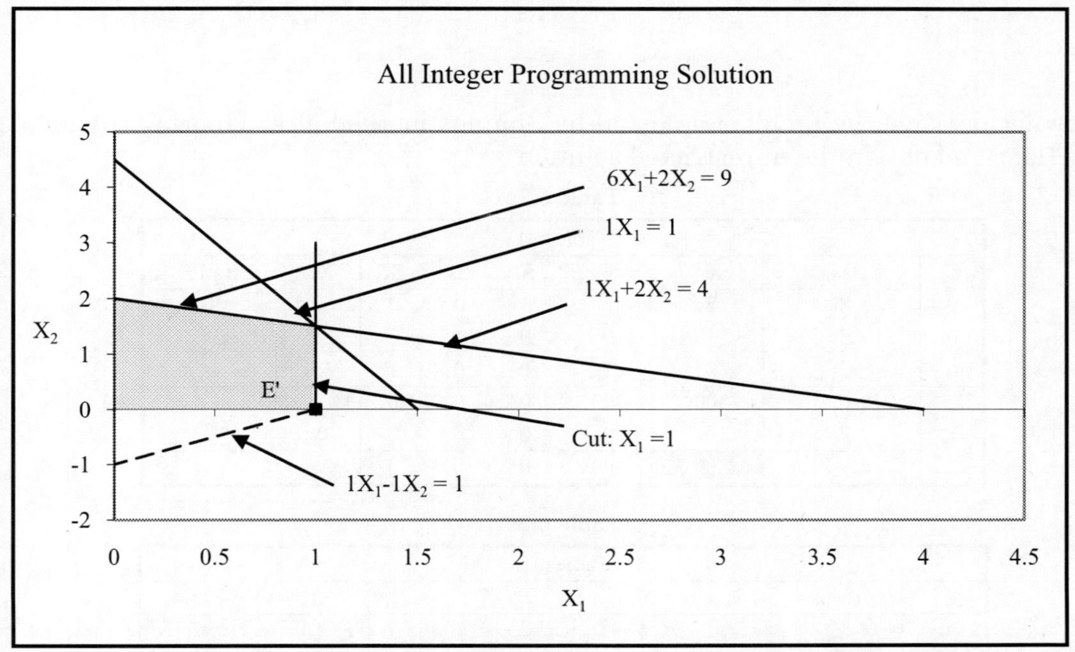

Figure 6.29

The integer solutions are $X_1 = 1$, $X_2 = 0$ and $Z = 1$.
(Remember that this is the cut we used in Simplex method for the same problem.)

■ Example 11

Solve the following IP for all integer solution by using Gomory's fractional cut. Also illustrate the cut geometrically.

$$\text{Maximize: } Z = 300X_1 + 200X_2$$
$$\text{Subject to: } 2X_1 + 2X_2 \le 60$$
$$X_1 \le 25$$
$$X_2 \le 35$$
$$X_1 ;\ X_2 \ge 0 \text{ and integer}$$

Unrestricted solution by Simplex method

As a first step let us write the problem in its standard form by adding appropriate slacks as shown below.

$$\text{Maximize: } Z = 300X_1 + 200X_2 + 0S_1 + 0S_2 + 0S_3$$
$$\text{Subject to: } 2X_1 + 1X_2 + S_1 = 60$$
$$X_1 + 0X_2 + \quad + S_2 = 25$$
$$0X_1 + X_2 + \quad + S_3 = 35$$
$$X_1 ;\ X_2 ;\ S_1 ;\ S_2 ;\ S_3 \ge 0$$

Now let us solve the problem as usual by Simplex method after ignoring the integer restrictions and obtain the unrestricted solution.

Table 6.24

Tableau 1

C_i	Basis	X_1	X_2	S_1	S_2	S_3	X_b	Ratio
0	S_1	2	1	1	0	0	60	30
0	S_2	1	0	0	1	0	25	25
0	S_3	0	1	0	0	1	35	–
	C_j	300	200	0	0	0		
	Z_j	0	0	0	0	0	0	
	$C_j–Z_j$	300	200	0	0	0		

Table 6.25

Tableau 2

C_i	Basis	X_1	X_2	S_1	S_2	S_3	X_b	Ratio
0	S_1	0	1	1	–2	0	10	10
300	X_1	1	0	0	1	0	25	–
0	S_3	0	1	0	0	1	35	35
	C_j	300	200	0	0	0		
	Z_j	300	0	0	300	0	7500	
	$C_j–Z_j$	0	200	0	–300	0		

Table 6.26

Tableau 3								
C_i	Basis	X_1	X_2	S_1	S_2	S_3	X_b	Ratio
200	X_2	0	1	1	−2	0	10	−5
300	X_1	1	0	0	1	0	25	25
0	S_3	0	0	−1	2	1	25	12.5
	C_j	300	200	0	0	0		
	Z_j	300	200	200	−100	0	9500	
	$C_j–Z_j$	0	0	−200	100	0		

Table 6.27

Tableau 4							
C_i	Basis	X_1	X_2	S_1	S_2	S_3	X_b
200	X_2	0	1	0	0	1	35
300	X_1	1	0	0.5	0	−0.5	12.5
0	S_2	0	0	−0.5	1	0.5	12.5
	C_j	300	200	0	0	0	
	Z_j	300	200	150	0	50	10,750
	$C_j–Z_j$	0	0	−150	0	−50	

The unrestricted solution is X_1= 12.5; X_2= 35 and Z = 10,750.

Fractional Gomory's cut

Since X_2 is already an integer let us apply Gomory's cut to X_1 and generate an additional constraint as follows. From Table 6.27:

$$X_1 + \frac{1}{2}S_1 - \frac{1}{2}S_3 = 12.5$$
$$(1+0)X_1 + (0+1/2)S_1 + (-1+1/2)S_3 = (12+1/2)$$

Now let us transfer the integer coefficients together and rearrange the equation as

$$1/2S_1 + 1/2S_3 = 1/2 + \{12 - X_1 + S_3\}$$
$$1/2S_1 + 1/2S_3 = 1/2 + \text{some integer}$$
$$1/2S_1 + 1/2S_3 \geq 1/2$$

This is indeed the needed additional Gomory's cut constraint for integer solution. Now to avoid the use of artificial variable let us convert the *greater than or equal to* type of inequality into a *less than or equal to* type of inequality by multiplying both the sides by −1.

$$-1/2S_1 - 1/2S_3 \leq -1/2$$
$$-1/2S_1 - 1/2S_3 + S_4 = -1/2 \text{ (slack } S_4 \text{ is added)}$$

Continuation of Simplex method after introducing the cut in Table 6.27

Now with this added constraint, our last Simplex tableau is rewritten in Table 6.28 and iteration for optimum solution is continued as usual.

In tableau 5 since all the entries in the last row are either zero or negative the contained solution looks like optimum. Clearly this is not so because one of the basic variables does not satisfy the non-negativity conditions ($S_4 = -0.5$).

Table 6.28

Tableau 5								
C_i	Basis	X_1	X_2	S_1	S_2	S_3	S_4	X_b
200	X_2	0	1	0	0	1	0	35
300	X_1	1	0	0.5	0	−0.5	0	12.5
0	S_2	0	0	−0.5	1	0.5	0	12.5
0	S_4	0	0	−0.5	0	−0.5	1	−0.5
	C_j	300	200	0	0	0	0	
	Z_j	300	200	150	0	50	0	10,750
	C_j-Z_j	0	0	−150	0	−50	0	

So to obtain the optimum solution let us use duality Simplex method as follows:

$$\text{Key column} = \text{Min}\left[(C_j - Z_j)/a_{ij}\right] : a_{ij} \leq 0$$

$$= \text{Min}\left[\left(\frac{-150}{-1/2}\right);\left(\frac{-50}{-1/2}\right)\right]$$

$$= \text{Min}[300, 100] = 100 \ \ (S_3 \text{ column})$$

$$\text{Key row}[35,\ 25/2,\ 25/2,\ -1/2] = -1/2 \ \ (S_4 \text{ row})$$

Thus, S_4 is outgoing and S_3 incoming in tableau 6. So the Simplex Table 6.28 is modified accordingly and Simplex calculations are carried away in the usual way. The result is reported in Table 6.29.

Table 6.29

Tableau 6								
C_i	Basis	X_1	X_2	S_1	S_2	S_3	S_4	X_b
200	X_2	0	1	−1	0	0	2	34
300	X_1	1	0	1	0	0	−1	13
0	S_2	0	0	−1	1	0	1	12
0	S_3	0	0	1	0	1	−2	1
	C_j	300	200	0	0	0	0	
	Z_j	300	200	100	0	0	100	10,700
	C_j-Z_j	0	0	−100	0	0	−100	

From the table the integer solutions are: $X_1 = 13$; $X_2 = 34$; $Z = 10{,}700$.

Geometric explanation of the cut

Unrestricted Solution

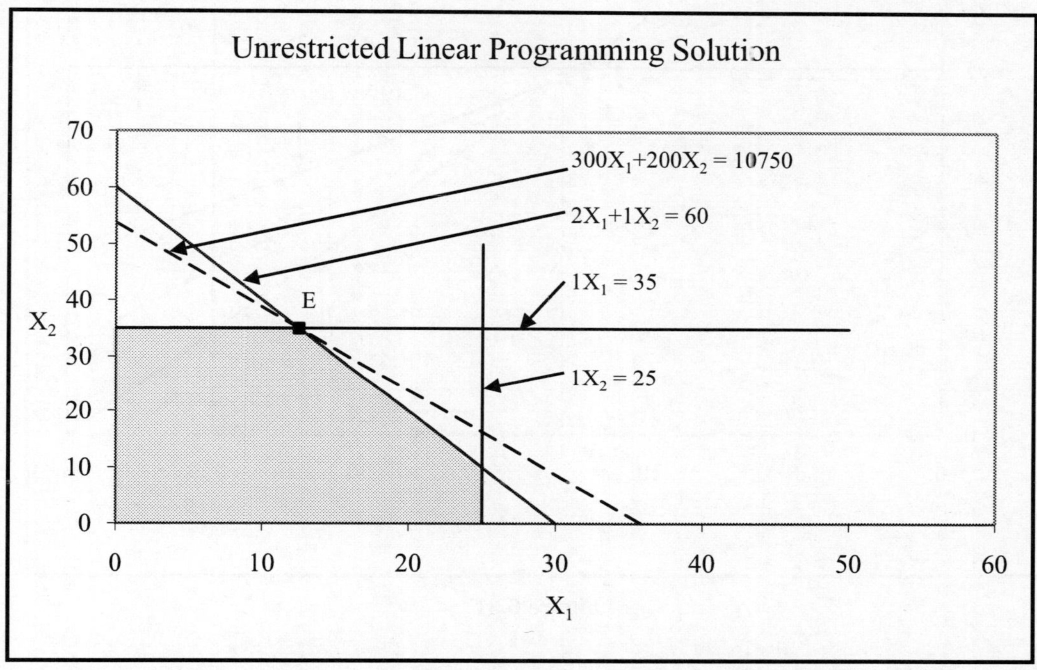

Figure 6.30

Unrestricted solution: $X_1 = 12.5$; $X_2 = 35$; $Z = 10{,}750$.

Integer Solution

Calculation of cut in terms of decision variables using the constants.

$$-\frac{1}{2}S_1 - 1/2S_3 \le -\frac{1}{2}$$

$$-\frac{1}{2}(60 - 2X_1 - 1X_2) - \frac{1}{2}(35 - X_2) \le -\frac{1}{2}$$

$$-30 + X_1 + \frac{1}{2}X_2 - 17.5 + \frac{1}{2}X_2 \le -\frac{1}{2}$$

$$X_1 + X_2 \le -0.5 + 17.5 + 30$$

$$X_1 + X_2 \le 47$$

Now let us introduce an additional constraint as cut, namely $X_1 + X_2 \le 47$ into the graph.

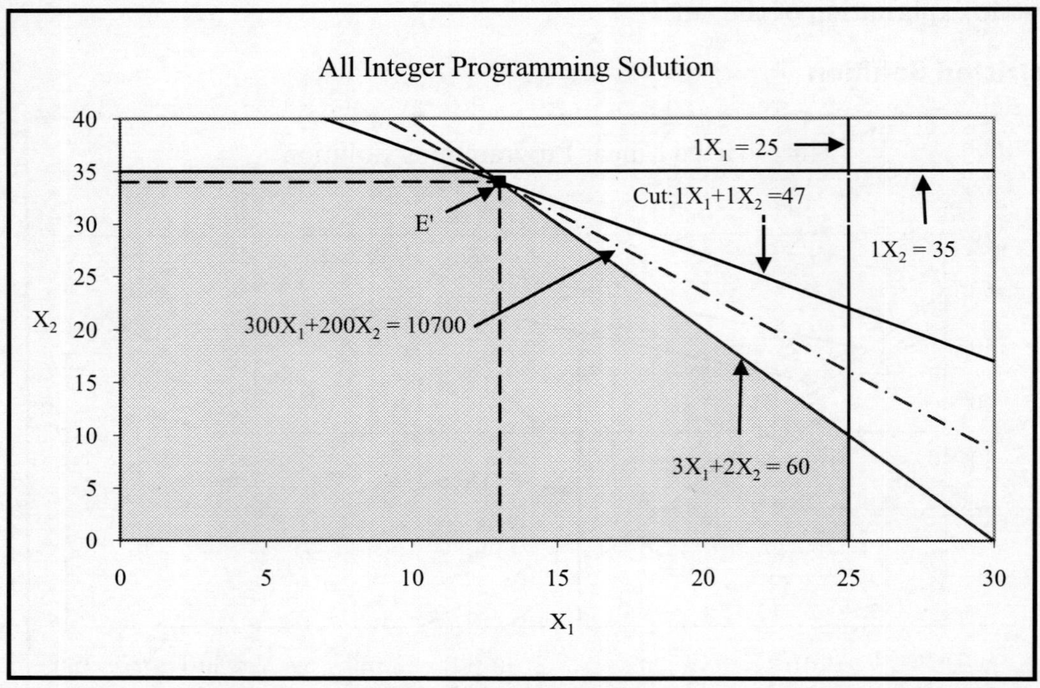

Figure 6.31

Integer solution: $X_1 = 13$; $X_2 = 34$; $Z = 10{,}700$.

■ Example 12

Solve the following IP by incorporating Gomory's fractional cut for complete integer solution.

$$\begin{aligned} \text{Maximize: } Z &= 4X_1 - 11X_2 \\ \text{Subject to: } & 2X_1 - X_2 \leq 4 \\ & 2X_1 + 5X_2 \leq 16 \\ & -1X_1 + 2X_2 \leq 4 \\ & X_1 ;\ X_2 \geq 0 \quad \text{and integer} \end{aligned}$$

Unrestricted solution by Simplex method

As a first step let us write the problem in its standard form by adding appropriate slacks as shown below.

$$
\begin{aligned}
\text{Maximize: } Z &= 4X_1 - 11X_2 + 0S_1 + 0S_2 + 0S_3 \\
\text{Subject to: } & 2X_1 - X_2 + S_1 = 4 \\
& 2X_1 + 5X_2 + S_2 = 16 \\
& -1X_1 + 2X_2 + S_3 = 4 \\
& X_1 ; X_2 ; S_1 ; S_2 ; S_3 \geq 0 \text{ and integer}
\end{aligned}
$$

Now let us solve the problem as usual by Simplex method ignoring the integer restriction.

Table 6.30

Tableau 1								
C_i	Basis	X_1	X_2	S_1	S_2	S_3	X_b	Ratio
0	S_1	2	−1	1	0	0	4	−4
0	S_2	2	5	0	1	0	16	3.2
0	S_3	−1	2	0	0	1	4	2
	C_j	4	11	0	0	0		
	Z_j	0	0	0	0	0	0	
	C_j-Z_j	4	11	0	0	0		

Table 6.31

Tableau 2								
C_i	Basis	X_1	X_2	S_1	S_2	S_3	X_b	Ratio
0	S_1	1.5	0	1	0	0.5	6	4
0	S_2	4.5	0	0	1	−2.5	6	1.33333333
11	X_2	−0.5	1	0	0	0.5	2	−4
	C_j	4	11	0	0	0		
	Z_j	−5.5	11	0	0	5.5	22	
	C_j-Z_j	9.5	0	0	0	−5.5		

Table 6.32

Tableau 3								
C_i	Basis	X_1	X_2	S_1	S_2	S_3	X_b	Ratio
0	S_1	0	0	1	−0.333	1.333	4	
4	X_1	1	0	0	0.222	−0.56	1.333333	
11	X_2	0	1	0	0.111	0.222	2.666667	
	C_j	4	11	0	0	0		
	Z_j	4	11	0	2.111	0.222	34.66667	
	C_j-Z_j	0	0	0	−2.111	−0.22		

So the unrestricted solutions are: $X_1 = 4/3$; $X_2 = 8/3$; $Z = 104/3$

Now since the fractional component for X_2 is more, we introduce cut only to this value at this stage.

$$(1+0)X_2 + (1/9)S_2 + (2/9)S_3 = 2 + 2/3$$
$$(1/9)S_2 + (2/9)S_3 = 2/3 + (2 - X_2)$$
$$(1/9)S_2 + (2/9)S_3 \geq 2/3$$
$$(-1/9)S_2 + (-2/9)S_2 \leq -(2/3)$$
$$(-1/9)S_2 + (-2/9)S_2 + S_4 = (2/3)$$

This cut constraint is included in Table 6.33 at the bottommost row. In it though the index row tells us that the optimum solution has been already reached the corresponding X_b column warns us that one of the basic variables S_4 does not satisfy the non-negativity condition ($S_4 = -2/3$).

Table 6.33

Tableau 4								
C_i	Basis	X_1	X_2	S_1	S_2	S_3	S_4	X_b
0	S_1	0	0	1	−0.333	1.333	0	4
4	X_1	1	0	0	0.222	−0.56	0	1.333333
11	X_2	0	1	0	0.111	0.222	0	2.666667
0	S_4	0	0	0	−0.111	−0.22	1	−0.666667
	C_j	4	11	0	0	0	0	
	Z_j	4	11	0	2.111	0.222	0	34.66667
	C_j-Z_j	0	0	0	−2.111	−0.22	0	

So to overcome this difficulty let us use duality Simplex method as described below.

$$\text{Key column} = \text{Min}\left[(C_j - Z_j)/a_{ij}\right] : a_{ij} \leq 0$$
$$= \text{Min}\left[\left(\frac{-2.111}{-0.111}\right); \left(\frac{-0.22}{-0.22}\right)\right]$$
$$= \text{Min}[19, 1] = 1 \ (S_3 \text{ column})$$

$$\text{Key row } [4,\ 1.333,\ 2.6667,\ -0.6667] = -0.6667 \ \ (S_4 \text{ row})$$

So we replace S_4 by S_3 and do the necessary modifications in Table 6.33. Once this is done we make Simplex calculations as usual. The resulting table is reported in Table 6.34.

Table 6.34

Tableau 5								
C_i	Basis	X_1	X_2	S_1	S_2	S_3	S_4	X_b
0	S_1	0	0	1	−1	0	6	0
4	X_1	1	0	0	0.5	0	−2.5	3
11	X_2	0	1	0	0	0	1	2
0	S_3	0	0	0	0.5	1	−4.5	3
	C_j	4	11	0	0	0	0	
	Z_j	4	11	0	2	0	1	34
	C_j-Z_j	0	0	0	−2	0	−1	

From the table the integer solutions are: $X_1 = 3$; $X_2 = 2$; $Z = 34$.

6.8 THE GOMORY'S CUT FOR MIXED INTEGER PROBLEMS

In the previous section the proposed fractional cut was initiated to have integer values for all the basic variables including slack and surplus variables. By this we mean that the fractional cut does not yields feasible optimum solution unless otherwise all the variables are integers.

In this section we propose to insist the integer values only for some selective variables leaving others for non-integer values. In such cases the optimum value of the objective function will be much more than corresponding all integer solution and still less than the corresponding unrestricted solution. For solving such partial integer restricted problem we use special cut called Gomory's mixed-integer cut. The following example illustrates the method of creating cuts.

Example 13

Solve the following IP by applying Gomory's cut method. Also illustrate the cut graphically.

$$\text{Maximize: } Z = 3X_1 + 4X_2$$
$$\text{Subject to: } 3X_1 - X_2 \le 12$$
$$3X_1 + 11X_2 \le 66$$
$$X_1 ; \quad X_2 \ge 0 \quad \text{and } X_2 \text{ integer}$$

Unrestricted solution by Simplex method

Step 1: Let us solve the problem for unrestricted solution first by using the Simplex method in the usual manner.

Table 6.35

Simplex Tableau 1							
C_i	Basis	X_1	X_2	S_1	S_2	X_b	Ratio
0	S_1	3	−1	1	0	12	−12
0	S_2	3	11	0	1	66	6
	C_j	3	4	0	0		
	Z_j	0	0	0	0	0	
	C_j-Z_j	3	4	0	0		

Table 6.36

Simplex Tableau 2							
C_i	Basis	X_1	X_2	S_1	S_2	X_b	Ratio
0	S_1	3.2727	0	1	0.0909	18	5.5
4	X_2	0.2727	1	0	0.0909	6	22
	C_j	3	4	0	0		
	Z_j	1.0909	4	0	0.3636	18	
	C_j-Z_j	1.9091	0	0	−0.364		

Table 6.37

Simplex Tableau 3							
C_i	Basis	X_1	X_2	S_1	S_2	X_b	Ratio
3	X_1	1	0	0.3056	0.0278	5.5	
4	X_2	0	1	−0.0833	0.0833	4.5	
	C_j	3	4	0	0		
	Z_j	3	4	0.5833	0.4167	34.5	
	C_j-Z_j	0	0	−0.5833	−0.417		

The unrestricted solutions to the problem are: $X_1 = 5.5$; $X_2 = 4.5$; $Z = 34.5$

Step 2: Since X_2 is the integer restricted variable in our problem we introduce the Gomory's fractional cut only to this variable. From Table 6.37 we write the row corresponding to X_2 as:

$$\begin{pmatrix}\text{Basic}\\ \text{Varible(s)}\end{pmatrix}+\begin{pmatrix}\text{Non-basic with}\\ \text{+ve coefficients}\end{pmatrix}+\begin{pmatrix}\text{Non-basic with}\\ \text{−ve coefficients}\end{pmatrix}=I+f$$

$$0X_2 + 1/12S_2 - 1/12S_1 \geq 4 + 1/2$$

Where I is the integer component and f is the fractional component of the RHS. Now the cut is obtained as

$$\begin{pmatrix}\text{Non-basic with}\\ \text{+ve coefficients}\end{pmatrix}+\frac{f}{f-1}\begin{pmatrix}\text{Non-basic with}\\ \text{−ve coefficients}\end{pmatrix}\geq f$$

Step 3: So from above

$$1/12S_2 + \frac{1/2}{(1/2)-1}(-1/12)S_1 \geq 1/2$$

$$1/12S_2 + 1/12S_1 \geq 1/2$$

$$-1/12S_2 - 1/12S_1 \leq -1/2$$

$$-1/12S_2 - 1/12S_1 + S_3 = -1/2$$

So our new modified Simplex table after entering the new constraint at the bottommost row is written in Table 6.38.

Table 6.38

Tableau 4							
C_i	Basis	X_1	X_2	S_1	S_2	S_3	X_b
3	X_1	1	0	0.3056	0.0278	0	5.5
4	X_2	0	1	−0.0833	0.0833	0	4.5
0	S_3	0	0	−0.0833	−0.083	1	−0.5
	C_j	3	4	0	0	0	
	Z_j	3	4	0.5833	0.4167	0	34.5
	C_j-Z_j	0.0000	0.0000	−0.5833	−0.4167		

In Table 6.38 though the index row entries are non-positive one of the basic variables is negative ($S_3 = -0.5$). So to obtain the optimum non-negative solution we use the duality Simplex method to Table 6.38.

$$\text{Key column} = \text{Min}\left[(C_j - Z_j)/a_{ij}\right] : a_{ij} \leq 0$$

$$= \text{Min}\left[\left(\frac{-0.5833}{-0.0833}\right);\left(\frac{-0.4167}{-0.083}\right)\right]$$

$$= \text{Min}[7,5] = 5 \;\; (S_2 \text{ column})$$

$$\text{Key row}[5.5, 4.5, -0.5] = -0.5 \;\; (S_3 \text{ row})$$

$$S_3 = \frac{-0.5833}{-0.0833} = 7; \;\; S_2 = \frac{-0.4167}{-0.083} = 5$$

So the new entrant is S_2 in the place of S_3.

Solution by graphical method: Linear Programming Solution

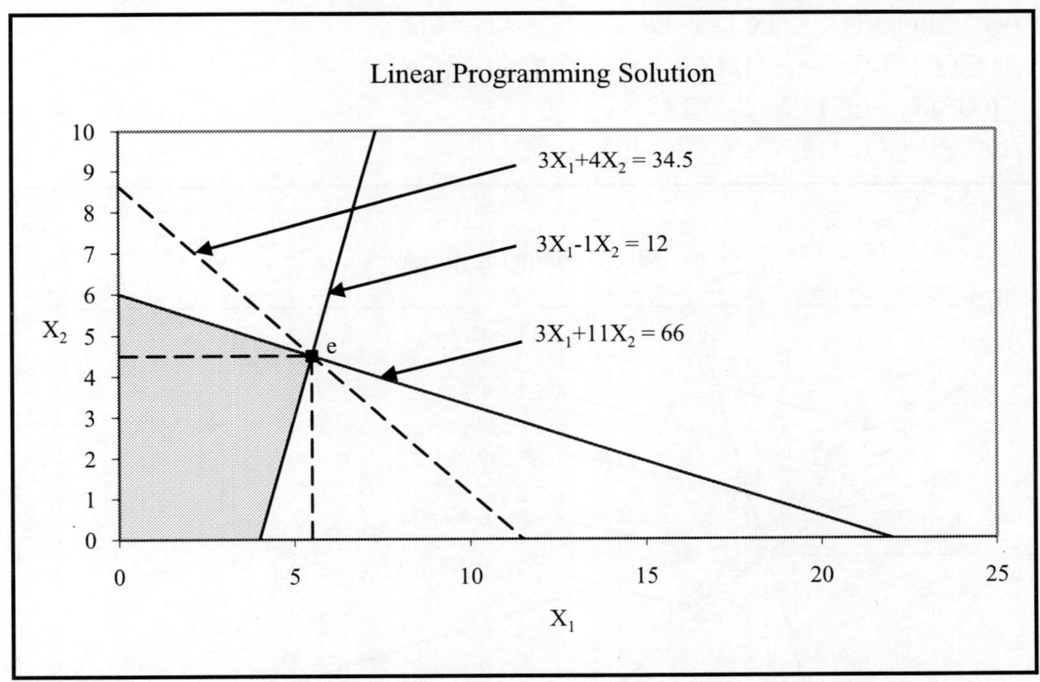

Figure 6.32

So the solution becomes $X_1 = 5.5$; $X_2 = 4.5$; and $Z = 34.5$.

Optimal tableau for integer solution follows from Table 6.38

Table 6.39

Tableau 5								
C_i	Basis	X_1	X_2	S_1	S_2	S_3	X_b	Ratio
3	X_1	1	0	0.2778	0	0.3333	5.333	198
4	X_2	0	1	−0.1667	0	1	4	54
0	S_2	0	0	1	1	−12	6	6
	C_j	3	4	0	0	0		
	Z_j	3	4	0.1667	0	5	32	
	C_j-Z_j	0	0	−0.1667	0	−5		

So the solution becomes $X_1 = 5.5$; $X_2 = 4.5$; $Z = 34.5$.

Mixed integer solution after the cut:

$$-(1/12)S_2 - (1/12)S_1 \le -1/2$$
$$(-1/12)(66 - 11X_2 - 3X_1) - 1/12(12 - 3X_1 + X_2) \le -1/2$$
$$-5.5 + (11/12)X_2 + (3/12)X_1 - 1 + (3/12)X_1 - (1/12)X_2 \le -1/2$$
$$(10/12)X_2 + (6/12)X_1 \le -0.5 + 5.5 + 1$$
$$(10/12)X_2 + (6/12)X_1 \le 6$$

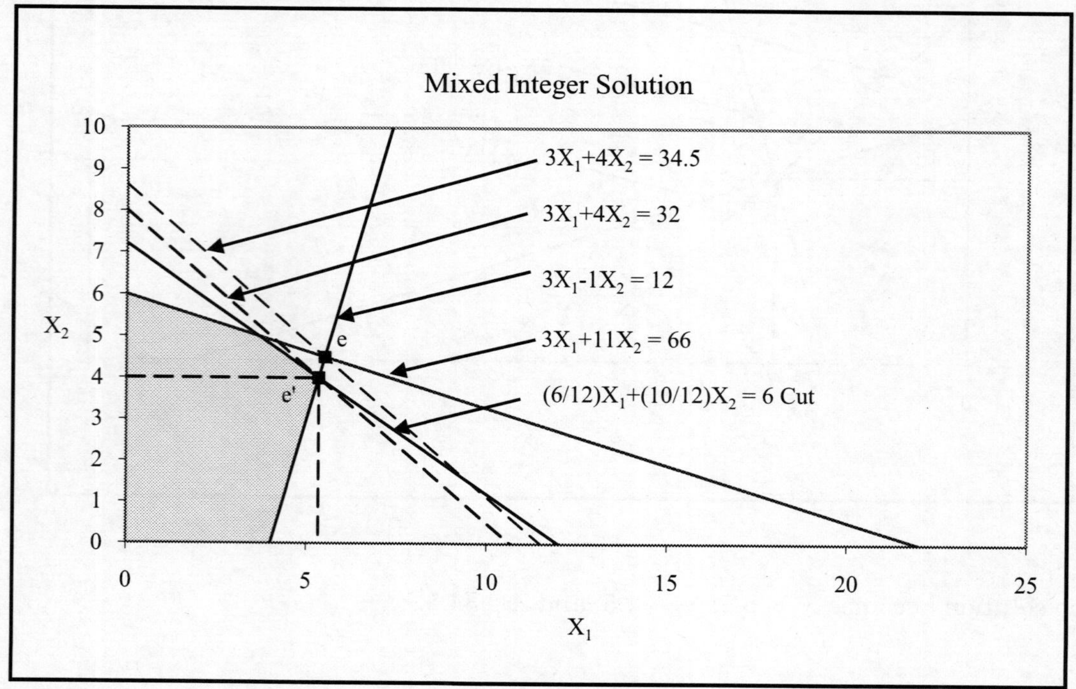

Figure 6.33

So the solution becomes $X_1 = 16/3$; $X_2 = 4$; $Z = 32$.

Note 1: In this problem we get the answer by using single cut only. This need not be the case all the time. In such cases this process will have to be repeated till we get the optimum solution.

Note 2: The optimum value has come down to 32 from 34.5.

Example 14

Solve the following IP by applying Gomory's cut method. Also obtain the corresponding graphical solution by using the cut.

$$\begin{aligned} \text{Maximize: } Z &= 4X_1 + 3X_2 \\ \text{Subject to: } & 2X_1 + X_2 \leq 4 \\ & 3X_1 + 5X_2 \leq 15 \\ & 2X_1 + 5X_2 \leq 10 \\ & X_1 \,;\; X_2 \geq 0 \quad \text{and } X_2 \text{ integer} \end{aligned}$$

Simplex Solution

Table 6.40

Tableau 1								
C_i	Basis	X_1	X_2	S_1	S_2	S_3	X_b	Ratio
0	S_1	2	1	1	0	0	4	2
0	S_2	3	5	0	1	0	15	5
0	S_3	2	5	0	0	1	10	5
	C_j	4	3	0	0	0		
	Z_j	0	0	0	0	0	0	
	C_j-Z_j	4	3	0	0	0		

Table 6.41

Tableau 2								
C_i	Basis	X_1	X_2	S_1	S_2	S_3	X_b	Ratio
4	X_1	1	0.5	0.5	0	0	2	4
0	S_2	0	3.5	−1.5	1	0	9	2.5714
0	S_3	0	4	−1	0	1	6	1.5
	C_j	4	3	0	0	0		
	Z_j	4	2	2	0	0	8	
	C_j-Z_j	0	1	−2	0	0		

Table 6.42

Tableau 3							
C_i	Basis	X_1	X_2	S_1	S_2	S_3	X_b
4	X_1	1	0	0.625	0	−0.125	1.25
0	S_2	0	0	−0.625	1	−0.875	3.75
3	X_2	0	1	−0.25	0	0.25	1.5
	C_j	4	3	0	0	0	
	Z_j	4	3	1.75	0	0.25	9.5
	C_j-Z_j	0	0	−1.75	0	−0.25	

Since X_2 is not an integer we develop a cut accordingly

$$\begin{pmatrix}\text{Basic}\\\text{Variable(s)}\end{pmatrix}+\begin{pmatrix}\text{Non-basic with}\\\text{+ve coefficients}\end{pmatrix}+\begin{pmatrix}\text{Non-basic with}\\\text{−ve coefficients}\end{pmatrix}=I+f$$

Where 'I' is the integer component and f is the fractional component of the RHS. So the above equation can be written as

$$X_2+(1/4)S_3+(-1/4)S_1=1+1/2$$

Now the cut is written as

$$\begin{pmatrix}\text{Non-basic with}\\\text{+ve coefficients}\end{pmatrix}+\frac{f}{f-1}\begin{pmatrix}\text{Non-basic with}\\\text{−ve coefficients}\end{pmatrix}\geq f$$

$$1/4S_3+\frac{1/2}{1/2-1}(-1/4S_1)\geq 1/2$$

$$1/4S_3+1/4S_1\geq 1/2$$

$$-1/4S_3-1/4S_1\leq -1/2$$

$$-1/4S_3-1/4S_1+S_4=-1/2$$

Now we introduce this cut in the last row as shown in Table 6.43.

Table 6.43

Tableau 4								
C_i	Basis	X_1	X_2	S_1	S_2	S_3	S_4	X_b
4	X_1	1	0	0.625	0	−0.125	0	1.25
0	S_2	0	0	−0.625	1	−0.875	0	3.75
3	X_2	0	1	−0.25	0	0.25	0	1.5
0	S_4	0	0	−0.25	0	−0.25	1	−0.5
	C_j	4	3	0	0	0	0	
	Z_j	4	3	1.75	0	0.25	0	9.5
	C_j-Z_j	0	0	−1.75	0	−0.25		

Now to identify the new entrant into basis in the place of S_4 we divide the negative coefficient S_4 row by corresponding values in $C_j - Z_j$th row and take least one for replacement.

$$S_3 = \frac{-7/4}{-3/4} = 2.33 \quad S_3 = \frac{-1/4}{-1/4} = 1$$

So the new entrant is S_3 in the place of S_4.

Table 6.44

Tableau 5								
C_i	Basis	X_1	X_2	S_1	S_2	S_3	S_4	X_b
4	X_1	1	0	0.75	0	0	−0.5	1.5
0	S_2	0	0	0.25	1	0	−3.5	5.5
3	X_2	0	1	−0.5	0	0	1	1
0	S_3	0	0	1	0	1	−4	2
	C_j	4	3	0	0	0	0	
	Z_j	4	3	1.5	0	0	1	9
	$C_j - Z_j$	0	0	−1.5	0	0	−1	

So from the table the solution values are: $X_1 = 1.5$; $X_2 = 1$; $Z = 9$.

Graphical Solution: Linear programming solution

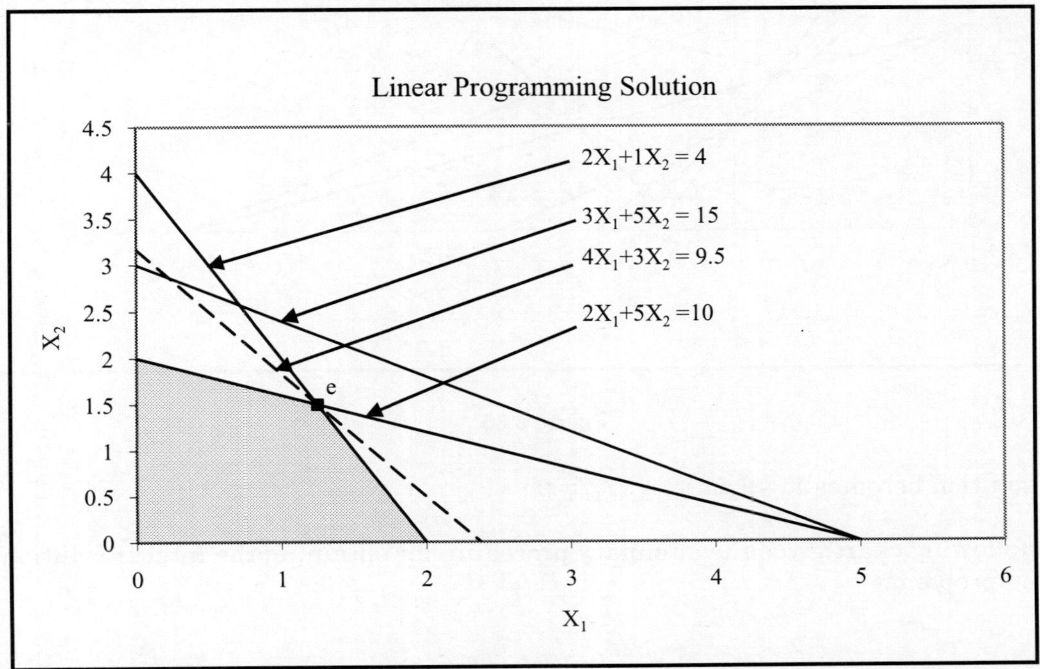

Figure 6.34

So from the figure the solution values are: $X_1=1.5$; $X_2= 1$; $Z = 9$.

Mixed Integer Solution: The Cut

$$-(1/4)S_3 - (1/4)S_1 \leq -1/2$$
$$-1/4(10 - 2X_1 - 5X_2) - 1/4(4 - X_2 - 2X_1 \leq -1/2$$
$$-10/4 + 1/2X_1 + 1.25X_2 - 1 + 0.25X_2 + 0.5X_1 \leq -0.5$$
$$X_1 + 1.5X_2 \leq -0.5 + 2.5 + 1$$
$$X_1 + 1.5X_2 \leq 3$$

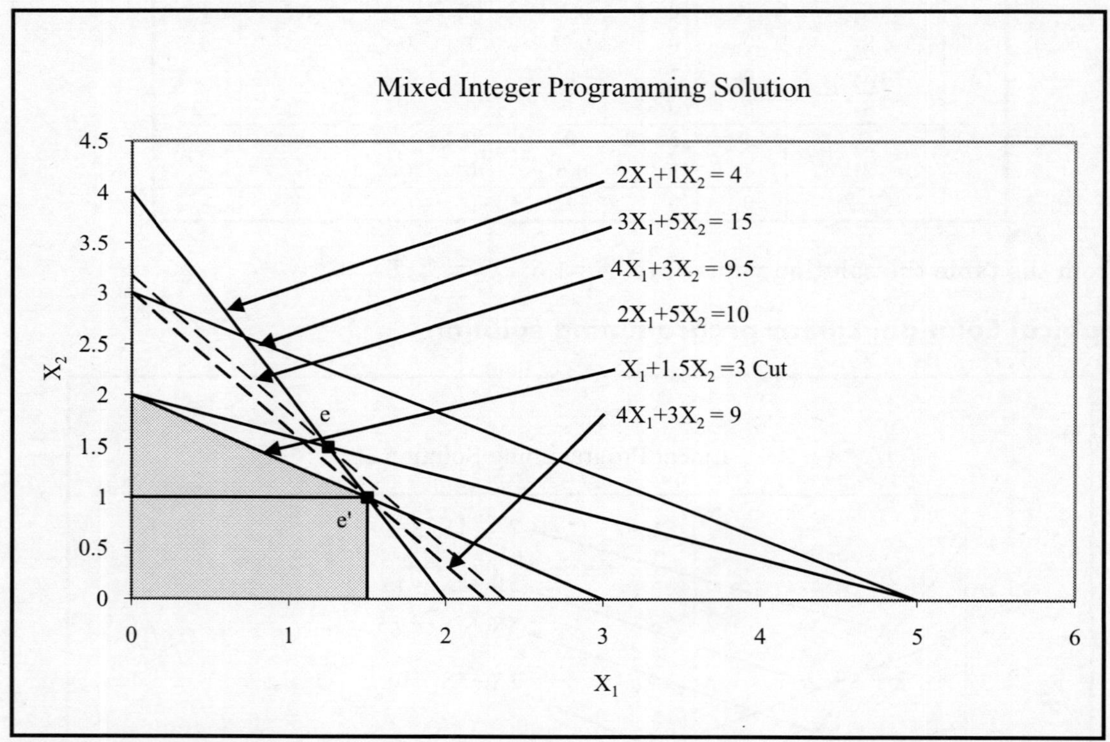

Figure 6.35

So the solution becomes $X_1= 1.5$; $X_2= 1$; $Z = 9$

The following chart gives the summary procedure for obtaining the integer solution to the given problem.

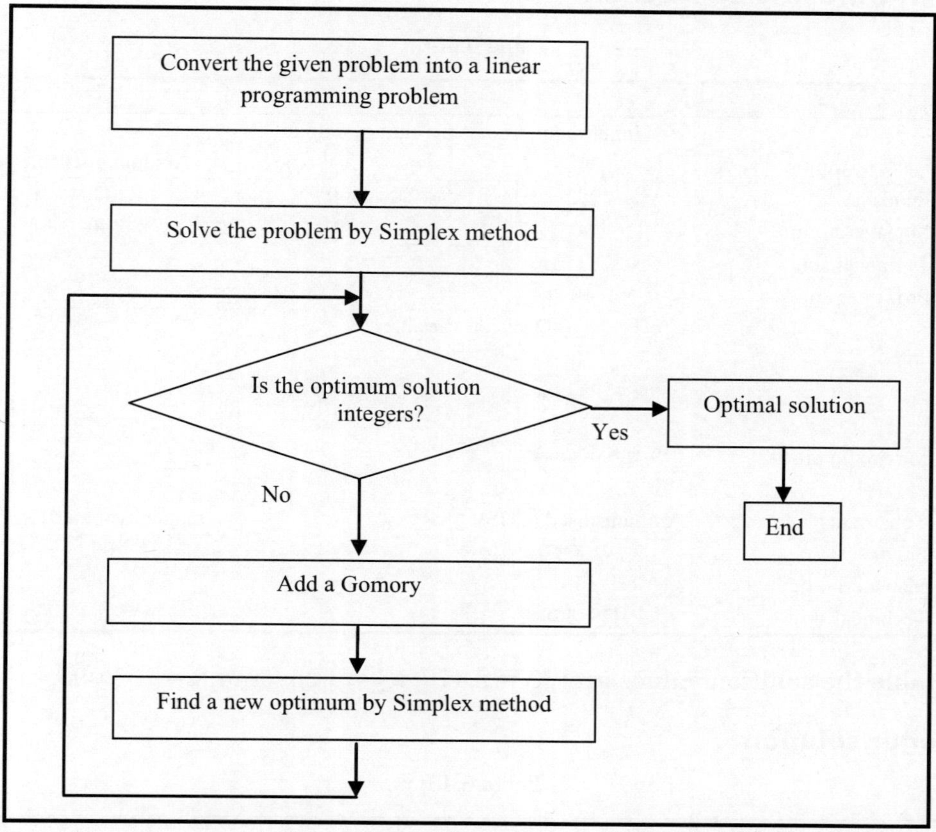

Figure 6.36

6.9 EXCEL SOLUTION TO INTEGER PROGRAMMING

■ Example 15

Solve the following linear programming problem.

a) Unrestricted solution
b) All integer solution
c) X_1 integer solution
d) X_2 integer solution

$$\text{Maximize: } Z = 10X_1 + 25X_2$$

$$\begin{aligned} \text{Subject to:} \quad & 282X_1 + 400X_2 \leq 2000 \quad \text{(Funds)} \\ & 4X_1 + 40X_2 \leq 140 \quad \text{(Managerial Time)} \\ & X_1 \leq 5 \quad \text{(Technical Skill)} \\ & X_1 ;\ X_2 \geq 0 \quad \text{and integer} \end{aligned}$$

Solution: a) Unrestricted solution

Table 6.45

	A	B	C	D
1		Input requirements per unit of output		
2	Inputs	X_1	X_2	Amount available
3	Funds	282	400	2000
4	Managerial time	4	40	140
5	Technical skill	1	0	5
6	Profit per unit	10	25	
7		Decision variables		
8		X_1	X_2	
9		2.479338842	3.25206116	
10				
11	Maximum profit	106.0950413		
12				
13	Constraints	Amounts used (LHS)		Amount available (RHS)
14	Funds	2000	<=	2000
15	Managerial time	140	<=	140
16	Technical skill	2.479338842	<=	5

From the table the solution values are: $X_1 = 2.479$, $X_2 = 3.252$ and R = 106.095.

b) All integer solution

Table 6.46

	A	B	C	D
1		Input requirements per unit of output		
2	Inputs	X_1	X_2	Inputs available
3	Funds	282	400	2000
4	Managerial time	4	40	140
5	Technical skill	1	0	5
6	Profit per unit	10	25	
7		Decision variables		
8		X_1	X_2	
9		4	2	
10				
11	Maximum profit	90		
12				
13	Constraints	Inputs used		Inputs available
14	Funds	1928	<=	2000
15	Managerial time	96	<=	140
16	Technical skill	4	<=	5

From the table the solution values are: $X_1 = 4$, X = 2 and value $=10\times4+25\times2 = 90$

c) X_1 integer solution

Table 6.47

	A	B	C	D
1		Input requirements per unit of output		
2	Inputs	X_1	X_2	Inputs available
3	Funds	282	400	2000
4	Managerial time	4	40	140
5	Technical skill	1	0	5
6	Profit per unit	10	25	
7		Decision variables		
8		X_1	X_2	
9		3	2.885	
10				
11	Maximum profit	102.125		
12				
13	Constraints	Inputs used		Inputs available
14	Funds	2000	<=	2000
15	Managerial time	127.4	<=	140
16	Technical skill	3	<=	5

From the table the solution values are: $X_1 = 3$, $X = 2.655$, value $=10 \times 3 + 25 \times 2.885 = 102.125$.

d) X_2 integer solution

Table 6.48

	A	B	C	D
1		Input requirements per unit of output		
2	Inputs	X_1	X_2	Inputs available
3	Funds	282	400	2000
4	Managerial time	4	40	140
5	Technical skill	1	0	5
6	Profit per unit	10	25	
7		Decision variables		
8		X_1	X_2	
9		2.836879433	3	
10				
11	Maximum profit	103.3687943		
12				
13	Constraints	Inputs used		Inputs available
14	Funds	2000	<=	2000
15	Managerial time	131.3475177	<=	140
16	Technical skill	2.836879433	<=	5

From the table the solution values are: $X_1 = 2.837$, $X_2 = 3$, $R = 10 \times 2.8369 + 25 \times 3 = 103.369$.

■ Example 16

Solve the following linear programming.

a) Unrestricted solution

b) All integer solution

c) X_1 integer solution

d) X_2 integer solution

$$\text{Maximize:} \quad Z = X_1 - X_2$$
$$\text{Subject to:} \quad X_1 + 2X_2 \le 4$$
$$6X_1 + 2X_2 \le 9$$
$$X_1 ;\ X_2 \ge 0 \text{ and integer}$$

Solution: a) Unrestricted solution

Table 6.49

	A	B	C	D
1		Input requirements per unit of output		
2	Inputs	X_1	X_2	Inputs available
3	Cash	1	2	4
4	Managerial time	6	2	9
5	Profit per unit	1	−1	
6		Decision variables		
7		X_1	X_2	
8		1.5	0	
9				
10	Maximum profit	1.5		
11	Constraints	Inputs used		Inputs available
12	Cash	1.5	<=	4
13	Managerial time	9	<=	9

From the table the solution values are: $X_1 = 1.5$, $X_2 = 0$, $R = 1 \times 1.5 - 1 \times 0 = 1.5$

b) All integer solution

Table 6.50

	A	B	C	D
1		Input requirements per unit of output		
2	Inputs	X_1	X_2	Inputs available
3	Cash	1	2	4
4	Managerial time	6	2	9
5	Profit per unit	1	−1	
6		Decision variables		
7		X_1	X_2	
8		1	0	
9				
10	Maximum profit	1		
11	Constraints	Inputs used		Inputs available
12	Cash	1	<=	4
13	Managerial time	6	<=	9

From the table the solution values are: $X_1 = 1$, $X_2 = 0$, $R = 1 \times 1 - 1 \times 0 = 1$

c) X_1 integer solution

Table 6.51

	A	B	C	D
1		Input requirements per unit of output		
2	Inputs	X_1	X_2	Inputs available
3	Cash	1	2	4
4	Managerial time	6	2	9
5	Profit per unit	1	−1	
6		Decision variables		
7		X_1	X_2	
8		1	0	
9				
10	Maximum profit	1		
11				
12	Constraints	Inputs used		Inputs available
13	Cash	1	<=	4
14	Managerial time	6	<=	9

From the table the solution values are: $X_1 = 1$, $X_2 = 0$, $R = 1 \times 1 - 1 \times 0 = 1$

d) X_2 integer solutions

Table 6.52

	A	B	C	D
1		Input requirements per unit of output		
2	Inputs	X_1	X_2	Amount available
3	Cash	1	2	4
4	Managerial time	6	2	9
5	Profit per unit	1	−1	
6		Decision variables		
7		X_1	X_2	
8		1.5	0	
9				
10	Maximum profit	1.5		
11				
12	Constraints	Amounts used (LHS)		Amount available (RHS)
13	Cash	1.5	<=	4
14	Managerial time	9	<=	9

From the table the solution values are: $X_1 = 1.5$, $X_2 = 0$, $R = 1 \times 1.5 - 1 \times 0 = 1.5$

■ Example 17

Solve the following linear programming for all integer solution by Excel solver.

$$\text{Maximize:} \quad Z = 5X_1 + 3X_2$$
$$\text{Subject to:} \quad 4X_1 + 2X_2 \le 25$$
$$X_1 \le 5$$
$$X_2 \le 8$$
$$X_1;\ X_2 \ge 0 \text{ and integers}$$

Solution

Table 6.53

	A	B	C	D
1		Input requirements per unit of output		
2	Inputs	X_1	X_2	Amount available
3	X	4	2	25
4	Y	1	0	5
5	Z	0	1	8
6	Profit per unit	5	3	
7		Decision variables		
8		X_1	X_2	
9		2	8	
10				
11	Maximum profit	34		
12				
13	Constraints	Amounts used (LHS)		Amount available (RHS)
14	X	24	<=	25
15	Y	2	<=	5

■ Example 18

Solve the following linear programming for X_2 integer value.

$$\text{Maximize:} \quad Z = 3X_1 + 4X_2$$
$$\text{Subject to:} \quad 3X_1 - X_2 \le 12$$
$$3X_1 - 11X_2 \le 66$$
$$X_1;\ X_2 \ge 0 \text{ and } X_2 \text{ integer}$$

Table 6.54

	A	B	C	D
1		Input requirements per unit of output		
2	Inputs	X_1	X_2	Amount available
3	A	3	−1	12
4	B	3	11	66
5	Profit per unit	3	4	
6		Decision variables		
7		X_1	X_2	
8		5.333333333	4	
9				
10	Maximum profit	32		
11				
12	Constraints	Amounts used (LHS)		Amount available (RHS)
13	A	12	<=	12
14	B	60	<=	66

6.10 USE OF '0–1' INTEGER PROGRAMMING

Several real life problems require only integer 0 or 1 for the decision variables. Such problems revolve around 0, 1 integer values depicting a 'yes' or 'no' situation. The assignment problem discussed in the previous section is one such typical example. Some other areas suitable for these types of binary variables are capital budgeting, project selection machines / job sequencing etc. Such problems are often solved by branch-and-bound method outlined in the previous section. Not only this method needs large number of iterations but also becomes inefficient when the problem is large. So in this section we propose to use Balas' implicit enumeration method to solve such problems.

1. Application in capital budgeting

■ Example 19

The Prestige, a real estate company in Bangalore, wants to invest in several projects under consideration over the next four years. However, with limited funds availability over the said period the management would like to select the most profitable projects only by keeping the yearly availability of funds in mind. The estimated net present value of the future stream of income, the annual capital requirements and the annual capital availability and other details are given below.

Table 6.55

	Plant expansion	Warehouse expansion	New machinery	Research	Total capital available
Net Present value	90,000	40,000	10,000	37,000	–
PV of I year Capital	15,000	10,000	10,000	15,000	40,000
PV of 2 year Capital	20,000	15,000		10,000	50,000
PV of 3 year Capital	20,000	20,000		10,000	40,000
PV of 4 year Capital	15,000	5,000	4,000	10,000	35,000

Solve the problem for optimum solution.

Solution

The four 0–1 decision variables are defined as follows

P = 1 if the plant expansion project is executed; 0 not executed.

W = 1 if the warehouse expansion project is executed; 0 not executed.

M = 1 if the machinery expansion project is executed; 0 not executed.

R = 1 if the research expansion project is executed; 0 not executed.

The company will try to maximize the net present value of the capital budgeting project. So the objective function to this problem may be written as:

$$\text{Max: } 90P+40W+10M+37R \text{ (in thousands)}$$

The relevant constraints are

$$\begin{aligned} \text{Maximize:} \quad & Z = 90P + 40W + 10M + 37R \text{ (in thousands)} \\ \text{Subject to:} \quad & 15P + 10W + 10M + 15R \le 40 \\ & 20P + 15W + 0M + 10R \le 50 \\ & 20P + 20W + 0M + 10R \le 40 \\ & 15P + 5W + 4M + 10R \le 35 \\ & P;\ W;\ M;\ R = 0,1 \end{aligned}$$

Excel Solution

Table 6.56

	A	B	C	D	E	F
1		Requirements				
2	Origin	P	W	M	R	Supply
3	Year 1	15	10	10	15	40
4	Year 2	20	15		10	50
5	Year 3	20	20		10	40
6	Year 4	15	5	4	10	35
7	Present value	90	40	10	37	
8		Decision variables				
9		1	1	1	0	
10						
11	Objective max	=SUMPRODUCT(B7:E7,B9:E9)				
12						
13	Constraints	Used				Available
14	Year 1	=SUMPRODUCT(B3:E3,$B$9:$E$9)		<=		=F3
15	Year 2	=SUMPRODUCT(B4:E4,$B$9:$E$9)		<=		=F4
16	Year 3	=SUMPRODUCT(B5:E5,$B$9:$E$9)		<=		=F5
17	Year 4	=SUMPRODUCT(B6:E6,$B$9:$E$9)		<=		=F6

Table 6.57

	A	B	C	D	E	F
1		Requirements				
2	Origin	P	W	M	R	Supply
3	Year 1	15	10	10	15	40
4	Year 2	20	15		10	50
5	Year 3	20	20		10	40
6	Year 4	15	5	4	10	35
7	Present value	90	40	10	37	
8		Decision variables				
9		1	1	1	0	
10						
11	Objective max	140				
12						
13	Constraints	Used				Available
14	Year 1	35		<=		40
15	Year 2	35		<=		50
16	Year 3	40		<=		40
17	Year 4	24		<=		35

Note: This time remember b_{in} restriction is to be imposed to B9:E9.

The integer programming solution is shown in Table 6.57. The optimum solution is P = 1, W = 1, M = 1, R = 0 with a total estimated present value of Rs. 1,40,000. Thus the company will fund the plant expansion, the warehouse expansion, and the machinery projects without any problem. The new research project is to be put off unless additional funds are made available. The value of the slack (excess supply) in the first year is Rs. 5000. The value of the slack in the second year is Rs. 15,000. The value of the slack in the fourth year is Rs. 11,000. Since the first year surplus is Rs. 5000 and first year requirement for research project is Rs.15,000 one will have to find an additional amount Rs. 10,000 in the very first year if at all they want to have the research project also on line in addition.

Similarly, the surplus available during the third year is Rs. 10,000. So the entire amount of Rs. 10,000 will have to be fed into the system during the third year from some other source. The following revised calculations after adjusting the needs in both the first and the third year show that the research project is also feasible (Table 6.58).

Table 6.58

	A	B	C	D	E	F
1				Requirements		
2	Origin	P	W	M	R	Supply
3	Year 1	15	10	10	15	50
4	Year 2	20	15		10	50
5	Year 3	20	20		10	50
6	Year 4	15	5	4	10	35
7	Present value	90	40	10	37	
8				Decision variables		
9		1	1	1	1	
10						
11						
12	Objective max	177				
13						
14	Constraints	Used				Available
15	Year 1	50		<=		50
16	Year 2	45		<=		50
17	Year 3	50		<=		50
18	Year 4	34		<=		35

P = 1, W = 1, M = 1, R = 1 and Z = 177.

Example 20

Infosys, a software company in Bangalore, wants to invest in several projects under consideration over the next two years. However, with limited fund availability over the said period the management would like to select the most profitable projects only. The estimated net present value, the capital requirements and capital availability and other details are given below. Obtain the optimum solution.

Table 6.59

	Projects									Total capital available
	1	2	3	4	5	6	7	8	9	
Net present value	16	15	17	15	40	12	13	9	12	–
Present value of the funds required in year 1	13	50	7	6	36	6	46	38	18	50
Present value of the funds required in year 2	3	5	6	2	34	6	4	7	3	20

Maximize: $16X_1 + 15X_2 + 17X_3 + 15X_4 + 40X_5 + 12X_6 + 13X_7 + 9X_8 + 12X_9$

Subject to: $13X_1 + 50X_2 + 7X_3 + 6X_4 + 36X_5 + 6X_6 + 46X_7 + 38X_8 + 18X_9 \leq 50$

$3X_1 + 5X_2 + 6X_3 + 2X_4 + 34X_5 + 6X_6 + 4X_7 + 7X_8 + 12X_9 \leq 20$

$X_1;\ X_2;\ X_3;\ X_4;\ X_5;\ X_6;\ X_7;\ X_8;\ X_9 = 0.1$

Solution

Table 6.60

	A	B	C	D	E	F	G	H	I	J	K
1		Requirements									
2	Origin	X_1	X_2	X_3	X_4	X_5	X_6	X_7	X_8	X_9	Supply
3	Year 1	13	50	7	6	36	6	46	38	18	50
4	Year 2	3	8	6	2	34	6	4	7	3	20
5	Present value	16	15	17	15	40	12	13	9	12	
6		Decision variables									
7		1	0	1	1	0	1	0	0	1	
8											
9	Objective max	72									
10											
11	Constraints	Used				Available					
12	Year 1	50		<=		50					
13	Year 2	20		<=		20					

Thus the optimal projects are 1, 3 ,4 ,6, 9. Such a choice makes use of all the available resources fully. The associated maximum profit is Rs.72 lakhs.

General Formulation of Capital Budgeting

$$\text{Maximize: } \sum_{j=1}^{n} b_j X_j$$

$$\text{Subject to: } \sum_{j=1}^{n} c_{ij} X_j \leq B \quad i=1,2,.....m$$

$$X_j = 0 \text{ or } 1$$

6.11 APPLICATIONS IN FIXED COST MODELLING

In many situations, the cost of production consists of two components, namely a set up cost (fixed cost) and a variable cost. The use of 0–1 variable approach is more ideal for this type of problem.

■ Example 21

Suppose A, B, C are the three products, with the per unit input requirements given below. If the respective sales profit are Rs. 40, Rs. 30, Rs. 50 obtain the maximum revenue.

Table 6.61

Inputs	Material need per unit of output			Supply
	A	B	C	
Material 1	0.4	0.5	0.6	20
Material 2		0.2	0.1	5
Material 3	0.6	0.3	0.3	21

Solution

From the data given the corresponding linear programming problem may be written as

$$\begin{aligned} \text{Maximize: } R &= 40A + 30B + 50C \\ \text{Subject to: } & 0.4A + 0.5B + 0.6C \leq 20 \\ & 0.0A + 0.2B + 0.1C \leq 5 \\ & 0.6A + 0.3B + 0.3C \leq 21 \\ & A\ ;\quad B\ ;\quad C \geq 0 \end{aligned}$$

The spreadsheet solution for this problem is shown below.

Table 6.62

	A	B	C	D	E
1	Material	A	B	C	Amount available
2	Material 1	0.4	0.5	0.6	20
3	Material 2		0.2	0.1	5
4	Material 3	0.6	0.3	0.3	21
5	Profit per unit	40	30	50	
6		Decision variables			
7		A	B	C	
8		27.5	0	15	
9					
10	Maximum profit	1850			
11					
12	Constraints	Amounts used (LHS)			Amount available (RHS)
13	Material 1	20	<=		20
14	Material 2	1.5	<=		5
15	Material 3	21	<=		21

The optimum values are A =27.5, B = 0, C =15, Maximum Profit = Rs. 1850.

■ Example 22

Suppose the following are the setup cost with their respective capacities for the products A, B, C, formulate a suitable linear programming problem and obtain its solution.

Table 6.63

Product	Setup cost	Maximum output
A	200	50
B	50	25
C	400	40

Solution

Let us incorporate this additional information by using 0–1 variable into our model.

S_A = 1 if the product A is produced, 0 if not.

S_B = 1 if the product B is produced, 0 if not.

S_C = 1 if the product C is produced, 0 if not.

So our new objective function becomes: Max. $= 40A+30B+50C-200S_A-50S_B-400S_C$.

Now we must go for production capacity constraints by using 0–1 variables. If the setup variable is 1 production of the said output is permitted to the given maximum capacity. On the other hand if the setup variable is 0 then we say that the said product is not permissible. For example, if $S_A = 1$ then we say that the product A is permissible to its maximum capacity, namely 50. So the constraints for all the three setup variables may be written as

$$A \le 50\ S_A$$
$$B \le 25\ S_B$$
$$C \le 40\ S_C$$

After moving all the RHS to LHS in the above constraints the complete model may be written as

$$\begin{aligned}
\text{Maximize: } R = &\ 40A + 30B + 50C - 200S_A - 50S_B - 400S_C \\
\text{Subject to: } &\ 0.4A + 0.5B + 0.6C \le 20 \\
&\ 0.0A + 0.2B + 0.1C \le 5 \\
&\ 0.6A + 0.3B + 0.3C \le 21 \\
&\ A - 50S_A \le 0 \\
&\ B - 25S_B \le 0 \\
&\ C - 400S_C \le 0 \\
&\ A\ ;\ B\ ;\ C \ge 0 \quad S_A\ ;\ S_B\ ;\ S_C = 1,0
\end{aligned}$$

For this problem the spreadsheet solution is shown below.

Table 6.64

	A	B	C	D	E	F	G
1		Requirements					
2	Constraints	A	B	C	Supply		
3	1	0.4	0.5	0.6	20		
4	2		0.2	0.1	5		
5	3	0.6	0.3	0.3	21		
6	Profit	40	30	50	−200	−50	−400
7	Setup cost						
8	Model						
9		A	B	C	S_A	S_B	S_C
10		25	20	0	1	1	0
11							
12	Max profit	1350					
13							
14	Constraints	Used			Available		
15	1	20		<=	20		
16	2	4		<=	5		
17	3	21		<=	21		
18	4	−25		<=	0		
19	5	−5		<=	0		
20	6	0		<=	0		

6.12 APPLICATION IN PRODUCTION AND DISTRIBUTION PROBLEMS

Example 23

The MTR foods in Bangalore operate with an annual capacity of 30,000 units Jamoon mix and supply the same to Madurai, Pune and Agra. Because of anticipated demand in future, the company wants to setup four new productive capacities at Chennai, Mumbai, Kolkata, and New Delhi. The estimated annual fixed cost and the respective capacities are shown below.

Table 6.65

Location	Setup cost	Maximum capacity
Chennai	1,75,000	10,000
Kolkata	3,00,000	20,000
Mumbai	3,75,000	30,000
New Delhi	5,00,000	40,000

The demands at the three destinations are listed below.

Table 6.66

Destination	Demand
Agra	30,000
Pune	20,000
Madurai	20,000

From the old and new production centres the per unit transport cost are given below:

Table 6.67

Location	Chennai	Pune	Agra
Chennai	5	2	3
Kolkata	4	3	4
Mumbai	9	7	5
New Delhi	10	4	2
Bangalore	8	4	3

Solution

With these given data let us use the 0–1 variable to obtain the best possible plant locations among the four proposals in hand.

If $p_1 = 1$ then plant is constructed in Chennai, 0 if not

If $p_2 = 1$ then plant is constructed in Kolkata, 0 if not

If $p_3 = 1$ then plant is constructed in Mumbai, 0 if not

If $p_4 = 1$ then plant is constructed in New Delhi, 0 if not

If X_{ij} = the transport cost from i[th] plant to j[th] destination then our objective function becomes

$$\text{Max: } X_{11}+2X_{12}+3X_{13}+4X_{21}+3X_{22}+4X_{23}+9X_{31}+7X_{32}+5X_{33}+10X_{41}+4X_{42}+2X_{43}+8X_{51}+4X_{52}+3X_{33}$$

The annual fixed cost of operation of all the four new plants in thousand rupees is written as

$$175p_1+300p_2+375p_3+500p_4$$

In the above formulation if $p_i = 1$ only such plants are taken for the cost calculation. If $p_i = 0$ these plants are not going to be executed and hence no cost is going to be incurred. Now let us construct capacity constraints for all the four new plants as follows:

$$X_{11}+X_{12}+X_{13}+X_{14} \leq 10p_1$$

If $p_1 = 1$ then the plant at Chennai will be taken up and supply is going to be augmented to the maximum of 10,000 units. On the other hand if $p_1 = 0$ then the plant will not be taken up. So $X_{11}= X_{12} = X_{13} = X_{14}= 0$ by definition. Now after transferring the RHS term into LHS the relevant constraint pertaining to Chennai production centre is written as

$$X_{11}+X_{12}+X_{13}+X_{13}-10p_1=0 \text{ (Chennai)}$$

Similarly for other centres the relevant constraints are

$$X_{21}+X_{22}+X_{23}+X_{24}-20p_2=0 \text{ (Kolkata)}$$
$$X_{33}+X_{32}+X_{33}+X_{34}-30p_3=0 \text{ (Mumbai)}$$
$$X_{41}+X_{42}+X_{43}+X_{44}-40p_2=0 \text{ (New Delhi)}$$

Note that the plant already exists in Bangalore hence we do not use 0–1 variable for this plant. So the relevant constraint is written as

$$X_{51}+X_{52}+X_{53} \leq 30$$

Now let us write the demand constraints for all the three destinations.

$$X_{11}+X_{21}+X_{31}+X_{41}+X_{51}=30$$
$$X_{12}+X_{22}+X_{32}+X_{42}+X_{52}=20$$
$$X_{13}+X_{23}+X_{33}+X_{43}+X_{51}=20$$

Now we write the complete model as follows

Maximize: $5X_{11} + 2X_{12} + 3X_{13} + 4X_{21} + 3X_{22} + 4X_{23} + pX_{31} + 7X_{32} + 5X_{33} + 10X_{41}$

Subject to: $+4X_{42} + 2X_{43} + 8X_{51} + 4X_{52} + 3X_{53} + 175p_1 + 300p_2 + 375p_3 + 500p_4$

$$X_{11} + X_{12} + X_{13} - 10p_1 \leq 0$$
$$X_{21} + X_{22} + X_{23} - 20p_2 \leq 0$$
$$X_{31} + X_{32} + X_{33} - 30p_3 \leq 0$$
$$X_{41} + X_{42} + X_{43} - 40p_4 \leq 0$$
$$X_{51} + X_{52} + X_{53} \leq 30$$
$$X_{11} + X_{21} + X_{31} + X_{41} + X_{51} = 30$$
$$X_{12} + X_{22} + X_{32} + X_{42} + X_{52} = 20$$
$$X_{13} + X_{23} + X_{33} + X_{43} + X_{53} = 20$$
$$X_{ij} \geq 0 \text{ for all i and j} \qquad p_1 \;;\; p_2 \;:\; p_3 \;;\; p_4 = 0,1$$

From the spreadsheet in Table 6.68 the following are the sclutions.

X_{41}=20; X_{42}=20; X_{51} = 30, p_4 = 1.The value of the objective fcnction = Rs. 860

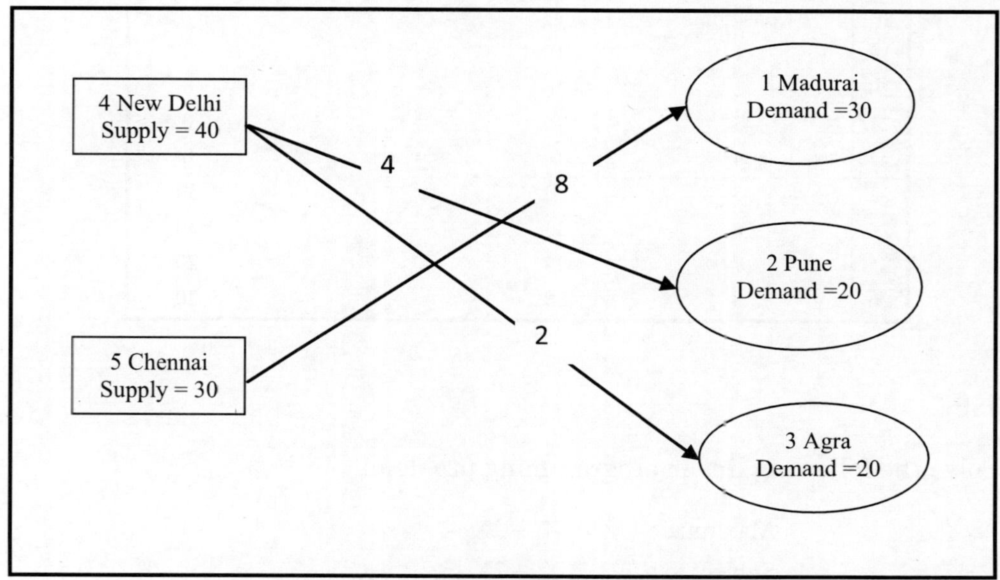

Figure 6.37

Table 6.68

	A	B	C	D	E
1			Destination		
2	Production node	1	2	3	Supply
3	1	5	2	3	10
4	2	4	3	4	20
5	3	9	7	5	30
6	4	10	4	2	40
7	5	8	4	3	30
8	Demand	30	20	20	
9			Destination		
10	Production node	1	2	3	P_i
11	1	0	0	0	0
12	2	0	0	0	0
13	3	0	0	0	0
14	4	0	20	20	1
15	5	30	0	0	
16	Model:				
17	Minimum cost:	860			
18	Constraints				
19	1	0	<=		0
20	2	0	<=		0
21	3	0	<=		0
22	4	0	<=		0
23	5	30	<=		30
24	6	30	=		30
25	7	20	=		20
26	8	20	=		20

EXERCISE

1. Solve the following linear programming problem.

$$\begin{aligned} \text{Maximize}: \quad & Z = X_1 + 2X_2 \\ \text{Subject to}: \quad & X_1 + 2X_2 \le 12 \\ & 4X_1 + 3X_2 \le 14 \\ & X_1 ; \quad X_2 \ge 0 \quad \text{and integers} \end{aligned}$$

2. Solve the following linear programming problem.

$$\begin{aligned} \text{Maximize}: \quad & Z = 3X_1 + 4X_2 \\ \text{Subject to}: \quad & 3X_1 + 2X_2 \leq 8 \\ & X_1 + 4X_2 \geq 10 \\ & X_1 ;\ X_2 \geq 0 \text{ and integers} \end{aligned}$$

3. Solve the following linear programming problem

$$\begin{aligned} \text{Maximize}: \quad & Z = 7X_1 + 9X_2 \\ \text{Subject to}: \quad & -X_1 + 3X_2 \leq 6 \\ & 7X_1 + X_2 \leq 35 \\ & X_1 ;\ X_2 \geq 0 \text{ and integers} \end{aligned}$$

4. Solve the following linear programming problem

$$\begin{aligned} \text{Maximize}: \quad & Z = X_1 + 2X_2 \\ \text{Subject to}: \quad & 2X_2 \leq 7 \\ & X_1 + X_2 \geq 7 \\ & 2X_1 \geq 11 \\ & X_1 ;\ X_2 \geq 0 \text{ and integers} \end{aligned}$$

5. The owner of a readymade garments store makes two types of shirt known as T-shirts and XL-shirts. He makes a profit of Re.1 and Rs. 4 per shirt on Zee and Button-down shirts respectively. He has two tailors (Tailor A and Tailor B) at his disposal to stitch these shirts. Tailor A and tailor B can devote at the most 7 hours/day and 15 hours/day respectively. Both these shirts are to be stitched by both the tailors. Tailor A and tailor B spend 2 hrs and 5 hrs respectively in stitching a T-shirt and 4 hrs and 3 hrs respectively in stitching a XL-shirt. How many shirts of both the types should be stitched in order to maximize daily profit?

   (i) Set-up and solve the linear programming problem.
   (ii) If the optimum solution is not integer valued, use Gomory's technique to derive the optimum integer solution.

6. A manufacturer of toys make two types of toy A and B. Processing these toys is done on two machines X and Y. The toy A requires 2 hrs on machine X and 6 hrs, on machine Y. Toy B requires 4 hrs. On machine X and 5 hrs on machine Y. The profit obtained on both the toys is the same i.e., Rs. 5/toy. What should be the daily production of each of the two toys? (A non-integer solution for this problem will not be accepted.)

CHAPTER

7

Assignment Problems

- *Introduction*
- *Methods for solving assignment problem*
- *Maximization problems*
- *Variants of assignment problems*
- *Branch-and-bounds method for solving assignment problems*
- *Solution to travelling salesmen problem by Hungarian method*
- *Solution to travelling salesmen problem by branch-and-bound method*
- *Linear programming formulation to the assignment problem*
- *Production and inventory applications*
- *Exercises*

7.1 INTRODUCTION

Assignment of workers to different machines, assignment of clerical staff to different work desks, assignment of salesmen to different sales areas, assignment of service crews to different aircraft are some of the examples of the practical assignment problems. In all these cases we assign a particular job or machine to a particular worker depending upon the work efficiency. In this section we propose to discuss the assignment of n specialists to n jobs on one-to-one basis so that the total time cost of the entire project is minimized.

If C_{ij} represents the time cost of assigning i^{th} individual to j^{th} job then the time cost for the entire model could be stated in a matrix form as shown in Table 7.1.

From the given cost matrix, one could easily visualize the assignment problem as a special case of the transportation problem discussed in the previous section. In transportation models the supplies from origins are represented by s_i in the last column.

Table 7.1

Individuals	Jobs						s_i
	1	2	. . .	j	. . .	n	
1	c_{11}	c_{12}	. . .	c_{1j}	. . .	c_{1n}	1
2	c_{21}	c_{22}	. . .	c_{2jj}	. . .	c_{2n}	1
.	.	.	. . .	.	. . .	.	.
.	.	.	. . .	.	. . .	.	.
i	c_{i1}	c_{i2}	. . .	c_{ij}	. . .	c_{in}	1
.	.	.	. . .	.	. . .	.	.
.	.	.	. . .	.	. . .	.	.
n	c_{n1}	c_{n2}	. . .	.	. . .	c_{nn}	1
d_j	1	1	. . .	1	. . .	1	

Such s_i's can take any value depending upon the supply conditions from the respective origins. In assignment problems since only one person is available in each type, all s_i's are unity as shown in the Table 7.1. Similarly, in transportation models the demand from each destination can take any value and commonly denoted by d_j. In assignment models these entries also takes the unit values because of the one-to-one rule of assignment. Thus, the assignment problem is a special type of transportation problem where m = n and $s_i = d_j = 1$ for all i and j values running from 1 to n.

The associated transportation equivalent assignment problem in its standard form is written as

$$\text{Minimize: } \sum_{i=1}^{n}\sum_{j=1}^{n} c_{ij}X_{ij}$$

$$\text{Subject to: } \sum_{j=1}^{n} X_{ij}=1 \quad i=1, 2, 3 \text{ .... } n \text{ (Supply constraints)}$$

$$\sum_{i=1}^{n} X_{ij}=1 \quad j = 1, 2, 3, 4 \text{ .... } n \text{ (Demand constraints)}$$

$$X_{ij} \geq 0 \text{ and integer.}$$

7.2 METHODS FOR SOLVING ASSIGNMENT PROBLEM

There are four methods of solving assignment problem. They are:

a) Complete Enumeration Method

b) Transportation Method

c) Hungarian Method

d) Branch-and-Bounds Method

e) Linear Programming Method

7.2.1 Complete Enumeration Method

In this method exhaustively all possible assignments are listed out first and the associated time cost for each and every possibility is worked out. The final solution clearly is the one having the minimum total time cost. This method works well provided few jobs are to be assigned to few individuals. For examples, if 3 individuals are to be assigned to 3 jobs then clearly the number of possibilities is given by the formula $^{3}P_{3} = 3! = 3\times2\times1=6$. The following example illustrates the utility of complete enumeration method.

Example 1

Bangalore Development Authority (BDA) received orders for the construction of three flyovers in Bangalore at three locations. It has three specialized engineers for this specific job. The BDA is to assign one job to one individual so that the total operational time is minimized. The following table gives the associated time cost in months for all possible assignments.

Table 7.2

Sl. No	Project leader	Flyovers 1	Flyovers 2	Flyovers 3
1	Ravi	10	15	9
2	Shiva	9	18	5
3	Roopa	6	14	3

Draw a network chart and obtain the optimum assignment that minimizes total time cost for BDA.

Solution

To arrive at the needed optimum solution, the BDA must consider all possible project leader-client combinations and select one that minimizes the total time cost. Since there are three engineers to be allotted for three flyovers there are $^{3}P_{3} = 3! = 3\times2\times1 = 6$ ways of job allotment. Table 7.3 summarizes all the six possibilities and their associated cost calculations. Figure 7.1 illustrates the problem with their respective time cost in months.

Table 7.3

Possibilities	Flyovers 1	Flyovers 2	Flyovers 3	Completion time
1	1	2	3	10+18+3 = 31
2	1	3	2	10+14+5 = 29
3	2	1	3	9+15+3 = 27
4	2	3	1	9+14+9 =32
5	3	1	2	6+15+5 = 26
6	3	2	1	9+18+6 =33

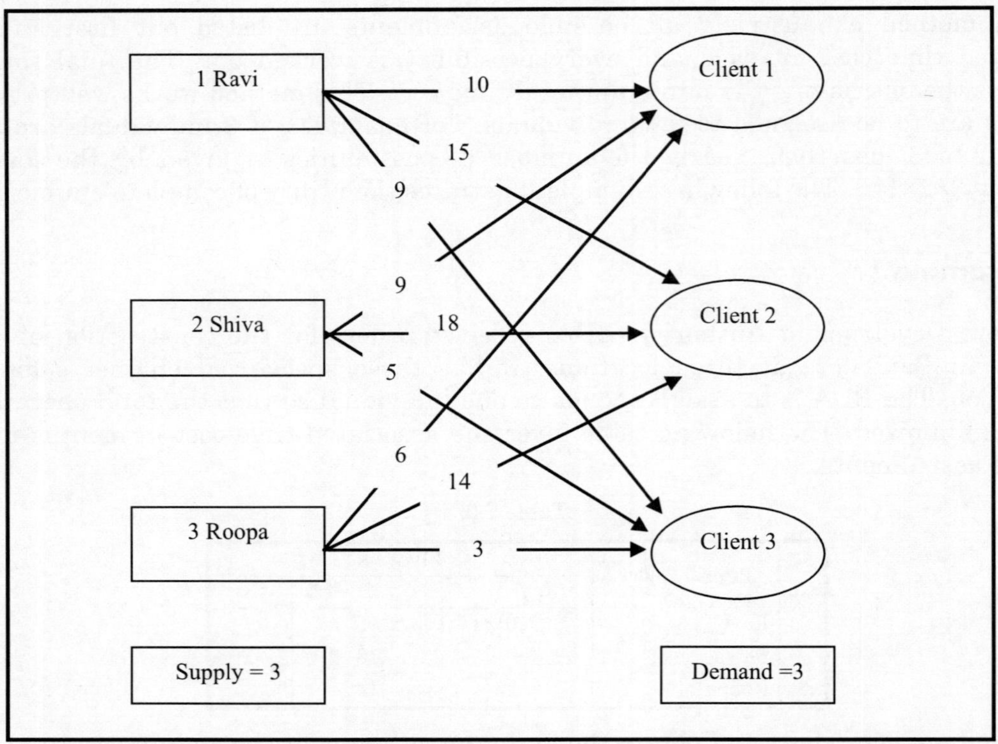

Figure 7.1

Among all assignments, assignment possibility 5 gives the least time cost of 26 months for doing job. Accordingly, Shiva (2) could be allotted to flyover 3, Roopa (3) could be allotted to flyover 1 and Ravi (1) could be allotted to flyover 2. The total minimum time cost to complete all the three projects = 26 months.

Example 2

A production engineer is considering the rational assignment of four jobs to four workers so that the aggregate time cost to perform all the four jobs by all the four workers is minimized. Based on the previous experience he has the following time cost table data.

Table 7.4

Worker	Jobs			
	A	B	C	D
1	45	40	51	67
2	57	42	63	55
3	49	52	48	64
4	41	45	60	55

Solution

In this method all the 24 exhaustive possibilities are being explored and the minimum time consuming assignment is highlighted in Table 7.5.

Assignments:

Worker 4 to job A

Worker 1 to job B

Worker 3 to job C

Worker 2 to job D

Minimum Time Cost = 184 units of time

Table 7.5

Sl. No	Jobs				Time	Sl. No	Jobs				Time
	A	B	C	D			A	B	C	D	
1	1	2	3	4	190	13	3	1	2	4	207
2	1	2	4	3	211	14	3	1	4	2	204
3	1	3	2	4	215	15	3	2	4	1	218
4	1	3	4	2	212	16	3	2	1	4	197
5	1	4	2	3	217	17	3	4	1	2	200
6	1	4	3	2	193	18	3	4	2	1	224
7	2	1	3	4	200	19	4	1	2	3	208
8	2	1	4	3	221	20	4	1	3	2	184
9	2	3	1	4	215	21	4	2	1	3	198
10	2	3	4	1	236	22	4	2	3	1	198
11	2	4	3	1	207	23	4	3	1	2	199
12	2	4	1	3	217	24	4	3	2	1	223

7.2.2 Transportation Method

Since the stated assignment problem is a special case of transportation problem often we will be tempted to use the transportation techniques discussed in the previous section to solve the assignment problems. While making such an effort unfortunately in initial solutions stage itself the solution degenerate. For a transportation solution, as per the optimality conditions we must have at least $n + n - 1 = 2n - 1$ ($m = n$) basic allotted cells. In assignment problem we will be having only n assignments to n jobs on one to one basis. As a consequence $n - 1$ basic variables will take zero values. Thus, if we want to use the transportation method for solving assignment problem we must use $n - 1$ dummy (epsilon) allotments to least cost independent cells in our initial allotment. Because these limitations normally we do not opt this technique for getting the optimum solution. However, to illustrate the utility of this method we propose to work out the problem discussed in Example 2 again by transportation method.

Table 7.6

Worker	Jobs			
	A	B	C	D
1	45	40	51	67
2	57	42	63	55
3	49	52	48	64
4	41	45	60	55

Solution

First let us obtain the initial solution by VAM method as described in Table 7.7.

Table 7.7

Workers		Jobs				s_i	Row penalty			
		A	B	C	D		I	II	III	IV
1		45	40	51	**67** 1	1	5	6	16	**67**
2		57	**42** 1	63	55	1	**13**	-	-	-
3		49	52	**48** 1	64	1	1	1	**16**	-
4		**41** 1	45	60	55	1	4	**14**	-	-
d_j		1	1	1	1					
Column penalty	I	4	2	3	0					
	II	4	-	3	3	Total Cost = 198				
	III	-	-	-	3					
	IV	-	-	-	67					

Clearly the solution degenerates because in the solution table we make only 4 highlighted allotments. Normally for a non-degenerated solution we need 2n – 1 = 7 allotted cells. So to eliminate the degeneracy in our initial solution we place three epsilon (very small quantity) allotments in three independent cells as shown in Table 7.8 and proceed further for the final solution by MODI method. Now after such placements remember that we are having the needed 7 allotted cells altogether.

Table 7.8

Workers		Jobs				Supply
		A	B	C	D	
		V_1	V_2	V_3	V_4	
1	U_1	45 ε	40 ε	51	67 **1**	1
2	U_2	57	42 **1**	63	55	1
3	U_3	49 ε	52	48 **1**	64	1
4	U_4	41 **1**	45	60	55	1
Demand		1	1	1	1	

MODI Method

Now from Table 7.8, for the assigned cells we know that:

$U_1 + V_1 = 45$, $U_1 + V_2 = 40$, $U_1 + V_4 = 67$, $U_2 + V_2 = 42$, $U_3 + V_1 = 49$, $U_3 + V_3 = 48$, $U_4 + V_1 = 41$

The solutions are

$U_1 = 0$, $U_2 = 2$, $U_3 = 4$, $U_4 = -4$, $V_1 = 45$, $V_2 = 40$, $V_3 = 44$, $V_4 = 67$

Now after incorporating these solution values into the Table 7.9 we obtain $C_{ij} - U_j - V_j$ for all the empty cells. The so obtained results are reported in Table 7.9 for all the empty cells.

Table 7.9

Workers		Jobs A	Jobs B	Jobs C	Jobs D	Supply
		$V_1 = 45$	$V_2 = 40$	$V_3 = 44$	$V_4 = 67$	
1	$U_1 = 0$	45 ε	40 ε	51	67 1	1
2	$U_2 = 2$	57 10	42 1	63 17	55 (−14)	1
3	$U_3 = 4$	49 ε	52	48 1	64 1	1
4	$U_4 = -4$	41 1	45 9	60 20	55 −8	1
Demand		1	1	1	1	

In Table 7.9 since –14 is the largest cost saving value for C_{2D}, it deserves for allotment to improve the solution. For such a transfer the associated closed loop is shown in Table 7.9. After effecting the transfer from cell C_{1D} to this cell under reference the results are reported in Table 7.10.

Table 7.10

Workers		Jobs A	Jobs B	Jobs C	Jobs D	Supply
		V_1	V_2	V_3	V_4	
1	U_1	45 ε	40 1	51	67	1
2	U_2	57	42 ε	63	55 1	1
3	U_3	49 ε	52	48 1	64	1
4	U_4	41 1	45	60	55	1
Demand		1	1	1	1	

Now to check for optimality of new assigned cells we know that for allotted cells

$U_1 + V_1 = 45$, $U_1 + V_2 = 40$, $U_2 + V_2 = 42$, $U_2 + V_4 = 55$, $U_3 + V_1 = 49$, $U_3 + V_3 = 48$, $U_4 + V_1 = 41$

The solutions are

$U_1 = 0$, $U_2 = 2$, $U_3 = 4$, $U_4 = -4$, $V_1 = 45$, $V_2 = 40$, $V_3 = 44$, $V_4 = 53$

Table 7.11

Workers		Jobs								Supply
		A		B		C		D		
		$V_1 = 45$		$V_2 = 40$		$V_3 = 44$		$V_4 = 53$		
1	$U_1 = 0$	45	(ε)	40	1	51	7	67	14	1
2	$U_2 = 2$	57	10	42	ε	63	17	55	**1**	1
3	$U_3 = 4$	49	ε	52	8	48	**1**	64	15	1
4	$U_4 = -4$	41	1	45	9	60	20	55	6	1

Now let us incorporate these values into Table 7.11 and calculate the $C_{ij} - U_j - V_j$ for all the empty cells. The results are reported in Table 7.11.

Since all the calculated values for empty cells are positive no further improvement is possible, the obtained solution is the optimum solution.

Assignments:

Worker 4 to job A

Worker 1 to job B

Worker 3 to job C

Worker 2 to job D

The associated minimum time cost for the entire project = 40+55+48+41 = 184 days. This is exactly the same result we got by complete enumeration method.

7.2.3 Hungarian Method

Kuhn first developed a computational algorithm for solving the assignment problem by bringing in a theorem. However, since a Hungarian mathematician Konig has proved the associated theorem in 1961, this method is often called Hungarian Method. Since reduction methodology is the key in this method, it is also called Reduced Matrix Method.

Theorem

If the elements of a matrix are divided into two classes by a property R then the minimum number of lines that contain all the elements with the property R is equal to the maximum number of elements with the property R, with no two elements on the same line.

Property

If a constant is added or subtracted to any row and/or column of the cost matrix of an assignment problem, the resulting assignment problem has the same optimal solution as that of the original problem.

New algorithm for the assignment problem is illustrated in the following example

1. In a cost matrix if some elements are negative, then we convert the problem into an equivalent assignment problem by adding a suitable constant to the relevant column or row so that all the elements are positive in the resulting matrix. As per the theorem such an alteration will leave the original solution unaltered.
2. To assist the final assignment on the basis of zero time cost we bring in as many zeros as possible into the cost matrix by aspirate row and column operations to various rows and columns. Note that such a move will leave the optimum solution unaltered as per the property stated above.
3. As a third step we cover all resulting zeros by minimum number of column/row lines.
4. If the numbers of such minimum lines are equal to n, the number of assignments, we start our assigning work giving priority to zero assignment costs.
5. If the numbers of such lines are less than n we repeat the earlier steps till the condition 4 is satisfied for assignment.

Solution

Table 7.12

	Jobs			
Workers	A	B	C	D
1	45	40	51	67
2	57	42	53	55
3	49	52	48	64
4	41	45	60	55

Now according to Hungarian method we try to bring in more and more zero cost into action by appropriate row and column operations. We know that such a transformation will leave the original solution to the original problem unaltered.

Table 7.13

Workers	A	B	C	D	Row Min.
1	45	40	51	67	40
2	57	42	53	55	42
3	49	52	48	64	48
4	41	45	60	55	41

To bring in more zeros let us obtain the row minimums in Table 7.13 and subtract these minimums from the respective row elements. Such an operation will bring in at least one zero per row as shown in Table 7.14.

In the row operated results in Table 7.14 we cover all the zeros with a minimum of three lines. Since 3 < 4 we continue our effort to bring in some more zeros into operation by the column operation.

Table 7.14

Workers	A	B	C	D
1	5	0	11	27
2	15	0	21	13
3	1	4	0	16
4	0	4	19	14
Column Min.	0	0	0	13

To bring in some more zeros into action this time let us obtain the column minimums Table 7.14 and subtract these minimums from the respective column elements. Such an operation will bring in at least one zero per column as shown in Table 7.15.

Now to cover all zeros in Table 4.15 we need a minimum of four lines as shown in the Table 7.15. Since the minimum number of lines equal to the number of allotments that is to be made in our problem our iteration ends now.

Now for final allotment we search row by row till we get a single zero rows in Table 7.16.

Table 7.15

Workers	A	B	C	D
1	5	0	11	14
2	15	0	21	0
3	1	4	0	3
4	0	4	19	1

Since the very first row is having single zero corresponding to job B and worker 1, let us assign worker 1 to job B and honour the assignment by putting a square over this zero as shown in the Table 7.16. Once this is done we cancel all other zeros in this column corresponding to B and avoid allotment of any other worker to job B in future.

Table 7.16

Workers	A	B	C	D
1	5	[0]	11	14
2	15	0	21	[0]
3	1	4	[0]	3
4	[0]	4	19	1

After effecting this step we continue our search for a single zero row afresh. This time the very second row is found to have the needed single zero corresponding to job D and to worker 2. Accordingly, allot worker 2 to job D and honour it by putting a square over this zero as usual. Also cross it out the leftover zeros in column D if any to avoid any further allotment. After effecting this step we continue our search for a single zero row afresh. This time the third row is having single zero corresponding to column C. So allot worker 3 to job C and honour it by putting the square over this zero as usual. Also cross out the leftover zeros in that column if any.

After effecting this step we continue our search for a single zero row afresh. This time the last row is having the single zero in the very first column. So, allot worker 4 to job A, and complete the assignment. Once this is done, bring in the original cost table and mark the assignment squares as shown in Table 7.17.

Table 7.17

Workers	Jobs A	B	C	D
1	45	[40]	51	67
2	57	42	53	[55]
3	49	52	[48]	64
4	[41]	45	60	55

From Table 7.17 the assignments follows:

Worker 4 to job A

Worker 1 to job B

Worker 3 to job C

Worker 2 to job D

The minimum total time cost for completion = 41+40+55+48= 184 time units.

■ Example 3

Given the following cost matrix, obtain the optimum assignment and the associated minimum operational time cost.

Table 7.18

Operators	Machines 1	2	3	4	5
A	160	130	175	190	200
B	135	120	130	160	175
C	140	110	155	170	185
D	50	50	80	80	110
E	55	35	70	80	105

Solution

Table 7.19

Operators	Machines 1	2	3	4	5	Row Min.
A	160	130	175	190	200	130
B	135	120	130	160	175	120
C	140	110	155	170	185	110
D	50	50	80	80	110	50
E	55	35	70	80	105	35

Step 1: Since all the cost matrix elements are positive there is no need for converting them all into positive numbers.

Step 2: To bring in zeros into action to help our final assignment, first we obtain the row minimums as listed out in the last column of Table 7.19.

Step 3: We subtract these row minimums from their respective rows elements. The resultant matrix is shown in Table 7.20 with at least one zero for each row. Now shade the columns and rows starting from the column or row having the highest number of zeros so that all the zeros in the cost matrix are covered. There are many ways of doing this trick. Among the many ways remember that we consider only the minimum number of lines (shade) scheme for consideration. In our illustration a minimum of 2 lines, one across the second column and another across the fourth row, are needed two cover all zero elements as shown in Table 7.20. Since we are supposed to make 5 (n = 5) assignments in our problem 2 line coverage is not enough for optimum assignment.

Step 4: To bring in some more zeros into action this time we obtain the column minimums and subtract these minimums from their respective columns. Such an action in addition will bring at least one zero for each column. In this modified resultant matrix shown in Table 7.21 we need a minimum of two rows and one column to cover all the zeros. Since once again $(n_1 = 3) < (n = 5)$ the obtained matrix is not enough for final assignment.

Table 7.20

Operators	Machine time 1	2	3	4	5
A	30	0	45	60	70
B	15	0	10	40	55
C	30	0	45	60	75
D	0	0	30	30	60
E	20	0	35	45	70
Column Min.	0	0	10	30	55

Table 7.21

Operators	Machine time 1	2	3	4	5
A	30	0	35	30	15
B	15	0	0	10	0
C	30	0	35	30	20
D	0	0	20	0	5
E	20	0	25	15	15

Step 5: To bring some more zeros in action we identify the minimum value 15 among all the uncovered elements in Table 7.21 and subtract it from all such uncovered elements. At the same time we add this 15 to all the intersection elements of our shades. The final resultant matrix is shown Table 7.22. Now we need at least five lines (three rows and two columns) to cover all the zeros. Since ($n_1 = 5$) = ($n = 5$) we arrive at the final stage of our iteration.

Step 6: Now to make the correct assignment we examine the rows one by one until a row with only one zero element is found. In our illustrative case the third row has a single zero in column 2. As a token of assigning operator C to machine 2, block this zero by a square as shown in the table. Once this is done cross out all other zeros in column 2 to avoid any further allotment to machine 2 as shown in Table 7.23.

Table 7.22

Operators	Machine time 1	2	3	4	5
A	15	0	20	15	0
B	15	15	0	10	0
C	15	0	20	15	5
D	0	15	20	0	5
E	5	0	10	0	0

Table 7.23

Operators	Machine time 1	2	3	4	5
A	15	0	20	15	0
B	15	15	0	10	0
C	15	[0]	20	15	5
D	0	15	20	0	5
E	5	0	10	0	0

Step 7: One again we continue our search for single zero row. This time it is the very first row which has a single zero in the fifth column. As a token of allotting operator A to machine 5 block this zero by a square as shown in Table 7.24. At the same time to avoid any more allotment to machine 5 crosses out other zeros in the fifth column.

Table 7.24

Operators	Machine time 1	2	3	4	5
A	15	0	20	15	[0]
B	15	15	0	10	0
C	15	[0]	20	15	5
D	0	15	20	0	5
E	5	0	10	0	0

Step 8: One again we continue our search for single zero row. This time it is the second row having a single zero in the third column. As a token of allotting operator B to machine 3 block this zero by a square as shown in the Table 7.25. At the same time to avoid any more allotment to machine 3 crosses out other zeros in the third column.

Table 7.25

Operators	Machine time 1	2	3	4	5
A	15	0	20	15	[0]
B	15	15	[0]	10	0
C	15	[0]	20	15	5
D	0	15	20	0	5
E	5	0	10	0	0

Step 9: This time it is the fifth row having a single zero in the fourth column. As a token of allotting operator E to machine 4 block this zero by a square as shown in Table 7.26. At the same time to avoid any more allotment to machine 4, cross out other zeros in the fourth column.

Table 7.26

Operators	Machine time 1	2	3	4	5
A	15	0	20	15	[0]
B	15	15	[0]	10	0
C	15	[0]	20	15	5
D	0	15	20	0	5
E	5	0	10	[0]	0

Step 10: This time it is the fourth row having a single zero in the first column. As a token of allotting operator D to machine 1 block this zero by a square as shown in Table 7.27. At the same time to avoid any more allotment to machine 1 cross out other zeros in the first column. With this allotment we reach the final stage as shown in Table 7.27.

Table 7.27

Operators	Machine time 1	2	3	4	5
A	15	0	20	15	[0]
B	15	15	[0]	10	0
C	15	[0]	20	15	5
D	[0]	15	20	0	5
E	5	0	10	[0]	0

Final Summary Solution:

Operator A is assigned to machine 5

Operator B is assigned to machine 3

Operator C is assigned to machine 2

Operator D is assigned to machine 1

Operator E is assigned to machine 4

Now we use the following original table for our minimum cost calculation.

Table 7.28

Operators	Machine time 1	2	3	4	5
A	160	130	175	190	[200]
B	135	120	[130]	160	175
C	140	[110]	155	170	185
D	[50]	50	80	80	110
E	55	35	70	[80]	105

Now from the Table 7.28 the minimum time spent by all the 5 operators to complete all the five assignments = 50+110+130+200+80 = 570 days.

Example 4

Given the following cost matrix obtain the minimum operation time cost. Also state the optimum assignment.

Table 7.29

Operators	Machine time 1	2	3	4	5	6
A	41	72	39	52	25	51
B	22	29	49	65	81	50
C	27	39	60	51	32	32
D	45	50	48	52	37	43
E	29	40	39	26	30	33
F	82	40	40	60	51	30

Solution

In our problem since all the cost elements are positive we go straight to the second step. By applying the theorem we reduce the given matrix by subtracting a constant to various rows and columns until a complete assignment matrix with zeros in appropriate places are obtained.

Table 7.30

	Machine time					
Operators	1	2	3	4	5	6
A	41	72	39	52	25	51
B	22	29	49	65	81	50
C	27	39	60	51	32	32
D	45	50	48	52	37	43
E	29	40	39	26	30	33
F	82	40	40	60	51	30
Column Min.	22	29	39	26	25	30

In Table 7.30 we obtain the column minimum for all the six columns. Subtract the column minimum from the respective column elements. By doing so we will bringing at least one zero in every column in Table 7.31.

Shade the columns and rows starting from the column or row having the highest number of zeros. In practice this could be done in many ways, we prefer the one having less number of column and rows. In our illustration three rows and one column are shaded. Clearly, the number of rows ($n_1 = 4$) < ($n = 6$). So the final allotment cannot be done at this stage.

In the Table 7.31, this time let us get row minimum for all the six rows and subtract them from the respective rows. This will bring at least one zero more for every row. Now in a minimum number of ways we shade all the zeros.

Table 7.31

Operators	Machine time 1	2	3	4	5	6	Row Min.
A	19	43	0	26	0	21	0
B	0	0	10	39	56	20	0
C	5	10	21	25	7	2	2
D	23	21	9	26	12	13	9
E	7	11	0	0	5	3	0
F	60	11	1	34	26	0	0

In a minimum way we can shade three rows and two columns and cover all the resulting zeros as shown in Table 7.32. But still we have shortfall of one row or column shading ($n_1 = 5$) < ($n = 6$).

Table 7.32

Operators	Machine time 1	2	3	4	5	6
A	19	43	0	26	0	21
B	0	0	10	39	56	20
C	3	8	19	23	5	0
D	14	12	0	17	3	4
E	7	11	0	0	5	3
F	60	11	1	34	26	0

Now we choose the smallest element among the uncovered numbers in the above table. Such a number is 3 in our problem. Subtract this number 3 from all the uncovered elements. Also add this 3 to the elements at the intersections of shades and get the resultant Table 7.33. Now we are in need of at least three columns and three rows totalling 6 shades to cover all zeros in the resulting Table 7.33.

Table 7.33

Operators	1	2	3	4	5	6
	Machine time					
A	19	43	3	26	0	24
B	0	0	13	39	56	23
C	0	5	19	20	2	0
D	11	9	0	14	0	4
E	7	11	3	0	5	6
F	57	8	1	31	23	0

The final solution and the allotments are shown by squared boxes in the Table 7.34.

Table 7.34

Operators	1	2	3	4	5	6
	Machine time					
A	19	43	3	26	[0]	24
B	0	[0]	13	39	56	23
C	[0]	5	19	20	2	0
D	11	9	[0]	14	0	4
E	7	11	3	[0]	5	6
F	57	8	1	31	23	[0]

Assignment:

Operator A is assigned to machine 5

Operator B is assigned to machine 2

Operator C is assigned to machine 1

Operator D is assigned to machine 3

Operator E is assigned to machine 4

Operator F is assigned to machine 6

The minimum cost is obtained from the original table as shown in Table 7.35.

Table 7.35

Operators	Machine time 1	2	3	4	5	6
A	41	72	39	52	[25]	51
B	22	[29]	49	65	81	50
C	[27]	39	60	51	32	32
D	45	50	[48]	52	37	43
E	29	40	39	[26]	30	33
F	82	40	40	60	51	[30]

Now from the starting table the minimum time spent by all the 6 to complete all the six assignments = 25+29+27+48+26+30 = 185 days.

■ Example 5

Solve the following assignment problem.

Table 7.36

Individuals	Jobs 1	2	3	4
A	120	100	80	90
B	80	90	110	70
C	110	140	120	100
D	90	90	80	90

Let us take row minimums and subtract them all from their respective rows. All the emerging zeros are covered with a minimum of three columns in Table 7.38.

Table 7.37

Individuals	Jobs 1	2	3	4	Row Min.
A	120	100	80	90	80
B	80	90	110	70	70
C	110	140	120	100	100
D	90	90	80	90	80

Table 7.38

Individuals	Jobs 1	2	3	4
A	40	20	0	10
B	10	20	40	0
C	10	40	20	0
D	10	10	0	10
Column Min.	10	10	0	0

Now since 3 < 4 we take column minimums and subtract them all from their respective columns to bring in some more zeros into action. Now we need at least three columns and one row to cover all zeros in Table 7.39.

Table 7.39

Individuals	Jobs 1	2	3	4
A	30	10	0	10
B	0	10	40	0
C	0	30	20	0
D	0	0	0	10

Now since number of covered columns and rows equals the number of assignment (4 = 4) we continue further to our final allotment step. Table 7.40 shows the final assignment.

Table 7.40

Individuals	Jobs 1	2	3	4
A	30	10	[0]	10
B	[0]	10	40	0
C	0	30	20	[0]
D	0	[0]	0	10

The needed minimum time cost is obtained from the original cost table as follows:

Table 7.41

Individuals	Jobs 1	2	3	4
A	120	100	[80]	90
B	[80]	90	110	70
C	110	140	120	[100]
D	90	[90]	80	90

Final Assignment:

Individual A is assigned Job 3

Individual B is assigned Job 1

Individual C is assigned Job 4

Individual D is assigned Job 2

Minimum time cost for the allotment = 80 +80 + 100 + 90 = 350.

Example 6

Solve the following assignment problem to minimize the total layover time of drivers. The drivers should have a minimum layover of 3 hours between two trips when they stay away from their home city. Obtain the pair of driver's route that minimizes the total layover time that they stay away from home city. Also mention the driver's base city.

Table 7.42

City A			City B		
Route no.	Departure from A	Arrival at B	Route no.	Departure from B	Arrival at A
101	10 A.M	11 A.M	201	11 A.M	12.15 P.M
102	11 A.M	12 NOON	202	1.30 P.M	2.45 P.M
103	3.30 P.M	4.30 P.M	203	6 P.M	7.15 P.M
104	7.30 P.M	8.30 P.M	204	9 P.M	10.15 P.M

Solution

Let us first obtain the layover time for all possible pairs of trips for drivers based in city A. The driver who starts from city A by route 101 at 10 A.M on a particular day will reach the city B by 11 A.M on the same day. If he wants to come back to city A on the same day by route 201 then he will have to start immediately at 11 A.M without enjoying the layoff. Since such a return trip is not permitted, he will have to wait in city B for another 24 hours and starts the journey at 11 A.M on the next day in route 201. In Table 7.43 the so calculated 24 hours is entered in the very first row and first column pertaining to route 101 and 201. Similarly, all other calculations are made and reported in Table 7.43 having prescribed three hours layoff in mind.

Table 7.43

Routes A to B	Routes B to A			
	201	202	203	204
101	24.00	26.50	7.00	10.00
102	23.00	25.50	6.00	9.00
103	18.50	21.00	25.50	4.50
104	14.50	17.00	21.50	24.50

Similarly, let us obtain the layover time for all possible pairs of trips for drivers based in city B as we have done in Table 7.43. The so obtained results are reported in Table 7.44.

Table 7.44

Routes A to B	Routes B to A			
	201	202	203	204
101	21.75	19.25	14.70	11.75
102	22.75	20.25	15.75	12.75
103	3.25	24.75	20.25	17.25
104	7.25	28.75	24.25	21.25

From Tables 7.43 and 7.44 we construct the minimum layover time Table 7.45 by only taking the minimum values in the respective cells. Remember that the highlighted layovers are the minimum layoffs of drivers based in city A originally and the un-highlighted layoffs are minimum layoffs of drivers based in city B originally.

Table 7.45

Routes	201	202	203	204
101	21.75	19.25	7.00	10.00
102	22.75	20.25	6.00	9.00
103	3.25	21.00	20.25	4.50
104	7.25	17.00	21.50	21.25

Now from this staring Table 7.45 we proceed further and obtain the needed optimum solution by using the Hungarian method in the usual manner. Table 7.46 gives the needed final allotment solution.

Table 7.46

Routes	201	202	203	204
101	13	0.75	0	0
102	15	2.75	0	0
103	0	8	18.8	0
104	0	0	16	15

In Table 7.47 we mark the optimum allotment of Table 7.46 with the original minimum layoffs arrived in Table 7.45.

Table 7.47

Routes	201	202	203	204
101	21.75	19.25	7.00	10.00
102	22.75	20.25	6.00	9.00
103	3.25	21.00	20.25	4.50
104	7.25	17.00	21.50	21.25

Final assignment:

Route 101 driver from city A will take route 203 from city B

Route 102 driver from city A will take route 204 from city B

Route 201 driver from city B will take route 103 from city A

Route 104 driver from city A will take route 202 from city B

Table 7.48

Routes	Driver Base
101–203	City A
102–204	City A
103–201	City B
104–202	City A

Total minimum layover time = 7.00+9.00+3.25+17.00 = 36.25 hours.

7.3 MAXIMIZATION PROBLEMS

Often we will come across assignment problems involving maximization. To obtain the solution for such problems we first convert the given maximization problem into an equivalent minimization by simply subtracting all data values from the maximum value of the given data matrix. Once the optimum minimum assignment is listed out in the usual way the need maximum value to the original problem is obtained by adding the corresponding cell values in the original data matrix given in the problem.

■ Example 7

The following table gives the profit contribution of the assigned machine to the assigned job. Obtain the maximum profit yielding assignment and the associated maximum profit.

Table 7.49

	Machines					
Operators	1	2	3	4	5	Max. 62
A	30	37	40	28	40	
B	40	24	27	21	36	
C	40	32	33	30	35	
D	25	38	40	36	36	
E	29	62	41	34	39	

Solution

The maximum data value from the given table is Rs. 62. So to convert this maximization problem into a minimization we subtract all the given elements in Table 7.49 from 62 and format the corresponding minimization problem as shown in Table 7.50. Once this task is accomplished we obtain the needed allotment in the usual manner as shown below from Tables 7.50 to 7.54.

Table 7.50

Operators	Machines 1	2	3	4	5	Row Min.
A	32	25	22	34	22	22
B	22	38	35	41	26	22
C	22	30	29	32	27	22
D	37	24	22	26	26	22
E	33	0	21	28	23	0

Table 7.51

Operators	Machine 1	2	3	4	5
A	10	3	0	12	0
B	0	16	13	19	4
C	0	8	.7	10	5
D	15	2	0	4	4
E	33	0	21	28	23
Column Min.	0	0	0	4	0

Table 7.52

Operators	Machine 1	2	3	4	5
A	10	3	0	8	0
B	0	16	13	15	4
C	0	8	7	6	5
D	15	2	0	0	4
E	33	0	21	24	23

Table 7.53

Operators	Machine 1	2	3	4	5
A	14	3	0	8	0
B	0	12	9	11	0
C	0	4	3	2	1
D	19	2	0	0	4
E	37	0	21	24	23

Table 7.54

Operators	Machine 1	2	3	4	5
A	14	3	[0]	8	0
B	0	12	9	11	[0]
C	[0]	4	3	2	1
D	19	2	0	[0]	4
E	37	[0]	21	24	23

Final Assignment:

Operator A is assignment machine 3

Operator B is assignment machine 5

Operator C is assignment machine 1

Operator D is assignment machine 4

Operator E is assignment machine 2

Table 7.55

Operators	Machine 1	2	3	4	5
A	30	37	[40]	28	40
B	40	24	27	21	[36]
C	[40]	32	33	30	35
D	25	38	40	[36]	36
E	29	[62]	41	34	39

The associated optimum profit due to this assignment from the original table = 40 + 36 + 40 + 36 + 62 = Rs. 214.

7.4 VARIANTS OF ASSIGNMENT PROBLEMS

7.4.1 Excess of Individuals than the Jobs

Sometimes we come across problems with more managers than the jobs to be allotted. Under such circumstances the supply exceeds the demand. To do such problems we create dummy jobs so that m = n and solve the problem in the usual manner.

■ Example 8

Solve the following assignment problem and get the minimum time cost.

Table 7.56

Spots	Trucks 1	2	3	4
A	3	6	2	6
B	7	1	4	4
C	3	8	5	8
D	6	4	3	3
E	5	2	4	3
F	5	7	6	2

Since the spots are 2 in excess than the trucks we create 2 more dummy trucks and continue the problem as follows in the usual manner.

In this problem note that row minimum step is unnecessary because all row minimums are zeros. Subtracting zeros from the respective row elements leaves the original matrix unaltered. So only the column minimum and other operations are used in this problem.

Table 7.57

Spots	Trucks 1	2	3	4	5	6
A	3	6	2	6	0	0
B	7	1	4	4	0	0
C	3	8	5	8	0	0
D	6	4	3	7	0	0
E	5	2	4	3	0	0
F	5	7	6	2	0	0
Column Min.	3	1	2	2	0	0

Table 7.58

Spots	Trucks 1	2	3	4	5	6
A	0	5	0	4	0	0
B	4	0	2	2	0	0
C	0	7	3	6	0	0
D	3	3	1	5	0	0
E	2	1	2	1	0	0
F	2	6	4	0	0	0

Table 7.59

Spots	Trucks 1	2	3	4	5	6
A	0	5	[0]	4	0	0
B	4	[0]	2	2	0	0
C	[0]	7	3	6	0	0
D	3	3	1	5	0	[0]
E	2	1	2	1	[0]	0
F	2	6	4	[0]	0	0

Table 7.60

Spots	Trucks 1	2	3	4	5	6
A	3	6	[2]	6	0	0
B	7	[1]	4	4	0	0
C	[3]	8	5	8	0	0
D	6	4	3	7	0	[0]
E	5	2	4	3	[0]	0
F	5	7	6	[2]	0	0

Final assignment:

Spot A is assigned to Truck 3

Spot B is assigned to Truck 2

Spot C is assigned to Truck 1

Spot D is assigned to Truck 6

Spot E is assigned to Truck 5

Spot F is assigned to Truck 4

The associated total time cost = 2+1+3+0+0+2 = 8 units of time.

Example 9

Solve the following assignment problem and get the minimum time cost.

Table 7.61

Operators	Machine time 1	2	3	4	5
A	6	2	5	3	6
B	2	5	8	7	7
C	7	8	6	9	8
D	6	2	3	4	5
E	9	3	8	9	7
F	4	7	4	6	8

Since the operators are 1 in excess than the machines we create 1 more dummy machine and continue the problem as shown in Table 7.62.

Table 7.62

Operators	Machine time 1	2	3	4	5	6
A	6	2	5	3	6	0
B	2	5	8	7	7	0
C	7	8	6	9	8	0
D	6	2	3	4	5	0
E	9	3	8	9	7	0
F	4	7	4	6	8	0

Once again the row minimum step is avoided because all row minimums are zeros. Subtracting zeros from the respective row elements leaves the original matrix unaltered. So, only column minimum and other operations are highlighted.

Table 7.63

Operators	Machine time 1	2	3	4	5	6
A	6	2	5	3	6	0
B	2	5	8	7	7	0
C	7	8	6	9	8	0
D	6	2	3	4	5	0
E	9	3	8	9	7	0
F	4	7	4	6	8	0
Column Min.	2	2	3	3	5	0

Table 7.64

Operators	Machine time 1	2	3	4	5	6
A	4	0	2	0	1	0
B	0	3	5	4	2	0
C	5	6	3	6	3	0
D	4	0	0	1	0	0
E	7	1	5	6	2	0
F	2	5	1	3	3	0

Table 7.65

Operators	Machine time 1	2	3	4	5	6
A	4	0	2	0	1	1
B	0	3	5	4	2	1
C	4	5	2	5	2	0
D	4	0	0	1	0	1
E	6	0	4	5	1	0
F	1	4	0	2	2	0

Table 7.66

Operators	Machine time 1	2	3	4	5	6
A	4	0	2	[0]	1	1
B	[0]	3	5	4	2	1
C	4	5	2	5	2	[0]
D	4	0	0	1	[0]	1
E	6	[0]	4	5	1	0
F	1	4	[0]	2	2	0

Table 7.67

Operators	Machine time 1	2	3	4	5	6
A	6	2	5	[3]	6	0
B	[2]	5	8	7	7	0
C	7	8	6	9	8	[0]
D	6	2	3	4	[5]	0
E	9	[3]	8	9	7	0
F	4	7	[4]	6	8	0

Final assignment:

Operator A is assigned to Machine 4

Operator B is assigned to Machine 1

Operator C is assigned to Machine 6

Operator D is assigned to Machine 5

Operator E is assigned to Machine 2

Operator F is assigned to Machine 3

Total optimum time cost = 3+2+0+5+3+4 =17

7.4.2 Excess of Jobs than the Individuals: One-to-One Solution

Sometimes we come across problems with more jobs than the individuals available to do the job. Under such circumstances the demand exceeds the supply. To do such problems we create dummy individuals so that m = n and solve the problem in the usual manner.

■ Example 10

Solve the following assignment problem and get the minimum time cost.

Solution

Table 7.68

Operators	Machine time 1	2	3	4	5	6
A	3	6	2	6	4	6
B	7	1	4	4	2	3
C	3	8	5	8	2	4
D	8	4	3	7	2	2

This time since the machines are more than operators by 2 we create 2 dummy operators and continue the problem as usual as shown in Table 7.69.

Table 7.69

Operators	Machine time 1	2	3	4	5	6
A	3	6	2	6	4	6
B	7	1	4	4	2	3
C	3	8	5	8	2	4
D	8	4	3	7	2	2
E	0	0	0	0	0	0
F	0	0	0	0	0	0

Table 7.70

Operators	1	2	3	4	5	6	Row Min.
	Machine time						
A	3	6	2	6	4	6	2
B	7	1	4	4	2	3	1
C	3	8	5	8	2	4	2
D	8	4	3	7	2	2	2
E	0	0	0	0	0	0	0
F	0	0	0	0	0	0	0

In this problem column minimum step is unnecessary because all column minimums are zeros. Subtracting zeros from the respective column elements leaves the original matrix unaltered. So only row minimum and other operations are highlighted.

Table 7.71

Operators	1	2	3	4	5	6
	Machine time					
A	1	4	0	4	2	4
B	6	0	3	3	1	2
C	1	6	3	6	0	2
D	6	2	1	5	0	0
E	0	0	0	0	0	0
F	0	0	0	0	0	0

Table 7.72

Operators	Machine time 1	2	3	4	5	6
A	0	3	[0]	3	2	4
B	6	[0]	4	3	2	2
C	0	5	3	6	[0]	2
D	6	2	2	5	1	[0]
E	[0]	0	1	0	1	0
F	0	0	1	[0]	1	0

Table 7.73

Operators	Machine time 1	2	3	4	5	6
A	3	6	[2]	6	4	6
B	7	[1]	4	4	2	3
C	3	8	5	8	[2]	4
D	8	4	3	7	2	[2]
E	[0]	0	0	0	0	0
F	0	0	0	[0]	0	0

Final assignment:

Operator A is assigned to Machine 3

Operator B is assigned to Machine 2

Operator C is assigned to Machine 5

Operator D is assigned to Machine 6

Operator E is assigned to Machine 1

Operator F is assigned to Machine 4

Thus the total optimum time cost = 2+1+2+2+0+0= 7 units of time

7.4.3 Excess of Jobs than the Individuals: Multiple Assignment

In the previous illustration there were two excess jobs. We created two dummy managers to take care of these two additional jobs. Under multiple assignments we will introduce only one dummy and assign both the excess jobs to that dummy. The solution remains one and the same as shown below.

■ Example 11

Solve the following assignment problem and get the minimum time cost.

Table 7.74

	Machine time					
Operators	1	2	3	4	5	6
A	3	6	2	6	4	6
B	7	1	4	4	2	3
C	3	8	5	8	2	4
D	8	4	3	7	2	2

Solution

This time remember that we create only one dummy operator E as shown in Table 7.75.

Table 7.75

	Machine time					
Operators	1	2	3	4	5	6
A	3	6	2	6	4	6
B	7	1	4	4	2	3
C	3	8	5	8	2	4
D	8	4	3	7	2	2
E	0	0	0	0	0	0

Table 7.76

Operators	Machine time 1	2	3	4	5	6	Row Min.
A	3	6	2	6	4	6	2
B	7	1	4	4	2	3	1
C	3	8	5	8	2	4	2
D	8	4	3	7	2	2	2
E	0	0	0	0	0	0	0

Table 7.77

Operators	Machine time 1	2	3	4	5	6
A	1	4	0	4	2	4
B	6	0	3	3	1	2
C	1	6	3	6	0	2
D	6	2	1	5	0	0
E	0	0	0	0	0	0

Table 7.78

Operators	Machine time 1	2	3	4	5	6
A	1	4	[0]	4	2	4
B	6	[0]	3	3	1	2
C	1	6	3	6	[0]	2
D	6	2	1	5	0	[0]
E	[0]	0	0	[0]	0	0

Table 7.78(a)

Operators	Machine time 1	2	3	4	5	6
A	3	6	[2]	6	4	6
B	7	[1]	4	4	2	3
C	3	8	5	8	[2]	4
D	8	4	3	7	2	[2]
E	[0]	0	0	[0]	0	0

Final assignment:

Operator A is assigned to Machine 3
Operator B is assigned to Machine 2
Operator C is assigned to Machine 5
Operator D is assigned to Machine 6
Operator E is assigned to Machine 1
Operator E is assigned to Machine 4

Total time cost = 2+1+2+2+0+0 = 7 units of time

7.4.4 Prohibited Jobs

Sometimes, due to some legal constraints, certain jobs are not assigned to certain individuals. Often certain clients do not want to have certain individuals for their concerns. Under such cases we handle the situation by imposing hefty penalties to these cells and proceed further in the usual way. In the final assignment matrix these cells remain un-allotted. The following example is designed to illustrate this type of problem.

Example 12

Solve the following assignment problem and get the minimum time cost. By X we mean the prohibited jobs.

Table 7.79

	Client				
Engineers	1	2	3	4	5
A	80	40	X	70	40
B	X	80	60	40	40
C	70	X	60	80	70
D	70	80	30	50	X
E	40	40	50	X	80

For unwanted assignment we put a hefty penalty 2000 hours and solve the problem as usual.

Table 7.80

	Client					
Engineers	1	2	3	4	5	Row Min
A	80	40	2000	70	40	40
B	2000	80	60	40	40	40
C	70	2000	60	80	70	60
D	70	80	30	50	2000	30
E	40	40	50	2000	80	40

Table 7.81

Engineers	Client 1	2	3	4	5
A	40	0	1960	30	0
B	1960	40	20	0	0
C	10	1940	0	20	10
D	40	50	0	20	1970
E	0	0	10	1960	40
Column Min.	0 0	0 0	0	0	0

Table 7.82

Engineers	Client 1	2	3	4	5
A	40	0	1960	30	0
B	1960	40	20	0	0
C	10	1940	0	20	10
D	40	50	0	20	1970
E	0	0	10	1960	40

Table 7.83

Engineers	Client 1	2	3	4	5
A	40	0	1960	30	0
B	1960	40	20	0	0
C	10	1940	0	20	10
D	40	50	0	20	1970
E	0	0	10	1960	40

Table 7.84

Engineers	Client 1	2	3	4	5
A	80	40	X	70	[40]
B	X	80	60	[40]	40
C	[70]	X	60	80	70
D	70	80	[30]	50	X
E	40	[40]	50	X	80

Assignment:

Manager A to Client 5
Manager B to Client 4
Manager C to Client 1
Manager D to Client 3
Manager E to Client 2

The minimum total cost = 40+40+70+30+40 =220

■ Example 13

Solve the following assignment problem and get the minimum time cost. Here by X we mean prohibited jobs.

Table 7.85

Engineers	Client 1	2	3	4	5
A	9	11	15	10	11
B	12	9	X	10	9
C	X	11	14	11	7
D	14	8	12	7	8

Solution

Table 7.86

	Client				
Engineers	1	2	3	4	5
A	9	11	15	10	11
B	12	9	X	10	9
C	X	11	14	11	7
D	14	8	12	7	8
E	0	0	0	0	0

This time we put a very big M as penalty for prohibited jobs in Table 7.87.

Table 7.87

	Client					
Engineers	1	2	3	4	5	Row Min.
A	9	11	15	10	11	9
B	12	9	M	10	9	9
C	M	11	14	11	7	7
D	14	8	12	7	8	7
E	0	0	0	0	0	0

Table 7.88

	Client				
Engineers	1	2	3	4	5
A	0	2	6	1	2
B	3	0	M	1	0
C	M	4	7	4	0
D	7	1	5	0	1
E	0	0	0	0	0

Table 7.89

Engineers	Client 1	2	3	4	5
A	[0]	2	6	1	2
B	3	[0]	M	1	0
C	M	4	7	4	[0]
D	7	1	5	[0]	1
E	0	0	[0]	0	0

Table 7.90

Engineers	Client 1	2	3	4	5
A	[9]	11	15	10	11
B	12	[9]	X	10	9
C	X	11	14	11	[7]
D	14	8	12	[7]	8
E	0	0	[0]	0	0

Final Assignment:

Engineer A is assigned to Client 1

Engineer B is assigned to Client 2

Engineer C is assigned to Client 5

Engineer D is assigned to Client 4

Engineer E is assigned to Client 3

Total time cost = 9+9+7+7+0=32 units of time.

7.4.5 Unique Vs. Multiple Optimum Solutions

In all the previous illustrations at the final stage we select a row/column containing only one zero for allotment. However, a situation may arise with various rows and columns have multiple zeros. The following example illustrates one such case.

■ Example 14

Solve the following assignment problem.

Table 7.91

	Client				
Engineers	1	2	3	4	5
A	25	18	32	20	21
B	34	25	21	12	17
C	20	17	20	32	16
D	20	28	20	16	27

Since the clients are more than engineers we create a dummy engineer E in Table 7.92. To bring in zeros in action we obtain the row minimums in Table 7.92. These minimums are subtracted from the respective row elements and the results are produced in Table 7.93. In it all the zeros are covered with three rows and one column as shown. Since 4 < 5 optimality is not reached yet. So bring in some zeros we subtract the smallest 4 from the uncovered elements. We also add this 4 to the elements at the intersection points. The resulting table is presented in 7.94.

Table 7.92

	Client					
Engineers	1	2	3	4	5	Row Min.
A	25	18	32	20	21	18
B	34	25	21	12	17	12
C	20	17	20	32	16	16
D	20	28	20	16	27	16
E	0	0	0	0	0	0

Table 7.93

Engineers	Client 1	2	3	4	5
A	7	0	14	2	3
B	22	13	9	0	5
C	4	1	4	16	0
D	4	12	4	0	11
E	0	0	0	0	0

Solution 1

Table 7.94

Engineers	Client 1	2	3	4	5
A	7	[0]	14	6	3
B	18	9	5	[0]	1
C	4	1	4	20	[0]
D	[0]	8	0	0	7
E	0	0	[0]	4	0

Solution 2

Table 7.95

Engineers	Client 1	2	3	4	5
A	7	[0]	14	6	3
B	18	9	5	[0]	1
C	4	1	4	20	[0]
D	0	8	[0]	0	7
E	[0]	0	0	4	0

The first optimum solution follow from Table 7.94 and the second optimum solution follow from the Table 7.95 the associated cost calculations are shown in Table 7.96.

Table 7.96

Solution 1		
Engineer	Client	Cost
A	2	18
B	4	12
C	5	16
D	1	20
E	3	0
Total		66

Solution 2		
Engineer	Client	Cost
A	2	18
B	4	12
C	5	16
D	3	20
E	1	0
Total		66

■ Example 15

A medical care centre plans to assign five doctors at its command to five hospitals at its command. Given the following revenue matrix (Rs. in thousands) obtain the best assignment that brings the maximum revenue to the centre.

Table 7.97

	Hospitals				
Doctors	1	2	3	4	5
A	40	46	48	36	48
B	48	32	36	29	44
C	49	35	41	38	45
D	30	46	49	44	44
E	37	41	48	43	47

Since it is a maximization problem let us first convert this into a minimization problem by subtracting all the cells values in Table 7.97 from the highest value 49 in it. The resulting matrix is shown in Table 7.98.

Table 7.98

Doctors	Hospitals 1	2	3	4	5	Row Min.
A	9	3	1	13	1	1
B	1	17	13	20	5	1
C	0	14	8	11	4	0
D	19	3	0	5	5	0
E	12	8	1	6	2	1

To bring in some more zeros in action this time we subtract the respective row minimums from their respective row entries and rewrite the resulting matrix as shown in Table 7.99.

Table 7.99

Doctors	Hospitals 1	2	3	4	5
A	8	2	0	12	0
B	0	16	12	19	4
C	0	14	8	11	4
D	19	3	0	5	5
E	11	7	0	5	1
Column Min.	0	2	0	5	0

Since 3 < 5 we subtract the column minimum from the respective columns and report it in Table 7.100. In Table 7.100 since the minimum number of lines equal to the number of doctors (5 = 5) our iteration ends here.

Table 7.100

Doctors	Hospitals 1	2	3	4	5
A	12	0	0	7	0
B	0	10	8	10	0
C	0	8	4	2	0
D	23	1	0	0	5
E	14	5	0	0	1

There exists more than one optimal assignment to this problem because of existence of multiple zeros in different rows and columns.

The multiple assignments are:

1. A-2, B-5, C-1, D-3, E-4

Doctors	Hospitals 1	2	3	4	5
A	12	[0]	0	7	0
B	0	10	8	10	[0]
C	[0]	8	4	2	0
D	23	1	[0]	0	5
E	14	5	0	[0]	1

2. A-2, B-1, C-5, D-3, E-4

Doctors	Hospitals 1	2	3	4	5
A	12	[0]	0	7	0
B	[0]	10	8	10	0
C	0	8	4	2	[0]
D	23	1	[0]	0	5
E	14	5	0	[0]	1

3. A-2, B-5, C-1, D-4, E-3

Doctors	Hospitals 1	2	3	4	5
A	12	[0]	0	7	0
B	0	10	8	10	[0]
C	[0]	8	4	2	0
D	23	1	0	[0]	5
E	14	5	[0]	0	1

4. A-2, B-1, C-5, D-4, E-3

Doctors	Hospitals 1	2	3	4	5
A	12	[0]	0	7	0
B	0	10	8	10	[0]
C	[0]	8	4	2	0
D	23	1	[0]	0	5
E	14	5	0	[0]	1

Note that all the 4 assignments yield the same revenue of Rs. 321 thousands. The following chart illustrates the steps that are to be followed while solving the assignment problem.

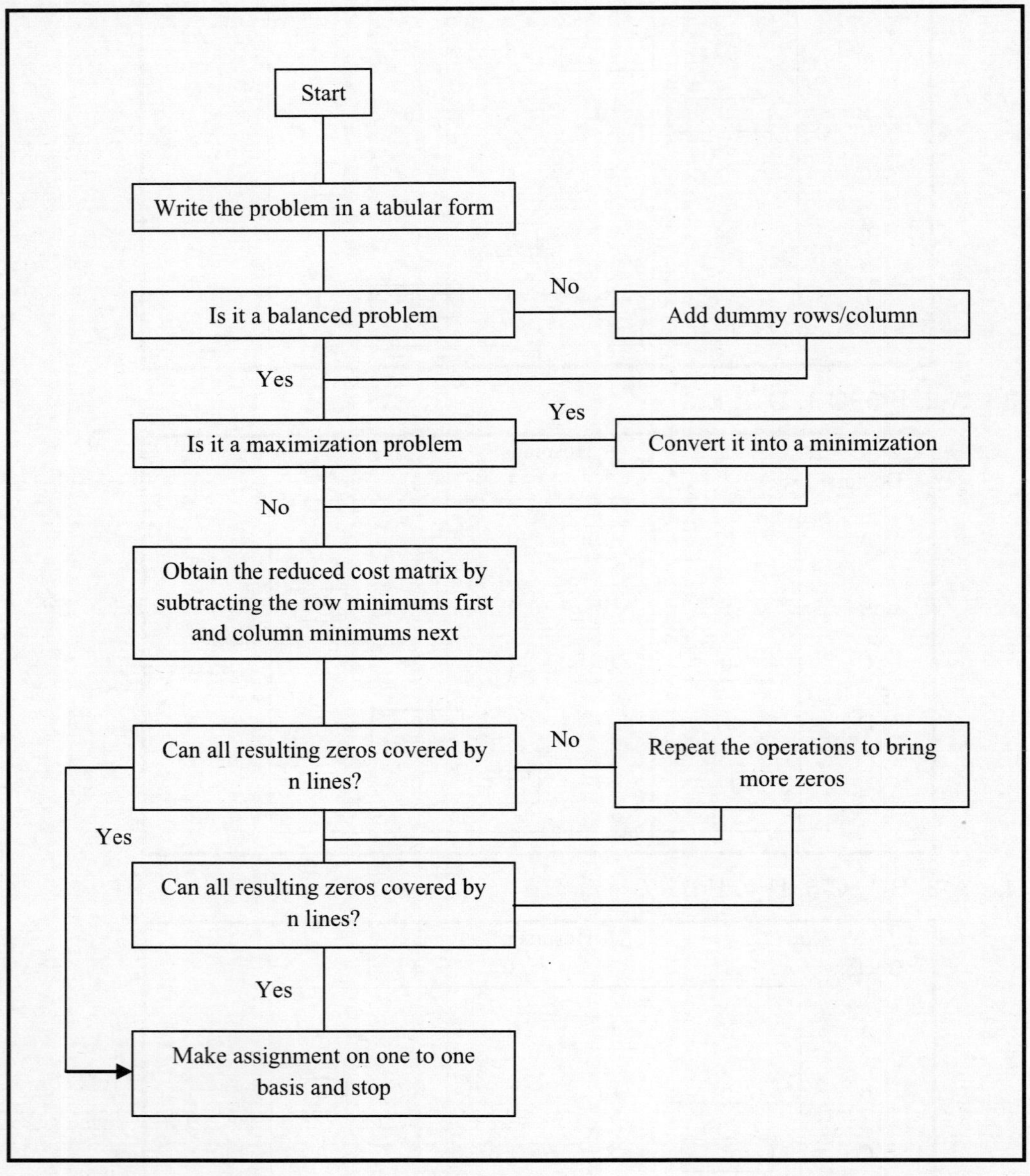

Figure 7.2

7.5 BRANCH-AND-BOUNDS METHOD FOR SOLVING ASSIGNMENT PROBLEMS

Since assignment problems are having finite number of alternative assignments we often prefer to go for complete enumeration method for finding the optimum solution. Remember, that the example 2 in this section was solved by using complete enumeration method. In it since we had 4 jobs to be allotted to 4 workers we explored all the possible 4! = 24 outcomes to find the least cost optimum outcome. Instead of 4 if it were to be a 10 job 10 workers model, definitely we would not have been ventured this method of complete enumeration, because 10! = 36,28,800 possibilities are to be explored for finding the least cost optimum assignment.

As an alternative, in this section we propose to suggest branch-and-bounds method for enumeration. In a way this method also employ enumeration but certainly not for all the possibilities. By applying some rule or the other this method splits the entire set of possibilities into two sets—one in which most probably the optimum solution lies and the other in which in all probability the solution does not lie and hence needs no further examination. Thus, the branching and bounding essentially keeps on dividing the feasible region into smaller and smaller segments until the final optimum solution is encountered in the selected smaller subset. We illustrate the use of this method in solving the following assignment problem.

Example 16

Solve the following assignment problem by branch-and-bounds method.

Table 7.101

Worker	Job			
	1	2	3	4
A	90	18	48	50
B	72	28	85	80
C	53	92	12	78
D	20	70	70	25

Solution

Step 1: First we obtain the lower bound for the total assignment cost to the given problem. Here by lower bound we mean the total assignment cost cannot fall below this level under any circumstances even if we take all the possibilities under consideration. Remember, that such an assignment necessarily need not yield a feasible solution in the sense that one job to one individual. It may be the situation in which two or more jobs assigning situation to a single individual it may also be the case in which two or more individuals are assigned a single job. To get one such minimum start, we simply observe the highlighted least costs in each column and add them all to get the total. Thus in our illustration this lower bound works out to be 20+18+12+25 = 75. Remember, that such a choice allots D to job 1, A to job 2, C to job 3 and again D to job 4, a repeated assignment. Definitely, even we explore all the 24 feasible possibilities the total cost cannot be less than this.

Step 2: In this step we device the process of looking for solutions. We begin by assigning job 1 to individuals A, B, C, and D in turn and then consider the smallest value assignments for the leftover jobs among leftover individuals as described in step 1 after ignoring the concerned column and row in the original table.

Step 3 (a): Assignment of job 1 to worker A: We assign job 1 to worker A with an associated cost of 90 stated by the first row first column element in Table 7.101. Assign the remaining jobs 2, 3, 4 to remaining workers B, C, and D on the basis of minimum cost principle as described in step 1 after ignoring the row 1 and column 1 in Table 7.101 for which assignment has been made. The remaining part of the table is written in Table 7.102.

Table 7.102

Worker	Job		
	2	3	4
B	28	85	80
C	92	12	78
D	70	70	25

So the resulting total cost of this assignment is summarized in Table 7.103.

Table 7.103

Job	Ind.	Cost
1	A	90
2	B	28
3	C	12
4	D	25
	Total	155

Step 3 (b): Assignment of job 1 to worker B

This time we assign job 1 to worker B with an associated cost of 72 stated by the second row first column element in Table 7.101. Assign the remaining jobs 2, 3, 4 to remaining workers A, C, and D on the basis of minimum cost principle as described in step 1 after ignoring the row 2 and column 1 of Table 7.101 for which assignment has been made. The leftover table is reported in Table 7.104.

Table 7.104

Worker	Job		
	2	3	4
A	18	48	50
C	92	12	78
D	70	70	25

So the resulting total cost of this assignment is summarized in Table 7.105.

Table 7.105

Job	Ind.	Cost
1	B	72
2	A	18
3	C	12
4	D	25
	Total	127

Step 3 (c): Assignment of job 1 to worker C: This time we assign job 1 to worker C with an associated cost of 53 stated by the third row first column element in Table 7.101. Assign the remaining jobs 2, 3, 4 to remaining workers A, B, and D on the basis of minimum cost principle as described in step 1 after ignoring the row 3 and column 1 in Table 7.101 for which assignment has been made. The leftover table is reported in Table 7.106.

Table 7.106

Worker	Job		
	2	3	4
A	18	48	50
B	28	85	80
D	70	70	25

So the resulting total cost of this assignment is summarized in Table 7.107.

Table 7.107

Job	Ind.	Cost
1	C	53
2	A	18
3	A	48
4	D	25
	Total	144

Step 3 (d): Assignment job 1 to worker D: This time we assign job 1 to worker D with an associated cost of 20 stated by the third row first column element in Table 7.101. Assign the remaining jobs 2, 3, 4 to remaining workers A, B, and C on the basis of minimum cost principle as described in step 1 after ignoring the row 4 and column 1 in Table 7.101 for which assignment has been made. The leftover table is reported in Table 7.108.

Table 7.108

Worker	Job		
	2	3	4
A	18	48	50
B	28	85	80
C	92	12	78

So the resulting total cost of this assignment is summarized in Table 7.109.

Table 7.109

Job	Ind.	Cost
1	D	20
2	A	18
3	C	12
4	A	50
	Total	100

Figure 7.3 shows the tree diagram summarizing all the four results. It may be noted that the first two options are feasible in the sense that both satisfy the one job to one worker rule. In the third case A is allotted job 2 and 3. In the fourth option A is assigned job 2 and 4.

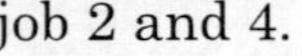

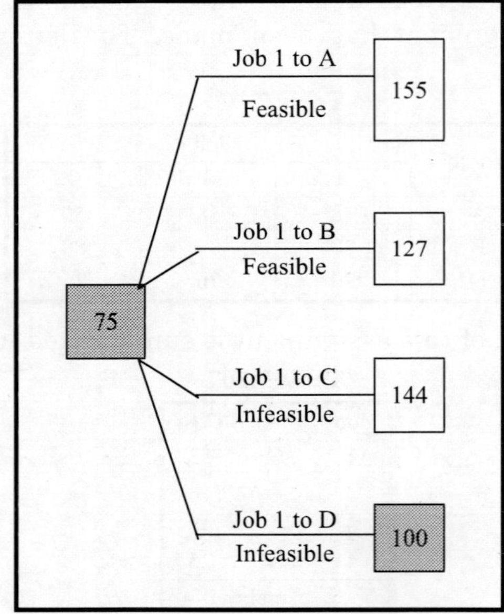

Figure 7.3

Step 4: Among the four possibilities, since 100 is the least cost allotment our next branching out starts from this branch as shown in the Figure 7.4 irrespective of the solutions feasibility or not. Since job 1 is already allotted to D at this stage we consider the next job 2 for allotment among the leftover workers A, B, and C in turn. After allotting the job 1 to D the leftover cost summary is shown in Table 7.110.

Table 7.110

Worker	Jobs		
	2	3	4
A	18	48	50
B	28	85	80
C	92	12	78

The resulting allotments are summarized in Table 7.111 and the associated tree diagram is shown in Figure 7.4.

Table 7.111

Job	Ind.	Cost
1	D	20
2	A	18
3	C	12
4	A	78
	Total	128

Job	Ind.	Cost
1	D	20
2	B	28
3	C	12
4	A	50
		110

Job	Ind.	Cost
1	D	20
2	C	92
3	A	48
4	A	50
		210

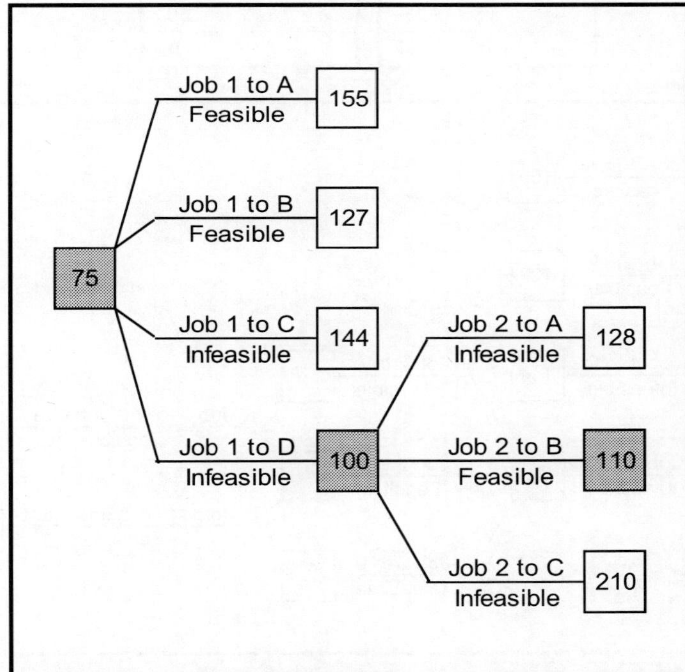

Figure 7.4

Step 5: Among the three possibilities, since 110 is the least cost allotment our next branching out starts from this branch as shown in the Figure 7.5 irrespective of the solutions feasibility or not. Since job 1 is already allotted to D at this stage and job 2 is allotted to B at this stage we consider the next job 3 for allotment among the leftover workers A, and C in turn. After allotting the job 1 to D and job 2 to B the leftover cost summary is shown in Table 7.112.

The resulting allotments are summarized in Table 7.113 and the associated tree diagram is shown in Figure 7.5.

Table 7.112

Worker	Jobs	
	3	4
A	48	50
C	12	78

Table 7.113

Job	Ind.	Cost
1	D	20
2	B	28
3	A	48
4	C	78
	Total	174

Job	Ind.	Cost
1	D	20
2	B	28
3	C	12
4	A	50
		110

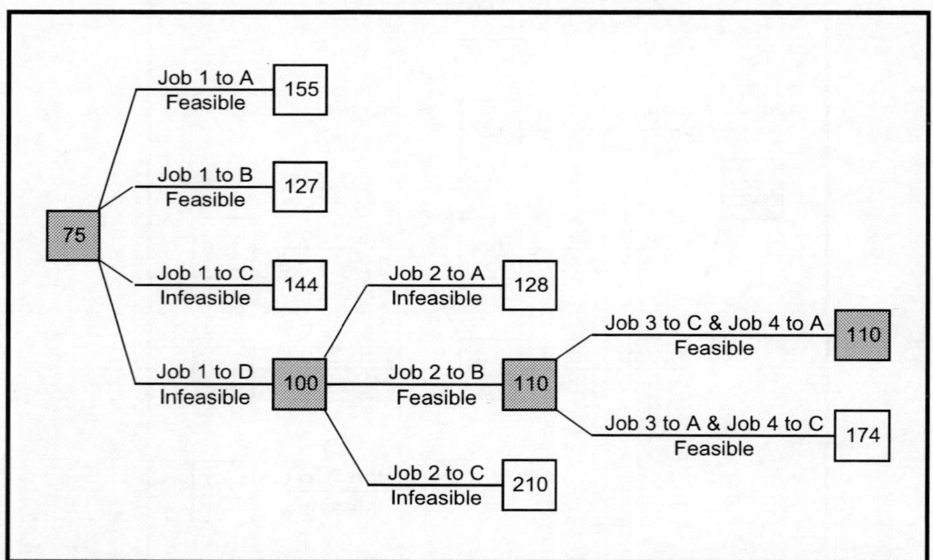

Figure 7.5

Thus, the solution values are summarized in Table 7.114.

Table 7.114

Worker	Job	Cost
A	4	50
B	2	28
C	3	12
D	1	20
	Total	110

7.6 SOLUTION TO TRAVELLING SALESMEN PROBLEM BY HUNGARIAN METHOD

A salesperson is assigned n cities to visit. He is given with the distance between all possible pairs of cities and instructed to visit each of the cities only once in a continuous trip and return to starting city so that the total distance travelled is minimized. Since complete cycle is involved, it makes no difference which city is the starting city.

This type of travelling salesman problem is a special type of assignment problem in which only one sale personal is assigned to all the sales destinations one after another with an additional condition that he starts from his home town, visits each city only once and comes back to his home town. The following examples illustrate such a business tour related problems.

■ Example 17

A salesmen travels from his home city A to four cities B, C, D, and E and comes back to his home city A. Given the inter-city distance matrix below, obtain the route map for effective minimum distance visit.

Table 7.115

From \ To	A	B	C	D	E
A	X	375	600	150	190
B	375	X	300	350	175
C	600	300	X	350	500
D	160	350	350	X	300
E	190	175	500	300	X

Solution

Note that in the given problem travelling from A to A, B to B, C to C, D to D, and E to E has no meaning. So to avoid such allotment we put a hefty penalty of M as shown in Table 7.116 so that such meaningless allotments get automatically eliminated in the final solution.

Table 7.116

From	To A	B	C	D	E	Row Min.
A	M	375	600	150	190	150
B	375	M	300	350	175	175
C	600	300	M	350	500	300
D	160	350	350	M	300	160
E	190	175	500	300	M	175

Row minimum is taken and subtracted from the respective row elements and the result is reported in Table 7.117.

Table 7.117

From	To A	B	C	D	E
A	M	225	450	0	40
B	200	M	125	175	0
C	300	0	M	50	200
D	0	190	190	M	140
E	15	0	325	125	M
Column Min.	0	0	125	0	0

Column minimum is taken and subtracted from the respective column elements and reported in Table 7.118.

Table 7.118

From \ To	A	B	C	D	E
A	M	225	325	0	40
B	200	M	0	175	0
C	300	0	M	50	200
D	0	190	65	M	140
E	15	0	200	125	M

The highlighted minimum 15 from uncovered elements is subtracted from all the uncovered elements including itself, and the same 15 is added to intersectional elements and the resulting result is reported in Table 7.119.

Table 7.119

From \ To	A	B	C	D	E
A	M	275	325	0	40
B	235	M	0	175	0
C	285	0	M	0	150
D	0	205	30	M	105
E	0	0	150	75	M

Again minimum 30 from uncovered elements is subtracted from all the uncovered elements including itself, and the same 30 is added to intersectional elements as shown in Table 7.120.

Table 7.120

From \ To	A	B	C	D	E
A	M	275	295	0	10
B	265	M	0	205	0
C	285	0	M	0	120
D	0	205	0	M	85
E	0	0	120	75	M

The minimum number of lines = n = 5. So the Table 7.120 is ready for optimum allotment. In Table 7.120(a) the final optimum allotment is reported in usual manner.

Table 7.120(a)

From \ To	A	B	C	D	E
A	M	275	295	[0]	10
B	265	M	0	205	[0]
C	285	[0]	M	0	120
D	0	205	[0]	M	85
E	[0]	0	120	75	M

In Table 7.121 with the allotment made in Table 7.120(a) the total minimum cost calculation is highlighted using the original table data given.

Table 7.121

From	A	B	C	D	E
A	X	375	600	[150]	190
B	375	X	300	350	[175]
C	600	[300]	X	350	500
D	160	350	[350]	X	300
E	[190]	175	500	300	X

Optimum Solution: A to D, B to E, C to B, D to C, E to A.

This may be sequenced as A to D, D to C, C to B, B to E, and E to A.

Total distance =150+350+300+175+190 = 1165 km.

■ Example 18

A salesman wants to visit five cities to do his business. His visit starts from his home city A and invariably ends in A. The inter-city transport costs are listed in the following matrix. Obtain the route map so that the travel cost is minimized.

Table 7.122

From	A	B	C	D	E
A	X	12	24	25	15
B	6	X	16	18	7
C	10	11	X	18	12
D	14	17	22	X	16
E	12	13	23	25	X

It may be noted that in this problem the matrix of distances between different cities is not symmetrical. The distance from A to B is 12 km while the distance between B to A is only 6 km. Such situation may arise in modern one-way traffic system.

Solution

In this problem to avoid prohibited routes like A to A, B to B etc., we put a hefty penalty of Rs. 200. Once this is done the problem is solved in the usual manner by using Hungarian method.

Table 7.123

From \ To	A	B	C	D	E	Row Min.
A	200	12	24	25	15	12
B	6	200	16	18	7	6
C	10	11	200	18	12	10
D	14	17	22	200	16	14
E	12	13	23	25	200	12

Row minimum is taken and subtracted from the respective row elements and reported in Table 7.124.

Table 7.124

From \ To	A	B	C	D	E
A	188	0	12	13	3
B	0	194	10	12	1
C	0	1	190	8	2
D	0	3	8	186	2
E	0	1	11	13	188
Column Min.	0	0	8	8	1

Column minimum is taken and subtracted from the respective row elements and reported in Table 7.125.

Table 7.125

From	To A	B	C	D	E
A	188	0	4	5	2
B	0	194	2	4	0
C	0	1	182	0	1
D	0	3	0	178	1
E	0	1	3	5	187

Since the number of line = n = 5, the final allotment made as usual by searching single zero rows and columns in Table 7.126.

Table 7.126

From	To A	B	C	D	E
A	188	[0]	4	5	2
B	0	194	2	4	[0]
C	0	1	182	[0]	1
D	0	3	[0]	178	1
E	[0]	1	3	5	187

Table 7.127

From	To A	B	C	D	F
A	X	[12]	24	25	15
B	6	X	16	18	[7]
C	10	11	X	[18]	12
D	14	17	[22]	X	16
E	[12]	13	23	25	X

The optimum minimum distance route solution is obtained as: (A to B) + (B to E) + (C to D) + (D to C) + (E to A) = 12 +7 + 18 + 22 +12 = Rs. 71. This is not possible because he has to move from A to B, B to E and E to A without visiting C and D. Thus in this problem we have two sub-tours as described below.

Sub-Tour 1: A to B, B to E and E to A

Sub-Tour 2: D to C and C to D

So to achieve this task with the obtained minimum expense of Rs. 71, two salespersons are needed at the rate of one for each per tour. If we have only one sale person to look after the affairs then the tour could be inevitably more costlier as shown in Table 7.128.

Table 7.128

From	To A	To B	To C	To D	To E
A	188	0	4	5	2
B	0	194	2	4	0
C	0	1	182	0	1
D	0	3	0	178	1
E	0	1	3	5	187

To get the next best single tour solution we consider entries like 1 and 2 as well in addition to zero's as shown in Table 7.128.

Table 7.129

From	To A	To B	To C	To D	To E
A	X	12	24	25	15
B	6	X	16	18	7
C	10	11	X	18	12
D	14	17	22	X	16
E	12	13	23	25	X

Accordingly, the salesman travels from A to B, B to C, C to D, D to E and E to A. The associated distance cost = 12 + 16 + 18 + 16 +12 = Rs. 74.

In Tables 7.130 and 7.131 an equivalent solution is considered by considering 3 in addition to zeros.

Equivalent Solution

A to B; B to E; E to C; C to D; D to A.

Table 7.130

From \ To	A	B	C	D	E
A	188	[0]	4	5	2
B	0	194	2	4	[0]
C	0	1	182	[0]	1
D	[0]	3	0	178	1
E	0	1	[3]	5	187

Table 7.131

From \ To	A	B	C	D	E
A	X	[12]	24	25	15
B	6	X	16	18	[7]
C	10	11	X	[18]	12
D	[14]	17	22	X	16
E	12	13	[23]	25	X

The distance travelled = 12+7+18+14+23 = Rs. 74.

Note that both these assignments are costlier by Rs. 3 than the original two sub-tour solution cost of Rs. 71.

Example 19

An American sales executive wants to visit five major cities in India. To do this job he has booked both the way flight ticket to city D in India. Given the local travelling expenses in thousands of rupees based on distance travelled obtain the least cost travel route.

Table 7.132

From \ To	A	B	C	D	E
A	X	4	7	3	4
B	4	X	6	3	4
C	7	6	X	7	5
D	3	3	7	X	7
E	4	4	5	7	X

Solution

Note that in this problem the distance travelled matrix is symmetric in the sense that the distance from A to B (4) is equal from B to A (4). To bring more zeros into action let us get the row minimums first in Table 7.133.

Table 7.133

From \ To	A	B	C	D	E	Row Min.
A	200	4	7	3	4	3
B	4	200	6	3	4	3
C	7	6	200	7	5	5
D	3	3	7	200	7	3
E	4	4	5	7	200	4

Now let us subtract these row minimums from their respective row values and report the result in Table 7.134. Now as per Hungarian rule we cover all the resulting zeros with a minimum of four columns as shown in Table 7.134.

Table 7.134

From	A	B	C	D	E
A	197	1	4	0	1
B	1	197	3	0	1
C	2	1	195	2	0
D	0	0	4	197	4
E	0	0	1	3	196
Column Min.	0	0	1	0	0

Since 5 < 4, the reduced Table 7.134 is not enough for optimum allotment. So to bring in some more zeros into action this time let us obtain the column minimums and subtract these minimums from the respective columns and report the result in Table 7.135.

Table 7.135

From	A	B	C	D	E
A	197	1	3	0	1
B	1	197	2	0	1
C	2	1	194	2	0
D	0	0	3	197	4
E	0	0	0	3	196

This time again we were able to cover all the zeros by two columns and two rows only. Again since 5 < 4 we identify the minimum value among the uncovered elements as 1 in Table 7.135 and subtract this value from all the uncovered elements. Also add this element to all the intersectional elements. Such a result is reported in Table 7.136.

Now we need a minimum of five columns to cover all zeros as shown in Table 7.136. So let us begin optimum allotment as usual.

Table 7.136

From \ To	A	B	C	D	E
A	196	0	2	0	1
B	0	196	1	0	1
C	1	0	193	2	0
D	0	0	3	198	5
E	0	0	0	4	197

Table 7.137

From \ To	A	B	C	D	E
A	196	0	2	[0]	1
B	[0]	196	1	0	1
C	1	0	193	2	[0]
D	0	[0]	3	198	5
E	0	0	[0]	4	197

Table 7.138

From \ To	A	B	C	D	E
A	X	4	7	[3]	4
B	[4]	X	6	3	4
C	7	6	X	7	[5]
D	3	[3]	7	X	7
E	4	4	[5]	7	X

Thus the least cost assignment is obtained A to D, B to A, C to E, D to B, and E to C

Total expense: 3000+4000+5000+3000+5000 = Rs. 20,000

However, the tour cannot be managed by one salesperson. The resulting two sub-tours are:

Sub-Tour 1: D to B, B to A and A to D

Sub-Tour 2: C and E and E to C.

So to get the next best result let us consider unity element as well in addition to zeros for allotment and complete the allotment as shown below in Table 7.139.

Table 7.139

From \ To	A	B	C	D	E
A	196	0	2	[0]	1
B	0	196	[1]	0	1
C	1	0	193	2	[0]
D	0	[0]	3	198	5
E	[0]	0	0	4	197

Table 7.140

From \ To	A	B	C	D	E
A	X	4	7	[3]	4
B	4	X	[6]	3	4
C	7	6	X	7	[5]
D	3	[3]	7	X	7
E	[4]	4	5	7	X

Route: D to B, B to C, C to E, E to A and A to D

Travel Expense: 3000+6000+5000+3000+4000 = Rs. 21,000

Example 20

Given the inter-city distance matrix below, obtain the route map for effective minimum distance visit. Also state the minimum number of salesperson necessary to complete task at the stated minimum distance.

Table 7.141

From \ To	A	B	C	D	E
A	X	103	188	136	38
B	103	X	262	176	52
C	187	262	X	85	275
D	136	176	85	X	162
E	38	52	275	162	X

Solution

Table 7.142

From \ To	A	B	C	D	E	Row Min.
A	M	103	188	136	38	38
B	103	M	262	176	52	52
C	187	262	M	85	275	85
D	136	176	85	M	162	85
E	38	52	275	162	M	38

Table 7.143

From \ To	A	B	C	D	E
A	M	65	150	98	0
B	51	M	210	124	0
C	102	177	M	0	190
D	51	91	0	M	77
E	0	14	237	124	M
Column Min.	0	14	0	0	0

Table 7.144

From \ To	A	B	C	D	E
A	M	51	150	98	0
B	51	M	210	124	0
C	102	163	M	0	190
D	51	77	0	M	77
E	0	0	237	124	M

Table 7.145

From \ To	A	B	C	D	E
A	M	0	150	98	0
B	0	M	210	124	0
C	51	112	M	0	190
D	0	26	0	M	77
E	0	0	288	175	M

Table 7.145 (a)

From \ To	A	B	C	D	E
A	M	0	150	98	0
B	0	M	210	124	0
C	51	112	M	0	190
D	0	26	0	M	77
E	0	0	288	175	M

Table 7.146

From \ To	A	B	C	D	E
A	X	103	188	136	38
B	103	X	262	176	52
C	187	262	X	85	275
D	136	176	85	X	162
E	38	52	275	162	X

Assignment: A to E, B to A, C to D, D to C, E to B.

Sub-Tour 1: A to E, E to B, B to A

Sub-Tour 2: D to C, C to D.

Distance: (38+52+103) + (85+52) = 363

Since the travel involves sub-tours for effective management at least two sales persons are essential.

7.7 SOLUTION TO TRAVELLING SALESMEN PROBLEM BY BRANCH-AND-BOUND METHOD

■ Example 21

Given the inter-city distance matrix below, obtain the route map for effective minimum distance visit by a single salesperson by using branch-and-bounds method.

Table 7.147

From \ To	1	2	3	4	5	6	Row Min.
1	M	25	18	35	50	39	18
2	21	M	28	16	30	13	13
3	22	28	M	14	16	20	14
4	35	12	14	M	12	12	12
5	50	30	16	12	M	8	8
6	39	15	20	12	7	M	7

Let us get the optimum solution to this problem by using Hungarian method as usual first. Row minimum operation has been performed and reported below in Table 7.148.

Table 7.148

From \ To	1	2	3	4	5	6
1	M	7	0	17	32	21
2	8	M	15	3	17	0
3	8	14	M	0	2	6
4	23	0	2	M	0	0
5	42	22	8	4	M	0
6	32	8	13	5	0	M
Column Min.	8	0	0	0	0	0

Column minimum operation has been performed and reported below in Table 7.149.

Table 7.149

From \ To	1	2	3	4	5	6
1	M	7	0	17	32	21
2	0	M	15	3	17	0
3	0	14	M	0	2	6
4	15	0	2	M	0	0
5	34	22	8	4	M	0
6	24	8	13	5	0	M

In it since the number of shading equals 6 we can readily start our optimum allotment.

Table 7.150

From \ To	1	2	3	4	5	6
1	M	7	[0]	17	32	21
2	[0]	M	15	3	17	0
3	0	14	M	[0]	2	6
4	15	[0]	2	M	0	0
5	34	22	8	4	M	[0]
6	24	8	13	5	[0]	M

As usual assignment has been done as per marking in Table 7.150.

Table 7.151

From	To 1	2	3	4	5	6	Row Min.
1	M	25	[18]	35	50	39	18
2	[21]	M	28	16	30	13	13
3	22	28	M	[14]	16	20	14
4	35	[12]	14	M	12	12	12
5	50	30	16	12	M	[8]	8
6	39	15	20	12	[7]	M	7

Assignment: 1 to 3, 2 to 1, 3 to 4, 4 to 2, 5 to 6 and 6 to 5.

Sub-Tour 1: 1 to 3, 3 to 4, 4 to 2, 2 to 1.

Sub-Tour 2: 5 to 6, 6 to 5.

Distance : 65+15 = 80 km.

Table 7.152

From	To 1	2	3	4	5	6	Row Min.
1	M	[25]	18	35	50	39	18
2	21	M	[28]	16	30	13	13
3	22	28	M	[14]	16	20	14
4	35	12	14	M	[12]	12	12
5	50	30	16	12	M	[8]	8
6	[39]	15	20	12	7	M	7

From original Table 7.147 the upper bound for the branch-and-bound approach is obtained by using the simple allotment as per the given order, namely 1 to 2 to 3 to 4 to 5 to 6 as shown in Table 7.152.

So the upper bound for the total distance = 25+28+14+12+8+39 =126 km. The lower bound of 80 km follows from the assignment that has been made by using Hungarian method in Table 7.151.

Now we proceed towards the final solution by reducing the gap between the upper bound and the lower bound by branching out method highlighted in the previous section. To achieve this task we break the smaller sub-tour 2 into two smaller sub-tours.

Tour 2(a): city 5 to 6

Tour 2(b): city 6 to 5

Once this is done we treat tour 5 to 6 as unaccepted prohibited tour. So accordingly we put a hefty penalty M for this tour in Table 7.153 and continue our assignment as shown below.

Table 7.153

From	1	2	3	4	5	6	Row Min.
1	M	7	0	17	32	21	0
2	0	M	15	3	17	0	0
3	0	14	M	0	2	6	0
4	15	0	2	M	0	0	0
5	34	22	8	4	M	M	4
6	24	8	13	5	0	M	0

Row minimum operation has been done and the resulting result is reported in the following Table 7.154.

Table 7.154

From	1	2	3	4	5	6
1	M	7	0	17	32	21
2	0	M	15	3	17	0
3	0	14	M	0	2	6
4	15	0	2	M	0	0
5	30	18	4	0	M	M
6	24	8	13	5	0	M

Table 7.154(a)

From	1	2	3	4	5	6
1	M	7	[0]	17	32	21
2	0	M	15	3	17	[0]
3	[0]	14	M	0	2	6
4	15	[0]	2	M	0	0
5	30	18	4	[0]	M	M
6	24	8	13	5	[0]	M

Assignments: 1 to 3, 2 to 6, 3 to 1, 4 to 2, 5 to 4 and 6 to 5.

Sub-Tour 1: 1 to 3, 3 to 1.

Sub-Tour 2: 2 to 6, 6 to 5, 5 to 4, and 4 to 2.

Distance: 40 + 44 = 84 km.

Thus due to this branching out the lower limit went up to 84 km. However, even now there are two sub-tours and hence one salesperson cannot manage the tour.

Now let us treat the other branch of sub-tour 2(b) as unacceptable by imposing a hefty penalty of big M. Let us redo the calculations once again starting from Table 7.150 by allotting a hefty penalty M to relevant cell as shown in Table 7.155.

Table 7.155

From	1	2	3	4	5	6	Row Min.
1	M	7	0	17	32	21	0
2	0	M	15	3	17	0	0
3	0	14	M	0	2	6	0
4	15	0	2	M	0	0	0
5	34	22	8	4	M	0	0
6	24	8	13	5	M	M	5

Table 7.156

From	1	2	3	4	5	6
1	M	7	0	17	32	21
2	0	M	15	3	17	0
3	0	14	M	0	2	6
4	15	0	2	M	0	0
5	34	22	8	4	M	0
6	19	3	8	0	M	M

Table 7.157

From	1	2	3	4	5	6
1	M	7	0	19	32	23
2	0	M	13	3	15	0
3	0	12	M	0	0	6
4	17	0	2	M	0	0
5	34	20	6	4	M	0
6	19	1	6	0	M	M

Table 7.157 (a)

From	1	2	3	4	5	6
1	M	7	[0]	19	32	23
2	[0]	M	13	3	15	0
3	0	12	M	0	[0]	6
4	17	[0]	2	M	0	0
5	34	20	6	4	M	[0]
6	19	1	6	[0]	M	M

Assignment: 1 to 3, 2 to 1, 3 to 5, 4 to 2 , 5 to 6, 6 to 4.

Tour: 1 to 3, 3 to 5, 5 to 6, 6 to 4, 4 to 2, 2 to 1.

Distance: 87 km.

Since 87 > 84 let us break the previous step sub-tour as 1 to 3 and 3 to 1 and work out the assignment under the assumption that 1 to 3 is unacceptable. The suitably modified matrix is presented in Table 7.158.

Table 7.158

From	1	2	3	4	5	6	Row Min.
1	M	7	M	17	32	21	7
2	0	M	15	3	17	0	0
3	0	14	M	0	2	6	0
4	15	0	2	M	0	0	0
5	30	18	4	0	M	M	0
6	24	8	13	5	0	M	0

Table 7.159

From	1	2	3	4	5	6
1	M	0	M	10	25	14
2	0	M	15	3	17	0
3	0	14	M	0	2	6
4	15	0	2	M	0	0
5	30	18	4	0	M	M
6	24	8	13	5	0	M

Table 7.160

From	1	2	3	4	5	6
1	M	0	M	10	25	14
2	0	M	13	3	17	0
3	0	14	M	0	2	6
4	15	0	0	M	0	0
5	30	18	2	0	M	M
6	24	8	11	5	0	M

Table 7.160 (a)

From	1	2	3	4	5	6
1	M	0	M	10	25	14
2	0	M	13	3	17	0
3	0	14	M	0	2	6
4	15	0	0	M	0	0
5	30	18	2	0	M	M
6	24	8	11	5	0	M

Assignment: 1 to 2, 2 to 6, 3 to 1, 4 to 3, 5 to 4, 6 to 5.

Tour: 1 to 2, 2 to 6, 6 to 5, 5 to 4, 4 to 3, 3 to 1.

Distance: 93 km.

Now before concluding let us explore the other branch as well. Let us consider tour 3 to 1 as unacceptable and redo the assignment again.

Table 7.161

From	1	2	3	4	5	6	Row Min.
1	M	7	0	17	32	21	0
2	0	M	15	3	17	0	0
3	M	14	M	0	2	6	0
4	15	0	2	M	0	0	0
5	30	18	4	0	M	M	0
6	24	8	13	5	0	M	0

Table 7.162

From	1	2	3	4	5	6
1	M	7	0	21	36	21
2	0	M	15	7	21	0
3	M	10	M	0	2	2
4	15	0	2	M	4	0
5	26	14	0	0	M	M
6	20	4	9	5	0	M

Table 7.163

From	1	2	3	4	5	6
1	M	5	0	21	34	19
2	0	M	17	9	21	0
3	M	8	M	0	0	0
4	15	0	2	M	4	0
5	24	12	0	0	M	M
6	20	4	11	7	0	M

Table 7.163 (a)

From	1	2	3	4	5	6
1	M	5	0	21	34	19
2	0	M	17	9	21	0
3	M	8	M	0	0	0
4	15	0	2	M	4	0
5	24	12	0	0	M	M
6	20	4	11	7	0	M

Assignment: 1 to 3, 2 to 1, 3 to 6, 4 to 2, 5 to 4 and 6 to 5.

Tour: 1 to 3, 3 to 6, 6 to 5, 5 to 4, 4 to 2, 2 to 1.

Distance: 90 km.

Clearly movement along both the branches of this node results in feasible solutions. But both of them result in longer distance coverage.

Thus 87 km tour discussed in the previous node gets confirmed as the best possible tour under the given situation. Remember, that this tour (1 to 3, 3 to 5, 5 to 6, 6 to 4, 4 to 2 and 2 to 1) is achievable by a single salesperson in the order specified with a distance of 87 km.

The following tree diagram describes the summary of all the discussions we had till now.

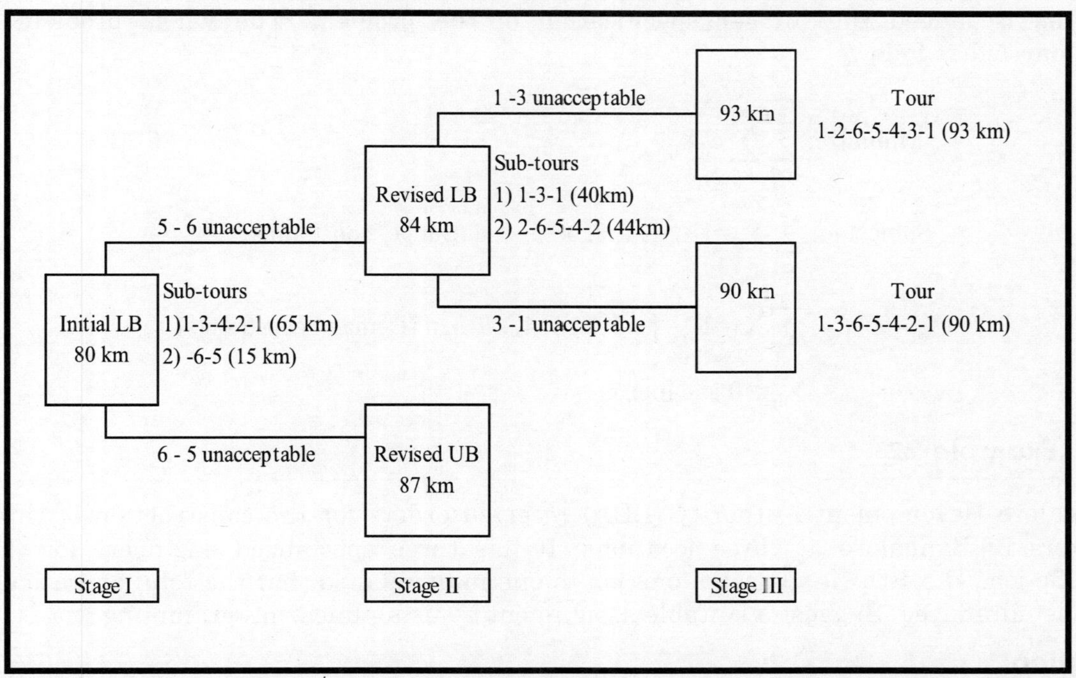

Figure 7.6

7.8 LINEAR PROGRAMMING FORMULATION TO THE ASSIGNMENT PROBLEM

If X_{ij} = Assignment of project leader i to client j then

$$X_{ij} = \begin{cases} 1 \text{ if project leader i assigned client j} \\ 0 \text{ otherwise} \end{cases}$$

where i = 1, 2, 3 ,.......n and j = 1, 2, 3, ...n

Table 7.164

Individuals	Jobs						s_i
	1	2	. . .	j	. . .	n	
1	c_{11}	c_{12}	. . .	c_{1j}	. . .	c_{1n}	1
2	c_{21}	c_{22}	. . .	c_{2j}	. . .	c_{2n}	1
.	.	.	. . .	.	. . .	.	.
.	.	.	. . .	.	. . .	.	.
i	c_{i1}	c_{i2}	. . .	c_{ij}	. . .	c_{in}	1
.	.	.	. . .	.	. . .	.	.
.	.	.	. . .	.	. . .	.	.
n	c_{n1}	c_{n2}	. . .	.	. . .	c_{nn}	1
d_j	1	1	. . .	1	. . .	1	

This is indeed the integer specification of the problem. The whole problem is summarized as follows

$$\text{Minimize: } \sum_{i=1}^{n}\sum_{j=1}^{n} c_{ij}X_{ij}$$

$$\text{Subject to: } \sum_{j=1}^{n} X_{ij} = 1 \qquad i = 1, 2, 3 \text{ .... } n \text{ (Supply constraints)}$$

$$\sum_{i=1}^{n} X_{ij} = 1 \qquad j = 1, 2, 3, 4 \text{ .... } n \text{ (Demand constraints)}$$

$$X_{ij} \geq 0 \text{ and integer.}$$

■ Example 22

Bangalore Development Authority (BDA) received orders for the construction of three flyovers in Bangalore at three locations. It has three specialized engineers for this specific job. The BDA is to assign one job to one individual so that the total operational time is minimized. Suggest a suitable assignment by using linear programming.

Solution

The BDA must consider all possible project leader–client combinations that minimize the total time spent. The alternatives and the estimated project completion time in months are given in Table 7.165.

Table 7.165

Project leader	Clients		
	1	2	3
1 Ravi	10	15	9
2 Shiva	9	18	5
3 Roopa	6	14	3

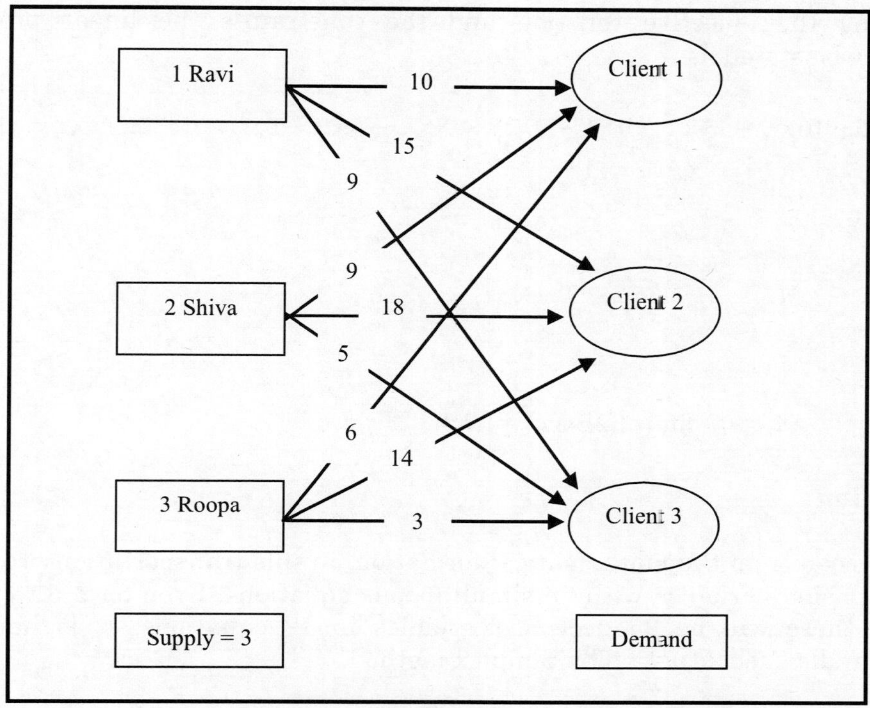

Figure 7.7

The chart illustrates this problem with their respect time cost in months for all 9 possible assignments. By combining this notation and the completion time data given let us compute the constraints as follows.

The objective is to minimize the total time required by all the three project managers. Thus the objective function is written as

$$\text{Min: } 10X_{11} + 15X_{12} + 9X_{13} + 9X_{21} + 18X_{22} + 5X_{23} + 6X_{31} + 14X_{32} + 3X_{33}$$

The conditions for the assignment should reflect that each project leader could be assigned to at the most one client.

$$X_{11} + X_{12} + X_{13} \leq 1$$
$$X_{21} + X_{22} + X_{23} \leq 1$$
$$X_{31} + X_{32} + X_{33} \leq 1$$

Similarly, each client should have one project leader.

$$X_{11} + X_{21} + X_{31} = 1$$
$$X_{12} + X_{22} + X_{32} = 1$$
$$X_{13} + X_{23} + X_{33} = 1$$

Combining the objective function and the constraints the linear programming problem may be stated as

$$\text{Min: } 10X_{11} + 15X_{12} + 9X_{13} + 9X_{21} + 18X_{22} + 5X_{23} + 6X_{31} + 14X_{32} + 3X_{33}$$

$$\begin{array}{llllllllll}
X_{11} + & X_{12} + & X_{13} & & & & & & & \leq 1 \\
 & & & X_{21} + & X_{22} + & X_{23} & & & & \leq 1 \\
 & & & & & & X_{31} + & X_{32} + & X_{33} & \leq 1 \\
X_{11} & & & + X_{21} & & & + X_{31} & & & = 1 \\
 & X_{12} & & & + X_{22} & & & + X_{32} & & = 1 \\
 & & X_{13} & & & + X_{23} & & & + X_{33} & = 1
\end{array}$$

$X_{ij} \geq 0$ for I=1,2,3 and j=1,2,3

Excel Solution

As could be seen from the mathematical formation, in this transportation problem there are n × n decision variables with 2n simultaneous equations. Even for a 5 worker and 5 job models there will be 25 decision variables in 12 equations. It is impossible to manually handle such situation by Simplex method.

Table 7.166

	A	B	C	D	E	F	G
1			Clients				
2	Engineers	1	2	3	Supply		
3	1	10	15	9	1		
4	2	9	18	5	1		
5	3	6	14	3	1		
6	Demand	1	1	1			
7							
8	Model						
9	Minimum cost	26					
10			Clients				
11	Engineers	1	2	3	Total		
12	1	0	1	0	1	<=	1
13	2	0	0	1	1	<=	1
14	3	1	0	0	1	<=	1
15	Total	1	1	1			
16		=	=	=			
17		1	1	1			

However, the excel spreadsheet is having a built in solver option. As in the case of the simple linear programming once again we propose use this technique irrespective of the number of decision variables.

Assignment:

Engineer 1 to client 2, Engineer 2 to client 3, Engineer 3 to client 1

Time minimum time cost = 15+6+5 = 26 months

Example 23

Given the following assignment matrix obtains the minimum operational time using excel solver.

Table 7.167

	Machine time					
Operators	1	2	3	4	5	6
A	41	72	39	52	25	51
B	22	29	49	65	81	50
C	27	39	60	51	32	32
D	45	50	48	52	37	43
E	29	40	39	26	30	33
F	82	40	40	60	51	30

Solution

Assignment:

Manager A to Client 5

Manager B to Client 2

Manager C to Client 1

Manager D to Client 3

Manager E to Client 4

Manager F to Client 6

Total minimum time cost = 185

Table 7.168

	A	B	C	D	E	F	G	H	I	J
1					Clients					
2	Manager	1	2	3	4	5	6	s_i		
3	A	41	72	39	52	25	51	1		
4	B	22	29	49	65	81	50	1		
5	C	27	39	60	51	32	32	1		
6	D	45	50	48	52	37	43	1		
7	E	29	40	39	26	30	33	1		
8	F	82	40	40	60	51	30	1		
9	d_j	1	1	1	1	1	1			
10										
11	Model									
12	Minimum cost		185							
13					Clients					
14	Manager	1	2	3	4	5	6	Total		
15	A	0	0	0	0	1	0	1	<=	1
16	B	0	1	0	0	0	0	1	<=	1
17	C	1	0	0	0	0	0	1	<=	1
18	D	0	0	1	0	0	0	1	<=	1
19	E	0	0	0	1	0	0	1	<=	1
20	F	0	0	0	0	0	1	1	<=	1
21	d_j	1	1	1	1	1	1			
22		=	=	=	=	=	=			
23		1	1	1	1	1	1			

Example 24

Solve the following assignment problem and get the minimum time cost.

Table 7.169

	Clients				
Manager	1	2	3	4	s_i
A	3	6	2	6	1
B	7	1	4	4	1
C	3	8	5	8	1
D	6	4	3	7	1
E	5	2	4	3	1
F	5	7	6	2	1
d_j	1	1	1	1	

Solution

Let us introduce to dummy jobs and make the number of jobs to number of managers and go ahead with the excel method in the usual manner and get the final solution.

Assignment:

Manager A to Client 3

Manager B to Client 2

Manager C to Client 1

Manager D to Client 6

Manager E to Client 5

Manager F to Client 4

The minimum total cost = 3+1+2+2 = 8

Table 7.170

	A	B	C	D	E	F	G	H	I	J
1					Clients					
2	Manager	1	2	3	4	5	6	s_i		
3	A	3	6	2	6	0	0	1		
4	B	7	1	4	4	0	0	1		
5	C	3	8	5	8	0	0	1		
6	D	6	4	3	7	0	0	1		
7	E	5	2	4	3	0	0	1		
8	F	5	7	6	2	0	0	1		
9	d_j	1	1	1	1	1	1			
10										
11	Model									
12	Min	8								
13					Clients					
14	Manager	1	2	3	4	5	6	Total		
15	A	0	0	1	0	0	0	1	<=	1
16	B	0	1	0	0	0	0	1	<=	1
17	C	1	0	0	0	0	0	1	<=	1
18	D	0	0	0	0	0	1	1	<=	1
19	E	0	0	0	0	1	0	1	<=	1
20	F	0	0	0	1	0	0	1	<=	1
21	d_j	1	1	1	1	1	1			
22		=	=	=	=	=	=			
23		1	1	1	1	1	1			

Example 25

Solve the following assignment problem and get the minimum time cost.

Table 7.171

	Clients						
Manager	1	2	3	4	5	6	s_i
A	3	6	2	6	4	6	1
B	7	1	4	4	2	3	1
C	3	8	5	8	2	4	1
D	6	4	3	7	2	2	1
d_j	1	1	1	1	1	1	

Solution: One-to-one assignment

Table 7.172

	A	B	C	D	E	F	G	H	I	J
1				Clients						
2	Manager	1	2	3	4	5	6	s_i		
3	A	3	6	2	6	4	6	1		
4	B	7	1	4	4	2	3	1		
5	C	3	8	5	8	2	4	1		
6	D	6	4	3	7	2	2	1		
7	E	0	0	0	0	0	0	1		
8	F	0	0	0	0	0	0	1		
9	d_j	1	1	1	1	1	1			
10										
11	Model									
12	Minimum cost		7							
13				Clients						
14	Manager	1	2	3	4	5	6	Total		
15	A	0	0	1	0	0	0	1	<=	1
16	B	0	1	0	0	0	0	1	<=	1
17	C	0	0	0	0	1	0	1	<=	1
18	D	0	0	0	0	0	1	1	<=	1
19	E	1	0	0	0	0	0	1	<=	1
20	F	0	0	0	1	0	0	1	<=	1
21	d_j	1	1	1	1	1	1			
22		=	=	=	=	=	=			
23		1	1	1	1	1	1			

Assignment:

Manager A to Client 3 Manager B to Client 2

Manager C to Client 5 Manager D to Client 6

Manager E to Client 1 Manager F to Client 4

The minimum total cost = 7

Solution: Multiple assignments

Table 7.173

	A	B	C	D	E	F	G	H	I	J
1					Clients					
2	Manager	1	2	3	4	5	6	s_i		
3	A	3	6	2	6	4	6	1		
4	B	7	1	4	4	2	3	1		
5	C	3	8	5	8	2	4	1		
6	D	6	4	3	7	2	2	1		
7	E	0	0	0	0	0	0	2		
8	d_j	1	1	1	1	1	1			
9										
10	Model									
11	Minimum cost		7							
12					Clients					
13	Manager	1	2	3	4	5	6	Total		
14	A	0	0	1	0	0	0	1	<=	1
15	B	0	1	0	0	0	0	1	<=	1
16	C	0	0	0	0	1	0	1	<=	1
17	D	0	0	0	0	0	1	1	<=	1
18	E	1	0	0	1	0	0	2	<=	2
19	d_j	1	1	1	1	1	1			
20		=	=	=	=	=	=			
21		1	1	1	1	1	1			

Assignment:

Manager A to Client 3 Manager B to Client 2

Manager C to Client 5 Manager D to Client 6

Manager E to Client 1 and 4

The minimum total cost = 7

E to 1, 4 we mean job 4 and 1 are assigned to F the dummy manager. Here the minimum cost = 7.

■ Example 26

Solve the following assignment problem and get the minimum time cost. By excel solver. By "X" we mean the prohibited jobs.

Table 7.174

Manager	Clients 1	2	3	4	5	s_i
A	80	40	x	70	40	1
B	x	80	60	40	40	1
C	70	x	60	80	70	1
D	70	80	30	50	x	1
E	40	40	50	x	80	1
d_j	1	1	1	1	1	

Table 7.175

	A	B	C	D	E	F	G	H	I
1					Clients				
2	Manager	1	2	3	4	5	s_i		
3	A	80	40	1000	70	40	1		
4	B	2000	80	60	40	40	1		
5	C	70	1000	60	80	70	1		
6	D	70	80	30	50	2000	1		
7	E	40	40	50	2000	80	1		
8	d_j	1	1	1	1	1			
9									
10	Model								
11	Minimum cost		220						
12					Clients				
13	Manager	1	2	3	4	5	Total		
14	A	0	0	0	0	1	1	<=	1
15	B	0	0	0	1	0	1	<=	1
16	C	1	0	0	0	0	1	<=	1
17	D	0	0	1	0	0	1	<=	1
18	E	0	1	0	0	0	1	<=	1
19	d_j	1	1	1	1	1			
20		=	=	=	=	=			
21		1	1	1	1	1			

Assignment:

Manager A to Client 5
Manager B to Client 4
Manager C to Client 1
Manager D to Client 3
Manager E to Client 2

The minimum total cost = 220

Example 27

Given the profit matrix of the assignment problem obtain the maximum profit.

Table 7.176

	Jobs				
Individuals	1	2	3	4	s_i
A	1	8	4	1	1
B	5	7	6	5	1
C	3	5	4	2	1
D	3	1	6	3	1
d_j	1	1	1	1	

Solution

Table 7.177

	A	B	C	D	E	F	G	H
1			Clients					
2	Individuals	1	2	3	4	s_i		
3	A	1	8	4	1	1		
4	B	5	7	6	5	1		
5	C	3	5	4	2	1		
6	D	3	1	6	3	1		
7	d_j	1	1	1	1			
8			Clients					
9	Individuals	1	2	3	4	s_i		
10	A	7	0	4	7	1		
11	B	3	1	2	3	1		
12	C	5	3	4	6	1		
13	D	5	7	2	5	1		
14	d_j	1	1	1	1			
15								
16	Model							
17	Max. return=		22					
18			Clients					
19	Individuals	1	2	3	4	Total		
20	A	0	1	0	0	1	<=	1
21	B	0	0	0	1	1	<=	1
22	C	1	0	0	0	1	<=	1
23	D	0	0	1	0	1	<=	1
24	d_j	1	1	1	1			
25		=	=	=	=			
26		1	1	1	1			

A to 2, B to 4, C to 1, D to 3, Profit = Rs. 22

7.9 PRODUCTION AND INVENTORY APPLICATIONS

■ Example 28

Suppose Sleepwell Company produces beds of a standard size and sells it throughout the four quarters in a year. The quarterly production capacity and the respective quarterly production are shown below. Since production and demand normally do not match for all quarters the company will have to keep the excess supply, if any, as a buffer stock, called inventories. Such an inventory costs the producer in the form of rent, interest and so on. These costs are given in the last column of the Table 7.178.

Table 7.178

Quarters	Production capacity (in square metres)	Demand (in square metres)	Production cost (per square metres)	Inventory cost (in square metres)
1	600	400	2	0.25
2	300	500	5	0.25
3	500	400	3	0.25
4	400	400	3	0.25

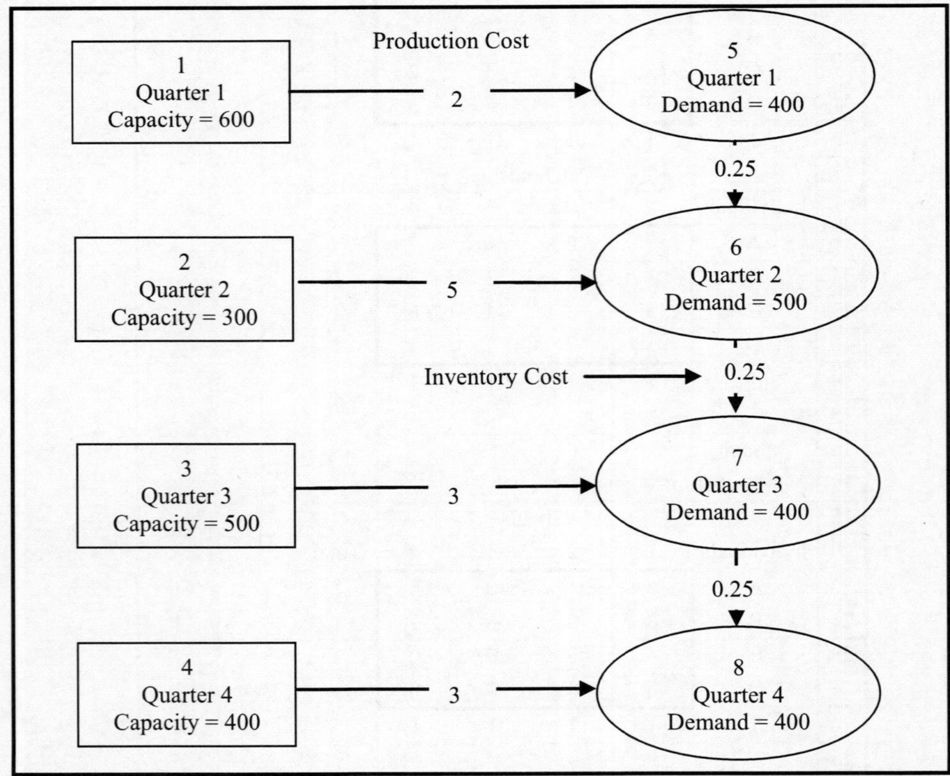

Figure 7.8

X_{15} represents the number of square metres of beds manufactured in quarter 1. The capacity pertaining to this quarter is 600. So the production constant is written as

$$X_{15} \leq 600$$

Similarly the constraints for other quarters are reported below:

$$X_{16} \leq 600$$
$$X_{26} \leq 300$$
$$X_{37} \leq 500$$
$$X_{48} \leq 400$$

Now let us consider the constraint for the demand node 5. For this the arrow X_{15} comes in and X_{56} out. So the relevant constraint may be written as:

$$X_{15} - X_{56} = 400$$

Similarly the other constraints are written as

$$X_{56} + X_{26} - X_{67} = 500$$
$$X_{67} + X_{37} - X_{78} = 400$$
$$X_{78} + X_{48} = 400$$

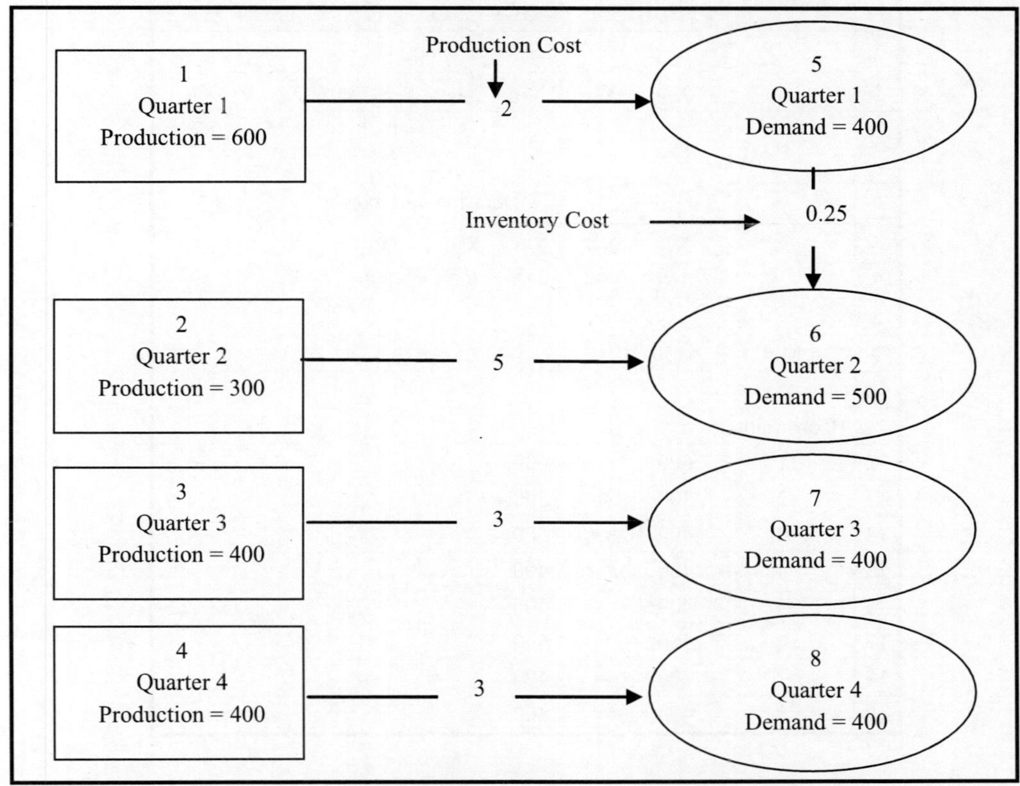

Figure 7.9

Now the complete linear programming model is written as

$$\text{Min: } 2X_{15} + 5X_{26} + 3X_{37} + 3X_{48} + 0.25X_{56} + 0.25X_{67} + 0.25X_{78}$$

$$\begin{aligned}
\text{S.to: } & X_{15} \le 600 \\
& X_{26} \le 300 \\
& X_{37} \le 500 \\
& X_{48} \le 400 \\
& X_{15} - X_{56} = 400 \\
& X_{26} + X_{56} - X_{67} = 500 \\
& X_{37} + X_{67} - X_{78} = 400 \\
& X_{48} + X_{78} = 400 \\
& X_{ij} \ge 0 \quad \text{for all i and j}
\end{aligned}$$

Excel Solution

The corresponding excel solution is obtained and reported in Table 7.179.

Table 7.179

	A	B	C	D	E	F	G	H
1		Data						
2		X_{15}	X_{26}	X_{37}	X_{48}	X_{56}	X_{67}	X_{78}
3		2	5	3	3	0.25	0.25	0.25
4								
5		Decision variables						
6		X_{15}	X_{26}	X_{37}	X_{48}	X_{56}	X_{67}	X_{78}
7		600	300	400	400	200	0	0
8								
9	Min	5150						
10								
11	Constraints							
12	1	600	<=	600				
13	2	300	<=	300				
14	3	400	<=	500				
15	4	400	<=	400				
16	5	400	=	400				
17	6	500	=	500				
18	7	400	=	400				
19	8	400	=	400				

EXERCISES

1. What do you mean by assignment problem? Discuss the merits and demerits of the various methods of solution seeking.

2. In which way the assignment problem is a special case of the transportation problem. Explain your answer with a suitable illustration.

3. Enumerate the similarities and differences between an assignment and transportation problem formulation.

4. Assignment problem gives degenerated solutions when it solved by transportation techniques. Explain.

5. Describe the Hungarian method with suitable illustration of your own.

6. What is meant by unbalanced assignment problem? How do you handle the situation to the solution?

7. A research department in management has four research assistants at its disposal in handling four research projects. Given the time cost across the project versus assistants, obtain the least time cost assignment on one-to-one basis.

	Projects			
Assistants	1	2	3	4
A	15	29	35	20
B	21	27	33	17
C	20	22	32	15
D	23	23	31	42

8. Five part-time engineers in a building concern are to be assigned different projects on one-to-one basis though all can do all the projects, technically speaking. However because of technical skill they possess and the projects needs often vary. Accordingly the time for completion across engineers and projects vary considerably as given in the Table. The contractual rates for A, B, C are Rs. 1000 per day and for D and E it is Rs. 500 per day. Obtain the least time for completing all the five units. Also get the least cost for completion.

	Project time cost in days				
Engineers	1	2	3	4	5
A	35	29	35	20	23
B	21	27	33	17	24
C	20	45	32	15	45
D	34	56	67	23	23
E	23	23	45	42	32

9. The director of a newly started management institute in Bangalore has enough preparations to start four new courses. In this context one professor, one associate professor and two assistant professors were appointed after thorough examination across the courses. Efficiency ratings were calculated for all the four teachers across the four courses. How should the director allot the teachers on one-to-one basis to realize his objective?

	Course ratings as percent			
Engineers	1	2	3	4
A	60	40	60	70
B	20	60	50	70
C	20	30	40	60
D	30	10	20	40

10. In a workshop producing a product, five lathes are to assigned five newly appointed operators on one-to-one basis. The lathes and operators across productivity are given below. If each part is sold at Rs. 25 obtain operator allotment that maximizes the total productivity and hence the total profit. In a particular week the operator C is sick and not in a position to attend how does the manager reassign the job. The next week though all the operators are available 3 lathes breakdown. How does the manager react this time?

	Output turnout per week			
Operators	1	2	3	4
A	18	20	23	25
B	17	21	20	20
C	21	26	22	22
D	19	22	20	23
E	22	25	23	26

11. Describe specialty of travelling salesman problem. Given the following across city distance table find the minimum distance root provided his home town is in A.

	To				
From	A	B	C	D	E
A	-	7	6	8	4
B	7	-	8	5	3
C	6	8	-	9	7
D	8	5	9	-	8
E	4	6	7	8	-

12. Find the optimal assignment for the following cost matrix.

	Areas			
Salesmen	A1	A2	A3	A4
S1	11	17	8	16
S2	9	7	12	10
S3	13	16	15	12
S4	14	10	12	11

(C.A. Final November 1980)

13. A firm wants to purchase three different types of equipment and five manufacturers have come forward to supply one or all the three machines. However, the firm's policy is not to accept more than one machine from any of the manufacturers. the data relating to the price (in lakhs of rupees) quoted by different manufacturers are given below:

	Machines		
Manufacturers	1	2	3
A	2.99	3.11	2.63
B	2.78	2.87	2.57
C	2.92	3.05	2.8
D	2.82	3.1	2.74
E	3.11	2.9	2.64

Determine how best the firm can purchase three machines.

(Delhi University, M.B.A. April 1985)

14. A salesman has five cities A, B, C, D and E. The distances (in hundred km) between the five cities are as follows.

	To city				
From city	A	B	C	D	E
A	-	2	1	6	5
B	2	-	2	5	4
C	1	2	-	7	5
D	6	5	7	-	2
E	5	4	2	5	-

If the salesman starts from A and has to come back to City A, which route should he select so that the total distance travelled by him is minimized?

(Delhi University, M.B.A April 1984)

15. Find the optimal solution to the assignment problem having the following cost matrix.

Salesman	Sales territories (cost in Rs. thousand)			
	I	II	III	IV
A	30	25	26	28
B	26	32	24	20
C	20	22	18	27
D	23	20	21	19

(Delhi University, M.Com. 1985)

16. Suggest optimum assignment of 4 workers A, B, C and D to 4 jobs I, II, III and IV. The time taken by different workers in completing the different jobs is given below:

Workers	Jobs			
	I	II	III	IV
A	8	10	12	16
B	11	11	15	8
C	9	6	5	14
D	15	14	9	7

Also indicate the total time taken in completing the jobs.

(Delhi University, M.Com. 1989)

17. A company has to assign four workers a, b, c and d to four w, x, y and z. The cost matrix is given below:

Workers	Jobs (Cost in Rs).			
	I	II	III	IV
A	1000	1200	400	900
B	600	500	300	800
C	200	300	400	500
D	600	700	300	1000

Suggest an optimal assignment schedule and total cost pertaining thereto.

(Delhi University, M.Com. 1985)

18. Suggest optimum solution to the following assignment problem and also the minimum cost:

Salesman	Markets (cost in lakhs of rupees)			
	I	II	III	IV
A	44	80	52	80
B	60	56	40	72
C	36	60	48	48
D	52	76	36	40

(Delhi University, M.Com. 1987)

19. Solve the following assignment problem for minimum cost. The cost structure is given below:

	M1	M2	M3	M4	M5	M6
T1	31	62	29	42	15	40
T2	12	20	40	55	70	38
T3	15	30	50	42	29	22
T4	35	40	37	44	27	33
T5	20	30	28	16	32	23
T6	65	30	32	50	42	20

(Delhi University, M.A. (Business Economics) 1982)

20. Solve the following assignment problem for minimum cost. The cost structure of the contractors against tenders is given below:

	C1	C2	C3	C4	C5
T1	16	22	13	21	25
T2	14	12	17	11	20
T3	18	21	20	17	21
T4	26	29	22	33	31
T5	19	15	17	16	20

(Delhi University, M.A. (Business Economics) 1981)

21. Solve the assignment problem to maximize the sales.

Salesman	Territory			
	1	2	3	4
A	92	78	64	50
B	68	58	48	38
C	68	58	48	38
D	56	48	40	32

(Bangalore University, M.B.A. July/August 2006)

22. Solve the following assignment problem to minimize the layover time of drivers. The drivers must have a minimum layover of 2.5 hours between trips. Obtain the pair of trips that minimizes the layover time.

Route	Bangalore - Hassan		Route	Hassan - Bangalore	
No	Dep.	Arr.	No	Dep.	Arr.
101	7:00 A.M	10:00 A.M	201	10:00 A.M	1:00 P.M
102	9:00 A.M	12 Noon	202	11:00 A.M	2:00 P.M
103	11:00 A.M	2:00 P.M	203	2:00 P.M	5:00 P.M
104	3:00 P.M	6:00 P.M	204	4:00 P.M	7:00 P.M

23. Solve the following assignment problem, giving details about region-wise profit for various products for a trading company, for profit maximization. Also find the maximum possible profit.

	Products				
Regions	A	B	C	D	E
P	30	37	40	28	40
Q	40	24	27	21	21
R	40	32	33	30	30
S	25	38	40	36	36
T	29	62	41	25	34
F	12	50	52	38	23

(Bangalore University, M.B.A July/August 2005)

24. Solve the assignment problem for minimizing the total cost.

	a	b	c	d	e	f
A	18	31	63	20	28	36
B	48	80	75	51	61	52
C	49	35	90	44	51	39
D	76	49	34	59	45	39
E	41	18	59	32	29	25
F	12	50	52	38	23	32

(Bangalore University, M.B.A October 2004)

25. Solve the assignment problem to minimize the layover time of drivers. The drivers should have a minimum layover of 3 hrs between two trips. Obtain the pair of drivers' route that minimizes the layover.

Route No.	Departure	Arrival	Route No.	Departure	Arrival
101	10:00 A.M	11:00 A.M	201	11:00 A.M	12:15 P.M
102	11:00 A.M	12 :00 Noon	202	1:30 P.M	2:45 P.M
103	3:30 A.M	4:30 P.M	203	6:00 P.M	7.15 P.M
104	7:30 A.M	8:30 P.M	204	9:00 P.M	10:15 P.M

(Bangalore University, M.B.A October 2004)

26. A company has four sales representatives who are to be assigned to four different sales territories. The monthly sale increase estimated for each sales representative for different scales territories (in lakhs of rupees) are shown in the following table.

Sales Representatives	Sales Territories			
	I	II	III	IV
A	200	150	170	220
B	160	120	150	140
C	190	195	190	200
D	180	175	160	190

Suggest optimal assignment and the total maximum sales increase per month. If for certain reasons, sales representative B cannot be assigned to territory III, will the optimum assignment schedule be different? If so find that schedule and the effect on the total sale. (M.Com, Delhi, 1997)

27. ABC air line, operating 7 days in a week, has given the following timetable. The crew must have a minimum layover of 5 hours between flights. Obtain the paring of flights that minimizes layover time away from home. For any given pairing the crew will be based at the city that results the smallest layover.

Hyderabad to Delhi			Delhi to Hyderabad		
Flight No.	Departure	Arrival	Flight No.	Departure	Arrival
A1	6 A.M	8 A.M	B1	8 A.M	10 A.M
A2	8 A.M	10 A.M	B2	9 A.M	11 A.M
A3	2 P.M	4 P.M	B3	2 P.M	4 P.M
A4	8 P.M	10 P.M	B4	7 P.M	9 P.M

28. A company has just developed a new item for which it proposes to undertake a national television promotional campaign. It has decided to schedule a series of one-minute commercials during peak audience viewing hours of 1 to 5 P.M. To reach the widest possible audience, the company wants to schedule one commercial on each of the networks and to have only one commercial appear during each of the four one-hour time blocks. The exposure ratings for each hour, which represents the number of viewers per Rs.10,000 spent are given below.

	Network			
Viewing hours	A	B	C	D
1-2 P.M	27.1	18.1	11.3	9.5
2-3 P.M	18.9	15.5	17.1	10.6
3-4 P.M	19.2	18.5	9.9	7.7
4-5 P.M	11.5	21.4	16.8	12.8

a) Which network should be scheduled each hour to provide the maximum audience exposure?

b) How would the schedule change if it is decided not to use network A between 1 and 3 P.M?

CHAPTER

8

Sequencing Problems

8.1 INTRODUCTION

By sequencing we mean the arrangement of given 'n' jobs to be performed at 'm' service facilities called machines in a predetermined order so that the total time elapsed for the whole sequence of jobs is minimized. Here by machines we mean the service facilities through which the jobs are being carried out. It is presumed that the processing time for each job on each machine is known in advance. Further, the job once processed in machine 1 will go to the next machine 2 for processing. If the second machine is free then it will receive job immediately and start processing it. If it is not free then the job will wait till machine 2 is made free. Sometimes machine 2 waits for job from machine as well. In the following section we propose to highlight some special type of sequencing and their associated problems. However, no job priority is considered in this section. Moreover, due date of delivery for the various jobs undertaken are not in the preview in this section. The only objective is to finish all the jobs under consideration at an early date.

8.2 PROCESSING N JOBS IN TWO MACHINES

This is the simplest case in which n jobs J1, J2...Jn are processed in two machines M1 and M2 in the order M1M2. Tailoring could be one typical example for this type of sequencing. The tailor cuts the peace of cloth in one machine (M1) and stitches in another machine (M2). Remember, only after cutting the job goes for stitching and not in the reverse order. Thus, the specified order is M1 M2 . The n jobs that the tailor gets may be a blouse, a shirt, a paint, a suit etc. Under the circumstances the problem faced by the tailor is to sequence the jobs so that the total time elapsed for stitching all these items is at its minimum. Remember, that in this section we do not consider the due date of delivery offered by the tailor.

This type of problem is described completely as follows:

1. Only two machines M1 and M2 are available.
2. All the jobs are processed in the unique order, namely M1 first and M2 next.
3. The estimated processing times, namely m_{11}, m_{12}, ...m_{1n} for M1; and m_{21}, m_{22}, ...m_{2n} for M2 are known constants as described in Table 8.1 below.

Table 8.1

Machine	Job processing time				
	J1	J2	J3	....	Jn
M1	m_{11}	m_{12}	m_{13}	....	m_{1n}
M2	m_{21}	m_{22}	m_{23}	....	m_{2n}

In the above table the job number namely, 1, 2, 3... n are only indicative names and thus does not indicate any priority rule unless otherwise specified. Given the time cost, our purpose is to find the cost effective sequence of these n jobs. To illustrate the method of sequencing consider the following example.

■ Example 1

The time taken to do three jobs J1, J2, and J3 in two machines M1 and M2 are given below. Obtain the cost effective sequence of these three jobs, in the given order M1M2.

Table 8.2

Machine	Job timings data in number of days		
	J1	J2	J3
M1	4	5	2
M2	8	6	8

Solution by complete enumeration

Since there are three jobs to be performed, there are 3! = 3×2×1 = 6 possible ways of doing it. Table 8.3 and the associated Gantt charts from Figure 8.1 to 8.6 illustrate all the six possibilities and their associated graphical representations.

Table 8.3

Possibility	1	2	3	4	5	6
Sequence	J1 J2 J3	J1 J3 J2	J2 J1 J3	J2 J3 J1	J3 J1 J2	J3 J2 J1

In Figure 8.1 the job J1 (blouse) is taken up first in machine M1 for cutting. It takes 4 minutes as described in the chart. So till the cutting is over for blouse the stitching in M2 can not start. So really the stitching of blouse starts immediately after cutting as described in the chart. This work of blouse stitching continues for 8 minutes. Meanwhile, in machine M1 cutting of pants (J2) starts immediately after the cutting of blouse and continues for 5 minutes. There after it takes up the cutting of shirt (J3). From 12 minute onwards the stitching of pants starts. This job will take 6 minutes for stitching. From 18 minute onwards the shirt stitching starts. It takes another 8 minutes for completion. Thus, altogether M1 works for 11 minutes and machine M2 works up to 26th minute from the start of cutting including the first four minute ideal time. So for completing all the three jobs both the operations take 26 minutes. Similarly, the total time elapsed for all the remaining sequencing are described in Gantt charts 8.1 to 8.6.

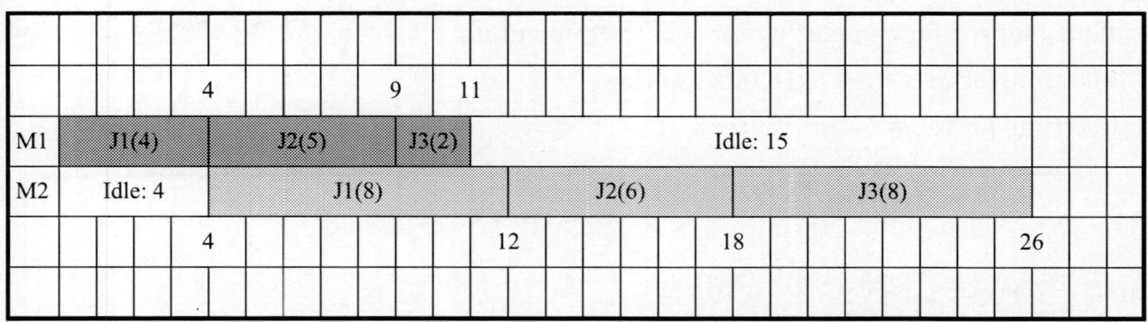

Figure 8.1 Possibility 1: J1 J2 J3

Time elapsed for completing the jobs: 26 time units

Idle time for M1=26−11=15 units of time

Idle time for M2 = 4 units of time

4 6 11
M1 J1(4) J3(2) J2(5) Idle: 15
M2 Idle: 4 J1(8) J3(8) J2(6)
4 12 20 26

Figure 8.2 Possibility 2: J1 J3 J2

Time elapsed for completing the jobs: 26 time units

Idle time for M1=26−11=15 units of time

Idle time for M2 = 4 units of time

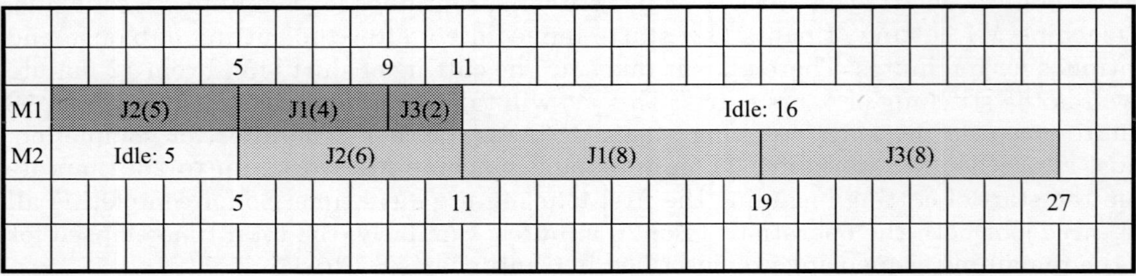

Figure 8.3 Possibility 3: J2 J1 J3

Time elapsed for completing the jobs: 27 time units

Idle time for M1=27−11=16 units of time

Idle time for M2 = 5 units of time

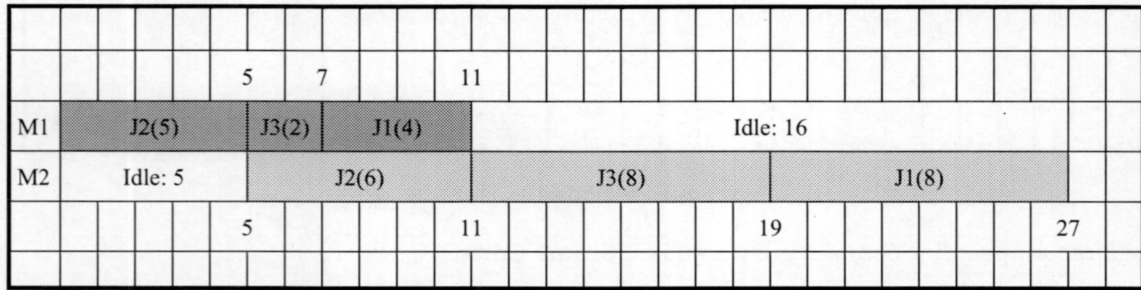

Figure 8.4 Possibility 4: J2 J3 J1

Time elapsed for completing the jobs: 27 time units

Idle time for M1 = 27−11=16 units of time

Idle time for M2 = 5 units of time

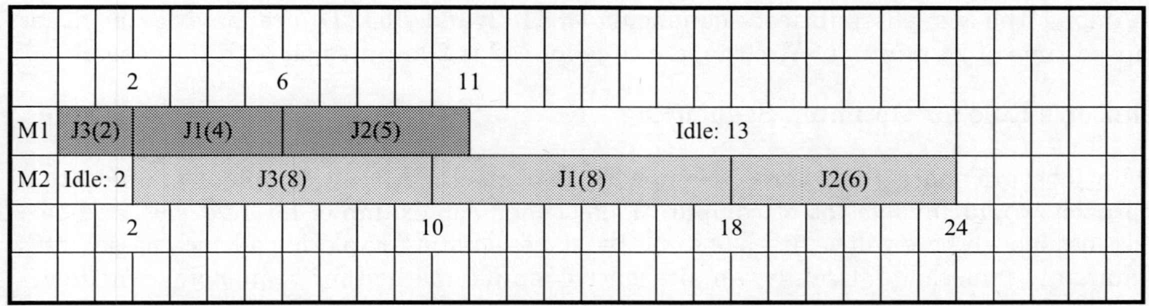

Figure 8.5 Possibility 5: J3 J1 J2

Time elapsed for completing the jobs: 24 time units

Idle time for M1=24−11=13 units of time

Idle time for M2 = 2 units of time

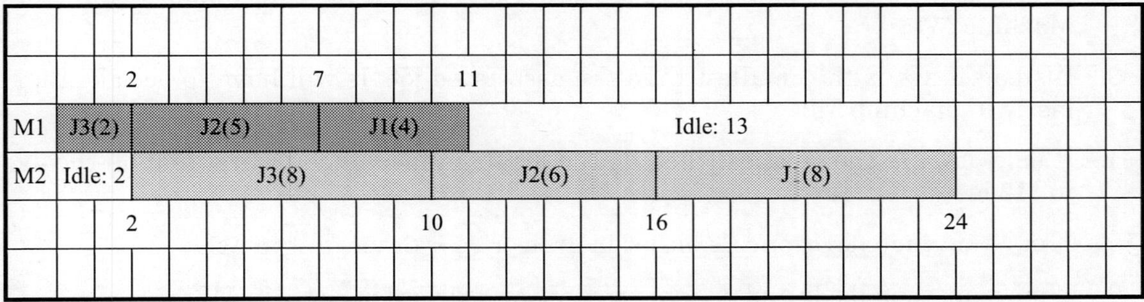

Figure 8.6 Possibility 6: J3 J2 J1

Time elapsed for completing the jobs: 24 time units

Idle time for M1=24−11=13 units of time

Idle time for M2 = 2 units of time

Table 8.4 summarizes the total time elapsed for all the six possibilities including the idle time for both M1 and M2.

Table 8.4

Alternatives	Total elapsed time	Idle time for M2	Idle time for M1
J1J2J3	26	4	16
J1J3J2	26	4	16
J2J1J3	27	5	15
J2J3J1	27	5	15
J3J1J2	24	2	13
J3J2J1	24	2	13

Among the six possibilities, sequencing J3J1J2 and J3J2J1 are having the least elapsed time of 24 minutes. Also the idle time for M2 is 2 minutes on both the cases.

Johnson's Rule for Optimum Sequencing

When jobs are more than three, even with two machines, the complete enumeration approach would involve lot of computational time. For example, for five jobs and two machines model there will be 5! = 5×4×3×2×1=120 possibilities. Exploring all such possibilities is humanly impossible. Luckily, an alternative simple method for sequencing was developed by Johnson and Bellman in the year 1953.

In this method we follow the following rules :

1. Look at the processing time through M1: m_{11}, m_{12} ...m_{1n} and M2: m_{21}, m_{22}............m_{2n} then select the smallest time consuming job.
2. If m_{1r} is the smallest then the associated job Jr will have to taken up first in Machine M1.
3. Instead if m_{2s} is the smallest then the associated job Js will have to be taken up lastly in machine M2.
4. If $m_{1r} = m_{2s}$ are the smallest then do the job Jr in machine M1 first and do the Js in M2 last.
5. If $m_{1r} = m_{1s}$ then do Jr and Js either in JrJs or JsJr order first in M1.
6. If $m_{2r} = m_{2s}$ then do Jr and Js ether in JrJs or JsJr order lastly in M2.
7. The remaining jobs are sequenced by merely repeating steps 1 to 6.
8. Continue this process by placing these jobs next to the first or proceeding to the last, as the case may be.
9. After placing all the jobs we simply calculate the overall time lapsed, as the time between start time of the job 1 in machine M1 and the finish time of the last job in machine M2.

Rules for the Calculation of Idle Time

Such sequencing does not necessarily keep both the machines fully engaged. Sometimes machine M2 is to be kept idle till the needed job is completed in M1 because of the order specification M1M2. Similarly, if all the jobs go for the second round M1 will have to wait for the completion of last job in M2.

1. Idle time of machine M1 = completion time of the last job in M2 – completion time of the last job in machine 1.
2. Idle time of machine M2 = sum of all idle times from the very bringing till the end.

Example 2

The time cost to do six jobs in two machines M1 and M2 are given below.

Table 8.5

Machine	Job timings in number of days					
	J1	J2	J3	J4	J5	J6
M1	5	9	4	7	8	6
M2	7	4	8	3	9	5

Obtain the optimum sequencing by using Johnson's Rule.

Solution

To get the optimum sequencing, we apply Johnson's rules 1 to 6 and obtain a new optimum sequence as follows. Since 3 hours is the minimum time for J4 in M2 in Table 8.5, according to Johnson's rule we push this job to the sixth place (refer Table 8.9) and also delete column pertains to this job from the data Table 8.5 to facilitate further iteration as shown in Table 8.6.

Table 8.6

Machine	Job timings data				
	J1	J2	J3	J5	J6
M1	5	9	4	8	6
M2	7	4	8	9	5

Since 4 hours is the minimum time for J3 in M1 and J2 in M2 we push J2 to the fifth and J3 to the first place and delete both the jobs to facilitate further iteration.

Table 8.7

Machine	Job timings data		
	J1	J5	J6
M1	5	8	6
M2	7	9	5

Since 5 is the minimum for J1 in M1 and J6 in M2 we push J6 to fourth place and J1 to the second place respectively and delete both the jobs to facilitate further iteration.

Table 8.8

Machine	Data
	J5
M1	8
M2	9

The leftover J5 is thus placed in the leftover third place. The final optimal sequential ordering is shown in Table 8.9.

Table 8.9

Machine	Optimum job sequencing					
	1	2	3	4	5	6
	J3	J1	J5	J6	J2	J4
M1	4	5	8	6	9	7
M2	8	7	9	5	4	3

Once the optimum sequencing is decided we obtain the time elapsed and the idle times in the usual manner either by using the Gantt chart 8.7 or tabular method described in Table 8.10.

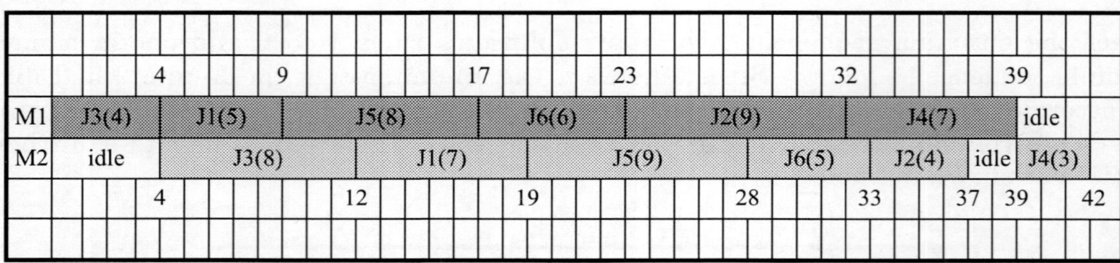

Figure 8.7

Table 8.10

Job sequence	M1				M2			
	Time in	Job time	Time out	Idle time	Time in	Job time	Time out	Idle time
J3	0	4	4	0	4	8	12	4
J1	4	5	9	0	12	7	19	0
J5	9	8	17	0	19	9	28	0
J6	17	6	23	0	28	5	33	0
J2	23	9	32	0	33	4	37	0
J4	32	7	39	3	39	3	42	2
Total				3				6

■ Example 3

The time to do given 6 jobs in two machine M1 and M2 are listed below.

Table 8.11

Machine	Job timings					
	J1	J2	J3	J4	J5	J6
M1	4	7	3	12	11	9
M2	11	7	10	8	10	13

Obtain the optimum job sequence that minimizes the total time elapsed.

Solution

Since 3 is minimum time to do the job J3 in machine M1, J3 is entered in the first place in the optimum sequencing Table 8.17 below. Once this is done block out J3 from the table and proceed for the next iteration as shown below.

Table 8.12

Machine	Job timings (Data table)					
	J1	J2	J3	J4	J5	J6
M1	4	7	3	12	11	9
M2	11	7	10	8	10	13

Since 4 is the minimum time for the job J1 in machine M1, J1 is entered in the second place in the optimum sequencing result table and block out J1 from the data table for next iteration.

Table 8.13

Machine	Job timings (Data table)					
	J1	J2	J3	J4	J5	J6
M1	4	7	3	12	11	9
M2	11	7	10	8	10	13

In the left out data table we encounter a tie to do the job J2 with processing time of 7 each in M1 and M2. The decision to put J2 either in the third or sixth does not pose any problem as far as the elapsed time is concerned. Given the priority to M2 we push this job to the sixth position in the result table and block out it from the data table.

Table 8.14

Machine	Job timings (Data table)					
	J1	J2	J3	J4	J5	J6
M1	4	7	3	12	11	9
M2	11	7	10	8	10	13

This time clearly 8 is least value to do J4 in M2. So we push this job to fifth position and block out it from the data table.

Table 8.15

Machine	Job timings (Data table)					
	J1	J2	J3	J4	J5	J6
M1	4	7	3	12	11	9
M2	11	7	10	8	10	13

Now J6 on M1 has the minimum time of 9. So we place J6 in the third place and block out J6 column.

Table 8.16

Machine	Job timings (Data table)					
	J1	J2	J3	J4	J5	J6
M1	4	7	3	12	11	9
M2	11	7	10	8	10	13

The last left out J5 invariably goes to the left out fourth place in the result table.

Table 8.17

Machine	Job timings result table					
	1	2	3	4	5	6
	J3	J1	J6	J5	J4	J2
M1	3	4	9	11	12	7
M2	10	11	13	10	8	7

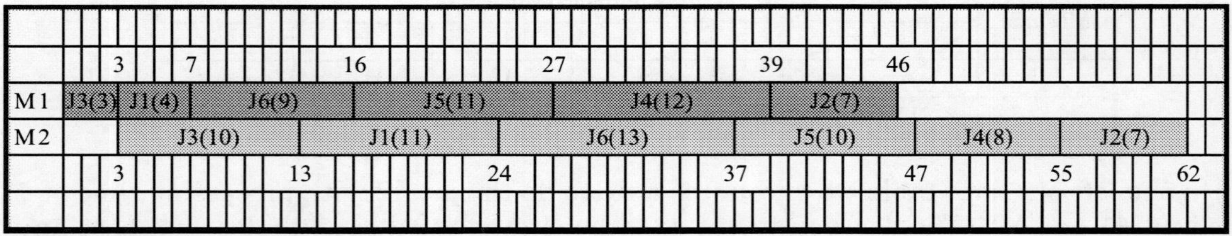

Figure 8.8

Time elapsed for completion is 62 hours. In between idle time for M2 is nil because in the chart above all the corresponding starting days for M2 are right side their corresponding finish dates pertaining to M1. Such a presentation shows the readiness of the jobs with M1 much before they are wanted in M2. The only idle time in the very beginning for M2 is 3 hours. The idle time for M1 is 16 hours.

Table 8.18

Job sequence	M1				M2			
	Time in	Job time	Time out	Idle time	Time in	Job time	Time out	Idle time
J3	0	3	3	0	3	10	13	3
J1	3	4	7	0	13	11	24	0
J6	7	9	16	0	24	13	37	0
J5	16	11	27	0	37	10	47	0
J4	27	12	39	0	47	8	55	0
J2	39	7	46	16	55	7	62	0
Total				16				3

Example 4

Given the following data obtain the optimum sequence of producing various types of bolts produced by a company. Also calculate time elapsed and the idle times for M1 and M2.

Table 8.19

Machine	Jobs time per unit in minutes					
	J1	J2	J3	J4	J5	J6
M1	4	12	14	20	8	18
M2	8	6	16	22	10	2
No. of piece per day	4	6	5	2	4	3

Solution

The smallest processing time is 2 minutes for the job J6 in machine M2. So let us push it to sixth place in our sequencing and delete it from the list to facilitate next iteration.

Table 8.20

Machine	Jobs time per unit in minutes				
	J1	J2	J3	J4	J5
M1	4	12	14	20	8
M2	8	6	16	22	10
No. of piece per day	4	6	5	2	4

Now the smallest processing time is 4 for J1 in M1. So let us push this job J1 to first place and then delete it from the list as usual.

Table 8.21

Machine	Jobs time per unit in minutes			
	J2	J3	J4	J5
M1	12	14	20	8
M2	6	16	22	10
No. of piece per day	6	5	2	4

Now the smallest processing time is 6 for J2 in M2. So we push this job to the fifth place and delete it from the list.

Table 8.22

Machine	Jobs time per unit in minutes		
	J3	J4	J5
M1	14	20	8
M2	16	22	10
No. of piece per day	5	2	4

Now the smallest processing time is 8 for J5 in M1. So we push this job to the second place and delete it from the list.

Table 8.23

Machine	Jobs time per unit in minutes	
	J3	J4
M1	14	20
M2	16	22
No. of piece per day	5	2

Now the smallest processing time is 14 for J3 in M1. So we push this job to the third place and delete it from the list. Also we place the left out to the left out place and complete the sequencing.

Table 8.24

Machine	Optimum jobs sequence					
	1	2	3	4	5	6
	J1	J5	J3	J4	J2	J6
M1	4	8	14	20	12	18
M2	8	10	16	22	6	2
No. of piece per day	4	4	5	2	6	3

Now the needed elapsed time and the idle time are obtained in the following table.

1. Total time lapsed = 286 minutes
2. Idle time on machine M2 = 48 minutes
3. Idle time machine M1 = 286–284 = 2 minutes

Table 8.25

Job	M1				M2			
	Time in	Job time	Time out	Idle time	Time in	Job time	Time out	Idle time
J1	0	4	4	0	4	8	12	4
J1	4	4	8	0	12	8	20	0
J1	8	4	12	0	20	8	28	0
J1	12	4	16	0	28	8	36	0
J5	16	8	24	0	36	10	46	0
J5	24	8	32	0	46	10	56	0
J5	32	8	40	0	56	10	66	0
J5	40	8	48	0	66	10	76	0
J3	48	14	62	0	76	16	92	0
J3	62	14	76	0	92	16	108	0
J3	76	14	90	0	108	16	124	0
J3	90	14	104	0	124	16	140	0
J3	104	14	118	0	140	16	156	0
J4	118	20	138	0	156	22	178	0
J4	138	20	158	0	178	22	200	0
J2	158	12	170	0	200	6	206	0
J2	170	12	182	0	206	6	212	0
J2	182	12	194	0	212	6	218	0
J2	194	12	206	0	218	6	224	0
J2	206	12	218	0	224	6	230	0
J2	218	12	230	0	230	6	236	0
J6	230	18	248	0	248	2	250	12
J6	248	18	266	0	266	2	268	16
J6	266	18	284	2	284	2	286	16
Total				2				48

Example 5

A soft drinks company has 6 manufacturing plants in six cities. It has the practice of overhauling the plants once in a year by two technicians A and B in the order AB. Since the plants are to be closed during the time of overhauling it cost Rs. 4000 per closure per plant. The idle time for A costs Rs. 1250 per day and the idle time for B costs Rs. 2150 per day. Obtain the optimum sequencing and the associated cost.

Table 8.26

Technician	Job timings					
	P1	P2	P3	P4	P5	P6
A	10	10	8	10	9	12
B	8	6	14	9	7	10

Solution

According to Johnson's the optimum job sequence is obtained as:

Table 8.27

Technician	Optimum jobs sequence					
	1	2	3	4	5	6
	P3	P6	P4	P1	P5	P2
A	8	12	10	10	9	10
B	14	10	9	8	7	6

Once the sequencing is obtained the total time elapsed, idle time and the associated down time are obtained and reported below.

Table 8.28

Plant	A				B				Down
sequence	Time in	Job time	Time out	Idle time	Time in	Job time	Time out	Idle time	time
P3	0	8	8	0	8	14	22	8	22
P6	8	12	20	0	22	10	32	0	24
P4	20	10	30	0	32	9	41	0	21
P1	30	10	40	0	41	8	49	0	19
P5	40	9	49	0	49	7	56	0	16
P2	49	10	59	6	59	6	65	3	16
Total				6				11	118

Minimum total cost = 118×4000+6×1250+11×2150 =Rs. 5,03,150.

8.3 PROCESSING N JOBS THROUGH THREE MACHINES

This type of problem has the following characteristics:

1. Only three machines are involved in the processing.
2. Job is done only in the prescribed order M1M2M3.
3. No surpassing of job is permitted.
4. The actual processing time on all the three machines are known in advance.

Table 8.29

Machines	Job processing time				
	1	2	3	....	N
M1	m_{11}	m_{12}	m_{13}	....	m_{1n}
M2	m_{21}	m_{22}	m_{23}	....	m_{2n}
M3	m_{31}	m_{32}	m_{33}	....	m_{3n}

As usual our objective is to find the optimal sequence of jobs that minimizes the elapsed time for completion. The previous method provided by Johnson is applicable if and only either one or both of the following additional conditions are fulfilled.

1. The smallest processing time on M1 is greater than or equal to the largest processing time on M2.
2. The smallest processing time on M3 is greater than or equal to the largest processing time on M2.

If these conditions are satisfied then we effectively convert the three-machine model into an equivalent two-machine model for easy handling. Once this is done we obtain the solution by applying the Johnson's method in the usual manner. If M4 and M5 are the two hypothetical machines then it can be shown M4i = M1i +M2i and M5i = M2i + M3i. Here the new order could be M4M5.

Example 6

Six jobs are to be performed by three machines in the order M1M2M3. If the order of the completion has no significance then obtain the optimum sequential ordering that minimizes the processing time and hence the time cost.

Table 8.30

	Machines		
Jobs	M1	M2	M3
J1	8	3	8
J2	3	4	7
J3	7	5	6
J4	2	2	9
J5	5	1	10
J6	1	6	9

Since the smallest processing time on M3 (6) is greater than or equal to the largest processing time on M2 (6) is fulfilled we can straightaway formulate for hypothetical machines M4 and M5 as shown below.

Table 8.31

	Machines			Hypothetical machines			
Jobs	M1	M2	M3	M4i = M1i+M2i	Rank	M5i = M2i+M3i	Rank
J1	8	3	8	11	5	11	
J2	3	4	7	7	3	11	
J3	7	5	6	12		11	6
J4	2	2	9	4	1	11	
J5	5	1	10	6	2	11	
J6	1	6	9	7	4	15	

Once the ranking (sequencing) is done for M4 and M5 the rest of the calculations are done in the routine manner using the original machines as shown below.

Table 8.32

Job sequence	M1				M2				M3			
	Time in	Job time	Time out	Idle time	Time in	Job time	Time out	Idle time	Time in	Job time	Time out	Idle time
J4	0	2	2	0	2	2	4	2	4	9	13	4
J5	2	5	7	0	7	1	8	3	13	10	23	0
J2	7	3	10	0	10	4	14	2	23	7	30	0
J6	10	1	11	0	14	6	20	0	30	9	39	0
J1	11	8	19	0	20	3	23	0	39	8	47	0
J3	19	7	26	27	26	5	31	25	47	6	53	0
Total				27				32				4

Time elapsed = 53 time units.

■ Example 7

Five jobs are to be performed by three machines in the order M1M2M3. If the order of the completion has no significance then obtain the optimum sequential ordering that minimizes the processing time and hence the cost.

Table 8.33

Jobs	Machines		
	M1	M2	M3
J1	16	10	8
J2	20	12	18
J3	12	4	16
J4	14	6	12
J5	22	8	10

Solution

Since the smallest processing time on M1 (12) is greater than or equal to the largest processing time on M2 (12) is fulfilled we can straightaway go for M4 and M5 combined method for sequencing as shown below.

Table 8.34

Jobs	Machines			Hypothetical machines	
	M1	M2	M3	M4i = M1i+M2i	M5i = M2i+M3i
J1	16	10	8	26	18
J2	20	12	18	32	30
J3	12	4	16	16	20
J4	14	6	12	20	18
J5	22	8	10	30	18

By using Johnson's optimum sequencing we obtain six optimum sequencings as follows.

Table 8.35

(i)	J3	J2	J4	J5	J1
(ii)	J3	J2	J4	J1	J5
(ii)	J3	J2	J1	J4	J5
(iv)	J3	J2	J5	J4	J1
(v)	J3	J3	J1	J5	J4
(vi)	J3	J2	J5	J1	J4

Once the ranking is done for M4 and M5 the rest of the calculations are done in the routine manner as shown below. Since all the six possibilities yield the same outcome we illustrate only the first two sequencings in the following tables.

Table 8.36

Job sequence	M1				M2				M3			
	Time in	Job time	Time out	Idle time	Time in	Job time	Time out	Idle time	Time in	Job time	Time out	Idle time
J3	0	12	12	0	12	4	16	12	16	16	32	16
J2	12	20	32	0	32	12	44	16	44	18	62	12
J4	32	14	46	0	46	6	52	2	62	12	74	0
J5	46	22	68	0	68	8	76	16	76	10	86	2
J1	68	16	84	18	84	10	94	16	94	8	102	8
Total				18				62				38

Table 8.37

Job sequence	M1				M2				M3			
	Time in	Job time	Time out	Idle time	Time in	Job time	Time out	Idle time	Time in	Job time	Time out	Idle time
J3	0	12	12	0	12	4	16	12	16	16	32	16
J2	12	20	32	0	32	12	44	16	44	18	62	12
J4	32	14	46	0	46	6	52	2	62	12	74	0
J1	46	16	62	0	62	10	72	10	74	8	82	0
J5	62	22	84	18	84	8	92	22	92	10	102	10
Total				18				62				38

8.4 PROCESSING N JOBS THROUGH M MACHINES

This type of problem is described as follows

1) There are n jobs described as J1, J2, J3,................Jn.
2) Each job is processed in m machine described as M1, M2, ...Mm.
3) No passing of job is permitted. It must be done M1M2M3.....Mm.
4) The actual processing time m_{11}, m_{12},..........m_{nm} are known in advance.

Table 8.38

Jobs	Machine times				
	M1	M1	M3	....	Mm
J1	m_{11}	m_{21}	m_{31}	....	m_{m1}
J2	m_{12}	m_{22}	m_{32}	....	m_{m2}
J3	m_{13}	m_{23}	m_{33}	....	m_{m3}
.	.	.	.	....	.
.	.	.	.	....	.
.	.	.	.	....	.
Jn	m_{1n}	m_{2n}	m_{3n}		m_{mn}

As usual this n-job by m-machine can be reduced to a two machine n job model provided one of the following condition is satisfied.

1) If the minimum time on M1 is greater than or equal to the maximum time on M2, M3, ... Mm−1.

2) If the minimum time on Mm is greater than or equal to the maximum time on M2, M3, ... Mm−1.

On satisfying the conditions the m machine problem is reduced to a two-machine problem as follows.

$$Di = m_{i1} + m_{i2} + \text{.........} + m_{im-1}$$

$$Ei = m_{i2} + m_{i3} + \text{.......} + m_{im}$$

Once the reduction is done we use the routine Johnson method for finding the optimum solution.

■ Example 8

Given the job timings obtain the optimum sequencings.

Table 8.39

Job	Machines				
	M1	M2	M3	M4	M5
J1	7	5	2	3	9
J2	6	6	4	5	10
J3	5	4	3	6	8
J4	8	3	5	2	6

Solution

Minimum processing time on M5 = 6 and Max (M2, M3, M4) are 6, 5, 6 respectively. Since minimum processing time on M5, namely 6 greater than or equal to the maximum of M2, M3, M4 the given problem can be converted into a two machine and four-job problem. In the following table the M6 is framed by adding the first four column elements for all the rows. Similarly the M7 is framed by adding the last four column elements for all the rows. Once this is done we allot the sequential ranking following the Johnson's rule. The respective sequence is obtained as J1, J3, J2 and J4.

Table 8.40

Job	Machines					Hypothetical machines			
	M1	M2	M3	M4	M5	M6	M7	Rank	Rank
J1	7	5	2	3	9	17	19	1	
J2	6	6	4	5	10	21	25	3	
J3	5	4	5	6	8	20	23	2	
J4	8	3	3	2	6	16	14		4

Once the sequencing is done we calculate the time in and time out in the usual manner using the given data in the original problem. The so obtained result is shown below.

Table 8.41

Job sequence	M1				M2				M3				M4				M5			
	Time in	Job time	Time out	Idle time	Time in	Job time	Time out	Idle time	Time in	Job time	Time out	Idle time	Time in	Job time	Time out	Idle time	Time in	Job time	Time out	Idle time
J1	0	7	7	0	7	5	12	7	12	2	14	12	14	3	17	14	17	9	26	17
J3	7	5	12	0	12	4	16	0	16	5	21	2	21	6	27	4	27	8	35	1
J2	12	6	18	0	18	6	24	2	24	4	28	3	28	5	33	1	35	10	45	0
J4	18	8	26	25	26	3	29	24	29	3	32	20	33	2	35	16	45	6	51	0
Total				25				33				37				35				18

Once the time in and time out table is completed we obtain the elapsed time and the associated idle times in the usual manner as in Table 8.41.

■ Example 9

Determine the best sequencing of 6 jobs on four machines given the operating sequence M1M2M3M3.

Table 8.42

Jobs	Processing time			
	M1	M2	M3	M4
A	6	5	3	4
B	7	2	5	5
C	9	6	3	3
D	8	5	5	4
E	8	3	4	3
F	9	5	5	4

Solution

Here we notice that Min (M1) = 6. Max (M2) = 6. Max (M3) = 5. So Min (M1) >Max (M2); Max (M3) So let us use Johnson's rule and obtain two new modified machines M5 and M6.

Table 8.43

	Processing time				New modified	
Jobs	M1	M2	M3	M4	M5	M6
A	6	5	3	4	14	12
B	7	2	5	5	14	12
C	9	6	3	3	18	12
D	8	5	5	4	18	14
E	8	3	4	3	15	10
F	9	5	5	4	19	14

Now by applying the Johnson's rule we get the following three optimum sequencings.

Table 8.44

1	F	D	C	B	A	E
2	D	F	C	A	B	E
3	D	F	A	B	C	E

Though we have three possible solutions the elapsed time for finishing the whole sequential job is going to be the same. So to illustrate let us work out for the tables for the calculation of the elapsed time for possibilities 1 and 2 only.

Table 8.45

Job sequence	M1				M2				M3				M4			
	Time in	Job time	Time out	Idle time	Time in	Job time	Time out	Idle time	Time in	Job time	Time out	Idle time	Time in	Job time	Time out	Idle time
F	0	9	9	0	9	5	14	9	14	5	19	14	19	4	23	19
D	9	8	17	0	17	5	22	10	22	5	27	3	27	4	31	4
C	17	9	26	0	26	6	32	4	32	3	35	5	35	3	38	4
B	26	7	33	0	33	2	35	1	35	5	40	0	40	5	45	2
A	33	6	39	16	39	5	44	4	44	3	47	4	47	4	51	2
E	39	8	47	10	47	3	50	10	50	4	54	6	54	3	57	3
Total				10				38				32				34

Table 8.46

Job sequence	M1				M2				M3				M4			
	Time in	Job time	Time out	Idle time	Time in	Job time	Time out	Idle time	Time in	Job time	Time out	Idle time	Time in	Job time	Time out	Idle time
D	0	8	8	0	8	5	13	8	13	5	18	13	18	4	22	18
F	8	9	17	0	17	5	22	11	22	5	27	4	27	4	31	5
C	17	9	26	0	26	6	32	4	32	3	35	5	35	3	38	4
A	26	6	32	0	32	5	37	0	37	3	40	2	40	4	44	2
B	32	7	39	15	39	2	41	2	41	5	46	1	46	5	51	2
E	39	8	47	10	47	3	50	13	50	4	54	7	54	3	57	3
Total				10				38				32				34

In both the cases the time elapsed = 57 hours.

8.5 SEQUENCING WITH DUE DATE COMMITMENTS: PROCESSING N JOBS IN ONE MACHINE

The simplest sequencing problem consists of 'n' jobs to be performed at one machine. In such models processing time and due dates for delivery committed on the very first day of the commencement of work are assumed to be known constants. Further, no new job is taken up until otherwise the preceding job is completed at one stretch and satisfactorily delivered to the customer. Remember, that in this section we consider only one machine the question of machine, ordering in carrying out the job does not arise.

Given this scenario, the completion or otherwise called the ***flow time*** of a given job, counted from the very start of the sequence, will differ depending upon its placement in the sequential order. The early placement definitely will have less flow time and the latter placement will have more flow time. However, the overall completion time otherwise called ***makespan*** time of all the jobs putting together will be a constant irrespective of the placement order of the individual jobs. ***Tardiness*** is a measure of delayed delivery and is obtained by subtracting the due date from the job completion date. If the job is completed well in advance than the delivery date, this measure is invariably becomes zero.

Even in this simple case of n jobs and one machine there is no unique sequencing rule that optimizes both processing efficiency and due date fulfilment for all the jobs. To illustrate the difficulties involved let us consider a simple five-job single process model. As a measure of efficiency, we propose to use only two important characteristics in this section. They are:

1. Average job completion time
2. Average delayed delivery.

■ Example10

Because of the fast approaching summer Mr. Ram wants to finish all the five jobs in hand early by rescheduling his six days work week to seven days work week. The five jobs A, B, C, D, and E, the associated job times and the respective due dates of delivery are given below.

Table 8.47

Job	Job time	Due date
A	5	10
B	10	15
C	2	5
D	8	12
E	6	8

Since there are 5 jobs on hand Mr. Ram can think of $5! = 5\times 4 \times 3 \times 2 \times 1 = 120$ options of sequencing them all. Clearly exploring all 120 options is humanly impossible for Mr. Ram. Hence in this section we propose to discuss only five simple rule based sequencing.

1. **FCFS**: In this method First Come First Serve is the rule of the sequencing. Hence, the associated sequencing order is given by A, B, C, D, and E.

2. **DUE DATE**: In this earliest due date sequencing method the due dates are arranged in the increasing order of magnitude. Accordingly, jobs with less due dates are taken up on priority.

3. **SLACK**: In this method we arrange the sequence in the increasing order of slacks otherwise called the time margin available for due date delivery.

4. **CRITICAL RATIO (CR)**: In this method we arrange the CR's in the increasing order of magnitude. The critical ratio is defined as the ratio of time remaining to work remaining.

5. **SEQUENCING ON THE BASIS OF PROCESSING TIME (SPT)**: In this method we arrange the sequence in the increasing order of processing time.

1. FCFS Solution

In this method the given original ordering is being restored in the very first column. The average completion and average tardiness are obtained and reported in the last row in Table 8.48.

Table 8.48

FCFS sequencing	Start time	Processing time	Completion time	Due date	Tardiness (delayed delivery)
A	0	5	5	10	0
B	5	10	15	15	0
C	15	2	17	5	12
D	17	8	25	12	13
E	25	6	31	8	23
Total			93		48
Average			93/5 = 18.6		48/5 =9.6

2. DUE DATE Solution

In this method due dates are rearranged in the increasing order of magnitude in column 5. Accordingly, in column 1 jobs are rescheduled. Also see that processing time in column three is also rearranged. After making these changes the rest of the calculations are done in the usual manner and reported in the Table 8.49.

Table 8.49

DUE DATE sequencing	Start time	Processing time	Completion time	Due date	Tardiness
C	0	2	2	5	0
E	2	6	8	8	0
A	8	5	13	10	3
D	13	8	21	12	9
B	21	10	31	15	16
Total			75		28
Average			15		5.6

3. SLACK Solution

In this method on the assumption that the job under reference is taken up first we calculate the remaining due date. These remaining due dates are often called slacks. Once this is done we arrange these slacks in the increasing order of magnitude as shown in column 2 and proceed further as shown in Table 8.50.

Definition of a SLACK

The slack for a job = time remaining after completing the job if at all it taken up on the very starting date = Due date – Processing time

Slack for Job A = 10–5 = 5

Slack for Job B = 15–10 = 5

Slack for Job C = 5–2 = 3

Slack for Job D = 12–8 = 4

Slack for Job E = 8–6 = 2

The rearranged job sequencing, slacks and the processing times are shown in the first, second and fourth column respectively in the following table. The rest of the calculations are carried out and reported as usual.

Table 8.50

SLACK sequencing	Slack	Start time	Processing time	Completion time	Due date	Tardiness
E	2	0	6	6	8	0
C	3	6	2	8	5	3
D	4	8	8	16	12	4
A	5	16	5	21	10	11
B	5	21	10	31	15	16
Total				82		34
Average				16.4		6.8

4. CR Solution

In this method the calculated CR is arranged in the increasing order of magnitude to decide about the sequencing of jobs in the first column in Table 8.51.

$$CR = \frac{\text{Due date}}{\text{Processing time}}$$

CR for Job A = 10/5 = 2

CR for Job B = 15/10 = 1.5

CR for Job C = 5/2 = 2.5

CR for Job D = 12/8 = 1.5

CR for Job E = 8/6 = 1.33

The rearranged job sequencing, CR and the processing times are shown in the first, second and fourth column respectively in the table. The rest of the calculations are carried out and reported as usual.

Table 8.51

CR sequencing	CR	Start time	Processing time	Completion time	Due date	Tardiness
E	1.33	0	6	6	8	0
D	1.50	6	8	14	12	2
B	1.50	14	10	24	15	9
A	2.00	24	5	29	10	19
C	2.50	29	2	31	5	26
Total				104		56
Average				20.8		11.2

5. SPT Solution

In this method for sequencing the jobs in column 1 the processing time is arranged in the increasing order of magnitude as shown in the Table 8.52.

Table 8.52

SPT sequencing	Start time	Processing time	Completion time	Due date	Tardiness
C	0	2	2	5	0
A	2	5	7	10	0
E	7	6	13	8	5
D	13	8	21	12	9
B	21	10	31	15	16
Total			74		30
Average			14.8		6

Summary

Table 8.53

Rule	Average completion time	Average tardiness	No. of jobs tardy	Maximum tardiness
FCFS	18.60	9.60	3	23
DUE DATE	15.00	5.60	3	16
SLACK	16.40	6.80	4	16
CR	20.80	11.20	4	26
SPT	14.80	6.00	3	16

Critical observations

1. All the rules finish the designed jobs in 31 days time.
2. However, no sequence rule completes all jobs on due date (with zero delivery delay).
3. The performance of FCFS is either met or exceeded by DUEDATE and SPT.

8.6 SEQUENCING WITH PRIORITY AND DUE DATE COMMITMENTS: PROCESSING 2 JOBS ON 2 MACHINES

Sometimes jobs done by a machine comes back again for second round for finishing. We illustrate one such problem in this section.

Forward Scheduling

This method of scheduling is used whenever there is no specific job delivery time in the entire sequence. Job is carried out and delivered on "as soon as possible" basis. But remember there can be propriety specification for delivery.

Backward Scheduling

This technique is useful in the assembly type operation where the job has a definite schedule of delivery. Since we will have to ensure their completion by a definite date, backward scheduling becomes handy. Here early delivery is not a problem. In such cases correct billing of needed materials and accurate processing estimates are essential. If schedule of delivery is not adhered to for whatsoever the reason, normally a penalty is levied either in the form of loss of profit or loss of creditability.

■ Example 11

Two jobs on two machines with sequence of operations are given below.

Table 8.54

Job J1		Job J2	
Sequence of operation	Processing time (days)	Sequence of operation	Processing time (days)
M1	4	M1	3
M2	5	M2	4
		M1	4

(a) Job J1 has been decided as priority job and will have to be delivered as early as possible. Draw a suitable Gannet chart by using forward pass and obtained time elapsed for completion of both the jobs.

(b) If J1 is the priority job still but this time J2 is to be delivered on 15th day evening at any cost. Draw a suitable Gannet chart by backward pass method.

Solution (a)

Forward Scheduling						
	0	4	7		13	17
M1	J1(4)	J2(3)			J2(4)	
M2		J1(5)		J2(4)		
		4		9	13	17

Figure 8.9

Forward Pass Method

Since job J1 is the priority job it is taken up first on the very first day in M1. Once the J1 is done in M1 it goes for M2 immediately as shown in Figure 8.9. Thus the completion of J1 in both M1 and M2 takes 4+ 5 = 9.

The first round of J2 starts on the fifth day morning and ends on the evening of day 7. But it waits for two days and goes for M2. Thus this phase starts on the morning of day 9 and ends on the evening of day 13. It comes back to M1 for finishing on the morning of day 13 and ends on the evening of day 17.

Thus, the priority job is done without break and delivered on the evening of day 9. The non-priority job J2 will be delivered at the most on the evening of day 17.

Solution (b)

Backward Scheduling							
	0	4	7		11		15
M1	J1(4)	J2(3)			J2(4)		
M2		J1(3)	J2(4)	J1(2)			
		4	7	11		13	

Figure 8.10

Backward Pass Method

This time still J1 is a priority job to be finished and delivered before J2, but the late emerging job must be completed and delivered on or before the evening of day 15. To accomplish such a task backward pass method is handy. So in our Gannet chart the job must be over at least on the evening of day 15. Accordingly, the second stretch of work on M1 is scheduled from day 11 to day 15. The job J1 in M2 is split into two parts. The second stretch is scheduled from day 11 to day 13 (two days). The first phase of job J2 in M1 is scheduled between day 4 and day 7. The second phase of J2 in M2 is scheduled from day 7 to day 11. The third phase of J2 is scheduled from day 11 to day 15 as committed.

8.7 SEQUENCING WITH PRIORITY AND DUE DATE COMMITMENTS: PROCESSING 2 JOBS ON N MACHINES

In all our previous sections we strictly assumed identical machine ordering for all jobs. In this section we bring in more realistic situation by allowing machine order to vary from job to job. For example, for Job 1 the machine ordering could be M1M2M3M4M5M6 and job 2 the machine ordering could be M2M1M3M6M4M5. Also we incorporate priority for early delivery as it has been done in the previous section.

■ Example 12

The following table gives the summary of machine ordering and the associated job completion timings.

Table 8.55

Job1	Machine order	M1	M2	M3	M4	M5	M6
	Time in hours	4	5	1	3	6	5
Job 2	Machine order	M2	M1	M3	M6	M4	M5
	Time in hours	6	3	2	4	3	5

1. By using the graphical method determine the earliest project completion time without any priority commitments.
2. If Job 1 is the priority job with as early as possible delivery condition, obtain the optimum project completion time.
3. If job 1 is the priority job with as early as possible delivery condition, obtain the optimum project completion time without insisting job ordering for both the jobs.

Solution (a)

The needed optimal sequence is determined by using graphical method. The job timings for job 1 is measured on the horizontal axis and the job timings for the job J2 is measured on the vertical axis. Having marked the job timings and the job sequence on both the axes we shade the intersectional rectangles for corresponding jobs as shown in the Figure 8.11. After having done the blocking as shown in Figure 8.11 we draw a 45° line from the origin to represent the uninterrupted smooth working path for both the jobs. Notice that our 45° line gets interrupted on day four because the job 2 with machine M2 is still on and will last for another two days.

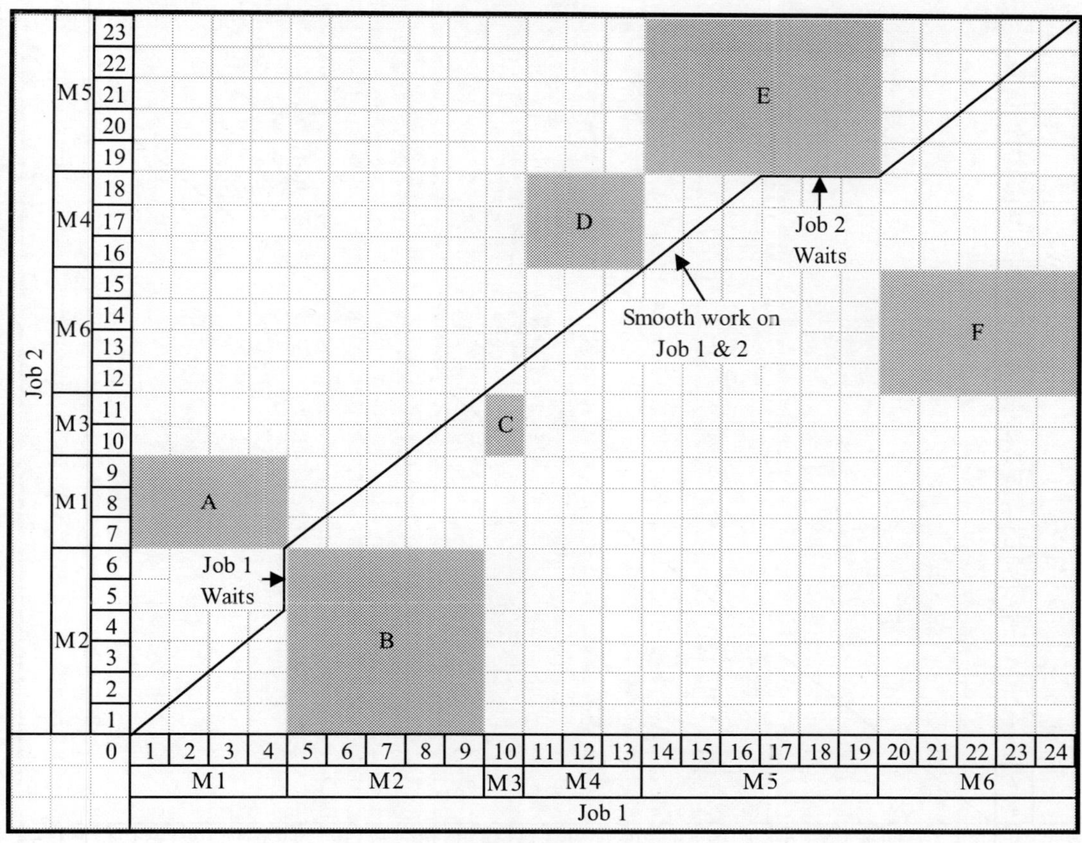

Figure 8.11

So the job 1 will have to wait for another two days to initiate the next machine M2 task. This is shown by a vertical movement of out 45° line. From this point onwards the uninterrupted smooth working of both the jobs continue up to day 16. This time job 2 will have to wait for machine M5's release from job 1 for three days. This is shown by the horizontal movement of our job line. From day 20 onwards the job 1 will be done on machine M6 and job 2 will be done on machine M5 without any interruption. Thus on 23rd working day job 2 will be over and on 24th working day job 1 will be over as shown in Figure 8.11. However, job 1 waits for 2 days due to interception literally the job will be over on (24 + 2) 26th day from starting. Similarly, since the job 2 waits for 3 days in between it will be over only on 26th day from the start. The entire summary is restated in Figure 8.12 after adjusting the waiting time in the respective axis. The same situation is illustrated in Gantt chart 8.13 also.

Note: *In this approach once job 1 waits and another time job 2 waits. Thus no priority of job is insisted. The only objective is to finish off both the projects early.*

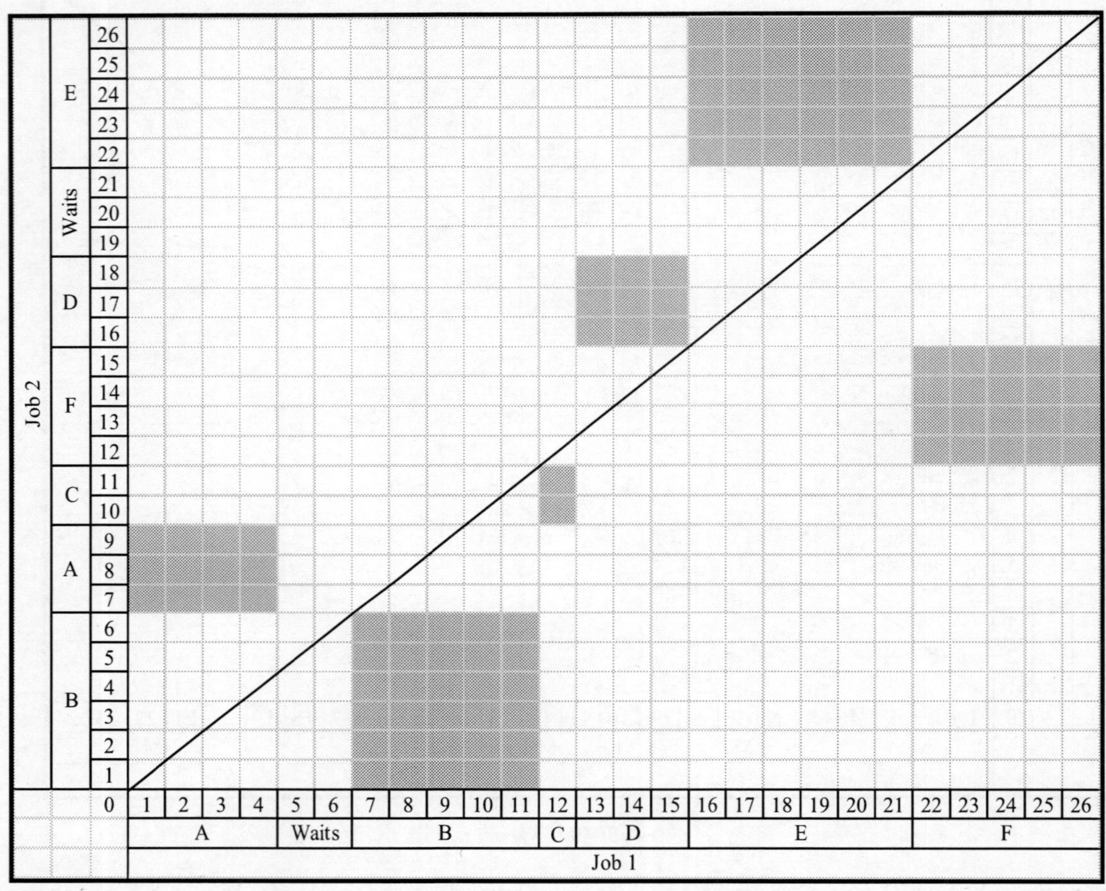

Figure 8.12

Day	1	2	3	4	5	6	7	8	9	10	11	12	13	14	15	16	17	18	19	20	21	22	23	24	25	26
J1	M1(4)				Wait		M2(5)					M3(1)	M4(3)			M5(6)						M6(5)				
J2	M2(6)						M1(3)			M3(2)		M6(4)				M4(3)			Wait			M5(5)				

Figure 8.13

Solution (b)

If the job 1 is our first priority with a due date of delivery on 24th evening then clearly job 1 must progress uninterruptedly as shown in Figure 8.14. If we still insist the prescribed order of doing job 2 then clearly job 2 will have to wait from day 5 to day 9 in the beginning and day 20 to day 24 in the second stage because of non-availability of relevant machines. Due to the 10 days delay the job 2 will be over only on 33rd day as shown in Gantt Chart 8.14.

Day	J2	J1
1	$M2_1(4)$	M1(4)
2		
3		
4		
5	Wait	M2(5)
6		
7		
8		
9		
10	$M2_2(2)$	M3(1)
11		M4(3)
12	M1(3)	
13		
14		M5(6)
15	M3(2)	
16		
17	$M6_1(3)$	
18		
19		
20	Wait	M6(5)
21		
22		
23		
24		
25	$M6_2(1)$	
26	M4(3)	
27		
28		
29	M5(5)	
30		
31		
32		
33		

Figure 8.14

Solution (c)

In Figure 8.15 though still job 1 is under priority consideration we do not insists any ordering pattern for both the jobs. Under this category in its interesting to note that both the jobs are delivered in time without even a day's delay.

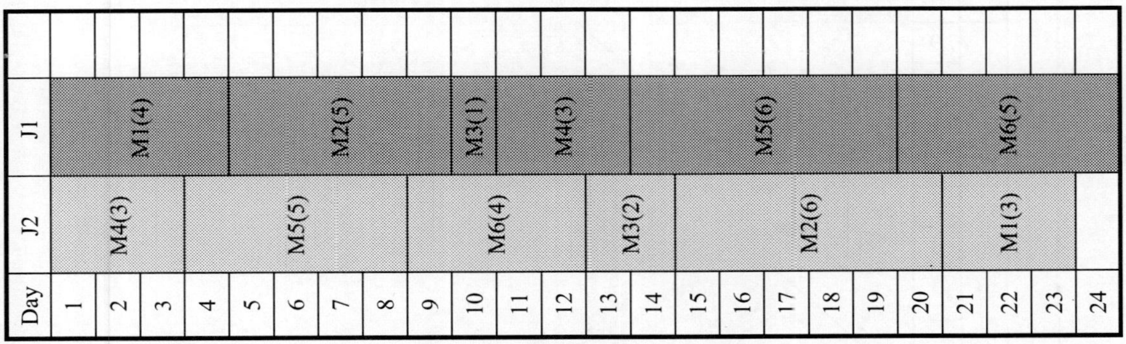

Figure 8.15

■ Example 13

The following table gives the summary of machine ordering and the associated job completion timings.

Table 8.56

Job1	Machine order	A	B	C	D	E
	Time in hours	3	4	2	6	2
Job 2	Machine order	B	C	A	D	E
	Time in hours	5	4	3	2	6

By using the graphical method determine the earliest project completion time without any priority commitments.

Solution 1

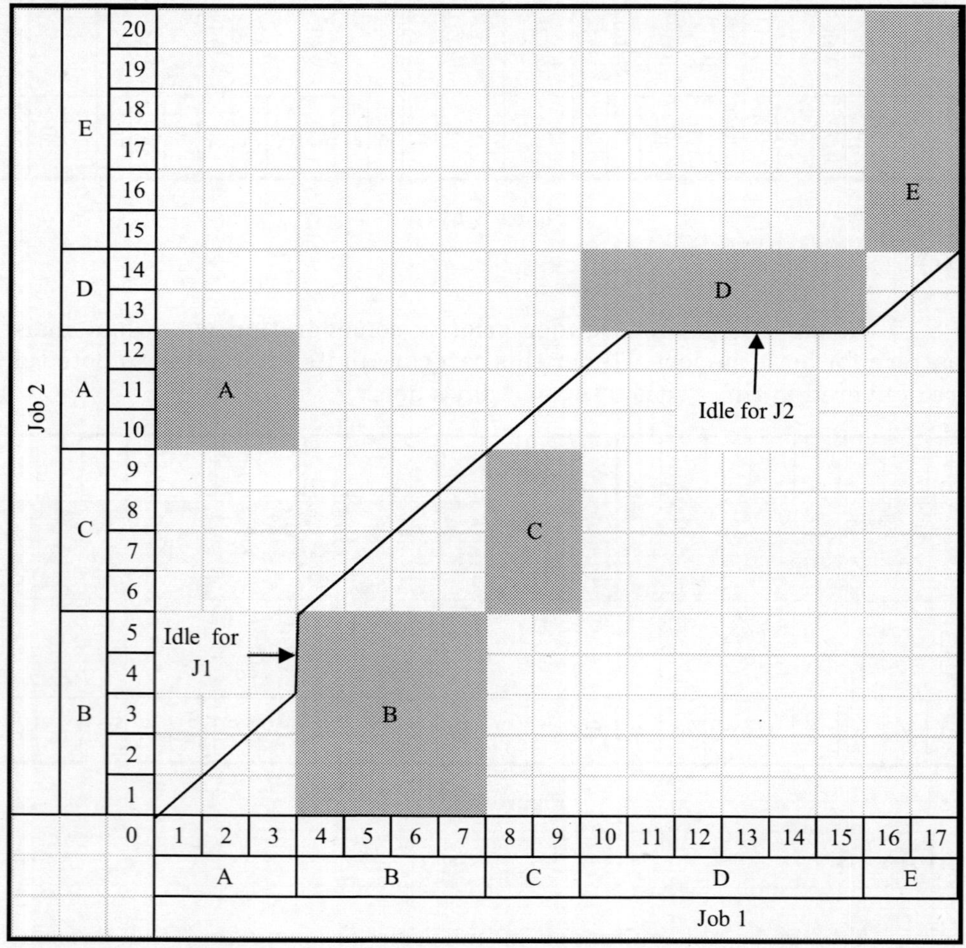

Figure 8.16

Thus, job 1 will be over by 17 working days and job 2 will be over by 20 working days.

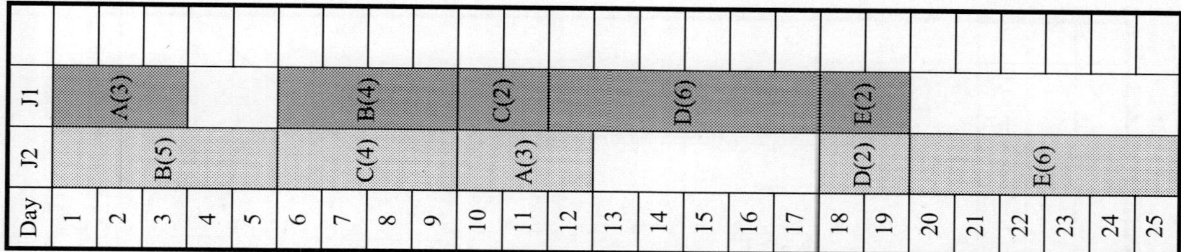

Figure 8.17

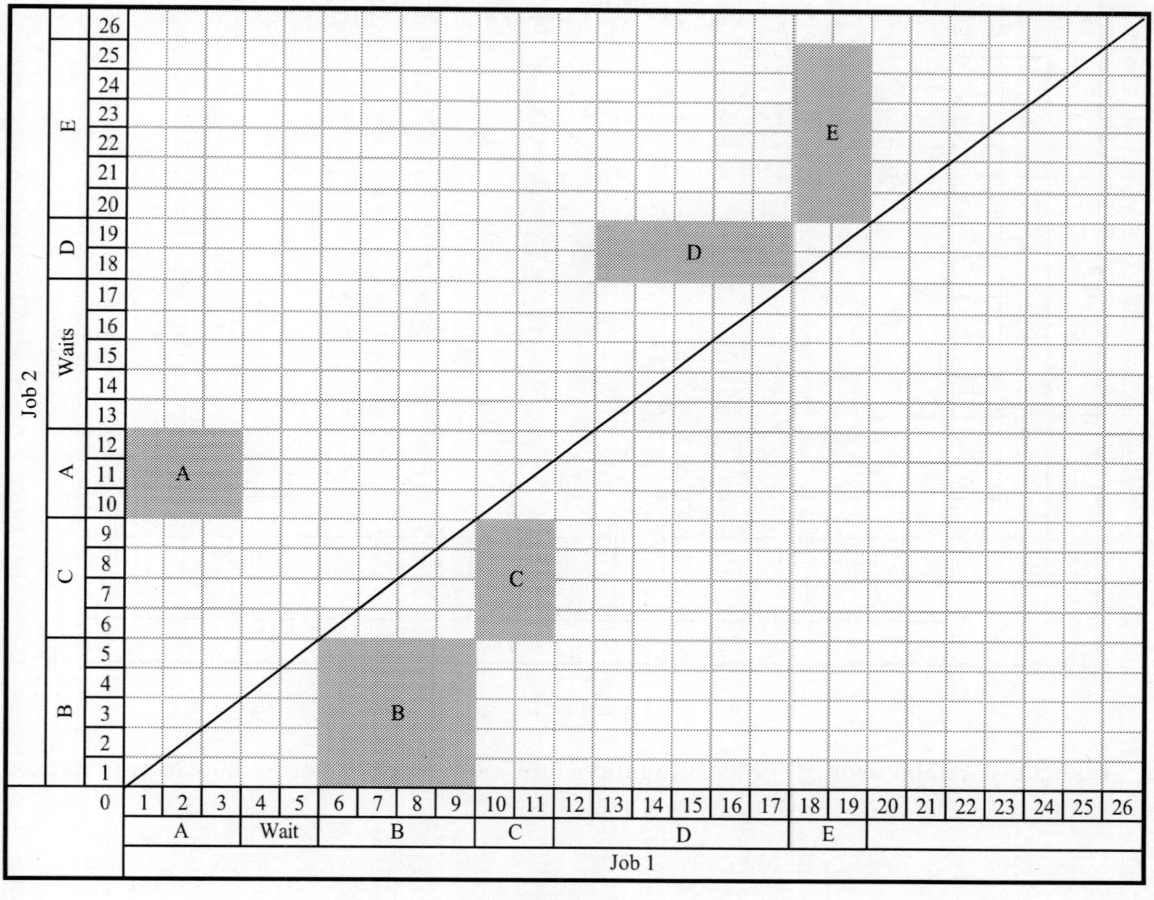

Figure 8.18

Thus, job 1 will be over by 19 days including the two days waiting time. Similarly, J2 will be over by 25 days including the five days waiting time.

Solution 2

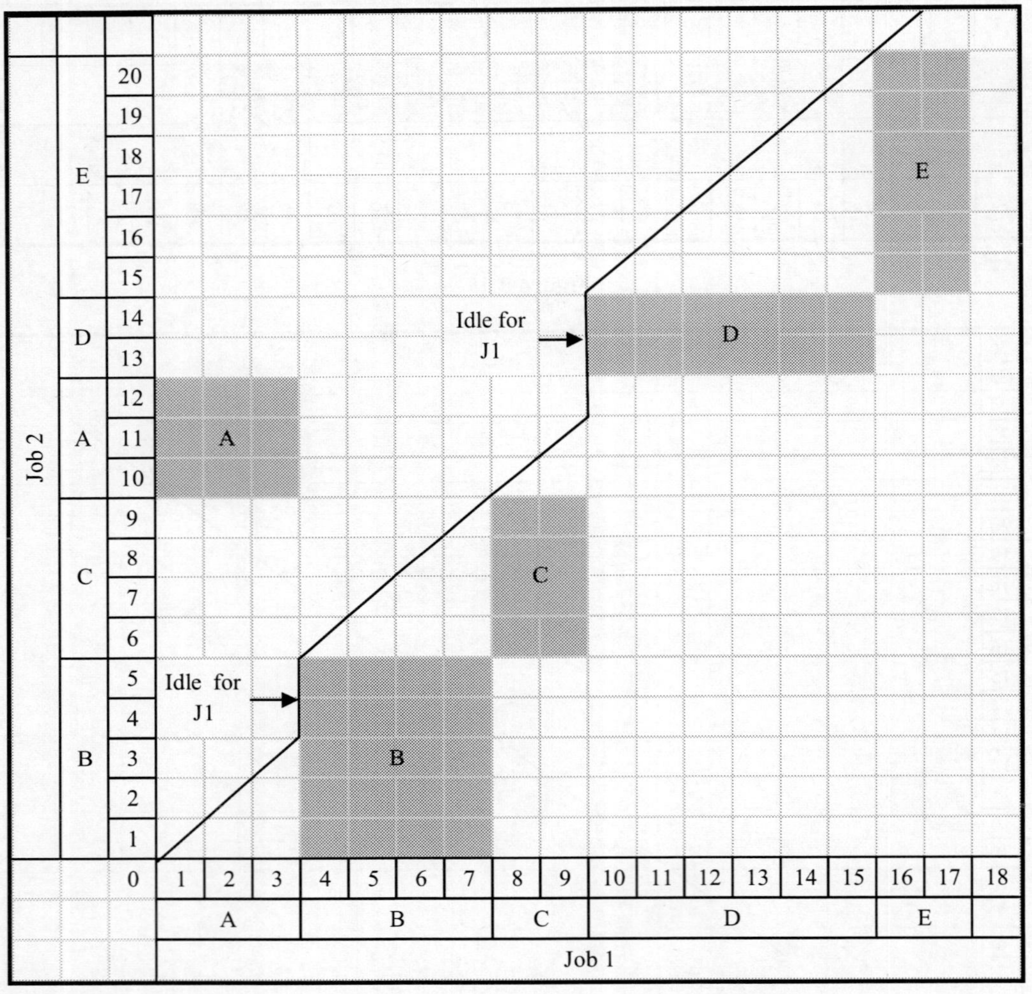

Figure 8.19

Thus, job 1 will be over by 17 working days and job 2 will be over by 20 working days.

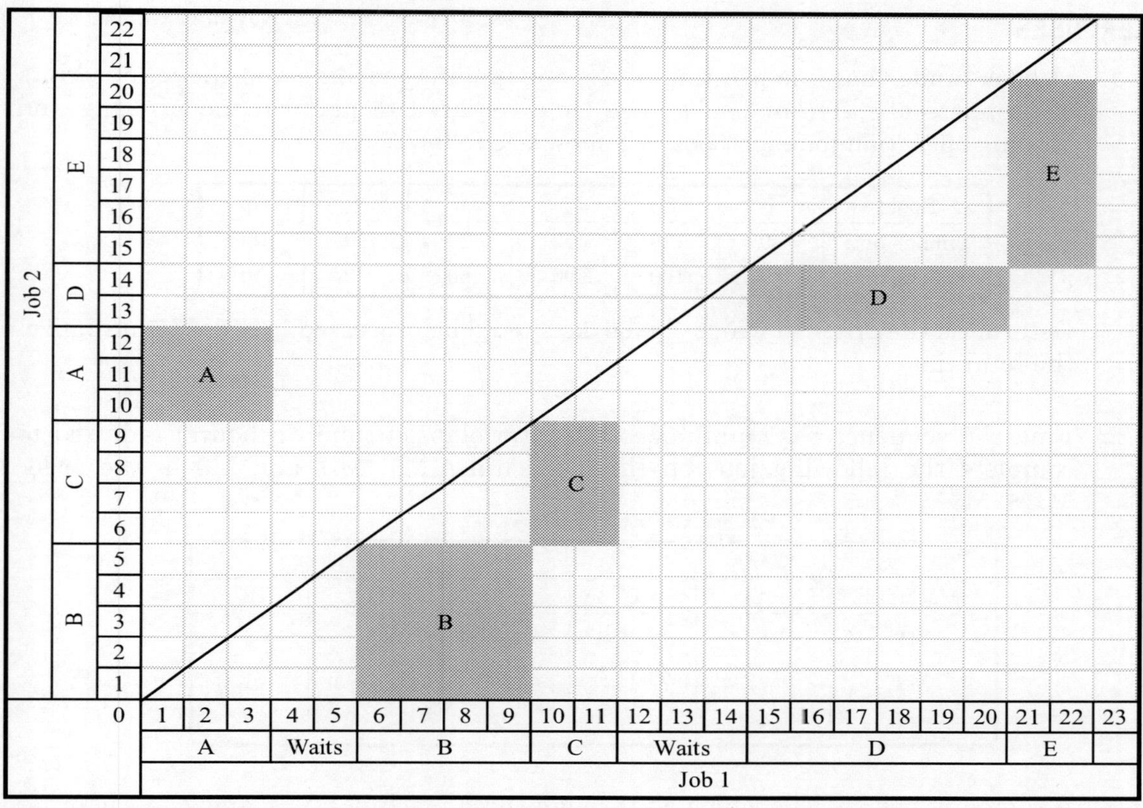

Figure 8.20

Thus job 1 will be over by 22 days including the five days waiting time. Similarly, J2 will be over by 20 days.

Day	1	2	3	4	5	6	7	8	9	10	11	12	13	14	15	16	17	18	19	20	21	22	23	24	25
J1	A(3)					B(4)				C(2)					D(6)						E(2)				
J2	B(5)					C(4)				A(3)			D(2)		E(6)										

Figure 8.21

EXERCISES

1. A book binder has one printing press, one binding machine and the manuscripts of numbers of different books. The time required to perform the printing and binding operation for each book are shown as follows:

Books	1	2	3	4	5	6
Printing time	30	120	50	20	90	100
Binding time	80	100	90	60	30	10

Determine the order in which the books should be processed in order to minimize the total time.

2. Find the sequence that minimizes the total elapsed time (in hours) required to complete the following jobs on three machines M1, M2, and M3 in the order M1M2M3.

	Jobs				
Machines	A	B	C	D	E
M1	5	7	6	9	5
M2	2	1	4	5	3
M3	3	7	5	6	7

3. A company has six jobs which go through three machines X, Y and Z in the order X,Y, Z. The processing time in minutes for each job for each machine is as follows:

	Jobs					
Machines	1	2	3	4	5	6
M1	18	12	29	35	43	37
M2	7	12	11	2	6	12
M3	19	12	23	47	28	36

What should be the sequence of the jobs?

(Bombay University, Diploma in Industrial Management 1980)

4. Four jobs are to be processed on three machines. All the jobs require machines in the same order i.e. Machine 1→Machine 2→Machine 3. The processing times are given below.

Jobs	Processing time		
	M1	M2	M3
A	5	6	7
B	8	7	8
C	7	2	10
D	3	5	9

Find the optimal solution so as to minimize make-span time.

(Delhi University, M.B.A April 1981)

5. Find the optimum sequence for the following sequencing problem of four jobs and five machines, when passing is not allowed, of which the processing time (in hours) is given below:

Jobs	Machines				
	A	B	C	D	E
1	7	5	2	3	9
2	6	6	4	5	10
3	5	4	5	6	8
4	8	3	3	2	6

Also find the total elapsed time. (Osmania University, M.B.A 1989)

6. A company has six jobs on hand, coded A to F. All the jobs have to go through two machines by scheduling the six jobs on the two machines. The time required for the job on each machine, in hours is given below.

Machines	Jobs					
	A	B	C	D	E	F
M1	1	4	6	3	5	2
M2	3	6	8	8	1	5

Draw a sequence table scheduling the six jobs on the two machines.

(Poona University, M.B.A 1982)

7. Find the sequence that minimizes the total elapsed time to complete the following tasks (clearly state any algorithm that you may use) and calculate the minimum elapsed time:

Machines	Jobs						
	A	B	C	D	E	F	G
Time on machine 1	3	8	7	4	9	8	7
Time on machine 2	4	3	2	5	1	4	3
Time on machine 3	6	7	5	11	5	6	12

(Bangalore University, M.B.A July/August 2005)

8. Six jobs are to be proceeded at three machines A, B, and C in the order ABC. The time taken by each job on each machine is indicated below. Each machine can process only one job at a time.

Machines	Jobs					
	J1	J2	J3	J4	J5	J6
A	12	8	7	11	10	5
B	7	10	9	6	10	4
C	3	4	2	5	1.5	4

Determine the sequence for the jobs so as to minimize the processing time.

(Delhi University, M.Com 1982)

9. Six jobs go first over machine I and then over machine II. The order of completion of the jobs has no significance. The following table gives the machine time in hours for six jobs and two machines.

Machines	Jobs					
	1	2	3	4	5	6
Machine I	5	9	4	7	8	6
Machine II	7	4	8	3	9	5

Find the sequence of jobs that minimizes the total time elapsed to complete the job. Find the minimum time by using Gantt chart or by any other method.

(Meerut University, M.Com 1979)

10. There are five jobs, each must go over through the machines A, B, and C in the order BAC. Each machine can process only one job at a time. Processing times (in hours) are given below:

Jobs	Machines		
	M1	M2	M3
A	5	8	4
B	6	10	6
C	2	6	8
D	3	7	6
E	4	11	5

Determine the sequence for jobs as to minimize total time required to complete them, and the total time that elapsed.

(Delhi University, M.Com 1985)

11. Five jobs pass through 3 machines X, Y, Z in the order XYZ. Processing times in each machine are given in minutes.

Find the best sequence that will minimize the total elapsed time. Find the idle time for each machine.

Jobs	Machines		
	X	Y	Z
1	11	7	6
2	12	6	11
3	9	6	9
4	8	5	8
5	6	6	5

(Bangalore University, July/August 2006)
(Osmania University, M.B.A 1982)

12. Provide the optimal job sequence involving three machines M1, M2, M3 (order M1, M2, M3) for the following:

Time on machines	Jobs				
	J1	J2	J3	J4	J5
M1	7	12	11	9	8
M2	8	9	5	6	7
M3	11	13	9	10	14

(Delhi University, M.Com 1983)

13. Seven jobs are to be processed on two machines A and B in the order A→B. Each machine can process only one job at a time. The processing times (in hours) are as follows:

Jobs	Machine A	Machine B
1	10	15
2	12	11
3	13	8
4	7	9
5	14	6
6	5	7
7	16	16

Suggest optimum sequence of processing the jobs and the total elapsed time.
(Delhi University, M.Com 1989)

14. Five jobs are performed first on machine X and then on machine Y. The time taken in hours by each job on each machine is given below:

Jobs	A	B	C	D	E
Time on machine X	6	2	10	4	11
Time on machine Y	3	7	8	9	5

Determine the optimal sequence of jobs that minimizes the total elapsed time to complete the jobs. Also compute the minimum time.

(Meerut University, M.Com 1983)

15. A book binder has one printing press, one binding machine and manuscripts of a number of different books. The time required to perform the printing and binding operations for each book is known. He wishes to determine the order to minimize the total time required to turn out all the books.

Books	Printing Time	Binding Time
1	30	80
2	120	100
3	50	90
4	20	60
5	90	30
6	110	10

An additional operation is added to the above process finishing. The time of the finishing operation is 20, 40, 60, 120, 30, 40 respectively. What should be the order of the books now? (Poona University, M.B.A 1988)

16. Using Graphical Method, find the minimum elapsed time to perform two jobs J1 and J2 by using machines A, D, C, and D. The sequence and the associated processing times are as follows:

Job1	Sequence	A	B	C	D
	Time in hours	2	4	5	1
Job 2	Sequence	B	C	D	A
	Time in hours	6	5	2	3

17. A machine shop has six machines A, B, C, D, E and F. Two jobs must be processed through each of these machines. The time spent on each machine and the necessary sequence of the jobs through the shop is given below.

Order	1	2	3	4	5	6
Job 1	A–25	C–15	D–15	B–35	E–30	F–30
Job2	B–30	C–35	A–15	D–15	F–20	E–25

In what sequence would the jobs be done on each of the machine in order to minimize total time necessary to finish the jobs? Also determine the total elapsed time.

18. The following table gives the summary of machine ordering and the associated job completion timings.

Job1	Machine order	A	B	C	D	E
	Time in hours	6	8	4	12	8
Job 2	Machine order	B	C	A	D	E
	Time in hours	10	8	6	4	12

By using the graphical method determine the earliest project completion time without any priority commitments.

CHAPTER

9

Replacement Models

- *Introduction*
- *Preventive maintenance cost*
- *Running cost and its discounted current value*
- *Price of the capital equipment*
- *Resale values of the equipment*
- *Replacement policy*
- *Replacement policy for items that deteriorate over time when time value of money is constant (alternatively rate of interest is zero)*
- *Replacement policy for items that deteriorate over time when time value of money is declining (alternatively rate of interest is positive)*
- *Group replacement of item that suddenly fails completely*
- *Excel calculation sheet for group replacement problems*
- *Exercises*

9.1 INTRODUCTION

In any business organization deliberate preventive maintenance of assets is essential for maintaining optimum production and healthy financial obligations. Although by improving the design and selecting appropriate ingredients one can optimize the efficiency, sudden random breakdown of equipment always hampers the business. In such cases the machines will have to be kept idle till the extensive repair work is carried out. Sometimes such failure may lead to safety hazards as in the case of traffic lights failure. To prevent such random break-downs the equipment are to be periodically switched off for maintenance work.

When an asset fails partly, replacement of failed parts by new ones could be the earliest remedy. Such replacements will involve only very little cost in the early stages. A good maintenance can postpone the time of complete replacement of the machine. As time passes the maintenance expenditure keep on increasing at an alarming rate. Around this period, if we look back the cumulative maintenance expenditure incurred in the past, it is going to be astonishingly very high. Continuing the policy of just maintenance will be a foolish proposition thereafter. At some point of time, going for a replacement of the old one by a new one seems to be more rational and economical. The correct and timely replacement of equipment not only saves the cost in the long run but also ensures the smooth supply chain management of the output. Well maintained plants and equipment with timely replacement policy will result in better capacity utilization, higher productivity and increased profit.

In general replacement policy identifies the best period for replacement from the financial point of view. If the item under reference is a heavy capital equipment with less resale value like electric generators, it must be used as long as possible though it involves heavy maintenance cost. On the other hand if the equipment is like cars having good resale value, early replacement could be the best policy. Inexpensive items like electric bulbs need to be replaced only after their complete failure.

Sometimes in technology driven economies like ours we do not wait till the due date for replacement. Early replacement by updated technology becomes inevitable due to technological obsolescence of the existing equipment. The updated technology mission definitively will be cost effective.

Other than financial consideration sometimes replacement is considered having the factors like available funds, policy of the company, attitudes of the employees etc. Thus each replacement policy can be considered unique as having some special features of its own. In this chapter we propose to discuss a few illustrative models for determining the correct replacement time.

9.2 PREVENTIVE MAINTENANCE COST

The following are some of the pre-planning steps normally known as Preventive Maintenance Schedule.

1. A schedule for daily or periodical inspection of equipment for day-to-day resetting or repairs.
2. A schedule for proper lubrication of moving parts of the machinery.
3. A plan for periodic monitoring of complex, crucial and expensive components for certain symptoms like noise, vibration, over-heating etc.
4. A plan for periodic replacement of certain essential moving parts as per the life span prescriptions of the concerned producer.
5. A plan for procurement of certain critical components for emergency use.
6. A plan for manpower upgradation for proper maintenance.
7. A plan for periodic shut down for check-up and overhauling.
8. A plan for predicting the possible breakdown using standard statistical tools like
    a. Normal distribution
    b. Negative exponential distribution
    c. Hyper-exponential distribution.

Objectives of Maintenance

Some of the important objectives of maintenance are listed below:

1. Minimizing the idle time of the plant
2. Minimize the repair time and cost
3. Minimize the production loss due to breakdown
4. Minimizing the accidents
5. Minimizing the inventory carrying cost
6. Improve the quality of the product
7. Improve the productivity
8. Extension of life span of the equipment
9. Optimize the production
10. Keep manpower and machines in good operational conditions.

Economics of Maintenance

By economics of maintenance we mean the optimum efficiency in the production at minimum maintenance. To know about the impact of maintenance on economy we must understand the various cost curves and their shapes.

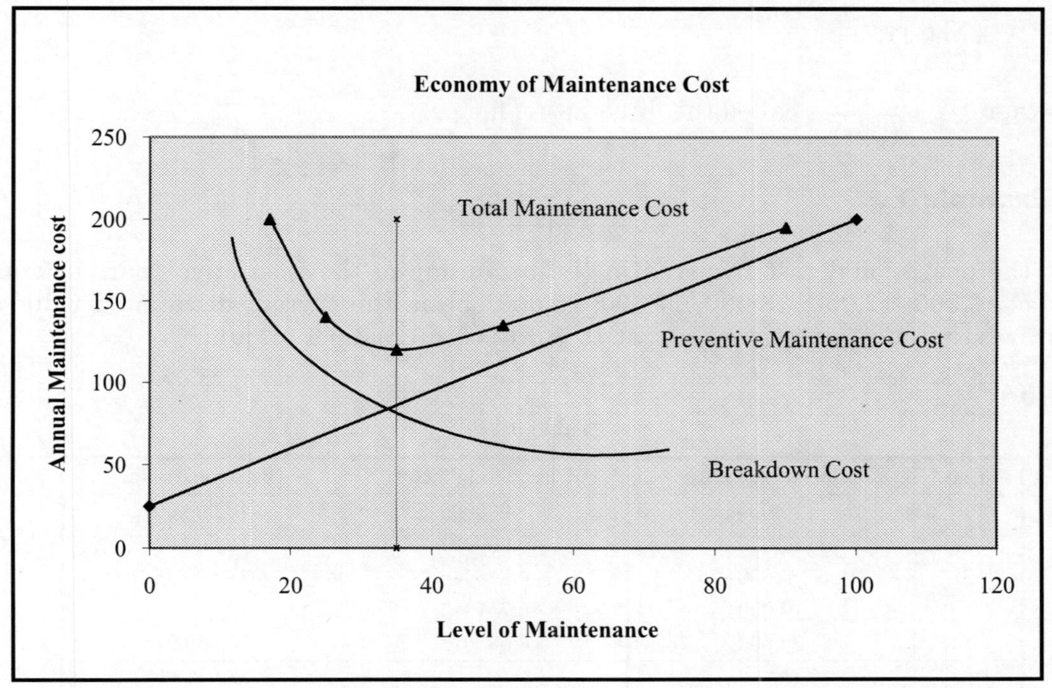

Figure 9.1

Figure 9.1 illustrates the relationships between the various cost curves and the associated optimum maintenance cost. The use of capital equipment in production lines needs a careful follow-up. As the age of the machine increases the maintenance cost per year also increases. At the same time the resale value of the used machine declines year after year. Thus, at some point of time it is more economical to discard the old machine and go for the new one. While doing so one must take the following facts into consideration.

9.3 RUNNING COST AND ITS DISCOUNTED CURRENT VALUE

While operating certain equipment, ingredients like lubricants, fuel, etc., are consumed. Such expenditures are called operation costs. Normally these costs increase over time due to variety of reasons like wear and tear of the capital equipment. Thus basic data on operational cost is necessary to make correct replacement decisions. When the value money undergo change over time then for correct decision making all the future operational costs are discounted and expressed as the current value worth at a given market rate of interest.

If $R_1, R_2 \ldots\ldots\ldots R_n$ are the operation costs (running cost) incurred in the beginning of years 1, 2, 3, ………………n respectively, then the discounted current value of the entire stream of operational cost denoted by R is obtained as:

$$R = R_1 + \frac{R_2}{(1+r)} + \frac{R_3}{(1+r)^2} \ldots\ldots\ldots\ldots + \frac{R_n}{(1+r)^{n-1}}$$

$$= R_1 + R_2 D^1 + R_3 D^2 \ldots\ldots\ldots\ldots + R_n D^{n-1}$$

$$= \sum_{i=1}^{n} R_i D^{i-1}$$

where $D = \dfrac{1}{(1+r)}$ is called the discounting factor.

■ Example 1

Given the operational cost incurred in the beginning of the respective years over five years as 10,000; 12,000; 13,000; 17,000; 20,000 obtain the current discounted value for the entire stream of operational cost at 10% rate of interest per annum.

Solution

Table 9.1

Year	Operational cost	Discounting factor	Discounted value
1	10,000	1.000000	10,000.00
2	12,000	0.909091	10,909.09
3	14,000	0.826446	11,570.25
4	17,000	0.751315	12,772.35
5	20,000	0.683013	13,660.27
Total	73,000	-	58,911.96

Example 2

A mineral processing has received offers for two types of dumper A and B. Dumper A has a pay load of 25 tons and is priced at Rs. 4,00,000. The dumper B also with a pay load of 25 tons is priced at Rs. 3,60,000. The estimated operational cost over the estimated life time of 5 years for both the types is given below.

Table 9.2

Year	1	2	3	4	5
Type A	8000	9000	10,000	11,000	12,000
Type B	14,000	16,000	18,000	20,000	22,000

Which dumper is to be preferred at 10% rate of interest per annum?

Solution

Table 9.3 Operational cost of A with and without discounting

Year	Operational cost	Discounting factor	Discounted value
1	8000	1.000000	8000.00
2	9000	0.909091	8181.82
3	10,000	0.826446	8264.46
4	11,000	0.751315	8264.46
5	12,000	0.683013	8196.16
Total	50,000		40,906.91

Table 9.4 Operational cost of B with and without discounting

Year	Operational cost	Discounting factor	Discounted value
1	14,000	1.000000	14,000.00
2	16,000	0.909091	14,545.45
3	18,000	0.826446	14,876.03
4	20,000	0.751315	15,026.30
5	22,000	0.683013	15,026.30
Total	90,000	-	73,474.08

Total cost of A without discounting = 4,00,000 +50,000 = 4,50,000

Total cost of B without discounting = 3,60,000 +90,000 = 4,50,000

Total cost of A with discounting = 4,00,000 +40,907 = 4,40,907

Total cost of B with discounting = 3,60,000 +73,474 = 4,13,474

When the time value of money is constant then (0% interest rate) one can go for either A or B type because the total expenditure on both comes to Rs. 4,50,000. If the expenses are discounted say at 10% then going for B costs less.

9.4 PRICE OF THE CAPITAL EQUIPMENT

The price of the machine that is under reference does not remain constant overtime. Normally it goes up in line with the rate of inflation. Thus while framing the replacement on a future date one must take the rate of inflation also into consideration. If inflation is at the rate I percent per annum then the price of the machine in n^{th} year, given the initial price C_0 in period 0, is obtained by using the compound interest formula.

If the compounding is done annually then the relevant formula is written as

$$C_n = C_0(1+I)^n$$

If the compounding is done m times in a year then this formula is written as

$$C_n = C_0\left(1+\frac{I}{m}\right)^{mn}$$

If compounding is done continuously then the above formula reduces to

$$C_n = C_0 e^{In}$$

To highlight the importance of inflation and the associated price rise, consider the following illustrative example.

■ Example 3

If 6000 is the price of a machine currently calculate it price over 10 years by all the three methods at 10% rate per annum.

Solution

Table 9.5

Year	Yearly compounding	Quarterly compounding	Continuous compounding
0	6000.00	6000.00	6000.00
1	6600.00	6622.88	6630.96
2	7260.00	7310.42	7328.26
3	7986.00	8069.33	8098.90
4	8784.60	8907.03	8950.58
5	9663.06	9831.70	9891.81
6	10,629.37	10,852.36	10,932.03
7	11,692.30	11,978.97	12,081.64
8	12,861.53	13,222.54	13,352.14
9	14,147.69	14,595.21	14,756.24
10	15,562.45	16,110.38	16,308.00

9.5 RESALE VALUES OF THE EQUIPMENT

It is generally believed that the value of the assets deteriorates with time. It is due to ageing effect on performance of the asset. Reduction in performance reduces the value of the assets. The rate of reduction in value over time is often called depreciation. The time value of the asset at a given point of time is called the resale value. To much depreciation is the healthy sign of replacement of the old by new. There are various methods of depreciation calculations available. Some of them are:

1. Straight line method
2. Some of years digit method
3. Diminishing balance method
4. Sinking fund method
5. Annuity charging method
6. Machine hour method
7. Insurance method
8. Revaluation method

Out of these eight methods we concentrate only the first three in this section.

(i) Straight line method

In this method the depreciation is assumed to be evenly distributed till the end of the lifetime of the concerned asset. Accordingly, the average depreciation per annum is obtained by simply dividing the total asset value at the time of purchase by the life time of the asset in years.

$$\text{Amount of depreciation per annum} = \frac{\text{Cost of the asset}}{\text{Life of the asset in years}}$$

This is the valid formula when the resale value of the concerned asset is zero on the expiry of the lifetime (scrap value = 0). For positive scrap value the asset, the modified formula may be written as

$$\text{Amount of depreciation per annum} = \frac{\text{Cost of the asset} - \text{Scrap value}}{\text{Life of the asset in years}}$$

■ Example 4

The cost of a machine on the purchase date is Rs. 25,000. If the life span of the machine is 10 years obtain the annual depreciation values by using straight line method by assuming zero resale value on the 10th year. If the scrap value on the 10th year is Rs. 5000 obtain annual rate of depreciation.

Solution

$$\text{Amount of depreciation per annum} = \frac{\text{Cost of the asset}}{\text{Life of the asset in years}} = \frac{25,000}{10} = \text{Rs. } 2500$$

$$\text{Amount of depreciation per annum} = \frac{\text{Cost of the asset} - \text{Scrap value}}{\text{Life of the asset in years}} = \frac{25,000 - 5000}{10} = \text{Rs. } 2000$$

(ii) Sum of the years digit value method

When the depreciation is maximum in the first year and steadily declines thereafter then the associated method of calculating depreciation is called sum of the digit value method. This is calculated using the following ratios.

$$r_1 = \frac{n}{d_1 + d_2 + \dots\dots\dots d_n} \quad \text{where } n = \text{life expectancy and } d_i = \text{sum of years digit}$$

$$r_2 = \frac{n-1}{d_1 + d_2 + \dots\dots\dots d_n}$$

$$r_3 = \frac{n-2}{d_1 + d_2 + \dots\dots\dots d_n}$$

....

$$r_n = \frac{1}{d_1 + d_2 + \dots\dots\dots d_n}$$

Using these ratios the depreciation for the respective years is obtained as

Depreciation for the first year $= \dfrac{C - R_s}{r_1}$

Depreciation for the second year $= \dfrac{C - R_s}{r_2}$

Depreciation for the third year $= \dfrac{C - R_s}{r_3}$ and so on.

Where C is the cost and R_s is the scrap value or resale value.

■ Example 5

A machine costing Rs. 25,000 has five years of life expectancy and Rs. 5000 scrap value. Obtain the rate of depreciation using sum of the year's digit method.

$$r_1 = \frac{5}{1+2+3+4+5} = \frac{5}{15}$$

$$r_2 = \frac{5-1}{d_1 + d_2 + \dots\dots\dots d_n} = \frac{4}{15}$$

$$r_3 = \frac{5-2}{d_1 + d_2 + ...d_n} = \frac{3}{15}$$

$$.r_4 = \frac{5-3}{d_1 + d_2 + ...d_n} = \frac{2}{15}$$

$$r_5 = \frac{5-4}{d_1 + d_2 + ...d_n} = \frac{1}{15}$$

Table 9.6

Year	r_i	Depreciation $(C-R_s)/r_i$
1	5/15	6666.67
2	4/15	5333.33
3	3/15	4000.00
4	2/15	2666.67
5	1/15	1333.33

(iii) Diminishing balance method

As time passes, the resale value of the capital equipment deteriorates due to its usage. The rate of such deterioration is often called depreciation. Often this re-sale value of the deterioration is calculated out of experience. Occasionally it is obtained by compounding formula stated above with negative rate called decay. If r is the rate at which the value declines per annum then the value of the machine after n years is obtained as

(i) $S_n = S_0(1-r)^n$

From this formula the rate of decay otherwise called depreciation is calculated as

$$S_n / S_0 = (1-r)^n$$

$$(S_n / S_0)^{\frac{1}{n}} = 1 - r$$

$$r = 1 - \left(\frac{S_n}{S_0}\right)^{\frac{1}{n}}$$

If the compounding is done m times in a year then this formula reduces to

(ii) $S_n = S_0\left(1 - \frac{r}{m}\right)^{nm}$

If compounding is done continuously then the above reduces to

(ii) $S_n = S_0 e^{-rn}$

■ Example 6

A machine costing Rs. 25,000 has five years of life expectancy and Rs. 5000 scrap value. Obtain the rate of depreciation using decay method on annual basis.

$$S_n = S_0(1-r)^n$$

$$S_n/S_0 = (1-r)^n$$

$$(S_n/S_0)^{\frac{1}{n}} = 1-r$$

$$r = 1-\left(\frac{S_n}{S_0}\right)^{\frac{1}{5}}$$

$$= 1-\left(\frac{5000}{25,000}\right)^{\frac{1}{5}}$$

$$= 1-(0.2)^{0.2}$$

$$= 1-0.72 = 0.28$$

■ Example 7

If 6000 is the price of a machine currently calculate its savage or otherwise called resale value over 10 years by all the three methods at 10% rate of decay per annum.

Solution

Table 9.7

Year	Yearly decay	Quarterly decay	Continuous decay
0	6000.00	6000.00	6000.00
1	5400.00	5422.13	5429.08
2	4860.00	4899.91	4912.49
3	4374.00	4427.99	4445.05
4	3936.60	4001.52	4022.09
5	3542.94	3616.13	3639.37
6	3188.65	3267.85	3293.07
7	2869.78	2953.12	2979.73
8	2582.80	2668.70	2696.20
9	2324.52	2411.67	2439.65
10	2092.07	2179.39	2207.51

9.6 REPLACEMENT POLICY

Normally replacement policy is categorized under the following four heads.

1. Replacement of items that deteriorate and loses it productive efficiency over time like machine tools, vehicles, equipment, buildings and so on.
2. Replacement of items which necessarily do not deteriorate and loses efficiency but fails suddenly like electric bulb, computer parts etc.
3. Replacement of items due to outdated technology like olden days radio valves.
4. Replacement of younger persons in the place of retired persons.

9.7 REPLACEMENT POLICY FOR ITEMS THAT DETERIORATE OVER TIME WHEN TIME VALUE OF MONEY IS CONSTANT (ALTERNATIVELY RATE OF INTEREST IS ZERO)

For finding the optimum replacement period under this category involves two items of costs. One item is the operational cost which normally tends to increase over time. The second item deserves attention is the wear and tear or otherwise called depreciation which normally decreases the resale value of the capital equipment over time. Having these two items in mind the correct optimum replacement cost is obtained by following two simple rules.

If we know the resale value, the operational cost and the cost of the equipment over time then the optimum replacement period is given by the period corresponding to the minimum average cost.

Let us consider a simple case of minimizing the annual average cost under the following assumptions.

(i) C = The total cost of the equipment that is to be replaced is constant over the time.

(ii) R(n) = The maintenance cost otherwise called the running cost of the equipment under reference in the year n. It is assumed to be an increasing function of time.

(iii) S(n) = The resale or otherwise called the scrap value of the equipment under reference in the year n. Normally it is a decreasing function of time.

(iv) n = The optimum replacement time.

Now the annual cost at any given point of time n

= Net machine repurchase cost during the nth year + cumulative running cost up to n^{th} year.

= Capital cost during the nth year – scrap value during the nth year + cumulative running cost up to n^{th} year.

Discrete Formation

Now the annual cost on both the accounts during n^{th} year denoted by TC(n) in its discrete form is written as

$$TC(n) = C - S(n) + \sum_{i=0}^{n} R_i$$

So the average annual cost per annum is written as

$$= AC(n) = \frac{TC(n)}{n} = \frac{1}{n}\left[C - S(n) + \sum_{i=1}^{n} R(i) \right]$$

$$= \frac{C - S(n)}{n} + \frac{\sum_{i=1}^{n} R(i)}{n}$$

In the above equation the average net capital cost is a decreasing function. Similarly, the average value of the cumulative running cost is an increasing function of time. The item under reference is to be replaced when the minimum average cost is equal to the current maintenance cost.

Continuous Formation

$$\text{So the annual } TC(n) = C - S(n) + \int_0^n R(n)\, dn$$

$$\text{So the average annual cost} = AC(n) = \frac{1}{n}\left[C - S(n) + \int_0^n R(n)\, dn \right]$$

Now for the minimum value of AC(n) we know that the first order derivative is equal to zero. ie., $\frac{d\{AC(n)\}}{dn} = 0$

$$\frac{d\{AC(n)\}}{dn} = -\frac{1}{n^2}\left[C - S(n) - \frac{1}{n^2} \int_0^n R(n)\, dn + \frac{1}{n} R(n) \right] = 0$$

$$= \left[C - S(n) - \frac{1}{n^2} \int_0^n R(n)\, dn + \frac{1}{n} R(n) \right] = 0$$

$$R(n) = \frac{C - S(n)}{n} + \frac{1}{n} \int_0^n R(n)\, dn = AC(n)$$

This is to say that the current running cost equals the current average cost, the replacement must be done. In other words, this is to say that the average cost is least then we go for replacement.

Example 8

The following table gives the running cost per year and resale value of a certain machinery whose purchase price is Rs. 65,000. Suggest a suitable replacement policy.

Table 9.8

Year	1	2	3	4	5	6	7
Running cost	14,000	15,000	17,000	20,000	24,000	28,000	33,000
Cost of new machine	65,000	65,000	65,000	65,000	65,000	65,000	65,000
Resale value	40,000	30,000	22,000	17,000	13,000	10,000	10,000

Solution

The total cost for the first year

= Running cost in the first year + purchase price of the capital equipment – resale value of the capital equipment.

= 14,000 + (65,000 – 40,000) = 39,000

= Therefore average cost per year during the first year = 39,000/1 = 39,000

The total cost for the second year

= Running for first year + running cost for the second year + purchase price of the capital equipment during the second year – resale value of the capital equipment during the second year.

= 14,000 +15,000 + (65,000 – 30,000) = 64,000

Therefore average cost per year = 64,000/2 = 32,000. Proceeding in this way the average costs are calculated and reported in the last column in Table 9.9. Note that this time the replacement machine cost found to be increasing over time.

Table 9.9

Year	Yearly running cost (Rs.)	Cumulative running cost (Rs.)	Cost of the new machine (Rs.)	Resale price (Rs.)	Net capital cost (Rs.)	Total cost (Rs.)	Average cost (Rs.)
	$R(t)$	$\sum R(t)$	$C(t)$	$S(t)$	$C(t)-S(t)$	$TC=C(t)-S(t)+\sum R(t)$	AC
1	14,000	14,000	65,000	40,000	25,000	39,000	39,000
2	15,000	29,000	65,000	30,000	35,000	64,000	32,000
3	17,000	46,000	65,000	22,000	43,000	89,000	29,667
4	20,000	66,000	65,000	17,000	48,000	1,14,000	28,500
5	24,000	90,000	65,000	13,000	52,000	1,42,000	28,400
6	28,000	1,18,000	65,000	10,000	55,000	1,73,000	28,833
7	33,000	1,51,000	65,000	10,000	55,000	206,000	29,429

In this table in the third column we represent cumulative running cost. In the fourth column the cost of the machine minus the resale value is given. This is simply the net capital cost of the new machine per year. In the fifth column we represent the total cost as the sum of column three and four. In the last column the average cost is being calculated and represented.

In the table the minimum average cost is Rs. 28,400 corresponding to the fifth year. Thus the machine should go for replacement once in five years. If he continues to use it for any more period without replacement he will have to incur more operational cost.

■ Example 9

A truck company has the following data for seven years. Obtain the optimum replacement policy.

Table 9.10

Year	1	2	3	4	5	6	7
Running cost	1500	2000	2500	3000	3750	4500	5750
Cost of new machine	5000	5250	5500	6000	6500	7250	8000
Resale value	2500	1250	750	500	400	250	0

Solution

The total cost for the first year

= Running cost in the first year + purchase price of the capital equipment – resale value of the capital equipment.

= 1500+ (5000–2500) = 4000

= Therefore average cost per year during the first year = 4000/1 = 4000

The total cost for the second year

= Running for first year + running cost for the second year + purchase price of the capital equipment during the second year – resale value of the capital equipment during the second year.

= 1500+2000+ (5250–1250) = 7500

Therefore average cost per year = 7500/2 = 3750. Proceeding in this way the average costs are calculated and reported in the last column. Note that this time the replacement machine cost found to be increasing over time.

In this table, in the third column we represent cumulative running cost. In the fourth column the cost of the truck minus the resale value is given. This is simply the net capital cost of the truck per year. In the fifth column we represent the total cost as the sum of column three and four. In the last column the average cost is being calculated and represented. In the table, the minimum average cost is Rs. 3583 corresponding to the third year.

Table 9.11

Year	Yearly running cost (Rs.)	Cumulative running cost (Rs.)	Cost of the new machine (Rs)	Resale price (Rs.)	Net capital cost (Rs.)	Total cost	Average cost
1	1500	1500	5000	2500	2500	4000	4000
2	2000	3500	5250	1250	4000	7500	3750
3	2500	6000	5500	750	4750	10,750	3583
4	3000	9000	6000	500	5500	14,500	3625
5	3750	12,750	6500	400	6100	18,850	3770
6	4500	17,250	7250	250	7000	24,250	4042
7	5750	23,000	8000	0	8000	31,000	4429
						Min =	3583

Thus the truck company should go for replacement once in three years. If he continues to use the truck for any more periods without replacement then definitely he will have to incur more operational cost.

Example 10

A certain machine has the purchase price of Rs. 16,600. Given its scrap value at any point of time only Rs. 1600 obtain the optimum replacement period.

Table 9.12

Year	1	2	3	4	5	6	7	8	9
Running cost	300	700	1000	1500	2100	2800	3600	4800	6200

Solution

Given the operation cost, the scrap value and the machine cost the optimum replacement corresponding to the minimum average cost is obtained in the following table.

Table 9.13

Year	Yearly running cost	Cumulative running cost	Cost of the new machine	Scrap or resale value	Net purchase price of new machine	Total cumulative cost	Average cost
1	300	300	16,600	1600	15,000	15,300	15,300
2	700	1000	16,600	1600	15,000	16,000	8000
3	1000	2000	16,600	1600	15,000	17,000	5667
4	1500	3500	16,600	1600	15,000	18,500	4625
5	2100	5600	16,600	1600	15,000	20,600	4120
6	2800	8400	16,600	1600	15,000	23,400	3900
7	3600	12,000	16,600	1600	15,000	27,000	3857
8	4800	16,800	16,600	1600	15,000	31,800	3975
9	6200	23,000	16,600	1600	15,000	38,000	4222
						Min =	3857

From the table above the average minimum cost is reached in the seventh year. So once in every seventh year optimum replacement must take place.

■ Example 11

A machine A costs Rs. 80,000. The annual operating cost for the first year is Rs. 2000, thereafter it increases by Rs. 15,000 per annum.

a) Find the optimum replacement period for Machine A.

b) Another machine B costing Rs. 1,00,000. The estimated operation cost for B is Rs. 4000 for the first year. Thereafter it increases by Rs. 7000. Can the firm having one year old machine A can go for replacement of machine B.

Solution

Table 9.14

Year	Yearly running cost (Rs.)	Cumulative running cost (Rs.)	Cost of a new machine (Rs.)	Resale value (Rs.)	Net capital cost (Rs.)	Total cumulative cost (Rs.)	Average cost (Rs.)
1	2000	2000	80,000	0	80,000	82,000	82,000
2	17,000	19,000	80,000	0	80,000	99,000	49,500
3	32,000	51,000	80,000	0	80,000	1,31,000	43,667
4	47,000	98,000	80,000	0	80,000	1,78,000	44,500
5	62,000	1,60,000	80,000	0	80,000	2,40,000	48,000
						Min=	43,667

In the above table since the minimum average cost is Rs. 43,667 it is recommended that the machine A should be replaced once in every three years.

Table 9.15

Year	Yearly running cost (Rs.)	Cumulative running cost (Rs.)	Cost of a new machine (Rs.)	Resale value (Rs.)	Net capital cost (Rs.)	Total cumulative cost (Rs.)	Average cost (Rs.)
1	4000	4000	1,00,000	0	1,00,000	1,04,000	1,04,000
2	11,000	15,000	1,00,000	0	1,00,000	1,15,000	57,500
3	18,000	33,000	1,00,000	0	1,00,000	1,33,000	44,333
4	25,000	58,000	1,00,000	0	1,00,000	1,58,000	39,500
5	32,000	90,000	1,00,000	0	1,00,000	1,90,000	38,000
6	39,000	1,29,000	1,00,000	0	1,00,000	2,29,000	38,167
						Min=	38,000

In the above table since the minimum average cost is Rs. 38,000 it is recommended the machine B should be replaced once in every five years.

Since the lowest average cost of B is less than the lowest average cost of A machine B clearly deserved to be replaced by machine A. To find the time of replacement we proceed as follows. The machine A deserved to be replaced by B provided A's next year running cost exceed the Rs. 38,000, the minimum average cost of B.

Table 9.16

Year	Total cumulative cost (Rs.)	Yearly total cost (Rs.)	Cost of one year old
1	82,000	82,000	17,000
2	99,000	17,000	32,000
3	131,000	32,000	47,000
4	178,000	47,000	62,000
5	240,000	62,000	

The third year running cost of the one-year old machine is Rs. 47,000. Since amount exceeds the minimum of Rs. 38,000, machine A needs to be replaced immediately by B.

Example 12

A truck owner currently has 2 two-year old and 1 one-year old mini trucks for operation. The purchase price per truck is Rs. 1,60,000. The following table shows the estimated running cost and resale value per truck.

Table 9.17

Year	1	2	3	4	5	6	7	8
Running cost	8000	9000	10,500	13,000	15,000	20,000	25,000	30,000
Resale value	1,45,000	1,32,000	1,22,000	1,14,000	1,09,000	90,000	70,000	50,000

Now he is considering two new bigger trucks with 50% excess capacity at Rs. 2,00,000 for replacement. The maintenance cost and resale value of the bigger truck is given below. Suggest a suitable replacement policy.

Table 9.18

Year	1	2	3	4	5	6	7	8
Running cost	10,000	12,000	14,000	16,500	19,000	25,000	30,000	40,000
Resale value	1,60,000	1,45,000	1,33,000	1,24,000	1,19,000	1,12,000	90,000	60,000

Solution

To begin with let us calculate the optimum replacement period for both mini and bigger trucks first in the usual way.

Table 9.19 Calculation of replacing time for smaller truck

Year	Yearly running cost (Rs.)	Cumulative running cost (Rs.)	Cost of a new machine (Rs.)	Resale value (Rs.)	Net capital cost (Rs.)	Total cumulative cost (Rs.)	Average cost (Rs.)
1	8000	8000	1,60,000	1,45,000	15,000	23,000	23,000
2	9000	17,000	1,60,000	1,32,000	28,000	45,000	22,500
3	10,500	27,500	1,60,000	1,22,000	38,000	65,500	21,833
4	13,000	40,500	1,60,000	1,14,000	46,000	86,500	21,625
5	15,000	55,500	1,60,000	1,09,000	51,000	1,06,500	21,300
6	20,000	75,500	1,60,000	90,000	70,000	1,45,500	24,250
7	25,000	1,00,500	1,60,000	70,000	90,000	1,90,500	27,214
8	30,000	1,30,500	1,60,000	50,000	1,10,000	2,40,500	30,063
						Min=	21,300

In the table the minimum average cost is Rs. 21,300 corresponding to the fifth year. Thus the truck company should go for replacement of old smaller one by the new smaller one in the fifth year.

Table 9.20 Calculation of replacing time for bigger truck

Year	Yearly running cost (Rs.)	Cumulative running cost (Rs.)	Purchase price (Rs.)	Resale price (Rs.)	Net capital cost (Rs.)	Total cumulative cost (Rs.)	Average cost (Rs.)
1	10,000	10,000	2,00,000	1,60,000	40,000	50,000	50,000
2	12,000	22,000	2,00,000	1,45,000	55,000	77,000	38,500
3	14,000	36,000	2,00,000	1,33,000	67,000	1,03,000	34,333
4	16,500	52,500	2,00,000	1,24,000	76,000	1,28,500	32,125
5	19,000	71,500	2,00,000	1,19,500	80,500	1,52,000	30,400
6	25,000	96,500	2,00,000	1,12,000	88,000	1,84,500	30,750
7	30,000	1,26,500	2,00,000	90,000	1,10,000	2,36,500	33,786
8	40,000	1,66,500	2,00,000	60,000	1,40,000	3,06,500	38,313
						Min=	30,400

In Table 9.20 the minimum average cost is Rs. 30,400 corresponding to the fifth year. Thus the truck company should go for replacement of old bigger one by new bigger one on the fifth year. Now from the tables given above

The average costs of three mini trucks =21,300 × 3 = 63,900

The average costs of two bigger ones = 30,400 ×2 = 60,800

Thus it is prudent to replace all the three mini ones by two bigger ones. Now the question is when such a decision is to be taken. To decide the time at which the replacement is to be made, we shall first find the total annual cost of running the three mini ones.

The year in which the average cost of the bigger trucks will be lower than the total cost of maintaining and running the mini trucks shall be the year of replacement.

Since in our problem, replacement of old smaller ones by big new ones is under consideration let us calculate the total cost for running all the three smaller ones (two 2-year old and one 1-year-old) in Table 9.21.

In the table the yearly additional cost for running one mini truck is reported as the 23,000 – 0 = 23,000; 45,000 – 23,000 = 22,000; 65,500 – 45,000 = 20,500 and so on in the second column. In the last column using second column we report the total cost for all the 3 machines with a breakup of 2 two-year-old and 1 one-year-old. For example, in the very first year since he is having 1 one-year-old and 2 two-year-old machines the total cost of running these = 22,000 + 2 × 20,500 = 63,000.

Table 9.21

Year	Total cumulative cost (Rs.)	Yearly cost (Rs.) per mini truck	Total yearly cost of 1 one-year-old and 2 two-years-old mini trucks (Rs.)
1	23,000	23,000 – 0 = 23,000	22,000 + 2 x 20,500 = 63,000
2	45,000	45,000 – 23,000 = 22,000	20,500 + 2 x 21,000 = 62,500
3	65,500	65,500 – 45,000 = 20,500	21,000 + 2 x 20,000 = 61,000
4	86,500	86,500 – 65,500 = 21,000	20,000 + 2 x 39,000 = 98,000
5	1,06,500	1,06,500 – 86,500 = 20,000	39,000 + 2 x 45,000 = 1,29,000
6	1,45,500	1,45,500 – 1,06,500 = 39,000	45,000 + 2 x 50,000 = 1,45,000
7	1,90,500	1,90,500 – 1,45,500 = 45,000	
8	2,40,500	2,40,500 – 1,90,500 = 50,000	

Similarly, for the second year he will be having 1 two-year-old and 2 three-year-old trucks. So the associated total running cost = 20,500 + 2 × 21,000 = 62,500. This way the cost for the rest of the years is obtained and reported in Table 9.21.

From Table 9.20 it follows that the average cost for two big trucks is 30,400 × 2 = 60,800. Since this cost is less than all the figures quoted in the last column in Table 9.21 we conclude that the truck owner should go for immediate replacement of all the three small ones by big ones.

■ Example 13

A mini bus transport company right now possesses three mini busses bought at the rate of Rs. 80,000. The cost of running a mini bus with its resale values over years are given below.

Table 9.22

Year	1	2	3	4	5	6	7	8
Running cost	3000	3600	4800	5000	8000	11,200	15,000	20,000
Resale value	70,000	61,000	55,000	49,000	32,000	20,000	10,000	5000

Currently among the three buses two of them are two-year-old and one is one-year-old. The company is planning for replacing them all by two bigger ones with 50% excess capacity at a cost of Rs. 1,20,000 per bus. The estimated per bus running costs and resale values of the bigger ones are given below.

Table 9.23

Year	1	2	3	4	5	6	7	8
Running cost	3400	3900	4700	5800	7200	9000	12,000	16,000
Resale value	1,00,000	92,000	86,000	81,000	76,000	66,000	54,000	40,000

Should the company venture for such replacement? If yes, when?

Solution

Using the data given, let us first calculate the minimum average costs for both the mini and bigger ones first.

Table 9.24 Calculation of replacing time for mini bus

Year	Yearly running cost (Rs.)	Cumulative running cost (Rs.)	Purchase price	Resale price (Rs.)	Net capital cost (Rs.)	Total cumulative cost	Average cost
1	3000	3000	80,000	70,000	10,000	13,000	13,000
2	3600	6600	80,000	61,000	19,000	25,600	12,800
3	4800	11,400	80,000	55,000	25,000	36,400	12,133
4	5000	16,400	80,000	49,000	31,000	47,400	11,850
5	8000	24,400	80,000	32,000	48,000	72,400	14,480
6	11,200	35,600	80,000	20,000	60,000	95,600	15,933
7	15,000	50,600	80,000	10,000	70,000	1,20,600	17,229
8	20,000	70,600	80,000	5000	75,000	1,45,600	18,200
						Min =	11,850

Table 9.25 Calculation of replacing time for bigger bus

Year	Yearly running cost (Rs.)	Cumulative running cost (Rs.)	Purchase price	Resale price (Rs.)	Net capital cost (Rs.)	Total cumulative cost	Average cost
1	3400	3400	1,20,000	1,00,000	20,000	23,400	23,400
2	3900	7300	1,20,000	92,000	28,000	35,300	17,650
3	4700	12,000	1,20,000	86,000	34,000	46,000	15,333
4	5800	17,800	1,20,000	81,000	39,000	56,800	14,200
5	7200	25,000	1,20,000	76,000	44,000	69,000	13,800
6	9000	34,000	1,20,000	66,000	54,000	88,000	14,667
7	12,000	46,000	1,20,000	54,000	66,000	1,12,000	16,000
8	16,000	62,000	1,20,000	40,000	80,000	1,42,000	17,750
						Min =	13,800

The average cost of three mini buses =11,850 × 3 = 35,550
The average cost of two bigger ones = 13,800 ×2 = 27,600

Thus it is prudent to replace all the three mini ones by two bigger ones. Now the question is when such a decision is prudent. To decide the time at which the replacement is to be carried out we shall first find the total annual cost of running the three mini ones.

The year in which the average cost of the new buses will be lower than the total cost of maintaining and running the mini ones shall be the year of replacement.

Since in our problem replacement of old smaller ones by big new ones is under consideration let us calculate the total cost for running all the three smaller ones (2 two-year-old and 1 one-year-old) as shown in Table 9.30.

In the Table 9.30 below the yearly additional costs are reported as the 13,000 – 0 = 13,000; 25,600 – 13,000 = 12,600, and 36,400 – 25,600 = 10,800 and so on in the seventh column. In the last column we report the total cost for all the 3 machines with a breakup of 2 two-year-old and 1 one-year-old. For example, in the very first year since he is having 1 one-year-old and 2 two-year-old machines the total cost = 12,600 + 2 × 10,800 = 34,200. Similarly, for the year 2, since he will be having 1 two-year-old and 2 three-year-old, the total cost =10,800 + 2 × 11,000 = 32,800. This way the cost for the rest of the years are obtained and reported in Table 9.26.

Table 9.26

Year	Total cumulative cost (Rs.)	Yearly cost (Rs.)					Total yearly cost of all the three (Rs.)				
1	13,000	13,000	–	0	=	13,000	12,600	+ 2 x	10,800	=	34,200
2	25,600	25,600	–	13,000	=	12,600	10,800	+ 2 x	11,000	=	32,800
3	36,400	36,400	–	25,600	=	10,800	11,000	+ 2 x	25,000	=	61,000
4	47,400	47,400	–	36,400	=	11,000	25,000	+ 2 x	23,200	=	71,400
5	72,400	72,400	–	47,400	=	25,000	23,200	+ 2 x	25,000	=	73,200
6	95,600	95,600	–	72,400	=	23,200	25,000	+ 2 x	25,000	=	75,000
7	1,20,600	1,20,600	–	95,600	=	25,000					
8	1,45,600	1,45,600	–	1,20,600	=	25,000					

From Table 9.25 the average cost for two big trucks is 13,800 × 2 = 27,600. Since this sum is less than all the future costs of running three small ones, immediate replacement all the three small ones by the two big ones is recommended.

9.8 REPLACEMENT POLICY FOR ITEMS THAT DETERIORATE OVER TIME WHEN TIME VALUE OF MONEY IS DECLINING (ALTERNATIVELY RATE OF INTEREST IS POSITIVE)

When the value of money changes over time at a constant rate reflected by the rate interest then all replacement decisions are to be revised based on the discounted values of the flow of various costs.

Use of discounting principle in replacement models

Let C be the purchase value (price) of the new equipment that is to be replaced. Suppose the machine is to be replaced at the beginning of the year 'n' every time with the associated operating costs R_1, R_2, ...R_n. Now K_n, the present value of the whole stream of replacement designed once in n years is obtained as:

$$K_n = \left((C+R_1)+R_2D^1+R_3D^2+.................+R_nD^{n-1}\right)$$
$$+\left((C+R_1)D^n+R_2D^{n+1}+R_3D^{n+2}+......+R_nD^{2n-1}\right)$$
$$+......\left((C+R_1)D^{2n}+R_2D^{2n+1}+R_3D^{2n+2}+......+R_nD^{3n-1}\right)+......$$
$$=(C+R_1)\left(1+D^n+D^{2n}+D^{3n}+......\right)+R_2D^1\left(1+D^n+D^{2n}+D^{3n}+......\right)$$
$$+......+R_nD^{n-1}\left(1+D^n+D^{2n}+D^{3n}+......\right)$$
$$=\left(C+R_1+R_2D^1+R_3D^2+......+R_nD^{n-1}\right)\left(1+D^n+D^{2n}+D^{3n}+......\right)$$
$$=\left[C+\sum_{i=1}^{n}R_iD^{i-1}\right]\left[1+D^n+D^{2n}+V^{3n}+.......\right]$$
$$K_n=\left[C+\sum_{i=1}^{n}R_iD^{i-1}\right]\left[\frac{1}{1-D^n}\right]........................(1) \text{ (Summation of geometric series)}$$

Similarly, if we plan for replacement once in (n+1) years then the present value of such a replacement scheme is obtained as

$$K_{n+1}=\left[C+\sum_{i=1}^{n+1}R_iD^{i-1}\right]\left[\frac{1}{1-D^{n+1}}\right]...................(2)$$

In the above expression K_n is the amount of money that is necessary in the beginning itself to pay all the future cost of acquiring and operating the equipment when it is replaced once in n years. Clearly no company can afford to have such a huge amount in the beginning itself. However whenever K_{n+1} is greater than K_n clearly replacement after n year is always recommended provided

$$K_n-K_{n-1}>0 \text{ and } K_{n-1}-K_n>0$$

$$K_{n+1}=\left[C+\sum_{i=1}^{n}R_iD^{i-1}+R_{n+1}D^n\right]\left[\frac{1}{1-D^{n+1}}\right]$$
$$=\frac{(1-D^n)K_n+R_{n+1}D^n}{1-D^{n+1}}$$
$$=\frac{(1-D^n)K_n}{1-D^{n+1}}+\frac{R_{n+1}D^n}{1-D^{n+1}}$$

$$K_{n+1} - K_n = \frac{(1-D^n)K_n}{1-D^{n+1}} - K_n + \frac{R_{n+1}D^n}{1-D^{n+1}}$$

$$= K_n\left[\frac{(1-D^n)}{1-D^{n+1}} - 1\right] + \frac{R_{n+1}D^n}{1-D^{n+1}}$$

$$= \frac{K_n\left[1-D^n-(1-D^{n+1})\right]+R_{n+1}D^n}{1-D^{n+1}}$$

$$= \frac{K_n\left[D^{n+1}-D^n)\right]+R_{n+1}D^n}{1-D^{n+1}} \text{.......................(3)}$$

If $K_{n+1} - K_n > 0$ then $K_n\left[D^{n+1} - D^n\right] + R_{n-1}D^n > 0$

$K_n\left[D^n(D-1)\right] + R_{n+1}D^n > 0$

Now let us divide both the sides by D^n

$K_n(D-1) + R_{n+1} > 0$

$R_{n+1} > K_n(1-D)$

$$\frac{R_{n+1}}{(1-D)} > K_n \text{..(4)}$$

Now let us multiply equation (3) by −1 it becomes

$$K_n - K_{n+1} = \frac{K_n\left[D^n - D^{n+1}\right] - R_{n+1}D^n}{1-D^{n+1}}$$

After replacing n by n−1 it becomes

$$K_{n-1} - K_n = \frac{K_{n-1}\left[D^{n-1} - D^n\right] - R_nD^{n-1}}{1-D^n}$$

Now if $K_{n-1} - K_n > 0$ then

$K_{n-1}\left[D^{n-1} - D^n\right] - R_nD^{n-1} > 0$

$K_{n-1}\left[D^{n-1}(1-D)\right] - R_nD^{n-1} > 0$

Now let us divide both the sides by $D^{(n-1)}$

$K_{n-1}[1-D] - R_n > 0$

$K_{n-1}[1-D] > R_n$

$$\frac{R_n}{(1-D)} < K_{n-1} \text{..(5)}$$

Inequalities (4) and (5) holds good provided the K_n is minimum.
Now from (5) above

$$\frac{R_n}{(1-D)} < K_{n-1}$$

$$R_n < (1-D)K_{n-1}$$

Now after substituting K_{n-1} value it is rewritten as

$$R_n < (1-D)\frac{(C+R_1)+R_2D+\ldots\ldots\ldots\ldots\ldots\ldots R_{n-1}D^{n-2}}{1-D^{n-1}}$$

$$R_n < \frac{(C+R_1)+R_2D+\ldots\ldots\ldots\ldots\ldots\ldots R_{n-1}D^{n-2}}{1-D^{n-2}}$$

$$\frac{(C+R_1)+R_2D+\ldots\ldots\ldots\ldots\ldots\ldots R_{n-1}D^{n-2}}{1+D+D^2+\ldots\ldots\ldots\ldots\ldots\ldots D^{n-2}} \quad \ldots\ldots\ldots(6)$$

The expression in the RHS is the weighted average of all cost components including period n−1. The associated weights 1, D, D^2, D^{n-1} simply the discounted current values of the respective operational costs.

Similarly, inequality (4) may be written as

$$\frac{R_{n+1}}{(1-D)} > K_n$$

$$R_{n+1} > (1-D)K_n$$

Now after substituting K_n value it is rewritten as

$$R_{n+1} > \frac{(1-D)(C+R_1)+R_2D+\ldots\ldots\ldots\ldots\ldots\ldots R_nD^{n-1}}{1-D^n}$$

$$R_{n+1} > \frac{(C+R_1)+R_2D+\ldots\ldots\ldots\ldots\ldots\ldots R_nD^{n-1}}{1-D^{n-1}}$$

$$R_{n+1} > \frac{(C+R_1)+R_2D+\ldots\ldots\ldots\ldots\ldots\ldots R_nD^{n-1}}{1+D+D^2+\ldots\ldots\ldots D^{n-1}} \quad \ldots\ldots\ldots(7)$$

Having these two inequalities in mind we can state the rules for optimum replacement as follows

1. ***Do not replace if the next period's cost is less than the weighted average of the previous cost.***
2. ***Replace if the next period's cost is greater than the weighted average of the previous costs.***

Example 14

Given the following data obtain the correct optimum replacement period for a given market rate of interest as 5% per annum when the scrap value of equipment is zero.

Table 9.27

Year	Price of equipment	Operating cost
1	5000	0
2	5000	500
3	5000	1000
4	5000	1500
5	5000	2000
6	5000	2500

Solution

Table 9.28

	Cost of machine	Running cost	Discounting factor	Discounted value of R	Cu. Discounted value of R	Total cost	Cumulative discounted value	Weighted average cost
Year	C	Ri	D^{i-1}	RiD^{i-1}	$Cu.RiD^{i-1}$	$C+RiD^{i-1}$	$Cu.D^{i-1}$	$(C+RiD^{i-1})/CuD^{i-1}$
1	5000	0	1.0000	0	0	5000	1.0000	5000
2	5000	500	0.9524	476	476	5476	1.9524	2805
3	5000	1000	0.9070	907	1383	6383	2.8594	2232
4	5000	1500	0.8638	1296	2679	7679	3.7232	2062
5	5000	2000	0.8227	1645	4324	9324	4.5460	2051
6	5000	2500	0.7835	1959	6283	11283	5.3295	2117

Step 1: Enter the cost of the machine in the second column of the table after entering the year in the first column. Enter the actual running cost the third column. Given the rate of interest obtain the discounted value per rupee and reported in the fourth column. For example, for the 0th year the current and discounted values are one and the same as reported as Re. 1.00.

Discounted value of a rupee receivable after one year at 5% $=1/(1+r)^{1-1}$
$=1/(1+0.05)^0 = 1.0000$

Discounted value of a rupee receivable after two years at 5% $=1/(1+r)^{2-1}$
$=1/(1+0.05)^1 = 0.9524$

Similarly, the discounted values for the remaining years are calculated and reported in the fourth column.

Step 2: In the fifth column using the discounted value in the fourth column we calculate the discounted worth the running cost. It is obtained simply by multiplying column three and four.

Step 3: In column number six the cumulative values of column number five is calculated and reported.

Step 4: In column number seven is obtained by simply adding the cost of replacement to column number six.

Step 5: The cumulative value of column number four is reported in column 8.

Step 6: In the last column we calculate the weighted average by dividing column 7 by column 8.

The minimum value in the last column corresponds to fourth year. Hence the optimum replacement should be once in four years.

■ Example 15

The purchase price of the equipment is Rs. 50. Given the following data on operation cost for 10 years obtain the correct optimum replacement period. Market rate of interest is 5% per annum.

Table 9.29

Year	1	2	3	4	5	6	7	8	9	10
Operation cost	0	10	20	30	40	50	60	70	80	90

Solution

Table 9.30

Year	Cost of machine C	Running cost Ri	Discounting factor D^{i-1}	Discounted value of R RiD^{i-1}	Cu. Discounted value of R $Cu.RiD^{i-1}$	Total cost $C+R^iD^{i-1}$	Cumulative discounted value $Cu.D^{i-1}$	Weighted average cost
1	50	0	1.0000	0	0	50	1.0000	50.00
2	50	10	0.9524	10	10	60	1.9524	30.49
3	50	20	0.9070	18	28	78	2.8594	27.16
4	50	30	0.8638	26	54	104	3.7232	27.82
5	50	40	0.8227	33	86	136	4.5460	30.02
6	50	50	0.7835	39	126	176	5.3295	32.96
7	50	60	0.7462	45	170	220	6.0757	36.28
8	50	70	0.7107	50	220	270	6.7864	39.81
9	50	80	0.6768	54	274	324	7.4632	43.46
10	50	90	0.6446	58	332	382	8.1078	47.16

The minimum value in the last column corresponds to third year. Hence the optimum replacement should be once in three years.

Example 16

Given the following data pertaining to two machines A and B obtain the correct optimum replacement period for each one of them and propose a best purchase policy, given market rate of interest as 5% per annum by assuming nil scrap value for both the machines.

Table 9.31

	Machine A		Machine B	
Year	Price of equipment	Operating cost	Price of equipment	Operating cost
1	5000	800	2500	1200
2	5000	800	2500	1200
3	5000	800	2500	1200
4	5000	800	2500	1200
5	5000	800	2500	1200
6	5000	1000	2500	1200
7	5000	1200	2500	1400
8	5000	1400	2500	1600
9	5000	1600	2500	1800
10	5000	1800	2500	2000

Solution

Table 9.32 Calculations for Machine A

Year	C	Ri	D^{i-1}	RiD^{i-1}	$Cu.RiD^{i-2}$	$C+R^iD^{i-1}$	$Cu.D^{i-1}$	W. average
1	5000	800	1.0000	800	800	5800	1.0000	5800.00
2	5000	800	0.9091	727	1527	6527	1.9091	3419.05
3	5000	800	0.8264	661	2188	7188	2.7355	2627.79
4	5000	800	0.7513	601	2789	7789	3.4869	2233.96
5	5000	800	0.6830	546	3336	8336	4.1699	1999.08
6	5000	1000	0.6209	621	3957	8957	4.7908	1869.59
7	5000	1200	0.5645	677	4634	9634	5.3553	1799.01
8	5000	1400	0.5132	718	5353	10,353	5.8684	1764.12
9	5000	1600	0.4665	746	6099	11,099	6.3349	1752.04
10	5000	1800	0.4241	763	6862	11,862	6.7590	1755.05

Table 9.33 Calculations for Machine B

Year	C	Ri	D^{i-1}	RiD^{i-1}	$Cu.RiD^{i-2}$	$C+R^{i}D^{i-1}$	$Cu.D^{i-1}$	W. average
1	2500	1200	1.0000	1200	1200	3700	1.0000	3700.00
2	2500	1200	0.9091	1091	2291	4791	1.9091	2509.52
3	2500	1200	0.8264	992	3283	5783	2.7355	2113.90
4	2500	1200	0.7513	902	4184	6684	3.4869	1916.98
5	2500	1200	0.6830	820	5004	7504	4.1699	1799.54
6	2500	1200	0.6209	745	5749	8249	4.7908	1721.83
7	2500	1400	0.5645	790	6539	9039	5.3553	1687.91
8	2500	1600	0.5132	821	7360	9860	5.8684	1680.22
9	2500	1800	0.4665	840	8200	10,700	6.3349	1689.04
10	2500	2000	0.4241	848	9048	11,548	6.7590	1708.56

The observations from Table 9.32 and 9.33 reveal that machine A needs replacement once in nine years and machine B needs replacement once in 8 years. The discounted current value of the operation cost for machine A is Rs. 1752.40 and Rs. 1680.22 for machine B. Thus it is more economical to go for B rather than A.

Importance of Discounting

Let us consider two machines A and B with expected operational cost to be incurred in three consecutive years. Both the machines have the same total of Rs. 5200 as operational cost for all the three years.

However, in discounted case the current discounted value for machine A is less than machine B at 5% rate of interest per annum.

Table 9.34

Year	Undiscounted operational cost for A	Undiscounted operational cost for 5
1	1900	2400
2	1600	1100
3	1700 .	1700
Total	5200	5200

Year	Discounted operational cost for A	Discounted operational cost for B
1	1900.00	2400.00
2	1523.81	1047.62
3	1541.95	1541.95
Total	4965.76	4989.57

Thus, in discounted terms machine A has less operational cost.

The following Table 9.35 gives the net replacement costs (purchase price – resale value) of two machines A and B. The machine A is to be replaced once in every three years and second machine is to be replaced once in every 6 years.

Table 9.35

Year	Original replacement cost of A	Discounted replacement cost of A	Original replacement cost of B	Discounted replacement cost of B
1	1000	1000.00	1700	1700.00
2	400	380.95	200	190.48
3	600	544.22	400	362.81
4			500	431.92
5			600	493.62
6			600	470.12
Total	2000	1925.17	4000	3648.94
Average	666.67	641.72	666.67	608.16

Now let us calculate the total discounted replacement cost for machine A for three years at 5% rate of interest as follows

$$=1000+\frac{400}{1+0.05}+\frac{600}{(1+0.05)^2}=1925.17$$

Therefore the average discounted cost for machine A is (1925.17)/3 = 641.72

Similarly, let us calculate the total discounted replacement cost for machine B for six years at 5% rate of interest as follows

$$=1700+\frac{200}{1+0.05}+\frac{400}{(1+0.05)^2}+\frac{500}{(1+0.05)^3}+\frac{600}{(1+0.05)^4}+\frac{600}{(1+0.05)^5}=3648.94$$

Therefore the average discounted cost for machine B is (3648.94)/3 = 608.16

Table 9.36

Year	Original replacement cost of A	Discounted replacement cost of A	Original replacement cost of B	Discounted replacement cost of B
1	1000	1000.00	1700	1700.00
2	400	380.95	200	190.48
3	600	544.22	400	362.81
4	1000	863.84	500	431.92
5	400	329.08	600	493.62
6	600	470.12	600	470.12
Total	4000	3588.20	4000	3648.94
Average	666.67	598.03	666.67	608.16

From Table 9.35 it is clear that the average original cost of replacement is Rs. 666.67 in both the cases. The discounted average replacement cost per year is less for machine B. However, this type of comparison is unfair because in one case we take three years and in another case we take 6 years while averaging. To overcome this difficulty let us consider one more cycle for machine A and make it also six years for a meaningful comparison. The Table 9.36 illustrates this modified method.

Clearly once again we get equal average costs per annum for original cost data without discounting. However, the discounted average cost gives a different picture now. Machine A has less average discounted cost per year than machine B.

9.9 GROUP REPLACEMENT OF ITEM THAT SUDDENLY FAILS COMPLETELY

In real life we come across many items like electric bulb works with full efficiency till it fails. Thus for such items the question of deterioration based replacement has no meaning. Till such complete failure occurs normally we do not go for replacement. Under such circumstances replacement poses practical problems and also loss of utility. If the bulbs are not available in the market for immediate replacement then the consumer will be at loss of the service. During office hours such instantaneous replacement would involve lot of inconvenience also. To overcome such difficulties, well planned precautionary pre-failure group replacement policy is often recommended irrespective of whether there is failure or not at the time of replacement. Such bulk replacement often costs less than the then and their replacement on the basis of complete failure.

The period between installation and possible complete failure in subsequent periods is obtained by using well known probability distributions. From a known probability function we derive the conditional probability of failure in a given small interval from t to t + Δt. This conditional probability may either decrease or increase or remain constant with time t. Here, in this section we consider only increasing probability situation only. The following illustration illustrates one such typical example.

■ Example 17

A company's establishment is fitted with 1000 bulbs of a certain standard. Over the entire lifetime of five years the cumulative rates of failure in percentage is given in Table 9.37. The company has two options for replacing the bulbs. In the first option each bulb costs Rs.10 per unit if it is replaced individually after its complete failure. The second option is to replace all the 1000 bulbs at a time periodically on preventive basis, irrespective of the condition of the individual bulbs. Such bulk replacement costs Rs. 2.50 per bulb. Suggest a suitable replacement policy for this company.

Table 9.37

Month	1	2	3	4	5
% of failure at the end of the month	10	25	50	80	100

Solution

Calculation of replacement cost per month under individual replacement policy

Let p_i stands for the probability of ith month's failure of new bulb installed in the beginning of the first month. This indeed is the conditional probability given by the difference between the proportion of "alive" bulbs in the beginning of the ith month and the proportion of "alive" bulbs at end of the ith month. In the following table we calculate the conditional probabilities for the entire lifetime of five months. Note that the bulb already served for four months surely fails in the fifth month so that the sum of the probabilities for all the five months equal to 1. In the last column we calculate the expected (weighted average) lifetime of the bulb.

Table 9.38

Months (X_i)	Probability	Calculation	Probability that a new bulb installed in the beginning of the first month o fails during i^{th} month (P_i)	X_iP_i
1	P1	10/100	0.10	0.10
2	P2	(25–10)/100	0.15	0.30
3	P3	(50–25)/100	0.25	0.75
4	P4	(80–50)/100	0.30	1.20
5	P5	(100–80)/100	0.20	1.00
Total			1.00	3.35

In a steady state the number of failure per day is obtained by dividing the total number of bulbs installed in the beginning. The expected lifetime is obtained from the Table 9.38. ie. 1000/3.35 = 299. Thus, the policy of replacing then and their on the day of failure is followed, and then at the rate of Rs. 10 per bulb it will cost Rs. 2990 to the company per month.

Calculation of replacement cost per month under group replacement policy

With the following two assumptions we calculate the number of failures and the associated replacements under group replacement.

1. The individual bulb that fails during a given month is replaced on then and their basis immediately.
2. The actual failure during a month for sub-population of bulbs with the same age is the same as the expected percentage of failures during the month for that sub-population.

Now let n_i denote the number of replacements at the end of ith month. If all the 1000 bulbs are newly installed in the beginning of the first month then on the basis of the assumptions listed above we arrive the following replacement month-wise series. In the very first month since the probability of failure of one-month-old bulb is p_1, the associated number of failure and hence replacement during the first month is simply $n_1 = n_0p_1$. Since the probability of failure of 2-month-old bulbs is p_2, the number of failures during the second month of those bulbs installed in the beginning of first month is given by n_0p_2. However, n_1 bulbs installed during the first month may also fail at the end of second month at the rate p_1. Such failure is given by n_1p_1. Thus the total number of failures at the end of the second month is given by $n_2 = n_0p_1+n_1p_1$. Similarly, for all the remaining months the number of failures of all ages of bulbs are calculated and reported in Table 9.39.

Table 9.39

$n_0 = n_0$	=	1000		
$n_1 = n_0\ p_1$	=	100	p_1	0.10
$n_2 = n_0\ p_2 + n_1\ p_1$	=	160	p_2	0.15
$n_3 = n_0\ p_3 + n_1\ p_2 + n_2\ p_1$	=	281	p_3	0.25
$n_4 = n_0\ p_4 + n_1\ p_3 + n_2\ p_2 + n_3\ p_1$	=	377	p_4	0.30
$n_5 = n_0\ p_5 + n_1\ p_4 + n_2\ p_3 + n_3\ p_2 + n_4\ p_1$	=	350	p_5	0.20
$n_6 = n_1\ p_5 + n_2\ p_4 + n_3\ p_3 + n_4\ p_2 + n_5\ p_1$	=	230		
$n_7 = n_2\ p_5 + n_3\ p_4 + n_4\ p_3 + n_5\ p_2 + n_6\ p_1$	=	286		

If all the bulbs (group) are replaced once in a month, then at the end of the first month the total replacement cost = 100 × 10 (individual replacement cost of bulbs failed in the first month) + (1000) × 2.5 = Rs. 2500 (bulk replacement cost on the last day of the first month) = 3500. So the average per month replacement cost is equal to Rs. 3500/1 = Rs. 3500.

If all the bulbs (group) are replaced once in two months, then at the end of the second month the total replacement cost = 100 × 10 (individual replacement cost of bulbs failed in the first month) + 160 × 10 (individual replacement cost of bulbs failed in the second month) + (1000) × 2.5 (bulk replacement cost on the last day of the second month) = 5100. So the average per month replacement cost is equal to Rs. 5100/2 = Rs. 2550.

If all the bulbs (group) are replaced once in three months, then at the end of the third month the total replacement cost = 100 × 10 (individual replacement cost of bulbs failed in the first month) + 160 × 10 (individual replacement cost of bulbs failed in the second month) + 281 × 10 (individual replacement cost of bulbs failed in the third month) + (1000) × 2.5 (bulk replacement cost on the last day of the third month) = 7910. So the average per month replacement cost is equal to Rs. 7910/3 = Rs. 2640.

If all the bulbs (group) are replaced once in four months, then at the end of the fourth month the total replacement cost = 100 × 10 (individual replacement cost of bulbs failed in the first month) + 160 × 10 (individual replacement cost of bulbs failed in the second month) + 281 × 10 (individual replacement cost of bulbs failed in the third month) + 377 × 10 (individual replacement cost of bulbs failed in the fourth month) + (1000) × 2.5 (bulk replacement cost on the last day of the fourth month) = 11,680. So the average per month replacement cost is equal to Rs. 11,680/4 = Rs. 2920.

If all the bulbs (group) are replaced once in five months, then at the end of the fifth month the total replacement cost = 100 × 10 (individual replacement cost of bulbs failed in the first month) + 160 × 10 (individual replacement cost of bulbs failed in the second month) + 281 × 10 (individual replacement cost of bulbs failed in the third month) + 377 × 10 (individual replacement cost of bulbs failed in the fourth month) + 350 × 10 (individual replacement cost of bulbs failed in the fifth month) + (1000) × 2.5 (bulk replacement cost on the last day of the fifth month) = 15,180. So the average per month replacement cost is equal to Rs. 15,180/5 = Rs. 3036.

The following table gives the entire summary calculations.

Table 9.40

End of month	No. of failures	Cumulative failures	Cumulative replacement cost after failure at Rs. 10	Total cost under group replacement at Rs. 2.50	Total cost	Average cost
1	100	100	1000	2500	3500	3500.00
2	160	260	2600	2500	5100	2550.00
3	281	541	5410	2500	7910	2636.67
4	377	918	9180	2500	11,680	2920.00
5	350	1268	12,680	2500	15,180	3036.00

Among all the five group replacement options, replacing once in two months cost least average (Rs. 2550). Further since this amount is less than the individual replacement cost per month obtained already (2550 < 2990) we recommend for group replacement of all the 1000 bulbs once in two months especially on the last day of the second month at the rate Rs. 2.50 per bulb coupled with individual replacement at the end of first and the second month on failure at Rs. 10 per bulb.

Example 18

An electrical distribution establishment is fitted with 1000 fuses of a certain standard. Over the entire lifetime of say five years the rates of failure in percentage over its life span is given hereunder. The company has two options of replacing the fuses. It costs Rs. 5 per fuse it replaces individually after its failure. On the other hand if it replaces all the 1000 fuses at a time it costs Rs. 2.50 per fuse. Suggest an optimum replacement policy.

Table 9.41

Year	1	2	3	4	5
% of failure at the end of the month	5	15	35	75	100

Solution

Let p_i stand for the probability of ith month's failure of new fuse installed in the beginning of the first month. This indeed is the conditional probability given by the difference between the proportion of "alive" fuses in the beginning of the ith month and the proportion of "alive" fuses at the end of the ith month. In the following table we calculate the conditional probabilities for all the five months. Note that the fuse already served four months surely fails in the fifth month so that the sum of the probabilities for all the five months is equal to 1. In the last column we calculate the expected (average) lifetime of the fuse.

Table 9.42

Months X_i	Probability	Calculation	Probability that a new bulb installed at time zero fails during ith month P_i	X_iP_i
1	P1	5/100	0.05	0.05
2	P2	(15 −5)/100	0.10	0.20
3	P3	(35 −15)/100	0.20	0.60
4	P4	(75 −35)/100	0.40	1.60
5	P5	(100 −75)/100	0.25	1.25
Total			1.00	3.70

In a steady state the number of failure per day is obtained as the ratio of the total number of fuses in the beginning to that of the expected failure as obtained in the above table. i.e.1000/3.7 = 270. Thus, the policy of replacing then and their, at the rate of Rs. 5 per fuse will cost Rs.1350.

Calculation of replacement cost per month under individual replacement policy

With the following two assumptions we calculate the number of failures and the associated replacements given the policy of group replacement.

1. The fuse that fails during a given month is replaced then and there with immediate effect.
2. That the actual failure during a month for sub-population of fuses with the same age is the same as the expected percentage of failures during the month for that sub-population.

Now let n_i denote the number of replacements at the end of ith month. If all the 1000 fuses are newly installed in the beginning of the first month then on the basis of the assumptions listed above we arrive the following replacement series month-wise. In the very first month since the probability of failure of one-month-old fuse is p_1, the associated number of replacements during the first month is simply $n_1 = n_0p_1$.

Since the probability of failure of 2-month-old fuse is p_2, the number of failures at the end of the second month of fuses installed in the beginning of first month is given by n_0p_2. However, n_1 fuses installed at during the first month may also fail at the end of second month at the rate p_1. Such failure is given by n_1p_1. Thus the total number of failures at the end of the second month is given by $n_2 = n_0p_1 + n_1p_1$. Similarly, for all the remaining months the number of failures for all ages of fuses are calculated and reported in the table below.

Table 9.43

$n_0 = n_0$	$= 1000$	
$n_1 = n_0\ p_1$	$= 50$	p_1 0.05
$n_2 = n_0\ p_2 + n_1\ p_1$	$= 103$	p_2 0.10
$n_3 = n_0\ p_3 + n_1\ p_2 + n_2\ p_1$	$= 210$	p_3 0.20
$n_4 = n_0\ p_4 + n_1\ p_3 + n_2\ p_2 + n_3\ p_1$	$= 431$	p_4 0.40
$n_5 = n_0\ p_5 + n_1\ p_4 + n_2\ p_3 + n_3\ p_2 + n_4\ p_1$	$= 333$	p_5 0.25
$n_6 = n_1\ p_5 + n_2\ p_4 + n_3\ p_3 + n_4\ p_2 + n_5\ p_1$	$= 155$	
$n_7 = n_2\ p_5 + n_3\ p_4 + n_4\ p_3 + n_5\ p_2 + n_6\ p_1$	$= 237$	

The calculation of average cost for all types of group replacement are worked out in the following table.

Table 9.44

End of year	No. of failures	Cumulative failures	Cumulative replacement cost after failure at Rs.5	Total cost due to group replacement at Rs.2.5	Total cost	Average cost
1	50	50	250	2500	2750	2750.00
2	103	153	765	2500	3265	1632.50
3	210	363	1815	2500	4315	1438.33
4	431	794	3970	2500	6470	1617.50
5	333	1127	5635	2500	8135	1627.00

Among all the five group replacements, replacing once in three months costs least (Rs.1438.33). Further since this amount is more than the individual replacement cost per month obtained already (1438.33 >1350) we recommend for the individual replacement at Rs. 5 per fuse.

■ Example 19

A large hotel in Bangalore city is having 500 rooms. Each room has 6 bulbs of certain standard. Over the entire lifetime of five years the cumulative rates of failure in percentage is given here under. The company has two options of replacing the bulbs. It costs Rs. 3 per bulb if it is replaced after its failure individually. On the other hand if it replaces all the 3000 bulbs at a time, it costs Rs. 1 per bulb. Which option is cost effective?

Table 9.45

Month	1	2	3	4	5
% of failure at the end of the month	10	25	50	80	100

Solution

Table 9.46

Months X_i	Probability that a new bulb installed at time zero fails during ith month P_i	X_i P_i
1	0.10	0.1
2	0.15	0.3
3	0.25	0.75
4	0.30	1.2
5	0.20	1
Total	1.00	3.35

In a study state the number of failure per day is obtained as the ratio of the total number of bulbs installed in the beginning to that of the expected failure as obtained in the above table 3000/3.35 = 896. Thus the present policy of replacing is followed then at the rate of Rs. 3 per bulb will cost 896 × 3 = Rs. 2688.

Now let n_i denote the number of replacements at the end of ith month. If all the 3000 bulbs are newly installed in the beginning then on the basis of the assumptions listed above we arrive the following results.

Table 9.47

$n_0 = n_0$	$= 3000$	
$n_1 = n_0\ p_1$	$= 300$	p_1 0.10
$n_2 = n_0\ p_2 + n_1\ p_1$	$= 480$	p_2 0.15
$n_3 = n_0\ p_3 + n_1\ p_2 + n_2\ p_1$	$= 843$	p_3 0.25
$n_4 = n_0\ p_4 + n_1\ p_3 + n_2\ p_2 + n_3\ p_1$	$= 1131$	p_4 0.30
$n_5 = n_0\ p_5 + n_1\ p_4 + n_2\ p_3 + n_3\ p_2 + n_4\ p_1$	$= 1050$	p_5 0.20
$n_6 = n_1\ p_5 + n_2\ p_4 + n_3\ p_3 + n_4\ p_2 + n_5\ p_1$	$= 689$	
$n_7 = n_2\ p_5 + n_3\ p_4 + n_4\ p_3 + n_5\ p_2 + n_6\ p_1$	$= 858$	

The calculation of average cost for all types of group replacement are worked out in the following table.

Table 9.48

End of month	Failures	Cost of failure at Rs. 3	Cumulative replacement cost	Total cost due to group replacement	Average cost
1	300	900	900	3000	3000
2	480	1440	2340	3900	1950
3	843	2529	4869	5340	1780
4	1131	3393	8262	7869	1967.25
5	1050	3150	11,412	11,262	2252.4

Among all the five group replacements, replacing once in three months cost least (Rs. 1780). Further since this amount is less than the individual replacement cost per month obtained already (1780 < 2688) we recommend the group replacement once in three months and individual replacement of bulbs failed during the first, second and third month.

9.10 EXCEL CALCULATION SHEET FOR GROUP REPLACEMENT PROBLEMS

Table 9.49

Month	1	2	3	4	5	Individual R. cost	5	Data
Failure at the end	5	15	35	75	100	Group R. Cost	2.5	1000

Month	Probability that a new bulb installed in 0 period fails during ith month	Expected Lifetime
1	0.05	0.05
2	0.10	0.20
3	0.20	0.60
4	0.40	1.60
5	0.25	1.25
Tot	1	3.70

Average individual replacement cost = 1351

n_0	=	n_0	=	1000		
n_1	=	$n_0\ p_1$	=	50	p_1	0.05
n_2	=	$n_0\ p_2 + n_1\ p_1$	=	102.5	p_2	0.10
n_3	=	$n_0\ p_3 + n_1\ p_2 + n_2\ p_1$	=	210.1	p_3	0.20
n_4	=	$n_0\ p_4 + n_1\ p_3 + n_2\ p_2 + n_3\ p_1$	=	430.8	p_4	0.40
n_5	=	$n_0\ p_5 + n_1\ p_4 + n_2\ p_3 + n_3\ p_2 + n_4\ p_1$	=	333.1	p_5	0.25
n_6	=	$n_1\ p_5 + n_2\ p_4 + n_3\ p_3 + n_4\ p_2 + n_5\ p_1$	=	155.3		
n_7	=	$n_2\ p_5 + n_3\ p_4 + n_4\ p_3 + n_5\ p_2 + n_6\ p_1$	=	236.9		

End of the month	No. of failures	Cumulative failure	Cumulative replacement cost on failure at Rs. 5	Group replacement cost at Rs. 2.5	Total cost	Average cost
1	50	50	250.00	2500	2750	2750.00
2	103	153	762.50	2500	3263	1631.25
3	210	363	1813.13	2500	4313	1437.71
4	431	793	3966.91	2500	6467	1616.73
5	333	1126	5632.16	2500	8132	1626.43

EXERCISES

1. What is replacement model? Describe some important replacement situations and replacement policies.

2. In a replacement problem the cost of the machine is Rs. 6100 and its scrap value is only Rs. 100. The maintenance costs are as follows.

Year	1	2	3	4	5	6	7	8
Costs	100	250	400	600	900	1200	1600	2000

When should the machine be replaced? (Delhi University, M.Com 1983)

3. An engineering company is offered a material handling equipment A. A is priced at Rs. 60,000 including the cost of installation and the cost of operation and maintenance are estimated to be Rs. 10,000 for each of the first five years, increasing every year by Rs. 3000 in the sixth and subsequent years. The company expects a return of 10% on all its investments. What is the optimum replacement period?

4. A firm considering the replacement of a machine, whose cost is Rs. 12,200 and the scrap value is Rs. 200. The maintenance costs are found from experience to be as follows:

Year	1	2	3	4	5	6	7	8
Costs	200	500	800	1200	1800	2500	3200	4000

When should the machine be replaced? (Delhi University, M.Com 1984)

5. A machine costs Rs. 60,000 and the running and maintenance costs, as well as salvage value over years are shown in the following table:

Age in Years	1	2	3	4	5	6	7	8
Running and maintenance cost in Rs.	6000	7000	9000	11,000	15,000	17,000	21,000	25,000
Salvage cost in Rs.	40,000	35,000	30,000	25,000	20,000	15,000	15,000	15,000

If the value is 10% per year and the salvage value is obtained at the end of the year; whereas the running and maintenance costs are incurred in the beginning of year, find the optimum age of replacement of this machine.

(Delhi University, M.B.A Dec 1980)

6. A manager finds from the past records that the cost per year of running an item of plant, whose purchase price is Rs. 30,000 is as given below:

Years	1	2	3	4	5	6	7	8
Running costs (Rs.)	5000	6000	7000	9000	11,000	14,000	17,000	20,000
Resale value (Rs.)	15,000	7500	3750	1850	1000	1000	1000	1000

How often should the plant be replaced?

(AIMA, Diploma in management, Jan. 1979)

7. The following failure rates have been observed for a certain type of light bulb.

Failure of week	1	2	3	4	5	6	7	8
Prob. of failure	0.05	0.13	0.25	0.43	0.68	0.88	0.96	1

The cost of replacing an individual failed bulb is Rs. 0.25. The decision is made to replace all bulbs simultaneously at fixed intervals and also to replace individual bulbs as they fail in service. If the cost of group replacement is 30 paise per bulb, what is the best interval between group replacements?

(IIM Calcutta 1982)

8. Perfect Machine Tools Ltd. has 40 lathes in service. The cost of repairing a breakdown is Rs. 1,00,000 per lathe. A preventive maintenance programme costing Rs. 300 per lathe per machine can prevent the breakdowns. Historical records indicate that the possibility of breakdown is as follows:

Months until breaks down	1	2	3	4	5	6	7	8
Problem of breakdown	0.1	0.05	0.1	0.15	0.2	0.25	0.1	0.05

a) What is the expected breakdown cost/month?

b) If preventive maintenance is followed, how often should the lath be serviced?

(Delhi University, M.B.A April 1982)

9. Machine A costs Rs. 45,000 and the operating costs are estimated at Rs. 1000 for the first year, increasing by Rs. 10,000 per year in the second and subsequent years. Machine B costs Rs. 50,000 and the operating costs are Rs. 2000 for the first year, increasing by Rs. 4000 per year in the second and subsequent years. If now you have a machine of A type, should you replace it with B? Assume that both the machines have no resale value, and the future costs are not discounted.

(Delhi University, M.Com, Dec. 1985)

10. A computer contains 20,000 resistors. When any resistor fails, it is replaced. The cost of replacing individually is Re. 1.00. If all the resistors are replaced at the same time, the cost per resistor is reduced to Re. 0.40. The present surviving at the end of the month t and the probability of failure during the month t are given below:

t	0	1	2	3	4	5	6
Percent surviving at the end of t	100	96	90	65	35	20	0
Probability of failure during the month t	-	0	0.1	0.3	0	0.2	0

What is the optimum replacement plan?

(Osmania University, M.B.A 1983)

11. An electronic equipment has 5000 IC's. If anyone IC fails it has to be replaced. Cost of replacing a single IC is Rs. 10. If all the IC's are replaced simultaneously at a time it would cost Rs. 3.5/IC. The data on surviving ones are given by

Month	0	1	2	3	4	5	6
Percentage of surviving at the end of each month	100%	97%	90%	70%	30%	15%	0%

Obtain the best replacement policy i.e replace as and when IC fails or replace all IC's at fixed intervals and replace the individual IC's which fail during the fixed interval.

12. The management of a large hotel is considering the periodic replacement of light bulbs fitted in rooms. There are 500 rooms in the hotel and each room has 6 bulbs. The management is now following the policy of replacing the bulbs as they fail at a total cost of Re. 1 by adopting periodic replacement policy. On the basis on information given below evaluate the alternatives and make a recommendation to the management:

Month of use	% of bulbs failing by that month
1	10
2	25
3	50
4	80
5	100

(Bangalore University, M.B.A, July/August 2005)

13. Mineral processing company has received offers for two types of dumper A and B. A has a payload of 25 tons and is priced at Rs. 4,00,000 while B also with a

payload of 25 tons is priced at Rs. 3,60,000. The operating costs over the estimated life of 5 years for both the types are as follows:

Year	1	2	3	4	5
Type(A) Rs.	8000	9000	10,000	11,000	12,000
Type(B) Rs.	14,000	16,000	18,000	20,000	22,000

Which dumper is to be preferred?

(Bangalore University, M.B.A July/August 2005)

14. A decorative series lamp set circuit contains 10,000 bulbs when any bulb fails, it is replaced. The cost of replacing a bulb individually is Re. 1 only. If all the bulbs are replaced at the same time, the cost per bulb would be reduced to 35 paise. The present surviving, say S(t), at the end of month t and P(t), the probabilities of failure during the month, are given below.

t	0	1	2	3	4	5	6
S(t)	100	97	90	70	30	15	0
P(t)	-	0.03	0.07	0.2	0.4	0.15	0.15

What is the optimal replacement plan?

(ICWA, December 1983)

15. Total number of bulbs:1000

Week	1	2	3	4	5
Percentage failing each week end	10	30	50	80	100

For each bulb replacement it will cost Rs. 2. If all bulbs have to be replaced it would cost Re. 0.5/bulb would group replacement help and at what interval?

(Bangalore University, M.B.A October 2004)

16. A pipeline is due for repairs. It will cost Rs. 10,000 and lasts for 3 years. Alternatively a new pipeline can be laid down at a cost of Rs. 30,000 and lasts for 10 years. Assuming cost of capital to be 10% and ignoring salvage value. What alternative should be chosen?

(C.A, May 1981)

CHAPTER

10

Waiting Line Models

- *Introduction*
- *Characteristics of the queuing system*
- *D. G. Kendall notations and symbols*
- *Steady state operating characteristics of queuing theory*
- *M/M/1 system*
- *M/M/1: ∞ /FCFO system*
- *M/M/k: ∞/FCFO model*
- *M/M/2: ∞/FCFO model*
- *M/M/3: ∞/FCFO model*
- *(M/G/1): (∞/FCFO) model*
- *(M/D/1): (∞/FCFO)*
- *(M/G/k): (∞/Blocked customers cleared)*
- *M/M/1: N/FIFO system*
- *Economic analysis of waiting line models*
- *Poisson arrival and Erlang service models*
- *Limitations queuing theory*
- *Exercises*

10.1 INTRODUCTION

Queues have become inevitable event in modern civilized life. We quite often stand in the long queue at the bus stand to book our bus tickets. In cinema house we stand in queue to buy cinema tickets. Sometimes automobiles wait in queue for service in service station, airplanes wait in queue for landing, and ships wait in the port for entry. Queue emerges wherever service rates are slower than the arrival rates. For example, if it takes ten minutes for serving a single customer and one customer arrives in every nine minutes then queue will form automatically and grow infinitely. If arrival rates are equal to service rates then there will be no queue formation. If service rates are larger than the arrival rates then again there will be no queue formation, instead the server will be idle for quite some time. For example, if the service rate is 10 minutes per customer and arrival rate is 11 minutes per customer then the service personal will be idle for one minute between every arrival and departure.

A waiting line problem is characterized by flow of customers arriving at random to get one or more service facilities made available. Sometimes the customers may be provided the service immediately on arrival or may be asked to wait until the facility is made available.

Sometimes, such queues are physically visible such as queue at the railway station for booking the ticket. On the other occasions like booking for cars, waiting for delivery is not visible physically. However, in both the cases the size of the queue is a stochastic variable linked to the waiting time. In the case of railway reservation more the number of booking counters more will be promptness of the service that too at lesser waiting time.

The queue and the associated waiting time are not only for human beings. In a manufacturing unit where several equipment are in use, there are occasional breakdowns. The broken down machine may immediately be attended or will have to wait before some other early arrivals are attended. In all these cases managers do not necessarily want the queue formation rather. In fact waiting queues are inevitable characteristics of modern life.

Given the situation, a customer has the option of either seeking the service after waiting in the queue or leaving the queue and getting the service elsewhere. When a customer leaves the queue and opts for service elsewhere then the opportunity of making profit by providing the service is being lost for the concerned organization. The option left to the organization is to balance this opportunity cost of losing the customer with that of the additional cost of providing an extra service facility. If the opportunity cost losing the customer is more than the extra service providing cost then the organization concerned will go for extra service facility.

10.2 CHARACTERISTICS OF THE QUEUING SYSTEM

1. **Input process (or arrival patterns):** Arrival is the very first component of the queuing system. The arrival pattern is a very important input to a satisfactory final solution. Customers do not come on regular intervals. Often arrivals occur at random. To handle such random arrivals we use well known statistical probability distribution depending upon the situation. In such a description the number of arrivals within a given time interval is a discrete random variable. Alternatively, sometimes we express the time between two successive arrivals by treating time as the random variable. This time the variable 'time' is a continuous variable. Analysis of the past data normally identifies the suitable type of probability distribution for arrivals. Many of the distributions found in practice can approximately fit into any one of the following well-known mathematical distributions. Uniform distribution, Poisson distribution, exponential distribution and Erlang distribution are some of the special types used in waiting line theories. A remarkable result in this context is that if a Poisson distribution describes the number of arrivals in a given time interval as a discrete random variable then the corresponding time between two successive arrivals is described by a continuous exponential random variable. This property is frequently used to derive certain important results in waiting line theories.

2. **Service:** Service time can again be expressed as a statistical variable. As in the case of arrival time, service can also be stated as number of persons served in a given interval of time or amount of time (or service time) for serving one person. The relevant probability distribution is to be identified again by using the past data. Once again it could be uniform, Poisson, Exponential or Erlang as the case may be.

3. **Service facility:** The essential service component can be either single channel type like a ticket counter at a bus station or several channels type like arrival platforms in a railway station. Sometimes the service is carried out in a sequence through several phases commonly known as multiple phase service. The handling of files in a government office is a good example for these types of multiple phase service. Similarly, most files must move through several service personnel following the official hierarchy till they arrive at the decision taking table. As the number of servers increase the number of customers in the waiting line decreases. The system with infinite number of servers, practically there will be no waiting line at all. Self-service fast food restaurants are good examples for infinite server pattern.

4. **Queue discipline:** By queue discipline we mean the order in which the service is offered to the customer. The commonly used method in the queue disposal is FIFO (First In First Out). Occasionally, some customers receive priority in service over others though they are late, on premium payment. Sometimes the last customer is served first and the first customer is served last. Such discipline is often referred by LIFO (Last In First Out). In lift service for example, the first customer goes interior and the last customer stands near the gate. So after receiving the service the last customer leaves the lift first and the first customer leaves lift last. On several occasions irrespective of order of arrival the server calls the customer randomly and provides the service. Such discipline is called SIRO (Service In Random Order). Since there is no queue formation, the only option left to the server is to call the customers at random and offer the service. Although several analytical methods are available for queue discipline, *First In First Out* discipline is commonly used in practice.

5. **Size of the population:** The population, which forms the queue, may be finite or infinite in size. If the arrival rate is not appreciably affected, then the source that generates the rate is said to be infinite. In a large city like Bangalore the potential ticket buyers at the railway booking counter is considered infinite for all practical purposes. This assumption is very convenient for analytical development of the waiting line theory. On the other hand, if the rate is influenced greatly by the source, then the referred source is finite. From practical purposes the population is considered as finite provided the rate of arrival varies considerably.

6. **Maximum size of the queue:** In general there is no maximum as such in queue discipline. However, sometimes only a finite number of customers alone are accommodative in the queue irrespective of the type of population. The service that is being provided by the city doctor on 'appointment basis' is a good example in this context. In this case although the population is infinite the maximum permissible number for service in the queue is finite.

7. **Channels:** When the number of server is one, then it is called single channel queuing system. Sometimes we will have single queue with multiple channels as well. In some cases multiple queues and multiple channels are also feasible.

8. **Traffic intensity:** If λ is rate of arrival per unit of time interval and μ is the rate of service per unit of time then the ratio $\rho = \lambda/\mu$ is often called traffic intensity. If this ratio is greater than one then the queue will grow without any limit overtime. If it is equal to one then clearly the original queue length will remain unaltered if it exists in the beginning. If it is less than one then the queue will diminish in its length gradually. In this section we consider only the special case of diminishing queue length situations for discussion. The following charts illustrate some of the queue disciplines used in practice.

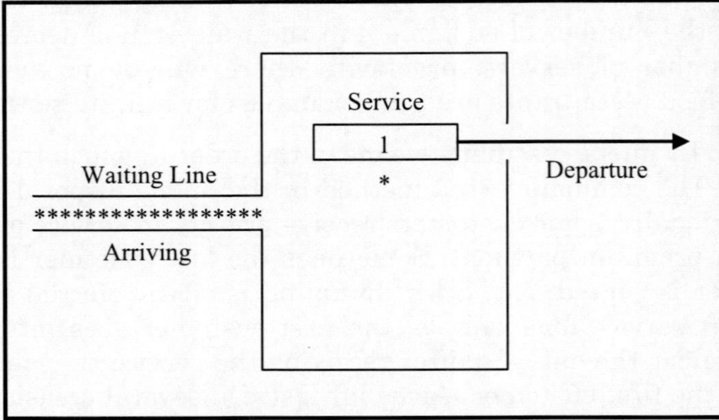

Figure 10.1 Single queue single service channel

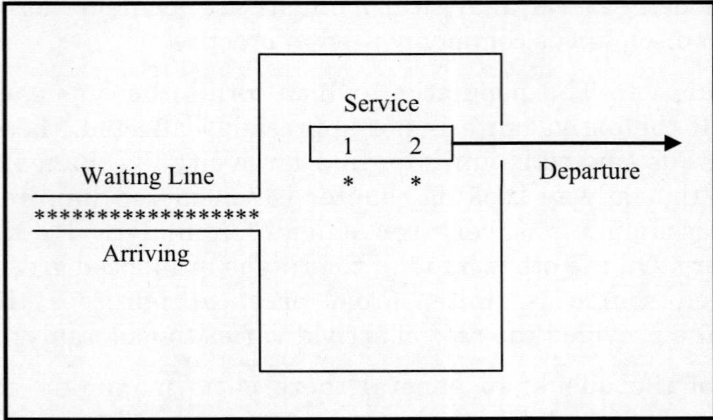

Figure 10.2 Single queue double service one after another

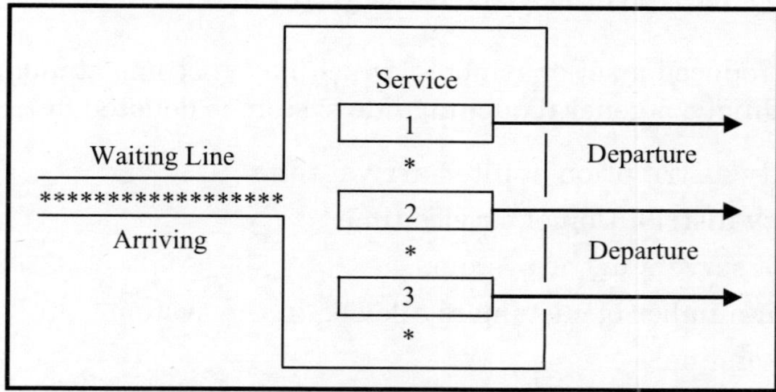

Figure 10.3 Single queue three channel service

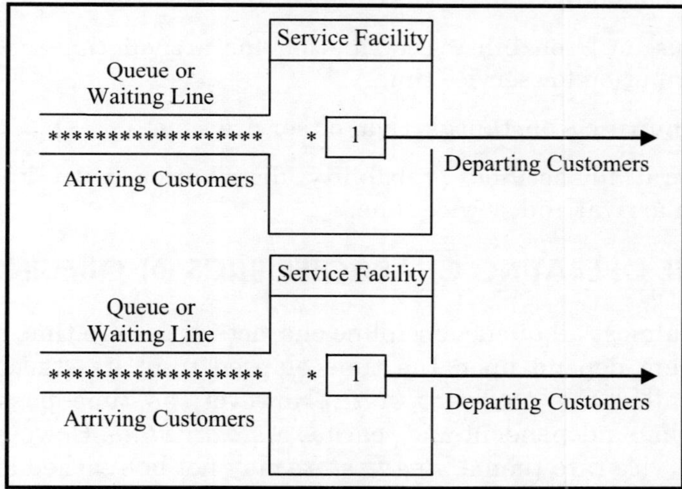

Figure 10.4 Double queue double channel independent service

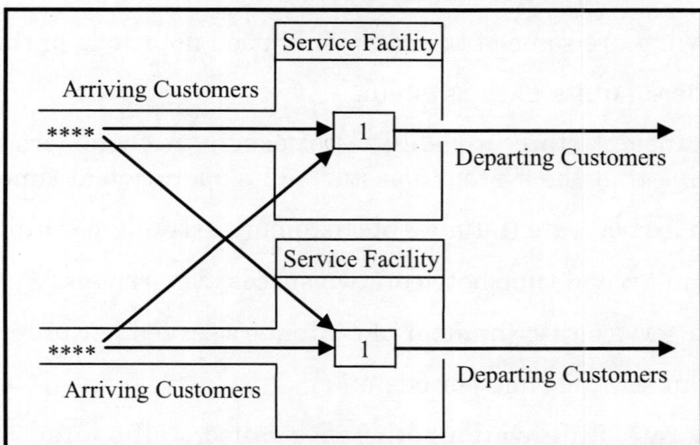

Figure 10.5 Double queue double channel dependent service

10.3 D.G. KENDALL NOTATIONS AND SYMBOLS

D.G. Kendall has introduced a set of symbols, which have become standard in the waiting line problems. Accordingly, a general queuing line system is denoted by $(a/b/c):(d/e)$.

where a = Probability distribution of inter-arrival time

b = Probability distribution of service time

c = Number of servers in the system

d = Maximum number of customers allowed in the system

e = Queue discipline.

Depending upon the letter appearing, varieties of waiting line models are being built. The letters that are commonly used in this context are given below.

- M denotes a Poisson Probability Distribution for arrival time or an Exponential Probability Distribution for service time.
- D denotes deterministic (constant) arrival or service time.
- G denotes general unspecified probability distribution with known mean and variance for both arrival and service time.

10.4 STEADY STATE OPERATING CHARACTERISTICS OF QUEUING THEORY

In the initial stages almost all queue discipline characteristics are time dependent in the sense that all of them depend upon the time elapsed from its start. Such an initial situation is often called the ***transient state***. However, as time passes the system's behaviour becomes time independent and reaches a ***steady state.*** However, if arrival rate is larger than the service rate then a ***steady state*** may not be reached at all irrespective of the time elapsed. In this section we propose to consider only ***steady state*** models with less arrival rates than the service rates for discussion.

Notations: The following are some of the commonly used notations in this section.

1. n = The number of units in the system.
2. $P_n(t)$ = The transient state probabilities of exactly n customers in the system at time t assuming that the system has started its operation at time zero.
3. λ = The mean arrival rate (number of customers arriving per unit of time).
4. $1/\lambda$ = The mean arrival time between two successive arrivals.
5. μ = The mean service rate (number of customers served per unit of time).
6. $1/\mu$ = The mean serving time per customer.
7. $\rho = \lambda/\mu$ = The rate of utilization, otherwise called traffic intensity, in the case of single channel model.

8. $\rho = \lambda / k\mu$ = The rate of utilization, otherwise called the traffic intensity, in the case of k channel models.
9. k = The number of parallel servers called channels.
10. P_o = The probability that no units are in the system.
11. L_q = The average number of units in the queue.
12. L_S = The average number of units in the system as a whole.
13. W_q = The average time that a given unit spends in the queue.
14. W_S = The average time that the given unit spends in the whole system.
15. P_W = The probability that an arriving unit has to wait in the queue.
16. P_n = The probability of having exactly n customers in the system.

Waiting line models consist of formulas and relationships describing the operational characteristics of the system under ***steady state***. Some of the commonly used characteristics are:

1. P_o = The probability that no units are in the system.
2. L_q = The average number of units in the queue.
3. L_S = The average number of units in the system.
4. W_q = The average time that a unit spends in the queue.
5. W_S = The average time that a given unit spends in the system.
6. P_W = The probability that an arriving unit has to wait in the queue.
7. P_n = The probability of having exactly n customers in the system.

With this information the manager concerned can bring in improvements in the system as a whole by affecting the necessary changes in parameters.

Structure of the single channel waiting line system

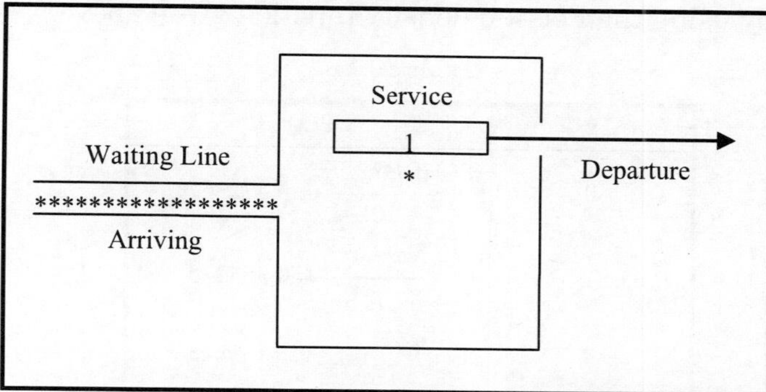

Figure 10.6 Single queue single service channel

1. The following chart exhibits the salient futures of the single channel waiting line system.
2. The customers assemble in a queue and get serviced from a single server and leave the system.

10.5 M/M/1 SYSTEM: QUEUING SYSTEM AS A STOCHASTIC PROCESS

It may be observed that the corresponding to any time t, there is a probability $P_n(t)$ indicating the probability that there are n customers in the system. The time t can vary from zero to infinite. Such a random phenomenon is known as stochastic process. Most elementary queuing systems, inputs (arriving units) and outputs (departing units) follow a special type of stochastic path called birth-death process. The term birth refers to the arrival of new units into the system and death refers to the departure of the served units.

Probability distribution of number of arrivals

For many waiting line systems the number of arrivals in a given interval occurs randomly and independently of other arrivals, and hence we cannot predict when the next arrival will occur. Thus, the number of arrival to the queue in a given interval is a stochastic variable. For such cases, the Poisson probability distribution provides a good approximation to the real life situation. It gives the probability of x arrivals in a given time interval. The mathematical form of the relevant Poisson distribution is given by

$$P(x) = \frac{\lambda^x e^{-\lambda}}{x!} \quad \text{for } x = 1, 2, 3, \ldots$$

Where x = The number of arrivals in a given time interval

λ = The mean number of arrivals in a given time interval

e = 2.71828.

Suppose from the past record the average number of arrivals in an hour = 45 then for a one minute period the number of average arrivals λ = 45/60 = 0.75 customers. Given this information, we can readily calculate the probability of any number of arrivals in the given one minute interval by using the Poisson density function given above.

Table 10.1

x	P(x)
0	$\frac{0.75^0 e^{-0.75}}{0!} = 0.472367$
1	$\frac{0.75^1 e^{-0.75}}{1!} = 0.354275$
2	$\frac{0.75^2 e^{-0.75}}{2!} = 0.132853$

$$P(x) = \frac{\lambda^x e^{-\lambda}}{x!} = \frac{0.75^x e^{-0.75}}{x!}$$

Thus, the probability of no customer in one minute interval is 0.472367, and the probability of one customer in the one minute interval is 0.354275 and so on.

Probability distribution of service time: Service time is the time that a customer spends in the service section from the start to the end. It is again a stochastic statistical variable similar to the number of arrivals discussed in the previous section. But this time, it is negative exponential distribution that suits the situation. Given the number of units served in the given minute interval we can always calculate the probability of being served by less than any given interval say 't' minutes. The probability of being served by less than t minutes is obtained by the negative exponential formula given below.

$$P\ (\text{service time} \leq t)\ 1 - e^{-\mu t}$$

Where μ = the mean number of units served in the given interval of one minute and $e = 2.71828$.

Suppose that the past data reveals that there are 60 customers served per hour. So the number of customers served per minute interval = 60/60 = 1. After having obtained this information, by using the above exponential function we can always calculate the probabilities of being served by less than any given time interval, may be half a minute, one minute etc. The following chart illustrates some such calculations.

$$P(\text{Service time} \leq 0.5 \text{ minutes}) = 1 - e^{-\mu t} = 1 - (2.71828)^{-1(0.5)} = 0.3935$$

$$P(\text{Service time} \leq 1 \text{ minutes}) = 1 - e^{-\mu t} = 1 - (2.71828)^{-1(1)} = 0.6321$$

$$P(\text{Service time} \leq 2 \text{ minutes}) = 1 - e^{-\mu t} = 1 - (2.71828)^{-1(2)} = 0.8647$$

10.6 M/M/1: ∞ /FCFO SYSTEM: SINGLE CHANNEL WAITING LINE MODEL WITH POISSON ARRIVAL AND EXPONENTIAL SERVICE TIME

Before exploring the salient features of the system let us restate our assumption as follows.

1. The arrival pattern of the customer follows Poisson distribution.
2. The service pattern follows exponential distribution.
3. A single queue is formed by the customer.
4. The service is based on *first come first serve* basis.
5. The queue length is infinite and the input population is also infinite.
6. The arrival rate is smaller than the service rate.
7. There is one single service facility that serves customer.

The mathematical methodology to derive formulas for the operating characteristics of this model is rather complex. Hence we propose to give only the final formulas and its usage in describing the steady state operational characteristics. Readers interested in the derivation of the results are advised to consult the specialized texts in this area. If λ refers to the mean number of arrivals and μ refers to the mean number of services offered per unit of time then the steady state operational characteristics are given by the following formulas.

1. The probability of no one in the system (service facility is idle): $P_0 = 1 - \frac{\lambda}{\mu}$
2. The average number of units waiting in the queue: $L_q = \frac{\lambda^2}{\mu(\mu - \lambda)}$
3. The average number of units in the system: $L_s = \frac{\lambda}{(\mu - \lambda)}$
4. The average time that an arriving unit spends in the queue: $W_q = \frac{\lambda}{\mu(\mu - \lambda)}$
5. The average time that an arriving unit spends in the system: $W_s = \frac{1}{\mu - \lambda}$
6. The probability that the arriving unit spends in the queue: $P_w = \frac{\lambda}{\mu}$
7. The probability of n units in the system: $P_n = \left(\frac{\lambda}{\mu}\right)^n \left(1 - \frac{\lambda}{\mu}\right)$
8. The probability of being in the system longer than t units of time: $e^{-(\lambda - \mu)t}$

To facilitate easy calculations one after another we can use the following alternative formulas as well.

1. The probability of no one in the system (service facility is idle): $P_0 = 1 - \frac{\lambda}{\mu}$
2. The average number of units waiting in the queue: $L_q = \frac{\lambda^2}{\mu(\mu - \lambda)}$
3. The average number of units in the system: $L_s = L_q + \frac{\lambda}{\mu}$
4. The average time that a unit spends in the queue: $W_q = \frac{L_q}{\lambda}$
5. The average time that a unit spends in the system: $W_s = W_q + \frac{1}{\mu}$
6. The probability that the arriving unit spends in the queue: $P_w = \frac{\lambda}{\mu}$

7. The probability of n units in the system: $P_n = \left(\frac{\lambda}{\mu}\right)^n P_0$

8. The probability of being in the system longer than t units of time: $e^{-(\lambda-\mu)t}$

Example 1

There is a toll arrangement in a newly opened flyover in Bangalore. There is an assistant at the toll gate to collect the fee. To the toll gate the mean arrival rate is $\lambda = 0.75$ vehicle per minute. The mean service rate of billing and collecting money is $\mu = 1$ vehicle per minute. If an arrival is described by Poisson and service time is described by exponential calculate all the seven steady state operational characteristics.

Solution

1. The probability that no one is in the system (the service is idle):

$$P_0 = 1 - \frac{\lambda}{\mu} = 1 - \frac{0.75}{1} = 0.25$$

2. The average number of units waiting in the queue:

$$L_q = \frac{\lambda^2}{\mu(\mu-\lambda)} = \frac{0.75^2}{1(1-0.75)} = 2.25 \text{ customers}$$

3. The average number of units in the system:

$$L_s = L_q + \frac{\lambda}{\mu} = 2.25 + \frac{0.75}{1} = 3 \text{ customers}$$

4. The average waiting time of an arriving unit in the queue:

$$W_q = \frac{L_q}{\lambda} = \frac{3.00}{0.75} = 3 \text{ minutes}$$

5. The average time that an arriving unit spends in the system:

$$W_s = W_q + \frac{1}{\mu} = 3 + \frac{1}{1} = 4 \text{ minutes}$$

6. The probability that the arriving unit spends in the queue:

$$P_w = \frac{\lambda}{\mu} = \frac{0.75}{1} = 0.75$$

7. The probability of a unit being in the system more than 3 units of time:

$$P(T > t) = e^{-(\mu-\lambda)t} = 2.817^{-(1-0.75)3} = 0.472367$$

8. The probability of n units in the system:

$$P_n = \left(\frac{\lambda}{\mu}\right)^n \left(1-\frac{\lambda}{\mu}\right) = \left(\frac{0.75}{1}\right)^n \left(1-\frac{0.75}{1}\right)$$

The last equation could be used to find out the probability of any number of customers in the system as follows.

$$P_0 = \left(\frac{\lambda}{\mu}\right)^0 P_0 = \left(\frac{0.75}{1}\right)^0 \times 0.25 = 0.250$$

$$P_1 = \left(\frac{\lambda}{\mu}\right)^1 P_0 = \left(\frac{0.75}{1}\right)^1 \times 0.25 = 0.1875$$

$$P_2 = \left(\frac{\lambda}{\mu}\right)^2 P_0 = \left(\frac{0.75}{1}\right)^2 \times 0.25 = 0.1406$$

Probability of 3 and above $= 1 - 0.250 - 0.1875 - 0.1406 = 0.4219$

Probability of not exceeding 2 $= 0.250 + 0.1875 + 0.1406 = 0.5781$

Excel Template

Table 10.2

	A	B	C	D	E
1		**M/M/1: ∞ /FCFO SYSTEM**			
2		**Assumptions**			
3		Poisson arrivals			
4		Exponential service time			
5		**Data**			
6		Number of channels	k	=	1
7		Mean arrival rate	λ	=	0.75
8		Mean service rate	μ	=	1
9		**Operating characteristics**			
10	1	Probability of no customers in the queue	P_o	=	=1-E7/E8
11	2	Average number of customers in the waiting line	L_q	=	=E7^2/(E8*(E8-E7))
12	3	Average number of customers in the system	L_s	=	=E11+E7/E8
13	4	Average time that a customer spends in the waiting line	W_q	=	=E11/E7
14	5	Average time that a customer spends in the system	W_s	=	=E13+1/E8
15	6	Probability of an arriving customer has to wait in queue	P_w	=	=E7/E8
16	7	Probability of unit in the system for more than 3 min.	P(t>3)	=	=EXP(-(E8-E7)*3)
17	8	Number of customers in the queue			Probability
18		0		=	=($E$7/$E$8)^B18*$E$10
19		1		=	=($E$7/$E$8)^B19*$E$10
20		2		=	=($E$7/$E$8)^B20*$E$10
21		3 and above		=	=1-SUM(E17:E20)
22		2 and below		=	=SUM(E17:E20)

To get all the characteristics we can develop ready to use excel worksheet template as shown in Table 10.2 by typing the formulas shown in the respective cells. The resulting answer display is shown in Table 10.3.

Excel Solution

Table 10.3

	A	B	C	D	E
1		**M/M/1: ∞ /FCFO SYSTEM**			
2		**Assumptions**			
3		Poisson arrivals			
4		Exponential service time			
5		**Data**			
6		Number of channels	k	=	1
7		Mean arrival rate	λ	=	0.75
8		Mean service rate	μ	=	1
9		**Operating characteristics**			
10	1	Probability of no customers in the queue	P_o	=	0.2500
11	2	Average number of customers in the waiting line	L_q	=	2.2500
12	3	Average number of customers in the system	L_s	=	3.0000
13	4	Average time that a customer spends in the waiting line	W_q	=	3.0000
14	5	Average time that a customer spends in the system	W_s	=	4.0000
15	6	Probability of an arriving customer has to wait in queue	P_w	=	0.7500
16	7	Probability of being in system for more than 3 minutes	P(t.>3)	=	0.4724
17	8	Number of customers in the queue			Probability
18		0		=	0.2500
19		1		=	0.1875
20		2		=	0.1406
21		3 and above		=	0.4219
22		2 and below		=	0.5781

Improving the waiting line operational characteristics

From Table 10.3 one could notice that the average waiting time is 3 minutes per customer. This is too long a period for a busy customer. In addition the average number of customers waiting in the line is 2.25. This means that about 75% of customers will have to wait for service. Further, the last observation in the table tells us that the probability of 3 or more customers in the system is 0.4219. All these observations tell us that there are still possibilities for improving the operational efficiency.

Thus, waiting line characteristics often indicate the desirable improvements. However, the decision of how to modify the waiting line configurations to improve the operational efficiency must be based on insight creativity. The following are some of the methods proposed to improve the characteristics.

- Increase the mean service rate by introducing new technology.

- Add a service personnel to the existing single channel system and improve the service rate.
- Add one or more service channels so that the service rate is improved.

Improvement by increasing mean service rate μ by merely adding service personnel

Example 2

Let the manager in example 1 appoint an additional person to assist the existing single lonely assistant. Such an arrangement definitely will increase the number of units served in a given time interval. Let the new service rate, after appointing the second service personnel into our single channel system, is 1.25 customers per minute. So let us redo the whole exercise with $\lambda = 0.75$ and $\mu = 1.25$.

Solution

Data

Mean arrival rate $\lambda = 0.75$; Mean service rate $\mu = 1.25$.

1. The probability of no one is in the system:

$$P_0 = 1 - \frac{\lambda}{\mu} = 1 - \frac{0.75}{1.25} = 1 - 0.6 = 0.4$$

2. The average number of units waiting in the queue:

$$L_q = \frac{\lambda^2}{\mu(\mu - \lambda)} = \frac{0.75^2}{1.25(1.25 - 0.75)} = \frac{0.75^2}{1.25 \times 0.5} = 0.9$$

3. The average number of units in the system:

$$L_s = L_q + \frac{\lambda}{\mu} = 0.9 + \frac{0.75}{1.25} = 1.5 \text{ customers per minute}$$

4. The average waiting time of an arriving unit in the queue:

$$W_q = \frac{L_q}{\lambda} = \frac{0.9}{0.75} = 1.2 \text{ customer per minutes}$$

5. The average time that an arriving unit spends in the system:

$$W_s = W_q + \frac{1}{\mu} = 1.2 + \frac{1}{1.25} = 2 \text{ minutes}$$

6. The probability that the arriving unit spends in the queue:

$$P_w = \frac{\lambda}{\mu} = 0.60$$

7. The probability of a unit being in the system more than 3 units of time:

$$P(T > t) = e^{-(\mu-\lambda)t} = 2.817^{-(1.25-0.75)3} = 0.2231$$

8. The probability of n units in the system

$$P_n = \left(\frac{0.75}{1.25}\right)^n P_0$$

$$P_0 = (0.6)^0 \times 0.4 = 0.4$$

$$P_1 = (0.6)^1 \times 0.4 = 0.24$$

$$P_2 = (0.6)^2 \times 0.4 = 0.144$$

Improved Excel solutions

Table 10.4

	A	B	C	D	E
1		**M/M/1: ∞ /FCFO SYSTEM**			
2		**Assumptions**			
3		Poisson arrivals			
4		Exponential service time			
5		**Data**			
6		Number of channels	k	=	1
7		Mean arrival rate	λ	=	0.75
8		Mean service rate	μ	=	1.25
9		**Operating characteristics**			
10	1	Probability of no customers in the queue	P_o	=	0.4000
11	2	Average number of customers in the waiting line	L_q	=	0.9000
12	3	Average number of customers in the System	L_s	=	1.5000
13	4	Average time that a customer spends in the waiting line	W_q	=	1.2000
14	5	Average time that a customer spends in the system	W_s	=	2.0000
15	6	Probability of an arriving customer has to wait in queue	P_w	=	0.6000
16	7	Probability of being in system more than 3 minutes	P(t.>3)	=	0.2231
17	8	Number of customers in the queue			Probability
18		0		=	0.4000
19		1		=	0.2400
20		2		=	0.1440
21		3 and above		=	0.2160
22		2 and below		=	0.7840

The above calculations reveal a remarkable improvement in both—in waiting time (1.2 min) and the average time spent in the system (2 min).

■ Example 3

If $\lambda = 5$ per hour and $\mu = 8$ in an M/M/1: ∞/FCFO discipline, then find out

(a) The probability the service is idle.

(b) The probability that at least 2 customers are in the queue.

(c) Expected time that the customer has to wait in the queue.

Solution

Data

Mean arrival rate $\lambda = 5$ per hour; Mean service rate $\mu = 8$ per hour

(a) The probability that no one in the system is simply the probability of the service idle:

$$P_0 = 1 - \frac{\lambda}{\mu} = 1 - \frac{5}{8} = \frac{3}{8} = 0.375$$

(b) The probability that at least 2 units are in the system:

$$P_n = \left(\frac{\lambda}{\mu}\right)^n P_0$$

$$P_0 = \left(\frac{5}{8}\right)^0 \times 0.375 = 0.375$$

$$P_1 = \left(\frac{5}{8}\right)^1 \times 0.375 = 0.234375$$

$$P_{n \geq 2} = 1 - 0.375 - 0.234375 = 0.390625$$

(c) The average number of units in the system:

$$W_q = \frac{\lambda}{\mu(\mu - \lambda)} = \frac{5}{8(8-5)} = \frac{5}{24} = 0.2083 \text{ customers.}$$

■ Example 4

In a drive-in ATM counter, only 5 cars can be parked inside the compound. In the case of more than 5 cars they will have to wait outside before getting into the inside parking queue. The arrivals follow Poisson with an average of 2 cars in every five minutes. The said counter can serve the customers at the rate of 1.5 minutes per customer and obeyed by exponential probability distribution. Find out:

(a) The probability the service is idle.

(b) Expected time that the customer has to spend in the system.

(c) The probability that a customer waits for inside queue but outside the compound.

Solution

Data

Mean arrival rate $\lambda = \frac{2}{5} \times 60 = 24$ per hour ; Mean service rate $\mu = \frac{60}{1.5} = 40$ per hour

(a) The probability of service facility remains idle means no one is in the queue:

$$P_0 = 1 - \frac{\lambda}{\mu} = 1 - \frac{24}{40} = 0.4$$

(b) The expected time that the customer spends in the system:

$$W_s = \frac{1}{\mu - \lambda} = \frac{1}{40 - 24} = \frac{1}{16} \text{ or } 0.0375 \text{ minutes}$$

(c) The probability of an incoming unit stays outside: Since the queue can accommodate only 5 cars, car number 6 and above will have to stay outside.

$$P(n \geq 6) = \left(\frac{\lambda}{\mu}\right)^{5+1} = \left(\frac{24}{40}\right)^{6} = 0.0467$$

■ Example 5

A child specialist in a government hospital takes an average 4 minutes per child for treatment. The patients arrive at an average rate of 10 per hour. Under M/M/1: ∞ /FIFO discipline, answer the following questions:

(a) What is the average length of the waiting line?

(b) What probability that the doctor is idle?

(c) What is the average time that a child has to wait in the queue?

(d) What is the probability that the child will have to wait more than 12 minutes?

Solution

Data

$\lambda = 10$ patients per hour; $\mu = 60/4 = 15$ patients per hour.

(a) Average number child waiting in the queue:

$$L_q = \frac{\lambda^2}{\mu(\mu - \lambda)} = \frac{10^2}{15(15 - 10)} = \frac{4}{3} = 1.3333$$

(b) Probability that the doctor is idle (no one is in the system):

$$P_0 = 1 - \frac{\lambda}{\mu} = 1 - \frac{10}{15} = \frac{1}{3} = 0.3333$$

(c) Average length of time that a child will have to wait in the queue:

$$W_q = \frac{\lambda}{\mu(\mu - \lambda)} = \frac{10}{15(15-10)} = \frac{10}{75} = 0.1333 \text{ hour}$$

(d) The probability that a child has to wait at least 12 minutes before treatment:

$$P(\text{waiting time}) \geq 1/5 \text{ hours} = \frac{\lambda}{\mu} e^{(-\lambda-\mu)1/5} = \frac{10}{15} 2.718^{-5} = 0.004492$$

■ Example 6

In a city nursing home the bench in the waiting room can accommodate only three patients. The lonely doctor takes an average of 5 minutes per patient for treatment. The arrival is at the rate of 10 patients per hour. Under M/M/1: ∞/FCFO discipline, answer the following questions:

(a) What is the probability that an arriving patient will have to wait outside the hall for want of accommodation?

(b) What is the probability that a patient directly enters and sits in the bench provided?

Solution

Data

From the data $\lambda = 10$ patients per hour; Mean service rate $\mu = 60/5 = 12$ patients per hour.

(a) Probability of arriving patients will have to wait outside.

$$P(n \geq 4) = \left(\frac{\lambda}{\mu}\right)^{3+1} = \left(\frac{10}{12}\right)^4 = 0.4823$$

(b) The probability that the arriving patient gets the seat accommodation

$$P(n \leq 3) = 1 - P(4 \geq 4) = 1 - 0.4823 = 0.5177$$

10.7 M/M/k: ∞/FCFO MODEL: k CHANNEL WAITING LINE MODELS WITH POISSON ARRIVAL AND EXPONENTIAL SERVICE TIME MODEL

In multiple channel models, two or more identical service capabilities are put into use. The arriving unit in this modified situation waits in the single queue as usual until one of the serving units is made available for servicing. Once the service unit is made available, irrespective of the service unit, the customer moves to the service unit and gets serviced. Because μ is the mean service rate per channel, the combined service rate for the multiple service system must be equal to kμ. As in the single channel the formulas for operating characteristics are valued only the combined service rate kμ is greater than the arrival rate of the single queue. The traffic intensity in the multiple channel system thus will be $\rho = \frac{\lambda}{k\mu}$.

Operational characteristics of a "k" channel model

1. The probability that no units in the system (the server is idle)

$$P_0 = \left[\sum_{n=0}^{k-1} \frac{(\lambda/\mu)^n}{n!} + \frac{(\lambda/\mu)^k}{k!}\left(\frac{k\mu}{k\mu-\lambda}\right)\right]^{-1}$$

$$= \left[\frac{(\lambda/\mu)^0}{0!} + \frac{(\lambda/\mu)^1}{1!} + \frac{(\lambda/\mu)^2}{2!} + \cdots\cdots + \frac{(\lambda/\mu)^k}{k!}\left(\frac{k\mu}{k\mu-\lambda}\right)\right]$$

2. The average number of units in the waiting line (length of the queue)

$$L_q = \frac{(\lambda/\mu)^k \lambda\mu}{(k-1)!(k\mu-\lambda)^2} P_0$$

3. The average number of units in the system

$$L_s = L_q + \frac{\lambda}{\mu}$$

4. The average time that a unit spends in the waiting line

$$W_q = \frac{L_q}{\lambda}$$

5. The average time that a unit spends in the system

$$W_s = W_q + \frac{1}{\mu}$$

6. The probability that the arriving unit has to wait in the queue

$$P_w = \frac{1}{k!}\left(\frac{\lambda}{\mu}\right)^k \left(\frac{k\mu}{k\mu-\lambda}\right) P_0$$

7. The probability of n units in the system

$$P_n = \frac{(\lambda/\mu)^n}{n!} P_0 \quad \text{for } n \le k$$

$$P_n = \frac{(\lambda/\mu)^n}{k!\,k^{n-k}} P_0 \quad \text{for } n > k$$

10.8 M/M/2: ∞/FCFO MODEL: TWO CHANNEL WAITING LINE MODELS WITH POISSON ARRIVAL AND EXPONENTIAL SERVICE TIME

■ Example 7

There is a toll arrangement in a newly constructed two-lane national highway bypass. There are two toll gates to collect the toll bill. The mean arrival rate is $\lambda = 0.75$ vehicle per minute. The mean service rate per channel is $\mu = 1$ vehicle per minute. If arrivals are described by Poisson and service time described by exponential calculate all the seven steady state operational characteristics.

Solution

1. The probability that no one in the system:

$$P_0 = \left[\sum_{n=0}^{2-1} \frac{(\lambda/\mu)^n}{n!} + \frac{(\lambda/\mu)^2}{2!}\left(\frac{k\mu}{k\mu-\lambda}\right)\right]^{-1}$$

$$= \left[\frac{(\lambda/\mu)^0}{0!} + \frac{(\lambda/\mu)^1}{1!} + \frac{(\lambda/\mu)^2}{2!}\left(\frac{2\mu}{2\mu-\lambda}\right)\right]^{-1}$$

$$= \left[1 + 0.75 + \frac{0.75^2}{2\times 1}\left(\frac{2\times 1}{2\times 1 - 0.75}\right)\right]^{-1} = 0.4545$$

2. The average number of units in the queue:

$$L_q = \frac{(\lambda/\mu)^2 \lambda\mu}{(2-1)!\,(k\mu-\lambda)^2} P_0$$

$$= \frac{(0.75/1)^2 \times 0.75 \times 1}{1! \times 1\,(2\times 1 - 0.75)^2} 0.4545 = 0.1227$$

3. The average number of units in the system:

$$L_s = L_q + \lambda/\mu$$

$$= 0.1227 + 0.75 = 0.8727$$

4. The average time a unit spends in the queue:

$$W_q = \frac{L_q}{\lambda}$$

$$= \frac{0.1227}{0.75} = 0.1636$$

5. The average time a unit spends in the system:

$$W_s = W_q + \frac{1}{\mu}$$

$$= 0.1636 + \frac{1}{1} = 1.1636$$

6. The probability that the arriving unit has to wait in the queue

$$P_w = \frac{1}{2!}\left(\frac{\lambda}{\mu}\right)^k \left(\frac{2\mu}{2\mu - \lambda}\right) P_0$$

$$= \frac{1}{2 \times 1}\left(\frac{0.75}{1}\right)^2 \left(\frac{2 \times 1}{2 \times 1 - 0.75}\right) 0.4545$$

$$= 0.2045$$

7. The probability of n units in the system

$$P_n = \frac{(\lambda / \mu)^n}{n!} P_0 \quad \text{for } n \le 2$$

$$P_0 = \frac{0.75^0}{0!} \times 0.4545 = 0.4545$$

$$P_1 = \frac{0.75^1}{1!} \times 0.4545 = 0.3409$$

$$P_2 = \frac{0.75^2}{2!} \times 0.4545 = 0.1278$$

$$P_n = \frac{(\lambda / \mu)^n}{2!\,2^{n-2}} P_0 \quad \text{for } n > 2$$

$$P_3 = \frac{0.75^3}{2 \times 1 \times 2^{3-2}} \times 0.4545 = 0.479$$

$$P_4 = \frac{0.753}{2 \times 1 \times 2^{4-2}} \times 0.4545 = 0.0180$$

Excel Template

Table 10.5

	A	B	C	D	E
1		**M/M/2: ∞ /FCFO SYSTEM**			
2		**Assumptions**			
3		Poisson arrivals time			
4		Exponential service time			
5		**Data**			
6		Number of channels	k	=	2
7		Mean arrival rate	λ	=	0.75
8		Mean service rate	μ	=	1
9		**Operating characteristics.**			
10	1		P_o	=	=(((E7/E8)^0)+((E7/E8)^1/FACT(1))+(((E7/E8)^2/FACT(2))*(E6*E8)/(E6*E8-E7)))^-1
11	2		L_q	=	=((E7/E8)^2*E7*E8)/(FACT(1)*(E6*E8-E7)^2)*E10
12	3		L_s	=	=E11+E7/E8
13	4		W_q	=	=E11/E7
14	5		W_s	=	=E13+1/E8
15	6		P_w	=	=(1/FACT(2))*((E7/E8)^2)*(E6*E8)/(E6*E8-E7)*E10
16	7	No. of units in queue			Probability
17		0			=(($E$7/$E$8)^B17)/FACT(B17)*$E$10
18		1			=(($E$7/$E$8)^B18)/FACT(B18)*$E$10
19		2			=(($E$7/$E$8)^B19)/FACT(B19)*$E$10
20		3			=(($E$7/$E$8)^B20)/(FACT(2)*2^(B20-2))*$E$10
21		4			=(($E$7/$E$8)^B21)/(FACT(2)*2^(B21-2))*$E$10
22		5 or more			=1-SUM(E17:E21)

To get all the characteristics we can develop ready to use excel worksheet as shown in Table 10.5 by typing the formulas shown in the respective cells. The resulting answer display is shown in Table 10.6.

Excel Solution

Table 10.6

	A	B	C	D	E
1		**M/M/2: ∞ /FCFO SYSTEM**			
2		**Assumptions**			
3		Poisson arrivals			
4		Exponential service time			
5		**Data**			
6		Number of channels	k	=	2
7		Mean arrival rate	λ	=	0.75
8		Mean service rate	μ	=	1
9		**Operating characteristics**			
10	1	Probability of no customers in the queue	P_o	=	0.4545
11	2	Average number of customers in the waiting line	L_q	=	0.1227
12	3	Average number of customers in the system	L_s	=	0.8727
13	4	Average time that a customer spends in the waiting line	W_q	=	0.1636
14	5	Average time that a customer spends in the system	W_s	=	1.1636
15	6	Probability of an arriving customer has to wait in queue	P_w	=	0.2045
16	7	Number of customers in the queue			Probability
17		0			0.4545
18		1			0.3409
19		2			0.1278
20		3			0.0479
21		4			0.0180
22		5 or more			0.0108

Clearly the two-channel delivery has improved the operational characteristics to a greater extent than the single channel model discussed already in example 1. By employing an order taker at each channel would further increase the mean service rate (say to 1.5), which in turn will improve all the operational characteristics.

■ Example 8

A company has two manufacturing units and two tool cribs one for each shop. Both the cribs are useable in both the units and handles identical tools. The service time was found to be described by exponential distribution with mean time of 2.5 minutes per workman. Arrivals of workmen are as usual described by Poisson with the mean 18 per hour. The production manager wants to reduce the waiting time in the queue by placing the both cribs at a common place useable by both the units. Justify the statement whether he is correct or not.

Solution

Two queue two channel models

As per the data made available $\lambda = 18$ per hour and $\mu = \frac{60}{2.5} = 24$ per hour

When the cribs are unit specific then the waiting time in the queue is obtained as

$$W_q = \frac{\lambda}{\mu(1-\lambda)} = \frac{18}{24(24-18)} = \frac{1}{8} = 0.125 \text{ hours}$$

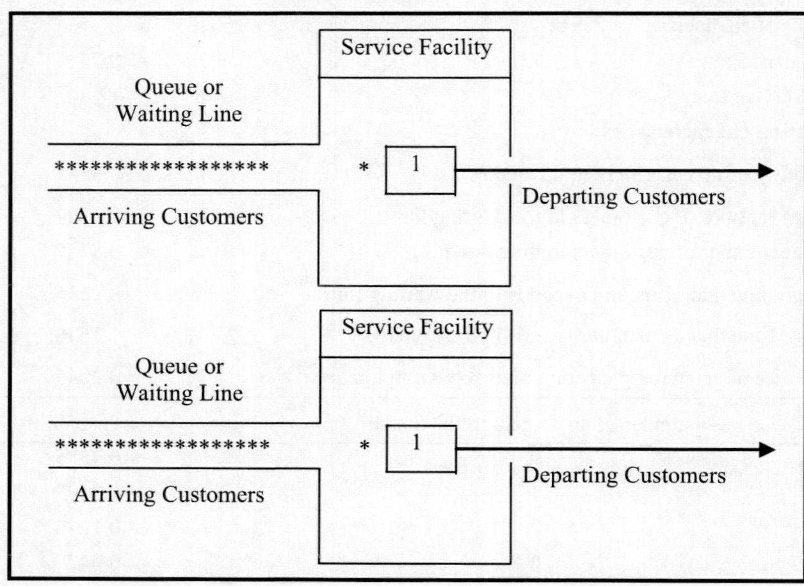

Figure 10.7

$$W_q = \frac{\lambda}{\mu(1-\lambda)} = \frac{18}{24(24-18)} = \frac{1}{8} = 0.125 \text{ hours}$$

Excel Solution

Table 10.7

	A	B	C	D	E
1		**M/M/1: ∞ /FCFO SYSTEM**			
2		**Assumptions**			
3		Poisson arrivals			
4		Exponential service time			
5		**Data**			
6		Number of channels	k	=	1
7		Mean arrival rate	λ	=	18
8		Mean service rate	μ	=	24
9		**Operating characteristics**			
10	1	Probability of no customers in the queue	P_o	=	0.2500
11	2	Average number of customers in the waiting line	L_q	=	2.2500
12	3	Average number of customers in the system	L_s	=	3.0000
13	4	Average time that a customer spends in the waiting line	W_q	=	0.1250
14	5	Average time that a customer spends in the system	W_s	=	0.1667
15	6	Probability of an arriving customer has to wait in queue	P_w	=	0.7500
16	7	Number of customers in the queue			Probability
17		0		=	0.2500
18		1			0.1875
20		2			0.1406
21		3		=	0.1055
22		4 and above		=	0.3164
23					

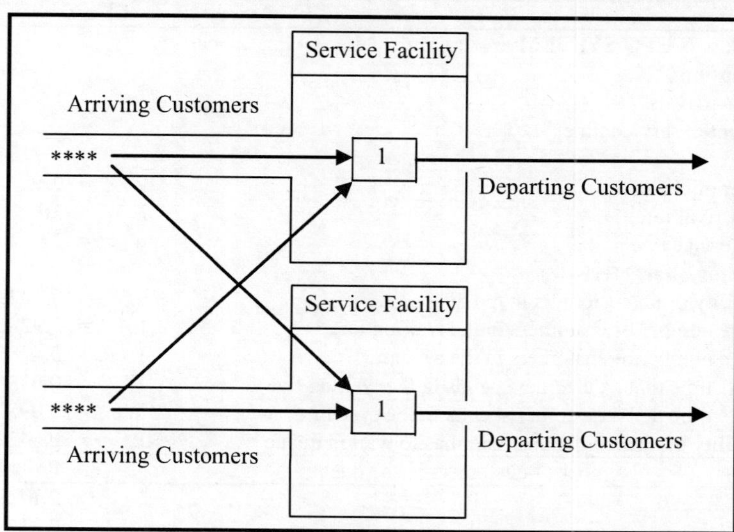

Figure 10.8

On the other hand if the cribs are assessable from both the units as shown in Figure 10.8 then the needed characteristics are obtained as:

$$\lambda = 18 + 18 = 36 \text{ per hour and } \mu = \frac{60}{2.5} = 24 \text{ hour}$$

$$P_0 = \left[\sum_{n=0}^{2-1} \frac{(\lambda/\mu)^n}{n!} + \frac{(\lambda/\mu)^2}{2!}\left(\frac{k\mu}{k\mu - \lambda}\right)\right]^{-1}$$

$$= \left[\frac{(\lambda/\mu)^0}{0!} + \frac{(\lambda/\mu)^1}{1!} + \frac{(\lambda/\mu)^2}{2!}\left(\frac{2\mu}{2\mu - \lambda}\right)\right]^{-1}$$

$$= \left[\frac{(36/24)^0}{0!} + \frac{(36/24)^1}{1!} + \frac{(36/24)^2}{2!}\left(\frac{2 \times 24}{2 \times 24 - 36}\right)\right]^{-1} = 0.1429$$

$$L_q = \frac{(\lambda/\mu)^2 \lambda\mu}{(2-1)!\,(k\mu - \lambda)^2} P_0 = \frac{(36/24)^2 \times 36 \times 24}{1! \times (2 \times 24 - 36)^2} 0.1429 = 1.9286$$

$$W_q = \frac{L_q}{\lambda} = \frac{1.9286}{36} = 0.0536$$

Since the new proposal reduces the waiting time substantially, it is time to advice him to go ahead with his proposal. The Table 10.8 shows these results after using the revised data in the template.

Excel Solution

Table 10.8

	A	B	C	D	E
1		**M/M/2: ∞ /FCFO SYSTEM**			
2		**Assumptions**			
3		Poisson arrivals			
4		Exponential service time			
5		**Data**			
6		Number of channels	k	=	2
7		Mean arrival rate	λ	=	36
8		Mean service rate	μ	=	24
9		**Operating characteristics**			
10	1	Probability of no customers in the queue	P_o	=	0.1429
11	2	Average number of customers in the waiting line	L_q	=	1.9286
12	3	Average number of customers in the system	L_s	=	3.4286
13	4	Average time that a customer spends in the waiting line	W_q	=	0.0536
14	5	Average time that a customer spends in the system	W_s	=	0.0952
15	6	Probability of an arriving customer has to wait in queue	P_w	=	0.6429
16	7	Number of customers in the queue			Probability
17		0			0.1429
18		1			0.2143
19		2			0.1607
20		3			0.1205
21		4			0.0904
22		5 or more			0.2712

10.9 M/M/3: ∞/FCFO MODEL: THREE CHANNEL WAITING LINE MODELS WITH POISSON ARRIVAL AND EXPONENTIAL SERVICE TIME MODEL

In multiple channel models, two or more identical service capabilities are put into action. The arriving unit in this modified situation waits in the single queue as usual until one of the serving units is made available for servicing. Once the service unit is made available, irrespective of the service unit, the customer moves to the service unit and gets serviced. Because μ is the mean service rate per channel the combined service rate for the multiple service system must be equal to 3μ. As in the single channel the formulas for operating characteristics are valued only the combined service rate kμ is greater than the arrival rate of the single queue.

Operational characteristics of a "3" channel model

1. The probability that no units in the system (the server is idle)

$$P_0 = \left[\sum_{n=0}^{2-1} \frac{(\lambda/\mu)^n}{n!} + \frac{(\lambda/\mu)^2}{2!}\left(\frac{k\mu}{k\mu-\lambda}\right)\right]^{-1}$$

$$= \left[\frac{(\lambda/\mu)^0}{0!} + \frac{(\lambda/\mu)^1}{1!} + \frac{(\lambda/\mu)^2}{2!} + \frac{(\lambda/\mu)^3}{3!}\left(\frac{3\mu}{3\mu-\lambda}\right)\right]^{-1}$$

2. The average number of units in the waiting line (length of the queue)

$$L_q = \frac{(\lambda/\mu)^3 \lambda\mu}{(3-1)!(3\mu-\lambda)^2} P_0$$

3. The average number of units in the system

$$L_s = L_q + \lambda/\mu$$

4. The average time that a unit spends in the waiting line

$$W_q = \frac{L_q}{\lambda}$$

5. The average time that a unit spends in the system

$$W_s = W_q + \frac{1}{\mu}$$

6. The probability that the arriving unit has to wait in the queue

$$P_w = \frac{1}{3!}\left(\frac{\lambda}{\mu}\right)^3\left(\frac{3\mu}{3\mu-\lambda}\right)P_0$$

7. The probability of n units in the system

$$P_n = \frac{(\lambda/\mu)^n}{n!} P_0 \quad \text{for } n \le 3$$

$$P_n = \frac{(\lambda/\mu)^n}{3!\ 3^{n-3}} P_0 \quad \text{for } n > 3$$

■ Example 9

In a Bangalore city government hospital, outpatients are treated by three equally efficient doctors. On an average it takes 20 minutes to treat a patient. The average time vary exponentially. The arrival rate is 6 per patients per hour and follows Poisson distribution. The medical superintendents are advised to investigate the following issues.

1. The average number of patients waiting in the queue.
2. The average time that the patient spends in the hospital.
3. The probability that a patient has to wait before the service starts.
4. The average percentage of idle time for each of the doctors.

Solution

1. The probability that no units in the system (the server is idle)

$$P_0 = \left[\sum_{n=0}^{2-1} \frac{(\lambda/\mu)^n}{n!} + \frac{(\lambda/\mu)^2}{2!}\left(\frac{k\mu}{k\mu-\lambda}\right)\right]^{-1}$$

$$= \left[\frac{(\lambda/\mu)^0}{0!} + \frac{(\lambda/\mu)^1}{1!} + \frac{(\lambda/\mu)^2}{2!} + \frac{(\lambda/\mu)^3}{3!}\left(\frac{3\mu}{3\mu-\lambda}\right)\right]^{-1}$$

$$= \left[\frac{(6/3)^0}{0!} + \frac{(6/3)^1}{1!} + \frac{(6/3)^2}{2!} + \frac{(6/3)^3}{3!}\left(\frac{3\times 3}{3\times 3-6}\right)\right]^{-1}$$

$$= 0.11111$$

2. The average number of patients waiting in the queue

$$L_q = \frac{(\lambda/\mu)^3\ \lambda\mu}{(3-1)!\ (3\mu-\lambda)^2} P_0$$

$$= \frac{(6/3)^3\ 6\times 3}{(3-1)!(3\times 3-6)^2} \times 0.1111 = 0.8889$$

3. The average number of units in the system

$$L_s = L_q + \lambda / \mu$$

$$= 0.8889 + \frac{6}{3} = 2.8889$$

4. The average time that the patient spends in the queue

$$W_q = \frac{L_q}{\lambda} = \frac{2.8889}{6} = 0.1481$$

5. The average time the patient spends in the hospital

$$W_s = W_q + \frac{1}{\mu} = 0.1481 + \frac{1}{3} = 0.4815$$

6. The probability that a patient has to wait in the queue before the service starts

$$P_w = \frac{1}{3!}\left(\frac{\lambda}{\mu}\right)^3 \left(\frac{3\mu}{3\mu - \lambda}\right) P_0$$

$$P_w = \frac{1}{3!}\left(\frac{\lambda}{\mu}\right)^3 \left(\frac{3\mu}{3\mu - \lambda}\right) P_0 = \frac{1}{3!}\left(\frac{6}{3}\right)^3 \left(\frac{3\times 3}{3\times 3 - 6}\right) \times 0.11111 = 0.6667$$

Excel Template

Table 10.9

	A	B	C	D	E
1		**M/M/3: ∞/FCFO SYSTEM**			
2		**Assumptions**			
3		Poisson arrivals			
4		Exponential service time			
5		**Data**			
6			k	=	3
7			λ	=	6
8			μ	=	3
9		**Operating characteristics**			
10	1		P_o	=	=(((E7/E8)^0/FACT(0))+((E7/E8)^1/FACT(1))+((E7/E8)^2/FACT(2))+(((E7/E8)^3/FACT(3))*(3*E8)/(3*E8-E7)))^-1
11	2		L_q	=	=((E7/E8)^3*E7*E8*E10)/(FACT(2)*(3*E8-E7)^2)
12	3		L_s	=	=E11+E7/E8
13	4		W_q	=	=E11/E7
14	5		W_s	=	=E13+1/E8
15	6		P_W	=	=(1/FACT(2))*((E7/E8)^2)*(E6*E8)/(E6*E8-E7)*E10
16	7	Units in the queue			Probability
17		0			=(($E$7/$E$8)^B17)/FACT(B17)*$E$10
18		1			=(($E$7/$E$8)^B18)/FACT(B18)*$E$10
19		2			=(($E$7/$E$8)^B19)/FACT(B19)*$E$10
20		3			=(($E$7/$E$8)^B20)/(FACT(2)*2^(B20-2))*$E$10
21					

7. The average percentage of idle time for each of the doctors.

$$P_n = \frac{(\lambda/\mu)^n}{n!} P_0 \quad \text{for } n \leq 3$$

$$P_0 = \frac{(6/3)^0}{0!} \times 0.11111 = 0.1111 \text{ (since no one in the queue all the doctors are free)}$$

$$P_1 = \frac{(6/3)^1}{1!} \times 0.1111 = 0.2222 \text{ (since one is in the queue two doctors are free)}$$

$$P_2 = \frac{(6/3)^2}{2!} \times 0.1111 = 0.2222 \text{ (since two are in the queue one doctor is free)}$$

Excel Solution

Table 10.10

	A	B	C	D	E
1		**M/M/3: ∞ /FCFO SYSTEM**			
2		**Assumptions**			
3		Poisson arrivals			
4		Exponential service time			
5		**Data**			
6		Number of channels	k	=	3
7		Mean arrival rate	λ	=	6
8		Mean service rate	μ	=	3
9		**Operating characteristics**			
10	1	Probability of no customers in the queue	P_o	=	0.1111
11	2	Average number of customers in the waiting line	L_q	=	0.8889
12	3	Average number of customers in the system	L_s	=	2.8889
13	4	Average time that a customer spends in the waiting line	W_q	=	0.1481
14	5	Average time that a customer spends in the system	W_s	=	0.4815
15	6	Probability of an arriving customer has to wait in queue	P_w	=	0.6667
16	7	Number of customers in the queue			Probability
17		0			0.1111
18		1			0.2222
19		2			0.2222
20		3			0.2222
21					

The calculation of P_0, the probability of no one in the queue for multiple channels is somewhat difficult indeed. Under such circumstances one can use the following readymade Table 10.11 for such calculations.

Table 10.11

λ / μ	Number of channels			
	2	3	4	5
0.05	0.9512	0.9512	0.9512	0.9512
0.10	0.9048	0.9048	0.9048	0.9048
0.15	0.8605	0.8607	0.8607	0.8607
0.20	0.8182	0.8187	0.8187	0.8187
0.25	0.7778	0.7788	0.7788	0.7788
0.30	0.7391	0.7407	0.7408	0.7408
0.35	0.7021	0.7046	0.7047	0.7047
0.40	0.6667	0.6701	0.6703	0.6703
0.45	0.6327	0.6373	0.6376	0.6376
0.50	0.6000	0.6061	0.6065	0.6065
0.55	0.5686	0.5763	0.5769	0.5769
0.60	0.5385	0.5479	0.5487	0.5488
0.65	0.5094	0.5209	0.5219	0.5220
0.70	0.4815	0.4952	0.4965	0.4966
0.75	0.4545	0.4706	0.4722	0.4724
0.80	0.4286	0.4472	0.4491	0.4493
0.85	0.4035	0.4248	0.4271	0.4274
0.90	0.3793	0.4035	0.4062	0.4065
0.95	0.3559	0.3831	0.3863	0.3867
1.00	0.3333	0.3636	0.3673	0.3678
1.05	0.3115	0.3451	0.3493	0.3499
1.10	0.2903	0.3273	0.3321	0.3328
1.15	0.2698	0.3103	0.3158	0.3165
1.20	0.2500	0.2941	0.3002	0.3011
1.25	0.2308	0.2786	0.2853	0.2863
1.30	0.2121	0.2638	0.2712	0.2723
1.35	0.1940	0.2496	0.2577	0.2590
1.40	0.1765	0.2360	0.2449	0.2463
1.45	0.1594	0.2230	0.2327	0.2343
1.50	0.1429	0.2105	0.2210	0.2228
1.55	0.1268	0.1986	0.2099	0.2118
1.60	0.1111	0.1872	0.1993	0.2014
1.65	0.0959	0.1762	0.1892	0.1915
1.70	0.0811	0.1657	0.1796	0.1821
1.75	0.0667	0.1556	0.1704	0.1731
1.80	0.0526	0.1460	0.1616	0.1646
1.85	0.0390	0.1367	0.1533	0.1565
1.90	0.0256	0.1278	0.1453	0.1487
1.95	0.0127	0.1193	0.1377	0.1413
2.00		0.1111	0.1304	0.1343
2.05		0.1032	0.1235	0.1276
2.10		0.0957	0.1169	0.1213
2.15		0.0884	0.1106	0.1152
2.20		0.0815	0.1046	0.1094
2.25		0.0748	0.0988	0.1039
2.30		0.0683	0.0933	0.0987
2.35		0.0621	0.0881	0.0937
2.40		0.0562	0.0831	0.0889

(Contd.)

λ / μ	Number of channels			
	2	3	4	5
2.45		0.0505	0.0783	0.0844
2.50		0.0449	0.0737	0.0801
2.55		0.0396	0.0693	0.0760
2.60		0.0345	0.0651	0.0721
2.65		0.0296	0.0612	0.0683
2.70		0.0249	0.0573	0.0648
2.75		0.0204	0.0537	0.0614
2.80		0.0160	0.0502	0.0581
2.85		0.0118	0.0469	0.0551
2.90		0.0077	0.0437	0.0521
2.95		0.0038	0.0406	0.0493
3.00			0.0377	0.0466
3.05			0.0350	0.0441
3.10			0.0323	0.0417
3.15			0.0297	0.0394
3.20			0.0273	0.0372
3.25			0.0250	0.0350
3.30			0.0227	0.0330
3.35			0.0206	0.0311
3.40			0.0186	0.0293
3.45			0.0166	0.0276
3.50			0.0148	0.0259
3.55			0.0130	0.0243
3.60			0.0113	0.0228
3.65			0.0096	0.0214
3.70			0.0081	0.0200
3.75			0.0066	0.0187
3.80			0.0051	0.0174
3.85			0.0038	0.0162
3.90			0.0025	0.0151
3.95			0.0012	0.0140
4.00			0.0000	0.0130
4.05				0.0120
4.10				0.0111
4.15				0.0102
4.20				0.0093
4.25				0.0085
4.30				0.0077
4.35				0.0070
4.40				0.0063
4.45				0.0056
4.50				0.0050
4.55				0.0043
4.60				0.0038
4.65				0.0032
4.70				0.0027
4.75				0.0022
4.80				0.0017

10.10 (M/G/1): (∞/FCFO) MODEL: SINGLE CHANNEL WAITING LINE MODEL WITH POISSON ARRIVALS AND ARBITRARY SERVICE TIME

In this model the arrival is described by Poisson as usual but the service time is described by unspecified probability distribution with known mean and S.D. For this unspecified service distribution the formulas for the seven operational characteristics are given by the following equations:

1. The probability of no one in the system:

$$P_0 = 1 - \frac{\lambda}{\mu}$$

2. The average number of units in the queue:

$$L_q = \frac{\lambda^2\sigma^2 + \left(\frac{\lambda}{\mu}\right)^2}{2\left(1 - \left(\frac{\lambda}{\mu}\right)\right)}$$

3. The average number of units in the system:

$$L_s = L_q + \frac{\lambda}{\mu}$$

4. The average number of units in the queue:

$$W_q = \frac{L_q}{\lambda}$$

5. The average time that a unit spends in the system:

$$W_s = W_q + \frac{1}{\mu}$$

6. The average time that a unit spends in the system:

$$P_w = \frac{\lambda}{\mu}$$

■ Example 10

A single seller handles the book sale in the company outlet at Indian Institute of Science campus. Customer arrivals are described by Poisson with the average arrival rate of 21 customers per hour ($\lambda = 21/60 = 0.35$ per min). A study shows that average service time is 2 minutes per customer with a S.D = 1.2. Calculate the operating characteristics:

Solution

1. The probability of no one in the system

$$P_0 = 1 - \frac{\lambda}{\mu} = 1 - \frac{0.35}{0.50} = 0.3$$

2. The average number of units in the queue

$$L_q = \frac{\lambda^2\sigma^2 + \left(\frac{\lambda}{\mu}\right)^2}{2\left(1 - \left(\frac{\lambda}{\mu}\right)\right)} = \frac{0.35^2 \times 1.2^2 + (0.35/0.50)^2}{2 \times (1 - 0.35/0.5)} = 1.1107$$

3. The average number of units in the system

$$L_s = L_q + \frac{\lambda}{\mu} = 1.1107 + (0.35/0.50) = 1.8107$$

4. The average time that a unit spends in the queue

$$W_q = \frac{L_q}{\lambda} = \frac{1.8107}{0.35} = 3.1733$$

5. The average time that a unit spends in the system

$$W_s = W_q + \frac{1}{\mu} = 3.1733 + \frac{1}{0.5} = 5.1733$$

6. The probability of the arriving unit has to wait in the queue

$$P_w = \frac{\lambda}{\mu} = \frac{0.35}{0.50} = 0.7$$

Excel Template

Table 10.12

	A	B	C	D	E
1		Book shop problem (M/G/1):(∞ /FCFO)			
2		**Assumptions**			
3		Poisson arrivals			
4		Arbitrary service			
5		**Data**			
6		Number of channels	k	=	1
7		Mean arrival rate	λ	=	0.35
8		Mean service rate	μ	=	0.5
9		S.D	σ	=	1.2
10		**Operating characteristics**			
11	1		P_o	=	=1-E7/E8
12	2		L_q	=	=((E7^2)*(E9^2)+(E7/E8)^2)/(2*(1-E7/E8))
13	3		L_s	=	=E12+E7/E8
14	4		W_q	=	=E13/E8
15	5		W_s	=	=E14+1/E8
16	6		P_w	=	=E7/E8
17					

Excel Solution

Table 10.13

	A	B	C	D	E
1		Book shop problem (M/G/1):(∞ /FCFO)			
2		**Assumptions**			
3		Poisson arrivals			
4		Arbitrary service			
5		**Data**			
6		Number of channels	k	=	1
7		Mean arrival rate	λ	=	0.35
8		Mean service rate	μ	=	0.5
9		S.D	σ	=	1.2
10		**Operating characteristics**			
11	1		P_o	=	0.3000
12	2		L_q	=	1.1107
13	3		L_s	=	1.8107
14	4		W_q	=	3.6213
15	5		W_s	=	5.6213
16	6		P_w	=	0.7000
17					

10.11 (M/D/1): (∞/FCFO): SINGLE CHANNEL WAITING LINE MODEL WITH POISSON ARRIVALS AND CONSTANT SERVICE TIME

In this specification the arrival is described by Poisson but the service time is a given constant with zero standard deviation. Such things happen in mechanized production lines. When the service is automatised the service time will be constant.

Operational characteristics

1. The probability of no one in the system

$$P_0 = 1 - \frac{\lambda}{\mu}$$

2. The average number of units in the queue

$$L_q = \frac{\left(\frac{\lambda}{\mu}\right)^2}{2\left(1 - \frac{\lambda}{\mu}\right)}$$

3. The average number of units in the system

$$L_s = L_q + \frac{\lambda}{\mu}$$

4. The average time that a unit spends in the queue

$$W_q = \frac{L_q}{\lambda}$$

5. The average time that a unit spends in the system

$$W_s = W_q + \frac{1}{\mu}$$

6. The probability of the arriving unit has to wait in the queue

$$P_w = \frac{\lambda}{\mu}$$

■ Example 11

In SBM the customers are provided ATM debit cards. Such customers can go to the ATM machine and draw the money with constant service time of 0.5 nineties per customer (S.D = 0). Customer arrivals are random and the average arrival rate is 21 customers per hour. ($\lambda = 21/60 = 0.35$ per min). Calculate the operating characteristics:

Excel Template

Table 10.14

	A	B	C	D	E	F
1		Book shop problem (M/G/1):(∞/FCFO)				
2		**Assumptions**				
3		Poisson arrivals				
4		Arbitrary service				
5		**Data**				
6		Number of channels	k	=	1	
7		Mean arrival rate	λ	=	0.35	
8		Mean service rate	μ	=	0.5	
9		S.D	σ	=	0	
10		**Operating characteristics**				
11	1		P_o	=	=1-E7/E8	
12	2		L_q	=	=((E7^2)*(E9^2)+(E7/E8)^2)/(2*(1-E7/E8))	
13	3		L_s	=	=E12+E7/E8	
14	4		W_q	=	=E13/E8	
15	5		W_s	=	=E14+1/E8	
16	6		P_w	=	=E7/E8	
17						

Excel Solution

Table 10.15

	A	B	C	D	E
1		Book shop problem (M/G/1):(∞/FCFO)			
2		**Assumptions**			
3		Poisson arrivals			
4		Arbitrary service			
5		**Data**			
6		Number of channels	k	=	1
7		Mean arrival rate	λ	=	0.35
8		Mean service rate	μ	=	0.5
9		S.D	σ	=	0
10		**Operating characteristics**			
11	1		P_o	=	0.3000
12	2		L_q	=	0.8167
13	3		L_s	=	1.5167
14	4		W_q	=	3.0333
15	5		W_s	=	5.0333
16	6		P_w	=	0.7000
17					

10.12 (M/G/k): (∞/BLOCKED CUSTOMERS CLEARED) MULTIPLE CHANNELS WITH POISSON ARRIVALS, ARBITRARY SERVICE TIME AND NO WAITING LINE

In this model the arriving units seek service from one or more service channels. If all channels are busy, the arriving units are denied access to the system in the sense no waiting is allowed. However, such blocked customers may be either lost forever or may attempt to return to the system later. These types of models are built with the following assumptions.

1. The system has k channels.
2. The arrival is described by Poisson distribution with mean λ.
3. The service time for each channel may have any unspecified distribution.
4. The mean service rate μ is same for all the channels.
5. The new arrival is entertained only at least one channel is free.
6. An arrival when all channels are busy is blocked in the sense that it is not allowed to wait.

The working of telephone system is the standing example for this type of model. A customer's call is attended provided at least one of the channels is free. If all the channels are busy then the customer is blocked from the access without keeping any waiting list. The only possibility left to be denied customer is to try again afresh. The question addressed in this situation is, how many channels or services should be planned?

Operational characteristics of M/G/k model with blocked customers cleared

We propose to find the best number of channels by computing the steady state probabilities that J of the k channels will be busy. These probabilities are obtained as

$$P_j = \frac{(\lambda/\mu)^j / j!}{\sum_{i=1}^{k} (\lambda/\mu)^1 / i!}$$

Where λ = the mean arrival rate

μ = the mean service rate for each channel

k = the number of channels

P_j = the probability of j of k channels is busy for j = 1, 2 ... k.

The most important is the value of P_k, the probability that all the k channels are busy and hence the access is blocked. P_k thus indicates the percentage of arrivals that are blocked. Another characteristic that deserves our attention in this context is the number of units in the system given by the formula

$$L = \frac{\lambda}{\mu}(1 - P_i)$$

Example 12

Wipro Software in Bangalore receives its orders through 800 telephones. The exponential arrival rate $\lambda = 12$ per hour. The time taken to process the order varies considerably from order to order. However let the average handling time $\mu = 6$ calls per hour. The calls received by all the 800 lines are channelled through 3 salesmen. Calls received on the 800 lines will be connected automatically to the any one of the free lines it exists, otherwise the call gets terminated hoping that the caller will contact later. However the research result shows that substantial terminated customers do not respond again, probably they all might have gone for some other software company. If the management's goal is to serve at least 90% of the callers, then how many telephones and sales personnel are ideal to this goal?

Solution

$$P_j = \frac{(\lambda/\mu)^j / j!}{\sum_{i=1}^{k} (\lambda/\mu)^1 / i!}$$

$$P_3 = \frac{(12/6)^3/3!}{(12/6)^0/0! + (12/6)^1/1! + (12/6)^2/2! + (12/6)^3/3!} = \frac{1.3333}{6.3333} = 0.2105$$

Thus 21% of the calls are being blocked in this case. So let us increase the sales persons by one more and recalculate P_4.

$$P_4 = \frac{(12/6)^4/4!}{(12/6)^0/0! + (12/6)^1/1! + (12/6)^2/2! + (12/6)^3/3! + (12/6)^4/4!}$$

$$= \frac{0.6667}{7} = 0.0952$$

Now only 9.52% are blocked. So the company should go for 4-channel call receiving system if at all it wants to handle 90% of the order calls.

The average number of sales personnel busy in our 4-line system is obtained as

$$L = \frac{\lambda}{\mu}(1 - P_i) = \frac{12}{6}(1.0.0952) = 1.8095$$

Though this average is less than 2 lines to have 90% target one must go for a four-line system only.

However, a final situation will emerge only on the basis of additional cost due to additional channel and opportunity cost of the orders lost. Thus economic analysis is imperative in all our models.

Excel Template

Table 10.16

	A	B	C	D
1	Book shop problem (M/G/1):(infinite/FCFO)			
2	**Assumptions**			
3	Poisson arrivals			
4	Exponential service time			
5	**Data**			
6	Number of channels	k	=	4
7	Mean arrival rate	λ	=	12
8	Mean service rate	μ	=	6
9	Utilization ratio	ρ	=	2
10	S.D	σ	=	0
11	Number of Channels	P_o		Probability
12	0			=($D$9^A12/FACT(A12))/($D$9^0/FACT(0)+$D$9^1/FACT(1)+$D$9^2/FACT(2)+$D$9^3/FACT(3)+($D$9^4/FACT(4)))
13	1			=($D$9^A13/FACT(A13))/($D$9^0/FACT(0)+$D$9^1/FACT(1)+$D$9^2/FACT(2)+$D$9^3/FACT(3)+($D$9^4/FACT(4)))
14	2			=($D$9^A14/FACT(A14))/($D$9^0/FACT(0)+$D$9^1/FACT(1)+$D$9^2/FACT(2)+$D$9^3/FACT(3)+($D$9^4/FACT(4)))
15	3			=($D$9^A15/FACT(A15))/($D$9^0/FACT(0)+$D$9^1/FACT(1)+$D$9^2/FACT(2)+$D$9^3/FACT(3)+($D$9^4/FACT(4)))
16	4			=($D$9^A16/FACT(A16))/($D$9^0/FACT(0)+$D$9^1/FACT(1)+$D$9^2/FACT(2)+$D$9^3/FACT(3)+($D$9^4/FACT(4)))

Type the following formula in cell D12 and copy by dragging down unto cell D16 and get the wanted results.

CellD12=($D$9^A12/FACT(A12))/($D$9^0/FACT(0)+$D$9^1/FACT(1)+$D$9^2/FACT(2)+$D$9^3/FACT(3)+($D$9^4/FACT(4)))

Excel Solution

Table 10.17

	A	B	C	D	E
1	Book shop problem (M/G/1):(∞/FCFO)				
2	**Assumptions**				
3	Poisson arrivals				
4	Exponential service time				
5	**Data**				
6	Number of channels	k	=	4	
7	Mean arrival rate	λ	=	12	
8	Mean service rate	μ	=	6	
9	Utilization ratio	ρ	=	2	
10	S.D	σ	=	0	
11	Number of busy lines in a n line system	P_n		Probability	
12	0			0.1428571	
13	1			0.2857143	
14	2			0.2857143	
15	3			0.1904762	
16	4			0.0952381	

10.13 M/M/1: N/FIFO SYSTEM

In this model the maximum size of the queue is finite in the sense there is no place in the queue for more than N. The car parking facility in a busy area is normally limited. In this situation the mean arrival rate λ can no longer be taken as constant. It varies depending upon the number of unit in the waiting line. Such models with constant population are called finite line calling population models. Such models have the following assumptions.

1. The arrival for each unit follows a Poisson distribution with mean arrival rate λ.
2. The service time follows an exponential distribution with mean μ.
3. The service seeking population is finite and denoted by N.

The mean arrival rate for this type of model is defined in terms of how often each unit arrives or seeks service. Note that in all the previous models with infinite population λ denotes the mean arrival rate. With the finite population the mean arrival rate for the system varies, depending upon the number of units in the system at the time of new entry. So here by λ we refer mean arrival rate for each unit (not for the system).

Operating characteristics

1. The probability that no one is in the system

$$P_0 = \frac{1}{\sum_{n=0}^{N} \frac{N!}{(N-n)!}\left(\frac{\lambda}{\mu}\right)^n}$$

2. The average number of units in the queue

$$L_q = N - \frac{\lambda+\mu}{\lambda}(1-P_0)$$

3. The average number of units in the system

$$L = L_q + (1-P_0)$$

4. The average time that a unit spends in the queue

$$W_q = \frac{L_q}{(N-L)\lambda}$$

5. The average time that a unit spends in the system

$$W_s = W_q + \frac{1}{\mu}$$

6. The probability that the arriving unit has to wait in the queue

$$P_w = (1-P_0)$$

7. The probability of n units in the system

$$P_n = \frac{N!}{(N-n)!}\left(\frac{\lambda}{\mu}\right)^n P_0$$

Example 13

In a bus station with one service personal the municipality collects entry tax from buses entering the bus stand. The mean arrival rate is λ = 120 buses per hour the service personal at an average takes 15 seconds per bus. If arrivals are described by Poisson and service times are described by exponential calculate:

1. The probability that the system has no vehicles.
2. The probability that the servant is busy.
3. Expected number of buses in the bus station.
4. Expected number of buses in the queue for service.

Solution

Data: $\lambda = 120$; $\mu = (60\times60)/15 = 240$; so $\rho = 1/2$

1. The probability that the system has no vehicles $P_0 = 1 - \rho = 1 - 0.5 = 0.5$
2. The probability that the servant is busy $P_W = \rho = 0.5$
3. Expected number of buses in the bus station at given time $L = \rho/(1 - \rho) = 0.5/(1 - 0.5) = 1$
4. Expected number of buses in the queue for service $L_S = \rho^2/(1 - \rho) = 0.5$

10.14 ECONOMIC ANALYSIS OF WAITING LINE MODELS

Queuing cost behaviour

In all the models discussed above the operating characteristics evaluated on the basis of subjective considerations. The manager may prefer to have one minute or less waiting to keep the incoming unit happy. But such a task is achieved only by adding one more service facility based on arrival rates.

In all cases if the service facility is inadequate a queue will always exist and continue to grow overtime, some of them may prefer to leave the queue permanently and go for alternative service provider. Some others after seeing the long queue may not join queue. In all cases the service provider not only loses the customers but also forego the respectable income from such units. This is often referred to as the cost of the waiting line. One way of reducing the queue length and the associated cost is to go for more number of service channels. Definitely such a move will bring down the length of the queue and the associated cost but at the same time increase the cost due to hiring of additional service personnel. Sometimes, too much of service channels will definitely keep the queue length zero but at the same time some service personnel or the other will be idle. A striking balance between these two types of cost is always desirable. The following diagram explains the cost of waiting and the cost of service.

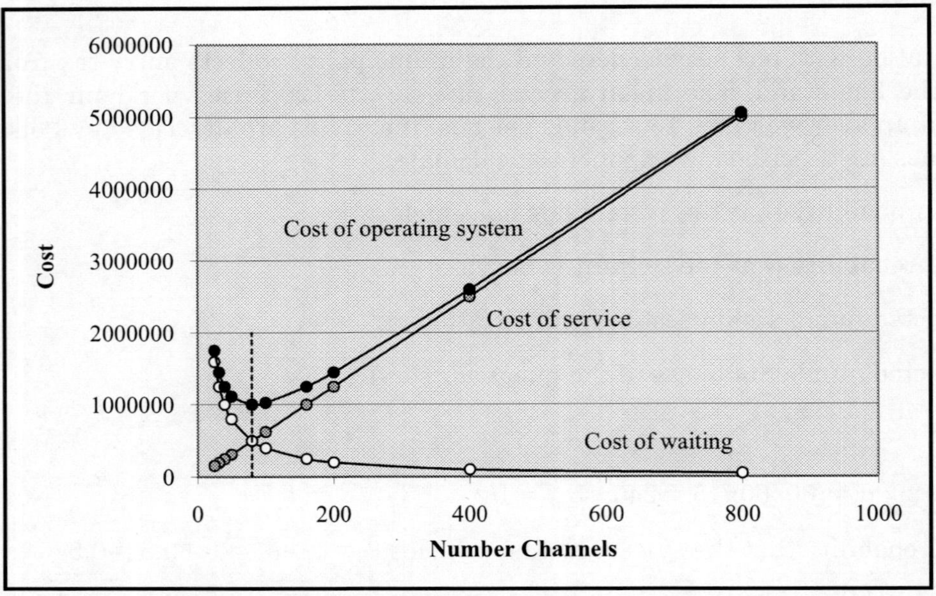

Figure 10.9

The service cost increases as the number of channels increases as shown by simple straight line cost function. But at the same time more the number of channels less and less will be the cost of waiting as shown by the rectangular hyperbolic cost curve. The number of channels that will provide a good approximation of the minimum total cost design can be found out by evaluating the total cost for several design alternatives. Such a minimum balancing total however necessarily need not coincide with zero queue length. In many practical situations the management would like to have an optimum queue size without risking loss of customers. To develop a total cost model we define the following additional new notations.

C_W = the waiting cost per unit time for each unit.

L = the average number of units in the system.

C_S = the service cost per unit time for each channel.

k = the number of channel in operation.

TC = total cost per unit of time.

The total cost is simply obtained by adding both the waiting and service costs.

$$TC = C_W L + C_S k$$

Out of the two cost components the waiting cost estimation is more difficult. This cost is really not the direct one and easily automatable. If care is not taken to reduce the queue length the frustrated customers will leave the queue and go elsewhere for similar service. Thus the waiting line cost is really an indirect cost of loss of business.

The service cost on the other hand follows directly. It is directly proportional to the number of service units provided. To explain the economic implications let us consider an illustration.

Example 14

In newly opened petrol pump with a single operator, the mean arrival rate is $\lambda = 0.75$ vehicle per minute. The mean service rate is $\mu = 1$ vehicle per minute. If arrival is described by Poisson and service time by exponential calculate the total operational cost. The waiting line cost per customer per minute is estimated as Rs. 5. The service cost to the operator of the machine is Rs. 10. To bring down the waiting time the bunk proposes to increase the service personnel, one for billing and another for putting petrol. Is it worthwhile to go for such a venture?

Solution

(a) The average number of units in the queue:

$$L_q = \frac{\lambda^2}{\mu(\mu-\lambda)} = \frac{0.75^2}{1(1-0.75)} = 2.25$$

$$TC = C_W L + C_S k = 10 \times 2.25 + 20 \times 1 = \text{Rs. } 42.25$$

(b) The average number of units in the queue

$$L_q = \frac{\lambda^2}{\mu(\mu-\lambda)} = \frac{0.75^2}{2(2-0.75)} = 0.225$$

$$TC = C_W L + C_S k = 10 \times 0.225 + 20 \times 1 = \text{Rs. } 22.25$$

So the punk owner must go for an additional service personal to reduce the queue pressure.

Example 15

In a machine shop, the rate of breakdown of machines was estimated as 4 per hour following a Poisson distribution. The cost of non-productive machine is found to be Rs. 9 per hour. The company has two options for appointing a service personnel. The first one does the job little slowly and repairs only 5 machines but at the same time charge only Rs. 3 per hour. The second one is little faster and repairs at the rate of 7 per hour and charges Rs. 5 per hour. Which proposal that one should opt?

Solution

The arrival rate $\lambda = 4$ per hour. The non-production cost due to waiting = Rs. 9 per hour.

Now the total cost for slow worker = Wage cost + Waiting line cost

$$6+\left(9\times\frac{\lambda}{\mu-\lambda}\right)=6+\left(9\times\frac{4}{7-4}\right)=\text{Rs. }18$$

Similarly, the total cost for slow worker = Wage cost + Waiting line cost

$$3+\left(9\times\frac{\lambda}{\mu-\lambda}\right)=3+\left(9\times\frac{4}{5-4}\right)=\text{Rs. }39$$

Since the total cost for the fast repairer is less the manager concerned should go for fast repairman.

■ Example 16

In an office the newly appointed typist receives at an average of 22 letters. She normally works 8 hours in a day and takes at an average of 20 minutes per letter. The opportunity cost of waiting letters is estimated as Rs. 80 paise per hour. The equipment operation cost plus labour cost work out to Rs. 40 per day.

a. What is the utilization ratio of the typist?

b. What is the average number of letters waiting?

c. What is the average typing time needed to type a letter?

d. What is the total cost of waiting?

e. Knowing fully the difficulty, the proprietor wants to computerize the whole processing of typing by hiring. The following table gives the additional cost as well as the improvement in the efficiency for two types of computer model.

Model 1	Additional cost per hour	Increase in efficiency
1	Rs. 37	50%
2	Rs. 39	75%

Propose the right choice for the model selection.

Solution

Data: Arrival rate λ = 22 letters. Service rate μ = (8 × 60)/20 = 24 letters per day.

a. Typist utilization rate:

$$\rho=\frac{\lambda}{\mu}=\frac{22}{24}=0.917$$

b. The average number of letters in the waiting line :

$$L_q = \frac{\lambda^2}{\mu(\mu-\lambda)} = \frac{22\times 22}{24(24-22)} = 10.8083$$

c. The average waiting time per letter :

$$W_s = \frac{1}{\mu-\lambda} = \frac{1}{24-22} = \frac{1}{2}\text{ day}$$

d. The average number of letters in the system:

$$L_S = \frac{\lambda}{\mu-\lambda} = \frac{22}{24-22} = 11$$

Total cost = Typist cost + Cost of 11 letters at 40 + 6.4 = 11 × 6.4 = Rs. 110.4

e. Computer model 1

Arrival rate λ = 22 letters. Service rate μ = 24 + 24 × 0.5 = 36 letters per day
Total cost = Typist cost + equipment cost + waiting line cost

$$= 40 + 37 + \frac{22}{36-22}\times 6.4 = \text{Rs. } 87.06$$

f. Computer model 2

Arrival rate λ = 22 letters. Service rate μ = 24 + (24 × 0.75) = 42 letters per day
Total cost = Typist cost + equipment cost + waiting line cost

$$= 40 + 39 + \frac{22}{42-22}\times 6.4 = \text{Rs. } 86.04$$

Compare all the options hiring type 2 computer is better.

Example 17

In a town bus terminal, customers arrive in queue to get their tickets. In this single channel reservation counter Poisson arrival and exponential service is hypothesized. The average arrival time between two successive arrivals is 5 minutes. The mean time to serve a unit is 2 minutes.

(a) Calculate the average waiting time in the queue.

(b) The management is willing to open a second service counter provided the consumer waits 5 minutes or more. By how much the flow of customers be increased to go for a second channel.

Solution

The average arrival rate λ = (1/5) × 60 = 12 customer per hour.

The average service rate μ = (1/2) × 60 = 30 customer per hour.

(a) The average waiting time in the queue is obtained as

$$W_q = \frac{\lambda}{\mu(\mu-\lambda)} = \frac{12}{30(30-12)} = 0.022 \text{ hours or } 1.333 \text{ minutes}$$

(b) Given the average waiting time as 5 minutes (1.12 hours) the appropriate arrival is obtained as follows.

$$W_q = \frac{1}{12} = \frac{\lambda}{\mu(\mu-\lambda)}$$

$$12\lambda = 30(30-\lambda)$$

$$12\lambda = 900 - 30\lambda$$

$$42\lambda = 900$$

$$\lambda = (900/42) = 21.43$$

So the average arrival rate should go up from the current 12 to 21.43 for the opening of the second reservation counter.

10.15 POISSON ARRIVAL AND ERLANG SERVICE MODELS

In Erlang model the service takes place in k different faces. Each face takes the average service time 1/μ. The actual service time however follows exponential distribution. For varying values of k we will have different distributions but the mean service time remains the same. The associated important characteristics are given below.

1. The average number of units in the queue

$$L_q = \frac{k+1}{2k} \times \frac{\lambda^2}{\mu(\mu-\lambda)}$$

2. The average number of units in the system

$$L_S = L_q + \frac{\lambda}{\mu} = \frac{k+1}{2k} \times \frac{\lambda^2}{\mu(\mu-\lambda)} + \frac{\lambda}{\mu}$$

3. The average waiting time of a unit in the queue

$$W_q = \frac{L_q}{\lambda} = \frac{k+1}{2k} \times \frac{\lambda}{\mu(\mu-\lambda)}$$

4. The average waiting time of a unit in the system

$$W_S = W_q + \frac{1}{\mu} = \frac{k+1}{2k} \times \frac{\lambda}{\mu(\mu-\lambda)} + \frac{1}{\mu}$$

Example 18

The repair of a certain machine involves four independent steps one after another. The time taken for each step follows exponential distribution with a mean of 5 minutes. Machine breakdown follows Poisson distribution with mean of 3 breakdowns per hour.

(a) What is the expected idle time when there is only one repairman in the workshop?

(b) What is the average waiting time of the breakdown machine in the queue?

(c) What is the expected number of broken down machines in the queue?

Solution

Data: $\lambda = 3$ per hour, $\mu = 12$ per hour, K = 4

1. The average waiting time of a unit in the queue

$$W_q = \frac{L_q}{\lambda} = \frac{K+1}{2K} \times \frac{\lambda}{\mu(\mu - \lambda)}$$

$$= \frac{4+1}{2 \times 4} \times \frac{3}{12(12-3)} = \frac{5}{8} \times \frac{3}{108} = 0.01736 \text{ or } 1.04 \text{ minutes}$$

2. The average waiting time of a unit in the system

$$W_S = \frac{K+1}{2K} \times \frac{\lambda}{\mu(\mu - \lambda)} + \frac{1}{\mu}$$

$$= \frac{4+1}{2 \times 4} \times \frac{3}{12(12-3)} + \frac{1}{12}$$

$$= \frac{5}{8} \times \frac{3}{108} + \frac{1}{12} = 0{,}10069 \text{ or } 6.0417 \text{ minutes}$$

3. The average number of units in the queue

$$L_q = \frac{K+1}{2K} \times \frac{\lambda^2}{\mu(\mu - \lambda)}$$

$$= \frac{3+1}{2 \times 4} \times \frac{3^2}{12(12-3)} = \frac{5}{8} \times \frac{9}{108} = 0.052 \text{ units}$$

10.16 LIMITATIONS OF QUEUING THEORY

1. Queuing problems are complex in the sense even in single queue single server models the characteristic results are obtained by using complex equations.

2. These models mostly assume Poisson arrivals and exponential service pattern. This need not be true in all the cases.

3. In multiple channel models the analysis becomes more difficult when departure from one queue becomes the arrival of another queue.

4. The arrivals and departures are difficult to handle by using simple formulas. Monte-Carlo technique could be of some use in such situation.

EXERCISES

1. What is queuing theory? In what areas of management can it be applied successfully? Give some examples.

2. A departmental store has one cashier. During the rush hours, customers arrive at the rate of 20 customers per hour. The average number of customers that can be processed by the cashier is 24 per hour. Assume that the conditions for the use of single-channel queuing model apply. What is the
   a) probability that the cashier is idle?
   b) average number of customers in the queuing system?
   c) average time that a customer spends in the system?
   d) average number of customers in the queue?
   e) average time a customer spends in the queue waiting for service?

   (Delhi University, M.Com 1986)

3. Telephone authorities studied a locality where public telephone booth has the following data.

   Arrivals (Mean arrival time) = 10 minutes

   The length of phone calls on an average = 3 minutes

   Arrival follows Poisson distribution and telephone calls negative exponential distribution.
   a) Find the probability that the phone is available when you go to the booth.
   b) The P&T department is prepared to install 2nd booth provided the customer has to wait for at least 3 minutes. By how much the flow of arrivals should be increased to justify the same.

4. At an oil refinery company, ships arrive to unload crude oil according to Poisson process with mean rate of 8 ships per week. Service time is exponentially distributed with mean service time of half day to unload one ship.
   a) What is the average number of ships waiting to deliver crude oil?
   b) What is the average time that a ship must wait before it is taken for service?
   c) What is the average total time that a ship spends at the refinery?
   d) What is the probability that a ship has to wait for more than 2 days before getting the service?

5. A TV repairman finds that the time spent on his job has an exponential distribution with mean 30 minutes. If he repairs sets in order in which they come and if the arrival of sets is approximately Poisson with an average rate of 10 per 8 hour day, what is his expected idle time each day? How many jobs are ahead of the set just brought in?

(Rajasthan University, M.Com 1986)

6. A bank has two tellers working on savings accounts. The first teller handles withdrawals only. The second teller handles deposits only. It has been found that the service time distribution of both deposits and withdrawals are exponential with mean service time of 3 minutes per customer. Depositors and withdrawers are found to arrive in a Poisson fashion throughout the day with mean arrival rate of 16 and 14 per hour. What would be the effect on the average waiting time for the depositors and withdrawers if each teller could handle both withdrawals and deposits?

(Delhi University, M.B.A 1987)

7. In a bank, cheques are cashed at a single "teller" counter. Customers arrive at the counter in a Poisson manner at an average rate of 30 customers per hour. The teller takes on an average a minute and a half to cash the cheque. The service time has been shown to be exponentially distributed.
   a) Calculate the percentage of time the teller is busy.
   b) Calculate the average time a customer is expected to wait.

(ICWA, Dec. 1986)

8. A steel fabrication plant was considering the installation of a second tool crib in the plant to save walking time for the skilled craftsmen who check out equipment at the tool cribs. The Poisson/ exponential assumption arrivals are justified in this case. The time of a craftsman is valued at Rs.20 per hour. The current facility receives an average of 10 calls per hour, with two cribs, each would average five calls per hour. Currently there are two attendants, each of whom service one craftsman at a time, each has a service rate of eight craftsmen per hour. Each could do just as well in a separate tool room. However, each craftsman would require six minutes less working time per cell. Evaluate the proposal to set up a new crib so that each attendant would run one crib.

(Delhi University, M.B.A 1992)

9. A bank withdrawer arrives with a mean inter-arrival time of 10 minutes. The service is at the rate of 12 customers/hr. Arrivals follow Poisson and service follows negative exponential distribution.
   a) What is the probability that a fresh arrival need not have to wait in the queue?
   b) The bank is prepared to open one more counter if the arrival has to wait at least 8 minutes. By how much must the flow of arrivals should be increased to justify the second counter?

(Bangalore University, M.B.A July/August 2006)

10. A post-office has two counters; the first one handles money orders and registration letters, while the second one handles all other business. It has been found that the service time distributions for both the counters are exponential with a mean service time of 4 minutes per customer. The customers are found to arrive at the first counter in a Poisson fashion with the mean arrival rate of 10 per hour while the customers arrive at the second counter in the Poisson fashion with the mean arrival rate of 12 per hour. What would be the effect on the average waiting time for the two types of customers? If each counter can handle all types of business.

11. In a super market the average arrival rate of customers is 5 per 30 minutes. The average time it takes to list and calculate the customer's purchases at the cash desk is 4.5 minutes and these times are exponentially distributed.

    a) How long will the customer have to wait for the service at cashier desk?

    b) What is the chance that the queue length will exceed 5?

    c) What is the probability that the cashier is working?

(Bangalore University, M.B.A July/August 2005)

12. A tax consulting firm has four service stations (counters) in its office to receive people, who have problems and complaints about their income, wealth and sales taxes. Arrivals average 80 persons in an 8-hour service per day. Each tax advisor spends an irregular amount of time servicing the arrivals, which have been found to have an exponential distribution. The average service time is 20 minutes. Calculate the average number of customers in the system, average number of customers waiting to be serviced, average time a customer spends in the system and the average waiting time for a customer. Calculate how many hours each week does a tax advisor spend performing his job. What is the probability that a customer has to wait before he gets service? What is the expected number of idle tax advisors at any specified time?

(Delhi University, M.B.A, Dec. 1984)

13. In a bank with a single server, there are 2 chairs for waiting customers. On an average, one customer arrives every 10 minutes and each customer takes 5 minutes for getting served. Making suitable assumptions, find

    a) The probability that an arrival will get a chair to sit down.

    b) The probability that an arrival will have to stand.

    c) Expected waiting time of a customer.

(C.A, May 1988)

14. Customers arrive at a sales counter manned by a single person according to a Poisson process with a mean rate of 20 per hour. The time required to serve a customer has an exponential distribution with a mean of 100 seconds. Find the average waiting time of a customer.

(C.A, May 1987)

15. An oil company is planning for a service station near Tumkur. Customer's arrival follows Poisson distribution with a mean of 18 automobiles/hr. Core pump could service with a mean of 6 automobiles/hr.

    a) If 3 pumps are installed what is the probability that an arrival would have to wait?

    b) What is the average waiting time in the queue?

(Bangalore University, M.B.A October 2004)

16. Four counters are being run on the frontier of a company to check passports and papers of tourists. The tourists choose the counter at random. If the arrivals at the frontier are Poisson at the rate λ and the service time is exponential with $\lambda/2$, what is the steady state average queue length?

(Bangalore University, M.B.A July/August 2005)

CHAPTER

11

Decision Theory

11.1 INTRODUCTION

In business, when a decision-maker faces several alternatives the decision analysis helps him to take appropriate decisions. For example, a global manufacturer might be interested in identifying the best location for his new plant. Suppose our manufacturer has identified six plant locations for his business in four different countries. Choosing one among the six locations is a difficult task indeed. The choice often depends upon factors like the political stability of the country concerned, extent of market, availability of raw material, its cost and so on. Under such circumstances, several scenarios could emerge by way of combining various factors. Accordingly, every scenario will have a probability of its own. Using profit or cost consequence in mind attempt will be made to identify the best possible plant location. With all due care often uncertain future complicates the selection process. In certain cases, the selected decision alternative may yield very good unexpected results even. In other cases it might be very fatal for the survival. Thus, the risk associated with any final decision is directly related to the risk associated with the respective individual factors taken into consideration. A good decision theory therefore should invariably relate to the theory of probability.

Let us begin the study of decision theory by considering a problem having reasonably few decision alternatives and reasonably few possible future events. To make a simple beginning, let us introduce ourselves to the influence diagrams and the concept of the associated payoff tables. In the second stage, we propose to introduce decision trees that will be simplify even if more complex real world problems.

11.2 FORMULATION OF THE PROBLEM

As usual, in any other field, the first step in the decision analysis is problem formulation. We begin with the verbal statement of the problem as follows. The Bangalore Development Authority (BDA) in its Civic Amenities site at HSR layout decides to construct a commercial complex. This location provides a spectacular view of the project. The BDA had preliminary architectural design for three alternative-sized projects. The alternatives are:

d_1 = a small complex with 100 shops

d_1 = a medium sized complex with 300 shops

d_3 = a large complex with 500 shops

Given these three alternatives, the BDA wants to identify the appropriate size that gives the maximum profit. A factor that affects the selection of the best alternative project is the uncertainty associated with the chance event otherwise called the demand for the shops in the proposed complex. In the decision analysis, the possible outcomes of a chance event are often referred as **states of nature**. The chance events are defined so that at a given point of time only one of the possible states of nature will occur. For the BDA problem let the two chance events be:

s_1 = Strong demand for the shops

s_2 = Weak demand for the shops

Thus, under the given circumstances the BDA must select the complex size having the state of nature that is going to follow in future in mind. Once the selection is made, depending upon the chance event that follows a final consequence will result.

11.3 INFLUENCE DIAGRAM

An influence diagram is a special graphical device describing the relationship between the decision variable, the chance events and the final consequence. In this presentation it is a common practice to use squares to represent decision **nodes**. Circles or ovals are used to depict the **chance nodes,** and diamonds are used to represent **consequence nodes**. The lines connecting these nodes are called **arcs.** These lines show the direction of influence that is being exerted.

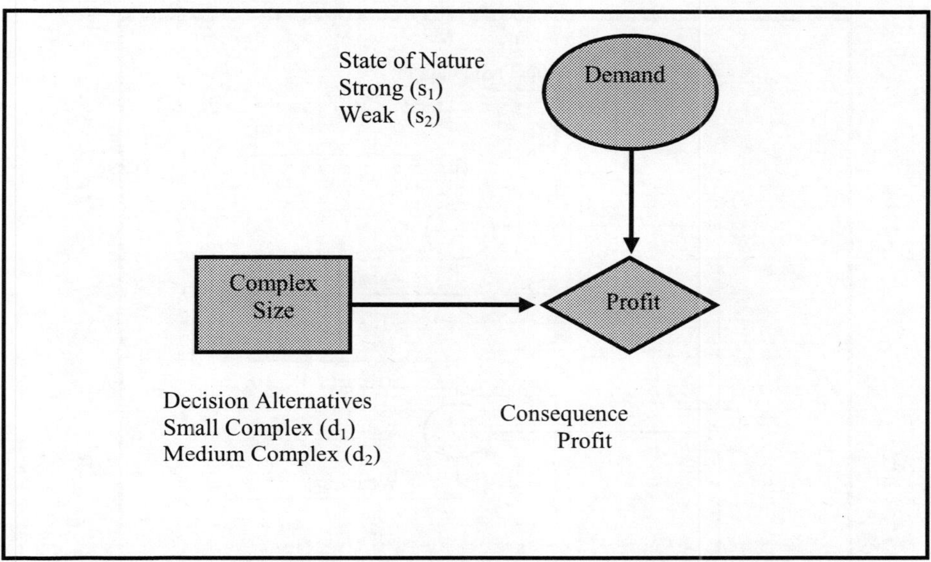

Figure 11.1 Influence diagram for the BDA problem

Payoff Table

Given the three decision alternatives and two states of nature, what size the BDA would choose? To arrive at an appropriate answer BDA must know the consequence associated with each decision alternative and each state on nature. In decision theory, we express the consequence resulting from a particular combination of a decision alternative and the state of nature as payoff.

Table 11.1

Payoff table for BDA

Decision alternatives	State of nature	
	Strong demand (s_1)	Weak demand (s_2)
Small complex (d_1)	8	7
Median complex (d_2)	14	5
Large complex (d_3)	20	−9

From the above table for example, if a medium complex is built and demand turns out to be weak then our BDA would receive a profit (payoff) of Rs. 5 crore. On the other hand, the same complex would fetch Rs. 14 crore provided the demand is strong. Thus, in general V_{ij} refers the payoff for i^{th} complex size and j^{th} demand pattern.

Decision Tree

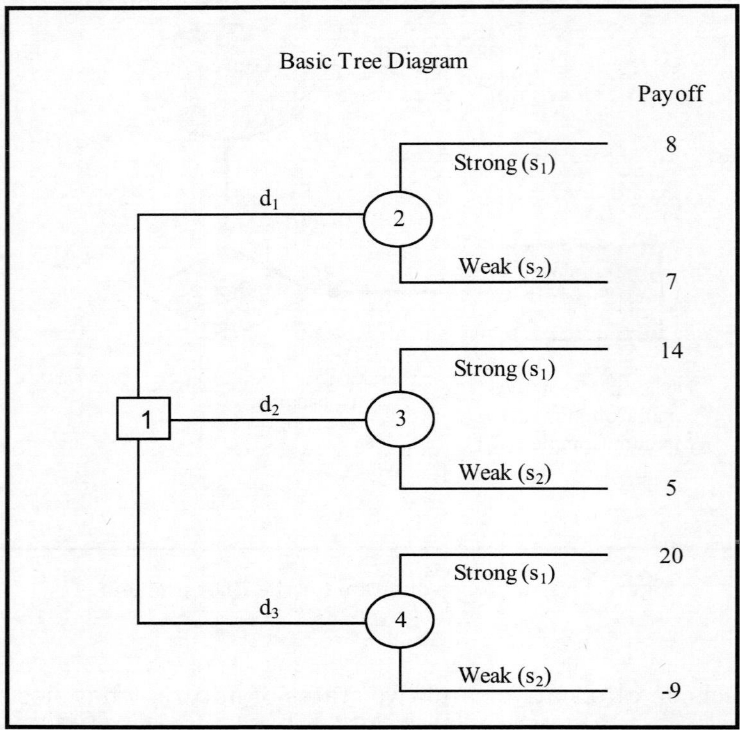

Figure 11.2

Decision tree is a novel diagrammatic representation of the decision problem. It is designed to enhance the logical sequence of the given problem. In our BDA problem since there are three decision alternatives, our decision tree will have three mutually exclusive branches as shown in the diagram to begin with. Further, since there are two states of nature associated, each branch drawn in the first step is further sub-divided into two minor branches leading to a total of six (3×2) branches finally. Once this is done we finally write the respective payoffs against their respective tree branches at the tail-end as shown in Figure 11.2. For example, if BDA wants to have the big complex and the associated demand turns out to be large, then it will receive a payoff of Rs. 20 crore, big boom indeed. On the other hand, the demand turns out to be small then the same old big complex decision will end up with a loss of Rs. 9 crore (−9).

In the diagram there are four junctions from which the branches emerge. Such points are called nodes and they are numbered from 1 to 4 as shown in the figure. However, to distinguish between decision nodes and chance nodes we use squares for decision nodes and circles or ovals for state of nature otherwise called chance nodes. Thus, in our diagram there is only one decision node [1] and three chance nodes from (2) to (4).

11.4 DECISION MAKING UNDER UNCERTAINTY

Under uncertain situations there is no way for the decision maker to assess the probabilities of the various states of nature. Our BDA illustrative example obviously does not have any probability statistics; instead it has only the pay off matrix under its command. Now given the tree diagram the real problem is to identify one best decision alternative among the three. To begin with, to have a simple model, we assume that BDA does not possess any probability statistics relating to the two chance events, namely strong and weak demand. Under the circumstances, to arrive at a decision alternative normally five basic approaches are being discussed in this section.

1) The Optimistic Approach: Maximum among Maximums

In this approach, as the name suggests, we propose to identify the best payoffs for all the three decision alternatives. Once this is done we chose the best as the final solution. Such an optimistic decision is often called maximum of the maximum payoff. To illustrate this approach let us go back to our BDA illustration once again.

i) If BDA constructs a small complex and the demand turns out to be weak then the corresponding payoff is Rs. 7 crore. On the other hand, if the demand is strong then our BDA will get Rs. 8 crore. Between the two payoffs since Rs. 8 crore is more, our BDA would construct small complex assuming a strong demand for it.

ii) If BDA constructs a medium complex and the demand turns out to be weak then the corresponding payoff is Rs. 5 crore. On the other hand, if the demand is strong then our BDA will get Rs. 14 crore. Between the two figures since Rs. 14 crore is more, our BDA would construct medium complex assuming a strong demand for it.

iii) If BDA constructs a large complex and the demand turns out to be weak then the corresponding payoff is Rs. −9 crore. On the other hand, if the demand is strong then our BDA will get Rs. 20 crore. Between the two figures since Rs. 20 crore is more, our BDA would construct large complex assuming a strong demand for it.

a) Tabular Method

The following Table 11.2 gives the summary statement of all the three alternatives discussed so far. The last column in the table gives the payoffs for all the three alternative decisions. Since we are interested in maximum of all these three maximums, the maximum value of Rs. 20 crore is reported in the last but one row in the table. The associated decision, namely d_3 (large complex) is reported in the last row. Thus, the final solution to this approach is to go for a large complex with full expectation of strong demand in future.

Table 11.2

Optimistic solution				
		State of nature		
		Strong demand	Weak demand	Maximum
Decision alternatives		s_1	s_2	payoff
Small complex	d_1	8	7	8
Medium complex	d_2	14	5	14
Large complex	d_3	20	−9	20
		Maximum of maximum payoff =		20
			Optimistic solution =	d_3

b) Decision Tree Method

In this method we take only the branch maximum payoffs and report against their respective chance nodes. For example, for the top most branch pertaining to small complex given in Figure 11.2 above, the branch payoffs are 8 (for strong demand) and 7 (for weak demand).

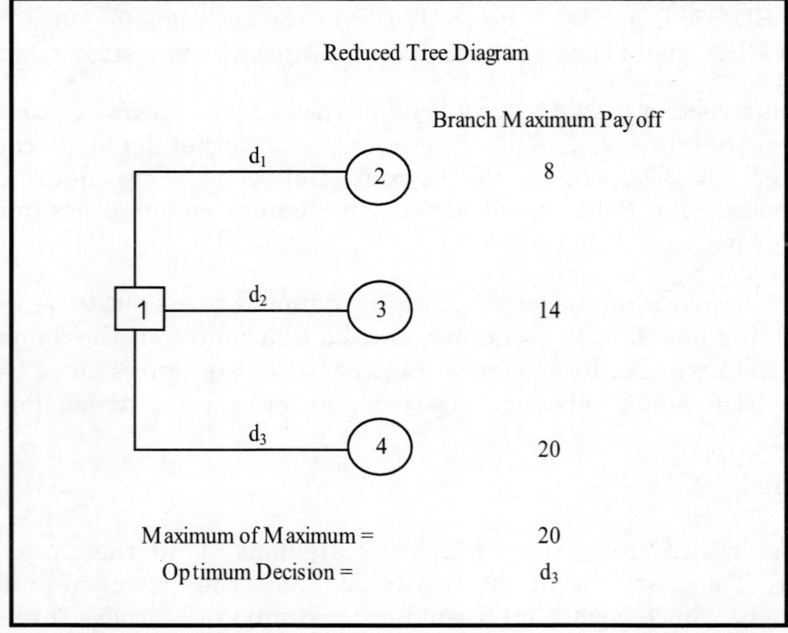

Figure 11.3

Among the two, since the payoff 8 is the maximum, we report this probability against the corresponding chance node (2) in Figure 11.3. In a similar way, the branch maximum payoffs are obtained and reported against node (3) and (4) in the figure. Among the three maximums reported, since 20 is 'the best' it is selected as the maximum of maximum and reported at the bottom of the figure. Since 20 is the payoff related to large complex we conclude that our BDA would go for the large complex (d_3) hoping to get Rs. 20 crore as the profit on completion.

2. Conservative Approach: Maximum among Minimums

This is indeed a defensive approach in the sense that our decision maker will try his best to avoid the worst thing that can happen to him in the business. Accordingly, he will obtain the minimum payoffs for all the three decision alternatives first. In the second stage he will choose the maximum among the three minimums obtained in the first stage.

i) To illustrate the approach, we go back to our BDA example. If BDA constructs a small complex and the demand turns out to be weak then the corresponding payoff is Rs. 7 crore. On the other hand, if the demand is strong, then our BDA will get Rs. 8 crore. Among the two figures since Rs. 7 crore is the minimum our BDA would construct small complex assuming a weak demand.

ii) If BDA constructs a medium complex and the demand turns out to be weak then the corresponding payoff is Rs. 5 crore. On the other hand, if the demand is strong, then our BDA will get Rs. 14 crore. Among the two figures since Rs. 5 crore is the minimum our BDA would construct medium complex assuming a weak demand.

iii) If BDA constructs a large complex and the demand turns out to be weak then the corresponding payoff is Rs. −9 crore. On the other hand, if the demand is strong, then our BDA will get Rs. 20 crore. Among the two figures since Rs. −9 crore is the minimum our BDA would construct large complex assuming a weak demand.

a) Tabular Method

The following table gives the summary statement of all the three discussions.

Table 11.3

Conservative solution

		State of nature		
		Strong demand	Weak demand	Minimum
Decision alternatives		s_1	s_2	payoff
Small complex	d_1	8	7	7
Medium complex	d_2	14	5	5
Large complex	d_3	20	−9	−9
		Maximum of minimum payoff =		7
		Optimistic solution =		d_1

Thus, the final solution to this approach is to go for a small complex with full expectation of weak demand in future.

b) Decision Tree Method

In this method we take only the branch minimum payoffs and report them all against their respective chance nodes. For example, for the top most branch pertaining to small complex given in Figure 11.2 above, the branch payoffs are 8 (for strong demand) and 7 (for weak demand).

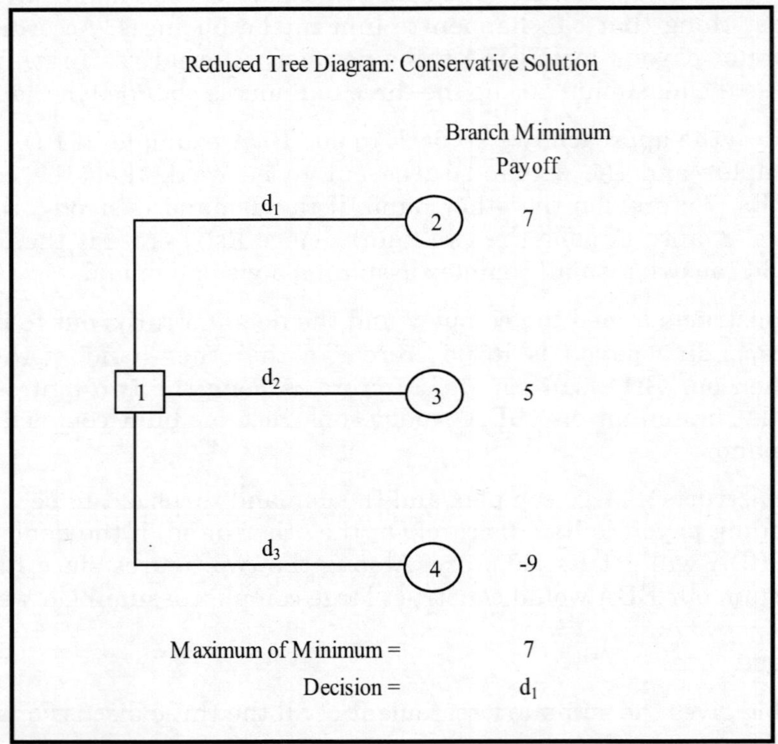

Figure 11.4

Among the two-since the payoff 7 is the minimum we report this probability against the corresponding chance node (2) in Figure 11.4. In a similar way the branch minimum payoffs are obtained and reported, against node (3) and (4) in the figure. Among the three minimums reported, since 7 is 'the best' it is selected as the maximum of minimum and reported at the bottom of the figure. Since 7 is the payoff related to small complex we conclude that our BDA would go for the small complex (d_1) hoping to get at least Rs. 7 crore as the profit on completion.

3. Regret Approach or Savage Approach: Minimum among Maximums

In this approach when every time the BDA calculates the payoff, the opportunity cost otherwise called the best alternative foregone is also obtained. This payoff due to foregone alternative is often called regret payoff.

i) To illustrate the approach, we go back to our BDA example. If BDA constructs a small complex and the demand turns out to be weak then the corresponding payoff is Rs. 7 crore. Under the circumstances (weak demand) the best alternative is going for the small complex yielding a payoff of Rs. 7 crore. Thus, the regret amount turns out to be (7–7) Rs. 0 crore.

ii) If BDA constructs a medium complex and the demand turns out to be weak then the corresponding payoff is Rs. 5 crore. Under the circumstances (weak demand) the best alternative is going for a small complex yielding a payoff of Rs. 7 crore. Thus, the regret amount turns out to be (7 – 5) Rs. 2 crore.

iii) If BDA constructs a large complex and the demand turns out to be weak then the corresponding payoff is Rs. –9 crore. Under the circumstances (weak demand) the best alternative is going for the small complex yielding a payoff of Rs. 7 crore. Thus, the regret amount turns out to be {7–(–9)} Rs. 16 crore.

iv) If BDA constructs a small complex and the demand turns out to be strong then the corresponding payoff is Rs. 8 crore. Under the circumstances (strong demand) the best alternative is going for large complex yielding a payoff of Rs. 20 crore. Thus, the regret amount turns out to be (20–8) Rs. 12 crore.

v) If BDA constructs a medium complex and the demand turns out to be strong then the corresponding payoff is Rs. 14 crore. Under the circumstances (strong demand) the best alternative is going for large complex yielding a payoff of Rs. 20 crore. Thus, the regret amount turns out to be (20–14) Rs. 6 crore.

vi) If BDA constructs a large complex and the demand turns out to be strong then the corresponding payoff is Rs. 20 crore. Under the circumstances (strong demand) the best alternative is going for large complex yielding a payoff of Rs. 20 crore. Thus, the regret amount turns out to be (20–20) Rs. 0 crore.

a) Tabular Method

All the six calculations stated above are reported in the regret payoff column in Table 11.4. Among the regrets the maximum regrets are reported in the last column. Since the minimum of all maximum regrets is our target it is reported in the last but one column.

Finally the decision-maker will minimize the maximum regrets. Thus, the final solution to this approach is to go for a medium complex with full expectation of strong demand in future.

Table 11.4

Mini max regret solution

	State of nature		Regret payoff		
	Strong demand	Weak demand	Strong demand	Weak demand	Maximum regret
Decision alternatives	s_1	s_2	s_1	s_2	payoff
Small complex d_1	8	7	12	0	12
Medium complex d_2	14	5	6	2	6
Large complex d_3	20	–9	0	16	16
		Minimum of maximum regrets =			6
		Minimum regret decision =			d_2

b) Decision Tree Method

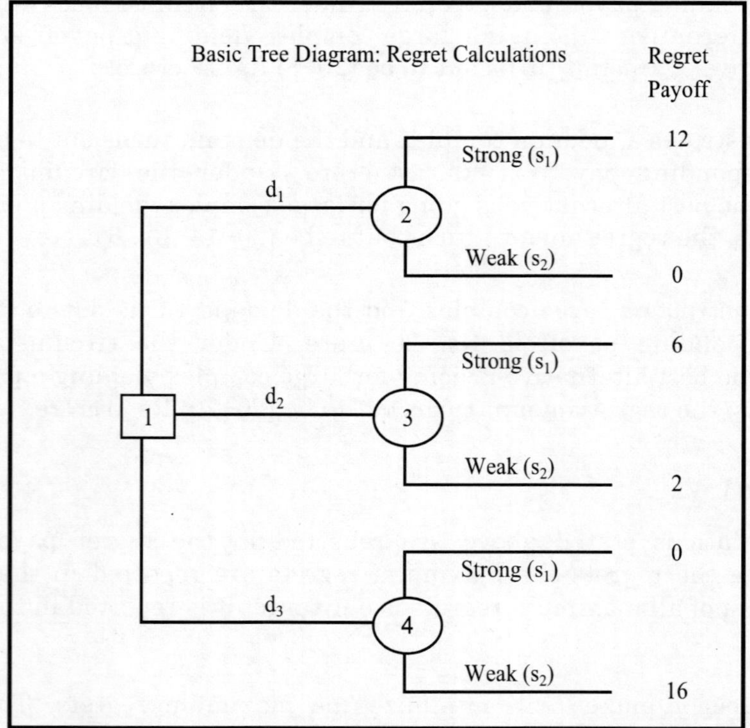

Figure 11.5

In this method all the associated six regret payoffs discussed earlier are written against the respective branches as shown in Figure 11.5.

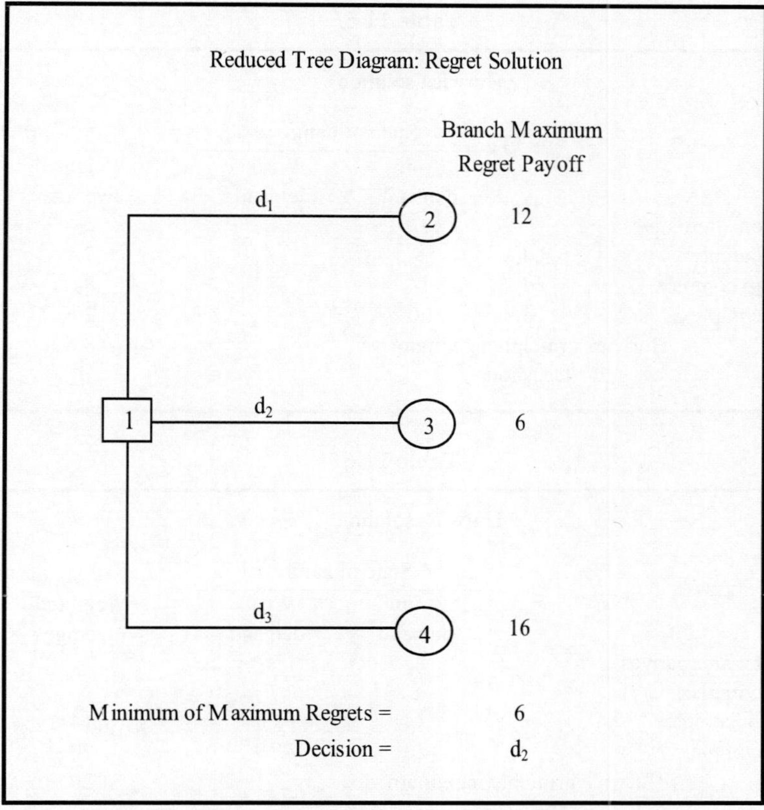

Figure 11.6

The maximum regrets of the concerned branches are obtained and reported against their respective chance nodes in Figure 11.6. The minimum of all these three regrets is the final solution to the problem. Since 6 is the minimum regret the associated decision is going for a medium complex.

4. Hurwicz Principle

In this method the needed optimum solution is obtained somewhere in between the optimistic Maximaxi principle and the conservative Maximini principle. Accordingly, the needed decision results are obtained as the weighted average of these two results. For this purpose a weight α is defined in the interval 0 to 1 both inclusive. According to this principle, for each decision alternative the weighted average is obtained as the sum of the maximum payoff multiplied by α and the minimum payoff multiplied by $(1-\alpha)$. Once such calculations are made, the best decision is taken based on the highest value among the values obtained. If α is 1 clearly our optimistic solution will be the result this time also. If it is 0 then the resulting result will be our conservative solution obtained in the previous section. The following tables illustrate this method.

Table 11.5

Hurwicz solution

		State of nature		$\alpha = 0.4$
		Strong demand	Weak demand	Weighted average
Decision alternatives		s_1	s_2	
Small complex	d_1	8	7	7.4
Medium complex	d_2	14	5	8.6
Large complex	d_3	20	–9	2.6
Hurwicz principle maximum =				8.6
Decision =				d_2

Table 11.6

Hurwicz solution

		State of nature		$\alpha = 1$
		Strong demand	Weak demand	Weighted average
Decision alternatives		s_1	s_2	
Small complex	d_1	8	7	8
Medium complex	d_2	14	5	14
Large complex	d_3	20	–9	20
Hurwicz principle maximum =				20
Decision =				d_3

Table 11.7

Hurwicz solution

		State of nature		$\alpha = 0$
		Strong demand	Weak demand	Weighted average
Decision alternatives		s_1	s_2	
Small complex	d_1	8	7	7
Medium complex	d_2	14	5	5
Large complex	d_3	20	–9	–9
Hurwicz principle maximum =				7
Decision =				d_1

5. Laplace Principle

This principle is based on the simple philosophy that the uncertain events are equally likely. Thus, it is again a weighted average with equal weights for all the uncertain events. In terms of our BDA it is simply $\alpha = 0.5$ case in Hurwicz principle.

Table 11.8

Laplace principle

		State of nature		α = 0.5
		Strong demand	Weak demand	Weighted average
Decision alternatives		s_1	s_2	
Small complex	d_1	8	7	7.5
Medium complex	d_2	14	5	9.5
Large complex	d_3	20	−9	5.5
Laplace principle maximum =				9.5
Decision =				d_2

11.5 DECISION MAKING UNDER RISK

The decision making situation wherein the decision maker knows the probability of each event is often called decision making under risk.

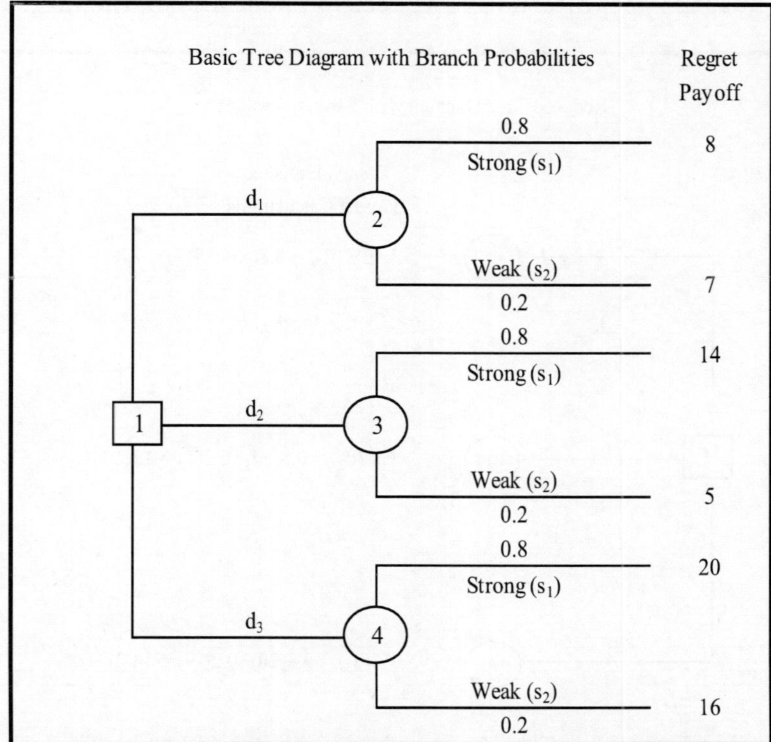

Figure 11.7

Such probabilities are obtained from the past data in the said business. Even in the absence of past data, on the basis of experience and judgment he can work out reasonably good probabilities. This additional information guides the decision maker to make a more accurate decision.

Expected Payoff Principle

The expected value concept developed in the probability theory is very handy in this context. Accordingly, the decision-maker calculates the expected value for all the possibilities and finalizes the decision on the basis of largest expected payoff. So let the subjective probability of relishing the strong demand is $P(s_1) = 0.8$ and weak demand is $P(s_2) = 0.2$. The calculations of expected payoffs could conveniently be done with the tree diagram. Accordingly, we redraw the tree diagram with the subjective branch probabilities as shown in Figure 11.7. Working backward through the tree, we compute the expected payoff values at each chance node as shown in Figure 11.8.

For example, at node (2) the branch expected value is obtained as $EV(d_1) = 0.8(8) + 0.2(7) = 7.8$. In the same way the expected values at node (3) and (4) are obtained and reported in the reduced diagram. From the diagram the profit of Rs. 14.2 is the highest expected value. Thus, the best alternative at decision node 1 is d_3, the large complex.

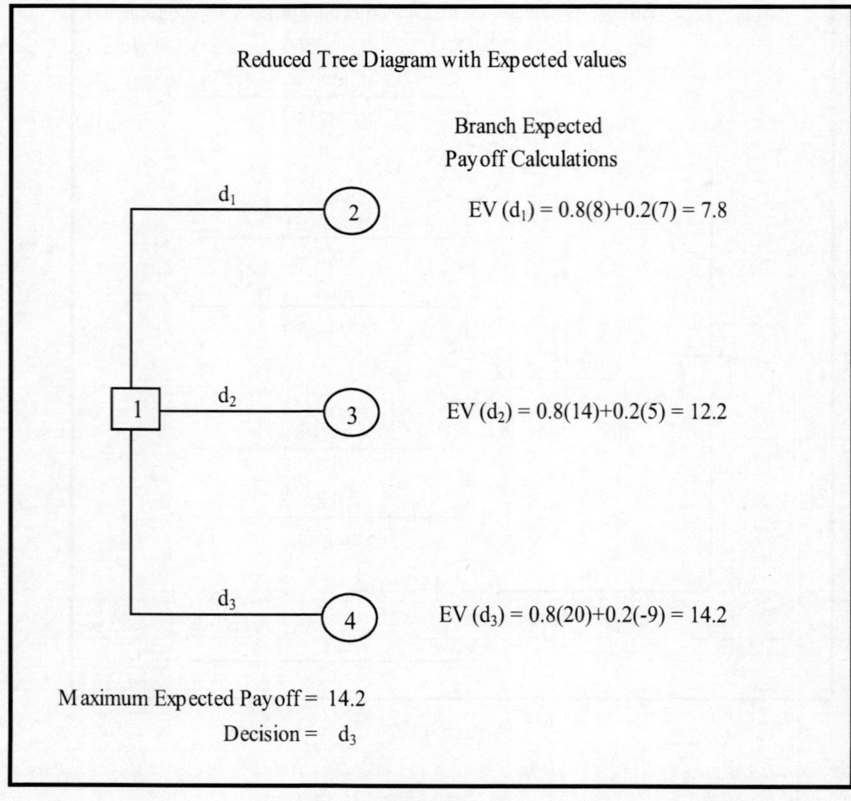

Figure 11.8

a) Tabular method for optimum decision by expected value approach

Table 11.9

Maximum expected value solution						
		State of nature		Prior probabilities		
		Strong demand	Weak demand	Strong demand	Weak demand	Expected value
Decision alternatives		s_1	s_2	s_1	s_2	payoff
Small complex	d_1	8	7	0.8	0.2	7.8
Medium complex	d_2	14	5	0.8	0.2	12.2
Large complex	d_3	20	−9	0.8	0.2	14.2
			Maximum expected value =			14.2
				Decision =		d_3

From Table 11.9, the profit of Rs. 14.2 is the highest expected value. Thus, the best decision is d_3, the large complex.

11.6 EXPECTED PAYOFF CALCULATIONS WITH PERFECT INFORMATION

Suppose BDA had the opportunity to conduct market research regarding the customer's tastes, the management could use this additional information for its advantage. Under the circumstances, BDA could determine, with certainty, prior to making decision, which state of nature is going to occur in the future. From payoff table discussed earlier note that,

i) If BDA knew with certainty that state of nature s_1 would occur, it would venture for large complex and knockout a hefty profit of Rs. 20 crore.

ii) If BDA knew with certainty that state of nature s_2 would occur, it would venture for small complex and earn a profit of Rs. 7 crore.

Now to compute the expected payoffs with perfect information, let us restore the probabilities 0.8 and 0.2 respectively for strong and weak demands respectively. Thus, there is a probability of 0.8 that this perfect information will indicate the state of nature s_1 and the resulting decision alternative d_3 will provide Rs. 20 crore profit. Similarly, with a probability of 0.2 for the state of nature s_2, the optimal decision d_1 will provide Rs. 7 crore profit. Thus, with perfect information the expected value is obtained as

$(EV)_{W.P.I} = 0.8(20) + 0.2(7) = 17.4$ where $(EV)_{W.P.I}$ refers expected value with perfect information.

In our earlier discussion without such perfect information the expected value was obtained as Rs. 14.2 crore. So hereafter we refer this value as the expected value without perfect information and denote it as $(EV)_{W.O.P.I.}$

Thus, $(EV)_{W.O.P.I} = 14.2$

It is worth to note that with perfect information our BDA could earn an additional profit of Rs. 3.2 crore (17.4 −14.2). Thus, to get this type of perfect market information, any market research costs less than or equal to Rs. 3.2 crore is rewarding to undertake.

Thus, the expected value of perfect information is given by

$$\begin{aligned}(EV)_{P.I} &= \left|(EV)_{W.P.I} - (EV)_{W.O.P.I}\right| \\ &= \left|17.4 - 14.2\right| \\ &= \text{Rs. } 3.2 \text{ crore}\end{aligned}$$

Since the expected value of the perfect information is Rs. 3.2 crore the final decision is to go for a market research and to assert the nature of the demand in future provided the market research costs anything less than Rs. 3.2 crore.

11.7 RISK PROFILES

The risk profile for a decision alternative shows the payoffs and their associated probabilities. In our illustrative example the expected value of Rs. 14.2 crore for the decision alternative d_3 is based on a 0.8 probability of getting Rs. 20 crore and a 0.2 probability of involving a loss of Rs. 9 crore. Here the 0.8 for Rs. 20 crore and 0.2 for Rs. −9 crore constitutes the needed risk profile.

The following diagram illustrates the risk profiles for all the three decision alternatives.

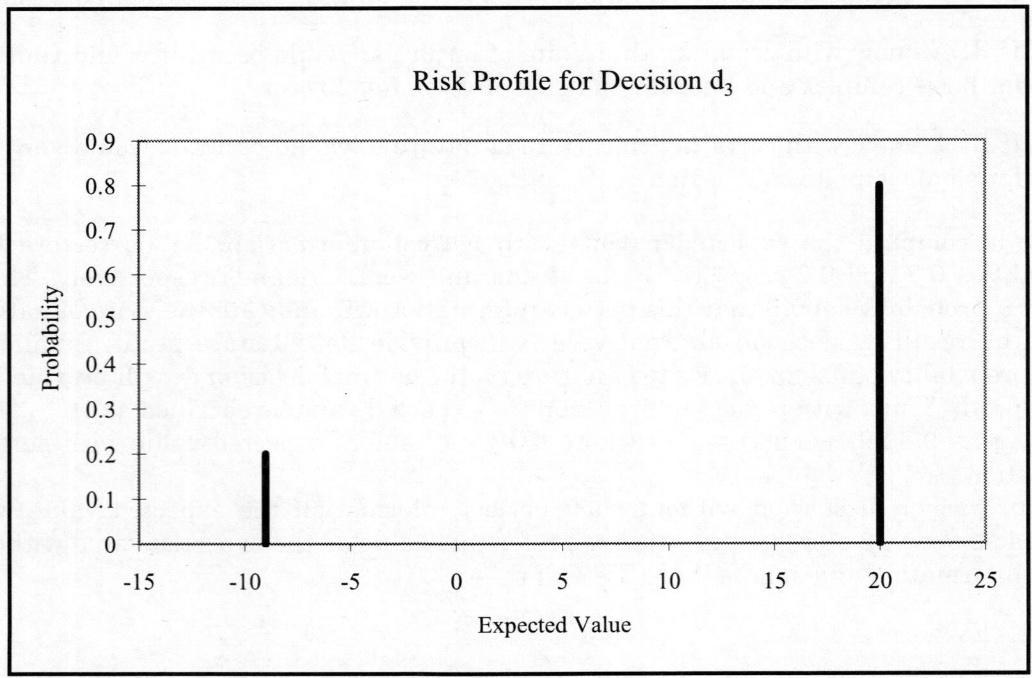

Figure 11.9

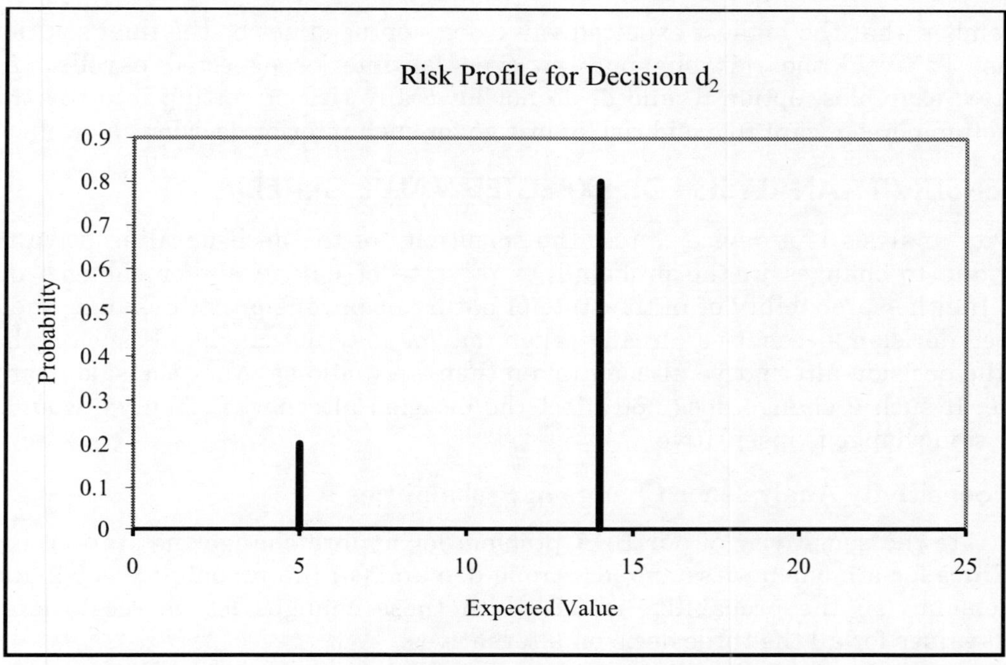

Figure 11.10

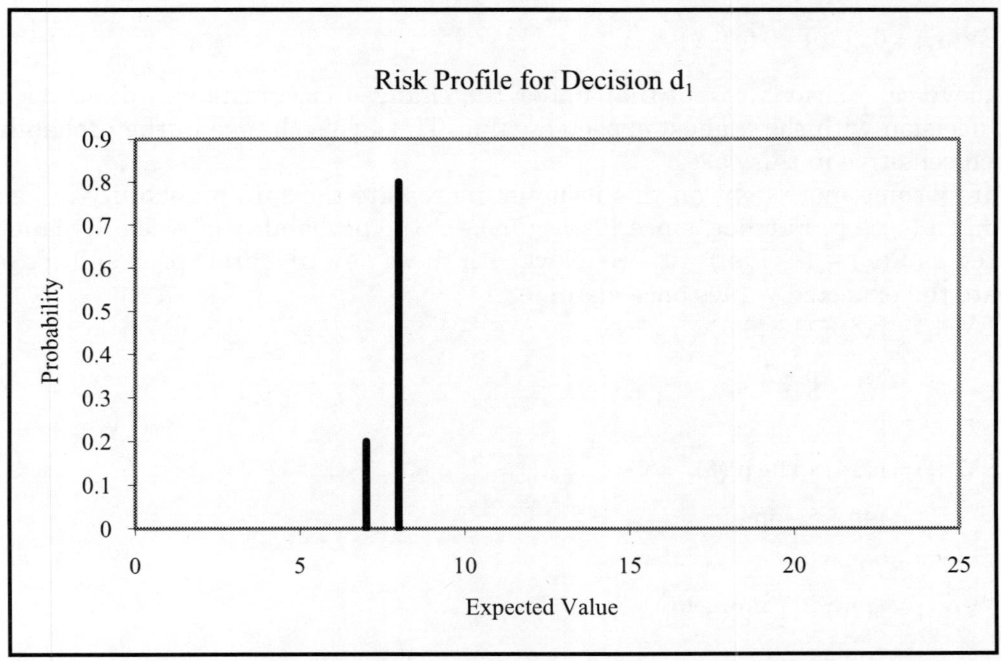

Figure 11.11

Remember, that the highest expected value decision need not be the final solution all the time. To avoid the risk one may go even for inferior expected payoffs. In our illustrative examples, option d_2 and d_3 do not have any risk of getting into the trap of loss. So often person want to avoid risk of loss go for such inferior decisions as well.

11.8 SENSITIVITY ANALYSIS FOR EXPECTED VALUE CRITERIA

Sensitivity analysis is a device to test the sensitivity of the decision alternative taken already due to changes in the probabilities of state of nature and/or changes in the payoffs. If either probability of in the state of nature or payoffs or both change, then the associated decision alternative already taken may or may not change. If such a change affects the decision alternative already taken then we would say that the said change is sensitive. If such a change does not affect the decision alternative then we would say, that the said change is insensitive.

11.8.1 Sensitivity Analysis for Change in Probabilities

To illustrate the sensitivity of our BDA problem let us interchange the state of nature probabilities for a moment. Assume for strong demand (s_1) the probability is 0.2 and for weak demand (s_2) the probability is 0.8. With these changes let us recalculate the expected values for all the three decision alternatives:

$$EV(d_1) = 0.2(8) + 0.8(7) = 7.2$$

$$EV(d_2) = 0.02(14) + 0.8(5) = 6.8$$

$$EV(d_3) = 0.2(20) + 0.08(-9) = -3.2$$

The above calculations reveal that under the changed circumstances d_1 seems to be the best decision with the highest expected value. Thus, this change in the probability is very much sensitive in this case.

To throw some more light on this issue let us restate the said probability in general as p so that $P(s_1) = p$. Further, since $P(s_1) + P(s_2) = 1$ the probability of weak demand may be restated as $P(s_2) = 1 - P(s_1) = 1 - p$. Now with these new algebraic probabilities let us recalculate the expected values once again.

$$EV(d_1) = p(8) + (1-p)(7)$$

$$= 8p - 7p + 7$$

$$= p + 7$$

$$EV(d_2) = p(14) + (1-p)(5)$$

$$= 14p + 5 - 5p$$

$$= 9p + 5$$

$$EV(d_3) = p(20) + (1-p)(-9)$$

$$= 20p - 9 + 9p$$

$$= 29p - 9$$

Thus, we are having three straight-line equations with the expected value as the dependent variable and the probability p as the independent variable. Figure 11.12 shows all the three straight lines and the range in which the decision is in sensitive.

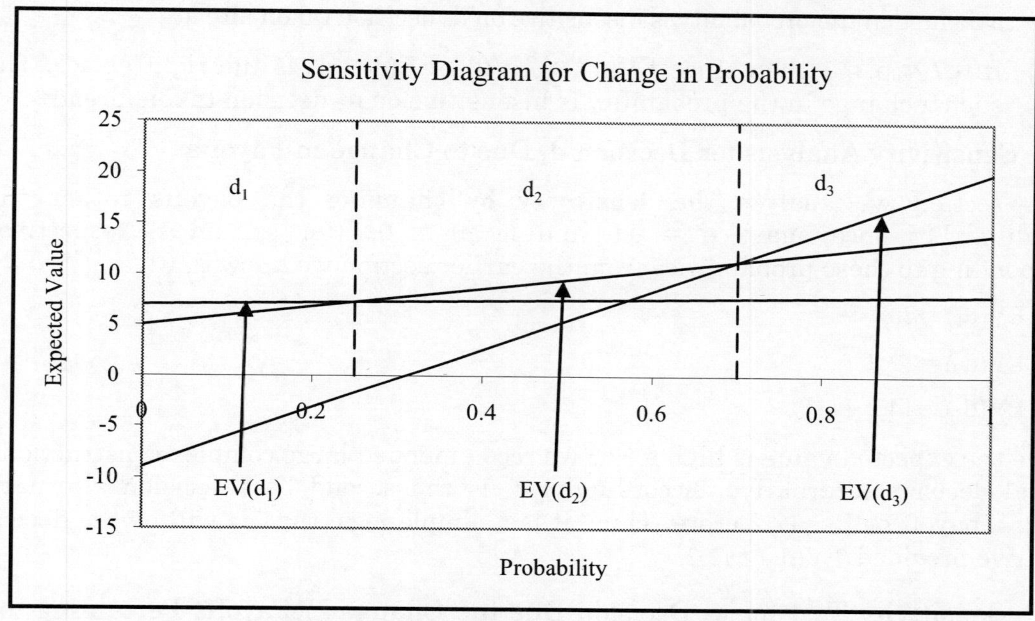

Figure 11.12

The diagram shows how the probability change affects the decision. On the vertical axis we measure the expected payoffs and on the horizontal axis we measure the probability p. Let us calculate the point of intersection of EV (d_1) and EV (d_2).

$$p + 7 = 9p + 5$$

$$9p = 2$$

$$p = 2/8 = 0.25$$

Similarly, the point of intersection between EV (d_2) and EV (d_3) is obtained as

$$29p - 9 = 9p + 5$$

$$20p = 14$$

$$p = 14/20 = 0.7$$

Now from the graph one can get the following five special results:

i) If $p < 0.25$ then since $EV(d_1) > EV(d_2) > EV(d_3)$ d_1 is the right choice. Thus such a change in the probability is sensitive on d_3 decision taken already.

ii) If $p = 0.25$, then $EV(d_1) = EV(d_2) > EV(d_3)$. So one can chose either d_1 or d_2. Thus such a change in the probability is sensitive on d_3 decision taken already.

iii) If $0.25 \leq p \leq 0.7$ then since $EV(d_2) > EV(d_1) > EV(d_3)$. d_2 choice is the right decision. Thus such a change in the probability is sensitive on d_3 decision taken already.

iv) If $p = 0.7$ then since $EV(d_2) = EV(d_3) > E(d_1)$ one can chose either d_2 or d_3. Thus such a change in the probability is insensitive on d_3 decision taken already.

v) If $0.7 < p < 1$ then since $EV(d_3) > EV(d_2) > E(d_1)$. d_3 is the right choice. Thus such a change in the probability is insensitive on d_3 decision taken already.

11.8.2 Sensitivity Analysis for Decision d_3 Due to Change in Payoffs

In this section we analyze the sensitivity by changing the payoffs rather than probability. Let once again $p = 0.8$ and $(1-p) = 0.2$ for s_1 and s_2 respectively. Corresponding to these probabilities from the earlier section we know that:

$$EV(d_1) = 7.8$$

$$EV(d_2) = 12.2$$

$$EV(d_3) = 14.2$$

Since the expected value is high for d_3 we recommended large complex construction as the final decision alternative. Accordingly, d_2 is the second best decision alternative yielding a payoff or Rs. 12.2 crore. Thus, it is a simple logic that d_3 will be the decision alternative provided $EV(d_3) \geq 12.2$.

11.8.3 Sensitivity Test for d_3 Decision Due the Change in Payoffs Pertaining to d_3 One at a Time

Now let A be the payoff for d_3 when the demand is strong and B be the payoff when the demand is weak.

So $EV(d_3) = 0.8A + 0.2B$

i) Assume that the payoff B is restored at the same old level of Rs. −9. Now let us try to calculate the minimum payoff A to restore the decision d_3.

Now d_3 holds good as the best payoff alternative provided.

$$EV(d_3) = 0.8A + 0.2(-9) \geq 12.2$$

$$0.8A \geq 14$$

$$A \geq 17.5$$

Recall that when demand is strong, d_3 has an estimated payoff of Rs. 20 crore. The preceding calculation shows that the decision alternative d_3 will hold good with strong demand provided the payoff is at least Rs. 17.5 crore.

ii) Assume that the payoff A is restored at the same old level of Rs. 20. Let us try to calculate the minimum payoff B to restore the decision d_3.

Now d_3 holds good as the best payoff alternative provided.

$$EV(d_3) = 0.8(20) + 0.2B \geq 12.2$$
$$0.2B \geq -3.8$$
$$B \geq -19$$

Recall that when demand was weak, d_3 has an estimated payoff of Rs. −9 crore. The preceding calculation shows that the decision alternative d_3 with weak demand will hold good provided the payoff B ≥ Rs. −19 crore.

However, in these cases decision alternative d_3 remains optimal only the chances in the expected payoffs for d_1 and d_2 meet the requirements $EV(d_1) \leq 14.2$ and $EV(d_2) \leq 14.2$.

11.8.4 Sensitivity Test for d_3 Decision Due the Change in Payoffs Pertaining to d_1 One at a Time

Now let C be the payoff for d_1 when the demand is strong and D be the payoff when the demand is weak.

So $EV(d_1) = 0.8C + 0.2D$

i) Assume that the payoff D is restored at the same old level of Rs. 7. Now let us try to calculate the minimum payoff C to restore the decision d_3.

Now d_3 holds good as the best payoff alternative provided.

$$EV(d_1) = 0.8C + 0.2(7) \leq 14.2$$
$$0.8C \leq 14.2 - 1.4$$
$$C \leq \frac{12.8}{0.8}$$
$$C \leq 16$$

ii) Assume that the payoff C is restored at the same old level of Rs. 8. Let us try to calculate the minimum pay off D to restore the decision d_3.

$$EV(d_1) = 0.8(8) + 0.2D \leq 14.2$$
$$0.2D \leq 14.2 - 6.4$$
$$D \leq \frac{7.8}{0.2}$$
$$D \leq 39$$

11.8.5 Sensitivity Test for d_3 Decision Due the Change in Payoffs Pertaining to d_2 One at a Time

Now let E be the payoff for d_2 when the demand is strong and F be the payoff when the demand is weak.

So $EV(d_2) = 0.8E+0.2F$

i) Assume that the payoff F is restored at the same old level of Rs. 5. Now let us try to calculate the minimum payoff E to restore the decision d_3.

Now d_3 holds good as the best payoff alternative provided.

$$EV(d_2) = 0.8E + 0.2(5) \leq 14.2$$

$$0.8E \leq 14.2 - 1.0$$

$$E \leq \frac{13.2}{0.8}$$

$$E \leq 16.5$$

ii) Assume that the payoff E is restored at the same old level of Rs. 14. Let us try to calculate the minimum payoff F to restore the decision d_3.

$$EV(d_2) = 0.8(14) + 0.2F \leq 14.2$$

$$0.2F \leq 14.2 - 11.2$$

$$F \leq \frac{3}{0.8}$$

$$F \leq 15$$

Table 11.10 Sensitivity analysis summary table for d_3 decision

		State of nature			
		Strong demand s_1		Weak demand s_2	
Decision alternatives		Original	Permissible change	Original	Permissible change
Small complex	d_1	8	≤ 16	7	≤ 39
Medium complex	d_2	14	≤ 16.5	5	≤ 15
Large complex	d_3	20	≥ 17.5	−9	≥ -19

11.9 DECISION ANALYSIS WITH SAMPLE INFORMATION

In the expected value approach discussed in the previous section we utilized the prior probability information at every stage. However, the decision-maker often wants to improve the prior probability information on the basis of actual field experiments. These revised posterior probabilities often change the decision alternatives already taken.

If the BDA is considering a six-month market research to learn more about the acceptance of its customers for the said complex project. On the basis of the study the BDA can expect any one of the following two results.

1. **Favourable report:** A significant number of customers conducted show interest in complex purchase.

2. **Unfavourable report:** Only very few contacted individuals show interest in complex purchase.

11.9.1 The Influence Diagram

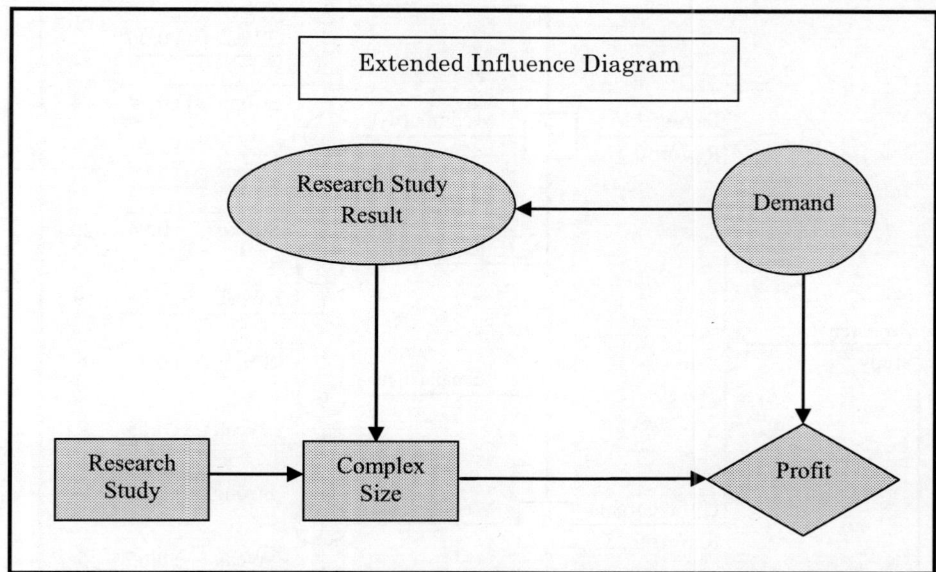

Figure 11.13

After introducing the market research our BDA problem becomes more complex. Note that, now we have two decision nodes one for research study and another for complex size. The chance node has also gone up to two one for market research and another for market demand. Finally, the consequence node is the profit as in the previous case. Figure 11.13 shows the extended influence diagram. Figure 11.14 shows the extended tree diagram.

11.9.2 Tree Diagram with Market Research Study

The tree diagram with the market study information given in Figure 11.4 illustrates the logical sequence for the decision and associated chance nodes. First BDA must decide whether to have the market study or not. Thus, at decision node [1] we have two branches, one for going to market research {chance node (2)} and another for not going for market research {decision node [5]}.

Once decision is taken to have the market study, the result may be favourable or not. Thus, our chance node (2) will have two branches once again, one for favourable market report {decision node [3]} and another for unfavourable market report {decision node [4]}.

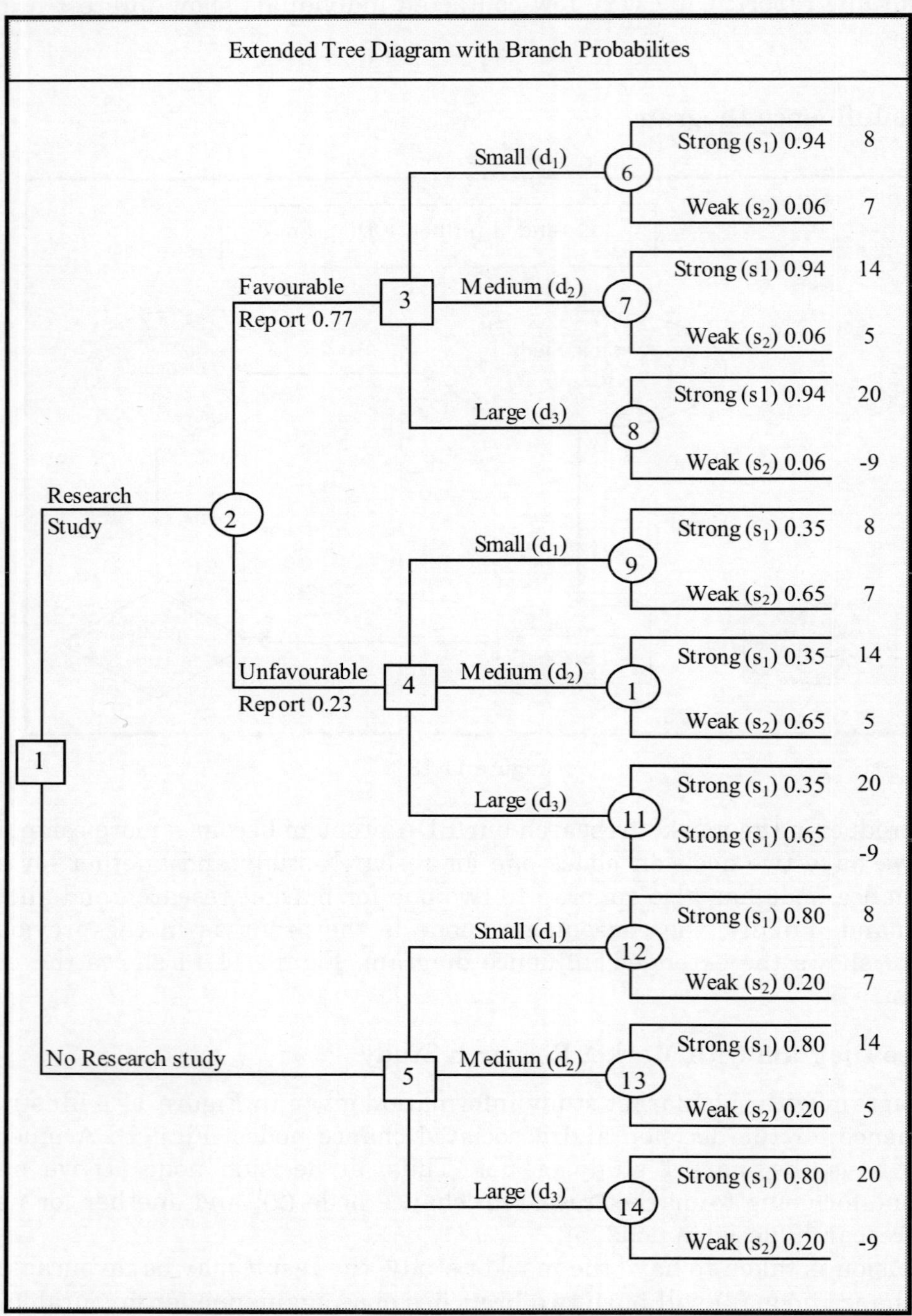

Extended Tree Diagram with Branch Probabilites
1
Research Study
2
Favourable Report 0.77
3
Small (d1)
6
Strong (s1) 0.94 8
Weak (s2) 0.06 7
Medium (d2)
7
Strong (s1) 0.94 14
Weak (s2) 0.06 5
Large (d3)
8
Strong (s1) 0.94 20
Weak (s2) 0.06 -9
Unfavourable Report 0.23
4
Small (d1)
9
Strong (s1) 0.35 8
Weak (s2) 0.65 7
Medium (d2)
1
Strong (s1) 0.35 14
Weak (s2) 0.65 5
Large (d3)
11
Strong (s1) 0.35 20
Strong (s1) 0.65 -9
No Research study
5
Small (d1)
12
Strong (s1) 0.80 8
Weak (s2) 0.20 7
Medium (d2)
13
Strong (s1) 0.80 14
Weak (s2) 0.20 5
Large (d3)
14
Strong (s1) 0.80 20
Weak (s2) 0.20 -9

Figure 11.14

If the report is favourable then again at decision node [3] we have three branches one for large size complex {chance node (6)} another for the medium size complex {chance node (7)} and a third one for the small size complex {chance node (8)}.

In this logical fashion the rest of the diagram is constructed and completed in Figure 11.14. Note that here the squares are used for decision nodes and circles or ovals are used for chance nodes.

To obtain the correct optimal decision alternatives at each decision node level we need the respective branch probabilities.

If the market research study is undertaken at decision node [1] then let us say

P (Favourable report) = 0.77

P (Unfavourable report) = 0.23

If market research report is favourable at decision node [3] then again let us say

P (Strong demand given a favourable report) = 0.94

P (Weak demand given a favourable report) = 0.06

If market research report is unfavourable at decision node [3] then again let us say

P (Strong demand given a unfavourable report) = 0.35

P (Weak demand given a unfavourable report) = 0.65

If market research is not taken then at decision node [5] we will have the associated prior probabilities discussed in the previous section

P (strong demand) = 0.8

P (weak demand) = 0.2

A decision strategy is a sequence of decisions taken at decision node on the basis of outcomes of the chance nodes. Here we propose to get the optimum strategy at each nodal level by using backward rolling rule as usual.

11.9.3 Rules

1. If you are at chance node, obtain the expected value by multiplying the payoffs at the tail end by the respective probabilities and add the results obtained. Write these expected values against the respective nodal levels after rolling back the relevant branches.

2. If you are at the decision node, select the branch which corresponds to the highest expected value and leaves the other branches. Write the so obtained branch expected value against the decision node after rolling back the relevant branches. Figure 11.15 shows the first stage of reduction.

   i) In Figure 11.15 after the first stage of reduction, the expected values for chance node (6) to (14) are reported using rule 1.

ii) In Figure 11.16 after the second stage of reduction, the best expected values for decision node [3] to [5] are reported.

iii) In Figure 11.17 after third stage of reduction the relevant expected values are reported for chance node (2) and decision node [5].

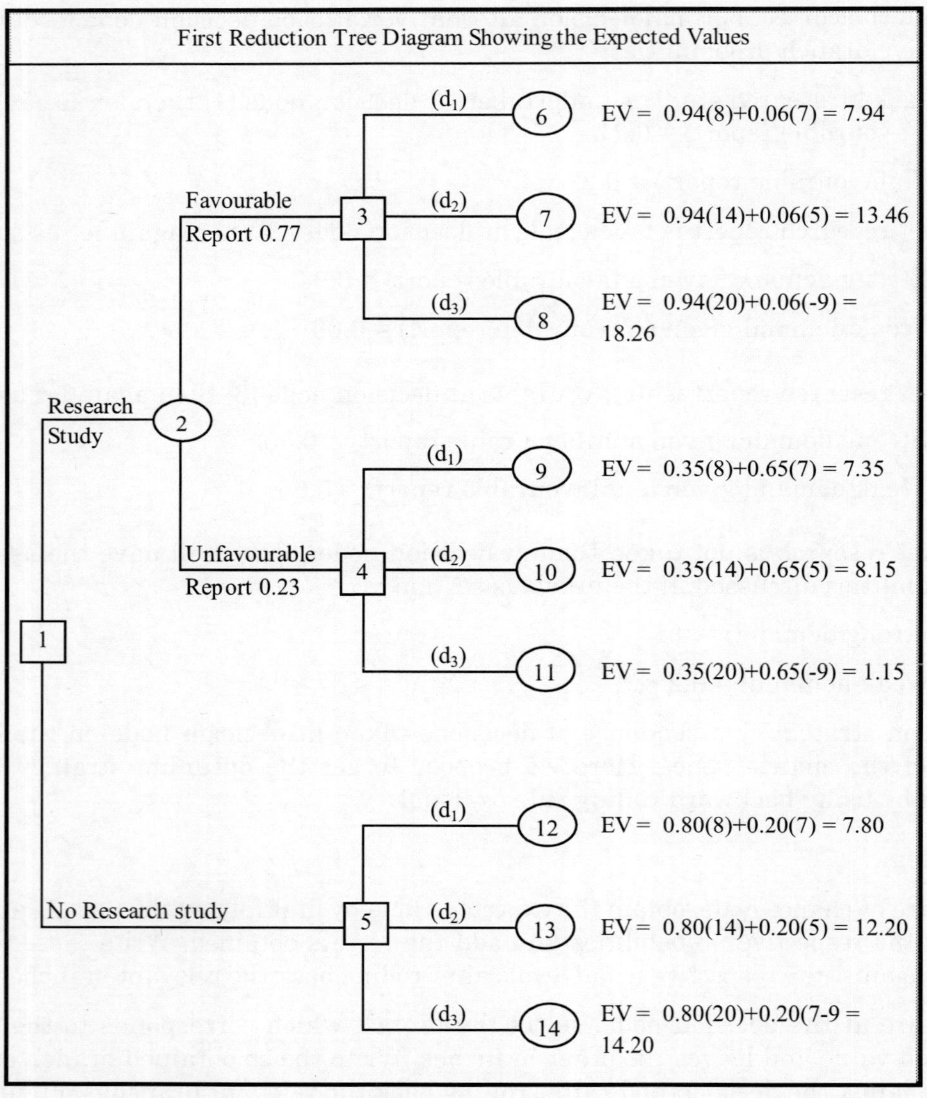

Figure 11.15

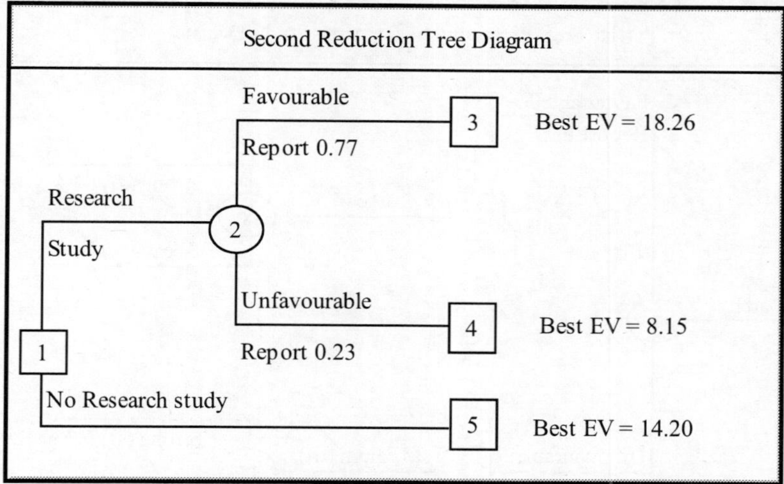

Figure 11.16

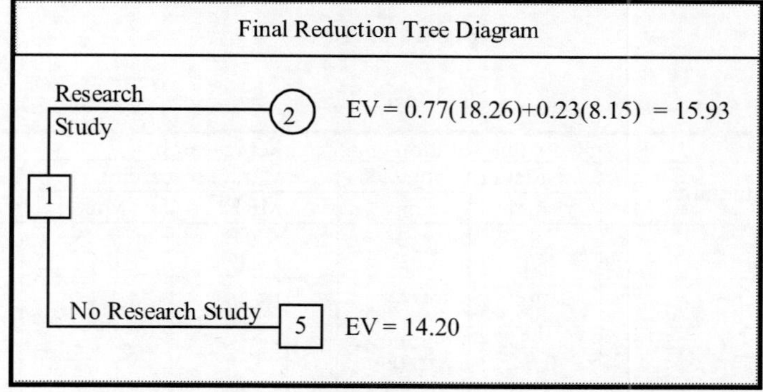

Figure 11.17

Since the expected value at chance node (2) is greater than the expected value at decision node number [5] our BDA would go for market research study finally. Once decision is taken for market research study the relevant part of the decision tree is reported in Figure 11.18 for ready reference.

Note: *MRF= Market research favourable*
MRU=Market research unfavourable
NMR= No market research

i) If market research study is favourable as per decision tree 11.18 it will go for a large complex.

ii) If market the study is unfavourable then as per the decision tree diagram 11.18 it would go for medium complex.

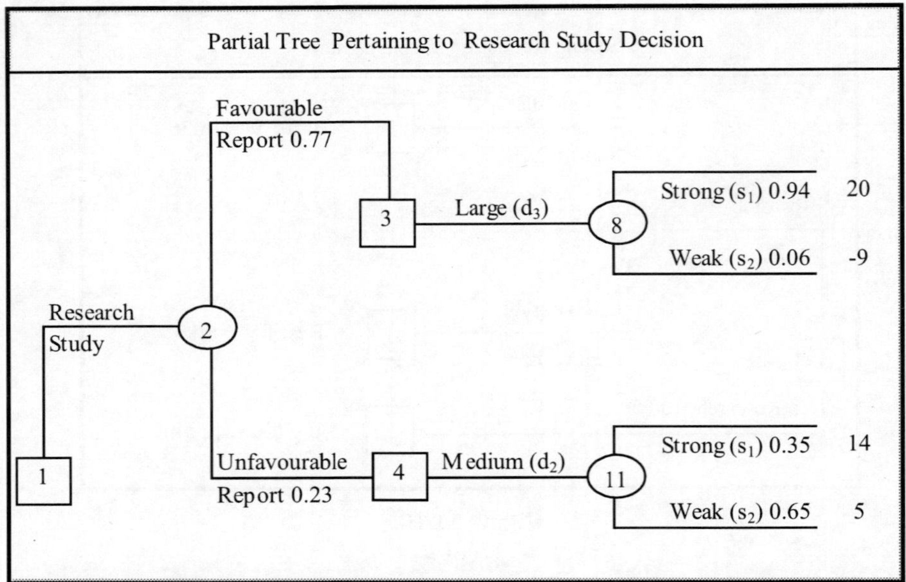

Figure 11.18

Table 11.11

Decision alternative	State of nature		Market research EV		EV
Expected value solution under market research					
	s_1	s_2	MRF	MRU	NMR
d_1	8	7	7.94	7.35	7.8
d_2	14	5	13.46	8.15	12.2
d_3	20	−9	18.26	1.15	14.2
$P(s_j)$	0.8	0.2			
$P(s_j/F)$	0.94	0.06			
$P(s_j/U$	0.35	0.65			
P(F)	0.77				
P(U)	0.23				
Maximum branch expected values			18.26	8.15	14.2
Reduced expected values (MR &NMR)			15.93		14.2
First level decision			MR		
Second level decision (MRF)			d_3		
Third level decision (MRU)			d_2		

11.10 RISK PROFILE FOR THE DECISION TO TAKE UP MARKET RESEARCH STUDY

Figure 11.18 above show the partial decision tree pertaining to the decision alternative for the market research study. Accordingly, BDA will get one of the payoffs depending upon the result of the research and the size of the demand shown at the terminal branches of the decision tree. Recall that a risk profile shows the possible payoffs with their associated probabilities.

i) The BDA will get Rs. 20 crore and go for large complex (d_3) provided the research study is favourable and the associated demand is strong. So the probability of getting Rs. 20 crore = 0.77 × 0.94 = 0.72.

ii) The BDA will get Rs. −9 crore and go for large complex (d_3) provided the research study is favourable and the associated demand is weak. So the probability of getting Rs. −9 crore = 0.77 × 0.06 = 0.05.

iii) The BDA will get Rs. 14 crore and go for medium complex (d_2) provided the research study is unfavourable and the associated demand is strong. So the probability of getting Rs. 14 crore = 0.23 × 0.35 = 0.08.

iv) The BDA will get Rs. 5 crore and go for large complex (d_2) provided the research study is favourable and the associated demand is weak. So the probability of getting Rs. 20 crore = 0.23 × 0.65 = 0.15.

The Table 11.12 and the associated Figure 11.19 show the risk profile for taking up the market research study.

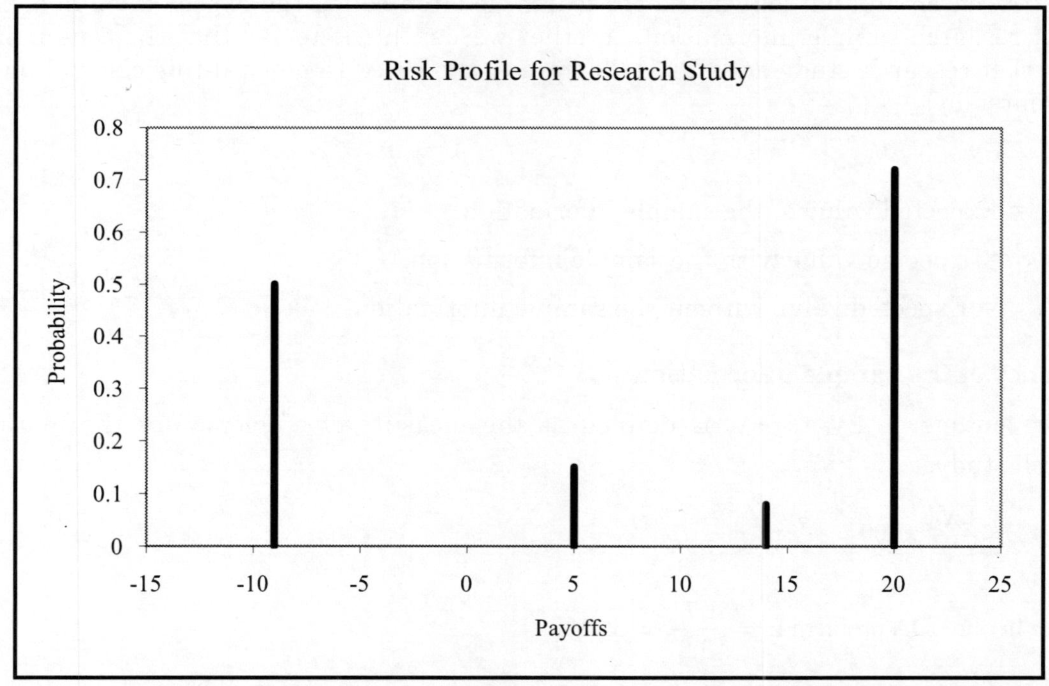

Figure 11.19

Table 11.12

Risk profile for market research	
Payoffs	Probability
–9	0.05
5	0.15
14	0.08
20	0.72
Total	1.00

Table 11.12 and the associated Figure 11.19 reveal that the probability getting Rs. 20 crore is very high at 72%.

Expected value of the market research sample information

The expected payoff when market research study is not undertaken is obtained as Rs. 14.2 crore in the previous section. Now with the market research the expected value is Rs. 15.93 crore. Thus, Rs. 1.73 crore, the difference between the expected value with market research and without market research, is often called the expected value of the market research sample information. In other words, this is to say that the conducting the market research study adds Rs. 1.73 crore to the BDA. In general this calculation is represented by:

$$EV_{SI} = |EV_{WSI} - EV_{WOSI}|$$

EV_{SI} = Expected value of the sample information.

EV_{WSI} = Expected value with the sample information.

EV_{WOSI} = Expected value without the sample information.

Efficiency of the sample information

The percentage of EV_{SI} to EV_{PI} is defined as the measure of efficiency for the market research study.

$$E = \frac{EV_{SI}}{EV_{PI}} \times 100$$

In our BDA problem $E = \frac{1.73}{3.2} \times = 54.1\%$

This is to say that the market research information is 54.1% efficient than the perfect information discussed in the beginning. Low efficiency often leads to search for better information by some other means. High efficiency information is as good as that of the perfect information.

11.11 COMPUTING THE BRANCH PROBABILITIES BY USING BAYE'S THEOREM

In the BDA problem discussed above the branch probabilities were specified in advance. No attempt has been made to calculate these probabilities.

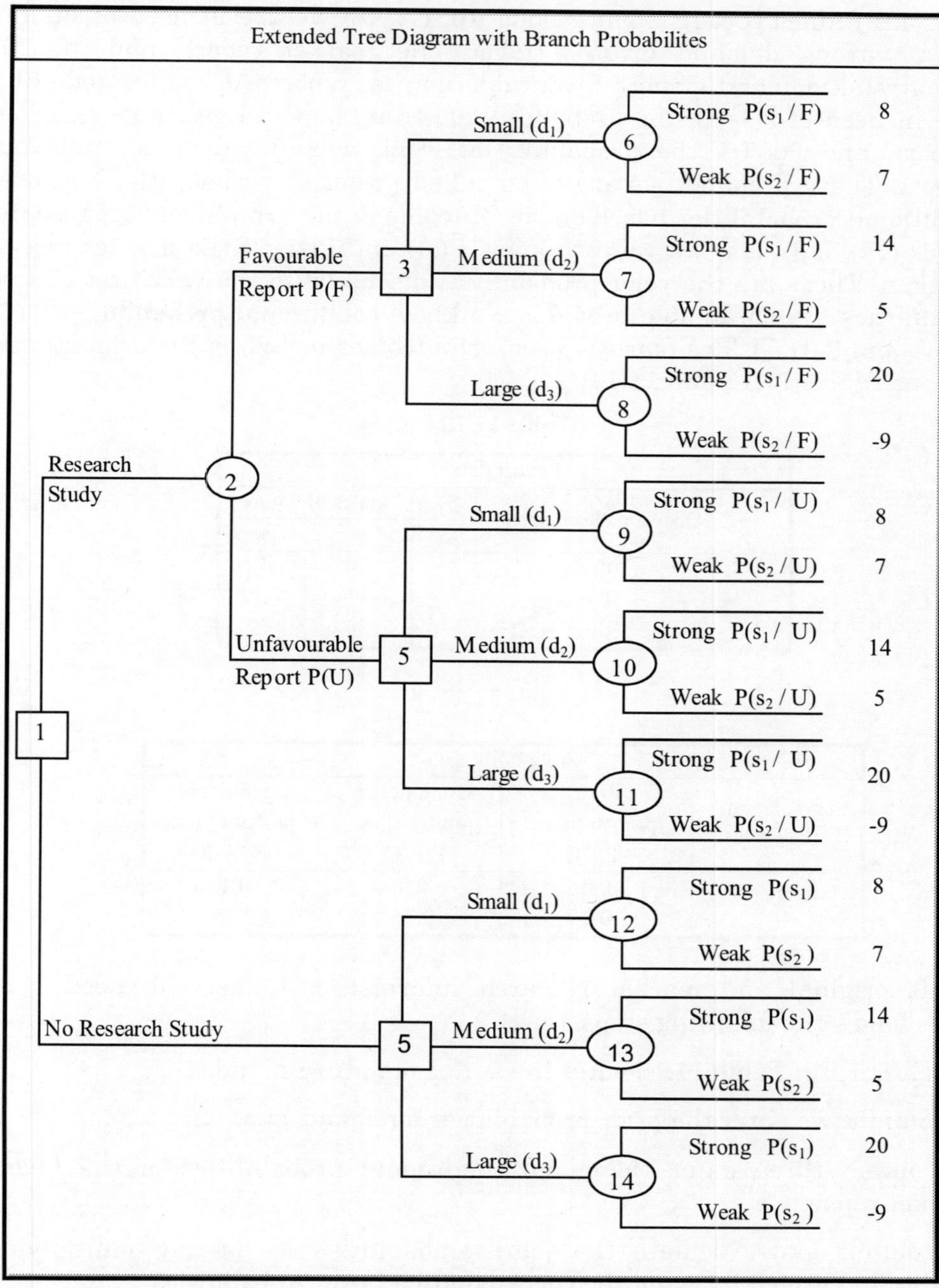

Figure 11.20

In this section we propose to use Baye's theorem to compute these branch probabilities. To highlight the decision tree, let F stand for favourable market research report and U stands for unfavourable market research report.

In the tree diagram in Figure 11.20, at chance node (2) we need to know the branch probabilities P(F) and P(U). At chance node (6), (7), (8) we are in need of $P(s_1/F)$, the probability of strong demand given a favourable market report, and $P(s_2/F)$, the probability of weak demand given a favourable market report. At chance node (9), (10), (11) we are in need of $P(s_1/U)$, the probability of strong demand given an unfavourable market report, and $P(s_2/U)$, the probability of weak demand given an unfavourable market report. These probabilities are often called posterior probabilities because they are all conditional probabilities based on the outcome of the sample market research. At chance nodes (12), (13), (14) we want to know $P(s_1)$ and $P(s_2)$, if the market research is not undertaken. These are the prior probabilities discussed in the very first section. All these probabilities are obtainable provided we know conditional probabilities $P(F/s_1)$, $P(F/s_2)$, $P(U/s_1)$, and $P(U/s_2)$. The market research outcomes including the original data are listed below in Tables 11.13 and 11.14.

Table 11.13

Payoff data		
Decision alternatives	States of nature	
	s_1	s_2
d_1	8	7
d_2	14	5
d_3	20	−9

Table 11.14

Probability data			
State of nature s_j	Prior probabilities $P(s_j)$	Conditional probabilities $P(F/s_j)$	Conditional probabilities $P(U/s_j)$
s_1	0.8	0.9	0.1
s_2	0.2	0.25	0.75

With this original and market research information to get the needed branch probabilities follow the following steps.

1. In column 1 in Table 11.15 enter the states of nature s_1 and s_2.
2. In column two enter the prior probabilities for s_1 and s_2.
3. In column three enter the given conditional probabilities of the favourable market report.
4. In column four calculate the joint probabilities by merely multiplying the conditional probability and the corresponding prior probabilities.

5. Add the probabilities in column 4 and obtain the needed probability of the favourable market research P(F).
6. Divide the respective joint probabilities by the P(F) = 0.77 and obtain the needed posterior probabilities, $P(s_1/F)$ and $P(s_2/F)$ in the last column.

To get the branch probabilities for the unfavourable market research replace the conditional probabilities in column 3 for unfavourable market research probabilities. Repeat all the steps listed above and get the needed probabilities as shown in Table 11.16.

Table 11.15

Calculation of posterior probabilities (F)				
State of nature s_j	Prior probabilities $P(s_j)$	Conditional probabilities $P(F/s_j)$	Joint probabilities $P(F \cap s_j)$	Posterior probabilities $P(s_j/F)$
s_1	0.8	0.9	0.72	0.94
s_2	0.2	0.25	0.05	0.06
		P(F) =	0.77	1.00

Table 11.16

Calculation of posterior probabilities (U)				
State of nature s_j	Prior probabilities $P(s_j)$	Conditional probabilities $P(U/s_j)$	Joint probabilities $P(U \cap s_j)$	Posterior probabilities $P(s_j/U)$
Strong s_1	0.8	0.1	0.08	0.35
Weak s_2	0.2	0.75	0.15	0.65
		P(U) =	0.23	1.00

Table 11.17

Summary of all the needed probabilities		
	States of nature	
Probabilities	s_1	s_2
$P(s_j)$	0.8	0.2
$P(s_j/F)$	0.94	0.06
$P(s_j/U)$	0.35	0.65
P(F)	0.77	
P(U)	0.23	

11.12 DECISION MAKING WITH EXPECTED UTILITY

In the preceding section we expressed the expected values although in money terms to arrive at the needed optimum profit solution. Often the decision alternative with best-expected money value need not necessarily be the best optimum solution. In certain special cases one will have to go beyond the mere expected money value to arrive at the correct optimum solution. Certainly the decision to go for fire insurance policy for our house does not provide a high money expected value than for not going for insurance. Similarly, people go for State Lotteries even though the expected money is very small. If we go by the rule of high expected money value no one will buy the state lotteries because not going for lottery certainly gives a high positive expected value always. Thus, the expected monetary value is not the only comfortable measure of the optimum decision under all circumstances. Often people find more joy in taking risks then avoiding it. Under such circumstances, one should go invariably for the highest expected utility rather than the mere highest expected money value.

11.13 THE CONCEPT UTILITY

Utility by definition is a measure of the wants satisfying capacity of the commodity when it is consumed. However, in this context we propose to construct utility numbers for all the payoffs expressed in money terms first. Once this is done we calculate the expected utility rather than the expected money value in the usual manner to arrive at the needed optimum solution. As an example, let us consider the problems faced by the Prestige group, relatively a new real estate firm, in Bangalore. The cash available with the firm currently is so manager that it cannot even think of investing more than one project at a time, though it is having two projects in hand. Under the circumstances, it can think of the following three decision alternatives.

d_1 = Make investment in the project A.

d_2 = Make investment in project B.

d_3 = Do not invest in either.

In Bangalore though the real estate is flourishing currently, the price of the real estate in future is very much uncertain, it may either go up, or come down or stable. Thus, the states of nature faced by the company are:

s_1 = Real estate prices go up in future.

s_2 = Real estate prices stay stable in future.

s_3 = Real estate prices come down in future.

With the best of information available the associated payoffs in money terms and their prior probabilities are reported in the following summary table.

Table 11.18

Money payoffs			
	State of nature		
Decision alternative	Prices up, s_1	Prices stable, s_2	Prices down, s_3
Invest in A, d_1	30,000	20,000	−50,000
Invest in B, d_2	50,000	−20,000	−30,000
Do not invest, d_3	0	0	0
Probability	0.3	0.5	0.2

From the table information given above we calculate the expected money values for all the three decision alternatives first.

$$EV(d_1) = 0.3(30,000) + 0.5(20,000) + 0.2(-50,000) = 9000$$

$$EV(d_2) = 0.3(50,000) + 0.5(-20,000) + 0.2(-30,000) = -1000$$

$$EV(d_1) = 0.3(0) + 0.5(0) + 0.2(0) = 0$$

The above calculations reveal that the expected money value is high in the case of d_1. So the optimum decision of the company is to go for the project A. However, it is high time to go for the risk profile before making such a decision.

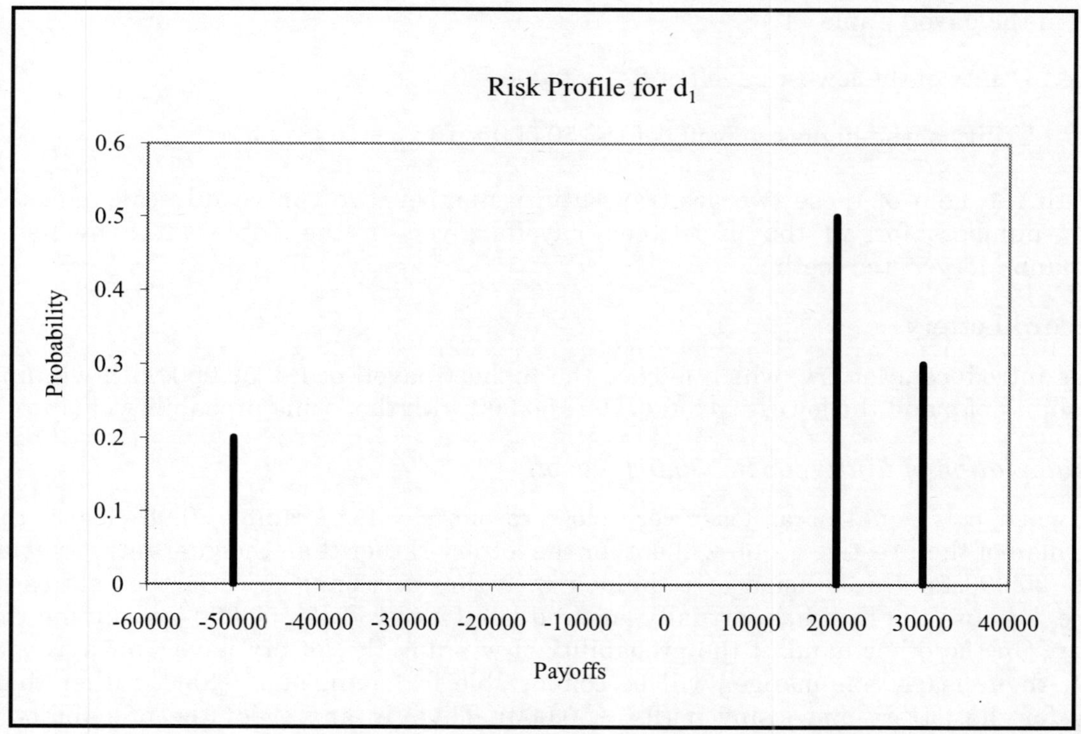

Figure 11.21

Though the expected value is high for d_1 it is very risky to take this venture. According to this decision the company will earn a profit of 30,000 provided price goes up in future. It will earn a profit of Rs. 20,000 if price level is going to be stable in future. If any of these two happens it is well and good for the new company. Unfortunately, if prices of the real estate come down in future (0.2 probability) the company will eventually incur a loss of Rs. 50,000, which is totally unacceptable at this stage of take-off. If it happens, it will have to shut down its operation in Bangalore once for all.

The second best solution of Project B also leads to a similar situation with a lose of Rs. 30,000, which is again unbearable for the firm. The best decision that the manager could take under the risky circumstances is to opt for d_3 and avoid the dear consequences of quitting the business.

The only way out to this tricky problem is to calculate the expected utility values rather than the expected money values. Such an optimum utility is capable of considering the risk factor as well into the analysis. If we succeed in getting such an optimum utility package the decision-making problem gets solved automatically.

11.14 ASSIGNMENT OF UTILITY INDEX FOR PAYOFFS

The Neumann and Morgenstern utility analysis is very much handy in solving this conflict. So let us assign two arbitrary utility levels to the highest and lowest payoffs given in the payoff Table 11.18.

Let, Utility of the lowest payoff of Rs. −50,000 = 0

Utility of the highest payoffs of Rs. 50,000 = 10

With the help of these two arbitrary utility numbers we can readily calculate the utility numbers for all the in-between payoffs given in the Table 11.18 by using Neumann Morgensten method.

Standard Lottery

Let us introduce a lottery, which fetches the highest payoff of Rs. 50,000 with winning probability of p and the lowest payoff of Rs. −50,000 with the losing probability of (1−p).

a) Calculation of utility for Rs. 30,000 payoff

Now when p is equal or at least very close to one say for example, 0.98 clearly the Chairman of the Prestige group will opt for the lottery rather than the guaranteed payoff of Rs. 30,000 for the decision d_1 combined with the increasing price of the estates in future, because the firm has virtually ensured itself a payoff Rs. 50,000 by winning the lottery. On the other hand, if the probability of winning the lottery is very meagre say 0.001, then clearly our manger will be comfortable in having Rs. 30,000, rather than going for the lottery and losing it (Rs. −50,000). Thus, in any event the probability p changes from 1 to 0 the Chairman's preference over the lottery goes on diminishing. Certainly at a point between 0 and 1, the manager will be in a confusing state of affair as whether to have the lottery or the ready cash payoff of Rs. 30,000. At this point the expected utility of the lottery must invariably equal to that of the ready cash payoff of Rs. 30,000. Let this critical probability be 0.9 as per the Chairman's opinion.

$$\begin{aligned} U(30{,}000) &= p[U(50{,}000)] + (1-p)[U(-50{,}000)] \\ &= 0.9(10) + 0.1(0) \\ &= 9 \end{aligned}$$

b) Calculation of utility for Rs. 20,000 payoff

Now since the guaranteed payoff is only Rs. 20,000 the Chairman will be indifferent between the lottery and the guaranteed payoff for a low probability only. Let this probability be p = 0.8.

$$\begin{aligned} U(20{,}000) &= p[U(50{,}000)] + (1-p)[U(-50{,}000)] \\ &= 0.8(10) + 0.2(0) \\ &= 8 \end{aligned}$$

c) Calculation of utility for Re. 0 payoff

Now since the guaranteed payoff is only Re. 0 the Chairman will be indifferent between the lottery and the guaranteed payoff for a still low probability only. Let this probability be p = 0.7.

$$\begin{aligned} U(20{,}000) &= p[U(50{,}000)] + (1-p)[U(-50{,}000)] \\ &= 0.7(10) + 0.3(0) \\ &= 7 \end{aligned}$$

d) Calculation of utility for Rs. –20,000 payoff

Now since the guaranteed payoff is only Rs. −20,000 the Chairman will be indifferent between the lottery and the guaranteed payoff for a still low probability only. Let this probability be p = 0.5.

$$\begin{aligned} U(-20{,}000) &= p[U(50{,}000)] + (1-p)[U(-50{,}000)] \\ &= 0.5(10) + 0.5(0) \\ &= 5 \end{aligned}$$

e) Calculation of utility for Rs. –30,000 payoff

Now since the guaranteed payoff is only Rs. −30,000 the Chairman will be indifferent between the lottery and the guaranteed payoff for a still low probability only. Let this probability be p = 0.4.

$$\begin{aligned} U(-30{,}000) &= p[U(50{,}000)] + (1-p)[U(-50{,}000)] \\ &= 0.4(10) + 0.6(0) \\ &= 4 \end{aligned}$$

The summary results of all these calculations are reported in Table 11.19.

Table 11.19

Utility payoffs			
	State of nature		
Decision alternative	Prices up, s_1	Prices stable, s_2	Prices down, s_3
Invest in A, d_1	9	8	0
Invest in B, d_2	10	5	4
Do not invest, d_3	7	7	7
Probability	0.3	0.5	0.2

Now from the Table 11.19 let us calculate the expected utility in the usual manner

$$EU(d_1) = 0.3(9) + 0.5(8) + 0.2(0) = 6.7$$

$$EU(d_2) = 0.3(10) + 0.5(5) + 0.2(4) = 6.3$$

$$EU(d_3) = 0.3(7) + 0.5(7) + 0.2(7) = 7$$

Note that the optimum decision with highest utility value is d_3, not to invest. The ranking assignment of the Chairman's utility index and the associated expected money payoffs are presented in Table 11.20.

Table 11.20

Expected utility and money values		
Decision alternatives ranking	Expected utility	Expected money value
Do not invest, d_3	7	0
Invest in A, d_1	6.7	9000
Invest in B, d_2	6.3	−1000

Note that even though the money expected value is high for d_1, the project A, the Chairman's decision is d_3, not to invest. The rationale behind the decision for not going for A is that the probability 0.2 for a loss of Rs. 50,000 is considered a very risky involvement. The seriousness of this risk is not brought out by the high-expected money value criterion.

Review Problems

■ Example1

A philanthropist wants to build a hospital in his village. He can build 100, 200, or 300 bed hospitals, depending upon the anticipated demand low, medium or high. The expected net profits and the prior probability probabilities regarding the states of nature is given in the following Table 11.21.

Table 11.21

State of nature	Decision nodes			Prior probability
	100 bed (d_1)	200 bed (d_2)	300 bed (d_3)	
Low (s_1)	20,000	−10,000	−30,000	0.2
Medium (s_2)	25,000	30,000	−5,000	0.3
High (s_3)	30,000	50,000	60,000	0.5

i) Compute EMV_{WOPI}, EMV_{WPI}, and EMV_{PI}.

ii) A researcher agrees to conduct a survey for Rs. 1500 for the philanthropist and provide him with information regarding the states of nature. Should the survey be conduced?

Solution

The following tree diagram exhibits the given problem.

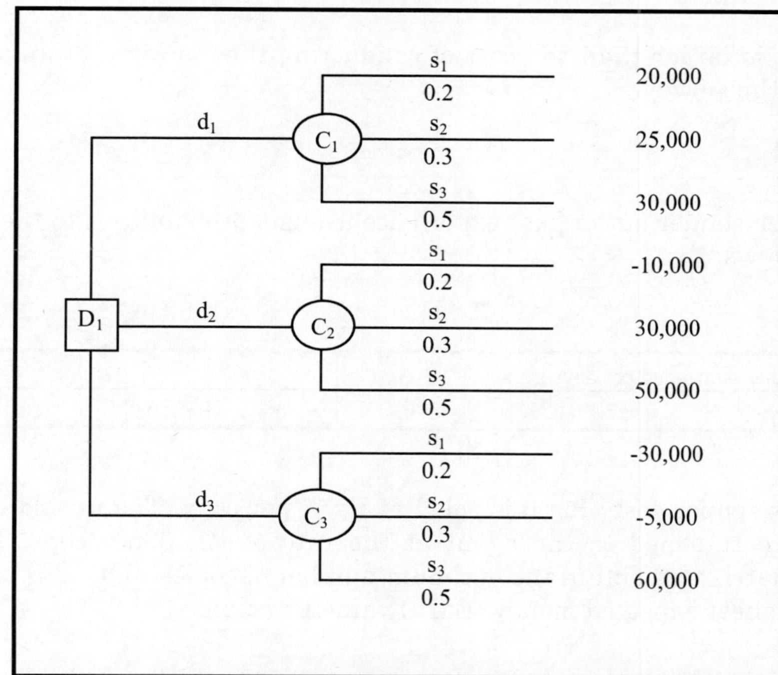

Figure 11.22

Let us first calculate the expected money values for all the nodes stated in the tree diagram by using the backward rolling principle.

$$EMV(C_1) = 20{,}000 \times 0.2 + 25{,}000 \times 0.3 + 30{,}000 \times 0.5 = 26{,}500$$
$$EMV(C_2) = -10{,}000 \times 0.2 + 30{,}000 \times 0.3 + 50{,}000 \times 0.5 = 32{,}000$$
$$EMV(C_3) = -30{,}000 \times 0.2 + -5000 \times 0.3 + 60{,}000 \times 0.5 = 22{,}500$$
$$EMV[D_1] = MAX(26{,}500, \quad 32{,}000, \quad 22{,}500) = 32{,}000$$

From the above computations it is clear that the highest expected payoff or profit is associated to decision node C_2. Thus $EMV_{W.O.P.I}$ = Rs. 32,000 and the associated decision is d_2.

For the calculation of EMV_{PI} we will have to calculate the highest payoff for each action under certainty. So for certainty of d_1 to occur the highest payoff is Rs. 20,000. For certainty of d_2 to occur the highest payoff is Rs. 50,000. For certainty of d_3 to occur the highest payoff is Rs. 60,000. Thus the expected pay off with perfect information is obtained as

$$EMV_{WPI} = 20{,}000 \times 0.2 + 30{,}000 \times 0.3 + 60{,}000 \times 0.5 = \text{Rs. } 43{,}000$$
$$EVM_{PI} = EMV_{WPI} - EMV_{WOPI} = \text{Rs. } 43{,}000 - 32{,}500 = 11{,}000$$

Since EMV_{PI} is larger than the cost of conducting the survey the philanthropist is advised to go for the survey.

■ Example 2

A daily newspaper vender out of past experience assigns probabilities to the demand for newspaper as follows.

Table 11.22

No. of copies demanded per day	1	2	3	4
Probability	0.4	0.3	0.2	0.1

A copy of the newspaper costs Rs. 6 is soled at Rs. 7 per copy. The unsold copies in the same evening are returned to the agent at the rate of Rs. 5 per copy. Formulate a suitable payoff matrix and obtain the optimum number of copies to be purchased for sale on the basis of highest expected money (EMV) value principle.

Solution

Profit per copy = Rs. 7 – Rs. 6 = Re. 1.

Loss per unsold copy = Rs. 6 – Rs. 5 = Re. 1.

If he orders only one copy per day and the same one copy sold in a day then there is no excess supply to return. Thus, he will make a profit of Re. 1 on that day as stated in the table on the top most corner corresponding to one copy order and one copy demand. Similarly, if the demand is one copy and he orders for two copies then the net gain is Rs. 2 – Re. 1 = Re. 1. This is stated in the table in the right place. If he orders for 4 copies and the demand is only one copy then after returning three copies the net profit = Re. 1 – Rs. 3 = Rs. –2. This figure is entered in the correct place in the Table 11.23. In a similar way all the entries are made in the table.

Table 11.23

Copies demanded	Copies ordered				Probability
	1	2	3	4	
1	1	0	–1	–2	0.4
2	1	2	1	0	0.3
3	1	2	3	2	0.2
4	1	2	3	4	0.1
EMV	1	1.2	0.8	0	

Given the probabilities in the last column the expected values are calculated and reported in the last row. For example, for the 2 copies order case the expected money value is obtained as:

$$\text{EMV}(2) = 0 \times 0.4 + 2 \times 0.3 + 2 \times 0.2 + 2 \times 0.1 = 1.2$$

In a similar way all other expected values are calculated and reported in the last row.

The maximum EMV = Rs. 1.2. Thus, the vender should order for 2 copies per day.

Example 3

A printing press has two machines new and old. The new machine gives good printing only when the paper quality is good. If the paper quality is bad then it gives only bad prints. Irrespective of the quality of the paper the slow going old machine gives good quality print all the time. The company is having 80% good quality and 20% of poor quality papers in its stock room. The good quality print in good quality paper gives him a net profit of Rs. 2000, at the same time the good print in bad quality paper gives him only Rs. 1600 profit. With new machine and good print the profit is Rs. 2400 and poor print it is Rs. 800. Suggest a suitable printing policy to earn more money as profit.

Solution

Figure 11.25 shows the associated tree diagram

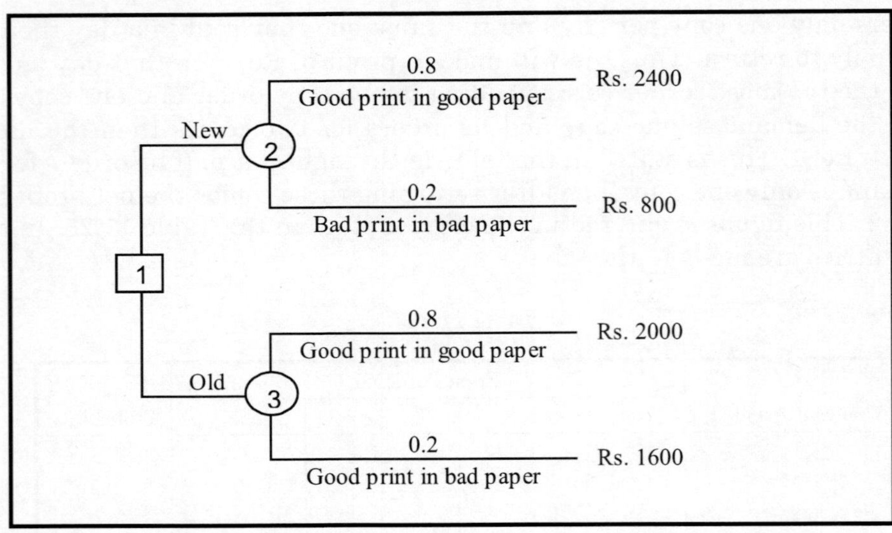

Figure 11.23

Expected profit at chance node (2) = 0.8 × 2400 + 0.2 × 800 = Rs. 2080

Expected profit at chance node (3) = 0.8 × 2000 + 0.2 × 1600 = Rs. 1920

Sine the expected profit is high in the case of new machine at action node [1] the printer should go for the new machine for printing.

Example 4

The XYZ investment bank is considering four options for investment; shares, bonds, real estate and saving certificates. The past data regarding the performance of the four options are given below.

Shares: There is 25% chance that these shares will decline by 10% of its value, a 30% chance that they will remain stable and a 45% chance that they will increase by 15%. The shares under consideration pay no dividends.

Bonds: The bond stand a 40% chance of increase in value by 5% and 60% chance of remaining stable. It's yield is at 12% per annum.

Real estate: This proposal has a 20% chance of increasing the value by 30%, a 25% chance of increasing the value by 20%, a 40% chance of increasing the value by 10% a 10% chance of being stable and a 5% chance of losing the value by 5%.

Saving certificates: These certificates yield 8.5% per annum with certainty.

Draw a suitable tree diagram and obtain the best possible investment option.

Solution

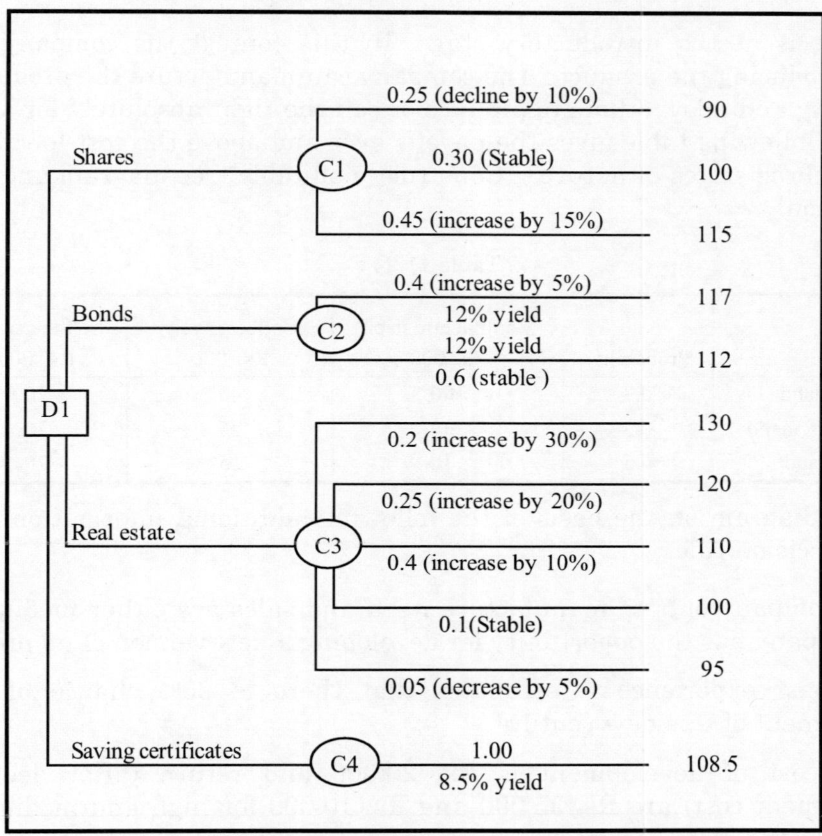

Figure 11.24

Assume that we have Rs. 100 to invest. The relevant decision tree is shown in Figure 11.24.

EMV at the chance node (C1) = $0.25 \times 90 + 0.30 \times 100 + 0.45 \times 115$ = Rs. 104.25

EMV at the chance node (C2) = $0.4 \times 117 + 0.60 \times 112$ = Rs. 114

EMV at the chance node (C3) = $0.2 \times 130 + 0.25 \times 120 + 0.40 \times 110 + 0.1 \times 100 + 0.05 \times 95 = 114.75$

EMV at chance node (C4) = $1 \times 108.50 = 108.50$

EMV at decision node [D1] = MAX(104.25, 114, 114.75, 108.50) = Rs. 114.75

The real estate alternative gives the highest EMV. Thus, real estate investment is the correct solution.

■ Example 5

A new product is at the introductory stage. In this context the company has three options for introducing the product. The company can manufacture the product by itself or collect royalty and allow others to produce or sell the right absolutely for a lump sum payment. The following table gives the payoffs over and above the cost for all the three options under three states of nature. Construct a suitable tree diagram and obtain the optimum solution:

Table 11.24

Event	Probability	Manufacture itself Rs. '000	Collect royalty Rs. '000	Sell the rights Rs. '000
High demand	0.2	100	40	20
Medium demand	0.3	30	25	20
Low demand	0.5	−10	15	20

Extend the diagram on the basis of the following additional information and obtain the optimum decision rule.

i) If the company opts to manufacture itself and sales are either medium or high, the company has the opportunity for developing a new version of its product.

ii) From past experience it estimates that there is 60% chance of successful development of this new venture.

iii) If the cost of development is Rs. 20,000 and return (after deducting the development cost) are Rs. 35,000 and Rs. 10,000 for high and medium demand respectively.

Solution

From Figure 11.25 we calculate the expected money values by backward rolling as follows

EVM at Chance Node (C1) = $\{1000(0.2 \times 100) + (0.3 \times 30) + (0.5 \times -10)\}$ = Rs. 24,000

EVM at Chance Node (C2) = $\{1000(0.2 \times 40) + (0.3 \times 25) + (0.5 \times 1.5)\}$ = Rs. 23,000

Decision at Decision node D1 = Max(Rs. 24,000, Rs. 20,000, Rs. 23,000) = Rs. 24,000.

Thus the correct decision is to manufacture the product by itself and make the expected profit of Rs. 24,000.

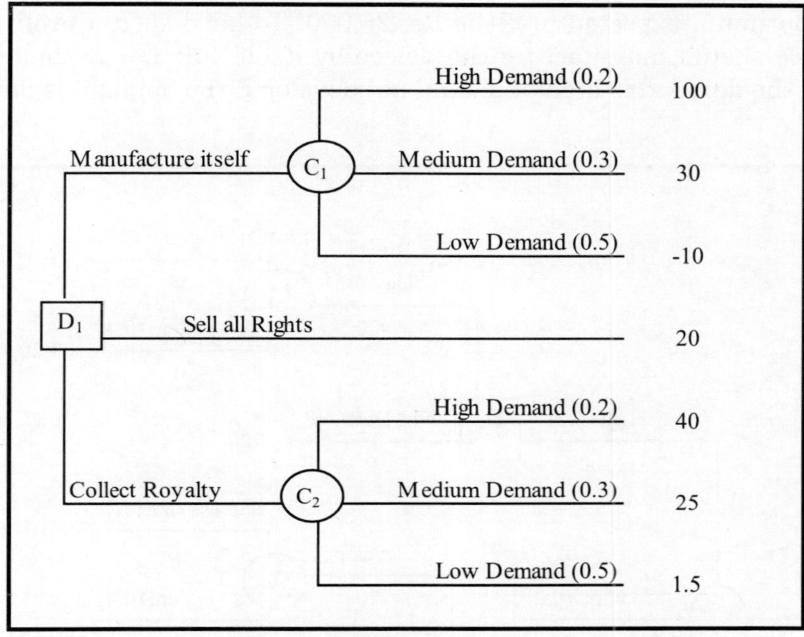

Figure 11.25

Solution for the Extended Tree Diagram shown

Note that in this diagram the tail end values are net pays values calculated from the bottom of the tree after making all necessary adjustments.

From Figure 11.26 we calculate the expected money values by backward rolling as follows

EMV at chance node (C3) = 1000{(0.6 × 135)+(0.4 × 80)} = 1,13,000

EMV at decision node [D2]= Max(1,13,000, 1,00,000) = 1,13,000

EMV at chance node (C4) = 1000{(0.6 × 40)+(0.4 × 10)} = 28,000

EMV at decision node [D3] = Max(28,000, 30,000) = 30,000

EMV at chance node (C1)= (0.2 × 1,13,000+0.3 × 30,000+0.5 × –10,000 = 26,000

EMV at chance node (C2)= (0.2 × 40,000+0.3 × 25,000+0.5 × 15,000 = 23,000

EMV at decision node [D1] = Max(26,000, 20,000, 23,000) = 26,000

Thus, the optimum expected profit is Rs. 26,000. To have such a profit the company under reference should manufacture the device by itself. Further it should develop the new version if the demand is high or should not develop if the demand is medium.

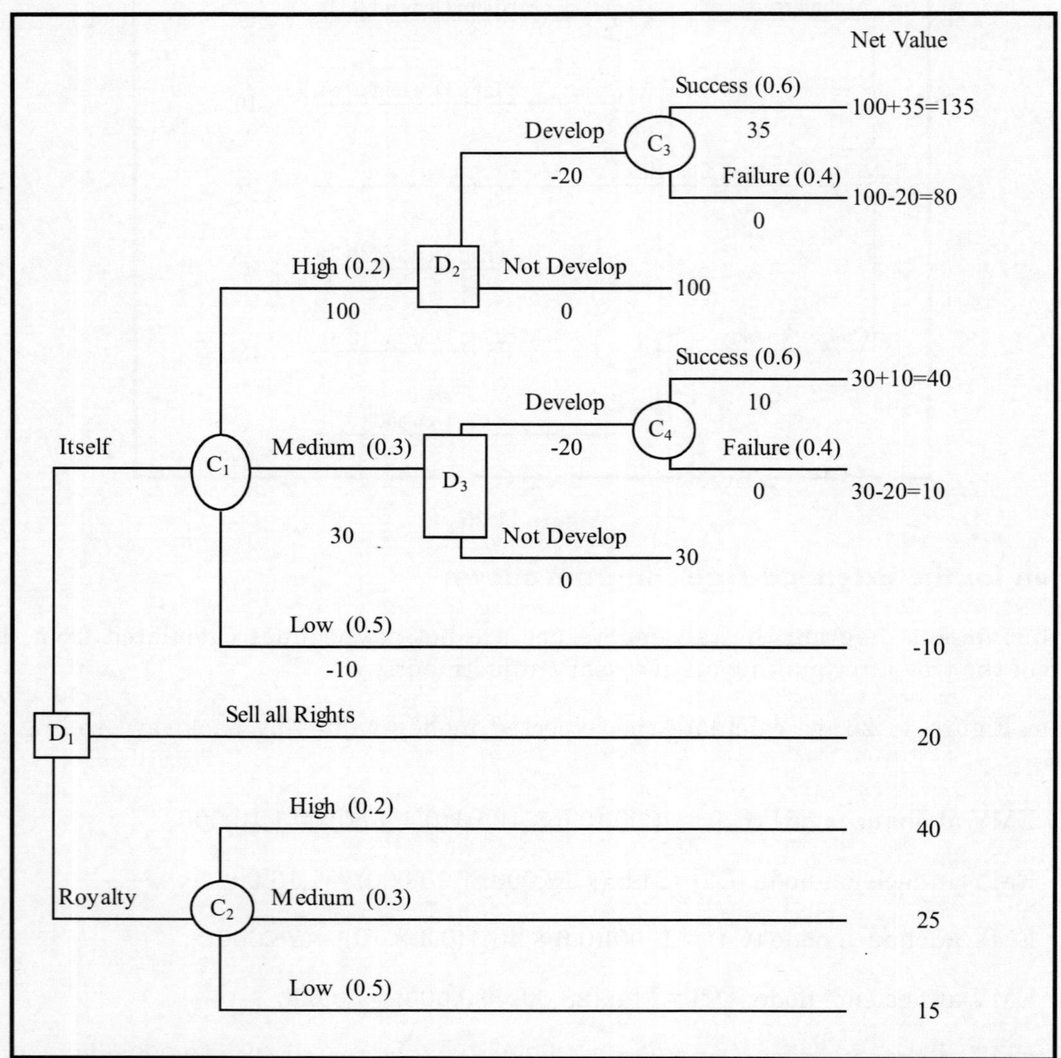

Figure 11.26

■ Example 6

A farmer is considering drilling a well in his farmland. In the past only 70% well drilled were successful at 20 metres depth. Moreover on failure up to 20 metres he prefers to go for another 25 metres. The probability of success this time is 20%. The prevailing cost of drilling is Rs. 500 per metre. In case if he does not succeed he will have to by water from his neighbour by paying Rs. 15,000 for the same period. Draw appropriate tree diagram and suggest the best decision for the farmer.

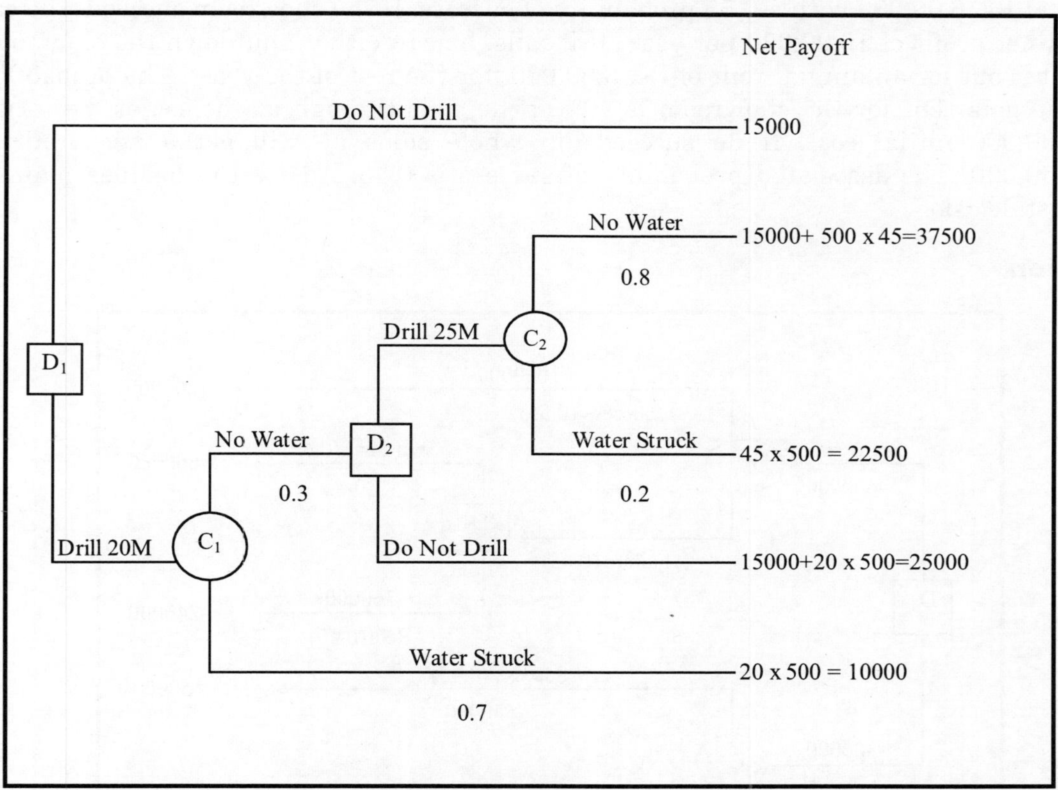

Figure 11.27

Solution by backward rolling from tree diagram in Figure 11.27

EMV(C2) = 0.8 × 37,500 + 0.2 × 22,500 = 34,500

EMV[D2] = MIN[25,000, 34,500] = 25,000

EMV(C1) = 0.3 × 25,000 + 0.7 × 10,000 = 14,500

EMV[D1] = MIN[14,500,15,000] = 14,500

Since the expected cost of drilling at both the stages is less than buying water the correct decision could be Drill 20M if no water struck then drill up to 25 M.

Example 7

A businessman from Chennai wishes to sell his products in Bangalore, he can set up a showroom in the city or can sell through a wholesaler. Setting up a show room will entail a cost of Rs. 6,00,000 with a 0.55 probability of success. If the showroom succeeds, he can gain a net profit of 10,00,000 per year. If it fails, he can either shutdown the showroom or rent it out for an annual rent of Rs. 3,60,000 (for the rest of the year). The probability that he gets rent for the showroom is 40%. If he sells through a wholesaler, he incurs Rs. 3,00,000 initial cost. If he succeeds in whole sales he will get a net profit of Rs. 5,50,000. The associated probability of success is 0.45. Advise the businessman on the best decision.

Solution

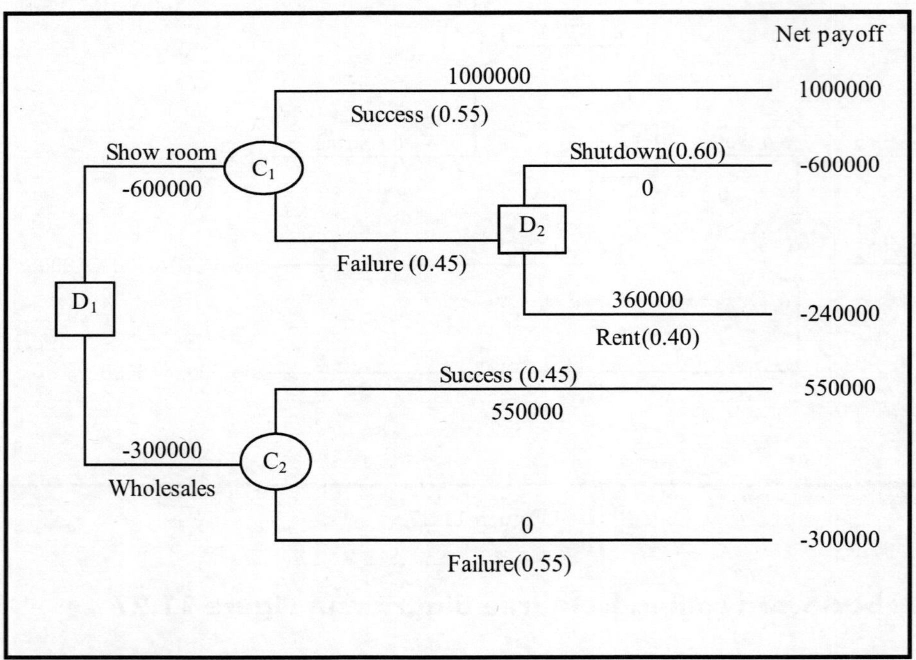

Figure 11.28

$EMV[D_2] = MAX(0.6 \times -6,00,000;\ 0.4 \times 2,40,000) = MAX(-3,60,000; -96,000) = -96,000$

$EMV(C_1) = 0.55 \times 10,00,000 + 0.45 \times -96,000 = 5,06,800$

$EMV(C_2) = 0.45 \times 5,50,000 + 0.55 + 0.55 \times -3,00,000 = 82,500$

$EMV[D_1] = MAX(82,500,\ 5,06,800) = 5,06,800$

Thus the businessman must opt for showroom at the first stage decision making.

Example 8

An organization has 3 alternative projects A, B and C under consideration. Which one it should choose based on decision tree approach. The data are given below for all the three projects.

Project A: For project A, a capital investment of Rs. 60,000 is required. If the project is completed on time, then the organization will receive reward of Rs. 1,00,000 as revenue. If not completed on time, a penalty of Rs. 5000 per day will be deducted from assured reward of 1,00,000 with a maximum penalty of Rs. 15,000. The following table gives the probabilities of delay.

Table 11.25

0 day	0.75
1 day	0.10
2 days	0.10
3 or more days	0.05
Total	1.00

Project B: Project B needs an initial investment of Rs. 75,000. After completion the company will get a reward of Rs. 75,000. If the said project is completed in 2 days the organization will get a follow up project that will require an expenditure of Rs. 20,000 and revenue reward of Rs. 50,000 with 70% chance and Rs. 75,000 with 30% chance. If more than 2 days are required for the first project, the follow up project will not be available. They have equal chance of completing the first project in two days.

Project C: Project C needs an initial investment of Rs. 1,00,000. A chance of success for this project is 70% the associated revenue reward will be 1,80,000. If the project fails then no revenue will be received. Suggest suitable optimum decision for the said investor.

Solution

EMV (C5) = $0.1 \times 35{,}000 + 0.1 \times 30{,}000 + 0.05 \times 25 = 7750$

EVM (C1) = $0.75 \times 40{,}000 + 7750 = 37{,}750$

EMV (C6) = $0.7 \times 30{,}000 + 0.3 \times 55{,}000 = 37{,}500$

EMV (C2) = $0.5 \times 37{,}500 + 0.5 \times 0 = 18{,}750$

EMV (C3) = $0.7 \times 80{,}000 + 0.3 \times -1{,}00{,}000 = 26{,}000$

EMV [d1] = MAX [37,750, 18,750, 26,000] = 37,500

Since EMV is high in the case of Project A the company should go for project A.

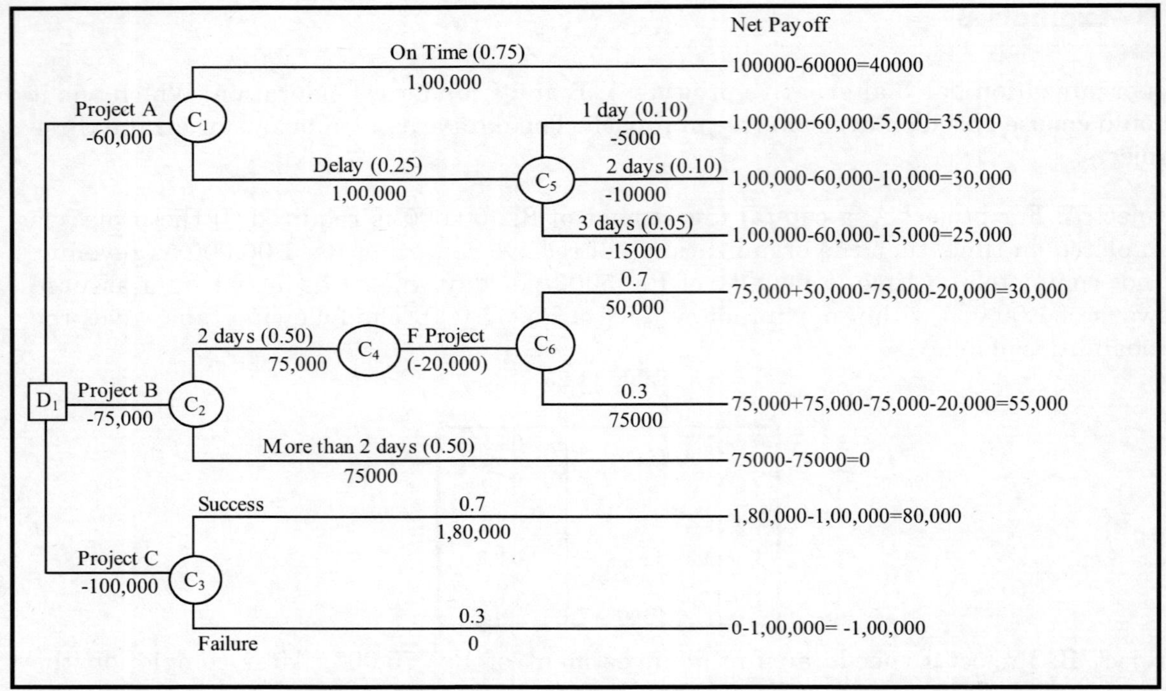

Figure 11.29

■ Example 9

A company is considering whether to go for the new product or not. If it decides to go further in producing the new product, it must either install a new division, which cost Rs. 4 crore or work overtime with existing division at an additional cost of Rs. 1.5 crore. Further in the case of new division before proceeding further it will have to get an approval from the government. It is believed that there is 70% chance of getting it approved from the government. The market survey has revealed the following facts regarding the quantity of sales.

Table 11.26

Amount of sales	Probability	Resulting profit
High	0.45	15
Medium	0.30	7
Low	0.20	3
Nil	0.05	-5

However, by overtime option the company knew that it will be able to meet up to the level of medium magnitude only. Suggest a suitable optimum decision.

Solution

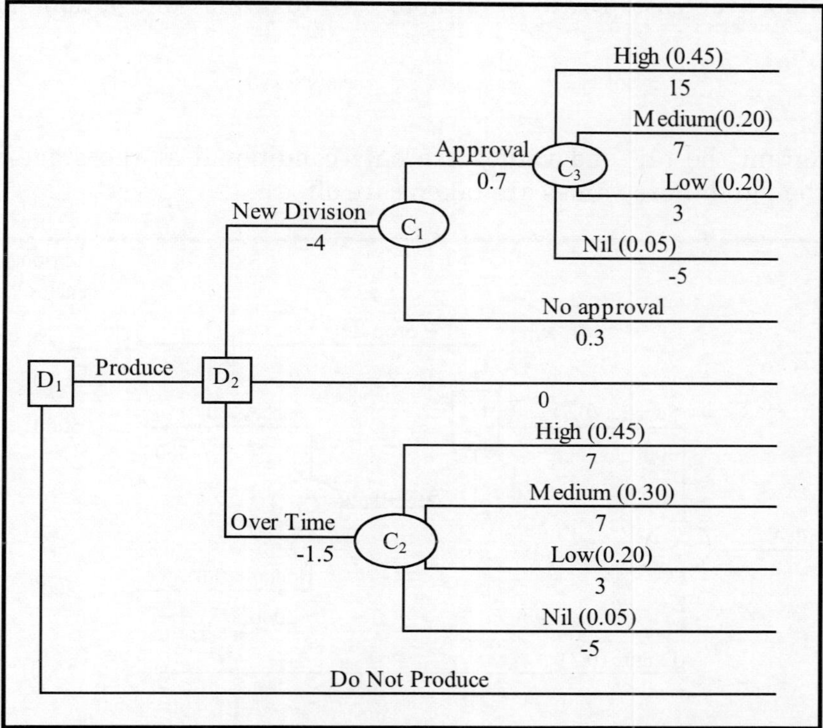

Figure 11.30

EMV (C3) = 0.45 ×15 +0.2 ×7 + 0.2 × 3 + 0.05 × −5 = 9.2

EMV (C1) = 0.7 × 9.2 + 0.3 × 0 = 6.44.

EMV (C2) =0.45 × 7 + 0.3 × 7 + 0.2 × 3 + 0.05 × −5 = 5.6

EMV [D2] = MAX [6.44 – 4; 5.6 – 1.5] = MAX [2.44; 4.1] = 4.1

EMV [D1] = 4.1

Thus the company must go for the new product by utilizing the over time.

■ Example 10

A businessman has two independent invest options A and B. However, he lacks funds for undertaking both simultaneously. He can undertake investment A and then stop or go for B on the event success in A. Alternatively, he can opt for B first and then stop or go for A provided he succeeds in B. The probability of success for A is 0.7 while for B it is 0.4. Both the investments need Rs. 2000 each and both pay back nothing when they fail.

Successful completion of A will pay Rs. 3000 over cost, and successful completion of B will pay Rs. 5000 over cost. Draw a suitable tree diagram and option the optimum solution.

Solution

In the tree diagram the tail end values are only conditional at the stage when move backward the then and there values are taken care off.

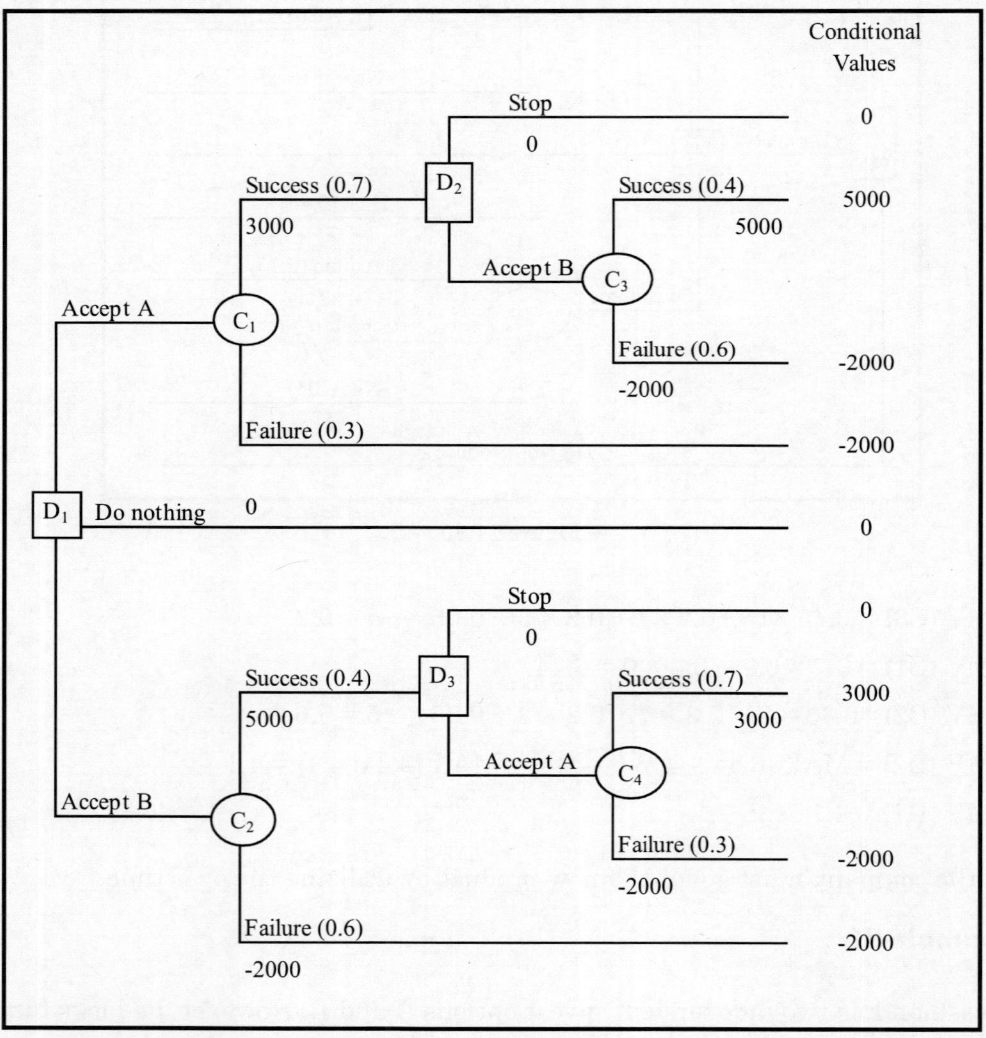

Figure 11.31

Evaluation of tree diagram by backward rolling

EMV (C3) = 0.4 ×5000+0.6 ×−2000 = 800

EMV(D2] = Max (800,0) = 800

EMV(C4) = 0.7 ×3000 +0.3 ×−2000 = 1500

EMV [D3] = MAX [1500,0] =1500

EMV (C1) = 0.7(3000+800)+0.3 ×−2000 = −1200

EMV (C2) = 0.4 (1500+5000)+0.6 ×−2000 =1400

EMV [D1] = MAX (−1200, 0, 1400) =1400

Thus, the best strategy is to accept the investment B first and if it is successful then accept investment A.

■ Example 11

There is 40% chance that a patient admitted to the hospital is suffering from cancer. The doctor has to decide whether a serious operation should be performed or not. If the patient is suffering from cancer and the serious operation is performed the chance that he will recover is 70%. Otherwise it is 35%. On the other hand if the patient is not suffering from cancer and the serious operation is performed the chance that he will recover is only 29% otherwise it is 100%. Assume that recovery and death are the only two alternatives. Construct an appropriate decision tree and find the best possible decision.

Solution

From the decision tree below we find that

Patient recovers when operation is done = 0.28+0.12 = 0.40

Patient recovers when no operation is done = 0.14+0.602 = 0.74

Since 0.74 is greater than 0.40 we conclude that doctor should not operate such a patient.

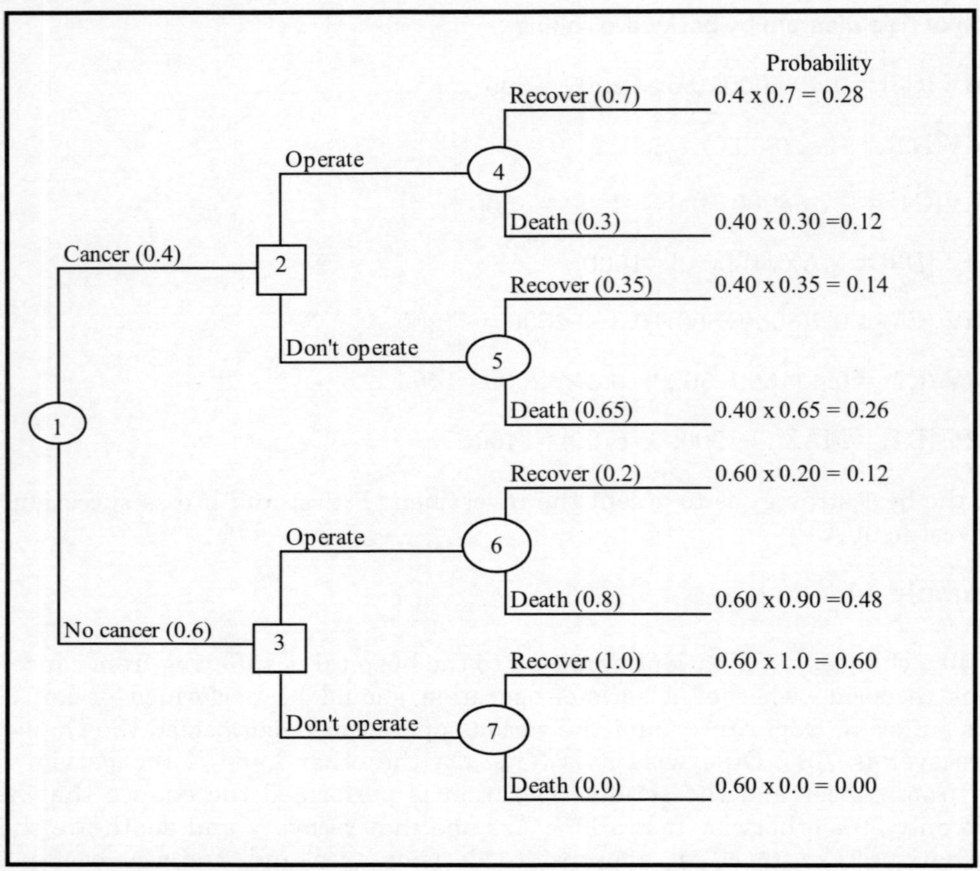

Figure 11.32

EXERCISES

1. Under an employment promotion program, it is proposed to allow sales of newspapers on the buses during off-peak hours. The vender can purchase the newspaper at a special discounted rate of 25 paise per copy against the selling price of 40 paise. Any unsold copies are, however a dead loss. A vender has estimated the following probability distribution for the number of copies demanded.

No. of copies demanded	15	16	17	18	19	20
Probability	0.04	0.19	0.33	0.26	0.11	0.07

How many copies should he order so that his expected profit is maximum?

2. A food company is contemplating the introduction of a revolutionary new product with new packaging to replace the existing product at same price (S_1) or a

moderate change in the composition of the existing product with a new packing at a small increase in price (S_2) or a small change in the composition of the existing except the word (new) with a negligible change in the price (S_3). The three possible states of nature of events are (i) high increase in sales (N_1) (ii) no change in sales (N_2) (iii) decrease in sales (N_3). The marketing department of the company worked out the payoffs in terms of yearly net profit for each course of action for these events (expected sales). This is represented in the following table.

States of nature	S_1	S_2	S_3
N_1	7,00,000	5,00,000	3,00,000
N_2	3,00,000	4,50,000	3,00,000
N_3	1,50,000	0	3,00,000

Which strategy should the company choose on the basis of (a) Maximini criterion? (b) Maximaxi criterion (c) Minimum regret criterion (d) Laplace criterion

3. A person has two independent investment options A and B available to him but he can undertake only one at a time due to certain constraints. He can choose a first and then stop, or if A is successful then take B and vice versa. The probability of success of A is 0.6, while for B it is 0.4. Both the investments require an initial capital outlay of Rs. 10,000 and both return nothing, if the venture is unsuccessful. Successful completion of A will return Rs. 20,000 (over cost) and successful completion of B will return Rs. 24,000 (over cost). Draw decision tree and determine the best strategy.

(C.A. May 1988)

4. Matrix Company is planning to launch a new product, which can be introduced initially in western India or in the entire country. If the product is introduced only in western India, the investment outlay will be Rs. 12 million. After two years, matrix can evaluate the project to determine, whether it should cover the entire country. For such expansion, it will have to incur an additional investment of Rs. 10 million. To introduce the product in the entire country right in the beginning would involve an outlay of Rs. 20 million. The product, in any case, will have a life of 5 years, after which the plant will have zero net value.

   If the product is introduced only in the western India, demand would be high or low with the probabilities of 0.8 and 0.2 respectively and annual cash inflow of Rs. 4 million and Rs. 2.5 million respectively.

   If the product is introduced in the entire country right from the beginning, the demand would be high or low with probabilities of 0.6 and 0.4 respectively and annual cash inflow of Rs. 5 million respectively.

   Based on the observed demand in the western India, if the product is introduced in the entire country, the following probabilities would exist for high and low demand on an all India basis.

Western India	Whole country	
	High demand	Low demand
High demand	0.9	0.1
Low demand	0.4	0.6

The hurdle rate applicable to this problem is 12 percent.

(a) Set up a decision tree for the investment situation.

(b) Advice the matrix company on the investment policy it should follow. Support your advice with appropriate reasoning. (ICWA, June 1990)

5. A company which operates a chain of lunch rooms, plan to install a unit in either of the two locations. The company feels that the probability of a unit being successful in location X is 3/4 and that, if it is successful it will make an annual profit of Rs. 4,00,000. If it is not successful, the company will lose Rs. 1,00,000 per year. The probability of a unit making success in location Y 0s only 1/2, but if it does succeed, the annual profit will be Rs. 6,00,000. If it does not succeed in location Y, the annual loss will be Rs. 1,20,000. Where should the company locate the new unit so as to maximize its expected gain?

(Banaras University, MMS 1984)

6. The demand for cakes made in a bakery is as follows:

No. of cakes	0	1	2	3	4	5
Probability	0.05	0.1	0.25	0.3	0.2	0.1

If the preparation cost is Rs. 2 and selling price is Rs. 4 per unit, how many should the baker make to maximize his profit?

7. A manager has a choice between (i) a risky contract promising Rs. 7 lakh with probability 0.6 and Rs. 4 lakh with probability 0.4 and (ii) a diversified portfolio consisting of two contracts with independent outcomes and each promising Rs. 3.5 lakh with probability 0.6 and Rs. 2 lakh with probability 0.4. Construct a decision tree using EMV criteria. Can you arrive at the decision using EMV criteria.

(Poona University, M.B.A 1982)

8. You have a new furnace installed. The dealer offers to sell you spare fuel pumps at Rs. 200 each, if you buy them during installation the pumps sell for Rs. 500 in retail. These pumps cannot be repaired if they fail. Manufacturer's records indicate the following probability of fuel pumps failures during the furnaces lifetime.

Failure	0	1	2	3	4
Probability	0.01	0.3	0.4	0.1	0.1

Ignoring installation and holding costs, how many spare fuel pumps should be purchased during installation? (Delhi University, M.B.A 1987)

9. A modern home appliances dealer finds that the cost of holding a mini cooking range in stock for a month is Rs. 200 (insurance, minor deterioration, interest on borrowed capital etc.). Customers who cannot obtain a cooking range immediately go to other dealers and he estimates that for every customer, who cannot get immediate delivery, he loses an average of Rs. 500. the probabilities of demand for 0, 1, 2, 3, 4, 5 mini cooking ranges in a month are 0.05, 0.1, 0.2, 0.3, 0.2 and 0.15 respectively. Determine the optimal stock level of cooking ranges. Also fin EVPI.

(Delhi University, M.B.A 1990)

10. A businessman has two independent investments A and B available to him, but he lacks the capital to undertake both of them simultaneously. He can choose to take A first and then stop, or if A is successful, then take B or vice-versa. The probability of success on A is 0.7, while for B, it is 0.4. Both investments require an initial capital outlay of Rs. 2000 and both return nothing, if the venture is unsuccessful. Successful completion of A will return Rs. 3000 (over cost) whereas successful completion of B will return Rs. 5000 (over cost). Draw the decision tree and determine the best strategy.

(C.A. May 1985)

11. The owner of a boat has estimated the following distribution of demand for a particular kind of boat.

No. demanded:	0	1	2	3	4	5	6
Probability:	0.14	0.27	0.27	0.18	0.09	0.04	0.01

Each boat costs him Rs. 7000 and he sells them for Rs. 10,000 each. Boats that are left unsold at the end of the season must be disposed off for Rs. 6000 each. How many boats should be stocked so as to maximize his expected profit?

(Delhi University, M.Com. 1988)

12. A mineral processing company wants to decide about the number of spare gear trains it has to order out at the time of placing order for a high horsepower gear-box connected to a grinding unit. Although the life of a gear can be as high as 30 years and more, sudden failures cannot be ruled out. In case of failure, it would be expensive and time consuming to get a spare gear-train. The cost would be Rs. 2,00,000 including the loss of production due to down time of the equipment. If ordered-out with gear-box, a gear train would cost only Rs. 10,000 per unit. The following data is based on an analysis of past experience of 100 gear-boxes.

Number of spare gear-trains required	0	1	2	3	4
Number of gear-boxes requiring the spares:	93	4	2	1	0

You are expected to advise the optimal order size.

(Poona University, M.B.A, 1982)

13. A milkman buys milk at Rs. 2 and sells it for Rs. 2.5 per litre. Unsold milk has to be thrown away. The daily demand in litres has the following probabilistic distribution.

Litres	46	48	50	52	54	58	60	62	64
Probability	0.01	0.03	0.06	0.1	0.2	0.25	0.15	0.1	0.1

If each day's demand is independent of the previous day's demand, how many litres should he order every day?

(Rajasthan University, M.B.A. 1982)

CHAPTER

12

Network Analysis

12.1 INTRODUCTION

At the time of implementing a large project one comes across variety of sub-activities. Given this scenario the job of the manager is to coordinate all the sub-activities and minimize the project completion time. The interdependency of the distinct sub-activities often needs a systematic approach to minimize the time cost. On some occasions the starting of one sub-activity depends upon the completion of some other preceding sub-activity. For example, in a house construction project, the electrical wiring of the building is prerequisite for plastering. Similarly, for painting, plastering is the prerequisite. To minimize the total time for project completion we often use 'Program Evaluation and Review Technique' (PERT). Sometimes we use 'Critical Path Method' (CPM) as well. Both these techniques use similar terminology and serve the same purpose though both of them were developed independently. During the late 1950's PERT technique was developed and used in connection with the planning and designing of the Polaris Submarine System.

The CPM was developed by the Du Pont Company and the Univac Division of Remington Rand Corporation as a device to control the maintenance of its chemical plants. However, the CPM technique is commonly used in projects with known constant activity timings. In this method not only the activity timings are known constants, but also the amounts of various resources required to perform the activities under reference are also constants.

PERT on the other hand, is preferably used in situation involving certain degree of uncertainty in activity completion timings. Thus, in essence, PERT is probabilistic in nature and hence used in research and development related projects. Since CPM is deterministic in nature it is used mostly in construction related projects. The following are some basic issues addressed in this section.

- What is the minimum time required to complete the project?
- What are the start and finish dates of each one of the sub-activities?
- Which are the critical activities that must be completed in time to keep the project moving?
- How long the non-critical activities can be delayed without disturbing the smooth functioning of the project?

12.2 PROJECT SCHEDULING WITH KNOWN ACTIVITY TIMINGS (CPM)

To begin with let us develop a list of inter-related activities that constitute the entire building construction project. The following example is designed to highlight some of the basic issues in the said construction project evaluation.

■ Example 1

The following table describes the steps involved in a new construction project initiated by Prestige constructions in Bangalore. In it nine certain activities are being described from A to I as stated in column 1. In column 2 the descriptions of the concerned activities are highlighted. In column 3 the immediate predecessor(s) to the activities under reference are described. It is clear that except the first two initial activities, all other activities are having predecessors. Such a schematic approach helps us to identify the sequence of activities. It is clear that unless A is completed activity E cannot started. Unless E is completed D and F cannot be started. The last column shows the time required for completing the respective activities.

Solution

As a first step, by using the information given in Table 12.1 we interlink the various activities as shown in Figure 12.1. In it the relevant activities are described by rectangles with their associated names. The associated inter-relationships are shown by the relevant arrows (arcs). In addition two activity nodes from A to I we deliberately add one start node, to highlight the start and one finish node, to highlight the finish in Figure 12.1.

Table 12.1

Activity	Description	Immediate predecessor	Activity time in months
A	Preparation plan	-	5
B	Identification of new customers	-	6
C	Develop prospectus	A	4
D	Selection of contractors	A	3
E	Prepare building documents	A	1
F	Approval of building plans	E	4
G	Construction	D,F	14
H	Finalize sales	B,C	12
I	Customers move in	G,H	2

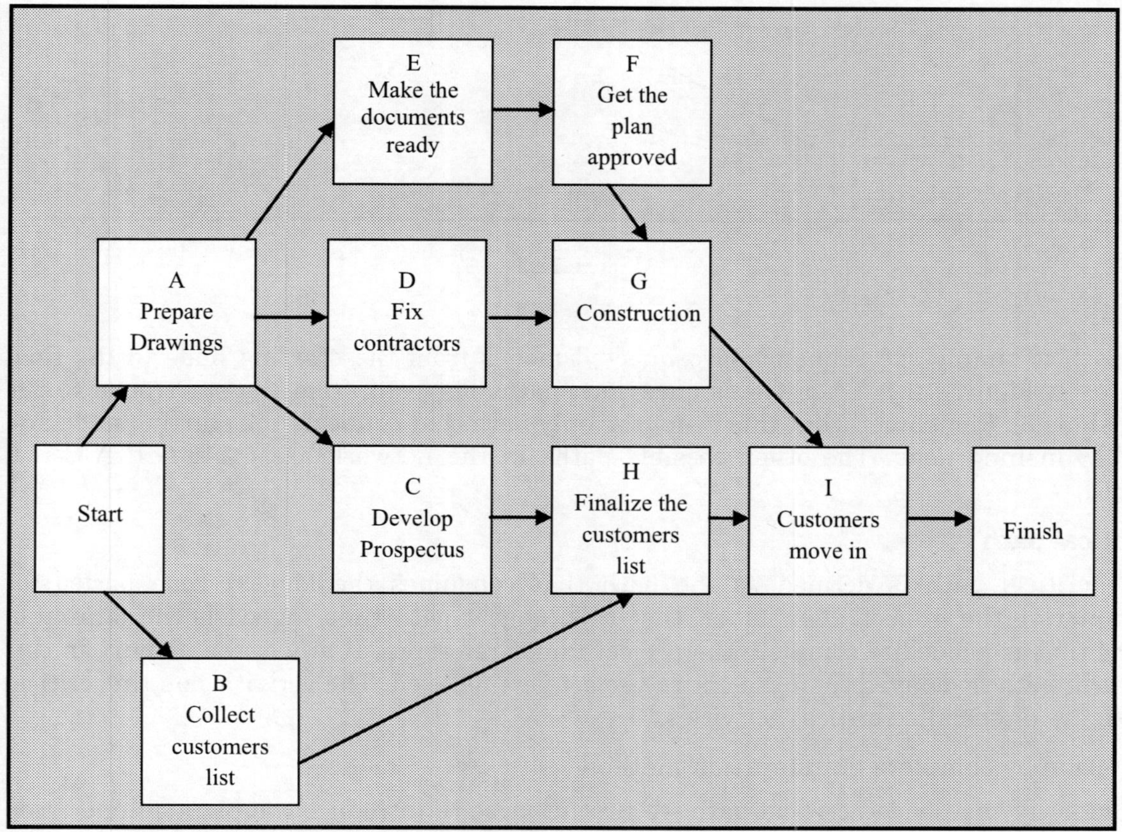

Figure 12.1

In Figure 12.2 we write the related activity names at the upper left-hand corners and the associated activity times just below these names. The leftover space is designed for further use.

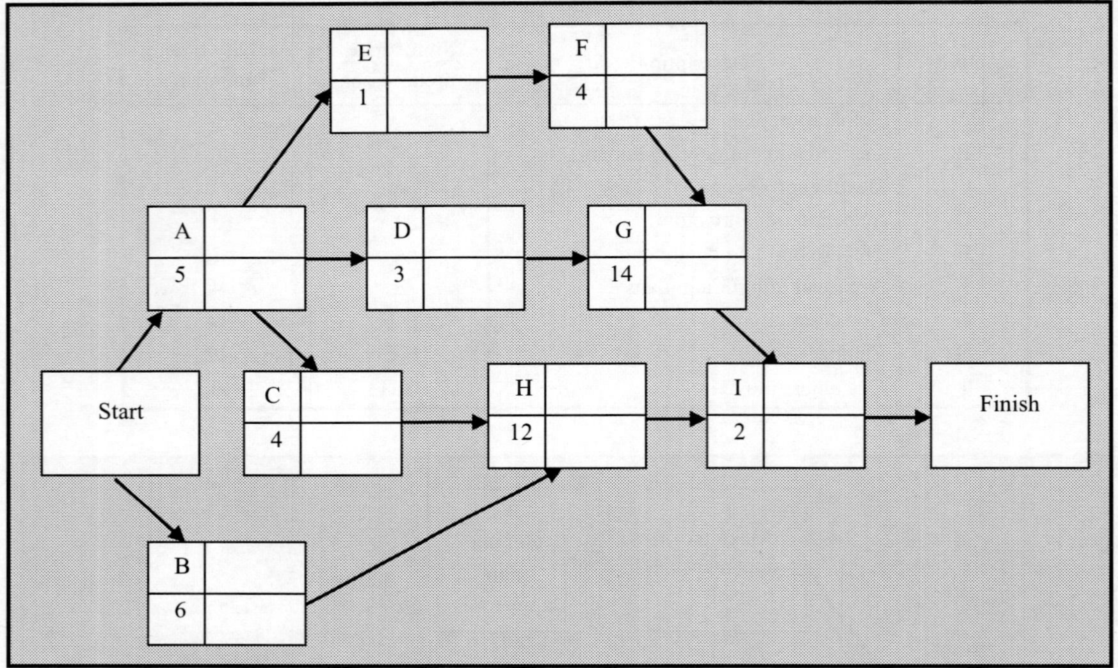

Figure 12.2

Path

A path is simply the sequence of connected modes from the starting node to the final node. In Figure 12.1 A-E-F-G-I is once such path that runs from the start node to the finish node. Remember, that this is not the only path that connects the starting node and final finishing node. The other possible paths in the network are: A-D-G-I; A-C-H-I; B-H-I.

Critical path

The critical path is defined as the path that consumes the longest time period for completing the project. Remember, that this longest path time indeed is the minimum time that is needed to complete the project under reference. If any of the activity in this critical path is delayed then the entire project gets delayed. The activities on this critical path are often called critical activities.

Solution by complete enumeration method

To identify the needed critical path we first propose to introduce complete enumeration method in Table 12.2. We list out all the possible paths. In column 2 we calculate the associated total time and report it.

Table 12.2

Path	Path time in months
A-E-F-G-I	5+1+4+14+2=26
A-D-G-I	5+3+14+1=23
A-C-H-I	5+4+12+2=23
B-H-I	6+12+2=20

The calculations in Table 12.2 yields the needed critical path A-E-F-G-I with the longest activity time of 26 months. Since the given problem is simple, we were able to identify all the possible paths and calculate the associated critical path timings. In bigger problem involving more number of diversified activities this method getting the critical path and the associated project completion time after exploring all possible paths is definitely time consuming and sometimes impossible even. To overcome this difficulty we often go for other short-cut methods. Before introducing such methods let us first discuss some basic concepts and rules.

Forward pass rule for early start and early finish of initial activities

For any activity **early start time** and **early finish time** are the two important concepts to be understood first. If ES stands for the earliest start time and EF stands for earliest finish time for an activity under reference then the earliest finish time is obtained by simply adding the activity time to the early starting time.

i.e., EF= ES + t (activity time)

In our illustration example, invariably activity A starts as soon as the project starts. So the earliest start time for activity A must be 0. With an activity time of 5 weeks the earliest finish time for completing this activity A is clearly 0 + 5 = 5 months. The following figure illustrates this process of making such calculations in a systematic way.

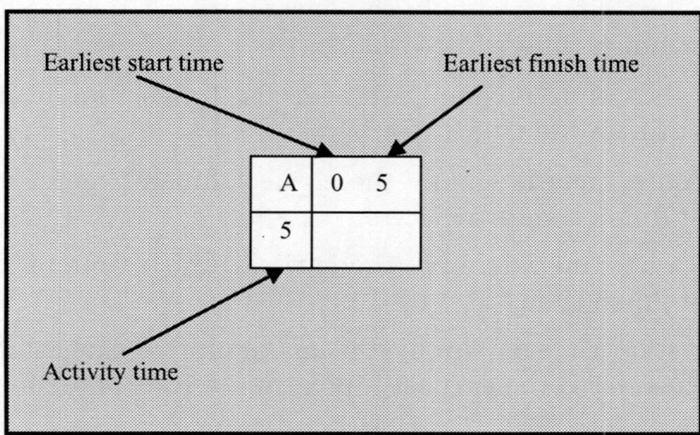

Figure 12.3

Similarly, activity B in our illustration also starts as soon as the project starts. So the earliest start time for B is also equal to 0. With an activity time of 6 months the earliest finish time for completing this activity B is 0 + 6 = 6 months. The following figure illustrates this process of making such calculations in a systematic way.

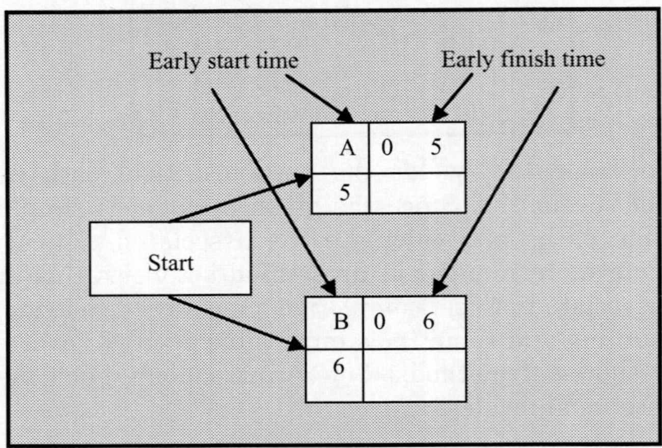

Figure 12.4

Forward pass rule for early start and early finish for succeeding activities

Since no succeeding activity can start unless otherwise all the preceding activities are completed the following stand-by rule could be used to obtain the earliest start time for all the subsequent activities.

The earliest start time for a succeeding activity is equal to the largest of the earliest finish time of all the preceding activities.

Now let us illustrate the application of this rule for the activities C, D, E, F, G, H, I.

For C the earliest start time is simply the earliest finish time of its predecessor A. Thus, for C the earliest finish time = 5 + 4 = 9.

For D the earliest start time is simply the earliest finish time of its predecessor A. Thus, for D the earliest finish time = 5 + 3 = 8.

For E the earliest start time is simply the earliest finish time of its predecessor A. Thus, for E the earliest finish time = 5 + 1 = 6.

For F the earliest start time is simply the earliest finish time of its predecessor E. Thus, for F the earliest finish time = 6 + 4 = 10.

For G the earliest start time is simply the maximum of earliest finish time of its predecessors D & F. Thus, for G the earliest finish time = 10 + 14 = 24.

For H the earliest start time is simply the maximum of earliest finish time of its predecessors C & G. Thus, for H the earliest finish time = 9 +12 = 21.

For I the earliest start time is simply the maximum of earliest finish time of its predecessors H & G. Thus, for H the earliest finish time = 24+ 2 =26.

This completes our whole exercise of forward pass as shown in Figure 12.5.

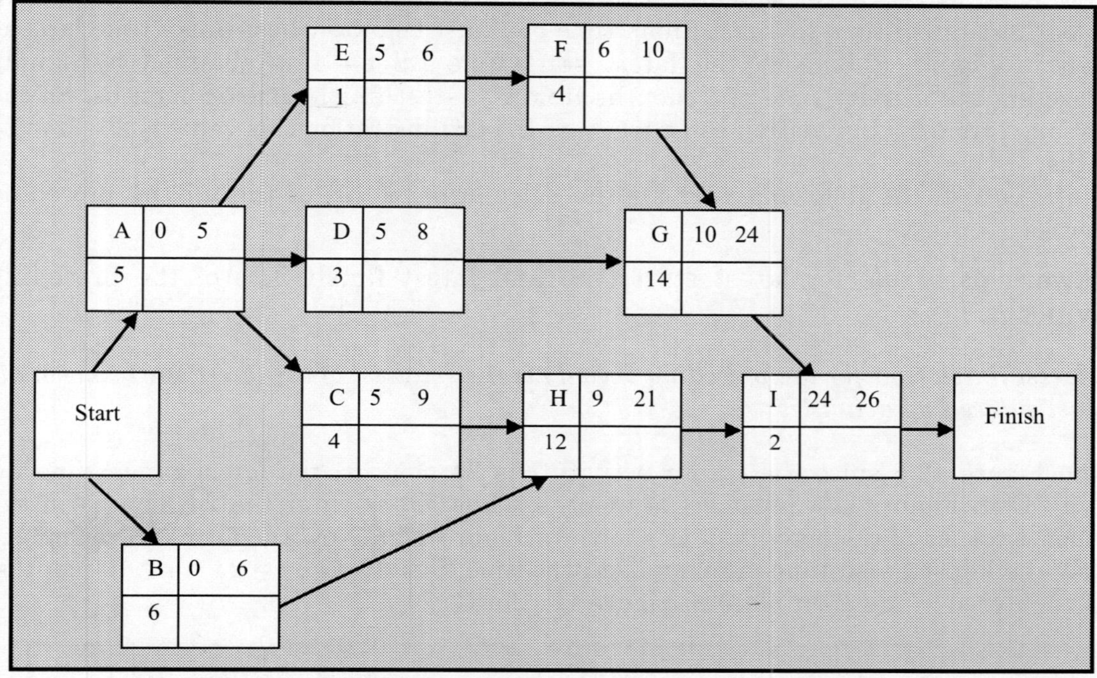

Figure 12.5

From the last activity I's entries one could read the earliest completion time for the project as 26 months.

Backward pass rule for latest start (LS) and latest finish (LF) of the last activity

To accomplish the final task of identifying the needed critical path in addition to early start and early finish time we must work out late start and late finish as well.

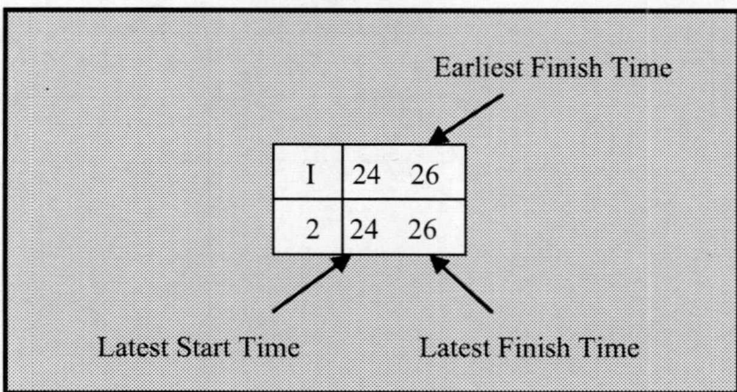

Figure 12.6

Such a task can be accomplished by backward pass method from the very last activity I. Now as far as I is concerned earliest finish (EF) is equal to the latest finish (LF) so we copy earliest finish time to latest finish time box just below earliest finish time box as shown in Figure 12.6. Now the latest start time LS for I is obtained by simply subtracting the activity time of 2 months from its latest finish time 26 months. So the latest start time for I is written as LS = LF – t. In our illustration this value is 26 – 2 = 24 months.

Now we use the following rule for the calculation of LF of the all the preceding activities.

Backward pass rule for latest start (LS) and latest finish (LF) of the preceding activities

The latest finish time for the preceding activity is the smallest of the latest start time of all the just succeeding activities.

In the Figure 12.6 since the LS for activity I is 24 months, the latest finish time for activity G in Figure 12.7 must be 24 months. Thus let us enter the 24 months in the southeast corner of the activity G as shown in Figure 12.7. Now the latest start for G is simply the latest finish time minus its activity time. So LS = LF– t = 24–14 = 10 months. This is entered in the right place in Figure 12.7 for G.

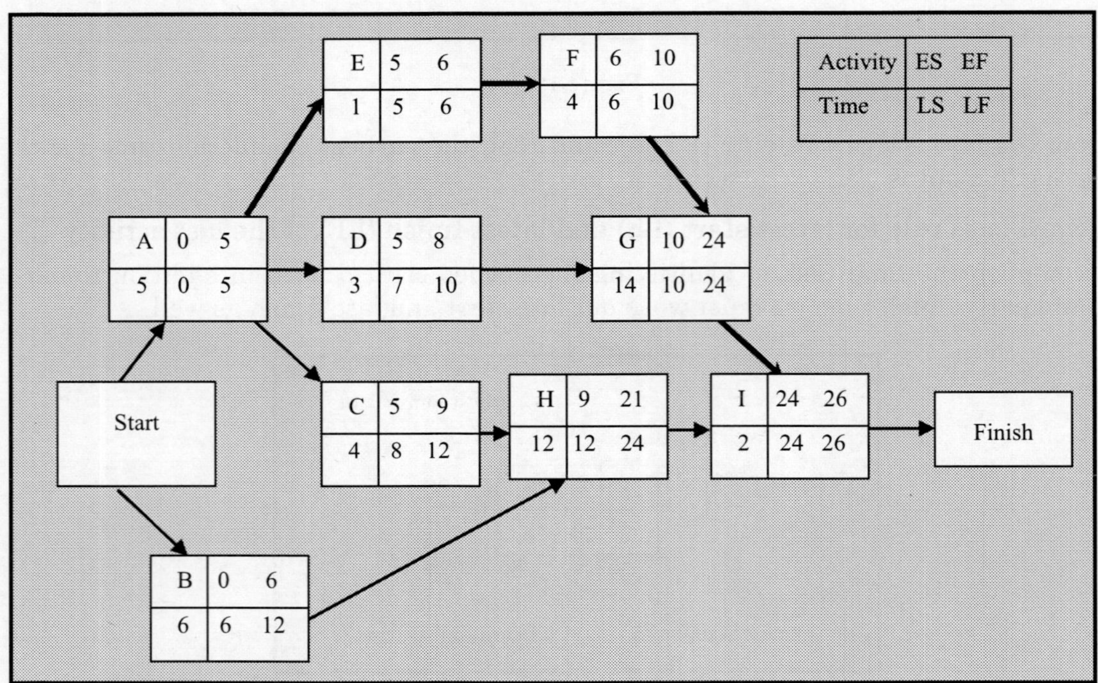

Figure 12.7

For H the latest finish time is simply the latest start time of its successor I. Thus, for H the latest start time 24 – 12 = 12.

For G the latest finish time is simply the latest start time of its successor I. Thus, for G the latest start time 24 – 10 = 14.

For F the latest finish time is simply the latest start time of its successor G. Thus, for F the latest start time 10 – 6 = 4.

For E the latest finish time is simply the latest start time of its successor FI. Thus, for E the latest start time 6 – 1 = 5.

For D the latest finish time is simply the latest start time of its successor G. Thus, for D the latest start time 10 – 7 = 3.

For C the latest finish time is simply the latest start time of its successor H. Thus, for C the latest start time 12 – 8 = 4.

For B the latest finish time is simply the latest start time of its successor H. Thus, for B the latest start time 12 – 6 = 6.

For A the latest finish time is simply the minimum of the latest start time of its successors E & D. Thus, for H the latest start time 5 – 5 = 0.

Float of an activity

The float of an activity may be defined as the extra cushion time available for its completion without affecting the overall project completion time. It is defined as *'the amount of time by which the completion of an activity could be delayed beyond the earliest completion time without affecting the overall project completion time'.* Accordingly, it is obtained by subtracting the earliest completion time from the latest completion time of the activity concerned.

Alternatively, it can also be defined as *'the amount of time by which the start of an activity could be delayed beyond the earliest start time without affecting the overall project completion time'.* Accordingly it is also obtained by subtracting the earliest start time from the latest start time of the activity concerned. The following table gives the summary of our network calculations.

Table 12.3

Activity	ES	LS	EF	LF	Float = LS – ES or LF – EF	Critical
A	0	0	5	5	0	Yes
B	0	6	6	12	6	-
C	5	8	9	12	3	-
D	5	7	8	10	2	-
E	5	5	6	6	0	Yes
F	6	6	10	10	0	Yes
G	10	10	24	24	0	Yes
H	9	12	21	24	3	-
I	24	24	26	26	0	Yes

Critical activity and critical path

The activity that does not have float is often called critical activity. In our illustration A, E, F, G, I are critical activities. The path connecting such critical activities is called the critical path. This path is indeed the longest path among all the paths available. The minimum time that is necessary to complete the project is obtained by adding the activity times of all the activities on the critical path. It is interesting to note that for a given project there can be more than one critical path.

Table 12.4

Path	Completion time in month	Nature of path
A-E-F-G-I	5+1+4+14+2 =26	Critical

In our illustration the longest path is none other than the only available critical path A-E-F-G-I. The associated time consumption on this path is 26 months. Thus, the minimum time needed for completing the entire project is 26 months for our illustrative example.

■ Example 2

Given the following data in months obtain the critical path and the associated minimum project completion time.

Table 12.5

Activity	Activity time in months	Preceding activities
A	2	-
B	15	A
C	6	B
D	16	A
E	6	D
F	2	B
G	2	D
H	4	C & E
J	2	G & H
K	4	F & H
L	1	J & K
M	10	L
N	20	L
O	2	M & N

Solution

As usual let us make both forward pass and backward pass calculations and report the result as shown Figure 12.8.

In the network the critical path is obtained by connecting all the highlighted critical (float less) activities.

Critical Path: A-D-E-H-K-L-N-O.

The associated minimum project time completion time = 2+16+6+4+4+1+20+2 = 55 months.

Note: *To get the minimum project completion time, in the above illustration we traced out only one critical path. Other non-critical path and other critical path (if exist) calculations are not necessary.*

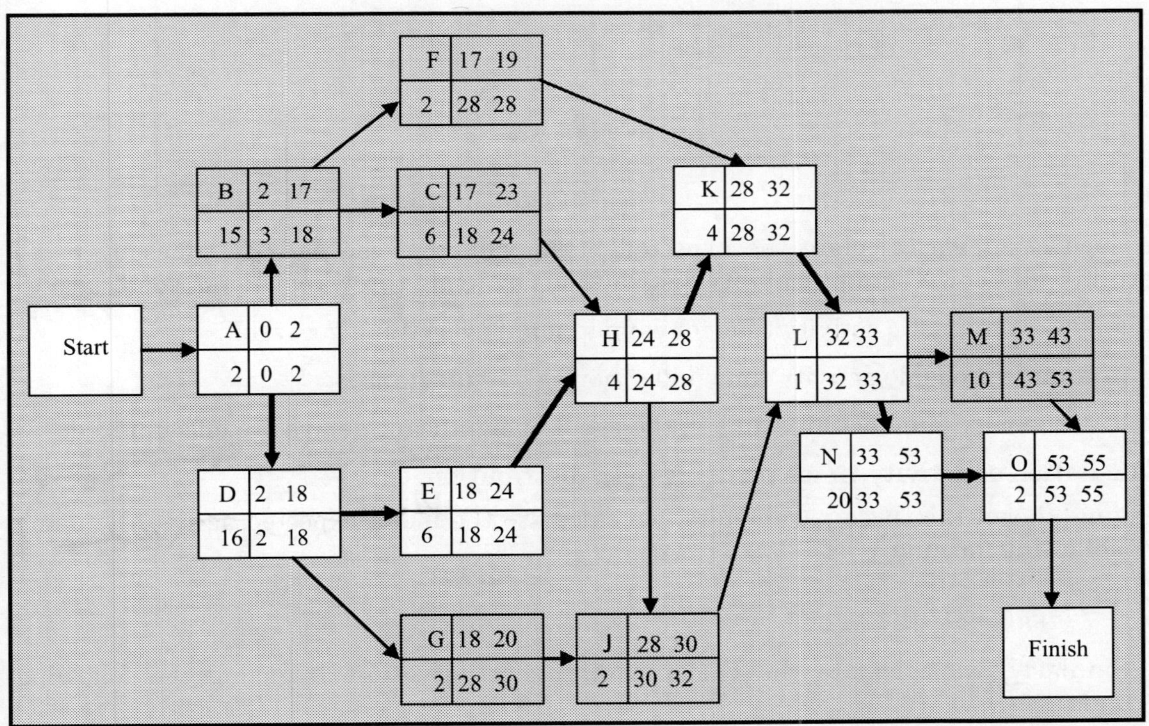

Figure 12.8

12.3 PROJECT SCHEDULING WITH UNCERTAIN ACTIVITY TIME

Till now activity times were assumed to be known constants. This need not be the case all the time. Very often activity time varies around a given mean time either way.

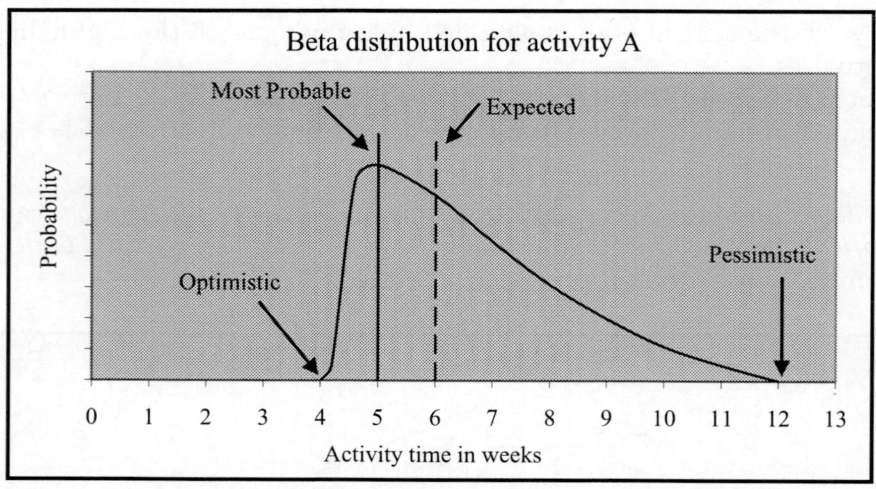

Figure 12.9

In such cases we calculate the expected mean values of the activity times by using certain well known beta probability distributions by using three known parameters.

a = the optimistic activity time for completion if everything goes well.

m = most probable activity time under normal circumstances.

b = the maximum activity time in the event of significant delays are encountered.

Calculation of activity times by using beta distribution

Having known the three parameters, we calculate the mean expected activity by using the following formula.

$$\text{Mean activity time} = t = \frac{a + 4m + b}{6}$$

Similarly, we calculate the variance of the activity time by using the following formula

$$\text{Variance of the activity time} = \left(\frac{b - a}{6}\right)^2$$

Greater the variance larger will be the probable deviation of the activity time from it estimated mean time and vice versa. Once the activity time is obtained for all the activities on the basis of Beta distribution by using the formula given above, we use either PERT or PCM method to get the solution in the usual manner.

■ Example 3

For the data given below draw a network diagram. Also obtain the critical path and the associated project completion time for the project under reference.

Table 12.6

Activity	Optimistic {a}	Most probable {m}	Pessimistic {b}
A	4	5	12
B	1	1.5	5
C	2	3	4
D	3	4	11
E	2	3	4
F	1.5	2	2.5
G	1.5	3	4.5
H	2.5	3.5	7.5
I	1.5	2	2.5
J	1	2	3

Table 12.7

Activity	Immediate predecessor
A	-
B	-
C	A
D	A
E	A
F	C
G	D
H	B,E
I	H
J	F,G,I

Solution

The activity time for the activity A in our illustration is obtained as

$$t = \frac{a + 4m + b}{6} = \frac{4 + 4(5) + 12}{6} = \frac{36}{6} = 6$$

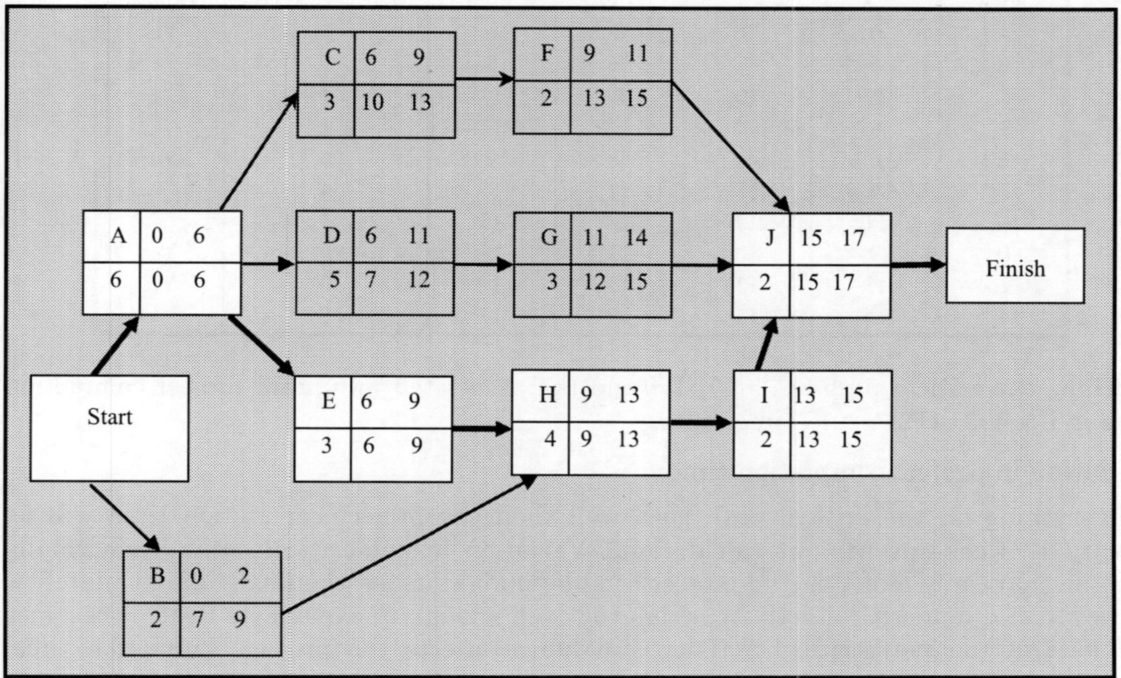

Figure 12.10

Table 12.8

Activity	Optimistic {a}	Most probable {m}	Pessimistic {b}	Activity time {t} estimate	Variance
A	4	5	12	6	1.78
B	1	1.5	5	2	0.44
C	2	3	4	3	0.11
D	3	4	11	5	1.78
E	2	3	4	3	0.11
F	1.5	2	2.5	2	0.03
G	1.5	3	4.5	3	0.25
H	2.5	3.5	7.5	4	0.69
I	1.5	2	2.5	2	0.03
J	1	2	3	2	0.11

Similarly, for all the activities the activity times are estimated and reported in Table 12.8. Once the activity times are obtained and reported for all activities, we use PERT/PCM method as usual and get the needed solution. The variance of the respective activities is obtained using the formula given and reported in the last column for later use.

Table 12.9

Activity	ES	LS	EF	LF	Float = LS−ES	Critical
A	0	0	6	6	0	Yes
B	0	7	2	9	7	-
C	6	10	9	13	4	-
D	6	7	11	12	1	-
E	6	6	9	9	0	Yes
F	9	13	11	15	4	-
G	11	12	14	15	1	-
H	9	9	13	13	0	Yes
I	13	13	15	15	0	Yes
J	15	15	17	17	0	Yes

Thus, the critical path is A-E-H-I-J and the associated minimum project completion time is T = 6+3+4+2+2 = 17 time units.

Variation in project completion time:

If everything on the critical path goes well then the project completion time will be exactly 17 weeks as per our calculations. Variation in non-critical activities normally does not have any effect on the project completion because of the float time availability. However, if a non-critical activity is delayed long enough to expand its float, it becomes the part of the new critical activity and hence affects the project completion time. Variation longer than the expected activity time on the critical path definitely will increase the project completion time. Similarly, shorter variation than the expected activity time on the critical path will reduce the completion time unless other activities become critical. Thus, the variance of the critical activities plays critical role in the calculation of the project completion time.

If the activities are independent then the variance of the completion time is the sum of the variance of the individual critical activities on the critical path. Therefore, in our illustration stated above the variance of the project completion time is obtained as

$$\sigma^2 = \sigma_A^2 + \sigma_E^2 + \sigma_H^2 + \sigma_I^2 + \sigma_J^2 = 1.78 + 0.11 + 0.69 + 0.03 + 0.11 = 2.72$$

The standard deviation of the project completion time: $\sigma = \sqrt{2.72} = 1.65$

If the project completion time T is morally distributed with the given standard deviation then we can readily use the standard normal distribution and obtain the probability of any given project completion time. For example, if the targeted project completion is 20 weeks then our standard normal variable Z is obtained as

$$Z = \frac{20-17}{1.65} = 1.82$$

Now from the standard normal table for Z = 1.82 the probability is 0.4656+0.5000 = 0.9656. Thus, there is a 96.65% chance for completing the project within given 20 weeks targeted time.

12.4 PROJECT CRASHING

Direct cost and desired level of crashing

Sometimes early completion of the project will give several advantages to the management. Such an outlook for early completion of the project is often called crashing. Such a task is achieved by crashing activities on the critical path by bringing additional resources into use. These additional resources like labour definitely will involve additional cost. This cost is often called direct cost of crashing. The worthiness of crashing must be decided by considering the additional benefit the management is going to get and the additional cost of crashing involved. Moreover, given the targeted project completion time and the activity-wise maximum permissible crash timings, one will work out the crashing by giving priority to the least crash cost activities. In this section we propose to discuss only the direct cost due to use of additional resources. More the resource we bring in more will be the additional cost of crashing. Thus, there are no concepts like optimum crashing in this section. We can at the most try to get the desired level of crashing and the associated costs. The following two examples illustrate such cases.

Linearity assumption of the direct crashing cost

We make the assumption that any portion of the given activity time within the permissible crash time is proportional to the given crashing cost as shown in Figure 12.11.

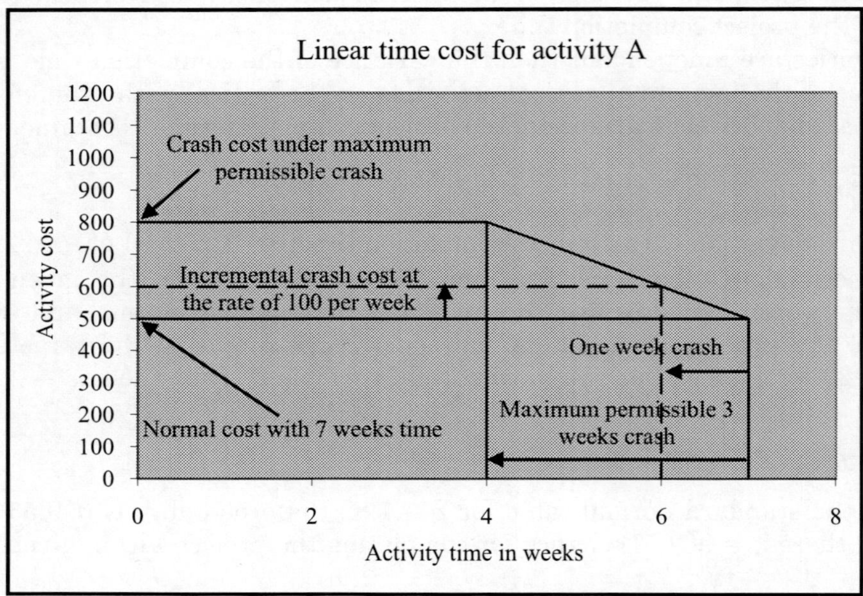

Figure 12.11

This assumption is often called linearity assumption of the crash time management. Table 12.10 and the associated linear graph 12.11 show the method of calculating the incremental weekly cost due to one week crashing of a given activity A.

Table 12.10

Activity	Normal time (in weeks)	Crash time (in weeks)	Maximum crash time (in weeks)	Normal cost before crashing (in Rs.)	Total cost after crashing (in Rs.)	Total incremental crash cost (in Rs.)	Average incremental crash cost per week (in Rs.)
A	7	4	3	500	800	300	100

■ Example 4

Given the data obtain the minimum project completion time. Also obtain crash solution and associated additional cost by shortening the project completion time by two weeks.

Table 12.11

Activity	Preceding activity	Normal time (in weeks)	Crash time (in weeks)	Normal cost (in Rs.)	Crash cost (in Rs.)
A	-	7	4	500	800
B	A	3	2	200	350
C	-	6	4	500	900
D	C	3	1	200	500
E_	B,D	2	1	300	550

No crash solution

Let us get the no crash solution to the problem by applying both forward and backward pass methods as shown in the following network chart 12.12.

(1) Critical Path: A-B-E

Project completion time: 12 weeks

The above chart shows that there are three critical activities A, B, and E along the critical path. The associated activity completion time is obtained as 12 weeks. The cost of the project with normal time is obtained by simply adding costs of all the activities in the project concerned.

So the total cost for completion = 500+200+500+200+300 = Rs. 1700.

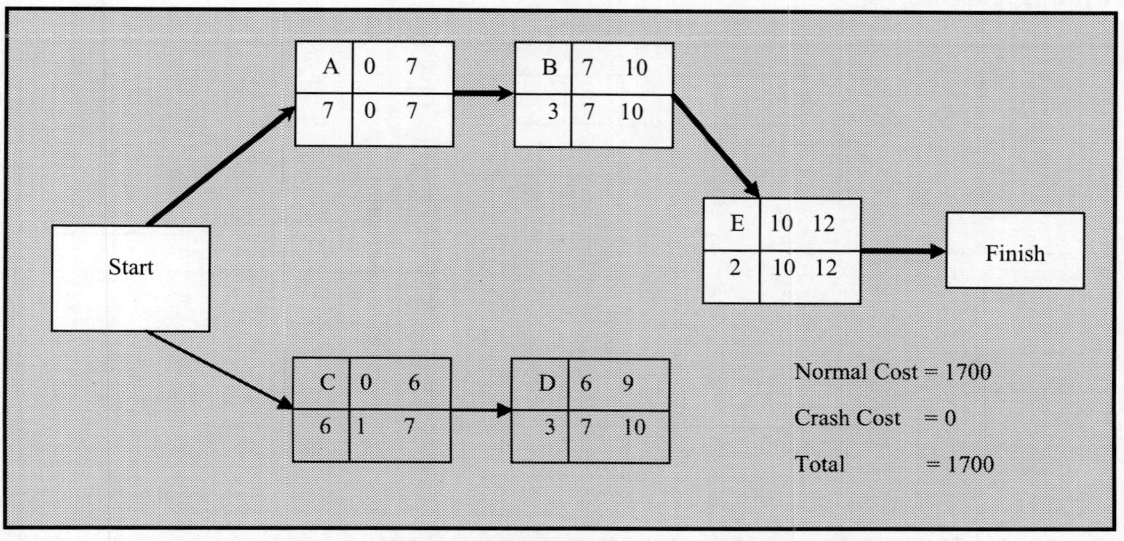

Figure 12.12 No crash solution

Calculation of weekly crash cost for all the activities on the basis of linearity assumption

In Table 12.12 for the activity A the maximum permissible crash cost is obtained as 800−500 =300. Such calculations are entered for all the activities in the seventh column. The average (incremental) crash cost for activity A is obtained by merely dividing 300 by the maximum permissible crash time 3. The so obtained average crash cost per week for all the activities are entered in the last column in Table 12.12.

Table 12.12

Activity	Normal time	Crash time	Maximum crash time	Normal cost	Crash cost	Total crash cost	Weekly incremental crash cost
A	7	4	3	500	800	300	100
B	3	2	1	200	350	150	150
C	6	4	2	500	900	400	200
D	3	1	2	200	500	300	150
E	2	1	1	300	550	250	250
Total normal project cost				1700			

Crash Solution

Now from Table 12.12 we notice that among the three critical activities the crash cost per day is less (100) for activity A. So let us crash activity A alone by one week and redo the whole exercise of forward and backward pass and obtain a critical path again as shown in Figure 12.13.

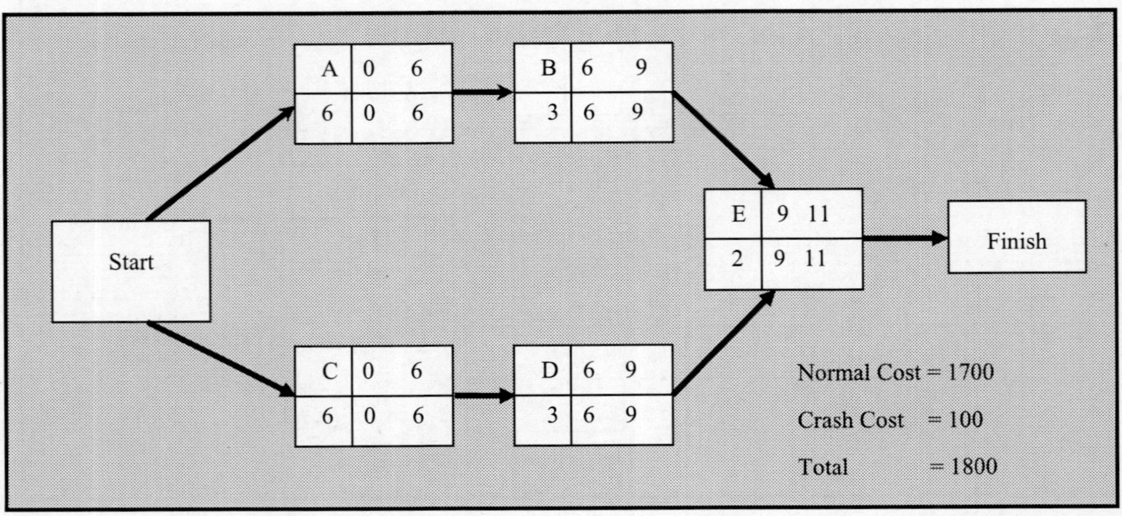

Figure 12.13 A is crashed by 1 week

(1) Critical Path: A-B-E

(2) Critical path: C-D-E

Project completion time: 11 weeks

From the above chart it is clear that the project completion time has come down by one week (from 12 to 11) with an additional crash cost of 100. But at the same time A-B-E and C-D-E are the two resulting critical paths this time. Now the minimum per week crashing cost for our new critical path for activity D is Rs. 150 from Table 12.17.

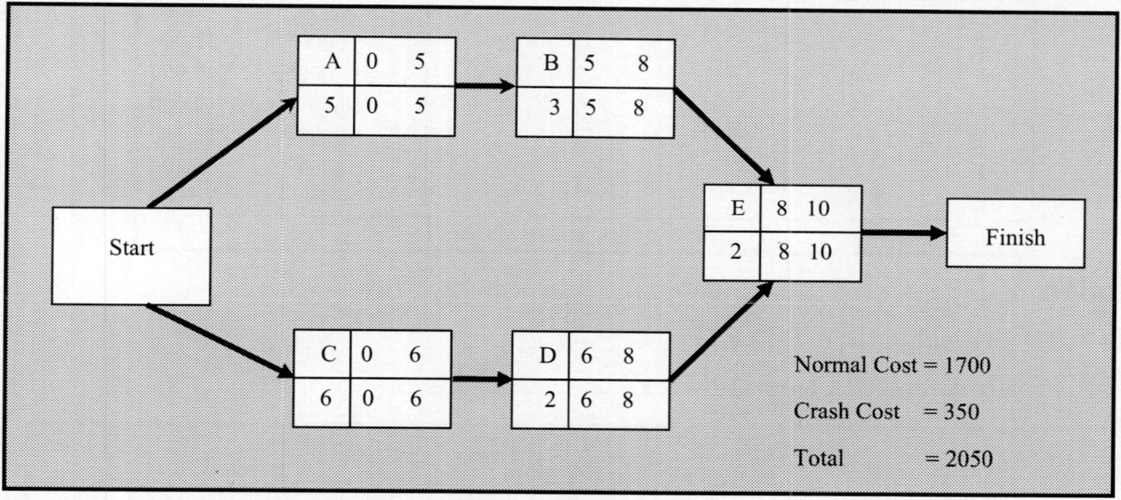

Figure 12.14 A is crashed by 2 weeks and D is crashed by 1 week

Therefore this time we crash both the activities A and D simultaneously by one week each and redo the forward and backward calculations once again as shown in Figure 12.14.

(1) Critical path: A-B-E

(2) Critical path: C-D-E

Project completion time: 10 weeks

The desired result of 10 days for project completion is obtained in chart 12.14 with a minimum of Rs. 350 additional crashing cost totalling to Rs. 2050. Table 12.13 gives the whole summary of crashing.

Table 12.13

Activity crashed	Direct project cost	Crash cost	Total cost	Project completion time
None	1700	0	1700	12
A(1)	1700	100	1800	11
A(2),D(1)	1700	350	2050	10

■ Example 5

For the data given below obtain the critical path and the associated project completion time. Crash the project by 4 weeks and obtain the associated crashing cost.

Table 12.14

Activity	Preceding activities	Normal time (in weeks)	Crash time (in weeks)	Normal cost (in Rs.)	Crash cost (in Rs.)
A	-	1	1	5000	5000
B	A	3	2	5000	12,000
C	A	7	4	11,000	17,000
D	B	5	3	10,000	12,000
E	B	8	6	8500	12,500
F	C	4	2	8500	16,500
G	E & F	1	1	5000	5000
Total project cost = 53,000					

Solution

Step 1: (1) Critical path: A-B-D-F-G

Project completion time: 14 weeks

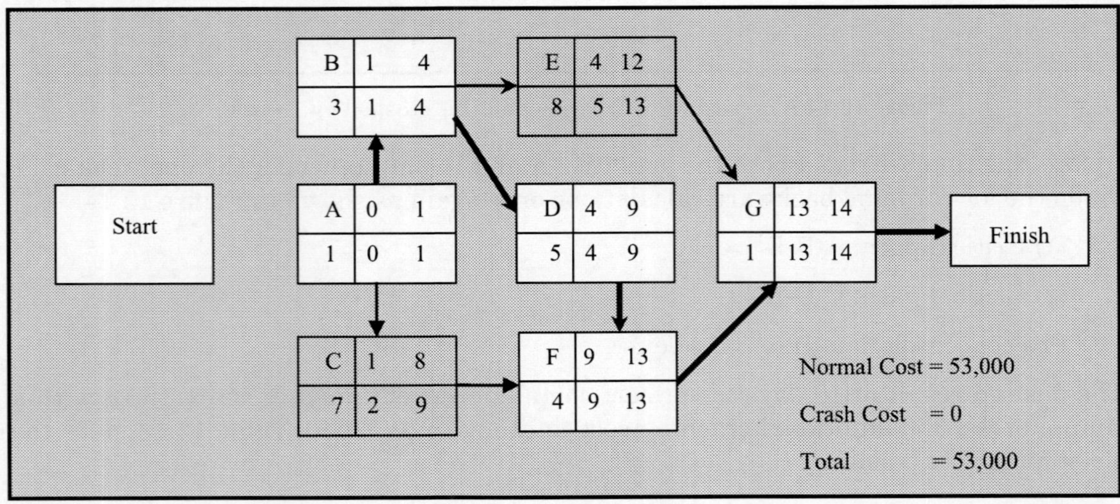

Figure 12.15 Solution without crashing

To begin with let us draw the network and do both forward and backward pass calculations and identify the critical path as shown in Figure 12.15. As per the critical path A-B-D-F-G shown the project completion needs 1+3+5+4+1=14 weeks time. At this stage since no crashing is considered, the total direct cost of the project will be

$$= 5000 + 85,000 + 5000 + 10,000 + 5000 + 11,000 + 85,000 = 53,000.$$

Step 2: Now let us calculate the weekly average (incremental) crash cost for all activities as shown in Table 12.15.

Table 12.15

Activity	Preceding activities	Normal time	Crash time	Maximum permissible crash	Normal cost	Crash cost	Total crash cost	Weekly incremental crash cost
A	-	1	1	0	5000	5000	0	-
B	A	3	2	1	5000	12,000	7000	7000
C	A	7	4	3	11,000	17,000	6000	2000
D	B	5	3	2	10,000	12,000	2000	1000
E	B	8	6	2	8500	12,500	4000	2000
F	C	4	2	2	8500	16,500	8000	4000
G	E & F	1	1	0	5000	5000	0	-
Total project cost = 53,000								

The above table gives the needed summary calculations. In this table we also calculate the maximum permissible crash for all the activities. Remember that we are planning to crash this project by 4 weeks in total. In other words we try to reduce project completion time from the present level of 14 weeks to just 10 weeks by bringing in additional resources into use and cutting short (crashing) the activity duration in some selective least cost activities on the critical path.

Among the five critical activities, namely A, B, D, F, and G the activity D is having the minimum of 1000 crashing cost per week. So let us crash this activity by one week to begin with and redo forward and backward pass calculations and identify the resulting critical path and the associated project completion time. The following chart 12.16 illustrates this step. With the additional crashing cost of 1000 per week the total cost of the project works out to be Rs. 54,000. The last earliest finish entry shows that the project completion time has come down by one week from 14 to 13 weeks.

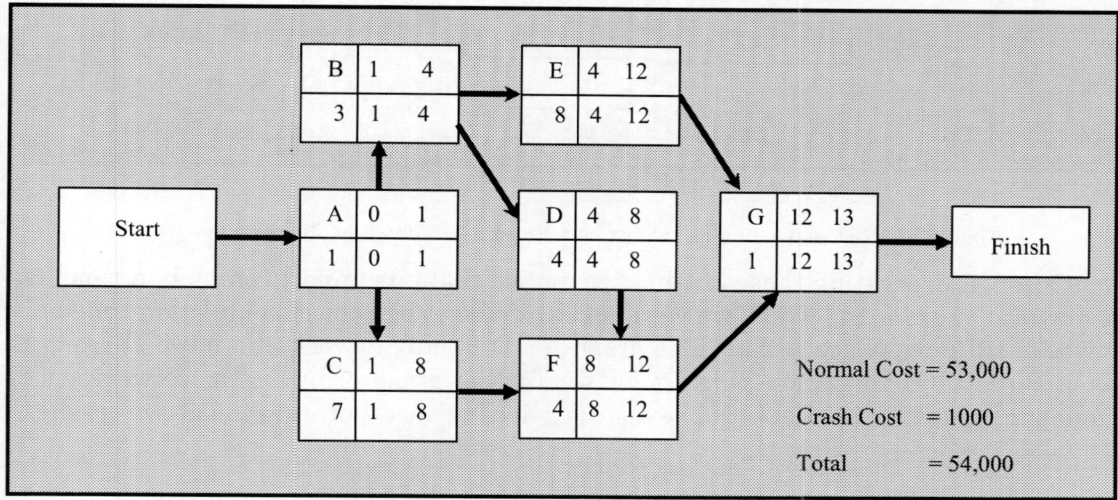

Figure 12.16 D is crashed by one week

(1) Critical path: A-B-E-G

(2) Critical path: A-B-D-E-G

(3) Critical path: A-C-F-G

Project completion time: 13 weeks

Under the new circumstances after incorporating the first crash altogether there are three critical paths and every activity is a critical activity. So, as per rule number 2, let as crash E on path (1) by one week, D on path (2) by one more week totalling to two weeks and C on path (3) by one week simultaneously. The final result of this crashing is shown in the following chart 12.17.

(1) Critical path: A-B-E-G

(2) Critical path: A-B-D-E-G

(3) Critical path: A-C-F-G

Project completion time: 12 weeks

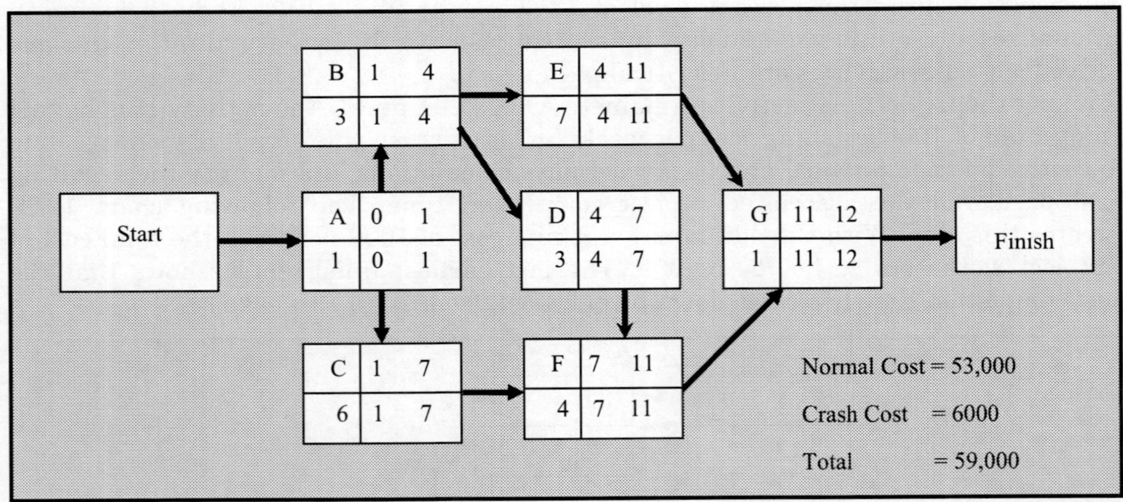

Figure 12.17 D is crashed by two weeks, E and C crashed by one week each

Now after incorporating these crashes, we notice that our project completion time has come down to 12 from 14. Thus, we were able to reduce the project completion time only by 2 weeks till now. Since permissible crash for D is only 2 weeks no more D crash is allowed. But still E and C are permissible for further crash of by one more week each. The following chart 12.18 shows the new situation after incorporating these two crashes.

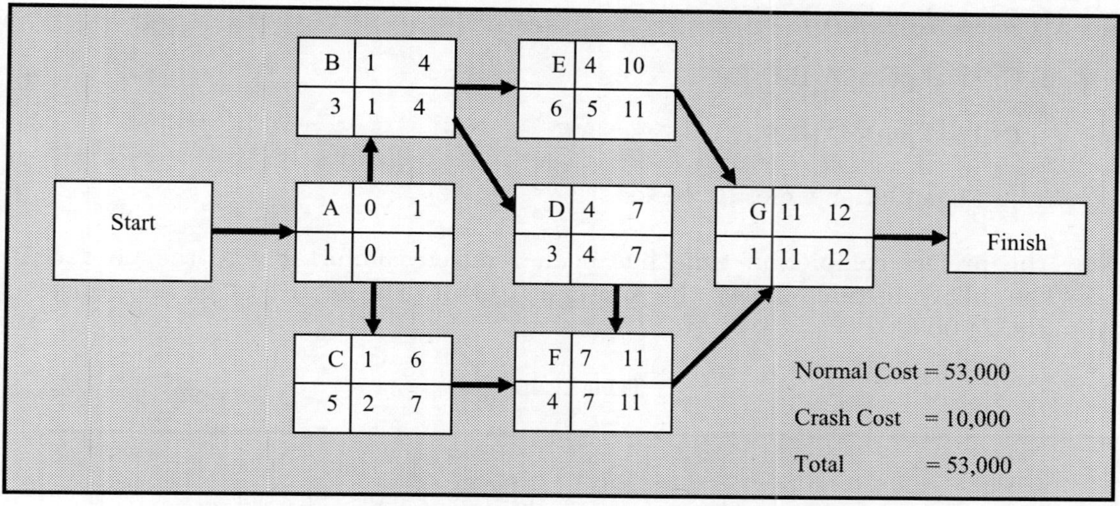

Figure 12.18 D is crashed by 2 weeks; E by 2 weeks C by 2 weeks

(1) Critical path: A-B-E-G

(2) Critical path: A-B-D-E-G

(3) Critical path: A-C-F-G

Project completion time: 12 weeks

However, project completion has not been affected this time. So this time let us crash C by one more week, F by one week and B by one week in addition to the crashes incorporated already. The following chart 12.19 illustrates this new situation.

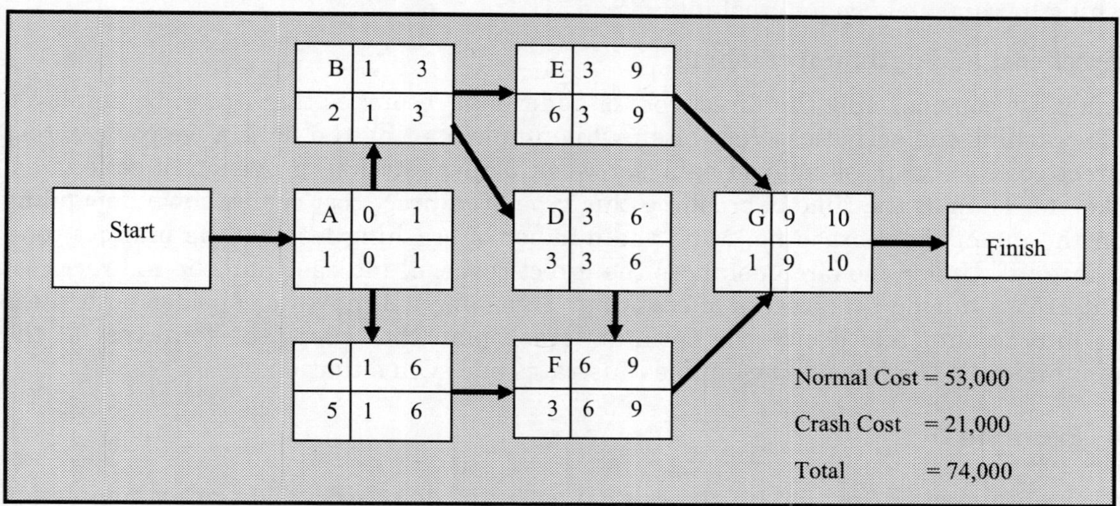

Figure 12.19 Crash D by two weeks, E by two weeks, and C by two week, F by one week and B by one week

(1) Critical path: A-B-E-G

(2) Critical path: A-B-D-E-G

(3) Critical path: A-C-F-G

Project completion time: 10 weeks

Now the project completion time has been reduced comfortably to the needed 10 weeks. The following table shows the summary result. The total cost of the project is equal to Rs. 74,000.

Table 12.16

Activity crashed with No. of weeks in brackets	Direct project cost	Crash cost	Total cost	Project completion time
None	53,000	0	53,000	14
D(1)	53,000	1000	54,000	13
E (1),D(2),C(1)	53,000	6000	59,000	12
E (2),D(2),C(2)	53,000	10,000	63,000	12
E (2),D(2),C(2),F(1),B(1)	53,000	21,000	74,000	10

The desired level of 4 weeks crashing has been achieved in the above table with the total cost of Rs. 74,000. Thus the additional cost of crashing in this case is Rs. 21,000. If the manager feels that this additional cost is less than the expected benefit out of crashing he can safely go for crashing.

Indirect cost and optimum crashing

Till now we consider only the direct cost of additional resources that are to be employed while crashing the activities. Very often while implementing the project we come across indirect costs as well, as a fixed sum per week or per day like provision of coffee, tea, snake etc. Thus at the time of crashing this type of indirect cost comes down depending upon the quantum of crashing. More the number of crashing less will be indirect total cost in total. Under the circumstances the direct cost will increase and the indirect cost will decrease during the process of crashing. Thus, one will have to consider both these costs into account and decide the optimum crash plan that keeps the total cost at the minimum level. The following example illustrates one such situation.

■ Example 6

For the data given below obtain the critical path and the associated project completion time under normal cost conditions. If there is an indirect cost of Rs. 3 per day obtain the optimum crash that keeps the total cost least.

Table 12.17

Activity	Preceding activities	Normal time (in days)	Crash time (in days)	Normal cost (in Rs.)	Crash cost (in Rs.)
A	-	2	2	10	10
B	A	12	10	30	35
C	A	9	6	10	13
D	B	3	2	35	55
E	B	5	3	60	84
F	A	12	10	50	58
G	C	12	10	45	50
H	D & E	1	1	10	10
I	G, F & H	1	1	10	10
Total project cost = 260					

Solution

The critical path: A-C-G-I

Project completion time: 24 Days

Now to work out the optimum crash let us first calculate the daily incremental cost of crashing as shown in Table 12.18.

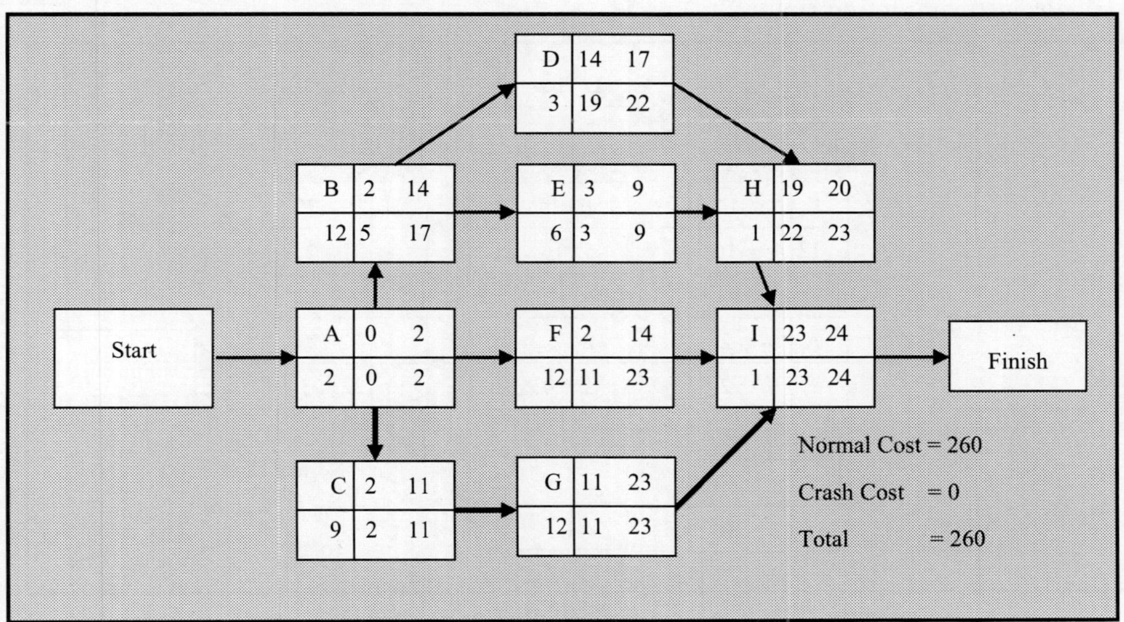

Figure 12.20 No crash solution

Table 12.18

Activity	Preceding activities	Normal time	Crash time	Normal cost	Maximum Permissible crash	Crash cost	Daily incremental cost
A	-	2	2	10	0	10	-
B	A	12	10	30	2	35	2.5
C	A	9	6	10	3	13	1
D	B	3	2	35	1	55	20
E	B	5	3	60	2	84	12
F	A	12	10	50	2	58	4
G	C	12	10	45	2	50	2.5
H	D & E	1	1	10	0	10	-
I	G, F & H	1	1	10	0	10	-
Total project cost = 260							

Since C is the least cost activity among the critical ones costing Re. 1, let us crash it first by one day and recalculate the network once again as shown below Figure 12.21.

The critical path: A-C-G-I

Project completion time: 23 Days

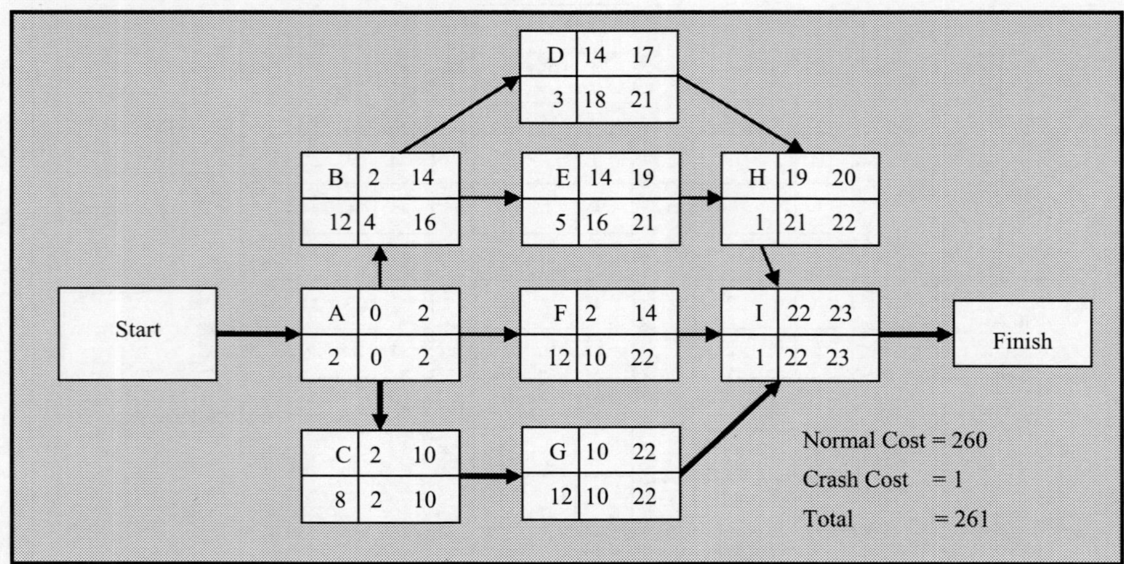

Figure 12.21 The least cost activity C is crashed by 1 day

Since the maximum permissible crash is 3 and 2 more crash is available for C let us crash it by one more day and redo the calculations as shown below in Figure 12.22.

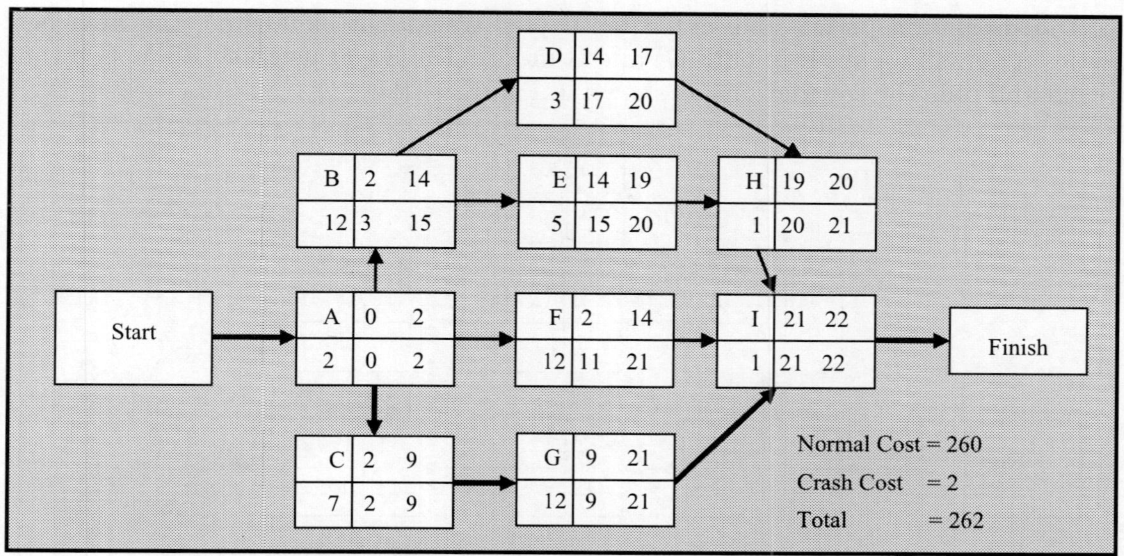

Figure 12.22 The least cost activity C is crashed by 2 days

The critical path: A-C-G-I

Project completion time: 22 Days

Since the maximum permissible crash is 3 and 1 more crash is available for C let us crash it by one more day and redo the calculations as shown in Figure 12.23.

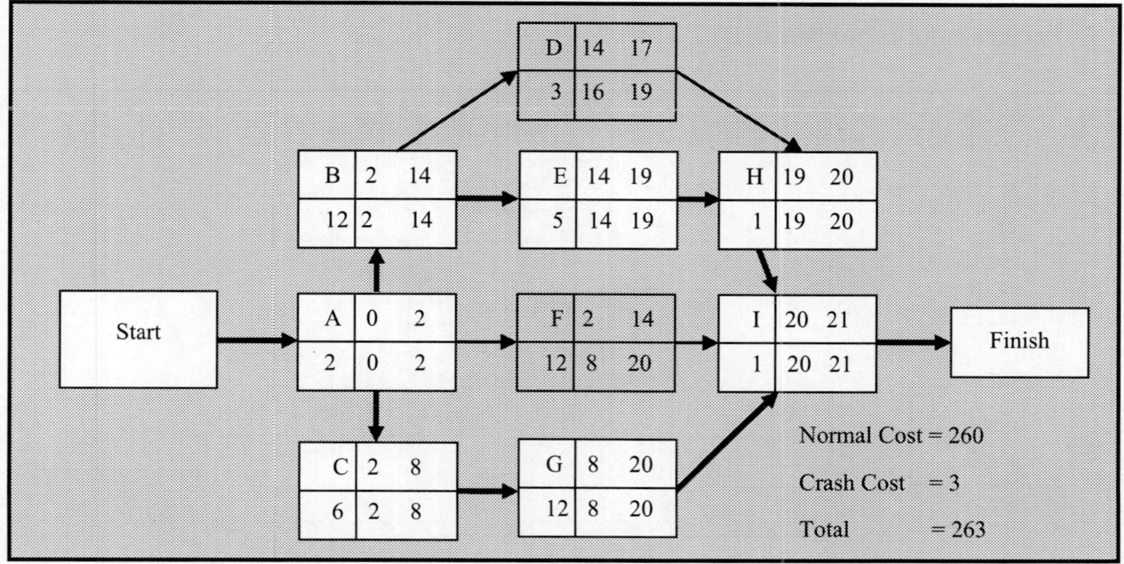

Figure 12.23 The least cost activity C is crashed by 3 days

(1) Critical path: A-C-G-I (2) Critical path: A-B-E-H-I

Project completion time: 21 Days

Since the maximum permissible crash for C of 3 weeks got exhausted, the next best activities for crashing are B in path (1) and G in Path (2). So let us crash B and G by one day each and redo the calculations as shown in Figure 12.24.

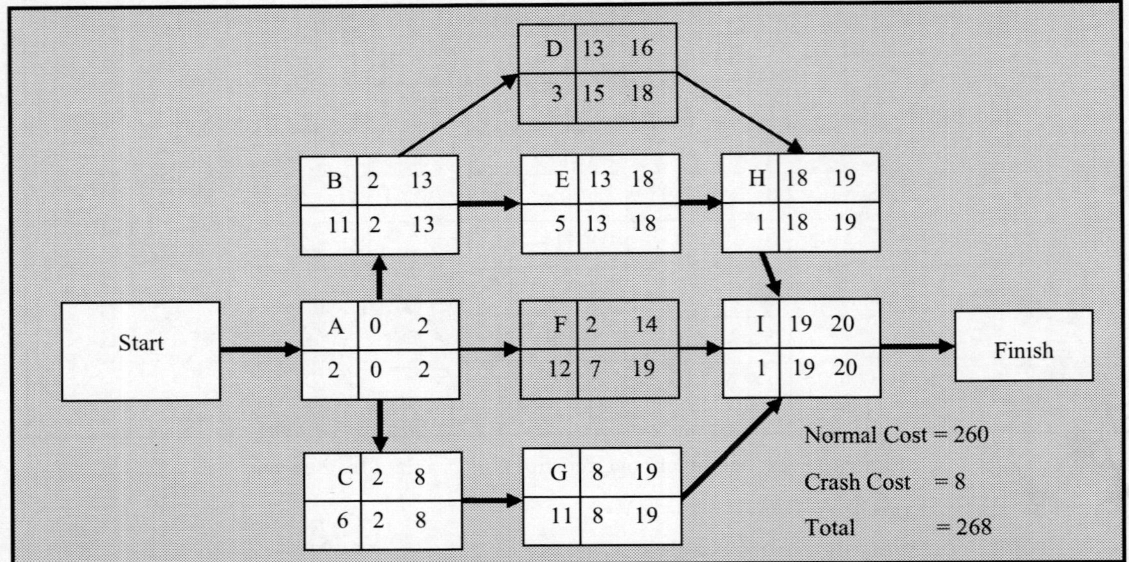

Figure 12.24 Activity C by 3 days, B & G by one day each

(1) Critical Path: A-C-G-I (2) Critical Path: A-B-E-H-I

Project completion time: 20 Days

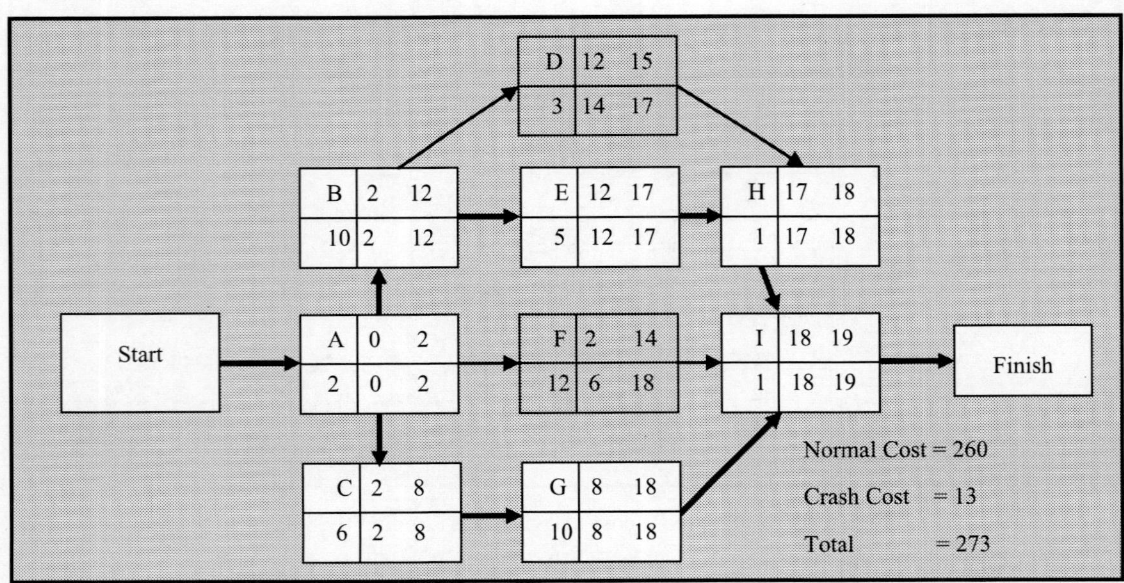

Figure 12.25 Activity C by 3 days, B & G by 2 days each

(1) Critical Path: A-C-G-I (2) Critical Path: A-B-E-H-I

Project completion time: 19 Days

The following table summarizes the whole proceeding with all the needed break ups. Clearly the optimum project completion is 21 days and the associated total project cost is Rs. 326.

Table 12.19

Project duration	Activity crashed with no. of days in bracket	Direct cost = Normal + Crash		Indirect cost		Total cost
24	No crash	260 + 0 =	260	3x24 =	72	332
23	C(1)	260 +1 =	261	3x23 =	69	330
22	C(2)	260 +2 =	262	3x22 =	66	328
21	C(3)	260 + 3 =	263	3x21 =	63	326
20	C(3) B(1) C(1)	260 + 8 =	268	3x20 =	60	328
19	C(3) B(2) C(2)	260 + 13 =	273	3x19 =	57	330

Since the minimum total cost Rs. 326, the optimum crashing could be the activity C for three days. The associated project completion time is 21 days.

12.5 LINEAR PROGRAMMING APPROACH TO CRASHING

Given the data obtain the minimum project completion time. Also obtain crash solution and associated additional cost by shortening the project completion time by two weeks.

Table 12.19(a)

Activity	Preceding activity	Normal time (in weeks)	Crash time (in weeks)	Normal cost (in Rs.)	Crash cost (in Rs.)
A	-	7	4	500	800
B	A	3	2	200	350
C	-	6	4	500	900
D	C	3	1	200	500
E	B,D	2	1	300	550

Let us redo Example 4 above by Linear Programming.

Let X_A = earliest finish time for activity A, and Y_A = amount of activity time crashed for the activity A. Now since the earliest start time for A is 0 in our model the earliest finish time depends upon the activity time minus the crash time

$$X_A \geq 0 + (7 - Y_A) \quad \text{or} \quad X_A + Y_A \geq 7 \qquad (1)$$

In particular of $Y_A = 0$ (no crashing) the earliest finish time for A is 0+7 = 7. Thus the earliest finish time for A depends upon the crash time of A.

Similarly, for the activity C the relevant constraint is obtained as

$$X_C \geq 0 + (6 - Y_C) \quad \text{or} \quad X_C + Y_C \geq 6 \qquad (2)$$

From forward pass we know that the earliest start time for B is simply the earliest finish time for A. Thus the constraint for the earliest finish time for B is written as

$$X_B \geq X_A + (3 - Y_B) \quad \text{or} \quad X_B + Y_B - X_A \geq 6 \qquad (3)$$

Similarly the co-starring for D is written as

$$X_D \geq X_C + (3 - Y_D) \quad \text{or} \quad X_D + Y_D - X_C \geq 6 \qquad (4)$$

Now for activity E, the preceding activities are B and D. So E's earliest start is the maximum value of the earliest finish time of both B and D. So to accommodate the double condition we write two constants as follows

$$X_E \geq X_B + (2 - Y_E) \quad \text{or} \quad X_E + Y_E - X_B \geq 2 \qquad (5)$$

$$X_E \geq X_D + (2 - Y_E) \quad \text{or} \quad X_E + Y_E - X_D \geq 2 \qquad (6)$$

Now since crashing is aimed at reducing the project completion time to 10 days the earliest finish time for E must be less than or equal to 10 days.

$$X_E \leq 10 \qquad (7)$$

Now from the data table the maximum allowable crash constraints are written as

$$Y_A \leq 3 \qquad (8)$$

$$Y_B \leq 1 \qquad (9)$$

$$Y_C \leq 2 \qquad (10)$$

$$Y_D \leq 2 \qquad (11)$$

$$Y_E \leq 1 \qquad (12)$$

Now it is time to formulate the objective function that minimizes the crashing cost

$$\text{Min: } 100Y_A + 150Y_B + 200Y_C + 150Y_D + 125Y_E$$

Excel Solution

To do this problem in excel worksheet open an excel worksheet with column and rows as shown in Table 12.20. Enter the formulas as given in Table 12.21 in the stated cells. Enter the other data entries as described in Table 12.20. Use solver option and solve the problem in the usual manner.

Cell F16	= SUMPRODUCT(A14:J14,A15:J15)
Cell C17	= SUMPRODUCT(A2:J2,$A$15:$J$15)
Cell C18	= SUMPRODUCT(A3:J3,$A$15:$J$15)
Cell C19	= SUMPRODUCT(A4:J4,$A$15:$J$15)
Cell C20	= SUMPRODUCT(A5:J5,$A$15:$J$15)
Cell C21	= SUMPRODUCT(A6:J6,$A$15:$J$15)
Cell C22	= SUMPRODUCT(A7:J7,$A$15:$J$15)
Cell C23	= SUMPRODUCT(A8:J8,$A$15:$J$15)
Cell C24	= SUMPRODUCT(A9:J9,$A$15:$J$15)
Cell C25	= SUMPRODUCT(A10:J10,$A$15:$J$15)
Cell C26	= SUMPRODUCT(A11:J11,$A$15:$J$15)
Cell C27	= SUMPRODUCT(A12:J12,$A$15:$J$15)
Cell C28	= SUMPRODUCT(A13:J13,$A$15:$J$15)

A is crashed by two weeks and activity D is crashed by one week given the projection completing time as 10 weeks. The minimum additional cost of crashing is Rs. 350.

Table 12.20

	A	B	C	D	E	F	G	H	I	J	K
1	X_A	Y_A	X_B	Y_B	X_C	Y_C	X_D	Y_D	X_E	Y_E	
2	1	1	0	0	0	0	0	0	0	0	7
3	–1	0	1	1	0	0	0	0	0	0	3
4	0	0	0	0	1	1	0	0	0	0	6
5	0	0	0	0	–1	0	1	1	0	0	3
6	0	0	–1	0	0	0	0	0	1	1	2
7	0	0	0	0	0	0	–1	0	1	1	2
8	0	0	0	0	0	0	0	0	1	0	10
9	0	1	0	0	0	0	0	0	0	0	3
10	0	0	0	1	0	0	0	0	0	0	1
11	0	0	0	0	0	1	0	0	0	0	2
12	0	0	0	0	0	0	0	1	0	0	2
13	0	0	0	0	0	0	0	0	0	1	1
14	0	100	0	150	0	200	0	150	0	250	
15	5	2	8	0	6	0	8	1	10	0	
16		Minimum crashing cost =				350					
17		1	7							>=	7
18		2	3							>=	3
19		3	6							>=	6
20		4	3							>=	3
21		5	2							>=	2
22		6	2							>=	2
23		7	10							<=	10
24		8	2							<=	3
25		9	0							<=	1
26		10	0							<=	2
27		11	1							<=	2
28		12	0							<=	1

12.6 ACTIVITY RESCHEDULING

Project scheduling is a methodical framing of a timetable giving the details of starting and finishing dates for each and every activity for the referred project. If all activities are critical in the final solution then this timetable has no flexibility at all. We simply go by what the network diagram says. However, if there are non-critical activities in the network we can always prepare modified schedules.

Table 12.21

Critical

E	12	17
5	12	17

Non-Critical

D	2	14
12	6	18

In Table 12.21 problem E is a critical activity for which early start and latest start are equal to 12th week. Similarly, the early finish and latest finish are equal to 17th weeks. So the job must start exactly at the end of 12th week (beginning of 13th week) and continue up to the end of 17th week. No rescheduling of whatsoever is possible without affecting the project completion time. The following chart illustrates this type of scheduling.

Time schedule for critical activity E																				
Activity																				
E																				
Day No.	1	2	3	4	5	6	7	8	9	10	11	12	13	14	15	16	17	18	19	20

Figure 12.26

However, in the case of non-critical activity like D having floats of 4 weeks in our above illustration we can reschedule the activity in five possible ways as shown below without affecting the project completion time.

Time schedule for non-critical activity D																				
Activity																				
D		Correct start												F						
D			One day late start												F					
D				Two days late start												F				
D					Three days late start												F			
D						Four days late start														
Day No.	1	2	3	4	5	6	7	8	9	10	11	12	13	14	15	16	17	18	19	20

Figure 12.27

12.7 ACTIVITY RESCHEDULING AND RECOURSE ALLOCATION

Till now we have considered only the time cost and money cost in all our network problems. In reality physical resources like labour, capital etc., are employed in the activities to complete the work in the stipulated time. More the resource used less will be time taken for completing the activity and vice versa. In fact only by bringing additional resources into use the stipulated time cost has been brought down in all our crashing illustrations discussed in the previous section.

In the technique CPM and PERT, implicitly we assume the unlimited availability of the physical resources so that we can increase them for crashing by spending extra money. Thus in this context the stipulated time for each activity depends upon the quantity of resources used. However when resources are limited in supply then our concept of critical path activity and floats has no significance in completing the project in time. Some of the critical activities may be delayed due to non-availability of resources. Delays in critical activity definitely delay the project completion time. Under such circumstances two alternative courses of action are highlighted in this section.

12.8 RESOURCE LEVELLING

In the first case, the non-critical activities are rescheduled from early start to late start by availing of the benefit of float. Often we will be forced to extent the activity time beyond the late start by reducing the resource use in earlier periods. Occasionally rescheduling of the critical activities become inevitable based on the rescheduling of its preceding activities. Due the rescheduling of the critical activities the project completion time also gets extended invariably. However, while rescheduling, priority is given to non-critical activities having the highest float first. This method is often called *"Resource Levelling"*.

12.9 RESOURCE SMOOTHING

Often we come across situations with lots of fluctuations in the quantum of resource use over time, though enough resources are available in the market. Take for example, if 10 workers on the first week, 2 workers in the second week and 15 workers in the third week are needed then managing the resource use is a difficult task. Eight workers will be jobless in the second week. Calling them back for the next week is unethical. Under such circumstances we opt for *resource smoothing* without affecting the overall projection time and the total resource use. The heuristic procedure for this type smoothing is described as follows.

1. After obtaining the network solution by critical path method to the given problem, an early start early finish schedule is prepared. Next we generate a resource loading for each day having the resource needs of the activities in action, starting from early start schedule.
2. A resource level, one unit below the peak requirement should be used to reschedule the jobs so that peak requirements do not exceed this reduced level commonly called trigger level. All jobs on the peak level need rescheduling. The activities having sufficiently large float could be used first for rescheduling so that the project is completed with the trigger level resources. Attempt must also be made to ensure uniform distribution of resources.
3. The trigger level is reduced by one time unit in the previous step is repeated with this reduced resource. This operation is repeated until the trigger level cannot be reduced further without increasing the project time.

Note: *Due to random choice of activities in the case of ties at various stages there may be more than one schedule for a given resource levelling.*

■ Example 7

For the data given below obtain the critical path and the associated project completion time. Also prepare a smooth Resource Levelling Schedule, given only 8 units of manpower available per day for the project under reference.

Table 12.22

Activity	Preceding activities	Activity time	Manpower consumption per day
A	-	2	7
B	A	2	4
C	A	2	4
D	A	2	4
E	B	2	5
F	C	3	3
G	D	1	2
H	E,F & G	1	6

Solution

From the data given we prepare the network chart and obtain the critical path and associated least project completion time as usual first. The critical activity is A-C-F-H and the associated project completion time is 8 days.

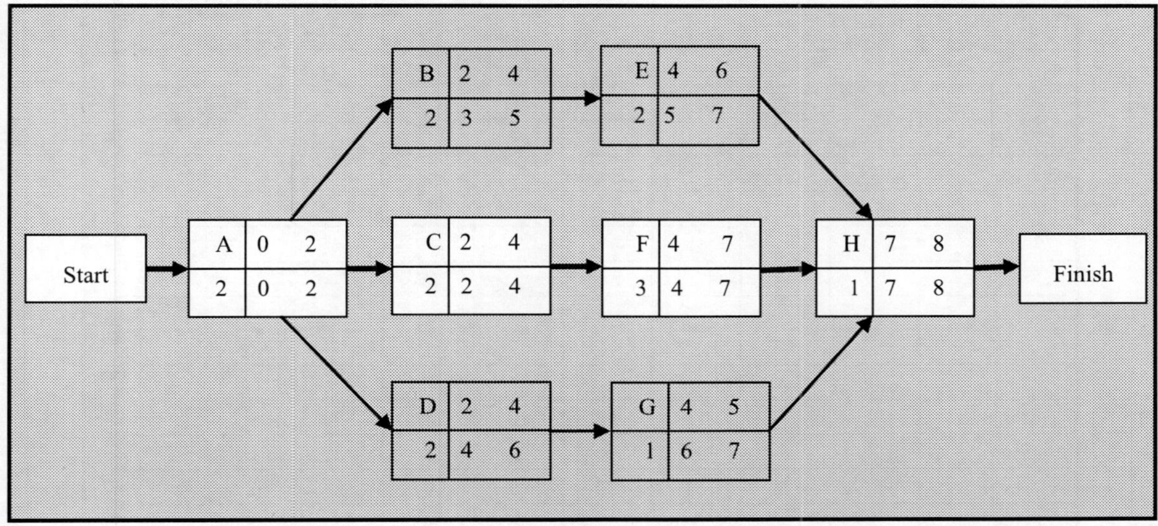

Figure 12.28

Now from the chart given above we prepare the early start-early finish and late start-late finish schedule as shown in Table 12.23 for all the 8 activities. In this table the critical activities are highlighted with zero float. The non-critical activities are shown with their associated levels of floats available in the last column.

Table 12.23

Activity	Preceding	Activity	ES	EF	LS	LF	Float
A	-	2	0	2	0	2	0
B	A	2	2	4	3	5	1
C	A	2	2	4	2	4	0
D	A	2	2	4	4	6	2
E	B	2	4	6	5	7	1
F	C	3	4	7	4	7	0
G	D	1	4	5	6	7	2
H	E,F,G	1	7	7	7	8	0

Using this table we next prepare the schedule for all the activities as shown below by using horizontal bars. The white bars represent the associated activity time. The green bars show the number of floats available.

Activity									
A	7	7							
B			4	4	Float				
C			4	4					
D			4	4	Float	Float			
E					5	5			
F					3	3	3		
G					2	Float			
H								6	
Demand	7	7	12	12	10	8	3	6	
Supply	8	8	8	8	8	8	8	8	
Surplus	1	1	−4	−4	−2	0	5	2	
Days No.	1	2	3	4	5	6	7	8	

Figure 12.29 Activity scheduling on early start and early finish basis

At the bottom rows we show the demand, supply and surplus/deficit of labour day-wise. Clearly the critical activities do not have any float as such and hence shown by white bars only.

The following chart shows the supply and demand for the labour resource before rescheduling.

Table 12.24

Day No.	Activity	Resource requirement		Resource available	Surplus / Deficit
1	A	7 =	7	8	1
2	A	7 =	7	8	1
3	B,C,D	4 + 4 + 4 =	12	8	−4
4	B,C,D	4 + 4 + 4 =	12	8	−4
5	E,F,G	5 + 3 + 2 =	10	8	−2
6	E,F	5 + 3 =	8	8	0
7	F	3 =	3	8	5
8	H	6 =	6	8	2
	Total		65		

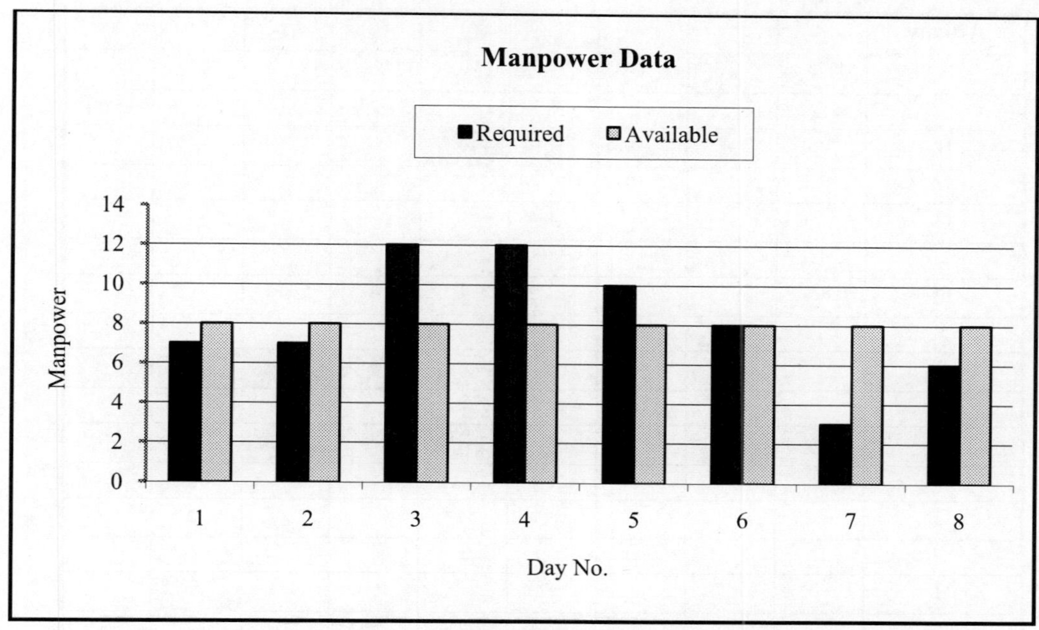

Figure 12.30

The Table 12.24 gives the summary of the pre-smoothing situation.

Day-by-day rescheduling and matching the demand and supply

Though the surplus does not pose any bottleneck to the production, the resource deficit must be avoided at every stage. By rescheduling the activities one can overcome such production bottlenecks.

Day 1

On day one, activity 1 is in operation by using 7 units of the labour resource. Since 8 units are available per day one unit will be inevitably surplus and hence no bottleneck to the production.

Day 2

Day two also poses no additional problem because the supply exceeds the demand.

Day 3 & 4

On day 3 & 4 three activities B, C, D are originally scheduled. Accordingly, the total resource requirement is equal to 4+4+4+12. Since only 8 units are available there will be a deficit of 4 units on each day, so one of the activities must be rescheduled to a future date. Clearly activity D deserves such a rescheduling because it has the highest float at this stage. Since it has a float of 2 days we simply reschedule this activity on 5th and 6th days by initiating a late start by two days as shown in Figure 12.31.

Activity									
A	7	7							
B			4	4	Float				
C			4	4					
D					4	4			
E					5	5	Float		
F					3	3	3		
G							2		
H								6	
Demand	7	7	8	8	12	12	5	6	
Supply	8	8	8	8	8	8	8	8	
Surplus	1	1	0	0	−4	−4	3	2	
Days No.	1	2	3	4	5	6	7	8	

Figure 12.31 Activity scheduling on early start and early finish basis

Once this rescheduling has been done, on 3rd & 4th days we will be left with only B and C activities in action totalling to 4 + 4 = 8 units of resource use on each day. Simply for D it is a late start and late finish situation. However, once this rescheduling for D is effected invariably its successor G is to be rescheduled to at least day 7 so that the predecessor D completes the task on the 6th day. The chart 12.31 shows the situation after both the rescheduling. Till now we have solved the problem up to fourth day.

Day 5

On 5th day now unfortunately we have three activities D, E, F in action totalling to 4+5+3 = 12 units of resource requirement. Clearly rescheduling of the non-critical activities is a must at this stage also. So let us delay E by one day and reschedule it on the 6th and 7th days as shown in Figure 12.32.

Activity										
A	7	7								
B			4	4	Float					
C			4	4						
D					4	4				
E						5	5			
F					3	3	3			
G							2			
H								6		
Demand	7	7	8	8	7	12	10	6	0	
Supply	8	8	8	8	8	8	8	8	8	
Surplus	1	1	0	0	1	−4	−2	2	8	
Days No.	1	2	3	4	5	6	7	8	9	

Figure 12.32 Activity scheduling on early start and early finish basis

At this stage on 5th problem is solved with one surplus unit of labour. It will have only D and F in action totalling to 4+3 = 7 units resource use.

Day 6

Now at this stage, on day 6 we will have activities D, E, F in action totalling to 4+5+3 = 12 units of resource requirement. At this stage rescheduling of E on 7th and 8th days will solve the problem. However, since H succeeds only after E's completion H (critical) is rescheduled to day 9. Now on 6th day we have only D and F totalling to 4+3 = 7 resource requirement. The situation is described in Figure 12.33.

Activity										
A	7	7								
B			4	4	Float					
C			4	4						
D					4	4				
E							5	5		
F					3	3	3			
G							2			
H									6	
Demand	7	7	8	8	7	7	10	5	6	
Supply	8	8	8	8	8	8	8	8	8	
Surplus	1	1	0	0	1	1	−2	3	2	
Days No.	1	2	3	4	5	6	7	8	9	

Figure 12.33 Activity scheduling on early start and early finish basis

Day 7

At this stage day 7 is having activities E, F, G totalling to 5 + 3 + 2 = 10 units of resource requirement. Now let us reschedule G to day 8 and solve the problem completely as shown in Figure 12.34.

On day 7 we will be left with only E and F totalling to 5 + 3 = 8 units of resource requirement.

Day 8

Only E and G are inaction totalling to 5 + 2 = 7 units of labour requirement.

Activity										
A	7	7								
B			4	4	Float					
C			4	4						
D					4	4				
E							5	5		
F					3	3	3			
G								2		
H									6	
Demand	7	7	8	8	7	7	8	7	6	
Supply	8	8	8	8	8	8	8	8	8	
Surplus	1	1	0	0	1	1	0	1	2	
Days No.	1	2	3	4	5	6	7	8	9	

Figure 12.34 Activity scheduling on early start and early finish basis

Day 9

Only H is left over with resource requirement of 6. Table 12.25 and the associated chart 12.35 gives the summary after rescheduling.

Table 12.25

Day No.	Activity	Resource requirement	Resource available	Surplus / Deficit
1	A	7 = 7	8	1
2	A	7 = 7	8	1
3	B,C	4+4 = 8	8	0
4	B,C	4+4 = 8	8	0
5	D,F	4+3 = 7	8	1
6	E,F	4+3 = 7	8	1
7	E,F	5+3 = 8	8	0
8	E,G	5+2 = 7	8	1
9	H	6 = 6	8	2
	Total	65		

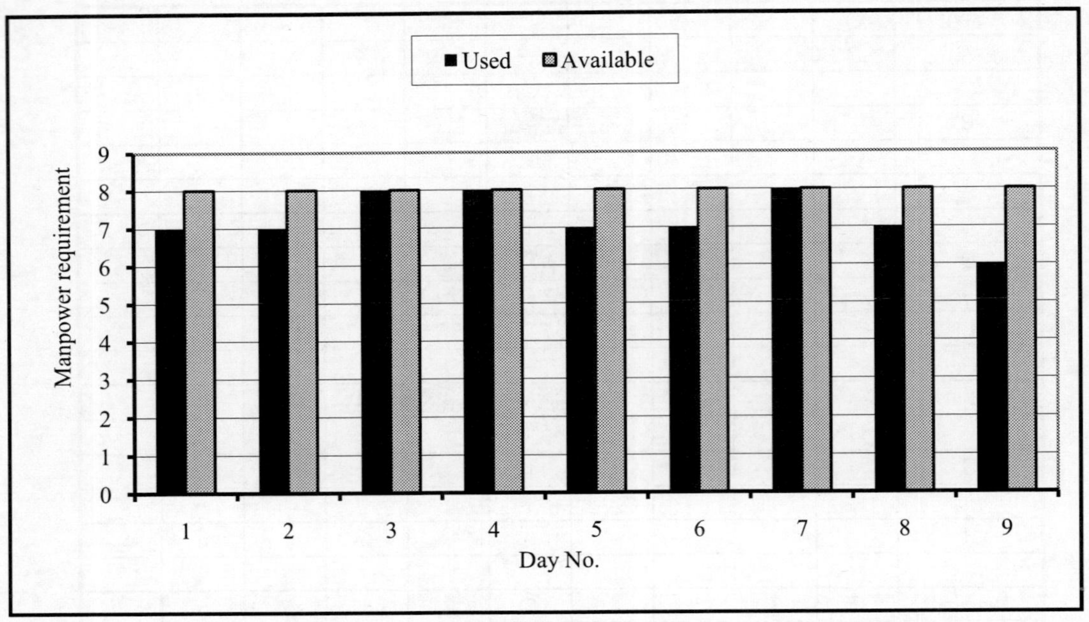

Figure 12.35

Note that such a smoothing leads to extension of project completion time by one day.

■ Example 8

For the data given below draw a suitable network diagram. Obtain the critical path and the associated project time. Given the manpower requirement per day for all the activities, obtain the total manpower necessary for the entire project compilation after smoothing in the range not less than 19 and not more then 21.

Table 12.26

Activity	Preceding activities	Activity time	No. of men/Per day
A	-	8	4
B	-	7	10
C	-	12	5
D	A	4	6
E	A	10	7
F	B,D	3	5
G	B,D	5	8
H	B,D	10	7
I	E,F	7	11
J	C,G	4	3

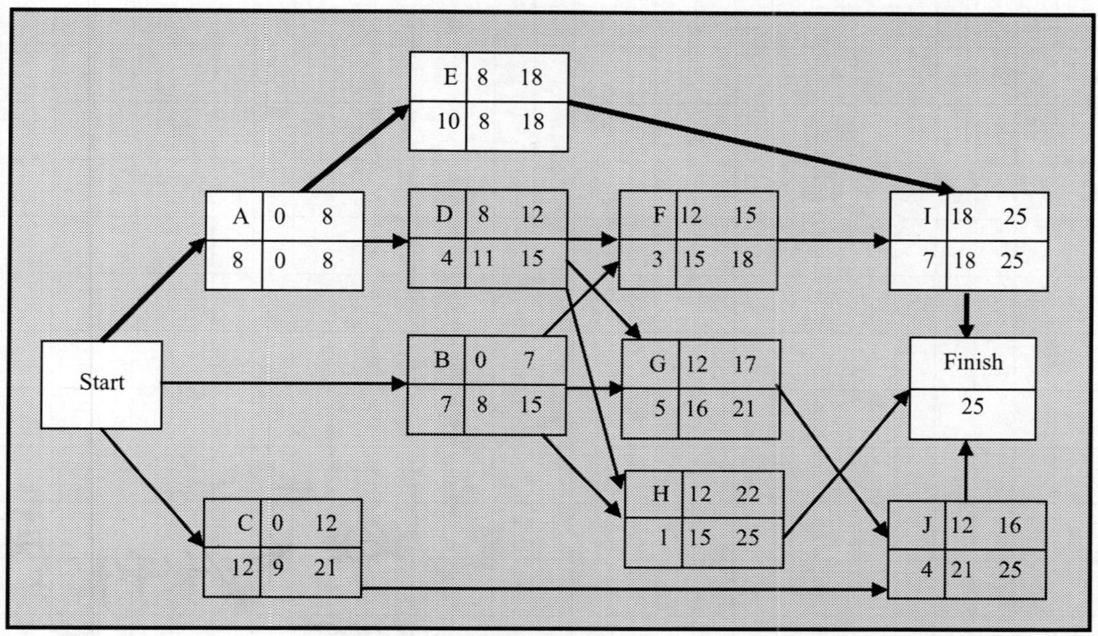

Figure 12.36

Computation of Critical path

Critical Path A-E-I: Activity time: 8 + 10 + 7 = 25

Table 12.27

Activity	Preceding activities	Normal time	ES	EF	LS	LF	Float
A	-	8	0	8	0	8	-
B	-	7	0	7	8	15	8
C	-	12	0	12	9	21	9
D	A	4	8	12	11	15	3
E	A	10	8	18	8	18	-
F	B,D	3	12	15	15	18	3
G	B,D	5	12	17	16	21	4
H	B,D	10	12	22	15	25	3
I	E,F	7	18	25	18	25	-
J	C,G	4	12	16	21	25	9

Note that the critical activities do not have floats and hence not possible for reschedule them. The non-critical activities have different floats. The activity time with early start are represented by light grey colour. Dark grey shows the associated floats.

Activity																										
A	4	4	4	4	4	4	4	4																		
B	10	10	10	10	10	10	10																			
C	5	5	5	5	5	5	5	5	5	5	5	5	F	F	F	F	F	F	F	F						
D									6	6	6	6	F	F	F											
E									7	7	7	7	7	7	7	7	7	7								
F													5	5	5	F	F	F								
G													8	8	8	8	8	F	F	F	F					
H													7	7	7	7	7	7	7	7	7	7	F	F	F	
I																			11	11	11	11	11	11	11	
J													3	3	3	3	F	F	F	F	F	F	F	F		
																										Total
Demand	19	19	19	19	19	19	19	9	18	18	18	18	30	30	30	25	22	14	18	18	18	18	11	11	11	470
Day No.	1	2	3	4	5	6	7	8	9	10	11	12	13	14	15	16	17	18	19	20	21	22	23	24	25	

Figure 12.37 Activity time schedule on early start early finish basis

Table 12.28

Day No.	Activity	Resource requirement		Day No.	Activity	Resource requirement	
1	A,B,C	4+10+5 =	19	14	E,F,G,H,J	7+5+8+7+3 =	30
2	A,B,C	4+10+5 =	19	15	E,F,G,H,J	7+5+8+7+3 =	30
3	A,B,C	4+10+5 =	19	16	E,G,H,J	7+8+7+3 =	25
4	A,B,C	4+10+5 =	19	17	E,G,H	7+8+7 =	22
5	A,B,C	4+10+5 =	19	18	E,H	7+7 =	14
6	A,B,C	4+10+5 =	19	19	H,I	7+11 =	18
7	A,B,C	4+10+5 =	19	20	H,I	7+11 =	18
8	A,C	5+4 =	9	21	H,I	7+11 =	18
9	C,D,E	5+6+7 =	18	22	H,I	7+11 =	18
10	C,D,E	5+6+7 =	18	23	H	11 =	11
11	C,D,E	5+6+7 =	18	24	H	11 =	11
12	C,D,E	5+6+7 =	18	25	H	11 =	11
13	E,F,G,H,J	7+5+8+7+3 =	30				
			244			G.Total	470

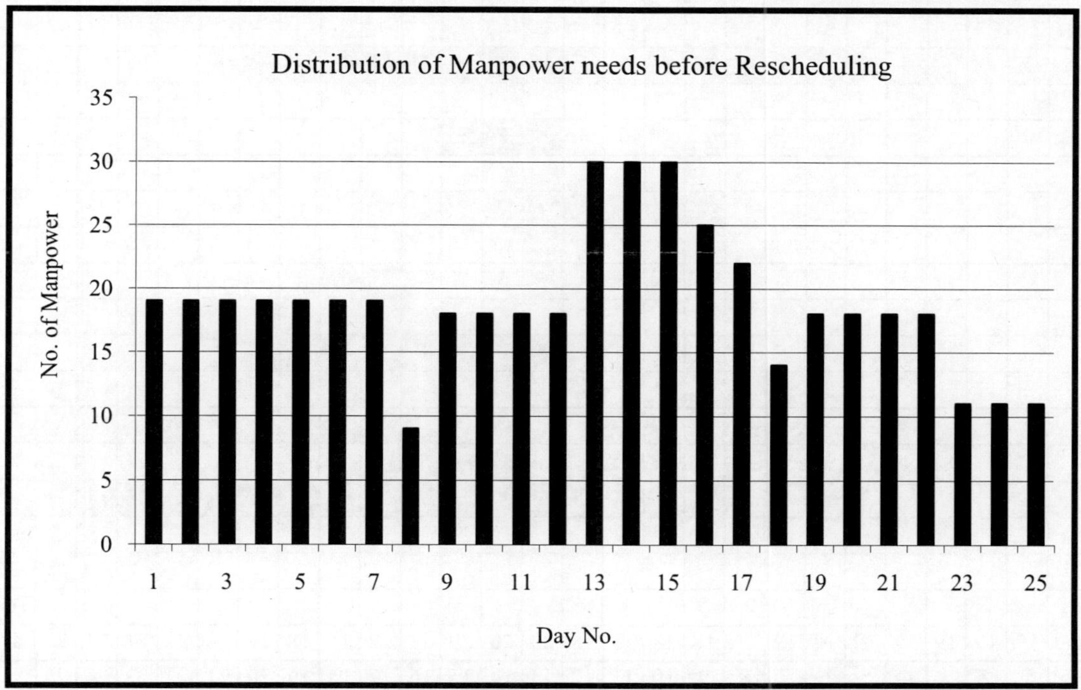

Figure 12.38

In Table 12.28 resource loading has been done for early start and early finish schedule shown in the table. For example, we notice that from day one today seven all the three activities, namely A, B, and C are in action by consuming 4 + 10 + 5 = 19 units of the resource under reference per day. On the 8th day only A and C are in action consuming 5+4 = 9 units of the resource.

Note that on this day D and E are not in action, they start the job only at the end of 8th day. By all means this is equal to 9th day morning. This procedure is repeated for all the 25 days available for the completion of the project. Note that the total resource requirement comes to 470 units as per the early start and early finish rule. The bar chart 12.38 is simply the graphical representation of the load scheduling.

Rescheduling

The peak level resource use is reduced by unit time to reschedule the loading giving priority to highest float in mind. The final rescheduling after repeated application of the rules laid down is reproduced in the Table 12.29. Note that the total resource use is kept at the original early start and early finish level of 470 units.

Activity																										
A	4	4	4	4	4	4	4	4																		
B	10	10	10	10	10	10	10																			
C	5	5	5	5	5	5	5	5	5	5	5	5	F	F	F	F	F	F	F	F	F					
D									6	6	6	6	F	F	F											
E									7	7	7	7	7	7	7	7	7	7								
F													5	5	5	F	F	F								
G													8	8	8	8	8	F	F	F	F					
H																7	7	7	7	7	7	7	7	7	7	
I																			11	11	11	11	11	11	11	
J																						3	3	3	3	
																										Total
Demand	19	19	19	19	19	19	19	9	18	18	18	18	20	20	20	22	22	14	18	18	18	21	21	21	21	470
Day No.	1	2	3	4	5	6	7	8	9	10	11	12	13	14	15	16	17	18	19	20	21	22	23	24	25	

Figure 12.39

Table 12.29

Day No.	Activity	Resource requirement	Day No.	Activity	Resource requirement
1	A,B,C	4+10+5 = 19	14	E,F,G	7+5+8 = 20
2	A,B,C	4+10+5 = 19	15	E,F,G	7+5+8 = 20
3	A,B,C	4+10+5 = 19	16	E,G,H	7+8+7 = 22
4	A,B,C	4+10+5 = 19	17	E,G,H	7+8+7 = 22
5	A,B,C	4+10+5 = 19	18	E,H	7+7 = 14
6	A,B,C	4+10+5 = 19	19	H,I	7+11 = 18
7	A,B,C	4+10+5 = 19	20	H,I	7+11 = 18
8	A,C	5+4 = 9	21	H,I	7+11 = 18
9	C,D,E	5+6+7 = 18	22	H,I,J	7+11+3 = 21
10	C,D,E	5+6+7 = 18	23	H,I,J	7+11+3 = 21
11	C,D,E	5+6+7 = 18	24	H,I,J	7+11+3 = 21
12	C,D,E	5+6+7 = 18	25	H,I,J	7+11+3 = 21
13	E,F,G	7+5+8 = 20			
		234			G. Total 470

Accordingly, the smoothened rescheduled bar chart is shown in the following chart 12.40.

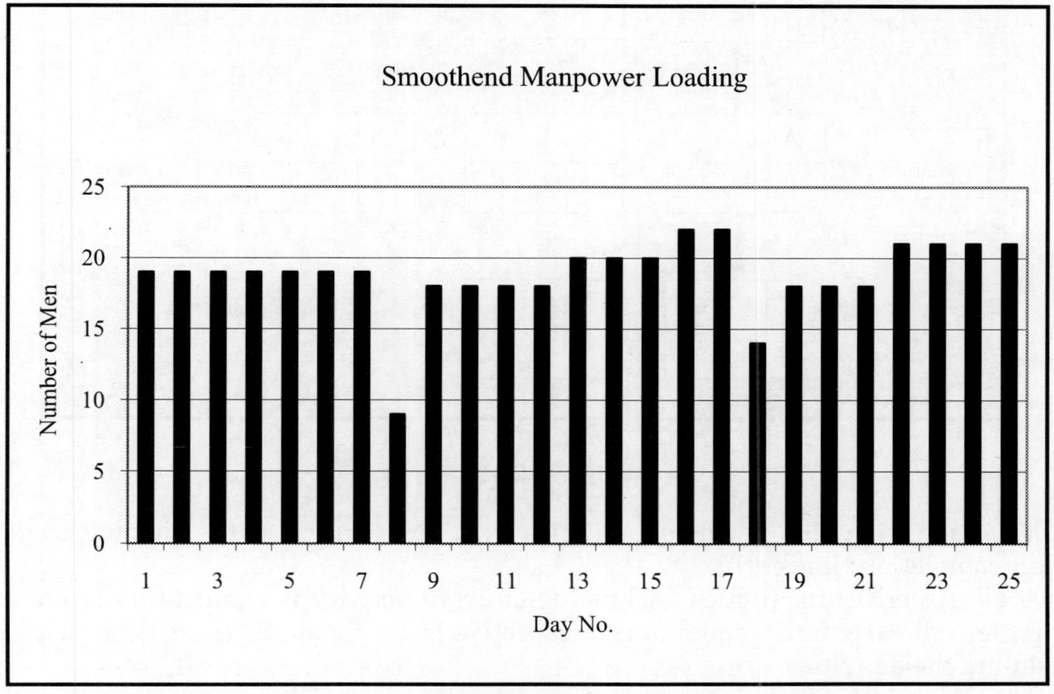

Figure 12.40

■ Example 9

For the data given below obtain the critical path and the associated project completion time. Also prepare a smooth resource-loading schedule without affecting the project completion time.

Table 12.30

Activity	Preceding activities	Activity time	Resource per day
A	-	1	4
B	A	2	10
C	A	5	5
D	B	3	6
E	B	6	7
F	C	3	5
G	E & F	1	8

Solution

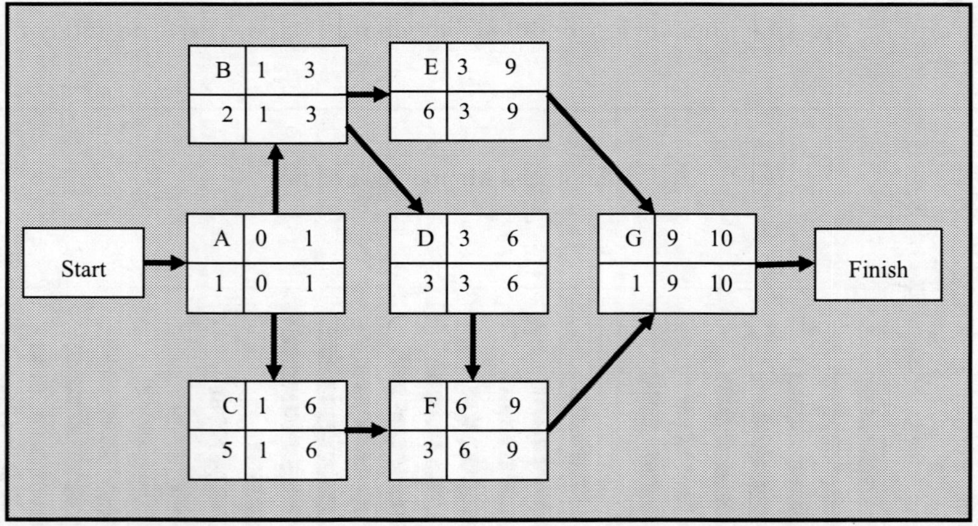

Figure 12.41

In this illustration all are critical activities with three critical paths yielding 10 days of project completion time each.

Since all are critical activities early start of every activity is equal to its late start. Similarly, for all early finish equal to the respective latest finish. So there is no scope for rescheduling the activities.

Table 12.31

Activity	Preceding activities	Activity time	ES	EF	LS	LF	Float
A	-	1	0	1	0	1	0
B	A	2	1	3	1	3	0
C	A	5	1	6	1	6	0
D	B	3	3	6	3	6	0
E	B	6	3	9	3	9	0
F	C	3	6	9	6	9	0
G	E & F	1	9	10	9	10	0

Activity											
A	4										
B		10	10								
C		5	5	5	5	5					
D				6	6	6					
E				7	7	7	7	7	7		
F							5	5	5		
G										8	
											Total
Demand	4	15	15	18	18	18	12	12	12	8	132
Day No.	1	2	3	4	5	6	7	8	9	10	11

Figure 12.42

As in the chart the following table summarizes resource loading totalling to 132 unit of resource.

Table 12.32

Day No.	Activity	Resource requirement
1	A	4 = 4
2	B,C	10+5 = 15
3	B,C	10+5 = 15
4	C.D,E	5+6+7 = 18
5	C.D,E	5+6+7 = 18
6	C.D,E	5+6+7 = 18
7	E,F	7+5 = 12
8	E,F	7+5 = 12
9	E,F	7+5 = 12
10	G	8 = 8
	Total	132

12.10 ALTERNATIVE APPROACH TO NETWORKING

In this alternative approach we propose to introduce a new concept called event in addition to activities described in the previous approach. However, here we describe the activities by arrows and the events by squares. In the network diagram activities described by arrows are linked to one another by events described by squares. An event by definition represents a specific accomplishment in the project and takes place at a particular instant of time and hence consumes no time as in the case of activity. It is simply a time oriented statistic reference point.

If there is only one preceding activity for the said event, then the event called 'end event' can occur only at the end of the activity under reference. If there are more than one preceding activities then the associated end event can occur only after completing all the preceding activities. At the same time it is important to note that the succeeding activities start immediately without any time break. Thus for events there is no separate starting and ending point in time scale. Such events are usually represented by either circle or squares. The following diagram shows an activity A with its two starting and ending events.

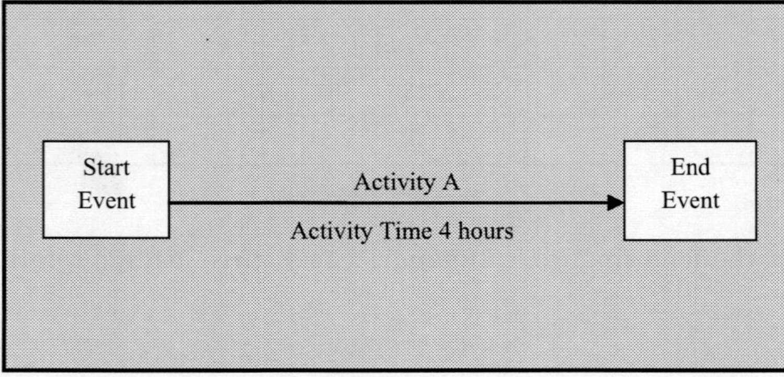

Figure 12.43

Merge, Bust and Merge & Bust Events

When one or more activity merges into an event then that event is called merge event. On the other hand more than one activity bust from an event then the associated event is called the bust event. An activity with both merging and busting activities is called merge & bust event. The following diagram shows all the three types.

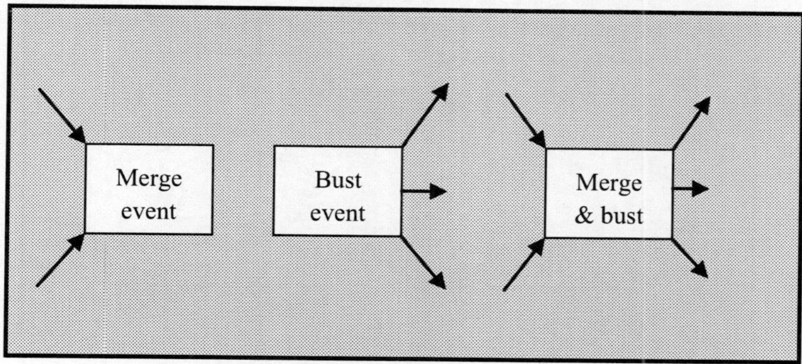

Figure 12.44

Dummy activity

In most projects many activities are performed concurrently with the same starting and ending events. Under such circumstances to maintain logical sequence of ordering the use of dummy activities is essential. The following figures show some the typical situations for the use of dummy activities.

1. Consider a situation in which A and B are the two concurrent activities with same starting and ending events. Further, let activity C succeed A, and D succeed both A and B. The following diagram describes the use of dummy in the correct place by dotted arrow.

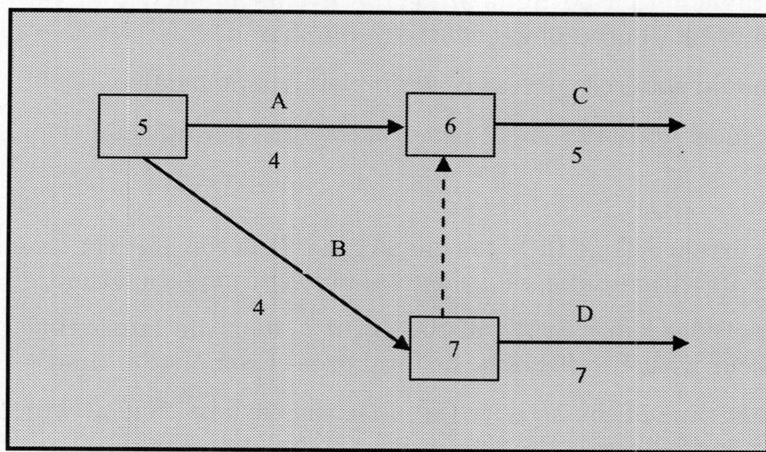

Figure 12.45

2. If activity C follows activity A and activity D follows both activities A and B then the following figure illustrates the use of dummy to reflect this logic sequence.

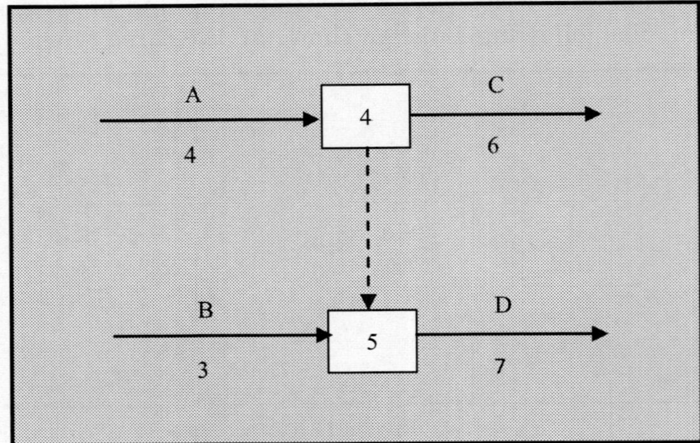

Figure 12.46

Guidelines for constructing network

1. Each activity is represented by only one arrow in the network.
2. No two or more activities identified by the same beginning and ending events.
3. The preceding and succeeding between the various activities are to be laid down clearly.
4. The flow of the diagram should be from the left to the right.
5. As per Flukerson's rule numbering the events is such that no event can be numbered until all preceding events are numbered.
6. Arrows should not cross unless it is completely unavoidable.
7. No two or more activities should have the same tail and head events.

Basic Notations

E_i = Earliest Occurrence Time (EOT_i) of the event i. It is the earliest time at which the event can occur without affecting the total project completion time.

L_i = Latest Occurrence Time (LOT_i) of the event i. It is the latest time at which the event can occur without affecting the total project completion time.

ES_{ij} = Earliest Start time of the activity (i,j). It is the earliest time at which the activity can start without affecting the total project completion time.

LS_{ij} = Latest Start time of the activity (i,j). It is the latest time at which the activity can start without affecting the total project completion time.

EF_{ij} = Earliest Finish time of the activity (i,j). It is the earliest time at which the activity can finish without affecting the total project completion time.

LF_{ij} = Latest Finish time of the activity (i,j). It is the latest time at which the activity can finish without affecting the total project completion time.

Earliest Occurrence Time (EOT) and Foreword Pass Calculations of Events

1) The computation starts from the very start node and ends with the end node. To accomplish this task we start an assumed Earliest Occurrence Time of 0 for the initial tail event number 1. Accordingly, $E_i = 0$ when $i = 1$.

2) The Earliest Start time of the activity (i,j) is eventually the earliest occurrence time of the tail event i. Accordingly, $ES_{ij} = E_i$.

3) The Earliest Finish time of the activity (i, j) is the earliest start time of the activity plus the activity time of the activity (i, j). Accordingly, $EF_{ij} = ES_{ij} + t_{ij} = E_i + t_{ij}$.

4) Earliest Occurrence Time of the head event j (j>i) is obtained as the maximum of the Earliest finish time of all the preceding activities. Accordingly, $E_j = \text{Max}(ES_{ij} + t_{ij}) = \text{Max}(E_i + t_{ij})$.

The so computed of EOT values are put into the lower left portion of the events concerned.

Latest Occurrence Time (LOT) and Backward Pass Calculations of Events

The Latest Occurrence Time (LOT) for any event represents the latest time by which the event can occur without affecting the project completion time. For such an occurrence all the preceding activities of the concerned event must be completed, without delaying the project completion. The calculation of LOT eventually starts form the very last event by setting EOT = LOT for it.

1) Latest finish time (LF_{ij}) of the activity (i,j) is simply the latest event time of event j. Accordingly, $LF_{ij} = L_j$.

2) Latest start time (LS_{ij}) of the activity (i,j) is the latest finish time of (i,j) minus the activity time of (i,j). Accordingly, $LS_{i,j} = LF_{ij} - t_{ij} = L_j - t_{ij}$.

3) LOT for the event i is the minimum of the latest start time of all the activities originating from that event. Thus $L_i = \text{Min}(LS_{ij}) = \text{Min}(LF_{ij} - t_{ij}) = \text{Min}(L_j - t_{ij})$.

Such computed LOT's are written into the lower right portion of the respective events.

Slack of an event

Slack is concerned with events. It is the cushion (extra) time available for the occurrence of the said event. Accordingly, the occurrence of the said event can safely be delayed up to the latest occurrence time without affecting the project completion time. Thus for any given event, the slack is defined as

Slack = LOT−EOT in general.

Head and Tail Slacks of events

If it is related to the event on the head side of the arrow pertaining to an activity (i,j) then it's called the head slack. If it is related to the event on the tail side of the arrow pertaining to an activity (i,j) then its called the tail slack.

Thus Tail Slack = $L_i - E_i$

Head Slack = $L_j - E_j$

Critical and Non-critical events

The events which do not have the slacks are called critical events. Such events should occur exaltedly at the specified point of time in the network. Eventually late or early occurrence of such events will extend the project completion time.

Float of an activity

Float relates to the activity. By float we refer the cushion (extra) time available for the said activity. This extra time availability depends upon the finish time of all the preceding activities, start time of all the succeeding activities and overall project completion time. Considering all the three cases, three types of floats are being defined in this section.

Total Float

This is the most important type of float concerned with the overall project completion time. It is defined as the time by which the starting time of the activity concerned could be delayed without affecting the project completion time on the assumption that preceding activity ends early (EOT_i) and succeeding activity starts late (LOT_j). It is obtained by subtracting the early start time from the latest start time.

Thus, the total float of the activity (i,j) $= LS_{ij} - ES_{ij}$

$$= LS_{ij} - E_i$$

$$= L_j - t_{ij} - E_i$$

Total float of activity (i, j) $= (L_j - E_i) - t_{ij} = LOT_j - EOT_i - t_{ij}$

Free Float

This is concerned with the commencement of the subsequent activity. For example, if the subsequent activity starts at EOT_j instead of LOT_j then the total float of the activity concerned is reduced by $LOT_j - EOT_j = (L_j - E_j)$. Such a float is called the free float. It is the cushion (extra) time available to complete the activity when the activity concerned finishes on EOT_j of the succeeding event.

Thus, Free Float $= \text{Total float} - (L_j - E_j)$

$= \text{Total float} - \text{Head slack}$

$= (L_j - E_i) - t_{ij} - (L_j - E_j)$

$= L_j - E_i - t_{ij} - L_j + E_j$

$= E_j - E_i - t_{ij}$

$= EOT_j - EOT_i - t_{ij}$

Independent Float

This float is concerned with both preceding and succeeding activities. It is the cushion (extra) time available to complete the activity concerned when the preceding activity ends at LOT_i and the succeeding activity starts at EOT_j. Thus,

Independent Float $= \text{Free Float} - (L_i - E_i)$

$= \text{Free Float} - \text{Tail slack}$

$= E_j - E_i - t_{ij} - (L_i - E_i)$

$= E_j - L_i - t_{ij}$

$= EOT_j - LOT_i - t_{ij}$

Critical path and Critical activities

After having obtained both LOT and EOT for all events we can work out the total time required for the project completion. All activities having zero float are called critical activities. Accordingly, these activities are to be taken up as specified by network calculations. The critical path is defined as the path connecting only critical events with zero slacks from the starting to the ending nodes. Such a path is obtained by connecting the events with zero slacks. This is indeed the longest path among all available paths. Moreover on this path all the involved activities are critical with zero total float. Eventually, such a path will have only critical activities on it. Moreover, though the critical path is the longest path, the time consumption for completion will be the least on this path. For a given network there can be more than one critical path. The associated activities on this critical path are called critical activities. The following example illustrates the method of getting the critical path and the associated project completion time.

■ Example 10

For the arrow diagram shown below.

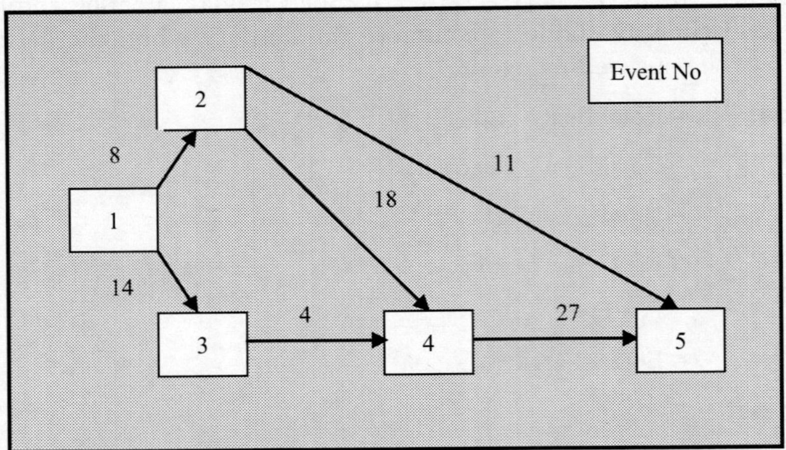

Figure 12.47 Network diagram

i) Obtain EOT, LOT and the associated slacks for all the five events.

ii) Find the critical path and the minimum project completion time.

iii) Calculate the ES, EF, LS, and LF for all the six activities.

iv) Calculate Total Float, Free Float and Independent Float for all the six activities.

i) Calculation of EOT for events

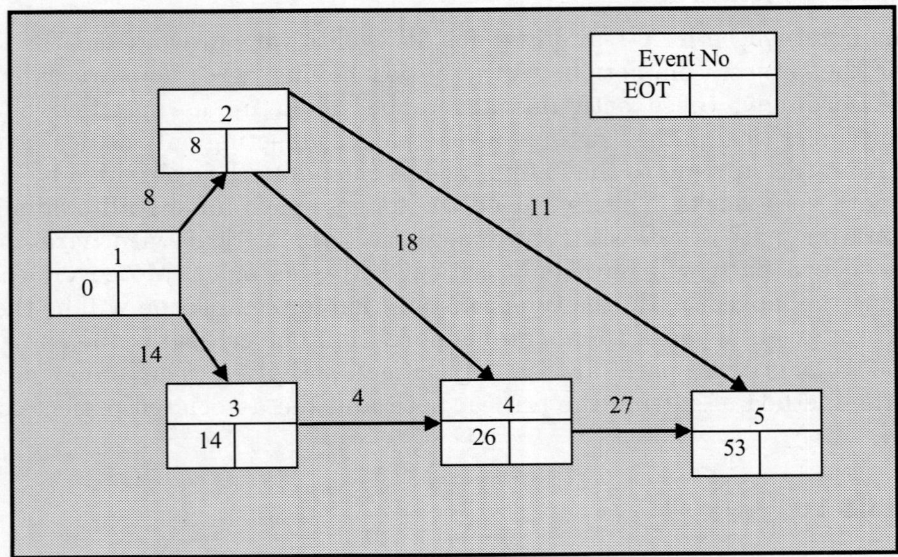

Figure 12.48

1. To calculate EOT and LOT for events the event nodal square is divided into three parts. The top half is reserved for naming the concerned event. The lower half is divided into two parts. The left side we represent the EOT and on the right side we state the LOT.

2. An event under reference will occur provided all the preceding activities have been accomplished. For example, in the above network for the start event 1 the earliest occurrence time ($EOT_1 = 0$) is the very beginning of the project under reference, with 0 staring time. Note that there is no preceding event for this initial event. Accordingly, $E_1 = 0$. This value of zero is reported in the left side of the bottom portion corresponding to event 1 in the above chart. Also note that when this event occurs the activity (1,2) starts. Accordingly, $ES_{1,2} = 0$ for activity (1-2).

3. Now the EOT_2 for the head [(for activity (i,j)] event 2 is simply the earliest finish time of the activity (1-2). Given the activity time 8 of units, the activity (1-2) will by over after 8 units of time from starting time of zero. Thus, the EOT_2 for event 2 is 8 units of time ($E_2 = 8$). This is stated again in the left side corner of the event 2 node. When event 2 starts invariably the next activities (2-5) and (2-4) also start. Accordingly, $ES_{2,4} = E_{2,5} = 8$.

4. For event 3 the EOT_3 is none other than the earliest finish time of the activity (1-3). Accordingly, $E_3 = EF_{13} = 0+14 = 14$ unit of time.

5. For EOT of event 4, the activities (1-2), (1-3), (3-4) and (2-4) must be over as per the chart. Route 1-2-4 will take 8+18 = 26 units of time. Similarly, route 1-3-4 will take 14+4 = 18 units of time. Thus the EOT for event 4 is equal to 26 units of time, the maximum of these two routes. In other words, it the maximum of early finish time of activity (2-4) and activity (3-4). This is stated in the respective left corner for event 4.

6. Similarly, EOT for event 5 = Max (8+11, 26+27) = 53 units of time.

ii) Calculation of LOT for events:

1. For the event 5 as per our assumption the latest occurrence time EOT5 = LOT5 = 53. This is shown in right side corner of the bottom space corresponding to event 5.

2. For the event 4 the latest occurrence time (LOT_4) is obtained by subtracting the activity time of activity (4-5) from LOT_5. Accordingly, $LOT_4 = 26$ (53−27 = 26). Also note that at this stage the latest finish ($LF_{4,5}$) for the activity (4-5) is 53 and the latest start time ($LS_{4,5}$) is = 53 – 27 = 26.

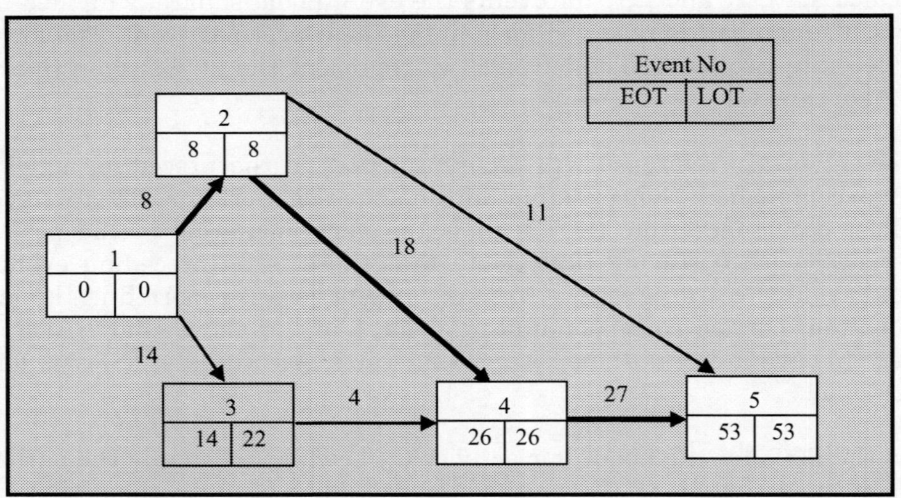

Figure 12.49

3. Similarly, $LOT_3 = 26-4 = 22$. Accordingly, $LF_{3,4} = 26$ and $LS_{3,4} = 26 - 4 = 22$.

4. LOT_2 for event 2 = Min (53−11, 26−18) = 8. At this time it is important to note that $LF_{2,5} = L_5 = 53$ and $LS_{2,5} = L_5 - t_{2,5} = 53 - 11 = 42$. Also, $LF_{2,4} = L_4 = 26$ and $LS_{2,4} = L_4 - t_{2,4} = 26 - 18 = 8$.

5. LOT_1 for event 1 = Min (22 −14, 8 −8) = 0. At this time it is important to note that $LF_{1,2} = L_2 = 8$ and $LS_{1,2} = L_2 - t_{1,2} = 8 - 8 = 0$. Also, $LF_{1,3} = L_3 = 22$ and $LS_{1,3} = L_3 - t_{1,3} = 22 - 14 = 8$.

iii) Calculation of slacks for events

In the following table both LOT, EOT and the associated slack for all the five events are summarized.

Table 12.33

Event	LOT	EOT	Slack = LOT−EOT
1	0	0	0
2	8	8	0
3	22	14	8
4	26	26	0
5	53	53	0

iv) Critical path for the project

From the network diagram shown above the critical path is obtained as:1-2-4-5

The associated project completion time: 8+18+27 = 53

v) Calculation of ES, EF, LS, LF for activities using event timings

Table 12.34 shows the method of getting Early Start (ES) and Early Finish (EF) timings for all the six activities.

Formula: $ES_{ij} = EOT_i = E_i$

$EF_{ij} = E_i + t_{ij}$

Table 12.34

Activity	$EOT_i = E_i$	$ES_{ij}=E_i$	Activity time	EF_{ij}
1-2	0	0	8	8
1-3	0	0	14	14
2-4	8	8	18	26
3-4	14	14	4	18
4-5	26	26	27	53
2-5	8	8	11	19

Table 12.35 shows the associated Late Start (LS) and Late Finish (LF) timings for all the six activities.

Formula: $LF_{ij} = LOT_j = L_j$

$LS_{ij} = L_j - t_{ij}$

Table 12.35

Activity	$LOT_j=L_J$	LF_{ij}	Activity time	LS_{ij}
1-2	8	8	8	0
1-3	22	22	14	8
2-4	26	26	18	8
3-4	26	26	4	22
4-5	53	53	27	26
2-5	53	53	11	42

vi) Calculation of Total float (TF), Free Float (FF) and Independent Float (IF) for activities using events timings

Table 12.36

	L_j	E_i	t_{ij}	Total Float = L_j–E_i–t_{ij}
1-2	8	0	8	0
1-3	22	0	14	8
2-4	26	8	18	0
3-4	26	14	4	8
4-5	53	26	27	0
2-5	53	8	11	34

	E_j	E_i	t_{ij}	Free = E_j–Ei–t_{ij}
1-2	8	0	8	0
1-3	14	0	14	0
2-4	26	8	18	0
3-4	26	14	4	8
4-5	53	26	27	0
2-5	53	8	11	34

	E_j	L_i	t_{ij}	Independently Float = E_j–L_i–t_{ij}
1–2	8	0	8	0
1-3	14	0	14	0
2-4	26	8	18	0
3-4	26	22	4	0
4-5	53	26	27	0
2-5	53	8	11	34

Example 11

For the network given below compute EOT, LOT, ES, EF, LS, LF, TF, FF, IF. Also obtain the critical path and the associated minimum project completion time.

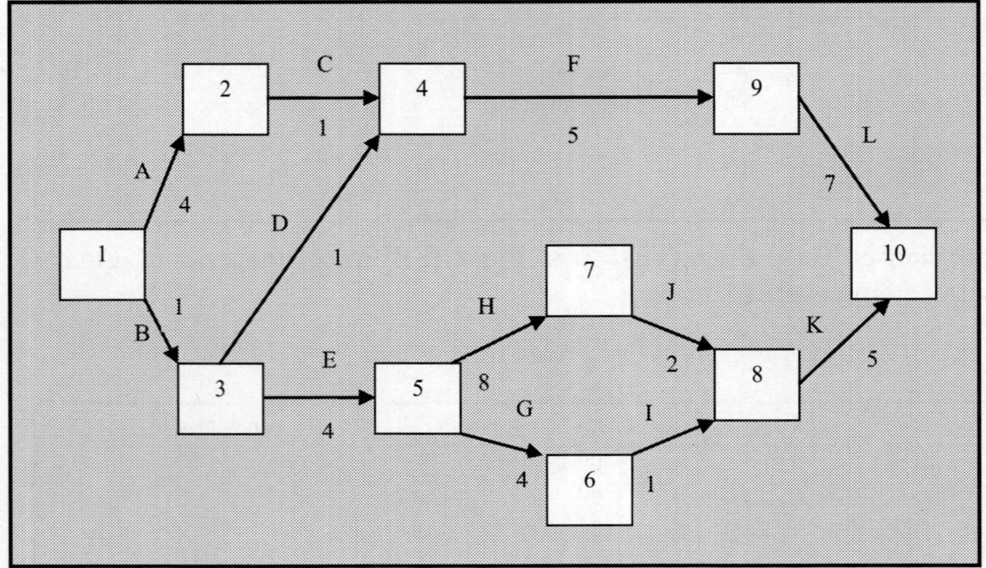

Figure 12.50

EOT calculations for events

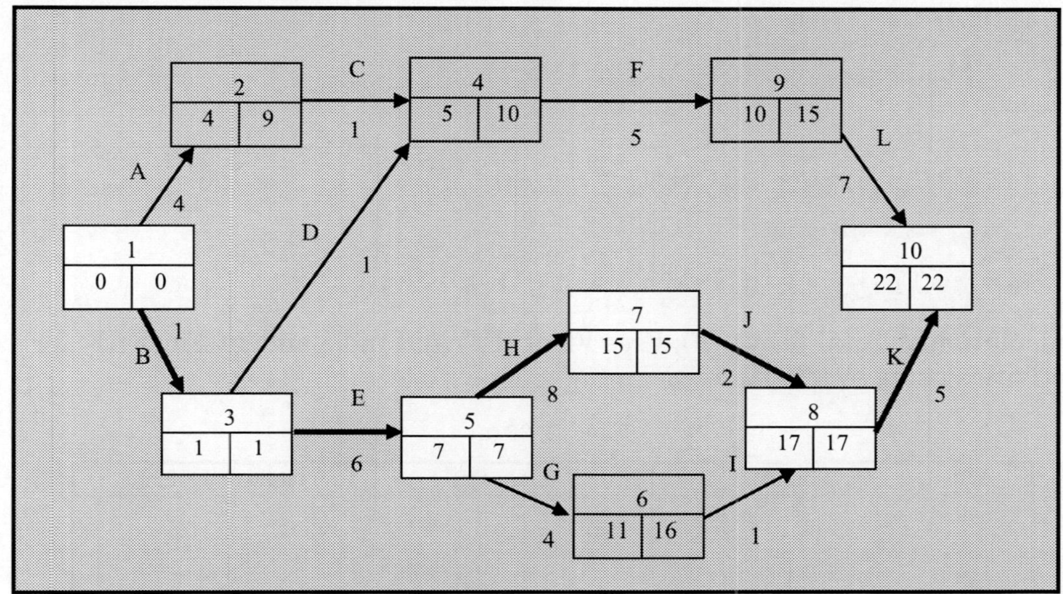

Figure 12.51

$E_1 = 0$

$E_2 = E_1 + t_{1,2} = 0+4 = 4$

$E_3 = E_1 + t_{1,3} = 0 + 1 = 1$

$E_4 = \text{Max}(E_2 + t_{2,4} ; E_3 + t_{3,4}) = \text{Max}(4+1 ; 1+1) = 5$

$E_5 = E_3 + t_{3,5} = 1+6 = 7$

$E_6 = E_5 + t_{5,6} = 7 + 4 = 11$

$E_7 = E_5 + t_{5,7} = 7 + 8 = 15$

$E_8 = \text{Max}(E_7 + t_{7,8} ; E_6 + t_{6,8}) = \text{Max}(15+2;11+1) = 17$

$E_9 = E_4 + t_{4,9} = 5+ 5 = 10$

$E_{10} = \text{Max}(E_9 + t_{9,10}; E_8 + t_{8,10}) = (10+7;17+5) = 22$

LOT calculations for events

$L_{10} = E_{10} = 22$

$L_9 = L_{10} - t_{9,19} = 22-7 = 15$

$L_8 = L_{10} - t_{8,10} = 22-5 = 17$

$L_7 = L_8 - t_{7,8} = 17 - 2 = 15$

$L_6 = L_8 - t_{6,8} = 17 - 1 = 16$

$L_5 = \text{Min}(L_6 - t_{5,6} \,;\, L_7 - t_{5,7}) = \text{Min}(16-4 \,;15-8) = 7$

$L_4 = L_9 - t_{4,9} = 15-5 = 10$

$L_3 = \text{Min}(L_5 - t_{3,5} \,;\, L_4 - t_{3,4}) = \text{Min}(7-6 \,;\, 10-1) = 1$

$L_2 = L_4 - t_{2,4} = 10-1 = 9$

$L_1 = \text{Min}(L_2 - t_{1,2} \,;\, L_3 - t_{1,3}) = \text{Min}(9-4 \,;\, 1-1) = 0$

iv) Calculation of Total float (TF), Free Float (FF) and Independent Float (IF) for activities using events timings

Table 12.36 (a)

Activity	L_j	E_i	t_{ij}	Total float $=L_j-E_i-t_{ij}$
1-2	9	0	4	5
1-3	1	0	1	0
2-4	10	4	1	5
3-4	10	1	1	8
3-5	7	1	6	0
4-9	15	5	5	5
5-6	16	7	4	5
5-7	15	7	8	0
6-8	17	11	1	5
7-8	17	15	2	0
8-10	22	17	5	0
9-10	22	10	7	5

Table 12.36(b)

Activity	E_j	E_i	t_{ij}	Free float $=E_j-E_i-t_{ij}$
1-2	4	0	4	0
1-3	1	0	1	0
2-4	5	4	1	0
3-4	5	1	1	3
3-5	7	1	6	0
4-9	10	5	5	0
5-6	11	7	4	0
5-7	15	7	8	0
6-8	17	11	1	5
7-8	17	15	2	0
8-10	22	17	5	0
9-10	22	10	7	5

Table 12.36(c)

Activity	E_j	L_i	t_{ij}	Ind. float $=E_j-L_i-t_{ij}$
1-2	4	0	4	0
1-3	1	0	1	0
2-4	5	9	1	−5
3-4	5	1	1	3
3-5	7	1	6	0
4-9	10	10	5	−5
5-6	11	7	4	0
5-7	15	7	8	0
6-8	17	16	1	0
7-8	17	15	2	0
8-10	22	17	5	0
9-10	22	15	7	0

Calculation of ES, EF, LS, LF, TF, FF, IF using activity times:

Table 12.37

	Activity	Duration	$ES_{ij} = (E_i)$	$EF_{ij} = (E_i+t_{ij})$	$LS_{ij} = (L_j-t_{ij})$	$LF_{ij} = (L_j)$	Total float $LS_{ij}-ES_{ij}$	Head slack	Free float	Tail slack	Ind. float
A	1-2	4	0	4	5	9	5	5	0	0	0
B	1-3	1	0	1	0	1	0	0	0	0	0
C	2-4	1	4	5	9	10	5	5	0	5	−5
D	3-4	1	1	2	9	10	8	5	3	0	3
E	3-5	6	1	7	1	7	0	0	0	0	0
F	4-9	5	5	10	10	15	5	5	0	5	−5
G	5-6	4	7	11	12	16	5	5	0	0	0
H	5-7	8	7	15	7	15	0	0	0	0	0
I	6-8	1	11	12	16	17	5	0	5	5	0
J	7-8	2	15	17	15	17	0	0	0	0	0
K	8-10	5	17	22	17	22	0	0	0	0	0
L	9-10	7	10	17	15	22	5	0	5	5	0

Critical Path = 1-3-5-7-8-10

Total project time = 1+6+8+2+5 = 22

■ Example 12

A small assembly plant assembles PC's through nine interlinked stages according to the adjoining precedence/process.

a) Draw an arrow diagram representing the above assembly unit.

b) Tabulate both EOT and LOT for all the events.

c) Hence obtain earliest start, earliest finish, latest start and largest finish time for all the activities.

d) Find the associated critical path and the project completion time.

e) Tabulate total float, free float and the independent float for all the activities.

Table 12.38

Activity	Duration in hours
1-2	2
1-3	2
1-4	1
2-5	4
3-6	8
3-7	5
4-6	3
5-8	1
6-9	5
7-8	4
8-9	3

Solution (a)

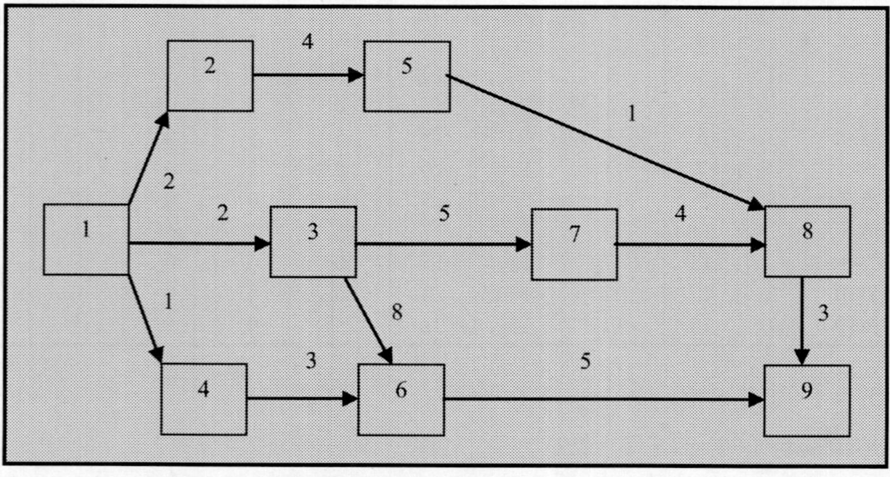

Figure 12.52 Arrow diagram

Solution (b)

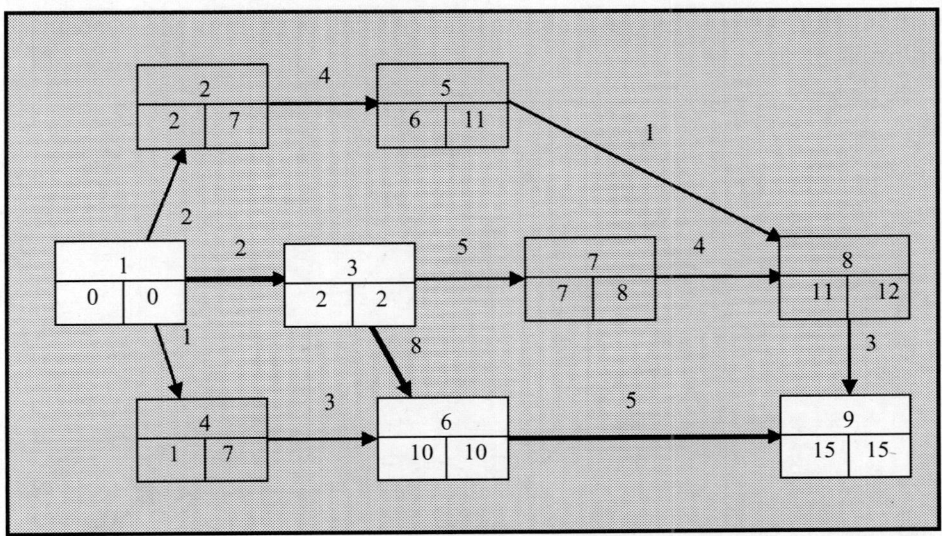

Figure 12.53 Network diagram

Solutions (c and e)

Table 12.39

Activity	Duration	ES	EF	LS	LF	Total float	Head slack	Free float	Tail slack	Ind. float
1-2	2	0	2	5	7	5	5	0	0	0
1-3	2	0	2	0	2	0	0	0	0	0
1-4	1	0	1	6	7	6	6	0	0	0
2-5	4	2	6	7	11	5	5	0	5	-5
3-6	8	2	10	2	10	0	0	0	0	0
3-7	5	2	7	3	8	1	1	0	0	0
4-6	3	1	4	7	10	6	0	6	6	0
5-8	1	6	7	11	12	5	1	4	5	−1
6-9	5	10	15	10	15	0	0	0	0	0
7-8	4	7	11	8	12	1	1	0	1	−1
8-9	3	11	14	12	15	1	0	1	1	0

Solution (d)

Critical path: 1-3-6-9

Project completion time: 2+8+5=15

Example 13

Given the following arrow diagram obtain the critical path and the associated project completion units of time.

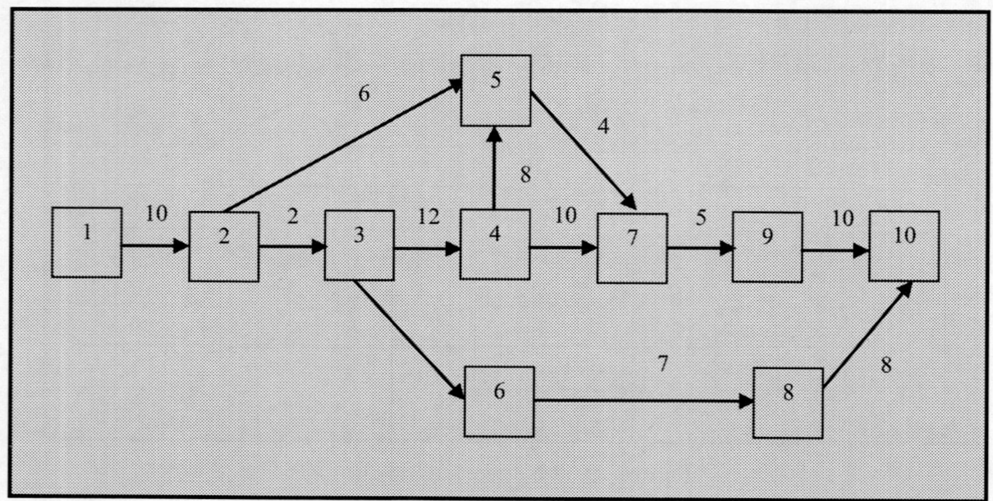

Figure 12.54

Solution

Calculation of EOT

Let the EOT for the very first event = 0

EOT for event 2 = 0 + 10 = 10

EOT for event 3 = 10 + 2 = 12

EOT for event 4 = 12 +1 2 = 24

EOT for event 5 = Max(10 + 6, 24+8) = 32

EOT for event 6 = Max(12+9, 24+5) = 29

EOT for event 7 = Max(24+10, 32+4) = 36

EOT for event 8 = 29+7 = 36

EOT for event 9 = 36 + 5 = 41

EOT for event 7 = Max(36+8, 41+10) = 51

Calculation of LOT

Let the EOT of event 10 = LOT of event 10 = 51

LOT for event 9 = 51 − 10 = 41

LOT for event 8 = 51 − 6 = 43

LOT for event 7 = 41 − 5 = 36

LOT for event 6 = 43 − 7 = 36

LOT for event 5 = 36 − 4 = 32

LOT for event 4 = Min (36 – 6, 36 –10, 32 – 8) = 24

LOT for event 3 = Min (24 –12, 32 – 6) =12

LOT for event 2 = Min (12 – 2, 32 – 6) = 10

LOT for event 1 = 10 − 10 = 0

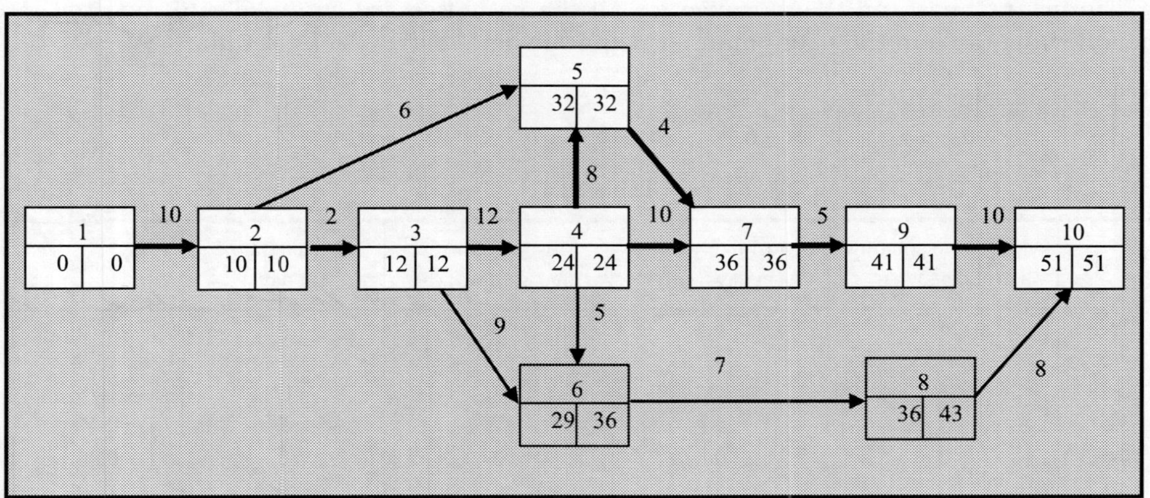

Figure 12.55 Network diagram

Critical Path: 1-2-3-4-5-7-9-10 =10

Project completion time =10+2+12+8+4+5+10 = 51 units of time.

■ Example 14

Various activities of a small project and other related information are shown in the following table. Draw a suitable network and obtain the minimum time for it effective completion. Also obtain the probability of completing this project in 48 days.

Table 12.40

Activity	Most optimistic time (in days)	Most likely time (in days)	Most pessimistic time (in days)
1-2	4	8	12
2-3	1	4	7
2-4	8	12	16
3-5	3	5	7
4-5	0	0	0
4-6	3	6	9
5-7	3	6	9
5-8	4	6	8
7-9	4	8	12
8-9	2	5	8
9-10	4	10	16
6-10	4	6	8

The mean times and the variance for all the activities are obtained and reported in the following table in the usual manner using the beta distribution formula.

$$t = \frac{a + 4m + b}{6} \text{ and } \sigma^2 = \left(\frac{b - a}{6}\right)^2$$

Table 12.41

Activity	Most optimistic time (in days) (a)	Most likely time (in days) (m)	Most pessimistic time (in days) (b)	Expected time (in days)	Variance
1-2	4	8	12	8	1.78
2-3	1	4	7	4	1.00
2-4	8	12	16	12	1.78
3-5	3	5	7	5	0.44
4-5	0	0	0	0	0.00
4-6	3	6	9	6	1.00
5-7	3	6	9	6	1.00
5-8	4	6	8	6	0.44
7-9	4	8	12	8	1.78
8-9	2	5	8	5	1.00
9-10	4	10	16	10	4.00
6-10	4	6	8	6	0.44

Critical Path: 1-2-4-5-7-9-10

Minimum time for completion: 8+12+0+6+8+10 = 44 days

Project completion time variance: 1.78+1.78+0+1+1.78+4 =10.34

Projection time standard deviation: $\sqrt{10.34} = 3.216$

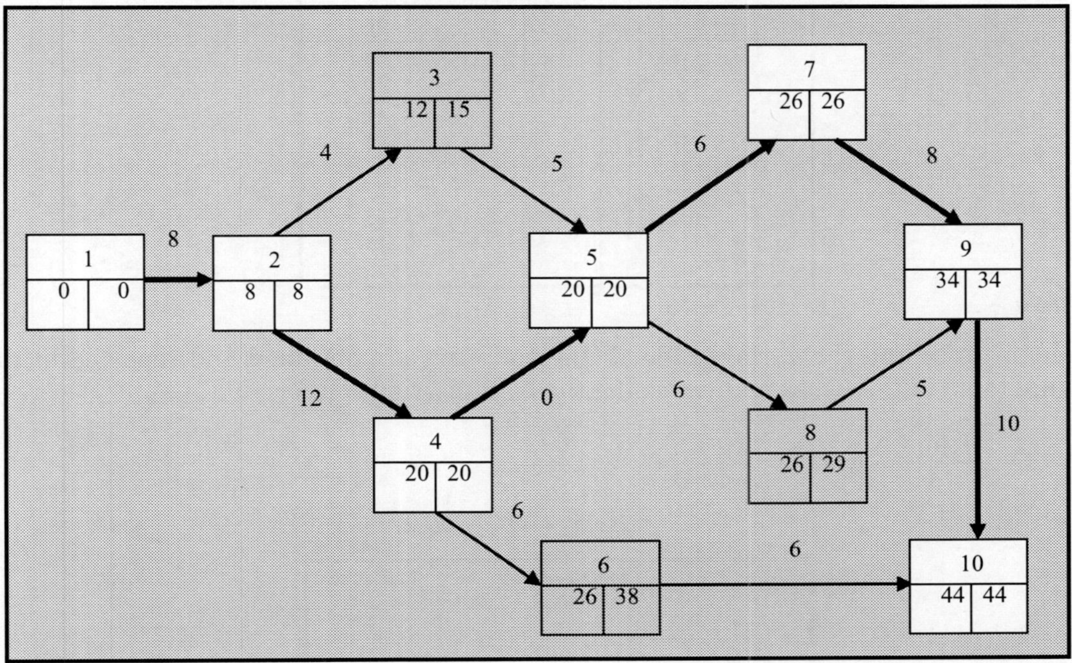

Figure 12.56 Network diagram

To find the probability of completing the project in 48 days we obtain the corresponding standard normal variant first using the following formula.

$$Z = \frac{X - \overline{X}}{\sigma} = \frac{48 - 44}{3.216} = 1.24$$

For Z = 1.24 the standard normal table value is 0.8925.

Thus there is the probability of completing the project in 48 days is 0.8925.

■ Example 15

The activities in a PERT project is shown below.

a) Draw a network diagram.

b) Find the critical path by using EOT and LOT calculations.

c) Find the probability of completing the project within 31 days.

d) What is the chance of the project duration exceeding 40 days?

Table 12.42

Job	Duration (in days)		
	a	m	b
1-2	3	6	15
2-3	6	12	30
3-5	5	11	17
7-8	4	19	28
5-8	1	4	7
6-7	3	9	27
4-5	3	6	15
1-6	2	5	14
2-4	2	5	8

Solution

The mean times and the variance for all the activities are obtained and reported in the following table in the usual manner using the beta distribution formula

$$t = \frac{a+4m+b}{6} \text{ and } \sigma^2 = \left(\frac{b-a}{6}\right)^2$$

Table 12.43

Activity	Expected time	Variance
1-2	7	4
2-3	14	16
3-5	11	4
7-8	18	16
5-8	4	1
6-7	11	16
4-5	7	4
1-6	6	4
2-4	5	1

a) PERT network is shown below

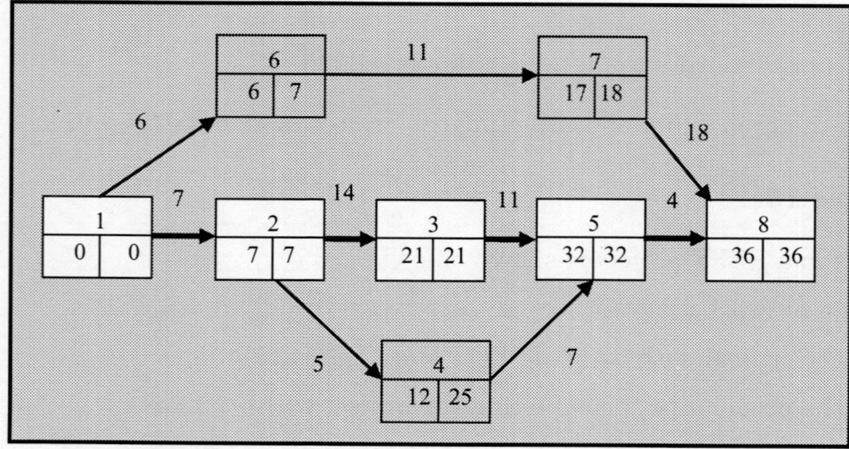

Figure 12.57

b) EOT and LOT are obtained and reported in the network shown as usual.

Expected time for completion obtained from the EOT or LOT of last event as 36 days.

Project variance is obtained by summing the variance of the critical activities on the critical path, i.e project variance = 4+16+4+1 = 25 days

c) Probability of completing the project before 31 days is given by:

$$P(T \le 31) = P\left(Z \le \frac{31-36}{5}\right) = P(Z \le -1) = 0.1587$$

d) Probability of completing the project exceeding 40 days is given by:

$$P(T > 40) = P\left(Z > \frac{40-36}{5}\right) = P(Z > 2) = 1 - 0.9772 = 0.0228$$

Example 16

A small 8-activity project has the following characteristics.

Table 12.44

Activity	Preceding activity	a	m	b
A	-	2	4	12
B	-	10	12	26
C	A	8	9	10
D	A	10	15	20
E	A	7	7.5	11
F	B,C	9	9	9
G	D	3	3.5	7
H	E,F,G	5	5	5

a) Draw PERT network for the given project

b) Prepare the activity schedule

c) Determine the critical path

d) If a 30-week dead line is imposed, what is the probability that the project will be completed within the time limit?

e) If the project manager wants to be 99% sure that the project is completed on the scheduled date, how many weeks before the date should he start the project work.

Solution

a) PERT network is shown below

The following is the relevant network for the project under reference

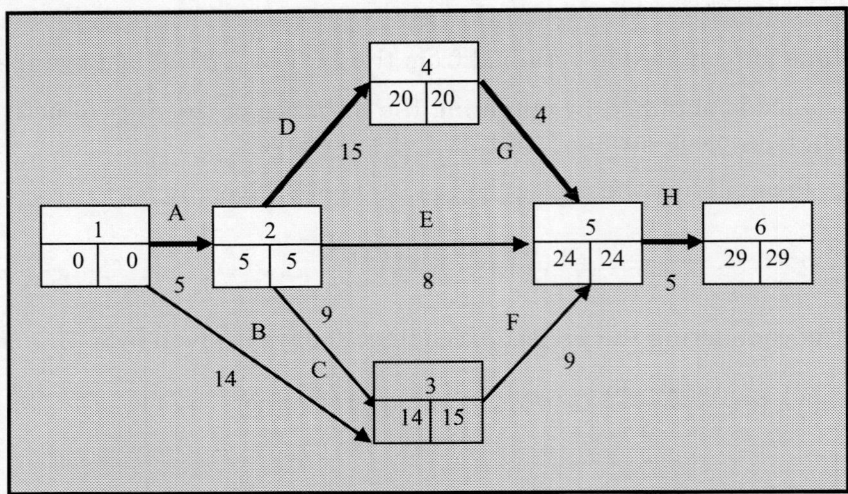

Figure 12.58

b) Calculations for the activity duration and scheduling times:

Table 12.45

Activity	Time			t	σ	Earliest time		Latest time	
	a	m	b			Start	Finish	Start	Finish
A	2	4	12	5	25/9	0	5	0	5
B	10	12	26	14	64/9	0	14	1	15
C	8	9	10	9	1/9	5	14	6	15
D	10	15	20	15	25/9	5	20	5	20
E	7	8	11	8	4/9	5	13	16	24
F	9	9	9	9	0	14	23	15	24
G	3	4	7	4	4/9	20	24	20	24
H	5	5	5	5	0	24	29	24	29

c) The critical path of the project is obtained as 1-2-4-5-6. The critical activities are A, D, G and H. The expected projection completion time = 29 weeks.

Project completion time variance $\frac{25}{9}+\frac{25}{9}+\frac{4}{9}+0=6$

d) The needed probability is obtained by finding the area of normal curve to the left of X = 30. Thus the relevant Z value $= Z = \frac{30-29}{\sqrt{6}} = 0.41$

$$P(X \leq 30) = 0.5 + P(0 \leq Z \leq 0.41) = 0.5 + 0.1591 = 0.6591$$

e) If the project starts T weeks before the project due date then X will represent the ordinate of the normal curve to the left of which 99% of the area lies.

The area between μ an X being 99−50 = 49%. From the table the associated Z value

$$2.33 = \frac{T - 29}{\sqrt{6}}$$

$$T = 29 + 2.33\sqrt{6} = 34.7 \quad \text{weeks}$$

Example 17

Consider the following schedule of activities and related information for the construction of a new plant. Assume that the cost and time needed for activities are independent to each other. Find

a) The critical path

b) Expected cost of construction of the plant

c) Expected time to build the plant

d) The standard deviation of the expected time.

Table 12.46

Activity	Expected time		Expected cost in Rs. (lakh)
	Months	Variance	
1-2	4	1	5
2-3	2	1	3
3-6	3	1	4
2-4	6	2	9
1-5	2	1	2
5-6	5	1	12
4-6	9	5	20
5-7	7	8	7
7-8	10	16	14
6-8	1	1	4

Solution

The following network shows the relevant network with LOT and EOT timings.

From the network the Critical Path is obtained as: 1-2-4-6-8.

Expected cost of construction is obtained by adding all the activity costs given in the problem: 5+3+4+9+2+12+20+7+14+4 = Rs. 80,00,000.

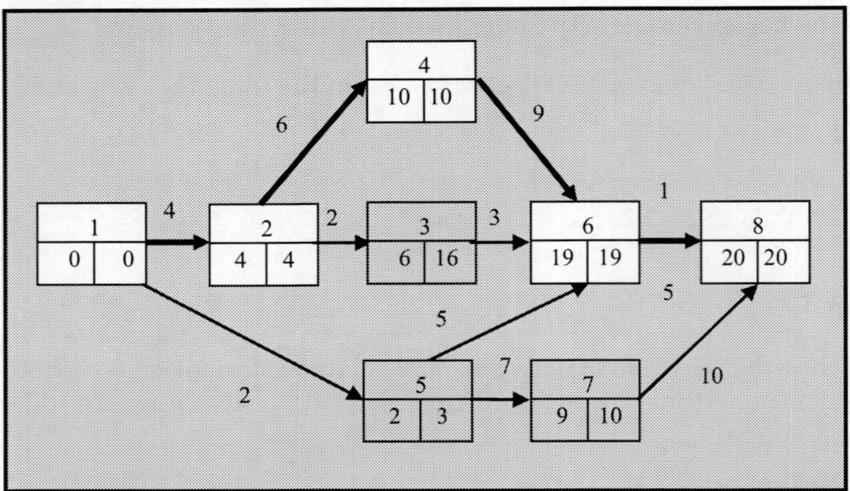

Figure 12.59

Expected construction time is obtained by adding the activity times along the critical path as = 4+6+9+1 = 20 months. Alternatively, it is obtained by the LOT or EOT of the last activity in the network.

Since the activities are independent as per the assumption given, the variance of the project completion time is obtained by simply summing the variance of the critical activities along the critical path.

Table 12.47

Critical activity	Variance
1-2	1
2-4	2
4-6	5
6-8	1
Total	9

The standard deviation of the project competition time of 20 months is obtained as

$$SD = \sqrt{9} = 3 \text{ months}$$

Example 18

XYZ company, planning a project to introduce a new product, has listed the following necessary activities.

i) Draw a critical path network for the project and determine the critical path and its duration.

ii) If the start of the activity B is delayed by three weeks, E is delayed by 2 weeks and G is delayed by 2 weeks how the total time for the projected is affected?

Table 12.48

Activity	Preceding activity	Activity time in weeks
A	-	6
B	-	3
C	A	5
D	A	4
E	A	3
F	C	3
G	D	5
H	B,D,E	5
I	H	2

iii) Assume that the time given in the above table are the expected times of the activities, the duration of which are normally distributed with the following standard deviations:

Activity:	A	B	C	D	E	F	G	H	I	J
SD	1	0.5	1	1	0.5	0.5	1	1	0.5	1

Ignoring the delays referred in (ii) above and the possible effect of uncertainty in non-critical activities, determine a 95% confidence.

Solution

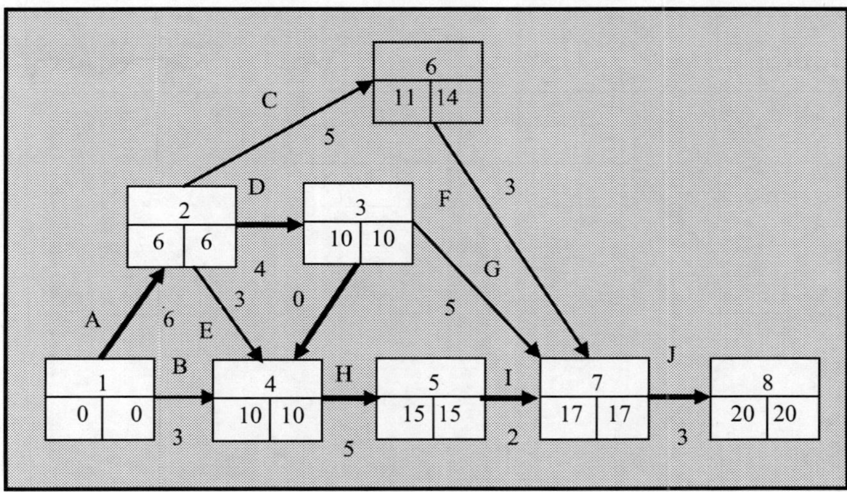

Figure 12.60

i) Critical path: A-D-H-I-J

Project Duration: 6+4+0+5+2+3=20 weeks

ii) From the network it is clear that since the activity B has 7 weeks float a 3-week delay will have no effect on the project completion time. E has a total float of 1 week; 2 week delay will increase the project completion time by one week. G has a total float of 2 weeks a 2 weeks delay will not affect the project completion time.

iii) Project S.D $\sqrt{\{1^2+1^2+1^2+(0.5)^2+1^2\}}=2.06$

95% confidence interval = $20 \pm 1.96.\ \text{S.D} = 20 \pm 1.96 \times 2.06 = (15.96 \text{ to } 24.04)$

Project crashing with direct cost only

■ Example 19

Given the data obtain the minimum project completion time. Also obtain crash solution and associated additional cost by shortening the project completion time by two weeks.

Table 12.49

Activity	Preceding activity	Normal time (in weeks)	Crash time (in weeks)	Normal cost (in Rs.)
A	-	7	4	500
B	A	3	2	200
C	-	6	4	500
D	C	3	1	200
E	B,D	2	1	300

No Crash Solution

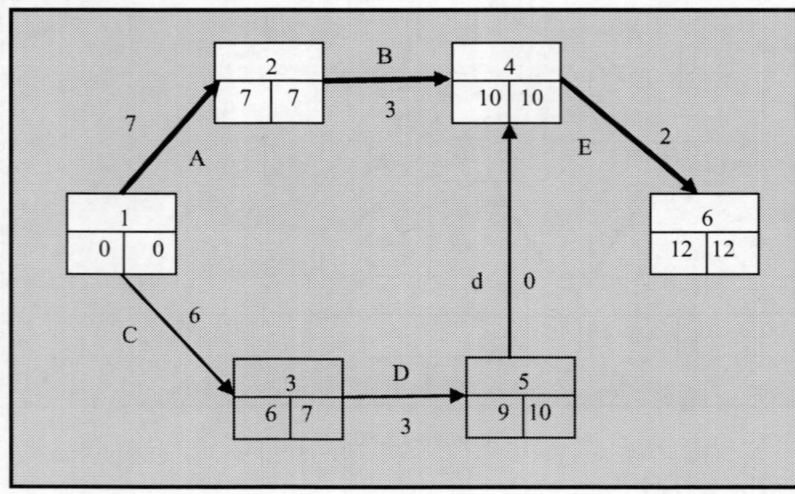

Figure 12.61

Let us get the no rash solution to the problem by applying both forward and backward pass methods as shown in network chart 12.61.

The above chart shows that there are three critical activities A, B, and E along the critical path. The associated activity completion time is obtained as 12 weeks. The cost of the project with normal time is obtained by simply adding costs of all the activities in the project concerned. So the total cost for completion = 500+200+500+200+300 = Rs. 1700.

Calculation of weekly crash cost for all the activities on the basis of linearity assumption

In the table given below for the activity A the maximum permissible crash cost is obtained as 800−500 =300. Such calculations are entered for all the activities in the seventh column. The average (incremental) crash cost for activity A is obtained by merely dividing 300 by the maximum permissible crash time 3. The so obtained average crash cost per week for all the activities are entered in the last column in Table 12.50.

Table 12.50

Activity	Normal time	Crash time	Maximum crash time	Normal cost	Crash cost	Total crash cost	Weekly incremental crash cost
A	7	4	3	500	800	300	100
B	3	2	1	200	350	150	150
C	6	4	2	500	900	400	200
D	3	1	2	200	500	300	150
E	2	1	1	300	550	250	250
Total normal project cost				1700			

Crash Solution

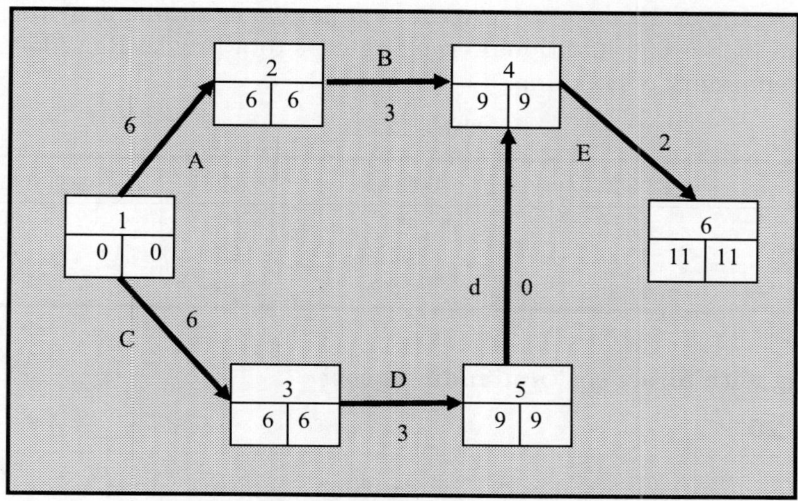

Figure 12.62 A is crashed by 1 week

Now from the Table 12.50 above we notice that among the three critical activities the crash cost per day is less (100) for activity A. So let us crash activity A alone by one week and redo the whole exercise of forward and backward pass and obtain a critical path again as shown in Figure 12.62.

From the above chart it is clear that the project completion time has come down by one week (from 12 to 11) with an additional crash cost of 100. But at the same time A-B-E and C-D-E are the two resulting critical paths this time. Now the minimum per week crashing cost for our new critical path for activity D is Rs. 150. Therefore this time we crash both activities A and D simultaneously by one week each and redo the forward and backward calculations once again as shown below in Figure 12.63.

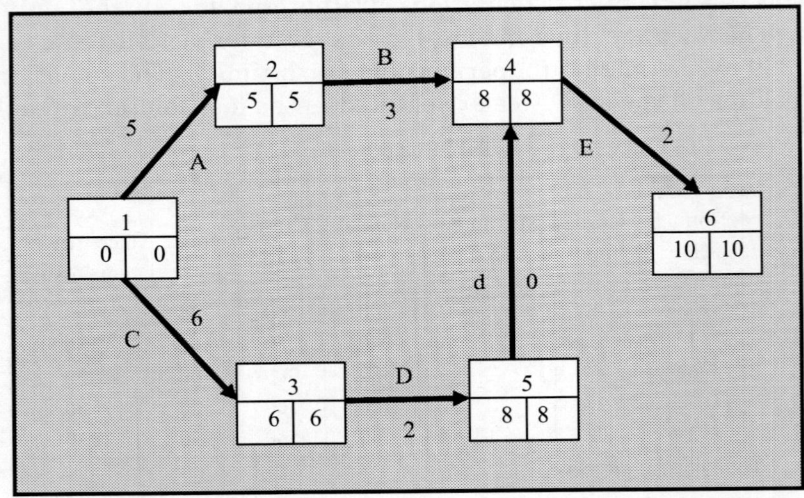

Figure 12.63 A is crashed by 2 week and D is crashed by 1 week

The desired result of 10 days for project completion is obtained in the above chart with a minimum of Rs. 350 additional crashing cost totalling to Rs. 2050. Table 12.51 gives the whole summary of crashing.

Table 12.51

Activity crashed	Direct project cost	Crash cost	Total cost	Project completion time
None	1700	0	1700	12
A (1)	1700	100	1800	11
A(2),D(1)	1700	350	2050	10

Project crashing with direct cost and indirect cost

■ Example 20

The activities, duration and direct activity time costs are given below. The indirect cost is Rs. 3000 per day. Obtain the optimum crash cost and the duration of the project.

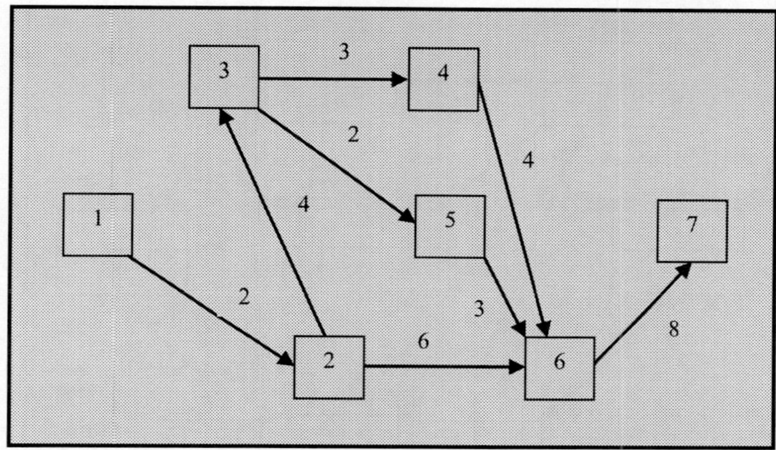

Figure 12.64 Network

Table 12.52

Activity	Time in days		Cost	
	Normal	Crash	Normal	Crash
1-2	2	2	3000	3000
2-3	4	3	4000	5000
2-6	8	8	6000	6000
3-4	3	2	2000	3500
3-5	2	2	2000	2000
4-6	4	3	4000	5000
5-6	3	3	4000	4000
6-7	8	5	8000	12,000

Solution

As a first step let us calculate the average cost of crashing on the basis of linearity assumption as shown below.

Table 12.53

Activity	Time in days		Maximum crash	Cost		Average crashing cost per day
	Normal	Crash		Normal	Crash	
1-2	2	2	0	3000	3000	-
2-3	4	3	1	4000	5000	1000
2-6	8	8	0	6000	6000	-
3-4	3	2	1	2000	3500	1500
3-5	2	2	0	2000	2000	-
4-6	4	3	1	4000	5000	1000
5-6	3	3	0	4000	4000	-
6-7	8	5	3	8000	12,000	1333.33333

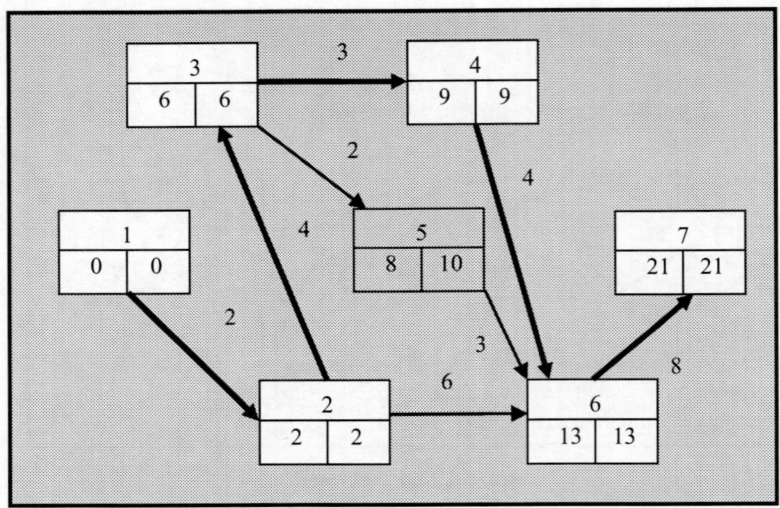

Figure 12.65

In Figure 12.65 we show both LOT and EOT calculations. Accordingly the critical path is obtained as1-2-3-4-6-7 and the associated project completion time cost is 21 days.

On the critical path, the least cost of 1000 per day is available on activities 2-3 and 4-6. So let us first crash the activity 4-6 by one day and recalculate both EOT and LOT as shown below in Figure 12.66.

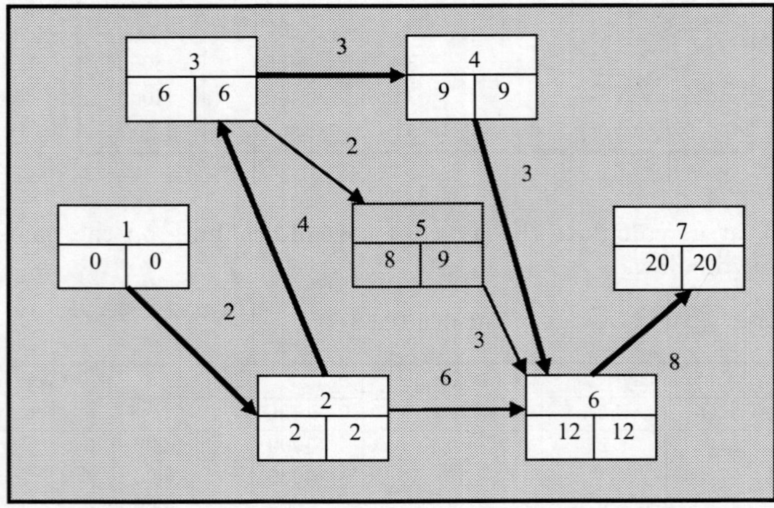

Figure 12.66

Note that because of this one day crash in 4-6 the project completion time has come down by one day from 21 to 20. However, the critical path remains the same: 1-2-3-4-6-7.

Now let us crash the other activity 2-3 by one day and observe the new situation as shown in Figure 12.67. Also note that no more crash on both 2-3 and 4-6 are permissible.

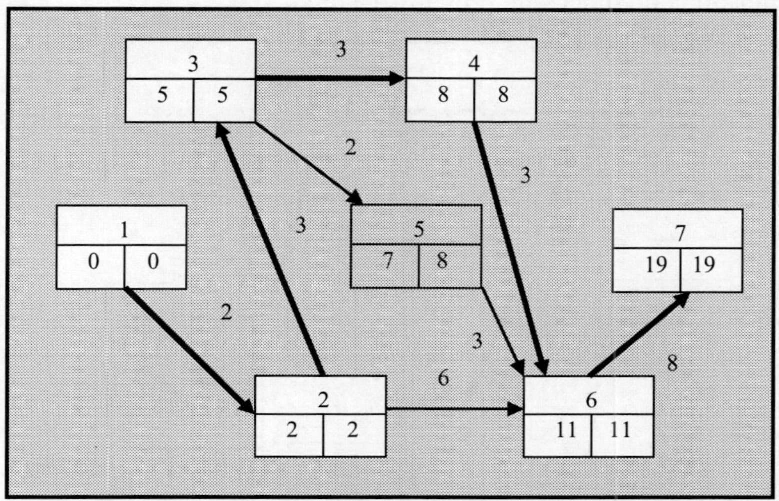

Figure 12.67

The next best alternative left to us is the activity 6-7 with average cost of Rs. 1333. Since the permissible crash is 3 for this activity, let us crash this activity by 3 days at one stroke. Figure 12.68 shows the result after applying this crash. There is no change in the critical path this time also. However, the project completion time has come down to 16 days.

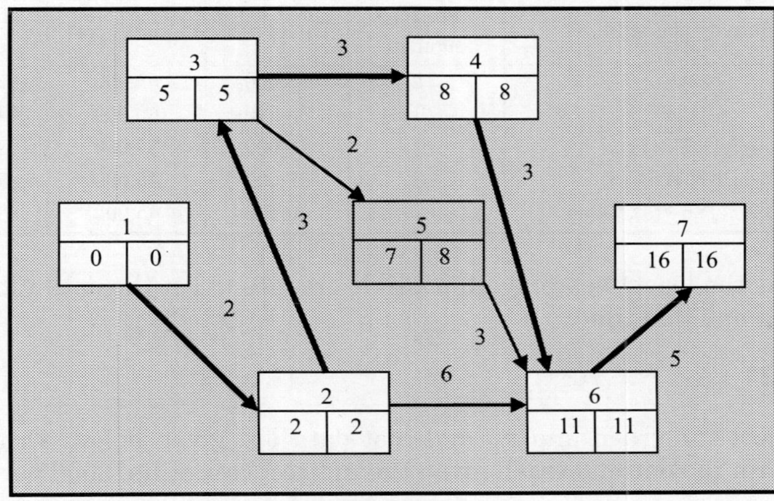

Figure 12.68

The next permissible crash is on the activity 3-4 by one day. After applying this crash the result is shown below in Figure 12.69. Though there is no change in the critical path the project completion time has come down to 15 days.

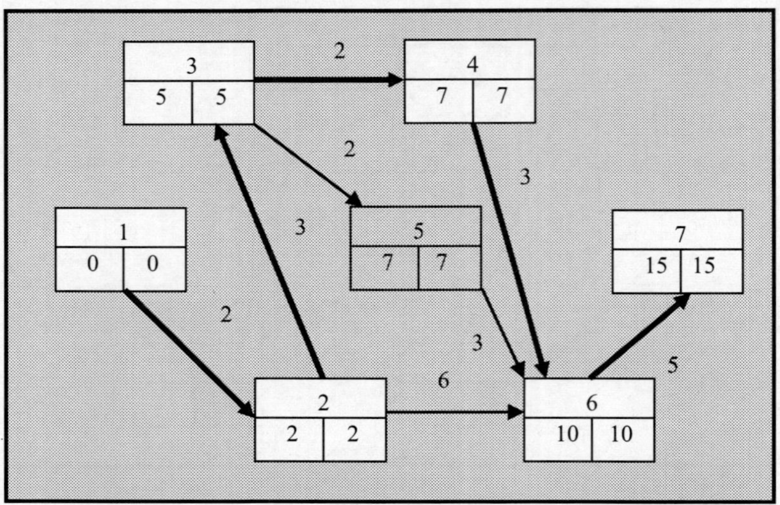

Figure 12.69

Since all the permissible crashes are exhausted let us calculate the involved costs in Table 12.54.

Table 12.54

Activity crashed	Project duration	Direct cost	Indirect cost	Total cost
None	21	33,000	63,000	96,000
4-6(1)	20	34,000	60,000	94,000
4-6(1), 2-3(1)	19	35,000	57,000	92,000
4-6(1),2-3(1),6-7(3)	16	39,000	48,000	87,000
4-6(1),2-3(1),6-7(3),3-4(1)	15	40,500	45,000	85,500

Since 85,000 is the least, we apply all possible crashes to get the optimum situation of 15 days project completion time.

■ Example 21

For a small project the crash and normal cost data are given below. Draw a network diagram and obtain the optimal crash given the indirect cost of Rs. 2000 per week.

Table 12.55

Activity	Preceding activities	Normal		Crash	
		Time in weeks	Cost in Rs.	Time in weeks	Cost in Rs.
A	-	9	12,000	6	18,000
B	A	14	14,000	4	24,000
C	A	4	2000	3	2400
D	C	6	44,000	4	56,000
E	-	14	16,000	13	1800
F	E	6	4000	6	4000
G	B,D	5	4000	3	4800
H	F,G	2	12,000	1	14,000

Solution

As usual let us get the network diagram and the associated critical path first without crash

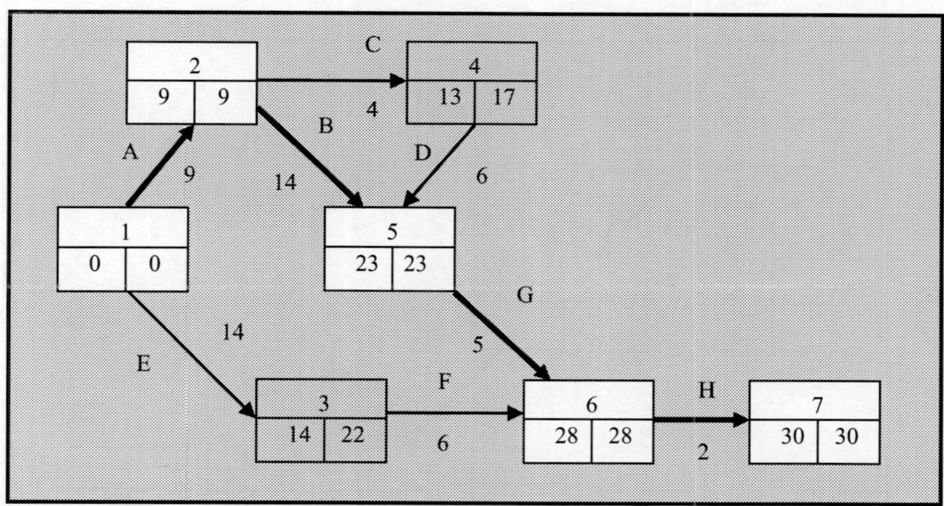

Figure 12.70 No crash solution

For the no crash network A-B-G-H is the critical path. The associated project completion is 30 weeks. Now let us calculate the average cost per week of crashing using the linearity assumption as in Table 12.56.

Now along the critical path since G is the cheapest activity with Rs. 400, let us crash it by one week first and redo LOT and EOT calculations once again as shown Figure 12.71.

Table 12.56

Activity	Time in days		Maximum crash	Cost		Average cost
	Normal	Crash		Normal	Crash	
A	9	6	3	12,000	18,000	2000
B	14	4	10	14,000	24,000	1000
C	4	3	1	2000	2400	400
D	6	4	2	44,000	56,000	6000
E	14	13	1	16,000	18,000	2000
F	6	6	0	4000	4000	-
G	5	3	2	4000	4800	400
H	2	1	1	12,000	14,000	2000
			Total	1,08,000		

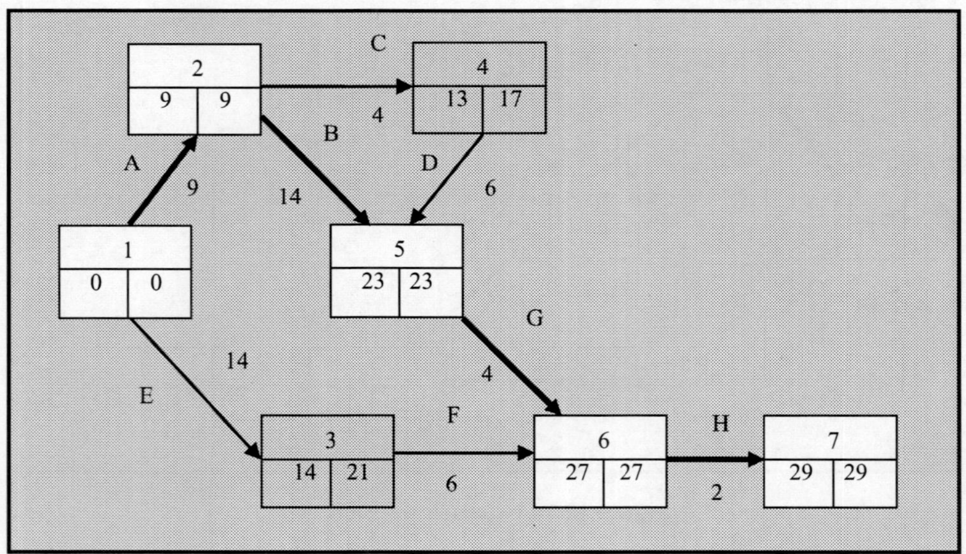

Figure 12.71 G by 1 week

Though the project completion has come down by one week there is no change in the critical path, so let us once again crash G by one more week and redo the needed calculations once again as shown below in Figure 12.72. This time the project completion time has been brought down by 2 weeks.

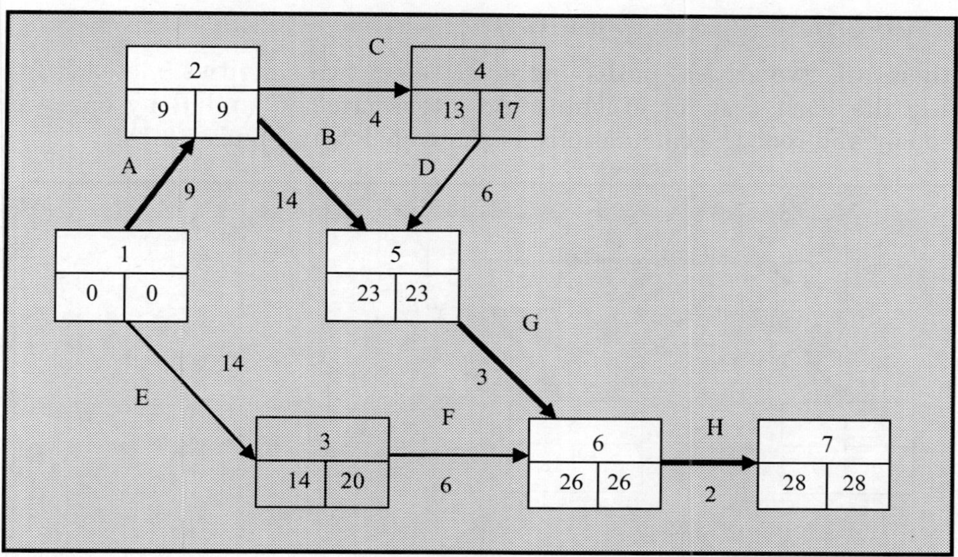

Figure 12.72 G by 2 weeks

Now we notice the maximum permissible crash for G is exhausted, the next best cheapest crashing could be activity B costing Rs. 1000 per week. It is to be noted that B could be crashed by all the available 4 weeks before other activities become critical So let us crash activity B by 4 weeks at one stroke and redo the necessary calculations as shown in Figure 12.73.

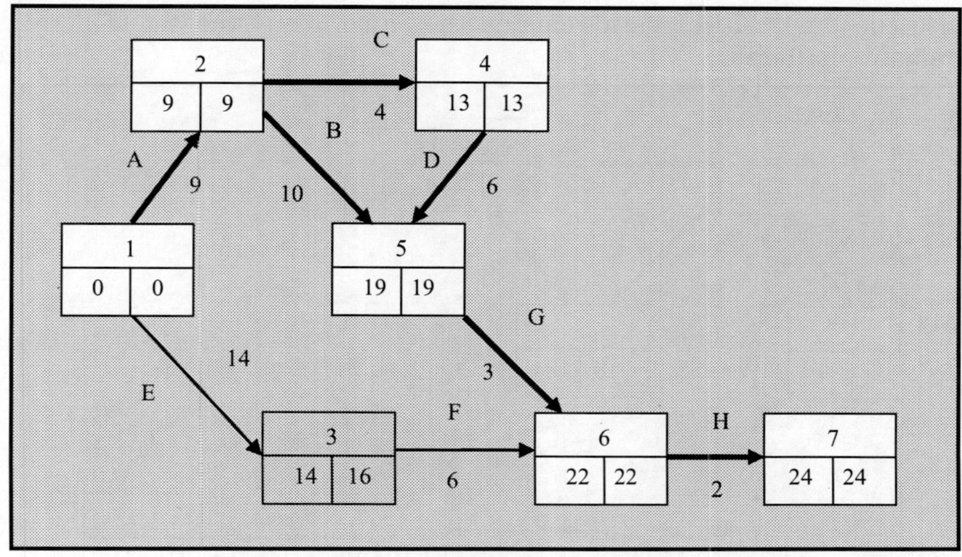

Figure 12.73 G by 2 weeks, B by 4 weeks

At this stage we have two critical paths

1. A-B-G-H 2. A-C-D-G-H.

Since there are two critical paths one must select two activities one each from each path having the least cost for crashing. So let us crash C and B by one week each simultaneously and redo the calculations as shown below in Figure 12.74.

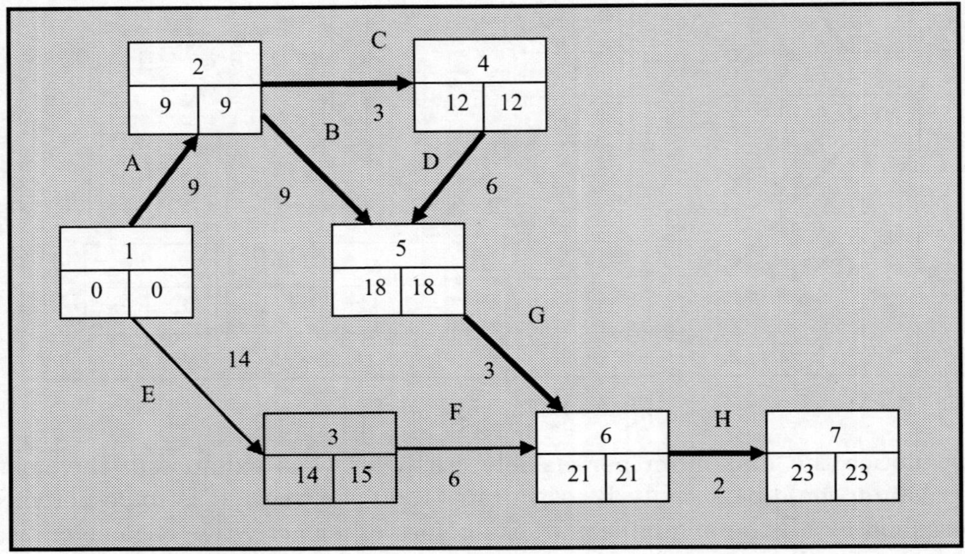

Figure 12.74 G by 2weeks, B by 5 weeks, C by 1 week

Now also the dual critical path remains the same. Thus the set of activities for crashing remain unaltered.

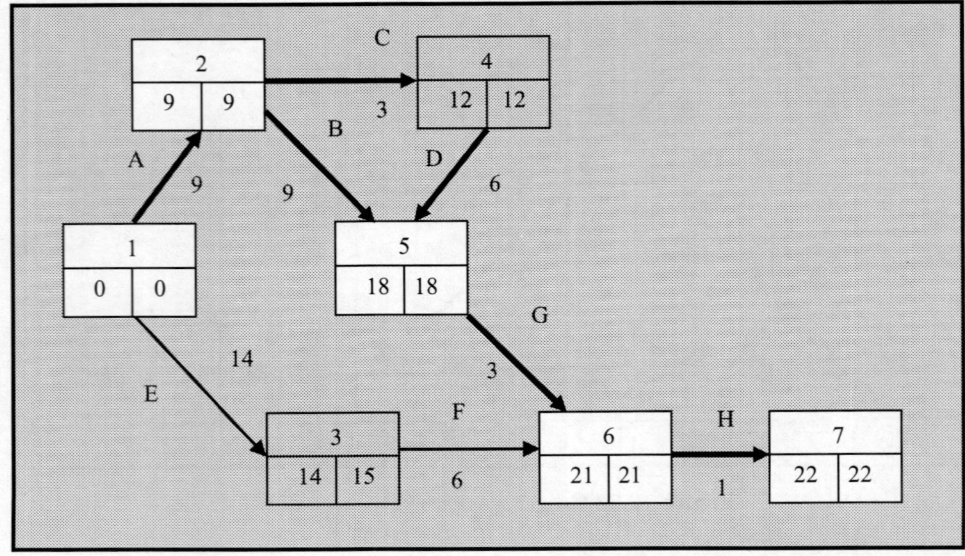

Figure 12.75 G by 2 week, B by 5 weeks, C by 1 week, H by 1 week

However, since activity C has been crashed fully to the permissible level it cannot be crashed further. Note that B in path 1 and D in path 2 costs 1000+6000 = 7000. However A and H are available on both the paths with 2000 cost each. So let us crash H by one week and redo the calculations again as shown in Figure 12.75. Now let us crash A by one week and redo the calculations as shown in Figure 12.76.

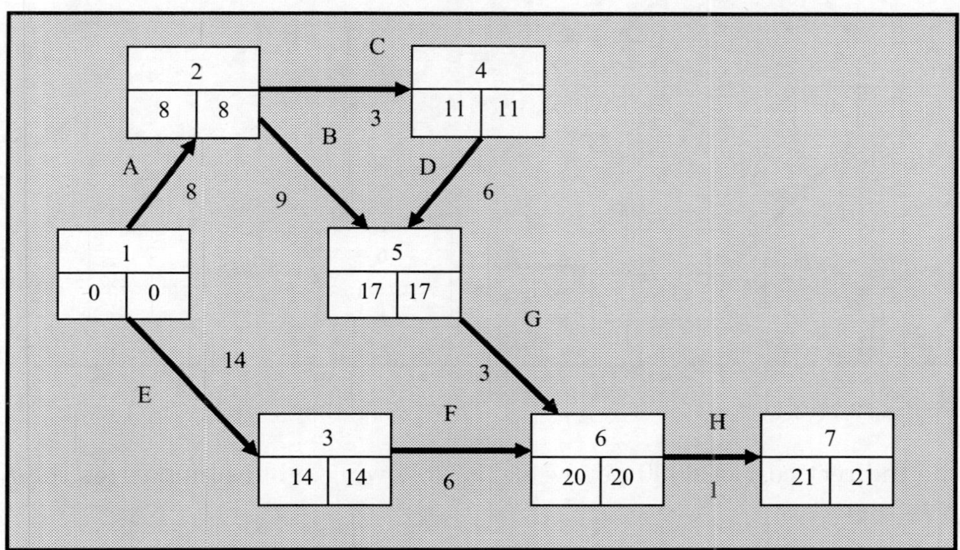

Figure 12.76 G by 2 weeks, B by 5 weeks, C by 1 week, H by 1 week, A by 1 week

At this stage we note that all the activities become critical yielding three distinct critical paths.

1. A-B-G*-H*
2. A-C*-D-G*-H*
3. E-F-H*

The starred ones are not permitted for crashing in the sense they all have reached their permissible limit already. Now as per the rule we must crash one activity each in all the three paths simultaneously.

The two alternatives are:

A in path 1 and 2 and E in path 3 costs: 1000 +1000 = 2000

B in path 1, D in path 2 and E in path 3 costs: 500 +3000 +1000 = 4500

Since 2000 is the least among the two, let as crash A and E by one week each and redo the needed EOT and LOT calculations as in Figure 12.77.

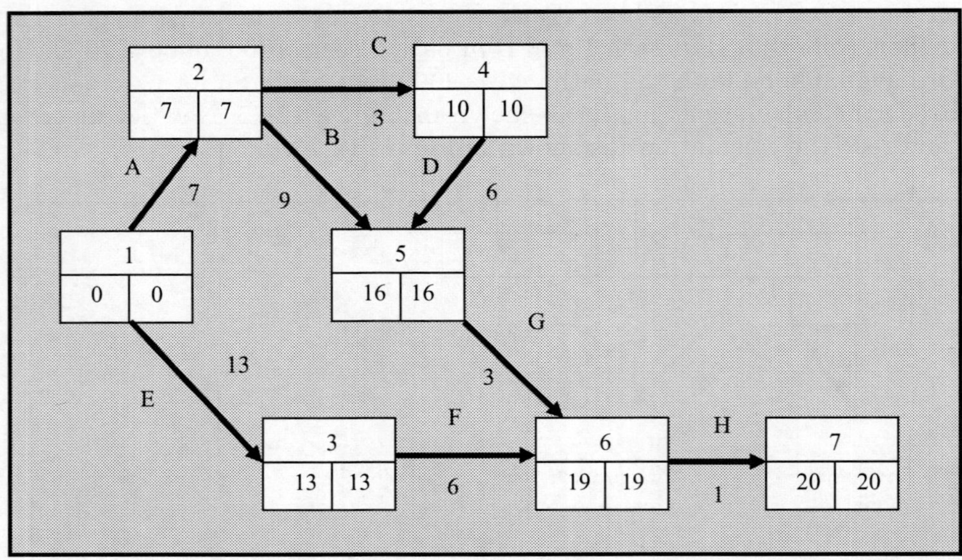

Figure 12.77 G by 2 weeks, B by 5 weeks, C by 1 week, H by 1 week, A by 1 week

Given the indirect cost of 2000 per week the following table summarizes the all the needed calculations.

Table 12.57

Step	Project duration (in weeks)	Crash cost (Rs.)	Direct cost (Rs.)	Indirect cost (Rs.)	Total cost (Rs.)
No crash	30	-	1,08,000	60,000	1,68,000
Crash G(1)	28	400	1,08,400	56,000	1,64,400
Crash G(1)	29	400	1,08,800	58,000	1,66,800
Crash B(4)	24	4000	1,12,800	48,000	1,60,800
Crash B(1),C(1)	23	1400	1,14,200	46,000	1,60,200
Crash H(1)	22	2000	1,16,200	44,000	1,60,200
Crash A(1)	21	2000	1,18,200	42,000	1,60,200
Crash A(1),E(1)	20	4000	1,22,200	40,000	1,62,200

From the above chart it is clear that 1,60,200 is the minimum total cost and the associated optimum crashings are 23 days, 22 days and 21 days.

Resource Levelling

■ Example 22

Consider the following problem

Table 12.58

Activity	Preceding activity	Duration in days	Men required
A	-	3	12
B	-	2	6
C	-	1	4
D	A	1	4
E	A	2	12
F	E	4	10
G	A	1	6
H	G	2	8
I	A	2	8

If the organization has only 18 men per day for use, could it get the project done without extending the time?

Solution

With the given data let us draw the network diagram and do all the needed forward and backward pass calculations.

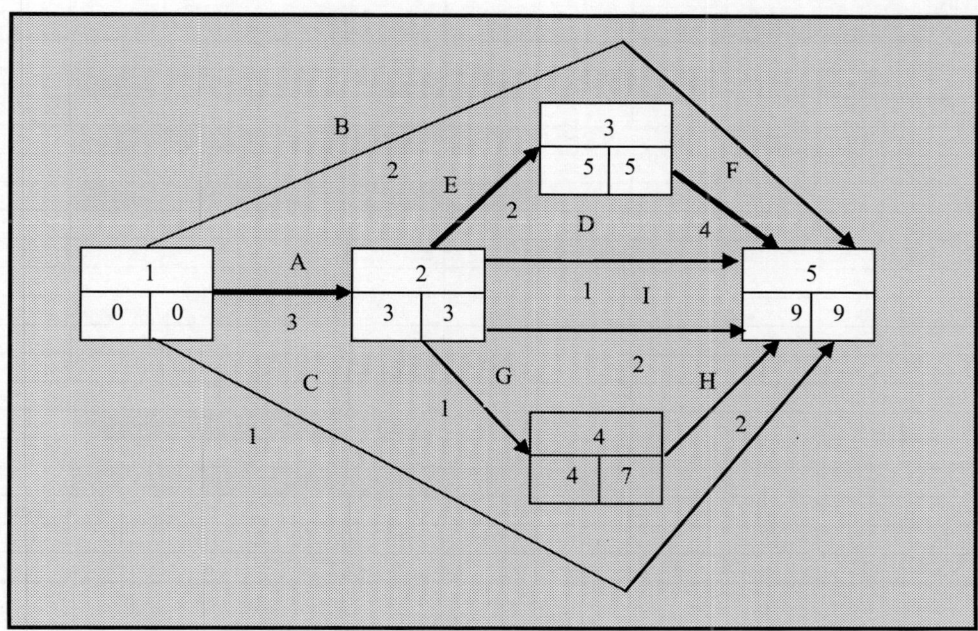

Figure 12.78

From the chart, the critical path is obtained as A-E-F with the minimum project completion time of 9 weeks. Now using the chart we calculate the ES, LS and the associated floats for all the activities.

Table 12.59

Activity	Duration in days	ES	EF	LS	Float
A	3	0	3	0	0
B	2	0	2	7	7
C	1	0	1	8	8
D	1	3	4	8	5
E	2	3	5	3	0
F	4	5	9	5	0
G	1	3	4	6	3
H	2	4	6	7	3
I	2	3	5	7	4

Now let us prepare time scale graph showing the activity time and their respective floats on the basis of early start and early finish.

Activity										
A	12	12	12							
B	6	6	F	F	F	F	F	F	F	
C	4	F	F	F	F	F	F	F	F	
D				4	F	F	F	F	F	
E				12	12					
F						10	10	10	10	
G				6	F	F	F			
H					8	8	F	F	F	
I				8	8	F	F	F	F	
										Total
Demand	22	18	12	30	28	18	10	10	10	158
Day No.	1	2	3	4	5	6	7	8	9	

Figure 12.79

On day one as per schedule we need 22 men though only 18 are available. Among the three activities in action C has the longest float of 8 days. So let as delay this C by two days and do it on the third day as shown below in Figure 12.80. So day 1, 2 and 3 problems are readily solved.

Activity										
A	12	12	12							
B	6	6	F	F	F	F	F	F	F	
C			4	F	F	F	F	F	F	
D				4	F	F	F	F	F	
E				12	12					
F						10	10	10	10	
G				6	F	F	F			
H					8	8	F	F	F	
I				8	8	F	F	F	F	
										Total
Demand	18	18	16	30	28	18	10	10	10	158
Day No.	1	2	3	4	5	6	7	8	9	

Figure 12.80

On day 4 we have D, E, G, and I are in action at this stage by consuming 30 men altogether. Since E is critical it cannot be disturbed. Among the remaining non-critical activities D, G, and I, D is having 5 days float and I is having 4 days float.

Activity										
A	12	12	12							
B	6	6	F	F	F	F	F	F	F	
C			4	F	F	F	F	F	F	
D					4	F	F	F	F	
E				12	12					
F						10	10	10	10	
G				6	F	F	F			
H					8	8	F	F	F	
I								8	8	
										Total
Demand	18	18	16	18	24	18	10	18	18	158
Day No,	1	2	3	4	5	6	7	8	9	

Figure 12.81

To have E and G on the fourth day totalling to 12 + 6 = 18 men let us crash activity D by one day and I by four days as shown below in Figure 12.81.

By doing so the fourth day's problem is amicably solved. For fifth day we need 24 men at this stage. On this day since E is critical once again we cannot disturb E. To tie over the situation we delay H by one day. The result is shown in Figure 12.82.

Activity										
A	12	12	12							
B	6	6	F	F	F	F	F	F	F	
C			4	F	F	F	F	F	F	
D					4	F	F	F	F	
E				12	12					
F						10	10	10	10	
G				6	F	F	F			
H						8	8	F	F	
I								8	8	
										Total
Demand	18	18	16	18	16	18	18	18	18	158
Day No.	1	2	3	4	5	6	7	8	9	

Figure 12.82

In the above chart we notice that the problem is completely solved without affecting the project completion time of 9 weeks.

■ Example 23

For the network given below obtain the critical path and the associated project completion time. Given the manpower need (bracket entries), reschedule the project without affecting the project completion time. The total number of manpower available per day is 50.

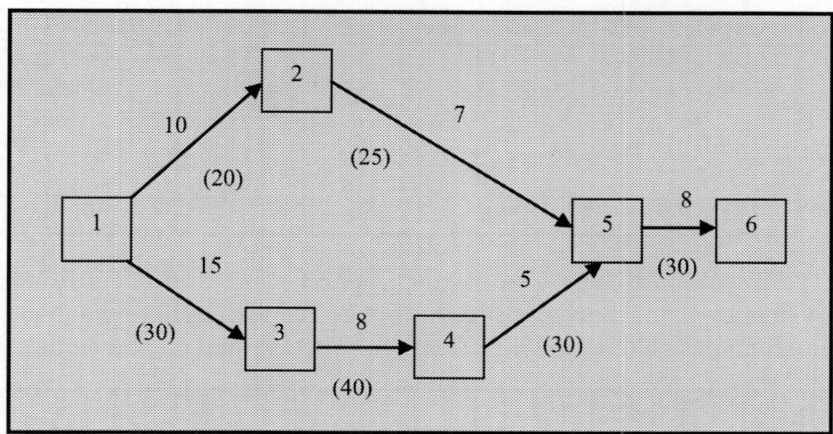

Figure 12.83

Solution

In Figure 12.86 we report both EOT and LOT calculations by using both forward and backward pass calculations. The resulting critical path is 1-3-4-5-6. The associated project completion time = 36 days.

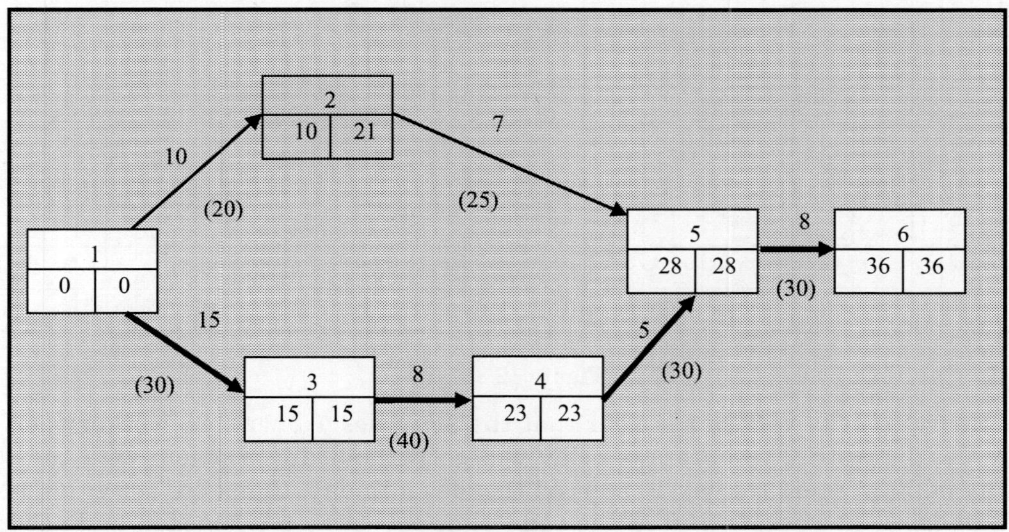

Figure 12.84

In the following table we summarize the resulting early start and early finish timings for all the six activities.

Table 12.60

Activity	ES	Activity time	EF	LS	Float
1-2	0	10	10	11	11
1-3	0	15	15	0	0
2-5	10	7	17	21	11
3-4	15	8	23	15	0
4-5	23	5	28	23	0
5-6	28	8	36	28	0

In Figure 12.85 we summarize the resulting early start and early finish timings for all the six activities by using horizontal bars. In the respective bars we also state the per day labour requirements. The respective floats are also shown on the right side.

Activity																																				
1-2	20	20	20	20	20	20	20	20	20	20	F	F	F	F	F	F	F	F	F	F	F															
1-3	30	30	30	30	30	30	30	30	30	30	30	30	30	30	30																					
2-5											25	25	25	25	25	25	25	F	F	F	F	F	F	F	F	F	F	F								
3-4																40	40	40	40	40	40	40	40													
4-5																								30	30	30	30	30								
5-6																													30	30	30	30	30	30	30	30
Demand	50	50	50	50	50	50	50	50	50	50	55	55	55	55	55	65	65	40	40	40	40	40	40	30	30	30	30	30	30	30	30	30	30	30	30	30
Day No.	1	2	3	4	5	6	7	8	9	10	11	12	13	14	15	16	17	18	19	20	21	22	23	24	25	26	27	28	29	30	31	32	33	34	35	36

Figure 12.85

From the chart it is very much clear that the activities 1-2 with 20 workers per day and activities 1-3 with 30 workers per day will go without any problem with the total availability of 50 workers till day 10. But on the eleventh day, thanks to activities 1-2, it ends and releases 20 workers. But at the same time activities 1-3 comes in and needs 25 workers. Such a move will lead to a shortage of 5 workers per day. To overcome this problem we use only the available 20 per day for another five days. From day 16 to day 23 we use the remaining 10 so that the total usage is 175 workers for this activity as shown in Figure 12.88. But at this stage the critical activities 3-4 are to be initiated with 40 workers per day. Such an arrangement will solve the problem completely as shown in Figure 12.86.

Activity																		
1-2	20	20	20	20	20	20	20	20	20	20	F	F	F	F	F	F	F	F
1-3	30	30	30	30	30	30	30	30	30	30	30	30	30	30	30			
2-5											20	20	20	20	20	10	10	10
3-4																40	40	40
4-5																		
5-6																		
Demand	50	50	50	50	50	50	50	50	50	50	50	50	50	50	50	50	50	50
Day No.	1	2	3	4	5	6	7	8	9	10	11	12	13	14	15	16	17	18

Activity																		
1-2	F	F	F															
1-3																		
2-5	10	10	10	10	10	F	F	F	F	F								
3-4	40	40	40	40	40													
4-5						30	30	30	30	30								
5-6											30	30	30	30	30	30	30	30
Demand	50	50	50	50	50	30	30	30	30	30	30	30	30	30	30	30	30	30
Day No.	19	20	21	22	23	24	25	26	27	28	29	30	31	32	33	34	35	36

Figure 12.86

■ Example 24

For the network given below, work out the project completion time and reschedule the same with the following resource constraints. The figures above the arrows indicate the activity time in months. For the project Rs. 54, Rs. 72 and Rs. 68 lakh are made available during the first, second and third year respectively. The money costs of all the activities are given below.

Table 12.61

Activity	Cost in lakh
1-2	20
1-3	45
2-5	39
3-4	40
4-5	30
5-6	24
Total	198

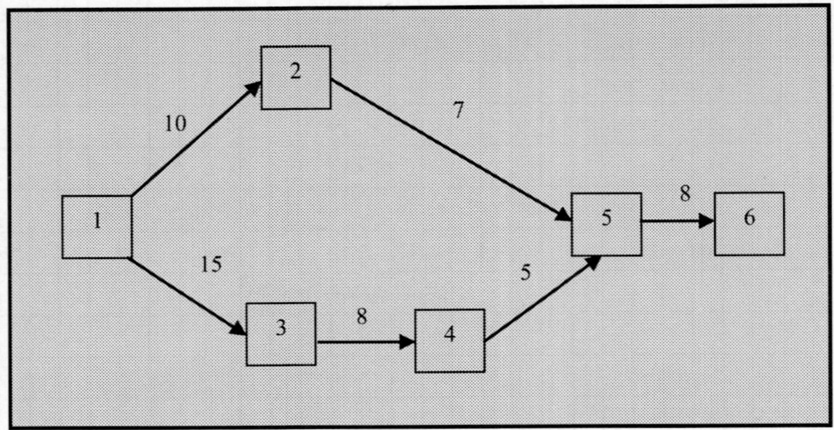

Figure 12.87

Solution

In the following chart we report both EOT and LOT calculations by using both reward and backward pass calculations. The resulting critical path is obtained as 1-3-4-5-6. The associated project completion time = 36 months.

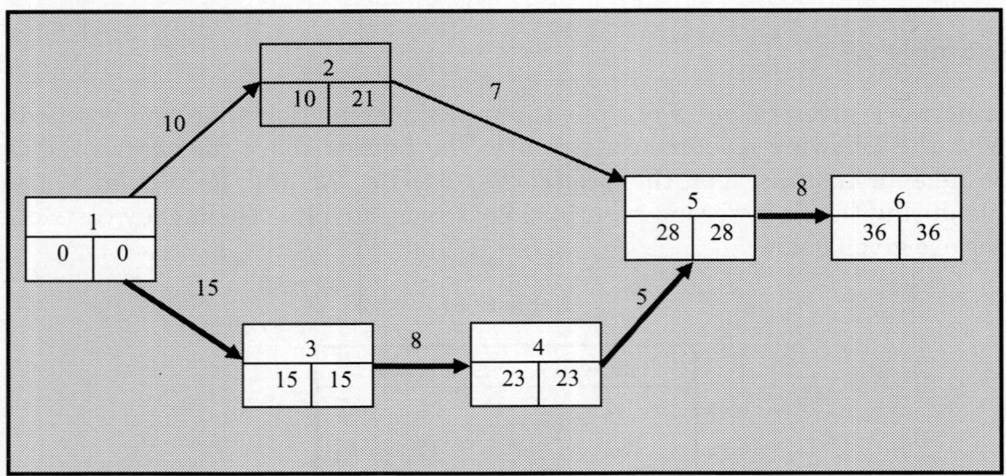

Figure 12.88

In the following table we summarize the resulting early start and early finish timings for all the six activities. In the chart 12.89 we summarize the resulting early start and early finish timings for all the six activities by using horizontal bars. The respective floats are also shown in the right side. Let us assume that the activities concerned use the funds uniformly over its entire period.

Table 12.62

Activity	ES	Activity time	EF	LS	Float
1-2	0	10	10	11	11
1-3	0	15	15	0	0
2-5	10	7	17	21	11
3-4	15	8	23	15	0
4-5	23	5	28	23	0
5-6	28	8	36	28	0

Table 12.63

Activity	Cost in lakh	Activity time in months	Cost per month in lakh
1-2	20	10	2
1-3	45	15	3
2-5	35	7	5
3-4	40	8	5
4-5	30	5	6
5-6	24	8	3
Total	194		

Activity																																				
1-2	2	2	2	2	2	2	2	2	2	2	F	F	F	F	F	F	F	F	F	F	F															
1-3	3	3	3	3	3	3	3	3	3	3	3	3	3	3	3																					
2-5											5	5	5	5	5	5	5	F	F	F	F	F	F	F	F	F	F	F								
3-4																5	5	5	5	5	5	5	5													
4-5																								6	6	6	6	6								
5-6																													3	3	3	3	3	3	3	3
												66												80												48
Demand	5	5	5	5	5	5	5	5	5	5	8	8	8	8	8	10	10	5	5	5	5	5	5	6	6	6	6	6	3	3	3	3	3	3	3	3
Day No.	1	2	3	4	5	6	7	8	9	10	11	12	13	14	15	16	17	18	19	20	21	22	23	24	25	26	27	28	29	30	31	32	33	34	35	36

Figure 12.89

From the chart above we obtain the funds required per year as follows

Total fund needed for the first year	:	Rs. 66 lakh
Total fund needed for the second year	:	Rs. 80 lakh
Total fund needed for the third year	:	Rs. 48 lakh
Total Project cost	:	**Rs. 194 lacks**

In the first year since we have only Rs. 54 lakh we will have a shortage of Rs. 12 lakh. This deficit could be adjusted by delaying the non-critical activities 1-2 by three months and non-critical activities 2-5 by two months as shown in Figure 12.90. The rescheduling will cost the first year by Rs. 54 lakh exactly.

Activity																																				
1-2				2	2	2	2	2	2	2	2	2	2	F	F	F	F	F	F	F	F															
1-3	3	3	3	3	3	3	3	3	3	3	3	3	3	3	3																					
2-5													5	5	5	5	5	5	5	F	F	F	F	F	F	F	F	F								
3-4																5	5	5	5	5	5	5	5													
4-5																								6	6	6	6	6								
5-6																													3	3	3	3	3	3	3	3
											54												92												48
Demand	3	3	3	5	5	5	5	5	5	5	5	5	10	8	8	10	10	10	10	5	5	5	5	6	6	6	6	6	3	3	3	3	3	3	3	3
Day No.	1	2	3	4	5	6	7	8	9	10	11	12	13	14	15	16	17	18	19	20	21	22	23	24	25	26	27	28	29	30	31	32	33	34	35	36

Figure 12.90

Still the second year will cost now Rs. 92 lakh though the funds are available only for Rs. 72 lakh. So we delay the non-critical activities 2-5 till whole of the float is exhausted as shown in the following chart. Such a rescheduling will cost only Rs. 72 lakh in the second year. Note that the remaining Rs. 68 lakh is enough for the third year. Figure 12.91 illustrates this final situation.

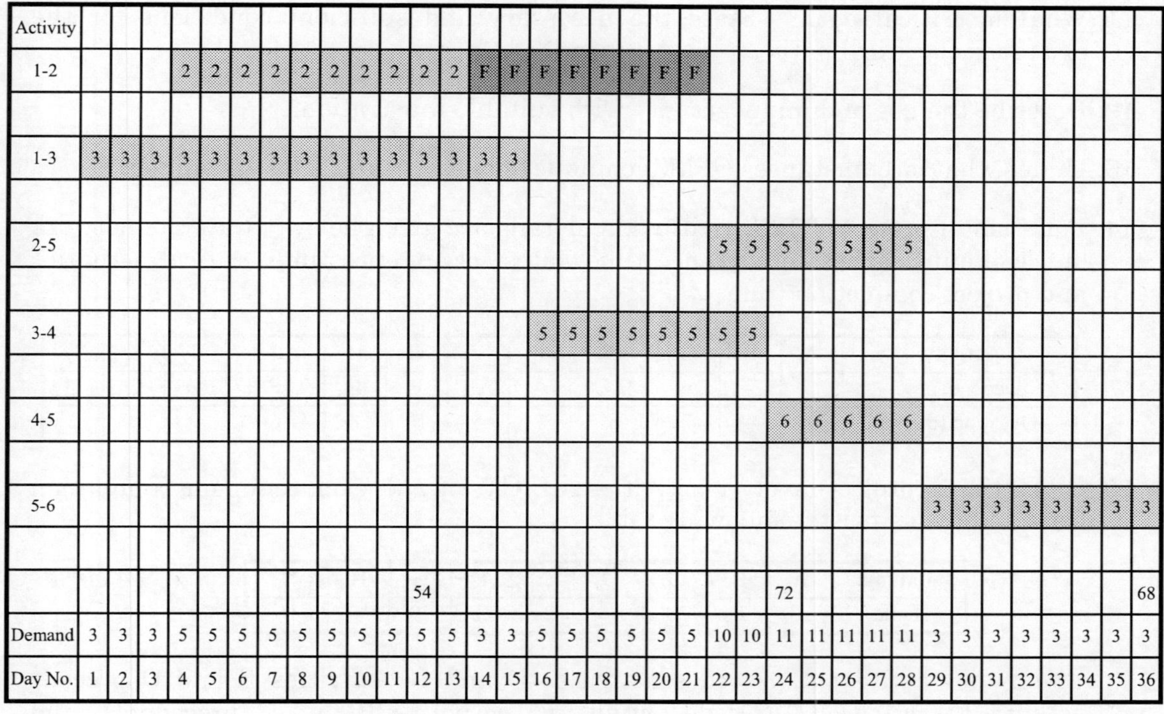

Activity																																				
1-2				2	2	2	2	2	2	2	2	2	2	F	F	F	F	F	F	F	F															
1-3	3	3	3	3	3	3	3	3	3	3	3	3	3	3	3																					
2-5																						5	5	5	5	5	5	5								
3-4																5	5	5	5	5	5	5	5													
4-5																								6	6	6	6	6								
5-6																													3	3	3	3	3	3	3	3
											54												72												68
Demand	3	3	3	5	5	5	5	5	5	5	5	5	5	3	3	5	5	5	5	5	5	10	10	11	11	11	11	11	3	3	3	3	3	3	3	3
Day No.	1	2	3	4	5	6	7	8	9	10	11	12	13	14	15	16	17	18	19	20	21	22	23	24	25	26	27	28	29	30	31	32	33	34	35	36

Figure 12.91

EXERCISES

1. Briefly state the area of application of network technique.
2. Distinguish between slack and float.
3. Distinguish between event and activity.
4. Define free float with illustration.
5. Interpret the meaning of negative independent float.
6. Compare and contrast CPM and PERT techniques.
7. Explain crashing with illustration.
8. What do you understand by resource levelling?
9. Distinguish between resource levelling and resource smoothing.
10. Illustrate the use of float in resolute levelling.

11. What is critical path? State the necessary and sufficient conditions for the existence of critical path.

12. Describe the use of dummy activity with suitable illustration.

13. Under what circumstances CPM technique is better than PERT technique.

14. For a small project of 12 activities the details are given below. Draw the network and compute earliest occurrence time, latest occurrence time, critical activities and project completion time.

Activity:	A	B	C	D	E	F	G	H	I	J	K	L
Dependency:	-	-	-	B,C	A	C	E	E	D,F,H	E	I,J	G
Duration (days):	9	4	7	8	7	5	10	8	6	9	10	2

15. Given that a project involves activities A,B,C,...........I; each requiring completion time, in day, as per the following table.

Activity:	A	B	C	D	E	F	G	H	I
Time	23	8	20	16	24	18	19	4	10

Given that activity A precedes activities D and C; activities B and D precede activity F; activity C precede activity C; activities B and G precede H, and activities F and G precede I, draw the network and calculate (i) total float, (ii) free float, (iii) independent float, (iv) critical path and (v) project completion time.

(M. Com, Delhi, 2005)

16. A project has fourteen activities A to M. The relationships which obtain among these activities are given here. Construct the network and number them.

A is the first operation

B and C can be performed in parallel and are immediate successors to A

D, E, and F follow B

G follows E

H follows D, but it cannot start until E is completed

I and J succeeds G

F and J precede K

H and I precede L

M succeeds L and K

The last operation N succeeds M and C

(CIWA, December, 1985)

17. A cabinet manufacturing company is planning to introduce a new model of cabinet which requires the following tasks.

Task	Description	Task time (in minutes)
A	Prepare the wheels	10
B	Mount the wheels	5
C	Assemble the sides	15
D	Attach the top	11
E	Attach the base	10
F	Insert the brackets	5
G	Insert the shelves	5
H	Attach the doors	10
I	Attach the back panel	10
J	Paint the unit	15

The wheels are mounted after they are prepared. The base cannot be attached until the sides are assembled and wheels mounted. The top cannot be attached nor the brackets installed. The back panel is attached after the base and top are attached. The doors are attached after the shelves are installed. The unit is painted after the back and doors are attached.

a) Identify the immediate predecessors of each task and draw the network.

b) Find the critical path(s), and list the critical activities.

c) Obtain the earliest and latest start and completion times of all the activities and their total, free and independent floats. (M.Com, Delhi, 2004)

18. A project has the following characteristics:

Activity	Preceding activity	Expected completion time (in weeks)
A	None	5
B	A	2
C	A	6
D	B	12
E	D	10
F	D	9
G	D	5
H	B	9
I	C,E	1
J	G	2
K	F,I,J	3
L	K	9
M	H,G	7
N	M	8

a) Draw a PERT network for this project.

b) Find the critical path and the project completion time.

c) Prepare the activity schedule showing the ES, EF, LS, LF and slack for each activity.

d) Will the critical path change if activity G takes 10 weeks instead of 5 weeks? If so what will be the new critical path? (M.Com, Delhi, 1987)

19. The following table gives the activities in a construction project and other relevant information.

Activity	Immediate predecessors	Time (in days)		Direct cost (in Rs.)	
		Normal	Crash	Normal	Crash
A	-	4	3	60	90
B	-	6	4	150	250
C	-	2	1	38	60
D	A	5	3	150	250
E	C	2	2	100	100
F	A	7	5	115	175
G	D, B, E	4	2	100	240

Indirect costs vary as follows:

Days:	15	14	13	12	11	10	9	8	7	6
Cost (in Rs.):	600	500	400	250	175	100	75	50	35	25

a) Draw an arrow diagram for the project.

b) Determine the project duration which will return in minimum total cost. (MBA, Delhi, 1985)

20. The data for a project are:

Activity	Preceding activity	Time (in weeks)		Direct cost (in Rs.)	
		Normal	Crash	Normal	Crash
A	none	3	2	18,000	19,000
B	none	8	6	600	1000
C	B	6	4	10,000	12,000
D	B	5	2	4000	10,000
E	A	13	10	3000	9000
F	A	4	4	15,000	15,000
G	F	2	1	1200	1400
H	C,E,G	6	4	3500	4500
I	F	2	1	7000	8000

a) Draw a project network diagram and find the critical path.

b) If the dead line of 15 weeks is imposed for completion of the project, what activities will be crashed, what would be the additional cost, what would be the critical activities of the network after crashing?

(M.Com, Delhi, 1987)

21. The following are the manpower requirements for each activity in a project.

Activity	Normal time (day)	Manpower required per day
1-2	10	2
1-3	11	3
2-4	13	4
2-6	14	3
3-4	10	1
4-5	7	3
4-6	17	3
5-7	13	5
6-7	9	8
7-8	1	11

a) Draw the network and find out total float and free float for each activity.

b) The contractor stipulates that during the first 26 days only 4 to 5 men and during the remaining days 8 to 11 men only can be made available. Rearrange the activities suitably for levelling the manpower resources satisfying the above conditions.

(ICEA, December, 1985)

22. The following table gives the list of various activities involved in the launch of a new CREDIT CARD service by a company, their immediate predecessors and their expected durations in days.

Where T_o = Most optimistic time,

T_m = most likely time,

T_p = most pessimistic time.

a) Draw an arrow diagram for the project

b) Find the expected project completion time for the project

c) Determine the probability of completing the project in 165 days.

Activity	Immediate predecessors	Expected durations in days		
		T_o	T_m	T_p
A	-	10	12	14
B	A	14	15	17
C	B	2	3	4
D	C	4	6	8
E	C	10	12	14
F	E	20	25	27
G	C	10	17	20
H	F	5	6	7
I	D	7	12	14
J	H,I	14	17	20
K	C	1	2	3
L	K	10	15	20
M	L	3	5	7
N	M,J	13	15	17
O	N	20	21	22
P	O	7	9	14
Q	P	2	3	4
R	Q	2	2	2
S	P	7	10	13
T	S	5	7	9
U	T,R,G	4	8	12

(MBA, Delhi, November, 1998)

CHAPTER 13

Inventory Control Models

13.1 INTRODUCTION

An inventory is defined as the idle stocking of tangible resources of any kind having economic value. Items carrying inventory includes raw materials, purchased parts, components, subassemblies, work-in-process, finished goods which are not delivered to the consumer etc. It is used as a buffer stock to bridge the gap the between the supply and demand. In the manufacturing industries the non-availability of certain crucial input in time often leads to its closure. It is therefore desirable to have physical stock of inputs for smooth running of the plant. Sometimes, procurement in large quantities and consuming a portion immediately and storing the balance as inventory for future use necessarily enjoys cost advantage under increasing price situations. Often bulk purchase may also result in possible large scale economy in transportation cost.

However, one should not forget the fact that though inventory is essential for business, the expenses associated with its financing and maintenance adds the cost substantially to the final product. Frequently money is borrowed from financial institutions for inventory maintenance. In all these accounts the business concerned has to incur various types of expenses like rent for the space, interest for the money borrowed, salary for the maintenance staff, insurance premium etc.

In addition to cost, inventories carry the risk of theft, spoilage, leakage etc. The most serious consequence of inventory is its obsolescence over time. In a dynamic world many items which are over-stocked for long period cannot be used due to technological improvement. Such items are to be disposed of only as scrap causing enormous waste of precious resources. The use of inventory control thus becomes an essential part of any modern business to strike a balance between conflicting objectives.

13.2 TYPES OF INVENTORY

In this context the inventory items are broadly classified into five categories.

a) Movement inventories
b) Buffer inventories
c) Anticipation inventories
d) Decoupling inventories
e) Cycle or lot size inventories

a) Movement inventories

Movement inventories are often called pipeline inventories. When coal is transported from coal fields to a certain industrial town, then the coal will be in its transits for some time and hence cannot provide any service to the final consumer till the time it reaches the work spot. Such items are called movement inventories.

b) Buffer inventories

Inventories are often held as buffer stock to meet the unforeseen demand and supply imbalances. The idea of keeping buffer stock is to render a high level of customer service and consequently reduce the number of stock outs and back orders. Such inventories are also called finished goods inventories. A stock out occurs when a customer is denied fulfilment of order because of no stock in the inventory. Some back orders are commonly used and the ordered quantity will be supplied in full from the next shipment of the item under reference is received.

c) Anticipatory inventories

In anticipation of future demand inventories are being kept. For example, crackers are produced and kept in inventory in anticipation of demand during the Diwali. Similarly, umbrellas are produced and kept in inventory with the anticipated demand during rainy season. In all such cases the idea of keeping inventory is to smoothen the production process for a longer duration on a continuous scale instead of producing it by working over time during the season.

d) Decoupling inventories

The idea behind these inventories is to decouple different parts of the production system. For example, in assembly line different parts are being assembled at different units. Since the work speed at different stages normally differs, for a smooth uninterrupted assembly at every stage certain amount of inventory is kept. In the absence of such inventories men and machine online cannot work smoothly. Longer the production cycle larger will be the semi-finished items that are kept in the inventories. Sometimes such inventories are also called work in progress inventories.

e) Cycle or lot size inventories

Cycle inventories are held for the simple reason that purchases in lots costs less. If the items are bought as and when they are needed the associated piecemeal purchase will cost more in total. This type of inventories are also called lot-size inventory.

13.3 INVENTORY COST PATTERNS

For determining the optimum inventory policy one must weigh the cost of inventory with the associated benefits. In this context we discuss five important cost components in this section.

1. **Ordering cost**

   While ordering for inventory experts have distinguished three types of cost.

   - **Order placing costs:** For placing orders one will have to incur cost. Such a cost is often called ordering cost. In government sectors for any type of work tenders inviting quotations, are mandatory. The lowest quoted quotation is honoured at the end. In this context tenders are advertised in leading newspapers. The expenditure incurred on this account is normally referred as ordering cost.

   - **Follow-up costs:** In certain cases the mere floating tenders is not enough, follow-up action is necessary. The cost of follow-up measures is commonly referred to as the follow-up cost.

   - **Receiving cost:** On arrival of the item under reference, the consignment is usually checked for the quoted quality and quantity. If it is a critical item, a thorough investigation is undertaken. The items must be stored properly and ledgers must be maintained all through till its final use. Such costs are clubbed together under the head 'receiving costs'.

   Putting all the items together, the ordering cost is expressed as a fixed sum per order in practice. Remember, that this cost is nothing to do with the ordered quantity and hence expressed as a lump sum amount per order.

2. Set-up cost

Here by set-up cost we mean the cost of preparing the production facility. If we produce the item under reference in large quantity once for all and keep the excess supply in inventory, then the set-up cost will be incurred only once in a given year. On the other hand, to keep the inventory cost at its minimum, sometimes we produce small quantities on repeated runs. In such cases the total set-up cost will be a certain multiple of single run set-up cost. It includes all expenses of dismantling the old production set-up and preparing for a new set-up for both men and material.

3. Inventory carrying cost

To hold on the inventories, a certain amount of costs will have to be incurred. Such costs are commonly referred under the head 'inventory carrying costs'.

- **Holding cost**: To hold on the inventories, funds are borrowed from the financial institutions, the interest on the borrowed amount is often called 'holding cost'.
- **Storage cost:** To store the inventory from the date of purchase to the date use, one will have to incur rental charges on buildings. On many occasions these items are insured for safety. All these costs are referred as 'storage costs'.
- **Operational cost:** Once items are stored, proper handling is another responsibility. Ledgers for outgoing and incoming items are to be kept updated. Thus, depending upon the quantum of inventory, adequate staffing is necessary. These costs are normally referred as 'operational cost'.
- **Shortage cost:** If the demand exceeds the stock in inventory shortage occurs. If the shortage is in semi-finished products, raw materials etc., then the production of final output will suffer. In the case of finished goods shortage means of lose potential customers, they may go to other competing firms to get their demand fulfilled. Sometimes shortages being voluntary in the company's policy in order to stimulate to demand. In all these cases the shortage cost is expressed per unit of lost sale.

4. Purchase cost

This refers to purchase price of the item under reference. If the unit cost is constant, it does not come in the way of inventory decision making because whether all the needed units are purchased at one go or piecemeal really does not matter. The purchase price become relevant wherever quantity discounts are being offered. In such situations bulk purchases will attract less unit price and hence become relevant in inventory models.

5. Stock out cost

This cost refers to the cost associated with not serving the customer. Stock out simply refers shortage. If it is internal then it would imply some production loss resulting in idle time for men and machine involved in the production. If it is external then it amounts to potential loss of sales. Such shortages evoke different reaction from customers. It would result either lose sale or back order. Among the two the types loss of sale should be viewed seriously.

13.4 IMPORTANT DEFINITIONS

- **Deterministic and probabilistic demand pattern:** The amount of inventory to be kept as stock depends upon the demand pattern for the commodity under consideration. The demand pattern may be either deterministic or probabilistic. In deterministic models the quantity needed in the subsequent periods are known in advance with certainty. It is usually expressed as the demand needs per period and normally referred as **demand rate**. However, this rate may be constant in the case of linear demand and variable in the case of nonlinear demand. Probabilistic or stochastic demand is the result of dynamism in the real world situation. To capture this dynamism we often express the demand pattern by a well known probability distribution.

- **Independent and dependent demand:** The demand for an item is independent of the demand for other products then the pattern is identified as independent demand. Demand for finished goods belongs to this category. The demand for components and sub-assemblies that are used as inputs for the final product are called dependent or derived demands. The demand for these items emerges provided the demand is there for the said finished products.

- **Lead-time:** In inventory models usually there will be a time gap between ordering and arrival of inventories. This time gap between ordering and delivery is often called the lead-time.

- **Time horizon:** It is the time over which the inventory level is controlled. It can be either finite or infinite depending upon the demand pattern.

- **Safety stock:** It is the level of inventory kept when the lead time demand is uncertain or demand is critical and storage cost is high. This type of inventory is necessary to cater to the needs for a short period.

- **Reorder quantity:** It refers to the number units of reordering. This quantity is determined based on a comprehensive inventory policy of the company concerned.

- **Reorder level (point):** It refers to the level of inventory stock in hand at the time of reordering. If the inventory level is equal or less than this prescribed amount, reordering is made.

- **Ordering cycle:** The time period between placement two successive orders is often called ordering cycle. The order may be placed on the basis of the following two types of inventory review system.

13.5 INVENTORY MANAGEMENT SYSTEMS

Basically we have two inventory management systems. They are:

- **Fixed order quantity system:** In this system the record of the inventory is checked continuously until a specified level (called reorder level) of inventory stock is reached. When inventory stock reaches this predefined level, reorder is placed for new arrivals in some fixed quantity. However, the reordering interval between two subsequent ordering may vary from order to another in these models.
- **Periodic review system:** In this system the inventory levels are checked periodically (not regularly as in the previous case) at regular time intervals and orders are placed for the balance having the available stock in hand in mind. This time though the ordering time intervals are equal to one another the quantity ordered each time depends upon the available inventory level at that point of time. Thus, in this method the quantity ordered varies from order to order but at the same time, the time interval between two consecutive orders remains the same.

13.6 INVENTORY DECISIONS

In inventory decision making, the following three basic questions are normally addressed. They are:

1. How much to order per order? This indeed refers to the optimum quantity that is to be ordered per order.
2. When should the order be placed? This refers to the date of ordering, given the lead time.
3. How much safety stock should be kept? This refers to the amount of buffer stock that is to be kept to manage the uncertain demand and supply conditions.

13.7 THE CLASSICAL ECONOMIC ORDER QUANTITY MODEL WITH UNIFORM DEMAND PATTERN

Economic order quantity model is based on deterministic demand pattern. By deterministic demand we mean the demand at a constant or nearly constant rate per period. This assumption simply means that the same number of units is taken from inventory for each equivalent period of time. For example, if 10 units are taken per day then 60 units will be taken per six day week and 240 units will be taken for 4 six day weeks. In this type of models the management is confronted with two opposing situations. When number of orders is few (order quantity is large) then the total ordering cost will be less, but at the same time, the carrying or holding cost of large quantity will be high. On the other hand, if number of orders are more (order quantity is less) then the total ordering cost will increase and carrying cost will diminish. Economic order quantity model finds a compromising solution to these two conflicting situations. The quantity corresponding to the minimum of the sum of these two cost components is often called economic order quantity (EOQ). Interestingly, such a compromising solution occurs at the point where these two cost components are equal to one another at their point of intersection as shown in Figure 13.1.

Assumptions of the EOQ model:

- The annual demand D is deterministic and occurs at a constant rate per period.
- The lead-time, that is, time between placing an order and receiving delivery, is a known constant. If this constant is zero then the delivery of items ordered is instantaneous.
- The order quantity Q per cycle is the same for all cycles.
- The inventory level increases by Q units when an order is received.
- The cost per order (C_O) is constant and does not depend on the quantity.
- The purchase price C per unit is a constant and thus no discount of any such is being offered on large scale purchase.
- The inventory holding cost C_h per unit is a constant. Sometimes this cost is expressed as a percentage of the total inventory purchase cost. On some other occasions it is expressed as a fixed sum per unit on annual basis.
- The inventory position is reviewed continuously and order is placed as soon as the inventory position reaches the reorder point so that the supply is received in time when the stock level becomes zero.
- Consequently shortages such as stock out or back orders do not exist.

The following example illustrates the method of getting such a solution.

■ Example 1

The annual demand for Maruti-800 car at an outlet in Bangalore is 800 units at the rate of Rs. 2,50,000 per unit. The ordering cost per order is Rs. 50,000. The cost of carrying (holding) inventory is 5% of the average inventory per unit per year. Given that the outlet works for 200 days in a year, find the economic order quantity (EOQ). Given the lead time of two days, obtain the ordering dates for the whole year. Illustrate the situation by a scaled diagram.

Solution

i) Tabular approach

The basic issue that is to be settled in the classical model is none other than the optimum quantity that is to be ordered. Once this issue is scheduled all other issues like—*when to order? How frequently to order*—are settled automatically. To find the optimum order quantity we shall consider two types of cost namely, the ordering cost on one side and the holding cost on the other side.

Ordering Cost

If only one order is placed for the whole year demand of 800 cars then the order cost will be Rs. 50,000 × 1 = Rs. 50,000. If the order is placed 2 times in a year at the rate of 400 cars per order then the ordering cost will be Rs. 50,000 × 2 = 1,00,000. In this way the ordering costs are obtained and reported in column number 5 in Table 13.1 for varying number of orders from order 1 to order 32 orders per year.

The Inventory Holding Cost

In column 1 the varying number of orders in a year is stated. Accordingly, in the second column the lot size per order is obtained and reported. So if only one order per year is placed then the associated lot size is stated as 800 units. Similarly, the second row shows the two orders per year situation with an order quantity of 400 units per order. For all other order numbers the associated lot sizes are obtained and reported in the Table 13.1.

Table 13.1

No. of order per year	No. of units per order	Average inventory	Carrying cost	Ordering cost	Total cost
1	800	400	50,00,000	50,000	50,50,000
2	400	200	25,00,000	1,0,0000	26,00,000
4	200	100	12,50,000	2,00,000	14,50,000
5	160	80	10,00,000	2,50,000	12,50,000
8	100	50	6,25,000	4,00,000	10,25,000
10	80	40	5,00,000	5,00,000	10,00,000
16	50	25	3,12,500	8,00,000	11,12,500
20	40	20	2,50,000	10,00,000	12,50,000
25	32	16	2,00,000	12,50,000	14,50,000
32	25	12.5	1,56,250	16,00,000	17,56,250

The third column shows the average inventory by simply dividing the second column by 2. The fourth column shows the inventory carrying cost calculations. For example, for two-order situation, the average inventory is 200 cars. So the corresponding inventory carrying cost = (200 × 2,50,000) × 0.05 = 25,00,000. Similar calculations are made and reported for all other rows in the table.

The last column gives the total of both the cost components. Note that as stated earlier the holding cost keep on falling as order increases. But at the same time the ordering cost is increasing as number of orders increases. The optimum minimum total cost necessarily occurs where the carrying cost and ordering cost are equal to one another at Rs. 5,00,000 level. In our illustration such a coincidence occurs when order per year is 10 and inventory level is 80 cars per cycle.

ii) Graphical Solution

In Figure 13.1 we measure the number of units to be ordered per cycle, as stated in the second column in the above table, on the X-axis. On the Y-axis we measure annual ordering cost, annual inventory holding cost and the consequent total cost. In the graph, the total cost reaches its minimum when the order quantity is 80 units per order. Also note that at that ordering level the annual ordering cost and the annual inventory holding costs are equal and hence intersect.

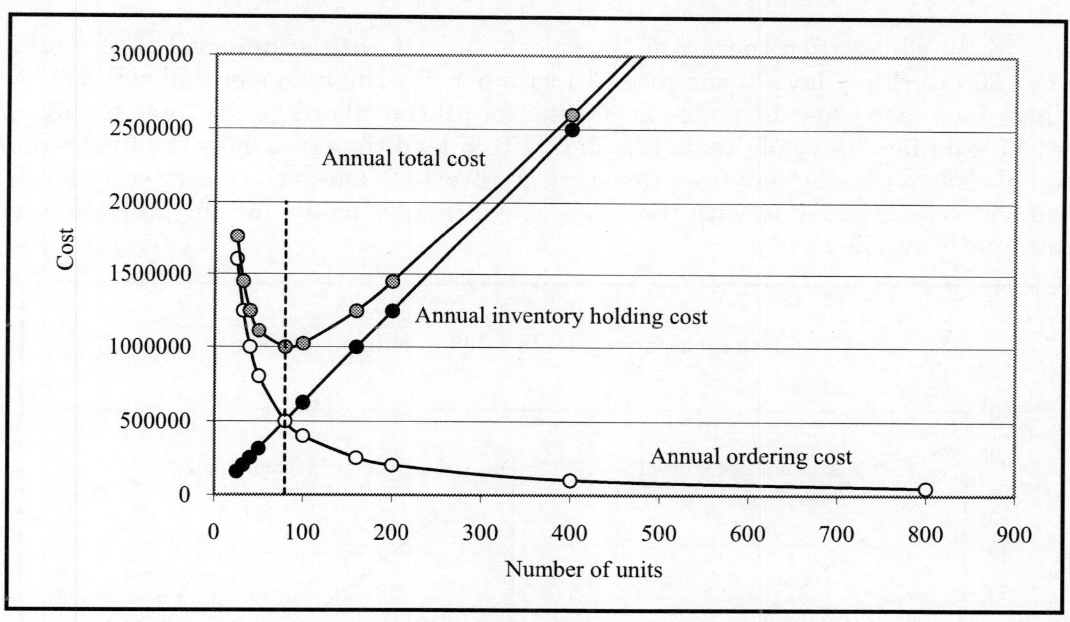

Figure 13.1

(b) When to order decision (Reorder Point) and how much to order (Reorder level)

Till now we only considered how much to order. The next logical question is when to order. To answer this question we must know the **inventory position** at the time of ordering and the average demand per day. When to order point is often called **reorder point.** Since our Maruti outlet works for 200 days in a year, at the uniform rate the demand per day is obtained as 800/200 = 4 cars per day. Suppose we begin at the origin with inventory of 80 cars then at the rate of 4 cars per day this stock will last for only 20 days. The 20 days time is called the order cycle. Given a **lead time** of 2 days from the date of order to delivery, the company must place its first reorder on 18th day itself so that the hand on stock of 8 cars gets exhausted on the evening of 20th. Remember, on 20th evening 80 new ordered cars will arrive. Here the 18th day is called reorder point. This amount of 8 cars is called the reorder level otherwise called **lead time demand.**

c) How frequently to order decision (inventory cycle)

The period between two consecutive reorders is normally referred as **inventory cycle**. If D is the annual demand and Q is the optimum reorder quantity the number of orders per year is obtained as n = D/Q. Since the number of working days in a year is 200 days the inventory cycle is obtained by using the following formula.

$$T = \frac{200}{D/Q} = \frac{200 \times Q}{D/Q} = \frac{200}{800/80} = 20 \text{ working days}$$

Thus, in our illustration there will 10 cycles in a year with an interval of 20 working days. For 200 working days at the rate of 4 cars per day the manager will sell 800 cars. Note that since each order is made for 80 cars, for all the 10 orders the total supply also will be 800 cars for the whole year. Further, with a lead-time of 2 days the first reorder must be placed on the 18th day from the start. Figure 13.2 shows the entire story for first 2 inventory cycles associated with their respective reorder points for our problem, given the lead time of two days.

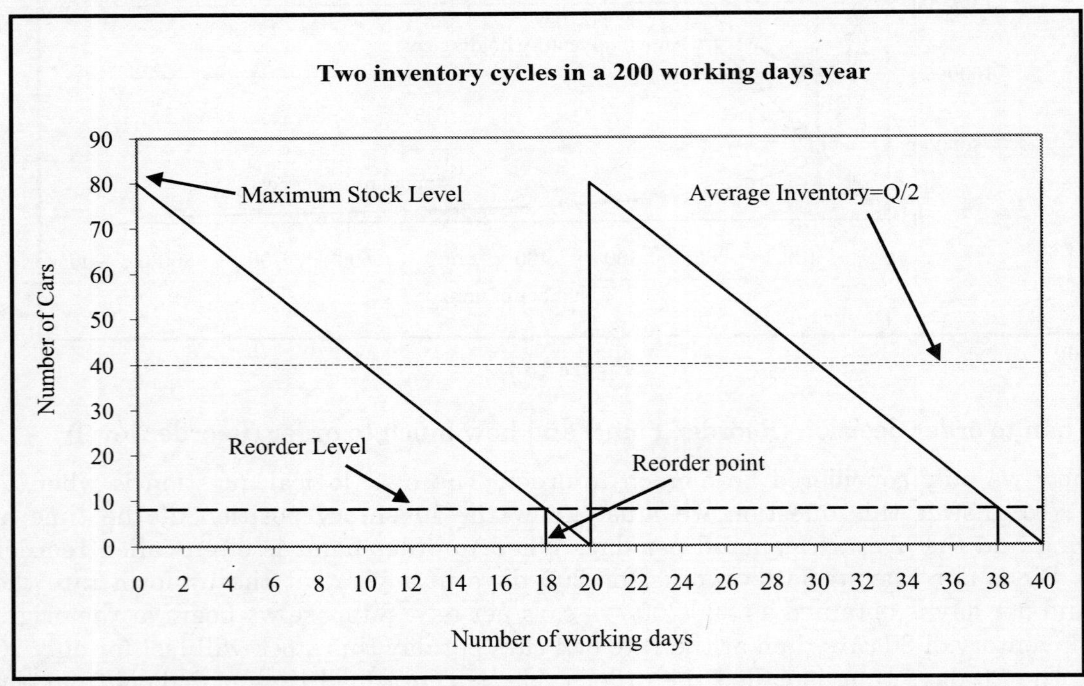

Figure 13.2

In the above graph the optimal inventory order per inventory cycle is 80 cars. Order is placed on 18, 38, 58, 78, 98, 118, 138, 158, 178, 198 working days respectively having the inventory cycle of 200/10 = 20 working days.

13.8 MATHEMATICAL MODELLING FOR EOQ MODEL

In this section we formulate an appropriate Mathematical Model to answer the following three major questions:

a) How much to order decision (EOQ)

b) When to order decision (reorder point)

c) How frequent to order decision (inventory cycle)

a) How much to order decision (EOQ)

Though yearly demand D is known in advance the quantity of inventory to be ordered at a point of time that brings down the total operational cost to the minimum is unknown in our problem. Let this unknown be Q units.

- Now the average inventory is Q/2 units.
- Let C be the cost per unit of purchase.
- Let I be inventory-carrying cost expressed as proportion of the value of the inventory.
- C_h = CI carrying or holding cost of one unit of inventory for the year.
- C_o = ordering cost per order.

Now the total inventory carrying cost per period is obtained as

$$\text{Total inventory carrying cost} = \begin{pmatrix}\text{Average}\\ \text{inventory}\\ \text{quantity}\\ \text{during the}\\ \text{period}\end{pmatrix} \times \begin{pmatrix}\text{Purchase}\\ \text{cost}\\ \text{per unit}\end{pmatrix} \times \begin{pmatrix}\text{Inventory}\\ \text{carrying}\\ \text{cost as}\\ \text{proportion}\end{pmatrix}$$

$$= \frac{Q}{2}CI$$

$$= \frac{Q}{2}C_h \qquad (CI = C_h)$$

Now the total ordering cost is obtained as

$$\text{Total ordering cost} = \begin{pmatrix}\text{Number of orders}\\ \text{per period}\end{pmatrix} \times \begin{pmatrix}\text{Ordering cost}\\ \text{per order}\end{pmatrix} = \frac{D}{Q}C_o$$

Now the total cost is obtained as the sum of these two components.

$$\begin{pmatrix}\text{Total annual cost}\\ \text{of inventory}\end{pmatrix} = \begin{pmatrix}\text{Annual holding}\\ \text{cost}\end{pmatrix} + \begin{pmatrix}\text{Annual ordering}\\ \text{cost}\end{pmatrix}$$

$$TC = \frac{Q}{2}C_h + \frac{D}{Q}C_o$$

For an optimum decision we are interested in getting the value of Q that minimizes the total operational cost of the inventory. From elementary calculus for such a minimum of TC we know that

$$\text{(i)}\ \frac{d}{dQ}(TC) = 0$$

$$\text{(ii)}\ \frac{d^2}{dQ^2}(TC) > 0$$

So from above

$$\frac{d}{dQ}(TC) = \frac{C_h}{2} - \frac{DC_o}{Q^2} = 0$$

$$\frac{C_h}{2} = \frac{DC_o}{Q^2}$$

$$Q^2 = \frac{2DC_o}{C_h}$$

$$Q^* = \sqrt{\frac{2DC_o}{C_h}}$$

Further,

$$\frac{d^2}{dQ^2}(TC) = \frac{2DC_o}{Q^3} > 0$$

So, $Q^* = \sqrt{\frac{2DC_o}{C_h}}$ Gives us the needed optimum cost minimizing level of Q.

Note 1: *In our Maruti Illustration in example 1 above the value of* Q^* *is obtained as*

$$Q^* = \sqrt{\frac{2DC_o}{C_h}} = \sqrt{\frac{2DC_o}{CI}} = \sqrt{\frac{2 \times 800 \times 50{,}000}{2{,}50{,}000 \times 0.05}} = 80$$

This is exactly the same order quantity that we obtained in the diagrammatic method. Once the EOQ is obtained other optimum decision parameters are obtained for the given data as usual as we have done it by graphical method.

Note 2: From above

$$Q^2 = \frac{2DC_o}{C_h}$$

$$Q^2C_h = 2DC_o$$

$$\frac{QC_h}{2} = \frac{2DC_o}{Q}$$

Total inventory carrying cost = Total ordering cost

Thus at the optimum inventory level total inventory cost = total ordering cost as discussed in the graphical and tabular solutions.

Spreadsheet solution to the Maruti illustration

Table 13.2

	A	B
1	**Data**	
2	Annual demand (D)	800.00
3	Ordering cost (C_0)	50000.00
4	Purchase cost per unit (C)	250000.00
5	Annual inventory holding rate in (I)	0.05
6	Inventory holding cost (C_h=IC)	12500.00
7	Working days per year	200.00
8	Lead time	2.00
9	**Optimum inventory policy parameters**	
10	Economic order quantity (Q)	=SQRT(2*B2*B3/B6)
11	Number of orders per year (D/Q)	=B2/B10
12	Average inventory (Q/2)	=B10/2
13	Annual inventory holding cost{(Q/2*C_h)	=B12*B6
14	Annual ordering cost	=B3*B11
15	Maximum inventory level (Q)	=B110
16	Cycle time (T)	=B7/B11
17	Demand per day (d)	=B10/B16
18	First reorder point	=B16-B8
19	Reorder level	=B17*B8

In the excel sheet displayed in Table 13.2 we reproduce the data given in the problem from cell B2 to B8 as described in the spreadsheet. To obtain the optimum policy parameters, type the formulas in the respective cells from B10 to B19 as shown in Table 13.2. Once this is done, the worksheet displays the answers as in Table 13.3.

Table 13.3

	A	B
1	**Data**	
2	Annual demand (D)	800.00
3	Ordering cost (C_0)	50000.00
4	Purchase cost per unit (C)	250000.00
5	Annual inventory holding rate in (I)	0.05
6	Inventory holding cost (C_h=IC)	12500.00
7	Working days per year	200.00
8	Lead time	2.00
9	**Optimum inventory policy parameters**	
10	Economic order quantity (Q)	80.00
11	Number of orders per year (D/Q)	10.00
12	Average inventory (Q/2)	40.00
13	Annual inventory holding cost (Q/2*C_h)	500000.00
14	Annual ordering cost	500000.00
15	Maximum inventory level (Q)	80.00
16	Cycle time (T)	20.00
17	Demand per day (d)	4.00
18	First reorder point	18.00
19	Reorder level	8.00

Table 13.4

	A	B
1	**Data**	
2	Annual demand (D)	1200.00
3	Ordering cost (C_0)	50000.00
4	Purchase cost per unit (C)	250000.00
5	Annual inventory holding rate in (I)	0.05
6	Inventory holding cost (C_h=IC)	12500.00
7	Working days per year	200.00
8	Lead time	2.00
9	**Optimum inventory policy parameters**	
10	Economic order quantity (Q)	97.98
11	Number of orders per year (D/Q)	12.25
12	Average inventory (Q/2)	48.99
13	Annual inventory holding cost (Q/2*C_h)	612372.44
14	Annual ordering cost	612372.44
15	Maximum inventory level (Q)	97.98
16	Cycle time (T)	16.33
17	Demand per day (d)	6.00
18	First reorder point	14.33
19	Reorder level	12.00

Note: *Once the spreadsheet solution is obtained, answers can be worked out for any new situation of the same type of problem. Only we will have to change the relevant data part of the worksheet. Once this is done since the formulas are already there in the spreadsheet automatically all the optimum parameter values are displayed instantaneously. For* *example, if the annual demand of our problem is changed from 800 units to 1200 units then the relevant optimum parameter values are displayed in Table 13.4.*

Moreover, the same worksheet could be used to do a different problem of the same type even. In Table 13.5 different problem is being done by merely changing the data values.

■ Example 2

The XYZ company has determined from an analysis of its accounting and production data for part number 625, its cost of purchase is Rs. 36 per order and Rs. 2 per part. Its inventory carrying charge is 9 percent of the average inventory. The demand for this part is 10,000 units per annum. Determine:

(i) What should be the economic order quantity?

(ii) What is the optimal number of orders?

Solution

Data

Demand per annum D = 10,000 units

Cost per order C_0 = Rs. 36

Purchase cost per part C = Rs. 2

Inventory carrying cost I = Re. 0.09 expressed as portion of the purchase price (9%)

(i) By definition total inventory holding cost = Average inventory × Purchase cost per unit × inventory carrying cost as portion of the value of the inventory

$$C_h = Q/2 \times C \times I = Q/2 \times 2 \times 0.09 = 0.09Q$$

The total Ordering cost C_0 = No. of orders per year × Ordering cost per order

$$= 10{,}000/Q \times 36$$

Now at EOQ we know that the total inventory holding cost = Total ordering cost

$$0.09Q^2 = 10{,}000Q \times 36$$

$$0.09Q^2 = 3{,}60{,}000$$

$$Q^2 = 3{,}60{,}000 / 0.09$$

$$Q^2 = 40{,}00{,}000$$

$$Q = 2000$$

(ii) Optimum number of orders $= \dfrac{D}{Q} = \dfrac{10{,}000}{2000} = 5$

Excel solution

In Table 13.5 we obtain the results by merely changing the relevant data values in Table 13.2.

Table 13.5

	A	B
1	**Data**	
2	Annual demand (D)	10000.00
3	Ordering cost (C_0)	36.00
4	Purchase cost per unit (C)	2.00
5	Annual inventory holding rate in (I)	0.09
6	Inventory holding cost (C_h=IC)	0.18
7	Working days per year	200.00
8	Lead time	2.00
9	**Optimum inventory policy parameters**	
10	Economic order quantity (Q)	2000.00
11	Number of orders per year (D/Q)	5.00
12	Average inventory (Q/2)	1000.00
13	Annual inventory holding cost (Q/2*C_h)	180.00
14	Annual ordering cost	180.00
15	Maximum inventory level (Q)	2000.00
16	Cycle time (T)	40.00
17	Demand per day (d)	50.00
18	First reorder point	38.00
19	Reorder level	100.00

Fixed lump sum inventory carrying cost models

Till now we were expressing the carrying cost as a portion of the purchase price per unit. Often we express this cost as a fixed sum per unit per year as well. The following example illustrates one such case.

■ Example 3

Western Electronics produces 2000 TV sets in a year. For this purpose it needs equal number of picture tubes of certain type. Each tube costs Rs. 10 and the cost of holding the tube in inventory is Rs. 2.40 per unit per year. In addition the cost of placing orders is Rs. 150 per order. Obtain the optimum order quantity and the number of orders, given the information that the company works for 200 days in a year with a lead time of 2 days.

Solution: Tabular presentation

Table 13.6

No. of orders per year	No. of units per order	Average inventory	Carrying cost	Ordering cost	Total cost
1	2000	1000	2400	150	2550
2	1000	500	1200	300	1500
4	500	250	600	600	1200
5	400	200	480	750	1230
8	250	125	300	1200	1500
10	200	100	240	1500	1740

Diagrammatic representation

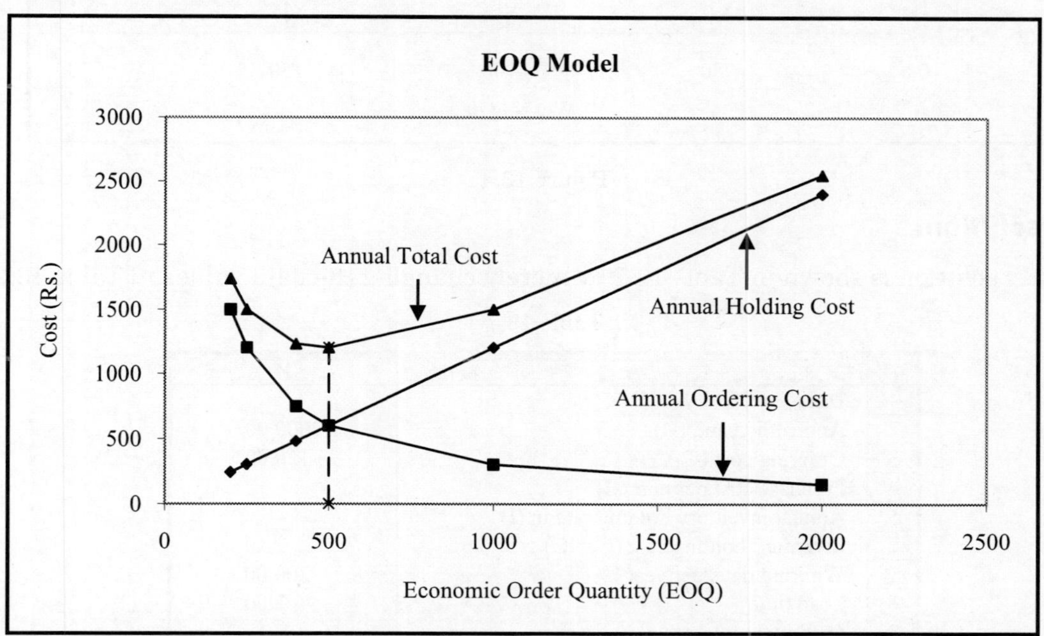

Figure 13.3

Thus, EOQ = 500 for our model. Since the total quantity needed for one year = 2000, the number of orders to be placed = 2000/500 = 4. Since the company works only for 200 days, the inventory cycle will be = 200/4 = 50 days. Further, since the order quantity is 500 and the cycle is 50 days. The demand per day = 500/50 = 10. Figure 13.4 illustrates the entire situation.

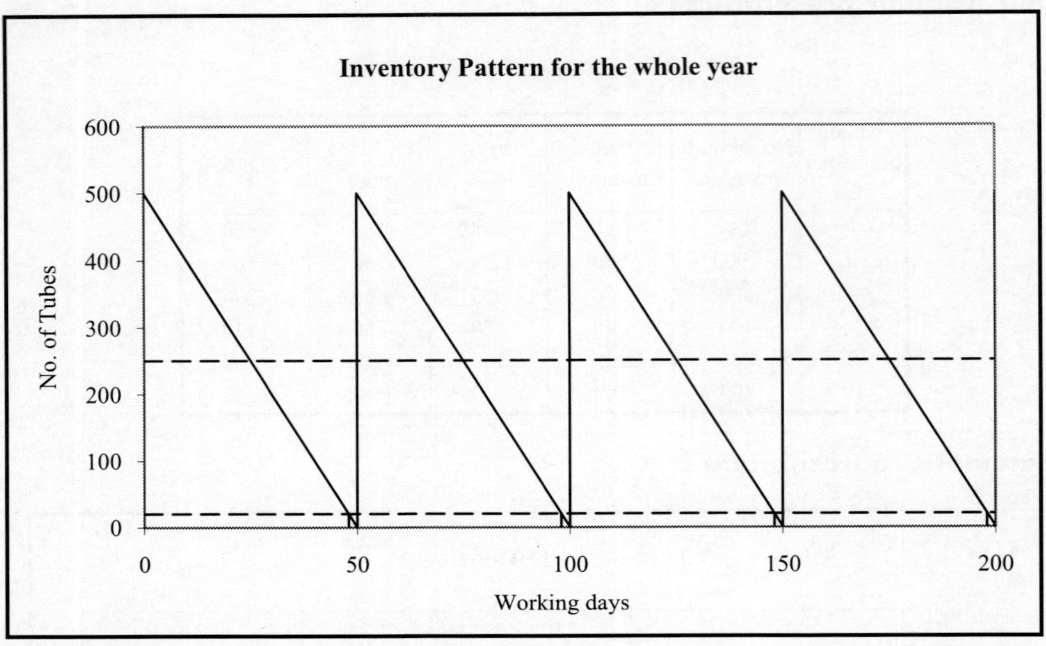

Figure 13.4

Excel solution:

The excel solution is shown in Table 13.7 by merely changing the data values in Table 13.2.

Table 13.7

	A	B
1	**Data**	
2	Annual demand (D)	2000.00
3	Ordering cost (C_0)	150.00
4	Purchase cost per unit (C)	-
5	Annual inventory holding rate in (I)	-
6	Inventory holding cost (C_h=IC)	2.40
7	Working days per year	200.00
8	Lead time	2.00
9	**Optimum inventory policy parameters**	
10	Economic order quantity (Q)	500.00
11	Number of orders per year (D/Q)	4.00
12	Average inventory (Q/2)	250.00
13	Annual inventory holding cost (Q/2*C_h)	600.00
14	Annual ordering cost	600.00
15	Maximum inventory level (Q)	500.00
16	Cycle time (T)	50.00
17	Demand per day (d)	10.00
18	First reorder point	48.00
19	Reorder level	20.00

Example 4

ABC plumbing supply company stocks thousands of plumbing items sold to regional plumbers, contractors, and retailers. Mr. X, the firm's general manager, wonders how much money could be saved annually if EOQ were used instead of the firm's present rules of thumb. He instructs Mr. Y to conduct an analysis to examine whether significant savings might result from using EOQ. Mr. Y develops the following estimates from accounting information.

$D = 10{,}000$

$Q = 400$

C_h = Rs. 4 per unit per year

C_o = Rs. 55 per order.

Solution

The total annual costs under present policy

$$= \frac{Q}{2} \times C_h + \frac{D}{Q} \times C_o$$

$$= \frac{400}{2} \times 4 + \frac{10{,}000}{400} \times 55$$

$$= \text{Rs. } 2175$$

$$\text{EOQ} = \sqrt{\frac{2DC_o}{C_h}}$$

$$= \sqrt{\frac{2 \times 1000 \times 55}{4}}$$

$$= \sqrt{2{,}75{,}000} = 524.4$$

The total annual cost when EOQ is followed

$$= \frac{Q}{2} \times C_h + \frac{D}{Q} \times C_o$$

$$= \frac{524.4}{2} \times 4 + \frac{10{,}000}{524.4} \times 55$$

$$= \text{Rs. } 2097.6$$

So the estimated cost savings = 2175 – 2097.6 = Rs. 77.4.

■ Example 5

XYZ Company buys in lots of 500 boxes enough to meet three months demand. The cost per box is Rs. 125 and the ordering cost is Rs. 150 per order. The inventory cost is estimated as 20% of the unit value of the product.

a) What is the total annual cost of the existing inventory policy?

b) How much money could be saved by employing the economic order quantity policy?

Solution: (a)

Ordering cost per order = Rs. 150

Current order quantity per order = 500 units

Annual demand = 500/3 ×12 = 2000 units

Inventory carrying cost per unit = 125 × 20/100 = Rs. 25

Total annual cost on existing inventory = Number of order × ordering cost + average inventory × inventory carrying cost per unit

$$= \frac{D}{Q} \times C_0 + \frac{Q}{2} \times C_h$$

$$= \frac{2000}{500} \times 150 + \frac{500}{2} \times 25$$

$$= \text{Rs. } 6850$$

Solution: (b)

$$\text{Economic order quantity (EOQ)} = \sqrt{\frac{2DC_0}{C_h}} = \sqrt{\frac{2 \times 2000 \times 150}{25}} = 155$$

$$\text{Total annual cost on EOQ level} = \frac{D}{Q} \times C_0 + \frac{Q}{2} \times C_h$$

$$= \frac{2000}{155} \times 150 + \frac{155}{2} \times 25$$

$$= \text{Rs. } 3873$$

Cost Savings = 6850 – 3873 = Rs. 2977.

13.9 SENSITIVITY ANALYSIS OF EOQ MODEL

We know that the sensitivity analysis is post-scenario status of EOQ solution. Accordingly, once the EOQ value is obtained we test for its sensitivity for varying cost conditions by changing both the carrying and ordering costs first by one by one and finally together. The following table shows such calculations for different levels of costs in the neighbourhood. As one could see from the table, the value of Q appears to be relatively stable, even with some variation in cost conditions.

Table 13.8

No. of orders per year	No. of units per order	Average inventory	Per unit carrying cost	Carrying cost	Ordering cost	Ordering cost	Total cost
4.00	500.00	250.00	2.40	600.00	150.00	600.00	1200.00
4.47	447.21	223.61	3.00	670.82	150.00	670.83	1341.64
3.87	516.40	258.20	2.40	619.68	160.00	619.67	1239.35
4.02	498.00	249.00	2.50	622.50	155.00	622.49	1244.99
3.98	502.17	251.09	2.30	577.50	145.00	577.49	1154.99
4.23	473.29	236.65	2.50	591.61	140.00	591.60	1183.22

One of the important characteristics of the EOQ model is its robustness at the minimum point of the total cost. The total cost curve is relatively flat. Due to this flatness the model gives reasonably good result even when the holding and carrying costs vary by small amount as shown in the Table 13.8.

13.10 EOQ MODEL WITH SINGLE PRICE BREAKS (QUANTITY DISCOUNTING MODELS)

The unit purchase price is assumed to be constant for all levels of purchase in our earlier version. Thus, there is no place for quantity discounts till now. In real life, it is not uncommon to find price discounts for large purchases. Usually, lower rates are quoted for large orders. On several occasions one could find more than one price breaks as well. Under such circumstances optimum quantity per order has to be decided with little more care. When the unit price is uniform then the purchase cost is totally irrelevant in determining the order quantity. However, this time in addition to order cost and holding cost we must include the purchase cost as well into our total cost equation. In it C_i refers to the purchase price at each bracket.

$$TC = \frac{Q}{2}C_h + \frac{D}{Q}C_0 + C_i D$$

Cost comparison approach

In this approach, as usual, the uniform demand pattern and the inventory acquisition lead-time are known constants. However, the total cost of annual inventory is calculated twice, once using EOQ values without discount and another time with discount. Among the two the decision is based on the least total cost. The following example illustrates one such model.

■ Example 6

Western Electronics Company produces 2000 TV sets in a year. For this purpose it needs equal number of a typical component. Each component costs Rs. 10 and the cost of holding the component in inventory is Rs. 2.40 per unit per year. In addition the cost of placing orders is Rs. 150 per order. Obtain the optimum order quantity. If the supplier offers a discounted price of Rs. 9 provided the order quantity is 1000 unit are more, test the viability of the company's claim for the discounted offer.

Solution

Let us first calculate the EOQ assuming that there is no discount of any type by using the formula

$$\text{EOQ} = Q = \sqrt{\frac{2DC_o}{C_h}} = \sqrt{\frac{2 \times 150 \times 2000}{2.4}} = 500$$

Clearly the economic order quantity of 500 units is not eligible for the discount because it is available only for purchase of more than 1000 units. So let us calculate the total cost for 500 units of EOQ at the usual rate of Rs. 10 per unit

$$TC = \frac{Q}{2}C_h + \frac{D}{Q}C_0 + C_iD$$

$$= \frac{500}{2} \times 2.40 + \frac{2000}{500} \times 150 + 10 \times 2000 = \text{Rs. } 21{,}200$$

Since the discount worthy minimum is 1000 units let us calculate the total cost one again for an order quantity of 1000 units at the discounted rate of Rs. 9.

$$TC = \frac{Q}{2}C_h + \frac{D}{Q}C_0 + C_iD$$

$$= \frac{1000}{2} \times 2.40 + \frac{2000}{1000} \times 150 + 9 \times 2000 = \text{Rs. } 19{,}500$$

To have one more feasibility check let as calculate the total cost for a higher order quantity for the whole year, namely 2000 units also as shown below at Rs. 9 rate.

$$TC = \frac{Q}{2}C_h + \frac{D}{Q}C_0 + C_i D$$

$$= \frac{2000}{2} \times 2.40 + \frac{2000}{2000} \times 150 + 9 \times 2000 = \text{Rs. } 20{,}550$$

The following Table 13.9 gives the summary calculations for all the three cases.

Table 13.9

No. of orders per year	No. of units per order	Average inventory	Carrying cost	Ordering cost	Purchase price	Purchase cost	Annual total cost
1	2000	1000	2400	150	9	18,000	20,550
2	1000	500	1200	300	9	18,000	19,500
4	500	250	600	600	10	20,000	21,200

Since for the order quantity of 1000 units the total cost is least, the order quantity must be 1000 for per order. Figure 13.5 illustrates this typical case.

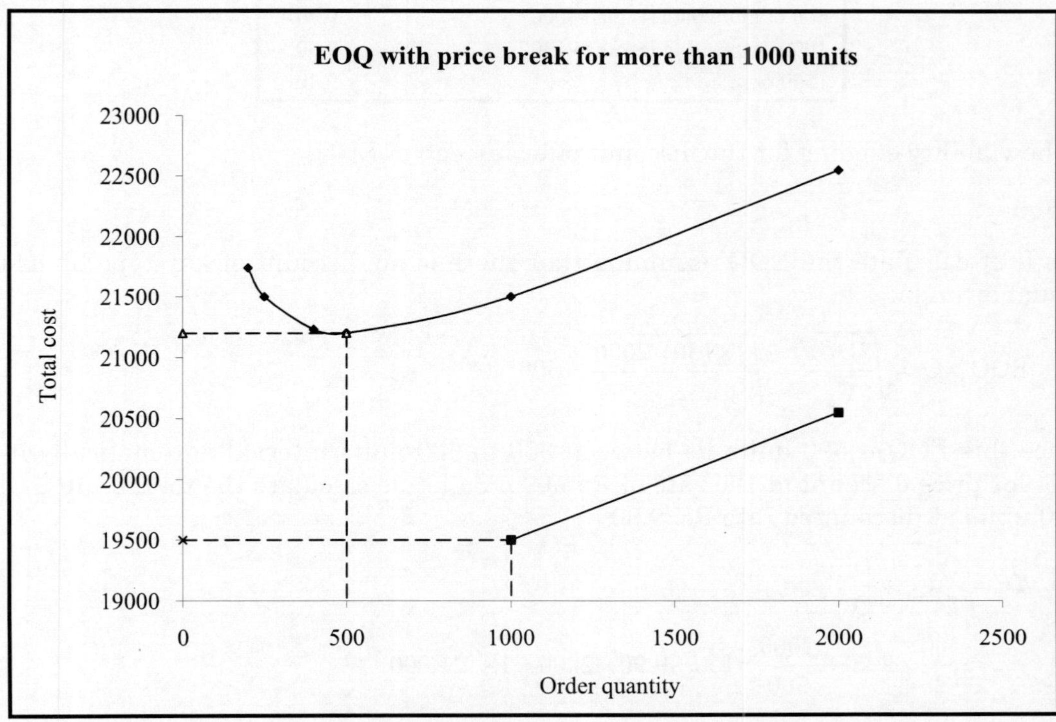

Figure 13.5

13.11 EOQ MODEL WITH MULTIPLE PRICE BREAKS (QUANTITY DISCOUNTING MODELS)

On many occasions price breaks are offered on multiple brackets. The following examples illustrate one such multiple break problems.

■ Example 7

Suppose the price breaks are described as follows Western Electronics Company produces 2000 TV sets in a year. For this purpose it needs equal number of picture tubes of certain type. Each tube costs Rs. 10 and the cost of holding the tube in inventory is Rs. 2.40 per unit per year. In addition the cost of placing orders is Rs. 150 per order. Obtain the optimum order quantity if the supplier offers multiple discounts for tubes as shown below.

Table 13.10

Order quantity	Price in Rs.
Less than 300	10.00
300 or more but less than 600	9.90
600 or more but less than 1000	9.70
1000 or more but less than 1500	9.50
1500 or more	9.40

Test the viability of going for the discount offer instead of EOQ.

Solution

Let us first calculate the EOQ assuming that there is no discount of any type by using the usual formula.

$$\text{EOQ} = Q = \sqrt{\frac{2DC_o}{C_h}} = \sqrt{\frac{2 \times 150 \times 2000}{2.4}} = 500$$

Since this EOQ = 500 units lie between 300 to 600 units interval automatically it is eligible for price discount at the rate of Rs. 9.90. So let us calculate the total annual cost for 500 units at discounted rate Rs. 9.90

$$TC = \frac{Q}{2}C_h + \frac{D}{Q}C_0 + C_iD$$

$$= \frac{500}{2} \times 2.40 + \frac{2000}{500} \times 150 + 9.90 \times 2000 = \text{Rs. } 21{,}000$$

Now to check the validity of this decision let us calculate the total costs for all the lower limit quantities of the higher discounts, namely 600, 1000, 1500, and 2000 as shown in Table 13.11.

Table 13.11

Category	Number of orders	Order quantity	Average inventory	Carrying cost	Ordering cost	Purchase price	Purchase cost	Annual total cost
2	4	500	250	600	600	9.90	19,800	21,000
3	3.3333333	600	300	720	500	9.70	19,400	20,620
4	2	1000	500	1200	300	9.50	19,000	20,500
5	1.3333333	1500	750	1800	200	9.40	18,800	20,800
6	1	2000	1000	2400	150	9.40	18,800	21,350

Since the total cost is least in the case of 1000 units order quantity, one must go two orders at the rate of 1000 units per order in a year.

Example 8

An electrical shop purchases and sells them in the market. In this regard the following information is available for a shopkeeper.

Expected annual sales = 8000 units

Ordering cost = Rs.180 per order

Holding cost = 10% of the average inventory

The items are available under the following conditions

Lot Size	Unit price
1-999	Rs. 22.00
1000-1499	Rs. 20.00
1500-1999	Rs. 19.00
2000 and above	Rs. 18.50

Find the best order size.

Solution

Let us calculate the EOQ having the lowest price of Rs. 18.50 under the assumption that there are no price breaks.

$$EOQ = Q = \sqrt{\frac{2DC_o}{C_h}} = \sqrt{\frac{2 \times 8000 \times 180}{0.10 \times 18.50}} = 1248 \text{ units}$$

From the given data, for the EOQ = 1248 units the associated price tag is Rs. 20. Thus the EOQ obtained is not a valued under the assumption of Rs. 18.50 price per unit.

Next let us calculate the EOQ having the next lowest price of Rs. 19.00 under the assumption that there are no price breaks.

$$EOQ = Q = \sqrt{\frac{2DC_o}{C_h}} = \sqrt{\frac{2 \times 8000 \times 180}{0.10 \times 19.00}} = 1231 \text{ units}$$

From the given data for EOQ = 1231 units the price tag is Rs. 20. Thus the EOQ obtained is not valued again under the assumption of Rs. 19 price per unit.

Next let us calculate the EOQ having the next lowest price of Rs. 20.00 under the assumption that there are no price breaks.

$$EOQ = Q = \sqrt{\frac{2DC_o}{C_h}} = \sqrt{\frac{2 \times 8000 \times 180}{0.10 \times 20.00}} = 1200 \text{ units}$$

From the given data for EOQ = 1200 units the price tag is Rs. 20. Thus the EOQ is valued at this price.

Now let us calculate the total cost for the order quantity of 1200 units at the rate Rs. 20 in the usual manner as shown in the Table 13.12.

Table 13.12

Category	Number of orders	Order quantity	Average inventory	Carrying cost	Ordering cost	Purchase price	Purchase cost	Annual total cost
1	6.6666667	1200	600	1200	1200	20.00	1,60,000	1,62,400
2	5.3333333	1500	750	1425	960	19.00	1,52,000	1,54,385
3	4	2000	1000	1850	720	18.50	1,48,000	1,50,570

In the second and the third row similar calculations are made for 1500 and 2000 units. Since the total cost is minimum at Q = 2000, it represents the optimum order quantity.

■ Example 9

Suppose the annual demand for a certain product is 5000 units. Let the holding cost be 20% and the ordering cost is Rs. 49 per order. However, the unit price for the product depends upon the order quantity due to discount offer given. The following data gives the discount details. Calculate the appropriate order decision.

Table 13.13

Discount category	Order size	Discount in %	Unit price
1	0 to 999	0	Rs. 5
2	1000 to 2499	3	Rs. 4.85
3	2500 and above	5	Rs. 4.75

Solution

5% discount and the associated minimum order of 2500 may be tempting to begin with. However, such a high level of ordering invariably involves higher holding cost as well. So before making decisions one must make cost calculations, one before one after availing of the discount for all the three categories.

For each discount category we calculate EOQ value having the respective discounted unit price in mind.

$$Q_1 = \sqrt{\frac{2DC_o}{C_h}} = \sqrt{\frac{2DC_o}{CI}} = \sqrt{\frac{2 \times 5000 \times 49}{5 \times 0.20}} = 700$$

$$Q_2 = \sqrt{\frac{2DC_o}{C_h}} = \sqrt{\frac{2DC_o}{CI}} = \sqrt{\frac{2 \times 5000 \times 49}{4.85 \times 0.20}} = 711$$

$$Q_3 = \sqrt{\frac{2DC_o}{C_h}} = \sqrt{\frac{2DC_o}{CI}} = \sqrt{\frac{2 \times 5000 \times 49}{4.75 \times 0.20}} = 718$$

To avail of 3% discount facility a minimum of 1000 units is to be bought. Since the EOQ = 711 < 1000 this quantity is not eligible for discount.

Similarly, to avail of 5% discount facility a minimum of 2500 units are to be bought. Since the EOQ = 718 < 2500 this quantity is also not eligible for discount.

Under the circumstances for the best decision we must make cost comparisons before and after availing of discounts.

$$TC = DC + \frac{Q}{2}CI + \frac{D}{Q}C_0$$

By using the above cost equation and the revised optimum order quantities $Q_1 = 700,\ Q_2 = 1000,\ Q_3 = 2500$. We calculate the total costs.

The whole of the needed calculations are made in Table 13.14.

Table 13.14

Category	Number of orders	Order quantity	Average inventory	Carrying cost	Ordering cost	Purchase price	Purchase cost	Annual total cost
1	7.1428571	700	350	350	350	5.00	25,000	25,700
2	5	1000	500	485	245	4.85	24,250	24,980
3	2	2500	1250	1187.5	98	4.75	23,750	25,035.5

From the table

$$TC_1 = 25{,}700$$
$$TC_2 = 24{,}980$$
$$TC_3 = 25{,}035.5$$

Among the three values the 3% discount offer cost is the minimum. Such a decision saves Rs. 720. The 5% discount offer total cost Rs. 25,035.5 is more than Rs. 25,700. So the company will go for 3% discount offer and order 1000 units per order. Note that, here mere EOQ order of 700 units without availing of the discount will cost more.

■ Example 10

Suppose the annual demand for a certain chemical is 3000 units. Let the holding cost be 20% of the purchase cost per unit per annum and the ordering cost Rs. 250 per order. However, the unit price for the product depends upon the order quantity due to discount offer given. The following data gives the discount details. Calculate the appropriate order decision.

Table 13.15

Discount category	Order size	Unit price
1	0 to 249	Rs. 20
2	250 to 499	Rs. 19
3	500 to 749	Rs. 17
4	750 to 999	Rs. 15
5	1000 and above	Rs. 13

Excel solution

To get the ready to use excel worksheet, type the given data and the formulas needed as shown in the Table 13.16. The corresponding display result sheet is shown in Table 13.17.

Table 13.16

	A	B	C	D	E	F
1		Category				
2		1	2	3	4	5
3	D	3000	3000	3000.00	3000.00	3000.00
4	Min. for discount	-	250	500.00	750.00	1000.00
5	(C)	20	19	17.00	15.50	13.50
6	(I)	0.2	0.2	0.20	0.20	0.20
7	I (C_h)	4	3.8	3.40	3.10	2.70
8	(C_o)	250	250	250.00	250.00	250.00
9	**Optimality**					
10	(Q*)/ minimum	=SQRT((2*B3*B8/B7))	=SQRT((2*C3*C8/C7))	=SQRT((2*D3*D8/D7))	=SQRT((2*E3*E8/E7))	=SQRT((2*F3*F8/F7))
11	**Costs**					
12	1) Purchase			=D3*D5	=E3*E5	=F3*F5
13	2) Inventory			=D6*D5*D10/2	=E6*E5*E4/2	=F6*F5*F4/2
14	3) Ordering			=D3/D10*D8	=E3/E4*E8	=F3/F4*F8
15	Total annual			=SUM(D12:D14)	=SUM(E12:E14)	=SUM(F12:F14)
16	Cost saving			Nil	=D15-E15	=D15-F15

Table 13.17

	A	B	C	D	E	F
1						
2		1	2	3	4	5
3	Yearly demand (D)	3000	3000	3000	3000	3000
4	Minimum purchase for Discount	-	250	500	750	1000
5	Purchase price per unit (C)	20	19	17	15.5	13.5
6	Inventory holding proportion per unit (I)	0.2	0.2	0.2	0.2	0.2
7	Inventory holding cost per unit (C_h)	4	3.8	3.4	3.1	2.7
8	Ordering cost (C_o)	250	250	250	250	250
9	**Optimality values**					
10	Optimum inventory (Q*)/ minimum	612.37	628.28	664.21	695.608	745.356
11	**Costs**					
12	1) Purchase cost			51,000	46,500	40,500
13	2) Inventory cost			1129.2	1162.5	1350
14	3) Ordering cost			1129.2	1000	750
15	Total annual cost			53,258	48,662.5	42,600
16	Cost saving			Nil	4595.82	10,658.3

In Table 13.17 category 1, and 2, and 3 the EOQ values are greater than the respective minimum eligible levels of discount. Thus EOQ ordering will qualify for discounts automatically at the respective rates. But for category 4 and 5, the respective EOQ values are less than the corresponding minimum discount eligibility levels (685.6 <750, 745.35 <1000). So, if at all the company is interested in really availing of the discount facility, it must make cost calculations for 4 and 5 categories and compare it with the cost corresponding to EOQ value for category 3. Because of this reason no cost calculations are made for Category 1 and 2 in the table.

Thus, the company should go for 1000 units and avail of the full discounted cost of Rs. 10,658.32.

13.12 ECONOMIC ORDER QUANTITY MODELS WITH PLANNED SHORTAGES

In general inventory situations, a shortage is considered undesirable. This is because shortages may mean loss of customer's goodwill, reduction of future orders, unfavourable reduction in the market share and so on. By shortage or stock out we mean the demand situation that cannot be honoured because of non-availability of stock in the inventory. Though generally such shortages are undesirable, there are occasions that permit planned shortages. Such shortages are permitted deliberately whenever the value of the inventory cost per unit is prohibitively high. In real life one will come across this type of situation with a new car dealer. Often the specific cars that a customer wants to have may not be in stock at the time of booking, but prepared to wait till next shipment arrives. Under such circumstances all such bookings after zero stock level will be cleared on the very first day of the subsequent inventory arrival. Thus, units going into the real inventory on the receiving date will be reduced by the number of advance booking. These types of model are often called *back order* models.

Really this is an extension of the EOQ model discussed in the previous section. As in the previous setting, all inventory ordering reaches the inventory in time on the zero stock evening and available for sale in the next morning. The constant linear demand assumption facilitates all such easy calculations well in time. Under heavy inventory cost situations, going for back ordering prove to be wise on certain grounds. If S indicates the number of optimum accumulated back orders then the relevant inventory will have the following characteristics.

- On arrival day itself the back order numbering to S units will be honoured on first priority and delivered to the customers. So only Q – S in number, will enter into the actual inventory on the next morning.
- In this type of models, the inventory cycle T is split into two parts. From the day one of fresh arrivals to the stock out day all the customers are honoured without any waiting. This period is referred as t_1. From the stock out day till next new arrival is really the waiting period for customers because of no stock in the inventory. This is referred as t_2 so that $T = t_1 + t_2$ as shown in Figure 13.6.

The inventory pattern with back orders is shown in the Figure 13.6. Here negative inventory represents the number of back orders. For the inventory model with back orders we encounter a new back order cost in addition to our usual holding and ordering costs. Here back order cost includes the labour and special delivery costs directly related to the handling of back orders. Another important component of this back order cost is the goodwill that the customers had although with the company. Longer the back order time larger will be the goodwill cost.

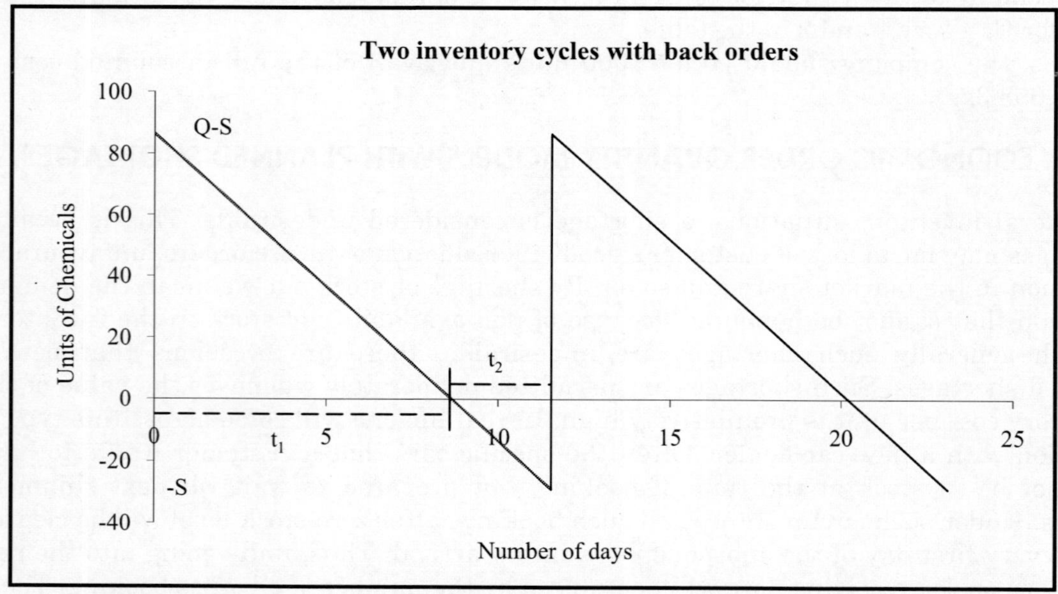

Figure 13.6

In the diagram the maximum inventory is Q – S because by assumption the back orders are cleared on the very first day of arrival itself. For t_1 days from the starting there is no back ordering and the units are delivered on demand at the constant rate specified. (If any booking is booked extra than the daily rate assumed, it is treated as back order booking.) From this point onwards for t_2 days no inventory is carried over; however back order cost is relevant for this period. Thus, the total cycle time in our model is $T = t_1 + t_2$ days. Under the defined conditions the average inventory is obtained as

$$\text{Average inventory} = \frac{1/2(Q-S)t_1 + 0t_2}{t_1 + t_2} = \frac{1/2(Q-S)t_1}{t_1 + t_2} = \frac{1/2(Q-S)t_1}{T}$$

If the demand rate per day is d as usual, we have

$$t_1 = \frac{Q-S}{d} \text{ days}$$

Further since Q units are ordered in every cycle, we have

$$T = \frac{Q}{d}$$

So the above average inventory equation may also be rewritten as

$$\text{Average inventory} = \frac{1/2(Q-S)t_1}{T} = \frac{1/2(Q-S)\left[(Q-S)/d\right]}{Q/d} = \frac{(Q-S)^2}{2Q}$$

Thus, in the above expression we express the average inventory as a function of the maximum order quantity Q and the maximum back order quantity S.

If D represents the annual demand as usual the number of orders per year is obtained as

$$\text{Number of orders per annum} = \frac{D}{Q}$$

Now it is time to develop an expression for average back orders. During t_1 period there is no back order in our problem. But in t_2 the maximum back order is S. So the average back order pertaining to this period = S/2. Now the average back order for the entire cycle is obtained as

$$\text{Average back order} = \frac{0t_1 + (S/2)t_2}{t_1 + t_2} = \frac{(S/2)t_2}{T}$$

Now the length of back order $t_2 = S/d$

$$\text{Therefore the average back order } = \frac{(S/2)t_2}{T} = \frac{(S/2)(S/d)}{Q/d} = \frac{S^2}{2Q}$$

Now let

C_h = per unit holding cost of inventory per annum.

C_o = Cost per order.

C_b = per unit holding cost of back orders per annum.

Now the total cost due to all the three items under reference is written as

TC = Average inventory × inventory holding cost per unit + number of orders in one year × order cost per order + average back order × back order cost per unit.

$$TC = \frac{(Q-S)^2}{2Q}C_h + \frac{D}{Q}C_o + \frac{S^2}{2Q}C_b$$

Now TC is a function of Q and S. So for the minimum TC, using the calculus technique the optimum order quantity Q and the back order quantity S are obtained as

$$Q = \sqrt{\frac{2DC_o}{C_h}\left(\frac{C_h + C_b}{C_b}\right)}$$

$$S = Q\left(\frac{C_h}{C_h + C_b}\right)$$

■ Example 11

The demand for a certain chemical per annum is 2000 kg. The cost of ordering is Rs. 25 per order. The purchase cost per kg of the chemical is Rs. 50. The inventory hoarding cost is 0.2 per kg per annum. The company works 200 days in a year. The lead time = 5 days.

(a) Obtain the EOQ solution for this model without allowing any back order. Also obtain the reorder level and other important parameters.

(b) The company is considering the possibility of allowing some back orders to occur. The back order cost has been estimated as Rs.30 per unit. Obtain the EOQ solution and other related parameters with this permitted shortage. Also find the cost saving due to beck ordering.

Solution (a)

Data

Annual demand = 2000 kg

Number of working days = 200

Demand per day = 2000/200 = 10 kg

Reordering cost per order C_0 = Rs. 25

Annual inventory holding rate I = 0.2

Purchase cost per kg = Rs.50

Inventory holding cost per kg per annum C_h = Rs. 10

Annul back order cost per unit C_b = Rs. 30

Economic order quantity $Q = \sqrt{\frac{2DC_o}{C_h}} = \sqrt{\frac{2DC_o}{CI}} = \sqrt{\frac{2 \times 2000 \times 25}{50 \times 0.2}} = 100$ kg

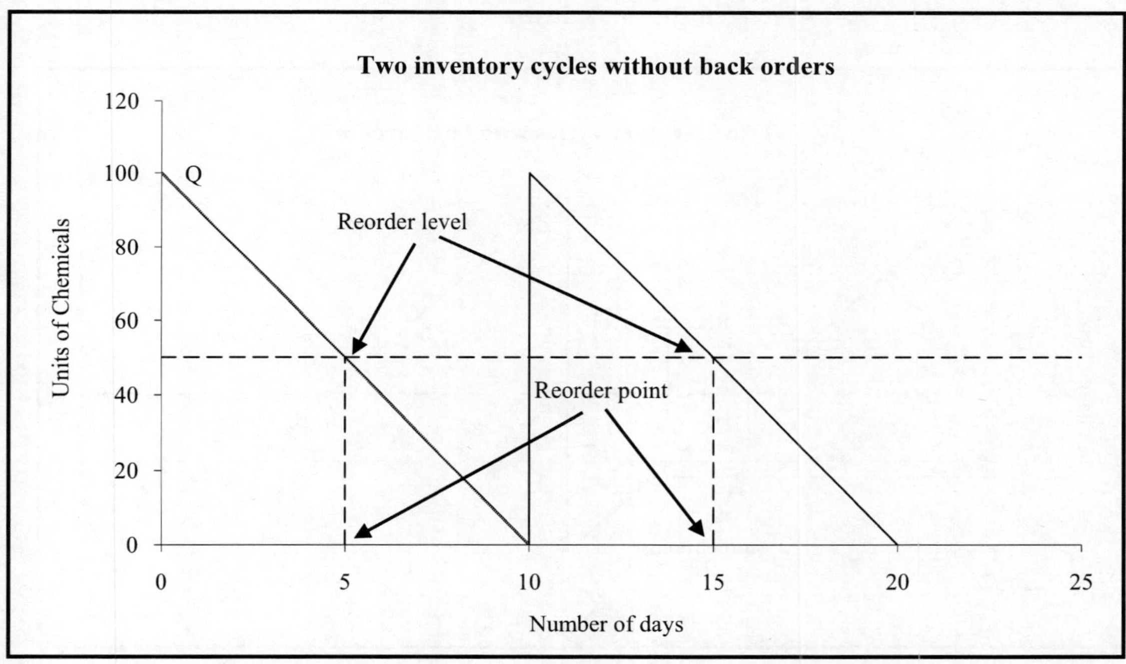

Figure 13.7

Maximum level of inventory = Q = 100 units

Number of cycle in a year given 200 working days = D/Q = 2000/100 = 20

Length of the cycle T = 200/20 = 10

Rate of demand per day = 100/10 = 2000/200 = 10

Reordering dates: 5th, 15th, 25th, 35th, 45th, 55th, 65th, 75th, 85th, 95th etc.

Reordering levels: = 5×10 = 50

Figure 13.7 shows all these parameters for two cycles.

Solution (b)

Economic Order Quantity with planned shortage $= Q = \sqrt{\frac{2DC_o}{C_h}\left(\frac{C_h + C_b}{C_b}\right)} = \sqrt{\frac{2 \times 2000 \times 25}{50 \times 0.2}\left(\frac{0.2 + 30}{30}\right)} = 115.47$ kg

Planned shortage $= S = Q\left(\frac{C_h}{C_h + C_b}\right) = 115.47\left(\frac{0.2}{0.2 + 30}\right) = 28.87$ kg

Maximum inventory = Q – S =115.47 – 28.87 = 86.60

Demand per day d = 2000/200 =10 kg

$$t_1 = \frac{Q - S}{d} = 86.6/10 = 8.66 \text{ days}$$

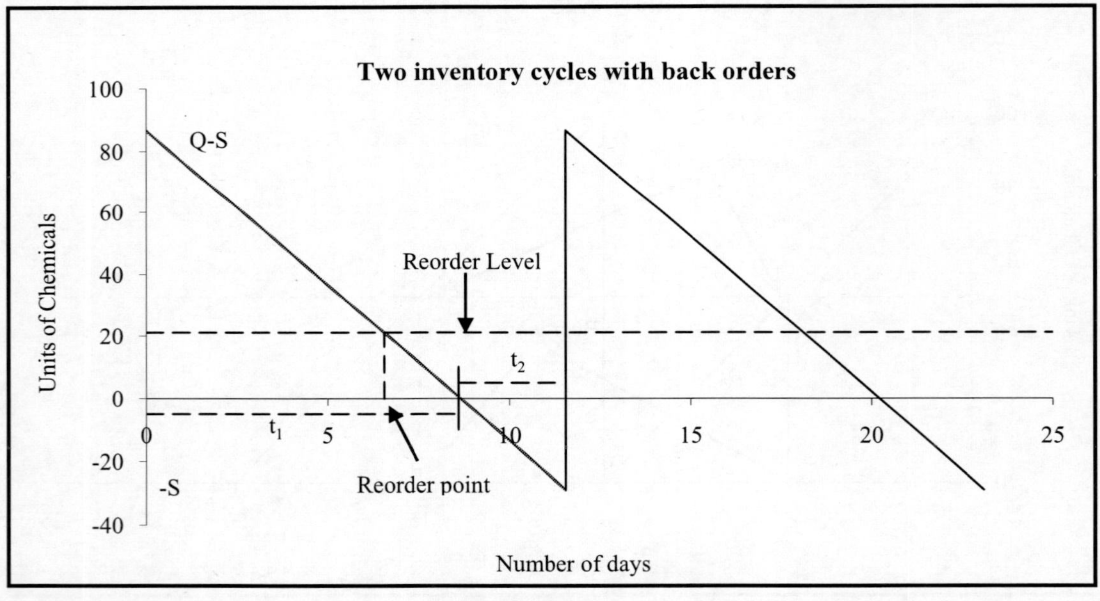

Figure 13.8

$$t_1 = \frac{S}{d} = \frac{28.87}{10} = 2.887 \text{ days}$$

Cycle time per cycle $T = \frac{200}{D/Q} = \frac{200}{2000/115.47} = 11.547$ days

Rate of demand per day d = 100/10 = 2000/200 = 10

Number of cycles in an year given 200 working days = D/Q = 2000/115.47 = 17.32

First reordering dates: 11.547 – 5 = 6.547 days

Reordering level = 50 – 28.87 = 21.13 units

Note that this reordering level + back ordering 21.13 +28.87 = 50 = 5×10 = 50

Note: *From the Figure 13.8 it is clear that on the day of booking the leftover stock 21.13 units will be enough for delivery up to the stock out period. For the remaining 2.89 day back order will be booked to the maximum 28.87 unit. On arrival 28.87 units will be cleared on the spot and the remaining 86.6 will go for the inventory.*

The EOQ without beck ordering = 100 units

The EOQ with back ordering = 115.47 units

Cost saving through back ordering = Rs. 133.97

Solution through excel worksheet

Table 13.18

	A	B	C
1	**DATA**		
2	Annual demand (D)	2000.00	
3	Ordering cost (C_O) per order	25.00	
4	Annual inventory holding rate (I)	0.20	
5	Cost per unit (C)	50.00	
6	Inventory hoarding cost C_h	10.00	
7	Annual back order cost per unit	30.00	
8	Working days per year	200.00	
9	Lead time	5.00	
10	**Optimality parameters**	**Without back order**	**With back order**
11	Economic order quantity Q*	=SQRT((2*B2*B3)/(B6))	=SQRT((2*B2*B3)/B6)*SQRT((B6+B7)/B7)
12	Back order quantity S*	0	=(C11*B6)/(B6+B7)
13	Maximum inventory Q*-S*	=B11-B12	=C11-C12
14	Average inventory level	=B13/2	=(C13*C13)/(2*C11)
15	Annual inventory holding cost	=(B11/2)*B6	=C14*B6
16	Number of orders per year	=B2/B11	=B2/C11
17	Annual ordering cost	=B16*B3	=C16*B3
18	Average back orders	0	=(C12*C12)/(2*C11)
19	Annual back order cost	0	=C18*B7
20	Total annual cost	=B15+B17	=C19+C17+C15
21	Demand per day d	=B2/B8	=B2/B8
22	No. of days with inventory t_1	=(B11-B12)/B21	=C13/B21
23	No. of days with back order t_2	=B24-B22	=C12/B21
24	Total cycle time T	=B8/B16	=C22+C23
25	Reorder level	=B9*B21	=B25-C12
27	Cost saving due to back ordering		=B20-C20

To get the excel solution for back order problem type the data pertaining to the problem from cell B2 to B9. To get the optimal parameter values before and after the back order type the formulas in the respective cells as shown in Table 13.18. The corresponding final result display is shown in Table 13.19.

Table 13.19

	A	B	C
1	**DATA**		
2	Annual demand (D)	2000	
3	Ordering cost (C_O) per order	25	
4	Annual inventory holding rate (I)	0.2	
5	Cost per unit (C)	50	
6	Inventory hoarding cost per unit per annum C_h	10	
7	Annual back order cost per unit	30	
8	Working days per year	200	
9	Lead time	5	
10	**Optimality parameters**	**Without back order**	**With back order**
11	Economic order quantity Q*	100	115.47
12	Back order quantity S*	0	28.87
13	Maximum inventory Q*-S*	100	86.60
14	Average inventory level	50	32.48
15	Annual inventory holding cost	500	324.76
16	Number of orders per year	20	17.32
17	Annual ordering cost	500	433.01
18	Average back orders	0	3.61
19	Annual back order cost	0	108.25
20	Total annual cost	1000	866.03
21	Demand per day d	10	10
22	Number of days with inventory t_1	10	8.66
23	number of days with back order t_2	0	2.89
24	Total cycle time T	10	11.55
25	Reorder level	50	21.13
27	Cost saving due to back ordering		133.97

■ Example 12

The Western Electronics produces 2000 TV sets for sale per annum. For the production it buys a certain component in the market. It costs Rs. 10 per unit. The inventory order cost is Rs. 150 per order. The inventory holding cost is Rs. 2.40 per unit per year. The shortage cost is Rs. 1.60 per unit. The lead time needed for delivery is 15 days. Obtain the back order cost savings.

Solution

Data

Annual demand = 2000 units

Number of working days = 250

Demand per day = 2000/250 = 8

Reordering cost per order C_0 = Rs. 150

Inventory holding cost per unit per annum C_h = Rs. 2.40

Annul back order cost per unit C_b = Rs. 1.60

Lead time 15 days

Excel Solution

The excel solution shown in Table 13.20. This time the reordering will take place after the stock being exhausted. Accordingly, the back order is placed only after the stock reaches –354.34 units. In other words this is to say that the reorder will be booked only when the back order alone reaches 354.34. If booking is done at this level for 790.57 then by the time of arrival the effective optimum back order will be 474.34 units. After honouring the entire bulk of back orders the leftover optimum inventory will be 316.23 as expected. The expected cost saving due to back ordering is Rs. 441.05.

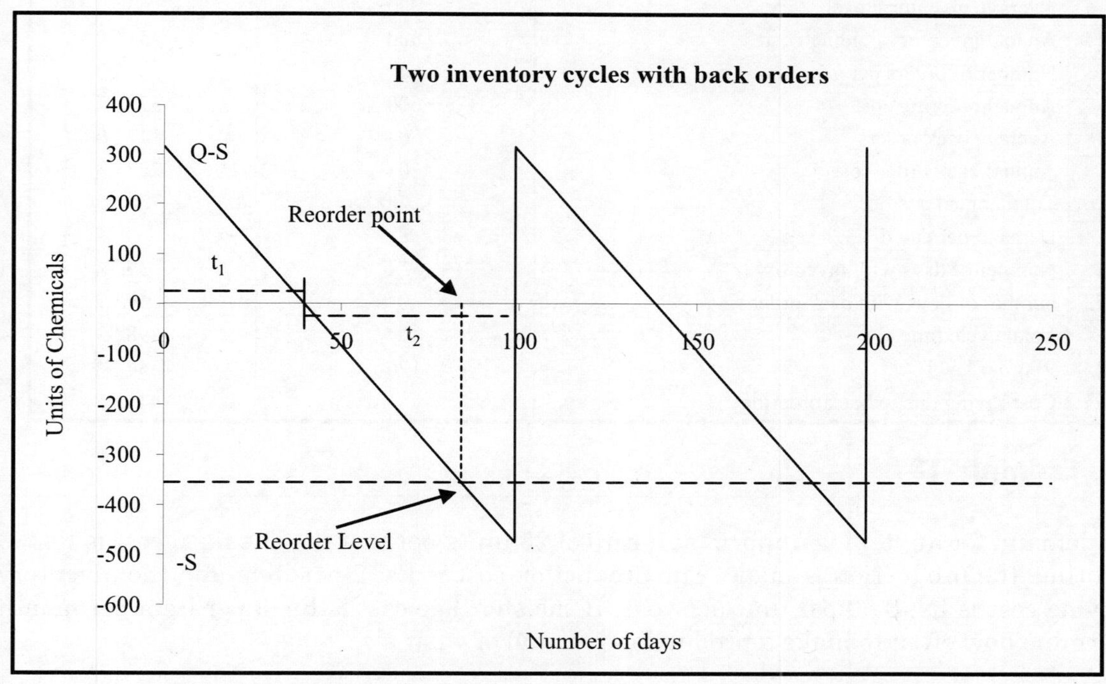

Figure 13.9

Table 13.20

	A	B	C
1	**DATA**		
2	Annual demand (D)	2000	
3	Ordering cost (C_O) per order	150	
4	Annual inventory holding rate (I)	-	
5	Cost per unit (C)	-	
6	Inventory hoarding cost per unit per annum C_h	2.4	
7	Annual back order cost per unit	1.6	
8	Working days per year	250	
9	Lead time	15	
10	**Optimality parameters**	**Without back order**	**With back order**
11	Economic order quantity Q*	500	790.57
12	Optimum back order quantity S*	0	474.34
13	Maximum inventory Q*-S*	500	316.23
14	Average inventory level	250	63.25
15	Annual inventory holding cost	600	151.79
16	Number of orders per year	4	2.53
17	Annual ordering cost	600	379.47
18	Average back orders	0	142.30
19	Annual back order cost	0	227.68
20	Total annual cost	1200	758.95
21	Demand per day d	8	8
22	Number of days with inventory t_1	62.5	39.53
23	number of days with back order t_2	0	59.29
24	Total cycle time T	62.5	98.82
25	Reorder level	120	-354.34
27	Cost saving due to back ordering		441.05

■ Example 13

The demand for an item is uniform at a rate of 25 units per month. The fixed cost is Rs. 30 each time the production is made. The production cost is Rs. 2 per item and the inventory carrying cost is Rs. 0.50 per unit per year. If the shortage cost is Rs. 3 per item per month determine how often to make a production run and of what size?

Solution

Data

$D = 25$ units

$C_h =$ Rs. 0.50

$C_s =$ Rs. 3 per item per month

Cost of production = Rs. 2 per unit

Production set-up cost = Rs. 30 per set-up

$$Q^* = \sqrt{\frac{2DC_o}{C_h}\left(\frac{C_h + C_b}{C_b}\right)} = \sqrt{\frac{2 \times 25 \times 30}{0.50}\left(\frac{0.5 + 3}{3}\right)} = 60$$

Optimum time =EOQ/D = 60/25 = 2.4 months.

13.13 ECONOMIC ORDER QUANTITY MODEL WITH GRADUAL ADDITION TO INVENTORY OR ECONOMIC PRODUCTION RUN LOT SIZE MODEL

This model is an application of EOQ model to production area. So it is assumed that the product is produced in the company itself and subsequently sold in the market. Both the production rate p and demand rate d are constants as usual. However, the arrival of goods to the inventory is not taking place at one stroke as in normal EOQ model. Instead, in this model the units are supplied to the inventory at a constant rate over entire production run starting from day one of to the last day run. For example, if in a production run of 10 days produces 50 units then, the inventory will receive 5 units per day from day one today ten totalling to 50 units. Thus, once order is placed for 50 units, the production begins, at a constant rate of 5 units per day and will be supplied for the whole production run totalling to 50 units. So from day one to day ten the inventory will increase at a constant rate given by the difference between the production rate and demand rate. However, once the production is stopped the inventory reaches its maximum and starts declining at the demand rate and becomes zero as in the case of EOQ model. Again another order is placed for 50 units another production run starts and ends after ten days. Here the total output 50 per production run is called the lot size.

This model is valid if and only if the production rate per day is more than the demand rate per day. If the demand rate is 2 units per day and production rate is 5 units per day then clearly the inventory will increase by three units (5-2) per day for the entire production of 10 days after fulfilling daily demand of 2 units. However, when the product run ends the inventory will have a maximum stock of 30 (3 × 10) units. Thereafter at the demand rate of 2 per day this stock will last for another 15 days (30/2). Given a lead time of 5 days, order will be placed for another lot of 50 units as soon as the inventory reaches 10 units (20th day) sufficient enough for another five days. During the lead time production facility will be set-up so the actual production starts on 25th day. Note that our lot size is 50 and our maximum inventory is only 30. In general if Q (50) is the lot size, the inventory will never have this maximum amount. The maximum inventory will always be less than order quantity Q. If the company works for 200 days in a year then there will be 200/25 = 8 inventory cycles altogether with the duration of 25 days per cycle. Remember, that the production run is only 10 days though the sales take place for the entire cycle period of 25 days. Total production = 10 × 8 × 5 = 400 units during the 8 production runs in a year. Similarly, the total sale = 200 × 2 = 400 for the whole year.

Generally in these models, an important added component is set-up cost in the place of ordering cost. To start the production run company will have to prepare the production facilities. Similarly, when production ends dismantling equipment takes place. Both the activities involve cost. Frequent production runs with less production per run will no doubt bring down the inventory holding cost but at the same time increase the set-up cost. Similarly, too few production runs with more production per run will bring down the set-up cost but definitely increase the inventory holding cost. The economic lot size (ELS) solution outlined in this section strikes a balance between these two conflicting situations and finds the optimum lot size. One the optimum lot size is fixed the remaining issues are settled automatically as in the case of simple EOQ model. Figure 13.10 illustrates two inventory cycles with all the characteristics explained.

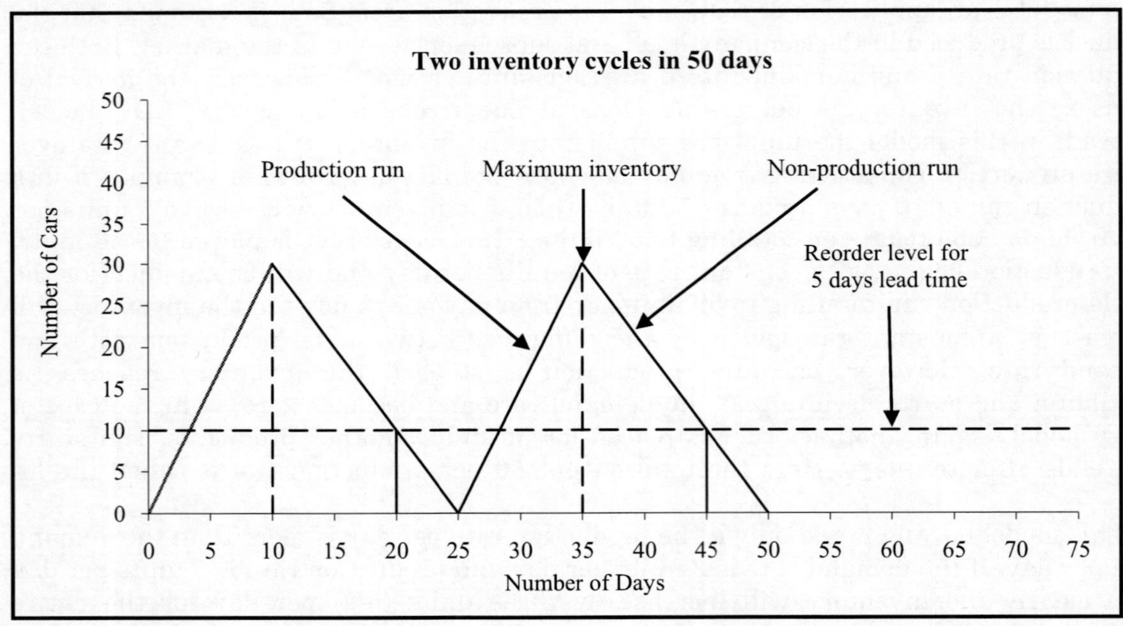

Figure 13.10

Assumptions

1. Annual demand, carrying cost, set-up, otherwise called, ordering cost are given constants as in simple EOQ model.
2. No safety stock is assumed.
3. Production takes place at p per day uniformly.
4. Demand is affected at d per day uniformly.
5. Volume discount does not exist.
6. There are no stock outs.
7. Supply rate p is greater than the demand rate d.

Computation of Maximum Inventory

Let, t = number of days for a production run.

Now during the production run, the rate per day inventory accumulation = p – d. So in the entire production run of t days the total maximum inventory will be = (p – d) t. If Q is the lot size per rum, then Q = pt. So the length of the production run t = Q/p days.

Therefore the maximum inventory = (p – d) t = (p – d) Q/p $= \left(1-\dfrac{d}{p}\right)Q$.

Average inventory $= \dfrac{1}{2}\left(1-\dfrac{d}{p}\right)Q$

Annual holding cost = Average inventory × Annual holding cost per unit $= \dfrac{1}{2}\left(1-\dfrac{d}{p}\right)QC_h$

If D is the annual demand and C_o (ordering cost) is the set-up cost per production run then

Annual set-up cost = number of production runs per year × set-up cost $= \dfrac{D}{Q}C_0$

Thus the total annual cost $= \dfrac{1}{2}\left(1-\dfrac{d}{p}\right)QC_b + \dfrac{D}{Q}C_0$

Suppose the sales take place for all the 200 working days in a year then the annual production D = 200 × d so that d = D/200.

Similarly, if P stands for the annual production corresponding to 200 working days then the annual production P = 200p so that p = P/200.

Therefore $\dfrac{d}{p} = \dfrac{D/200}{P/200} = \dfrac{D}{P}$

So we can rewrite the above total cost function as $TC = \dfrac{1}{2}\left(1-\dfrac{D}{P}\right)QC_b + \dfrac{D}{Q}C_o$

From elementary calculus for a minimum of TC we know that

$$\text{(i)} \frac{d}{dQ}(TC) = 0$$

$$\text{(ii)} \frac{d^2}{dQ^2}(TC) > 0$$

Therefore

$$\frac{d}{dQ}(TC) = \frac{1}{2}\left(1-\frac{D}{P}\right)C_b - \frac{D}{Q^2}C_o = 0$$

$$\frac{D}{Q^2}C_o = \frac{1}{2}\left(1-\frac{D}{P}\right)C_b$$

$$EOQ = Q = \sqrt{\frac{2DC_o}{(1-D/P)C_b}}$$

Since $\frac{d}{p} = \frac{D/200}{P/200} = \frac{D}{P}$ the EOQ value may also be obtained by using the formula:

$$EOQ = Q = \sqrt{\frac{2DC_o}{(1 - d/p)C_b}}$$

Other impotent results:

1. Optimum number of production runs in a year = N = D/Q
2. Optimum number of days in a production cycle = T = Q/d
3. Optimum number of days in a lot size production run t = Q/p
4. Maximum inventory = (p – d)t
5. Optimum total minimum inventory cost $= \frac{Q}{2}\left(\frac{p-d}{p}\right)C_h + \frac{D}{Q}C_0$

■ Example 14

A car company works for 200 days in a year and has the capacity to produce 5 units of a certain type of car per day. Their total sale for the 200 working days in a year is 400 units at the rate of 2 per day. The annual inventory holding cost is 1.667 per unit per year and the setup cost of Rs. 3.125 per run. Find the following parameters.

i) The optimum lot size

ii) The optimum number of production runs

iii) The optimum number of days in a production run

iv) The optimum number of days in an inventory cycle

v) The optimum inventory level

vi) The optimum reorder level give a lead time of 5 days

vii) Optimum inventory holding cost

viii) Optimum order cost.

Solution

i) The optimum lot size

$$Q = \sqrt{\frac{2DC_o}{(1 - d/p)C_b}} = \sqrt{\frac{2 \times 400 \times 3.125}{(1 - 2/5) \times 1.6667}} = 50$$

ii) The optimum number of production runs

N = D/Q = 400/50 = 8

iii) The optimum number of days in a production run

$t = Q/p = 50/5 = 10$ days

iv) The optimum number of days in an inventory cycle

$T = Q/d = 50/2 = 25$ days

v) The optimum inventory level

$(p - d)\ t = (5 - 2) \times 10 = 30$ units

vi) The optimum reorder level with a given lead time of 5 days

Lead time × demand rate = 5 × 2 =10 units

vii) Optimum inventory holding cost

Average inventory × holding cost per unit = (30/2) × 1.6667 = Rs. 25

viii) Optimum order cost = order cost × number of orders = 3.125 × 8 = Rs. 25

Excel Solution

Table 13.21

	A	B
1	**Data**	
2	Annual demand (D)	400.000
3	Ordering cost (C_0)	3.125
4	Purchase cost per unit (C)	-
5	Annual inventory holding rate in (I)	-
6	Inventory holding cost (C_h=IC)	1.667
7	Number of working t days in a year	200.000
	Production per day (p)	5.000
	Demand per day (d)	2.000
8	Lead time	5.000
9	**Optimum inventory policy parameters**	
10	Economic order quantity (Q)	50.000
11	Number of orders per year (D/Q)	8.000
12	Average inventory (Q/2)	25.000
13	Annual Inventory holding cost (Q/2(p-d)/p*C_h)	25.000
14	Annual ordering cost	25.000
15	Maximum inventory level (Q)	30.000
16	Cycle time (T)	25.000
17	Demand per day (d)	2.000
18	First reorder point	20.000
19	Reorder level	10.000

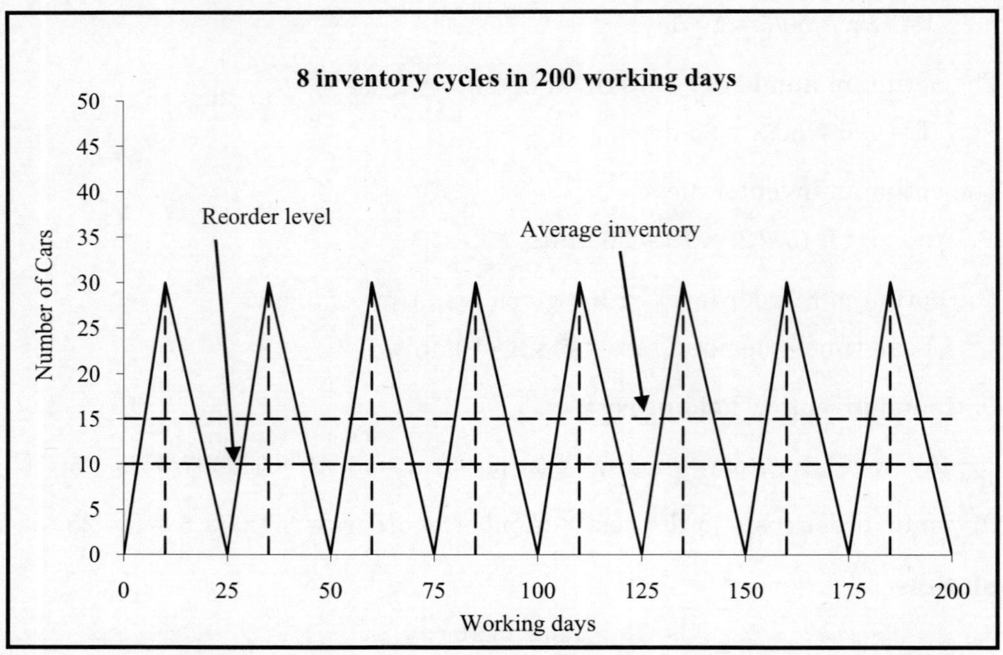

Figure 13.11

Example 15

XYZ company uses 10,000 units of a certain bulb per year. Each bulb costs Rs. 32. The set-up cost per production run is estimated as Rs. 55 per run. The inventory holding cost is obtained as 12.5 percent of the value of the inventory. Replenishment rate is uniform at 120 bulbs per day. The company is working for 250 days in a year.

Calculate

i) Optimal lot size order quantity

ii) Total optimum inventory cost

iii) Optimum number of set-ups.

Solution

i) Optimal lot size order quantity

$$Q = \sqrt{\frac{2DC_o}{(1-d/p)C_b}} = \sqrt{\frac{2\times 1000\times 55}{(1-40/120)\times 4}} = 642.26$$

ii) Total optimum inventory cost

$$\frac{Q}{2}\left(\frac{p-d}{p}\right)C_h + \frac{D}{Q}C_0 = \frac{642.26}{2}\left(\frac{120-40}{120}\right)\times 4 + \frac{10,000}{642.26}\times 55 = 1713.15$$

iii) Optimum number of set-ups.

$N = D/Q = 10,000/642.26 = 15.57$

Excel solution

Table 13.22

	A	B
1	**Data**	
2	Annual demand (D)	10000.000
3	Ordering cost (C_0)	55.000
4	Purchase cost per unit (C)	32.000
5	Annual inventory holding rate in (I)	0.125
6	Inventory holding cost (C_h=IC)	4.000
7	Number of working t days in a year	250.000
	Production per day (p)	120.000
	Demand per day (d)	40.000
8	Lead time	5.000
9	**Optimum inventory policy parameters**	
10	Economic order quantity (Q)	642.262
11	Number of orders per year (D/Q)	15.570
12	Average inventory (Q/2)	321.131
13	Annual inventory holding cost (Q/2(p-d)/p*C_h)	856.349
14	Annual ordering cost	856.349
15	Maximum inventory level (Q)	428.174
16	Cycle time (T)	16.057
17	Demand per day (d)	40.000
18	First reorder point	11.057
19	Reorder level	200.000

■ Example 16

The annual demand of a product is 1,00,000 units, the rate of production is 2,00,000 per year. The set-up cost per production run is Rs. 500 and variable production cost of each item is Rs. 10. The annual holding cost per unit is 20% of its value. Find the optimum production lot size and the length of the production run.

Solution

Data

$D = 1200$ units

$P = 2400$ units

$C_h = 10 \times 0.2 = \text{Rs. } 2.00$

$C_o = 500$

$$Q = \sqrt{\frac{2DC_o}{(1 - D/P)C_b}} = \sqrt{\frac{2 \times 1,00,000 \times 500}{(1 - 1,00,000/2,00,000) \times 2}} = 10,000$$

The length of the production run = Q/P = 10,000/2,00,000 = 0.05 year (0.6 month).

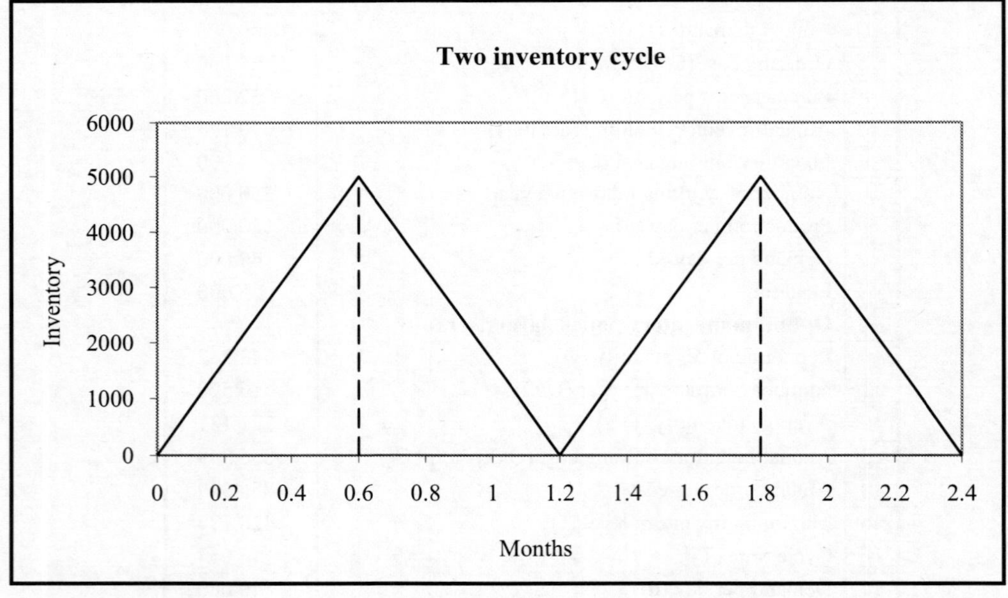

Figure 13.12

■ Example 17

A contractor has to supply 10,000 paper cones per day to a textile unit. He finds that when he starts a production run, he can produce 25,000 paper cones per day. The cost of holding a paper cone in stock for one year is 2 paise and set-up cost of production run is Rs.18. How frequently should the production run be made?

Solution

Let us assume that there are 300 working days in a year.

$D = 300 \times 10{,}000 = 10{,}00{,}000$ units

C_o = Rs. 16

C_h = 0.02 per unit per year

p = 25,000 units

d = 10,000 per day

$$Q = \sqrt{\frac{2DC_o}{(1 - d/p)C_b}} = \sqrt{\frac{2 \times 30{,}00{,}000 \times 500}{(1 - 10{,}000/25{,}000) \times 0.02}} = 95{,}000 \text{ units}$$

Number of days in a production cycle = EOQ/d = 95,000/25,000 = 3.8 days.

13.14 SINGLE PERIOD INVENTORY MODEL WITH PROBABILISTIC DEMAND

Till now we have discussed the inventory models based on the assumption that the demand rate is constant and deterministic. The EOQ was the only variable to be determined in such models. All other related values like demand rate, reorder point, reorder quantity were obtained by using obtained EOQ value. In this section we propose to discuss the probabilistic demand rate determination models. As a first step we consider only single period inventory model for discussion. In single period inventory model only one order is placed for period under reference. At the end of the period the unsold stocks will be sold for a salvage value and cannot be transferred to the next period for sale. Seasonal clothing like winter coat is good example for this type of models. Such items are sold at the end of the period as clearance sale. Newspaper sale could be considered another good example. The newspaper ordered for a particular day is not sold on that particular day, the sale of the same on the next day has no meaning. Since we order only once for the period, the only inventory decision we must make is how much of the product to be ordered at the beginning of the period. In most of the single period inventory models we will not know the exact demand. Even the forecasted demand will have wide variations. Under such circumstances we need information about the probabilities associated with the various demand values during lead time.

Let us consider a single period inventory model to find *how much to order* decision of a particular Bata shoe specially designed for summer months by a retail outlet in Bangalore. Since it is specially designed current summer it cannot be sold for the next summer. The retail buyer for such a Bata shoe buys it for Rs. 40 and sells it for Rs. 60 during the summer itself. The surplus shoes if any he will sell them at clearance sales counter for Rs. 30 during the last week of the summer. Here the real problem is to find out the right optimum quantity of shoes to be ordered. The quantity parched is in excess even by one pair he will have to sell it in the clearance sale at Rs. 30 with lose of Rs. 10 (40–30). On the other hand if there is a shortage by one pair, the retailer will lose the opportunity of making a profit of Rs. 20 (60–40). On either account he loses. The correct decision thus will avoid such uncomfortable situation. Such a dilemma is mainly due to the non-availability data pertaining to the demand for the said season. To overcome this problem let us assume a uniform probabilistic distribution for the demand pattern as shown in Figure 13.13.

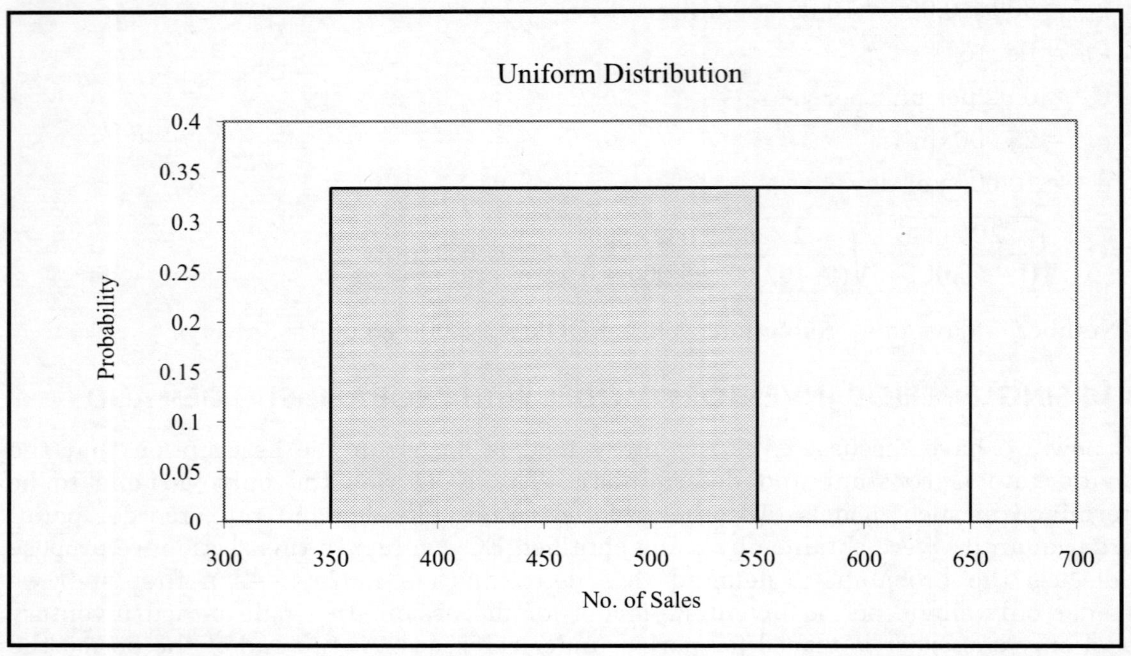

Figure 13.13

Accordingly, the seasonal demand could vary from 350 to 650 pairs with the average expected demand of 500 units.

Marginal Analysis

The incremental analysis to single period inventory model addresses the question *how much to order* decision by comparing the loss of ordering one unit less with the loss of ordering one unit more with the correct order cost. Let C_o be the loss due to over-estimating the demand by one more unit and C_u be the loss due to under estimating the demand by one unit. In our illustrative case thus C_o = Rs. 10 and C_u = Rs. 20.

Till now the exact level of demand is not known but the probabilistic demand pattern is assumed to be described by a uniform distribution as shown in Figure 13.13. Let us assume the retailer expects the average demand of 500 units will prevail during the next season. If this is the correct optimum decision then at the margin the loss due to not ordering one unit extra must be equal to the loss due to under order by one unit.

Order Quantity	Loss due to	Amount of Loss	Probability
Q = 501	Over-estimated demand	C_o = Rs.40–Rs.30 = Rs.10	P(demand ≤ 500)
Q = 500	Under-estimated demand	C_u = Rs.60–Rs.40 = Rs.20	P(demand > 500)

Since 500 units is assumed to be the demand, from the uniform probability distribution shown above clearly the probability of having more than 500 units demand = 0.5. Similarly, the probability of having the demand less than 500 units is also = 0.5.

Having known the respective probabilities of over-estimation and under-estimation we calculate the expected loss under both the situations as follows.

EL (Q = 501) = $C_o \times$ P(demand $\leq$500) = Rs. 10 × 0.5 = Rs. 5

EL (Q = 500) = $C_u \times$ P(demand >500) = Rs. 20 × 0.5 = Rs. 10

Given the situation the retailer will choose 501 units as the order quantity because the loss is least (Rs. 5 < Rs. 10) in this case. This is not the end of the story under this incremental approach. As next step we must assume a demand of 501 units and repeat this analysis for 502 units. This procedure should continue till the expected loss due to under-estimation is equal to the loss under over-estimation. If Q* is the optimum order quantity then clearly at this level.

EL (Q* +1) = EL (Q*)

Now we know that EL (Q* +1) = $C_o \times$ P(demand $\leq$ Q*)

and = EL (Q*)= $C_u \times$ P(demand >Q*)

So, $C_o \times$ P(demand <= Q*) = $C_u \times$ P(demand >Q*)

From the basic probability theory we know that

P(demand $\leq$Q*) + P(demand >Q*) =1

So P(demand $\leq$Q*)=1 – P(demand >Q*)

So, $c_o \times$ P(demand $\leq$ Q*) = $C_u \times$ 1 – P(demand $\leq$Q*)

Solving for P(demand $\leq$ Q*) we have

$$P(\text{demand} \leq Q^*) = \frac{C_u}{C_u + C_o}$$

Thus for the illustrative problem

$$P(\text{demand} \leq Q^*) = \frac{C_u}{C_u + C_o} = \frac{20}{20+10} = \frac{2}{3}$$

Now from the uniform distribution, the needed quantity Q* is obtained by moving 2/3 units from 350 to 650. So the correct optimum inventory order quantity Q* = 350 +200 = 550 units.

Example 18

Edward Chemicals retailing shop buys a certain chemical from Kumar Chemicals. It buys at Rs.15 per unit and sells it at Rs. 20 per unit. If the company purchases from Kumar Chemicals is less than the demand, to overcome the shortage a higher quality substitute is bought from Ram Chemicals at Rs. 19 per unit and sells it to the customers at the same rate of Rs. 20 and thus makes a profit of Re. 1. On the other hand, if its purchase from Kumar Chemicals is more than the need then the excess supply is reprocessed and sold for only Rs. 5 and hence makes a loss of Rs. 10 per unit. Out of past experience Edward company believes that the demand for its product follow a normal distribution with mean μ = 1000 units and standard deviation σ = 100 units. Obtain a single period optimum inventory order quantity.

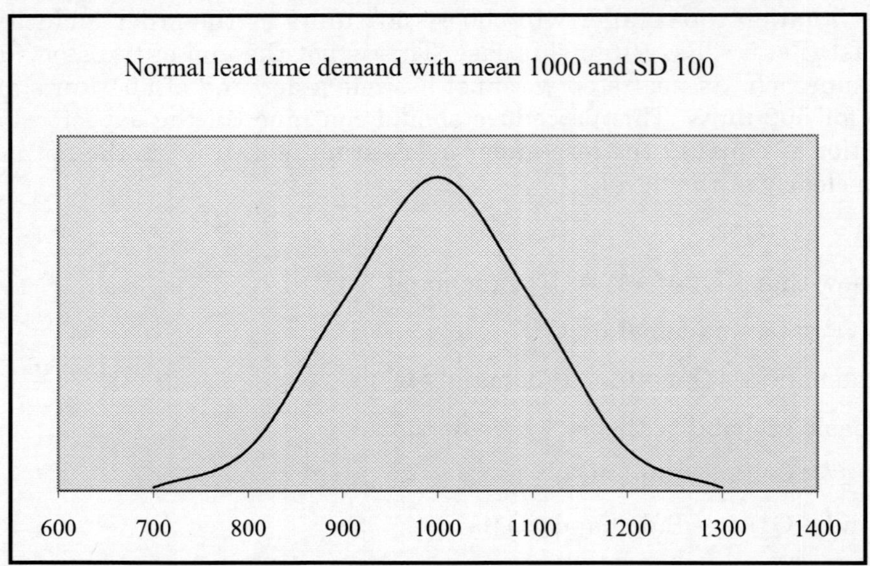

Figure 13.14

Solution

Let us first obtain the cost of under-estimation and over-estimation per unit of the said chemical purchased from Kumar Chemicals. Purchase from Kumar Chemicals cost only Rs. 15 per unit. The purchase of substitute to satisfy the excess demand from Ram Chemicals cost him Rs. 19. Thus the cost of under-estimating the demand is C_u = Rs. 4 (Rs. 19 – Rs. 15) per unit. Similarly, in the case of over-estimation the excess unit after reprocessing is sold only at Rs. 5 per unit. The loss due to this account is C_o = (Rs. 15 – Rs. 5) Rs. 10 per unit.

Thus the optimum quantity is obtained by using the formula as shown below:

$$P(\text{demand} \le Q^*) = \frac{C_u}{C_u + C_o} = \frac{4}{4+10} = 0.2857$$

Now from the normal distribution the area left of

$$Z = -0.57 = -0.50000 - 0.2143 = -0.2857.$$

Thus in original units the corresponding order quantity is obtained as

$$Q^* = 1000 - 0.57 \times 100 = 943 \text{ units}$$

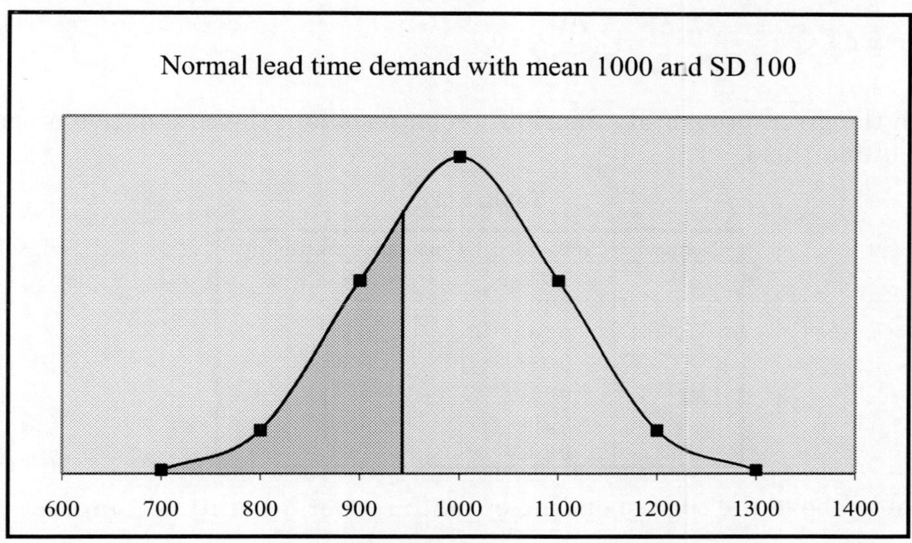

Figure 13.15

Thus with the assumed normal distribution Edward company should order for 943 units of the chemical in anticipation of the same demand to prevail. Note that in this case the under-estimation cost is less than the over-estimation cost. Accordingly, he is willing to risk a higher probability of under-estimation and hence a higher probability of stock out. If fact the probability of having a surplus purchase is 0.2857 and that of deficit purchase is 1.0000 – 0.2758 = 0.7143.

Example 19

For a TV dealer the cost of holding a TV in stock for a week is Rs. 30. The shortage cost per unit is Rs. 70. The probability distribution of the weekly demand is given by the following probability distribution.

Weekly Sales:	0	1	2	3	4	5	6
Probability:	0.05	0.10	0.20	0.25	0.20	0.15	0.05

How many units per week should the dealer order?

Solution

a) Marginal analysis method

The holding cost arises due to over-ordering. So $C_o = 30$. The shortage is due to under ordering. Thus $C_u = 70$.

The optimum quantity is obtained by using the formula as shown below:

$$P(\text{demand} \leq Q^*) = \frac{C_u}{C_u + C_o} = \frac{70}{70+30} = 0.7$$

Now from the given probability distribution let us obtain the cumulative probabilities as shown in Table 13.23.

Table 13.23

Demand	Probability	Cumulative probability
0	0.05	0.05
1	0.10	0.15
2	0.20	0.35
3	0.25	0.60
4	0.10	0.70
5	0.20	0.90
6	0.10	1.00

From the table above the corresponding optimum order quantity is 4 units.

a) Matrix method:

Since the weekly demand varies from 0 to 6 units the cost matrix for each strategy of holding 0 units to 6 units is worked out in Table 13.24. Once this is done the expected cost for each category of order is worked out and reported in the last row respective to each column.

Table 13.24

Demand	Probability	Order strategy						
		0	1	2	3	4	5	6
0	0.05	0	30	60	90	120	150	180
1	0.10	70	0	30	60	90	120	150
2	0.20	140	70	0	30	60	90	120
3	0.25	210	140	70	0	30	60	90
4	0.10	280	210	140	70	0	30	60
5	0.20	350	280	210	140	70	0	30
6	0.10	420	350	280	210	140	70	0
Expected cost =		227.5	162.5	107.5	72.5	62.5	62.5	82.5

In Table 13.24 since Rs. 62.5 is the least cost the corresponding weekly ordering could be either 4 units or 5 units.

13.15 MULTIPLE PERIOD ORDER QUANTITY MODELS WITH PROBABILISTIC DEMAND

a) Model which does not honour back order

In the previous section we considered only the once for all ordering situation with probabilistic demand pattern. In this section we propose to discuss the multi-period ordering situation with probabilistic demand. This time again whenever the inventory reaches a predetermined level, reorder is placed for Q units. However, the issues like the time by which reorder point will be reached, the time between orders, and the delivery time cannot be determined in advance. They all depend upon probabilistic demand pattern. The following diagram illustrates one such case.

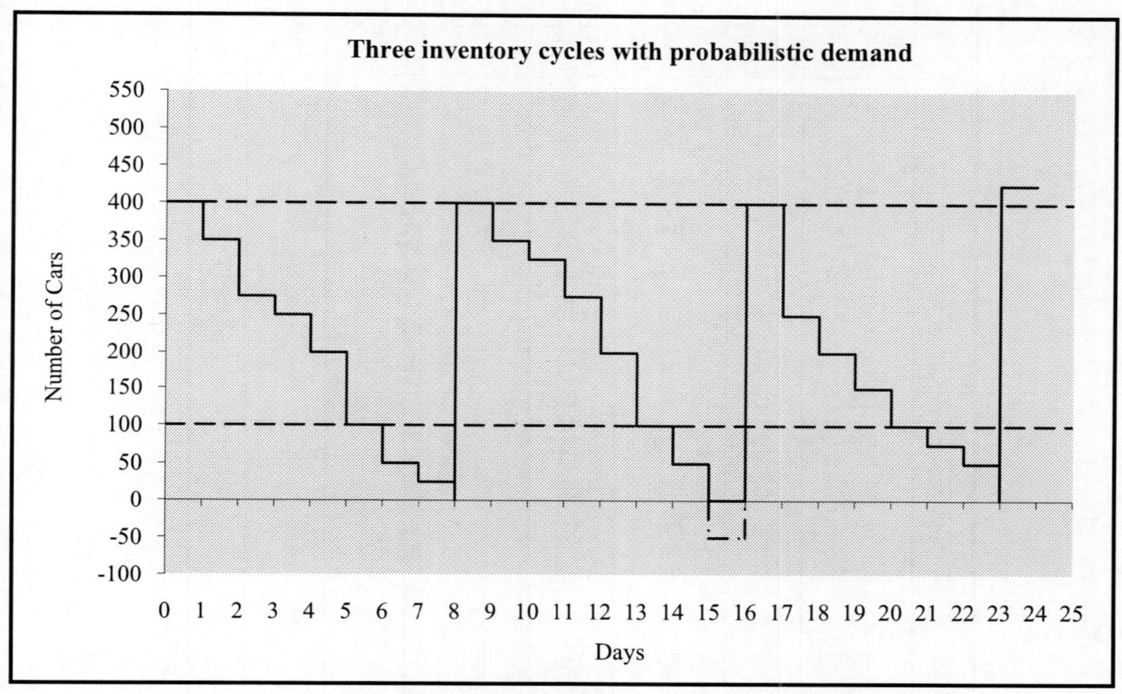

Figure 13.16

Initially there were 400 cars in the inventory. On the first day 50 cars were sold leading to a stock of 350 by the first day evening. On second day 75 cars were sold and the inventory came down to 275 units. On third day only 25 cars were sold leading to an inventory stock of 250 units. This way the day by day inventory position is shown in the graph and associated Table 13.25. On fifth day evening the inventory stock has come down to the reorder level of 100 units. So the order is booked for 400 cars immediately. On eighth day evening the stock has come down zero luckily the ordered quantity of 400 units come to the inventory on the same day and the inventory went up 400 units again. And the second cycle starts thereafter without any further problem.

On the evening of 13th day again the inventory reaches the reorder level of 100 units and the order for 400 units is placed immediately. The leftover stock of 100 units last only for 14th and 15th day sale. For 16th day sale no units are available in the stock and hence stock out occurs since stock outs are not allotted the customers visited on 16th were simply told no stock. On 16th evening the stock came and the inventory went up by 400 units and the next cycle starts as usual.

Table 13.25

Trail	Inventory at the beginning of the day	Daily demand	Inventory at the end of the day	Inventory cost	Lost sales	Shortage penalty	Stock received	Quantity ordered	Order cost
1	400	50	350	1750					
2	350	75	275	1375					
3	275	25	250	1250					
4	250	50	200	1000					
5	200	100	100	500				400	100
6	100	50	50	250					
7	50	25	25	125					
8	25	25	0	0			400		
9	400	50	350	1750					
10	350	25	325	1625					
11	325	50	275	1375					
12	275	75	200	1000					
13	200	100	100	500				400	100
14	100	50	50	250					
15	50	50	0	0					
16	0	50	0	0		5000	400		
17	400	150	250	1250					
18	250	50	200	1000					
19	200	50	150	750					
20	150	50	100	500				400	100
21	100	25	75	375					
22	75	25	50	250					
23	50	25	25	125			400		
24	425		Total	17,000		5000			300
				Grand Total Cost		22,300			

Once again the stock came down to 100 units on the evening of 20th day and immediate order has been placed for 400 units. But this time during the lead time of three days the company was able to sell only 75 units at the rate of 25 units per day. Thus on 23rd evening 25 units of unsolved units were there in the inventory. On the same evening 400 units new arrival also comes in and thus the total inventory becomes 425 units in total. If the order cost is Rs. 100 per order and holding cost is Rs. 5 per unit per day let us generate the cost data for 23 days as shown Table 13.25

Note: *Note that this is need not by the optimum minimum cost. The revision of reader quantity / and lead time will give a different picture altogether.*

b) Model which permits back order and honour

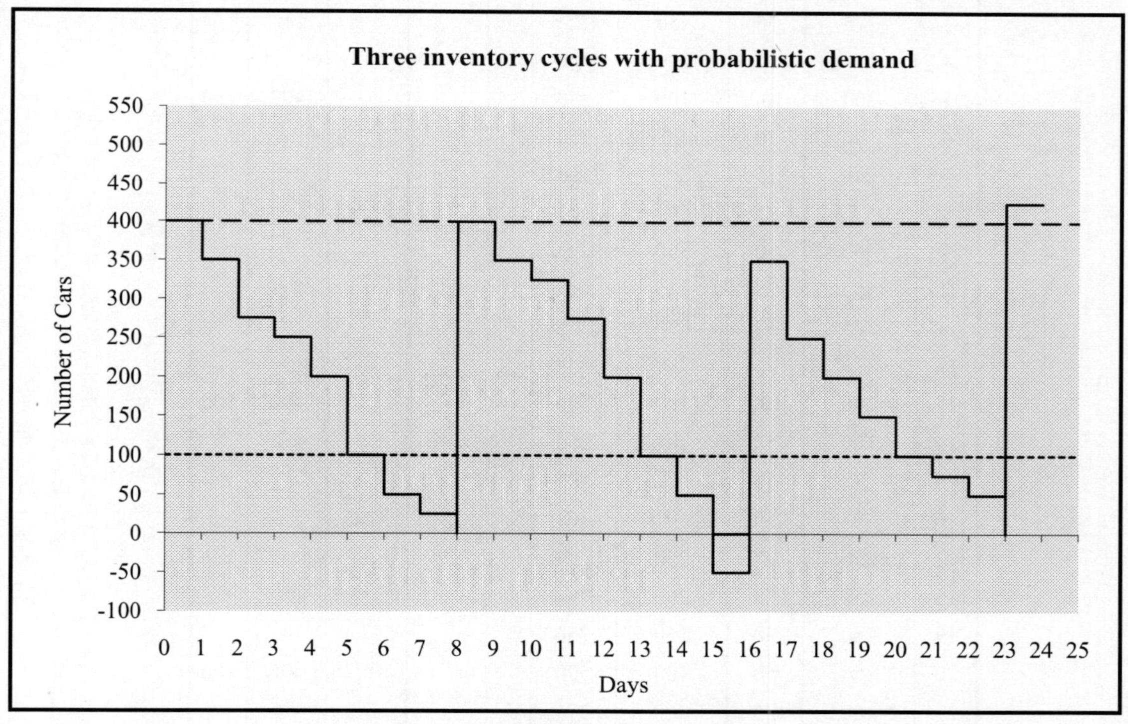

Figure 13.17

In our illustrative example above the question of back order arise in the second phase where the lead time demand is more than the reorder quantity. In such cases the excess quantity is booked but at the same time the allotment is made as soon as the next arrival comes in.

As a result, when the next arrival comes in the back order is cleared at one stroke and leftover will go for the inventory for the next cycle as shown in Figure 13.20 and the associated Table 13.26. See that this time the grand total cost has come down to Rs. 17,300.

Table 13.26

Trail	Inventory at the beginning of the day	Daily demand	Inventory at the end of the day	Inventory cost	Lost sales	Shortage penalty	Stock received	Quantity ordered	Order cost
1	400	50	350	1750					
2	350	75	275	1375					
3	275	25	250	1250					
4	250	50	200	1000					
5	200	100	100	500				400	100
6	100	50	50	250					
7	50	25	25	125					
8	25	25	0	0			400		
9	400	50	350	1750					
10	350	25	325	1625					
11	325	50	275	1375					
12	275	75	200	1000					
13	200	100	100	500				400	100
14	100	50	50	250					
15	50	50	0	0					
16	0	50	0	0			400		
17	350	100	250	1250					
18	250	50	200	1000					
19	200	50	150	750					
20	150	50	100	500				400	100
21	100	25	75	375					
22	75	25	50	250					
23	50	25	25	125			400		
24	425		Total	17,000					300
				Grand Total Cost		17,300			

■ Example 20

Suppose the daily demand for Maruti cars in a certain outlet in Bangalore has the following discrete probabilistic demand pattern. Inventory carrying cost is Rs. 5 per car per day in this showroom. Order placing cost is Rs. 100 per order irrespective of the order quantity. Loss in net revenue due to shortage of vehicle for delivery is Rs. 500 per unit. The reorder is placed whenever the evening closing stock is 27 units or less. The reordering quantity of 40 units is received on the evening of the 5th day and available for sale only on the morning of the 6th day from the date of ordering. The opening balance in the inventory is 45 units as on this day. Obtain the grand total cost for 25 days.

Solution

Table 13.27

Daily demand	Probability	Cumulative probability	Random number
3	0.17	0.17	00-16
4	0.25	0.42	17-41
5	0.25	0.67	42-66
6	0.20	0.87	67-86
7	0.13	1.00	87-99
Total	1.00	-	-

With this random numbers the stimulated result for 25 days is illustrated in Table 13.28. Since the first random number is 24 in Table 13.28(a), from Table 13.27 above the stimulated sales forecast is 4 cars. This information is reported in column number 4 in the table. Given the starting inventory as 45, the inventory in the evening of day one is 45 – 4 = 41. This is reported in column 5 in the very first row. At this stage no sales loss is being reported since no one moved out due to non-availability cars. The associated shortage penalty thus is zero. Further since the evening stock is greater than the prescribed minimum of 27 no reordering is needed at this stage and hence no reordering cost as well. All these columns are empty in the Table 13.28(a).

Now the evening inventory of 41 units becomes the beginning inventory for sales in the next morning as reported in row number 2. Now once again the random number for this day is 57, the simulated car sales could be 5 units. This sale leaves an inventory stock of 41 – 5 = 36 by the second day evening. This process of simulation is repeated for day three as well without any problem. When we move to fourth day the inventory stock gets depleted to 24 units by the evening. Since this is less than the reorder point of 27 units prescribed, reorder for 40 units is placed on the same evening. However, the arrival of this reorder will take 5 clear-cut days. So the new arrival could take place only on the evening of 8th day and available for sale only on 9th day as reported in the table. However, the stimulation table shows that the current stock meets the customers demand only up to the 8th day. On the 9th day because of non availability no sale takes place. This is reported as cost of lost sale (7 × 500 = Rs. 3500) in the corresponding column pertaining to 9th day.

From the next 10th day onwards the sales continue with new arrivals received on the ninth day evening. On 11th there is no problem, everything goes well. On 12th day once again the evening stock comes down to 27. Hence a second reorder is booked on this day and the corresponding order cost entries are made in the appropriate place in the table. On the 17th day though the predicted demand is 7, only one unit is available for sale from the inventory. The cost of 6 turnout customers (Rs. 3000) is entered in the corresponding column. The same evening second reordered 40 units are received.

On 18th and 19th days no additional problem is encountered. On 20th day the stock once again depleted to 27 units forcing a third reorder of 40 units. The whole summary results for all the 25 days simulation is given in Table 13.28. The calculations in the last column show that the cost due to the loss of sales due to non availability of stock seem to be greatest in the total of Rs. 9160. To avoid such situation a rethinking of reorder point is necessary.

Table 13.28 (a)

Trail	Random number	Inventory at the beginning of the day	Daily demand	Inventory at the end of the day	Inventory cost	Lost sales	Shortage penalty	Stock received	Quantity ordered	Order cost
1	24	45	4	41	205					
2	57	41	5	36	180					
3	77	36	6	30	150					
4	68	30	6	24	120				40	100
5	64	24	5	19	95					
6	88	19	7	12	60					
7	98	12	7	5	25					
8	48	5	5	0	0					
9	93	0	7	0	0	7	3500	40		
10	84	40	6	34	170					
11	11	34	3	31	155					
12	23	31	4	27	135				40	100
13	94	27	7	20	100					
14	83	20	6	14	70					
15	97	14	7	7	35					
16	83	7	6	1	5					
17	93	1	7	0	0	6	3000	40		
18	33	40	4	36	180					
19	49	36	5	31	155					
20	37	31	4	27	135				40	100
21	22	27	4	23	115					
22	27	23	4	19	95					
23	0	19	3	16	80					
24	43	16	5	11	55					
25	14	11	3	8	40			40		
Total			-		2360		6500			300
					Grand Total Cost		9160			

The Table 13.28 (b) is prepared for reordering level of 30 instead of 27 units with the same reorder quantity of 40 units each time. As a result the cost due to loss of sale is totally avoided. The total cost now is only Rs. 2995.

Table 13.28(b)

Trail	Random number	Inventory at the beginning of the day	Daily demand	Inventory at the end of the day	Inventory cost	Lost sales	Shortage penalty	Stock received	Quantity ordered	Order cost
1	24	45	4	41	205					
2	57	41	5	36	180					
3	77	36	6	30	150				40	100
4	68	30	6	24	120					
5	64	24	5	19	95					
6	88	19	7	12	60					
7	98	12	7	5	25					
8	48	5	5	0	0			40		
9	93	40	7	33	165					
10	84	33	6	27	135				40	100
11	11	27	3	24	120					
12	23	24	4	20	100					
13	94	20	7	13	65					
14	83	13	6	7	35					
15	97	7	7	0	0			40		
16	83	40	6	34	170					
17	93	34	7	27	135				40	100
18	33	27	4	23	115					
19	49	23	5	18	90					
20	37	18	4	14	70					
21	22	14	4	10	50					
22	27	10	4	6	30			40		
23	0	46	3	43	215					
24	43	43	5	38	190					
25	14	38	3	35	175					
Total					2695					300
					Grand Total Cost			2995		

The Table 13.28(c) is prepared for reordering level of same 27 units but with a reduced lead-time 4 days instead 5 days with the same reorder quantity of 40 units each time. Now the relevant total cost is Rs. 3395. Thus, in all our models, the optimality related issues are not discussed at all. The mathematical formulation for optimality model is beyond the scope of this book.

Table 13.28(c)

Trail	Random number	Inventory at the beginning of the day	Daily demand	Inventory at the end of the day	Inventory cost	Lost sales	Shortage penalty	Stock received	Quantity ordered	Order cost
1	24	45	4	41	205					
2	57	41	5	36	180					
3	77	36	6	30	150					
4	68	30	6	24	120				40	100
5	64	24	5	19	95					
6	88	19	7	12	60					
7	98	12	7	5	25					
8	48	5	5	0	0			40		
9	93	40	7	33	165					
10	84	33	6	27	135				40	100
11	11	27	3	24	120					
12	23	24	4	20	100					
13	94	20	7	13	65					
14	83	13	6	7	35			40		
15	97	47	7	40	200					
16	83	40	6	34	170					
17	93	34	7	27	135				40	100
18	33	27	4	23	115					
19	49	23	5	18	90					
20	37	18	4	14	70					
21	22	14	4	10	50			40		
22	27	50	4	46	230					
23	0	46	3	43	215					
24	43	43	5	38	190					
25	14	38	3	35	175					
Total			-		3095					300
					Grand Total Cost			3395		

Optimality model for probabilistic lead-time demand: An illustrative example

■ Example 21

Let us consider Deccan lighting system example. The company purchases special high density lighting system for National Lighting Corporation a well known light producer. One bulb costs Rs. 6 and per order the company spends Rs. 12. The inventory holding cost is 20% of the average inventory ($C_h = 6 \times 0.2 =$ Rs. 1.20 per unit). The number of units demanded varies considerably from day to day. Historical sales data tells that the demand with one week lead-time follow a normal distribution with mean value of 154 lights and standard deviation of 25 lights as shown in Figure 13.18. Since the mean demand is 154 per week it can expect for the whole year the expected demand could by 52 × 154 = 8008 units.

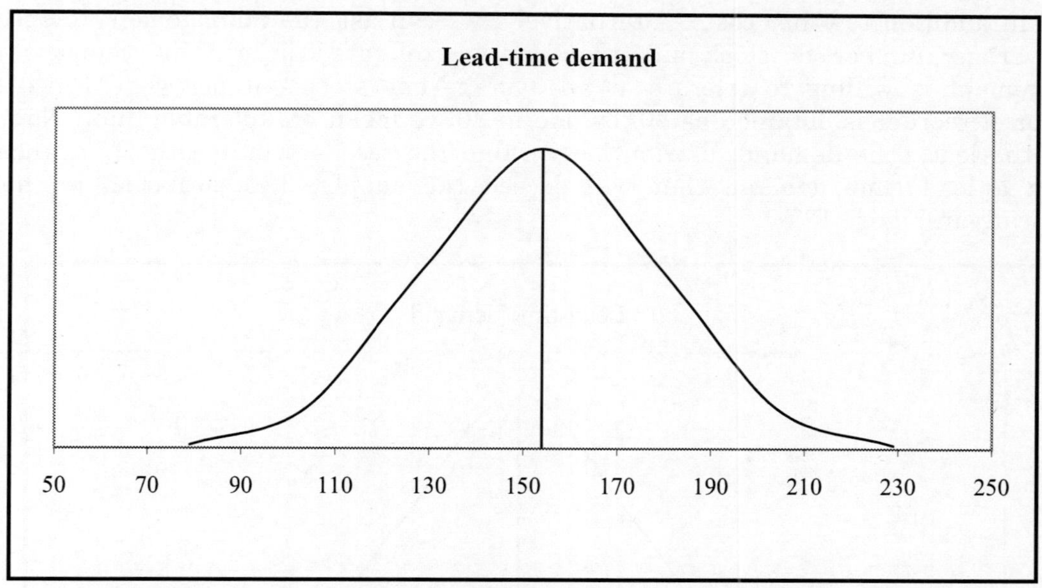

Figure 13.18

How much to order decision

Although we are in probabilistic demand pattern we can still get the approximately good inventory order quantity by applying our old EOQ formula as shown below.

$$Q = \sqrt{\frac{2DC_0}{C_h}} = \sqrt{\frac{2(8008)(12)}{1.20}} = 400 \text{ Units}$$

The so obtained 400 units could be considered an approximate value good enough for our purpose. This is mainly due to the reason that the total cost curve around the minimum point is very flat.

Number of orders decision

Since the total demand per annum is 8008 and the optimum order quantity is 400 the number order is obtained as 8008/400 = 20 orders approximately.

When to order decision

Since the mean of lead time demand is 154 units you may choose this level as the reordering level. However, considering the probability of demand the task becomes extremely important. If 154 is the mean lead-time demand symmetrically distributed about 154, then the lead-time demand will be more than 154 units roughly for about 50% of the time. Thus, if we use 154 units as the reorder level then about 50% of the time the demand will be in excess and the company will encounter shortage. This much of shortage is not likely to be accepted.

One way out is to approach the problem by bringing in the cost per stock out into the model in addition to other costs. Alternatively, we can ask the management to specify the average number of stock-outs that can be tolerated in a year. Suppose the management is willing to tolerate at an average one stock out per year. Since the number of reorder is obtained as 20, by one in 20 we mean 5% tolerable limit. Now by using the lead time demand distribution we find the reorder point with 5% chance of having a lead time demand that will exceed the supply. This situation is shown graphically in Figure 13.19.

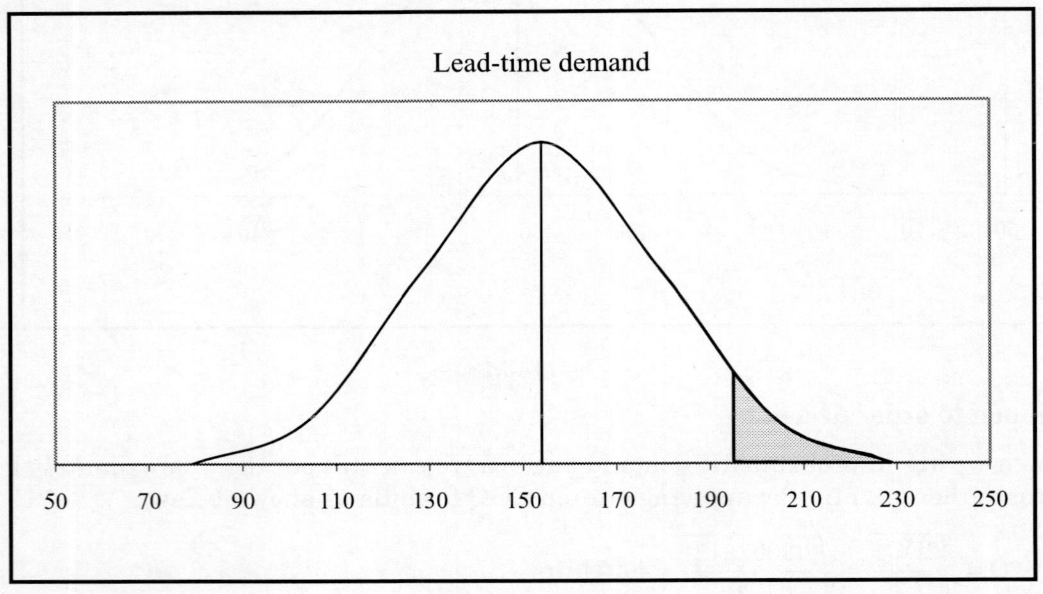

Figure 13.19

From the standard normal table for 5% right tail the corresponding Z = 1.645. So in original value term this amounts to $r = \mu + \sigma z = 154 + 1.645 \times 25 = 195$.

195 – 154 = 41 units serve here as the safety stock to meet any eventuality if any during the lead time. Roughly 95% percent of time this 195 units will be able to satisfy the lead-time demand. The anticipated annual cost is obtained as

Ordering cost = (D/Q) C_o	= (8008/400) × 12 = Rs. 240.24
Holding cost for normal inventory (Q/2) C_h	= (400/2) (1.20) = Rs. 240.00
Holding cost for safety stock = 41 × 1.20	= Rs. 49 20
Total cost	= Rs. 529.44

Note that if the demand is deterministic amounting to 8008 unit per year then Q = 400 and reorder level = 154. The corresponding total cost = Rs. 240.24 + Rs. 240 = Rs. 480.24

13.16 PERIODIC REVIEW MODEL WITH PROBABILISTIC DEMAND WITHOUT BACK ORDERS

In all the models discussed till now continuous monitoring was instated for placing reorders. Accordingly, once the reorder level is reached reordering takes place immediately. As an alternative to the continuous review system in this section we propose to introduce a periodic review system. Accordingly, the position of stock is reviewed periodically and reorder is placed wherever necessary to arrive at the desired stock level. Though, the review period is uniform, the order quantity varies due to the probabilistic pattern of demand assumed.

When a firm deals with multiple products the periodic review method is often recommended. Under the old method of continuous monitoring each and every product is to be monitored separately and order will have to be placed then and there as per the requirement. Such coordination from different sections of the firm under reference is very difficult indeed. If all the products are to be purchased for a single company then this periodic method is ideal because all the ordered items come in one shipment. Such an arrangement will save the transport cost a lot. Since no specific level of inventory for such models, on the day of the review the manager concerned note the existing stock and subtracts this level from a predetermined level of inventory and place order accordingly. Figure 13.20 and the associated Table 13.29 illustrate one such example.

$$Q = M - H$$

Where Q = the order quantity

H = the inventory on hand on the day of inspection

M = desired level of inventory stock.

The key issue in this type of model is to determine the optimum length of review period that keeps the total cost the least. Figure 13.20 illustrates one such model. The maximum inventory at the time of inspection is assumed to be 500 units in this representation. The inspection dates are highlighted as 6th, 12th and 18th days with equal interval between the successive inspections. On sixth evening the noticed stock is 50 units. So an order of 450 units is made in the same evening. With a lead time of two days, the inventory reaches the stores on the evening of the eighth day.

The stock of 50 units that was in existence on day sixth evening was found to be enough to satisfy the demand for the seventh and eighth days at the rate of 25 units per day.

Table 13.29

Trail	Inventory at the beginning of the day	Daily demand	Inventory at the end of the day	Inventory cost	Lost sales	Shortage penalty	Stock received	Quantity ordered	Order cost
1	400	50	350	1750					
2	350	75	275	1375					
3	275	25	250	1250					
4	250	50	200	1000					
5	200	100	100	500					
6	100	50	50	250				450	100
7	50	25	25	125					
8	25	25	0	0			450		
9	450	50	400	2000					
10	400	25	375	1875					
11	375	50	325	1625					
12	325	75	250	1250				250	100
13	250	100	150	750					
14	150	50	100	500					
15	100	100	0	0					
16	0	50	0	0		5000	250		
17	250	75	175	875					
18	175	25	150	750				350	100
19	150	50	100	500					
20	100	25	75	375					
21	75	25	50	250					
22	50	0	50	250					
23	50	0	50	250			350		
24	400		Total	17,500		5000			300
				Grand Total Cost			22,800		

Thus there is no stock out during this phase. The second phase starts from the morning of the 9th day with the arrived inventory of 450 units. On 12th day once again the stock is reviewed. The stock on the evening of that day was found to be 250 units. So an order of another 250 units is made on the same evening. But this time the inventory is received on the sixteenth evening with a lead time of 4 days. But on sixteenth, since there is no stock, back order is booked for 50 units but not honoured. So on 17th morning the inventory goes up to the full amount of booking, namely 250 units On 18th evening the stock was found to be only 150 units. Thus an order for 350 is made on the same evening. But this time the delivery takes 5 days to come in. On 23rd the unsold stock of 50 is also added to the inventory making it to 400 units on 24th day.

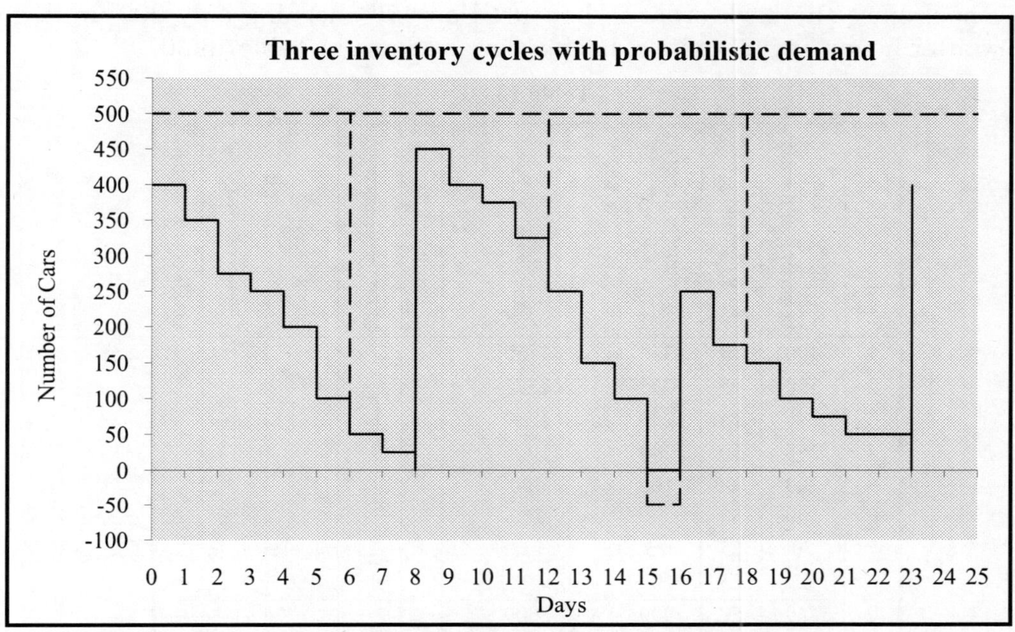

Figure 13.20 Back order orders are not honoured

13.17 PERIODIC REVIEW MODEL WITH PROBABILISTIC DEMAND WITH BACK ORDERS

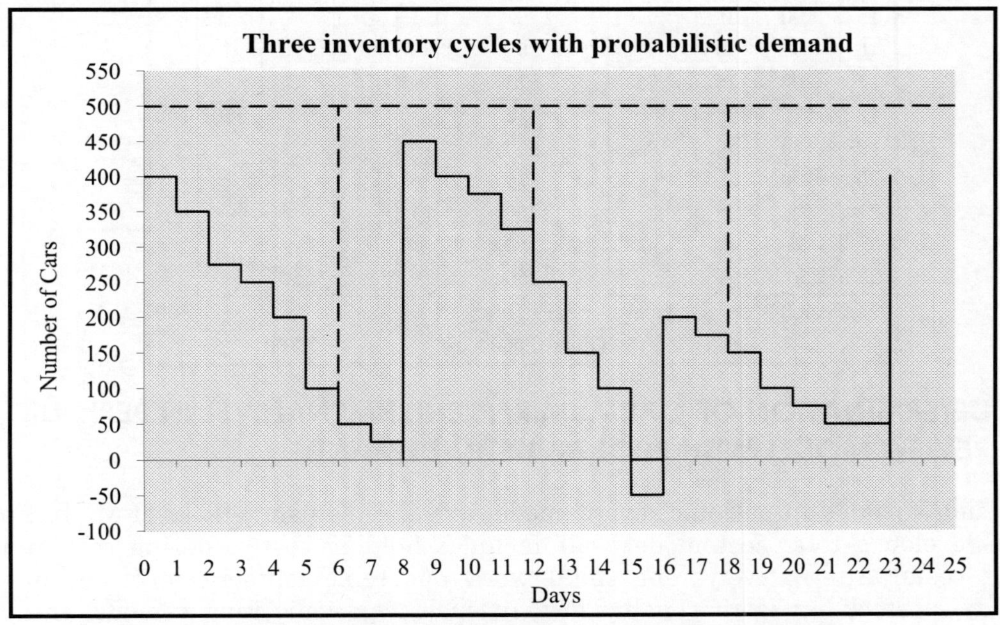

Figure 13.21 Back order orders are honoured

In this model since the back order is honoured on 17th morning only 200 units go for inventory after honouring 50 units of back order as shown in Table 13.30.

Table 13.30

Trail	Inventory at the beginning of the day	Daily demand	Inventory at the end of the day	Inventory cost	Lost sales	Shortage penalty	Stock received	Quantity ordered	Order cost
1	400	50	350	1750					
2	350	75	275	1375					
3	275	25	250	1250					
4	250	50	200	1000					
5	200	100	100	500					
6	100	50	50	250				450	100
7	50	25	25	125					
8	25	25	0	0			450		
9	450	50	400	2000					
10	400	25	375	1875					
11	375	50	325	1625					
12	325	75	250	1250				250	100
13	250	100	150	750					
14	150	50	100	500					
15	100	100	0	0					
16	0	50	0	0			250		
17	200	25	175	875					
18	175	25	150	750				350	100
19	150	50	100	500					
20	100	25	75	375					
21	75	25	50	250					
22	50	0	50	250					
23	50	0	50	250			350		
24	400		Total	17,500					300
				Grand Total Cost			17,800		

13.18 DETERMINATION OF OPTIMUM REPLENISHMENT LEVEL IN PERIODIC REVIEW MODEL WITH PROBABILISTIC DEMAND

The decision variable in periodic review model is the replenishment level M. To find M first we develop a total cost model that includes holding cost, ordering cost and the stock-out costs. Alternatively, one could work out replenishment level so that the probability of stock-out is reasonably low. In such a periodic review model the order quantity in each review period must be sufficient enough to cover the demand for the review period plus the following lead-time demand. In other words this is to say that the order quantity that brings the inventory position up to the level of M must last, until the order made in the next review period is received in the inventory. The length of time is equal to the review period plus the lead time. Figure 13.22 illustrates the normal probabilistic demand on both the accounts with means 250 units and standard deviation 45 units.

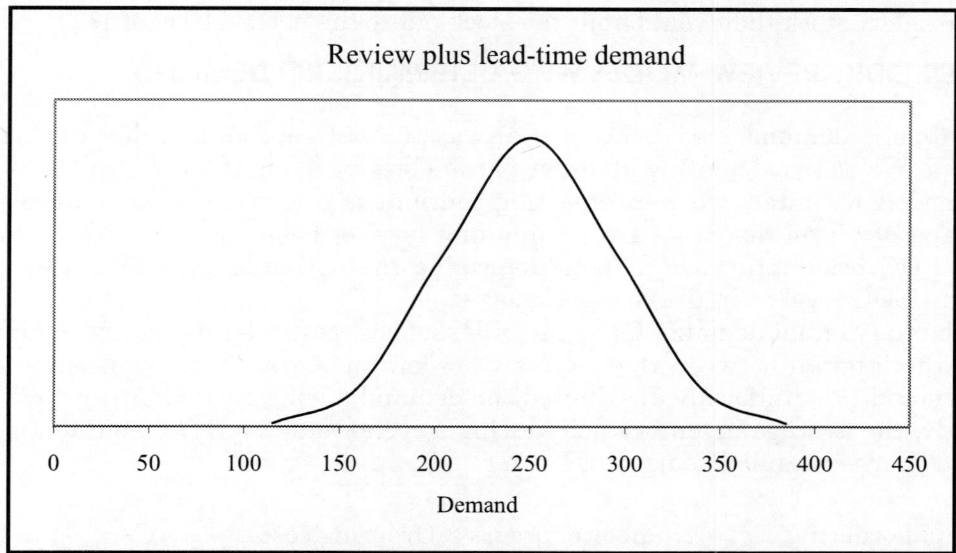

Figure 13.22

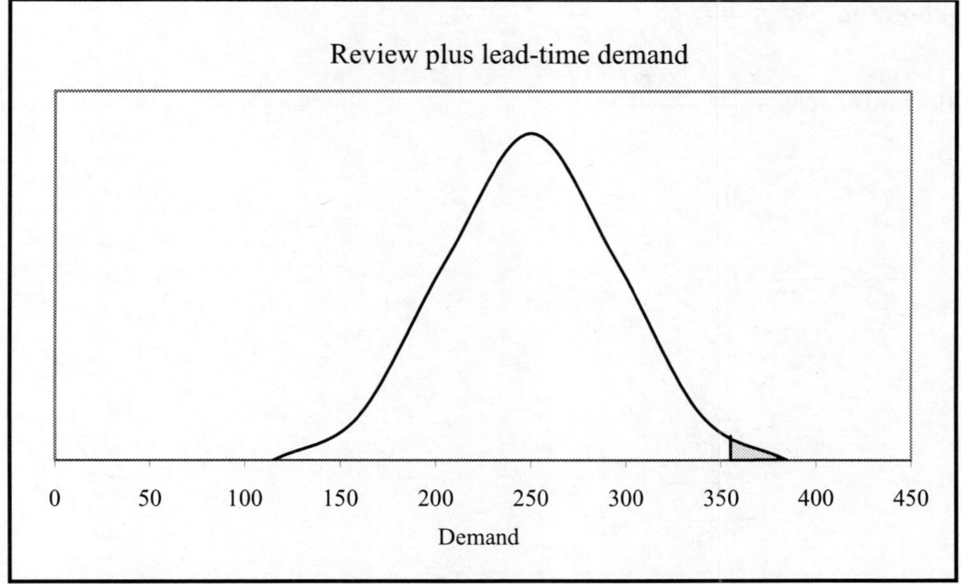

Figure 13.23

Figure 13.23 illustrates the corresponding replenishment level that allows 1% chance of a stock-out. From the standard normal table the value of M that is 2.33 times the standard deviation above the mean will allow stock-out with a given 1% probability.

If the demand had been deterministic, then again the replenishment demand would have been the demand during the review period plus the demand during the lead-time period. However, the replenishment level would have been only 250 units and no stock-out would have occurred. In the case of probabilistic demand the difference 355 – 250 = 105 is considered safety stock that would limit the stock-out to the needed level of 1%.

13.19 PERIODIC REVIEW MODEL WITH DETERMINISTIC DEMAND

In deterministic demand model, EOQ level was the only optimum value that is to be determined. We were calculating all other parameters by using the EOQ value obtained. In such models definitely we were insisting continuous reviewing. Orders were placed, provided reorder level reaches a given minimum level or below. In this section with the same old EOQ assumption we plan to determine the optimum periodic review period rather than EOQ level as in the classical model.

Let D be the annual demand, C_o be the ordering cost per order, C_h be the holding cost and T be the interval between the successive orders in years. Since by assumption the annual demand D is uniformly distributed the demand during one inventory cycle is DT. Accordingly, the average inventory held during a cycle would be DT/2. So the number of orders in a year is obtained as n = D/DT = 1/T

So annual ordering cost = Number of orders × Ordering cost $= \frac{D}{DT} \times C_o = \frac{C_o}{T}$

Annual holding cost $= \frac{DT}{2} \times C_h = \frac{DTC_h}{2}$

Total annual cost $= \frac{C_o}{T} + \frac{DTC_h}{2}$

$$\frac{d}{dT}\left(\frac{C_o}{T} + \frac{DTC_h}{2}\right) = 0$$

$$\frac{-C_o}{T^2} + \frac{DC_h}{2} = 0$$

$$\frac{DC_h}{2} = \frac{C_o}{T^2}$$

$$T^2 = \frac{2C_o}{DC_h}$$

$$T = \sqrt{\frac{2C_o}{DC_h}}$$

Example 22

Western Electronics company produces 2000 TV sets in a year. For this purpose it needs equal number of picture tubes of certain type. Each tube costs Rs. 10 and the cost of holding the tube in inventory is Rs. 2.40 per unit per year. In addition the cost of placing orders is Rs. 150 per order. Obtain the optimum order quantity and the number of orders given the information that the company works for 200 days in a year with a lead time of 10 days.

Solution

EOQ Solution : Tabular representation

Table 13.31

No. of orders per year	No. of units per order	Average inventory	Carrying cost	Ordering cost	Total cost
1	2000	1000	2400	150	2550
2	1000	500	1200	300	1500
4	500	250	600	600	1200
5	400	200	480	750	1230
8	250	125	300	1200	1500
10	200	100	240	1500	1740

Diagrammatic representation

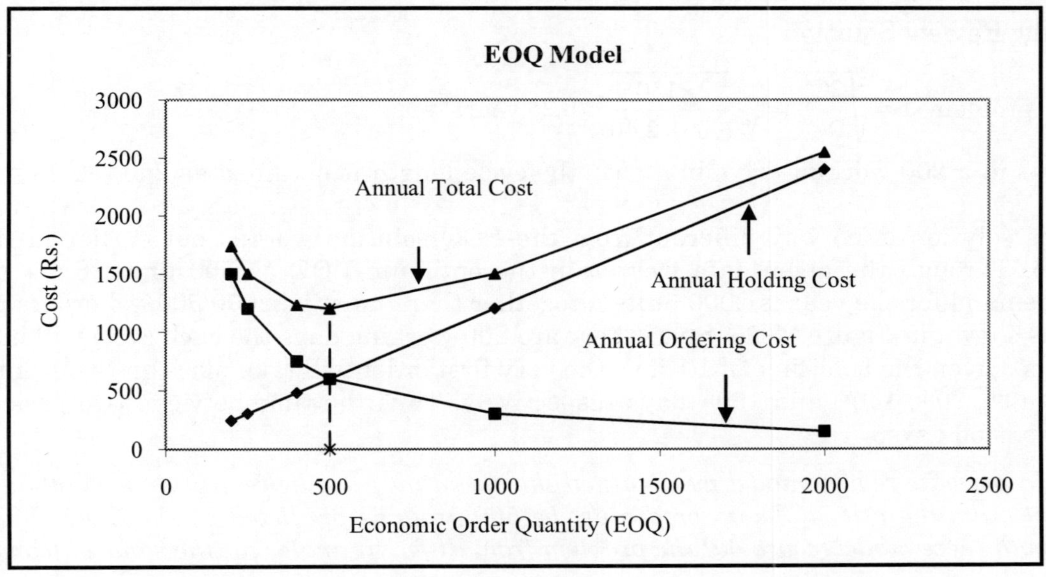

Figure 13.24

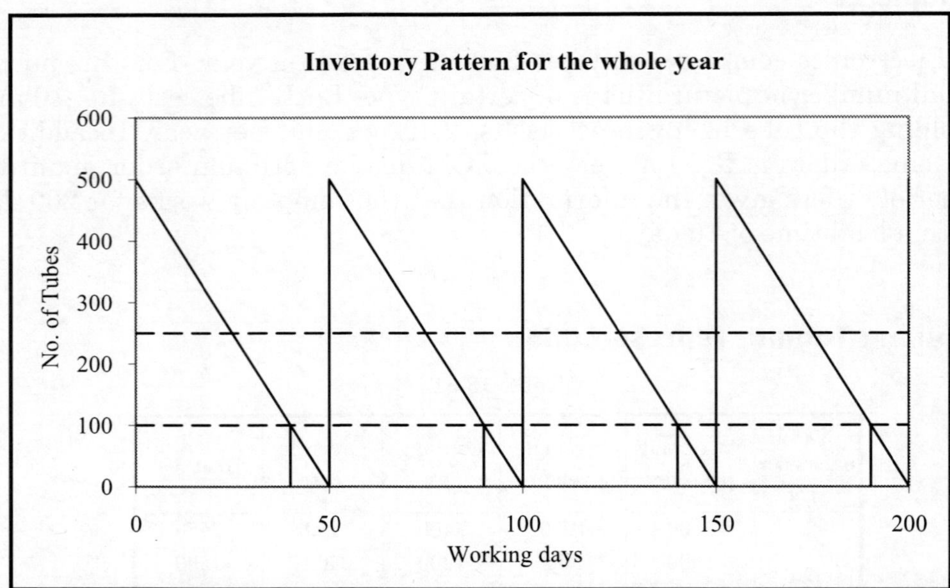

Figure 13.25

Thus the EOQ = 500 for our model. Since the total quantity needed for one year = 2000, the number of orders to be placed = 2000/500 = 4. Since the company works only for 200 days, the inventory cycle will be = 200/4 = 50 days. Further, since the order quantity is 500 and the cycle is 50 days the demand per day = 500/50 = 10. Cycle time = 90 – 40 = 50 days.

Periodic Review Solution

In our problem $T = \sqrt{\dfrac{2C_o}{DC_h}} = \sqrt{\dfrac{2 \times 150}{2000 \times 2.40}} = 0.25$ years.

Thus in a 200 working days in a year the cycle length is obtained as 200 × 0.25 = 50 days.

This solution is no way different from the EOQ solution worked out earlier in this section. Through the given data we obtain the optimum EOQ as 500 units. Since the total demand for the year is 2000 units altogether there must be 2000/500 = 4 orders per year as shown in Figure 13.26. Since there are 200 working days the each cycle will have 50 days. Given the lead time of 10 days the very first order will take place on the evening of the 40th, 90th, 140th and 190th days respectively. Thus the time between two periodic reviews is 50 days.

***Note:** In periodic review model we obtained the optimum periodic review time as 50 days. The optimum quantity to be ordered must be 500 units as predicted by the EOQ model. Thus both these models work out the problem from different angles for the same outcome.*

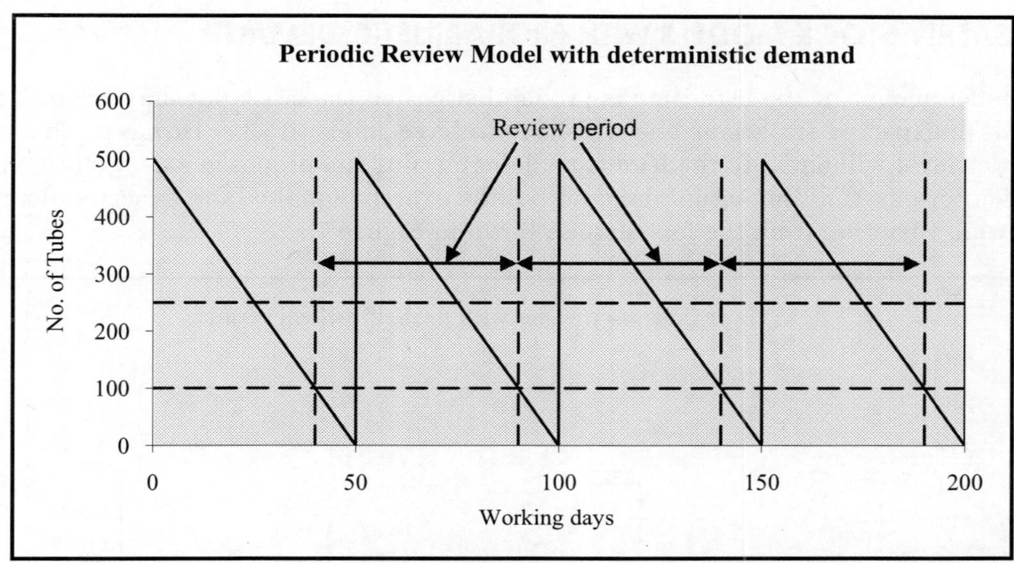

Figure 13.26

13.20 SAFETY STOCK MODELS WITH DETERMINISTIC DEMAND

In deterministic models, the stock is replenished as soon as the stock gets exhausted because of the known demand pattern and known lead time correctly. In such models, because ordered supply would reach by the moment the stock level reaches zero, there would be no stock outs unless otherwise it deliberately allowed as shown in Figure 13.37. In it a safety stock of 10 units is kept for safety.

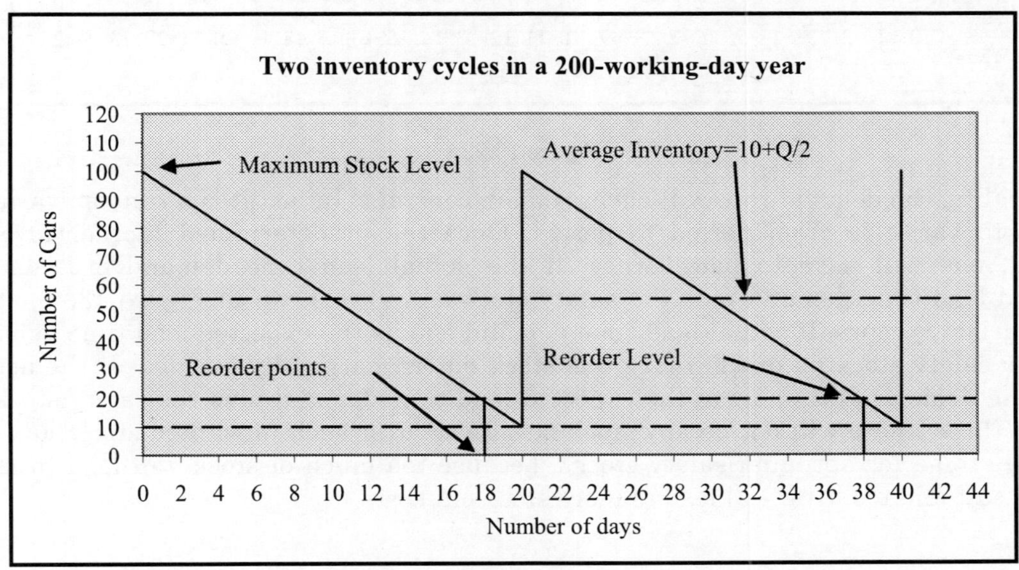

Figure 13.27

13.21 SAFETY STOCK MODELS WITH PROBABILISTIC DEMAND

Even in such models if the lead time is probabilistic then the stock out may occur. To tie over such unexpected situations one will have to have safety stocks. However, this time the reorder level will equal to the lead time demand plus the proposed safety stock. In an ideal inventory system the actual demand will be exactly equal to the expected demand and show no variation from the forecast as shown in Figure 13.28.

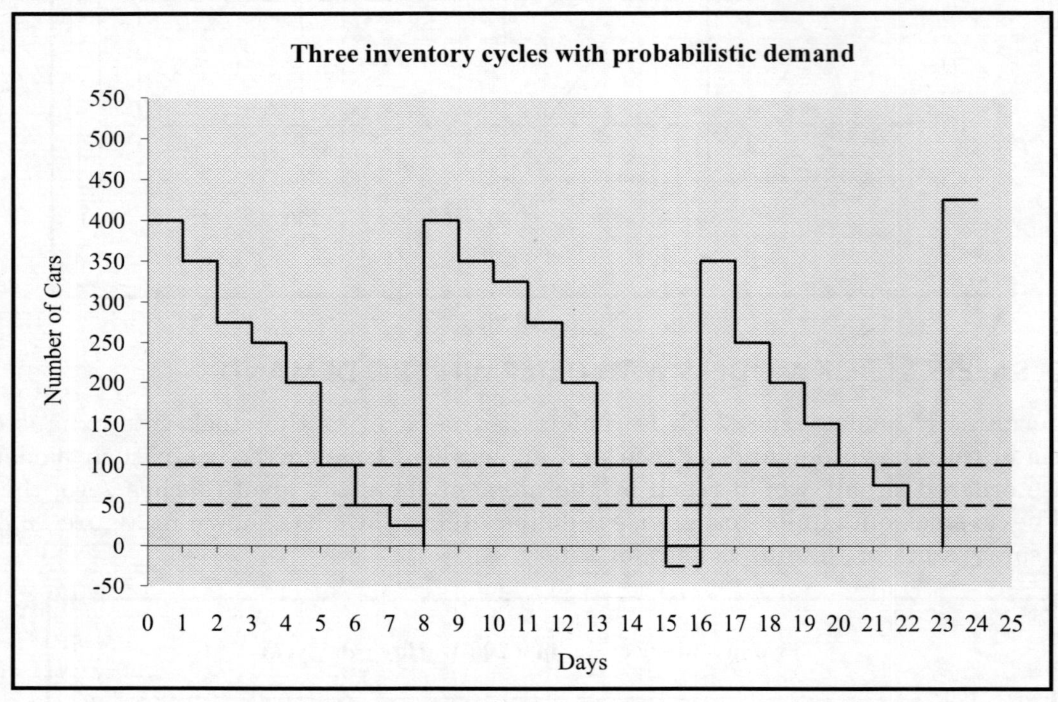

Figure 13.28

In reality, the demand rate will never remain constant. In fact it is a random variable subject to variation from period to period. Once the predetermined reorder level is reached, one will have to place order. If the actual lead time demand exceeds the expected lead time demand plus the safety stock as in cycle two in Figure 13.28, stock out invariably occurs. If actual lead time demand equals the expected lead time demand plus the safety stock as in the first cycle stock out is marginally avoided. If the actual lead time demand falls short of the expected lead time demand plus the safety stock as in cycle three, there will not be any stock out. To tie over such unwanted stock outs one will have to go for optimum safety stocks, because too much of stock will lead to more holding cost and too little will lead to more stock out costs.

Table 13.32

Trail	Inventory at the beginning of the day	Daily demand	Inventory at the end of the day	Inventory cost	Lost sales	Shortage penalty	Stock received	Quantity ordered	Order cost
1	400	50	350	1750					
2	350	75	275	1375					
3	275	25	250	1250					
4	250	50	200	1000					
5	200	100	100	500				400	100
6	100	50	50	250					
7	50	25	25	125					
8	25	25	0	0			400		
9	400	50	350	1750					
10	350	25	325	1625					
11	325	50	275	1375					
12	275	75	200	1000					
13	200	100	100	500				400	100
14	100	50	50	250					
15	50	50	0	0					
16	0	25	0	0	25	2500	400		
17	400	100	300	1500					
18	300	50	250	1250				350	100
19	250	50	200	1000					
20	200	50	150	750					
21	150	25	125	625					
22	125	25	100	500					
23	100	0	100	500			350		
24	450		Total	18,875		2500			300
				Grand Total Cost			21,675		

Determination of safety stock with constant lead time and probabilistic lead time demand when the stock out cost is known

To begin with let us assume that the lead time is constant and the probability distribution of the lead-time demand is known. Also assume that the stock out cost is known. Eventually when the lead time is constant then the question of safety stock arises only due to increase in lead-time demand. Given these inputs we can work out the optimum safety stock by considering the total least cost for different safety stock levels by trial-and-error method. The following example illustrates one such problem.

■ Example 23

Using the EOQ model an inventory manager has estimated the economic order quantity in respect of an item with annual usage of 24,000 units, let the average, to be equal to 4800 units, with a mean usage of 8 units per day. The lead time is known to be five working days. It is also known that the cost of carrying one unit of the item is Rs. 15 and the cost of being out of stock for this item is Rs. 50 per unit. From the past records the manager finds that the demand during the lead-time over the last 50 reorder period has been as follows:

Table 13.33

Lead time demand	250	300	350	400	450	500	550	600
Frequency	1	3	6	18	10	7	3	2

Help the manager to establish an appropriate level of safety stock and re-order point.

Solution

Let us first derive the respective probabilities by dividing the respective frequencies by the total of 50.

Table 13.34

Lead time demand	250	300	350	400	450	500	550	600
Frequency	1	3	6	18	10	7	3	2
Probability	0.02	0.06	0.12	0.36	0.2	0.14	0.06	0.04

Since the average per day demand is 80 units and the lead time is 5 days the expected lead-time demand must be 80 × 5 = 400 units. If the manager fixes the reorder level at 400 units without any safety stock then he will have the risk of stock out. From the table the associated risk is obtained as 0.2 + 0.14 + 0.06 + 0.04 = 0.44.

If he prefers to have 450 units with 50 units in the safety stock then the probability of facing out of stock will be equal to 0.14 + 0.06 + 0.04 = 0.25.

If he prefers to have 500 units with 100 units in the safety stock then the probability of facing out of stock will be equal to = 0.06 + 0.04 = 0.10.

If he prefers to have 550 units with 150 units in the safety stock then the probability of facing out of stock will be equal to = 0.04

If he prefers to have 600 units with 200 units in the safety stock then the probability of facing out of stock will be equal to = 0.

Now from the given data since the annual sale is 24,000 and the average sale is 4800 per cycle there will be 24,000/4800 = 5 inventory cycles. Thus, altogether the risk of running out of stock is five time. The cost of facing out-of-stock for various options are calculated and reported in the Table 13.35.

Table 13.35

Safety stock	Number of units short	Probability	Number of cycles	Stock out cost per unit	Expected annual cost	Total annual stock out cost
0	50	0.20	5	50	2500	
0	100	0.14	5	50	3500	
0	150	0.06	5	50	2250	
0	200	0.04	5	50	2000	10,250
50	50	0.14	5	50	1750	
50	100	0.06	5	50	1500	
50	150	0.04	5	50	1500	4750
100	50	0.06	5	50	750	
100	100	0.04	5	50	1000	1750
150	50	0.04	5	50	500	500
200	0	0.00	5	50	0	0

Let us calculate the total cost for all the five levels of safety stocks as shown in the Table 13.36.

Table 13.36

Level of safety stock	Stock out cost	Unit carrying cost	Total carrying cost	Total annual cost
0	10,250	15	0.00	10,250.00
50	4750	15	750.00	5500.00
100	1750	15	1500.00	3250.00
150	500	15	2250.00	2750.00
200	0	15	3000.00	3000.00

The total cost is least for 150 unit of safety stock.
So reorder level = Expected lead-time demand + safety stock = 400 +150 = 550.

Determination of safety stock with probabilistic lead time and probabilistic lead-time demand when the stock out cost is known

When discrete probability distribution for both lead time and lead-time demand are known then by using joint probability distribution we can get the needed optimum safety stock as usual. The following example illustrates one such case.

■ Example 24

As a part of review of the system of the company ABC, a typical stock item having the part No. Y 209 has been studied in detail and the results are reported below.

Table 13.37

Lead time	Probability
15 days	0.2
20 days	0.5
25 days	0.3

Table 13.38

Demand per day	Probability
5000 units	0.5
7000 units	0.5

The company has 240 working days in a year. The carrying cost per unit is Re. 0.15. the reorder level to this part is 1,50,000 units and the relevant reorder cost is Rs. 1000 per order. You are required to:

a) Calculate the level of buffer stock in a reorder level of 1,50,000 units.

b) Calculate the probability of the stock out.

c) Calculate the expected annual stock outs in units.

d) Calculate the stock out cost per unit at which it would be worthwhile raising the reorder level to 175.000 units.

Solution

a) From the data we generate the distribution of demand during the lead time as shown in Table 13.39.

Table 13.39

Lead time	Demand per day	Total lead time demand	Lead time probability	Demand probability	Joint probability	Expected demand
15	5000	75,000	0.2	0.5	0.10	7500
15	7000	1,05,000	0.2	0.5	0.10	10,500
20	5000	1,00,000	0.5	0.5	0.25	25,000
20	7000	1,40,000	0.5	0.5	0.25	35,000
25	5000	1,25,000	0.3	0.5	0.15	18,750
25	7000	1,75,000	0.3	0.5	0.15	26,250
				Total	1.00	1,23,000

Thus the expected demand during the lead time = 1,23,000 units

Safety stock = Reorder level – Expected demand = 1,50,000 – 1,23,000 = 27,000 units

b) The probability of stock out due to demand more than the reorder level 1,50,000 is given in the last row. If the lead-time demand is 1,75,000 then there will be a stock out because the reorder level is only 1,50,000.

c) $\text{EOQ} = \sqrt{\dfrac{2 \times 6000 \times 240 \times 1000}{0.15}} = 1{,}38{,}564$ units

Now from the data available expected daily demand

= 50,000 × 0.5 + 7000 × 0.05= 6000 units

Thus total annual demand = 6000 × 240 = 14,40,000 units

Thus the number of orders per annum = 14,40,000/1,38,564=10.39

Now the expected stock out per annum = 0.15(1,75,000–1,50,000) × 10.39 = 38,962 units

d) At the reorder level of 1,50,000 the stock out = 38,962 units. At 1,75,000 units level the stock out is nil. So additional cost = 25,000 × 0.15 = Rs. 3750

Unit cost at which the increased stock out is justified = 3750/38,962 = Re. 0.096

Determination of safety stock with probabilistic lead time and probabilistic lead-time demand when the stock-out cost is not known

In the absence of stock out cost the problem of safety stock is approached from a different angle by considering the service level. The safety level is defined as the ratio of number of units supplied without delay to that of the number of units denied. Accordingly a 90% service level we mean that 90% of customers are satisfied by offering the unit under reference and 10% of the customers are denied the service of offering the unit. The following example illustrates the method of getting safety stock in such cases.

■ Example 25

Given the probability distribution of lead time demand and the average lead time of 10 days, determine the safety stock if the company desires to have no stock out at all. Also determine the safety stock if it desires to have service level of 95%.

Solution

The safety stock at a given service level is equal to the difference between the maximum demand during the lead time at that service level and the expected demand during the lead time.

a) When no stock out is desired, the safety stock is obtained as

Safety stock = Maximum demand during lead time – average demand during lead time

Table 13.40

Lead-time demand	Mid value	Probability	Expected value	Cumulative frequency
81-100	95.5	0.06	5.73	0.06
102-120	110.5	0.19	21.00	0.25
121-140	130.5	0.28	36.54	0.53
141-160	150.5	0.40	60.20	0.93
161-180	170.5	0.05	8.53	0.98
181-200	190.5	0.02	3.81	1.00
	Total	1.00	135.80	

From the given distribution the maximum demand is 200 units. The expected demand = 135.80.

So the safety stock = 200 – 135.80 = 64.5 units

b) For 95% service level safety stock = maximum demand 95th percentile.

$$95^{\text{th}} \text{ percentile} = 160.5 + \frac{0.95 - 0.93}{0.05} = 168.5$$

Safety stock = 168.5 – 135.5 = 33 units

Safety stock with constant lead time and normal lead time demand distribution with known mean and standard deviation

Given the normal distribution with known mean μ and standard deviation σ, the 95% service level safety stock is obtained as Z σ. The following example illustrates the method of getting the safety stock.

■ Example 26

The demand for an item during the order period is known to be normally distributed with mean 600 units and standard deviation 50 units.

a) When it has 80 units of safety stock, find the service level it can offer.

b) When it has no safety stock find the service level

c) For 99% service level, find the safety stock

Solution

a) To find the needed probability we will find the area to the left of 680 in the standard normal curve.

$$Z = \frac{X - \mu}{\sigma} = \frac{680 - 600}{50} = 1.6$$

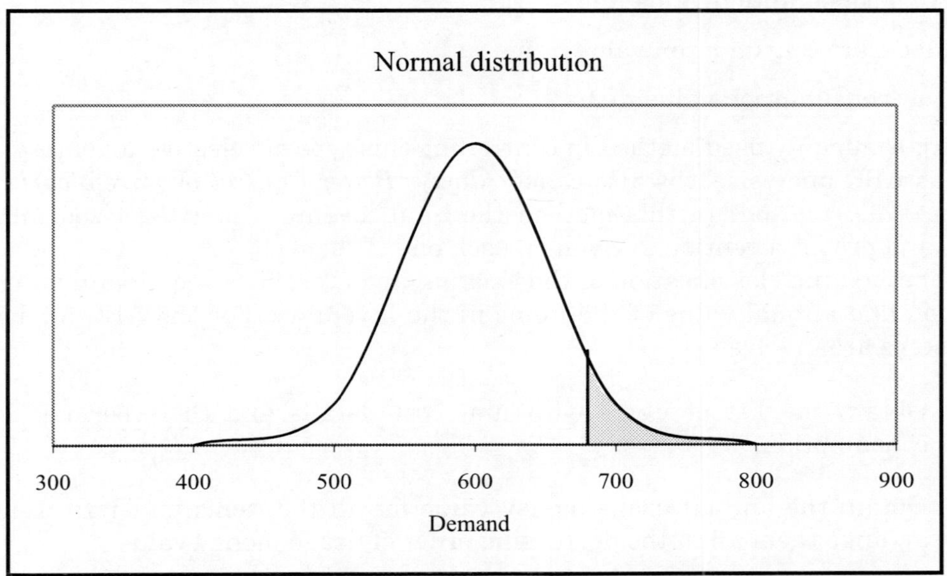

Figure 13.29

From the standard normal table the area corresponding to Z = 1.6 is 0.4452. Thus, the total area to the left of 680 = 0.5 + 0.4452 = 0.9452.

Thus, 80 units of safety stock the level of service offered is 94.52%.

b) When no safety stock is provided the service level will be only 50%.

c) For 99% service level for the table the corresponding Z = 2.33. Thus the level of safety stock = $\sigma Z = 2.33 \times 50 = 117$ units.

13.22 SELECTIVE APPROACHES TO INVENTORY CONTROL

Depending upon the type of business, organizations keep certain levels of inventories. When the number of items in the inventory is large definitely organization will have to mobilize large amount of money for inventory maintenance. Further, all items of inventory need not be controlled with equal importance. Some inventory items may be very important and some others may be too unimportant. Management often has laid priorities on items of importance. Some inventory items though consumed very little need an intensive control because of higher purchase cost. Similarly, some other items though used in plenty need very little or no control at all because of small amount money expenditure involved in procuring them. To have an effective control the inventory usage is broadly classified under three heads as shown below:

1. High consumption value items
2. Moderate consumption value items
3. Low consumption value items.

The most commonly used method in controlling this type of selective inventory control is known as **ABC** analysis. It is also called *Always Better Control* or *Pareto curve or the curve of mal distribution*. In this method the total inventory is categorized into three sub-heads and proper attention is given on each one of them.

In this method the classification of the existing inventory is based on annual physical consumption and annual value of the items in the inventory. For the ABC analysis the following steps are involved.

Step 1: Obtain the list of items with unit cost details and their periodic level of consumption.

Step 2: Obtain the annual usage money value for all the items under reference and arrange them all in the decreasing order of usage money value.

Step 3: Express the value for each item as a percentage of the aggregate usage value. Also get the cumulative value of the said percentage.

Step 4: Obtain the percentage value of each of the items. For example, if there are 20 items then the percentage value of each item 100/20 = 5%.

Step 5: Using the cumulative value of the items on the x-axis and cumulative usage value on the y-axis draw a cumulative curve.

Step 6: Determine the appropriate valve for A, B, and C categories. Normally the steep rising portion of the curve is classified as A. The moderately rising portion is classified as B and the very slow rising portion is put under the C category. After fixing these three classifications the items under category under A attracts a very strict monitoring. The protection against their stock out could be as high as 98% even. The C items may be put under open system without much botheration. For these items the reorder quantity may be even higher than the EOQ prescription. The following example illustrates the methodology ABC analysis.

Example 27

For the data given below perform ABC analysis.

Table 13.41

Item	No. of units	Unit price
1	700	5
2	2400	3
3	150	10
4	60	22
5	3800	1.5
6	4000	0.5
7	6000	0.2
8	300	3.5
9	30	8
10	2900	0.4
11	1150	1.1
12	410	6.2

Solution

In Table 13.42 we obtain the item-wise usage value percentage in the last but one column. In the last column according to the percentage we obtain the ranking in a decreasing order.

Table 13.42

Item	No. of units	Unit price	Usage value	Percentage	Rank
1	700	5	3500	9.8378166	4
2	2400	3	7200	20.237794	2
3	150	10	1500	4.2162071	7
4	60	22	1320	3.7102622	8
5	3800	1.5	5700	16.021587	3
6	4000	0.5	2000	5.6216095	6
7	6000	0.2	1200	3.3729657	9
8	300	3.5	1050	2.951345	11
9	30	8	240	0.6745931	12
10	2900	0.4	1160	3.2605335	10
11	1150	7.1	8165	22.950221	1
12	410	6.2	2542	7.1450656	5
Total			35,577	100	

Table 13.43

Item	Value	Cum. Value	% Number	Cum. Value	Rank
11	22.95	22.95	8.3	8.3	1
2	20.24	43.19	8.3	16.6	2
5	16.02	59.21	8.4	25	3
1	9.84	69.05	8.3	33.3	4
12	7.15	76.19	8.3	41.6	5
6	5.62	81.81	8.4	50	6
3	4.22	86.03	8.3	58.3	7
4	3.71	89.74	8.3	66.6	8
7	3.37	93.11	8.4	75	9
10	3.26	96.37	8.3	83.3	10
8	2.95	99.33	8.3	91.6	11
9	0.67	100.00	8.4	100	12

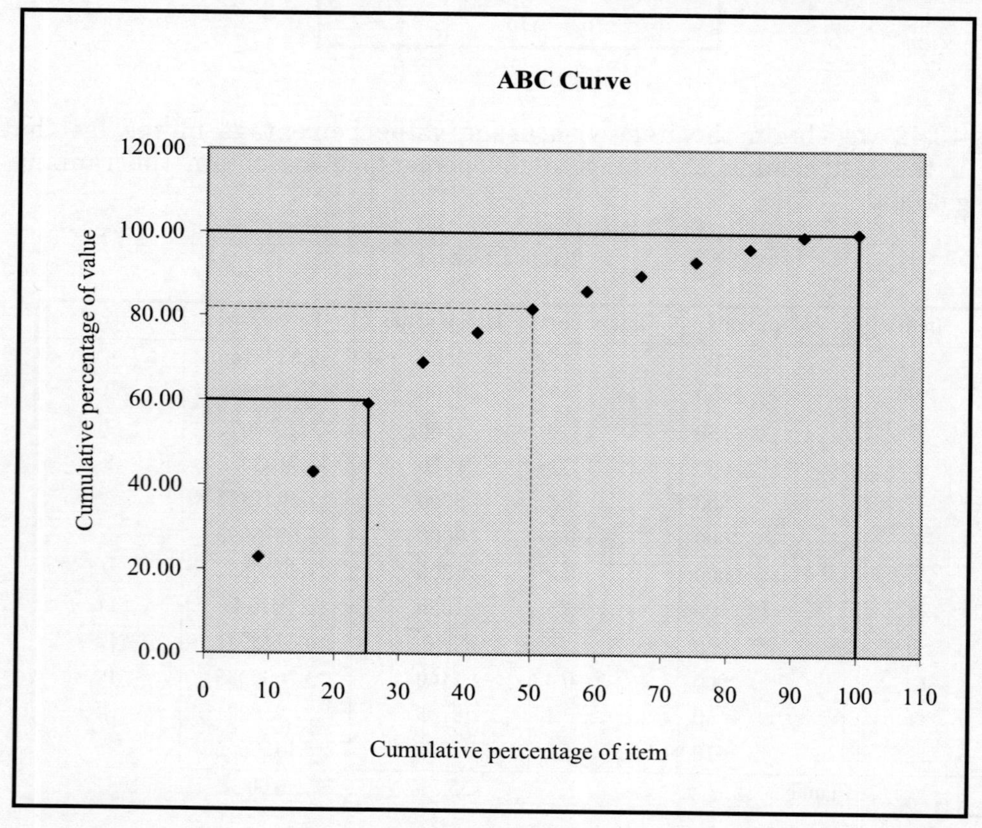

Figure 13.30

In Table 13.43 we rearrange the data rank-wise in the first and the second column. In column 3 we obtain the cumulative percentages. In Figure 13.30 we measure the cumulative percentage of items in x-axis and cumulative percentage of values on the y-axis for the graph highlighted. The desired three ranges are obtained in the graph.

13.23 OTHER INVENTORY CONTROL METHODS

1. **HML Method**

   To apply this method, the items are classified into three groups on the basis of unit price.

   H = High Price

   M = Medium Price

   L = Low price

2. **VED Method**

   In this method the classification is done on the basis of criticality of the items under reference.

   V = Vital

   E = Essential

   D = Desirable

3. **FSN Method**

   In this method the items are classified on the basis of level of consumption.

   F = Fast moving items

   S = Slow moving items

   N = Non-moving items

4. **SDE Method**

   In this method the classification is based on the availability of the item under reference.

   S = Scarce

   D = Difficult

   E = Easy to obtain

5. **SOS Method**

   In this method we classify the items on the basis of seasons.

   S =Seasonal

   OS = off-seasonal

6. GOLF Method

In this method the classification is based on the source of supply.

G = Government

O = Ordinarily available

L = Locally available

F = Foreign source of supply

13.24 ZERO INVENTORY SYSTEM (ZIN)

Often inventory holding cost may be exorbitantly high. Sometimes the space management for keeping the inventory may be difficult. In such cases zero inventories have been adopted by Japanese companies. Such methods were found to be worth consideration provided the materials needed are obtainable from local vendors.

Similar to this the Japanese also practised *Just In Time* (JIT) method as well. Accordingly, the supply will arrive just in time that too only the needed quantity for the day's work. Hence raw materials, work in progress as well as the finished goods inventories are kept at the minimum almost closer to zero level. This method is becoming more and more popular in India as well wherever possible.

13.25 INVENTORY MANAGEMENT FOR DEPENDENT DEMAND

The inventory management techniques discussed till now hold good only for finished good inventories normally held by retailing outlets. In this section we propose to consider the inventories in manufacturing industries. This time we are concerned with the online inventories like raw materials, parts, and sub-assemblies that give rise to finished product. The demand for these inputs heavily depends upon the demand for the final finished good. More the demand for finished good in the market more will be the demand for these inputs and vice versa. Under such circumstances the demand for the intermediary inputs are often called derived or dependent demands. Material requirement planning (MRP) is a technique that is being used to translate a set of period-by-period requirements of finished goods into a set of period-by-period requirement of raw materials, parts, and sub-assemblies.

In Figure 13.31 on zero day the finished goods in the inventory is 20 units. Since this is the reorder point, instruction is given to set-up production with a lead time of 5 days. Since no production is there on the zero day the raw material inventory of 100 units including a safety stock of 10 units remain unaltered. On 5th day the production run starts with a production of 6 units per day. Since demand is at 2 units per day the inventory incises at the rate of 4 units per day. On 20th the 90 units of input gets exhausted leaving only the safety stock of 10 units in inventory. Also, the production run ends on this day. However, on 15th day itself when the raw material reaches the reorder level of 40 units, with a given lead time of 10 days reorder is booked.

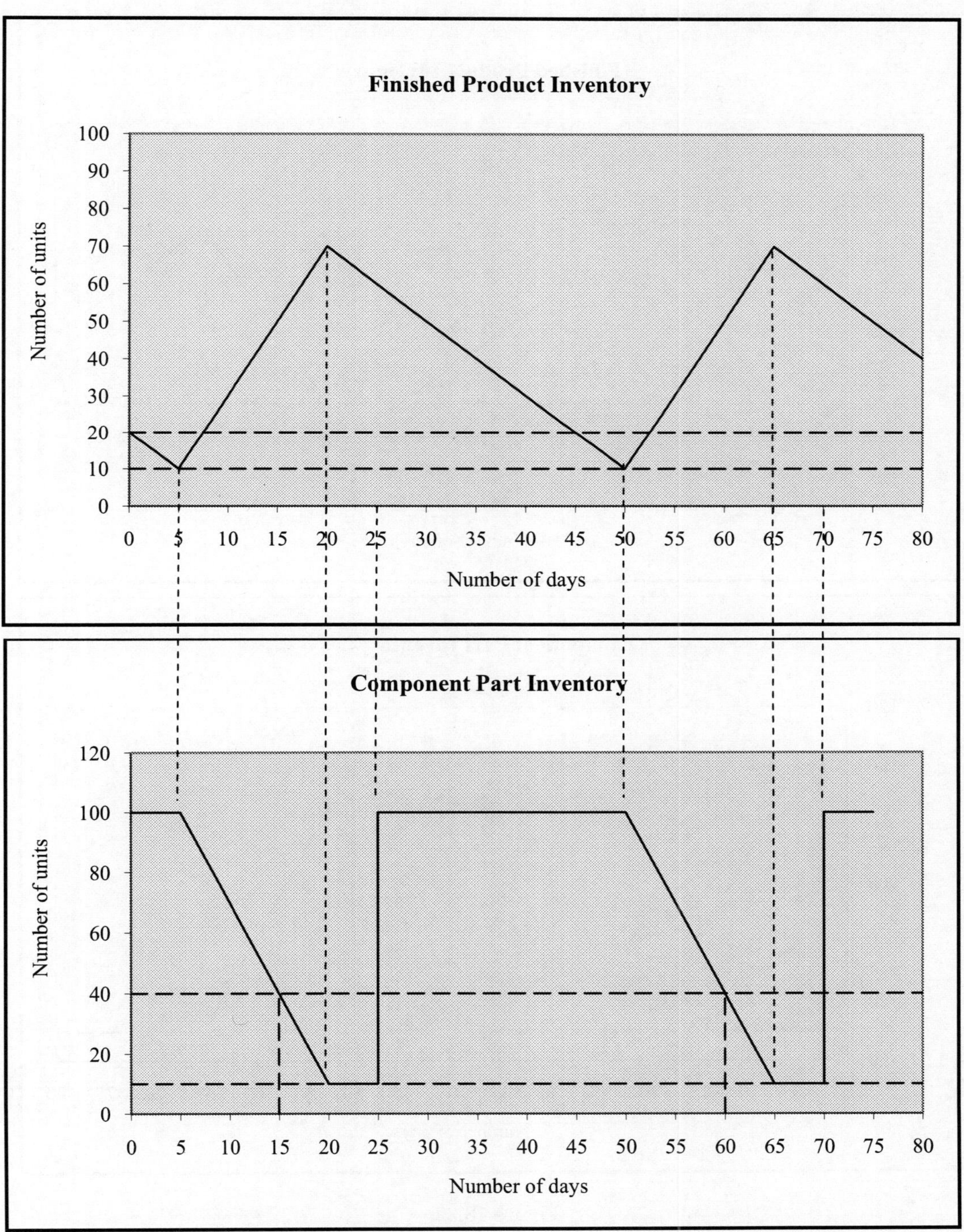
Finished Product Inventory
Number of units
100
90
80
70
60
50
40
30
20
10
0
0 5 10 15 20 25 30 35 40 45 50 55 60 65 70 75 80
Number of days
Component Part Inventory
Number of units
120
100
80
60
40
20
0
0 5 10 15 20 25 30 35 40 45 50 55 60 65 70 75 80
Number of days

Figure 13.31

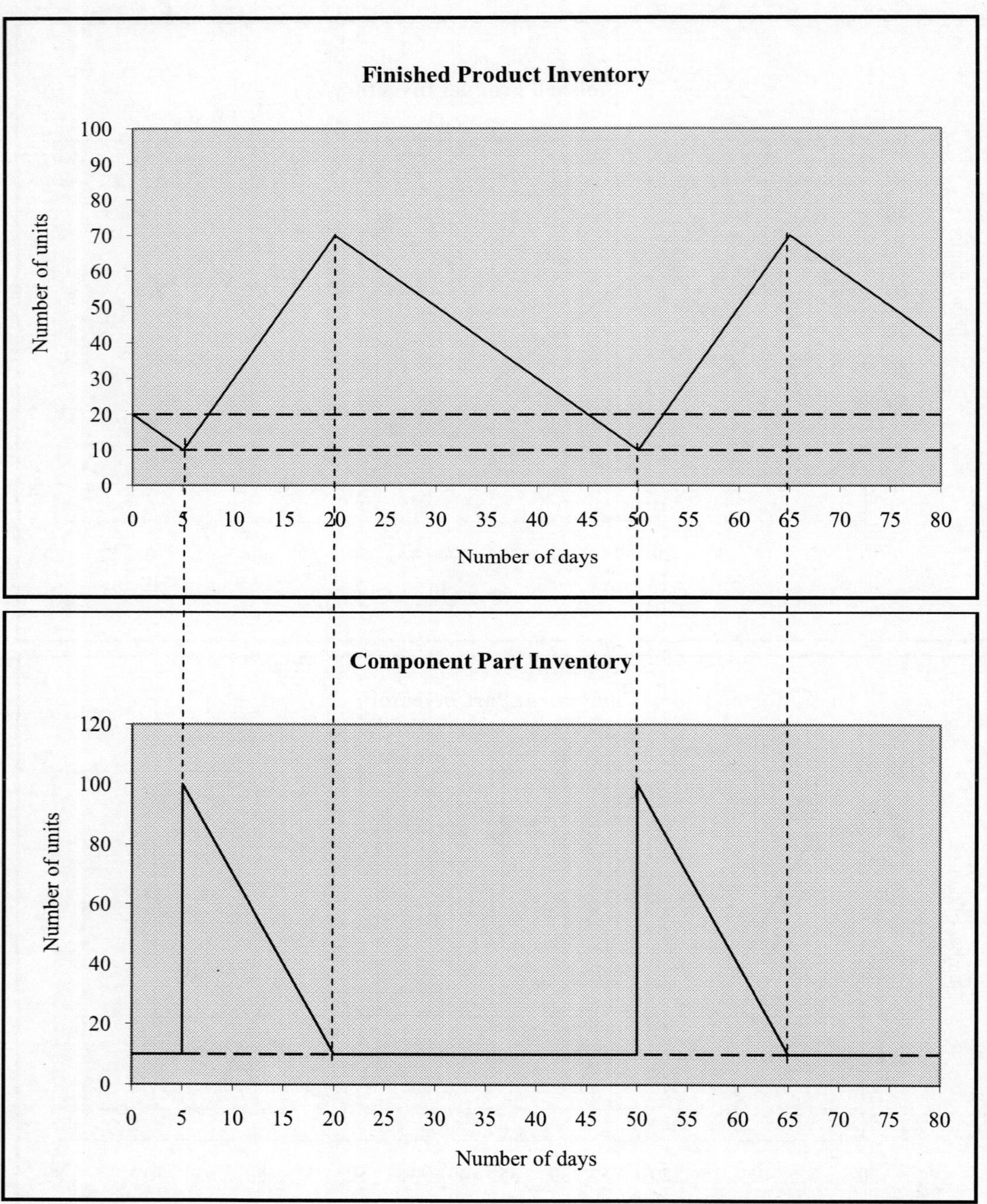
Finished Product Inventory
Number of units
100
90
80
70
60
50
40
30
20
10
0
0 5 10 15 20 25 30 35 40 45 50 55 60 65 70 75 80
Number of days
Component Part Inventory
Number of units
120
100
80
60
40
20
0
0 5 10 15 20 25 30 35 40 45 50 55 60 65 70 75 80
Number of days

Figure 13.32

Accordingly, the inventory increases by 100 units once again on the 25th day. Since the next production run starts only on 50th day, this inventory of raw material remain unused as shown in the graph. Another production run starts on this day and the cycle gets repeated in the same way. Clearly in this type of model the input inventory unnecessarily builds up on 25th day and remains unused fill the next production run starts on 50th day. Similarly, it happens in the later runs as well.

In the model shown in Figure 13.32, this unnecessary input inventory build-up is avoided. The input inventory order is booked only on the 40th day for 90 units. With lead time of 10 days the inventory stock rises to 100 units on the 50th day and available for production run.

EXERCISES

1. A contractor has to supply 10,000 paper cones per day to a textile unit. He finds that when he starts a production run, he can produce 25,000 paper cones per day. The cost of holding a paper cone in stock for one year is 2 paise and set-up cost of production is Rs. 18 how frequently the production run be made?

2. The annual requirements of a particular raw material are 2000 units costing Re.1 each to the manufacturer. The ordering costs is Rs. 10 per order and carrying cost 16% per annum of the average inventory value, find the economic ordering quantity and the total inventory cost per annum.

3. A company purchases 10,000 items per year for it use in its production shop. The unit cost is Rs. 10 per year, holding cost is Re. 0.80 per month and cost of making a purchase is Rs. 200. Determine the following if no shortages are allowed.

   i) The optimal order quantity

   ii) The optimal total yearly cost

   iii) The number of orders per year

   iv) The time between orders

4. The cost of placing an order is Rs. 150 per order. It is estimated that 1000 units will be used in the next 12 months. The carrying cost per month is Rs. 2.50. assuming that the demand is deterministic and continuous and that no stock-out are allowed, determine the optimal order quantity. What is the time between the placing of orders? The procurement lead time is one month. What is the reorder point based on the on-hand inventory level?

5. A company plans to consume 750 pieces of a particular component. Past records indicate that the purchasing department spent Rs. 14,555 for placing 20,500 purchase orders. The average inventory was valued at Rs. 1,45,000 and the total storage cost was Rs. 17,650. The company borrows capital at a rate of 15% a year. Fill the following table, if the price of a component is Rs. 10.

Lot size Purchase price/year Purchase expense/year	10	Storage cost/year Capital cost /year Total cost per year

6. A manufacturer is to supply 5000 bearings per day to an automobile dealer. He finds that when he starts a production run, he can produce 25,000 units per day. The cost of holding a bearing in stock for one year is 4 paise and the set-up cost for a production run is Rs. 18. find the economic production lot size. How frequently the production run be made?

7. A company has to purchase four items A, B, C and D for next year. The projected demand and unit price are as follows:

Item	Demand (units)	Unit price (Rs.)
A	60,000	6
B	40,000	4
C	1200	48
D	5000	2

If the company wants to restrict the total number of orders to 50 for all the four items, how many orders should be placed for each item?

(M.B.A, April 1985, Delhi University)

8. The cost of placing an order is Rs. 150 per order. It is estimated that 1000 units will be used in the next 12 months. The carrying cost per month is Rs. 250. Assuming that the demand is deterministic and continuous and that no stock-outs are allowed, determine the optimal order quantity. What is the time between the placing of orders? The procurement lead time is one month. What is the reorder point based on the on-hand inventory level?

(M.B.A, 1981, Delhi University)

9. A publishing house purchase 2000 units of a particular item per year at a unit cost of Rs. 20. The ordering cost per order is Rs. 50 and the inventory carrying cost is 25 percent. Find the optimal order quantity and minimum total cost including purchase cost. If a 3% discount is offered by the supplier for purchases in lot of 1000 or more, should the publishing house accept the proposal?

(ICWA, June 1986)

10. The probability distribution on monthly sales of an item is as follows:

Monthly sales (units)	0	1	2	3	4	5	6
Probability	0.01	0.06	0.25	0.30	0.22	0.10	0.06

The cost of carrying inventory (unsold during the month) is Rs. 30 per month and cost of unit storage is Rs. 70. Determine optimum stock to minimize expected cost.

11. BPL electronic company purchases transistors for its music system production. The yearly requirement is 400 units. The cost of transistor is Rs. 50. inventory carrying cost is 10% of the average inventory. The ordering cost per order is Rs. 40. The company received an attractive offer of 5% discount on purchase of 100 or more. What decision the BPL would take.

(C.A., May 1983)

12. Obtain:

    a) EOQ

    b) Number of orders

    c) ROL

    d) Safety stock for the following inventory problem:

    Annual demand = 36,000

    Cost per unit = 1 Re.

    Ordering cost = 25 Rs.

    Cost of capital = 15%

    Storage charge = 5%

    Lead time = 1/2month

    Safety stock = one month's consumption

(Bangalore University, M.B.A July/August 2005)

CHAPTER

14

Theory of Games-I Zero Sum Games

- *Introduction*
- *Two-person zero and positive sum games*
- *Conservative maximini and minimax approaches*
- *Solution by mixed strategy*
- *Solution by algebraic method*
- *Solution by sub-game method*
- *Dominance method: Dominance reduction rules*
- *Solution by graphical method*
- *Linear programming approach to the theory of games*
- *Exercises*

14.1 INTRODUCTION

All living beings on the mother earth compete with one another for survival all the time. The self-interest dominated rationality is the driving force for all such struggles. Playing carom board, with their own self-interest of winning is a typical example for pleasurable games because each player is interested in securing more points than others. Beside such pleasurable games we also come across materialistic out-come oriented games like winning the battle, gaining the market share, winning the election etc. In this section we do not consider pleasurable sportive games. Instead we propose to consider a kind of competitive conflict in which someone is declared the winner, the other is declared loser at the end.

If a business executive concerned is capable of predicting his competitor's plan of action in advance then the job of optimizing his business returns would not be a difficult task. Thus, the game theory is a special type of decision theory in which one's own choice of action, called strategy is determined after taking into account of all possible strategies of his opponents rather than mere consideration of several possible outcomes. Here, the mutual concern is the interdependency of each other's strategies. Each player is assumed to have a series of strategies to counter the strategies of his opponents. They play one or more combinations of strategies at a time. Thus, "if I do this, then he does something else than I thought off, so that I really ought to have done that instead of this" is the spirit of interdependency in the theory of game. The final outcome however, depends upon the quality of the strategies and skill of the player concerned.

Though the history of theory of games dates back to the 20th century, new turn towards its wider applicability is of recent origin. The work on "Theory and Practice of Games in Economics" by Jon Von Neumann published in the year 1944 added a new dimension to the theory of game. Neumann's approach uses Maximini and Minimax strategies in solving business related problems. In general, all the competitive situations with the following characteristics come under the purview of the theory of game.

1. There are finite numbers of competitors in the players' list.

2. For every player, there exists a finite number of actions called strategies.

3. All the players knew the rules of the game.

4. All the players play the strategies simultaneously so that no one knew the opponent's strategy until it is being played.

5. The final outcome of actions and reactions of all the participants' ends with a definite outcome, may be positive or negative or zero. Here by positive end result we mean a gain, by negative end result we mean a loss, and by zero end result we mean no loss no gain situation otherwise called a zero sum game.

14.2 TWO PERSON ZERO AND POSITIVE SUM GAMES

In a zero sum game one's own gain is equal to the loss of his opponent. Thus, zero will be the net result of the whole play. Suppose, there are two players A and B trying to win over the market share for each others products, then the gain in the market share to A from B is normally referred as A's payoff matrix. Positive entries in this A's payoff matrix refers a gain in the market share and negative entry refers a loss in the market share to the player A. However, equivalently we can also define payoff matrix for player B by merely converting all the positive entries into negative entries and all negative entries into positive entries in A matrix. Suppose firm A has three strategies of advertising in three leading newspapers and B has two strategies of advertising in two TV serials then the resulting game in both A and B's payoff format is shown in Table 14.1.

Example 1

Table 14.1

		A's Payoff	
		B's Strategies	
		1	2
A's Strategies	1	10	−6
	2	8	2
	3	−4	1

		B's Payoff	
		B's Strategies	
		1	2
A's Strategies	1	−10	6
	2	−8	−2
	3	4	−1

In the above representation since one is the mirror image of the other we can use either A's payoff or B's payoff, for problem solving. However, it is common practice to use A's payoff matrix all the time. Thus, unless otherwise specified, in our future endeavours we use only A's payoff matrix for all our discussion.

In positive sum games, one own gain need not be equal to his opponent loss. For example, if the strategies of advertisement of a cigarette company do not induce new customers into the cigarette smokers net then the total market will remain unaltered irrespective of the strategies of both A and B. Given this situation A's gain definitely means B's loss and vice versa. The appropriate game would be a zero sum type. On the other hand, advertisements given by the companies add new customers from non-smokers into smokers list then one market may increase without equivalent reduction in the other. Under the new situation A's gain may be entirely from the new customers. Under such circumstances one must use both A's and B's payoffs in solving problems. Clearly A's matrix and B's matrix are independent in such cases. In this section we propose to consider only the zero sum games.

From A's payoff matrix given in example 1 above we have the following six possible plays.

1. If A uses strategy 1 and B too uses strategy 1 then the game will yield a payoff of 10% market share to A. Accordingly, B will lose 10% of his market share to A.
2. If A uses strategy 1 and B uses strategy 2 then the game will yield a payoff of −6% market share to A. Accordingly, B will gain 6% of the market share from A.
3. If A uses strategy 2 and B uses strategy 1 then the game will yield a payoff of 8% market share to A. Accordingly, B will lose 8% of his market share to A.
4. If A uses strategy 2 and B too uses strategy 2 then the game will yield a payoff of 2% market share to A. Accordingly, B will lose 2% of his market share to A.
5. If A uses strategy 3 and B uses strategy 1 then the game will yield a payoff of – 4% market share to A. Accordingly, B will gain 4% of his market share from A.

6. If A uses strategy 3 and B uses strategy 2 then the game will yield a payoff of 1% market share to A. Accordingly, B will lose 1% of his market share to A.

In the above said game it is assumed that both A and B are aware of the entries in A's payoff matrix even before they start the game. The only thing that A does not know in advance is the strategy that B's is going to play for any given move by A. For example, if A plays strategy 1 then it is left to player B either to play strategy 1 or 2.

Similarly, the only thing that B does not know in advance is the strategy that A's is going to play for any given move by B. For example, if B plays strategy 1 then it is left to A either to play strategy 1 or 2 or strategy 3. Remember, that both the players are expected to play the strategies simultaneously and not one after another.

14.3 CONSERVATIVE MAXIMINI AND MINIMAX APPROACHES

When firms face strategic risky situations like this, normally they make cautious and defensive approaches all the time. Accordingly, for the play given in A's payoff matrix the following are the three possible defensive approaches to player A.

1. Let A play strategy 1, at the same time B can play either strategy 1 or strategy 2. If B plays 1 then the corresponding expected payoff to A is 10. On the other hand if B plays strategy 2 the corresponding expected payoff to A is –6. Among the two the worst thing that can happen to A is a payoff of –6.

2. Let A play strategy 2, at the same time B can play either strategy 1 or strategy 2. If B plays 1 then the corresponding expected payoff to A is 8. On the other hand if B plays strategy 2 the corresponding expected payoff to A is 2. Among the two the worst thing that can happen to A is a payoff of 2.

3. Let A play strategy 3, at the same time B can play either strategy 1 or strategy 2. If B plays 1 then the corresponding expected payoff to A is −4. On the other hand if B plays strategy 2 the corresponding expected payoff to A is 1. Among the two the worst thing that can happen is a payoff of −4.

Thus, the minimum payoff that A could expect from all the three strategies of his own are –6%, 2% and –1% respectively. Under the circumstances the best thing that A can do is to go for the maximum of all these three minimums. Accordingly, he will opt for strategy 2 and make the player B to part with either 8% or 2% of his market share. Because of this reasoning A is always called a Maximini player in the theory of game.

In a similar way B too will adopt a cautious approach of defence. Accordingly, for the play given in A's payoff matrix the following are the two possible defensive approaches to player B.

1. If B opts for strategy 1 then A can think of either strategy 1 or strategy 2 or strategy 3. If A play strategy 1 then the associated payoff from B to A is 0% of market share. Instead if A play strategy 2 then the associated payoff from B to A is 8%. As third

case, if A play strategy 3 then the associated pay from B to A is –4. Under the circumstances the worst thing that can happen to B is A's playing strategy 1. The associated payoff is 10% from B to A.

2. If B opts for strategy 2 then A can think of either strategy 1 or strategy 2 or strategy 3. If A play strategy 1 then the associated payoff from B to A is –6. Instead if A play strategy 2 then the associated payoff from B to A is 2%. As third case, if A play strategy 3 then the associated pay from B to A is 1. Under the circumstances the worst thing that can happen to B is A's playing strategy 2. The associated payoff is 2% from B to A.

Thus the maximum payoffs that B could offer to A from all the two strategies of his own are 10% and 2% respectively. Since B is going to make payment, it is logical to think of minimizing such payments. Among the two possibilities since 2 is the minimum, the right option for B is to go for the corresponding strategy 2 irrespective of A's play. Because of this reasoning B is always called a Minimax player in the theory of game. The whole set of arguments are summarised in the following matrix form.

Table 14.2

		A's payoff B's Strategies 1	2	Row Minimum	A's Maximini
A's Strategies	1	10	–6	–6	
	2	8	2	2	2
	3	–4	1	–4	
Col. Max		10	2		
B's Minimax		2			

Saddle Point Solution Exists
Value of the Game = 2
A's Strategy = 2
B's Strategy = 2

Now from the table above it is clear Player A wants to play the strategy 2 hoping to get a risk-less market share of 2%. At the same time B too wants to play strategy 2 and ready to part with a risk-less market share of 2%. Such a coincidental situation is often called the equilibrium solution to the model under reference. Since the solution in one way is the maximum and another way is the minimum it is often called the saddle point solution as well. So for a saddle point solution $(\text{Maximini})_A = (\text{Minimax})_B = 2$. Here 2 is called the value of the game and it is normally written as v = 2.

Example 2

For the given 2×3 game obtain the saddle point solution if it exists.

Table 14.3

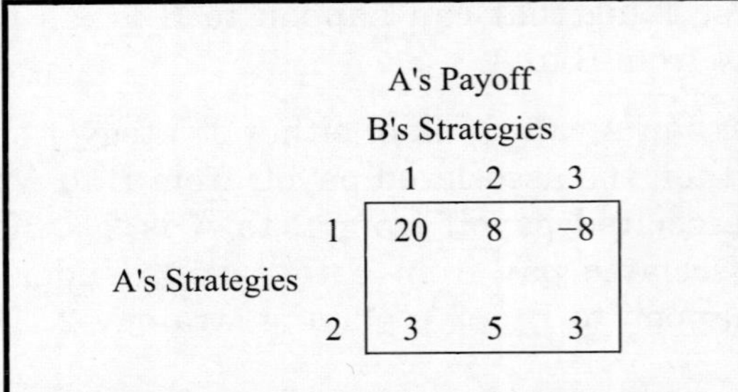

A's Payoff

A's Strategies	B's Strategies 1	2	3
1	20	8	−8
2	3	5	3

Solution

Table 14.4

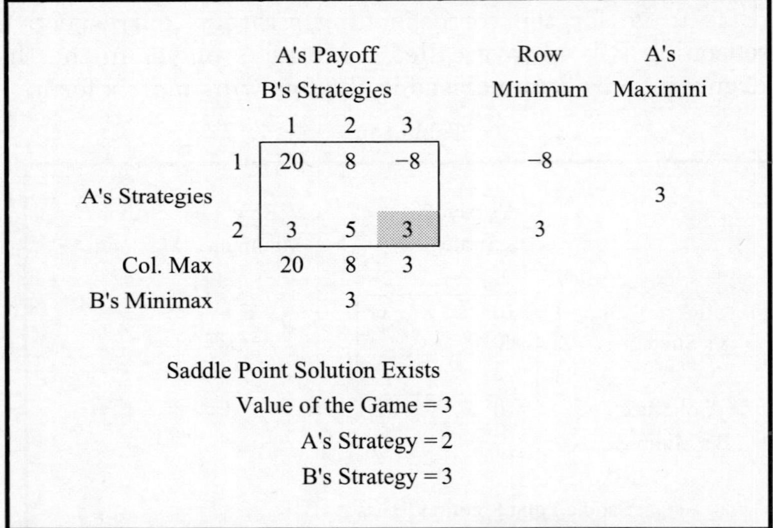

A's Payoff

A's Strategies	B's Strategies 1	2	3	Row Minimum	A's Maximini
1	20	8	−8	−8	3
2	3	5	3	3	
Col. Max	20	8	3		
B's Minimax		3			

Saddle Point Solution Exists

Value of the Game = 3

A's Strategy = 2

B's Strategy = 3

■ Example 3

For the given 2×2 game obtain the saddle point solution if it exists.

Table 14.5

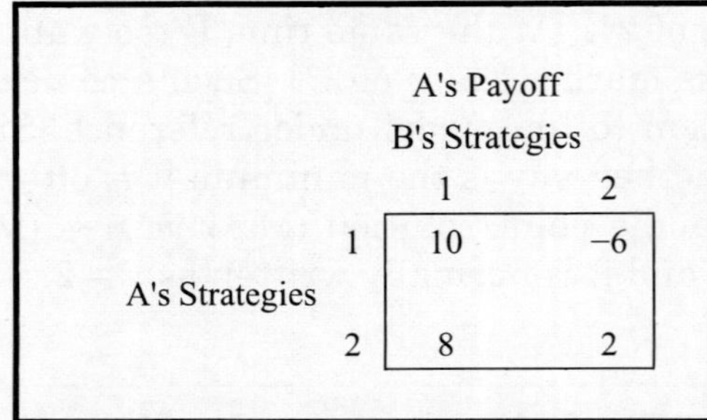

A's Payoff

A's Strategies	B's Strategies 1	2
1	10	−6
2	8	2

Solution

Table 14.6

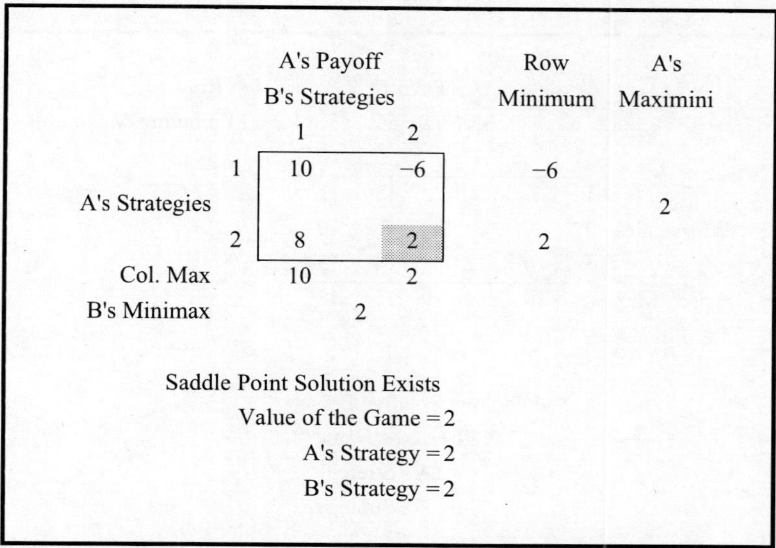

		A's Payoff B's Strategies 1	2	Row Minimum	A's Maximini
A's Strategies	1	10	−6	−6	2
	2	8	2	2	
Col. Max		10	2		
B's Minimax		2			

Saddle Point Solution Exists

Value of the Game = 2

A's Strategy = 2

B's Strategy = 2

■ Example 4

For the given 3×4 game obtain the saddle point solution if it exists.

Table 14.7

A's Payoff B's Strategies		1	2	3	4
A's Strategies	1	8	7	15	12
	2	9	14	8	10
	3	10	12	14	13

Solution

Table 14.8

		A's Payoff B's Strategies				Row Minimum	A's Maximini
		1	2	3	4		
	1	8	7	15	12	7	
A's Strategies	2	9	14	8	10	8	10
	3	10	12	14	13	10	
Col. Max		10	14	15	13		
B's Minimax			10				

Saddle Point Solution Exists
Value of the Game = 10
A's Strategy = 3
B's Strategy = 1

■ Example 5

For the given 3×4 game obtain the saddle point solution if it exists.

Table 14.9

		A's Payoff B's Strategies		
		1	2	3
	1	20	8	−8
A's Strategies	2	12	10	2
	3	3	5	3

Solution

Table 14.10

		A's Payoff B's Strategies 1	2	3	Row Minimum	A's Maximini
A's Strategies	1	20	8	−8	−8	
	2	12	10	2	2	3
	3	3	5	3	3	
Col. Max		20	10	3		
B's Minimax			3			

Saddle Point Solution Exists
Value of the Game = 3
A's Strategy = 3
B's Strategy = 3

■ Example 6

For the given 3×3 game obtain the saddle point solution if it exists.

Table 14.11

		A's Payoff B's Strategies 1	2	3
A's Strategies	1	20	8	−6
	2	12	10	2
	3	3	5	6

Solution

Table 14.12

		A's Payoff B's Strategies 1	2	3	Row Minimum	A's Maximini
A's Strategies	1	20	8	−6	−6	
	2	12	10	2	2	3
	3	3	5	6	3	
Col. Max		20	10	6		
B's Minimax			6			

Saddle Point Solution Does Not Exist
Value of the Game = Nil

14.4 SOLUTION BY MIXED STRATEGIES

From examples 1 to 5 we were able to get the saddle point solution exhibiting pure strategies for both the players. However, for the example 6 above there is no saddle point solution. For such models we can always find a mixed strategy solution by using theory of probabilities. Let player A play strategies 1, 2, 3 with probabilities x_1, x_2, x_3 respectively subject to $x_1 + x_2 + x_3 = 1$. Similarly, let the player B play strategies 1, 2, 3 with probabilities y_1, y_2, y_3 respectively subject to $y_1 + y_2 + y_3 = 1$. Once the probabilities are assigned, we can always calculate the expected payoffs called value of the game by mere substitution.

Table 14.13

			A's Payoff B's Strategies			Row Minimum	A's Maximini
			1	2	3		
			y_1	y_2	y_3		
	1	x_1	20	8	−6	−6	
A's Strategies	2	x_2	12	10	2	2	3
	3	x_3	3	5	6	3	
Col. Max			20	10	6		
B's Minimax				6			

Saddle Point Solution Does Not Exist

Value of the Game = Nil

Method of getting optimum strategic probabilities

Since A is a maximini player he will try to $\underset{Y}{\text{Maxi}}\,\underset{X}{\text{Mini}}\,E(x,y)$.

Similarly, player B is a Mini max player he will try to $\underset{X}{\text{Mini}}\,\underset{Y}{\text{Maxi}}\,E(x,y)$

Thus, for equilibrium $\underset{Y}{\text{Maxi}}\,\underset{X}{\text{Mini}}\,E(x_0,y_0) = \underset{X}{\text{Mini}}\,\underset{Y}{\text{Maxi}}\,E(x_0,y_0)$

According to Minimax theorem for any given problem there will always be at least one mixed strategy solution. Using any one of the following methods we can solve such games.

1. Algebraic Method
2. Sub-games Method
3. Dominance Method
4. Graphical Method
5. Linear Programming Method.

14.5 SOLUTION BY ALGEBRAIC METHOD

This method is often used for a very simple 2 × 2 type of games. The following example illustrates the method.

■ Example 7

For the given A's payoff, find the mixed strategy solution by algebraic method.

Table 14.14

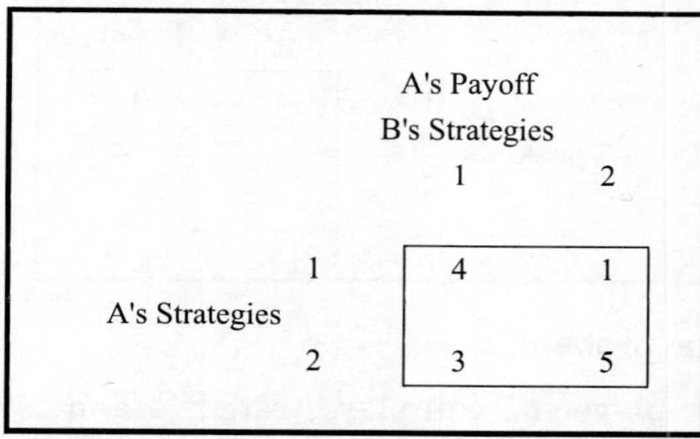

		A's Payoff B's Strategies	
		1	2
A's Strategies	1	4	1
	2	3	5

Solution

Step 1: First let us check whether the problem has a saddle point or not.

Table 14.15

		A's Payoff B's Strategies		Row Minimum	A's Maximini
		1	2		
A's Strategies	1	4	1	1	3
	2	3	5	3	
Col. Max		4	5		
B's Minimax		4			

Saddle Point Solution Does Not Exist

Value of the Game = Nil

Step 2: Since the problem does not have the saddle point, we proceed further for a strategic mixed solution. Let x_1 and x_2 are the two strategic probabilities for player A. Similarly, let y_1 and y_2 are the two strategic probabilities for the player B. Now, since $x_1 + x_2 = 1$ we can always replace $x_2 = 1 - x_1$. Similarly, since $y_1 + y_2 = 1$ we can always replace $y_2 = 1 - y_1$.

Table 14.16

A's Payoff

			B's Strategies	
			1	2
			y_1	y_2
A's Strategies	1	x_1	4	1
	2	x_2	3	5

Calculation of A's strategic probabilities

Since A is a maximini row player he will play strategy 1 and 2 in such a way that his expected earnings from both types are equal to one another irrespective of his opponent B's play.

If B plays his pure strategy 1 then the expected earnings by playing strategy 1 and 2 by A is obtained as

$$4x_1 + 3x_2 = v$$

$$\text{or } 4x_1 + 3(1 - x_1) = v$$

$$x_1 + 3 = v \quad (1)$$

If B plays his pure strategy 2 then the expected earnings by playing strategy 1 and 2 by A is obtained as

$$1x_1 + 5x_2 = v$$

$$\text{or } 1x_1 + 5(1 - x_1) = v$$

$$-4x_1 + 5 = v \quad (2)$$

Now from (1) and (2)

$$x_1 + 3 = -4x_1 + 5$$

$$5x_1 = 2$$

$$x_1 = 2/5$$

$$\text{So } x_2 = 1 - x_1$$

$$= 1 - 2/5 = 3/5$$

Hence, player A will play strategy 1 by 2/5 times of his total play and strategy 2 by 3/5 times of his total play.

Calculation of B's strategic probabilities

Since B is a column player he will play strategy 1 and 2 in such a way that his expected earnings from both types are equal to one another irrespective of his opponent A's play.

If A plays his pure strategy 1 then the expected earnings by playing strategy 1 and 2 by B is obtained as

$$4y_1 + 1y_2 = v$$
$$\text{or } 4y_1 + 1(1 - y_1) = v$$
$$3y_1 + 1 = v... \qquad (3)$$

Similarly, if A plays his pure strategy 2 then the expected earnings by playing strategy 1 and 2 by B is obtained as

$$3y_1 + 5y_2 = v$$
$$\text{or } 3y_1 + 5(1 - y_1) = v$$
$$-2y_1 + 5 = v... \qquad (4)$$

Now from (3) and (4)

$$3y_1 + 1 = -2y_1 + 5$$
$$5y_1 = 4$$
$$y_1 = 4/5$$
$$\text{So } y_2 = 1 - y_1$$
$$= 1 - 4/5 = 1/5$$

Hence, player B will play strategy 1 by 4/5 times of his total and strategy 2 by 1/5 times of his total play. The value of the game is obtained by merely substituting the strategic probabilities in any one of the equations given above.

$$v = 4y_1 + 1y_2 = 4 \times (4/5) + 1 \times (1/5) = 3.4$$

Formula method for a 2×2 game

In the algebraic method outlined, we obtain the needed mixed solution values by solving simultaneous systems of equations. To get such results one can use the following ready to use formula as well.

$$x_1 = \frac{a_{22} - a_{21}}{(a_{11} + a_{22}) - (a_{12} + a_{21})} \qquad y_1 = \frac{a_{22} - a_{12}}{(a_{11} + a_{22}) - (a_{12} + a_{21})}$$

$$V = \frac{a_{11}a_{22} - a_{12}a_{21}}{(a_{11} + a_{22}) - (a_{12} + a_{21})}$$

Table 14.17

			A's Payoff B's Strategies	
			1	2
			y_1	y_2
A's Strategies	1	x_1	a_{11}	a_{12}
	2	x_2	a_{21}	a_{22}

To check the correctness let us redo the example 7 by using the formula as shown below.

$$x_1 = \frac{a_{22} - a_{21}}{(a_{11} + a_{22}) - (a_{12} + a_{21})}$$

$$x_1 = \frac{5-3}{(4+5)-(1+3)} = \frac{2}{5}$$

$$x_2 = 1 - \frac{2}{5} = \frac{3}{5}$$

$$y_1 = \frac{a_{22} - a_{12}}{(a_{11} + a_{22}) - (a_{12} + a_{21})}$$

$$y_1 = \frac{5-1}{(4+5)-(1+3)} = \frac{4}{5}$$

$$y_2 = 1 - \frac{4}{5} = \frac{1}{5}$$

$$V = \frac{a_{11}a_{22} - a_{12}a_{21}}{(a_{11} + a_{22}) - (a_{12} + a_{21})}$$

$$V = \frac{4 \times 5 - 1 \times 3}{(4+5)-(1+3)} = \frac{17}{5} = 3.4$$

Warning: Do not use this method when there is a pure saddle point solution. Thus, before going for mixed strategy solution, check for pure strategy saddle point and stop once it is found. If not solution proceed for mixed strategy solution as suggested.

14.6 SOLUTION BY SUB-GAME METHOD

In this method first we divide the given big game into simple 2×2 games. The so obtained simple games are solved independently to arrive at the final solution. Now since B has three strategies in total he has better capacity to manipulate. So finally B will play that simple game that gives him the least payoff.

■ Example 8

For the given A's payoff, find the mixed strategy solution by algebraic method using sub-games.

Table 14.18

A's Payoff

		B's Strategies		
		1	2	3
A's Strategies	1	275	−50	−75
	2	125	130	150

Solution

Table 14.19

A's Payoff

		B's Strategies			Row Minimum	A's Maximini
		1	2	3		
A's Strategies	1	275	−50	−75	−75	125
	2	125	130	150	125	
Col. Max		275	130	150		
B's Minimax			130			

Saddle Point Solution Does Not Exist

Value of the Game = Nil

Since there is no saddle point solution to this problem let us split this 2×3 game into three 2×2 games first and try to solve them by algebraic method discussed.

Table 14.20

Sub-Game I			Sub-Game II			Sub-Game III		
	B's St.			B's St.			B's St.	
	1	2		1	3		2	3
	y_1	y_2		y_1	y_3		y_2	y_3
A's St. 1 x_1	275	−50	A's St. 1 x_1	275	−75	A's St. 1 x_1	−50	−75
A's St. 2 x_2	125	130	A's St. 2 x_2	125	150	A's St. 2 x_2	130	150

Solution for Sub-Game 1

Step 1: First let us check whether the sub-problem 1 has a saddle point solution or not.

Table 14.21

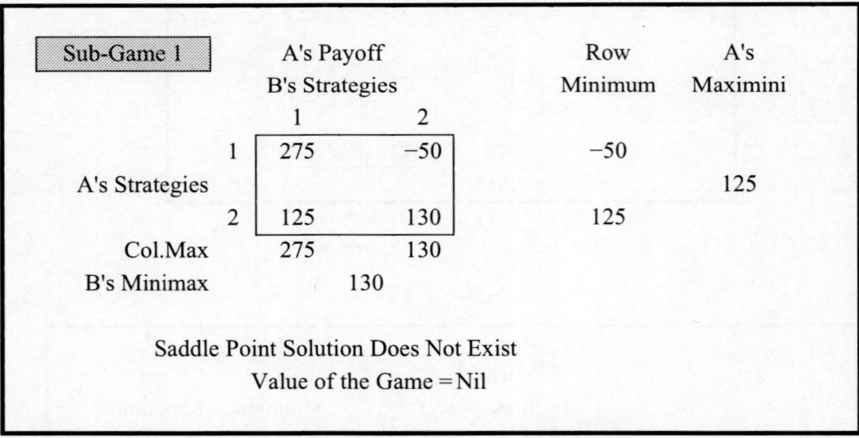

Sub-Game 1	A's Payoff B's Strategies 1	2	Row Minimum	A's Maximini
A's Strategies 1	275	−50	−50	125
A's Strategies 2	125	130	125	
Col.Max	275	130		
B's Minimax	130			

Saddle Point Solution Does Not Exist
Value of the Game = Nil

Step 2: Since the problem does not have the saddle point, we proceed further for a mixed strategy solution.

Table 14.22

A's Payoff	B's Strategies 1 (y_1)	2 (y_2)
A's Strategies 1 x_1	275	−50
A's Strategies 2 x_2	125	130

Let x_1 and x_2 are the two strategic probabilities for player A. Similarly, let y_1 and y_2 are the two strategic probabilities for the player B. Now, since $x_1 + x_2 = 1$ we can always replace $x_2 = 1 - x_1$. Similarly since $y_1 + y_2 = 1$ we can always replace $y_2 = 1 - y_1$.

Calculation of A's strategic probabilities

Since A is a minimax row player he will play strategy 1 and 2 in such a way that his expected earnings from both types are equal to one another irrespective of his opponent B's play.

If B plays his pure strategy 1 then the expected earnings by playing strategy 1 and 2 by A is obtained as

$$275x_1 + 125x_2 = v \quad \text{or}$$
$$275x_1 + 125(1 - x_1) = v$$
$$150x_1 + 125 = v... \qquad (1)$$

If B plays his pure strategy 2 then the expected earnings by playing strategy 1 and 2 by A is obtained as

$$-50x_1 + 130x_2 = v \quad \text{or}$$
$$-50x_1 + 130(1 - x_1) = v$$
$$-180x_1 + 130 = v... \qquad (2)$$

Now from (1) and (2)

$$150x_1 + 125 = -180x_1 + 130$$
$$330x_1 = 5$$
$$x_1 = 1/66$$
$$x_2 = 65/66$$

Hence, player A will play strategy 1 by 1/66 times of his total play and strategy 2 by 65/66 times of his total play. Value of the game v = (150×1/66) +125 =127.30

Calculation of B's strategic probabilities

Since B is a maximini column player he will play strategy 1 and 2 in such a way that his expected earnings from both types are equal to one another irrespective of his opponent A's play.

If a plays his pure strategy 1 then the expected earnings by playing strategy 1 and 2 by B is obtained as

$$275y_1 - 50y_2 = v \quad \text{or}$$
$$275y_1 - 50(1 - y_1) = v$$
$$325y_1 - 50 = v... \qquad (3)$$

Similarly, if A plays his pure strategy 2 then the expected earnings by playing strategy 1 and 2 by B is obtained as

$$125y_1 + 130y_2 = v \text{ or}$$
$$125y_1 + 130(1 - y_1) = v$$
$$-5y_1 + 130 = v \cdots \quad (4)$$

Now from (3) and (4)

$$325y_1 - 50 = -5y_1 + 130$$
$$330y_1 = 180$$
$$y_1 = 36/66$$
$$y_2 = 30/66$$

Hence, player B will play strategy 1 by 36/66 times of his total and strategy 2 by 30/66 times of his total play. The value of the game v = (325 × 36/66) − 50 = 127.30.

Solution for Sub-Game 2

Step 1: First let us check whether the problem has a saddle point or not.

Table 14.23

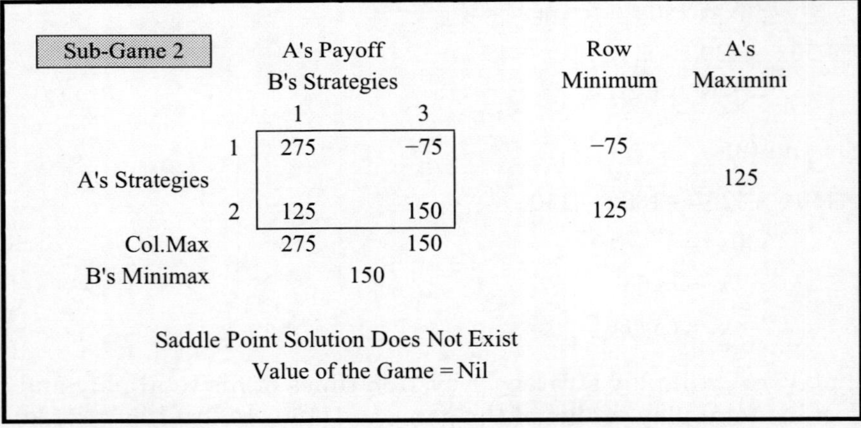

Sub-Game 2		A's Payoff B's Strategies 1	3	Row Minimum	A's Maximini
A's Strategies	1	275	−75	−75	125
	2	125	150	125	
Col.Max		275	150		
B's Minimax		150			

Saddle Point Solution Does Not Exist
Value of the Game = Nil

Step 2:

Table 14.24

A's Payoff			B's Strategies 1	2
			y_1	y_3
A's Strategies	1	x_1	275	−75
	2	x_2	125	150

Since the problem does not have the saddle point we proceed further for a mixed strategy solution. Let x_1 and x_2 are the two strategic probabilities for player A. Similarly, let y_1 and y_3 be the two strategic probabilities for the player B. Now, since $x_1 + x_2 = 1$ we can always replace $x_2 = 1 - x_1$. Similarly since $y_1 + y_3 = 1$ we can always replace $y_3 = 1 - y_1$.

Calculation of A's strategic probabilities

Since A is a maximini row player he will play strategy 1 and 2 in such a way that his expected earnings from both types are equal to one another irrespective of his opponent B's play.

If B plays his pure strategy 1 then the expected earnings by playing strategy 1 and 2 by A is obtained as

$$275x_1 + 125x_2 = v \quad \text{or}$$
$$275x_1 + 125(1 - x_1) = v$$
$$150x_1 + 125 = v... \qquad (1)$$

If B plays his pure strategy 2 then the expected earnings by playing strategy 1 and 2 by A is obtained as

$$-75x_1 + 150x_2 = v \quad \text{or}$$
$$-75x_1 + 150(1 - x_1) = v$$
$$-225x_1 + 150 = v... \qquad (2)$$

Now from (1) and (2)

$$150x_1 + 125x_2 = -225x_1 + 150$$
$$375x_1 = 25$$
$$x_1 = 1/15$$
$$x_2 = 14/15$$

Hence, player A will play strategy 1 by 1/66 times of his total play and strategy 2 by 65/66 times of his total play. The value of the game v = (150×1/15)+125 =135

Calculation of B's strategic probabilities

Since B is a minimax column player he will play strategy 1 and 2 in such a way that his expected earnings from both types are equal to one another irrespective of his opponent A's play.

If a plays his pure strategy 1 then the expected earnings by playing strategy 1 and 2 by B is obtained as

$$275y_1 - 75y_3 = v \quad \text{or}$$
$$275y_1 - 75(1 - y_1) = v$$
$$350y_1 - 75 = v... \qquad (3)$$

Similarly, if A plays his pure strategy 2 then the expected earnings by playing strategy 1 and 2 by B is obtained as

$$125y_1 + 150y_3 = v \text{ or}$$
$$125y_1 + 150(1 - y_1) = v$$
$$-25y_1 + 150 = v... \quad (4)$$

Now from (3) and (4)

$$350y_1 - 75 = -25y_1 + 150$$
$$375y_1 = 225$$
$$y_1 = 9/15$$
$$y_2 = 6/15$$

Hence, player B will play strategy 1 by 9/15 times of his total and strategy 3 by 6/15 times of his total play. The value of the game v = (350×9/15) −75 =135.

Solution for Sub-Game 3

Step 1: First let us check whether the problem has a saddle point or not.

Table 14.25

Sub-Game 3		A's Payoff B's Strategies 2	3	Row Minimum	A's Maximini
A's Strategies	1	−50	−75	−75	130
	2	130	150	130	
Col. Max		130	150		
B's Minimax		130			

Saddle Point Solution Exists

Value of the Game = 130

A's Strategy = 2

B's Strategy = 2

Remember since this sub-game has saddle point our search stops at this stage itself.

Final Solution: Now since B is having three strategies at his command he has better flexibility in handling the situation. Since the given payoff matrix is for A, as usual B will choose the sub-game that gives him the least payoff. In our example since sub-game 1 gives the least payoff (127.3) B will play always Sub−game I after ignoring the third strategy completely.

Note 1: Since in all the three games the expected payoff are positive A is the ultimate winner of the game.

Note 2: If it is a 3×2 game then invariably A has the better chance of handling the situation. So he will choose the sub-game that gives him the highest payoff.

14.7. DOMINANCE METHOD: Dominance Reduction Rules

1. If all the elements in a row of a given A's payoff matrix are greater than the corresponding elements of any other rows then player A would prefer to play the larger elements row and discard the smaller elements rows. Here the row that contains the larger elements is often called the dominant row and the associated strategy is called the dominant strategy for player A.
2. If all the elements in a column of a given A's payoff matrix are less then the corresponding elements of any other columns then player B would prefer to play the smaller elements column and discard the larger elements columns. Here the column that contains the smaller elements is often called the dominant column and the associated strategy is called the dominant strategy for player B.

■ Example 9

Reduce the following 3×2 game by using dominance rule. Also obtain the solution.

Table 14.26

		A's Payoff B's Strategies	
		1	2
A's Strategies	1	4	1
	2	−1	−2
	3	3	5

Solution

Step1:

Reduction by dominance

In the 3×2 Game stated above since A has 3 strategies, he has better capability of handling the situation. Further, since both the entries are negative, playing strategy 2 by A clearly involves payment to B irrespective of his opponent's play. Since in strategy 1 the corresponding entries are greater than strategy 2 entries, we say that strategy 1 dominates strategy 2 for A. By same reasoning we also notice that strategy 3 also dominates strategy 2 for A. So ultimately, player A will discard strategy 2 and play only strategy 1 and 3. The so reduced 2×2 game is shown below.

Table 14.27

		A's Payoff B's Strategies 1	2
A's Strategies	1	4	1
	3	3	5

Step 2:

Search for saddle point

Table 14.28

		A's Payoff B's Strategies 1	2	Row Minimum	A's Maximini
A's Strategies	1	4	1	1	3
	3	3	5	3	
Col. Max		4	5		
B's Minimax		4			

Saddle Point Solution Does Not Exist

Step 3:

Algebraic solution

Table 14.29

A's Payoff

		B's Strategies 1 y_1	2 y_2
A's Strategies	1 x_1	4	1
	3 x_3	3	5

Calculation of A's strategic probabilities

Since A is a maximini row player he will play strategy 1 and 2 in such a way that his expected earnings from both types are equal to one another irrespective of his opponent B's play.

If B plays his pure strategy 1 then the expected earnings by playing strategy 1 and 2 by A is obtained as

$$4x_1 + 3x_3 = v \text{ or}$$
$$4x_1 + 3(1 - x_1) = v$$
$$x_1 + 3 = v... \quad (1)$$

If B plays his pure strategy 2 then the expected earnings by playing strategy 1 and 2 by A is obtained as

$$1x_1 + 5x_3 = v \text{ or}$$
$$1x_1 + 5(1 - x_1) = v$$
$$-4x_1 + 5 = v... \quad (2)$$

Now from (1) and (2)

$$x_1 + 3 = -4x_1 + 5$$
$$5x_1 = 2$$
$$x_1 = 2/5$$
$$x_3 = 3/5$$

Hence, player A will play strategy 1 by 2/5 times of his total and strategy 3 by 3/5 times of his total play.

Calculation of B's strategic probabilities

Since B is a minimax column player he will play strategy 1 and 2 in such a way that his expected earnings from both types are equal to one another irrespective of his opponent A's play.

If a plays his pure strategy 1 then the expected earnings by playing strategy 1 and 2 by B is obtained as

$$4y_1 + 1y_2 = v \quad \text{or}$$
$$4x_1 + 1(1 - y_1) = v$$
$$3y_1 + 1 = v... \qquad (3)$$

Similarly, if A plays his pure strategy 2 then the expected earnings by playing strategy 1 and 2 by B is obtained as

$$3y_1 + 5y_2 = v \quad \text{or}$$
$$3x_1 + 5(1 - y_1) = v$$
$$-2y_1 + 5 = v... \qquad (4)$$

Now from (3) and (4)

$$3y_1 + 1 = -2y_1 + 5$$
$$5y_1 = 4$$
$$y_1 = 4/5$$
$$y_2 = 1/5$$

Hence, player B will play strategy 1 by 4/5 times of his total and strategy 3 by 1/5 times of his total play. The value of the game v = (3 × 4/5) +1 = 3.4.

■ Example 10

Reduce the following 2×4 game by using dominance rule. Also obtain it solution.

Table 14.30

A's Payoff

		B's Strategies			
		1	2	3	4
A's Strategies	1	0	2	–4	–7
	2	1	3	–4	–1

Solution

Step 1: Reduction by dominance

In the 2×4 Game stated above, since B has 4 strategies he has better capability of handling the situation. Further, since we are given with A's payoff negative entries refers gains to B. Clearly third and fourth strategies are dominated by B. So B will not play strategies 1 and 2 and make payments to A. So the reduced 2×2 game is obtained as

Table 14.31

		A's Payoff B's Strategies	
		3	4
A's Strategies	1	−4	−7
	2	−4	−1

Step 2: Search for saddle point

Table 14.32

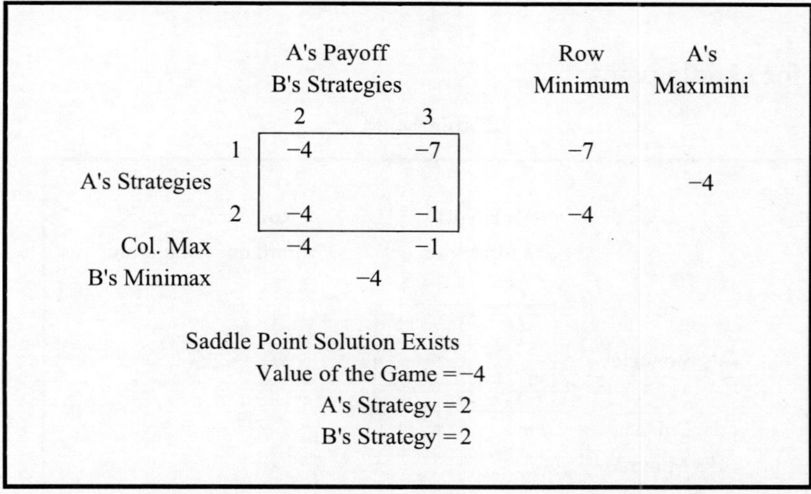

		A's Payoff B's Strategies		Row Minimum	A's Maximini
		2	3		
A's Strategies	1	−4	−7	−7	−4
	2	−4	−1	−4	
Col. Max		−4	−1		
B's Minimax		−4			

Saddle Point Solution Exists

Value of the Game = −4

A's Strategy = 2

B's Strategy = 2

14.8. SOLUTION BY GRAPHICAL METHOD

The graphical method is more appropriate two solve both (M×2) and (2×N) type of games. In the first type A have M strategies to play while B has only 2 strategies to counter. In the second type A have only two strategies while B has N strategies to counter. Often bigger games like (M×N) can be solved by this method after reducing the game into either (M×2) of (2×N) type. Deleting irreverent rows and columns by using dominance rule, often helps us a lot in solving the bigger problems by this method. The following example illustrates the method.

■ Example 11

Solve the following 2×3 game by graphical method.

Table 14.33

		A's Payoff B's Strategies		
		1	2	3
A's Strategies	1	1	3	11
	2	8	5	2

Solution

Step 1: Search for saddle point

Table 14.34

		A's Payoff B's Strategies			Row Minimum	A's Maximini
		1	2	3		
A's Strategies	1	1	3	11	1	2
	2	8	5	2	2	
Col. Max		8	5	11		
B's Minimax			5			

Saddle Point Solution Does Not Exist

Step 2: Graphical solution

Let x_1 and x_2 are the strategic probabilities for strategy 1 and 2 respectively to player A so that $x_1 + x_2 = 1$. Similarly, let y_1, y_2 and y_3 are the strategic probabilities for strategy 1, 2 and 3 of B so that $y_1 + y_2 + y_3 = 1$.

Table 14.35

A's Payoff

			B's Strategies		
			1	2	3
			y_1	y_2	y_3
A's Strategies	1	x_1	1	3	11
	2	x_2	8	5	2

Since our player A has only two strategies, in Figure 14.1 we prefer to measure his first strategy probability x_1 on the x-axis. At the origin since $x_1 = 0$ our player A invariably play strategy 2 with certainty so that $x_2 =1$. Similarly, if $x_1 =1$, the maximum, then we conclude that our player A play strategy 1 with certainty so that $x_2 = 0$. As a third case, if $x_1 = 1/3$ then clearly $x_2 = 2/3$ so that $x_1 + x_2 =1$. This time our player A will play first strategy 1/3rd times of his total and strategy 2 by 2/3rd times of his total.

On the vertical axis we measure resulting expected payoff from player B to player A. Suppose A plays strategy 1 and 2 with probabilities x_1 and x_2 respectively and B plays his pure strategy 1 with certainty (y_1=1) then the expected payoff for A is obtained as:

$$E(x_1) = 1x_1 + 8x_2 = 1x_1 + 8(1 - x_1) = 8 - 7x_1$$

Suppose A plays strategy 1 and 2 with probabilities x_1 and x_2 respectively and B plays his pure strategy 2 with certainty (y_2=1) then the expected payoff for A is obtained as:

$$E(x_1) = 3x_1 + 5x_2 = 3x_1 + 5(1 - x_1) = 5 - 2x_1$$

Suppose A plays strategy 1 and 2 with probabilities x_1 and x_2 respectively and B plays his pure strategy 3 with certainty (y_3=1) then the expected payoff for A is obtained as:

$$E(x_1) = 11x_1 + 2x_2 = 11x_1 + 2(1 - x_1) = 2 + 9x_1$$

Table 14.36 gives the summary of all the three expected payoff straight lines corresponding to the respective pure strategies of B.

Table 14.36

B's pure strategies	Expected payoff from B to A
1	$E(x_1) = 1x_1+8(1-x_1) = 8-7x_1$
2	$E(x_1) = 3x_1+5(1-x_1) = 5-2x_1$
3	$E(x_1) = 11x_1+2(1-x_1) = 2+9x_1$

Now we draw all these three pure strategy lines of B into the graph by identifying two points each as shown below.

Table 14.37

$E(x_1) = 8-7x_1$	
$E(x_1)$	x_1
8	0
1	1

$E(x_1) = 5-2x_1$	
$E(x_1)$	x_1
5	0
3	1

$E(x_1) = 2+9x_1$	
$E(x_1)$	x_1
2	0
11	1

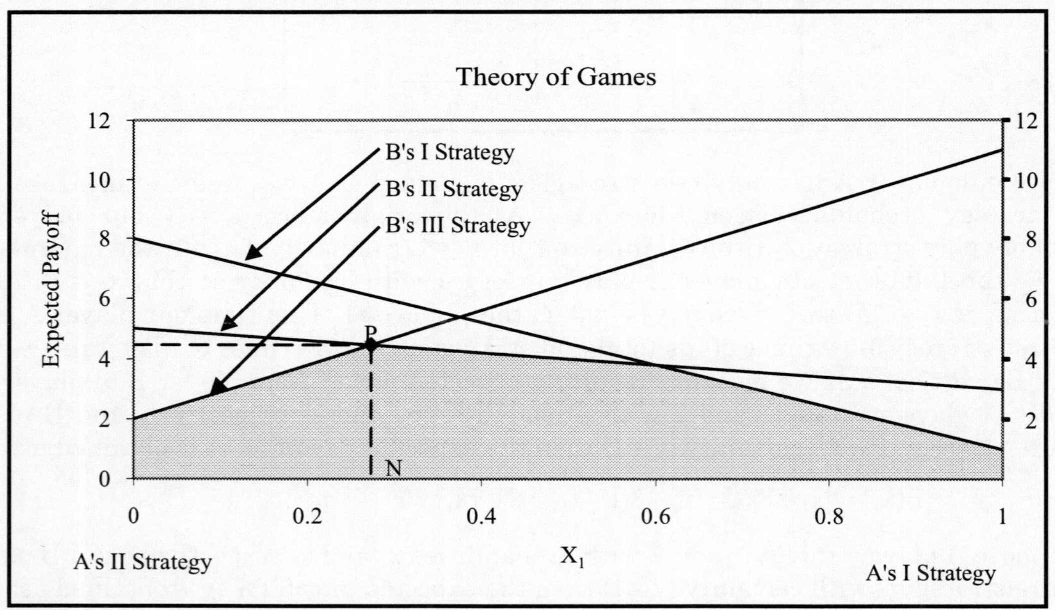

Figure 14.1

Remember, that our player A is a maximini player and thus will choose the maximum among the resulting minimums. The four corners of the shaded region constitute the four minimum points. Among the four minimums corners clearly the corner P is the maximum payoff yielding corner point. Once the Maximini point is identified graphically the corresponding x_1 value is obtained 3/11 from the graph. Similarly, from the vertical axis we read the corresponding value of the game as v = 49/11. However, this method fails to yield y_1, y_2 and y_3, values pertaining to player B.

However, since at the equilibrium point P the pure strategy 2 and 3 lines of B intersect, and strategy 1 line completely lie above the point P, we conclude that strategy 1 is inferior as for as B is concerned. After removing this strategically inferior strategy 1 we rewrite our resulting 2×2 game as shown below.

Table 14.38

A's Payoff

		B's Strategies	
		2	3
A's Strategies	1	3	11
	2	5	2

Step 3: Calculation of A's strategic probabilities

Since A is a maximini row player he will play strategy 1 and 2 in such a way that his expected earnings from both types are equal to one another irrespective of his opponent B's play.

If B plays his pure strategy 1 then the expected earnings by playing strategy 1 and 2 by A is obtained as

$$3x_1 + 5x_2 = v \quad \text{or}$$
$$3x_1 + 5(1 - x_1) = v$$
$$-2x_1 + 5 = v... \qquad (1)$$

If B plays his pure strategy 2 then the expected earnings by playing strategy 1 and 2 by A is obtained as

$$11x_1 + 2x_2 = v \quad \text{or}$$
$$11x_1 + 2(1 - x_1) = v$$
$$9x_1 + 2 = v... \qquad (2)$$

Now from (1) and (2)

$$-2x_1 + 5 = 9x_1 + 2$$
$$-11x_1 = -3$$
$$x_1 = 3/11$$
$$x_2 = 8/11$$

Hence, player A will play strategy 1 by 2/5 times of his total and strategy 2 by 3/5 times of his total play.

Calculation of B's strategic probabilities

Since B is a minimax column player he will play strategy 1 and 2 in such a way that his expected earnings from both types are equal to one another irrespective of his opponent A's play.

If a plays his pure strategy 1 then the expected earnings by playing strategy 1 and 2 by B is obtained as

$$3y_2 + 11y_3 = v \quad \text{or}$$
$$3y_2 + 11(1 - y_2) = v$$
$$-8y_2 + 11 = v... \qquad (1)$$

Similarly, if A plays his pure strategy 2 then the expected earnings by playing strategy 1 and 2 by B is obtained as

$$5y_2 + 2y_3 = v \quad \text{or}$$
$$5y_2 + 2(1 - y_2) = v$$
$$3y_2 + 2 = v... \qquad (2)$$

Now from (3) and (4)

$$-8y_2 + 11 = 3y_2 + 2$$
$$-11y_2 = -9$$
$$y_2 = 9/11$$
$$y_3 = 2/11$$

The value of the game v = (9×3/11) +2 = 49/11.

■ Example 12

Solve the following 2×4 game by graphical method.

Table 14.39

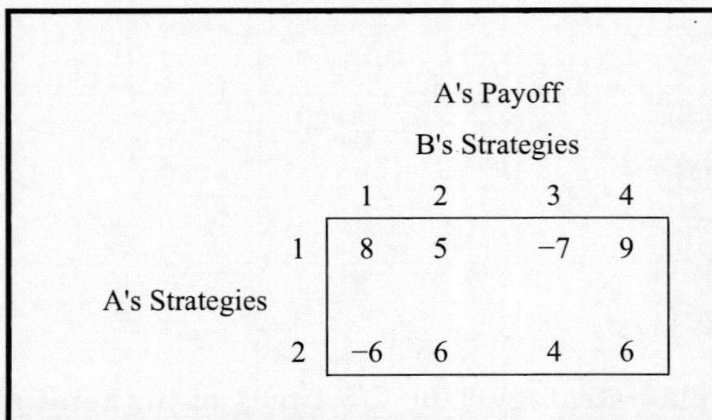

A's Payoff

		B's Strategies			
		1	2	3	4
A's Strategies	1	8	5	−7	9
	2	−6	6	4	6

Solution: Search for saddle point

Table 14.40

		A's Payoff B's Strategies				Row Minimum	A's Maximini
		1	2	3	4		
A's Strategies	1	8	5	−7	9	−7	−6
	2	−6	6	4	6	−6	
Col. Max		8	6	4	9		
B's Minimax				4			

Saddle Point Solution Does Not Exist

Graphical solution

Suppose A plays strategy 1 and 2 with probabilities x_1 and x_2 respectively and B plays his pure strategy 1 with certainty (y_1=1) then the expected payoff for A is obtained as:

$$E(x_1) = 8x_1 - 6x_2 = 8x_1 - 6(1-x_1) = -6 + 14x_1$$

Suppose A plays strategy 1 and 2 with probabilities x_1 and x_2 respectively and B plays his pure strategy 2 with certainty (y_2=1) then the expected payoff for A is obtained as:

$$E(x_1) = 5x_1 + 6x_2 = 5x_1 + 6(1-x_1) = 6 - 1x_1$$

Suppose A plays strategy 1 and 2 with probabilities x_1 and x_2 respectively and B plays his pure strategy 3 with certainty (y_3=1) then the expected payoff for A is obtained as:

$$E(x_1) = -7x_1 + 4x_2 = -7x_1 + 4(1-x_1) = 4 - 11x_1$$

Suppose A plays strategy 1 and 2 with probabilities x_1 and x_2 respectively and B plays his pure strategy 4 with certainty (y_4=1) then the expected payoff for A is obtained as:

$$E(x_1) = 9x_1 + 6x_2 = 9x_1 + 6(1-x_1) = 6 + 3x_1$$

Now we draw all the four pure strategy lines of B into the graph as shown below.

The following Table 14.41 gives the summery of all the three expected payoff straight lines corresponding to the respective pure strategies of B.

Table 14.41

1	$E(x_1) = -6+14x_1$
2	$E(x_1) = 6 - 1x_1$
3	$E(x_1) = 4-11x_1$
4	$E(x_1) = 6+3x_1$

Now we draw all these four pure strategy lines of B into the graph by identifying two points each as shown below.

Table 14.42

$E(x_1) = -6+14x_1$	
$E(x_1)$	x_1
−6	0
8	1

$E(x_1) = 6-1x_1$	
$E(x_1)$	x_1
6	0
5	1

$E(x_1) = 4-11x_1$	
$E(x_1)$	x_1
4	0
−7	1

$E(x_1) = 6+3x_1$	
$E(x_1)$	x_1
6	0
9	1

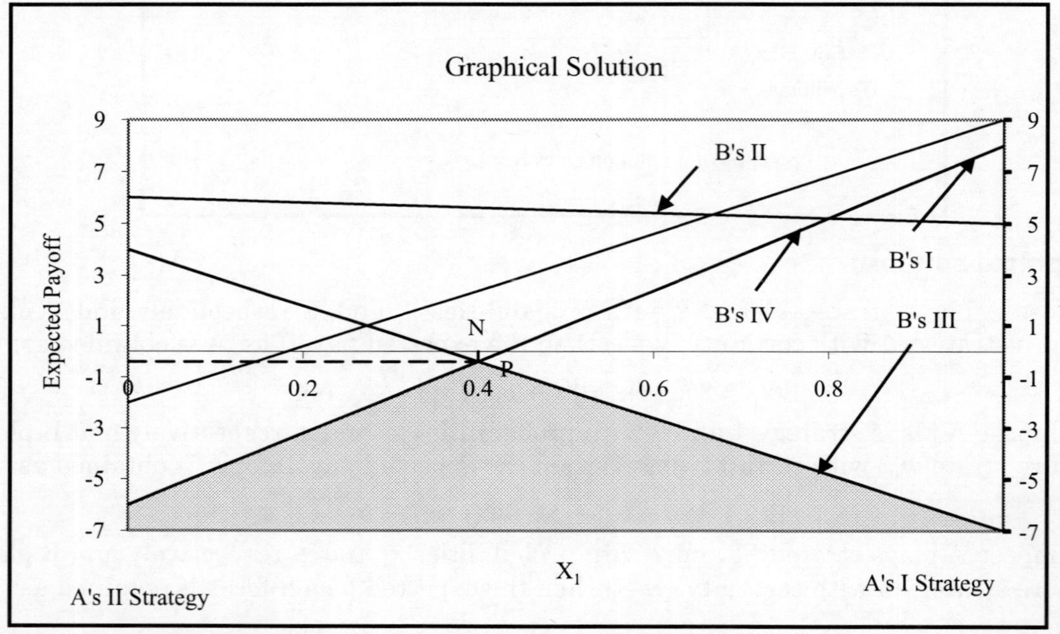

Figure 14.2

From the figure it is clear that though player A plays both the strategies, player B will play only strategy 1 and 3. Thus our corresponding 2×2 game may be rewritten as:

Table 14.43

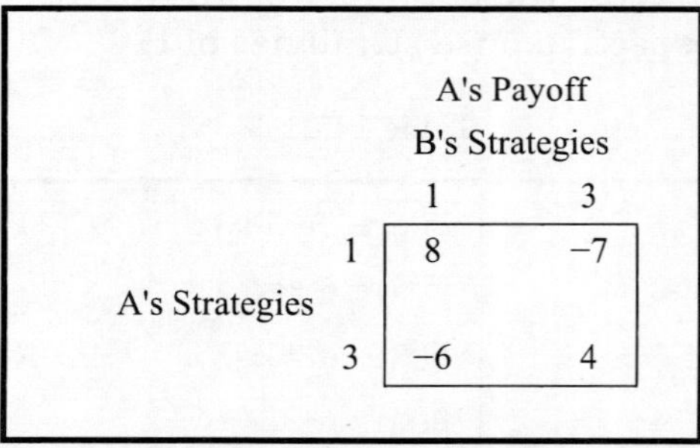

A's Payoff

		B's Strategies	
		1	3
A's Strategies	1	8	−7
	3	−6	4

Let us this time solve this game by using the formula

$$x_1 = \frac{a_{22} - a_{21}}{(a_{11} + a_{22}) - (a_{12} + a_{21})}$$

$$= \frac{4-(-6)}{(8+4)-(-7-6)} = \frac{2}{5}$$

$$x_2 = 1 - \frac{2}{5} = \frac{3}{5}$$

$$y_1 = \frac{a_{22} - a_{12}}{(a_{11} + a_{22}) - (a_{12} + a_{21})}$$

$$= \frac{4-(-7)}{(8+4)-(-7-6)} = \frac{11}{25}$$

$$y_3 = 1 - \frac{11}{25} = \frac{14}{25}$$

$$v = \frac{a_{11}a_{22} - a_{12}a_{21}}{(a_{11} + a_{22}) - (a_{12} + a_{21})}$$

$$= \frac{8\times 4 - (-7)\times(-6)}{(8+4)-(-7-6)}$$

$$= \frac{17}{5} = -2/5 = -0.4.$$

(2×N) Type Game with Multiple Solutions

Even in a (2×N) game multiple solutions are also possible. The following example illustrates one such case.

■ Example 13

Solve the following 2×4 game for multiple solutions by graphical method.

Table 14.44

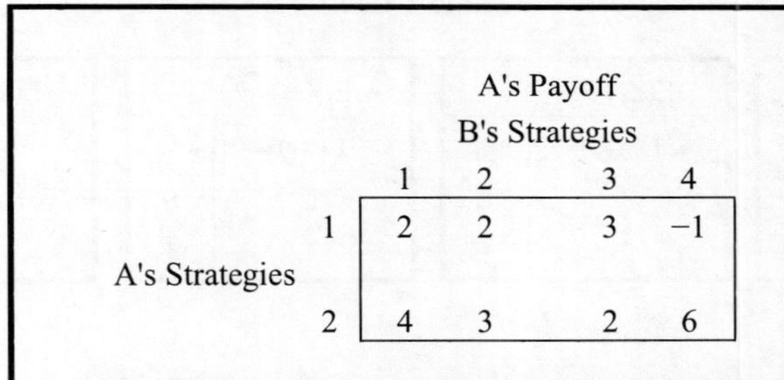

A's Payoff

		B's Strategies			
		1	2	3	4
A's Strategies	1	2	2	3	−1
	2	4	3	2	6

Solution

Step 1: Search for saddle point

Table 14.45

		A's Payoff B's Strategies				Row Minimum	A's Maximini
		1	2	3	4		
A's Strategies	1	2	2	3	−1	−1	2
	2	4	3	2	6	2	
Col. Max		4	3	3	6		
B's Minimax			3				

Saddle Point Solution Does Not Exist

Step 2: Graphical solution

The needed expected payoff lines for all the four pure strategies of B are found and reported in Table 14.46 and inserted in the graph.

Table 14.46

B's pure strategies	Expected payoff
1	$E(x_1) = 2x_1+4(1-x_1) = -2x_1+4$
2	$E(x_1) = 2x_1+3(1-x_1) = -x_1+3$
3	$E(x_1) = 3x_1+2(1-x_1) = x_1+2$
4	$E(x_1) = -1x_1+6(1-x_1) = -7x_1+6$

As usual we identify four lines as shown in Table 14.47.

Table 14.47

$E(x_1) = -2+4x_1$	
$E(x_1)$	x_1
−2	0
2	1

$E(x_1) = 3-1x_1$	
$E(x_1)$	x_1
3	0
2	1

$E(x_1) = 2+x_1$	
$E(x_1)$	x_1
2	0
3	1

$E(x_1) = 6-7x_1$	
$E(x_1)$	x_1
6	0
−1	1

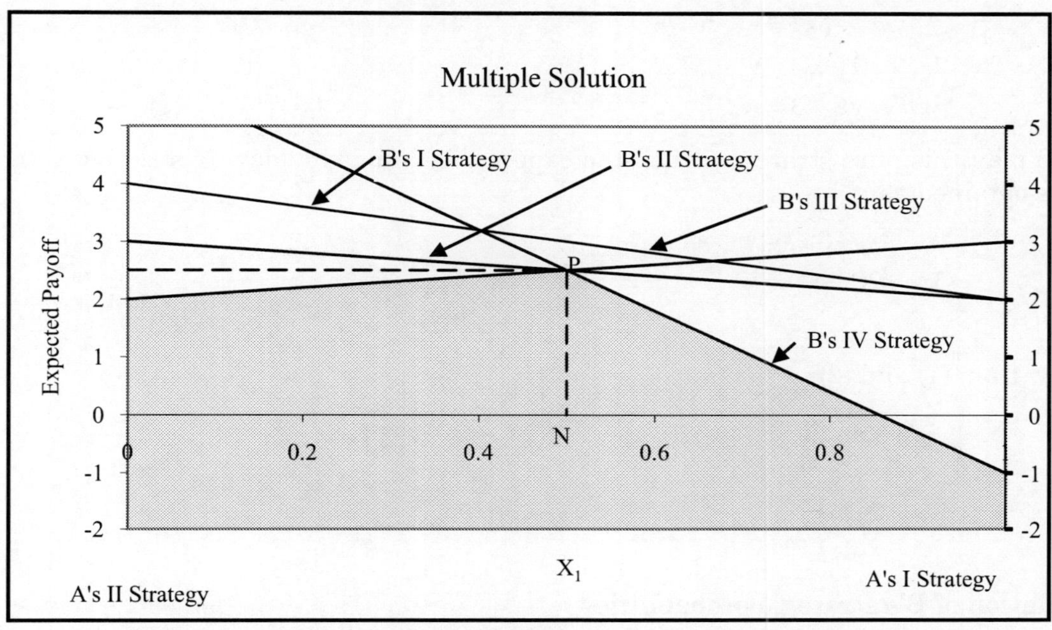

Figure 14.3

At the Maximini equilibrium point P we notice that B's strategy lines 2, 3, 4 pass through. So strategy 1 can be discarded as an inefficient strategy for B. With the remaining three strategies the equilibrium point P can be thought of as the intersection of either strategies 2&3, or 2&4, or 3&4 as described by the following three games.

Table 14.48

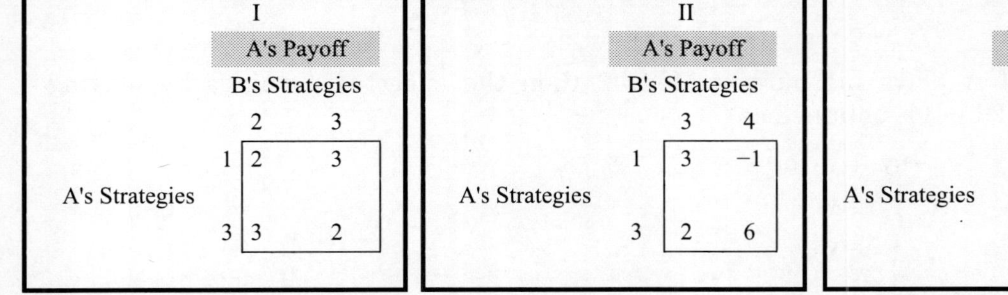

I — A's Payoff

A's Strategies	B's Strategies 2	B's Strategies 3
1	2	3
3	3	2

II — A's Payoff

A's Strategies	B's Strategies 3	B's Strategies 4
1	3	−1
3	2	6

III — A's Payoff

A's Strategies	B's Strategies 2	B's Strategies 4
1	2	−1
3	3	6

Solution for first possibility

Calculation of A's strategic probabilities

Since A is a maximini row player he will play strategy 1 and 2 in such a way that his expected earnings from both types are equal to one another irrespective of his opponent B's play.

If B plays his pure strategy 1 then the expected earnings by playing strategy 1 and 2 by A is obtained as

$$2x_1 + 3x_2 = v \quad \text{or}$$
$$2x_1 + 3(1 - x_1) = v$$
$$-x_1 + 3 = v... \qquad (1)$$

If B plays his pure strategy 2 then the expected earnings by playing strategy 1 and 2 by A is obtained as

$$3x_1 + 2x_2 = v \quad \text{or}$$
$$3x_1 + 2(1 - x_1) = v$$
$$x_1 + 2 = v... \qquad (2)$$

Now from (1) and (2)

$$-x_1 + 3 = x_1 + 2$$
$$-2x_1 = -1$$
$$x_1 = 1/2$$
$$x_2 = 1/2$$

Calculation of B's strategic probabilities

Since B is a minimax column player he will play strategy 1 and 2 in such a way that his expected earnings from both types are equal to one another irrespective of his opponent A's play.

If A plays his pure strategy 1 then the expected earnings by playing strategy 1 and 2 by B is obtained as

$$3y_3 - y_4 = v \quad \text{or}$$
$$3y_3 - 1(1 - y_3) = v$$
$$4y_3 - 1 = v... \qquad (3)$$

Similarly, if A plays his pure strategy 2 then the expected earnings by playing strategy 1 and 2 by B is obtained as

$$2y_2 - 6y_3 = v \quad \text{or}$$
$$2y_2 - 1(1 - y_2) = v$$
$$y_2 + 2 = v... \qquad (4)$$

Now from (3) and (4)

$$-y_2 + 3 = y_2 + 2$$
$$-2y_2 = -1$$
$$y_2 = 1/2$$
$$y_3 = 1/2$$

The value of the game v = (1/2) +2 = 5/2.

Solution for second possibility

Calculation of A's strategic probabilities

Since A is a maximini row player he will play strategy 1 and 2 in such a way that his expected earnings from both types are equal to one another irrespective of his opponent B's play.

If B plays his pure strategy 1 then the expected earnings by playing strategy 1 and 2 by A is obtained as

$$3x_1 + 2x_2 = v \quad \text{or}$$
$$3x_1 + 2(1 - x_1) = v$$
$$x_1 + 2 = v... \qquad (1)$$

If B plays his pure strategy 2 then the expected earnings by playing strategy 1 and 2 by A is obtained as

$$-1x_1 + 6x_2 = v \quad \text{or}$$
$$-1x_1 + 6(1 - x_1) = v$$
$$-7x_1 + 6 = v... \qquad (2)$$

Now from (1) and (2)

$$x_1 + 2 = -7x_1 + 6$$
$$8x_1 = 4$$
$$x_1 = 1/2$$
$$x_2 = 1/2$$

Calculation of B's strategic probabilities

Since B is a minimax column player he will play strategy 1 and 2 in such a way that his expected earnings from both types are equal to one another irrespective of his opponent A's play.

If A plays his pure strategy 1 then the expected earnings by playing strategy 1 and 2 by B is obtained as

$$3y_3 - y_4 = v \quad \text{or}$$
$$3y_3 - 1(1 - y_3) = v$$
$$4y_3 - 1 = v... \qquad (3)$$

Similarly, if A plays his pure strategy 2 then the expected earnings by playing strategy 1 and 2 by B is obtained as

$$2y_3 + 6y_4 = v \quad \text{or}$$
$$2y_3 - 6(1 - y_3) = v$$
$$-4y_3 + 6 = v... \qquad (4)$$

Now from (3) and (4)

$$4y_3 - 1 = -4y_3 + 6$$
$$8y_3 = 7$$
$$y_3 = 7/8$$
$$y_4 = 1/8$$

The value of the game v = (1/2) +2 = 5/2.

Solution for third possibility

Calculation of A's strategic probabilities

Since A is a maximini row player he will play strategy 1 and 2 in such a way that his expected earnings from both types are equal to one another irrespective of his opponent B's play.

If B plays his pure strategy 1 then the expected earnings by playing strategy 1 and 2 by A is obtained as

$$2x_1 + 3x_2 = v \quad \text{or}$$
$$2x_1 + 3(1 - x_1) = v$$
$$-x_1 + 3 = v... \qquad (1)$$

If B plays his pure strategy 2 then the expected earnings by playing strategy 1 and 2 by A is obtained as

$$-1x_1 + 6x_2 = v \quad \text{or}$$
$$-1x_1 + 6(1 - x_1) = v$$
$$-7x_1 + 6 = v... \qquad (2)$$

Now from (1) and (2)

$$-x_1 + 3 = -7x_1 + 6$$
$$6x_1 = 3$$
$$x_1 = 1/3$$
$$x_2 = 2/3$$

Calculation of B's strategic probabilities

Since B is a minimax column player he will play strategy 1 and 2 in such a way that his expected earnings from both types are equal to one another irrespective of his opponent A's play.

If A plays his pure strategy 1 then the expected earnings by playing strategy 1 and 2 by B is obtained as

$$2y_2 - 1y_4 = v \quad \text{or}$$
$$2y_2 - 1(1 - y_2) = v$$
$$3y_2 - 1 = v... \qquad (3)$$

Similarly, if A plays his pure strategy 2 then the expected earnings by playing strategy 1 and 2 by B is obtained as

$$3y_2 + 6y_4 = v \quad \text{or}$$
$$3y_2 + 6(1 - y_2) = v$$
$$-3y_2 + 6 = v... \qquad (4)$$

Now from (3) and (4)

$$3y_2 - 1 = -3y_2 + 6$$
$$0y_2 = 7$$
$$y_2 = \text{infinite}$$

Thus, there is no optimum solution to this third model. Thus, possibility 1 and 2 yields optimum solutions.

■ Example 14

Solve the following 2×3 game by graphical method.

Table 14.49

		A's Payoff B's Strategies		
		1	2	3
A's Strategies	1	6	4	3
	2	2	4	8

Solution: Search for saddle point

Table 14.50

		A's Payoff B's Strategies			Row Minimum	A's Maximini
		1	2	3		
A's Strategies	1	6	4	3	3	3
	2	2	4	8	2	
Col. Max		6	4	8		
B's Minimax			4			

Saddle Point Solution Does Not Exist

Graphical solution

Table 14.51

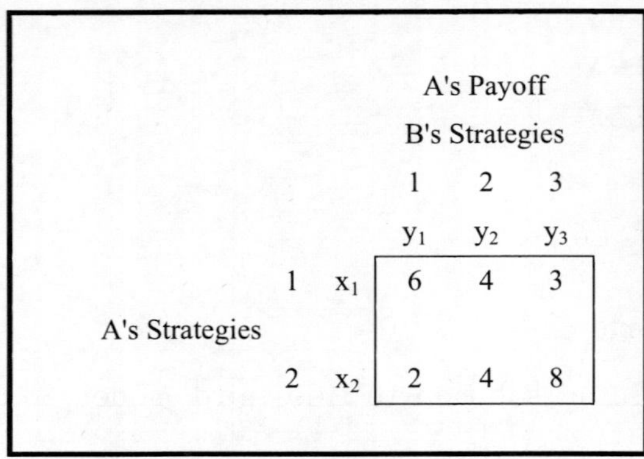

		A's Payoff B's Strategies		
		1	2	3
		y_1	y_2	y_3
A's Strategies	1 x_1	6	4	3
	2 x_2	2	4	8

Suppose A plays strategy 1 and 2 with probabilities x_1 and x_2 respectively and B plays his pure strategy 1 with certainty (y_1=1) then the expected payoff for A is obtained as:

$$E(x_1) = 6x_1 + 2x_2 = 6x_1 + 2(1 - x_1) = 2 + 4x_1$$

Suppose A plays strategy 1 and 2 with probabilities x_1 and x_2 respectively and B plays his pure strategy 2 with certainty (y_2=1) then the expected payoff for A is obtained as:

$$E(x_1) = 4x_1 + 4x_2 = 4x_1 + 4(1 - x_1) = 4$$

Suppose A plays strategy 1 and 2 with probabilities x_1 and x_2 respectively and B plays his pure strategy 3 with certainty (y_3=1) then the expected payoff for A is obtained as:

$$E(x_1) = 3x_1 + 8x_2 = 3x_1 + 8(1 - x_1) = 8 - 5x_1$$

Now let us identify two points on each one these three lines as shown in the following table and draw them on the graph sheet as shown in Figure 14.4.

Table 14.52

$E(x_1) = 2+4x_1$	
$E(x_1)$	x_1
2	0
6	1

$E(x_1) = 4$	
$E(x_1)$	x_1
4	0
4	1

$E(x_1) = 8-5x_1$	
$E(x_1)$	x_1
8	0
3	1

Clearly, this time there is no single maximums among the minimums. In the interval AB all the values of x_1 will yield the same 4 unit payoff to player A. For example if A plays at D then the corresponding solution values are obtained by using the formula as usual.

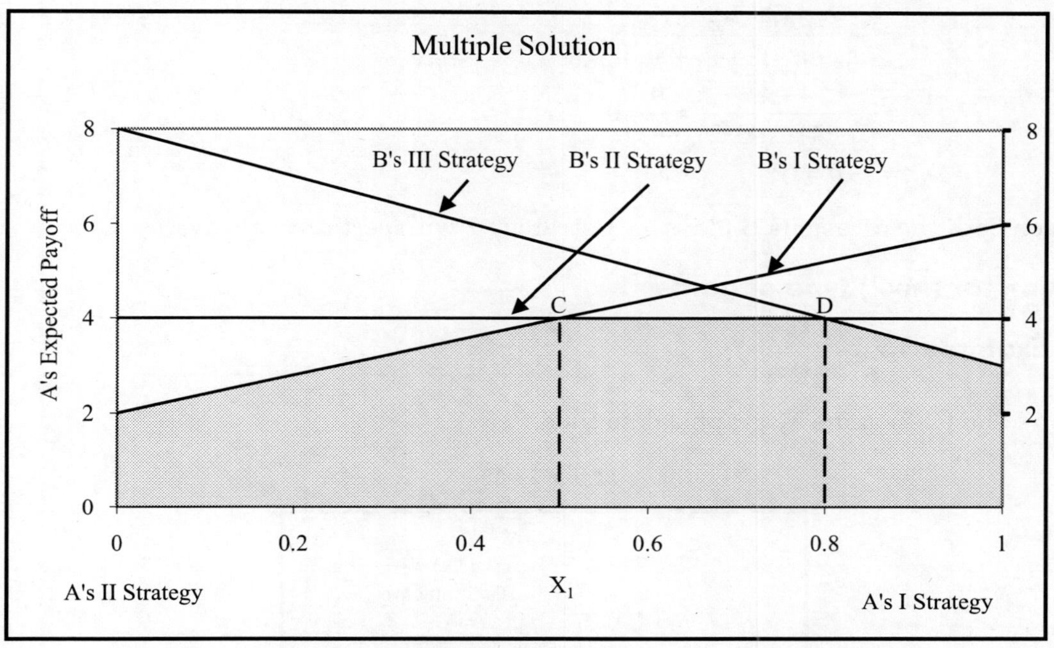

Figure 14.4

$$x_1 = \frac{a_{22} - a_{21}}{(a_{11} + a_{22}) - (a_{12} + a_{21})}$$

$$= \frac{8-4}{(4+8)-(3+4)} = \frac{4}{5} = 0.8$$

$$x_2 = 1 - \frac{4}{5} = \frac{1}{5} = 0.2$$

$$y_2 = \frac{a_{22} - a_{12}}{(a_{11} + a_{22}) - (a_{12} + a_{21})}$$

$$y_2 = \frac{8-3}{(4+8)-(3+4)} = \frac{5}{5} = 1$$

$$y_3 = 1 - 1 = 0$$

If A play at C then the corresponding solution values are obtained by using the formula as usual.

$$x_1 = \frac{a_{22} - a_{21}}{(a_{11} + a_{22}) - (a_{12} + a_{21})}$$

$$x_1 = \frac{4-2}{(6+4)-(4+2)} = \frac{2}{4} = 0.5$$

$$x_2 = 1 - 0.5 = 0.5$$

$$y_1 = \frac{a_{22} - a_{12}}{(a_{11} + a_{22}) - (a_{12} + a_{21})}$$

$$y_1 = \frac{4-4}{(6+4)-(4+2)} = \frac{0}{4} = 0$$

$$y_2 = 1 - 0 = 1$$

So on both the occasions B plays only strategy 2 with certainty always.

Solution for (M×2) type of game

■ Example 15

Solve the (4×2) game by graphical method.

Table 14.53

A's Payoff

		B's Strategies 1	B's Strategies 2
A's Strategies	1	2	4
	2	2	3
	3	3	2
	4	−2	6

Solution: Search for saddle point

Table 14.54

		B's Strategies 1	B's Strategies 2	Minimum	Maximini
A's Strategies	1	2	4	2	
	2	2	3	2	2
	3	3	2	2	
	4	−2	6	−2	
Col. Max		3	6		
B's Minimax			6		

Saddle Point Solution Does Not Exist

In this game since B is having two strategies we measure y_1, the probability of strategy 1 of player B on the X-axis. When $y_1 = 1$ ($y_2 = 0$) then clearly our player B will use strategy 1 with certainty. Similarly when $y_1 = 0$ ($y_2 = 1$) B will use strategy with certainty. As usual the expected Payoff is measured on the vertical axis. Player A's pure strategy lines are obtained as usual and reported in the following table and reproduced in the graph.

Table 14.55

A's Pure strategies	Expected Payoff
1	$E(y_1) = 2y_1+4(1-y_1) = -2y_1+4$
2	$E(y_1) = 2y_1+3(1-y_1) = -y_1+3$
3	$E(y_1) = 3y_1+2(1-y_1) = y_1+2$
4	$E(y_1) = -1x_1+6(1-y_1) = -8y_1+6$

Table 14.56

$E(y_1) = 4-2y_1$	
$E(y_1)$	y_1
4	0
2	1

$E(y_1) = 3-1y_1$	
$E(y_1)$	y_1
3	0
2	1

$E(y_1) = 2+y_1$	
$E(y_1)$	y_1
2	0
3	1

$E(y_1) = 6-8x_1$	
$E(y_1)$	y_1
6	0
−2	1

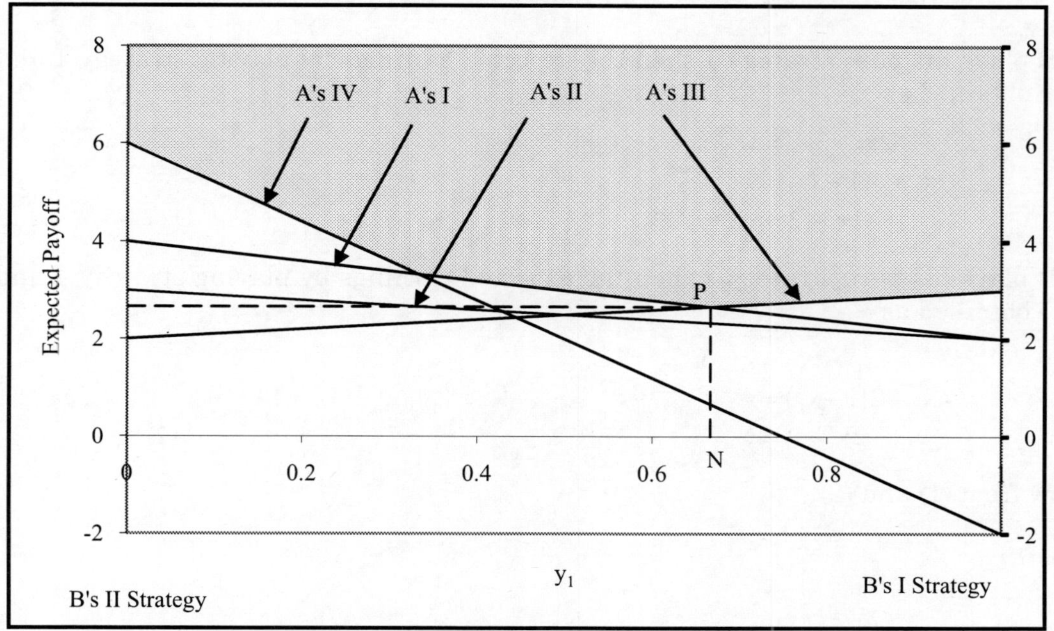

Figure 14.5

This time remember our player B is a minimax player. Accordingly, he will choose the minimum payoff among all the four maximum payoffs given by the corner points of shaded region. At the equilibrium point P since only the third and first strategy lines intersect the strategies 2 and 4 are discarded as inferior strategies. Thus, the (4×2) game ultimately reduces to a 2×2 game with strategies 1 and 3 of A as shown below. Here onwards the algebraic solution is obtained in the usual manner.

Table 14.57

		A's Payoff B's Strategies 1	2
A's Strategies	1	2	4
	3	3	2

Calculation of A's strategic probabilities

Since A is a maximini row player he will play strategy 1 and 2 in such a way that his expected earnings from both types are equal to one another irrespective of his opponent B's play.

If B plays his pure strategy 1 then the expected earnings by playing strategy 1 and 2 by A is obtained as

$$2x_1 + 3x_3 = v \quad \text{or}$$
$$2x_1 + 3(1 - x_1) = v$$
$$-1x_1 + 3 = v... \qquad (1)$$

If B plays his pure strategy 2 then the expected earnings by playing strategy 1 and 2 by A is obtained as

$$4x_1 + 2x_3 = v \quad \text{or}$$
$$4x_1 + 2(1 - x_1) = v$$
$$2x_1 + 2 = v... \qquad (2)$$

Now from (1) and (2)

$$-x_1 + 3 = 2x_1 + 2$$
$$-3x_1 = -1$$
$$x_1 = 1/3$$
$$x_3 = 2/3$$

Calculation of B's strategic probabilities

Since B is a minimax column player he will play strategy 1 and 2 in such a way that his expected earnings from both types are equal to one another irrespective of his opponent A's play.

If A plays his pure strategy 1 then the expected earnings by playing strategy 1 and 2 by B is obtained as

$$2y_1 + 4y_2 = v \quad \text{or}$$
$$2y_1 + 4(1 - y_1) = v$$
$$-2y_2 + 4 = v... \qquad (3)$$

Similarly, if A plays his pure strategy 2 then the expected earnings by playing strategy 1 and 2 by B is obtained as

$$3y_1 + 2y_2 = v \quad \text{or}$$
$$3y_1 + 2(1 - y_1) = v$$
$$y_1 + 2 = v... \qquad (4)$$

Now from (3) and (4)

$$-2y_1 + 4 = y_1 + 2$$
$$-3y_1 = -2$$
$$y_1 = 2/3$$
$$y_2 = 1/3$$

The value of the game v = 2/3 +2 = 8/3.

■ Example 16

Solve the following 4×4 game by using dominance rule.

Table 14.58

		A's Payoff			
		B's Strategies			
		1	2	3	4
A's Strategies	1	19	6	7	5
	2	7	3	14	6
	3	12	8	18	4
	4	8	7	13	−1

Solution

Step 1: Search for saddle point solution

Table 14.59

		A's Payoff B's Strategies 1	2	3	4	Row Minimum	A's Maximini
A's Strategies	1	19	6	7	5	5	
	2	7	3	14	6	3	5
	3	12	8	18	4	4	
	4	8	7	13	−1	−1	
Col. Max		19	8	18	6		
B's Minimax			6				

Saddle Point Solution Does Not Exist

Step 2: Dominance rule and removal of inefficient strategies

Since all the elements of column 2 are less than the corresponding elements of column 1 and 3 we remove column 1 and 3 from B's point of view. Similarly, all the elements of row 3 are greater than row 4 elements we remove row 4 from A's point of view. So the final reduced pay of becomes

Table 14.60

		A's Payoff B's Strategies 2	4
A's Strategies	1	6	5
	2	3	6
	3	8	4

Now let us solve this (3×2) game by graphical method.

Table 14.61

A's pure strategies	B's expected payoff
1	$E(y_2) = 6y_2+5(1-y_2) = y_2+5$
2	$E(y_2) = 3y_2+6(1-y_2) = -3y_2+6$
3	$E(y_2) = 8y_2+4(1-y_2) = 4y_2+4$

Table 14.62

$E(y_2) = 5+y_2$	
$E(y_2)$	y_2
5	0
6	1

$E(y_2) = 6-3y_2$	
$E(y_2)$	y_2
6	0
3	1

$E(y_2) = 4+4y_2$	
$E(y_2)$	y_2
4	0
8	1

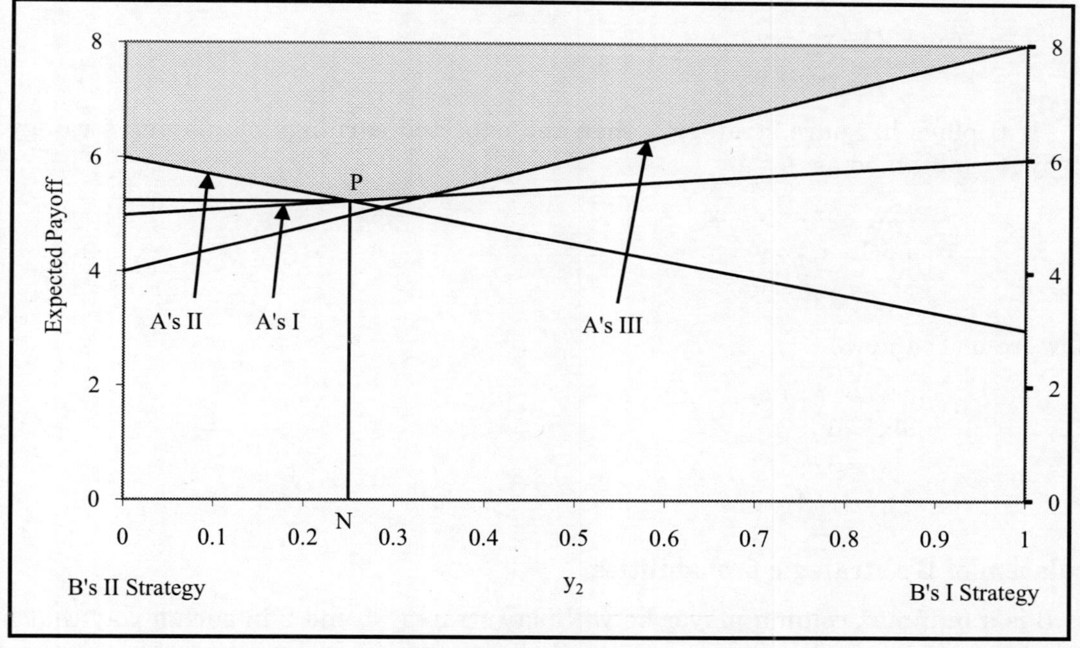

Figure 14.6

In the above graph at P, since only strategy 1 and 2 are intersecting we remove the inefficient strategy 3 and restate the problem as follows.

Table 14.63

A's Payoff

A's Strategies	B's Strategies 2	4
1	6	5
2	3	6

Calculation of A's strategic probabilities

Since A is a maximini row player he will play strategy 1 and 2 in such a way that his expected earnings from both types are equal to one another irrespective of his opponent B's play.

If B plays his pure strategy 1 then the expected earnings by playing strategy 1 and 2 by A is obtained as

$$6x_1 + 3x_2 = v \quad \text{or}$$
$$6x_1 + 3(1 - x_1) = v$$
$$3x_1 + 3 = v \ldots \qquad (1)$$

If B plays his pure strategy 2 then the expected earnings by playing strategy 1 and 2 by A is obtained as

$$5x_1 + 6x_2 = v \quad \text{or}$$
$$5x_1 + 6(1 - x_1) = v$$
$$-x_1 + 6 = v \ldots \qquad (2)$$

Now from (1) and (3)

$$3x_1 + 3 = -x_1 + 6$$
$$4x_1 = 3$$
$$x_1 = 3/4$$
$$x_2 = 1/4$$

Calculation of B's strategic probabilities

Since B is a minimax column player he will play strategy 1 and 2 in such a way that his expected earnings from both types are equal to one another irrespective of his opponent A's play.

If A plays his pure strategy 1 then the expected earnings by playing strategy 1 and 2 by B is obtained as

$$6y_2 + 5y_4 = v \quad \text{or}$$
$$6y_2 + 5(1 - y_2) = v$$
$$y_2 + 5 = v \ldots \qquad (3)$$

Similarly, if A plays his pure strategy 2 then the expected earnings by playing strategy 1 and 2 by B is obtained as

$$3y_2 + 6y_4 = v \quad \text{or}$$
$$3y_2 + 6(1 - y_2) = v$$
$$-3y_2 + 6 = v \ldots \qquad (4)$$

Now from (3) and (4)

$$y_2 + 5 = -3y_2 + 6$$
$$4y_2 = 1$$
$$y_2 = 1/4$$
$$y_4 = 3/4$$

The value of the game v = 1/4 +5 = 5.25.

14.9 LINEAR PROGRAMMING APPROACH TO THE THEORY OF GAMES

The existence of saddle point gave us pure strategies for both the players. In the absence of saddle point we opt for mixed strategy solution by using probabilities. In all these illustrations at the final stage we were left with very simple 2×2 game. In reality such a reduction is not possible in all cases. For a more general (M×N) type of game, none of the above methods is applicable. Under such circumstances linear programming technique will be handy to handle the situation. To apply the linear programming technique, we must convert the given game into a linear programming problem first.

Let us consider (M×N) game with the following payoff matrix for A.

$$A = \begin{bmatrix} a_{11} & a_{12} & \cdots & a_{1j} & \cdots & a_{1n} \\ a_{21} & a_{22} & \cdots & a_{2j} & \cdots & a_{2n} \\ . & . & \cdots & . & \cdots & . \\ a_{i1} & a_{i2} & \cdots & a_{ij} & \cdots & a_{in} \\ . & . & \cdots & . & \cdots & . \\ a_{m1} & a_{m2} & \cdots & a_{mj} & \cdots & a_{mn} \end{bmatrix}$$

Let A have M strategies with probabilities x_1, x_2, ...x_m so $x_1 + x_2 + ...x_m = 1$ that and B is having N strategies with probabilities y_1, y_2 ....y_n so that $y_1 + y_2 + ...y_n = 1$. With these probabilities we rewrite the game in its standard form as shown below with respective probabilities thereon.

B's Strategy

			1	2	...	j	...	n
			y_1	y_2	...	y_j	...	y_n
A's Strategy	1	x_1	a_{11}	a_{12}	...	a_{1j}	...	a_{1n}
	2	x_2	a_{21}	a_{22}	...	a_{2j}	...	a_{2n}
	.	.	.	.	...	.	...	.
	i	x_i	a_{i1}	a_{i2}	...	a_{ij}	...	a_{in}
	.	.	.	.	...	.	...	.
	m	x_m	a_{m1}	a_{m2}	...	a_{mj}	...	a_{mn}

Conversion of A's maximini strategies into a linear programming problem

Since A is a maximini player in our game he will select the strategic probabilites in such a way that maximises the minimum expected payoffs across columns. The following table summarizes the expected payoffs constraints across columns.

Table 14.64

B's pure strategies	A's expected payoffs
1	$a_{11}x_1 + a_{21}x_2 + \cdots + a_{m1}x_m \geq v$
2	$a_{12}x_1 + a_{22}x_2 + \cdots + a_{m2}x_m \geq v$
.	.
.	.
.	.
n	$a_{1n}x_1 + a_{2n}x_2 + \cdots + a_{mn}x_m \geq v$

Thus, the needed linear programming problem in its standard form may be written as

Maximize: v

Subject to:

$$a_{11}x_1 + a_{21}x_2 + \ldots\ldots a_{m1}x_m \geq v$$
$$a_{12}x_1 + a_{22}x_2 + \ldots\ldots a_{m2}x_m \geq v$$
$$\ldots\ldots\ldots\ldots\ldots\ldots$$
$$a_{1n}x_1 + a_{2n}x_2 + \ldots\ldots a_{mn}x_m \geq v$$
$$x_1 + x_2 + \ldots\ldots x_m = 1$$
$$x_1, x_2 \ldots\ldots x_m \geq 0$$

Now let us simplify this game by dividing all the constraints by the value of the game provided it is positive, without affecting the inequalities. If it is negative then we can always add an approprate constant to all the payoff entries and make the value of the game positive without affecting the optimum solution.

$$a_{11}\frac{x_1}{v} + a_{21}\frac{x_2}{v} + \ldots\ldots a_{m1}\frac{x_m}{v} \geq 1$$
$$a_{12}\frac{x_1}{v} + a_{22}\frac{x_2}{v} + \ldots\ldots a_{m2}\frac{x_m}{v} \geq 1$$
$$\ldots\ldots\ldots\ldots\ldots\ldots$$
$$a_{1n}\frac{x_1}{v} + a_{2n}\frac{x_2}{v} + \ldots\ldots a_{mn}\frac{x_m}{v} \geq 1$$
$$\frac{x_1}{v} + \frac{x_2}{v} + \ldots\ldots \frac{x_m}{v} = \frac{1}{v}$$
$$x_1, x_2 \ldots\ldots x_m \geq 0$$

Now let $\frac{x_1}{v} = X_1, \frac{x_2}{v} = X_2 .... \frac{x}{v} = X_m$ and $\frac{1}{v} = V$.

Thus, from the last constraint our objective function becomes

$$\text{Minimize: } \frac{1}{v} = V = X_1 + X_2 + ...X_m$$

$$\begin{aligned} \text{Subject to: } \quad & a_{11}X_1 + a_{21}X_2 + ...a_{m1}X_m \geq 1 \\ & a_{12}X_1 + a_{22}X_2 + ...a_{m2}X_m \geq 1 \\ & .... \\ & a_{1n}X_1 + a_{2n}X_2 + ...a_{mn}X_m \geq 1 \\ & X_1, X_2 ....X_m \geq 0 \end{aligned}$$

Conversion of B's minimax strategies into a linear programming problem

Since B is a minimax player in our game he will select the strategic probabilities in such a way that minimizes the maximum expected payoffs across rows. The following table summarizes the expected payoffs constraints across rows.

Table 14.65

A's pure strategies	B's expected payoffs
1	$a_{11}y_1 + a_{12}y_2 + \cdots + a_{1n}y_n \leq v$
2	$a_{21}y_1 + a_{22}y_2 + \cdots + a_{2n}y_n \leq v$
.	.
.	.
.	.
m	$a_{m1}y_1 + a_{m2}y_2 + \cdots + a_{mn}y_n \leq v$

Thus, the linear programming problem for B is written as

$$\text{Minimize: } v$$

$$\begin{aligned} \text{Subject to: } \quad & a_{11}y_1 + a_{12}y_2 + ...a_{1n}y_n \leq v \\ & a_{21}y_1 + a_{22}y_2 + ...a_{2n}y_n \leq v \\ & .... \\ & a_{m1}y_1 + a_{m2}y_2 + ...a_{mn}y_n \leq v \\ & y_1 + y_2 + ...y_n = 1 \\ & y_1, y_2 ....y_n > 0 \end{aligned}$$

As usual to simplify the problem let us divide all the constraints by v, the value of the game

$$a_{11}\frac{y_1}{v} + a_{12}\frac{y_2}{v} + \ldots\ldots\ldots\ldots\ldots a_{1n}\frac{y_n}{v} \le 1$$

$$a_{21}\frac{y_1}{v} + a_{22}\frac{y_2}{v} + \ldots\ldots\ldots\ldots\ldots a_{2n}\frac{y_n}{v} \le 1$$

$$\ldots\ldots\ldots\ldots\ldots\ldots\ldots\ldots\ldots\ldots\ldots\ldots\ldots\ldots\ldots\ldots$$

$$a_{m1}\frac{y_1}{v} + a_{m2}\frac{y_2}{v} + \ldots\ldots\ldots\ldots\ldots a_{mn}\frac{y_n}{v} \le 1$$

$$\frac{y_1}{v} + \frac{y_2}{v} + \ldots\ldots\ldots\ldots\ldots\ldots\ldots\ldots\ldots\frac{y_n}{v} = \frac{1}{v}$$

Now let $\frac{y_1}{v} = Y_1, \frac{y_2}{v} = Y_2 \ldots\ldots\ldots\ldots\ldots\ldots \frac{Y}{v} = Y_m$ and $\frac{1}{v} = V$.

Thus, from the last constraint our objective function becomes

Maximize: $\frac{1}{v} = V = Y_1 + Y_2 + \ldots\ldots\ldots\ldots\ldots\ldots\ldots Y_m$

Subject to:

$$a_{11}Y_1 + a_{12}Y_2 + \ldots\ldots\ldots\ldots\ldots\ldots a_{1n}Y_n \le 1$$
$$a_{21}Y_1 + a_{22}Y_2 + \ldots\ldots\ldots\ldots\ldots\ldots a_{2n}Y_n \le 1$$
$$\ldots\ldots\ldots\ldots\ldots\ldots\ldots\ldots\ldots\ldots\ldots\ldots\ldots\ldots\ldots\ldots$$
$$\ldots\ldots\ldots\ldots\ldots\ldots\ldots\ldots\ldots\ldots\ldots\ldots\ldots\ldots\ldots\ldots$$
$$a_{m1}Y_1 + a_{m2}Y_2 + \ldots\ldots\ldots\ldots\ldots a_{mn}Y_n \le 1$$
$$Y_1, Y_2, \ldots\ldots\ldots\ldots\ldots\ldots\ldots\ldots\ldots\ldots\ldots Y_n \ge 0$$

Note: A's Maximini and B's Minimax are dual to one another. Thus solving one problem facilitate the other and vice versa. Thus, Maximini = Minimax = V in all problems.

■ Example 17

Solve the following (3×3) game by linear programming method.

Table 14.66

A's Payoff		B's Strategies		
		1	2	3
A's Strategies	1	3	−1	−3
	2	−3	3	−1
	3	−4	−3	3

Solution

Search for saddle point

In Table 14.67 A's maximini solution is –3 and B's minimax value is 3 our strategic solution will yield a value in between these two values. So it can be negative as well. To avoid the negative value let us add 5 to all the payoff entries and make all of them positive so that v is positive. So our new payoff matrix is shown in Table 14.68.

Table 14.67

		A's Payoff B's Strategies 1	2	3	Row Minimum	A's Maximini
A's Strategies	1	3	−1	−3	−3	
	2	−3	3	−1	−3	−3
	3	−4	−3	3	−4	
Col. Max		3	3	3		
B's Minimax			3			

Saddle Point Solution Does Not Exist

Table 14.68

		A's Payoff B's Strategies 1	2	3
A's Strategies	1	8	4	2
	2	2	8	4
	3	1	2	8

In the above payoff matrix since all the entries are positive the value of the game must be positive. Now in standard notation let us write the minimax linear programming problem of B as follows

$$\text{Maximize:} \quad V = Y_1 + Y_2 + Y_3$$

$$\text{Subject to:} \quad 8Y_1 + 4Y_2 + 2Y_3 \leq 1$$

$$2Y_1 + 8Y_2 + 4Y_3 \leq 1$$

$$1Y_1 + 2Y_2 + 8Y_3 \leq 1$$

$$Y_1 ; \quad Y_2 ; \quad Y_3 \geq 0$$

Now we are having two constraints let us introduce three slacks and make the constraints as equalities.

$$\text{Maximize:} \quad Y_1 + Y_2 + Y_3$$

$$\text{Subject to:} \quad 8Y_1 + 4Y_2 + 2Y_3 + 0Y_4 + 0Y_5 + 0Y_6 = 1$$

$$2Y_1 + 8Y_2 + 4Y_3 + Y_5 = 1$$

$$1Y_1 + 2Y_2 + 8Y_3 + Y_6 = 1$$

$$Y_1 ;\ Y_2 ;\ Y_3 ;\ Y_4 ;\ Y_5 ;\ Y_6 \geq 0$$

Solution

1) Simplex solution to our modified problem

Table 14.69

Tableau 1

	A	B	C	D	E	F			G	H
1	C_j	Basis	Y_1	Y_2	Y_3	Y_4	Y_5	Y_5	X_b	M.R
2	0	Y_4	8	4	2	1	0	0	1	0.125
3	0	Y_5	2	8	4	0	1	0	1	0.5
4	0	Y_6	1	2	8	0	0	1	1	1
5		C_j	1	1	1	0	0	0	0	
6		Z_j	0	0	0	0			0	
7		$C_j - Z_j$	1	1	1	0				

Table 14.70

Tableau 2

	A	B	C	D	E	F			G	H
1	C_j	Basis	Y_1	Y_2	Y_3	Y_4	Y_5	Y_5	Output	Mini
2	1	Y_1	1	1/2	1/4	1/8	0	0	1/8	1/2
3	0	Y_5	0	7	7/2	−1/4	1	0	3/4	3/14
4	0	Y_6	0	3/2	31/4	−1/8	0	1	7/8	7/62
5		C_j	1	1	1	0	0	0		
6		Z_j	1	1/2	1/4	1/8	0	0	1/8	
7		$C_j - Z_j$	0	1/2	3/4	−1/8	0	0		

Table 14.71

Tableau 3										
	A	B	C	D	E	F	G	H	I	J
1	C_j	Basis	Y_1	Y_2	Y_3	Y_4	Y_5	Y_5	X_b	M.R
2	1	Y_1	1	14/31	0	4/31	0	−1/31	3/31	3/14
3	0	Y_5	0	196/31	0	−6/31	1	−14	11/196	11/196
4	1	Y_3	0	6/31	1	−1/62	0	31	7/12	7/12
5		C_j	1	1	1	0	0	4/31		
6		Z_j	1	20/31	1	−7/62	0	−3/21	13/62	
7		$C_j - Z_j$	0	11/31	0	−7/62	0	−3/21		

Table 14.72

Tableau 4										
	A	B	C	D	E	F	G	H	I	J
1	C_j	Basis	Y_1	Y_2	Y_3	Y_4	Y_5	Y_5	X_b	M.R
2	1	Y_1	1	0	0	1/7	1/14	0	1/14	
3	1	Y_2	0	1	0	−3/98	31/196	−1/14	11/196	
4	1	Y_3	0	0	1	−1/98	−3/98	1/7	5/49	
5		C_j	1	1	1	0	0	0		
6		Z_j	1	1	1	−5/49	−11/196	−1/14	45/196	
7		$C_j - Z_j$	0	0	0	−5/49	−11/196	−1/14		

Thus, the equilibrium solutions to our modified problem are:

$Y_1 = 1/14 = 0.071428571$, $Y_2 = 11/196 = 0.05612245$, $Y_3 = 5/49 = 0.10204082$,

$V = 45/196 = 0.0229591837$.

Now we obtain the solution to our original problem for the modified solution as follows.

$$y_1 = \frac{Y_1}{V} = \frac{1/14}{45/196} = \frac{1}{14}\frac{196}{45} = \frac{14}{45} = 0.311111$$

$$y_2 = \frac{Y_2}{V} = \frac{11/196}{45/196} = \frac{11}{196}\frac{196}{45} = \frac{11}{45} = 0.244444$$

$$y_3 = \frac{Y_3}{V} = \frac{5/49}{45/196} = \frac{5}{49}\frac{196}{45} = \frac{20}{45} = 0.4444444$$

$$v = \frac{1}{V} = \frac{1}{45/196} = \frac{196}{45} = 4.3555556$$

Remember, that we have added 5 to all elements in the original payoff matrix to convert them all into positive entries; we subtract this 5 from the value of the game to arrive at the correct value

So the corrected value $= \frac{196}{45} - 5 = \frac{196-225}{45} = \frac{-29}{45}$

The final solution for A as per the duality rule is obtained from the last Simplex table. From the last Simplex table the answer for our primal is obtained as

$$X_1 = 5/49,\ X_2 = 11/196,\ X_3 = 1/14$$

Therefore

$$x_1 = \frac{X_1}{V} = \frac{5/49}{45/196} = 20/45$$

$$x_2 = \frac{X_2}{V} = \frac{11/49}{45/196} = 11/45$$

$$x_3 = \frac{X_3}{V} = \frac{1/14}{45/196} = 14/45$$

2) Spreadsheet solution to the modified problem

Table 14.73

	Material requirements			
Material	Y_1	Y_2	Y_3	Amount available
1	8	4	2	1
2	2	8	4	1
3	1	2	8	1
Profit per unit	1	1	1	
Model	Decision Variables			
	Y_1	Y_2	Y_3	
	0.071428571	0.05612245	0.10204082	
Maximum	0.229591837			
Constraints	Amounts Used (LHS)			Amount Available (RHS)
1	1	<=		1
2	1	<=		1
3	1	<=		1

EXERCISES

1. Define the following
   (a) Competitive game.
   (b) Pure strategies.
   (c) Mixed strategies.
   (d) Two person zero sum game.
   (e) Two person positive sum game.
   (f) Payoff matrix.
   (g) Saddle point.
   (h) Dominance and strategic dominance.
   (i) Strategic probabilities.
2. With suitable illustration explain Maximini and Minimax principles in the theory of game.
3. With suitable example illustrate the method of converting a game theory into a linear programming.
4. Discuss the algebraic solution for a 2×2 game by giving an illustration.
5. How do you find the strategic dominance by graphical method?
6. Solve the following game, using the concept of dominance.

		Player B			
		1	2	3	4
Player A	1	8	10	9	14
	2	10	11	8	12
	3	13	12–Jan	14	13

(C.A, May 1984)

7. For the following payoff tables transform the zero-sum game into an equal linear programming problem and solve it by using simplex method.

		Player B		
		1	2	3
Player A	1	1	−1	3
	2	3	5	−3
	3	6	2	−2

(Madurai University, B.Sc (Maths),1983)

8. There is a famous Italian game called Two-Tiger Morra. The game is played by two people and each of who shows one or two fingers and simultaneously guesses the number of fingers his opponent will show. If just one player guesses correctly, he wins an amount equal to the sum of the fingers shown by himself and by his opponent, otherwise the game is considered as a draw. Obtain the payoff matrix corresponding to it.

9. Can there be a two person zero sum game which has a value zero. If so write the payoff matrix of one such game and find out the optimal strategies of the player. (M. Com, Delhi, 1999)

10. In a small town, there are only two stores that handle Sunday goods—ABC and XYZ. The total number of customers is equally divided between the two, because price and quality of goods sold are same. Both the stores have good reputation in the community, and they render good customer service. Assume that a gain of customer by ABC is a loss to XYZ and vice versa. Both stores plan to run annual pre-Diwali sales during the first week of November. Sales are advertised through a local newspaper, radio, and television media. With the aid of an advertising firm, store ABC constructed the game matrix given below. (Figures in the matrix represent a gain or loss of customers)

		Strategy of XYZ		
		Newspaper	Radio	Television
Strategy of ABC	Newspaper	30	40	−80
	Radio	0	15	−20
	Television	90	20	50

Determine optimal strategies and worth of such strategies by both ABC and XYZ. (M.B.A, Delhi, November, 1999)

11. Reduce the following two person zero sum game to 2×2 order and obtain the optimum strategies for each player and the value of the game.

		Player B			
		B_1	B_2	B_3	B_4
Player A	A_1	3	2	4	0
	A_2	3	4	2	4
	A_3	4	2	4	0
	A_4	0	4	0	8

(M.Com, Delhi, 1984)

12. A company is currently involved in negotiations with its union on the upcoming wage contract. With the aid of an outside mediator, the table below was constructed by the management group. The pluses are to be interpreted as proposed wage increases while minus figure indicates that a wage reduction. The mediator informs the management group that he has been in touch with the union and that they have constructed a table that is comparable to the table developed by the management. Both the company and the union must decide on the overall strategy before negotiations begin. The management group understands the relationship of the company strategies to the union strategies in the following table but lacks specific knowledge of the game theory to select the best strategy (or strategies) for the firm. Assist the management on this problem. What game value and strategies are available to the opposing groups?

Conditional cost to company

		Union Strategies			
		U_1	U_2	U_3	U_4
Company Strategies	C_1	0.25	0.27	0.05	−0.02
	C_2	0.20	0.16	0.08	0.08
	C_3	0.14	0.12	0.15	0.13
	C_4	0.30	0.14	0.19	0.00

(M.B.A, Delhi, 1994)

13. For the following payoff matrix find the value of the game and strategies of player A and B using linear programming.

		Player B		
		1	2	3
Player A	1	3	−1	4
	2	6	7	−2

14. Two companies are competing for similar product. The pay of matrix in terms of their advertising plan is shown below.

		Competitor B		
		Large advt.	Medium advt.	Small advt.
Competitor A	Large advt.	70	80	50
	Medium advt.	90	60	95
	Small advt.	105	90	65

(M.B.A, Delhi, November, 1997)

15. The following matrix represents player A's payoff in a two person, zero sum game.

		Player B		
		1	2	3
Player A	1	30	40	−80
	2	0	15	−20
	3	90	20	50

Find the optimal strategies for the two players and the value of the game.

(M.Com, Delhi, 1985)

CHAPTER

15

Theory of Games - II Non-Zero Sum Games

- *Introduction*
- *Non-cooperative and cooperative pure strategy games*
- *Pure strategy non-cooperative prisoner's dilemma model*
- *Application of prisoners' dilemma in oligopolistic market models, advertising models , common resource sharing models*
- *A revisit of advertising models: Games with co-operation*
- *Nash equilibrium and its applications in advertising, choice of products, prisoner's dilemma a revisit, Investing*
- *Nash equilibrium and Honeymoon trip*
- *Application of mixed strategies to investment: Maximizing the expected payoff*
- *Application of mixed strategies: Matching the rupee*
- *Repeated games*
- *Tit-for-tat strategy*
- *Sequential game: Advantage of moving first*
- *Threats, commitments and creditability*
- *Exercises*

15.1 INTRODUCTION

Game theory is the study of how people behave in strategic situations. By 'strategic' we mean a situation in which each participant, before deciding his action, must consider how others might respond to his action. In a competitive market since there is large number sellers, selling a homogeneous product, strategic interaction with other sellers is not so important. Similarly, in a monopolist market also the question of strategic interaction does not arise because except himself no one will be there to interact. However, under oligopolistic situations since the market is shared by few, each firm must act strategically when others act. In this context the theory of game is quite useful in understanding the underlying behavioural characters of the oligopolistic market.

The key objective of the theory of game is to determine the optimal strategy for underlying game between two or more participants. A strategy is simply a rule or plan of action for playing the game. For example, for a price setting firm, a strategy might be:

"I will keep my price high as long as all my competitors do the same, but once my competitors go for price reduction, I too will lower my price by more percentage." In an action the bidder's strategy might be: "I will make a first bid of Rs. 20,000 to convince others that I am very serious about winning the contract, but I will drop myself out if other bidders push the rate to Rs. 50,000."

"Your competitor is as rational as you are" is the spirit of the game that we are going to discuss in this section. Taking the competitors' behaviour into account is the main theme of the game. Even under conditions of complete symmetry and perfection of each other's information, determining the optimum strategies is a difficult task indeed.

In this section we plan to introduce non-zero sum game. Unlike zero sum games where in one firms loss is exactly equal to the other firms gain, in the non-zero sum game the net effect can be either positive or net negative in the final. Some occasion both players will gain in the final solution. In some other cases both will lose in the final solution. In certain other case one may gain and other may lose but not with zero net. In this section we plan to bring in such interesting games into play.

Thus, this time player A will have his own payoff matrix and player B will have another payoff matrix of his own. Clearly this time B's matrix is not the obtainable from A's matrix by simply multiplying by a minus one as the case of zero sum game. Both the matrices must be taken into consideration simultaneously for the final equilibrium solution. In the following table we represent both A's and B's payoff matrices.

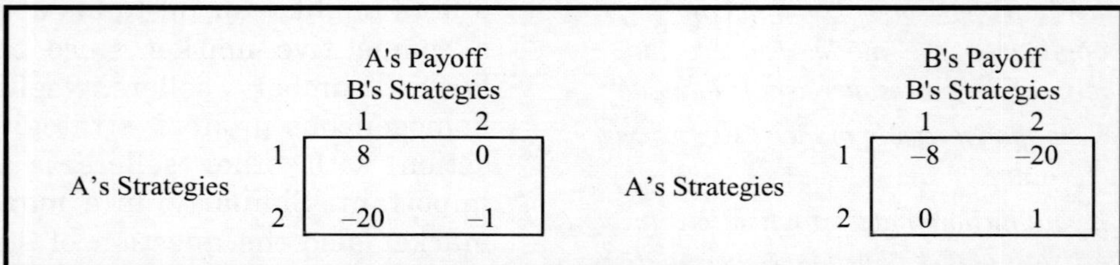

A's Payoff

		B's Strategies 1	B's Strategies 2
A's Strategies	1	8	0
	2	–20	–1

B's Payoff

		B's Strategies 1	B's Strategies 2
A's Strategies	1	–8	–20
	2	0	1

Figure 15.1

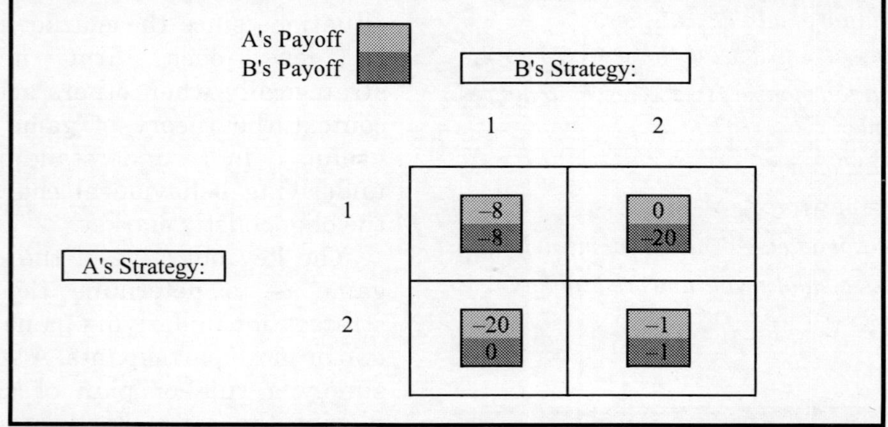

Figure 15.2

However, to handle the situation simultaneously we put both the matrices one over the other as shown in Figure 15.2. In it the top light grey colour cell highlights the A's matrix and the bottom dark grey colour highlighted cells constitute the B's payoff matrix.

Definition of dominance in non-zero sum game

1. If all the elements in a row of a given A's payoff matrix are greater than the corresponding elements of any other row/rows, then player A would prefer to play the larger elements row and discard the smaller elements row/rows. Here the row that contains the larger elements is often called the dominant row and the associated strategy is called the dominant strategy for player A.
2. If all the elements in a column of a given B's payoff matrix are greater than the corresponding elements of any other column/columns then player B would prefer to play the larger elements column and discard the smaller elements column/columns. Here the column that contains the larger elements is often called the dominant column and the associated strategy is called the dominant strategy for player B.

Thus, for the definition of dominance remember the quote all the time.

I am doing the best I can no matter what you do.
You are doing the best you can no matter what I do.

Here it is important to note that in the definition of dominance for B, we consider only the biggest elements column because he has his own payoff matrix this time in mind and not A's matrix as in the case of zero sum game.

Example 1

For the game given below obtain dominant strategy for both A and B.

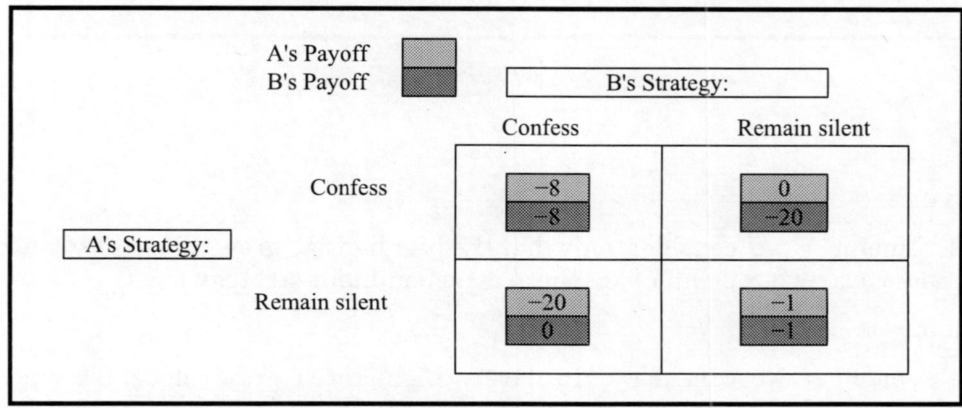

Figure 15.3

Solution:

A's Dominance

This time remember we consider only lightly highlighted grey colour A's matrix for consideration. Clearly –8 > –20 and 0 > –1, thus for A, the first row reference "confess" is the dominating row for player A. Thus player A will play strategy 1 with certainty all the time.

B's Dominance

This time remember we consider only darkly highlighted grey colour B's matrix for consideration. Clearly –8 > –20 and 0 > –1, thus for B the first column reference "confess" is the dominating column. Thus, B will play strategy 1 with certainty all the time.

■ Example 2

For the game given below obtain dominant strategy for both A and B.

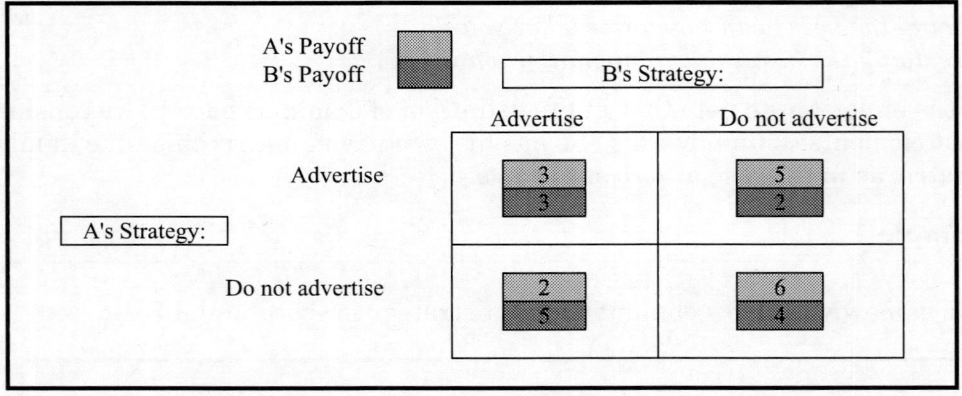

Figure 15.4

Solution:

A's Dominance

This time remember we consider only lightly highlighted grey colour A's matrix for consideration. Clearly 3 > 2 and 5 < 6, there is no dominant strategy for A.

B's Dominance

This time remember we consider only darkly highlighted grey colour B's matrix for consideration. Clearly 3 > 2 and 5 > 4, thus for B the first column (advertise) is a dominant strategy. Thus B will play strategy advertise with certainty all the time.

Example 3

For the game referred hereunder obtain the dominant strategies for both A and B if exists.

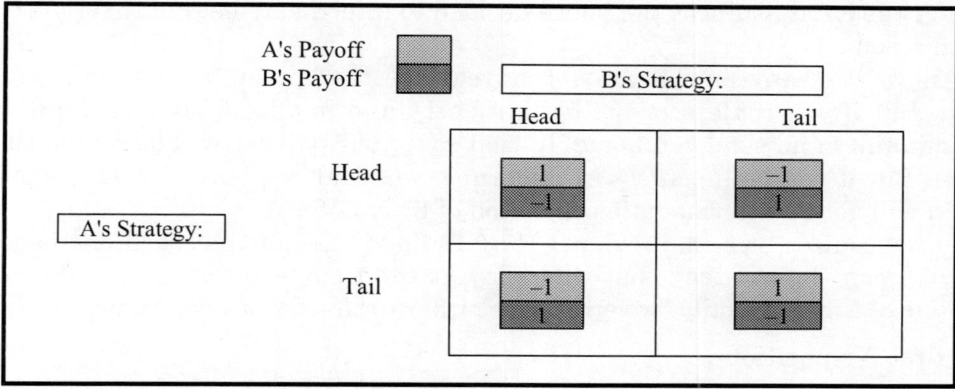

Figure 15.5

Solution:

A's Dominance

This time remember we consider only lightly highlighted grey colour A's matrix for consideration. Clearly 1 = 1 and –1 < 1, there is no dominant strategy for A.

B's Dominance

This time remember we consider only darkly highlighted grey colour B's matrix for consideration, clearly, –1 < 1 and 1 > –1. Thus, B also does not have dominant strategy this time.

15.2 NON-COOPERATIVE AND COOPERATIVE PURE STRATEGY GAMES

In cooperative game players often negotiate binding contracts for mutual benefits and make joint strategies. In non-cooperative games, negotiation and binding contracts are totally absent. If a coat is produced at the cost of Rs. 2000 and the customer is willing to pay Rs. 3000 the cooperative bargaining may fetch the seller Rs. 2500 at the end. Under non-cooperative game each firm knows that by undercutting its price will fetch more market share but at the same time initiate a price war.

15.3 PURE STRATEGY NON-COOPERATIVE PRISONER'S DILEMMA MODEL

Many times in real life people fail to cooperate with each other though knowing fully well that such an act is mutually beneficial. In this context, to stress the difficulties involved in maintaining co-operation, often *prisoner's dilemma* is being highlighted in game theory.

It is simply a tale about two criminals behaviour. Accordingly, the two criminals A and B were caught by the police for minor crimes committed by them. The police have enough evidence to convict both A and B for a year or so for the minor crimes that they had committed. The police also suspect both A and B for their involvement in the big bank robbery that took place recently in that area, but they lack evidence to convict them. Under the circumstances the police decided to interrogate both of them in separate rooms with a deal.

"Right now, we can convict you for a year for the minor crimes that you have committed. But if you confess to the bank robbery and implicate your partner, we will give you immunity and send you home. In that case your partner will be convicted for 20 years term. But if you both confess to the crime, we would not care for your testimony; instead you will get an intermediate conviction of 8 years term."

Would they confess or remain silent? Here the punishment that each one is going to get depends upon the strategy that they play at the time of interrogation. Remember, that each one of them has only two strategies either to confess or keep silent.

Consider first A's position

I do not know what B is going to do at the time of interrogation. If he remains silent, my best strategy is to confess and go free rather than keep myself silent and get convicted for one year for the minor crime that I have committed.

Instead, if B confesses my best strategy could be still to confess and get 8 years of jail instead of keeping silent and get a capital punishment of 20 years. Here, irrespective of his opponent's strategy A decides to confess at all cost. Thus A's strategy to confess is a dominant strategy.

Now consider B's decision

I do not know what A is going to do at the time of interrogation. If he remains silent, my best strategy is to confess and go free rather than keep myself silent and get convicted for one year for the minor crime that I have committed.

Instead, if A confesses my best strategy could be still to confess and get 8 years of jail instead of keeping silent and get a capital punishment of 20 years. Here, irrespective of his opponent's strategy B is decided to confess at all cost. Thus, B's strategy to confess is also a dominant strategy.

So under the said situation both will confess and get 8 years punishment rather than keeping silent and get only one year punishment under co-operation. By each pursuing his own self-interest in the absence of co-operation leads to this painful situation of 8 years punishment. Instead, if both had remained silent, as a predetermined strategy under cooperative discussion, they would have had only one-year punishment for the minor crimes that they have really committed. Since they are being questioned separately, the destructive self-interest dominated the scene. Co-operation between the two is difficult to maintain, and also is individually irrational.

i) Solution by dominant strategy

Figure 15.6 represents the payoff matrix for the situation discussed. In the payoff matrix it is interesting to note that for A, the punishment after confessing is less than the corresponding punishment by remaining silent (−8<−20, 0<−1). Thus confessing is a dominant (good) strategy for A. Similarly, for B punishment after confessing is less than the corresponding punishment by remaining silent (−8<−20, 0<−1). Thus, confessing is a dominant strategy for B also. Thus, ultimately both confess and get 8 years punishment.

Dominant Strategy:

I am doing the best I can, no matter what you do.
You are doing the best you can, no matter what I do.

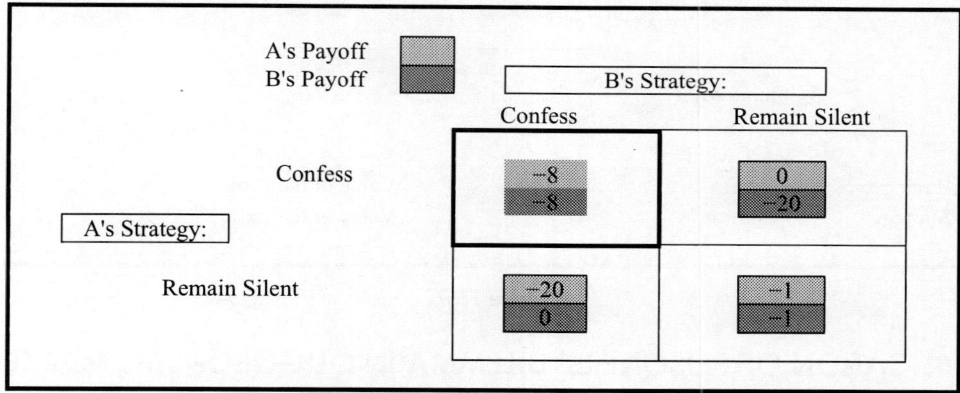

Figure 15.6

Note that in Figure 15.6, the darkly bordered box exhibits the final solution.

ii) Solution by maximini strategies (conservative solution)

In this game, both A and B are conservative in the sense both of them choose the best among the worst thing that can happen. As far as A is concerned the worst thing that can happen is given by row minimum. Among these two worst things the best is given by maximum of these two minimums. This is reported in the right side of the in Figure 15.7.

Similarly, B is also conservative. The column minimums are reported at the bottom of the figure. Among these two minimums the maximum is reported at the bottom as a maximini strategy. Thus both the players apply maximini strategies and end up with a conservative 'confess-confess' solution. Thus our dominant solution is nothing but the conservative maximini solution.

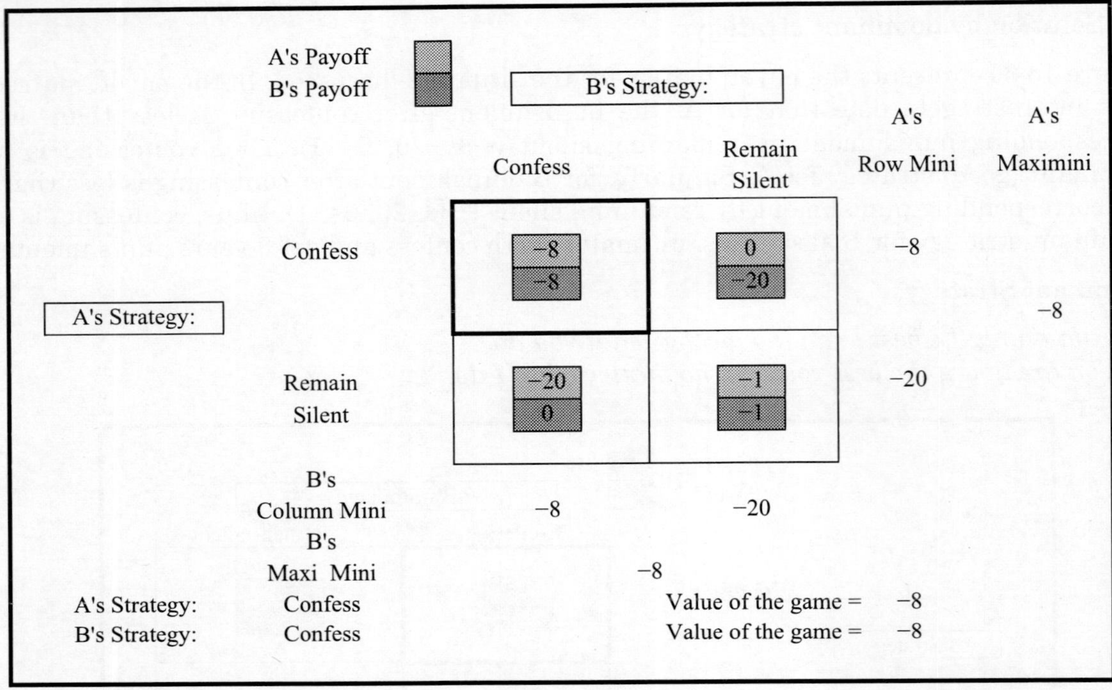

Figure 15.7

15.4 APPLICATION OF PRISONERS' DILEMMA IN OLIGOPOLISTIC MARKET MODELS

Consider the two oil producing oligopolies Iran and Iraq. Both the countries produce and sell crude oil to the rest of the world. After prolonged discussion they decided to produce oil at low level to keep the oil price high in the world market. After having made such a collusive agreement, each one of them must keep up the allotted quota to optimize the total revenue. However, individually still both of them can ignore the agreement and try to produce more hoping to get more revenue. The following are the four possible outcomes.

(a) If Iran keeps up to the agreement and goes for low production then Iraq will get \$60 billion with its own high production and \$50 billion with its own low production. In any case since \$60 > \$50 billion Iraq is better of with high production.

(b) If Iran violates the agreement and goes for high production then Iraq will get \$40 billion with its own high production and \$30 billions with its own low production. Once again since \$40 > \$30 billion Iraq once again will prefer to go for high production.

Thus, irrespective of Iran's low or high strategy Iraq will follow the strategy of high production. Producing at high level is a dominant strategy for Iraq all the time.

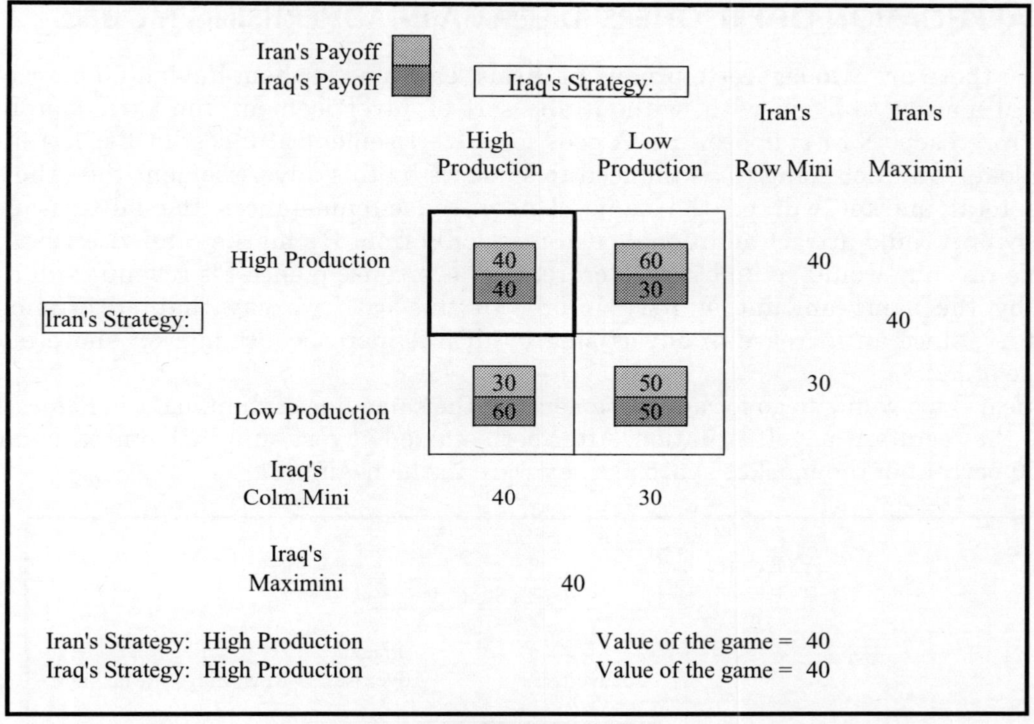

Figure 15.8

(c) If Iraq lives up to the agreement and goes for low production then Iran will get $60 billion with its own high production and $50 billion with its own low production. In any case since $60 > $ 50 billion Iran is better of with high production.

(d) If Iraq violates the agreement and goes for high production then Iran will get $40 billion with its own high production and $30 billion with its own low production. Once again since $40 > $30 billion Iran once again will prefer to go for high production.

Thus, irrespective of Iraq's low or high strategy Iran will follow the strategy of high production. Producing at high level is a dominant strategy for Iran for all the time.

Thus, ultimately both of them will opt for violation and produce high and get a reward of $40 billion each. It is important to note that producing less by both yields $50 billion. This example illustrates why oligopolies have the trouble in maintaining monopoly profit by agreeing to produce less. Thus, though the monopoly outcome is jointly rational when they are left free, self-interested cheating seems to be rational. Just as self-interest drives the prisoners in the prisoners' dilemma to confess, self-interest makes it difficult for our oligopolies to maintain the cooperative outcome with low production, high prices, and a high monopoly profit. Note that the solution in Figure 15.8 is conservative as well.

15.5 APPLICATION OF PRISONERS' DILEMMA IN ADVERTISING MODELS

Suppose there are two cigarette producing firms, each one of them having 50 percent of the total market to begin with, without any sort of advertisement and earn a profit of Rs. 4 crore each. Now suppose firm A goes for advertisement at a cost of Rs. 1 crore. If non-smokers are not going to be induced to smokers by this advertisement then the size of the total market will be the same. Under the circumstances the advertisement strategy of A could attract additional customers only from B's market and vice versa. So firm A's revenue would go up by Rs. 2 crore, and as a consequence B's revenue will come down by the same amount of Rs. 2 crore. In this case we say that advertising is predatory, since an increase in advertising by firm A increases its market share at the expense of B.

If firm B too wants to go for advertisement, on the same line of thinking, the Figure 15.9 will be the resultant payoff situation. After both started advertising each one will end up with 50 percent of the market share as they were in the beginning.

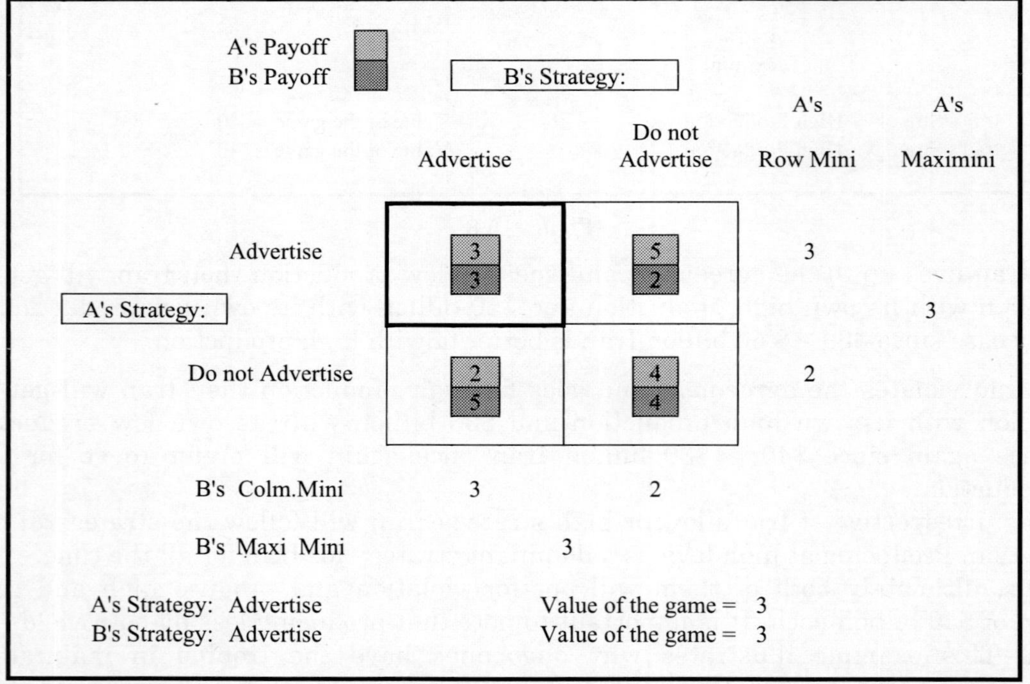

Figure 15.9

Now, each one of them will incur an additional advertising cost Rs. 1 crore. So each one them will receives a net profit of Rs. 3 crore.

15.6 APPLICATION OF PRISONERS' DILEMMA IN COMMON RESOURCE SHARING MODELS

The prisoners' dilemma explains many real-life situations. It explains the difficulty involved in maintaining co-operation, even though co-operation would make both the players better off. In some situations non-cooperative equilibrium can be fatal to the society. Imagine that two farmers A and B own adjacent farm land. Clearly the groundwater available is a common resource for both the farmers. Drilling a bore well to get water for irrigating costs Rs. 1 lakh in that area. If each of them drills one well, they hope to get one half of the water from the common resource and earn a profit of Rs. 5 lakh each. Suppose one of them goes for a second well without the knowledge of the other, he will tap two-thirds of the common resource and make a profit of Rs. 6 lakh. However, if each one of them goes for the second well without the knowledge of the other than the said common resource gets split into two equal parts once again. In this case each one of them will have an additional cost of 1 lakh and earn a profit of Rs. 4 lakh each. Figure 15.10 shows the relevant payoff. Drilling two wells is the dominant strategy for both of them. Once again the self-interest leads to inferior outcome.

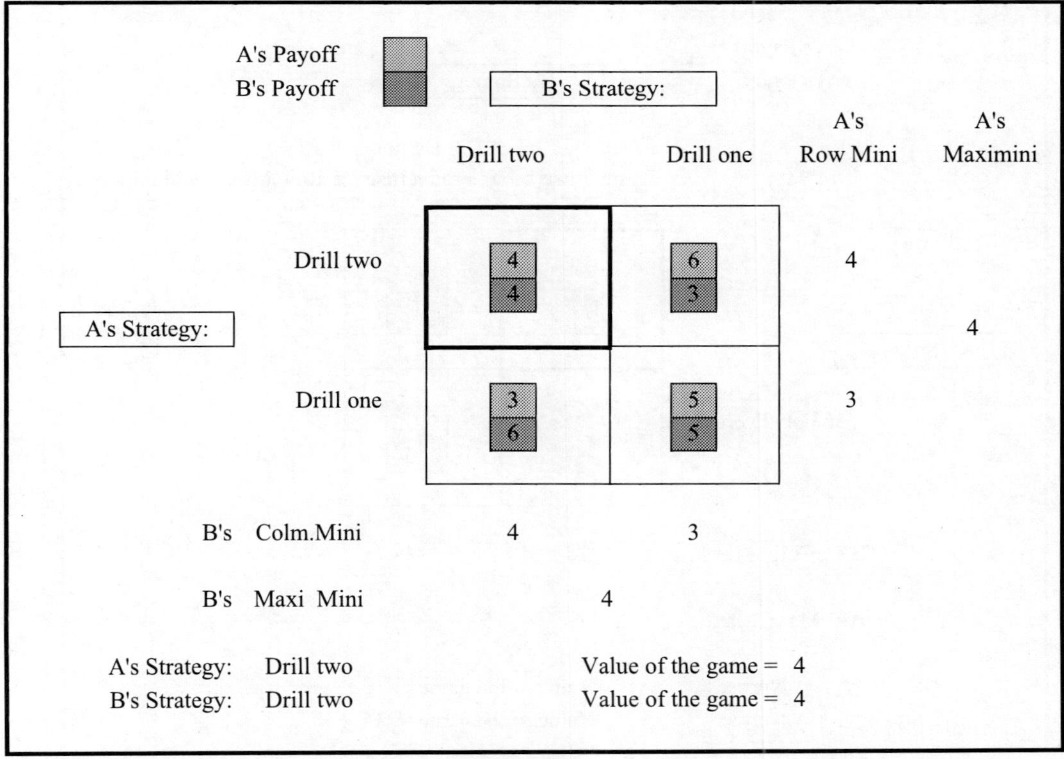

Figure 15.10

15.7 A REVISIT OF ADVERTISING MODELS: GAMES WITH COOPERATION

Till now we have discussed enough illustrations for non-cooperative games. Often advertising takes place under co-operation. Consider two farms, one grows grapes and another grows pineapple. To begin with neither of them advertises and each one of them gets Rs. 10 lakh as profit. Now suppose grape grower wants to put an advertisement stating that the fresh fruits are healthier, at the cost of Rs. 3 lakh.

Such an advertisement is cooperative in the sense that such an advertisement not only will increase the market for grapes but also for pineapple because pineapple is also a fresh healthy food. Thus, the pineapple grower, the unintended gainer, will be happy with the advertisement and the cooperative. Now let this advertisement add Rs. 4 crore revenue to both. The resulting outcome is reported in the payoff matrix. The advertised grape grower will get a profit of Rs. 11 lakh. The happiest pineapple grower will get a profit of Rs. 14 lakh.

If our pineapple grower also wants to reciprocate the act of his good friend, he too will go for an advertisement in similar line stressing the importance of fresh fruit in healthcare. If he spends another Rs. 3 crore on advertisement an additional market of Rs. 4 lakh each will be generated.

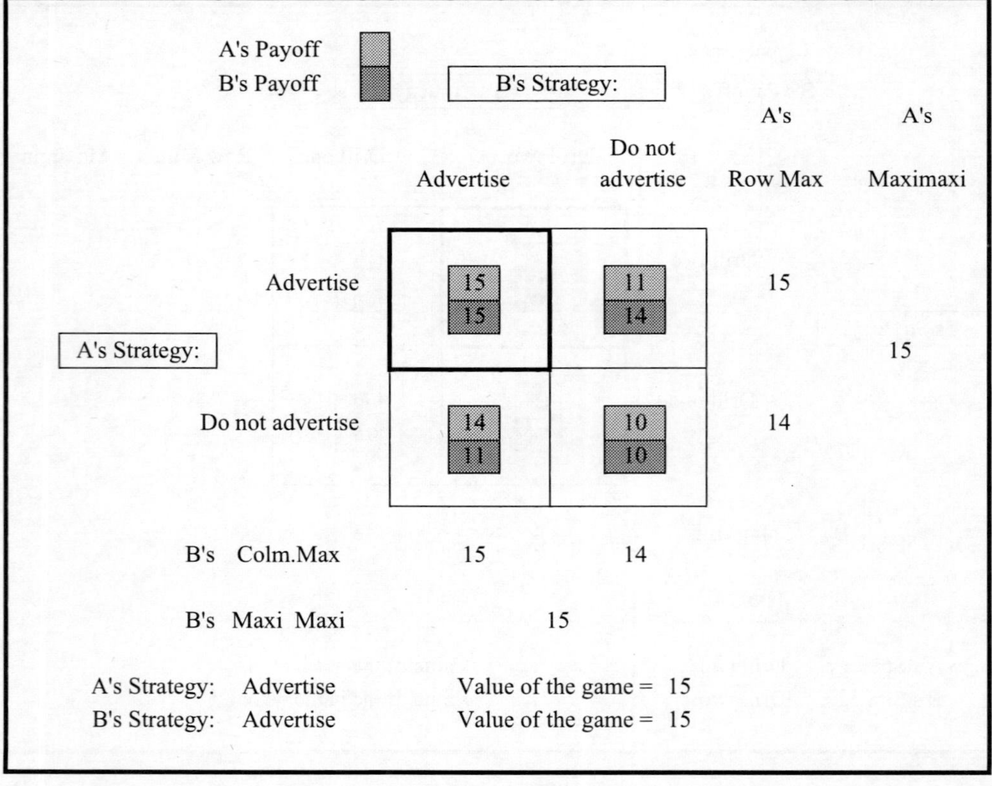

Figure 15.11

The final result could be a win-win situation of Rs. 12 crore profits for each one of them. However, in market where advertising is cooperative, no producer will opt for advertisement, because such advertisement not only benefits him but also benefits his competitor. Note that this game is based on dominance but at the same time optimistic in the sense both the strategies are Maximaxi type.

15.8 NASH EQUILIBRIUM AND ITS APPLICATIONS IN ADVERTISING

To determine the likely outcome of a game we have been searching "self enforcing" or "stable" strategies. The equilibrium under dominance rule that we have discussed so far are no doubt stable and self-enforcing, but in many games one or more players will not have the dominant strategies to play. We therefore need a little relaxed and more general equilibrium concept to deal such situations.

Nash equilibrium is a set of strategies such that each player is doing the best it can, given the actions of his opponents. Because each player has no tendency to deviate from its Nash strategies, the strategies are stable.

Nash Equilibrium

I am doing the best I can, given what you're doing
You're doing the best you can, given what I am doing

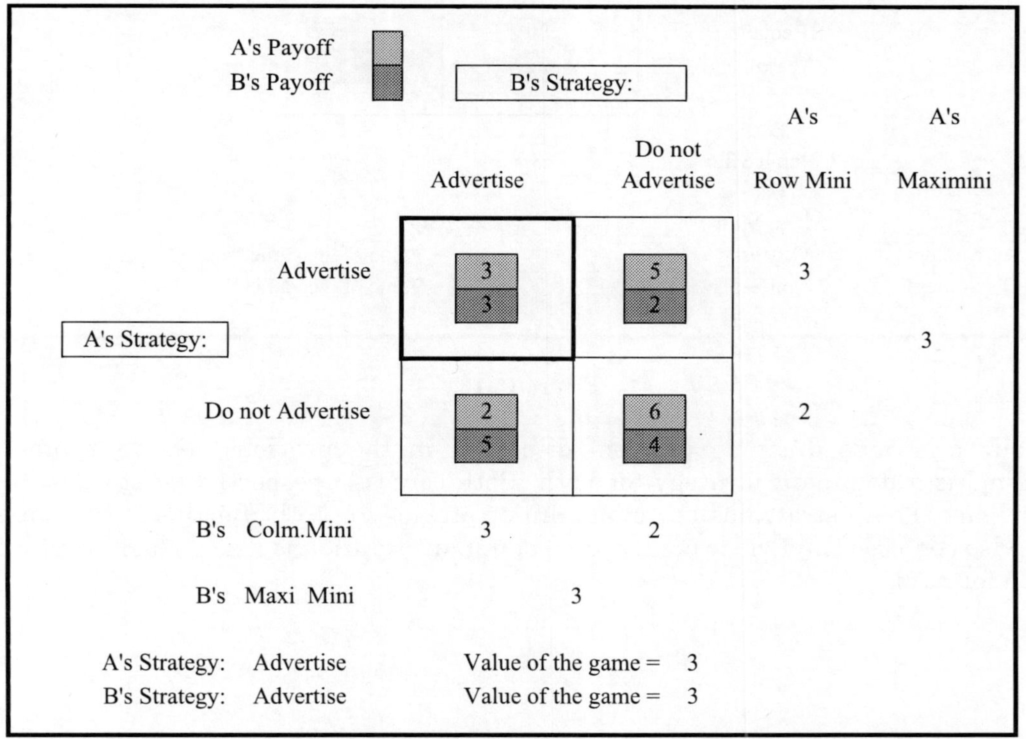

Figure 15.12

In this modified game described in Figure 15.12 clearly A does not have dominant strategy. However, B still has a dominant strategy of going for advertisement. Under the new circumstances A must put himself into B shoes and act on the precondition that B would go for advertisement. When B's choice is restricted to "Advertise", A too will go for advertising because that only gives him more profit of Rs. 3 crore (3>2). Thus "advertise-advertise" will be the final solution. Here clearly B's solution is based on dominance and A's is based on Nash equilibrium. Note that the equilibrium is conservative as well.

15.9 PRISONERS' DILEMMA: A REVISIT UNDER NASH EQUILIBRIUM

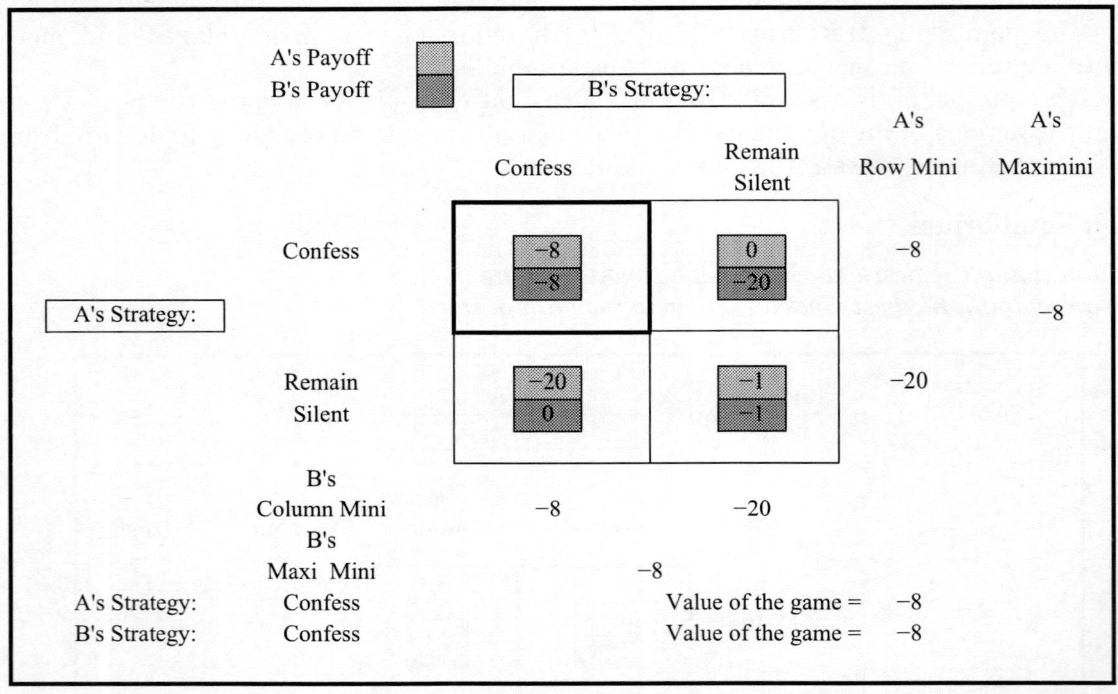

Figure 15.13

In our prisoner's dilemma problem discussed in the previous section 'confessing-confessing' is a dominant strategy for both. Both confess irrespective of strategy of his counter part. These strategies are maximini strategies for both. Further, if A confesses first B also confesses and vice versa. So the dominant solution is also a Nash equilibrium solution for both.

15.10 NASH EQUILIBRIUM IN INVESTING

The concepts of Nash equilibrium heavily relay upon individual rationally. Each player's strategy depends upon not only on his own strategy but also the rationality of his opponent. Consider the following illustration. From the payoff matrix in Figure 15.14 it is clear that investing is a dominant strategy for firm B because by doing so it will be better irrespective of what firm A does. Under the given conditions firm A also prefer to invest (20 > –100) and make a profit of Rs. 20 crore. Clearly the outcome (invest-invest) is a Nash equilibrium.

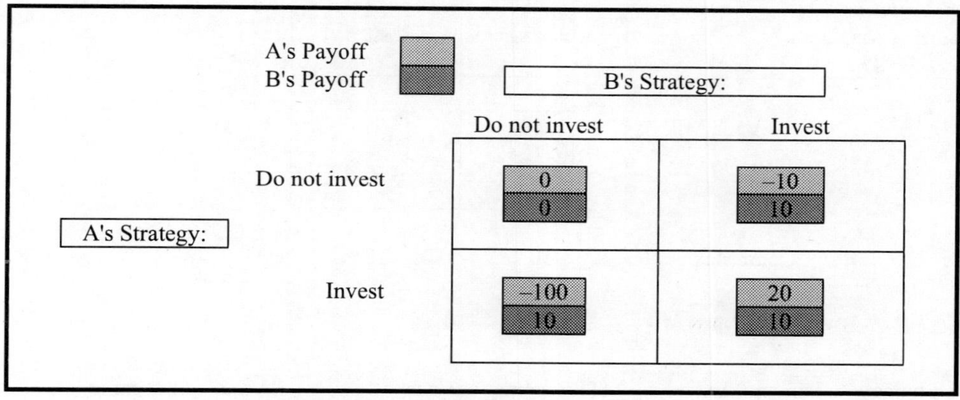

Figure 15.14

15.11 NASH EQUILIBRIUM IN CHOICE OF PRODUCTS

Two ready mix food processing companies MTR and TTK in Bangalore are interested in introducing a new product into their market. The proposed products may be either Jamoon mix or Vada mix for the coming Diwali season. Remember, that resources permit only one item for the said season. If MTR introduces Jamoon mix, then TTK can think of only Vada mix and vice versa. If both the products are introduced simultaneously then the respective payoffs are going to be negative as shown in Figure 15.15.

The payoff matrix clearly shows that neither of them has dominant strategies in this case. Let MTR take the lead and introduce Jamoon mix into the market first. Under the rule "I can do the best given what you are doing" the TTK can think of only introducing Vada mix. Under Nash Equilibrium each of them will get Rs. 10 crore as profit. On the other hand if TTK leads first and introduces Jamoon mix then MTR can think of introducing Vada mix. This is again another Nash equilibrium fetching Rs. 10 crore profit each. Clearly, the solution is optimistic type.

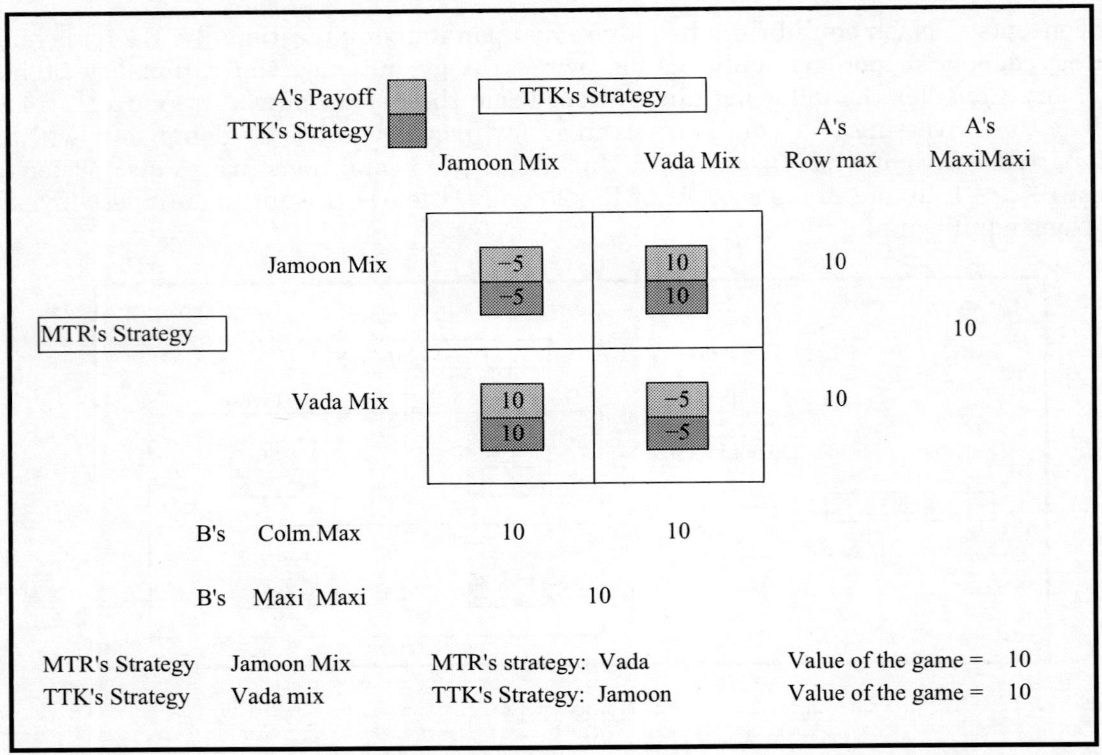

Figure 15.15

15.12 NASH EQUILIBRIUM AND HONEYMOON TRIP

In the Honeymoon Trip payoff shows Jim is more interested in Ooty and John is more interested in Simla. In this approach it is interesting to note that neither of them has dominant strategies to drag their partner to their place of choice. The maximini strategies show one is interested in Ooty and another is interested in Simla. This type of assignment could be of some interest for a holiday trip and not for a honeymoon trip. Luckily, we have two Nash equilibriums situations for honeymoon trip in this case. If John reveals his choice to Ooty first then Jim can also do the same. Instead if Jim were to put up her claim Simla first, then the best that our poor John can do is to agree for Simla.

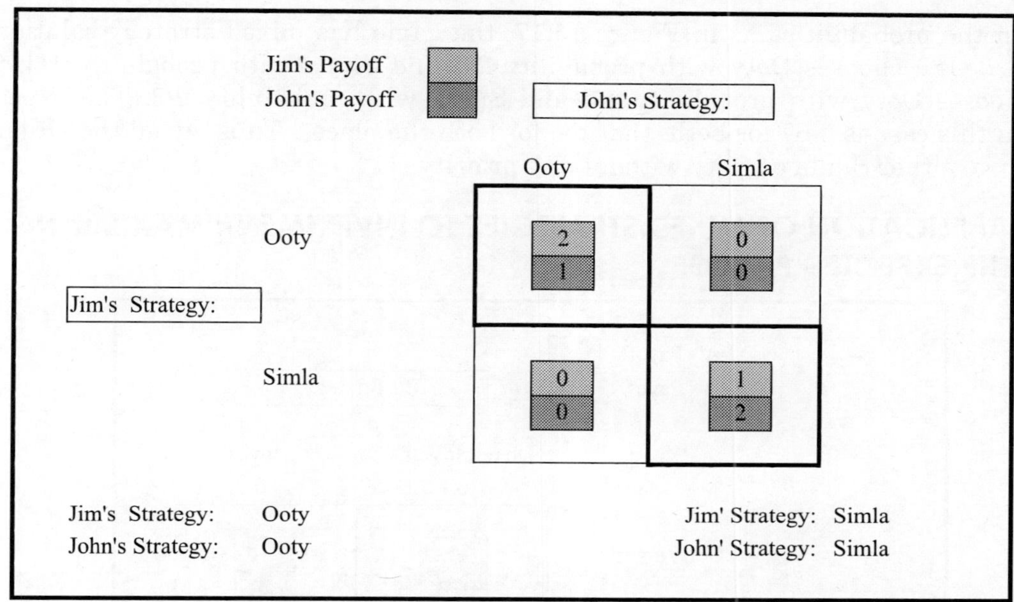

Figure 15.16

15.13 A MIXED STRATEGIES: EXPECTED PAYOFF APPROACH TO HONEYMOON MODEL

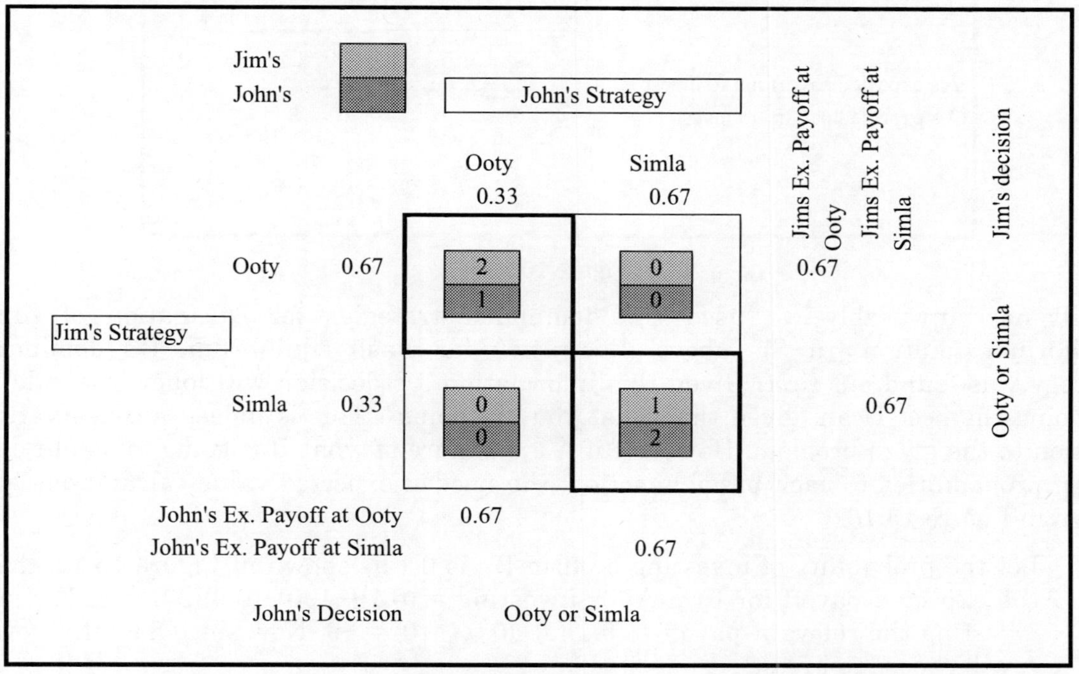

Figure 15.17

Given the probabilities as in Figure 15.17, the game has mixed strategy solution as well. In it Jim chooses Ooty with probability 2/3 and Simla with probability 1/3. Our John chooses Ooty with probability 1/3 and Simla with probability 2/3. The expected payoff in this case is 0.67 for both, that too for both the places. Thus, given the situation they can go either Simla or Ooty without any priority.

15.14 APPLICATION OF MIXED STRATEGIES TO INVESTMENT: MAXIMIZING THE EXPECTED PAYOFF

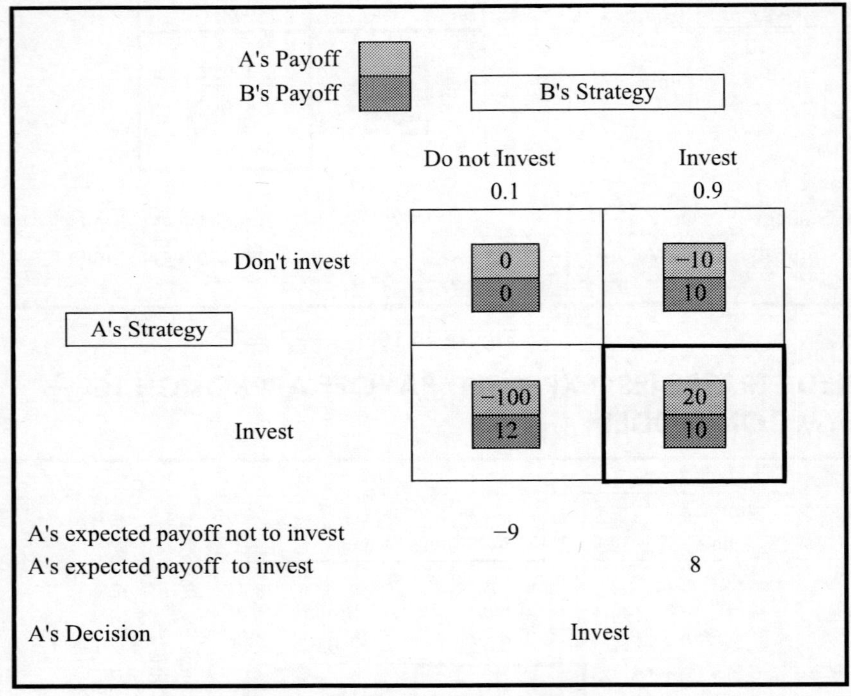

Figure 15.18

Till now invariably we used the dominant strategies for calculation of final equilibrium solution. In the above illustration for Nash equilibrium B's dominant strategy was found out first. Given this information A's decision will follow the rule "I am doing the best I can, given that what you are doing". This is indeed a conservative solution to the given problem. However, if A is not sure of what B is going to do but can assign probabilities to each possible action, the needed expected value calculations are shown in Figure 15.18.

i) Let the probability of investing by firm B's is 0.1 as shown in Figure 15.12, then the expected payoff for firm A for investing = (0.1)(−100)+(0.9)(20) = 8. For not investing the relevant payoff = (0.1)(0)+(0.9)(−10) = −9. Now since 8>−9 the form A will choose to invest.

ii) On the other hand, is firm A believes that the probability of investing by firm B is 0.5 then the expected payoff for firm A for investing is (0.5)(−100)+(0.5)(20) = −40. For not investing the relevant payoff = (0.5)(0)+(0.5)(−10) = −5. Since −5 > −40 the firm B will not invest.

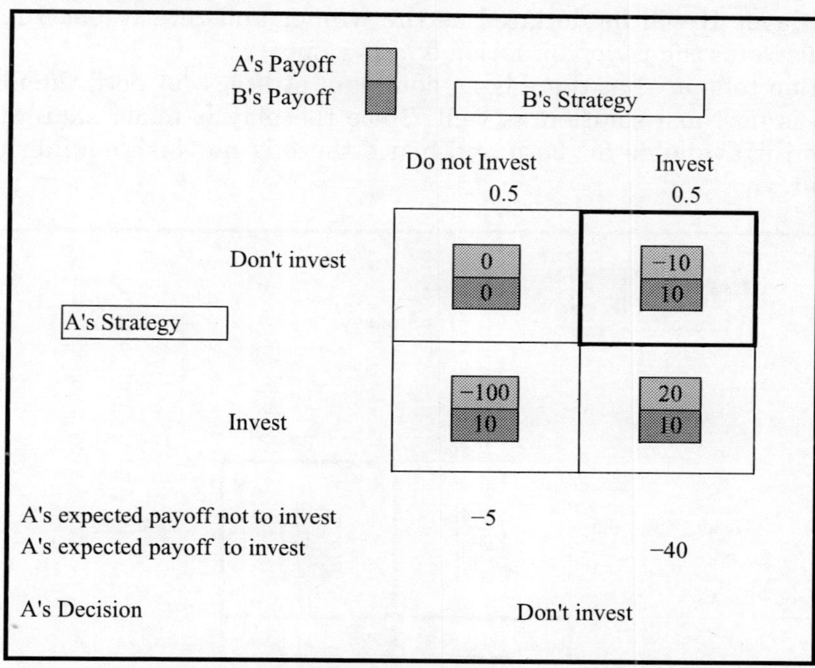

Figure 15.19

15.15 APPLICATION OF MIXED STRATEGIES: MATCHING THE RUPEE

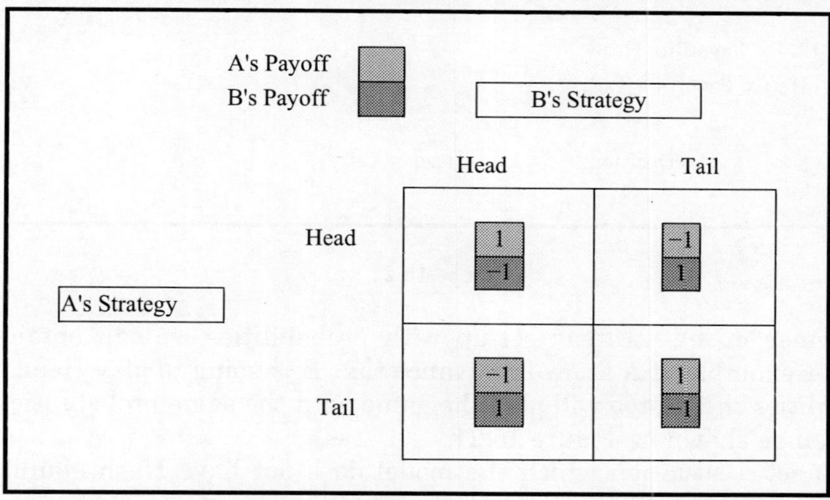

Figure 15.20

In this game both the players toss the coin independently but reveal the outcome simultaneously. If the outcome were to match either as "head-head" or "tail-tail" then player A is declared as winner of the play and will receive one rupee from B. On the other hand, if the coins do not match in the sense that the outcome is either head-tail or tail-head then player B will be declared as the winner and receive one rupee form A. Figure 15.20 illustrates the payoff situation of this game.

Mere inspection tells us that there is no dominant strategy for both the players. Also note that there is no Nash solution as well. Since the play is made simultaneously no prior information is available for both and hence there is no Nash equilibrium in pure strategies format.

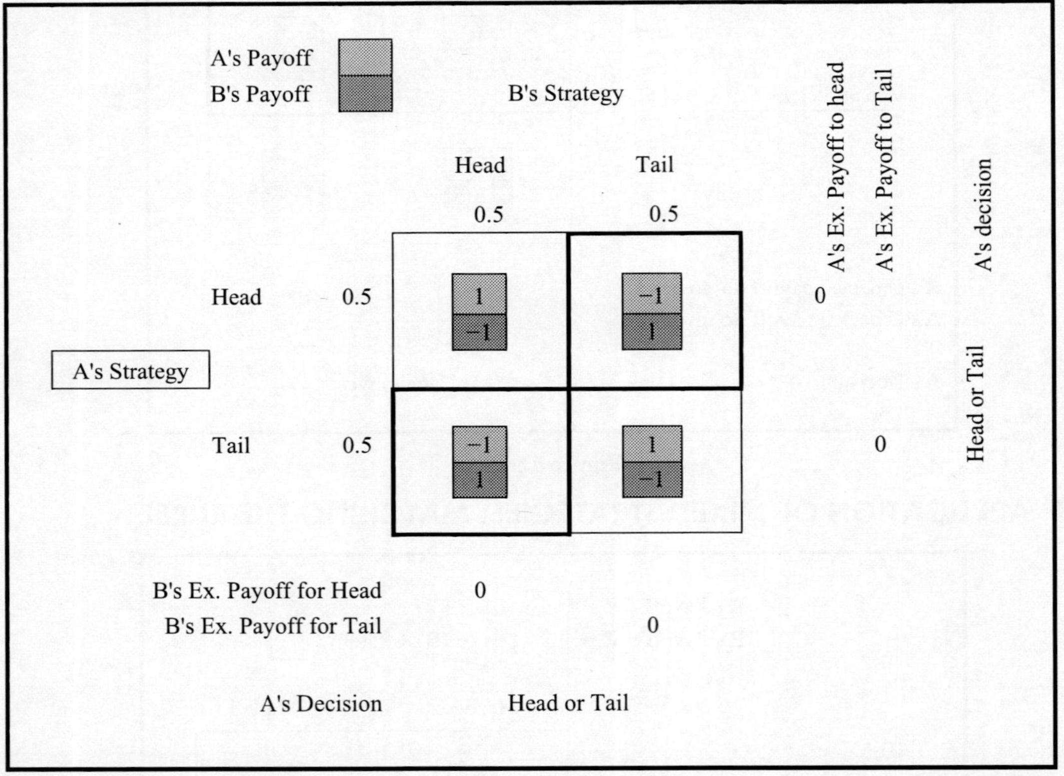

Figure 15.21

However, under mixed strategy set-up with probabilities we can obtain the Nash equilibrium. For example, if A knew in advance that B is going to play Head and Tail at 1/2, 1/2 probabilities then A too will play the game with the same probabilities yielding a 0 expected payoff as shown in Figure 15.21.

We can also set a situation which the model does not have Nash equilibrium even under mixed strategy play. Figure 15.22 illustrates one such case.

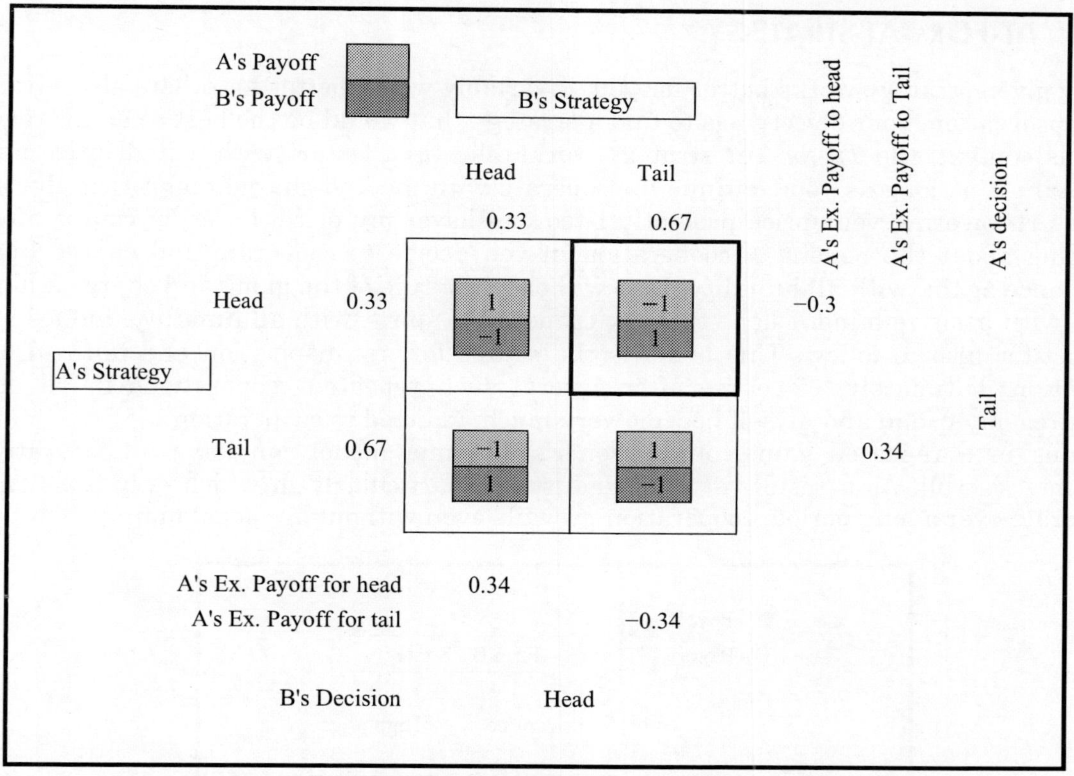

Figure 15.22

15.16 REPEATED GAMES

Prisoner's dilemma types of games discussed so far are of non-cooperative in nature and based on individual self interest. Also the final solutions are inefficient in such models. Is there any way out to this dilemma, so that ogolopolistic coordination and cooperation could prevail? Really in our prisoner's dilemma the opportunity to confess may be once or twice in their entire lifetime. But in oligopolistic markets firms set both output and price again and again for many more times. Such games are often called repeated games. In such repeated games, strategies become more and more complex. During such repetitions they not only develop reputation of their own but also study the reputation of its competitors.

If both firms A and B charge a very high price for their products then both of them are entailed to a very high profit. But definitely they are afraid of charging such a high price because of the fear that the competitor may go for a low price at any time and loot a very hefty profit by himself. But if this game is repeated again and again many times both of them notify price decisions of the competitor on every first day of the respective months. Should one play the game differently, perhaps changing price over and again in line with his competitor?

15.17 TIT-FOR-TAT STRATEGY

If any given strategy works better for you, invariably works better for others also. Given this observation, your objective is to find a strategy that would be the best at an average. In this context, the *Tit-for-Tat* strategy works like this. I start with a high price and maintain it as long as you continue to cooperate with me and charge high price. But as you start lowering your price promptly I too will lower my price. Later, of course after learning about the benefit of cooperation, if you decide to cooperate and charge high price once again, with all humility, I too will obey the rule of the game and charge a high price with positive mind. I hope that my cooperative spirit, with all humility, induce my competitor also to follow. This belief is the source for my inspiration and faith in co-operation. Ultimately, of course after several such repeated cooperations in all our ventures, my friend and myself became very much addicted to cooperation.

Thus in a repeated game, our prisoner's dilemma model can have a cooperative outcome as well. As a result, in some industries, particularly in which only few firms cooperate over a long period, cooperation prevails even without any legal biding.

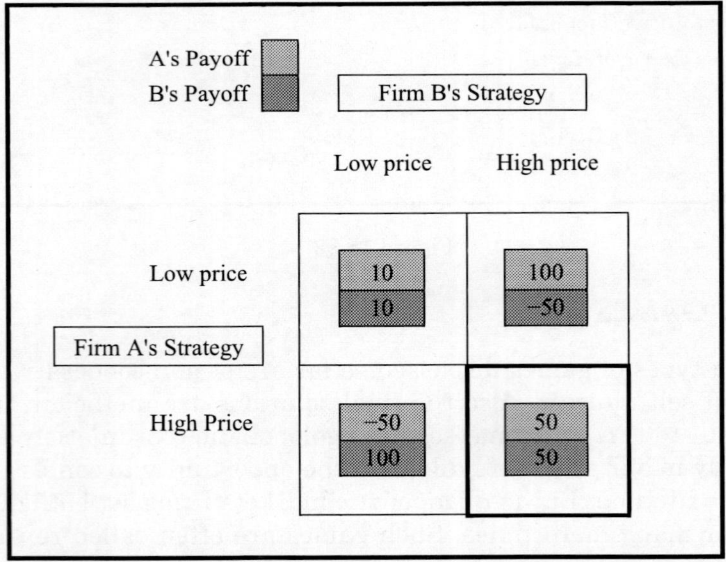

Figure 15.23

15.18 SEQUENTIAL GAME: ADVANTAGE OF MOVING FIRST

In most of the games discussed so far, both the players are assumed to take the strategic decisions simultaneously. For example, the Cournot spring water problem is a sequential model in which one firm set output first and the second one sets the output later by taking the first firm's output in mind by assuming him as his business leader.

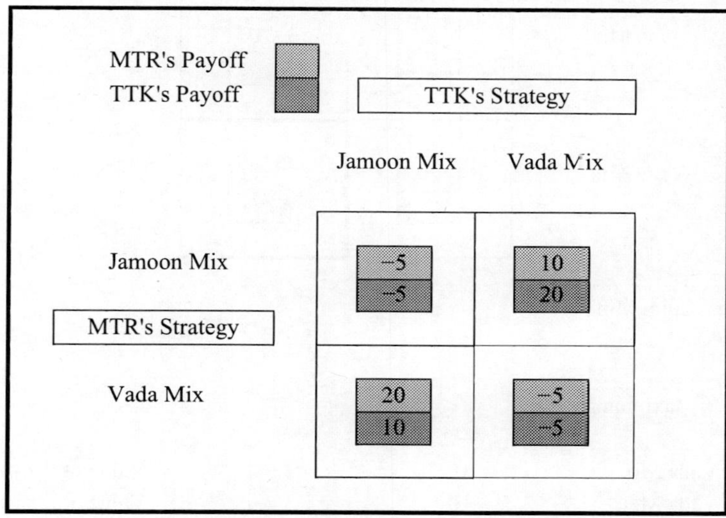

Figure 15.24

On the basis of second firms output decision the very first firm revises his output decision by assuming the second duopoly as his business leader. This process of revision continues till the final equilibrium is obtained.

As a simple illustration let us have relooked at our Jamoon mix and Vada mix problem once again with a slight change in the payoff as shown in Figure 15.24.

Clearly this time Vada mix is having a clear cut advantage Rs. 20 crore profit. Thus, the first incoming person to this new venture is going to be the gainer. Suppose both the firms, in ignorance of each other's interest, enters the Vada mix business to take the advantage of Rs. 20 crore profit, the result would by pathetic in the sense that both of them will end up with a loss of Rs. 5 crore as shown Figure 15.25.

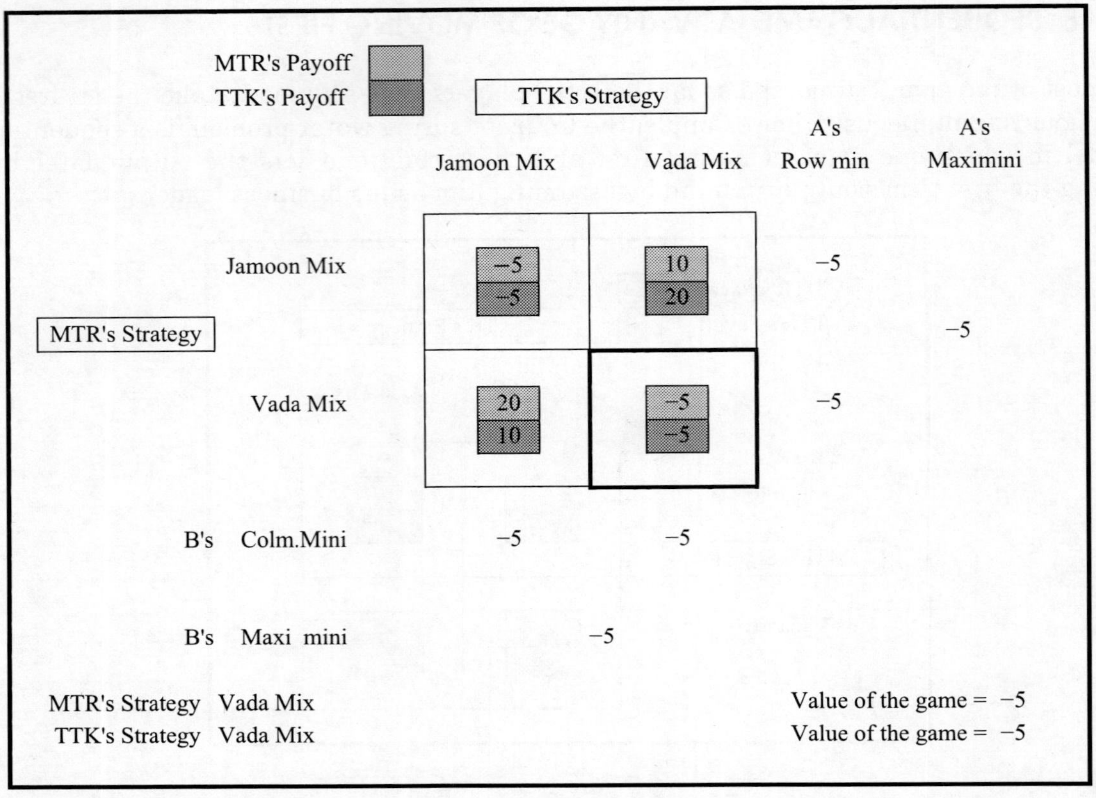

Figure 15.25

Advantage of Moving First

Both the new ventures are still profitable under sequential model. For example, in our illustration if TTK introduces Vada mix first, it will get a profit of Rs. 20 crore in a year. The late MTR at best can go for Jamoon mix and get a profit of Rs. 10 crore. If MTR introduces Vada mix first it will get a profit of Rs. 20 crore in a year. The late TTK at best can go for Jamoon mix and get a profit of Rs. 10 crore.

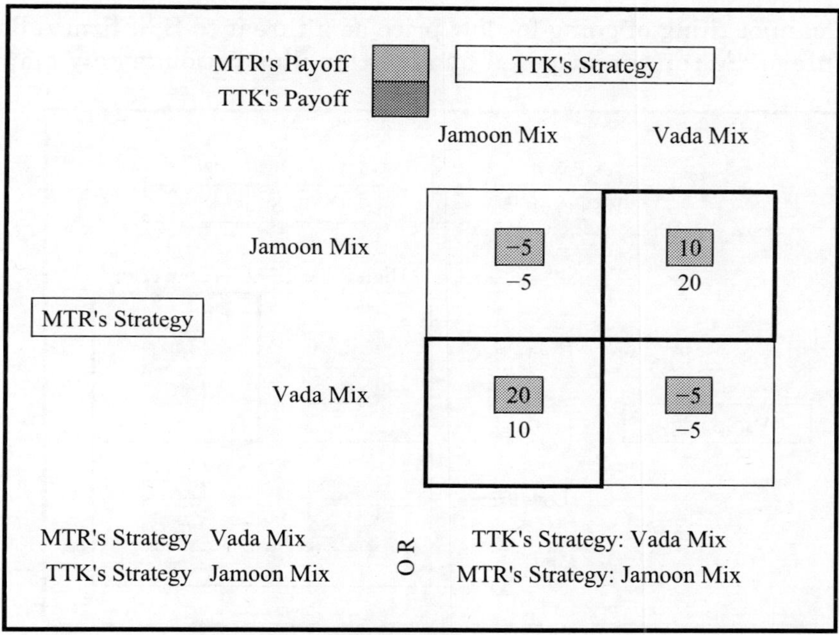

Figure 15.26

15.19 THREATS, COMMITMENTS AND CREDITABILITY

In the sequential game described above, who will make strategic move to enter first into the business and grasp the huge profit is typical question. A strategic move is one that influences the other persons choice in a manner favourable to one's self. If for example, if MTR simply announces its decision to go for Vada mix to knockout the benefit there is no guarantee that such a threat will work. TTK can make still big announcement in a different tone expressing its readiness to introduce the Vada mix. In this context the threat must be creditable and believable. For example, at the time of announcing its readiness MTR sign a big contract for the forward purchase of raw materials for Vada making then there is every possibility of our TTK to believe the readiness of MTR. Thus the thread is useful only if it is creditable.

Empty threat

Let firm A produce a computer that can be used for word processor and other uses. Let firm B produce a computer especially for word processor alone. Given the payoff matrix in Figure 15.27, firm A using his dominant strategy of high price will try to restore top left corner square for its advantage for harvesting Rs. 100 crore profit hoping that B too will follow high price strategy.

However, for firm B clearly low price is a dominant strategy. So ultimately the top right square will be the equilibrium square for him given the A's high price strategy. If A's threat is real it must be in a position to induce firm B to have a high price strategy.

Clearly firm A cannot thing of going for low price as a threat to B, if firm A does so, it is for its disadvantage. So the threat posed by A in this illustration is only empty and not real.

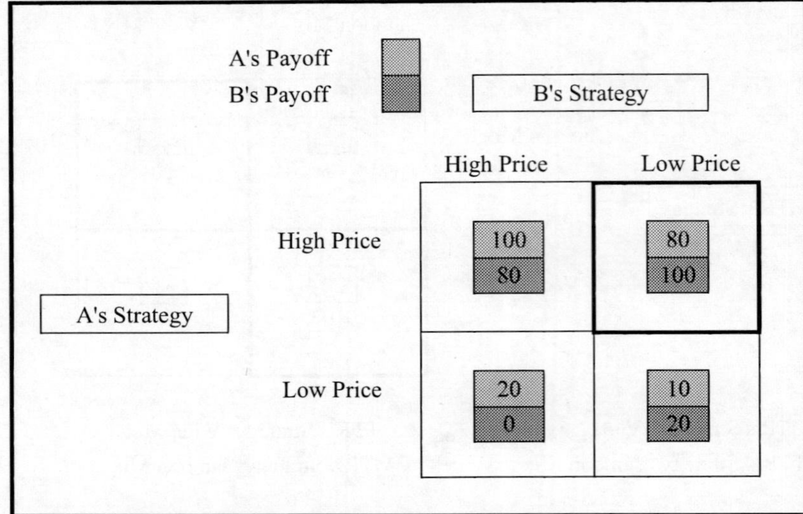

Figure 15.27

Real threat

Many times, a firm can send real threat signals to other firms. HAL in India is producing small commercial airplane. It buys small engines for its use from NAL. However, the supplier NAL is capability of producing both small and big engines. The associated payoffs are shown in Figure 15.16.

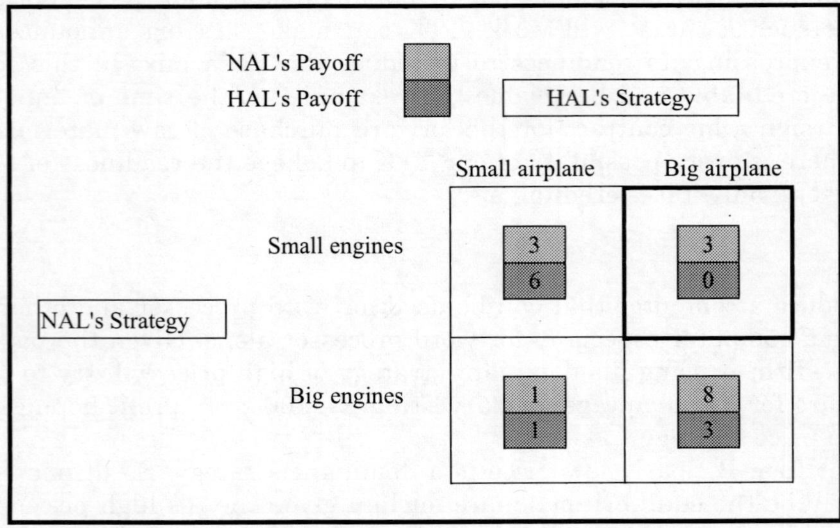

Figure 15.28

Here we have a sequential game in which HAL is the leader in the sense that it will first decide whether to go for a small pane or a big plane. From the payoff matrix it is clear that HAL will go for a small airplane that gives her Rs. 6 thousand crore profit. With this first decision, it is for NAL to go for small engines and have a profit of Rs. 3 thousand crore.

However, it looks that the bottom right square is more attractive for NAL hopping to get Rs. 8 thousand crore by producing big engines. If it does so it will force HAL to go for big planes. But this threat is not creditable. Once the HAL decision to go for small plane is notified NAL will have no interest to carry out it threat.

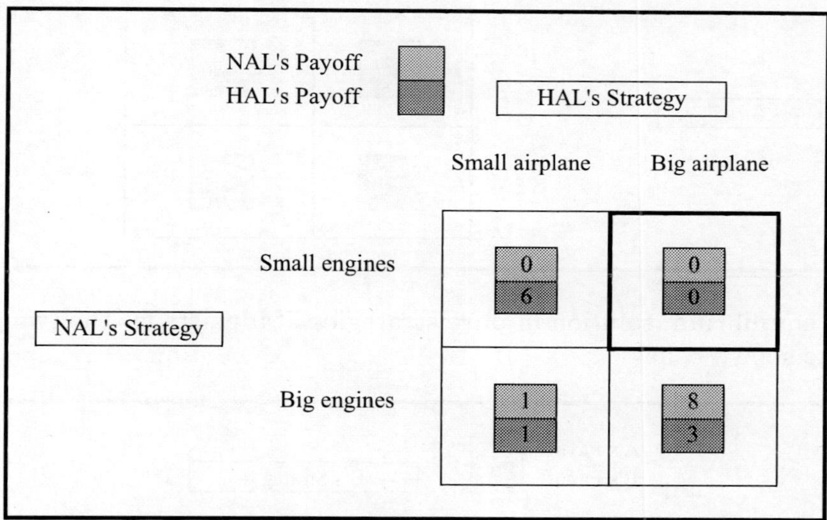

Figure 15.29

If NAL wants to make her threat creditable it should reduce its reliability of profit making by small engines. It can do this by shutting down or destroying some of it small engine producing capabilities. One such situation is shown in Figure 15.16. Under the new situation if HAL is to go for small planes our NAL will sell its big engines to other plane producers and settle for Rs. 1 thousand crore. But this is better than going for small engine and getting nothing. Clearly NAL has improved it affairs now.

EXERCISES

1. Find the equilibrium solution in pure strategies, if it exists for the two people zero sum game shown below.

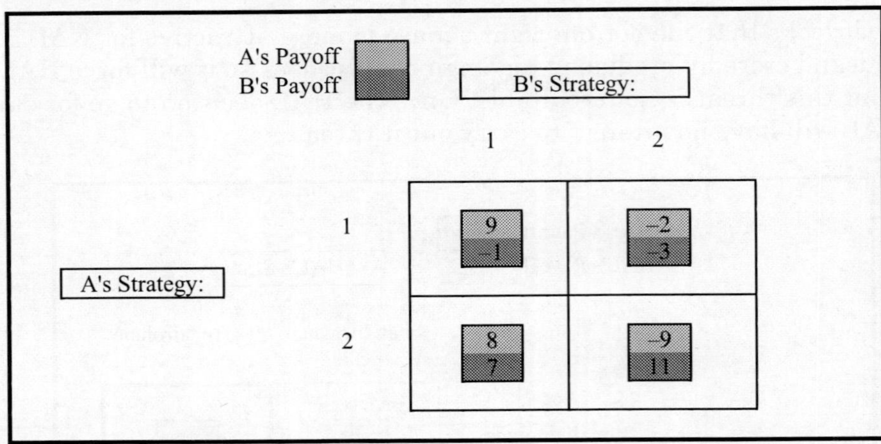

2. Find the equilibrium solution in pure strategies, if it exists for the two person zero sum game shown below.

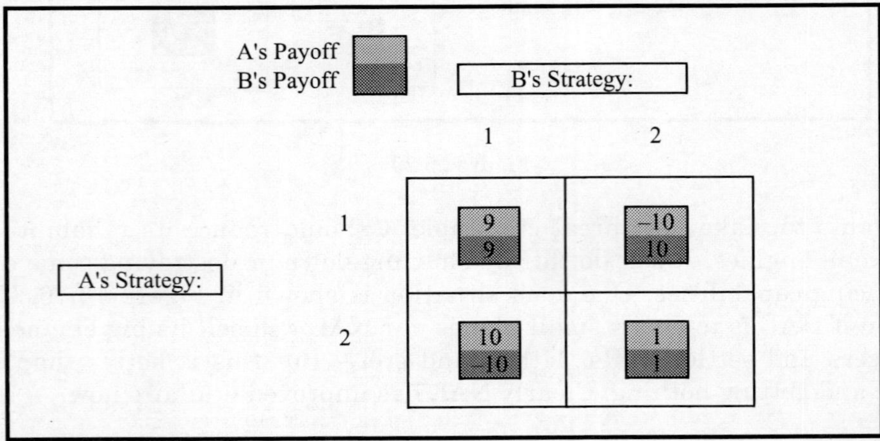

3. Given that each player role is to maximize the expected reward, show that for the game shown below show that under equilibrium the probabilities for each player under mixed strategy are (1/2, 1/2).

CHAPTER

16

Simulation Models

16.1 INTRODUCTION

Simulation is an experimentation oriented learning process of certain uncertain real-life situations. Though computerized mathematical models of simulation are readily available, it is often recommended for its ability in handling complicated and uncertain real-life situations that does not permit mathematical modelling. For effective handling, the inputs of simulation models are classified into two types. Controllable inputs and probabilistic inputs are two types of inputs commonly used in simulation models. In the case of controllable input models of simulation the analysts selects the value or values of these controllable inputs. However, in the case of probabilistic inputs the analysts randomly generate the inputs by using random numbers. Thus, the insertion of these two types of input into the model generate the needed outputs. By using different sets of controllable and probabilistic inputs the analysts stimulate different sets of outputs. After reviewing all such stimulated outputs the analysts recommends the suitable values of the controllable inputs that will give the suitable output for the real model. The following are some of the areas in which the simulation method could be of great help.

1. ***New product development:*** The very purpose of this model is to verify the profitability of the new venture. For this purpose a profit model is developed with probabilistic inputs like demand, labour cost, parts cost, etc. The only controllable input in such a modelling is whether the new product is to be introduced or not.
2. ***Over-booking in Indian Railway:*** Correct number booking according to the number of seats available is no longer a valid proposition for optimizing the profit. Especially the last minute cancellations will leave unfilled vacant seats and hence involve loss to the railway. To avoid such situations and the associated loss, railways reserve tickets even beyond the seat availability under the head 'Reservation Against Cancellation' category. However, one should always know the number of probabilistic cancellations so that the 'Reservation Against Cancellation' bookings is limited to only that number. Remember, too many 'Reservation Against Cancellation' bookings unnecessarily block the customers from other modes of transportation and loss in faith in railways. In such models the reservations are probabilistic inputs and the actual number of available seats are controllable inputs.
3. ***Inventory controls:*** Inventory definitely involves cost but at the same time ensures a smooth flow of the commodity under reference by considering both demand and supply. However, too much of inventory is too bad for the company's profitability. The variables like product demand, delivery lead-time are the two probabilistic inputs in such models. Order quantity and reorder points are the two controllable inputs.

16.2 METHODS OF SIMULATION

Simulation can be initiated by either of the following two methods:

1. Monte Carlo method
2. System stimulation method

Monte Carlo method uses the random numbers to generate the needed probabilistic inputs. Since random numbers are random by definition, the so generated probabilistic inputs are also random in these models. Wherever the situation is uncertain and not amenable for mathematical modelling, Monte Carlo technique is recommended.

System stimulation, on the other hand, is one in which there is a miniature reproduction of the operating environment. Testing of an airplane for its plying worthiness is actually tested with the miniature models in an artificially created real-life environment before the real models are being put into use.

Monte Carlo method: Although simulation can be done in many ways we will adopt probabilistic simulation by using *Monte Carlo* method in the section to follow. The chance element is a very significant feature of *Monte Carlo* simulation and this approach can be used when the given process has a *random* or *chance* component in it. In this method a given problem is solved by simulating the data with the help of random number generator. What is significant here is that the variables under consideration may not explicitly follow well known theoretical distributions like Poisson, Normal, Binomial, etc. Normally the relevant probability distribution is obtained directly from the past record. The following examples illustrate such a formulation.

16.3 DEMAND SUPPLY RELATED SIMULATIONS

■ Example 1

A newspaper agent in a local area keeps the demand for news paper over 200 working days in a year as follows. Given the random numbers 61, 74, 24, 03, 59, 16, 84, 92, 52, 07 simulate the demand for the newspaper for 10 days.

Table 16.1

Demand:	5	6	7	8	9	10	11	12	Total
No. of days:	5	10	16	50	62	38	12	8	200

Solution

If this past experience is adequate enough to describe the distribution then we can effectively convert the data into a frequency distribution as follows.

Table 16.2

Demand:	5	6	7	8	9	10	11	12	Total
No. of days:	0.02	0.05	0.08	0.25	0.31	0.19	0.06	0.04	1

Since we have probabilities up to two decimals which will all add up to 1.00 we need 100 random numbers of two digits to represent each point of the probability distribution. Table 16.3 illustrates the said assignment.

Table 16.3

Demand	Probability	Cumulative probability	Random number interval
5	0.02	0.02	00-01
6	0.05	0.07	02-06
7	0.08	0.15	07-14
8	0.25	0.40	15-39
9	0.31	0.71	40-70
10	0.19	0.90	71-89
11	0.06	0.96	90-95
12	0.04	1.00	96-99

Now since the probability for the demand 5 is 0.02 we attach random numbers 00-01 to this interval. Since the next level of 6 units demand is having the probability 0.05 we attach five random numbers 02-06. In this way we attach random numbers in proportion to their probability interval as shown in Table 16.3. The calculation of the cumulative probability in the third column will assist us in such an allotment.

Once this is done we are ready to simulate the demand pattern for any number of days by choosing as many random numbers as needed from a random number table. Given the 10 random numbers 61, 74, 24, 03, 59, 16, 84, 92, 52, 07, the associated simulated demand pattern for all the ten days run are obtained and reported in Table 16.4. Our first random number is 61 in Table 16.4. From Table 16.3 the associated level of demand is obtained 9 and reported in Table 16.4. Similarly, we simulate the demand for all other random numbers and report them all in Table 16.4.

Table 16.4

Day	1	2	3	4	5	6	7	8	9	10
Random No:	61	74	24	03	59	16	84	92	52	07
Demand:	9	10	8	6	9	8	10	11	9	7

In the above table the last row shows the stimulated demand for newspaper for 10 days.

■ Example 2

The daily production of mopeds in a factory varies from 146 to 154 depending upon the availability of raw material and other working conditions. The associated probabilities for various levels of production are highlighted in Table 16.5.

Table 16.5

Production per day	146	147	148	149	150	151	152	153	154
Probability	0.04	0.09	0.12	0.14	0.11	0.10	0.20	0.12	0.80

The finished mopeds are transported in a specially designed lorry accommodating only 150 mopeds to the maximum. If there is any excess moped produced over and above 150, they can not find space in the lorry and hence can be carried over to the next day lorry for transport. Using the following random numbers 80, 81, 76, 64, 43, 18, 26, 10, 12, 65, 68, 69, 61, 57 simulate the production levels for 15 days and obtain the answer to the following questions

i) What will be the average number of mopeds waiting in the factory?

ii) What will be the average number of empty spaces on the lorry?

Solution

As a first step we allot random numbers from 00 to 99 in proportion to the associated probabilities with the production of mopeds as shown in Table 16.6.

Table 16.6

Production	Probability	Cumulative probability	Random number interval
146	0.04	0.04	00-03
147	0.09	0.13	04-12
148	0.12	0.25	13-24
149	0.14	0.39	25-38
150	0.11	0.50	39-49
151	0.10	0.60	50-59
152	0.20	0.80	60-79
153	0.12	0.92	80-91
154	0.08	1.00	92-99

Based on the given random numbers we stimulate the production on daily basis as shown in Table 16.7 using Table 16.6.

Table 16.7

Sl. No.	Random number	Production	No. waiting in the factory	No. of empty space in the lorry
1	80	153	3	0
2	81	153	6	0
3	76	152	8	0
4	75	152	10	0
5	64	152	12	0
6	43	150	12	0
7	18	148	10	0
8	26	149	9	0
9	10	147	6	0
10	12	147	3	0
11	65	152	5	0
12	68	152	7	0
13	69	152	9	0
14	61	152	11	0
15	57	151	12	0
		Average	8.2	0

The average number of mopeds waiting in the factory = 8.2

Average number of empty space in the lorry = 0

■ Example 3

The daily production of scooters in a factory varies from 146 to 154 depending upon the availability of raw material and other working conditions. The associated probabilities for the varying levels of output are highlighted in Table 16.8 based on past experience. Obtain the stimulated production levels for 10 days.

Table 16.8

Production per day	97	98	99	100	101	102	103
Probability	0.05	0.10	0.20	0.25	0.20	0.15	0.05

Produced scooters should be sent through a tempo having the space only for 100 scooters. If more than 100 scooters are produced in a day then extra scooters produced are to be kept and clubbed with next day's production for transport. Find the average number of empty space in the tempo and the average number of scooters waiting for want of space in the tempo. Random number for 10 days: 1010, 0730, 7040, 6080, 9130, 4870, 5170, 2180, 8750, 9833.

Solution

As a first step we allot random numbers from 00 to 99 in proportion to their associated probabilities with the production of scooters as shown in Table 16.9.

Table 16.9

Production	Probability	Cumulative probability	Random number interval
97	0.05	0.05	00-04
98	0.10	0.15	05-14
99	0.20	0.35	15-34
100	0.25	0.60	35-59
101	0.20	0.80	60-79
102	0.15	0.95	80-94
103	0.05	1.00	95-99

Since the given random numbers contain four digits and we need only two-digit random numbers we consider only the first two digits for our problem. So the needed random numbers are: 10, 07, 70, 60, 91, 48, 51, 21, 87, and 98. Based on these random numbers we stimulate the production on daily basis as shown in the Table 16.10 for 10 days.

The average number of scooters waiting in the factory = 3.1

Average number of empty space in the lorry = 0.4

Table 16.10

Sl. No.	Random number	Production	No. waiting in the factory	No. of empty space in the lorry
1	10	98	0	2
2	7	98	0	2
3	70	101	1	0
4	60	101	2	0
5	91	102	4	0
6	48	100	4	0
7	51	100	4	0
8	21	99	3	0
9	87	102	5	0
10	98	103	8	0
		Average	3.1	0.4

■ Example 4

The profitability of a certain project depends upon its promptness in completing the job on the agreed up on date. The delayed completion attracts a penalty reduction of Rs. 5000 per day subject to a maximum of Rs. 15,000. The expected profit per day is Rs. 40,000. The following table gives the probabilities of delayed delivery. By using the random numbers 363, 125, 490, 780, 691, 911, 712, 489, 514, 418 simulate profit for 10 days and obtain the average profit.

Table 16.11

Number of days delay	Probability
0	0.70
1	0.15
2	0.10
3 and above	0.05
Total	1.00

Solution

As a first step we allot random numbers from 00 to 99 in proportion to their associated probabilities with the days of delayed delivery as shown in Table 16.12.

Table 16.12

No. of days delay	Probability	Cumulative probability	Random number interval
0	0.70	0.70	00-69
1	0.15	0.85	70-84
2	0.10	0.95	85-94
3 and above	0.05	1.00	95-99

Based on these random numbers entered in the second column let us simulate the number days of delayed for 10 projects as shown in Table 16.13. Remember, since the given random numbers contain three digits and we need only two digits random numbers we consider only the first two-digit for our problem. So the needed random numbers are: 36, 12, 49, 78, 69, 91, 71, 48, 51, and 41.

Table 16.13

Sl. No.	Random number	No. of days delay	Profit per day
1	36	0	40,000
2	12	0	40,000
3	49	0	40,000
4	78	1	35,000
5	69	0	40,000
6	91	2	30,000
7	71	1	35,000
8	48	0	40,000
9	51	0	40,000
10	41	0	40,000
		Average	38,000

Average profit per project = Rs. 38,000

■ Example 5

An agricultural farm manufactures 30 kg of a perishable product in a day. The varying levels of demand and their associated probabilities are given in Table 16.14.

Table 16.14

Demand	Probability
27	0.10
28	0.15
29	0.20
30	0.35
31	0.15
32	0.05

The production costs and sales price per unit are Rs. 40 and Rs. 50 respectively. Unsold quantity, if any, is to be disposed of at a loss of Rs. 15. Also there is a penalty of Rs. 5 per unit if the demand is not met. Using the following random numbers 10, 99, 65, 99, 95, 01, 79, 11, 16, 20 estimate the total profit/loss for the farm for the next 10 days. If the farm decides to produce 29 units per day instead of 30 units will there be any improvement in the profitability?

Solution

As a first step we allot random numbers from 00 to 99 in proportion to their associated probabilities as shown in Table 16.15.

Table 16.15

Demand	Probability	Cumulative probability	Random number interval
27	0.10	0.10	00-09
28	0.15	0.25	10-24
29	0.20	0.45	25-44
30	0.35	0.80	45-79
31	0.15	0.95	80-94
32	0.05	1.00	95-99

Based on the given random numbers, we stimulate the demand on daily basis as shown in Table 16.16.

Table 16.16 Demand simulation when the production is 30 units per day

Sl. No.	Random number	Estimated demand	Profit	Disposal loss	Penalty for shortage	Net profit
1	10	28	280	30	0	250
2	99	32	300	0	10	290
3	65	30	300	0	0	300
4	99	32	300	0	10	290
5	95	32	300	0	10	290
6	01	27	270	45	0	225
7	79	30	300	0	0	300
8	11	28	280	30	0	250
9	16	28	280	30	0	250
10	20	28	280	30	0	250
					Total profit	2695

Table 16.17 Demand simulation when the production is 29 units per day

Sl. No.	Random number	Estimated demand	Profit	Disposal loss	Penalty for shortage	Net profit
1	10	28	280	15	0	265
2	99	32	290	0	15	275
3	65	30	290	0	5	285
4	99	32	290	0	15	275
5	95	32	290	0	15	275
6	01	27	270	30	0	240
7	79	30	290	0	5	285
8	11	28	280	15	0	265
9	16	28	280	15	0	265
10	20	28	280	15	0	265
					Total profit	2695

The total profit remains unaltered even though the output is reduced by one unit.

16.4 WAITING LINE SIMULATION MODELS

In the waiting line models arrival time between two successive customers need not be constant all the time. In reality the inter-arrival time between two successive customers varies randomly. Similarly, the service time also cannot be a constant. Normally it also varies randomly depending upon the type of service. The following example illustrates the usefulness of simulation in this area.

■ Example 6

Dr. Strong is a dentist who schedules all her patients for 30 minutes appointment. Accordingly, the inter-arrival time between two consecutive arrivals is scheduled as 30 minutes. However, for some of his patients he takes more than 30 minutes and some others he takes less than 30 minutes depending upon the type of dental work to be under taken. The following table shows the various categories of dental work with their associated time for completion with their probabilities as well.

Table 16.18

Category	Time required	Probability of category
Filling	45 minutes	0.40
Crown	60 minutes	0.15
Cleaning	15 minutes	0.15
Extracting	45 minutes	0.10
Check-up	15 minutes	0.20

Simulate the dentist's clinic for four hours and determine the average waiting time for the patents and also the idle time for the doctor. Use the following random numbers for handling the above problem. 40, 82, 11, 34, 25, 66, 17, 79.

Solution

As a first step we allot random numbers from 00 to 99 in proportion to their associated probabilities for varying types of work as shown in Table 16.19.

Table 16.19

Category	Time	Probability	Cumulative probability	Random number interval
Filling	45 m	0.40	0.40	00-39
Crown	60 m	0.15	0.55	40-54
Cleaning	15 m	0.15	0.70	55-69
Extracting	45 m	0.10	0.80	70-79
Check-up	15 m	0.20	1.00	80-99

Based on the given random numbers, we simulate the service time starting from 8 AM as shown Table 16.20.

Table 16.20

Sl. No.	Arrival time	Random number	Service time in minutes	Service begins	Service ends	Waiting time for patient	Idle time for doctor
1	8.00 AM	40	60	8.00 AM	9.00 AM	0	0
2	8.30 AM	82	15	9.00 AM	9.15 AM	30	0
3	9.00 AM	11	45	9.15 AM	10.00 AM	15	0
4	9.30 AM	34	45	10.00 AM	10.45 AM	30	0
5	10.00 AM	25	45	10.45 AM	11.30 AM	45	0
6	10.30 AM	66	15	11.30 AM	11.45AM	60	0
7	11.00 AM	17	45	11.45AM	12.30 PM	45	0
8	11.30 AM	79	45	12.30 PM	1.15 PM	60	0
					Average	35.625	0

Thus the patient will have to wait at an average 35.625 minutes.

The doctor is to be busy all the time.

■ Example 7

The Regional Manager of Janatha Cooperative Bank sent a proposal to start a drive in facility banking in a commercial area in Bangalore. The research term of the bank has projected inter-arrival times and their associated probabilities as shown in Table 16.21.

Table 16.21

Time between two consecutive arrivals (in minutes)	Probability
3	0.17
4	0.25
5	0.25
6	0.20
7	0.13
Total	1.00

The bank wants to set only one teller's counter in this branch to begin with. The service timing and the associated probabilities are reported in Table 16.22.

Table 16.22

Service time (in minutes)	Probability
3	0.10
4	0.30
5	0.40
6	0.15
7	0.05
Total	1.00

Before going for actual construction work of the branch, the manager wants to know the number of car parking facilities to be provided. He also wants to know the mean waiting time by the customers from a sample of 25 customers. Suppose the location has facility to park only two cars, how many of them will move away from the proposed counter for want of parking space to some other nearby bank.

Solution

To begin with let us allot random numbers to the probability distribution of inter-arrival time given in Table 16.21. The Table 16.23 shows this allotment in the usual way.

Table 16.23

Time between two consecutive arrivals (in minutes)	Probability	Cumulative probability	Random number
3	0.17	0.17	00-16
4	0.25	0.42	17-41
5	0.25	0.67	42-66
6	0.20	0.87	67-86
7	0.13	1.00	87-99
Total	1.00	-	0

Similarly, for the service time also by using the Table 16.22 we allot the random numbers as shown in Table 16.24.

Table 16.24

Service time (in minutes)	Probability	Cumulative probability	Random number
3	0.10	0.10	00-09
4	0.30	0.40	10-39
5	0.40	0.80	40-79
6	0.15	0.95	80-94
7	0.05	1.00	95-99
Total	1.00	-	

In Table 16.25 we obtain the simulated work sheet for first 25 trials. In it, column 2 and 6, random numbers generated from a random number table are presented for inter-arrivals times and service times separately. Since for the first customer arrival, the random number is 24, the associated random arrival time is 4 minutes from Table 16.23. So if the car carriage starts functioning from 11 A.M, then the first customer will arrive at 11.04 A.M and park his car comfortably. Similarly, for service, the random number is 39 and the associated random service time from Table 16.24 is 4 minutes. Though the carriage starts at 11 A.M, it will have to wait up to the arrival of the first customer at

11.04 A.M. This 4 minute waiting time for the server is stated in the 10th column in Table 16.25. However, the customer need not wait in this case and this is being reported as 0 in the 9th column of the table. The service for the first customer will be over by 11.08 A.M.

Table 16.25

Trail	Random number	Inter- arrival time	Arrival time	Service begins	Random number	Service time	Service ends	Weighting time (consumer)	Idle time (server)	Queue length
(1)	(2)	(3)	(4)	(5)	(6)	(7)	(8)	(9)	(10)	(11)
1	24	4	11.04	11.04	39	4	11.08		4	
2	57	5	11.09	11.09	45	5	11.14		1	
3	77	6	11.15	11.15	27	4	11.19		1	
4	68	6	11.21	11.21	29	4	11.25		2	
5	64	5	11.26	11.26	35	4	11.30		1	
6	88	7	11.33	11.33	68	5	11.38		3	
7	98	7	11.40	11.40	51	5	11.45		2	
8	48	5	11.45	11.45	93	6	11.51		0	
9	93	7	11.52	11.52	61	5	11.57		1	
10	84	6	11.58	11.58	2	3	12.01		1	
11	11	3	12.01	12.01	56	5	12.06		0	
12	23	4	12.05	12.06	72	5	12.10	1	0	1
13	94	7	12.12	12.12	83	6	12.18		2	
14	83	6	12.18	12.18	2	3	12.21		0	
15	97	7	12.25	12.25	99	7	12.32		4	
16	83	6	12.31	12.32	10	4	12.35	1	0	1
17	93	7	12.38	12.38	34	4	12.42		3	
18	33	4	12.42	12.42	53	5	12.47		0	
19	49	5	12.47	12.47	94	6	12.53		0	
20	37	4	12.51	12.53	97	7	1.00	2	0	1
21	22	4	12.55	1.00	12	4	1.04	5	0	2
22	27	4	12.59	1.04	87	6	1.10	5	0	2
23	0	3	1.02	1.10	79	5	1.15	8	0	2
24	43	5	1.07	1.15	13	4	1.19	8	0	2
25	14	3	1.10	1.19	49	5	1.24	9	0	-
							Total	39		

The second customer random number for inter-arrival is 57. So from Table 16.23 the associated inter-arrival time is obtained as 5 minutes. So the second customer will arrive at 11.04 + 05 = 11.09 A.M as reported. Since the service of the first customer was over by 11.08 A.M, the second customer can be served immediately. But this time also the server will have to wait for 1 minute as reported. Remember, that at this point of time his car alone is parked in the two car parking spaces provided. Similarly other arrivals calculations are made and reported in the table up to the 11th customer.

In the case of 12th customer the arrival time is obtained as 12.05. But the 11th customer will be served up to 12.06. So this time the customer will have to wait for 1 minute. Accordingly, there will be 1 person in the queue. This is reported in the last column. This procedure is followed for all the 25 trials and entries are made accordingly. In the last column the maximum number in the queue is reported as 2. Therefore the maximum number of car parking that the bank building can have is two in the manager's building plan. Under such circumstances no customer will move out for want of car parking space. The total waiting time for all the 25 customers is 39 minutes. Thus the average waiting time per customer is 34 minute in our illustrative problem.

16.5 INVENTORY SIMULATION MODELS

In the inventory system, uncertainties bound to occur both in demand and lead time. However, in certain realistic situations, lead time is considered constants but the demand is considered stochastic variable based on certain well known probability distribution. In example 8 we take an illustrative case with known constant lead time and uncertain stochastic demand pattern. In example 9 we take both demand and lead time as uncertain stochastic variables.

■ Example 8

Suppose the daily demand for Maruti cars in a certain outlet in Bangalore has the discrete probability distribution as shown in Table 16.26. The inventory carrying cost is Rs. 5 per car per day in this showroom. Order placing cost is Rs. 100 per order irrespective of the order quantity. Loss in net revenue due to shortage of vehicle for delivery is Rs. 500 per unit. Reorder is placed whenever the evening closing stock reaches 27 units or less. The reordered quantity of 40 units is received only on the evening of the 5th day and available for sale only on the morning of the 6th day from the date of ordering. The opening balance in the inventory is 45 units. Obtain the grand total cost for a stimulated run of 25.

Table 16.26

Daily demand	Probability
3	0.17
4	0.25
5	0.25
6	0.20
7	0.13
Total	1.00

Solution

Let us assign random numbers to the given probability distribution in the usual way from 00 to 99 in proportion to their respective probabilities.

Table 16.27

Daily demand	Probability	Cumulative probability	Random number
3	0.17	0.17	00-16
4	0.25	0.42	17-41
5	0.25	0.67	42-66
6	0.20	0.87	67-86
7	0.13	1.00	87-99
Total	1.00	-	-

With this random number the stimulated result for 25 days is illustrated in Table 16.28. Since the first random number is 24 in Table 16.28, from Table 16.27 the simulated sales forecast is 4 cars. This information is reported in column number 4 in the Table 16.28. Given the starting inventory as 45, the inventory in the evening of day one is 45 – 4 = 41 units. This is reported in column 5 in the very first row. At this stage no sales loss is being reported since no one moved out due to non-availability cars. The associated shortage penalty thus is zero. Further since this evening stock is greater than the prescribed minimum of 27 no reordering is needed and hence no reordering cost is being entered. All these column entries are empty in the very first row in the Table 16.28.

Now the evening inventory of 41 units becomes the beginning inventory for sales in the next morning as reported in row number 2. Now once again the random number for this day is 57 and the simulated sale is 5 units. This leaves an inventory stock of 41−5 = 36 by the second day evening. This process of simulation is repeated for day three as well without any problem. When we move to fourth day the inventory stock gets depleted to 24 units by the evening. Since this is less than the reorder point of 27 units prescribed, reorder for 40 units is placed on the same evening. However, the arrival of this reorder will take 5 clear-cut days. So the new arrival could take place only on the evening of 8th day and available for sale only on 9th day morning as reported in the table. However, the simulation table shows that the current stock meets the customers demand only up to the 8th day. On the 9th day, because of non-availability inventory stock no sale takes place. This is reported as cost of lost sale (7 × 500 = Rs. 3500) in the corresponding column pertaining to 9th day.

From the next 10th day onwards the sales continue with new arrivals received on the ninth day evening. On 11th there is no problem, everything goes well. On 12th day once again the evening stock comes down to 27. Hence a second reorder is booked on this day and the corresponding order cost entries are made in the appropriate place in the table. On the 17th day though the predicted demand is 7 units only one unit is available for sale from the inventory. The cost of 6 turnout customers (Rs. 3000) is entered in the corresponding column. The same evening second reordered 40 units are received.

On 18th and 19th days no additional problem is encountered. On 20th day the stock one again depletes to 27 units forcing a third reorder of 40 units. The whole summary results for all the 25 days simulation is given in the Table 16.28. The calculations in the last column show that the cost due to the loss of sales seems to be the greatest in the total of cost of Rs. 9100.

Table 16.28

Trail	Random number	Inventory at the beginning of the day	Daily demand	Inventory at the end of the day	Inventory cost	Lost sales	Shortage penalty	Stock received	Quantity ordered
1	24	45	4	41	205				
2	57	41	5	36	180				
3	77	36	6	30	150				
4	68	30	6	24	120				40
5	64	24	5	19	95				
6	88	19	7	12	60				
7	98	12	7	5	25				
8	48	5	5	0	0				
9	93	0	7	0	0	7	3500	40	
10	84	40	6	34	170				
11	11	34	3	31	155				
12	23	31	4	27	135				40
13	94	27	7	20	100				
14	83	20	6	14	70				
15	97	14	7	7	35				
16	83	7	6	1	5				
17	93	1	7	0	0	6	3000	40	
18	33	40	4	36	180				
19	49	36	5	31	155				
20	37	31	4	27	135				40
21	22	27	4	23	115				
22	27	23	4	19	95				
23	0	19	3	16	80				
24	43	16	5	11	55				
25	14	11	3	8	40			40	
Total					2360		6500		300
						Grand Total Cost			9160

To avoid such situation a rethinking of reorder point is necessary. Table 16.29 is prepared for reordering level of 30 units in the place of 27 units with the same reorder quantity of 40 units each time. As a result the cost due to loss of sale is totally avoided. The total cost now is only Rs. 2765 this time as shown in Table 16.29.

Table 16.29

Trail	Random number	Inventory at the beginning of the day	Daily demand	Inventory at the end of the day	Inventory cost	Lost sales	Shortage penalty	Stock received	Quantity ordered	Order cost
1	24	45	4	41	205					
2	57	41	5	36	180					
3	77	36	6	30	150				40	
4	68	30	6	24	120					100
5	64	24	5	19	95					
6	88	19	7	12	60					
7	98	12	7	5	25					
8	48	5	5	0	0			40		
9	93	40	7	33	165					
10	84	33	6	27	135				40	
11	11	27	3	24	120					
12	23	24	4	20	100					100
13	94	20	7	13	65					
14	83	13	6	7	35					
15	97	7	7	0	0			40		
16	83	40	6	34	170					
17	93	34	7	27	135					
18	33	27	4	23	115				40	
19	49	23	5	18	90					
20	37	18	4	14	70					100
21	22	14	4	10	50					
22	27	10	4	6	30					
23	0	6	3	3	15			40		
24	43	40	5	35	175					
25	14	35	3	32	160					
Total			-		2465					300
		Grand Total Cost						2765		

Example 9

Suppose the daily demand for TVS two wheeler in an outlet has the discrete probability distribution as shown in Table 16.30.

Table 16.30

Daily demand	Probability
3	0.17
4	0.25
5	0.25
6	0.20
7	0.13
Total	1.00

Inventory carrying cost is Rs. 5 per car per day. Order placing cost is Rs. 100 per order. The loss in net revenue due to shortage of the vehicle is Rs. 500 per unit. The reorder takes place when the evening closing stock is 30 units or less. The lead time for the delivery of 40 units of reordering follows the discrete probability distribution as shown in Table 16.31. The opening balance in the inventory is 45 units. Obtain the grand total cost for first 25 days sales plan.

Table 16.31

Number of days from reorder to delivery	Probability
3	0.20
5	0.60
7	0.20
Total	1.00

Solution

As usual we attach random numbers to both the demand and the reorder lead time. These results are reported in Tables 16.32 and 16.33 respectively.

Table 16.32

Daily demand	Probability	Cumulative probability	Random number
3	0.17	0.17	00-16
4	0.25	0.42	17-41
5	0.25	0.67	42-66
6	0.20	0.87	67-86
7	0.13	1.00	87-99
Total	1.00	-	-

Table 16.33

Number of days from reorder to delivery	Probability	Cumulative probability	Random number
3	0.20	0.20	00-19
5	0.60	0.80	20-79
7	0.20	1.00	80-99
Total	1.00	-	-

Having both the demand and reorder delivery forecast in mind the simulation record is shown in Table 16.34.

Table 16.34

Days	Random number for demand	Random number for ordering	Order arrival in days	Inventory at the beginning of the day	Daily demand	Inventory at the end of the day	Inventory cost	Lost sales	Shortage penalty	Stock received	Quantity ordered	Order cost
1	24			45	4	41	205					
2	57			41	5	36	180					
3	77	85	7	36	6	30	150				40	100
4	68			30	6	24	120					
5	64			24	5	19	95					
6	88			19	7	12	60					
7	98			12	7	5	25					
8	48			5	5	0	0					
9	93			0	7	0	0	7	3500			
10	84	70	5	40	6	34	170			40		
11	11			34	3	31	155					
12	23			31	4	27	135				40	
13	94			27	7	20	100					
14	83			20	6	14	70					
15	97			14	7	7	35					
16	83			0	6	0	0	6	3000			
17	93			40	7	33	165			40		
18	33	19	3	33	4	29	145				40	
19	49			29	5	24	120					
20	37			24	4	20	100					
21	22			60	4	56	280			40		
22	27			56	4	52	260					
23	0			52	3	49	245					
24	43			49	5	44	220					
25	14			44	3	41	205					
Total					-	648	3240		6600			300
						Grand Total Cost			10,140			

Thus the grand total cost for 25 runs is Rs. 10,140.

■ Example 10

Suppose the daily demand for four-wheelers in a sales outlet follows the following distribution as shown in Table 16.35.

Table 16.35

Daily demand	Probability
0	0.10
1	0.40
2	0.30
3	0.20
Total	1.00

The delivery time in number of days follows the following probability distribution as shown in Table 16.36.

Table 16.36

Number of days from reorder to delivery	Probability
2	0.20
3	0.60
4	0.20
Total	1.00

Inventory carrying cost is Rs. 5 per unit per day. Order placing cost is Rs. 10 per order. Loss in revenue due to shortage is Rs. 50 per unit. Using Monte Carlo simulation for 20 days obtain the grand total cost given the starting inventory 8 units. Use reorder point as 5 units and reorder quantity as 4 units.

Solution

As usual we attach random numbers to both the demand and the reorder delivery time. These results are reported in Tables 16.37 and 16.38 respectively.

Table 16.37

Daily demand	Probability	Cumulative probability	Random number
0	0.10	0.10	00-09
1	0.40	0.50	10-49
2	0.30	0.80	50-79
3	0.20	1.00	80-99
Total	1.00	-	-

Table 16.38

Number of days from reorder to delivery	Probability	Cumulative probability	Random number
2	0.20	0.20	00-19
3	0.60	0.80	20-79
4	0.20	1.00	80-99
Total	1.00	-	-

Having both the demand and reorder delivery forecast in mind the simulation record is shown in Table 16.39.

Table 16.39

Trail	Random number for demand	Random number for delivery time	Simulated delivery time in days	Inventory at the beginning of the day	Daily demand	Inventory at the end of the day	Inventory cost	Lost sales	Shortage cost	Stock received	Quantity ordered	Order cost
1	44			8	1	7	35					
2	84	82	4	7	3	4	20				4	10
3	50			4	2	2	10					
4	85			2	3	0	0	1	50			
5	40			0	1	0	0	1	50			
6	96	88	4	4	3	1	5			4	4	10
7	16			1	1	0	0					
8	16			0	1	0	0	1	50			
9	97			0	3	0	0	3	150			
10	92	39	3	4	3	1	5			4	4	10
11	33			1	1	0	0					
12	83			0	3	0	0	3	150			
13	42	16	2	4	1	3	15			4	4	10
14	7			3	0	3	15					
15	77	66	3	7	2	5	25			4	4	10
16	50			5	2	3	15					
17	20			3	1	2	10					
18	50	95	4	6	2	4	20			4	4	10
19	58			4	2	2	10					
20	44			2	1	1	5					
Total					36		190	9	450	20	24	60
						Grand Total Cost				700		

Thus the grand total cost for 25 runs is Rs. 700.

16.6 NETWORK SIMULATION MODELS

■ Example 11

A project consists of seven activities. The time for performance of each activity is random variable with the respective probability distribution given in Table 16.40.

Table 16.40

Activity	Immediate predecessor	Time in days / Probability				
A	-	3	4	5		
		0.2	0.6	0.2		
B	-	4	5	6	7	8
		0.1	0.3	0.3	0.2	0.1
C	A	1	3	5		
		0.15	0.75	0.1		
D	B,C	4	5			
		0.8	0.2			
E	D	3	4	5	6	
		0.1	0.3	0.3	0.3	
F	D	5	7			
		0.2	0.8			
G	F	2	3			
		0.5	0.5			

a) Draw the diagram and identify the critical path using the expected activity time.

b) Simulate the project using random numbers to determine the activity times. Find the critical path and the project duration.

c) Repeat the simulation four more times. Is the same path critical for all the simulations.

a) The following table gives the summary of expected time for all the given activities

Table 16.40 (a)

Activity	Immediate predecessor	Time in days / Probability					Expected time
A	-	3	4	5			4
		0.2	0.6	0.2			
B	-	4	5	6	7	8	5.9
		0.1	0.3	0.3	0.2	0.1	
C	A	1	3	5			4.2
		0.15	0.75	0.1			
D	B,C	4	5				4.2
		0.8	0.2				
E	D	3	4	5	6		4.8
		0.1	0.3	0.3	0.3		
F	D	5	7				6.6
		0.2	0.8				
G	F	2	3				2.5
		0.5	0.5				

The following chart gives the needed network calculations obtained by using the expected time for activities.

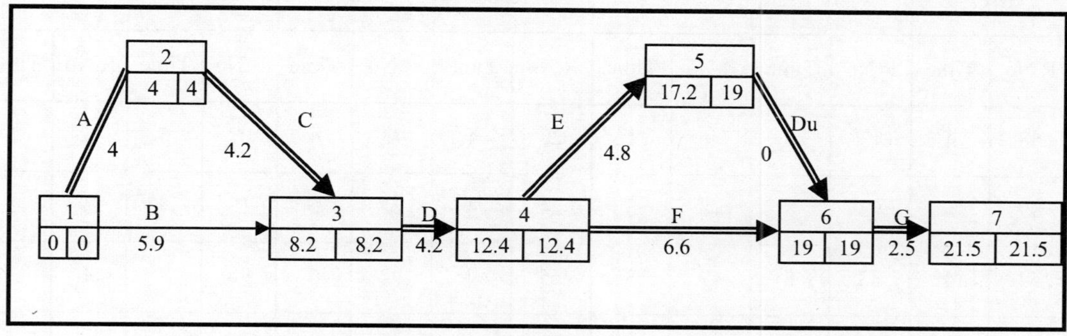

Figure 16.1

So the critical Path: 1-2-3-4-6-7. The project completion time = 21.5 days

Table 16.41 Random number assignment

Activity	Time	Probability	Cu. Probability	Random number interval
A	3	0.20	0.20	00-19
	4	0.60	0.80	20-79
	5	0.20	1.00	80-00
B	4	0.10	0.10	00-09
	5	0.30	0.40	10-39
	6	0.30	0.70	40-69
	7	0.20	0.90	70-89
	8	0.10	1.00	90-99
C	1	0.15	0.15	00-14
	3	0.75	0.90	15-89
	5	0.10	1.00	90-99
D	4	0.80	0.80	00-79
	5	0.20	1.00	80-99
E	3	0.10	0.10	00-09
	4	0.30	0.40	10-39
	5	0.30	0.70	40-69
	6	0.30	1.00	70-99
F	5	0.20	0.20	00-19
	7	0.80	1.00	20-99
G	2	0.50	0.50	00-49
	3	0.50	1.00	50-99

Table 16.42 Simulation work sheet for five runs

Run	A		B		C		D		E		F		G	
	R.No	Time	R.No	Time	R.No	Time	R.No	Time	R.No	Time	R.No	Time	R.No	Time
1	68	4	13	5	9	1	20	4	73	4	07	5	92	3
2	99	5	93	8	18	3	24	4	22	6	10	5	29	2
3	54	4	33	5	49	3	65	4	92	6	98	7	00	2
4	57	4	12	5	31	3	96	5	85	6	92	7	91	3
5	77	4	37	5	34	3	11	4	27	4	10	5	50	3

The following charts illustrate all the five runs and the associated results.

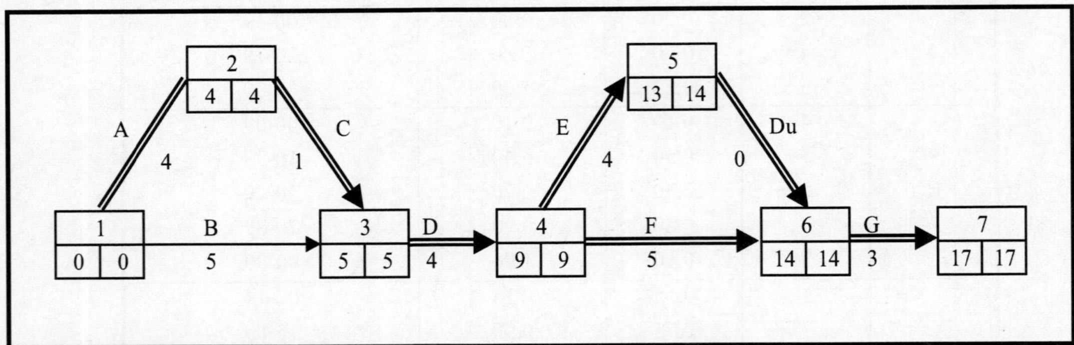

Figure 16.2 Run 1

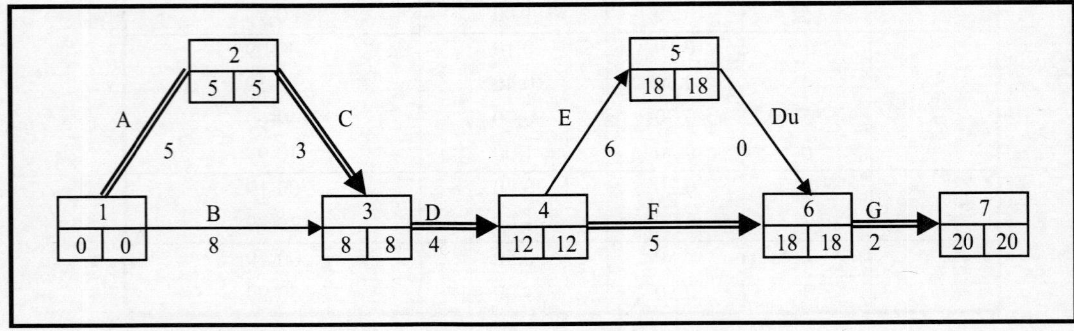

Figure 16.3 Run 2

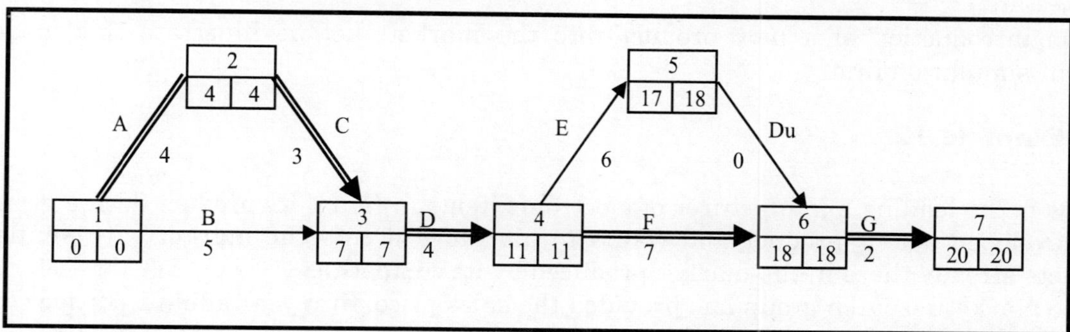

Figure 16.4 Run 3

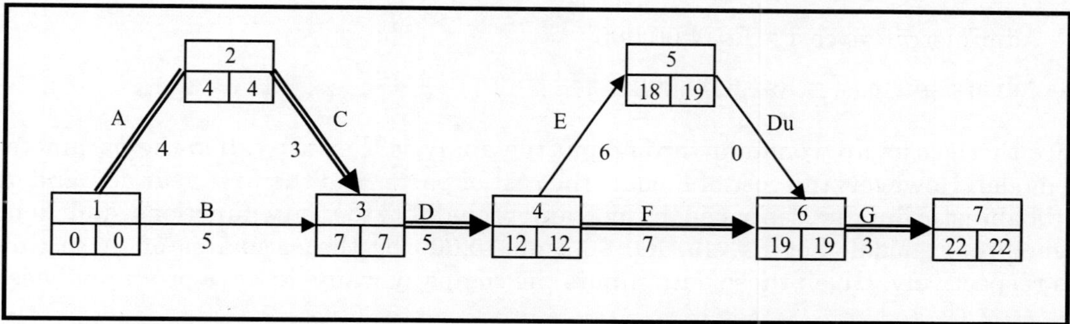

Figure 16.5 Run 4

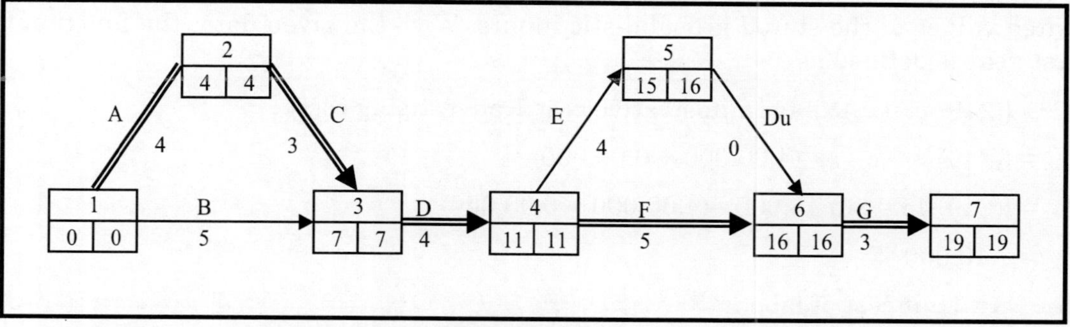

Figure 16.6 Run 5

16.7 SIMULATION BY USING COMPUTER GENERATED RANDOM NUMBERS: RISK ANALYSIS

In all our previous illustrations we used manual simulation for 10 to 15 runs at the most. If we want to simulate for about 5000 runs then definitely one must go for computerized simulation. In this section we propose to simulate any number of runs by using excel worksheet. In the event of uncertainty we use risk analysis for predicting the outcome of the concerned decision. In this section we illustrate the use of risk analysis for the introduction of a new product into the market. Let us illustrate this model without simulation first.

■ Example 12

Philips is the leading manufacturer of electronic items in India. Its product design group has developed a new product and wants to introduce it into the market with similar products already there in the market produced by its competitor.

The market research group has provided the sales price, first year administrative cost and advertising expenditure as follows.

- Selling price per unit = Rs. 249
- Administrative cost = Rs. 4,00,000
- Advertising cost = Rs. 6,00,000

Since these costs are constant throughout the analysis they are called the parameters of the model. However, the cost of labour, the cost of parts, and the first year demand are the probabilistic inputs described by certain probability density functions and hence random in our model. Let Rs. 45, Rs. 90, and 15,000 be the estimates of these three inputs respectively. Given these data inputs the company wants to have profit analysis.

WHAT-IF-ANALYSIS

What-if-analysis is the basic approach to risk theory. It simply calculates the profit for generated values of the stated probabilistic inputs. With the given data, the function for the first year is defined as

$$= \{(249 - c_1 - c_2)\, x\} - (\text{administrative cost} + \text{advertising cost})$$

$$= \{(249 - c_1 - c_2)\, x\} - (4{,}00{,}000 + 6{,}00{,}000)$$

$$= \{(249 - 45 - 90)\, 15{,}000\} - (4{,}00{,}000 + 6{,}00{,}000)$$

$$= 7{,}10{,}000$$

Where c_1 = Unit cost of labour

c_2 = Unit cost of parts

x = The first year demand

Recollect that the direct labour cost per unit, parts cost per unit and first year demand describes the *base-case scenario* of the model under reference.

Thus the base year scenario gave us an anticipated profit of Rs. 7,10,000.

Though the base-case scenario is impressive for the estimated data there is no guarantee that these estimates are correct. For instance the company believes that labour cost could range anywhere from Rs. 43 to Rs. 47.

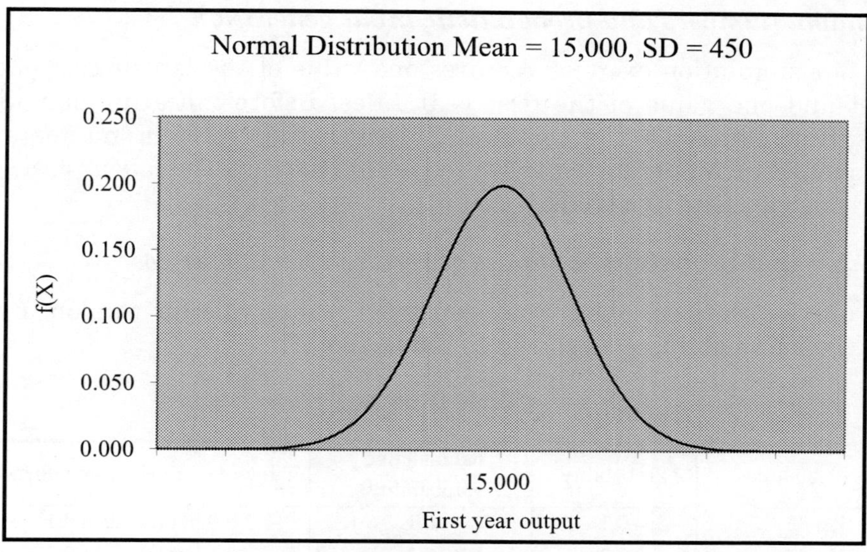

Figure 16.9

Computer generated random numbers

In all our previous illustrative examples random numbers were generated using random number tables. Computer generated random numbers are nothing but randomly selected decimals in the interval from 0 to 1 excluding 1 as shown in Table 16.44. The generated numbers are equally likely and uniformly distributed in the said interval. These numbers are obtainable in excel spreadsheet by using built-in functions available in the computer. To get such numbers one must place the formula = RAND() in any one of the cell and copy it to any number of shells depending upon the number of random numbers necessary. In the above table, 100 three-digit random number are being generated and reported.

Table 16.44

0.517	0.371	0.490	0.728	0.848	0.020	0.742	0.091	0.870	0.397
0.240	0.664	0.572	0.289	0.319	0.154	0.522	0.298	0.955	0.965
0.033	0.267	0.634	0.497	0.439	0.995	0.497	0.027	0.156	0.642
0.202	0.966	0.919	0.063	0.516	0.285	0.106	0.912	0.885	0.592
0.975	0.774	0.011	0.904	0.814	0.568	0.427	0.075	0.382	0.192
0.727	0.145	0.896	0.151	0.171	0.707	0.744	0.468	0.830	0.950
0.164	0.094	0.761	0.966	0.282	0.364	0.365	0.584	0.653	0.940
0.663	0.975	0.459	0.914	0.057	0.856	0.991	0.293	0.956	0.955
0.997	0.955	0.316	0.055	0.901	0.893	0.603	0.476	0.023	0.190
0.471	0.008	0.251	0.522	0.842	0.290	0.117	0.240	0.879	0.953

Computer random numbers and probabilistic input generation

Each trial in our simulation exercise requires one value of the labour cost one value of the parts cost and one value of the first year sales. Before initiating our stimulating model let us first illustrate the methods of generating random numbers and the associated probabilistic inputs for all the three inputs namely, the labour cost, parts cost and first year sales for first five trials.

Random number and probabilistic inputs generation for labour cost

Since we have five values of labour cost, construct 5 intervals for random numbers in proportion to their probabilities as shown in Table 16.45.

Table 16.45

Direct labour cost	Probability	Cumulative probability	Interval of random numbers
Rs. 43	0.1	0.1	0.0 but less than 0.1
Rs. 44	0.2	0.3	0.1 but less than 0.3
Rs. 45	0.4	0.7	0.3 but less than 0.7
Rs. 46	0.2	0.9	0.7 but less than 0.9
Rs. 47	0.1	1	0.9 but less than 1.0
Total	1	-	-

Table 16.46

	A	B	C
1	Direct labour cost data		
2	Lower	Upper	Per unit
3	random no.	random no.	labour cost
4	0.0	0.1	43
5	0.1	0.3	44
6	0.3	0.7	45
7	0.7	0.9	46
8	0.9	1	47
9			
10	Simulation of probabilistic inputs for labour cost		
11	Trail	Random number	Direct labour cost
12	1	0.7258	=VLOOKUP(B12,$A$4:$C$8,3)
13	2	0.3234	=VLOOKUP(B13,$A$4:$C$8,3)
14	3	0.4795	=VLOOKUP(B14,$A$4:$C$8,3)
15	4	0.1948	=VLOOKUP(B15,$A$4:$C$8,3)
16	5	0.7451	=VLOOKUP(B16,$A$4:$C$8,3)

First let us create random number for the proposed five trials. To do this enter the formula in cell B12 as =RAND() as shown in Table 16.46 and copy this formula up to cell

B16. This will generate 5 needed random numbers as shown in column two in the table. Once the random numbers are ready, by using the guideline Table 16.45 we can readily obtain the probabilistic inputs manually as we were doing in the previous illustrations.

Table 16.47

	A	B	C
1		Direct labour cost data	
2	Lower	Upper	Per unit
3	random no.	random no.	labour cost
4	0.0	0.1	43
5	0.1	0.3	44
6	0.3	0.7	45
7	0.7	0.9	46
8	0.9	1	47
9			
10		Simulation of probabilistic inputs for labour cost	
11	Trial	Random number	Direct labour cost
12	1	0.7258	46
13	2	0.3234	45
14	3	0.4795	45
15	4	0.1948	44
16	5	0.7451	46

However, in this section we propose to assign this job to the computer also. To do this job we type the formula = VLOOKUP(B12,$A$4:$C$8,3) in cell C12 and copy it to the desired five cells up to C16 as shown in Table 16.46. The complete resulting result is shown in Table 16.47.

Table 16.48

	A	B	C
1		Direct labour cost data	
2	Lower	Upper	Per unit
3	random no.	random no.	labour cost
4	0.0	0.1	43
5	0.1	0.3	44
6	0.3	0.7	45
7	0.7	0.9	46
8	0.9	1	47
9			
10		Simulation of probabilistic inputs for labour cost	
11	Trail	Direct labour cost	
12	1	=VLOOKUP(RAND(),$A$4:$C$8,3)	
13	2	=VLOOKUP(RAND(),$A$4:$C$8,3)	
14	3	=VLOOKUP(RAND(),$A$4:$C$8,3)	
15	4	=VLOOKUP(RAND(),$A$4:$C$8,3)	
16	5	=VLOOKUP(RAND(),$A$4:$C$8,3)	

If we do not want to have the random number on display in the Table 16.47 then the random number generating augment could be included in the very formula itself to generate the final probabilistic labour input. This time, type the following formula in the cell B12 in Table 16.48 = VLOOKUP(RAND(),$A$4:$C$8,3). The final result is displayed in, Table 16.49.

Table 16.49

	A	B	C
1		Direct labour cost data	
2	Lower	Upper	Per unit
3	random no.	random no.	labour cost
4	0.0	0.1	43
5	0.1	0.3	44
6	0.3	0.7	45
7	0.7	0.9	46
8	0.9	1	47
9			
10		Simulation of probabilistic inputs for labour cost	
11	Trial	Direct labour cost	
12	1	45	
13	2	46	
14	3	44	
15	4	45	
16	5	45	

Note: *But note that the new results are quite different from the previous ones because this time a new set of random numbers is being generated and accordingly a new but still random probabilistic inputs are generated in Table 16.49.*

Random number and probabilistic inputs generation for parts cost

However, since our probability distribution is uniform we will have to adopt a different method for generating probabilistic inputs for parts cost input as illustrated Tables 16.50 and 16.51.

$$\text{Part cost} = a + r\,(b-a)$$

where r = random number from 0 to 1

a = smallest value of the part cost

b = largest value of the part cost

Table 16.50

	A	B	C
1	Parts cost (uniform distribution) data		
2	Smallest value	80	
3	Largest value	100	
4			
5	Simulated part cost probabilistic input		
6	Trial number	Random number	Parts cost
7	1	0.8473	=$B$2+B7*($B$3−$B$2)
8	2	0.6605	=$B$2+B8*($B$3−$B$2)
9	3	0.4529	=$B$2+B9*($B$3−$B$2)
10	4	0.7935	=$B$2+B10*($B$3−$B$2)
11	5	0.9535	=$B$2+B11*($B$3−$B$2)

Table 16.51

	A	B	C
1	Parts cost (uniform distribution) data		
2	Smallest value	80	
3	Largest value	100	
4			
5	Simulated part cost probabilistic input		
6	Trial number	Random number	Parts cost
7	1	0.8473	96.95
8	2	0.6605	93.21
9	3	0.4529	89.06
10	4	0.7935	95.87
11	5	0.9535	99.07

First the random numbers are generated as discussed above in the second column as usual. Once this is done we enter the formula =$B$2+B7*($B$3−$B$2) in cell C7 and copy it down the line as shown in Table 16.50. Accordingly, the computer calculates the part cost. For example, for the random number 0.8473 the associated probabilistic part cost = 80+0.8473(100−80) = 96.95. In the Table 16.51 such probabilistic parts costs are obtained and reported for 5 trials.

In Table 16.52 the random number generation is not shown explicitly. This time the appropriate built-in formula is =$B$2+RAND()*($B$3−$B$2). The resulting results are displayed in Table 16.53.

Table 16.52

	A	B	C
1	Parts cost (uniform distribution) data		
2	Smallest value	80	
3	Largest value	100	
4			
5	Simulated part cost probabilistic input		
6	Trial number	Parts cost	
7	1	=$B$2+RAND()*($B$3−$B$2)	
8	2	=$B$2+RAND()*($B$3−$B$2)	
9	3	=$B$2+RAND()*($B$3−$B$2)	
10	4	=$B$2+RAND()*($B$3−$B$2)	
11	5	=$B$2+RAND()*($B$3−$B$2)	

Table 16.53

	A	B	C
1	Parts cost (uniform distribution) data		
2	Smallest value	80	
3	Largest value	100	
4			
5	Simulated part cost probabilistic input		
6	Trial number	Parts cost	
7	1	97.58	
8	2	91.63	
9	3	99.98	
10	4	86.16	
11	5	82.73	

Random number and probabilistic inputs generation for first year sales

Since the demand is assumed to follow a normal distribution with mean 15,000 and SD of 4500 we need a different procedure for generating probabilistic inputs. Once again the excel worksheet is used and the needed random numbers are generated. In Table 16.54 the needed five random numbers are generated in the second column as usual from B7 to B11. The associated probabilistic demand input is obtained by typing the formula = NORMINV(B7,$B$2,$B$3) in C7. For the rest of the cell this formula is simply copied down the line as shown in Table 16.54. The result is reported in Table 16.55.

Table 16.54

	A	B	C
1	First year sales (normal distribution) data		
2	Mean	15,000	
3	Standard deviation	4500	
4			
5	Simulated probabilistic inputs for demand		
6	Trial number	Random number	Demand
7	1	0.8473	=NORMINV(B7,$B$2,$B$3)
8	2	0.6605	=NORMINV(B8,$B$2,$B$3)
9	3	0.4529	=NORMINV(B9,$B$2,$B$3)
10	4	0.7935	=NORMINV(B10,$B$2,$B$3)
11	5	0.9535	=NORMINV(B11,$B$2,$B$3)

Table 16.55

	A	B	C
1	First year sales (normal distribution) data		
2	Mean	15,000	
3	Standard deviation	4500	
4			
5	Simulated probabilistic inputs for demand		
6	Trial number	Random number	Demand
7	1	0.8473	19612
8	2	0.6605	16863
9	3	0.4529	14467
10	4	0.7935	18683
11	5	0.9535	22557

Table 16.56

	A	B
1	First year sales (normal distribution)	
2	Mean	15,000
3	Standard deviation	4500
4		
5	Simulated probabilistic inputs for demand	
6	Trial number	Demand
7	1	=NORMINV(RAND(),$B$2,$B$3)
8	2	=NORMINV(RAND(),$B$2,$B$3)
9	3	=NORMINV(RAND(),$B$2,$B$3)
10	4	=NORMINV(RAND(),$B$2,$B$3)
11	5	=NORMINV(RAND(),$B$2,$B$3)

In Table 16.56 the random number creation in column 2 is avoided. This time we straightaway type the formula = NORMINV(RAND(),$B$2,$B$3) in B7 and achieve the desired result as shown in Table 16.56 and 16.57.

Table 16.57

	A	B
1	First year sale (normal distribution)	
2	Mean	15,000
3	Standard deviation	4500
4		
5	Simulated probabilistic inputs for demand	
6	Trial number	Demand
7	1	18,436
8	2	19,328
9	3	4338
10	4	17,216
11	5	16,757

Having known the method of generating the probabilistic inputs of various kinds of input we initiate the needed profit simulation as shown in Table 16.58 for about 500 trials. In Table 16.58, trial number 6 to trial number 495 have been hidden for want of space for presentation. The data relating to Philips new venture is presented on the top of the excel sheet in the first 14 rows. The selling price, administrative cost, and advertising costs are presented in cells C3, C4, C5 respectively.

The relevant data entries for the three inputs are presented from row 7 to row 14 in Table 16.58. The direct labour cost and the related discrete probability are entered in a tabular form to facilitate easy generation of probabilistic labour cost inputs. For example, 0.0 in cell A10 and 0.1 in cell B10 enable the computer to assign 43 as labour cost provided the computer generated random number is in the interval 0.0 to 0.1. Thus approximately 10% of the stimulated random numbers will be assigned the labour cost Rs. 43.

For the uniform probability distribution pertaining to parts cost with a smallest value of 80 and highest value of 100 enable the computer to allot the parts cost in accordance with random number it generates by using the formula 'smallest cost + random number × largest cost – smallest cost'. Finally, the normal distribution with mean 15,000 entered in cell E13 and standard deviation 4500 entered in cell E14 will enable the computer to describe the needed probabilistic inputs for first year demand.

Now our simulation exercises start from line number 21 in our excel sheet. Type the formulas for row 21 as follows

Table 16.58

	A	B	C	D	E
1	PHILIPS RISK ANALYSIS (Data)				
2					
3	Selling price per unit		249.00		
4	Administrative cost		400000.00		
5	Advertising cost		600000.00		
6					
7	Direct cost of labour			Part costs (uniform distribution)	
8	Lower	Upper	Cost per unit	Smallest value	80.00
9	random no.	random no.		Largest value	100.00
10	0	0.1	43.00		
11	0.1	0.3	44.00	Demand (normal distribution)	
12	0.3	0.7	45.00		
13	0.7	0.9	46.00	Mean	15000.00
14	0.9	1	47.00	Standard dev.	4500.00
15					
16					
17			Simulation Table		
18					
19		Direct	Parts cost	First-Year	
20	Trials	labour cost	per unit	demand	Profit
21	1	45	83.56	7832.18	−56715.83
22	2	46	89.93	12587.27	423217.58
23	3	44	81.76	20501.92	1526571.73
24	4	45	82.43	12532.88	523621.33
25	5	44	87.67	12477.67	464048.81
26	496	44	85.98	18718.33	1227872.46
27	497	44	86.76	17415.35	1059216.14
28	498	45	80.82	9293.21	144776.29
29	499	44	96.96	18882.04	1039998.91
30	500	43	93.50	14187.47	596092.16
31					
32			Summary Statistics		
33			Mean Profit		680079.10
34			Standard Deviation		521577.95
35			Minimum Profit		−683924.35
36			Maximum Profit		2227700.02
37			Number of Losses		51.00
38			Probability of Loss		0.10
39					

Cell A21&A22 Enter 1 & 2 and copy this up to 500 and create the needed 500 trial numbers.

Cell B21 Write the formula =VLOOKUP(RAND(),$A$10:$C$14,3)

This will generate the necessary probabilistic labour cost inputs to the model

Cell C21 Enter the formula =$E$8+RAND()*($E$9−$E$8)

This will generate the necessary probabilistic parts cost inputs to the model

Cell C21 Enter the formula =NORMINV(RAND(),$E$13,$E$14)

This will generate the necessary probabilistic first year demand inputs to the model

Cell E21 Enter the formula =($C$3−B21−C21)*D21−$C$4−$C$5

This will generate the needed profit calculations for the first trial

Now to have 500 trial results copy cells A21: E21 down the line till row number 520

Simulation Summary Results

Cell E523 Enter the formula =AVERAGE(E21:E520)

This will calculate the mean profit per trial

Cell E524 Enter the formula =STDEV(E21:E520)

This will calculate the standard deviation of the profit distribution

Cell E525 Enter the formula =MIN(E21:E520)

This will calculate the minimum profit among the 500 trials

Cell E526 Enter the formula = Max(E21:E520)

This will calculate the maximum profit among the 500 trials

Cell E527 Enter the formula =COUNTIF(E21:E520, "<0")

This will count the number of negative profits in 500 trials

Cell E528 E527/E528

This will calculate the proportion of loss in 500 trials

16.8 INVENTORY SIMULATION: A REVISIT

■ Example 13

Usha, a leading fan manufacturer in India, manufactures a new mini table-fan. A dealer in Bangalore receives it at a cost of Rs. 75. He sells it at Rs. 125 and makes a profit of Rs. 50 per unit. The monthly demand for fans follows a normal probability distribution with mean 100 units and standard deviation of 20 units. The dealer receives Q units in the beginning of every month and replenishes its inventory. The inventory cost per unit is Rs. 15. In the case of stock outs he loses Rs. 50, from the profit margin per unit. It wants to have a simulation and arrive at a net average profit per month. It also wants to know the service level for a given level of replenishment Q.

In this mode the controllable input is Q and the probabilistic input is the demand D.

When the demand is less than the supply then

Gross profit = Rs. 50×D

The inventory cost = Rs. 15 (Q–D)

Therefore net profit = Gross profit – Holding Cost = Rs. 50×D – Rs. 15 (Q–D)

When the demand exceeds the supply then

Gross profit = Rs. 50×D

The shortage cost = Rs. 50 (D–Q)

Therefore net profit = Gross profit – Shortage cost = Rs. 50×D – Rs. 50 (D–Q)

The Simulation Model

As usual the rows 1 to 12 in Table 16.53 all the data with description are entered. Real simulation starts from row 17.

Cell A17 Enter 1 to refer first simulation

Cell B17 Write the formula =NORMINV(RAND()<$C$10:$C$11,3)

This will generate the necessary probabilistic demand input to the model

Now if the demand is less than or equal to 100, the replenishment level, then the sales will be as in A17. If the demand is greater than 100 then the sales will be 100. Accordingly write the formula in cell C17 as follows:

Table 16.59

	A	B	C	D	E	F	G
1			Inventory model				
2							
3	Gross profit per unit		50				
4	Holding cost per unit		15				
5	Shortage cost per unit		50				
6							
7	Replacement Demand		100				
8							
9	Demand (Normal)						
10	Mean		100				
11	SD		20				
12							
13							
14				Simulation			
15							
16	Month	Demand	Sales	Gross Profit	Holding Cost	Shortage Cost	Net Profit
17	1	88	88	4388	184	0	4204
18	2	81	81	4069	279	0	3790
19	3	139	100	5000	0	1951	3049
20	4	94	94	4680	96	0	4584
21	294	82	82	4117	265	0	3852
22	295	148	100	5000	0	2408	2592
23	296	93	93	4650	105	0	4545
24	297	71	71	3540	438	0	3102
25	298	104	100	5000	0	185	4815
26	299	109	100	5000	0	446	4554
27	300	138	100	5000	0	1900	3100
28							
29	Total	30440	27763		Summary Statistics		
30					Mean Profit		4069
31					Standard Deviation		713.93
32					Minimum Profit		612
33					Maximum Profit		4998
34					Service Level		0.9121
35							

Cell C17 =IF(B17<=$C$7,B17,$C$7)

This will generate the necessary sales

Cell D17 =$C$3*C17

This will calculate the profit for first trial

Cell E17 =IF(B17<=$C$7,$C$4*($C$7–B17,0)

This will calculate the holding cost if Q exceeds D

Cell F17 =IF(B17>$C$7,$C$5*(B17–$C$7,0)

This will calculate the shortage cost if D exceeds Q

Cell G17 =D17–E17–F17

This will calculate the net profit

Now copy A17:G17 down the line till row 316 and generate the complete the simulation

Summary Results

Cell B318 =SUM(B17:B316)

This will give the total demand

Cell C318 =SUM(C17:C316)

This will give the total sales

Cell G319 =AVERAGE((G17:G316)

This will give the mean profit

Cell G320` =STDEV(G17,G316)

This will give the SD for mean profit

Cell G321 =MIN(G17:G316)

This will give the minimum net profit

Cell G322 =MAX(G17:G316)

This will give the maximum net profit

Cell G323 =C318/B318

This will give the service level.

16.9 WAITING LINE SIMULATION: A REVISIT

Example 14

The State Bank of Mysore Branch at Jnanabharati in Bangalore university campus wants to open one ATM counter. The average waiting at ATM should not be more than one minute. The customer arrival is described by a uniform distribution in the interval 0-5 minutes. The customer service by the machine is described by a normal distribution with mean 2 minutes and standard deviation 0.5 minutes. Table 16.54 illustrates the simulation process for 300 trials.

Simulation Model

As usual the rows 1 to 9 all the data with description are entered. Real simulation starts from row 16.

Cell A16 Enter 1 to refer first simulation

Cell B16 =$C$4+RAND()*($C$5−$C$4)

This will generate the necessary probabilistic inter-arrival time input to customer 1

Cell C16 =B16

This will generate arrival time for customer 1.

Cell D16 =C16

This will start time for customer 1

Cell E16 =D16−C16

This will calculate the waiting time for customer 1

Cell F16 =NORMINV(RAND(),$C$8,$C$9)

This will calculate the service time for customer 1

Cell G16 =D16+F16

This will calculate the completion time for customer 1

Cell H16 =G16−C16

This will calculate the time in the system for customer 1.

Simulation for the second customer starts in row 17.

Table 16.60

	A	B	C	D	E	F	I	G
1	Waiting Line Model							
2								
3	Inter-arrival times (uniform Distribution)							
4	Smallest value		0					
5	Largest value		5					
6								
7	Service time (Normal Distribution)							
8	Mean		2					
9	Standard deviation		0.5					
10								
11								
12	Simulation							
13								
14–15	Customer	Inter-arrival time	Arrival time	Service start time	Waiting time	Service time	Completion time	Time in system
16	1	1.4	1.4	1.4	0.0	1.9	3.4	1.9
17	2	3.7	5.1	5.1	0.0	1.4	6.5	1.4
18	3	1.4	6.5	6.5	0.0	2.7	9.2	2.7
19	4	1.0	7.5	9.2	1.7	0.9	10.1	2.6
20	495	4.6	1258.9	1258.9	0.0	1.2	1260.1	1.2
21	496	0.6	1259.5	1260.1	0.7	0.9	1261.0	1.6
22	497	1.4	1260.8	1261.0	0.2	1.5	1262.5	1.7
23	498	4.2	1265.1	1265.1	0.0	2.1	1267.2	2.1
24	499	4.6	1269.7	1269.7	0.0	2.6	1272.3	2.6
25	500	1.7	1271.4	1272.3	0.9	2.8	1275.0	3.6
26								
27				Summary statistics				
28				Number waiting			253	
29				Probability of waiting			0.6325	
30				Average waiting			1.5	
31				Maximum waiting time			7.9	
32				Utilization of ATM			0.8	
33				Number waiting >1 minute			192	
34				Probability of waiting > 1 min.			0.48	
35								

Cell A17 Enter 2 to refer the second simulation

Cell B17 =$C$4+RAND()*($C$5−$C$4)

This will generate the necessary probabilistic inter-arrival time input to customer 2

Cell C17 =C16+B17

This will generate arrival time for customer 2.

Cell D17 =IF(C17>G16,C17,G16)

This will calculate start time for customer 2

Cell E17 =D17−C17

This will calculate the waiting time for customer 2

Cell F17 =NORMINV(RAND(),$C$8,$C$9)

This will calculate the service time for customer 2

Cell G17 =D17+F17

This will calculate the completion time for customer 2

Cell H17 =G17−C17

This will calculate the time in the system for customer 2

Now to complete the simulation Copy Cells A17:H17 down the line till row 515

Summary Results

Cell G518 =COUNTIF(E116:E515, ">0")

This will count the number of customers made to wait

Cell G519 =E518/400

This will give the probability of waiting

Cell G520 =AVERAGE(E116:G515)

This will give the mean waiting time

Cell G521 =MAX(G116:G515)

This will give the maximum waiting time

Cell G522 =SUM(F116:F515/G515−G115)

This will give the utilization time of ATM

Cell G523 =COUNTIF(E116:E515, ">1")

This will count the number of customers who have to wait more than one minute

Cell G524 =G523/400

This will give the probability of waiting more than 1 minute.

EXERCISES

1. What is simulation? Describe the simulation process with suitable illustration.

2. "When it becomes difficult to use an optimization technique for solving a problem, one has to resort to simulation". Discuss.

   (M.Com. Rajasthan, 1987)

3. "Simulation is typically the process of carrying out sampling experiments on the model of the system rather than the system itself." Elucidate this statement by taking some example.

   (MBA Delhi, April 1994)

4. Discuss Monte Carlo simulation. Illustrate its usage through an illustration.

5. Customers arrive at a milk booth for required service. Assume that inter-arrival and service times are constants and given by 1.8 and 4 time units respectively. Simulate the system by hand computation for 14 time units. What is the average waiting time per customer? What is the percentage idle time of the facility? (Assume that the system starts at t = 0).

   (Andhra Productivity Council,1978)

6. A bakery keeps stock of a particular brand of cakes. Previous experience shows the daily demand pattern for the item with associated probabilities, as given.

Daily demand	0	10	20	30	40	50
Probability	0.1	0.2	0.15	0.5	0.12	0.02

   Use the following sequence of random numbers to stimulate the demand for next 10 days. Also find the average demand per day. Random numbers: 25, 39, 65, 76, 12, 05, 73, 89, 19, 49.

   (ICWA, December, 1986)

7. A confectioner sells confectionery items. Past data on demand per week in hundred kilograms with frequency is given below.

Demand/week	0	5	10	15	20	25
Frequency	2	11	8	21	5	3

Using following sequence of random numbers, generate the demand for next 15 weeks. Also find the average demand per week. Random numbers: 35, 52, 90, 13, 23, 73, 34, 57, 35, 83, 94, 56, 67, 66, 60.

(ORSI, May, 1983)

8. Inter-arrivals and service duration studied over past five years for a single channel queuing system revealed the following patterns:

9.

Inter-arrival time		Service time	
Minutes	Probability	Minutes	Probability
2	0.19	1	0.15
4	0.22	3	0.28
6	0.32	5	0.30
8	0.17	7	0.18
10	0.10	9	0.11

Using the random number table given above, simulate the queue behaviour for a period of 60 minutes. 1908, 3227, 5974, 8196, 2748, 4507, 2665, 5278, 7792, 4649, 8512, 8363, 3410, 4206, 4397, 0753, 1748, 2375.

Use the above simulation to estimate:

i) the probability of the server being idle

ii) the mean time spent by the customer waiting for service.

(DMS, Mumbai,1996)

10. A retailer deals in a perishable commodity. The daily demand and supply are variable. The data for the past 500 days show the following demand and supply.

Supply		Demand	
Availability	No. of days	Demand (Kg.)	No. of days
10	40	10	50
20	50	20	110
30	190	30	200
40	150	40	100
50	70	50	40

The retailer buys the commodity at Rs. 20 per kg and sells it at Rs. 30 per kg. Any commodity remaining at the end of the day has no sale value. Moreover, the loss (unearned profit) on any unsatisfied demand is Rs. 8 per kg. Given the following pairs of random numbers, simulate six days sales, demand, and profit.

(31,18), (63, 84), (15,79), (07,32), (43,75), 81,27).

The first random number in the pair is for supply and the second random number is for the demand, viz, in the first pair (31,18) use 31 to simulate supply and 18 to stimulated demand.

(CA, November, 2000)

11. An investment company wants to study the investment projects based on market demand, profit, and the investment required, which are independent of each other. Following probability distribution has been estimated for each of the three factors.

Annual demand (in 1000 Units)	25	30	35	40	45	50	55
Probability	0.05	0.1	0.2	0.3	0.2	0.1	0.05

Profit per unit	3	5	7	9	10
Probability	0.1	0.2	0.4	0.2	0.1

Investment (in Thousands)	2750	3000	3500
Probability	0.25	0.5	0.25

Using simulation process, repeat the trial 10 times, compute the return on investment on each trial taking these factors into consideration. What is the most likely return?

(30, 12, 16), (59, 09, 69), (63, 94, 26), (27, 08, 74), (64, 60, 61):

(28, 28, 72), (31, 23, 57), (54, 85, 20), (64, 68, 18), (32, 31, 87)

In each bracket above, the first random number is for annual demand, the second one is for profit, and the last one is for investment required.

12. A company has problems with the stock outs for one of its components and contemplating making alternations either to reorder quantity or to the reorder level or to both. Before making any decision, the company wishes to explore whether any guidance can be obtained by simulating the operation of the system. The pattern of weekly demand over the past few years has been as follows:

Weekly demand	500	525	550	575	600	625	650	675	700	725
Frequency	10	15	30	50	55	60	40	20	10	5

Ordering cost is Rs. 20 per order and the carrying costs is Rs. 5 per unit. The estimated loss when the demand cannot be met is Rs. 12 per unit. When the stock reaches the pre-set order point, a replenishment order is issued.

i) You are required to describe, using a flow chart other means, how a simulation model for this problem might work.

Using a reorder point of 2500 and an order quantity of 2000, a simulation of 20 weeks' operations has been run on a computer and the following summary produced.

Simulation summary			
Number of periods	20 weeks	Carrying cost	Rs. 1984.13
Average inventory	1031.8	Ordering cost	Rs. 120
Orders	6	Stock out cost	Rs. 20,580
Average demand	606.3 units	Total cost	Rs. 22684.13
Lead time	4 weeks		

ii) You are required to interpret the above summary and suggest what could be done next to make the simulation more realistic.

(ICMA, May, 1992)

CHAPTER

17

Goal Programming

- *Introduction*
- *Linear programming with single goal of maximum revenue*
- *Goal programming with single goal of targeted revenue*
- *Pre-emptive and non-pre-emptive models goal programming with multiple goals*
- *Exercises*

17.1 INTRODUCTION

In traditional linear programming models we optimize the single quantifiable objective such as profit, output, cost etc. However, setting clear-cut targets instead of simple maximization or minimization of objectives is always advantages in the pursuit of business problems.

In modern business world each and every business unit, however small may be, definitely will have multiple goals. Achievements of multiple goal objectives often create difficulties especially when goals are conflicting. For example, a company interested in a new venture will have to select the site of its operation first. This choice of site depends upon the construction cost which varies from place to place. If the cost is the only criteria then it must go for the place where the land cost plus the cost construction is minimum. However, before finalizing such decisions the company should consider the availability of inputs, transport, the energy sources and its cost, the price of skilled labourers etc. In such a situation with diversified objectives the complexity of the problem gets multiplied. Goal programming techniques are often useful in solving such problems with multiple and diversified goals. In such models the solution heavily depends upon the listed out priorities to the diversified objectives. Given the clearly established priority goals, the goal programming technique tries to minimize the deviation of each one of them from their respective target levels according to their listed out priorities. In such models, the goals are satisfied in a given sequential order, higher goals taking priorities, lower goals are pushed to lower levels. In the final solution, some goals may be over-satisfied and some others may be under-satisfied. Thus compromising rather than optimizing is the correct approach in goal programming. In the model formation, therefore, we incorporate all goals and try to solve them having the priority in mind.

Thus, in goal programming, all the objects are assigned to specific target levels for achievement. Goal programming treats these targets as *goals to aspire for* and not as absolute constraints. It then attempts to find an optimal solution that comes as "close as possible" to the targets in the order of specified priorities.

The concept of Goal Programming was introduced by Charles and Cooper in 1961 by incorporating a method for solving infeasible linear programming problem involving various conflicting constraints. In 1965 Ijiri developed the model based on appropriate priorities for various goals and in the form of weightage for the same priority goals. Thus, it is very much clear that the objective function formulation is not for optimum, as in the case of linear programming, but as far as possible very near to it. The latter work of Lee (1972) and Ignizio shows several applications of goal programming business world. The following example illustrates one such problem.

Goal programming is solved either by graphical method or by a suitably modified Simplex method.

17.2 LINEAR PROGRAMMING WITH SINGLE GOAL OF MAXIMUM REVENUE

■ Example 1

Consider the following linear programming problem we have dealt with in chapter 2.

$$\text{Maximize: } R = 10X_1 + 15X_2$$

$$\text{Subject to: } \quad 4X_1 + 2X_2 \geq 100 \quad \text{(Constraint K)}$$

$$1X_1 + 2X_2 \geq 70 \quad \text{(Constraint L)}$$

$$X_1 \;,\; X_2 \geq 0$$

Graphical Solution

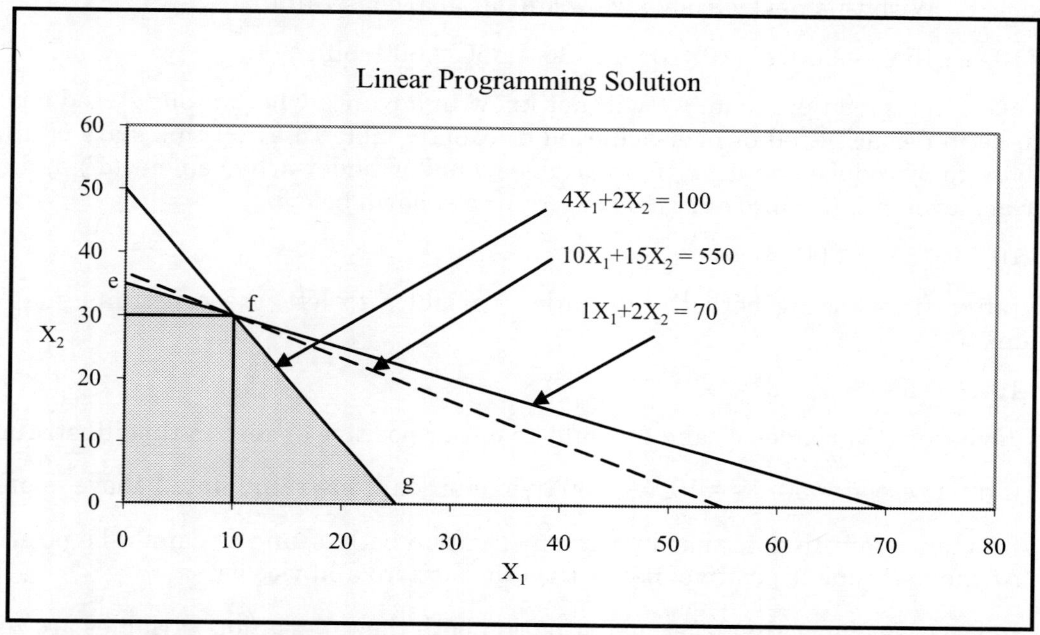

Figure 17.1

The optimum results are $X_1 = 10$, $X_2 = 30$ and maximum R = Rs. 550.

17.3 GOAL PROGRAMMING WITH SINGLE GOAL OF TARGETED REVENUE

Example 2

Now, suppose the manager, instead of this unknown optimum, has the set goal of achieving a known revenue target of at least Rs. 600 or more. If we return to the graphical solution to this problem in example 1, we observe that the maximum revenue obtainable is only Rs. 550. Thus, the objective revenue of Rs. 600 is not achievable for this problem by simple linear programming. However, goal programming recognizes that any goal stated may be either over-achieved or under achieved or exactly met. Given this goal programming terminology, the goal of Rs. 600 is under-achieved by Rs. 50 (d^-) in this problem. (Instead if he sets the target at Rs. 500 then the target will be over achieved by Rs. 50 (d^+). If the manager sets the target of Rs. 550 then clearly the target is met exactly ($d^- = d^+ = 0$). Thus an infeasible solution in the case of simple linear programming is converted into a feasible solution model in the goal programming with an under-achieved target value of Rs. 50.

Accordingly, we now modify the objective function in the original linear programming problem into a goal constraint. In this illustrative case we know that the goal of Rs. 600

will be under-achieved by Rs. 50. Let this deviation in the under-achieved case be denoted as d^-. With this new variable we write our goal constraint as

$$10X_1 + 15X_2 = 600 - d^- \ (10\times10 + 15\times30 = 550 = 600-50)$$

However, in a given problem, we will not know in advance whether our stated goal is going to be under-achieved or over-achieved or exactly met. To avoid this state of affair we propose to introduce two deviation variables, one for under-achievement (d^-) and one for over-achievement (d^+) into our goal constraint as shown below.

$$10X_1 + 15X_2 = 600 - d^- + d^+$$

Now after transferring both the deviation variables to left, the constraint may be rewritten as

$$10X_1 + 15X_2 + d^- - d^+ = 600$$

The deviational variables d^- and d^+ could be either positive or zero in this illustration.

- If d^+ is zero, $d^- = 50$, $X_1 = 10$, $X_2 = 30$ then our goal constraint stated above is met.
- It is important to note that in a given situation both d^- and d^+ cannot be positive. At the most one is positive means the other is zero and vice versa.
- In the case of exactly target met situation both these deviation variables are zero.
- In the case of an under-achievement d^- is positive and $d^+ = 0$.
- In the case of an over-achievement d^+ is positive and $d^- = 0$.

Now we have a total of three constraints, two resource constraints stated in the very problem and one goal constraint obtained in this problem formulation from the goal function given. To find the needed optimum solution, we first convert this problem into a standard minimization problem by including appropriate objective goal function. It is stated that the targeted revenue is Rs. 600. Clearly, over-achievement of this targeted Rs. 600 is not at all a problem. It is a well come move indeed. If we are going to under achieve this target of Rs. 600 then clearly the under achievement deviation variable d^- must be kept at its minimum. Accordingly, the problem may be restated as

Minimize: $Z = d^-$

Subject to:

$$10X_1 + 15X_2 + d^- - d^+ = 600$$
$$4X_1 + 2X_2 \leq 100$$
$$1X_1 + 2X_2 \leq 70$$
$$X_1 ;\ X_2 \quad d^- \quad d^+ \geq 0$$

Graphical Solution

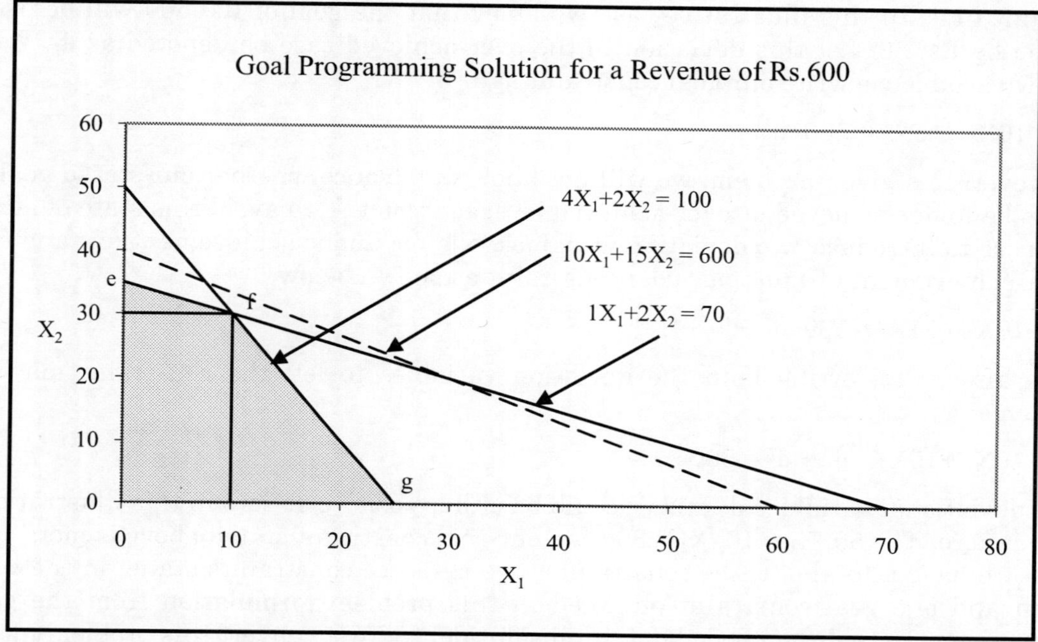

Figure 17.2

Clearly our goal constraint for Rs. 600 is not reachable because this constraint shown by the dotted line in Figure 17.2 excludes our feasible region completely fixed by resource constraints. Under such circumstances, the best possible solution is obtained at f in the diagram by utilizing both the resources fully. At f the values of outputs are: $X_1 = 10$, $X_2 = 30$. The relevant revenue achievable and the corresponding deviation variables are obtained by substituting these values in our goal constraint.

$$10X_1 + 15X_2 + d^- - d^+ = 600$$

$$10 \times 10 + 15 \times 30 + d^- - 0 = 600$$

$$d^- = 50$$

$$R = \text{Rs.}550$$

■ Example 3

Now, suppose the manager, instead of an unknown optimum, has the set goal of achieving a known revenue target of at least Rs. 500 or more. If we return to the graphical solution to this problem in example 1, we observe that the maximum revenue obtainable is Rs. 550. Given this goal programming terminology, the goal of Rs. 500 is over achieved by Rs. 50 (d^+) in this problem.

Accordingly, we now modify the objective function in the original problem into a goal constraint first. In this illustrative case we know that the goal of Rs. 500 will be over-achieved by Rs. 50. Let this deviation in the over achieved case be denoted as d^+. With this new variable we write our goal constraint as

$$10X_1 + 15X_2 = 500 + d^+$$

However, in a given problem, we will not know in advance whether our stated goal is going to be under-achieved or over-achieved or exactly met. Two avoid this state of affair we propose to introduce two deviation variables, one for under-achievement (d^-) and one for over-achievement (d^+) into our goal constraint as shown below

$$10X_1 + 15X_2 = 500 + d^+ - d^-$$

Now after transferring both the deviation variables to left the constraint may be rewritten as

$$10X_1 + 15X_2 - d^+ + d^- = 500$$

The deviational variables d^- and d^+ could be either positive or zero in this illustration. If d^- is zero, $d^+ = 50$, $X_1 = 10$, $X_2 = 30$ then our goal constraint stated above is met.

Now we have a total of three constraints, two resource constraints stated in the very problem and one goal constraint obtained in this problem formulation from the goal function given. To find the needed optimum solution, we first convert this problem into a standard minimization problem by including appropriate objective goal function. It is stated that the targeted revenue is Rs. 500. Clearly, over-achievement of this targeted Rs. 500 is not at all a problem. It is a well come move indeed. Thus once again the underachievement deviation variable d^- is to be minimized. Accordingly, the problem may be restated as

$$\begin{aligned}
\text{Minimize:} \quad & Z = d^- \\
\text{Subject to:} \quad & 10X_1 + 15X_2 + d^- - d^+ = 500 \\
& 4X_1 + 2X_2 \leq 100 \\
& 1X_1 + 2X_2 \leq 70 \\
& X_1 ; \ X_2 \quad d^- \ d^+ \geq 0
\end{aligned}$$

Graphical Solution

Clearly our goal constraint for Rs. 500 is over-achievable because this constraint shown by the dotted line is below f in Figure 17.3. Under the circumstances, the best possible solution is obtained at f in the diagram. At the point f the values of outputs are: $X_1 = 10$, $X_2 = 30$. The relevant revenue achievable and the corresponding deviation variables are obtained by substituting these values in our goal constraint.

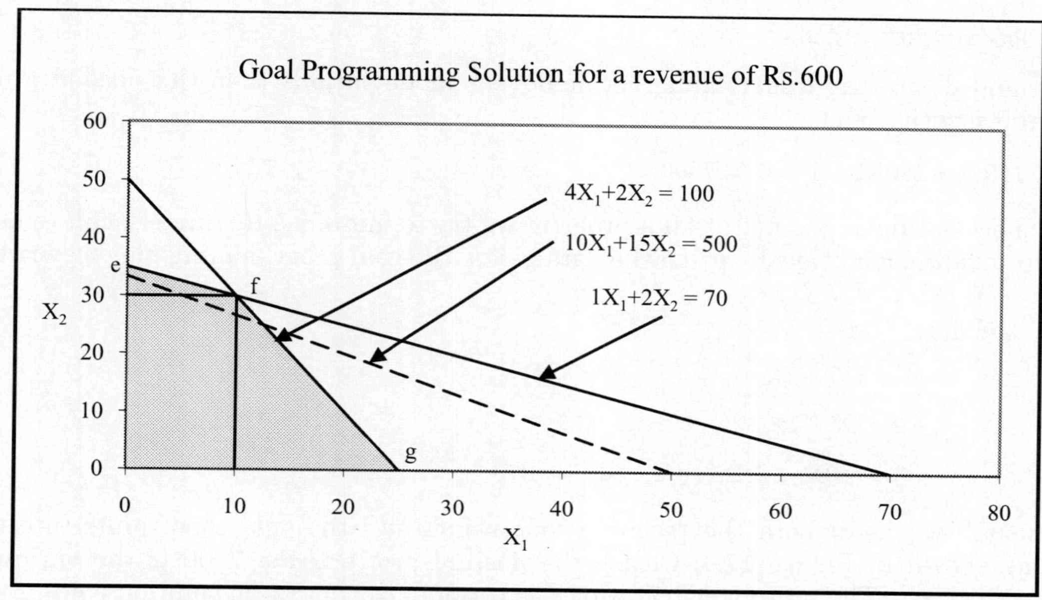

Figure 17.3

$$10X_1 + 15X_2 + d^- - d^+ = 500$$

$$10 \times 10 + 15 \times 30 + 0 - 50 = 500$$

$$d^+ = 50$$

$$R = Rs.550$$

Example 4

An office equipment manufacturing company manufactures chairs and tables. The company earns a profit of Rs. 150 per chair and Rs. 250 per table by selling them in the market. Both the products uses common assembly department and common finishing department as stated in Table 17.1.

Table 17.1

Product type	Hours of time required per unit		Profit per unit
	Assembly department	Finishing department	
Chair	1	1	150
Table	3	1	250
Hours available	60	40	

The company facing a transition period wants to have fair profit of at least Rs. 7000. Formulate a suitable goal programming and obtain the best possible solution.

Solution

Let X_1 and X_2 stands for the chairs and tables produced by the company. Now using the data one profit and the two resource constraints are obtained as

Profit $=150X_1 + 250X_2$

If d^- and d^+ are the under- and over-achievement deviations then the needed profit constraint is written as

$$150X_1 + 250X_2 + d^- - d^+ = 7000$$

Since over-achieving is not at all a problem we try to minimize the under-achievement if any in this problem. Now by using the data given, the goal programming may be written as

Minimize: d^-

$$1X_1 + 3X_2 \leq 60$$

$$1X_1 + 1X_2 \leq 40$$

$$150X_1 + 250X_2 + d^- - d^+ = 7000$$

As usual we insert both the resource constraints and the goal constraints into the graph as shown in Figure 17.4. Cleary the desired profit of Rs. 7000 is the optimum profit in this case. The profit constraint passes through the optimum solution point given by the resource constraints. At equilibrium point f our optimum solutions are X_1=30 and X_2=10. Now let us substitute these values in our profit constraint as shown below.

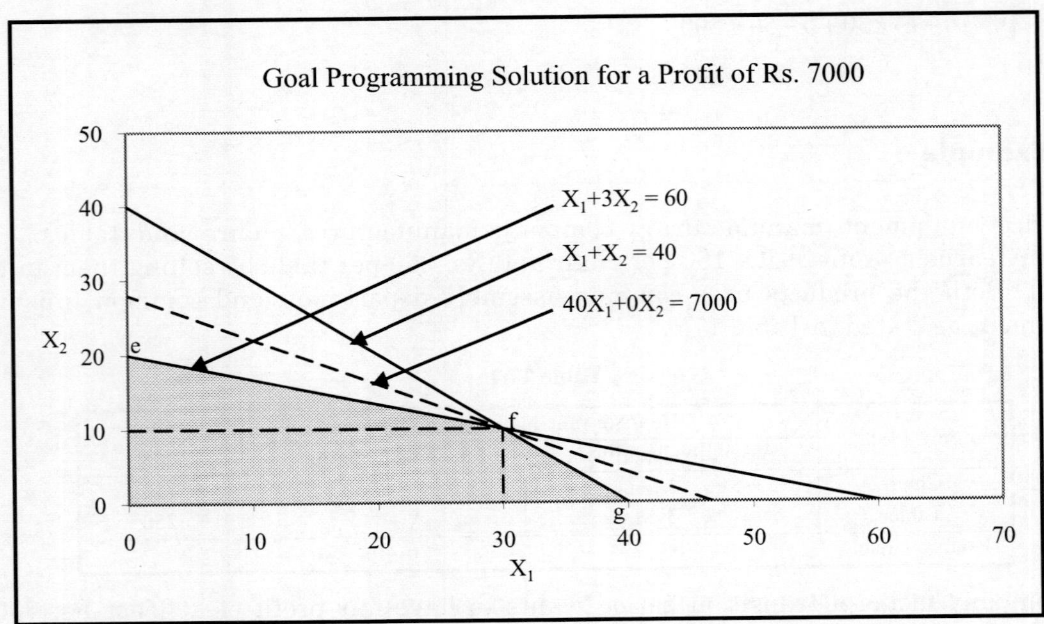

Figure 17.4

$$150X_1 + 250X_2 + d^- - d^+ = 7000$$

$$150 \times 30 + 250 \times 10 + 0 - 0 = 7000$$

$X_1 = 30$, $X_2 = 10$, $d^- = 0$, $d^+ = 0$ and Profit = 7000.

Thus in this case the target is neither over-achieved nor under-achieved. In fact the achieved result is simply a linear programming solution.

17.4 PRE-EMPTIVE AND NON-PRE-EMPTIVE MODELS OF GOAL PROGRAMMING WITH MULTIPLE GOALS

In this section we consider the real goal programming by introducing multiple goals into the model. Instead of seeking optimum goal values for each every goal, in goal programming we fix specific numerical target levels for each and every goal and look for a solution by minimizing the negative deviation variables. The goal set can be one sided of two sided. In one sided goal models sometimes we set a lower limit below which we do not accept it all. For example, if our target is to get profit from a certain business then we set a lower limit for profit below which does not accept. On some other occasions, we set upper limit for targets beyond which we do not prefer to have that venture. For example, if we undertake a risky venture of investing in share market, we fix a maximum level of risk index beyond which we do not opt for investment. These multiple goals may be either complementary to each other or competitive to one another. Moving closer to one target may lead to moving closer to the other in the complementary goal programming. Moving closer to one may leads to moving away from another in competitive goal programming. In two-sided goal models we set both lower and upper limits for acceptable range. Thus in such cases missing from either side is not desirable.

Depending upon the importance of goals, we classify the goal programming into two broad categories as pre-emptive on non-pre-emptive types. In pre-emptive goal programming the priorities are being listed out in the beginning. Once the priority is listed we focus our attention to first priority goal and try to achieve its optimum solution or result as close as possible. After achieving the 'as close as possible solution' to the first priority goals we proceed to the next priority goals and solve it for optimum solution or 'as close as possible' solution. After solving the second priority goal we proceed further to leftover goals in the order specified in the same way and obtain the final solution.

Non-pre-emptive goal programming involves situations where all the goals are roughly of equal importance. In such models first we establish a specified goal for each and every objective and then seek a solution that minimizes the weighted sum of the specified deviations from their respective goals.

17.4.1 Pre-emptive Goal Programming by Graphical Method

■ Example 5

Mr. A has Rs. 80,000 to make investments either in Infosys or Wipro shares. Wipro share costs Rs. 50 per share and Infosys share costs Rs. 25 per share. The return per share in the case of Infosys is Rs. 3 per share per annum and the Wipro it is Rs. 5 per share per annum. Thus, though the return per share is high in the case of Infosys the associated risk index is 0.5. However, the low return yielding Wipro risk index is 0.25 only. Formulate a linear programming problem and obtain the maximum return, given that Mr. A does not want to have a combined risk of more than 700.

Solution

Table 17.1(a)

Investments	Share price	Returns per share	Annual return per Rs. 100	Risk index
Infosys	25	3	12%	0.5
Wipro	50	5	10%	0.25

Simple Linear Programming Problem Formulation with Single Goal of Returns Maximization

From the summary data Table 17.1(a) we formulate the simple linear programming problem as follows

$$\text{Maximize: } R = 3X_1 + 5X_2$$

$$\text{Subject to: } 25X_1 + 50X_2 \leq 80{,}000 \text{ (Resource)}$$

$$0.5X_1 + 0.25X_2 \leq 700 \quad \text{(Risk)}$$

Graphical Solution

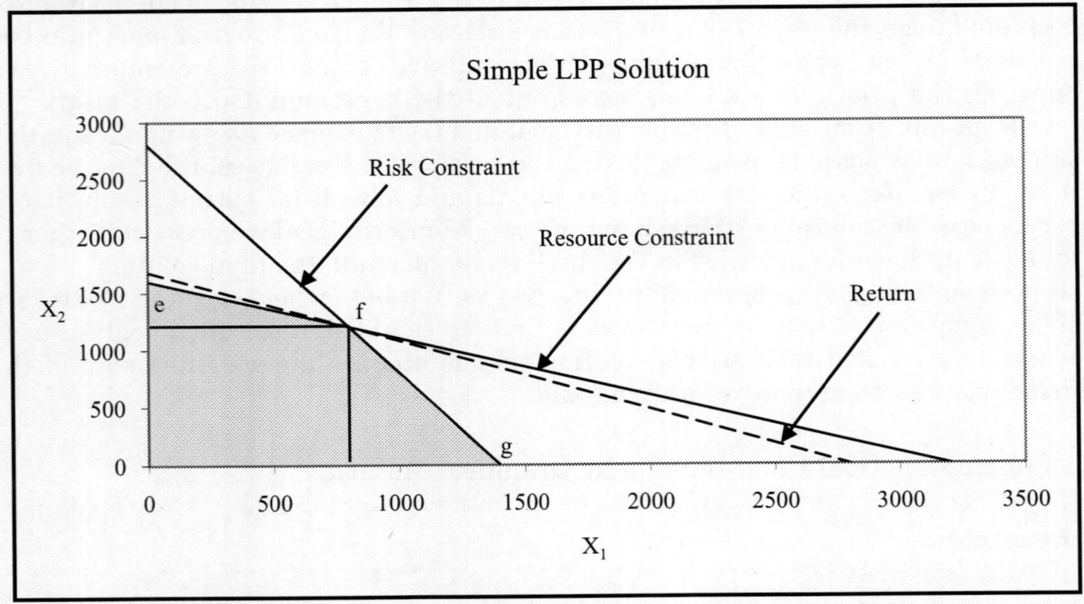

Figure 17.5

$X_1 = 800$, $X_2 = 1200$, Return = 8400.

Example 6

In our simple linear programming problem we optimized Mr. A's return given his combined risk index of 700 or less. Perhaps, if he prefers to have more risk of greater 700, then the associated returns could have been higher. From the data Table 17.1(a) it is clear that the return per Rs. 100 investment is more in the case of Infosys than Wipro, but at the same time the risk factor is more in Infosys than Wipro. Thus, the Mr. A's investment plan is a complicated one in the sense more the return means more risky as well. If he wants to have less risky then returns are also low. To overcome this problem let us propose a method for the calculation of risk index per share as follows.

Suppose Mr. A's investment is only at Infosys at the rate of Rs. 25 he can obtain a total of (80,000/25) = 3200 shares. This portfolio will have a risk index of 3200×0.5 = 1600, and associated return of 80,000× 0.12 = Rs. 9600. On the other hand only Wipro shares of 1600 (80,000/50) units are punched, then the associated returns is 8000 × 0.10 = Rs. 8000, and the associated risk index = 1600 × 0.25 = 400.

Table 17.2

I	W	Total investment	Risk index	Returns
3200	0	3200 x 25 = 80,000	3200 x 0.5 =1600	80,000 x 0.12=9600
0	1600	1600 x 50 = 80,000	1600 x 0.25=400	80,000 x 0.10=8000

Table 17.2 reveals that the complete investment in Infosys is more return yielding but at the same time with the highest risk index of 1600. Similarly, complete investment in Wipro yields only Rs. 8000 but at the same time with the low risk index of 400.

Table 17.3 shows 33 possible combinations of purchases and their associated summary results. If Mr. A invests all the money in Infosys, then the associated risk index is 1600 and the related profit is the highest at Rs. 9600 as shown in highlighted row 1. On the other hand, if the whole money is invested in Wipro, then the associated return is Rs. 8000 with lowest risk index of just 400 as shown by highlighted row 33.

As a compromise to both the extreme cases Mr. A wants to have a minimum of at least Rs. 9000 as return and at the same time the combined risk index of not more than 700. Once again we are in trouble. The table reveals that for an assured return of Rs. 9000 the associated risk index is 1150 as shown by the highlighted row13. Similarly, for an assured risk index of 700 the associated return is only Rs. 8400 as highlighted row 25. Thus no situation exists satisfying both the goals simultaneously in our illustration.

Thus, this portfolio selection problem is a multi-criteria decision problem involving two conflicting goals. The one is concerned with risk and the other is concerned with returns. The goal programming approach is precisely developed to tackle such situations. In this programming method we try to achieve a result closest to both the goals in the event of absence of exact goal achievements. However, for such modelling sometimes we choose the goal priorities as well. Suppose Mr. A proposes priority to risk and do not want to compromise anything more than 700. Then this goal is often called first priority goal. But at the same time as long as the first priority goal is fulfilled he will try to optimize the second goal of high returns. Thus, our problem may be restated as follows.

Table 17.3

S.N0	X_1	Risk	X_2	Risk	Total risk	Total return	Investment	d_1^+	d_1^-	d_2^+	d_2^-
1	3200	0.5	0	0.25	1600	9600	80,000	900	0	600	0
2	3100	0.5	50	0.25	1562.5	9550	80,000	863	0	550	0
3	3000	0.5	100	0.25	1525	9500	80,000	825	0	500	0
4	2900	0.5	150	0.25	1487.5	9450	80,000	788	0	450	0
5	2800	0.5	200	0.25	1450	9400	80,000	750	0	400	0
6	2700	0.5	250	0.25	1412.5	9350	80,000	713	0	350	0
7	2600	0.5	300	0.25	1375	9300	80,000	675	0	300	0
8	2500	0.5	350	0.25	1337.5	9250	80,000	638	0	250	0
9	2400	0.5	400	0.25	1300	9200	80,000	600	0	200	0
10	2300	0.5	450	0.25	1262.5	9150	80,000	563	0	150	0
11	2200	0.5	500	0.25	1225	9100	80,000	525	0	100	0
12	2100	0.5	550	0.25	1187.5	9050	80,000	488	0	50	0
13	2000	0.5	600	0.25	1150	9000	80,000	450	0	0	0
14	1900	0.5	650	0.25	1112.5	8950	80,000	413	0	0	50
15	1800	0.5	700	0.25	1075	8900	80,000	375	0	0	100
16	1700	0.5	750	0.25	1037.5	8850	80,000	338	0	0	150
17	1600	0.5	800	0.25	1000	8800	80,000	300	0	0	200
18	1500	0.5	850	0.25	962.5	8750	80,000	263	0	0	250
19	1400	0.5	900	0.25	925	8700	80,000	225	0	0	300
20	1300	0.5	950	0.25	887.5	8650	80,000	188	0	0	350
21	1200	0.5	1000	0.25	850	8600	80,000	150	0	0	400
22	1100	0.5	1050	0.25	812.5	8550	80,000	113	0	0	450
23	1000	0.5	1100	0.25	775	8500	80,000	75	0	0	500
24	900	0.5	1150	0.25	737.5	8450	80,000	37.5	0	0	550
25	800	0.5	1200	0.25	700	8400	80,000	0	0	0	600
26	700	0.5	1250	0.25	662.5	8350	80,000	0	37.5	0	650
27	600	0.5	1300	0.25	625	8300	80,000	0	75	0	700
28	500	0.5	1350	0.25	587.5	8250	80,000	0	113	0	750
29	400	0.5	1400	0.25	550	8200	80,000	0	150	0	800
30	300	0.5	1450	0.25	512.5	8150	80,000	0	188	0	850
31	200	0.5	1500	0.25	475	8100	80,000	0	225	0	900
32	100	0.5	1550	0.25	437.5	8050	80,000	0	263	0	950
33	0	0.5	1600	0.25	400	8000	80,000	0	300	0	1000

First priority goal: Determine a portfolio that has a risk index 700 or less.

Second priority goal: Determine a portfolio that yields a return of at least Rs. 9000 or more.

Resource constraint: Let X_1 stand for Infosys shares and X_2 stand for Wipro shares bought by Mr. A. Obviously, the first constraint could be the availability of funds namely, Rs. 80,000 as usual. Since the Infosys share costs Rs. 25 and Wipro share costs Rs. 50 the resource constraint is obtained as

$$25 X_1 + 50 X_2 \leq 80,000 \ldots \quad (1)$$

Goal constraints for first priority goal: Since the Infosys has the risk index of 0.50 and Wipro has the risk index of 0.25 the combined portfolio risk index is obtained by the equation $0.50X_1 + 0.25X_2$. However, depending upon X_1 and X_2 this index can be either equal or less or greater than 700 as highlighted in Table 17.4. By incorporating the necessary deviation variables the first priority goal constraint is written as

$$0.50X_1 + 0.25X_2 = 700 + d_1^+ - d_1^- \ldots \tag{2}$$

d_1^+ = the amount by which the portfolio index exceeds the target value of 700

d_1^- = the amount by which the portfolio index is less than the target value of 700

For example, in Table 17.4 shown above if $X_1 = 900$ and $X_2 = 1150$ then the associated risk index is 0.5(900)+0.25(1150) = 737.5. Hence, the relevant $d_1^+ = 37.5$. Similarly, if $X_1 = 800$ and $X_2 = 1200$, then 0.5(800)+0.25(1200) = 700. In this case, $d_1^+ = d_1^- = 0$. Further if $X_1 = 700$ and $X_2 = 1250$, then 0.5(700)+0.25(1250) = 662.5. So the relevant $d_1^- = 37.5$. Note that in this type of model specification both the deviations cannot take positive values. If negative deviation is zero then the positive deviation must be zero or positive. Similarly, if the positive deviation is zero then the negative deviation must be either zero or positive. In any case clearly both the variables cannot take positive values. These values are often called deviation values in goal programming. By bring all the deviation variables to the left the equation (2) may rewrite as

$$0.5X_1 + 0.25X_2 - d_1^+ + d_1^- = 700 \ldots \tag{3}$$

Goal constraints for second priority goal

In a similar way the annual return goal equation is written as

$$3X_1 + 5X_2 = 9000 + d_2^+ - d_2^- \ldots \tag{4}$$

$$3X_1 + 5X_2 - d_2^+ + d_2^- = 9000 \ldots \tag{5}$$

Objective function formulation: In goal programming models with priorities, the first priority objective function will have only first priority deviation variables. In our model since the total risk index is not to exceed the value 700, under-achievement like 400 units of risk is readily acceptable. So we need not bother ourselves with d_1^- deviation variable in our illustration. Under the circumstances the relevant positive deviation variable d_1^+ must be kept at its minimum. Thus, the complete model to deal with the first priority goal in its goal programming format may be written as

First priority model

Minimize: d_1^+

Subject to: $25X_1 + 50X_2 \leq 80,000$

$0.50X_1 + 0.25X_2 - d_1^+ + d_1^- = 700$ (First Goal)

$3X_1 + 5X_2 - d_2^+ + d_2^- = 9000$ (Second Goal)

$X_1; X_2; d_1^+; d_1^-, d_2^+, d_2^- \geq 0$

Graphical solution to the first priority goal

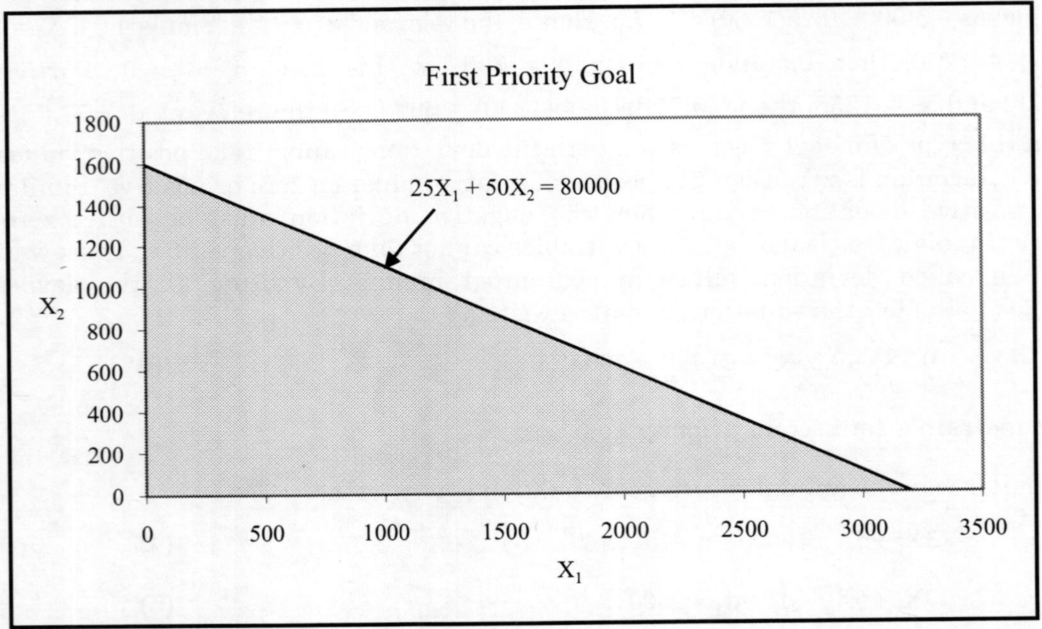

Figure 17.6

To begin with we obtain the portfolio selection graph as shown in Figure 17.6 by making use of our first funds availability constraint in our problem. The relevant straight-line equation is written as

$$25\,X_1 + 50\,X_2 = 80,000$$

In Figure 17.6 we show this straight line. Accordingly, any point on or below this line is a feasible portfolio.

Now we write the relevant first goal constraint as

$$0.5X_1 + 0.25X_2 - d_1^+ + d_1^- = 700$$

When priority goal is exactly met then in the above equation there cannot be either positive or negative deviations. So the constant reduces to

$$0.50X_1 + 0.25X_2 = 700$$

Now let us superimpose this first goal constraint into this graph and obtain the relevant feasible region and the solution as shown in Figure 17.7.

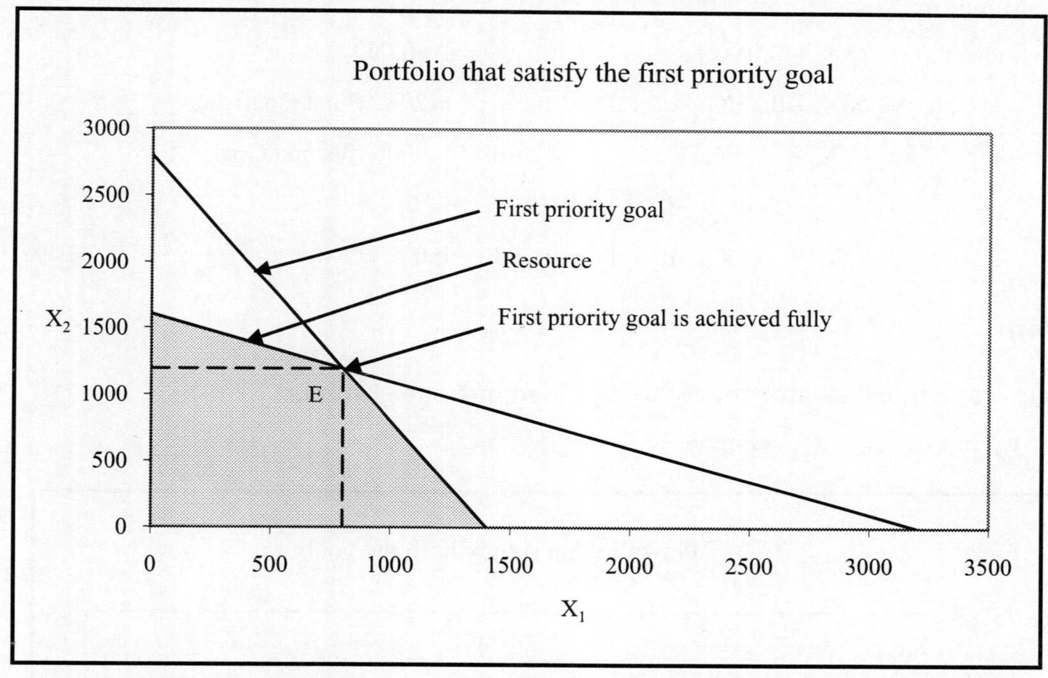

Figure 17.7

At all points on this first priority goal line both the deviations are zero and the target is fully met. All points below this line the risk index will be less than 700. Remember 700 or less risk index is always preferred by the investor. More than 700 risk index is not permitted. Thus, all the points on or below this line satisfy the first priority goal constraint. The shaded region in Figure 17.7 identifies all the solution points satisfying both the fund constraint and the risk index of our first priority goal. Clearly, E is the point at which both the constraints are exactly fulfilled and the corresponding deviations are zero. So the solution values are $X_1 = 800$, $X_2 = 1200$ and $R = 8400$ $d_1^- = 0$, $d_1^+ = 0$.

Second priority constraint

At this stage we are not certain about the fulfilment of our second priority goal relating to portfolio returns. The second priority goal constraint is written as

$$3X_1 + 5X_2 - d_2^+ + d_2^- = 9000$$

In this case over-fulfilment is not at all a problem because over-fulfilment means getting more than Rs. 9000 as returns. So our concern is to minimize the under fulfilment deviation variable d_2^- without undermining the result already achieved in our first priority goal diagram. So after adding the second priority goal constraint, the priority 2 level linear programming model is written as

$$\begin{aligned}
\text{Minimize:} \quad & d_2^- \\
\text{Subject to:} \quad & 25X_1 + 50X_2 \leq 80{,}000 \\
& 0.50X_1 + 0.25X_2 - d_1^+ + d_1^- = 700 \quad \text{(First Goal)} \\
& 3X_1 + 5X_2 - d_2^+ + d_2^- = 9000 \quad \text{(Second Goal)} \\
& d_1^+ = 0 \\
& X_1\ ;\ X_2\ ;\ d_1^+\ ;\ d_1^-\ ;\ d_2^+\ ;\ d_2^- \geq 0
\end{aligned}$$

Solution

Now our relevant constraint that is to be added into the graph is

$$3X_1 + 5X_2 - d_2^+ + d_2^- = 9000$$

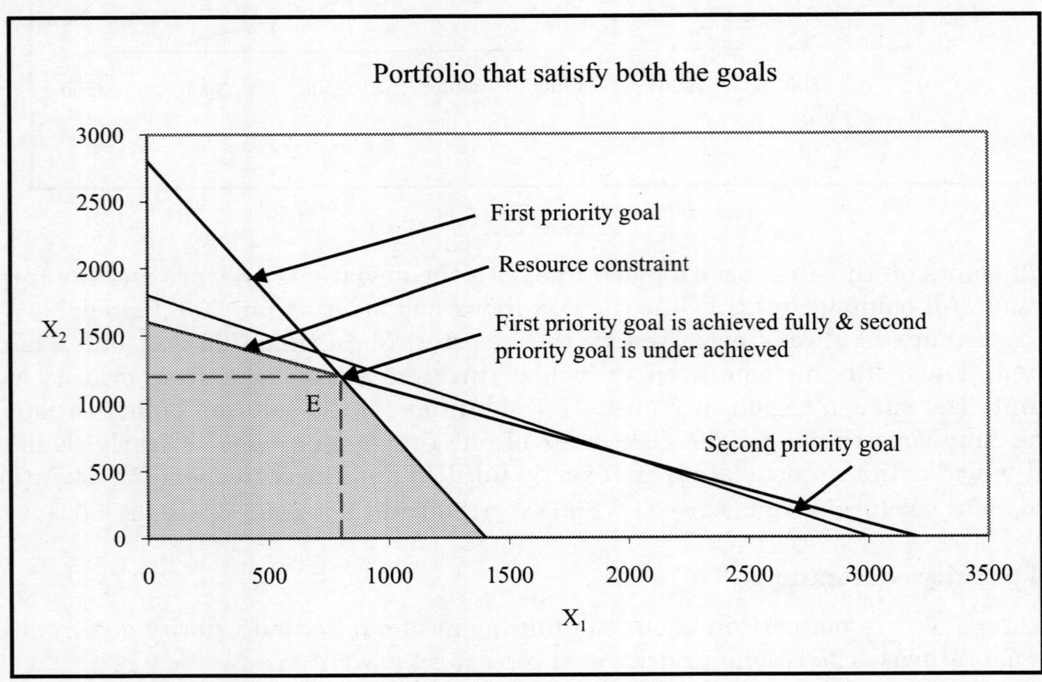

Figure 17.8

If this constraint is fulfilled exactly then our reduced straight line becomes

$$3X_1 + 5X_2 = 9000$$

Let us continue our graphical method by superimposing our new second goal constraint also in the graph as shown in Figure 17.8.

Clearly our new second priority goal constraint is completely above the feasible region marked during our first priority goal insertion. Clearly, all possible first priority fulfilling solutions points under-fulfils the second priority. So at the most one can think of getting the nearest best possible solution. Clearly, one such point is again at E. At this point the goal achievement positions are obtained by substituting the coordinates (800, 1200) of this point in the respective goal constraints as shown below.

First goal

$$0.50X_1 + 0.25X_2 - d_1^+ + d_1^-$$

$$0.5 \times 800 + 0.25 \times 1200 - 0 + 0 = 700$$

So $d_1^+ = 0$; $d_1^- = 0$ and this goal is fully achieved as desired

Second goal

$$3X_1 + 5X_2 - d_2^+ + d_2^- = 9000$$

$$3 \times 800 + 5 \times 1200 - 0 + 600 = 9000$$

So $d_2^+ = 0$; $d_1^- = 600$ and this goal is under-fulfilled than the desired minimum of Rs. 9000 by Rs. 600. Thus the best possible return is only Rs. 8400.

At this point the first priority goal is fully met at 700 index level but the corresponding returns are only Rs. 8400 with under fulfilment of Rs. 600.

Note: *The goal programming problem 4 is solved as sequence of linear programming problems, one linear program for each priority. However, for the notational convenience we write the entire goal programming problem in one concise statement. While doing so we write the overall objective function for our portfolio selection problem having the listed hierarchy of priority in mind. We can do so by writing the objective function as* Min: $P_1(d_1^+) + P_2(d_2^-)$. *Here the priority levels P_1 and P_2 are not the numerical weights of the priority levels but simply reminds us the levels of priority.*

Now we write the complete goal programming as

$$\text{Min: } P_1(d_1^+) + P_2(d_2^-)$$

$$\text{S.t : } 25X_1 + 50X_2 \leq 80{,}000$$

$$0.50X_1 + 0.25X_2 - d_1^+ + d_1^- = 700 \quad \text{(First Goal)}$$

$$3X_1 + 5X_2 - d_2^+ + d_2^- = 9000 \quad \text{(Second Goal)}$$

$$X_1, X_2, d_1^+, d_1^-, d_2^+, d_2^- \geq 0$$

Priority Reversed: A Revisit

Example 7

Now let us reverse the priority goals so that the desired level of Rs. 9000 or more as the first priority goal and preferable risk index of 700 or less as our second priority goal and redo the above problem once again.

Resource constraint

Let X_1 stand for Infosys shares and X_2 stand for Wipro shares bought by Mr. A. Obviously, the first constraint could be availability of funds namely Rs. 80,000 as usual. Since the Infosys share costs Rs. 25 and Wipro share costs Rs. 50 the resource constraint is obtained as

$$25\,X_1 + 50\,X_2 \leq 80{,}000 \ldots \qquad (1)$$

Goal constraints for first priority goal

Since the Infosys has the risk index of 0.5 and Wipro has the risk index of 0.25 the combined portfolio risk index is obtained by the equation $0.50X_1 + 0.25X_2$. However, depending upon X_1 and X_2 this index can be either equal or less or greater than 700 as highlighted in Table 17.4. By incorporating the necessary deviation variables the first priority goal constraint is written as

$$3X_1 + 5X_2 = 9000 + d_1^+ - d_1^- \ldots \qquad (2)$$

d_1^+ = the amount by which the portfolio index exceeds the target value of 9000

d_1^- = the amount by which the portfolio index is less than the target value of 900

Thus (2) may be rewritten as

$$3X_1 + 5X_2 - d_1^+ + d_1^- = 9000 \ldots \qquad (3)$$

Goal constraints for second priority goal

In a similar way the annual return goal equation is written as

$$0.5X_1 + 0.25X_2 = 700 + d_2^+ - d_2^- \ldots \qquad (4)$$

$$0.5X_1 + 0.25X_2 - d_2^+ + d_2^- = 700 \ldots \qquad (5)$$

Objective function formulation

In our model according to first priority goal the return out of investment must be at last Rs. 9000. So we need not bother ourselves with over-achievement deviation d_1^+ deviation variable in our illustration. Under the circumstances the relevant negative deviation variable d_1^- must be kept at its minimum. Thus, the complete model to deal with the first priority goal in its linear programming format may be written as

First priority model

$$\text{Minimize: } d_1^-$$

$$\begin{aligned} \text{Subject to: } & 25X_1 + 50X_2 \leq 80{,}000 \\ & 3X_1 + 5X_2 - d_1^+ + d_1^- = 9000 \quad \text{(First Goal)} \\ & 0.5X_1 + 0.25X_2 - d_2^+ + d_2^- = 700 \quad \text{(Second Goal)} \\ & X_1 ; X_2 ; d_1^+ ; d_1^- ; d_2^+ ; d_2^- \geq 0 \end{aligned}$$

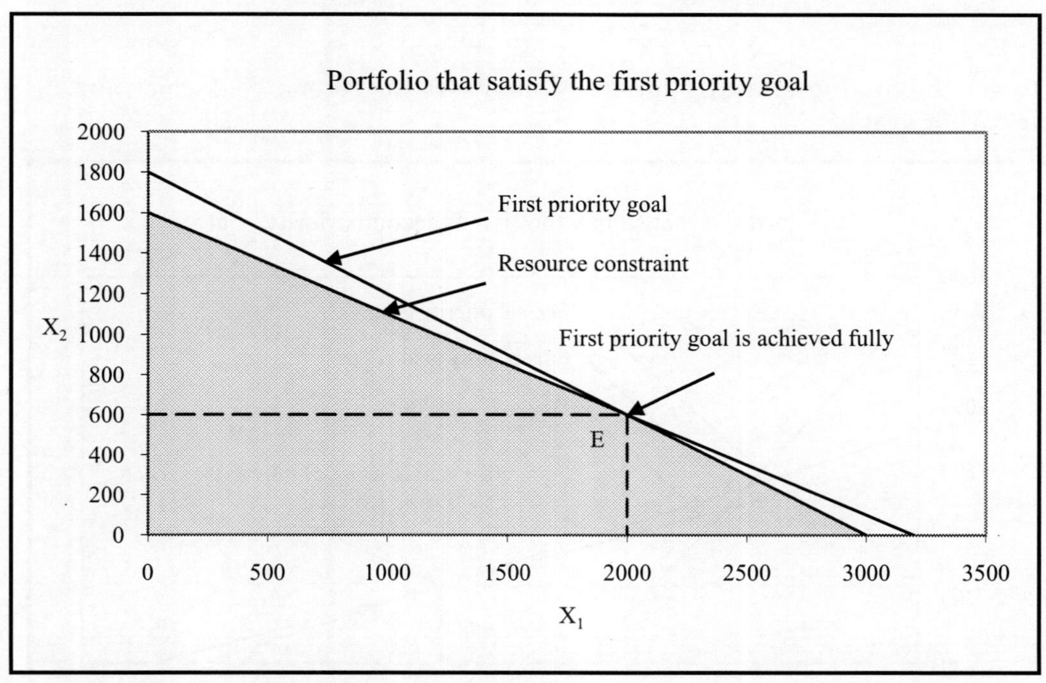

Figure 17.9

Clearly the first priority goal is met at E. So the solution values are X_1 = 2000, X_2 = 600 and R = 9000.

Second priority goal model

$$0.5X_1 + 0.25X_2 - d_2^+ + d_2^- = 700\ldots \qquad (5)$$

If this goal is to be fully satisfied then both the positive and negative deviation coefficient must be zero. So the above constant may be written as

$$0.5X_1 + 0.25X_2 = 700$$

In this goal since 700 or less risk is preferred we need not bother ourselves with under-achievement. So in case of over-achievement the coefficient of over-achievement d_2^+ must be kept at its minimum.

Minimize: d_2^+

Subject to:

$$25X_1 + 50X_2 \leq 80{,}000$$

$$3X_1 + 5X_2 - d_1^+ + d_1^- = 9000 \quad \text{(First Goal)}$$

$$0.5X_1 + 0.25X_2 - d_2^+ + d_2^- = 700 \quad \text{(Second Goal)}$$

$$d_1^- = 0$$

$$X_1\ ;\ X_2\ ;\ d_1^+;\ d_1^-;\ d_2^+\ ;\ d_2^- \geq 0$$

So to get the final graphical solution let us insert the second goal constraint as well into our earlier graph.

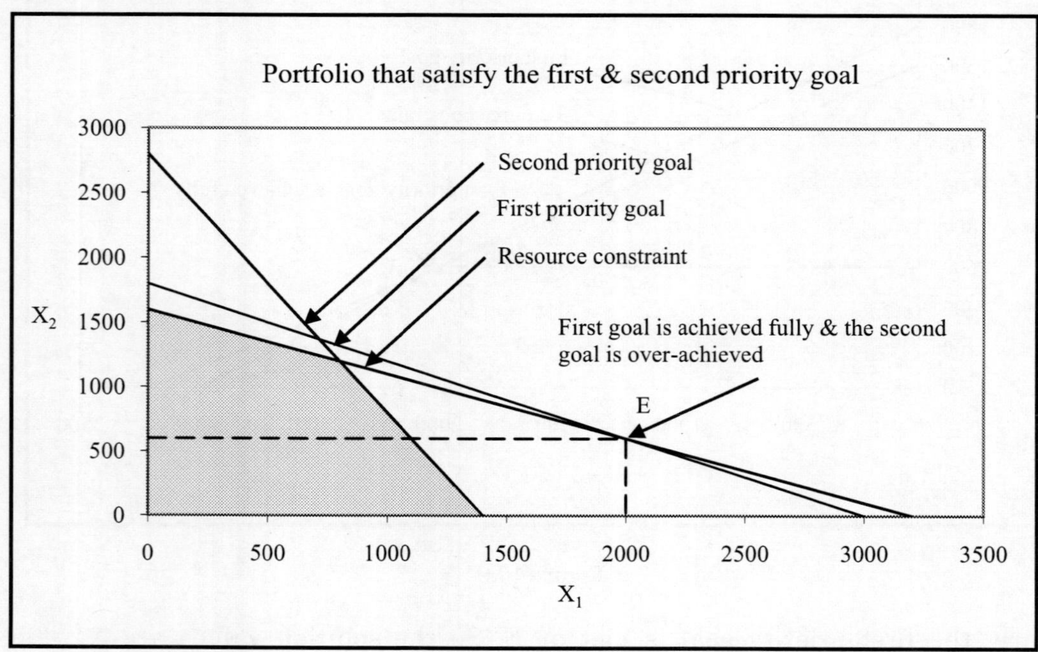

Figure 17.10

From the graph the coordinate of E is obtained as

First goal

$$3X_1 + 5X_2 - d_2^+ + d_2^- = 9000$$

$$3 \times 2000 + 5 \times 600 - 0 + 0 = 9000$$

So $d_2^+ = 0$; $d_2^- = 0$ and this goal is fully achieved as desired.

Second goal

$$0.50X_1 + 0.25X_2 - d_1^+ + d_1^-$$

$$0.5 \times 2000 + 0.25 \times 600 - 450 + 0 = 700$$

So $d_1^+ = 450$; $d_1^- = 0$ and this goal is over-achieved as desired by 450 index points leading to a total of 1150 points. Thus the profit target of Rs. 9000 is achievable only with the risk index of 1150 points.

Example 8

Solve the following pre-emptive goal programming problem graphically.

$$\text{Minimize:} \quad Z = P_1 d_1^- + P_2 d_2^+ + P_3 d_3^+ + P_4 d_4^-$$

$$\begin{aligned}
\text{Subject to:} \quad & 25X_1 + 30X_2 - d_1^+ + d_1^- = 150 \\
& 15X_1 + 5X_2 - d_2^+ + d_2^- = 45 \\
& 5X_1 + 10X_2 - d_3^+ + d_3^- = 60 \\
& X_1 - d_4^+ + d_4^- = 5 \\
& X_1, X_2, d_1^+, d_1^-, d_2^+, d_2^-, d_3^+, d_3^-, d_4^+, d_4^- \geq 0
\end{aligned}$$

Solution

In the above graph as far as the first constraint is concerned we need not worry about positive deviation, only under-achievement is to be minimized. So on or above the line is a welcome move. As far as goal two is concerned positive deviation is to be kept at its minimum. So on or below the line is a welcome move. In the case of third constraint positive deviation is to be minimized. On or below the line is a welcome move. Thus the shaded areas ABCD clearly satisfy all our first three constraints. For the fourth constraint the negative deviation is to be minimized.

Thus the area on the line and above the lines deserve our attention. Thus the fourth constraint is not at all possible to achieve fully. However, D is the closest best point so that the negative deviation is minimized.

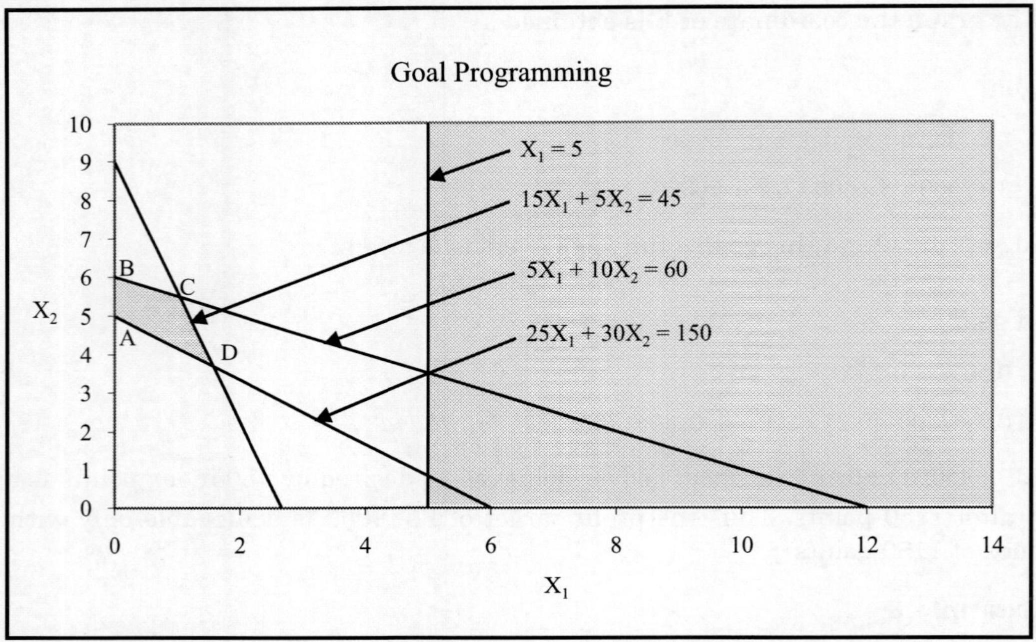

Figure 17.11

The coordinate of point D is obtained from the graph as X_1 = 2.2 and X_2 = 2.4 substituting these values in the respective goal constraints we verify their fulfilment or otherwise.

Goal 1

$$25X_1 + 30X_2 - d_1^+ + d_1^- = 150$$

$$25 \times 2.2 + 30 \times 2.4 - 0 + 0 = 150$$

Thus this goal is fulfilled fully.

Goal 2

$$15X_1 + 5X_2 - d_2^+ + d_2^- = 45$$

$$15 \times 2.2 + 5 \times 2.4 - 0 + 0 = 45$$

Thus this goal is fulfilled fully.

Goal 3

$$5X_1 + 10X_2 - d_3^+ + d_3^- = 60$$

$$5 \times 2.2 + 10 \times 2.4 - 0 + 25 = 60$$

Thus $d_3^- = 25$ and $d_3^+ = 0$. So this goal is achieved more than expected.

Goal 4

$$X_1 - d_4^+ + d_4^- = 5$$

$$2.2 - 0 + 2.8 = 5$$

So $d_4^- = 2.8$ and $d_4^+ = 0$.

Thus only fourth goal is under-achieved with a minimum negative deviation of 2.8.

■ Example 9

A manufacturing firm produces two products A and B. The production of these products uses one hour each of a certain machine time per unit. The total machine hour at the firm's disposal is 400 hours. As per the market survey the firm can cell a maximum of 240 units of A and 300 unit of B. The net profit from the sale of A and B are Rs. 800 and Rs. 400 respectively. The managers priority are described as:

1. He wants to avoid any under-utilization of the normal production capacity.
2. He wants to sell the maximum units of A and B in the market. Since the net profit from A is twice the net profit from B, the manager has twice desire to sell product A over B.
3. He wants to minimize the over time operation of the plant as much as possible.

Solution

First goal constraint

If X_1 and X_2 are the number of units of A and B produced then under the assumption of permitted over-production the capacity constant may be written as

$$X_1 + X_2 + d_1^- - d_1^+ = 400$$

Since the sales goals are fixed at maximum levels we need not bother our selves with the over-achieving positive deviations.

Thus our sales constraints may be written as:

Second goal constraint

$$X_1 + d_2^- = 240$$

$$X_2 + d_3^- = 300$$

With these constraints the relevant goal programming mode is written as

$$\text{Maximize:} \quad P_1 d_1^- + P_2(2d_2^- + d_3^-) + P_3 d_1^+$$

$$\text{Subject to:} \quad X_1 + X_2 + d_1^- - d_1^+ = 400$$

$$X_1 + d_2^- = 240$$

$$X_2 + d_3^- = 300$$

$$X_1; X_2; d_1^-; d_2^-; d_3^-; d_1^+ \geq 0$$

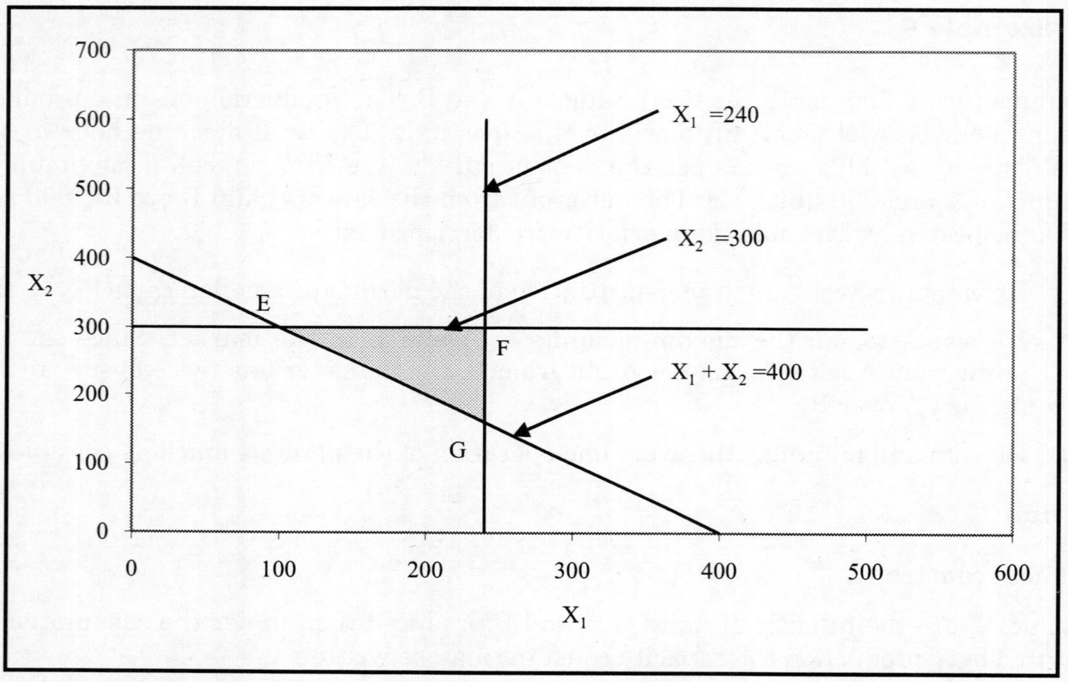

Figure 17.12

Since under-utilization of capacity is not permitted the feasible region of capacity constraint could be on the right side of the capacity line EG. Since as per the second goal for both the products the maximum sale is 240 and 300 the relevant feasible region could lie below these two lines. Thus the feasible region satisfying all the goal constraints is shown by the shaded region EFG. Clearly at F goal I and 2 are satisfied. In fact goal 1 is over fulfilled. Goal 2 is correctly fulfilled as per the prescribed maximum. The goal 3 aiming at minimizing over time operation is not achievable in this problem. The final solution values are

$$X_1 = 240; X_2 = 300; d_1^- = 0,\ d_2^- = 0,\ d_3^- = 0,\ d_1^+ = 140$$

Example 10

A firm produces two products A and B and sells them in the market for a profit of Rs. 400 and Rs. 300 respectively. Both these products use a certain raw material whose supply is limited to 300 units per production period. The raw material need is 1 kg per unit for each product. Further each unit of A needs 2 hours of processing time and each unit of B needs 1 hour of processing time. The firm has 400 hours of processing time at its disposal per month. The marketing department predicts a maximum sale of 150 unit of A and 350 unit of B per month. The manager of the concerned set the goals in their order of importance as follows.

Goal 1: Avoid under-utilization of normal productive capacity.

Goal 2: See that the maximum number of units of product A and B are sold. However, since the profit ratio of the two products is 400: 300, the manager wants to give the same weightage to the sales of these two products.

Goal 3: He wants to minimize the over time operation of the plant.

Convert given data into a goal programming and obtain it solution graphically.

Solution

Set up constraints and goals as follow.

Raw material constraint

$$X_1 + X_2 \leq 300$$

Processing capacity goal constraint

If d_1^- = the number of hours of unutilized processing time and d_1^+ = the number of hours of over utilization of the processing time then the process utilization goal constraint is written as

$$2X_1 + X_2 + d_1^- - d_1^+ = 400$$

Market goal constraints

$$X_1 + d_2^- = 150$$

$$X_2 + d_3^- = 350$$

Note that in the above constraints since the sale levels are maximum, deviational variables are provided only for their under achievements.

With P_1, P_2 and P_3 being the pre-emptive priority factors, then the relevant goal programming may be summarized as

$$\text{Maximize: } Z = P_1 d_1^- + P_2(4d_2^- + 3d_3^-) + P_3 d_1^+$$

$$\text{Subject to: } X_1 + X_2 \leq 300$$

$$2X_1 + X_2 + d_1^- - d_1^+ = 400$$

$$X_1 + d_2^- = 150$$

$$X_2 + d_3^- = 350$$

$$X_1, X_2, d_1^-, d_1^+, d_3^- \geq 0$$

1. Graphical Solution

Let us draw the raw material constraint and shade the feasible region first as shown in the Figure 17.13. Next we introduce all the goal constraints by setting the respective deviation variables to zero. The first goal of avoiding the under-utilization of productive capacity is achieved by setting $d_1^- = 0$. Since in the second goal the manager insists the maximum sale as 150 and 350 units respectively our feasible region is reduced to the triangular area given by ATO.

Thus now we shift our focus to the second goal. Since $X_1 = 150$ is the maximum permitted sale, the area DET represents undesirable (excess sale) deviation as for this goal is concerned. Thus our feasible region is restricted to CBE in the figure.

The line depicting $X_2 = 350$ lie outside the feasible region. This amount of maximum sale is not at all possible. So we shift this goal line till it touches the feasible region at the corner at A. Since the product A yields a greater profit of Rs. 400 over the product B which yields Rs. 300, a hefty penalty is stipulated for deviating from the output target of product A. between points A, B, and E we may obtain the optimum solution by computing the values of d_2^- and d_3^- and the value of Z.

Given the resource constraint by the shaded area in Figure 17.13, the first priority goal of non-under-utilization of processing time insists all points on or above line segment BC but on or below the resource constraint line AT. The second goal constraint $X_1 = 150$ or less insists all the points on or below the line segment EC. Thus ultimately the triangular area BEC becomes the best restricted area.

Table 17.4

Point	X_1	X_2	d_2^-	d_3^-	$Z = 4d_2^- + 3d_3^-$
A	0	300	150	50	750
B	100	200	50	150	650
E	150	150	0	200	600 (minimum)

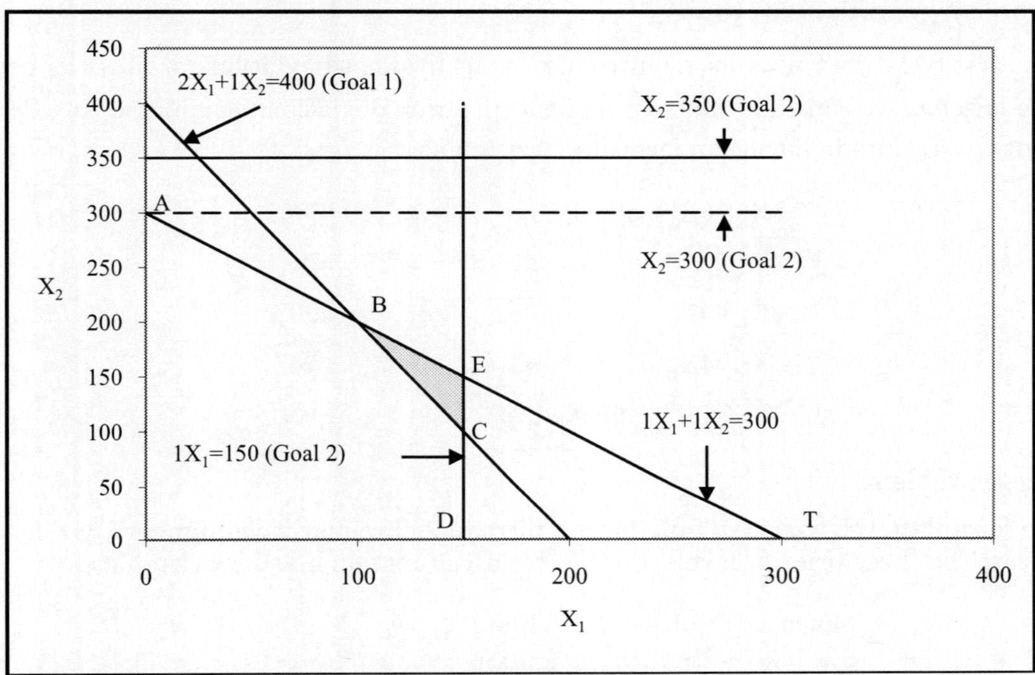

Figure 17.13

Thus the best possible solution at E is written as: $X_1 = 150$, $X_2 = 150$. Substituting these values in the goal 1 constraint yields $2 \times 150 + 150 - d_1^+ = 400$ or $d_1^+ = 500$, implying that 50 hours of processing capacity shall be overtime work.

■ Example 11

Suppose a producer is producing two products 1 and 2 by using two inputs labour and capital. Each unit of the product 1 requires 2 hours of labour and 4 units of capital. Each unit of the product 2 products requires 3 hours of labour and 4 units of capital. For everyday use, 20 units of labour and 24 units of capital are made available by the company. These products are sold at Rs. 8 and Rs. 6 respectively in the market. Formulate a goal programming problem given the goal priorities as follows and obtain the best possible solution both graphical and simplex method.

Goal 1: The profit per day should be at least Rs. 36.

Goal 2: Product 2 must be two times of the product 1.

Goal 3: The labour should be fully utilized.

Goal 4: The capital should be fully utilized.

Solution by graphical method

For the first two constraints let us introduce $d_1^+, d_1^-, d_2^+, d_2^-$ as deviation variables as usual. For the last two we consider only two under utilization deviation variables d_3^-, d_4^-. So the relevant constraints in deviation forms are written as

$$\begin{aligned} 8X_1 + 6X_2 - d_1^+ + d_1^- &= 36 \\ X_1 - 2X_2 \quad - d_2^+ + d_2^- &= 0 \\ 2X_1 + 4X_2 \quad + d_3^- &= 20 \\ 3X_1 + 4X_2 \quad + d_4^- &= 24 \\ X_1, X_2, d_1^+, d_1^-, d_2^+, d_2^-, d_3^-, d_4^- &\geq 0 \end{aligned}$$

Complete problem

Now to formulate the objective function we introduce three positive numbers $P_1 > P_2 > P_3$ to indicate the three priority levels. So the objective function may be written as

$$\begin{aligned} \text{Minimize:} \quad & P_1 d_1^- + P_2(d_2^+ + d_2^-) + P_3 d_3^- \\ \text{Subject to:} \quad & 8X_1 + 6X_2 - d_1^+ + d_1^- = 36 \\ & X_1 - 2X_2 \quad - d_2^+ + d_2^- = 0 \\ & 2X_1 + 4X_2 \quad + d_3^- = 20 \\ & 3X_1 + 4X_2 \quad + d_4^- = 24 \\ & X_1, X_2, d_1^+, d_1^-, d_2^+, d_2^-, d_3^-, d_4^- \geq 0 \end{aligned}$$

1. Graphical Solution

First we draw both labour and capital constraints first and shade the feasible area as highlighted in the Figure 17.14. It is clear that our production line intersect the labour constraint outside the feasible region.

Therefore, the nearest best solution lies at E in the graph with under utilization of labour.

So the optimum solutions are $X_1 = 24/5 = 4.8$ $X_2 = 12/5 = 2.4$.

Goal 1

$$8X_1 + 6X_2 - d_1^+ + d_1^- = 36$$

$$8 \times 4.8 + 6 \times 2.4 - 16.8 + 0 = 36$$

Thus goal constraint is over achieved by Rs. 16.80 leading to a total of Rs. 52.8.

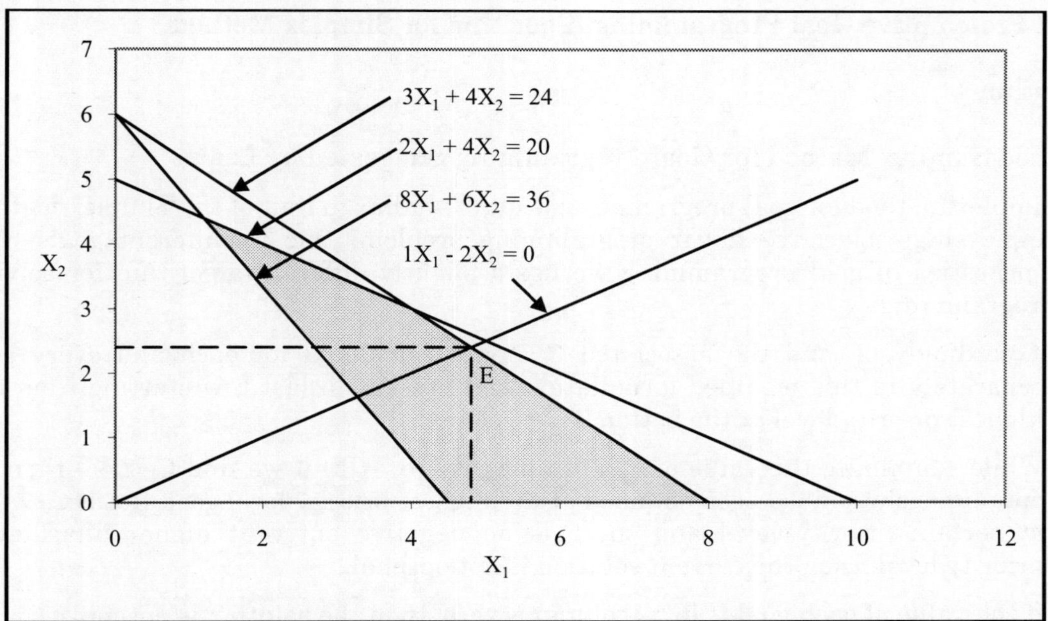

Figure 17.14

Goal 2

$$X_1 - 2X_2 - d_2^+ + d_2^- = 0$$
$$4.8\text{-}2 \times 2.4 - 0 + 0 = 0$$

Thus the second goal is fully achieved with zero deviation on either side.

Goal 3

$$2X_1 + 4X_2 + d_3^- = 20$$
$$2 \times 4.8 + 4 \times 2.4 + 0.8 = 20$$

This target is under-achieved by 0.8 units of labour unemployed.

Goal 4

$$3X_1 + 4X_2 \;\; + d_4^- = 24$$
$$3 \times 4.8 + 4 \times 2.8 + 0 = 24$$

Thus at equilibrium the capital is fully utilized.

17.4.2 Pre-emptive Goal Programming Algorithm for Simplex Method

Algorithm 1

Modified Simplex Method for Goal Programming Suggested by Lee

The Simplex method for goal programming is quite similar to that of the Simplex method for simple single objective linear programming problem. Due to different goals with listed priorities in goal programming we use a slightly different algorithm for solving goal programming.

1. Accordingly, Z_j and the associated $C_j–Z_j$ are calculated for each and every goal separately in this modified formulation and present in last row starting from the highest priority level at the bottom.
2. While examining the value of $C_j–Z_j$ in priority row (Pi) if we find $C_j–Z_j \leq 0$ then an optimum solution has been obtained at the highest priority level goal. But if $C_j–Z_j \geq 0$ at certain priority level and there is no negative entry at higher unachieved priority level, then the current solution is not optimal.
3. If the value of each goal is in X_b column is zero, then the solution is optimal.
4. In order to determine the entering variable, we examine $C_j–Z_j$ at the highest priority p_1, we select the largest negative value or else move to the next highest priority p_2 and select the largest negative value.
5. Apply usual procedure for minimum positive ratio, to select the leaving variable.
6. The negative values of $C_j–Z_j$ row under lower priority rows are ignored.

■ Example 12

Solve the following Goal Programming by using modified Simplex method

$$\text{Minimize: } P_1 d_1^- + P_2(2d_2^- + d_3^-) + P_3 d_1^+$$

$$\begin{aligned}\text{Subject to: } & X_1 + X_2 - d_1^+ + d_1^- = 400 \\ & X_1 + d_2^- = 240 \\ & X_2 + d_3^- = 300 \\ & X_1 ;\ X_2 ;\ d_1^+ ;\ d_1^- ;\ d_2^- ;\ d_3^- \geq 0\end{aligned}$$

In our modified deviation introduced statement of the goal programming all the negative deviation variables are having positive unity coefficients. Such a presentation allows us to start the Simplex tableau 1 with all the negative deviation variables having unit coefficients in the basis column as shown in Tableau 1. Since this is minimization problem we calculate $Z_j–C_j$ in the index row for each and every goal separately and present them all as shown in Tableau 1 starting from the first priority goal at the bottom. The needed calculations are shown below.

$$Z_1 - C_1 = (P_1 \times 1 + 2P_2 \times 1 + P_2 \times 0) - 0 = P_1 + 2P_2$$

$$Z_2 - C_2 = (P_1 \times 1 + 2P_2 \times 0 + P_2 \times 1) - 0 = P_1 + P_2$$

$$Z_3 - C_3 = (P_1 \times 1 + 2P_2 \times 0 + P_2 \times 0) - p_1 = 0$$

$$Z_4 - C_4 = (P_1 \times 0 + 2P_2 \times 1 + P_2 \times 0) - 2p_2 = 0$$

$$Z_5 - C_5 = (P_1 \times 0 + 2P_2 \times 0 + P_2 \times 1) - p_2 = 0$$

$$Z_6 - C_6 = (P_1 \times -1 + 2P_2 \times 0 + P_2 \times 0) - p_3 = -P_3 - P_1$$

Table 17.5

Tableau 1									
C_i	Basis	X_1	X_2	d_1^-	d_2^-	d_3^-	d_1^+	X_b	MR
P_1	d_1^-	1	1	1	0	0	−1	400	400
$2P_2$	d_2^-	1	0	0	1	0	0	240	240
P_2	d_3^-	0	1	0	0	1	0	300	–
C_j		0	0	P_1	$2P_2$	P_2	P_3	0	
$Z_j–C_j$	P_3	0	0	0	0	0	−1	0	
	P_2	2	1	0	0	0	0	780	
	P_1	1	1	0	0	0	−1	400	

Now let us give priority to our first priority goal index row given at the bottom in Tableau 1. In it there are two positive unity entries. So arbitrarily we chose the highlighted first unity entry for identifying the key column. Accordingly, the pivot element is identified as unity. So X_1 enters in and d_1^- leaves the basis. The necessary Simplex calculations are reported in Tableau 2.

Table 17.6

Tableau 2									
C_i	Basis	X_1	X_2	d_1^-	d_2^-	d_3^-	d_1^+	X_b	MR
P_1	d_1^-	0	1	1	−1	0	−1	160	160/1
0	X_1	1	0	0	1	0	0	240	
P_2	d_3^-	0	1	0	0	1	0	300	300/1
C_j		0	0	P_1	$2P_2$	P_2	P_3	0	
$Z_j–C_j$	P_3	0	0	0	0	0	−1		
	P_2	0	1	0	−2	0	0		
	P_1	0	1	0	−1	0	−1		

In Tableau 2 the relevant pivot element is identified as highlighted unity in the first priority row. So this time X_2 enters in and d_1^- leaves the basis. The necessary Simplex calculations are reported in Tableau 3.

In Tableau 3 in the first priority index row all the entries are zero or negative. So let us move to the second priority index row.

Table 17.7

Tableau 3									
C_i	Basis	X_1	X_2	d_1^-	d_2^-	d_3^-	d_1^+	X_b	MR
0	X_2	0	1	1	−1	0	−1	400	-
0	X_1	1	0	0	1	0	0	240	-
P_2	d_3^-	0	0	−1	1	1	1	300	300/1
C_j		0	0	P_1	$2P_2$	P_2	P_3	0	
$Z_j–C_j$	P_3	0	0	0	0	0	−1		
	P_2	0	0	−1	−1	0	1		
	P_1	0	0	−1	0	0	0		

In it the only positive unit entry constitutes the key column. After identifying the highlighted unit pivot element we make the necessary simplex calculations and report it in Tableau 4. This time, remember d_1^+ enters in and d_3^- leaves the basis.

Table 17.8

Tableau 4									
C_i	Basis	X_1	X_2	d_1^-	d_2^-	d_3^-	d_1^+	X_b	MR
0	X_2	0	1	1	0	1	0	300	
0	X_1	1	0	0	1	0	0	240	
P_3	d_1^+	0	0	−1	1	1	1	140	
C_j		0	0	P_1	$2P_2$	P_2	P_3	0	
$Z_j–C_j$	P_3	0	0	−1	1	1	0		
	P_2	0	0	0	−2	−1	0		
	P_1	0	0	−1	0	0	0		

In Tableau 4 the first and the second priority index rows all the entries are either zero or negative. So we confirm that the first and the second priority goals are fully achieved. However, as far as the third priority row is concerned there are two positive unity entries in the index row suggesting the inclusion of either d_2^- or d_3^-. Such an inclusion is not possible because at higher priority level (P_2) these deviation variables are already positive. So the solution values are: $X_1 = 240$, $X_2 = 300$, $d_1^- = d_2^- = d_3^- = 0$, $d_1^+ = 140$.

Example 13

Solve the following Goal Programming by using modified Simplex method

$$\text{Minimize:} \quad P_1 d_1^- + P_2(2d_2^- + d_3^-) + P_3 d_1^+$$

$$\text{Subject to:} \quad X_1 + X_2 - d_1^+ + d_1^- = 300$$

$$X_1 + d_2^- = 200$$

$$X_2 + d_3^- = 200$$

$$X_1\ ;\ X_2\ ;\ d_1^+\ ;\ d_1^-\ ;\ d_2^-\ ;\ d_3^- \geq 0$$

In our modified deviation introduced statement of the goal programming all the negative deviation variables have positive unity coefficients. Such a presentation allow us to start the simplex Tableau 1 with all the negative deviation variables having unit coefficients in the basis column as shown in Tableau 1. Since this is a minimization problem we calculate $Z_j–C_j$ in the index row for each and every goal separately and present them all as shown in Tableau 1 starting from the first priority goal at the bottom. The needed calculations are shown below.

$$Z_1 - C_1 = (P_1 \times 1 + 2P_2 \times 1 + P_2 \times 0) - 0 = P_1 + 2P_2$$

$$Z_2 - C_2 = (P_1 \times 1 + 2P_2 \times 0 + P_2 \times 1) - 0 = P_1 + P_2$$

$$Z_3 - C_3 = (P_1 \times 1 + 2P_2 \times 0 + P_2 \times 0) - p_1 = 0$$

$$Z_4 - C_4 = (P_1 \times 0 + 2P_2 \times 1 + P_2 \times 0) - 2p_2 = 0$$

$$Z_5 - C_5 = (P_1 \times 0 + 2P_2 \times 0 + P_2 \times 1) - p_2 = 0$$

$$Z_6 - C_6 = (P_1 \times -1 + 2P_2 \times 0 + P_2 \times 0) - p_3 = -P_3 - P_1$$

Table 17.9

Tableau 1									
C_i	Basis	X_1	X_2	d_1^-	d_2^-	d_3^-	d_1^+	X_b	MR
P_1	d_1^-	1	1	1	0	0	−1	300	300
$2P_2$	d_2^-	1	0	0	1	0	0	200	200
P_2	d_3^-	0	1	0	0	1	0	240	–
C_j		0	0	P_1	$2P_2$	P_2	P_3	0	
$Z_j–C_j$	P_3	0	0	0	0	0	−1	0	
	P_2	2	1	0	0	0	0	640	
	P_1	1	1	0	0	0	−1	300	

Now let us give priority to our first priority goal index row given at the bottom in Tableau 1. In it there are two positive unity entries. So arbitrarily we choose the highlighted first unity entry for identifying the key column. Accordingly, the pivot element is identified as unity. So X_1 enters in and d_1^- leaves the basis. The necessary simplex calculations are reported in Tableau 2.

Table 17.10

Tableau 2									
C_i	Basis	X_1	X_2	d_1^-	d_2^-	d_3^-	d_1^+	X_b	MR
P_1	d_1^-	0	1	1	−1	0	−1	100	100
0	X_1	1	0	0	1	0	0	200	–
P_2	d_3^-	0	1	0	0	1	0	240	240
C_j		0	0	P_1	$2P_2$	P_2	P_3	0	
$Z_j–C_j$	P_3	0	0	0	0	0	−1	0	
	P_2	0	1	0	−2	0	0	240	
	P_1	0	1	0	−1	0	−1	100	

In Tableau 2 the relevant pivot element is identified as highlighted unity in the first priority row. So this time X_2 enters in and d_1^- leaves the basis. The necessary simplex calculations are reported in Tableau 3.

Table 17.11

Tableau 3									
C_i	Basis	X_1	X_2	d_1^-	d_2^-	d_3^-	d_1^+	X_b	MR
0	X_2	0	1	1	−1	0	−1	100	−100
0	X_1	1	0	0	1	0	0	100	–
P_2	d_3^-	0	0	−1	1	1	1	240	240
C_j		0	0	P_1	$2P_2$	P_2	P_3	0	
Z_j-C_j	P_3	0	0	0	0	0	−1	0	
	P_2	0	0	−1	−1	0	1	240	
	P_1	0	0	−1	0	0	0	0	

In Tableau 3 in the first priority index row all the entries are zero or negative. So let us move to the second priority index row. In it the only positive unit entry constitutes the key column. After identifying the highlighted unit pivot we make the necessary simplex calculations are reported in Tableau 4. This time remember d_1^+ enters in and d_3^- leaves the basis.

Table 17.12

Tableau 4									
C_i	Basis	X_1	X_2	d_1^-	d_2^-	d_3^-	d_1^+	X_b	MR
0	X_2	0	1	1	0	1	0	100	
0	X_1	1	0	0	1	0	0	100	
P_3	d_1^+	0	0	−1	1	1	1	240	
C_j		0	0	P_1	$2P_2$	P_2	P_3	0	
Z_j-C_j	P_3	0	0	−1	1	1	0	240	
	P_2	0	0	0	−2	−1	0	0	
	P_1	0	0	−1	0	0	0	0	

In Tableau 4 the first and the second priority index rows all the entries are either zero or negative. So we confirm that the first and the second priority goals are fully achieved. However, as far as the third priority row is concerned there are two positive unity entries in the index row suggesting the inclusion of either d_2^- or d_3^-. Such an inclusion is not possible because at higher priority level (P_2) these deviation variables are already positive. So the solution values are: $X_1 = 100$, $X_2 = 100$, $d_1^- = d_2^- = d_3^- = 0$, $d_1^+ = 240$.

Example 14

Solve the following Goal Programming by using modified Simplex method.

Minimize: $P_1d_1^- + P_2(d_2^+ + d_2^-) + P_3d_3^-$

Subject to:
$$8X_1 + 6X_2 - d_1^+ + d_1^- = 36$$
$$X_1 - 2X_2 - d_2^+ + d_2^- = 0$$
$$2X_1 + 4X_2 + d_3^- = 20$$
$$3X_1 + 4X_2 + d_4^- = 24$$
$$X_1 \; ; \; X_2 \; ; d_1^+ \; ; d_1^- \; ; d_2^+ \; ; d_2^- \; ; d_3^-; \; d_4^- \geq 0$$

Solution

In our modified deviation introduced statement of the goal programming all the negative deviation variables are having positive unity coefficients. Such a presentation allows us to start the Simplex Tableau 1 with all the negative deviation variables having unit coefficients in the basis column as shown in Tableau 1. Since this is minimization problem we calculate $Z_j - C_j$ in the index row for each and every goal separately and present them all as shown in Tableau 1 starting from the first priority goal at the bottom. The needed calculations are shown below.

Table 17.13

Tableau 1											
C_i	Basis	X_1	X_2	d_1^+	d_1^-	d_2^+	d_2^-	d_3^-	d_4^-	X_b	M.R
P_1	d_1^-	8	6	−1	1	0	0	0	0	36	4.5
P_2	d_2^-	1	−2	0	0	−1	1	0	0	0	0
P_3	d_3^-	2	4	0	0	0	0	1	0	20	10
0	d_4^-	3	4	0	0	0	0	0	1	24	8
C_j		0	0	0	P_1	P_2	P_2	P_3	0	0	
$Z_j - C_j$	P_3	2	4	0	0	0	0	0	0		
	P_2	1	−2	0	0	−1	0	0	0		
	P_1	8	6	−1	0	0	0	0	0		

In Tableau 1, $Z_1 - C_1 = 8P_1 + P_2 + 2P_2 - 0 = 8P_1 + P_2 + 2P_2$. The coefficients are P_1, P_2 and P_3 are shown in Tableau 1 in the index row according to their priorities. Similarly, all other index row calculations are made and reported in the Tableau 1. Since in the first goal priority row, 8 is the largest positive number we insert X_1 into the basis displacing d_2^- and carry out the next iteration calculations as usual. The so obtained results are shown in Tableau 2.

Table 17.14

Tableau 2											
C_i	Basis	X_1	X_2	d_1^+	d_1^-	d_2^+	d_2^-	d_3^-	d_4^-	X_b	M.R
P_1	d_1^-	0	22	−1	1	8	−8	0	0	36	1.636364
0	X_1	1	−2	0	0	−1	1	0	0	0	0
P_3	d_3^-	0	8	0	0	2	−2	1	0	20	2.5
0	d_4^-	0	10	0	0	3	−3	0	1	24	2.4
C_j		0	0	0	P_1	P_2	P_2	P_3	0	0	
$Z_j - C_j$	P_3	0	8	0	0	2	−2	0	0		
	P_2	0	0	0	0	−1	−1	0	0		
	P_1	0	22	−1	0	8	−8	0	0		

Since P_1 is the first priority in the last row the coefficient 22, we insert X_2 into the basis displacing d_1^- and carry out the iteration as usual. The so obtained results are shown in Tableau 3.

Table 17.15

Tableau 3										
C_i	Basis	X_1	X_2	d_1^+	d_1^-	d_2^+	d_2^-	d_3^-	d_4^-	X_b
P_1	X_2	0	1	−1/22	1/22	8/22	−8/22	0	0	36/22
P_2	X_1	1	0	−2/22	2/22	−6/22	6/22	0	0	72/22
P_3	d_3^-	0	0	8/22	−8/22	−20/22	20/22	1	0	152/22
0	d_4^-	0	0	10/22	−10/22	−14/22	14/22	0	1	168/22
C_j		0	0	0	P_1	P_2	P_2	P_3	0	0
$Z_j - C_j$	P_3	0	0	8/22	−1	−20/22	20/22	0	0	
	P_2	0	0	0	0	−1	−1	0	0	
	P_1	0	0	0	−8/22	0	0	0	0	

Since P_1 and P_2 rows the results are zero or negative this time we look row corresponding to P_3. In this row since the coefficient 8/22 is the largest we insert d_1^+ into the basis displacing d_4^- and carry out the iteration as usual. The so obtained results are shown in Tableau 4.

Table 17.16

Tableau 4										
C_i	Basis	X_1	X_2	d_1^+	d_1^-	d_2^+	d_2^-	d_3^-	d_4^-	X_b
P_1	X_2	0	1	0	0	3/20	−3/20	0	1/10	12/5
P_2	X_1	1	0	0	0	−2/5	2/5	0	1/5	24/5
P_3	d_3^-	0	0	0	0	−2/5	2/5	1	−4/5	4/5
0	d_4^-	0	0	1	−1	−14/5	14/10	0	22/10	84/5
C_j		0	0		P_1	P_2	P_2	P_3	0	0
$Z_j - C_j$	P_3	0	0	0	0	−2/5	2/5	0	−4/5	
	P_2	0	0	0	0	−1	−1	0	0	
	P_1	0	0	0	1	0	0	0	0	

Thus $Z_j - C_j$ are less than or equal to zero at the third iteration. Note that for the vector d_2^- the $Z_j - C_j = -P_2 + 2/5P_3$ because P_2 dominates the final value will be negative. So the optimum solutions are $X_1 = 24/5 = 4.8$, $X_2 = 12/5 = 2.4$. The profit is $264/5 = 52.8$. Thus the profit goal of 36 is over-achieved by Rs. 16.8 (84/5). Goal 2 is achieved fully so that $d_2^+ = d_2^- = 0$. Goal 3 has not been achieved in the sense that $d_3^- = 0.8$. In the diagrammatic solution discussed earlier, the highest achievable point E reveals the same result.

Excel Solution Pre-Emptive Goal Models

Table 17.17

Inputs	X_1	X_2	d_1^+	d_1^-	d_2^+	d_2^-	d_3^-	d_4^-	
1	8	6	−1	1	0	0	0	0	36
2	1	−2	0	0	−1	1	0	0	0
3	2	4	0	0	0	0	1	0	20
4	3	4	0	0	0	0	0	1	24
			400	40	40	1			
	Decision Variables								
	X_1	X_2	d_1^+	d_1^-	d_2^+	d_2^-	d_3^-	d_4^-	
	4.8	2.4	16.8	0	0	0	0.8	0	
Min:	0								
Constraints	Used (LHS)								Available (RHS)
1	36	=							36
2	0	=							0
3	20	=							20
4	24	=							24

17.5 NON-PRE-EMPTIVE GOAL PROGRAMMING WITH MULTIPLE GOALS

In non-pre-emptive goal programming we do not order the objectives on priority basis. Instead we state each and every objective by numerical numbers. Thus the question of considering the goals on priority basis does not arise in this case. Once the goals are stated in numerical numbers, by giving them equal importance to all the goals under reference, we seek a solution by minimizing the sum of the deviations of the goal targets from their respective objectives. For this type of non-pre-emptive programming it is assumed that the decision maker has a linear utility function with respect to the objective functions.

Sometimes instead of giving equal importance we give varying weights called *penalties* in accordance with their relative importance. In such cases we minimize the weighted sum of the deviations. The weights used in a goal programming are indicative of the decision maker's utility on various objectives listed. The following example illustrates a non-pre-emptive goal programming with equal importance.

Example 15

The production manager of a company wants to schedule a week's production run for two products A and B. Each one of them requires labour and two raw materials in a fixed proportion as given in the following Table 17. 10.

Table 17.18

Inputs	Product		Maximum resource availability
	A	B	
Labour	2	4	600
Material 1(kg)	4	5	1000
Material 2(kg)	5	4	1200

These two products A and B are newer version of certain old varieties and will be priced at Rs. 20 and Rs. 32 respectively in the market. The manager is no doubt interested in maximizing the profit of the new venture but at the same also interested in maintaining the current of work force of different kind intact having the long run in mind. The research team tells the manager that one unit production of A will retain 0.3 unit of labour and one unit production of B will retain 0.75 units of labour.

Had the production manager considered only profit maximization under simple linear programming model he could have earned Rs. 5486.67 and restored only 100.3 units of labour force. To avoid this situation the manager is ready to limit his profit to Rs. 5400, At the same time he wants to retain 108 units of the labour force intact. Formulate a suitable goal program and solve it by graphical method. Also confirm the so obtained results by using Simplex method.

Solution

Let X_1 and X_2 stand for the output levels of the products A and B respectively. The relevant goals and constants of the problem are formulated as shown below.

$$20X_1 + 32X_2 = 5400 \quad \text{(Goal 1)}$$

$$0.3X_1 + 0.75X_2 = 108 \quad \text{(Goal 2)}$$

$$2X_1 + 4X_2 \leq 600 \quad \text{(Labour)}$$

$$4X_1 + 5X_2 \leq 1000 \quad (M_1)$$

$$5X_1 + 4X_2 \leq 1200 \quad (M_2)$$

If d_1^- = the number of rupees below the profit goal of Rs. 5400 and d_1^+ = the number rupees above the profit goal of Rs. 5400 the relevant profit constraint may be written as

$$20X_1 + 32X_2 + d_1^- - d_1^+ = 5400$$

If d_2^- = the number of people below the workforce goal of 108 and d_2^+ = the number of people above the workforce goal of 108 the relevant workforce constraint may be written as

$$0.3X_1 + 0.75X_2 + d_2^- - d_2^+ = 108$$

Now, considering the nature of goals, it may be presumed that over-achievement of either of them would attract no penalty. Thus we propose to find solution which minimizes the under-achievement variables for both the goals. Thus the suitable goal programming problem in its complete form may be written as

Minimize: $Z = d_1^- + d_2^-$

Subject to:
$$20X_1 + 32X_2 + d_1^- - d_1^+ = 5400$$
$$0.3X_1 + 0.75X_2 + d_2^- - d_2^+ = 108$$
$$2X_1 + 4X_2 + S_1 = 600$$
$$4X_1 + 5X_2 + S_2 = 1000$$
$$5X_1 + 4X_2 + S_3 = 1200$$
$$X_1; X_2; S_1; S_2; S_3; d_1^-; d_1^+; d_2^-; d_2^+ \geq 0$$

In goal programming remember the objective functions are also written as constraints and introduced in the graph. Remember, in such converted constraints the needed unit coefficients to start the simplex method are provided by under-achievement deviation variables d_1^- and d_2^- hence no need to bring in artificial variables. So in our first simplex tableau 1 three slacks relating to resource constraints and two negative deviation variables relating to objective constraints are entered in the basis column.

Minimize: $Z = d_1^- + d_2^-$

Subject to:
$$20X_1 + 32X_2 + d_1^- - d_1^+ = 5400$$
$$0.3X_1 + 0.75X_2 + d_2^- - d_2^+ = 108$$
$$2X_1 + 4X_2 + S_1 = 600$$
$$4X_1 + 5X_2 + S_2 = 1000$$
$$5X_1 + 4X_2 + S_3 = 1200$$
$$X_1, X_2, S_1, S_2, S_3, d_1^-, d_1^+, d_2^-, d_2^+ \geq 0$$

Table 17.19

Tableau 1												
C_i	Basis	X_1	X_2	d_1^-	d_1^+	d_2^-	d_2^+	S_1	S_2	S_3	X_b	M.Ratio
1	d_1^-	20	32	1	−1	0	0	0	0	0	5400	168.75
1	d_2^-	0.3	0.75	0	0	1	−1	0	0	0	108	144
0	S_1	2	4	0	0	0	0	1	0	0	600	150
0	S_2	4	5	0	0	0	0	0	1	0	1000	200
0	S_3	5	4	0	0	0	0	0	0	1	1200	300
	C_j	0	0	1	0	1	0	0	0	0		
	Z_j	20.3	32.8	1	−1	1	−1	0	0	0	5508	
	$Z_j–C_j$	20.3	32.8	0	−1	0	−1	0	0	0		

Table 17.20

Tableau 2												
C_i	Basis	X_1	X_2	d_1^-	d_1^+	d_2^-	d_2^+	S_1	S_2	S_3	X_b	M.Ratio
1	d_1^-	7.2	0	1	−1	−42.7	42.7	0	0	0	792	18.5625
0	X_2	0.4	1	0	0	1.33	−1.33	0	0	0	144	−108
0	S_1	0.4	0	0	0	−5.33	5.33	1	0	0	24	4.5
0	S_2	2	0	0	0	−6.67	6.67	0	1	0	280	42
0	S_3	3.4	0	0	0	−5.33	5.33	0	0	1	624	117
	C_j	0	0	1	0	1	0	0	0	0		
	Z_j	7.2	0	1	−1	−42.7	42.7	0	0	0	792	
	$Z_j–C_j$	7.2	0	0	−1	−43.7	42.7	0	0	0		

Table 17.21

Tableau 3												
C_i	Basis	X_1	X_2	d_1^-	d_1^+	d_2^-	d_2^+	S_1	S_2	S_3	X_b	M.Ratio
1	d_1^-	4	0	1	−1	0	0	−8	0	0	600	150
0	X_2	0.5	1	0	0	0	0	0.25	0	0	150	300
0	d_2^+	0.075	0	0	0	−1	1	0.19	0	0	4.5	60
0	S_2	1.5	0	0	0	0	0	−1.25	1	0	250	166.6667
0	S_3	3	0	0	0	0	0	−1	0	1	600	200
	C_j	0	0	1	0	1	0	0	0	0		
	Z_j	4	0	1	−1	0	0	−8	0	0	600	
	$Z_j–C_j$	4	0	0	−1	−1	0	−8	0	0		

Table 17.22

Tableau 4												
C_i	Basis	X_1	X_2	d_1^-	d_1^+	d_2^-	d_2^+	S_1	S_2	S_3	X_b	M.Ratio
1	d_1^-	0	0	1	−1	53.3	−53.3	−18	0	0	360	6.75
0	X_2	0	1	0	0	6.67	−6.67	−1	0	0	120	18
0	X_1	1	0	0	0	−13.3	13.3	2.5	0	0	60	−4.5
0	S_2	0	0	0	0	20	−20	−5	1	0	160	8
0	S_3	0	0	0	0	40	−40	−8.5	0	1	420	10.5
	C_j	0	0	1	0	1	0	0	0	0		
	Z_j	0	0	1	−1	53.3	−53.3	−18	0	0	360	
	$Z_j–C_j$	0	0	0	−1	52.3	−53.3	−18	0	0		

Table 17.23

Tableau 5												
C_i	Basis	X_1	X_2	d_1^-	d_1^+	d_2^-	d_2^+	S_1	S_2	S_3	X_b	M.Ratio
1	d_2^-	0	0	0.02	−0.02	1	−1	−0.34	0	0	6.75	
0	X_2	0	1	−0.13	0.13	0	0	1.25	0	0	75	
0	X_1	1	0	0.25	−0.25	0	0	−2	0	0	150	
0	S_2	0	0	−0.38	0.38	0	0	1.75	1	0	25	
0	S_3	0	0	−0.75	0.75	0	0	5	0	1	150	
	C_j	0	0	1	0	1	0	0	0	0		
	Z_j	0	0	0.02	−0.02	1	−1	−0.34	0	0	6.75	
	$Z_j–C_j$	0	0	−0.98	−0.02	0	−1	−0.34	0	0		

So the solution values are:

$X_1 = 150$; $X_2 = 75$; $d_2^- = 6.75$; employment $= 108 - 6.75 = 101.25$; $R = 5400$.

In the example illustrated above both the goals were given equal importance. A rupee under-achievement in the profit goal treated equal to one person under-achievement in employment goal clearly, this is an unfair method indeed. A 5 rupees under-achievement would imply 5/5400 = 0.09% while under-achievement of 5 person in workforce would imply a 5/108 = 4.63%. To avoid such glaring mistake often we use weightage while deciding the goal priorities. For example, if the manager wants to give up a sum of Rs. 100 for restoring one person in workforce then the relevant weighted objective function may be written as

Minimize: $Z = d_1^- + 100d_2^-$

Now to see the effect of this modification let us work out the problem once again by considering our new weighted objective function.

Minimize: $Z = d_1^- + 100d_2^-$

Subject to:

$$20X_1 + 32X_2 + d_1^- - d_1^+ = 5400$$
$$0.3X_1 + 0.75X_2 + d_2^- - d_2^+ = 108$$
$$2X_1 + 4X_2 + S_1 = 600$$
$$4X_1 + 5X_2 + S_2 = 1000$$
$$5X_1 + 4X_2 + S_3 = 1200$$
$$X_1; X_2; S_1; S_2; S_3; d_1^-; d_1^+; d_2^-; d_2^+ \geq 0$$

Table 17.24

Tableau 1												
C_i	Basis	X_1	X_2	d_1^-	d_1^+	d_2^-	d_2^+	S_1	S_2	S_3	X_b	M.Ratio
1	d_1^-	20	32	1	−1	0	0	0	0	0	5400	168.75
100	d_2^-	0.3	0.75	0	0	1	−1	0	0	0	108	144
0	S_1	2	4	0	0	0	0	1	0	0	600	150
0	S_2	4	5	0	0	0	0	0	1	0	1000	200
0	S_3	5	4	0	0	0	0	0	0	1	1200	300
	C_j	0	0	1	0	100	0	0	0	0		
	Z_j	50	107	1	−1	100	−100	0	0	0	16200	
	$Z_j–C_j$	50	107	0	−1	0	−100	0	0	0		

Table 17.25

Tableau 2												
C_i	Basis	X_1	X_2	d_1^-	d_1^+	d_2^-	d_2^+	S_1	S_2	S_3	X_b	M.Ratio
1	d_1^-	7.2	0	1	−1	−42.7	42.7	0	0	0	792	18.5625
0	X_2	0.4	1	0	0	1.33	−1.33	0	0	0	144	−108
0	S_1	0.4	0	0	0	−5.33	5.33	1	0	0	24	4.5
0	S_2	2	0	0	0	−6.67	6.67	0	1	0	280	42
0	S_3	3.4	0	0	0	−5.33	5.33	0	0	1	624	117
	C_j	0	0	1	0	100	0	0	0	0		
	Z_j	7.2	0	1	−1	−42.7	42.7	0	0	0	792	
	$Z_j–C_j$	7.2	0	0	−1	−143	42.7	0	0	0		

Table 17.26

Tableau 3													
C_i	Basis	X_1	X_2	d_1^-	d_1^+	d_2^-	d_2^+	S_1	S_2	S_3	X_b	M.Ratio
1	d_1^-	4	0	1	−1	0	0	−8	0	0	600	150
0	X_2	0.5	1	0	0	0	0	0.25	0	0	150	300
0	d_2^+	0.075	0	0	0	−1	1	0.19	0	0	4.5	60
0	S_2	1.5	0	0	0	0	0	−1.25	1	0	250	166.667
0	S_3	3	0	0	0	0	0	−1	0	1	600	200
	C_j	0	0	1	0	100	0	0	0	0		
	Z_j	4	0	1	−1	0	0	−8	0	0	600	
	$Z_j–C_j$	4	0	0	−1	−100	0	−8	0	0		

Table 17.27

Tableau 4													
C_i	Basis	X_1	X_2	d_1^-	d_1^+	d_2^-	d_2^+	S_1	S_2	S_3	X_b	M.Ratio
1	d_1^-	0	0	1	−1	53.3	−53.3	−18	0	0	360	
0	X_2	0	1	0	0	6.67	−6.67	−1	0	0	120	
0	X_1	1	0	0	0	−13.3	13.3	2.5	0	0	60	
0	S_2	0	0	0	0	20	−20	−5	1	0	160	
0	S_3	0	0	0	0	40	−40	−8.5	0	1	420	
	C_j	0	0	1	0	100	0	0	0	0		
	Z_j	0	0	1	−1	53.3	−53.3	−18	0	0	360	
	$Z_j–C_j$	0	0	0	−1	−46.7	−53.3	−18	0	0		

The new solution values are:

$X_1 = 60$, $X_2 = 120$, $d_1^- = 360$, total profit = 5400–360 = Rs. 5040. Since $d_2^- = 0$ the desired employment goal is fully achieved now.

■ Example 16

Suppose a producer is producing two products in X_1 and X_2 quantities by using two inputs–labour and capital. Each unit of the product one requires 2 hours of labour and 3 units of capital. Each unit of the second product requires 3 hours of labour and 2 units of capital. For everyday use 1500 hours of labour and 1500 units of capital are made available. The sum of these two products should not exceed 600 units. These products are sold in the market at Rs. 10 and Rs. 12 respectively. Considering this as a simple linear programming problem find X_1 and X_2 that maximizes the profit.

Solution

$$\text{Maximize}: \quad 10X_1 + 12X_2$$

$$2X_1 + 3X_2 \le 1500 \quad \text{(Labour)}$$

$$3X_1 + 2X_2 \le 1500 \quad \text{(Capital)}$$

$$X_1 + X_2 \le 600 \quad \text{(Output)}$$

Graphical solution

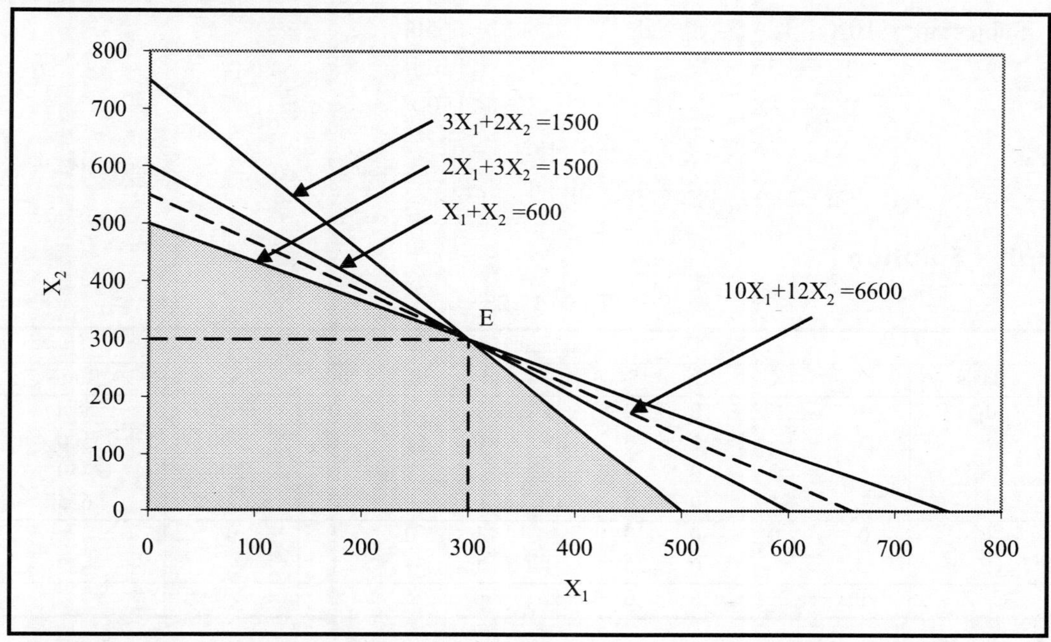

Figure 17.15

$X_1 = 300; X_2 = 300: R = 6600$

■ Example 17

In the above example 16 if the management wants to have a profit of Rs. 6650 and joint output level of 625 units as first and second equally ranked goals obtain the best possible solution.

Solution

For the first two goal constraints let us introduce $d_1^+, d_1^-, d_2^+, d_2^-$ as deviation variables. So the relevant constraints in deviation forms are written as.

$$
\begin{aligned}
10X_1 + 12X_2 - d_1^+ + d_1^- &= 6650 \\
2X_1 + 3X_2 &\leq 1500 \\
3X_1 + 2X_2 &\leq 1500 \\
X_1 + X_2 \quad - d_2^+ + d_2^- &= 625 \\
X_1 \; ; \; X_2 \; ; \; d_1^+ \; ; \; d_1^- \; ; d_2^+ \; ; d_2^- &\geq 0
\end{aligned}
$$

Objective function: Inclusive of the objective function the complete problem may be written as

$$\text{Minimize: } d_1^- + d_1^+ + d_2^- + d_2^+$$

$$\begin{aligned} \text{Subject to: } & 10X_1 + 12X_2 - d_1^+ + d_1^- && = 6650 \\ & 2X_1 + 3X_2 && \leq 1500 \\ & 3X_1 + 2X_2 && \leq 1500 \\ & X_1 + X_2 \quad - d_2^+ + d_2^- && = 625 \\ & X_1 ; X_2 ; d_1^+ ; d_1^- ; d_2^+ ; d_2^- && \geq 0 \end{aligned}$$

1. Simplex Solution

Table 17.28

Tableau 1											
C_i	Basis	X_1	X_2	S_1	S_2	d_1^+	d_1^-	d_2^-	d_2^-	X_b	Ratio
1	d_1^-	10	12	0	0	−1	1	0	0	6650	554.17
0	S_1	2	3	1	0	0	0	0	0	1500	500
0	S_2	3	2	0	1	0	0	0	0	1500	750
1	d_2^-	1	1	0	0	0	0	−1	1	625	625
	C_j	0	0	0	0	1	1	1	1	0	
	Z_j	11	13	0	0	−1	1	−1	1		
	$C_j–Z_j$	−11	−13	0	0	2	0	2	0		

Table 17.29

Tableau 2											
C_i	Basis	X_1	X_2	S_1	S_2	d_1^+	d_1^-	d_2^-	d_2^-	X_b	Ratio
1	d_1^-	2	0	−4	0	−1	1	0	0	650	325
0	X_2	2/3	1	1/3	0	0	0	0	0	500	750
0	S_2	5/3	0	−2/3	1	0	0	0	0	500	300
1	d_2^-	1/3	0	−1/3	0	0	0	−1	1	125	375
	C_j	0	0	0	0	1	1	1	1		
	Z_j	7/3	0	−13/3	0	−1	1	−1	1		
	$C_j–Z_j$	−7/3	0	13/3	0	2	0	2	0		

Table 17.30

Tableau 3											
C_i	Basis	X_1	X_2	S_1	S_2	d_1^+	d_1^-	d_2^-	d_2^-	X_b	Ratio
1	d_1^-	0	0	−16/5	−6/5	−1	1	0	0	50	
0	X_2	0	1	3/5	−2/5	0	0	0	0	300	
0	X_1	1	0	−2/5	3/5	0	0	0	0	300	
1	d_2^-	0	0	−1/5	−1/5	0	0	1	1	25	
	C_j	0	0	0	0	1	1	1	1		
	Z_j	0	0	−17/5	−7/5	−1	1	−1	1		
	$C_j–Z_j$	0	0	17/5	7/5	2	0	2	0		

$X_1 = 300$, $X_2 = 300$, $d_1^- = 50$, $d_2^- = 25$; R = 6650−50 = 6600; Output = 625−25 = Rs. 600.

2. Excel Solver Solution

Table 17.31

	Data						
Constraints	X_1	X_2	d_1^+	d_1^-	d_2^+	d_2^-	RHS
1	10	12	−1	1	0	0	6650
2	2	3	0	0	0	0	1500
3	3	2	0	0	0	0	1500
4	1	1	0	0	−1	1	625
	Decision Variables						
	X_1	X_2	d_1^+	d_1^-	d_2^+	d_2^-	
	300	300	0	50	0	25	
Minimize	75						
Constraints	LHS						RHS
1	6650	=					6650
2	1500	<=					1500
3	1500	<=					1500
4	625	=					625

Thus, the final solutions are $X_1 = 300$, $X_2 = 300$, $d_1^- = 50$, $d_2^- = 25$.

3. Graphical Solution

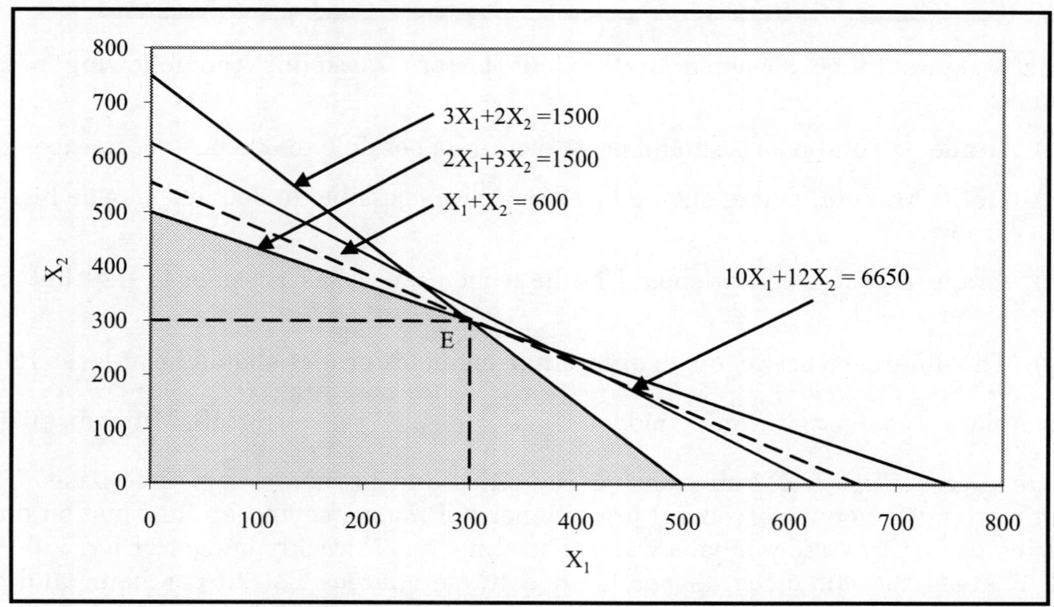

Figure 17.16

$X_1 = 300$, $X_2 = 300$, $d_1^- = 50$, $d_2^- = 25$; $R = 6650-50 = 6600$; Output $= 625-25 = 600$. Note that this is indeed our simple linear programming solution.

EXERCISES

1. "In goal programming, we accept to 'satisfy' or come as close as possible to satisfying, the various goals." Discuss.

2. "Goal programming appears to be the most appropriate, flexible and powerful quantitative technique for complex decision problems involving conflicting objectives." Elucidate.

3. Distinguish between non-pre-emptive and pre-emptive goal programming problems with illustrations.

4. You have been retained as a consultant to formulate a model for job evaluation in an organization. The data collected for 5 representative benchmark jobs across the hierarchy of all the blue collar jobs evaluated on a 5 point scale four factors viz. skill, effort, working condition, and responsibility.

Job grade	Skill	Effort	Work conditions	Responsibility
A	4	5	5	5
C	4	4	4	4
M	5	4	3	3
D	2	3	2	2
Z	1	2	2	1

Find weights to be assigned to the four factors satisfying the following broad guidelines:

(i) Grade A: Total points should be as nearer as possible to 600 but not in excess.

(ii) Grade M: Total points should be as nearer as possible to 360 but may be higher or lower.

(iii) Grade Z: Total points should be at least as near as possible to 120 but not lower.

(iv) The difference between two different benchmark grades should be at least 120.

Formulate a goal programming model. (MMS, Mumbai, 2000)

5. A rural clinic hires its staff from nearby cities and towns on part time basis. The clinic attempts to have a general practitioner (GP), a nurse and an internist on duty during at last a portion of each week. The clinic has a weekly budget of Rs. 1200. A GP charges the clinic Rs. 49 per hour, a nurse charges Rs. 20 per hour, and an internist charges Rs. 150. The clinic has established the following goals in order of priority.

(i) A nurse should be available for at least for 30 hours per week.

(ii) The weekly budget of Rs. 1200 should not be exceeded.

(iii) A GP or internist should be available at least 20 hours per week

(iv) An internist should be available at least 6 hours per week

Formulate a goal programming model for determining the number of hours to hire each staff member in order to satisfy the various goals. (MBA, Delhi, March 2004)

6. A farmer has 1000 acres of land on which he intends to plant corn, wheat, and soybean. Each acre of corn costs Rs. 10,000 for preparation, requires 7 working days of labour, and yields a profit of Rs. 3000. An acre of wheat costs Rs. 12,000 to prepare, requires 10 working days of labour, and yields Rs. 4000 profit. An acre of soybean costs Rs. 7000 to prepare, requires 8 working days of labour, and yields Rs. 2000 profit. The farmer has taken a loan of Rs. 80,000 for crop preparation and has contracted with a union for 6000 work days of labour. A granary has agreed to purchase 200 acres of corn, 500 acres of wheat, and 300 acres of soybean the farmer has established the following goals, in order of their importance.

(i) In order to maintain good relation with the union, the labour contract must be honoured: that is the 6000 must be used.

(ii) Preparation cost should not exceed the loan amount so that additional finance will not be secured.

(iii) The farmer desires a profit of at least Rs. 1,05,00,000 in order to remain in good financial position.

(iv) Contracts for excess labour should be avoided.

(v) The farmer would like to use as much of the available acreage as possible.

(vi) The farmer would like to meet the sales with the granary. However, the goal should be weighted according to the profit returned by each group.

Formulate a goal programming model for determining the number of acres of each crop the farmer should plant in order to satisfy the goals in the best possible way. (MBA, Delhi, March 2003)

7. Consider the goal programming model:

$$\text{Minimize: } Z = P_1 + d_1^+ + P_1 d_2^+ + 30P_2 d_3^+ + 40P_2 d_4^- + P_3 d_5^-$$

$$\begin{aligned} \text{Subject to: } & 2X_1 + 4X_2 + d_1^- - d_1^+ = 80 \\ & 3X_1 + 3X_2 + d_2^- - d_2^+ = 80 \\ & X_1 \quad d_3^- - d_3^+ = 10 \\ & X_2 \quad +d_4^- - d_4^+ = 10 \\ & 30X_1 + 40X_2 + d_5^- - d_5^+ = 1200 \end{aligned}$$

All variables are to be non-negative.

The modified Simplex tableau giving IBFS is shown below. Carry out the Simplex iterations until an optimum solution is reached. State clearly the optimal solution and to what extent the goals have been achieved.

			0	0	0	P_1	0	P_1	$30P_2$	0	$40P_3$	0	P_3	0
C_b	X_b	b	X_1	X_2	d_1^-	d_1^+	d_2^-	d_2^+	d_3^-	d_3^+	d_4^-	d_4^+	d_5^-	d_5^+
0	d_1^-	80	2	4	1	−1	0	0	0	0	0	0	0	0
0	d_2^-	80	3	3	0	0	1	−1	0	0	0	0	0	0
$30P_2$	d_3^-	10	1	0	0	0	0	0	1	−1	0	0	0	0
$40P_2$	d_4^-	10	0	1	0	0	0	0	0	0	1	−1	0	0
P_1	d_5^-	1200	30	40	0	0	0	0	0	0	0	0	1	−1
	P_3	1200	−30	−40	0	0	0	0	0	0	0	0	0	1
C_j-Z_j	P_2	700	−30	−40	0	0	0	0	0	30	0	40	0	0
	P_1	0	0	0	0	1	0	1	0	0	0	0	0	0

(MMS, Mumbai, 2001)

8. A super star company produces three models of microcomputers–SS100, SS150, and SS250. Most of the components are imported from Far East, and the company only assembles the microcomputers. The operation hours required to produce one unit of the microcomputers are 2 hours for SS100, 3 hours for SS150, and 4 hours for SS250. The normal capacity of the assembly line is 400 hours per month. The profit per unit is Rs. 10,000 for SS100, Rs. 15,000 for SS150 and Rs. 25,000 for SS250. The president of the company has set the following goals, according to their order of importance:

   (i) Avoid under-utilization of production capacity

   (ii) Meet the outstanding orders–30 units for SS100, 20 units for SS150, and 50 units for SS250.

   (iii) Avoid the over-utilization of the production capacity.

   (iv) Maximize total profit as much as possible.

   Formulate a goal programming model for the problem and solve it (up to three iterations only).

9. A manufacturing company produces two products A and B. In one hour time 100 units of A are produced. Similarly in an hour 80 units are produced. The total working hours made available is 8 hours per day. If the company had the order to supply 1600 units per day, and has the following constraints.

P_1: Delivery 1600 unit per day.

P_2: Overtime per day restricted to 2 hours.

P_3: Minimize the under-utilization of production and fix weightage as per productivity rate.

P_4: Minimize over-utilization of overtime and fix weightage as per over-time cost.

10. Solve the following GP problem by Simplex method.

Minimize: $Z = p_1 d_1^- + p_2(2d_2^- + d_3^-)p_3 d_1^+ d_4^+$

Subject to: $X_1 + X_2 + d_1^- - d_1^+ = 300$

$X_1 + d_2^- = 200$

$X_2 + d_3^- = 200$

$X_1, X_2, d_1^-, d_1^+, d_2^-, d_2^+, d_3^- \geq 0$

CHAPTER 18

Markov Process

- *Introduction*
- *Markov process*
- *Tree diagram approach for calculation of state probabilities*
- *Systematic approach for calculation of state probabilities*
- *Calculation of steady probabilities using formula*
- *Calculation of state and steady probabilities using excel worksheet*
- *Finite and absorbing states*
- *Absorbing chain and its application in accounts*
- *Exercises*

18.1 INTRODUCTION

On several occasions, we will be interested in knowing the time path of a given random variable over time. For example, the manager concerned may be interested in knowing the way in which the demand for Maruti cars will change over time. Similarly, an investor in capital market may be interested in knowing the behaviour of the price of a certain shares that he is having. Such a time path of the random variables over time is often called the *stochastic process*.

In this section we plan to introduce one such stochastic process often called Markov process, in analyzing the evolution of a given system overtime by using repeated trials. In such an evaluation process, the state of the system at a particular period on a future date is obtained by using initial otherwise called transition probabilities. Markov process could be of great help in identifying the probability that a machine which is functioning very well in the current period will continue to do so in the future period. Similarly, it can also be used to find the probability that a particular customer shopping at Food World currently will switch over to Big Bazaar on a later period.

18.2 MARKOV PROCESS

1. State of the system

Suppose we are interested in analyzing the customer's loyalty to the Food World and Big Bazaar in Bangalore city. In this illustration we propose to analyze the sequence of shopping patterns by a customer who normally goes for shopping once in a week regularly. Given the choice, in a weekly trip he can shop either at Food World or at Big Bazaar. A particular store selected in a given week is often referred as the *state of the system.* In the illustration, because our typical customer has two alternatives at each trial, at a given point of time he may be either in state 1 or 2.

State 1: The customer shops at Food World.

State 2: The customer shops at Big Bazaar.

With this state of the system definition, if we say that the system is in state 1 in trip two, we simply say that our customer shops at Food World in the second trip. Given the initial state (state 1 or state 2) in period 0, by using Markov Chain process we will be in a position to find the probability of his shopping either at Big Bazaar or Food World on any future date. To find the probability of the various states at successive trials by using Markov process, we need the probability that a customer remains with the same shop in the subsequent trial.

2. Transitional probability

Suppose the market research study reveals that out of 100 customers at Food World during the current trip, only 90 come for Food World in the subsequent trip. This amounts to a shift over of 10 customers from Food World to Big Bazaar in the subsequent trip. Similarly, suppose that the same market research study reveals that out of 100 customers at Big Bazaar during the current trip, only 80 come for Big Bazaar again for shopping in the subsequent trip. This amounts to a shift over of 20 customers from Big Bazaar to Food World in the subsequent trip. Thus, among the current Food World Customers 90% are loyal to Food World and come again to Food World in the subsequent trial. Only 10% of current Food World customers go for Big Bazaar in the subsequent trial. In a similar way among the current Big Bazaar customers, only 80% come again to Big Bazaar in the subsequent trial and 20% move to Food World in the Subsequent trial. These probabilities observed in the beginning are only transitional in character and hence called transitional probabilities. For developing the Markov process we assume that these transitional probabilities are constants for all the trials to come. Table 18.1 gives the whole summary this discussion.

Table 18.1

Transition probabilities		
Current week shopping	Next week shopping	
	Food World	Big Bazaar
Food World	0.9	0.1
Big Bazaar	0.2	0.8

Since in the above table, there are two rows and two columns, it is common practice to represent these transitional probabilities in a matrix form as shown below.

$$P = \begin{bmatrix} P_{11} & P_{12} \\ P_{21} & P_{22} \end{bmatrix} = \begin{bmatrix} 0.9 & 0.1 \\ 0.2 & 0.8 \end{bmatrix}$$

where P_{ij} = Probability of making transition from state i in a given period to state j in the subsequent period. In Figure 18.1 we show the corresponding brand switching transition diagram.

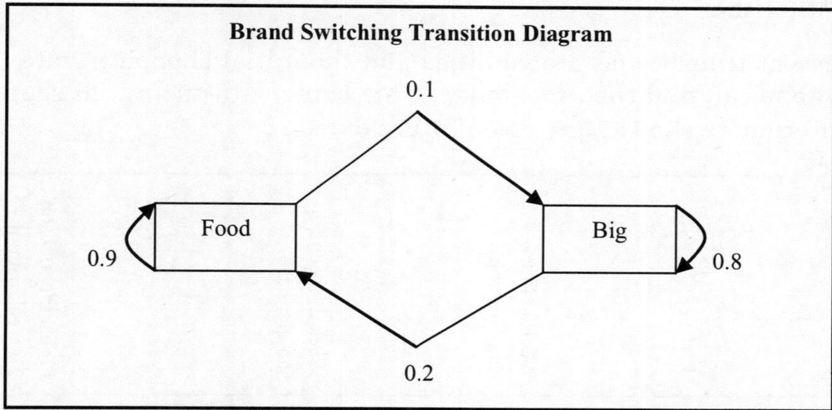

Figure 18.1

Assumption Underlying Markov Chain Process

1. **First Order Condition:** In our loyalty switching problem, the customer's shopping choice for any given period depends upon his choice in the just preceding period and the associated transition probability matrix. The process with this assumed first order characteristic is often called *First order Markov process.* If the customer's current shopping choice depends upon two just consecutive preceding shopping choices then it is called *second order Markov process.* In this text only the *first order Markov* models are being addressed.

2. **Stationary or Homogenous Models:** The transitional probabilities remain unaltered from time to time for the entire sequence in Markov process. Thus, such models are often called *stationary or homogenous* Markov Chain models. In addition if it satisfies the first order condition as well then it is called first *order stationary* Markov Chain process.

3. **Uniform Time Period Models**: The change from one state to another takes place only once during the given time period. Thus the time intervals are equal for all periods. The uniform time could be one week or one month or one year depending upon the situation.

State and steady state probabilities

By *state probability* we refer the probability of our Food World customer, at different future states under reference, like the first week, second week etc. After crossing several such states these state probabilities themselves remain unaltered in all the future periods. Such an altered state probabilities are called *steady state probabilities*. In the following Figure 18.2 we illustrate the method of calculating the state 1 and state 2 probabilities of our Food World customer by using tree diagram approach.

18.3 TREE DIAGRAM APPROACH FOR CALCULATION OF STATE PROBABILITIES

Given the constant transitional probabilities and the initial shopping state (Food world or Big Bazaar) one can find the probability of his being with the initial shop as well his being in the alternative shop at any given future date.

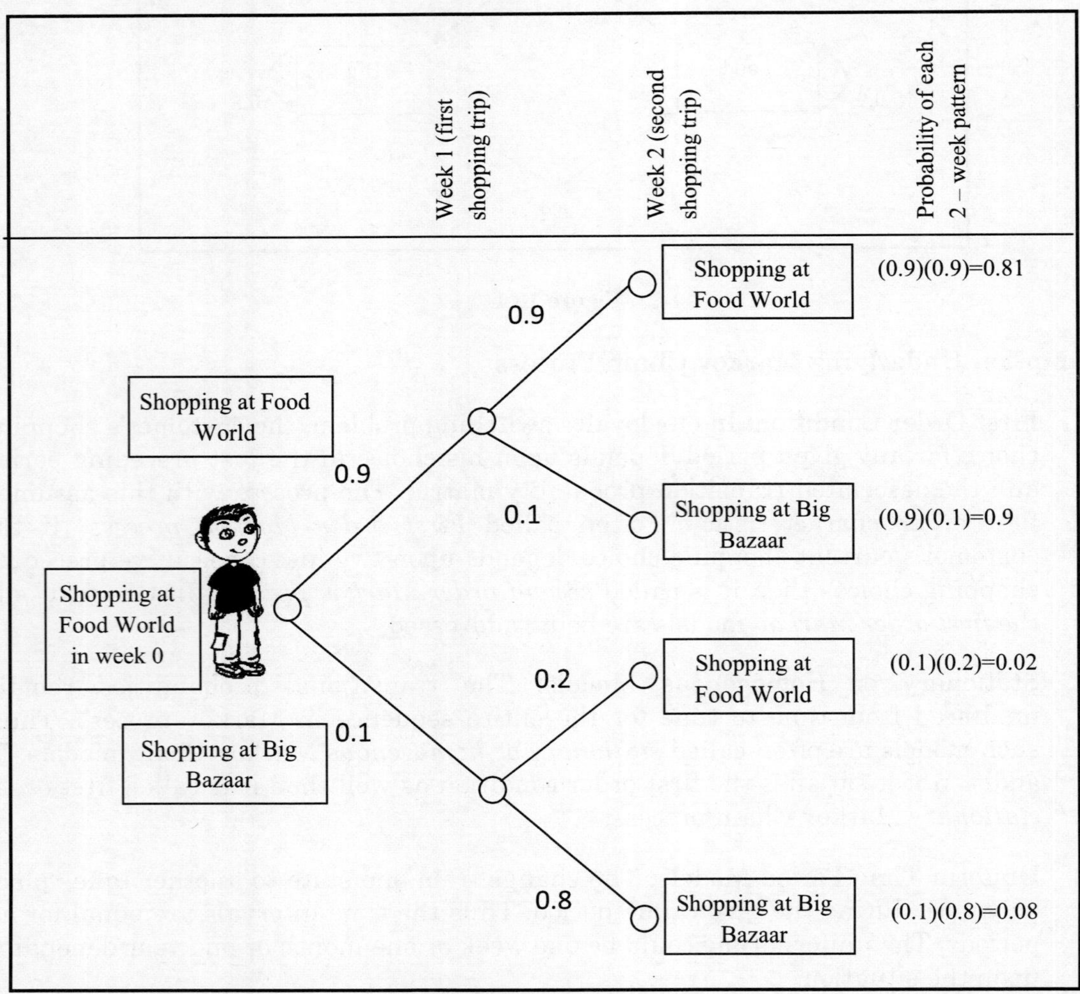

Figure 18.2

To illustrate this process let our customer's current shopping be in Food World (state 1) in week 0 as shown in the tree diagram in Figure 18.2. Now for our customer, the probability of visiting Food World in the first week is given here as $P_{11} = 0.9$. Thus, given this probability, the probability of restoring our customer at Food World (state1) in second week is obtained as (0.9) × (0.9) = 0.81.

In a similar way the probability of our customer switching over to Big Bazaar in the first week is given in the Figure 18.2 as $P_{12} = 0.1$. So the probability of this customer who has switched over to Big Bazaar in the first week, coming back to Food World in the second week is given by (0.1) × (0.2) = 0.02. So in week 2 the probability of our customer being with Food World is obtained as (0.81) + (0.02) = 0.83.

Similarly, the probability of our customer who has switched over to Big Bazaar in week 1 and continues to be there in the second week is given by (0.1) × (0.8) = 0.08. The probability of the customer who was loyal to Food World in the first week but loyal to Big Bazaar in the second week is given by (0.9) × (0.1) = 0.09. Thus, the probability of customer being with Big Bazaar at the second week is given by (0.08) + (0.09) = 0.17.

Thus, ultimately the probability of our typical customer, who was found in the Food World in the zero week, to be seen in the Food World again for the second week purchase is only 0.83. The probability of his being in the Big Bazaar during the second week is 0.17. Thus it is important to note that the probability of our customer being in Food World and Big Bazaar in week 2 is = 0.83+0.17 = 1.00.

In this way, given the initial shopping information of our typical customer, one can find the probability of being with Food World as well as with Big Bazaar on any future date by merely repeating these calculations up to the concerned week. Such probabilities are called state probabilities because they keep on changing during the week under reference.

18.4 SYSTEMATIC APPROACH FOR CALCULATION OF STATE PROBABILITIES

Whatever may be the desirability and simplicity, the tree diagram approach is cumbersome and time consuming, if we are interested in many more such calculations with longer period under reference. To make the task simple, let us proceed in a systematic way as shown below.

Let $\pi_i(n) =$ the probability that our system is in state i (1, or 2) in period n. For example, $\pi_1(1)$ denotes the probability of the system under consideration in state 1 (Food World) in period 1. Similarly, $\pi_2(1)$ refers to the probability of the system in state 2 (Big Bazaar) in period 1.

With these notational features, we use $\pi_1(0)$ and $\pi_2(0)$ to refer to the probabilities of the system being in state 1 and 2 respectively in period 0, the starting point of the system. In particular, if $\pi_1(0) = 1$ or $\pi_2(0) = 0$, then with certainty we say that our customer is in Food World to begin within the period 0 as described in Figure 18.1. Alternatively, if $\pi_2(0) = 1$ or $\pi_1(0) = 0$ we would say that our customer is in Big Bazaar in period 0.

Thus, $\prod(0)=[\pi_1(0) \quad \pi_2(0)]=[1 \quad 0]$ refers the initial probability vector of the system in period zero as in Figure 18.2. Similarly, $\prod(n)=[\pi_1(n) \quad \pi_2(n)]$ refers to the probability vector of this system in period n. Using this notation, one can find the state probabilities for the period (n+1) by merely multiplying the known state probabilities of the previous period (n) by the given transition probability matrix.

Thus, in general $\prod(n+1)=\prod(n)\begin{bmatrix} P_{11} & P_{12} \\ P_{21} & P_{22} \end{bmatrix}$

In particular, $\prod(1)=\prod(0)\begin{bmatrix} P_{11} & P_{12} \\ P_{21} & P_{22} \end{bmatrix}$

Similarly

$$\prod(2)=\prod(1)\begin{bmatrix} P_{11} & P_{12} \\ P_{21} & P_{22} \end{bmatrix}$$

$$=\prod(0)\begin{bmatrix} P_{11} & P_{12} \\ P_{21} & P_{22} \end{bmatrix}\begin{bmatrix} P_{11} & P_{12} \\ P_{21} & P_{22} \end{bmatrix}$$

$$=\prod(0)\begin{bmatrix} P_{11} & P_{12} \\ P_{21} & P_{22} \end{bmatrix}^2$$

$$\prod(3)=\prod(2)\begin{bmatrix} P_{11} & P_{12} \\ P_{21} & P_{22} \end{bmatrix}$$

$$=\prod(0)\begin{bmatrix} P_{11} & P_{12} \\ P_{21} & P_{22} \end{bmatrix}^2\begin{bmatrix} P_{11} & P_{12} \\ P_{21} & P_{22} \end{bmatrix}$$

$$=\prod(0)\begin{bmatrix} P_{11} & P_{12} \\ P_{21} & P_{22} \end{bmatrix}^3$$

$$\prod(n)=\prod(n-1)\begin{bmatrix} P_{11} & P_{12} \\ P_{21} & P_{22} \end{bmatrix}$$

$$=\prod(0)\begin{bmatrix} P_{11} & P_{12} \\ P_{21} & P_{22} \end{bmatrix}^{n-1}\begin{bmatrix} P_{11} & P_{12} \\ P_{21} & P_{22} \end{bmatrix}$$

$$=\prod(0)\begin{bmatrix} P_{11} & P_{12} \\ P_{21} & P_{22} \end{bmatrix}^n$$

Calculation of state probabilities for week 1

Let us begin with our system is in state 1 in period 0 as in Figure 18.2. So $\prod(0)=[1 \quad 0]$ follows from Figure 18.2. Now we can compute the probabilities for period 1 as follows

$$\Pi(1) = \Pi(0)\begin{bmatrix} P_{11} & P_{12} \\ P_{21} & P_{22} \end{bmatrix}$$

$$\begin{aligned}
[\pi_1(1) \quad \pi_2(1)] &= [\pi_1(0) \quad \pi_2(0)] \begin{bmatrix} P_{11} & P_{12} \\ P_{21} & P_{22} \end{bmatrix} \\
&= [1 \quad 0] \begin{bmatrix} 0.9 & 0.1 \\ 0.2 & 0.8 \end{bmatrix} \\
&= [1 \times 0.9 + 0 \times 0.2 \quad 1 \times 0.1 + 0 \times 0.8] \\
&= [0.9 \quad 0.1]
\end{aligned}$$

Calculation of state probabilities for week 2

The stated probabilities $\pi_1(1) = 0.9$ and $\pi_2(1) = 0.1$ are the state probabilities for period 1 of our customer who was shopping at Food World during the 0 period. Here $\pi_1(1) = 0.9$ refers to the state probability of our Food World customer shopping at Food World again in period 1. Similarly, $\pi_2(1) = 0.1$ refers to the state probability of our Food World customer in period 0 found in Big Bazaar in period 1.

Similarly,

$$\Pi(2) = \Pi(1)\begin{bmatrix} P_{11} & P_{12} \\ P_{21} & P_{22} \end{bmatrix}$$

$$\begin{aligned}
[\pi_1(2) \quad \pi_2(2)] &= [\pi_1(1) \quad \pi_2(1)] \begin{bmatrix} P_{11} & P_{12} \\ P_{21} & P_{22} \end{bmatrix} \\
&= [0.9 \quad 0.1] \begin{bmatrix} 0.9 & 0.1 \\ 0.2 & 0.8 \end{bmatrix} \\
&= [0.9 \times 0.9 + 0.1 \times 0.2 \quad 0.9 \times 0.1 + 0.1 \times 0.8] \\
&= [0.83 \quad 0.17]
\end{aligned}$$

(Note that the same values were obtained in tree diagram approach as well.)

18.5 CALCULATION OF STEADY PROBABILITIES USING FORMULA

From the preceding discussion we know that

$$\Pi(n+1) = \Pi(n)\begin{bmatrix} P_{11} & P_{12} \\ P_{21} & P_{22} \end{bmatrix}$$

$$[\pi_1(n+1) \quad \pi_2(n+1)] = [\pi_1(n) \quad \pi_2(n)] \begin{bmatrix} P_{11} & P_{12} \\ P_{21} & P_{22} \end{bmatrix}$$

Because of the steady and constant probabilities in the long run, the time reference 'n' could conveniently be omitted while representing the steady state probabilities. So let us rename the associated steady state probabilities merely by π_1 and π_2 for state 1 and 2 respectively without any time reference.

$$\pi_1(n+1) = \pi_1(n) = \pi_1$$
$$\pi_2(n+1) = \pi_2(n) = \pi_2$$

So the above equation may be rewritten as

$$[\pi_1 \quad \pi_2] = [\pi_1 \quad \pi_2]\begin{bmatrix} P_{11} & P_{12} \\ P_{21} & P_{22} \end{bmatrix}$$
$$= [\pi_1 \quad \pi_2]\begin{bmatrix} 0.9 & 0.1 \\ 0.2 & 0.8 \end{bmatrix}$$
$$= [\pi_1 \times 0.9 + \pi_2 \times 0.2 \quad \pi_1 \times 0.1 + \pi_2 \times 0.8]$$

Thus

$$\pi_1 = 0.9\pi_1 + 0.2\pi_2$$
$$\pi_2 = 0.1\pi_1 + 0.8\pi_2$$

Further we also know that the sum of these two steady state probabilities is always equal to 1. So

$$\pi_1 + \pi_2 = 1$$
$$\pi_2 = 1 - \pi_1$$

Substituting this value of π_2 in the above equation it becomes

$$\pi_1 = 0.9\pi_1 + 0.2(1 - \pi_1)$$
$$\pi_1 = 0.9\pi_1 + 0.2 - 0.2\pi_1$$
$$\pi_1 - 0.7\pi_1 = 0.2$$
$$0.3\pi_1 = 0.2$$
$$\pi_1 = 2/3$$

$$\pi_2 = 1 - \pi_1 = 1 - 2/3 = 1/3$$

18.6 CALCULATION OF STATE AND STEADY PROBABILITIES USING EXCEL WORKSHEET

The vectors $\Pi(1)$, $\Pi(2)$, $\Pi(3)$, ... show the *state* probabilities that our Food World customer with Food World or Big Bazaar overtime. In Table 18.2 we generate 30 sets of state probabilities of our Food World customer. To prepare the work sheet type the formula in Cell B6 =B5*$E$6+C5*$E$8 and C6 = B5*$F$6+C5*$F$8 and copy the formulas down the line up to 35th row as shown in Table 18.2. The result is displayed in Table 18.3.

In Table 18.3, from period 17 onwards we notice that these state probabilities do not change and become stationary. These stationary probabilities are also often called *steady-state probabilities*. Accordingly, if there were 1000 customers to begin with in Food World then there will be only 667 customers left with Food World after 17th period and thereafter. The remaining 333 of our original Food World customers by now have already moved to Big Bazaar.

Table 18.2

	A	B	C	D	E	F	G
1	Period	Initial Shopping at Food World					
2							
3							
4		$\pi_1(n)$	$\pi_2(n)$				
5	0	1	0				
6	1	=B5*$E$6+C5*$E$8	=B5*$F$6+C5*$F$8		0.9	0.1	
7	2	=B6*$E$6+C6*$E$8	=B6*$F$6+C6*$F$8	P =			
8	3	=B7*$E$6+C7*$E$8	=B7*$F$6+C7*$F$8		0.2	0.8	
9	4	=B8*$E$6+C8*$E$8	=B8*$F$6+C8*$F$8				
10	5	=B9*$E$6+C9*$E$8	=B9*$F$6+C9*$F$8				
11	6	=B10*$E$6+C10*$E$8	=B10*$F$6+C10*$F$8				
12	7	=B11*$E$6+C11*$E$8	=B11*$F$6+C11*$F$8				
13	8	=B12*$E$6+C12*$E$8	=B12*$F$6+C12*$F$8				
14	9	=B13*$E$6+C13*$E$8	=B13*$F$6+C13*$F$8				
15	10	=B14*$E$6+C14*$E$8	=B14*$F$6+C14*$F$8				
16	11	=B15*$E$6+C15*$E$8	=B15*$F$6+C15*$F$8				
17	12	=B16*$E$6+C16*$E$8	=B16*$F$6+C16*$F$8				
18	13	=B17*$E$6+C17*$E$8	=B17*$F$6+C17*$F$8				
19	14	=B18*$E$6+C18*$E$8	=B18*$F$6+C18*$F$8				
20	15	=B19*$E$6+C19*$E$8	=B19*$F$6+C19*$F$8				
21	16	=B20*$E$6+C20*$E$8	=B20*$F$6+C20*$F$8				
22	17	=B21*$E$6+C21*$E$8	=B21*$F$6+C21*$F$8				
23	18	=B22*$E$6+C22*$E$8	=B22*$F$6+C22*$F$8				
24	19	=B23*$E$6+C23*$E$8	=B23*$F$6+C23*$F$8				
25	20	=B24*$E$6+C24*$E$8	=B24*$F$6+C24*$F$8				
26	21	=B25*$E$6+C25*$E$8	=B25*$F$6+C25*$F$8				
27	22	=B26*$E$6+C26*$E$8	=B26*$F$6+C26*$F$8				
28	23	=B27*$E$6+C27*$E$8	=B27*$F$6+C27*$F$8				
29	24	=B28*$E$6+C28*$E$8	=B28*$F$6+C28*$F$8				
30	25	=B29*$E$6+C29*$E$8	=B29*$F$6+C29*$F$8				
31	26	=B30*$E$6+C30*$E$8	=B30*$F$6+C30*$F$8				
32	27	=B31*$E$6+C31*$E$8	=B31*$F$6+C31*$F$8				
33	28	=B32*$E$6+C32*$E$8	=B32*$F$6+C32*$F$8				
34	29	=B33*$E$6+C33*$E$8	=B33*$F$6+C33*$F$8				
35	30	=B34*$E$6+C34*$E$8	=B34*$F$6+C34*$F$8				

Table 18.3

Period	Initial Shopping at Food World		
	$\pi_1(n)$	$\pi_2(n)$	
0	1	0	
1	0.900	0.100	$P = \begin{vmatrix} 0.9 & 0.1 \\ 0.2 & 0.8 \end{vmatrix}$
2	0.830	0.170	
3	0.781	0.219	
4	0.747	0.253	
5	0.723	0.277	
6	0.706	0.294	
7	0.694	0.306	
8	0.686	0.314	
9	0.680	0.320	
10	0.676	0.324	
11	0.673	0.327	
12	0.671	0.329	
13	0.670	0.330	
14	0.669	0.331	
15	0.668	0.332	
16	0.668	0.332	
17	0.667	0.333	
18	0.667	0.333	
19	0.667	0.333	
20	0.667	0.333	
21	0.667	0.333	
22	0.667	0.333	
23	0.667	0.333	
24	0.667	0.333	
25	0.667	0.333	
26	0.667	0.333	
27	0.667	0.333	
28	0.667	0.333	
29	0.667	0.333	
30	0.667	0.333	

■ Example 1

Suppose Big Bazaar is initiating an advertising program to attract customers from Food World. Such a programme is expected to increase the probability of Food World customers switching over to Big Bazaar from the present 0.1 to 0.15. Obtain the revised steady state probabilities.

Solution

The revised transient matrix is obtained as

Table 18.4

Current month shopping	Next month shopping	
	Food World	Big Bazaar
Food World	0.85	0.15
Big Bazaar	0.2	0.8

Table 18.5

Period	Initial Shopping at Food World	
	$\pi_1(n)$	$\pi_2(n)$
0	1	0
1	0.850	0.150
2	0.753	0.248
3	0.689	0.311
4	0.648	0.352
5	0.621	0.379
6	0.604	0.396
7	0.592	0.408
8	0.585	0.415
9	0.580	0.420
10	0.577	0.423
11	0.575	0.425
12	0.574	0.426
13	0.573	0.427
14	0.572	0.428
15	0.572	0.428
16	0.572	0.428
17	0.572	0.428
18	0.572	0.428
19	0.572	0.428
20	0.572	0.428
21	0.571	0.429
22	0.571	0.429
23	0.571	0.429
24	0.571	0.429
25	0.571	0.429
26	0.571	0.429
27	0.571	0.429
28	0.571	0.429
29	0.571	0.429
30	0.571	0.429

$$P = \begin{vmatrix} 0.85 & 0.15 \\ 0.2 & 0.8 \end{vmatrix}$$

With this new transitional probability matrix we repeat the calculation made in Table 18.2 and report it in Table 18.5. To achieve this result copy the worksheet in

Table 18.2 and change the probabilities in the matrix as shown Table 18.5. The result will be automatically displayed as shown in Table 18.5.

Alternatively, we recalculate the steady state probabilities once again under the new circumstances.

$$\pi_1 = 0.85\pi_1 + 0.2\pi_2$$
$$\pi_2 = 0.15\pi_1 + 0.8\pi_2$$

Further we also know that the sum of these two steady state probabilities is always equal to one. So

$$\pi_1 + \pi_2 = 1$$
$$\pi_2 = 1 - \pi_1$$

Substituting this value of π_2 in the above equation it becomes

$$\pi_1 = 0.85\pi_1 + 0.2(1 - \pi_1)$$
$$\pi_1 = 0.85\pi_1 + 0.2 - 0.2\pi_1$$
$$\pi_1 - 0.65\pi_1 = 0.2$$
$$0.35\pi_1 = 0.2$$
$$\pi_1 = 0.571$$

$\pi_2 = 1 - \pi_1 = 1 - 0.571 = 0.429$. These values π_1 and π_2 are in conformity with the steady state probabilities obtained in Table 18.5 after period 21.

■ Example 2

Instead of advertisement let this time the Big Bazaar initiate a gift programme to all its loyal customers to prevent switching over to Food World. Given the new transition matrix, obtain the revised steady state probabilities.

Table 18.6

Current month shopping	Next month shopping	
	Food World	Big Bazaar
Food World	0.9	0.1
Big Bazaar	0.15	0.85

Now let us recalculate the state and steady state probabilities once again under the new circumstances and find the resulting steady state as shown in Table 18.7.

Table 18.7

Period	Initial Shopping at Food World		
	$\pi_1(n)$	$\pi_2(n)$	
0	1	0	
1	0.900	0.100	$P = \begin{vmatrix} 0.9 & 0.1 \\ 0.15 & 0.85 \end{vmatrix}$
2	0.825	0.175	
3	0.769	0.231	
4	0.727	0.273	
5	0.695	0.305	
6	0.671	0.329	
7	0.653	0.347	
8	0.640	0.360	
9	0.630	0.370	
10	0.623	0.377	
11	0.617	0.383	
12	0.613	0.387	
13	0.610	0.390	
14	0.607	0.393	
15	0.605	0.395	
16	0.604	0.396	
17	0.603	0.397	
18	0.602	0.398	
19	0.602	0.398	
20	0.601	0.399	
21	0.601	0.399	
22	0.601	0.399	
23	0.601	0.399	
24	0.600	0.400	
25	0.600	0.400	
26	0.600	0.400	
27	0.600	0.400	
28	0.600	0.400	
29	0.600	0.400	
30	0.600	0.400	

Alternatively, the result is also obtained as follows:

$$\pi_1 = 0.9\pi_1 + 0.15\pi_2$$
$$\pi_2 = 0.1\pi_1 + 0.85\pi_2$$

Further we also know that the sum of these two steady state probabilities is always equal to 1. So

$$\pi_1 + \pi_2 = 1$$
$$\pi_2 = 1 - \pi_1$$

Substituting this value of π_2 in the above equation it becomes

$$\pi_1 = 0.9\pi_1 + 0.15(1-\pi_1)$$
$$\pi_1 = 0.9\pi_1 + 0.15 - 0.15\pi_1$$
$$\pi_1 - 0.75\pi_1 = 0.15$$
$$0.25\pi_1 = 0.15$$
$$\pi_1 = 0.600$$

$$\pi_2 = 1 - \pi_1 = 1 - 0.600 = 0.400$$

These values π_1 and π_2 are in conformity with the steady state probabilities obtained in Table 18.7 after 24 periods.

Out of 1000 customers only 600 will be with the Food World leaving 400 to Big Bazaar by this time.

Example 3

Given the transitional matrix

$$P = \begin{bmatrix} P_{11} & P_{12} \\ P_{21} & P_{22} \end{bmatrix} = \begin{bmatrix} 0.9 & 0.1 \\ 0.2 & 0.8 \end{bmatrix}$$

Obtain the state and steady state probabilities by using excel worksheet, given initial condition that this time our customer is seen in Big Bazaar in period 0.

Solution

Table 18.8 shows the state probability calculation for 30 periods for our Big Bazaar customer. This time note that the proximities 0 and 1 are suitably interchanged in the table.

This time the steady state is reached from 24th period onwards. Accordingly, out of 1000 Big Bazaar customers after 24th period only 667 will be with the Big Bazaar. The remaining 333 will have a switch over to Food World by this time.

Table 18.8

Period	Initial Shopping at Big Bazaar		
	$\pi_1(n)$	$\pi_2(n)$	
0	0	1	
1	0.200	0.800	$P = \begin{vmatrix} 0.9 & 0.1 \\ 0.2 & 0.8 \end{vmatrix}$
2	0.340	0.660	
3	0.438	0.562	
4	0.507	0.493	
5	0.555	0.445	
6	0.588	0.412	
7	0.612	0.388	
8	0.628	0.372	

(Contd.)

9	0.640	0.360	
10	0.648	0.352	
11	0.653	0.347	
12	0.657	0.343	
13	0.660	0.340	
14	0.662	0.338	
15	0.664	0.336	
16	0.664	0.336	
17	0.665	0.335	
18	0.666	0.334	
19	0.666	0.334	
20	0.666	0.334	
21	0.666	0.334	
22	0.666	0.334	
23	0.666	0.334	
24	0.667	0.333	
25	0.667	0.333	
26	0.667	0.333	
27	0.667	0.333	
28	0.667	0.333	
29	0.667	0.333	
30	0.667	0.333	

***Note:** In Tables 18.3 and 18.8 the steady state probabilities are found to be one and the same. Thus the steady state probabilities are independent of the initial shopping pattern.*

■ Example 4

In a certain market only two brands of soaps A and B are available. A consumer was found to purchase A brand last month. There is 80% chance that he will buy the same brand this month also. Similarly, the customer who purchased B brand last month, there is 90% for restoring it now also.

a) Calculate the probability for a currently A brand purchaser moving to B brand after two purchases from now.

b) Calculate the probability for a currently B brand purchaser moving to A brand after two purchases from now.

c) Calculate the steady state probability of a currently A brand purchaser.

Solution (a)

Table 18.9

From brand (this month)	To brand (next month)	
	A	B
A	0.80	0.20
B	0.10	0.90

$$P = \begin{bmatrix} P_{11} & P_{12} \\ P_{21} & P_{22} \end{bmatrix} = \begin{bmatrix} 0.8 & 0.2 \\ 0.1 & 0.9 \end{bmatrix}$$

To begin with our system is in state 1 (brand A) in period 0. So $\Pi(0) = [1 \quad 0]$ as follows from the definition. Now we can compute the probabilities for period 1 as follows

$$\Pi(1) = \Pi(0) \begin{bmatrix} P_{11} & P_{12} \\ P_{21} & P_{22} \end{bmatrix}$$

$$[\pi_1(1) \quad \pi_2(1)] = [\pi_1(0) \quad \pi_2(0)] \begin{bmatrix} P_{11} & P_{12} \\ P_{21} & P_{22} \end{bmatrix}$$

$$= [1 \quad 0] \begin{bmatrix} 0.8 & 0.2 \\ 0.1 & 0.9 \end{bmatrix}$$

$$= [1 \times 0.8 + 0 \times 0.1 \quad 1 \times 0.2 + 0 \times 0.9]$$

$$= [0.8 \quad 0.2]$$

The state probabilities $\pi_1(1) = 0.8$ and $\pi_2(1) = 0.2$ are the state probabilities of our customer who was buying brand A during the 0 period. Here $\pi_1(1) = 0.8$ refers to the state probability of our brand A customer buying brand A again in period 1. Similarly, $\pi_2(1) = 0.2$ refers to the state probability of our brand A customer moving to brand B in period 1. Now,

$$\Pi(2) = \Pi(1) \begin{bmatrix} P_{11} & P_{12} \\ P_{21} & P_{22} \end{bmatrix}$$

$$[\pi_1(2) \quad \pi_2(2)] = [\pi_1(1) \quad \pi_2(1)] \begin{bmatrix} P_{11} & P_{12} \\ P_{21} & P_{22} \end{bmatrix}$$

$$= [0.8 \quad 0.2] \begin{bmatrix} 0.8 & 0.2 \\ 0.1 & 0.9 \end{bmatrix}$$

$$= [0.8 \times 0.8 + 0.2 \times 0.1 \quad .0.8 \times 0.2 + 0.2 \times 0.9]$$

$$= [0.66 \quad 0.34]$$

So the probability for a currently A brand purchaser moving to B brand after two purchases from now is 0.34.

Excel Solution

Table 18.10

Period	Initial A brand purchase				
	$\pi_1(n)$	$\pi_2(n)$			
0	1	0			
1	0.800	0.200		0.8	0.2
2	0.660	0.340	P =		
3	0.562	0.438		0.1	0.9
4	0.493	0.507			

Solution (b)

To begin with our system is in state 2 (B Brand) in period 0. So $\Pi(0)=\begin{bmatrix}0 & 1\end{bmatrix}$ follows from the definition. Now we can compute the probabilities for period 1 as follows

$$\Pi(1)=\Pi(0)\begin{bmatrix}P_{11} & P_{12}\\ P_{21} & P_{22}\end{bmatrix}$$

$$\begin{bmatrix}\pi_1(1) & \pi_2(1)\end{bmatrix}=\begin{bmatrix}\pi_1(0) & \pi_2(0)\end{bmatrix}\begin{bmatrix}P_{11} & P_{12}\\ P_{21} & P_{22}\end{bmatrix}$$

$$=\begin{bmatrix}0 & 1\end{bmatrix}\begin{bmatrix}0.8 & 0.2\\ 0.1 & 0.9\end{bmatrix}$$

$$=\begin{bmatrix}0\times 0.8+1\times 0.1 & 0\times 0.2+1\times 0.9\end{bmatrix}$$

$$=\begin{bmatrix}0.1 & 0.9\end{bmatrix}$$

The state probabilities $\pi_1(1)=0.9$ and $\pi_2(1)=0.1$ are the transitional probabilities of our customer who was buying brand A during the 0. Here $\pi_1(1)=0.9$ refers to the transitional probability of our brand B customer buying brand B again in period 1. Similarly, $\pi_2(1)=0.2$ refers to the transitional probability of our brand B customer moving to brand A in period 1. Now,

$$\Pi(2)=\Pi(1)\begin{bmatrix}P_{11} & P_{12}\\ P_{21} & P_{22}\end{bmatrix}$$

$$\begin{bmatrix}\pi_1(2) & \pi_2(2)\end{bmatrix}=\begin{bmatrix}\pi_1(1) & \pi_2(1)\end{bmatrix}\begin{bmatrix}P_{11} & P_{12}\\ P_{21} & P_{22}\end{bmatrix}$$

$$=\begin{bmatrix}0.1 & 0.9\end{bmatrix}\begin{bmatrix}0.8 & 0.2\\ 0.1 & 0.9\end{bmatrix}$$

$$=\begin{bmatrix}0.1\times 0.8+0.9\times 0.1 & 0.1\times 0.2+0.9\times 0.9\end{bmatrix}$$

$$=\begin{bmatrix}0.17 & 0.83\end{bmatrix}$$

Excel Solution

Table 18.11

Period	Initial B brand purchase		
	$\pi_1(n)$	$\pi_2(n)$	
0	0	1	
1	0.100	0.900	
2	0.170	0.830	$P = \begin{vmatrix}0.8 & 0.2\\ 0.1 & 0.9\end{vmatrix}$
3	0.219	0.781	
4	0.253	0.747	

So the probability for a currently B brand purchaser moving to A brand after two purchases from now is 0.17.

Solution (c)

Excel Solution

Table 18.12

Period	Initial Shopping at Food World		
	$\pi_1(n)$	$\pi_2(n)$	
0	1	0	
1	0.800	0.200	$P = \begin{vmatrix} 0.8 & 0.2 \\ 0.1 & 0.9 \end{vmatrix}$
2	0.660	0.340	
3	0.562	0.438	
4	0.493	0.507	
5	0.445	0.555	
6	0.412	0.588	
7	0.388	0.612	
8	0.372	0.628	
9	0.360	0.640	
10	0.352	0.648	
11	0.347	0.653	
12	0.343	0.657	
13	0.340	0.660	
14	0.338	0.662	
15	0.336	0.664	
16	0.336	0.664	
17	0.335	0.665	
18	0.334	0.666	
19	0.334	0.666	
20	0.334	0.666	
21	0.334	0.666	
22	0.334	0.666	
23	0.334	0.666	
24	0.333	0.667	
25	0.333	0.667	
26	0.333	0.667	
27	0.333	0.667	
28	0.333	0.667	
29	0.333	0.667	
30	0.333	0.667	

Steady State Probability of a customer currently buying A brand. From the preceding discussion we know that

$$[\pi_1(n+1) \quad \pi_2(n+1)] = [\pi_1(n) \quad \pi_2(n)] \begin{bmatrix} P_{11} & P_{12} \\ P_{21} & P_{22} \end{bmatrix}$$

Now because of the observed steady state property

$$\pi_1(n+1) = \pi_1(n) = \pi_1$$

$$\pi_2(n+1) = \pi_2(n) = \pi_2$$

So the above equation may be rewritten as

$$\begin{aligned}[\pi_1 \quad \pi_2] &= [\pi_1 \quad \pi_2]\begin{bmatrix} P_{11} & P_{12} \\ P_{21} & P_{22} \end{bmatrix} \\ &= [\pi_1 \quad \pi_2]\begin{bmatrix} 0.8 & 0.2 \\ 0.1 & 0.9 \end{bmatrix} \\ &= [\pi_1 \times 0.8 + \pi_2 \times 0.1 \quad \pi_1 \times 0.2 + \pi_2 \times 0.9]\end{aligned}$$

Thus

$$\pi_1 = 0.8\pi_1 + 0.1\pi_2$$
$$\pi_2 = 0.2\pi_1 + 0.9\pi_2$$

Further we also know that the sum of these two steady state probabilities is always equal to 1. So

$$\pi_1 + \pi_2 = 1$$
$$\pi_2 = 1 - \pi_1$$

Substituting this value of π_2 in the above equation it becomes

$$\pi_1 = 0.8\pi_1 + 0.1(1-\pi_1)$$
$$\pi_1 = 0.8\pi_1 + 0.1 - 0.1\pi_1$$
$$\pi_1 - 0.7\pi_1 = 0.1$$
$$0.3\pi_1 = 0.1$$
$$\pi_1 = 1/3$$

$$\pi_2 = 1 - \pi_1 = 1 - 1/3 = 2/3.$$

These values π_1 and π_2 are in conformity with the steady state probabilities obtained in Table 18.12.

■ Example 5

Suppose that there are three brands of detergent D_1, D_2, and D_3. It is observed that every month customers are being drawn from brand D_1 to brand D_2 to the extent of 30%; another 10% drawn to D_3 brand and as a result only 40% is restored by the brand D_1. Similarly, 20% of those using D_2 brand in a given month, 20% are found to move to D_1 brand, another 30% move to brand D_3. Only 50% are loyal to D_2 brand. Moreover, as far as D_3 is concerned only 80% are loyal to D_3, 15% move to D_1 and another 5% move to D_2.

a) Formulate a suitable transitional probability matrix and interpret the result.

b) Calculate the probability for a currently D_1 brand purchaser restores to D_1 brand after two purchases from now.

c) Calculate the probability for a currently D_1 brand purchaser moving to D_2 brand after two purchases from now.

d) Calculate the probability for a currently D_1 brand purchaser moving to D_3 brand after two purchases from now.

e) Calculate the steady state for a currently D_1 brand purchaser.

Solution (a)

Table 18.13 Transition Probability Matrix

From Brand (this month)	To Brand (next month)		
	D_1	D_2	D_3
D_1	0.60	0.30	0.10
D_2	0.20	0.50	0.30
D_3	0.15	0.05	0.80

Alternatively,

$$P = \begin{bmatrix} 0.60 & 0.30 & 0.10 \\ 0.20 & 0.50 & 0.30 \\ 0.15 & 0.05 & 0.80 \end{bmatrix}$$

The rows of this matrix tell us the restoration as well as loss of market to other brands. For example, the first row convey the message that as far as D_1 is concerned 60% is restored, 30% move to D_2 and 10% move to D_3.

Similarly, the columns tell us the respective restoration and gain from others, for example the first column shows a 60% of its own, 20% from D_2 and 15% from D_3. Note that the sums of the row probabilities are always equal to unity.

Solution (b)

We know that

$$\begin{bmatrix} \pi_1 & \pi_2 & \pi_3 \end{bmatrix} = \begin{bmatrix} \pi_1 & \pi_2 & \pi_3 \end{bmatrix} \begin{bmatrix} P_{11} & P_{12} & P_{13} \\ P_{21} & P_{22} & P_{23} \\ P_{31} & P_{32} & P_{33} \end{bmatrix}$$

$$= \begin{bmatrix} \pi_1 & \pi_2 & \pi_3 \end{bmatrix} \begin{bmatrix} 0.60 & 0.30 & 0.10 \\ 0.20 & 0.50 & 0.30 \\ 0.15 & 0.05 & 0.80 \end{bmatrix}$$

$$\pi_1 = 0.60\pi_1 + 0.20\pi_2 + 0.15\pi_3$$
$$\pi_2 = 0.30\pi_1 + 0.50\pi_2 + 0.05\pi_3$$
$$\pi_3 = 0.10\pi_1 + 0.30\pi_2 + 0.80\pi_3$$

$$0 = -0.40\pi_1 + 0.20\pi_2 + 0.15\pi_3$$
$$0 = \quad 0.30\pi_1 - 0.50\pi_2 + 0.05\pi_3$$
$$0 = \quad 0.10\pi_1 + 0.30\pi_2 - 0.20\pi_3$$

Further we know that $\pi_1 + \pi_2 + \pi_3 = 1$.

So to get the solution we take the first two equations in addition to the probability conditions stated above. Now by using the Cramer's Rule the results are obtained reported below.

Excel Template

Table 18.14

	A	B	C	D
1–2	Period	Initial purchase		
3–4		$\pi_1(n)$	$\pi_2(n)$	$\pi_3(n)$
5	0	1	0	0
6	1	=B5*$B$38+C5*$B$39+D5*$B$40	=B5*$C$38+C5*$C$39+D5*$C$40	=B5*$D$38+C5*$D$39+D5*$D$40
7	2	=B6*$B$38+C6*$B$39+D6*$B$40	=B6*$C$38+C6*$C$39+D6*$C$40	=B6*$D$38+C6*$D$39+D6*$D$40
8	3	=B7*$B$38+C7*$B$39+D7*$B$40	=B7*$C$38+C7*$C$39+D7*$C$40	=B7*$D$38+C7*$D$39+D7*$D$40
9	4	=B8*$B$38+C8*$B$39+D8*$B$40	=B8*$C$38+C8*$C$39+D8*$C$40	=B8*$D$38+C8*$D$39+D8*$D$40
10	5	=B9*$B$38+C9*$B$39+D9*$B$40	=B9*$C$38+C9*$C$39+D9*$C$40	=B9*$D$38+C9*$D$39+D9*$D$40
11	6	=B10*$B$38+C10*$B$39+D10*$B$40	=B10*$C$38+C10*$C$39+D10*$C$40	=B10*$D$38+C10*$D$39+D10*$D$40
12	7	=B11*$B$38+C11*$B$39+D11*$B$40	=B11*$C$38+C11*$C$39+D11*$C$40	=B11*$D$38+C11*$D$39+D11*$D$40
13	8	=B12*$B$38+C12*$B$39+D12*$B$40	=B12*$C$38+C12*$C$39+D12*$C$40	=B12*$D$38+C12*$D$39+D12*$D$40
14	9	=B13*$B$38+C13*$B$39+D13*$B$40	=B13*$C$38+C13*$C$39+D13*$C$40	=B13*$D$38+C13*$D$39+D13*$D$40
15	10	=B14*$B$38+C14*$B$39+D14*$B$40	=B14*$C$38+C14*$C$39+D14*$C$40	=B14*$D$38+C14*$D$39+D14*$D$40
16	11	=B15*$B$38+C15*$B$39+D15*$B$40	=B15*$C$38+C15*$C$39+D15*$C$40	=B15*$D$38+C15*$D$39+D15*$D$40
17	12	=B16*$B$38+C16*$B$39+D16*$B$40	=B16*$C$38+C16*$C$39+D16*$C$40	=B16*$D$38+C16*$D$39+D16*$D$40
18	13	=B17*$B$38+C17*$B$39+D17*$B$40	=B17*$C$38+C17*$C$39+D17*$C$40	=B17*$D$38+C17*$D$39+D17*$D$40
19	14	=B18*$B$38+C18*$B$39+D18*$B$40	=B18*$C$38+C18*$C$39+D18*$C$40	=B18*$D$38+C18*$D$39+D18*$D$40
20	15	=B19*$B$38+C19*$B$39+D19*$B$40	=B19*$C$38+C19*$C$39+D19*$C$40	=B19*$D$38+C19*$D$39+D19*$D$40
21	16	=B20*$B$38+C20*$B$39+D20*$B$40	=B20*$C$38+C20*$C$39+D20*$C$40	=B20*$D$38+C20*$D$39+D20*$D$40
22	17	=B21*$B$38+C21*$B$39+D21*$B$40	=B21*$C$38+C21*$C$39+D21*$C$40	=B21*$D$38+C21*$D$39+D21*$D$40
23	18	=B22*$B$38+C22*$B$39+D22*$B$40	=B22*$C$38+C22*$C$39+D22*$C$40	=B22*$D$38+C22*$D$39+D22*$D$40
24	19	=B23*$B$38+C23*$B$39+D23*$B$40	=B23*$C$38+C23*$C$39+D23*$C$40	=B23*$D$38+C23*$D$39+D23*$D$40
25	20	=B24*$B$38+C24*$B$39+D24*$B$40	=B24*$C$38+C24*$C$39+D24*$C$40	=B24*$D$38+C24*$D$39+D24*$D$40
26	21	=B25*$B$38+C25*$B$39+D25*$B$40	=B25*$C$38+C25*$C$39+D25*$C$40	=B25*$D$38+C25*$D$39+D25*$D$40
27	22	=B26*$B$38+C26*$B$39+D26*$B$40	=B26*$C$38+C26*$C$39+D26*$C$40	=B26*$D$38+C26*$D$39+D26*$D$40
28	23	=B27*$B$38+C27*$B$39+D27*$B$40	=B27*$C$38+C27*$C$39+D27*$C$40	=B27*$D$38+C27*$D$39+D27*$D$40
29	24	=B28*$B$38+C28*$B$39+D28*$B$40	=B28*$C$38+C28*$C$39+D28*$C$40	=B28*$D$38+C28*$D$39+D28*$D$40
30	25	=B29*$B$38+C29*$B$39+D29*$B$40	=B29*$C$38+C29*$C$39+D29*$C$40	=B29*$D$38+C29*$D$39+D29*$D$40
31	26	=B30*$B$38+C30*$B$39+D30*$B$40	=B30*$C$38+C30*$C$39+D30*$C$40	=B30*$D$38+C30*$D$39+D30*$D$40
32	27	=B31*$B$38+C31*$B$39+D31*$B$40	=B31*$C$38+C31*$C$39+D31*$C$40	=B31*$D$38+C31*$D$39+D31*$D$40
33	28	=B32*$B$38+C32*$B$39+D32*$B$40	=B32*$C$38+C32*$C$39+D32*$C$40	=B32*$D$38+C32*$D$39+D32*$D$40
34	29	=B33*$B$38+C33*$B$39+D33*$B$40	=B33*$C$38+C33*$C$39+D33*$C$40	=B33*$D$38+C33*$D$39+D33*$D$40
35	30	=B34*$B$38+C34*$B$39+D34*$B$40	=B34*$C$38+C34*$C$39+D34*$C$40	=B34*$D$38+C34*$D$39+D34*$D$40
37				
38		0.6	0.3	0.1
39	P =	0.2	0.5	0.3
40		0.15	0.05	0.8
41				

$$A = \begin{vmatrix} -0.4 & 0.2 & 0.15 \\ 0.3 & -0.5 & 0.05 \\ 1 & 1 & 1 \end{vmatrix} = 0.29$$

$$A_1 = \begin{vmatrix} 0 & 0.2 & 0.15 \\ 0 & -0.5 & 0.05 \\ 1 & 1 & 1 \end{vmatrix} = 0.085 \qquad \pi_1 = \frac{A_1}{A} = 0.2931034$$

$$A_2 = \begin{vmatrix} -0.4 & 0 & 0.15 \\ 0.3 & 0 & 0.05 \\ 1 & 1 & 1 \end{vmatrix} = 0.065 \qquad \pi_2 = \frac{A_2}{A} = 0.2241379$$

$$A_3 = \begin{vmatrix} -0.4 & 0.2 & 0 \\ 0.3 & -0.5 & 0 \\ 1 & 1 & 1 \end{vmatrix} = 0.14 \qquad \pi_3 = \frac{A_3}{A} = 0.4827586$$

Thus in the long run the steady state market shares will be as follows,

$$\pi_1 = 0.2931; \qquad \pi_2 = 0.2241; \qquad \pi_3 = 0.4828$$

Solution (c, d, e)

$$\Pi(2) = \Pi(0)P^2$$

$$= \begin{bmatrix} 1 & 0 & 0 \end{bmatrix} \begin{bmatrix} 0.60 & 0.30 & 0.10 \\ 0.20 & 0.50 & 0.30 \\ 0.15 & 0.05 & 0.80 \end{bmatrix}^2$$

$$= \begin{bmatrix} 1 & 0 & 0 \end{bmatrix} \begin{bmatrix} 0.60 & 0.30 & 0.10 \\ 0.20 & 0.50 & 0.30 \\ 0.15 & 0.05 & 0.80 \end{bmatrix} \begin{bmatrix} 0.60 & 0.30 & 0.10 \\ 0.20 & 0.50 & 0.30 \\ 0.15 & 0.05 & 0.80 \end{bmatrix}$$

$$= \begin{bmatrix} 1 & 0 & 0 \end{bmatrix} \begin{bmatrix} 0.435 & 0.335 & 0.235 \\ 0.265 & 0.325 & 0.410 \\ 0.220 & 0.110 & 0.670 \end{bmatrix}$$

$$\begin{bmatrix} \pi_1(2) & \pi_2(2) & \pi_3(2) \end{bmatrix} = \begin{bmatrix} 0.435 & 0.335 & 0.230 \end{bmatrix}$$

c) $\pi_1(2) = 0.435$

d) $\pi_2(2) = 0.335$

e) $\pi_3(2) = 0.230$

Excel Solution:

Excel Result Sheet

Table 18.14(a)

Period	Initial Shopping at Food World		
	$\pi_1(n)$	$\pi_2(n)$	$\pi_3(n)$
0	1	0	0
1	0.600	0.300	0.100
2	0.435	0.335	0.230
3	0.363	0.310	0.328
4	0.329	0.280	0.392
5	0.312	0.258	0.430
6	0.303	0.244	0.453
7	0.299	0.236	0.466
8	0.296	0.231	0.473
9	0.295	0.228	0.477
10	0.294	0.226	0.480
11	0.294	0.225	0.481
12	0.293	0.225	0.482
13	0.293	0.225	0.482
14	0.293	0.224	0.482
15	0.293	0.224	0.483
16	0.293	0.224	0.483
17	0.293	0.224	0.483
18	0.293	0.224	0.483
19	0.293	0.224	0.483
20	0.293	0.224	0.483
21	0.293	0.224	0.483
22	0.293	0.224	0.483
23	0.293	0.224	0.483
24	0.293	0.224	0.483
25	0.293	0.224	0.483
26	0.293	0.224	0.483
27	0.293	0.224	0.483
28	0.293	0.224	0.483
29	0.293	0.224	0.483
30	0.293	0.224	0.483

$$P = \begin{vmatrix} 0.6 & 0.3 & 0.1 \\ 0.2 & 0.5 & 0.3 \\ 0.15 & 0.05 & 0.8 \end{vmatrix}$$

Table 18.15

Period	Initial Shopping at Food World		
	$\pi_1(n)$	$\pi_2(n)$	$\pi_3(n)$
0	1	0	0
1	0.600	0.300	0.100
2	0.435	0.335	0.230
3	0.363	0.310	0.328
4	0.329	0.280	0.392

$$P = \begin{vmatrix} 0.6 & 0.3 & 0.1 \\ 0.2 & 0.5 & 0.3 \\ 0.15 & 0.05 & 0.8 \end{vmatrix}$$

Example 6

Recently, a market research unit has conducted a survey of customers buying habits three brands of toothpaste in Bangalore. It estimates that, at present, 20% of the customers buy brand A, 50% of the customers buy brand B and 30% of the customers buy brand C. In addition the research unit reported the following switching brand matrix.

Table 18.16

Brand just bought	Brand next bought		
	A	B	C
A	0.6	0.3	0.1
B	0.4	0.5	0.1
C	0.2	0.1	0.7

What will be the expected distribution of market shares in two time periods latter?

Solution

The initial condition indicates the present market share.

$$\Pi(0) = [\pi_1(0) \quad \pi_2(0) \quad \pi_3(0)] = [0.20 \quad 0.50 \quad 0.30]$$

Further from the table given above the transitional probability matrix is obtain as:

$$P = \begin{bmatrix} 0.6 & 0.3 & 0.1 \\ 0.4 & 0.5 & 0.1 \\ 0.2 & 0.1 & 0.7 \end{bmatrix}$$

The expected distribution of market share after 2 periods from row $\Pi(2)$ is obtained as $\Pi(0) \times P^2$.

$$\Pi(2) = [\pi_1(2) \quad \pi_2(2) \quad \pi_3(2)]$$
$$= [\pi_1(0) \quad \pi_2(0) \quad \pi_3(0)]\, P^2$$
$$= [\pi_1(0) \quad \pi_2(0) \quad \pi_3(0)] \begin{bmatrix} 0.6 & 0.3 & 0.1 \\ 0.4 & 0.5 & 0.1 \\ 0.2 & 0.1 & 0.7 \end{bmatrix}^2$$
$$= [0.20 \quad 0.50 \quad 0.30] \begin{bmatrix} 0.6 & 0.3 & 0.1 \\ 0.4 & 0.5 & 0.1 \\ 0.2 & 0.1 & 0.7 \end{bmatrix}^2$$
$$= [0.420 \quad 0.312 \quad 0.268]$$

Thus, after 2 periods the expected market shares will be 42%, 31.2%, 26.8% respectively for three brands A, B, and C.

Excel Solution

Table 18.17

Period	Initial Shopping at Food World		
	$\pi_1(n)$	$\pi_2(n)$	$\pi_3(n)$
0	0.2	0.5	0.3
1	0.380	0.340	0.280
2	0.420	0.312	0.268
3	0.430	0.309	0.261
4	0.434	0.310	0.256
P =	0.6	0.3	0.1
	0.4	0.5	0.1
	0.2	0.1	0.7

■ Example 7

A research team surveyed the shoppers in an area to discover the brand switching habits among customers in three brands A, B, and C type of detergents. Finally they were able to get the following information from the shoppers.

Table 18.18

Current purchase	Previous purchase	Number
A	A	200
B	B	150
C	C	100
B	A	50
C	A	25
A	B	80
C	b	45
A	C	130
B	C	20

a) By using the given data obtain the transition matrix.

b) Calculate the market share for each of the brands after two periods from now.

Solution

Let us tabulate the given data first as shown below.

Table 18.19

From brand	To brand			
	A	B	C	Total
A	200	50	25	275
B	80	150	45	275
C	130	20	100	250
Total	410	220	170	800

a) Now the needed transition matrix is obtained by dividing each element in the above table by the respective row total.

$$P = \begin{bmatrix} 8/11 & 2/11 & 1/11 \\ 16/55 & 30/55 & 9/55 \\ 13/25 & 2/25 & 10/25 \end{bmatrix}$$

b) The current market share vector is obtained by dividing the column total by the grand total in the data summary table given above.

$$\Pi(0) = [410/800 \quad 220/800 \quad 170/800]$$

Now the needed $\Pi(2)$ is obtained as

$$\Pi(2) = \Pi(0) \times P^2$$

$$= \begin{bmatrix} 41/80 & 22/80 & 17/80 \end{bmatrix} \begin{bmatrix} 8/11 & 2/11 & 1/11 \\ 16/55 & 30/55 & 9/55 \\ 13/25 & 2/25 & 10/25 \end{bmatrix}^2$$

$$= \begin{bmatrix} 0.571 & 0.258 & 0.164 \end{bmatrix}$$

Thus, the expected market share after 2 periods latter will be: 57.71%, 25.84%, and 16.44% respectively for A, B, and C brands.

Excel Solution

Table 18.20

Period	Initial Shopping at Food World		
	$\pi_1(n)$	$\pi_2(n)$	$\pi_3(n)$
0	0.5125	0.275	0.2125
1	0.559	0.260	0.177
2	0.571	0.258	0.164
3	0.572	0.257	0.160
4	0.571	0.257	0.158
	0.73	0.18	0.09
P =	0.29	0.55	0.16
	0.52	0.08	0.40

■ Example 8

Suppose that new razor blades were introduced in the market by three companies at the same time. When they were introduced, each company had an equal share of the market, but during the first year the following changes took place.

i) Company A retained 90% of its customers, lost 3% to B and 7% to C

ii) Company B retained 70% of the customers, lost 10% to A and 20% to C

iii) Company C retained 80% of the customers, lost 10% to A and 10% to B

Assume that no changes in the buying habits of the consumer occur.

1. What is the market share of these companies at the end of the second year?

2. What are the long run market shares of these three companies?

Solution

1) From the data given we obtain the initial condition and the transition matrix obtained as

$$\Pi(0) = \left[\pi_1(0) \quad \pi_2(0) \quad \pi_3(0)\right] = \left[0.333 \quad 0.333 \quad 0.333\right]$$

$$P = \begin{bmatrix} 0.90 & 0.03 & 0.07 \\ 0.10 & 0.70 & 0.20 \\ 0.10 & 0.10 & 0.80 \end{bmatrix}$$

$$\Pi(1) = \Pi(0) \times P$$

$$= \left[0.333 \quad 0.333 \quad 0.333\right] \begin{bmatrix} 0.90 & 0.03 & 0.07 \\ 0.10 & 0.70 & 0.20 \\ 0.10 & 0.10 & 0.80 \end{bmatrix}$$

$$= \left[0.367 \quad 0.277 \quad 0.356\right]$$

Market share after one year will be: 36.7%, 27.7%, and 35.6% respectively.

$$\Pi(2) = \Pi(0) \times P^2$$

$$= \left[0.333 \quad 0.333 \quad 0.333\right] \begin{bmatrix} 0.90 & 0.03 & 0.07 \\ 0.10 & 0.70 & 0.20 \\ 0.10 & 0.10 & 0.80 \end{bmatrix} \begin{bmatrix} 0.90 & 0.03 & 0.07 \\ 0.10 & 0.70 & 0.20 \\ 0.10 & 0.10 & 0.80 \end{bmatrix}$$

$$= \left[0.3933 \quad 0.2403 \quad 0.3664\right]$$

Market share at the end of second year: 39.33%, 24.03%, and 36.64% respectively.

2) Let π_1, π_2 and π_3 be the expected market share in the long run.

$$\left[\pi_1 \quad \pi_2 \quad \pi_3\right] = \left[\pi_1 \quad \pi_2 \quad \pi_3\right] \left[P\right]$$

$$= \left[\pi_1 \quad \pi_2 \quad \pi_3\right] \begin{bmatrix} 0.90 & 0.03 & 0.07 \\ 0.10 & 0.70 & 0.20 \\ 0.10 & 0.10 & 0.80 \end{bmatrix}$$

$$\pi_1 = 0.90\pi_1 + 0.10\pi_2 + 0.10\pi_3$$
$$\pi_2 = 0.03\pi_1 + 0.07\pi_2 + 0.10\pi_3$$
$$\pi_3 = 0.07\pi_1 + 0.20\pi_2 + 0.80\pi_3$$

Now from the first two equations and the probability condition we obtain

$$0.10\pi_1 - 0.10\pi_2 - 0.10\pi_3 = 0.... \qquad (1)$$
$$-0.03\pi_1 + 0.3\pi_2 - 0.10\pi_3 = 0.... \qquad (2)$$
$$\pi_1 + \pi_2 + \pi_3 = 1.... \qquad (3)$$

Now as Cramer's rule we obtain the results as shown below

Excel Solution

Table 18.21

Period	Initial Shopping at Food World		
	$\pi_1(n)$	$\pi_2(n)$	$\pi_3(n)$
0	0.33333	0.33333	0.33333
1	0.367	0.277	0.357
2	0.393	0.240	0.366
3	0.415	0.217	0.369
4	0.432	0.201	0.367
5	0.445	0.190	0.364
6	0.456	0.183	0.361
7	0.465	0.178	0.357
8	0.472	0.174	0.354
9	0.478	0.171	0.351
10	0.482	0.169	0.348
11	0.486	0.168	0.346
12	0.489	0.167	0.345
13	0.491	0.166	0.343
14	0.493	0.165	0.342
15	0.494	0.165	0.341
16	0.495	0.164	0.341
17	0.496	0.164	0.340
18	0.497	0.164	0.339
19	0.498	0.163	0.339
20	0.498	0.163	0.339
21	0.498	0.163	0.338
22	0.499	0.163	0.338
23	0.499	0.163	0.338
24	0.499	0.163	0.338
25	0.499	0.163	0.338
26	0.499	0.163	0.338
27	0.500	0.163	0.338
28	0.500	0.163	0.338
29	0.500	0.163	0.338
30	0.500	0.163	0.338

$$P = \begin{vmatrix} 0.50 & 0.03 & 0.07 \\ 0.10 & 0.70 & 0.20 \\ 0.10 & 0.10 & 0.80 \end{vmatrix}$$

$$A = \begin{vmatrix} 0.1 & -0.1 & -0.1 \\ -0.03 & 0.3 & -0.1 \\ 1 & 1 & 1 \end{vmatrix} = 0.08$$

$$A_1 = \begin{vmatrix} 0 & -0.1 & -0.1 \\ 0 & 0.3 & -0.1 \\ 1 & 1 & 1 \end{vmatrix} = 0.04 \qquad \pi_1 = \frac{A_1}{A} = 0.5$$

$$A_2 = \begin{vmatrix} 0.1 & 0 & -0.1 \\ -0.03 & 0 & -0.1 \\ 1 & 1 & 1 \end{vmatrix} = 0.013 \qquad \pi_2 = \frac{A_2}{A} = 0.1625$$

$$A_3 = \begin{vmatrix} 0.1 & -0.1 & 0 \\ -0.03 & 0.3 & 0 \\ 1 & 1 & 1 \end{vmatrix} = 0.027 \qquad \pi_1 = \frac{A_3}{A} = 0.3375$$

So in the long run the steady equilibrium market share will be: 50%, 16% and 34% respectively.

18.7 FINITE AND ABSORBING STATES

In our illustrative example of three detergents there are only three finite numbers of states in the sense that the customer under reference must be using either D_1 or D_2 or D_3 and no other brands. Note that all the 100% of the customers are absorbed by the three states.

Further, none of the states are completely absorbing in the sense having the probability of 1.00. Such states are called non-observing states. For such non-observing states the following properties hold good.

- For two states i and j, the sequence of transition that begins at i and ends at j is called a *path* from i to j.
- A state j is *reachable* from the state i provided there is a defined path from i to j.
- The states i and j are said to be *communicative* provided there exists a path from i to j.
- A state i is known as *transient* state if there exists a state j reachable from i, and at the same time i is not reachable from j. Thus, a state i is transient provided there is a way to leave i and joint j but no way to come back from j to i again.
- If a state is not transient then it is called a recurrent state.

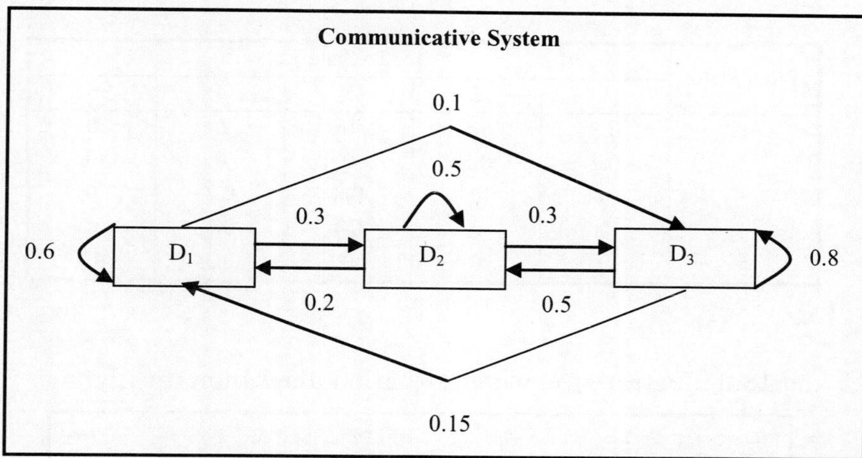

Figure 18.3

In Figure 18.3, it could be seen that each of the three stages is reachable from all others and hence becomes commutative as well. Also none of these three states are in absorbing stage and hence become recurrent and suitable for Markov process analysis.

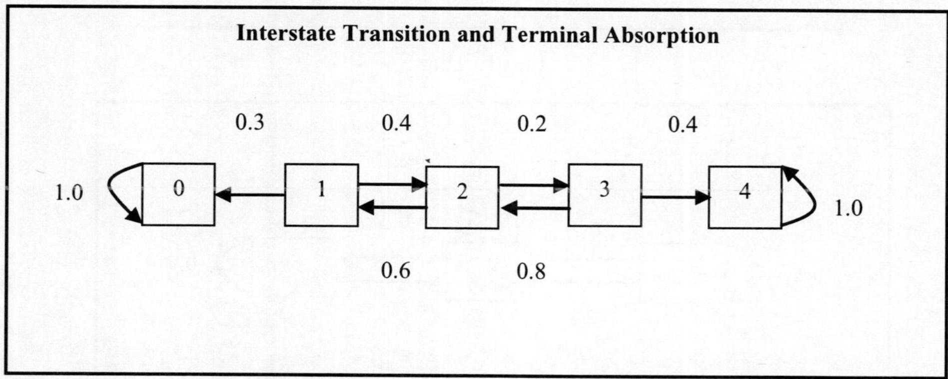

Figure 18.4

In Figure 18.4, states 1, 2 and 3 are in transition state and hence reachable from any other, but state 0 and 4 are in absorbing state with probability 1.00. In such a state once reached no way to come out it.

■ Example 9

Consider the following Probability Transition matrix.

a) Provide an example of a path in it.

b) Give examples of states which communicate.

c) Provide an example of recurrent state.

d) Identify the absorbing states if any.

Table 18.22

From State	To State				
	1	2	3	4	5
1	0.3	0.5	0	0	0
2	0.7	0.5	0	0	0
3	0	0	0.8	0.2	0
4	0	0	0.4	0.3	0.3
5	0	0	0	0.3	0.9

Solution

Figures 18.5 and 18.6 illustrate the whole situation diagrammatically.

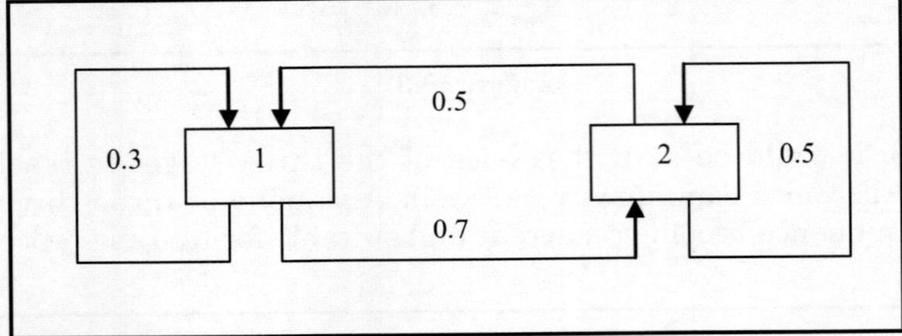

Figure 18.5

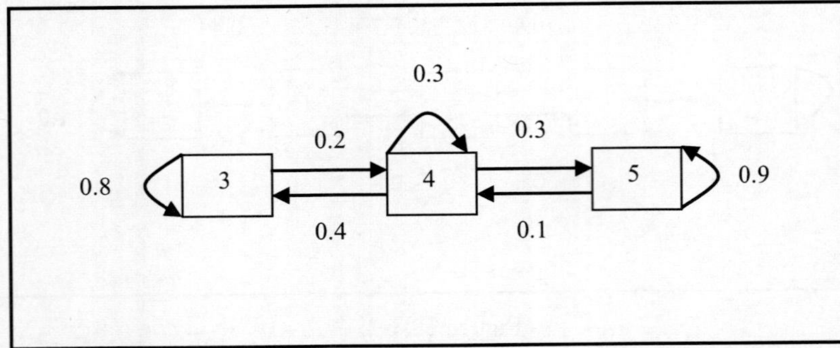

Figure 18.6

a) A path could be seen from state 3 to state 5 via state 4. Similarly a path could be traced from state 1 to 2 also.

b) The state i and j are communicative provided from state i one can reach state j and from state j one can reach i. For example, state 1 and 2 are communicative because state 1 is reachable from state 2 and vice versa.

c) Each state in this matrix is recurrent, because for each i there exist a state j that is reachable from it.

d) None of the states is transient because all of them are recurrent.

e) None of the states is absorbing because no state is having probability 1.

18.8 ABSORBING CHAIN AND ITS APPLICATION IN ACCOUNTS

Till now in all our problems none of the states were found to be absorbing. In this section we consider an application of Markov process with two observing states. In accounting related problems an amount receivable is either received within the overdue time of 90 days or written off as bad debt beyond 90 days. Thus, whatever may be the transient state, finally the said account ends either as paid category or written off category. These two categories are often called observing category. Any account beyond 90 days default is definitely going to be written off with the probability of 1. Similarly, even the paid category within 90 days will not be considered again. Thus the fully paid and fully written cases are not at all in the preview of our analysis. Further, for absorbing type of problems there is no long run steady state probability as such, in the sense that no account will prolong after 90 days. All probability calculations are state probabilities only in this section.

Given this scenario, let us illustrate the Food World's accounting department for illustration. Food World departmental store has two aging categories for its amount receivable: (1) accounts that are classified as 0-30 days old and (2) accounts that are classified as 31-90 days old. If any portion of an account exceeds 90 days, it is written off as bad debt. To illustrate this method of classification let us concentrate on Mr. A's account Suppose customer Mr.A's balance as on April 30 is as follows.

Table 18.23

Date of purchase	Amount billed
August 15	Rs. 25
September 18	Rs. 10
September 28	Rs. 50
Total balance to be paid	Rs. 85

An aging account receivable on September 30 would assign the balance of Rs. 85 to 31 to 90 days category because the oldest unpaid bill of August 15 is 46 days old. Suppose our customer pays the August 15th bill of Rs. 25 on October 7. Now on this day the remaining balance of Rs. 60 deserves a 0-30 day category and will be recorded accordingly, because the oldest unpaid September 18th bill is only 12 days old on September 30. This method of obtaining aging account receivable is called total balance method because the total balance till the aging day is placed in this account. Note that in this illustration the balance portion of Rs. 60 will be shifted from 31-90 days category to 0-30 day's category once the August bill of Rs. 25 is paid on October 7.

Let, on December 31 the departmental store show a total of Rs. 3000 in its accounts receivable from all the customers. Out of this due amount how much the Food World will eventually be collect and how much will eventually treated, as bad debt on December 31 while submitting the year ending financial statement for notification is the real problem.

Now let us see how we can evaluate the account receivable using Markov process. First let us concentrate on imaginary amount 1 rupee which is there on the account receivable account instead of the actual amount of Rs. 3000. As the business concern the manager would be interested in monitoring the situation on weekly basis. At a given point of time the said one rupee will lie in any one of the following four mutually exhaustive states of the system.

1. Paid category
2. Bad debt and hence written off category
3. 0-30 days category
4. 31-90 days category

Now let us check track record of the said one rupee account on weekly basis by using Markov process.

Let p_{ij} = the probability of a rupee in state i in current week moving to the state j in the next week. Based on historical data let the obtained transition probability matrix by Food World be

$$P = \begin{bmatrix} p_{11} & p_{12} & p_{13} & p_{14} \\ p_{21} & p_{22} & p_{23} & p_{24} \\ p_{31} & p_{32} & p_{33} & p_{34} \\ p_{41} & p_{42} & p_{43} & p_{44} \end{bmatrix} = \begin{bmatrix} 1.0 & 0.0 & 0.0 & 0.0 \\ 0.0 & 1.0 & 0.0 & 0.0 \\ 0.4 & 0.0 & 0.3 & 0.3 \\ 0.4 & 0.2 & 0.3 & 0.1 \end{bmatrix}$$

Now let us interpret the entries in the transition matrix by taking the third row as example. From the transaction matrix third row it is clear that the probability of a rupee moving from state 3 (0-30 days) to state 1 (paid category) in the next week is 0.4. Also notice that the probability of this rupee to be there in state 3 itself (0-30 days) in the next week is 0.3, and moving to category 4 (31-90 days) in the next week is also 0.3. But definitely this rupee cannot become a bad debt in the next week because of the zero state in the matrix. Once a rupee is moved into paid category (state 1) the probability of moving to any other state in later weeks is invariably shown by zero entries as shown in the first row of the matrix.

Similarly, once the rupee enters into state 2 bad debt will not be converted into any other state in later weeks as revealed by the second row. Here state 1 and 2 are called absorbing states since once the rupee moves into these states it gets absorbed and hence will not move to any other state. Also note, in the above table the sum of all the probabilities added up to unity for all the four rows as expected.

In Markov process, since we have two definite paid up and bad debt observing states we do not go for the calculation of steady state probability as stated earlier. Irrespective of the states they belong in the transitory period they all ultimately end up in any one of these two observing states.

We will be interested only in

1. Getting the probabilities of a 0-30 age category ending with the paid category as well as bad debt category.
2. Getting the probabilities of a 31-90 age category ending with the paid category as well as bad debt category.

The computation of the said absorbing state probabilities requires the use of what is called *fundamental matrix*–the mathematical logic underlying the fundamental matrix is beyond the scope of this text. However, we derive the fundamental matrix from the transition probability matrix. To do the task we begin the computation by partitioning the given transition matrix as shown below.

$$P = \left[\begin{array}{c} \left\langle\begin{array}{cc|cc} 1.0 & 0.0 & 0.0 & 0.0 \\ 0.0 & 1.0 & 0.0 & 0.0 \end{array}\right\rangle \\ \hline \left\langle\begin{array}{cc|cc} 0.4 & 0.0 & 0.3 & 0.3 \\ 0.4 & 0.2 & 0.3 & 0.1 \end{array}\right\rangle \end{array}\right] = \left[\begin{array}{c} \left\langle\begin{array}{cc|cc} 1.0 & 0.0 & 0.0 & 0.0 \\ 0.0 & 1.0 & 0.0 & 0.0 \end{array}\right\rangle \\ \hline \langle R|Q\rangle \end{array}\right]$$

where, $R = \begin{bmatrix} 0.4 & 0.0 \\ 0.4 & 0.2 \end{bmatrix} \qquad Q = \begin{bmatrix} 0.3 & 0.3 \\ 0.3 & 0.1 \end{bmatrix}$

Now the needed fundamental matrix is defined as

$$N = (I - Q)^{-1}$$

$$I - Q = \begin{bmatrix} 1 & 0 \\ 0 & 1 \end{bmatrix} - \begin{bmatrix} 0.3 & 0.3 \\ 0.3 & 0.1 \end{bmatrix} = \begin{bmatrix} 0.7 & -0.3 \\ -0.3 & 0.9 \end{bmatrix}$$

$$I - Q = \begin{bmatrix} 0.7 & -0.3 \\ -0.3 & 0.9 \end{bmatrix}$$

$$\text{Cofactor of } I - Q = \begin{bmatrix} 0.9 & 0.3 \\ 0.3 & 0.7 \end{bmatrix}$$

$$\text{adj of } I - Q = \begin{bmatrix} 0.9 & 0.3 \\ 0.3 & 0.7 \end{bmatrix}$$

$$\text{Determinant } I - Q = 0.54$$

$$N = (I - Q)^{-1} = \begin{bmatrix} 1.67 & 0.56 \\ 0.56 & 1.3 \end{bmatrix}$$

$$N \times R = \begin{bmatrix} 1.67 & 0.56 \\ 0.56 & 1.3 \end{bmatrix} \times \begin{bmatrix} 0.4 & 0 \\ 0.4 & 0.2 \end{bmatrix} = \begin{bmatrix} 0.89 & 0.11 \\ 0.74 & 0.26 \end{bmatrix}$$

The first row in NR above gives us the probabilities of a 0-30 category will end up with respective observing states. With probability of 0.89 this category will end with paid state. Similarly, with probability of 0.11 this category will end with bad debt state. The second row in NR gives us the probabilities of a 31-90 category will end up with respective observing states. With probability of 0.74 this category will end with paid state. Similarly, with probability of 0.26 this category will end with Bad Debt State.

■ Example 10

Consider the receivable account of a firm. Suppose that, for this firm, amounts receivable turns into a bad debt provided the account is more than three months old.

Now at the beginning of each month, each of the accounts may be classified into one of the following states.

1. New account.
2. Account is one month overdue for payment.
3. Account is two months overdue for payment.
4. Account is three months overdue for payment.
5. Account has been paid.
6. Account is written off as bad debt.

Suppose further that on the basis of past experience, the bank evaluated the following transition matrix for the six states stated.

Table 18.24

State	State					
	1	2	3	4	5	6
1	0	0.7	0	0	0.3	0
2	0	0	0.6	0	0.4	0
3	0	0	0	0.4	0.6	0
4	0	0	0	0	0.8	0.2
5	0	0	0	0	1	0
6	0	0	0	0	0	1

i) What is the probability of a new account shall eventually be paid?

ii) What is the probability that a new account shall be written off as bad debt?

iii) What is the probability that a one month overdue account shall eventually be paid?

iv) What is the probability that a one month overdue account shall eventually be written off?

v) What is the probability that a two months overdue account shall eventually be paid?

vi) What is the probability that a two months overdue account shall eventually be written off?

vii) If the bank makes a credit sale of Rs. 4,00,000 per month then what amount it can expect to lose per annum by way bad debt.

Solution

To answer these questions let us we first partition the given transition matrix as shown below with the states listed as transient states as one part and absorbing states as another part.

Table 18.25

State	State					
	t_1	t_2	t_3	t_4	a_1	a_2
t_1	0	0.7	0	0	0.3	0
t_2	0	0	0.6	0	0.4	0
t_3	0	0	0	0.4	0.6	0
t_4	0	0	0	0	0.8	0.2
a_1	0	0	0	0	1	0
a_2	0	0	0	0	0	1

Now let $Q = \begin{bmatrix} 0 & 0.7 & 0 & 0 \\ 0 & 0 & 0.6 & 0 \\ 0 & 0 & 0 & 0.4 \\ 0 & 0 & 0 & 0 \end{bmatrix}$ $R = \begin{bmatrix} 0.3 & 0 \\ 0.4 & 0 \\ 0.6 & 0 \\ 0.8 & 0.2 \end{bmatrix}$

Now the needed fundamental matrix is defined as

$$NR = (I - Q)^{-1} R$$

$$I - Q = \begin{bmatrix} 1 & 0 & 0 & 0 \\ 0 & 1 & 0 & 0 \\ 0 & 0 & 1 & 0 \\ 0 & 0 & 0 & 1 \end{bmatrix} - \begin{bmatrix} 0 & 0.7 & 0 & 0 \\ 0 & 0 & 0.6 & 0 \\ 0 & 0 & 0 & 0.4 \\ 0 & 0 & 0 & 0 \end{bmatrix}$$

$$= \begin{bmatrix} 1 & -0.7 & 0 & 0 \\ 0 & 1 & -0.6 & 0 \\ 0 & 0 & 1 & -0.4 \\ 0 & 0 & 0 & 1 \end{bmatrix}$$

$$(I - Q)^{-1} = \begin{bmatrix} 1 & 0.7 & 0.42 & 0.168 \\ 0 & 1 & 0.6 & 0.24 \\ 0 & 0 & 1 & 0.4 \\ 0 & 0 & 0 & 1 \end{bmatrix}$$

$$(I - Q)^{-1} R = \begin{bmatrix} 1 & 0.7 & 0.42 & 0.168 \\ 0 & 1 & 0.6 & 0.24 \\ 0 & 0 & 1 & 0.4 \\ 0 & 0 & 0 & 1 \end{bmatrix} \times \begin{bmatrix} 0.3 & 0 \\ 0.4 & 0 \\ 0.6 & 0 \\ 0.8 & 0.2 \end{bmatrix}$$

$$= \begin{bmatrix} 0.9664 & 0.0336 \\ 0.9520 & 0.0480 \\ 0.9200 & 0.0800 \\ 0.8000 & 0.2000 \end{bmatrix}$$

The ij[th] element of this fundamental matrix gives the expected number of periods that will have to be spent in the transient state before reaching any one of the absorbing states. Similarly, if the chain begins in a transient state say t_i then the probability that it shall eventually absorbed in absorbing state a_j is given by the ij[th] element of the matrix $(I-Q)^{-1}R$.

i) The probability of a new account holder (t_1) shall eventually make payment (a_1) = 0.9664

ii) The probability that a new account holder (t_1) shall be written off (a_2) as bad debt = 0.0336

iii) The probability that a one month overdue account holder (t_2) shall eventually make payment (a_1) = 0.9520

iv) The probability that a one month overdue account holder (t_2) shall eventually written off (a_2) = 0.0480

v) The probability that a two months overdue account holder (t_3) shall eventually make payment (a_1) = 0.9200

vi) What is the probability that a two months overdue account holder (t_3) shall eventually written off (a_2) = 0.0800

vii) For a credit sale of Rs. 4,00,000 per month the corresponding annual sales will by 12 × 4,00,000 = Rs. 48,00,000. Thus the amount that the bank is likely to lose as bad debt = 48,00,000×0.0336 = Rs. 1,61,280.

EXERCISES

1. What do you understand by Markov process? In what areas of management can they be applied?

2. What do you understand by transition probabilities? Is the assumption of stationary transition probabilities realistic, in your opinion? Why or why are not?

3. Distinguish between recurrent and an absorbing state.

4. Discuss the importance of the following assumption in Markov Chain

   a) Finite state

   b) First-order process

   c) Stationary and transition probabilities

   d) Uniform time period

5. How do we calculate the steady state probabilities? Do you think they are independent of initial conditions?

6. Explain the use of probability tree in Markov process with suitable illustration.

7. (a) The purchase patterns of two brands of toothpaste can be expressed as a Markov process with the following probabilities:

	Brand X	Brand Y
Brand X	0.90	0.10
Brand Y	0.05	0.95

(i) Which brand appears to have more loyal customers?

(ii) What are the projected market shares for the two brands?

(b) Suppose that in part (a) a new toothpaste brand enters the market such that the following transitional probabilities exist

	Brand X	Brand Y	Brand Z
Brand X	0.80	0.10	0.10
Brand Y	0.05	0.75	0.20
Brand Z	0.40	0.30	0.30

What are the new long run market shares? Which brand will suffer from introduction of the new brand of toothpaste?

8. A market survey is made on three brands of breakfast foods X, Y and Z. Every time the customer purchases a new package, he may buy the same brand or switch over to other brands. The following estimates are obtained, expressed as decimal fractions:

$$\text{Present Brand}\begin{matrix} \\ X \\ Y \\ Z \end{matrix}\overset{\text{Next Brand}}{\begin{matrix} X & Y & Z \\ \end{matrix}}\begin{pmatrix} 0.7 & 0.2 & 0.1 \\ 0.3 & 0.5 & 0.2 \\ 0.3 & 0.3 & 0.4 \end{pmatrix}$$

At this time, it is estimated that 30 percent of the people buy brand X, 20 percent brand Y and 50 percent brand Z. What will the distribution of customers by two times periods later and at equilibrium?

(Delhi University, M.B.A, Dec. 1984)

9. On January 1, (this year), Bakery A had 40% of its market share while the other two bakeries B and C has 40% and 20% respectively of the market share. Based upon a study by a marketing research firm, the following facts were compiled. Bakery A retains 90% of its own customers while gaining 5% of B's customers and 10% of C's customers. Bakery B retains 85% of its own customers while gaining 5% of A's customers and 7% of C's customers. Bakery C retains 83% of its own customers and gains 5% of A's customers and 10% of B's customers. What will each firm share be on January 1, next year and what will each firm's market share be at equilibrium?

(Delhi University, M.B.A, November 1996)

10. A manufacturing company has a certain piece of equipment that is inspected at the end of each day and classified as overhauled, good, fair or inoperative. If the piece is inoperative, it is overhauled, a procedure that takes one day. Assume that the working condition of the equipment follows a Markov process with the following transitional matrix.

From state	To state			
	1	2	3	4
1	0.00	0.75	0.25	0.00
2	0.00	0.50	0.50	0.00
3	0.00	0.00	0.50	0.50
4	1.00	0.00	0.00	0.00

It costs Rs. 125 a machine (including last time) on the average and Rs. 75 in production is in lost if the machine is fund inoperative. Using the steady state probability, compute the expected per day cost of maintenance.

11. A credit card company is attempting to determine a more effective set of credit control polices. It has traditionally classified all its accounts receivable into five distinct categories listed below.

Account receivable category	Status of account receivable
1	0-30 days late
2	31-90 days late
3	91 plus date late
4	Paid in full
5	Defaulted

Based on the research, the company has derived the following transition matrix describing the behaviour of various categories on a weekly basis.

From category	To category				
	1.00	2.00	3.00	4.00	5.00
1.00	0.40	0.20	0.10	0.20	0.10
2.00	0.30	0.40	0.10	0.10	0.10
3.00	0.20	0.40	0.10	0.10	0.20
4.00	0.00	0.00	0.00	1.00	0.00
5.00	0.00	0.00	0.00	0.00	1.00

Currently the company has the following accounts in various categories.

Category	Amount
1	Rs. 2 Million
2	Rs. 4 Million
3	Rs. 3 Million

You are required to calculate the expected amounts which may eventually be (i) collected, and (ii) defaulted.

CHAPTER

19

Business Forecasting

19.1 APPROACHES TO FORECASTING

Under uncertain business environment, forecasting the future is very much essential. Augmenting inventory without knowledge of the future demand is really a wasteful exercise. If the future demand is correctly forecasted then to the meet the eventuality the decision maker can plan and have the correct inventory. In case of flexible production one can think of even zero inventories. In all such cases the accuracy of prediction that matters much. Two basic approaches are available for forecasting future outcomes. They are:

1. Qualitative
2. Quantitative

Qualitative forecasting uses various judgement techniques based on experience. For example, forecasting the future demand for a new product/service, cannot have past data and therefore has to relay on one's own experience or experience of experts in the concerned field. Even in the case of existing products, with all the relevant past data, the business enterprise often prefer to go for qualitative judgement method in the event of rapid chances in the environmental conditions that makes the past data totally irreverent. However, since the qualitative models use personal judgements they are always subjective.

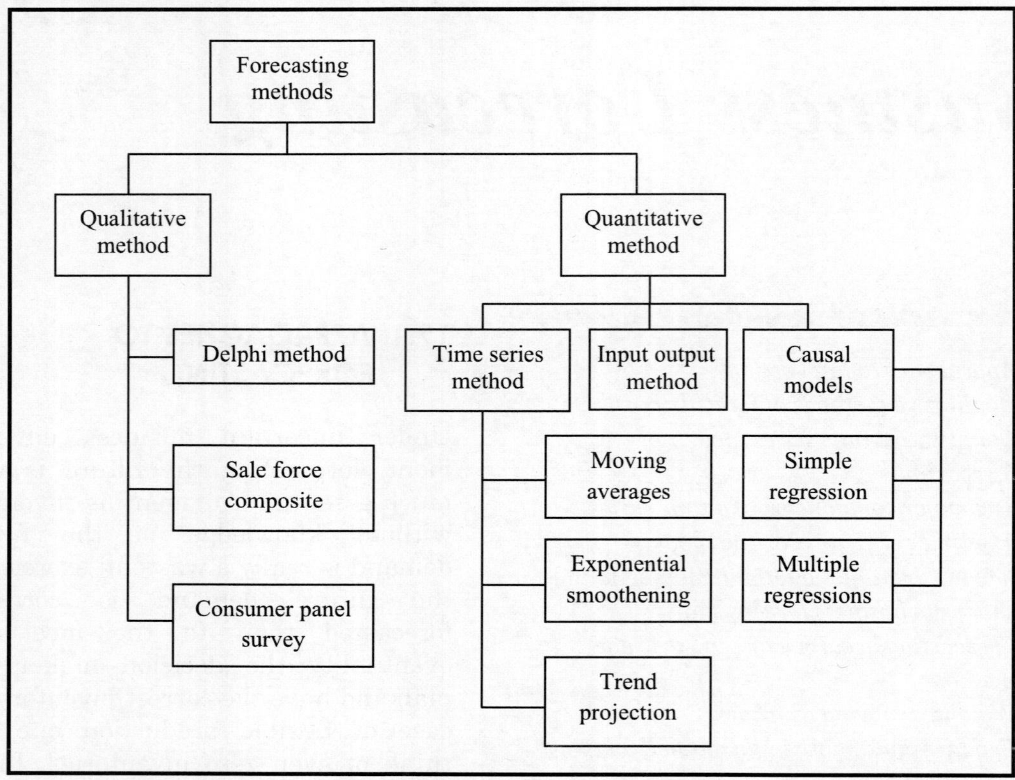

Figure 19.1

Availability of reliable past data is the essential precondition for effective use of the quantitative techniques for business forecasting. In addition to the availability of reliable past data, there must be sufficient evidence that the future will look like the past. In such models the past is taken as the guide for predicting the future. The above chart illustrates the various methods which are used in business forecasting.

19.2 QUALITATIVE METHODS OF FORECASTING

1. Delphi Method

One of the most commonly used qualitative forecasting techniques is the Delphi Method. This technique, originally was developed by the research group at the Rant Corporation, through "group consensus". The invited experts are not known to each other and their interaction takes place through a coordinator. The other participants other than experts are the staff and respondents. The personal staff assists the experts by preparing, collecting, tabulating and summarizing the discussion. The experts are asked to respond a series questionnaire. The coordinator obtains the experts forecast and assimilates them. Each of the experts whose estimate is not included in the middle 50% of the forecast is asked to explain the reasoning underlying the forecast. Then the coordinator disseminates to each expert a summary of forecasts by all, and the opinion of those whose forecast were found to be outside the middle 50%. With this additional information about all other's views each one of them is permitted to revise their estimate if it is necessary. This procedure is repeated many more times till consensus is arrived. Remember by consensus we do not mean a single answer output, it only expects relatively narrow spread opinions within which the majority of experts concur. In the year 1940 Olaf Helmer, Delphi and Cordon developed this method for Rand Corporation of USA in the area of technological change.

Merits:

- Maintenance of anonymity of respondent's identity is the plus point of this approach.
- Since the respondents are physically polled together at the time of investigation the result obtained could be the best.

Care to be taken:

- The panelist must be really rich in their experience.
- It presupposes the organizers ability in conceptualizing the problem for discussion, generate considerable thinking, and simulate dialogue.

2. Collective Opinion or Sales force composite

In this method all the personnel involved in the sales, from top to bottom, are asked to estimate the likely sales in their respective areas. The so obtained estimates are reviewed about their reliability. Once reliable review is completed in the respective sales area, the so obtained area level estimates are clubbed together to get the district level estimates. Similarly by combining the district levels, regional level estimates are obtained. Ultimately the national level estimates are obtained by combining the regional level estimates.

Merits:

- Since the salesmen are always closer to the final consumer, the result obtained at the lower level is more reliable.
- Such estimates are reviewed to eliminate personal bias if any before the use in the next stage.
- Top to bottom collective opinion is the master key for superiority of this approach.
- It is simple and does not use any high level mathematical and statistical tools.
- The forecast is entirely based on first hand information. Secondary data is totally absent in this approach.
- It suits well even for the introduction of a new product into the market.

Demerits:

- Completely subjective.
- It can be used only for short run predictions.
- At the bottom level, salesmen are normally unaware of economic changes taking place during the time of forecasting. In such cases the predictions may go wrong.

3. Consumer Panel Survey

This is the most direct method of forecasting. Accordingly, the final consumers themselves are asked to estimate their future demand. Thus this method has faith fully in consumers who are really buying the products.

Demerits:

- Whenever shortage is expected on a future date, the consumer normally over-states the demand estimate.
- This method is not useful in the case of household purchases.
- When multiple alternatives are available, their estimate for a particular item under reference need not be so accurate.
- Sometimes the consumers may express their desire and not the actual demand based on their buying power.
- Since the consumer list is inexhaustible the practical utility of this approach is very much low.

19.3 QUANTITATIVE TIME SERIES MODELS OF FORECASTING

Since the past data is related to data in different time periods, it is often referred to as time series data forecasting. Time series refers to statistical data collected at regular intervals of time. The regular interval may be an hour, a day, a month, a year, a decade etc. The statistical method used to analyze the time related data is called the time series analysis. An essential aspect of managing any organisation is planning for its future development. Its long run success is closely related to the organizational ability in predicting the future events correctly.

Good reasoning and awareness of the past and current global economic environment are some of the basic essentials for future planning. However, converting the economic environment into facts like the number of units of sale in the next quarter, amount of profit in the next year, market share in the decade to come and so on, is definitely a difficult task for the manager concerned. In this chapter we propose to introduce several such statistical forecasting methods to assist the decision-makers in the process of prediction and planning.

If the manager's task is to predict the sales figures for his product during the next festival season, invariably he will have to look into the festive sales of at least few past years. Using the past data one can visualize the trend in the expected volume of sales during the arrival of the next festive season. A further review of the said data might reveal a definite seasonal pattern within a year such as peek sales in festive seasons and weak sales non-festive seasons. Such resourceful historical information is often called the time series data.

'Past is the guide for the future' is the slogan of this crucial chapter. The data on variables like price, output, employment etc., are collected at regular time intervals all over the world. The data classified on the basis of time constitute vital information for the control of business activity. A scientific management of business enterprises thus requires continuous decision making process.

The following Table 19.1 shows the hypothetical per capita consumption of cotton from 1980 to 1995. To make the data more appealing often we use pictorial representations like pie chart, bar chart, line charts etc. The following line graph shows the scattered diagram of our hypothetical data pertaining to the per capita consumption of cotton. Such graphs depict the economic history of the concerned variable over the past. In spite of some ups and downs from time to time an overall downward trend in per-capita consumption is noticeable in Figure 19.2.

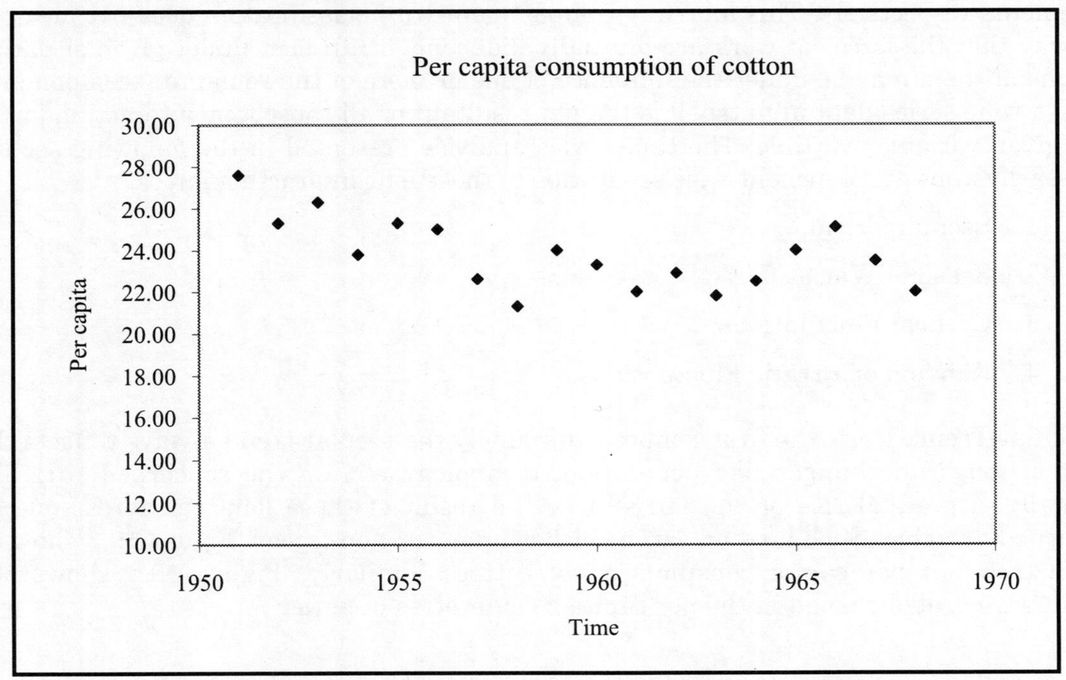

Figure 19.2

Table 19.1

Consumption of cotton in India			
Year	Per capita consumption	Year	Per capita consumption
1951	27.60	1960	23.30
1952	25.30	1961	22.00
1953	26.30	1962	22.90
1954	23.80	1963	21.80
1955	25.30	1964	22.50
1956	25.00	1965	24.00
1957	22.60	1966	25.10
1958	21.30	1967	23.50
1959	24.00	1968	22.00

19.4 THE MAJOR COMPONENTS OF A TIME SERIES

The forces affecting the time series data are classified and grouped into a number of ways depending upon the need. However, to have a common yardstick, we normally classify the data into four major groups to exhibit long-run movements, periodic movements within a year, recurring changes that do not occur necessarily with fixed time frame, and random movements respectively. This four-way compartment-wise classification does not mean in any way that the forces at work are mutually independent. In fact, under given situation any one of them may be under the influence of one or more of the remaining components. Thus, a time series data in nutshell is the net resultant of all these four interacting forces on a given variable overtime. The time series analysis presented in the following section necessarily aims at component-wise separation of these four interacting forces.

1. Secular Trend.
2. Seasonal Variations.
3. Cyclical Fluctuations.
4. Random or Erratic Fluctuations.

1. Secular Trend: Here, the first component, namely the secular trend always reflects the effect of long run changes like population, technology etc. on the concerned variable. Normally, a gradual rise or the fall will be the result of these long run forces on the concerned variable. Such a trend can be either linear or non-linear. Figure 19.3 shows a linear trend in per capita consumption of cotton. Similarly, Figure 19.4 shows the non-linear parabolic trend in the per capita consumption of power.

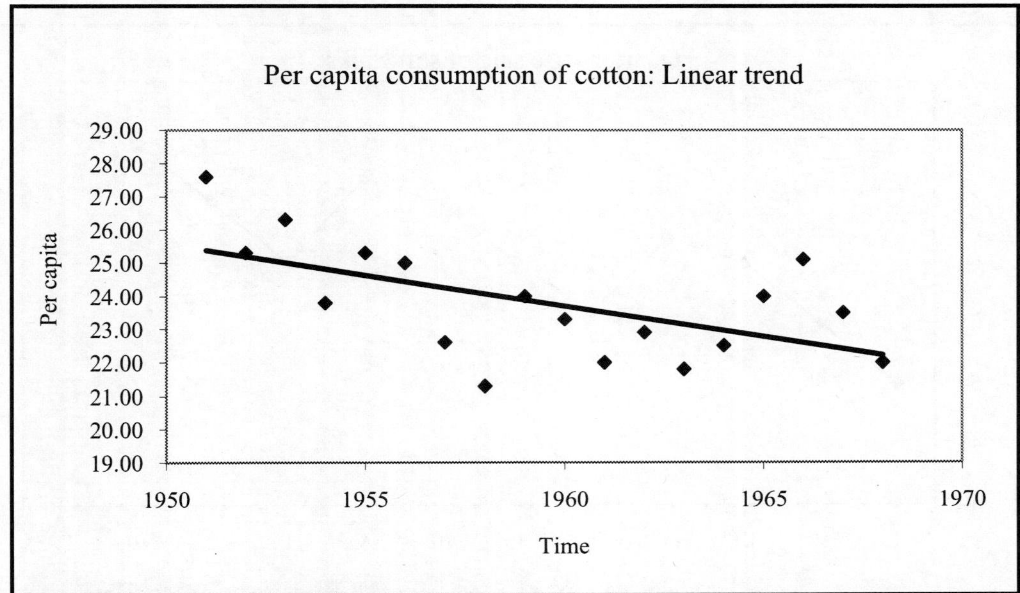

Figure 19.3

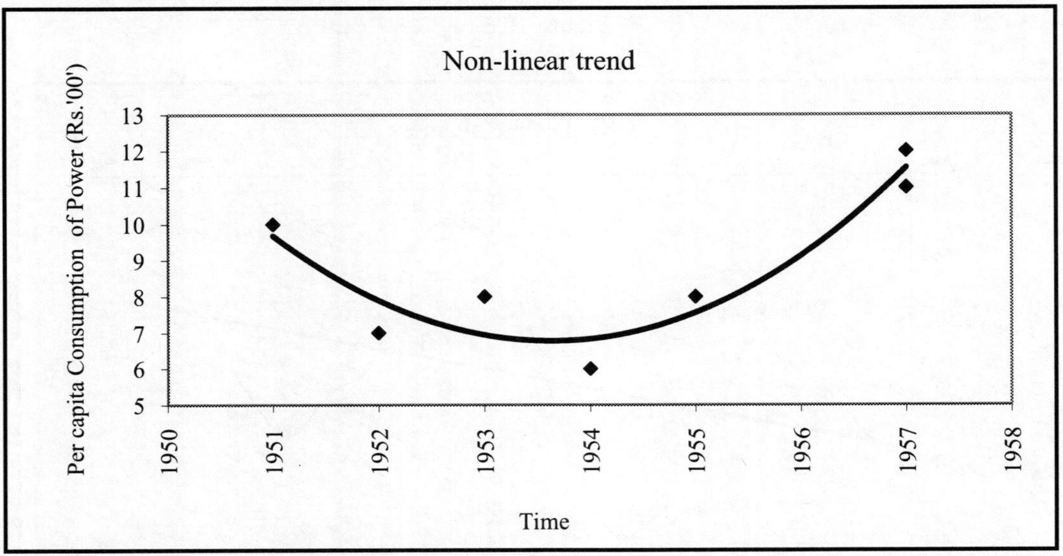

Figure 19.4

2. Seasonal Variations: Seasonal variation normally shows a regular recurring pattern from season to season in a year. Rainfall and temperature are some of the examples for this type of fluctuation. In these variables year after year the said pattern is repeated almost in a similar manner. Figure 19.5 depicts one such seasonal pattern in addition to the trend in the sale of softdrinks for three consecutive years on quarterly basis.

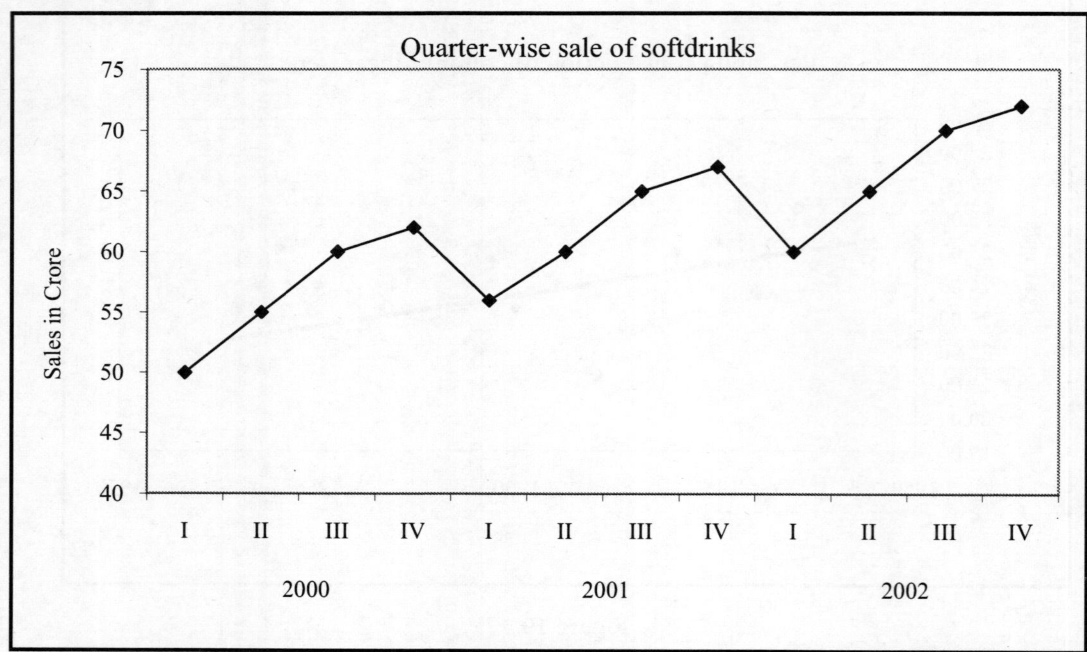

Figure 19.5

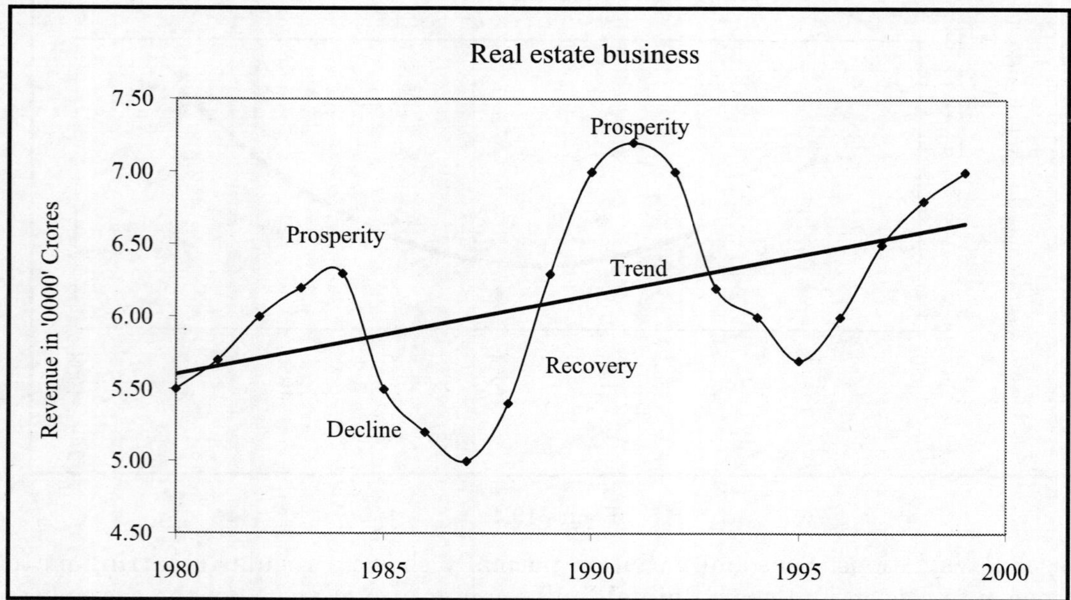

Figure 19.6

3. Cyclical Fluctuations: The third type of fluctuation is again recurring but not with regular uniform intervals. Figure 19.6 illustrates one such cyclical variation in addition to the trend. These cyclical changes indeed are long run movements and can have both rising and falling phases. Here the successive recurrence may occur for every three or four or more years as the case may be depending upon the type of variable concerned. Figure 19.6 depicts one such hypothetical situation relating to real estate business.

4. Random or Erratic Fluctuations: Intermingled with the above three well-defined types, there are numerable small variations with random characteristic. The fluctuations of the given series due to such random forces are called erratic fluctuations. The day-to-day irregularities in business activity could be one such example for this type. When the amounts of sales are classified by days, the figures for different days would expect to show random fluctuations. However, if we represent the same data of sales against weeks then some of these irregularities will cancel themselves and resulting data gets smoothened to some extent. If we repeat sales against months then the data will be still better and smother. Hence, larger the time interval more will be the smoothness and vice versa. Figure 19.7 illustrates one such hypothetical data with random fluctuations.

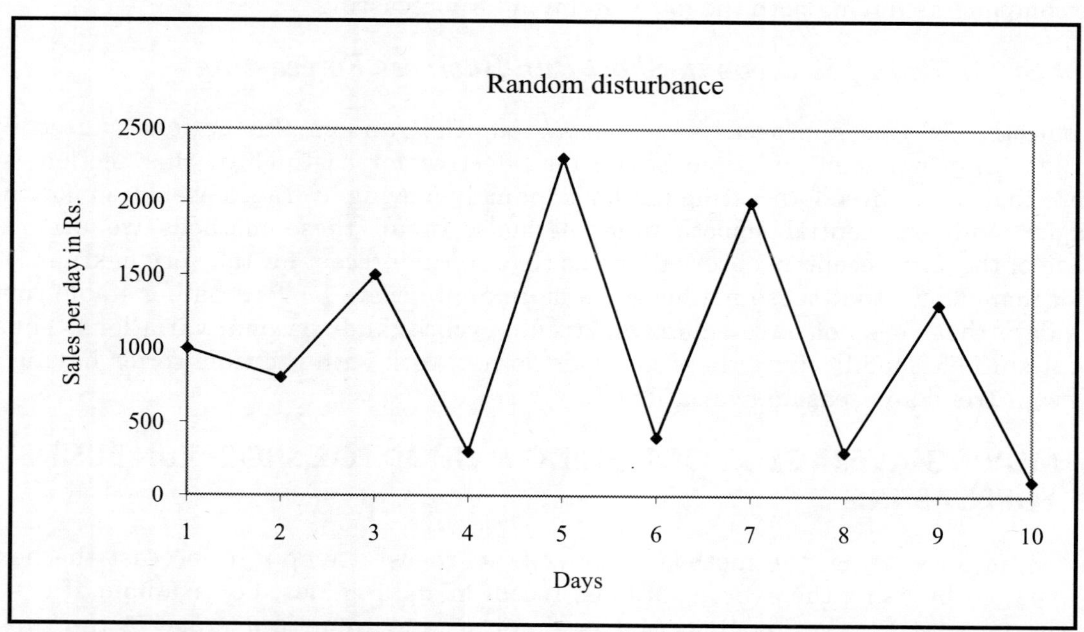

Figure 19.7

19.5 THE TIME SERIES FORECASTING MODELS

A time series normally is assumed to contain one or more types of component discussed in the previous section. Under such circumstances, the given time series would represent only the net effect of all these forces. For example, the volume of production of an industry might show a net increase that was due to a sharp seasonal rise that offsets the cyclical drop. Thus unless these individual effects are separated, the given time series information will not be of much use in the short run predictions. In other words, this is to say that the seasonal budget of an industry invariably should take the seasonal elements into account while budgeting. Similarly, the recent heavy investment decisions of NRI's in India cannot ignore the long run trends of the Indian economic development. Thus, in this sense, the time series generally requires more extensive component-wise analysis instead of concentrating on the net effect alone.

The relationships between these four components are normally expressed either in the multiplicative form or additive form.

Multiplicative form: $X = T \times S \times C \times R$.

Additive form: $X = T + S + C + R$.

In the following section we propose to analyze the various methods of separating these components having both the forms in mind for forecasting.

Use of Smoothening Methods in Short Run Business Forecasting

The objective of this approach is to *smooth out* the random fluctuations caused by irregular components of the time series for effective forecasting. In this section we propose to discuss three forecasting methods, namely moving averages, weighted moving averages, and exponential smoothening methods. In all these methods we use the average of the very recent 'n' observations as the typical forecast for the very next $(n+1)^{th}$ period. Remember, that these methods are more appropriate for forecasting stable time series data that does not have significant trend, cyclical and seasonal variations. Thus, without suitable modification, these methods do not work with the time series having a significant trend and/seasonal variations.

19.6 MOVING AVERAGE SMOOTHENING METHOD FOR SHORT RUN BUSINESS FORECASTING

As a tool of forecasting, the method of moving averages attempts to forecast the next $(n+1)^{th}$ period by using the average of most recent 'n' data values. For example, if n = 3 then the forecast for the fourth period is simply the arithmetic average of the three recent past data values. In this method, the successive average values are calculated by including the next new value in the place of one oldest value in the previous set. While doing so, since we move from one set of data to another set of data till the end, the technique is often called moving average method. The following example illustrates the calculation procedure for predicting the sales for the 13th week by using three-period moving average.

■ Example 1

The demand for an item is observed for 12 weeks and recorded below.

Table 19.2

Weeks	1	2	3	4	5	6	7	8	9	10	11	12
Sales	17	21	19	23	18	16	20	18	22	20	15	22

Calculate 3-weekly moving average and forecast the demand for week 13 in each case.

Solution

Three-period moving average

$$\text{Moving average for weeks 1 to 3} = \frac{17+21+19}{3} = 19$$

$$\text{Moving average for weeks 2 to 4} = \frac{21+19+23}{3} = 21$$

- -

- -

$$\text{Moving average for weeks 10 to 12} = \frac{20+15+22}{3} = 19$$

In a similar way all the 9 moving averages are obtained and reported in the fourth column of Table 19.3. Now, the moving average corresponding to the first three weeks is taken as the forecast for the fourth week and entered accordingly. Similarly, the second moving average from second to fourth week is taken as the forecast value for the fifth week. This procedure is repeated till the end corresponding to the 12th week. The needed forecast for the unknown 13th week is simply the average of 10th, 11th and 12th week data values.

Table 19.3

Three-period moving average forecasting		
Weeks	No. of units sold Y_t	Three-period moving average/forecast F_t
1	17	
2	21	
3	19	
4	23	19
5	18	21
6	16	20
7	20	19
8	18	18
9	22	18
10	20	20
11	15	20
12	22	19
13	Forecast	**19**

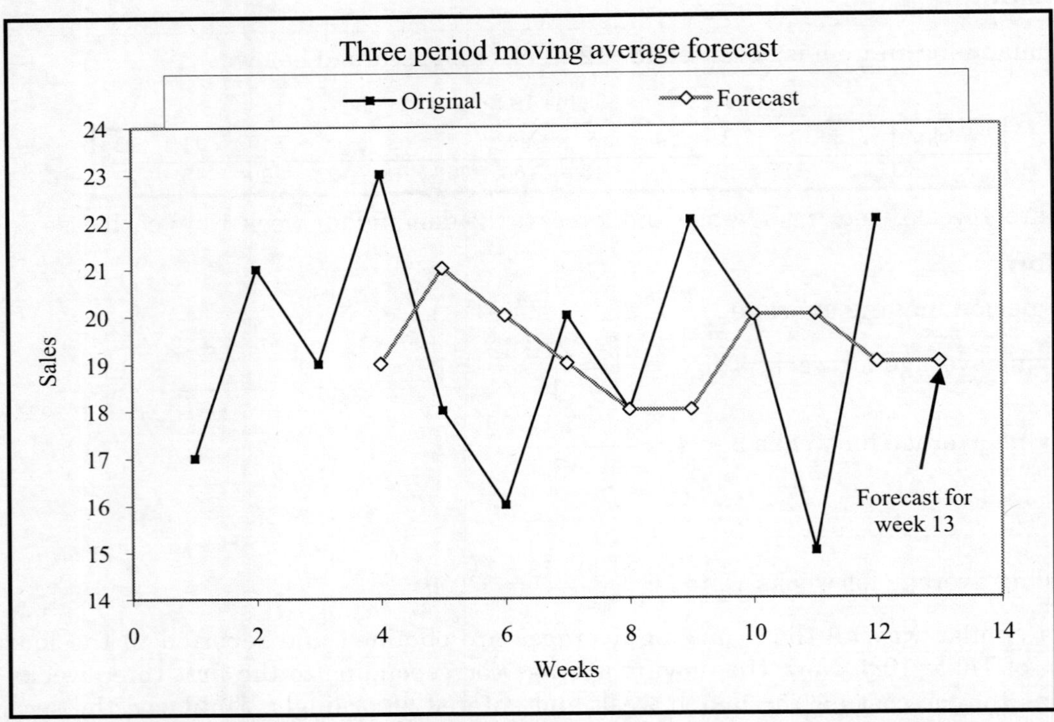

Figure 19.8

19.7 FORECASTING ACCURACY MOVING AVERAGE METHOD

While selecting the period for moving average, the reliability of the forecast must be taken into consideration. Undoubtedly, the choice of period must keep the error at minimum. The following are some of the standard forecasting errors normally used.

1. Mean Absolute Deviation (MAD) $= \dfrac{\sum |Y_t - F_t|}{n}$

2. Mean Square Error (MSE) $= \dfrac{\sum (Y_t - F_t)^2}{n}$

3. Mean Absolute Percentage Error (MAPE) $= \sum \dfrac{(Y_t - F_t)}{Y_t} \times 100$

Where, Y_t = actual value of the data pertaining to period t, and F_t = forecasted value of the data pertaining to period t.

One possible approach for choosing the number of periods to be included while averaging is to use trial and error method and identify the periods that are to be averaged so that the desired error is minimum. As a first step, let us redo example 1 again and calculate the above stated error measurements by taking three periods for averaging. Table 19.2 gives the summary of the needed calculations.

Table 19.4

Three-period moving average forecasting				
Weeks	No. of units sold Y_t	Three-period moving average/forecast F_t	Forecast error $\lvert Y_t - F_t \rvert$	Squared forecast error $(Y_t - F_t)^2$
1	17			
2	21			
3	19			
4	23	19	4	16
5	18	21	3	9
6	16	20	4	16
7	20	19	1	1
8	18	18	0	0
9	22	18	4	16
10	20	20	0	0
11	15	20	5	25
12	22	19	3	9
13	Forecast	19		
n =	9	Total =	24	92
MAD =	2.666666667	MSE =	10.22	

$$\frac{\sum \lvert Y_t - F_t \rvert}{n} = \frac{24}{9} = 2.6667$$

$$\frac{\sum (Y_t - F_t)^2}{n} = \frac{92}{9} = 10.22$$

***Note:** While calculating the errors we notice that we have lost 3 data values. So in the calculation the total error is divided only by 9 and not 12.*

Four-period moving average:

The four-period moving average forecast is reported in Table 19.5. From it we calculate the MAD and MSE as follows.

$$\text{MAD} = \frac{\sum \lvert Y_t - F_t \rvert}{n} = \frac{21.75}{8} = 2.72$$

$$\text{MSE} = \frac{77.1875}{8} = 9.65$$

***Note:** If MAD is considered three-year moving average is better because MAD is less only in that case. If MSE is the choice, then four-year moving average is better.*

Table 19.5

Four-period moving average forecasting				
Weeks	No. of units sold Y_t	Four-period moving average/forecast F_t	Forecast error $\|Y_t - F_t\|$	Squared forecast error $(Y_t - F_t)^2$
1	17			
2	21			
3	19			
4	23			
5	18	20.00	2.00	4.00
6	16	20.25	4.25	18.06
7	20	19.00	1.00	1.00
8	18	19.25	1.25	1.56
9	22	18.00	4.00	16.00
10	20	19.00	1.00	1.00
11	15	20.00	5.00	25.00
12	22	18.75	3.25	10.56
13	Forecast	19.75		
n =	8	Total =	21.75	77.1875
MAD =	2.71875	MSE =	9.65	

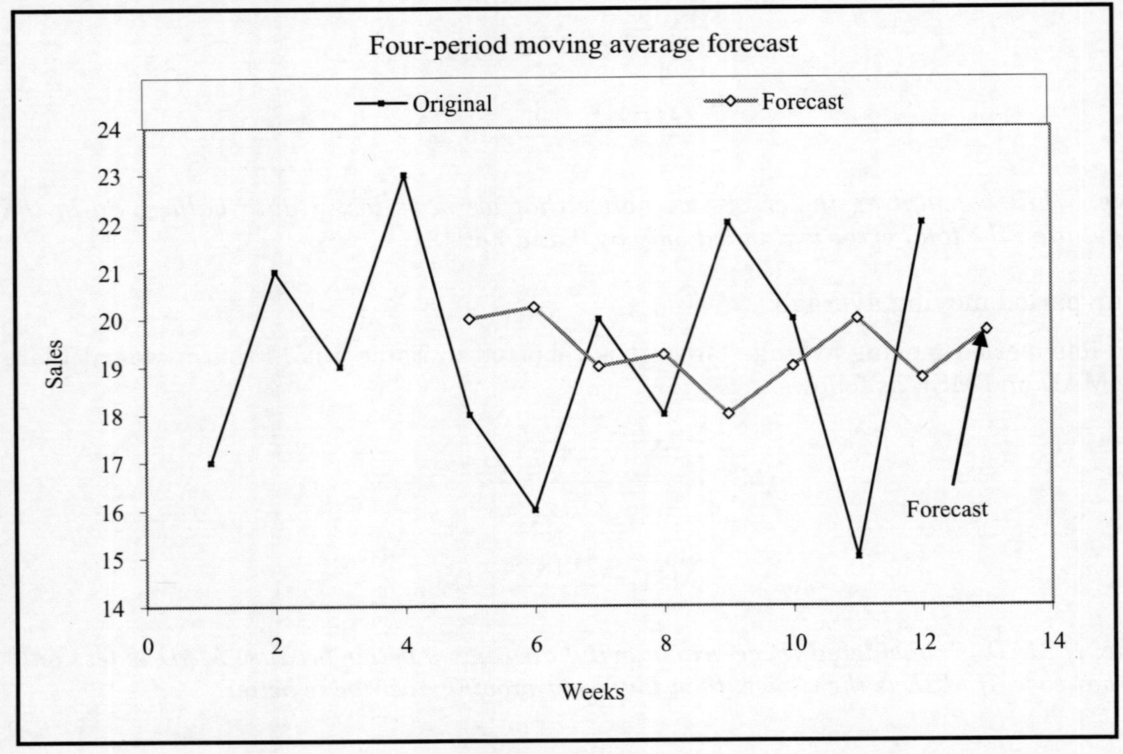

Figure 19.9

19.8 WEIGHTED MOVING AVERAGE

In all the previous calculations all the selected observations were given equal importance in the calculation of the moving average. For example, in the three-year moving average case, each of the three observations was given the weightage of 1/3 totalling to one. Similarly, in the four-period moving average case, every time all the four observations were given 1/4 weights, totalling to one again. However, experience tells us that recent past data is more reliable than the faraway period data for current prediction. Accordingly, for our three-year moving average problem 1 let us calculate the MSE values by giving different weighing schemes based on the *recent past* principle.

■ Example 2

For the data given below forecast the sales for the thirteenth week by using weighted moving average method.

Table 19.6

Weeks	1	2	3	4	5	6	7	8	9	10	11	12
Sales	17	21	19	23	18	16	20	18	22	20	15	22

Solution

i) Weighted three-period moving average with the weighing scheme of 0.7 + 0.2+ 0.1=1

With these weights the needed weighted forecast for the fourth week = 0.7(19) + 0.2(21) + 0.1(17) = 19.20. For the remaining periods, in a similar way the calculated results are reported in Table 19.7. From the table the MSE is obtained as MSE $= \frac{120.03}{9} = 13.33667$.

Table 19.7

Three period weighted moving average				
Weights	0.7	0.2	0.1	
Week	No. of units sold Y_t	Moving average forecast F_t	Forecast error $(Y_t - F_t)$	Squared forecast error $(Y_t - F_t)^2$
1	17			
2	21			
3	19			
4	23	19.20	3.80	14.44
5	18	22.00	-4.00	16.00
6	16	19.10	-3.10	9.61
7	20	17.10	2.90	8.41
8	18	19.00	-1.00	1.00
9	22	18.20	3.80	14.44

(Contd.)

Three period weighted moving average				
Weights	0.7	0.2	0.1	
Week	No. of units sold Y_t	Moving average forecast F_t	Forecast error $(Y_t - F_t)$	Squared forecast error $(Y_t - F_t)^2$
10	20	21.00	-1.00	1.00
11	15	20.20	-5.20	27.04
12	22	16.70	5.30	28.09
13	-	20.40		
Total	-	-	-	120.03
n =	9	MSE =	13.336667	

ii) Weighted three-period moving average with the weighing scheme of 0.5 + 0.3+ 0.2=1

Table 19.8

Three-period weighted moving average				
Weights	0.5	0.3	0.2	
Week	No. of units sold Y_t	Moving average forecast F_t	Forecast error $(Y_t - F_t)$	Squared forecast error $(Y_t - F_t)^2$
1	17			
2	21			
3	19			
4	23	19.20	3.80	14.44
5	18	21.40	-3.40	11.56
6	16	19.70	-3.70	13.69
7	20	18.00	2.00	4.00
8	18	18.40	-0.40	0.16
9	22	18.20	3.80	14.44
10	20	20.40	-0.40	0.16
11	15	20.20	-5.20	27.04
12	22	17.90	4.10	16.81
13	-	19.50		
Total	-	-	-	102.30
n =	9	MSE =	11.366667	

$$\text{MSE} = \frac{102.30}{9} = 11.36667$$

iii) Weighted three-period moving average with the weighing scheme of 0.4 + 0.4+ 0.2=1

$$\text{MSE} = \frac{97.92}{9} = 10.88$$

From the above illustrations it is very much clear that MES depends upon both the weighing scheme and the number of data points taken into consideration for averaging. So, for best result one must find both the number of data points and the weighing scheme that keeps the MSE at the minimum.

Table 19.9

Three-period weighted moving average				
Weights	0.4	0.4	0.2	
Week	No. of units sold Y_t	Moving average forecast F_t	Forecast error $(Y_t - F_t)$	Squared forecast error $(Y_t - F_t)^2$
1	17			
2	21			
3	19			
4	23	19.40	3.60	12.96
5	18	21.00	–3.00	9.00
6	16	20.20	–4 .20	17.64
7	20	18.20	1.80	3.24
8	18	18.00	0.00	0.00
9	22	18.40	3.60	12.96
10	20	20.00	0.00	0.00
11	15	20.40	–5.40	29.16
12	22	18.40	3.60	12.96
13	-	18.80		
Total	-	-	-	97.92
n =	9	MSE =	10.88	

■ Example 3

Given the demand for an item for 15 months forecast the demand for the month 16 by using a 3-monthly moving average method. Also test its reliability by using MAD measure.

Table 19.10

Month	Demand	Month	Demand
1	280	9	309
2	288	10	315
3	266	11	320
4	295	12	332
5	302	13	310
6	310	14	308
7	303	15	320
8	328		

Solution

Table 19.11

Three-period moving average forecast			
Weeks	No. of units sold Y_t	Three-period moving average/ forecast F_t	Absolute forecasting error $\lvert Y_t - F_t \rvert$
1	280		
2	288		
3	266		
4	295	278.00	17.00
5	302	283.00	19.00
6	310	287.67	22.33
7	303	302.33	0.67
8	328	305.00	23.00
9	309	313.67	4.67
10	315	313.33	1.67
11	320	317.33	2.67
12	332	314.67	17.33
13	310	322.33	12.33
14	308	320.67	12.67
15	320	316.67	3.33
16	-	**312.67**	-
Total	-	-	136.67
n =	12	MAD =	11.39

$$\text{MAD} = \frac{\sum \lvert Y_t - F_t \rvert}{n} = \frac{136.67}{12} = 11.39$$

■ Example 4

Given the demand for an item for 15 months forecast the demand for the month 16 by using a 3-monthly moving average method. Also obtain its MAPE (Mean Absolute Percentage Error) measure.

Table 19.12

Month	Demand	Month	Demand
1	280	9	309
2	288	10	315
3	266	11	320
4	295	12	332
5	302	13	310
6	310	14	308
7	303	15	320
8	328		

Solution

Table 19.13

Three-period moving average forecast				
Weeks	No. of units sold Y_t	Three-period moving average/forecast F_t	Absolute forecasting error $\lvert Y_t - F_t \rvert$	$\frac{\lvert Y_t - F_t \rvert}{Y_t} \times 100$
1	280			
2	288			
3	266			
4	295	278.00	17.00	5.76
5	302	283.00	19.00	6.29
6	310	287.67	22.33	7.20
7	303	302.33	0.67	0.22
8	328	305.00	23.00	7.01
9	309	313.67	4.67	1.51
10	315	313.33	1.67	0.53
11	320	317.33	2.67	0.83
12	332	314.67	17.33	5.22
13	310	322.33	12.33	3.98
14	308	320.67	12.67	4.11
15	320	316.67	3.33	1.04
16	-	312.67	-	
Total	-	-	136.67	43.72
n =	12	MAPE =	11.39	

$$\text{MAPE} = \sum \frac{\lvert Y_t - F_t \rvert}{Y_t} \times 100 = \frac{43.72}{12} = 11.39$$

19.9 EXPONENTIAL SMOOTHENING METHOD FOR SHORT RUN BUSINESS FORECASTING

As in the previous case, this method also uses the weighted moving average by giving priority to the recent past data. However, in the previous method the weights were chosen arbitrarily for all the three recent past data values in a declining order of magnitude so that the total is I always. In exponential smoothening method we choose the weight for the very recent past data and allow the system to determine the remaining weights in a decreasing order of magnitude totalling to one again. Also this time the past data values are not limited to three as in the previous case. In fact, this method accommodates all the past data values in the prediction. The suitably designed declining weighing exponential scheme is normally written as

$$F_{t+1} = \alpha Y_t + \alpha(1-\alpha)Y_{t-1} + \alpha(1-\alpha)^2 Y_{t-2} + \alpha(1-\alpha)^3 Y_{t-3} + \ldots\ldots\ldots\ldots \quad (1)$$

Where F_{t+1} is the proposed forecast for the next period, Y_t is the actual value pertaining to current period, and Y_{t-1}, Y_{t-2} and so on are the values of the successive preceding periods, α is a fractional weight chosen for the very recent past in the interval $0 \le \alpha \le 1$ inclusive. Here the so chosen α is often called the smoothening agent.

Points to be remembered:

Weights are declining: Since α is less than 1 the successive weights are declining, since each successive weight is obtained by multiplying the pervious weight by (1 – α), which is also a fraction.

Sum of the weights equal to unity: Like the weighted moving average method, the sum of the weights here also work out to be unity. This can be proved as follows.

$$\sum W_1 = \alpha + \alpha(1-\alpha) + \alpha(1-\alpha)^2 + \alpha(1-\alpha)^3 + \dots\dots\dots$$

$$= \alpha\left[1 + (1-\alpha) + (1-\alpha)^2 + (1-\alpha)^3 + \dots\dots\dots\right]$$

$$= \alpha\left[\frac{1}{1-(1-\alpha)}\right] = \alpha\left[\frac{1}{\alpha}\right] = 1$$ (since the term in side of the square bracket is a geometric progression)

Despite the fact that the exponential smoothening provides a forecast for the period t+1 as the weighted average of all the past observations of the series, for practical calculations one need not have to reserve all the previous period data values intact. For the t+1th period forecast, namely F_{t+1} we need only Y_t, the actual observation of the series in period t, and F_t, the forecasted value for the period t, in addition to α. Now let us demonstrate this highly useful information by appropriate simplification of our lengthy equation (1).

Remember, the estimation of equation (1) is beyond anybody's comprehension because it could run backwards to any length of period. Since the above equation holds good for all t values let us write the corresponding forecast for the current period t as in equation (2).

$$F_{t+1} = \alpha Y_t + \alpha(1-\alpha)Y_{t-1} + \alpha(1-\alpha)^2 Y_{t-2} + \alpha(1-\alpha)^3 Y_{t-3} + \dots \quad (1)$$

$$F_t = \alpha Y_{t-1} + \alpha(1-\alpha)Y_{t-2} + \alpha(1-\alpha)^2 Y_{t-3} + \alpha(1-\alpha)^3 Y_{t-4} + \dots \quad (2)$$

Now let us multiply equation (2) by $(1-\alpha)$ on either side

$$(1-\alpha)F_t = \alpha(1-\alpha)Y_{t-1} + \alpha(1-\alpha)^2 Y_{t-2} + \alpha(1-\alpha)^3 Y_{t-3} + \alpha(1-\alpha)^4 Y_{t-4} + \dots \quad (3)$$

Now subtract (3) from (1)

$$F_{t+1} - (1-\alpha)F_t = \alpha Y_t$$

$$F_{t+1} = \alpha Y_t + (1-\alpha)F_t \dots \quad (4)$$

As stated earlier, by using equation (4) one can forecast for the period t+1 as the weighted average of the actual value of the series in period t and the forecast value of the series for the period t by using α and (1 – α) as the respective weights.

However, in effect the exponential smoothening forecast given by equation (4) for any given period t+1 can easily be demonstrated as the weighted average of all the past actual values. Since under our declining weighting scheme the weighted associated to the very first period is definitely going to be very small. In such a situation we can easily assume that forecast value for the very first period is equal to the actual value of the series in period 1. i.e., $F_1 = Y_1$.

Now the forecast for the period 2 is obtained as

$$\begin{aligned} F_2 &= \alpha Y_1 + (1-\alpha)F_1 \\ &= \alpha Y_1 + (1-\alpha)Y_1 \qquad \because (Y_1 = F_1) \\ &= Y_1 \\ F_3 &= \alpha Y_2 + (1-\alpha)F_2 \\ &= \alpha Y_2 + (1-\alpha)Y_1 \\ F_4 &= \alpha Y_3 + (1-\alpha)F_3 \\ &= \alpha Y_3 + (1-\alpha)\{\alpha Y_2 + (1-\alpha)Y_1 \\ &= \alpha Y_3 + \alpha(1-\alpha)Y_2 + (1-\alpha)^2 Y_1 \end{aligned}$$

and so on.

Thus, the forecast for the fourth period is the weighted sum of all the three previous observations as expected. Further, the sum of all the three weights

$$\begin{aligned} &= \alpha + \alpha(1-\alpha) + (1-\alpha)^2 \\ &= \alpha + \alpha - \alpha^2 + \alpha^2 - 2\alpha + 1 = 1 \end{aligned}$$

To draw some more useful information let us rewrite the equation (4) in slightly different format as shown below.

$$\begin{aligned} F_{t+1} &= \alpha Y_t + (1-\alpha)F_t \\ &= \alpha Y_t + F_t - \alpha F_t \\ &= F_t + \alpha(Y_t - F_t) \ldots \qquad (5) \end{aligned}$$

In this format the forecast for the period t+1 is equal to the forecast for the period t plus α times the difference between the actual and forecast for the period t. Here the second term is often called the correction term in exponential smoothening. Here again the smoothening constant α plays a crucial role in determining MSE value. Invariably one must try for α that keeps the MSE value at a bare minimum.

■ Example 5

For the data given below, forecast the sales for the thirteenth week by method of exponential smoothening given $\alpha = 0.2$. Also obtain MSE.

Table 19.14

Weeks	1	2	3	4	5	6	7	8	9	10	11	12	13
Demand	17	21	19	23	18	16	20	18	22	20	20	15	22

Solution

Since for the first data 17 we attach very little weight to begin our iteration, let us assume $Y_1 = F_1$.

$$Y_1 = F_1 = 17$$
$$F_2 = \alpha Y_1 + (1-\alpha)F_1$$
$$= 0.2 \times 17 + (1-0.2)17 = 17$$
$$F_3 = \alpha Y_2 + (1-\alpha)F_2$$
$$= 0.2 \times 21 + (1-0.2)17 = 17.80$$
$$F_4 = \alpha Y_3 + (1-\alpha)F_3$$
$$= 0.2 \times 19 + (1-0.2)18.23 = 18.04$$
$$= \ldots\ldots\ldots\ldots\ldots\ldots\ldots\ldots$$
$$= \ldots\ldots\ldots\ldots\ldots\ldots\ldots\ldots$$
$$F_{13} = \alpha Y_{12} + (1-\alpha)F_{12}$$
$$= 0.2 \times 22 + (1-0.2)18.48 = 19.18$$

In the same way the rest of the calculations are obtained and reported in Table 19.15.

From Table 19.14 the forecast for week 13 is obtained as 19.18 units. The accuracy of this estimate will be known only at the end of week 13, on the day the final actual sale occurs. Figure 19.9 shows the corresponding smoothening effect of this method.

Table 19.15

$\alpha = 0.2$		Exponential smoothening		
Week	Actual units sold	Exponential smoothening forecast	Forecast error	Squared forecast error
	Y_t	F_t	$(Y_t - F_t)$	$(Y_t - F_t)^2$
1	17			
2	21	17.00	4.00	16.00
3	19	17.80	1.20	1.44
4	23	18.04	4.96	24.60
5	18	19.03	-1.03	1.07
6	16	18.83	-2.83	7.98
7	20	18.26	1.74	3.03
8	18	18.61	-0.61	0.37
9	22	18.49	3.51	12.34
10	20	19.19	0.81	0.66
11	15	19.35	-4.35	18.94
12	22	18.48	3.52	12.38
13	-	19.18		
Total			10.92	98.80
n =	11	MSE =	8.98	

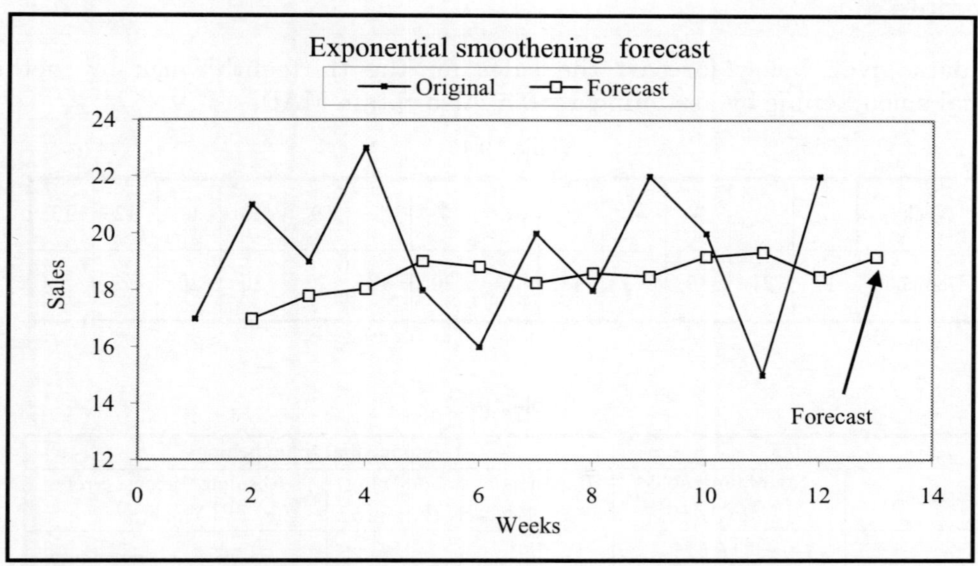

Figure 19.10

MSE = 8.98 gives us the measure of efficiency. This efficiency in exponential smoothening method wholly depends upon the choice of α, the smoothening factor. To illustrate the importance of the smoothening coefficient α, let us calculate the efficient measure MSE once again by assuming α = 0.3 in Table 19.16.

Table 19.16

α = 0.3		Exponential smoothening		
Week	Actual units sold Y_t	Exponential smoothening forecast F_t	Forecast error $(Y_t\text{-}F_t)$	Squared forecast error $(Y_t\text{-}F_t)^2$
1	17			
2	21	17.00	4.00	16.00
3	19	18.20	0.80	0.64
4	23	18.44	4.56	20.79
5	18	19.81	−1.81	3.27
6	16	19.27	−3.27	10.66
7	20	18.29	1.71	2.94
8	18	18.80	−0.80	0.64
9	22	18.56	3.44	11.83
10	20	19.59	0.41	0.17
11	15	19.71	−4.71	22.23
12	22	18.30	3.70	13.69
13	-	19.41		
Total			8.03	102.86
n =	11	MSE =	9.35	

Among the two values α = 0.2 seems to be the better choice if we go by MSE criterion.

■ Example 6

For the data given below forecast the sales for the thirteenth week by method of exponential smoothening by assuming $\alpha = 0.5$. Also obtain MAD.

Table 19.17

Weeks	1	2	3	4	5	6	7	8	9	10	11	12	13
Demand	17	21	19	23	18	16	20	18	22	20	20	15	22

Solution

Table 19.18

$\alpha = 0.5$		Exponential smoothening	
Week	No. of units sold (Y_t)	Exponential smoothening forecast (F_t)	Absolute forecast error $\lvert Y_t - F_t \rvert$
1	17		
2	21	17.00	4.00
3	19	19.00	0.00
4	23	19.00	4.00
5	18	21.00	3.00
6	16	19.50	3.50
7	20	17.75	2.25
8	18	18.88	0.88
9	22	18.44	3.56
10	20	20.22	0.22
11	15	20.11	5.11
12	22	17.55	4.45
13	-	19.78	
Total			30.96
n =	11	MAD =	2.81

MAD = 2.81

■ Example 7

For the data given below forecast the sales for the thirteenth week by method of exponential smoothening. Also obtain MAPE.

Table 19.19

Weeks	1	2	3	4	5	6	7	8	9	10	11	12	13
Demand	17	21	19	23	18	16	20	18	22	20	20	15	22

Solution

Table 19.20

$\alpha = 0.2$		Exponential smoothening		
Week	No. of units sold Y_t	Exponential smoothening forecast F_t	Absolute forecast error $\lvert Y_t - F_t \rvert$	$\frac{\lvert Y_t - F_t \rvert}{Y_t} \times 100$
1	17			
2	21	17.00	4.00	19.05
3	19	17.80	1.20	6.32
4	23	18.04	4.96	21.57
5	18	19.03	1.03	5.73
6	16	18.83	2.83	17.66
7	20	18.26	1.74	8.70
8	18	18.61	0.61	3.38
9	22	18.49	3.51	15.97
10	20	19.19	0.81	4.05
11	15	19.35	4.35	29.01
12	22	18.48	3.52	15.99
13	-	19.18		
Total				147.43
n =	11		MAPE =	13.40

MAPE = 13.40.

19.10 USE OF LINEAR TREND PROJECTION FOR LONG-RUN BUSINESS FORECASTING

19.10.1 Method of Least Squares

The least squares method outlined in this section identifies both 'a' and 'b' uniquely without leaving any scope for subjectivity in our calculation. In this method both 'a' and 'b' values are obtained by minimizing the sum of the squared deviations of the actual values from their respective trend values. For a detailed discussion on least squares method readers are advised to refer the corresponding section in the regression analysis in any statistics text.

Method: 1 (Both for odd and even number of years)

If $Y = a + bX$ is the proposed line of regression then the associated normal equations are written as

$$\sum Y = na + b\sum X... \quad (1)$$

$$\sum XY = a\sum X + b\sum X^2... \quad (2)$$

By using these two normal equations we can get both 'a' and 'b'. After having obtained these values, the needed linear trend line is obtained by mere substitution.

Method: 2

When the number of years is even then our natural choice of origin would be midpoint of the two middle years. Under such circumstances we attach obituary odd numbers to the respective years so that the sum of such number will be equal to zero once again. However, this time one unit on the x-axis measures only half of a year. If we shift the origin to the middle of the time period, then the sum of arbitrarily assigned X values will become zero. This choice really saves a lot of calculations in the formula for both 'a' and 'b' values in equation (1) and (2).

Now by choice $\sum X = 0$

Further

$$\overline{X} = \frac{\sum X}{n} = \frac{0}{n} = 0$$

So equation (1) and (2) above reduces to

$$\sum Y = na + b\sum X$$

$$\sum Y = na + b \times 0$$

$$a = \frac{\sum Y}{n}$$

$$\sum XY = a \times 0 + b\sum X^2$$

$$b = \frac{\sum XY}{\sum X^2}$$

The following examples illustrate the use of both the methods.

Trend by OLS Method for Odd Number of Years

■ **Example 8**

For the data given below obtain the linear trend line by using OLS method and project the profit for the year 1956.

Table 19.21

Profit data					
Year	1951	1952	1953	1954	1955
Profit (Y)	6	7	10	6	12

Solution: Method 1

If $Y = a + bX$ is the proposed line of regression then the associated normal equations are written as

$$\sum Y = na + b\sum X... \quad (1)$$

$\sum XY = a\sum X + b\sum X^2 ...$ (2)

$41 = 5a + 10b...$ (3)

$93 = 10a + 30b...$ (4)

Table 19.22

Least squares method				
Year	X	Y	XY	X^2
1951	0	6	0	0
1952	1	7	7	1
1953	2	10	20	4
1954	3	6	18	9
1955	4	12	48	16
Total	10	41	93	30

Now from equation (3) and (4) as per Cramer's rule

$$a = \frac{\begin{vmatrix} 41 & 10 \\ 93 & 30 \end{vmatrix}}{\begin{vmatrix} 5 & 10 \\ 10 & 30 \end{vmatrix}} = \frac{30 \times 41 - 93 \times 10}{5 \times 30 - 10 \times 10} = \frac{300}{50} = 6$$

$$b = \frac{\begin{vmatrix} 5 & 41 \\ 10 & 93 \end{vmatrix}}{\begin{vmatrix} 5 & 10 \\ 10 & 30 \end{vmatrix}} = \frac{5 \times 93 - 41 \times 10}{5 \times 30 - 10 \times 10} = \frac{55}{50} = 1.1$$

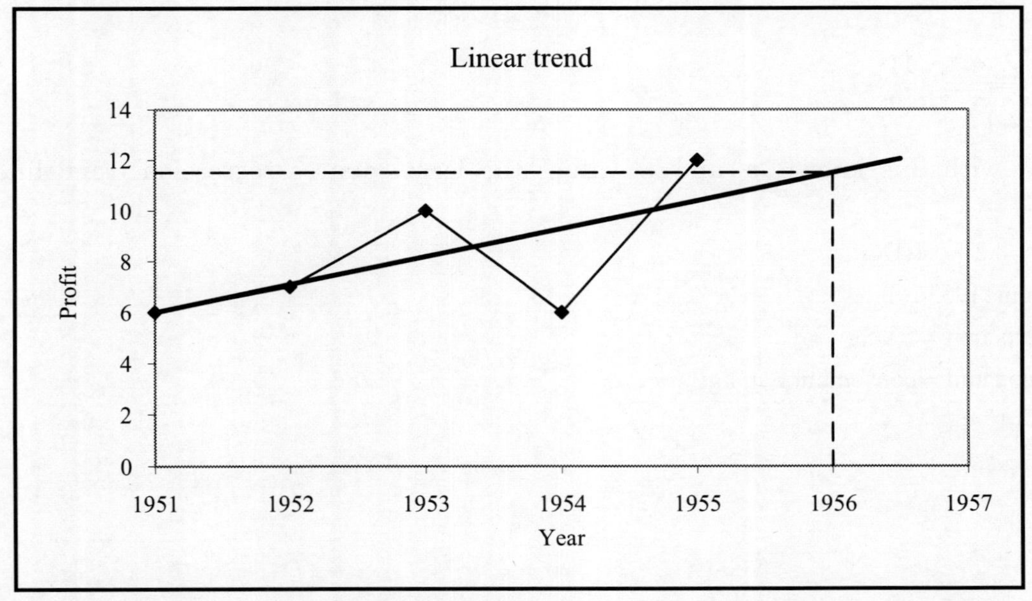

Figure 19.11

Thus, the needed regression may be written as

$Y_c = 6 + 1.1\,X...$

Origin : 1951

X-axis unit = 1 Year

Y-axis unit = Profit per unit in Rs.

Projection for 1956 $Y_{1956} = 6 + 1.1 \times 5 = 11.5$

Solution: Method 2

For this method 1953, the middle year, is taken as the origin.

Table 19.23

Least squares method				
Year	X	Y	XY	X^2
1951	−2	6	−12	4
1952	−1	7	−7	1
1953	0	10	0	0
1954	1	6	6	1
1955	2	12	24	4
Total	0	41	11	10

Now from Table 19.7 above

$$a = \frac{\sum Y}{n} = \frac{41}{5} = 8.2$$

$$b = \frac{\sum XY}{\sum X^2} = \frac{11}{10} = 1.1$$

Now with this known 'a' and 'b' values the least squares regression trend line is written as

$Y_C = 8.2 + 0.1.1X...$

Origin : 1953

X-axis unit = 1 Year

Y-axis unit = consumption in kg.

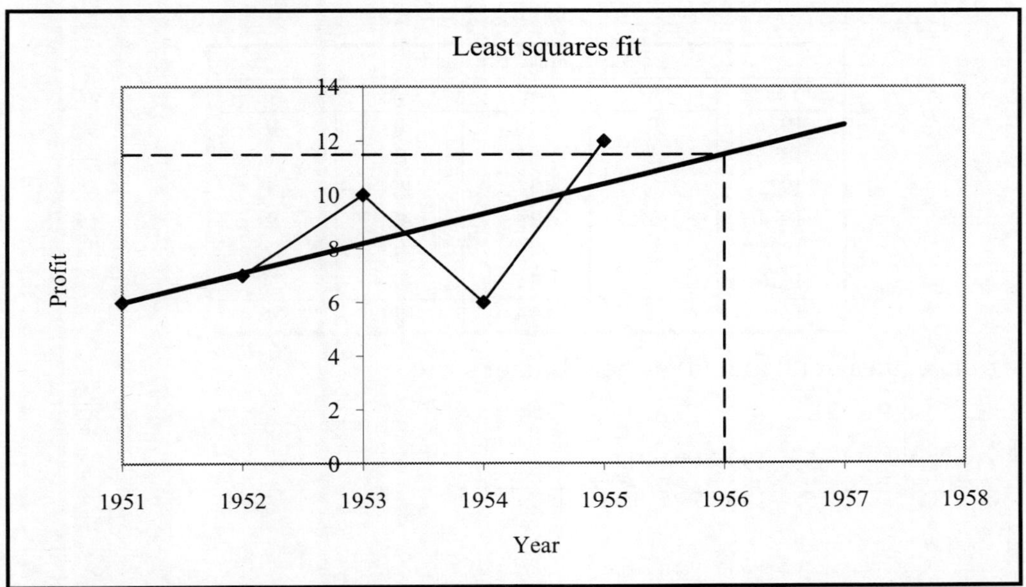

Figure 19.12

Projection for 1956 $Y_{1956} = 8.2 + 1.1 \times 3 = 11.5$

Note: *Note that the intercept 'a' differs in both the methods though the forecast is one and the same.*

Trend by OLS Method for Even Number of Years

■ Example 9

For the data given below obtain the linear trend by using OLS

Table 19.24

Profit data						
Year	1971	1972	1973	1974	1975	1976
Profit (Y)	10	12	11	14	12	15

Method 1

If $Y = a + bX$ is the proposed line of regression then the associated normal equations are written as

$$\sum Y = na + b\sum X ... \quad (1)$$

$$\sum XY = a\sum X + b\sum X^2 ... \quad (2)$$

$$74 = 6a + 15b ... \quad (3)$$

$$199 = 15a + 55b ... \quad (4)$$

Table 19.25

Least squares method				
Year	X	Y	XY	X^2
1971	0	10	0	0
1972	1	12	12	1
1973	2	11	22	4
1974	3	14	42	9
1975	4	12	48	16
1976	5	15	75	25
Total	15	74	199	55

Now from equation (3) and (4) as per Cramer's rule

$$a = \frac{\begin{vmatrix} 74 & 15 \\ 199 & 55 \end{vmatrix}}{\begin{vmatrix} 6 & 15 \\ 15 & 55 \end{vmatrix}} = \frac{74 \times 55 - 199 \times 15}{6 \times 55 - 15 \times 15} = \frac{1085}{105} = 10.3333$$

$$b = \frac{\begin{vmatrix} 6 & 74 \\ 15 & 199 \end{vmatrix}}{\begin{vmatrix} 6 & 15 \\ 15 & 55 \end{vmatrix}} = \frac{6 \times 199 - 74 \times 15}{6 \times 55 - 15 \times 15} = \frac{84}{105} = 0.8$$

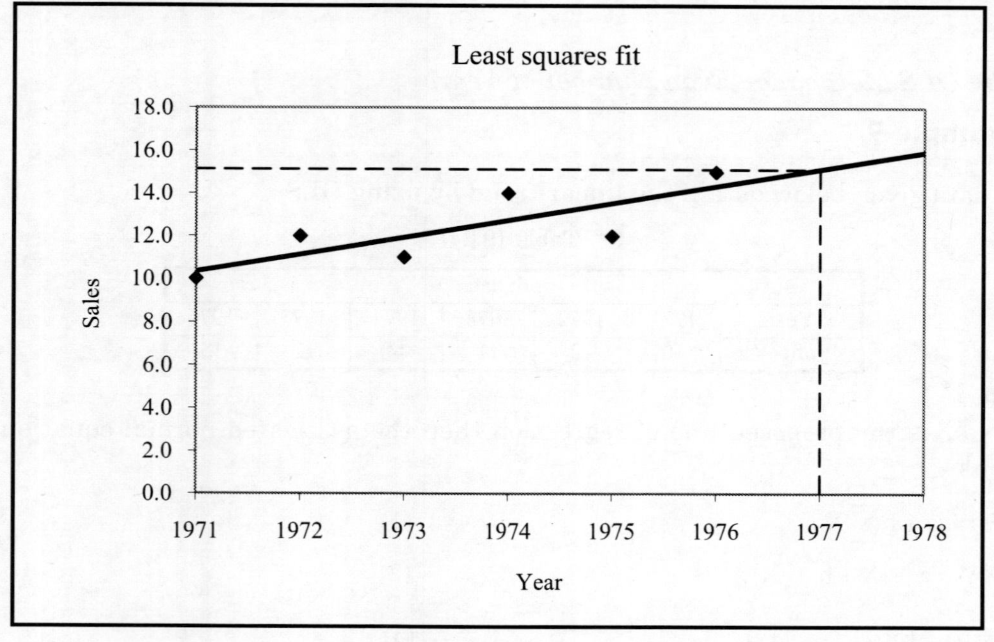

Figure 19.13

Thus, the needed regression may be written as

$Y_C = 10.3333 + 0.8X$

Origin : 1971

X-axis unit = One Year

Y-axis unit = Output in kg.

Projection for 1977 $Y_{1977} = 10.333 + 0.8 \times 6 = 15.13$

Method 2

Now from Table 19.26,

$$a = \frac{\sum Y}{n} = \frac{74}{6} = 12.3333$$

$$b = \frac{\sum XY}{\sum X^2} = \frac{28}{70} = 0.4$$

Table 19.26

Least squares method				
Year	X	Y	XY	X^2
1971	−5	10	−50	25
1972	−3	12	−36	9
1973	−1	11	−11	1
1974	1	14	14	1
1975	3	12	36	9
1976	5	15	75	25
Total	0	74	28	70

So the equation of the fitted line may be written as

$Y_c = 12.3333 + 0.4X...$

Origin : Jan 1974

X-axis unit = 1 / 2 Year

Y-axis unit = Output in kg.

Projection for 1977 $Y_{1977} = 12.3333 + 0.4 \times 7 = 15.13$

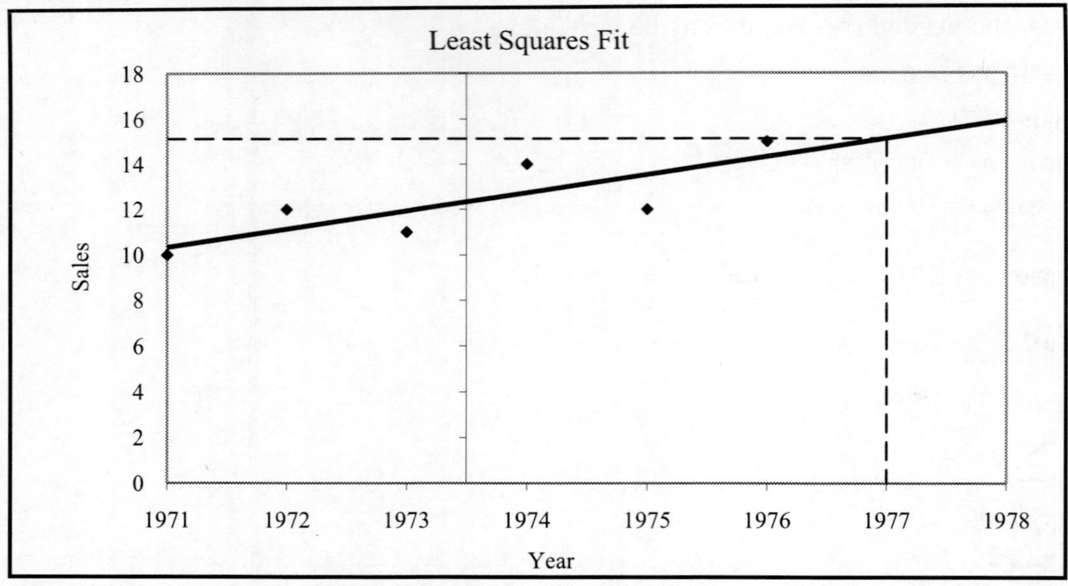

Figure 19.14

19.11 INDEXING THE SEASONALS FOR FORECASTING

The trend projection method discussed in the previous section is useful only for the long run prediction based on annual data. Thus, when yearly data are given, we obtain the forecast for the nearby future years by trend method. However, projections are to be made on, say monthly or weekly basis, then clearly the recurring seasonal variations in the concerned year must be incorporated into the projected trend. Sales of softdrinks, woollen garments, cooking gas and so on are certainly not expected to be the same in all the months in a year. Figure 19.15 illustrates one such quarterly seasonal pattern. The second year quarter-wise sales pattern is exactly the same as that of the first year. This pattern is followed in the third year as well. The consideration of monthly or quarterly data makes it easy to study seasonal patterns and make appropriate monthly or quarterly projections. The measures of seasonal variations are usually referred to as indices of seasonal variation rather than absolute numbers as in the case of trend values discussed in the previous section. Some of the commonly used methods are discussed in the following illustrations.

i) Construction for Seasonal index for Monthly Data: Probably the simplest method of showing the seasonal variation is to express the figure for each month as percentage of the total for all the twelve months. If there were no seasonal variation in the data, then each of these monthly percentages would have been 8 1/3 % for all the twelve months, which is just one-twelfth of the total 100% for the whole year. The amount that each of them differs from 8 1/3% reveals the amount of seasonal factor present in the said data.

Another method that perhaps makes it easier to see the effect of seasonal variation on the different months is to multiply each of the percentages in the previous method by 12.

Once again if there were no seasonal variations in the given data then each of the monthly indexes would be 100 with a total of 1200 for the whole year. Thus, 100 is again one-twelfth of the total 1200. The amount of difference from 100 reveals the amount of seasonal factor in the relevant data.

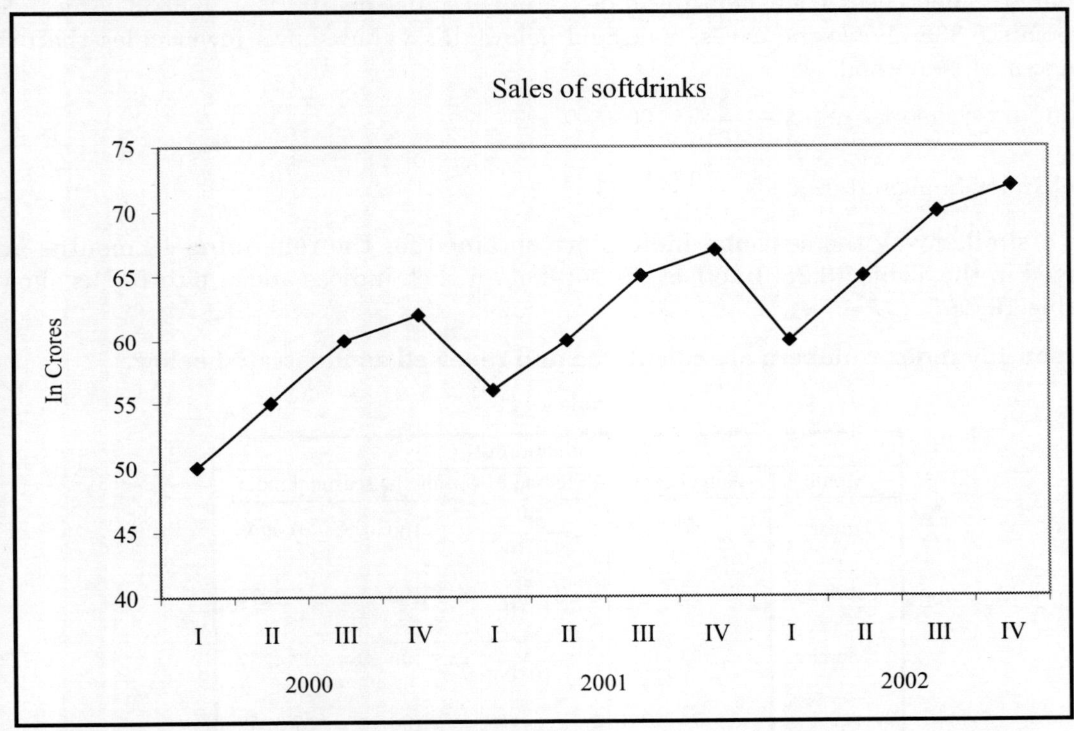

Figure 19.15

In this method third method to get the monthly indices for all the 12 months, we simply multiply the indices obtained in method 1 by 12 instead of 100. Thus, the annual total in this method would be 12 instead of 100 as in method 1. Thus here, each month is assigned an index of 1 to represent non-seasonal pattern totalling to 12. The deviation of the actual index from 1 thus represents the seasonal pattern. The following two examples are designed to highlight all the three methods.

■ Example 10

For the data given below obtain the seasonal index numbers for all the 12 months.

Table 19.27

Sales of softdrinks													
Month	Jan	Feb	Mar	Apr	May	Jun	Jul	Aug	Sep	Oct	Nov	Dec	Total
Sales	80	73	75	74	85	87	94	99	90	88	85	80	1010

Solution: Method 1

In this method the seasonal indices are expressed as percentage of the total for the whole year so that the total of all the 12 index numbers is equal to 100 as shown in Table 19.28. Thus, if at all there is no seasonal then all the twelve monthly sales must be equal to one another. In such cases each monthly index number must be 100/12 = 8.3333. Any index above this 8.333 shows the excess sale and below this value shows fewer sales than the average in this method.

$$\text{January Seasonal index} = \frac{80}{1010} \times 100 = 7.9$$

$$\text{February Seasonal index} = \frac{73}{1010} = 7.23$$

In a similar way the seasonal indices are obtained for the remaining 10 months and reported in the Table 19.28. In effect the total of all such indices adds up to 100 as shown in Table 19.28.

The monthly index numbers are calculated and reported as illustrated below.

Table 19.28

Sales of softdrinks		
Month	Actual data	Method of getting the seasonal index
January	80	$\frac{80}{1010}$ x 100 = 7.9208
February	73	$\frac{73}{1010}$ x 100 = 7.2277
March	75	$\frac{75}{1010}$ x 100 = 7.4257
April	74	$\frac{74}{1010}$ x 100 = 7.3267
May	85	$\frac{85}{1010}$ x 100 = 8.4158
June	87	$\frac{87}{1010}$ x 100 = 8.6139
July	94	$\frac{94}{1010}$ x 100 = 9.3069
August	99	$\frac{99}{1010}$ x 100 = 9.8020
September	90	$\frac{90}{1010}$ x 100 = 8.9109
October	88	$\frac{88}{1010}$ x 100 = 8.7129
November	85	$\frac{85}{1010}$ x 100 = 8.4158
December	80	$\frac{80}{1010}$ x 100 = 7.9208
Total	1010	- 100.00

Method 2

In this method, to get the monthly indices for all the 12 months, we simply multiply the indices obtained in method 1 by 12. Thus, the annual total in this method would be 1200 instead of 100 in method 1. Thus here, each month is assigned an index of 100 to represent non-seasonal pattern totalling to 1200. The deviation of the actual index from 100 thus represents the seasonal pattern. Table 19.29 illustrates the underling method of calculation.

Table 19.29

Sales of softdrinks		
Month	Actual data	Method of getting the seasonal index
January	80	$\frac{80}{1010}$ x 1200 = 95.0495
February	73	$\frac{73}{1010}$ x 1200 = 86.7327
March	75	$\frac{75}{1010}$ x 1200 = 89.1089
April	74	$\frac{74}{1010}$ x 1200 = 87.9208
May	85	$\frac{85}{1010}$ x 1200 = 100.9901
June	87	$\frac{87}{1010}$ x 1200 = 103.3663
July	94	$\frac{94}{1010}$ x 1200 = 111.6832
August	99	$\frac{99}{1010}$ x 1200 = 117.6238
September	90	$\frac{90}{1010}$ x 1200 = 106.9307
October	88	$\frac{88}{1010}$ x 1200 = 104.5545
November	85	$\frac{85}{1010}$ x 1200 = 100.9901
December	80	$\frac{80}{1010}$ x 1200 = 95.0495
Total	1010	- 1200.00

For example

January Seasonal index = $\frac{80}{1010} \times 1200 = 95.0495$

February Seasonal index = $\frac{73}{1010} \times 1200 = 86.7327$

In a similar way the seasonal indices are obtained for the remaining 10 months and reported in Table 19.29.

Method 3

In this method to get the monthly indices for all the 12 months, we simply multiply the indices obtained in method 1 by 12 instead of 100. Thus, the annual total in this method would be 12 instead of 100 as in method 1. Thus here, each month is assigned an index of 1 to represent non-seasonal pattern totalling to 12. The deviation of the actual index from 1 thus represents the seasonal pattern. Table 19.30 illustrates the underlying method of calculation.

Table 19.30

Sales of softdrinks		
Month	Actual data	Method of getting the seasonal index
January	80	$\frac{80}{1010}$ x 12 = 0.9505
February	73	$\frac{73}{1010}$ x 12 = 0.8673
March	75	$\frac{75}{1010}$ x 12 = 0.8911
April	74	$\frac{74}{1010}$ x 12 = 0.8792
May	85	$\frac{85}{1010}$ x 12 = 1.0099
June	87	$\frac{87}{1010}$ x 12 = 1.0337
July	94	$\frac{94}{1010}$ x 12 = 1.1168
August	99	$\frac{99}{1010}$ x 12 = 1.1762
September	90	$\frac{90}{1010}$ x 12 = 1.0693
October	88	$\frac{88}{1010}$ x 12 = 1.0455
November	85	$\frac{85}{1010}$ x 12 = 1.0099
December	80	$\frac{80}{1010}$ x 12 = 0.9505
Total	1010	- 12.00

For example,

January seasonal index = $\frac{80}{1010} \times 12 = 0.950495$

February seasonal index = $\frac{73}{1010} \times 12 = 0.867327$

In a similar way the seasonal indices are obtained for the remaining 10 months and reported in Table 19.30.

ii) Construction of Seasonal Indices for Quarterly Data

Example 11

For the quarterly data given below obtain the seasonal index numbers by all the three methods.

Table 19.31

Data table				
Quarters	I	II	III	IV
Y_t	200	150	250	300

Solution: Method 1

In the first method as usual we obtain the seasonal indices as percentages of the total for all the four quarters as shown below in table. All the four indices add up to 100 so that for non-seasonal pattern each month index deserves only 25.

Table 19.32

Sales of softdrinks		
Quarters	Actual data	Method of getting the seasonal index
I	200	$\frac{200}{900}$ x 100 = 22.2222
II	150	$\frac{150}{900}$ x 100 = 16.6667
III	250	$\frac{250}{900}$ x 100 = 27.7778
IV	300	$\frac{300}{900}$ x 100 = 33.3333
Total	900	- 100.00

Method 2

In the second method we obtain the seasonal indices by merely multiplying the indices in Table 19.32 above by 4, the number of quarter in our data. Here the total adds up to 400 with associated number of 100 for each quarter.

Table 19.33

Sales of softdrinks		
Quarters	Actual data	Method of getting the seasonal index
I	200	$\frac{200}{900}$ x 400 = 88.8889
II	150	$\frac{150}{900}$ x 400 = 66.6667
III	250	$\frac{250}{900}$ x 400 = 111.1111
IV	300	$\frac{300}{900}$ x 400 = 133.3333
Total	900	- 400.00

Method 3

Table 19.34

Sales of softdrinks		
Quarters	Actual data	Method of getting the seasonal index
I	200	$\frac{200}{900}$ x 4 = 0.8889
II	150	$\frac{150}{900}$ x 4 = 0.6667
III	250	$\frac{250}{900}$ x 4 = 1.1111
IV	300	$\frac{300}{900}$ x 4 = 1.3333
Total	900	- 4.0000

In this method instead of 400 in the previous method we multiply by only 4 to get a total of 4. In this method each month is assigned an index of 1 do depict non-seasonal pattern.

19.12 USE OF SEASONAL INDICES IN BUSINESS FORECASTING

In the illustrative example pertaining to monthly data given above we took the data for only 12 months in a particular year. Hence these indices cannot be taken as representative values for future. For better real life example one must take more number of years into consideration and obtain an average representative values. For example, if we take 10 years for consideration, then we will be having 10 January data, 10 February data etc. To average all the ten observations for each month we propose to use the arithmetic mean method in this section. Normally, three methods are used to obtain the seasonal index.

1. Method of averages
2. Ratio to trend method
3. Link relative method

Calculation of seasonal indices by method of averages

This technique is useful for forecasting provided we have only seasonal variation the given data series. Hence this method is not appropriate in the presence of trend in addition to seasonal variations. In such cases the ratio to trend method is used to obtain the needed seasonal indices. The following example illustrates the method of quarterly projection for seasonal dominant data.

Example 12

The following table shows the quarterly break-up of yearly sale of a certain seasonal items for 8 years. By using least square method obtain the trend line for the yearly data and project the trend sale for the year 2008. Using the seasonal break-up obtain the seasonal indices for all the four quarters and forecast the quarterly sale for the year 2008.

Table 19.35

Year	Quarters				Total
	I	II	III	IV	
1999	18	27	23	14	82
2000	17	30	20	13	80
2001	21	26	23	20	90
2002	21	29	26	16	94
2003	19	28	22	14	83
2004	22	30	26	16	94
2005	24	32	24	19	99
2006	21	29	29	13	92

Solution

Let us first forecast the sale for the year 2007 by using the annual data.

Trend projection for 2008 on annual data basis

If $Y = a + bX$ is the proposed line of regression then the associated normal equations are written as

$$\sum Y = na + b\sum X... \qquad (1)$$

$$\sum XY = a\sum X + b\sum X^2... \qquad (2)$$

Table 19.36

Least squares method				
Year	X	Y	XY	X^2
1999	0	82	0	0
2000	1	80	80	1
2001	2	90	180	4
2002	3	92	276	9
2003	4	83	332	16
2004	5	94	470	25
2005	6	99	594	36
2006	7	92	644	49
Total	28	712	2576	140

$$712 = 8a + 28b... \qquad (3)$$

$$2576 = 28a + 140b... \qquad (4)$$

Now from equation (3) and (4) as per Cramer's rule

$$a = \frac{\begin{vmatrix} 712 & 28 \\ 2576 & 140 \end{vmatrix}}{\begin{vmatrix} 8 & 28 \\ 28 & 140 \end{vmatrix}} = \frac{712 \times 140 - 28 \times 2576}{8 \times 140 - 28 \times 28} = 82$$

$$b = \frac{\begin{vmatrix} 8 & 712 \\ 28 & 2576 \end{vmatrix}}{\begin{vmatrix} 8 & 28 \\ 28 & 140 \end{vmatrix}} = \frac{8 \times 2576 - 28 \times 712}{8 \times 140 - 28 \times 28} = 2.00$$

Thus the trend equation may be written as $Y_c = 82 + 2X$.

Now the needed forecast for the year is obtained by substituting X = 9 in the trend equation.

So $Y_{2008} = 82 + 2 \times 9 = 100$ units

Now to break-up this annual forecast quarter-wise we construct the seasonal indices first for all the four quarters by method of averages.

Construction of quarterly seasonal index by method of averages

In this method we take the average values of all the four quarterly values as the typical representative for the respective quarters. Once this is done we obtain the seasonal indices in the usual manner by attaching index 1 for all the four quarters to represent non-seasonal pattern. Remember that this method is applicable if and only if seasonal pattern alone dominates the given data. Thus, we assume the absence of both trend and cyclical variations in the given series over the years. The needed seasonal indices calculations are highlighted in Table 19.37 below.

Table 19.37

Sales of softdrinks										
Quarters	1999	2000	2001	2002	2003	2004	2005	2006	Average	Method of getting the seasonal index
I	18	17	21	21	19	22	24	21	20.375	$\frac{20.38}{89.00}$ x 4 = 0.9157
II	27	30	26	29	28	30	32	29	28.875	$\frac{28.88}{89.00}$ x 4 = 1.2978
III	23	20	23	26	22	26	24	29	24.125	$\frac{24.13}{89.00}$ x 4 = 1.0843
IV	14	13	20	16	14	16	19	13	15.625	$\frac{15.63}{89.00}$ x 4 = 0.7022
Total	82	80	90	92	83	94	99	92	89.000	- 4.0000

In Table 19.37 the annual projection, namely 100 units is equally spread over (100/4 = 25) to all the four quarters is shown. The respective quarter-wise indices are reported in the third column. In the fourth column the needed quarter-wise projection for the year 2008 is obtained by multiplying the respective quarterly projection by the respective seasonal indices.

Table 19.38

Projected quarters-wise sale for the year 2008			
Quarters	Quarterly projection on the basis of yearly projection	Seasonal index	Quarterly projection
1	25	0.9157	22.8925
2	25	1.2978	32.445
3	25	1.0843	27.1075
4	25	0.7022	17.555
Total	100	4.0000	100

Example 13

The following table shows the monthly break-up of yearly sale of certain seasonal items for 8 years. Obtain the seasonal indices for all the twelve months and forecast the monthly sales for the year 2007.

Table 19.39

	Sales							
Months	1999	2000	2001	2002	2003	2004	2005	2006
Jan	18	17	21	21	19	22	24	21
Feb	27	30	26	29	28	30	32	29
Mar	23	20	23	26	22	26	24	29
Apr	21	23	20	19	20	21	22	23
May	23	24	21	20	19	19	23	21
Jun	21	24	25	23	24	21	20	19
Jul	20	21	22	23	19	23	22	21
Aug	23	24	25	21	21	22	24	21
Sep	24	23	21	20	19	21	20	22
Oct	20	21	20	21	23	21	19	21
Nov	23	21	24	25	23	22	21	22
Dec	21	23	20	21	20	22	23	21
Total	264	271	268	269	257	270	274	270

Solution

Least square fit and projection for the year 2007.

Table 19.40

Least squares method				
Year	X	Y	XY	X^2
1999	0	264	0	0
2000	1	271	271	1
2001	2	268	536	4
2002	3	269	807	9
2003	4	257	1028	16
2004	5	270	1350	25
2005	6	274	1644	36
2006	7	270	1890	49
Total	28	2143	7526	140

If $Y = a + bX$ is the proposed line of regression then the associated normal equations are written as

$$\sum Y = na + b\sum X \ldots \qquad (1)$$

$$\sum XY = a\sum X + b\sum X^2 \ldots \qquad (2)$$

$$2143 = 8a + 28b \ldots \qquad (3)$$

$$7526 = 28a + 140b \ldots \qquad (4)$$

Now from equation (3) and (4) as per Cramer's rule

$$a = \frac{\begin{vmatrix} 2143 & 28 \\ 7526 & 140 \end{vmatrix}}{\begin{vmatrix} 8 & 28 \\ 28 & 140 \end{vmatrix}} = \frac{2143 \times 140 - 28 \times 7526}{8 \times 140 - 28 \times 28} = 265.75$$

$$b = \frac{\begin{vmatrix} 8 & 2143 \\ 28 & 7526 \end{vmatrix}}{\begin{vmatrix} 8 & 28 \\ 28 & 140 \end{vmatrix}} = \frac{8 \times 7526 - 28 \times 2143}{8 \times 140 - 28 \times 28} = 0.607$$

Thus the trend equation may be written as $Y_c = 265.75 + 0.607X$.

Now the needed forecast for the year 2007 is obtained by substituting X = 8 in the trend equation.

So $Y_{2007} = 265.75 + 0.607 \times 8 = 270.607$

Now to have monthly break-up of this annual data we obtain seasonal index for all the 12 months first by using simple average.

Construction of monthly seasonal index by method of averages

Table 19.41

Months	Sales									
	1999	2000	2001	2002	2003	2004	2005	2006	Average	Calculation of seasonal indices
Jan	18	17	21	21	19	22	24	21	20.38	$\frac{20.4}{268}$ x 12 0.9127
Feb	27	30	26	29	28	30	32	29	28.88	$\frac{28.9}{268}$ x 12 1.2935
Mar	23	20	23	26	22	26	24	29	24.13	$\frac{24.1}{268}$ x 12 1.0807
Apr	21	23	20	19	20	21	22	23	21.13	$\frac{21.1}{268}$ x 12 0.9463
May	23	24	21	20	19	19	23	21	21.25	$\frac{21.3}{268}$ x 12 0.9519
Jun	21	24	25	23	24	21	20	19	22.13	$\frac{22.1}{268}$ x 12 0.9911
Jul	20	21	22	23	19	23	22	21	21.38	$\frac{21.4}{268}$ x 12 0.9575
Aug	23	24	25	21	21	22	24	21	22.63	$\frac{22.6}{268}$ x 12 1.0135
Sep	24	23	21	20	19	21	20	22	21.25	$\frac{21.3}{268}$ x 12 0.9519
Oct	20	21	20	21	23	21	19	21	20.75	$\frac{20.8}{268}$ x 12 0.9295
Nov	23	21	24	25	23	22	21	22	22.63	$\frac{22.6}{268}$ x 12 1.0135
Dec	21	23	20	21	20	22	23	21	21.38	$\frac{21.4}{268}$ x 12 0.9575
Total	264	271	268	269	257	270	274	270	267.9	12.0000

In this method we take the average values of eight January's as the typical representative for January. Similarly, for February we take the average of all the eight February values as the typical representative value. In a similar way the averages for all the 12 months are obtained and reported in the last column of Table 19.41. Once this is done we obtain the seasonal indices in the usual manner by attaching index 1 for all the twelve months to represent non-seasonal pattern. Remember that this method is applicable if and only if seasonal pattern alone dominates the given data. Thus, we assume the absence of both trend and cyclical variations in the given series over the years.

Projection of monthly sales for 2007 using the seasonal indices

In Table 19.42 we write the non-seasonal projection as 270.607/12 = 22.550595 for each month. These figures are multiplied by the corresponding seasonal indices to obtain the needed monthly seasonal projections for the year 2007 as shown in Table 19.42.

Table 19.42

Projected monthly sales for the year 2007			
Months	Yearly projection	Seasonal index	Monthly projection
Jan	22.550595	0.9127	20.581928
Feb	22.550595	1.2935	29.169195
Mar	22.550595	1.0807	24.370428
Apr	22.550595	0.9463	21.339628
May	22.550595	0.9519	21.465911
Jun	22.550595	0.9911	22.349895
Jul	22.550595	0.9575	21.592195
Aug	22.550595	1.0135	22.855028
Sep	22.550595	0.9519	21.465911
Oct	22.550595	0.9295	20.960778
Nov	22.550595	1.0135	22.855028
Dec	22.550595	0.9575	21.592195
Total	270.60714	12.000	270.598120

19.13 FORECASTING BY USING SEASONAL INDICES OBTAINED BY RATIO TO TREND METHOD

In the previous section we have discussed the method of forecasting when the time series is having only seasonal and random variables. In this section we propose to discuss the method of forecasting when the given time series is having both seasonal and trend components in addition to random component. Invariably, the data must be monthly based or at the most quarterly based for this type of forecasting.

Many situations in economics and business involve period to period comparison. One must be interested in knowing the implications of a fall in the cement production by 5% compared to last month, or in the rise in the production of electric power by 7% this month over last month. Care must be taken in using such information because the presence of seasonal influence might make such comparisons meaningless. For instance, the fact that electric power consumption is down by 4% in September to October might be only due to the seasonal effect associated with a decrease in air-conditioning and not because of long run decline in the power consumption. Indeed after adjusting the seasonal effect, one might even find the use of more electric power by certain percentage. Removing the seasonal effect from the given time series is often called deseasonalizing the time series. After doing so one could make period to period meaningful comparison. To do this we first compute the seasonal indices and use them to deseasonalize the given series.

The multiplicative model:

If T stands for trend, S stands for seasonal and I stands for the irregular random variation in the given time series data then the relevant multiplicative model is written as

$$Y_t = T_t \times S_t \times I_t$$

In such a presentation T_t is the trend measured in units in which the items are being measured. However, I_t and S_t are measured in relative terms, with values above 1.00 indicating the respective seasonal effects above the trend, and values below 1.00 indicating the seasonal effects below the trend. We illustrate the use of the multiplicative model in isolating the seasonals by using the following numerical example.

■ Example 14

For the quarterly data given below, obtain the seasonal indices by method of ratio to trend method. Also get the deseasonalized sales and forecast the sales figures for all the four quarters in 1984.

Table 19.43

Sales of sweet				
Year	Quarters			
	I	II	III	IV
1980	4.8	4.1	6.0	6.5
1981	5.8	5.2	6.8	7.4
1982	6.0	5.6	7.5	7.8
1983	6.3	5.9	8.0	8.4

Solution

Calculation of seasonal indices by ratio to trend method:

Figure 19.16 clearly indicates the presence of seasonals. Every second quarter the sales is minimum and every fourth quarter the sales is maximum. Since the data available is on quarterly basis to, disassociate the trend from seasonal we use four-quarter moving average method as illustrated in Table 19.44.

$$\text{First moving average} = \frac{4.8 + 4.1 + 6.0 + 6.5}{4} = \frac{21.4}{4} = 5.35$$

$$\text{Second moving average} = \frac{4.1 + 6.0 + 6.5 + 5.8}{4} = \frac{22.4}{4} = 5.6$$

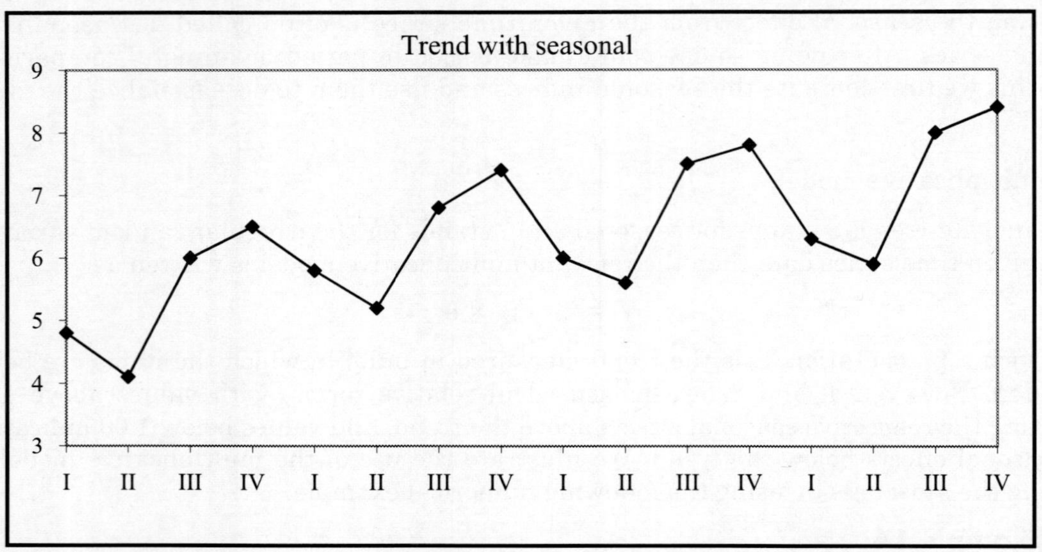

Figure 19.16

Clearly the first moving average 5.35 represents average quarterly sales for the very first year. This figure will eventually be placed in the mid-way between second and the third quarters. Similarly the second moving average 5.6 is to be placed in the middle of third and fourth and so on. Recall that the very purpose of computing moving average is to isolate the combined seasonal and irregular components. Since the moving average values are placed in the middle of two consecutive periods they are really not comparable with the actual data value. To overcome this difficulty we take the average of the first and the second moving averages and amicably place the result in the middle corresponding to the third quarters as shown in the Table 19.44. The same principle is followed for the later periods as well as shown in the table. Effectively we use a two-period moving average to fix the values correctly. Once the centring is done we divide the original value by the corresponding centred value and report it in the last column of the table. The first entry in this table 1.096 reveals the excessive sale of 0.096 over and above average 1 in the third quarter. Similarly 1981 first quarter entry is found to be less than one. Such an observation reveal the under sale during that quarter. Altogether in the last column we have 12 seasonal indices from the third quarter 1980 to second quarter 1983.

Table 19.44

Years	Quarters	Y_t	4 Quarters moving total	4 quarters moving average	Two-period cantered average	Calculation of seasonal indices
1980	I	4.800				
	II	4.100				
			21.400	5.350		
	III	6.000			5.475	6.000 / 5.475 = 1.096
			22.400	5.600		
	IV	6.500			5.738	6.500 / 5.738 = 1.133
			23.500	5.875		
1981	I	5.800			5.975	5.800 / 5.975 = 0.971
			24.300	6.075		
	II	5.200			6.188	5.200 / 6.188 = 0.840
			25.200	6.300		
	III	6.800			6.325	6.800 / 6.325 = 1.075
			25.400	6.350		
	IV	7.400			6.400	7.400 / 6.400 = 1.156
			25.800	6.450		
1982	I	6.000			6.538	6.000 / 6.538 = 0.918
			26.500	6.625		
	II	5.600			6.675	5.600 / 6.675 = 0.839
			26.900	6.725		
	III	7.500			6.763	7.500 / 6.763 = 1.109
			27.200	6.800		
	IV	7.800			6.838	7.800 / 6.838 = 1.141
			27.500	6.875		
1983	I	6.300			6.938	6.300 / 6.938 = 0.908
			28.000	7.000		
	II	5.900			7.075	5.900 / 7.075 = 0.834
			28.600	7.150		
	III	8.000				
	IV	8.400				

Averaging the quarterly seasonal:

In Table 19.45, the obtained 12 seasonal indices are averaged by using arithmetic mean method. After obtaining the needed averages they are adjusted to make the total 4, the number of quarters in a year.

Table 19.45

Quts.	Years				Average S index	Adjusted seasonal index S.I
	1980	1981	1982	1983		
I	-	0.971	0.918	0.908	0.932	$\frac{0.932}{4.004} \times 4.000 = 0.931$
II	-	0.840	0.834	0.834	0.836	$\frac{0.836}{4.004} \times 4.000 = 0.835$
III	1.096	1.075	1.109	-	1.093	$\frac{1.093}{4.004} \times 4.000 = 1.092$
IV	1.133	1.156	1.141	-	1.143	$\frac{1.143}{4.004} \times 4.000 = 1.142$
Total	-	-	-	-	4.004	4.00

Depersonalizing the original sales by using seasonal indices

Table 19.46

Years	Quarters	Original data (Y_t)	Seasonal index (S.I)	Deseasonalised sales $Y_d = Y_t/S.I$	Seasonal sales
1980	I	4.800	0.931	5.156	-0.356
	II	4.100	0.835	4.910	-0.810
	III	6.000	1.092	5.495	0.505
	IV	6.500	1.142	5.692	0.808
1981	I	5.800	0.931	6.230	-0.430
	II	5.200	0.835	6.228	-1.028
	III	6.800	1.092	6.227	0.573
	IV	7.400	1.142	6.480	0.920
1982	I	6.000	0.931	6.445	-0.445
	II	5.600	0.835	6.707	-1.107
	III	7.500	1.092	6.868	0.632
	IV	7.500	1.142	6.567	0.933
1983	I	6.300	0.931	6.767	-0.467
	II	5.900	0.835	7.066	-1.166
	III	8.000	1.092	7.326	0.674
	IV	8.400	1.142	7.356	1.044

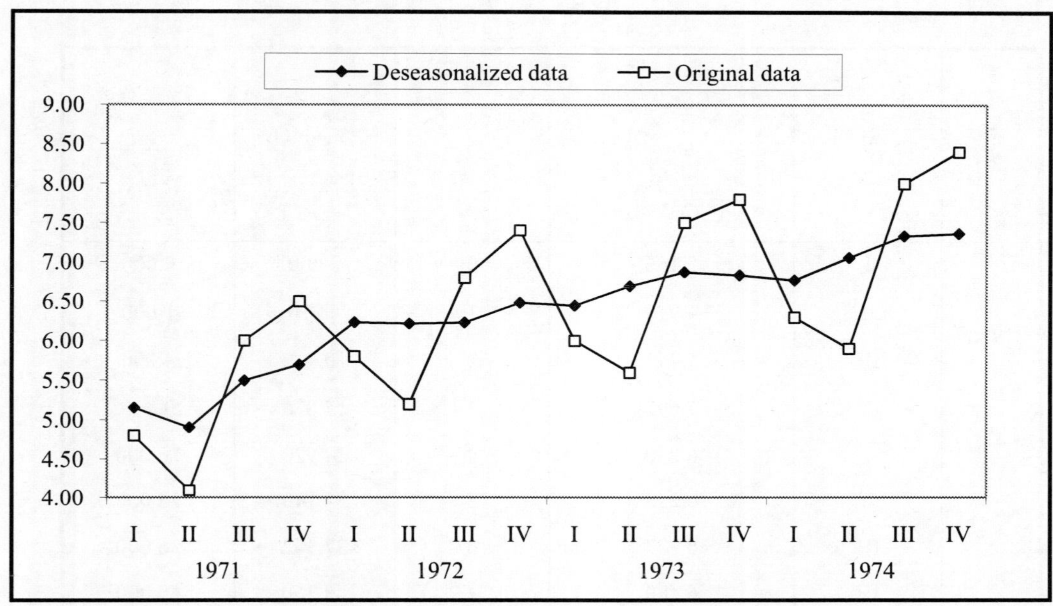

Figure 19.17

In Table 19.46 the original data values are divided by their respective seasonal indices and needed deseasonalized sales are obtained and reported in the table. The resulting deseasonalized and original data are presented in Figure 19.17.

Fitting a trend line for the deseasonalized data:

From Table 19.47,

$$101.52 = 16a + 120b... \qquad (1)$$

$$810.564 = 120a + 1240b... \qquad (2)$$

Now from equation (1) and (2) as per Cramer's rule

$$a = \frac{\begin{vmatrix} 101.52 & 28 \\ 810.564 & 140 \end{vmatrix}}{\begin{vmatrix} 16 & 120 \\ 120 & 1240 \end{vmatrix}} = 5.261$$

$$b = \frac{\begin{vmatrix} 16 & 101.52 \\ 120 & 810.564 \end{vmatrix}}{\begin{vmatrix} 16 & 120 \\ 120 & 1240 \end{vmatrix}} = 0.145$$

We fit a linear trend line for the deseasonalized data in Table 19.47.

Table 19.47

Year	Quarters	Deseasonalized sales (Y_d)	X	XY_d	X^2
1980	I	5.156	0	0.000	0.000
	II	4.910	1	4.910	1.000
	III	5.495	2	10.990	4.000
	IV	5.692	3	17.076	9.000
1981	I	6.230	4	24.920	16.000
	II	6.228	5	31.140	25.000
	III	6.227	6	37.362	36.000
	IV	6.480	7	45.360	49.000
1982	I	6.445	8	51.560	64.000
	II	6.707	9	60.363	81.000
	III	6.868	10	68.680	100.000
	IV	6.567	11	72.237	121.000
1983	I	6.767	12	81.204	144.000
	II	7.066	13	91.858	169.000
	III	7.326	14	102.564	196.000
	IV	7.356	15	110.340	225.000
Total	-	101.520	120.000	810.564	1240.000
n =	16	a =	5.261	b =	0.145

Thus the trend equation may be written as $Y_c = 5.261 + 0.145X$.

Now the needed quarter-wise forecasts for the year 2007 are obtained by substituting X = 16, 17, 18, 19 in the trend equation obtained.

$Y_{1984\ I} = 5.261 + 0.145 \times 16 = 7.575$

$Y_{1984\ II} = 5.261 + 0.145 \times 17 = 7.719$

$Y_{1984\ III} = 5.261 + 0.145 \times 18 = 7.863$

$Y_{1984\ IV} = 5.261 + 0.145 \times 19 = 8.008$

The so obtained trend line and the deseasonalized data are shown in Figure 19.18.

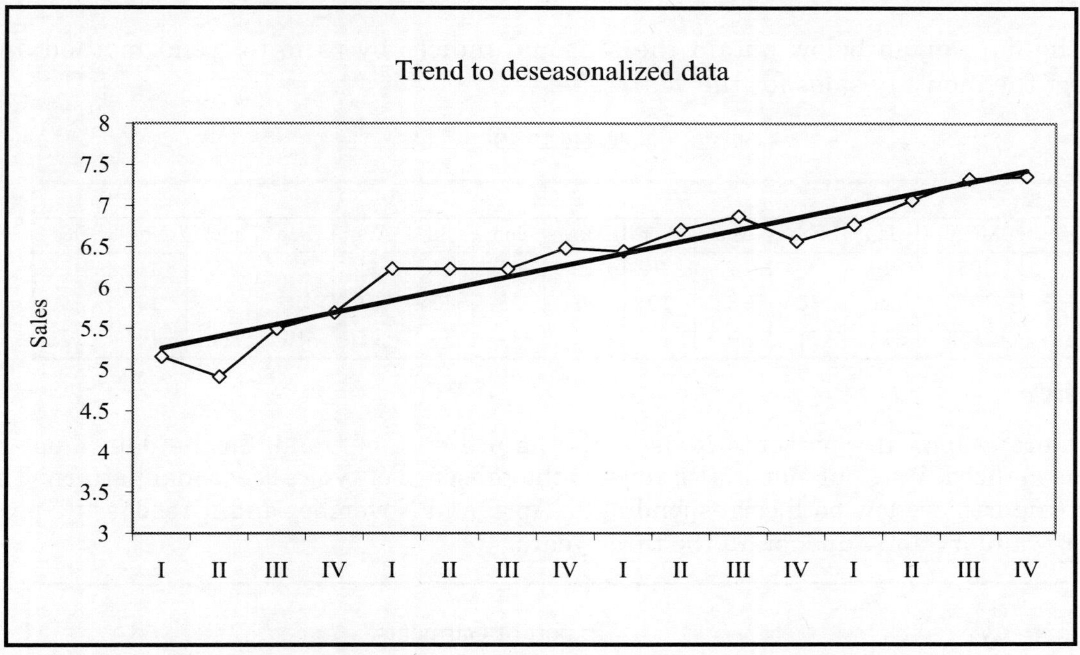

Figure 19.18

Quarter-wise forecasting for the needed quarters of 1984

In the following table the trend for all the four quarters for the next year 1984 is obtained by the regression equation $Y_c = 6.34+0.072X$. The so obtained trend values are multiplied by the respective seasonal indices obtained in the previous section and the needed final projected results are reported in the last column.

Table 19.48

Quarter-wise final forecast					
Year	Quts	x	Trend forecast (Y_c)	S_t	Quarterly forecast (Y_c x S_t)
1984	I	16	7.574	0.931	7.051
	II	17	7.719	0.835	6.445
	III	18	7.863	1.092	8.586
	IV	19	8.008	1.142	9.145

Example 15

For the data given below obtain the seasonal indices by ratio to trend method and project the monthly sales for the year 1954.

Table 19.49

Sales													
	Month	Jan	Feb	Mar	Apr	May	Jun	Jul	Aug	Sep	Oct	Nov	Dec
Sales	1951	10	9	7	9	10	12	11	9	7	9	11	12
	1952	11	10	8	10	11	13	12	10	9	10	13	14
	1953	12	11	9	11	12	14	13	11	10	11	14	15

Solution

The mere graphical representation reveals the presence of trend. So the data must be deseasonalized first. Further it also reveals the presence of typical seasonal pattern. The export figures are low both corresponding to April and November and it reaches its peak during January and June for all the three years.

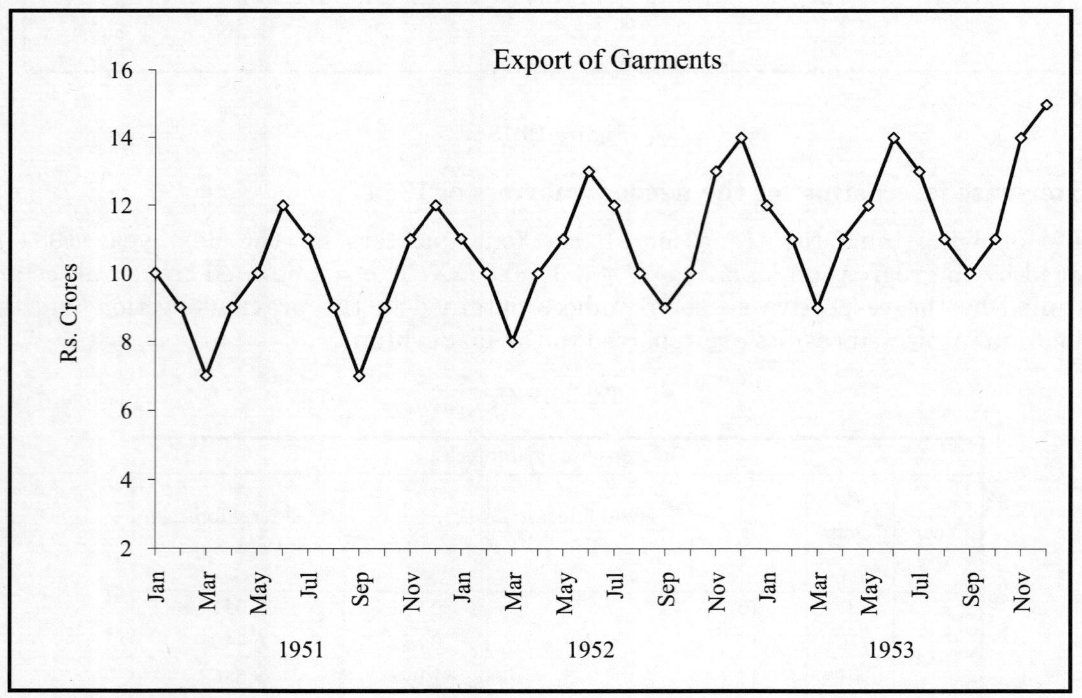

Figure 19.19

Calculation of seasonal indices by ratio to trend method: Table 19.50 illustrates the method of obtaining the seasonals by ratio to term by applying 12 month moving average.

Table 19.50

		Y_t	12 month moving total	12 period moving average	Two-period centred average	Seasonal index calculations
1951	Jan	10				
	Feb	9				
	Mar	7				
	Apr	9				
	May	10				
	Jun	12				
			116	9.67		
	Jul	11			9.71	11.00 / 9.71 = 1.13
			117	9.75		
	Aug	9			9.79	9.00 / 9.79 = 0.92
			118	9.83		
	Sep	7			9.88	7.00 / 9.88 = 0.71
			119	9.92		
	Oct	9			9.96	9.00 / 9.96 = 0.90
			120	10.00		
	Nov	11			10.04	11.00 / 10.04 = 1.10
			121	10.08		
	Dec	12			10.13	12.00 / 10.13 = 1.18
			122	10.17		
1952	Jan	11			10.21	11.00 / 10.21 = 1.08
			123	10.25		
	Feb	10			10.29	10.00 / 10.29 = 0.97
			124	10.33		
	Mar	8			10.42	8.00 / 10.42 = 0.77
			126	10.50		
	Apr	10			10.54	10.00 / 10.54 = 0.95
			127	10.58		
	May	11			10.67	11.00 / 10.67 = 1.03
			129	10.75		
	Jun	13			10.84	13.00 / 10.84 = 1.20
			131	10.92		

(Contd.)

		Y_t	12 month moving total	12 period moving average	Two-period centred average	Seasonal index calculations
	Jul	12			10.96	12.00 / 10.96 = 1.09
			132	11.00		
	Aug	10			11.04	10.00 / 11.04 = 0.91
			133	11.08		
	Sep	9			11.13	9.00 / 11.13 = 0.81
			134	11.17		
	Oct	10			11.21	10.00 / 11.21 = 0.89
			135	11.25		
	Nov	13			11.29	13.00 / 11.29 = 1.15
			136	11.33		
	Dec	14			11.38	14.00 / 11.38 = 1.23
			137	11.42		
	Jan	12			11.46	12.00 / 11.46 = 1.05
			138	11.50		
	Feb	11			11.54	11.00 / 11.54 = 0.95
			139	11.58		
	Mar	9			11.63	9.00 / 11.63 = 0.77
			140	11.67		
	Apr	11			11.71	11.00 / 11.71 = 0.94
			141	11.75		
	May	12			11.79	12.00 / 11.79 = 1.02
			142	11.83		
	Jun	14			11.88	14.00 / 11.88 = 1.18
1953			143	11.92		
	Jul	13				
	Aug	11				
	Sep	10				
	Oct	11				
	Nov	14				
	Dec	15				

Averaging the Seasonals

Table 19.51

Month	Years			S. index average	Adjusted seasonal index calculations
	1951	1952	1953		
Jan	-	1.08	1.05	1.06	$\frac{12.00}{11.95} \times 1.06 = 1.0644$
Feb	-	0.97	0.95	0.96	$\frac{12.00}{11.95} \times 0.96 = 0.9640$
Mar	-	0.77	0.77	0.77	$\frac{12.00}{11.95} \times 0.77 = 0.7732$
Apr	-	0.95	0.94	0.94	$\frac{12.00}{11.95} \times 0.94 = 0.9439$
May	-	1.03	1.02	1.02	$\frac{12.00}{11.95} \times 1.02 = 1.0243$
Jun	-	1.20	1.18	1.19	$\frac{12.00}{11.95} \times 1.19 = 1.1950$
Jul	1.13	1.09	-	1.11	$\frac{12.00}{11.95} \times 1.11 = 1.1146$
Aug	0.92	0.91	-	0.91	$\frac{12.00}{11.95} \times 0.91 = 0.9138$
Sep	0.71	0.81	-	0.76	$\frac{12.00}{11.95} \times 0.76 = 0.7632$
Oct	0.90	0.89	-	0.90	$\frac{12.00}{11.95} \times 0.90 = 0.9038$
Nov	1.10	1.15	-	1.12	$\frac{12.00}{11.95} \times 1.12 = 1.1247$
Dec	1.18	1.23	-	1.21	$\frac{12.00}{11.95} \times 1.21 = 1.2151$
Total				11.95	12.0000

Once the seasonal indices are obtained as ratio to trend we average them in the usual manner as shown in Table 19.51. Here it is important to note that at the time of removing the trend we have lost one data for each month in our illustration. There will be only two indices for each month.

Deseasonalizing original data and trend calculations

Once the adjusted seasonal index is obtained for all the 12 months we can always remove the seasonal effects and get the deseasonalized sales for the entire data. In Table 19.52 the needed deseasionlizing is done. Figure 19.20 presents the graph for the deseasonalized data.

For example the deseasonalized sales for January 1951

$$= \frac{\text{The original data for January 1951}}{\text{Seasonal index for January 1951}} = \frac{10}{1.0652} = 9.39$$

In a similar way the deseasonalized sales are obtained and reported for the entire data in Table 19.52. (The corresponding seasonal sales are obtained by simply subtracting the corresponding deseasonalized sales from the original data. For example the seasonal sales for January 1951 $= 10 - 9.3387 = 0.6613$)

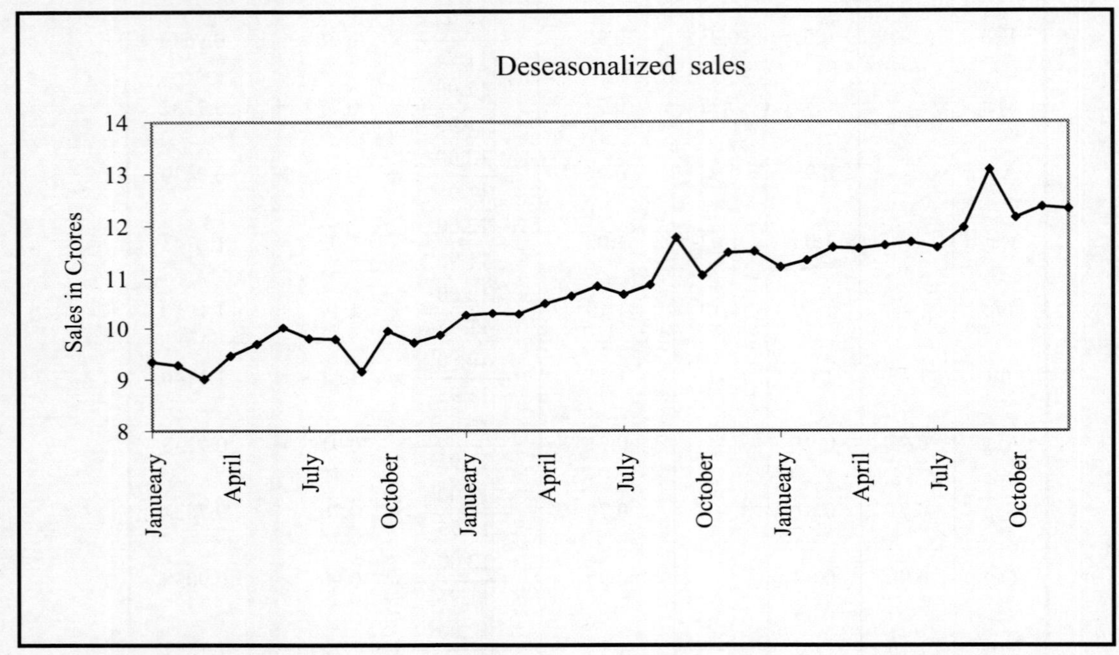

Figure 19.20

Using the so obtained deseasonalized sales we obtain the trend values by using the OLS method in Table 19.52.

Now from the table

$$a = \frac{\sum Y_d}{N} = \frac{391.01}{36} = 10.861$$

$$b = \frac{\sum XY_d}{\sum X^2} = \frac{734.28}{15540} = 0.5$$

Table 19.52

Year	Month	Y_t	S.I	D. Sales Yd $=Y_t / S.I$	X_t	X_t^2	X_tY_d
1951	Jan	10	1.0644	9.39	-35	1225.00	-328.81
	Feb	9	0.9640	9.34	-33	1089.00	-308.09
	Mar	7	0.7732	9.05	-31	961.00	-280.64
	Apr	9	0.9439	9.53	-29	841.00	-276.50
	May	10	1.0243	9.76	-27	729.00	-263.60
	Jun	12	1.1950	10.04	-25	625.00	-251.05
	Jul	11	1.1146	9.87	-23	529.00	-226.98
	Aug	9	0.9138	9.85	-21	441.00	-206.83
	Sep	7	0.7632	9.17	-19	361.00	-174.27
	Oct	9	0.9038	9.96	-17	289.00	-169.29
	Nov	11	1.1247	9.78	-15	225.00	-146.71
	Dec	12	1.2151	9.88	-13	169.00	-128.39
1952	Jan	11	1.0644	10.33	-11	121.00	-113.68
	Feb	10	0.9640	10.37	-9	81.00	-93.36
	Mar	9	0.7732	11.64	-7	49.00	-81.48
	Apr	10	0.9439	10.59	-5	25.00	-52.97
	May	11	1.0243	10.74	-3	9.00	-32.22
	Jun	13	1.1950	10.88	-1	1.00	-10.88
	Jul	12	1.1146	10.77	1	1.00	10.77
	Aug	10	0.9138	10.94	3	9.00	32.83
	Sep	9	0.7632	11.79	5	25.00	58.96
	Oct	10	0.9038	11.06	7	49.00	77.45
	Nov	13	1.1247	11.56	9	81.00	104.03
	Dec	14	1.2151	11.52	11	121.00	126.74
1953	Jan	12	1.0644	11.27	13	169.00	146.56
	Feb	11	0.9640	11.41	15	225.00	171.16
	Mar	9	0.7732	11.64	17	289.00	197.87
	Apr	11	0.9439	11.65	19	361.00	221.41
	May	12	1.0243	11.72	21	441.00	246.03
	Jun	14	1.1950	11.72	23	529.00	269.46
	Jul	13	1.1146	11.66	25	625.00	291.57
	Aug	11	0.9138	12.04	27	729.00	325.01
	Sep	10	0.7632	13.10	29	841.00	379.99
	Oct	11	0.9038	12.17	31	961.00	377.31
	Nov	14	1.1247	12.45	33	1089.00	410.78
	Dec	15	1.2151	12.35	35	1225.00	432.08
Total		391		391.01	0.00	15540.00	734.28
n =	36	a =	10.861	b =	0.05		

So the equation of the fitted line may be written as

$\hat{Y} = 10.8613 + 0.5X...$

Origin: Jan 1974

X – axis unit = 1 / 2 Year

Y – axis unit = Output in kg.

The trend line is obtained as Y_c = 10.861+0.05 X. The needed trend forecasts for all the 12 months for 1954 is obtained by substituting 37, 39, 41, 43, 45. 47, 49, 51, 53, 55, 57, 59 for X in the trend equation and the resulting results are reported in Table 19. 53. In the same Table the monthly seasonally adjusted forecasts are obtained and reported in the last column.

Table 19.53

Year	Month	X_t	Y_c	S_i	Forecast $Y_c S_i$
1954	Jan	37	12.609	1.06	13.42
	Feb	39	12.704	0.96	12.25
	Mar	41	12.798	0.77	9.90
	Apr	43	12.893	0.94	12.17
	May	45	12.987	1.02	13.30
	Jun	47	13.082	1.19	15.63
	Jul	49	13.176	1.11	14.69
	Aug	51	13.271	0.91	12.13
	Sep	53	13.365	0.76	10.20
	Oct	55	13.46	0.90	12.16
	Nov	57	13.554	1.12	15.24
	Dec	59	13.649	1.22	16.58

19.14 FORECASTING BY USING SEASONAL INDICES OBTAINED BY METHOD OF LINK RELATIVE

Method of Link Relatives: In this method again we try to smoothening the data but by using a different method. The logic behind this method is that for any consecutive data with short time intervals may not exhibit trend.

Step: 1 In this method each monthly or quarterly observations are expressed as a ratio of the preceding observation. The so obtained relative is often called the link relative. It is defined as

$$Y_t = \left(\frac{Y_t}{Y_{t-1}}\right) \times 100$$

Step 2: In the second stage we obtain the average of the link relatives over the years under consideration.

Step 3: In the third step, we take the seasonal index for the month of January as hundred and calculate the index for the month of February. For February the seasonal index is given by the product of February link relative and the seasonal index of January. Similarly, the seasonal index for March is obtained as the product of seasonal index for February and the link relative for March, etc. The following example illustrates the method of calculations.

■ Example 16

For the data given below obtain the seasonal index by link relative method and project the quarter-wise sales for the year 1979.

Table 19.54

Original data				
Year	I	II	III	IV
1980	6.83	6.26	6.11	6.33
1981	6.54	5.79	5.64	6.15
1982	6.81	6.27	6.28	6.70
1983	7.01	5.91	5.63	6.16
1984	5.95	5.48	6.11	6.80

Solution

In the table above the top left corner we represent the original data. In the adjacent right side the link relative are obtained and reported. For example, the link relative for the second quarter 1974 is obtained as

$$\text{Link relative for the second quarter 1974} = \frac{Y_t}{Y_{t-1}} \times 100 = \frac{6.26}{6.83} \times 100 = 91.65$$

$$\text{Link relative for the third quarter 1974} = \frac{Y_t}{Y_{t-1}} \times 100 = \frac{6.11}{6.26} \times 100 = 97.60$$

In a similar manner for entire data the link relatives are obtained and reported in Table 19.55. In the 8th row we average all the link relatives quarter-wise. In the 9th row we obtain the chain relative for all the three-quarters from quarter II to IV after assuming 100 for the first quarter.

For example,

$$\text{The second quarter chain relative} = \frac{89.73}{100} \times 100 = 89.73$$

Table 19.55

Seasonal indices by link relative method									
	Original data				Link relatives				
Year	I	II	III	IV	I	II	III	IV	Total
1980	6.83	6.26	6.11	6.33	-	91.65	97.60	103.60	
1981	6.54	5.79	5.64	6.15	103.32	88.53	97.41	109.04	
1982	6.81	6.27	6.28	6.70	110.73	92.07	100.16	106.69	
1983	7.01	5.91	5.63	6.16	104.63	84.31	95.26	109.41	
1984	5.95	5.48	6.11	6.80	96.59	92.10	111.50	111.29	
Average					103.82	89.73	100.39	108.01	
Relative chain					100.00	89.73	90.08	97.29	
Corrected chain relatives					100.00	89.48	89.58	96.54	375.60
Corrected seasonal indices					106.50	95.30	95.40	102.81	400.00
Correction term					0.25				

The third quarter chain relative $= \dfrac{100.39}{89.73} \times 100 = 90.08$

The fourth quarter chain relative $= \dfrac{108.01}{90.08} \times 100 = 97.29$

Chain relative of the first quarter by assuming the last quarter to 100 is $= \dfrac{103.3}{100} \times 93.73 = 101.0$

Difference =101.0 – 100.00 = 1.00

Difference per quarter = 1.0/4 = 0.25

Second quarter corrected chain relative = 89.73 – 0.25 = 89.48

Third quarter corrected chain relative = 90.08 – 2 × 0.25 = 89.58

Fourth corrected chain relative = 97.29 – 3 × 0.25 = 89.6 = 96.54

Deseasonalizing the original sales by using seasonal indices obtained by link relative methods

In Table 19.56 the original data values are divided by their respective seasonal indices and needed deseasonalized sales are obtained.

Table 19.56

Years	Quts	Original data (Yt)	Seasonal index (St)	Deseasonalised sales
1980	I	6.830	1.065	6.413
	II	6.260	0.953	6.569
	III	6.110	0.954	6.405
	IV	6.330	1.028	6.158
1981	I	6.540	1.065	6.141
	II	5.790	0.953	6.076
	III	5.640	0.954	5.912
	IV	6.150	1.028	5.982
1982	I	6.810	1.065	6.394
	II	6.270	0.953	6.579
	III	6.280	0.954	6.583
	IV	6.700	1.028	6.518
1983	I	7.010	1.065	6.582
	II	5.910	0.953	6.201
	III	5.630	0.954	5.901
	IV	6.160	1.028	5.992
1984	I	5.950	1.065	5.587
	II	5.480	0.953	5.750
	III	6.110	0.954	6.405
	IV	6.800	1.028	6.615

Trend for deseasonalized data by using least squares method

We fit a linear trend line for the deseasonalized data in Table 19.57

$$124.673 = 20a + 190b... \quad (1)$$

$$2470 = 190a + 1178.242b... \quad (2)$$

Now from equation (1) and (2) as per Cramer's rule

Table 19.57

Year	Quarters	Deseasonalized sales (Y_d)	X	XY_d	X^2
1980	I	6.413	0	0.000	0.000
	II	6.569	1	6.569	1.000
	III	6.405	2	12.810	4.000
	IV	6.158	3	18.474	9.000
1981	I	6.141	4	24.564	16.000
	II	6.076	5	30.380	25.000
	III	5.912	6	35.472	36.000
	IV	5.982	7	41.874	49.000
1982	I	6.394	8	51.152	64.000
	II	6.579	9	59.211	81.000
	III	6.583	10	65.830	100.000
	IV	6.518	11	71.698	121.000
1983	I	6.582	12	78.984	144.000
	II	6.201	13	80.613	169.000
	III	5.901	14	82.614	196.000
	IV-	5.992	15	89.880	225.000
1984	I	5.587	16	89.392	256.000
	II	5.750	17	97.750	289.000
	III	6.405	18	115.290	324.000
	IV	6.615	19	125.685	361.000
Total	-	124.763	190.000	1178.242	2470.000

$$a = \frac{\begin{vmatrix} 124.673 & 120 \\ 2470 & 1240 \end{vmatrix}}{\begin{vmatrix} 20 & 190 \\ 190 & 1178.242 \end{vmatrix}} = 6.338$$

$$b = \frac{\begin{vmatrix} 20 & 124.673 \\ 190 & 2470 \end{vmatrix}}{\begin{vmatrix} 20 & 190 \\ 190 & 1178.242 \end{vmatrix}} = -0.011$$

Trend Equation $Y_c = 6.338 - 0.011X$. The trend projection and seasonally adjustment projection are shown in Table 19.58.

Table 19.58

Year	Quts	X	Trend forecast (Y_c)	SI	Quarterly forecast ($Y_c \times SI$)
1985	I	20	6.128	1.065	6.526
	II	21	6.117	0.953	5.830
	III	22	6.106	0.954	5.825
	IV	23	6.096	1.028	6.267

19.15 BAROMETRIC FORECASTING

Trend projection and exponential smoothening methods for projecting the future is based on past time series data. On several occasions there may not be clear pattern in the past time series relating to the variable under consideration. To overcome this problem we find a closely related second series with needed clear-cut pattern. By observing the changes in the related series it may be possible to predict the changes in the original series. For example, suppose that brick company sales exhibited very large yearly fluctuations over the last decade so that our old method of trend projection forecast is useless. However, it is observed during the said period the firm's sale of bricks is highly correlated with the housing units started. Thus housing starts can be predicted, This information can be used to predict the demand for bricks.

Use of leading indicators in business forecasting

If two series of data either increase or decrease at the same time, then one series may be regarded as a coincident indicator of the other series. In the following Figure 19.21 series 1 is the coincident indicator of the series 2 and vice versa because the two series have same packs and troughs periods.

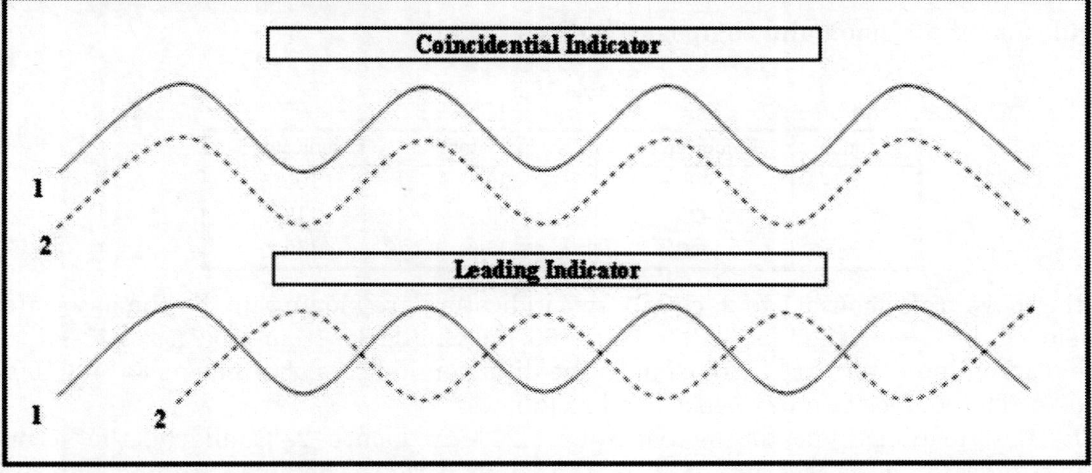

Figure 19.21

If change one series consistently occur prior to chances in the other a leading indicator has been identified. In Figure 19.21 a leading indicator has been identified. In Figure 19.21 series 1 is a leading indicator of the series 2.

For purposes of forecasting leading indicators are of primary interest. Meteorologist uses changes in the barometric pressure to predict the weather. Similarly, leading indicators can be used to forecast the changes in general economic conditions.

Leading indicators of some selected variables	
Leading indicators	Predicted variable
Average work week	Manufacturing output
New orders for durables	Sales of durables
New order for consumables	Sales of consumables
New housing plan sanction	Private housing
Petrol price	Consumer goods price

Use of composite and diffusion indices in business forecasting

Although a time series showing the stock prices may be useful in predicting general economic conditions, no single indicator is identified yet to having a perfect forecasting record. Even when a single leading indicator correctly indicates variations in economic activity, the lead time contains random fluctuations.

In such a situation one can think of constructing an index of several leading indicators to increase the accuracy of forecasting. United States of America uses 11 important leading indicators to obtain the composite index. Among the 11 if 8 indicators increase from June to July then diffusion index is obtained as (8/11) × 100 = 72.6. When the so obtained index is 50% or more for several months, it forecasts, improve economic conditions. Clearly as the index approaches 100% then there is a clear-cut improvement in economic conditions.

Calculation of diffusion and composite indices

Table 19.59

Month	L. indicator 1	L. indicator 2	L. indicator 3
1	400	30	100
2	425	29	110
3	460	33	135

For the move from month 1 to 2, clearly two indicators are moving up. So the associated diffusion index is obtained as (2/3) × 100 = 66.7%. Similarly, from month 2 to 3 all the three leading indicators are increasing so the diffusion index is obtained as (3/3) × 100 = 100.00%. The composite index is obtained as follows.

The first leading indicator increases by 6.25% in month 2. Similarly, the second leading indicator decreases by –0.33 in the same period. The third leading indicator increases by 10%. So at an average the composite increase is by 4.31%. Thus by assuming an index of 100% for the base period the composite index for the second period is reported as 104.31%. By similar calculations the composite index for the third month is obtained and reported as 120%.

Table 19.60

Month	Diffusion index	Composite index
1	-	100
2	66.7	104.31
3	100	120

19.16 INPUT-OUTPUT FORECASTING MODELS

The present-day economic system may be thought of as a composite of mutually interrelated industries. In such a system all the industries are related to one another. In general each industry buys it inputs from other industries and sells its output to other industries and final consuming sectors like households, governments etc. There may be some industries buying only primary inputs from the household sectors. Similarly, some industries sell their outputs only to final consumption sectors. With all such varieties of business, circular interdependence remains a common feature in all the economies. Such a comprehensive system cannot be handled by usual methods developed in the previous sections.

There are two sections in the theory of programming in this regard. The first section of the theory deals with internal consistency of the various programs under-taken simultaneously. Here by consistency we mean the co-ordination of various inter-related decisions. In the second part of the theory of programming we aim at optimizing the various interrelated decisions. This part of programming has been discussed in detail in the previous section under the heading "Linear programming". In this section we take up, the problem of internal consistency or otherwise called the input-output analysis without much emphasis on optimality.

19.17 INPUT-OUTPUT MODEL

The input-output model, sometimes called the Leontief model, in the name of its originator W.W. Leontief, provides a simplified version of the general multi-market equilibrium analysis of Leon Walras. To quote Henderson and Quant," The input-output system is an empirically oriented multi-market analysis. In this the assumptions represent a considerable amount of simplification of the general multi-market equilibrium analysis (Lean Walras Model). Utility functions are omitted and consumer demands are taken on the basis of outside information without regard to the equilibrium of the individual consumers. The industry, rather than the firm, is the unit of production. The production function of each industry is of the constant coefficient type, and there are no optimization problems in the production side".

Assumption of the model: The input-output model is based on few important assumptions.

1. The whole of the economy is divided into number of interrelated sectors.
2. In each industry only one homogeneous product is produced.
3. In all the industries inputs are employed in a fixed proportion. In other words, it assumes the presence of constant returns to sale.
4. Techniques of production are assumed to be static.
5. There are no external economies and diseconomies.

The interdependence in input-output analysis is due to the fact that every industry employs the outputs of other industries as their inputs apart from its own product. For example, wheat is used as input by the very agricultural sector itself as seen. It is also used as input in bread making in bakery industry. Similarly, steel is used as an input in both railways and petroleum industry. Similarly, the steel industry uses both railway transport and petrol as inputs in its own production. Therefore, in the presence of such an inter-industry relationship, the real problem in input-output analysis is to find out the level of outputs in all the industries simultaneously that confirms the given inter-industry consumption levels as inputs and the leftover for the final consumption by the sectors like household, government, business etc.

19.17.1 Input-Output Matrix in Physical Units

The inter-industry matrix normally is classified into two broad categories as producing and consuming sectors. While constricting the input-output table often we divide production sector into two or more sub-sectors like agriculture, manufacturing etc. Similarly, the final consumption sector is also subdivided into households sector, government sector, export sector etc.

In Table 19.61 we represent the sector-wise outputs along the rows. For example, in the first row we show the distribution of the agricultural output among the three sectors listed. Out of 160 units, thirty units are consumed as input in the same agricultural sector, 40 units are sold to power sector as input. The leftover 90 units are sold to household sector as final demand. Similarly, second row shows the distribution of power output to its own and the remaining two sectors, namely agriculture and household. However, in the third row we show the distribution of household sectors output, namely labour to the remaining two sectors. The specialty of this sector is that it does not consume its own output. Thus, the corresponding entry in the table occupies zero. Similarly, column-wise we show the sector-wise input requirements. For example, to produce 160 units of agricultural output we need 30 units of agricultural output, 5 units of power and 60 units labour from household sector as inputs.

Table 19.61 Inter-industry flow matrix in physical numerical units

Input	Output			
	Agriculture	Power	Household	Total output
Agriculture	30	40	90	160
Power	5	20	40	65
Household	60	25	0	85

Symbolic representation of the input-output table in physical terms:

In symbolic terms the above matrix may be reproduced as

Table 19.62 Inter-industry flow matrix in physical algebraic units

Input	Output			
	Agriculture	Power	Household	Total output
Agriculture	X_{11}	X_{12}	D_1	X_1
Power	X_{21}	X_{22}	D_2	X_2
Household	L_1	L_2	D	N

The basic supply = demand row relations of the above table

$$X_1 = X_{11} + X_{12} + D_1$$
$$X_2 = X_{21} + X_{22} + D_2$$

19.17.2 Open and Closed Input-Output Models

In the above hypothetical illustration we introduced only two production sectors. The consumption of final goods by consumers is treated as autonomous and hence is not related to the amount of labour supplied to the process of production. Such a treatment is often called the open model. In the alternative closed model approach household is included as a third sector in the third row similar to any other sector. Accordingly, the final goods it consumes are considered inputs to produce the labour output. In this section we concentrate only on the open models.

19.17.3 Technical Input Coefficient Matrix

The technical coefficient matrix shows the level of input requirements per unit of the output produced. For example, the total output of the agricultural sector in our illustration is 160. To produce this much of output it needs 30 units of inputs from agricultural, 5 units from power. Therefore, for producing one unit worth of output the industry needs 0.1875(30/160) units of inputs from the agriculture, 0.03125 (5/160) units of inputs from power. In a similar manner we calculate inputs coefficients for power sector also. The following table shows the input-output coefficient matrix for the illustration given in the previous table. Note that the third column pertaining to household sector in excluded in the table.

Table 19.63

Input	Output			
	Agriculture		Power	
Agriculture	(30/160) =	0.18750	(40/65) =	0.61538
Power	(5/160) =	0.03125	(20/65) =	0.30769
Household	(60/160) =	0.37500	(25/65) =	0.38462

Symbolic representation of the technical coefficient matrix

In symbolic terms the above matrix may be reproduced as

Table 19.64

Input	Output	
	Agriculture	Power
Agriculture	$\frac{X_{11}}{X_1} = a_{11}$	$\frac{X_{12}}{X_2} = a_{12}$
Power	$\frac{X_{21}}{X_1} = a_{21}$	$\frac{X_{22}}{X_2} = a_{22}$

Now let us rewrite the basic row identities stated earlier in terms of the technical coefficients as

$$X_1 = a_{11}X_1 + a_{12}X_2 + D_1$$
$$X_2 = a_{21}X_1 + a_{22}X_2 + D_2$$

The above equations may be rewritten as

$$(1 - a_{11})X_1 - a_{12}X_2 = D_1 \ldots \quad (1)$$
$$-a_{21}X_1 + (1 - a_{22})X_2 = D_2 .. \quad (2)$$

Now, given the technical coefficients and the final demand, we can calculate the needed production levels X_1 and X_2 by using equation (1) and (2) with the help of Cramer's rule.

Numerical illustration: Let us use the data given in example 1 above and calculate the needed output levels as follows

$$X_1 = 0.1875X_1 + 0.61538X_2 + 90$$
$$X_2 = 0.03125X_1 + 0.30769X_2 + 40$$

The above equations may be rewritten as

$$(1-0.1875)X_1 - 0.61538X_2 = 90$$

$$0.8125X_1 - 0.61538X_2 = 90... \quad (1)$$

$$-0.03125X_1 + (1-0.30769)X_2 = 40$$

$$-0.03125X_1 + 0.69231X_2 = 40... \quad (2)$$

Now from (1) and (2) as per Cramer's rule

$$X_1 = \frac{\begin{vmatrix} 90 & -0.61538 \\ 40 & 0.69231 \end{vmatrix}}{\begin{vmatrix} 0.8125 & -0.61538 \\ -0.03125 & 0.69231 \end{vmatrix}} = \frac{62.3079 + 24.6152}{0.5625 - 0.01923} = \frac{86.9231}{0.54327} = 160$$

$$X_2 = \frac{\begin{vmatrix} 0.8125 & 90 \\ -0.03125 & 40 \end{vmatrix}}{\begin{vmatrix} 0.8125 & -0.61538 \\ -0.03125 & 0.69231 \end{vmatrix}} = \frac{32.5 + 2.8125}{0.5625 - 0.01923} = \frac{35.3125}{0.54327} = 65$$

Thus to meat both inter industry and final desired amount of demand 160 units of X_1 and 65 units of X_2 are to be produced.

Matrix representation of the input output problem:

$$\begin{bmatrix} X_1 \\ X_2 \end{bmatrix} = \begin{bmatrix} a_{11} & a_{12} \\ a_{21} & a_{22} \end{bmatrix} \begin{bmatrix} X_1 \\ X_2 \end{bmatrix} + \begin{bmatrix} D_1 \\ D_2 \end{bmatrix}$$

If A denotes the technical coefficient matrix and X denotes the output vector and D denotes the final demand vector then

$$X = AX + D$$

$$D = X - AX$$

$$D = (I - A)X$$

$$X = (I - A)^{-1} D$$

Using this relation we can always calculate the level of outputs for any given level of final demand vector by simply pre-multiplying the given final demand vector by $(I-A)^{-1}$. The matrix $(I-A)^{-1}$ is often called Leontief Matrix.

Pertaining to our illustrative example let us calculate $(I-A)^{-1}$ first

$$A = \begin{bmatrix} 0.1875 & 0.61538 \\ 0.03125 & 0.30769 \end{bmatrix}$$

$$\text{So } (I-A) = \begin{bmatrix} 1 & 0 \\ 0 & 1 \end{bmatrix} - \begin{bmatrix} 0.8175 & 0.61538 \\ 0.03125 & 0.30769 \end{bmatrix}$$

$$= \begin{bmatrix} 0.8125 & -0.61538 \\ -0.03125 & 0.69231 \end{bmatrix}$$

$$\text{By definition } (I-A)^{-1} = \frac{\text{adj } (I-A)}{|I-A|}$$

$$\text{adj } (I-A) = \begin{bmatrix} 0.69231 & 0.61538 \\ 0.03125 & 0.8125 \end{bmatrix}$$

$$|I-A| = \begin{vmatrix} 0.8125 & -0.61538 \\ -0.03125 & 0.69231 \end{vmatrix}$$

$$= (0.8125)(0.69231) - (-0.03125)(-0.61538)$$

$$= 0.54327$$

$$(I-A)^{-1} = \frac{\text{adj } (I-A)}{|I-A|}$$

$$= \frac{1}{0.54327} \begin{bmatrix} 0.69231 & 0.61538 \\ 0.03125 & 0.8125 \end{bmatrix}$$

$$= \begin{bmatrix} 1.27434 & 1.1327 \\ 0.0575 & 1.4956 \end{bmatrix}$$

In the above $(I–A)^{-1}$ matrix the first column reveals that to produce one unit of agricultural product for final consumption we need 1.27434 units of agricultural product and 0.0575 units of power. Similarly, the second column reveals that to produce one unit of power for final consumption we need 1.1327 units of agricultural product and 1.4956 units of power.

In our illustration since $D = \begin{bmatrix} 90 \\ 40 \end{bmatrix}$, the needed output vector is obtained by pre-multiplying D by $(I–A)^{-1}$

$$\text{Therefore} \quad X = (I-A)^{-1}D$$

$$X = \begin{bmatrix} 1.27434 & 1.1327 \\ 0.0575 & 1.4956 \end{bmatrix} \begin{bmatrix} 90 \\ 40 \end{bmatrix}$$

$$= \begin{bmatrix} 90 \times 1.27434 + 1.1327 \times 40 \\ 0.0575 \times 90 + 1.4955 \times 40 \end{bmatrix}$$

$$= \begin{bmatrix} 160 \\ 65 \end{bmatrix}$$

19.17.4 Equilibrium Price Determination

Given the exogenous wage rate we can always determine the equilibrium output prices. From the given technical coefficient matrix with given wage rate $P_w = 10$ we obtain the following equations

$$P_1 = 0.1875P_1 + 0.03125P_2 + 0.375P_w$$

$$0.8125P_1 - 0.03125P_2 = 0.375 \times 10 = 3.75... \qquad (1)$$

$$P_2 = 0.61538P_1 + 0.30768P_2 + 0.3846P_w$$

$$-0.61538P_1 + 0.6923P_2 = 0.3846 \times 10 = 3.846... \qquad (2)$$

Now from equation (1) and (2) as per Cramer's rule

$$P_2 = \frac{\begin{vmatrix} 3.75 & -0.03125 \\ 3.846 & 0.69231 \end{vmatrix}}{\begin{vmatrix} 0.8125 & -0.03125 \\ -0.61538 & 0.6923 \end{vmatrix}}$$

$$= \frac{2.5995375 + 0.1089375}{0.56249 - 0.01923}$$

$$= \frac{2.708475}{0.54326} = 5$$

$$P_2 = \frac{\begin{vmatrix} 0.8125 & 3.75 \\ -0.61538 & 3.846 \end{vmatrix}}{\begin{vmatrix} 0.8125 & -0.03125 \\ -0.61538 & 0.6923 \end{vmatrix}}$$

$$= \frac{3.12487 + 2.307675}{0.56249 - 0.01923}$$

$$= \frac{5.432545}{0.54326} = 10$$

Thus under equilibrium P_1 = Rs.5; $P_2 = 10$; $P_w = 10$

19.17.5 Input-Output Coefficient Matrix: A More Generalized Case

Table 19.65

Output / Input	Industry 1	Industry 2	...	Industry n	Household	Total output
Industry 1	X_{11}	X_{12}	...	X_{1n}	D_1	X_1
Industry 2	X_{21}	X_{22}		X_{2n}	D_2	X_2
...	...	...	...	...	...	...
...	...	...	...	...	...	...
Industry n	X_{n1}	X_{n2}		X_{nn}	D_n	
Household	L_1	L_2	...	L_n	D	N

The input-output coefficients, as explained earlier, give us the input requirements per unit of output of all the industries. It is obtained by simply dividing the total requirements of input by the total output of the concerned industry. For example, since X_{21} is the total input requirement of the first industry from the second, for an output level X_1, per unit requirement of the first industry from the second denoted by a_{21} is obtained as X_{21}/X_1. Similarly, $a_{31} = X_{31}/X_1$ give us per unit input requirement of the first industry from the third and so on. Thus, for the whole of the economy the input-output coefficient matrix is given by:

$$A=\begin{bmatrix} a_{11} & a_{12} & a_{13} & a_{14} & - & - & a_{1n} \\ a_{21} & a_{22} & a_{23} & a_{24} & - & - & a_{2n} \\ - & - & - & - & - & - & - \\ - & - & - & - & - & - & - \\ - & - & - & - & - & - & - \\ - & - & - & - & - & - & - \\ a_{n1} & a_{n2} & a_{n3} & a_{n4} & - & - & a_{nn} \end{bmatrix}$$

Now interns of input-output coefficients the whole system of inter-industry relations is rewritten as

$$X_1 = a_{11}X_1 + a_{12}X_2 + \dots\dots\dots\dots a_{1n}X_n + D_1$$
$$X_2 = a_{21}X_1 + a_{22}X_2 + \dots\dots\dots\dots a_{2n}X_n + D_2$$

$$X_n = a_{n1}X_1 + a_{n2}X_2 + \dots\dots\dots\dots a_{nn}X_n + D_{n1}$$

The system in matrix form:

Since we have a system of simultaneous equations, let us restate the whole system in its matrix form as shown below.

$$\begin{bmatrix} X_1 \\ X_2 \\ - \\ - \\ - \\ - \\ X_n \end{bmatrix} = \begin{bmatrix} a_{11} & a_{12} & a_{13} & a_{14} & - & - & a_{1n} \\ a_{21} & a_{22} & a_{23} & a_{24} & - & - & a_{2n} \\ - & - & - & - & - & - & - \\ - & - & - & - & - & - & - \\ - & - & - & - & - & - & - \\ - & - & - & - & - & - & - \\ a_{n1} & a_{n2} & a_{n3} & a_{n4} & - & - & a_{nn} \end{bmatrix} \times \begin{bmatrix} X_1 \\ X_2 \\ - \\ - \\ - \\ - \\ X_n \end{bmatrix} + \begin{bmatrix} D_1 \\ D_2 \\ - \\ - \\ - \\ - \\ D_n \end{bmatrix}$$

Now let us specify the various components of the system in a matrix form as shown below. Now the whole system in its matrix form may be restated as:

$X = A X + D$

i.e., $X (I - A) = D$

The actual question which is to be answered now is: Given a certain vector of final demands $D = (D_1, D_2, D_a \dots D_n)$, what quantity of goods should be produced by each of the n industries, bearing in mind the inter-industry requirements? Now to get the needed X we pre-multiplying both the sides of the above equation by $[I-A]^{-1}$

i.e., $X = [I - A]^{-1} D$

Therefore, to find out the vector of requirements we must first subtract the technical coefficient matrix A from the unit matrix I and then find its inverse. Now by pre-multiplying the given autonomous demand vector D by the inverse so obtained we get the needed vector X, composed of the output levels for all the 'n' sectors.

■ Example 17

Given the following technical coefficient matrix, obtain the gross outputs for a final demand of 50 and 400 units. If the demand for the industry 1's output is projected at 60 units in the next year.

Table 19.66

Industry	Input to	
	1	2
1	0.40	0.25
2	0.20	0.50

Solution

$$A = \begin{bmatrix} 0.40 & 0.25 \\ 0.20 & 0.50 \end{bmatrix}$$

$$I = \begin{bmatrix} 1.00 & 0.00 \\ 0.00 & 1.00 \end{bmatrix}$$

$$I - A = \begin{bmatrix} 0.60 & -0.25 \\ -0.20 & 0.50 \end{bmatrix}$$

$$D = \begin{bmatrix} 50 \\ 400 \end{bmatrix}$$

$$\text{Co-factor of } I - A = \begin{bmatrix} 0.50 & 0.20 \\ 0.25 & 0.60 \end{bmatrix}$$

$$\text{adj of } I - A = \begin{bmatrix} 0.50 & 0.25 \\ 0.20 & 0.60 \end{bmatrix}$$

$$\text{Determinant } I - A = 0.25$$

$$[I - A]^{-1} = \begin{bmatrix} 2.00 & 1.00 \\ 0.80 & 2.40 \end{bmatrix}$$

$$\begin{bmatrix} X_1 \\ X_2 \end{bmatrix} = \begin{bmatrix} 2.00 & 1.00 \\ 0.80 & 2.40 \end{bmatrix} \times \begin{bmatrix} 50 \\ 400 \end{bmatrix} = \begin{bmatrix} 50 \\ 1000 \end{bmatrix}$$

■ Example 18

Given the following technical coefficient matrix obtain the gross outputs for a final demand of 300 and 100 units of final demand.

Table 19.67

Industry	Input to	
	1	2
1	0.10	0.50
2	0.20	0.25

Solution

$$A = \begin{bmatrix} 0.10 & 0.50 \\ 0.20 & 0.25 \end{bmatrix}$$

$$I = \begin{bmatrix} 1.00 & 0.00 \\ 0.00 & 1.00 \end{bmatrix}$$

$$I - A = \begin{bmatrix} 0.90 & -0.50 \\ -0.20 & 0.75 \end{bmatrix}$$

$$D = \begin{bmatrix} 300 \\ 100 \end{bmatrix}$$

$$\text{Co-factor of } I - A = \begin{bmatrix} 0.75 & 0.20 \\ 0.50 & 0.90 \end{bmatrix}$$

$$\text{adj of } I - A = \begin{bmatrix} 0.75 & 0.50 \\ 0.20 & 0.90 \end{bmatrix}$$

$$\text{Determinant } I - A = 0.58$$

$$[I - A]^{-1} = \begin{bmatrix} 1.30 & 0.87 \\ 0.35 & 1.57 \end{bmatrix}$$

$$\begin{bmatrix} X_1 \\ X_2 \end{bmatrix} = \begin{bmatrix} 1.30 & 0.87 \\ 0.35 & 1.57 \end{bmatrix} \times \begin{bmatrix} 300 \\ 100 \end{bmatrix} = \begin{bmatrix} 478 \\ 261 \end{bmatrix}$$

19.17.6 Hawkins–Simon Condition

In all our illustrations a certain portion of the own output is used as inputs in almost all the sectors. The remaining portion is used by other sectors including the household sector. The portion of the output, called net output, sold to other sectors constitutes the source of revenue for the concerned sector. If this net output is zero or negative it is of no use in producing such an item. The net output would be positive provided the $a_{ii} < 1$ in our technical coefficient matrix. The existence of net output is only necessary and certainly not sufficient. Even if the net output is positive, it may not be available for final consumption. So to have positive net final demand two more conditions are to be satisfied.

1. $|I - A| > 0$

2. $(I-A)^{-1}$ must exist

Thus to have a meaningful positive final output in all the sectors the following three Hawkins-Simon conditions must be satisfied.

(a) $a_{ii} < 1$ for all i

(b) $|I - A| > 0$

(c) $(I-A)^{-1}$ must exist

EXERCISES

1. What is meant by Semi-logarithmic chart? How can this chart be used to compute the growth rates of two countries?

2. Why is it logical to plot two curves representing two series in different units (such as output and price) on the same semi-logarithmic chart?

3. What conditions might make a chain relative as an unsatisfactory measure of fluctuations in a time series? Explain.

4. The data on the life insurance and disposable income per family is given below

Year	1960	1961	1962	1963	1964	1965	1966
L.I	Rs.10,200	Rs.10,800	Rs.11,400	Rs.12,200	Rs.13,300	Rs.14,700	Rs.15,900
D.Y	6100	6300	6500	6800	7300	7700	8200

   a) Plot both the time series on the same arithmetic chart using the same scale.

   b) Plot both the time series data on the same semi-log chart as close as possible by using appropriate scale. Does this chart tell the same story of the relative growth, as did the first one?

5. What are the advantages and disadvantages of semi-averages for computing a straight-line trend as compared with the method of least squares?

6. The following table shows the number of workers in a factory for a period of nine years.

Year	1961	62	63	64	65	66	67	68	69
Workers	40	38	42	44	41	50	60	55	70

   With the help of least square method compute the straight-line trend and estimate the number of workers for the year 1975.

7. For the following data compute the equation of the straight-line by the method of semi-averages. Hence calculate the trend value corresponding to the year 1960.

Year	Consumption	Year	Consumption
1947	7.7	1957	11.0
1948	7.8	1958	8.20
1949	7.5	1959	11.4
1950	9.5	1960	10.1
1951	7.3	1961	10.3
1952	9.0	1962	12.1
1953	9.6	1963	11.5
1954	7.5	1964	9.40
1955	10.7	1965	10.4
1956	12.7	1966	14.8
		1967	11.6

8. The machine with an expected life period of 10 years is found to be depreciating at a uniform rate. The original cost of the machine was Rs.1,00,000 and it is estimated that it will worth only Rs.10,000 after five years. Compute the trend line and calculate the worth of the machine after 7 years.

9. Why the measurement of seasonal variation is called the problem of averages?

10. Define the term 'specific seasonals' and propose a method for the calculation of such seasonals.

11. The following table shows the sales of a company by quarters.

Quarters	1965	1966	1967	1968	1969
1	15	17	18	20	21
2	22	23	27	27	30
3	24	26	29	30	32
4	35	38	40	45	46
Total	96	104	114	122	129

a) Compute an index of seasonal variations by quarters by averaging quarterly sales.

b) Compute an index of seasonal variation by quarters by using a ratio to 2 items average of a four-quarter moving average.

CHAPTER

20

Investment and Break-Even Analysis

20.1 INTRODUCTION

Investment is defined as the acquisition of durable productive facilities in the expectation of future gain. In can be either tangible asset in the form of physical capital like the plant, equipment, buildings, machinery etc. It may also include the intangible assets like training of personnel. Capital expenditure normally refers huge amount of money expenditures incurred at a point of time may be in the beginning or some intermediate period. It represents long run commitment of funds for a better profitability. Thus investment decisions become vital for all business firms.

In today's business environment, financial management occupies more importance. For example, an increase in the plant and equipment expenditure would affect the firm's profitability, its cash position, its borrowing ability, its dividend payments and so forth. Capital expenditures have such widespread ramifications because the various activities of the firm are so interrelated. The investment analysis involves three essential steps.

- Making the list of worthy alternative investment projects.
- Estimation of the funds availability for investment.
- Selection of suitable evaluation technique for the selection of some of the viable projects keeping the financial commitments in mind.

The present chapter is designed to address some of these financial matters and related issues in detail. The first part is designed to deal with the time value of money. The second part is designed to deal with the interrelations between cost, revenue, volume and profit by using the break-even analysis.

20.2 ANNUAL COMPOUNDING

If P_0 stands for the principal amount, r stands for the rate of interest per annum and n stands for number of years then the amount at the end of any given year under annual compounding is calculated as

After one year

$$P_1 = P_0 + P_0 r$$
$$= P_0(1 + r)$$

After two years

$$P_2 = P_0(1 + r) + P_0(1 + r)r$$
$$= P_0(1 + r)(1 + r)$$
$$= P_0(1 + r)^2$$

After three years

$$P_3 = P_0(1 + r) + P_0(1 + r)r$$
$$= P_0(1 + r)^2 + P_0(1 + r)^2 r$$
$$= P_0(1 + r)^2(1 + r)$$
$$= P_0(1 + r)^3$$

After n years

$$P_n = P_0(1 + r)^n$$

20.3 QUARTERLY COMPOUNDING

If interest is compounded quarterly (4 times in a given year) then the total number of periods for n years = 4n. Similarly, if r is the rate of interest per annum then interest per quarter = r/4. So the above formula is modified as

$$P_n = P_0\left(1+\frac{r}{4}\right)^{4n}$$

20.4 MONTHLY COMPOUNDING

If compounding is done 12 times then the above formula becomes

$$P_n = P_0\left(1+\frac{r}{12}\right)^{12n}$$

20.5 k TIMES COMPOUNDING

In general, if compounding is done k times in a year then the general compounding formula is written as

$$P_n = P_0\left(1+\frac{r}{k}\right)^{kn}$$

20.6 INFINITE OR CONTINUOUS COMPOUNDING

In particular, if the number of times compounding, namely k tends to infinity, amounting to continuous compounding, then the corresponding formula is given by

$$P_n = \underset{k\to\infty}{\text{Lt}}\ P_0\left(1+\frac{r}{k}\right)^{kn} = P_0e^{rn}$$

Here the money grows continuously at exponential rate.

■ Example 1

Find the value of Rs. 1000 at 12% interest rate after 2 years by

a) annual compounding
b) bi-annual compounding
c) quarterly compounding
d) monthly compounding
e) continuous compounding

Solution

a) Compounding annually

$$P_n = P_o(1+r)^n$$

$$P_2 = 1000(1+0.12)^2$$
$$= 1000(1.12)^2$$
$$= 1254.4$$

b) Compounding bi-annually

$$P_n = P_o(1+r/2)^{2n}$$
$$P_2 = 1000(1+0.12/2)^{2\times2}$$
$$= 1000(1+.06)^4$$
$$= 1000(1.06)^4$$
$$= 1262.48$$

c) Compounding quarterly

$$P_n = P_o(1+r/4)^{4n}$$
$$P_2 = 1000(1+0.12/4)^{4\times2}$$
$$= 1000(1+0.03)^8$$
$$= 1000(1.03)^8$$
$$= 1266.77$$

d) Continuous compounding

$$P_n = P_o e^{rn}$$
$$= 1000\times(2.718281828)^{0.12\times2}$$
$$= 1271.24915$$

Note: At annual rate, Rs. 1000 becomes Rs. 1254.4. If compounding is done bi-annually then the same amount after two years becomes Rs. 1262.48. If compounding is done quarterly then the said investment after two years becomes Rs. 1266.77. At continuous compounding it becomes Rs. 1271.2915. Thus we notice that the money grows faster and faster when the number of compounding in a given year increases. However, it cannot grow more than Rs. 1271.24915 at the given interest rate of 12%.

20.7 NOMINAL VS. EFFECTIVE ANNUAL INTEREST RATES

The interest rate specified on annual basis is normally known as *nominal interest rate.* If the compounding more than once in a year the actual annualised interest rate would be higher than the specified nominal interest rate and such a rate is called *effective interest rate* denoted by j.

Effective interest rate when compounding is done k times in a year

When compounding is done k times in a year then we know that

$$P_n = P_o\left(1+\frac{r}{k}\right)^{kn} \ldots \quad (1)$$

If i is the effective annualized interest rate then on the basis of annual compounding

$$P_n = P_o(1+i)^n \ldots \quad (2)$$

Now we obtain the annualized effective interest rate we equate (1) and (2)

$$P_o(1+i)^n = P_o\left(1+\frac{r}{k}\right)^{kn}$$

$$(1+i)^n = \left(1+\frac{r}{k}\right)^{kn}$$

$$(1+i) = \left(1+\frac{r}{k}\right)^{k}$$

$$i = \left(1+\frac{r}{k}\right)^{k} - 1$$

Example 2

An investment of Rs. 1000 grows at 12% per annum compounded bi-annually. Calculate the maturity value after two years. Also get its annualized effective interest rate.

Solution

$$P_n = P_o\left(1+\frac{r}{k}\right)^{kn}$$

$$P_n = 1000\left(1+\frac{0.12}{2}\right)^{2\times 1}$$

$$= 1000(1+0.06)^2$$

$$= 1000(1.06)^2$$

$$= 1123.6$$

Annualized effective interest rate on the assumption that the compounding is done bi-annually

$$i = \left(1 + \frac{r}{k}\right)^k - 1$$

$$= \left(1 + \frac{0.12}{2}\right)^2 - 1$$

$$= (1.06)^2 - 1$$

$$= 1.1236 - 1$$

$$i = 0.1236$$

Thus, the effective interest rate per annum is $0.1236 \times 100 = 12.36\%$.

■ Example 3

An investment of Rs. 1000 grows at 12% per annum compounded quarterly. Calculate the maturity value after two years. Also get its effective interest rate.

Solution

$$P_n = P_o\left(1 + \frac{r}{k}\right)^{kn}$$

$$P_n = 1000\left(1 + \frac{0.12}{4}\right)^{4\times 2}$$

$$= 1000(1 + .03)^8$$

$$= 1000(1.03)^8$$

$$= 1266.77$$

The effective interest rate on annual basis is given by

$$i = \left(1 + \frac{r}{k}\right)^k - 1$$

$$= \left(1 + \frac{0.12}{4}\right)^4 - 1$$

$$= (1.03)^4 - 1$$

$$= 1.1255088 - 1$$

$$= 0.1255088$$

Thus, the effective interest rate per annum is $0.1255088 \times 100 = 12.55088\%$.

To verify the validity of this result let us calculate the maturity value compounded annually at the effective rate $j = 0.1255088$.

$$P_n = P_o(1+i)^n$$
$$= 1000(1+0.1255088)^2$$
$$= 1000(1.1255088)^2$$
$$= 1266.77$$

Note that both the maturity values are equal.

Example 4

An investment of Rs. 1000 grows at 12% per annum compounded monthly. Calculate the maturity value after two years. Also get its effective interest rate

Solution

$$P_n = P_o\left(1+\frac{r}{k}\right)^{kn}$$
$$P_n = 1000\left(1+\frac{0.12}{12}\right)^{12\times 2}$$
$$= 1000(1+.01)^{24}$$
$$= 1000(1.01)^{24}$$
$$= 1269.735$$

The effective interest rate on annual basis is given by

$$i = \left(1+\frac{r}{k}\right)^{k} - 1$$
$$= \left(1+\frac{0.12}{12}\right)^{12} - 1$$
$$= (1.01)^{12} - 1$$
$$= 1.126825 - 1$$
$$= 0.126825$$

Thus, the effective interest rate per annum is 0.126825 × 100 = 12.6825%.

Example 5

Investment in NSC deposits grows at 12% per annum compounded quarterly. Calculate the effective interest rate i.

Solution

$$i = \left(1 + \frac{r}{k}\right)^k - 1$$
$$= \left(1 + \frac{0.12}{4}\right)^4 - 1$$
$$= (1.03)^4 - 1$$
$$= 1.1255088 - 1$$
$$= 0.1255088$$

Thus, the effective interest rate per annum is 0.1255088 × 100 = 12.55088%.

Annualized effective interest rate when compounding is done continuously

$$P_o(1+i)^n = P_o e^{kn}$$
$$(1+i)^t = ekn$$
$$(1+i) = e^r$$
$$i = e^r - 1$$

■ Example 6

For the NSC problem given above calculate the effective interest rate by assuming continuous compounding.

Solution

$$i = e^r - 1$$
$$= (2.718281828)^{0.12} - 1$$
$$= 1.1275 - 1$$
$$= 0.1275$$

Thus, the effective interest rate is 0.1275 × 100 = 12.75%.

20.8 THE TIME VALUE OF MONEY (DISCOUNTING PRINCIPLE)

Most financial decisions, such as purchase of assets or procurement of funds, affect the firm's cash flows in different points of time. For example, if fixed assets is purchased, it will require an immediate cash outflow. Such outflows invariably generate cash inflows in the form of income in future. Similarly, if the firm borrows funds from banks and other individuals, it enjoys cash inflow under the obligation of returning same on future dates with interest. Similarly, when the firm under reference raises funds by issuing equity shares, the firm's cash balance increases immediately. However, at the time of issuing dividend its cash balance automatically decreases. Under the circumstances a sound decision making requires that the cash inflows and outflows must be comparable over a given period of time. In fact, the nominal cash inflows and outflows are not comparable. Cash flows are comparable provided they are adjusted for time and risk.

The recognition of time value of money and risk is extremely important in financial decision making. In the absence of such adjustment mere money comparison in absolute term often proves to be fatal. The firm under reference makes decisions that maximize the wealth of its owner by using the concept called net present value of all the cash flows.

Having Rs. 100 right now in hand and getting Rs. 112 after one year are equivalent provided the prevailing market rate of interest is 12% per annum. This statement amounts to that one hundred you have right now is invested in a bank it will become 112 after one year at 12% rate of interest per annum. Thus the present discounted value of Rs. 112 is equal to Rs. 100. Instead of Rs. 112 someone is ready to give Rs. 120 after one year then clearly the option of Rs. 120 after one year is preferable than Rs. 100 right now. This phenomenon is often called the time value of money.

Discounting on annual basis

If P_0 is the investment in period 0 then the value after n years denoted by P_n, at interest rate of r per annum is given by the compound interest formula

$$P_n = P_0(1+r)^n \ldots \quad (1)$$

If n = 1 then the above formula reduces to

$$P_1 = P_o(1+r)$$

Now from this formula the present value of P_1 at the discount rate of r is obtained as

$$P_o = \frac{P_1}{(1+r)}$$

Similarly, if n = 2 then equation (1) reduces to

$$P_2 = P_0(1+r)^2$$

$$P_0 = \frac{P_2}{(1+r)^2}$$

Similarly, the present discounted value of the income receivable after n^{th} years is given by the formula

$$P_o = \frac{P_n}{(1+r)^n}$$

Discounting by k times in a year

If compounding is done at k times in a given year then we know that

$$P_n = P_0\left(1+\frac{r}{k}\right)^{kn}$$

So the present discounted value is obtained as

$$P_0 = \frac{P_n}{\left(1+\frac{r}{k}\right)^{kn}}$$

Continuous discounting

If k tends to infinity under continuous discounting then we know that

$$P_n = P_0 e^{rn}$$

So the current discounted value at continuous discounting is obtained as:

$$P_0 = \frac{P_n}{e^{rn}}$$

■ Example 7

Calculate the present value for an expected income of Rs. 5000 after 5 years given the market interest rate of 9% per annum.

a) Under Annual Discounting

b) Under Quarterly Discounting

c) Under Continuous Discounting

Solution

a) Under Annual Discounting

$$P_o = \frac{P_n}{(1+r)^n}$$

$$= \frac{5000}{(1+.09)^3}$$

$$= \frac{5000}{1.09^3}$$

$$= 3249.66$$

b) Under Quarterly Discounting

We know that

$$P_0 = \frac{P_n}{\left(1+\frac{r}{k}\right)^{kn}}$$

$$= \frac{5000}{\left(1+\frac{0.09}{4}\right)^{4\times5}}$$

$$= \frac{5000}{1.0225^{20}}$$

$$= 3204.08$$

c) Under Continuous Discounting

Under continuous compounding we know that

$$P_0 = \frac{P_n}{e^{rn}}$$

$$= \frac{5000}{2.178281828^{0.09\times5}}$$

$$= 3188.141$$

i) Present Value (PV) of a stream of income receivable on future dates on annual basis:

If A_1 , A_2 , ...A_n are the stream of income receivable over n years respectively then the current value of all the income streams is obtained as.

$$PV = \frac{A_1}{(1+r)} + \frac{A_2}{(1+r)^2} + \dots\dots\dots + \frac{A_n}{(1+r)^n}$$

■ Example 7(a)

In a business the following table gives the stream of income receivable at the end of each year for five years is given. Calculate the current worth of the stream of income at the rate of interest 12% per annum.

Months	Jan	Feb	Mar	Apr	May
Income	2000	3000	5000	4000	6000

$$PV = \frac{A_1}{(1+r)} + \frac{A_2}{(1+r)^2} + \frac{A_3}{(1+r)^3} + \frac{A_4}{(1+r)^4} + \frac{A_5}{(1+r)^5}$$

$$= \frac{2000}{(1+0.12)} + \frac{3000}{(1+0.12)^2} + \frac{5000}{(1+0.12)^3} + \frac{4000}{(1+0.12)^4} + \frac{6000}{(1+0.12)^5}$$

$$= \frac{2000}{(1.12)} + \frac{3000}{(1.12)^2} + \frac{5000}{(1.12)^3} + \frac{4000}{(1.12)^4} + \frac{6000}{(1.12)^5}$$

$$= 1785.741 + 2391.582 + 3558.901 + 2542.072 + 3404.561 = 13682.83$$

ii) Present Value (PV) of a stream of income receivable on future dates on quarterly basis:

If A_1, A_2, ...A_n are the stream of income receivable over n quarters respectively then at the interest rate of r per annum the current value of all the income streams over n quarters is obtained as.

$$PV = \frac{A_1}{\left(1+\frac{r}{4}\right)} + \frac{A_2}{\left(1+\frac{r}{4}\right)^2} + \dots + \frac{A_n}{\left(1+\frac{r}{4}\right)^n}$$

■ Example 7(b)

In a business the following table gives the stream of income receivable at the end of each quarter for five quarters is given. Calculate the current worth of the stream of income at the rate of interest 12% per annum.

Quarters	1	2	3	4	5
Income	2000	3000	5000	4000	6000

$$PV = \frac{A_1}{\left(1+\frac{r}{4}\right)} + \frac{A_2}{\left(1+\frac{r}{4}\right)^2} + \dots + \frac{A_n}{\left(1+\frac{r}{4}\right)^n}$$

$$= \frac{2000}{\left(1+\frac{0.12}{4}\right)} + \frac{3000}{\left(1+\frac{0.12}{4}\right)^2} + \frac{5000}{\left(1+\frac{0.12}{4}\right)^3} + \frac{4000}{\left(1+\frac{0.12}{4}\right)^4} + \frac{6000}{\left(1+\frac{0.12}{4}\right)^5}$$

$$= \frac{2000}{(1+0.03)} + \frac{3000}{(1+0.03)^2} + \frac{5000}{(1+0.03)^3} + \frac{4000}{(1+0.03)^4} + \frac{6000}{(1+0.03)^5}$$

$$= 1941.748 + 2827.788 + 4575.948 + 3553.948 + 5175.653 = 18074.84$$

iii) Present Value (PV) of a stream of income receivable on future dates on monthly basis:

■ Example 7(c)

In a business the following table gives stream of income receivable at the end of each month for five months is given. Calculate the current worth of the stream at the rate of interest 12% per annum.

Year	Jan	Feb	Mar	Apr	May
Income	2000	3000	5000	4000	6000

$$PV = \frac{A_1}{\left(1+\frac{r}{12}\right)} + \frac{A_2}{\left(1+\frac{r}{12}\right)^2} + \ldots\ldots\ldots\ldots\ldots + \frac{A_n}{\left(1+\frac{r}{12}\right)^n}$$

$$= \frac{2000}{\left(1+\frac{0.12}{12}\right)} + \frac{3000}{\left(1+\frac{0.12}{12}\right)^2} + \frac{5000}{\left(1+\frac{0.12}{12}\right)^3} + \frac{4000}{\left(1+\frac{0.12}{12}\right)^4} + \frac{6000}{\left(1+\frac{0.12}{12}\right)^5}$$

$$= \frac{2000}{(1+0.01)} + \frac{3000}{(1+0.01)^2} + \frac{5000}{(1+0.01)^3} + \frac{4000}{(1+0.01)^4} + \frac{6000}{(1+0.01)^5}$$

$$= 1980.198 + 2840.888 + 48.52.951 + 3843.921 + 5708.794 = 19326.75$$

20.9 ANNUITIES

An annuity refers to a series of payments of a certain fixed amount of money for a designated period of time at equal intervals. Such periodical payments occur in many practical situations like the instalment payments for a TV purchased at home, equated monthly instalments (EMI) for the house building loan taken etc. The time interval between two successive payments is called the *payment interval,* and the time between the first and the last payment is called the *term* of the annuity concerned. On the basis of term, the annuity may be classified as follows:

- **Annuity Certain:** When the term of an annuity begins and ends on a certain fixed date then the referred annuity is called annuity certain.
- **Perpetuity:** The annuity that begins on a particular date but never terminates is called perpetuity. Without touching the principal if interest alone is paid regularly then the said annuity never terminates. It terminates if and only if the whole of the principal is paid back at one instalment.
- **Contingent Annuity:** It begins on a particular date as usual but termination is contingent up of certain future happenings. For example, the payment of LIC premium starts on a particular date but ends as soon as the concern person dies.

Annuity can also be defined on the basis of the time of payments

- **Ordinary annuity:** In the case of ordinary annuity the payment is made at the end of each period.
- **Annuity Due:** When the payment is made at the beginning of each period then it is called annuity due.
- **Deferred annuity:** No doubt the payment takes place at the end of each period but the annuity does not begin till some time to come. It begins only after some specified period.

In the present discussion we assume only ordinary annuity unless otherwise specified.

Terminal worth of annuity certain

By amount of an ordinary annuity we mean the terminal value worth of the series of payments starting from the beginning to the end. Let a personal loan be taken from a bank and repayment is made in equated instalments. Eventually such instalments will have a principal component as well as interest component in it. Let

A = the periodic payment

n = the number of payments

r = the rate of interest per period.

Since the first payment is made at the end of the first period, its terminal value is obtained by using the compound interest formula for the remaining n–1 period. Similarly, since the second instalment is paid at the end of the second month, its terminal value is obtained by using the compound interest formula for the remaining n–2 periods. This way we calculate the terminal worth for all the n payments. The total of all such terminal worth values is the needed annuity amount. At this stage note that the last payment would not attract any interest since it is paid at the end of the last period.

If M is terminal worth amount of annuity and A is the instalment payment amount then by using the compound interest formula for each payment the annuity amount M is written as

$$M = A(1+r)^{n-1} + A(1+r)^{n-2} + \dots\dots\dots + A(1+r)^2 + A(1+r) + A \dots \qquad (1)$$

Now to get the value of M easily we multiply both the sides of above equation (1) by (1+r) and generate equation (2) as shown below.

$$M(1+r) = A(1+r)^n + A(1+r)^{n-1} + \dots\dots + A(1+r)^3 + A(1+r)^2 + A(1+r)\dots \qquad (2)$$

Now let us subtract equation (2) from (1)

$$M(1+r) - M = A(1+r)^n - A$$

$$M + Mr - M = A(1+r)^n - A$$

$$Mr = A(1+r)^n - A$$

$$M = \frac{A\left[(1+r)^n - 1\right]}{r}$$

If (1+r) = R then the above formula reduces to

$$M = \frac{A\left[(1+r)^n - 1\right]}{(1+r) - 1}$$

$$= A\left[\frac{R^n - 1}{R - 1}\right]$$

Thus, the total terminal worth or otherwise called the annuity is obtained simply by multiplying the instalment amount A by the term $\frac{R^n - 1}{R - 1}$. This term is often called the amount factor. For easy reference this amount factor is obtained from the readymade Table always. The following example illustrates the method of calculating the annuity amount M. This amount is nothing but the effective total money paid by the borrower to the bank.

■ Example 8

A company wants to establish a sinking (or depreciation) fund for a period of 10 years for replacing Rs. 8,00,000 worth of a machine at the end of 9th year. Calculate the amount that is to be retained out of its profits every year if the market rate of interest is 8% per annum.

Solution

Data

M = 8,00,000, r = 0.08 and n = 10, A =?

$$M = \frac{A\left[(1+r)^n - 1\right]}{r}$$

$$8,00,000 = \frac{A\left[(1+0.08)^{10} - 1\right]}{0.08}$$

$$8,00,000 = A\left[\frac{2.156925 - 1}{0.08}\right]$$

$$8,00,000 = A\left[\frac{1.156925}{0.08}\right]$$

$$8,00,000 = A(14.48656)$$

$$A = \frac{8,00,000}{14.48656} = 55,223.59$$

Hence an amount of Rs. 55,223.59 is to be kept away from the profit each month.

■ Example 8(a)

A company wants to establish a sinking (or depreciation) fund for a period of 10 months for replacing Rs. 8,00,000 worth of a machine at the end of 9th month. Calculate the amount that is to be retained out of its profits every month if the market rate of interest is 12% per annum.

Solution

Data

M = 8,00,000, rate of interest per month r = 0.012/12 =0.01 and n = 10 A = ?

$$M = \frac{A\left[(1+r)^{n}-1\right]}{r}$$

$$8{,}00{,}000 = \frac{A\left[(1+0.01)^{10}-1\right]}{0.01}$$

$$8{,}00{,}000 = A\left[\frac{1.104622-1}{0.01}\right]$$

$$8{,}00{,}000 = A\left[\frac{0.104622}{0.01}\right]$$

$$8{,}00{,}000 = A(10.46221)$$

$$A = \frac{8{,}00{,}000}{10.46221} = 76{,}465.66$$

Hence an amount of Rs.76,465.66 is to be kept away from the profit each month.

The present value of the annuity certain under annual compounding

For an annuity, the present value is simply the sum total of the present values of all the periodic payments. The present value of the first payment is obtained as A/(1+r).The present value of the second payment would be $A/(1+r)^2$ and so on. If V is the present value of the annuity certain and A is the instalment payment then:

$$\text{Present value (PV)} = \frac{A}{(1+r)} + \frac{A}{(1+r)^2} + \ldots\ldots \frac{A}{(1+r)} \ldots \quad (1)$$

Now let us multiply equation (3) by (1+r) on both the sides

$$PV(1+r) = A + \frac{A}{(1+r)} + \frac{A}{(1+r)^2} + \ldots\ldots \frac{A}{(1+r)^{n-1}} \ldots \quad (2)$$

Now let us subtract (2) from (1) and get

$$PV(1+r) - PV = A - \frac{A}{(1+r)^n}$$

$$PV + rPV - PV = A - \frac{A}{(1+r)^n}$$

$$rPV = A\left[1 - \frac{1}{(1+r)^n}\right]$$

$$PV = \frac{A\left[1 - \frac{1}{(1+r)^n}\right]}{r}$$

$$PV = \frac{A\left[1 - (1+r)^{-n}\right]}{r}$$

If R = (1+r) then the above result is restated as

$$PV = \frac{A\left[1 - (1+r)^{-n}\right]}{(1+r) - 1}$$

$$PV = A\left[\frac{1 - R^{-n}}{R - 1}\right]$$

Thus again the present value of the annuity certain is obtained by merely multiplying the instalment payment A by the term $\left[\frac{1 - R^{-n}}{R - 1}\right]$. This ready to use factor is called the present value factor *(PVF)*. This factor is available in ready to usable tabular form for all combinations.

■ Example 9

Determine the present value of an annuity certain, given the instalment payment of Rs. 2000 payable at the end of each year for next 4 years. The market interest rate is 6% per annum.

Solution

Data A = 2000, r = 0.06 and n = 4, PV=?

$$PV = \frac{A\left[1 - \frac{1}{(1+r)^n}\right]}{r}$$

$$= \frac{2000\left[1 - \frac{1}{(1+0.06)^4}\right]}{0.06}$$

$$= \frac{2000\left[1 - \frac{1}{1.06^4}\right]}{0.06}$$

$$= \frac{2000\left[1 - 0.792094\right]}{0.06}$$

$$PV = 2000(0.207969)/0.06$$

$$= \text{Rs. } 6930.2$$

Example 10

Mr. X purchased a colour TV from a ONIDA outlet for Rs. 12,800. He has made a down payment of Rs. 4000 and agreed to pay the balance in an equated monthly payment of 24 instalments. If the market rate of interest is 12% compounded monthly, how much Mr. X should pay every month?

Solution

The present value of the annuity PV = Price of TV – Down payment = 12,800–4000 = 8800

Further, the monthly interest = Annual interest (r) / 12 = 12%/12 = 1% and n = 24 months.

$$PV = \frac{A\left[1-(1+r)^{-n}\right]}{(1+r)-1}$$

$$PV = A\left[\frac{1-R^{-n}}{R-1}\right]$$

$$8800 = A\left[\frac{1-1.01^{-24}}{1.01-1}\right]$$

$$= A(21.2434)$$

$$A = \frac{8800}{21.2434} = 414.25$$

So our customer must pay Rs. 414.25 per instalment so that the present value of all the future payments equals to the present due of Rs. 8800.

Terminal worth of the perpetuity

We know that in the case of perpetuity though the first payment starts as usual it never ends with n^{th} period payment, the payment goes on and on forever. So there is no concept like terminal worth for perpetuity in such cases.

The present value of the perpetuity

Now we know that the formula for an annuity certain

$$PV = \frac{A\left[1-(1+r)^{-n}\right]}{(1+r)-1}$$

$$PV = A\left[\frac{1-R^{-n}}{R-1}\right]$$

In the case of perpetuity we know that the payment continues indefinitely. Thus in the above formula n tends to ∞. If n tends to ∞ then the term R^{-n} tends to zero. So the above formula reduces to

$$\underset{n\to\infty}{\text{Lt}}\ PV = \underset{n\to\infty}{\text{Lt}}\ A\left[\frac{1-R^{-n}}{R-1}\right]$$

$$= A\left[\frac{1-0}{1+r-1}\right]$$

$$= \frac{A}{r}$$

Thus, in the case of perpetuity the needed present value factor (PVF) is simply the reciprocal of the interest rate r.

Example 11

A freehold estate was bought for Rs. 4,00,000. At what annual rent should it be let so that the owner gets 12% return on the purchase value?

Solution

Here we will simply find the periodic payment A with the given V = 4,00,000 and r = 0.12

$$PV = \frac{A}{r}$$

$$4{,}00{,}000 = \frac{A}{0.12}$$

$$A = 4{,}00{,}000 \times 0.12 = \text{Rs. } 48{,}000 \text{ per annum}$$

Example 11(a)

A free hold estate was bought for Rs.4,00,000.At what rent month should it be let so that the owner gets 12% return per annum on the purchase value.

Solution

Here we will simply find the periodic payment A with the given V = 4,00,000 and interest per month r = 0.12/12 = 0.01

$$PV = \frac{A}{r}$$

$$4{,}00{,}000 = \frac{A}{0.0.01}$$

$$A = 4{,}00{,}000 \times 0.01 = \text{Rs. } 4000 \text{ per month}$$

The present value of a deferred annuity

If the annuity begins after p periods from now and continue up to n periods then the present value of the successive payments would be

$$PV = AR^{-(p+1)} + AR^{-(p+2)} + AR^{-(p+3)} \ldots\ldots AR^{-(p+n)} \ldots \qquad (5)$$

To obtain PV as usual we multiply both the sides of equation (5) by R first

$$PV \times R = AR^{-(p)} + AR^{-(p+1)} + AR^{-(p+2)} \ldots\ldots AR^{-(p+n-1)} \ldots \quad (6)$$

Now let us subtract equation (5) from (6)

$$PV \times R - PV = AR^{-p} - AR^{-(p+n)}$$

$$PV \times (R-1) = AR^{-p} - AR^{-(p+n)}$$

$$PV = \frac{AR^{-p}}{R-1} - \frac{AR^{-(p+n)}}{R-1}$$

$$= A\left[\frac{R^{-p}}{R-1} - \frac{R^{(p+n)}}{R-1}\right]$$

Thus this time the present value factor (PVF) of a deferred annuity is $\left[\frac{R^{-p}}{R-1} - \frac{R^{(p+n)}}{R-1}\right]$

■ Example 12

The instalment of a deferred annuity is Rs. 3000 per annum. Its first instalment starts 3 years from now. If the current market interest rate is 8% per annum obtain it present value.

Solution

Data: $A = \text{Rs. } 3000$, $R = 1 + r = 1 + 0.08$, and $p = 3$

$$PV = \frac{AR^{-p}}{R-1} - \frac{AR^{-(p+n)}}{R-1}$$

$$= \frac{3000 \times 1.08^{-3}}{1.08-1} - \frac{3000 \times 1.08^{-(3+8)}}{1.08-1} = \text{Rs. } 9508.62$$

The present value of deferred perpetuity

If perpetuity is to begin after a certain period from now, its present value is obtained by the formula.

$$PV = \frac{AR^{-p}}{R-1}$$

Since the second term $\frac{AR^{-p-n}}{R-1}$ tends to zero as n tends to infinity in the deferred annuity for formula. So the relevant present value factor is given by the formula $\frac{R^{-p}}{R-1}$.

Example 12(a)

The instalment of a deferred annuity is Rs.10,000 per month. Its first instalment starts 3 months from now. If the current market interest rate is 12% per annum obtain it present value.

Solution

Data

A = Rs.10,000, the monthly interest r = 0.12/12 =0.01 R = 1+ r = 1+.01, and p = 3

$$PV = \frac{AR^{-p}}{R-1} - \frac{AR^{-(p+n)}}{R-1}$$

$$= \frac{10000 \times 1.01^{-3}}{1.01-1} - \frac{3000 \times 1.01^{-(3+8)}}{1.01-1}$$

$$= \frac{9705.901}{0.01} - \frac{8963.237}{0.01}$$

$$= 970590.1 - 896323.7$$

$$= \text{Rs. } 74266.43$$

Example 13

Determine the present value of perpetuity of Rs. 1000 starting 4 years from now, carrying an interest rate of 5% per annum.

Solution

$$PV = A\left(\frac{R^{-p}}{R-1}\right) = \frac{1000 \times 1.05^{-4}}{1.05-1} = \text{Rs. } 16,454.04$$

Example 13(a)

Determine the present value of perpetuity of Rs. 2000 starting 4 months from now, carrying an interest rate of 6% per annum.

Solution

A = 1000, the monthly interest rate r = 0.06/12 = 0.005

$$PV = A\left(\frac{R^{-p}}{R-1}\right) = \frac{2000 \times 1.005^{-4}}{1.005-1} = \text{Rs. } 392099.0087$$

20.10 CAPITAL BUDGETING UNDER CERTAINTY

Companies make investments overtime in variety of real assets. The major objective of investment decision is to identify the current worth of real assets in comparison to their current costs. In this context, for correct decision making one must have the following sets of information.

- Estimates of capital outlays required for a project and the estimates of future promising cash inflows are known constants.
- A correct set of standards to select the projects for execution so that the long term economic benefits to the owners are maximized.

In selecting the suitable criteria for investment the following two basis principles are being used.

- ***The bigger the better principle:*** Accordingly, other things being equal, bigger the benefits more will be the preference.
- ***A bird in hand is better than two in bush principle:*** Accordingly, other things being equal, early benefits are preferable than later benefits.

The following are the four methods of assessing the investment worthiness in comparison with their future cash flows.

20.11 PAYBACK METHOD

The payback method is a very simple popular method of assessing the investment worthiness. It is based on the second principle highlighted above. Thus, it is a measure of rapidity of the investment earning capacity. The time period that is needed to recover the initial cost of investment proposal through the future stream of income flows is called the payback period. If C_0 is the initial investment and C_i is the cash inflow in the ith year then the relevant payback condition is written as

$$C_0 = \sum_{i=1}^{n} C_i$$

In this expression the period n is called the payback period. If this calculated pay-back period is less than certain maximum acceptable limit fixed by the company, the proposed investment is accepted, otherwise it is rejected.

Similarly, if there are two or more mutually exclusive projects then the project having less payback period is always preferred. While calculating the payback period, the nature of cash inflows must be taken into account. All the annual cash inflows may be either equal or unequal. If the annual cash inflows are equal to one another than the necessary payback period is obtained by simply dividing the initial investment by the amount of cash inflow per year. The following example illustrates one such situation.

■ Example 14

The cost of a project is Rs. 2,00,000. The annual earnings from this project is Rs. 20,000 per year. Obtain the payback period for this project.

Solution

$$\text{Payback period} = \frac{\text{Initial Investment}}{\text{Yearly cash flow}} = \frac{2,00,000}{20,000} = 10 \text{ years}$$

If the cash flows differ from year to year then the payback period is obtained by equating the cumulative cash inflows to the initial investment. The following example illustrates one such case.

■ Example 15

An investment proposal put forward by a company needs Rs. 48,000. The expected cash flows over five years of its lifetime is given in Table 20.1.

Table 20.1

Year	1	2	3	4	5
Cash Flow	15,000	20,000	10,000	6000	2000

Solution

Table 20.2

Year	Cash flow	Cumulative cash flow
1	15,000	15,000
2	20,000	35,000
3	10,000	45,000
4	6000	51,000
5	2000	53,000

Clearly the cumulative cash inflow Rs. 45,000 is less than the initial investment of Rs. 48,000. At the same time the cumulative cash flow of Rs. 51,000 is greater than our initial investment. So the needed payback lies in between third and the fourth year. Over and above the cumulative payback of Rs. 45,000 we need another Rs. 3000 to cover the initial cost of Rs. 48,000. Notice that the fourth year full cash back is Rs. 6000. So needed Rs. 3000 payback could be attributed to first six months at the rate of Rs. 500 per month. Thus, the ultimate payback period equals to 3.5 years.

Example 16

A company XYZ is having two investment proposals A and B with two different initial investment packages. The expected cash flows and the initial capital needs are given below. By using the payback period criteria rank the priority of these two proposals.

Table 20.3 Proposal A

Year	0	1	2	3	4	5
Cash Flow	−50,000	20,000	20,000	10,000	10,000	5000

Table 20.4 Proposal B

Year	0	1	2	3	4	5
Cash Flow	−100,000	40,000	40,000	10,000	10,000	10,000

Clearly the payback period for proposal A is three years (20,000 + 20,000 + 10,000 = 50,000). The payback period for proposal B is four years (40,000 + 40,000 + 10,000 + 10,000 = 1,00,000). Thus, the proposal A is the first priority investment and proposal B is the second priority investment.

Example 17

A project costs Rs. 10,00,000 and yields a profit of Rs. 1,60,000 per year. Such a cash flow attracts a tax of 50% on cash inflow and depreciation of 12% on capital investment. Calculate the payback period for this project.

Solution

Let us calculate the net cash flow after deducting the tax and depreciation.

The annual cash flow before the tax and depreciation	= Rs. 1,60,000
(−) Tax at the rate of 50% on Rs. 1, 60,000	= Rs. 80,000
(+) Depreciation at the rate of 12% on 10,00,000	= Rs. 1,12,000
Net cash flow per annum	= Rs. 2,00,000

$$\text{Payback period} = \frac{\text{Rs. } 10{,}00{,}000}{\text{Rs. } 2{,}00{,}000} = 5 \text{ Years}$$

20.12 RELEVANCE OF POST-PAYBACK METHOD PERIOD

In addition to payback period, investments yield till its lifetime. So in addition to payback period cash flows, one will have to keep in mind the cash inflows beyond the payback period as well for correct decision making. The following example illustrates the relevance of post payback period cash inflow for decision making.

Example 18

A company is proposing to expand its production and can go in either for an automatic machine costing Rs. 2,24,000 with an estimated lifetime of five and half years or an ordinary machine costing Rs. 60,000 having an estimated life of 8 years. The annual sales and costs are estimated as follows:

Table 20.5

Particulars	Automatic machine Rs.	Ordinary machine Rs.
Sales	1,50,000	1,50,000
Material Cost	50,000	50,000
Labour Cost	12,000	60,000
Variable overhead Cost	24,000	20,000

Compute the comparative profitability under the payback period method. Ignore income tax.

(I.C.W.A Final)

Solution

Table 20.6 Profit calculations

Particulars	Automatic machine Rs.	Ordinary machine Rs.
Sales	1,50,000	1,50,000
Material Cost	−50,000	−50,000
Labour Cost	−12,000	−60,000
Variable overhead Cost	−24,000	−20,000
Annual Profit	64,000	20,000

Table 20.7 Payback calculations

Particulars	Automatic machine Rs.	Ordinary machine Rs.
Cost	2,24,000	60,000
Annual profit	64,000	20,000
Payback period	3.50	3.00

Clearly ordinary machine is better because it has less payback period.

20.13 RELEVANCE OF POST-PAYBACK PROFIT

In the following table we show the post-payback profit calculations

Table 20.8

Particulars	Automatic machine Rs.	Ordinary machine Rs.
Annual Profit	64,000	20,000
Estimated lifetime	5.5 years	8 years
Payback period	3.5 years	3 years
Balance lifetime	2 years	5 years
Profit for the remaining period	Rs. 1,28,000	Rs. 1,00,000

Clearly this time post-payback period profit is more for the automatic machine.

■ Example 19

A company has to choose one of the following two mutually exclusive projects. Investment required for each project is Rs. 30,000. Both the projects have to be depreciated on straight line basis. The tax rate is 50%.

Table 20.9

Year	Profit before depreciation	
	Project A Rs.	Project B Rs.
1	8400	8400
2	9600	9000
3	14,000	8000
4	14,000	10,000
5	4000	20,000

Calculate the payback period and decide the project priority.

Solution

Let us first obtain the depreciation by straight line method:

Depreciation for project A on straight light basis = Initial investment/ Lifetime
= 30,000/5 = Rs. 6000

Depreciation for project B on straight light basis = Initial investment/ Lifetime
= 30,000/5 = Rs. 6000

Payback period for project A

From the table it is clear that an amount of Rs. 25,000 is recovered from the first 3 years cash flow. The fourth year cash flow is Rs. 10,000. So the remaining amount Rs. 5000 is recovered in 6 months period. So the total payback period = 3.5 years.

Table 20.10

Year	Profit	Less depreciation	Profit after depreciation	Less tax 50%	Profit after tax	Add depreciation	Cash inflow	Cumulative cash inflow
1	8400	6000	2400	1200	1200	6000	7200	7200
2	9600	6000	3600	1800	1800	6000	7800	15,000
3	14,000	6000	8000	4000	4000	6000	10,000	25,000
4	14,000	6000	8000	4000	4000	6000	10,000	35,000
5	4000	6000	−2000	0	−2000	6000	4000	39,000

Payback period for project B

From Table 20.11 it is clear that an amount of Rs. 29,700 is recovered from the first 4 years cash flow.

The fifth year cash flow is Rs. 13,000. So the remaining amount Rs. 300 is recovered in 8 days as shown below.

Table 20.11

Year	Profit	Less depreciation	Profit after depreciation	Less tax 50%	Profit after tax	Add depreciation	Cash inflow	Cumulative cash inflow
1	8400	6000	2400	1200	1200	6000	7200	7200
2	9000	6000	3000	1500	1500	6000	7500	14,700
3	8000	6000	2000	1000	1000	6000	7000	21,700
4	10,000	6000	4000	2000	2000	6000	8000	29,700
5	20,000	6000	14,000	7000	7000	6000	13,000	42,700

Cash flow per day = Rs. 13,000/365 = Rs. 35.6144.

Number of days to have a cash flow of Rs. 300 = 300/35.61= 8.423.

So the total payback period = 4 years 8 days approximately.

Thus project-A is better than project-B in this case.

Merits of Payback Method

1. Simple to understand and easy to compute.
2. In the event of shortage of funds for investment it permits the firm to choose the quick return yielding investment.
3. Enables quick ascertainment of degree of risk.
4. This method is handy in the event of obsolescence due to rapid technical change.

Demerits of Payback Method

1. It ignores the time value of money and thus the interest concept totally.
2. Does not consider the whole lifetime of the investment while calculating the payback period. The income after the payback period is completely ignored in this approach.
3. The profitability is totally ignored under liquidity consideration.

20.14 ACCOUNTING OR AVERAGE RATE OF RETURN METHOD

This is a simple investment criterion that is being commonly used in practice. In this method we first arrive at the average rate of profit per annum from the book of accounts after adjusting for tax and depreciation.

Average profit = (Total profit for all the years / Lifetime of the investment) $\times 100$

In the next step we arrive at the average investment per annum by assuming linear depreciation.

Average investment = (original investment – scrap)/2

Now ARR = Average profit/ average investment

Finally the accounting rate of return (ARR) is obtained as the ratio of the average annual profit (after tax deductions) to that of the average annual investment in the project concerned. The following examples illustrate this method.

Example 20

Let the average annual book profit per year is Rs. 4000 for a five years of lifetime cash flow investment proposal. Further, let the initial investment be Rs. 32,000 for the said project. Given the desired return rate of 20%, assess the worthiness of this project by using the average return rate principle.

Solution

The average investment by assuming straight line depreciation = 32,000/2 = Rs. 16,000.

The adjusted average return per annum is given in this problem as Rs. 4000.

Therefore the needed ARR = 4000/16,000 = 0.25 or 25%.

Since this percentage is more than the desired profit rate of 20% it is worthwhile to go for the project.

Example 21

A company X is having two investment proposals A and B with two different initial investment packages having the life time of 5 years. The expected cash flows and the initial capital needs are given below. By using the Average rate of return criteria rank the priority of these two proposals.

Table 20.12 Proposal A

Year	0	1	2	3	4	5
Cash Flow	−50,000	20,000	20,000	10,000	10,000	5,000

Table 20.13 Proposal B

Year	0	1	2	3	4	5
Cash Flow	−1,00,000	40,000	40,000	10,000	10,000	10,000

Solution

For proposal A, the total profit for its entire lifetime of 5 years = 20,000 + 20,000 + 10,000 + 10,000 + 5,000 = 65,000.

So the average return per year = 65,000/5 =Rs. 13,000.

The average investment per annum assuming linear rate of depreciation = 50,000/2 = 25,000. So the average rate of return (ARR) = 13,000/25,000 = 0.52 (52%)

For proposal B the total profit for its entire lifetime of 5 years = 40,000+40,000 10,000+10,000+10,000 = Rs. 1,10,000. So the average rate of return per year = 1,00,00/5 = Rs. 20,000.

The average investment per annum assuming linear depreciation is 1,00,000/2 = 50,000.

So the average rate of return (ARR) = 20,000/50,000 = 0.4 (40%).

Thus, the proposal A is the first priority investment and proposal B is the second priority investment.

Merits of this method

1. This an easily understandable method based on readily available book information.
2. The entire lifetime accounting profit flows are considered for calculation.
3. In the case ranking the projects, this method is very much simple to use it.

Demerits of this method

1. It is based on accounting profit and not on cash flows.
2. It ignores the time value of money and hence the interest concept altogether.
3. This method is inadequate in ranking project with different lifetime.
4. Returns on investment are calculated on the basis of post tax book profits.

20.15 NET PRESENT VALUE METHOD

The net present value (NPV) is obtained by subtracting the initial investment from the sum of the discounted values of the cash flows. If the so obtained NPV is positive then the project is accepted. When we compare two projects then we naturally prefer the one having the highest NPV.

$$\text{NPV} = \sum_{i-1}^{n} \frac{C_i}{(1+r)^n} - C_0$$

or

$$\text{NPV} = \sum_{i-0}^{n} \frac{C_i}{(1+r)^n}$$

■ Example 22

A company X has Rs. 2,00,000 to make investment. It has five project proposals under consideration. The following table gives the summary of the initial investments, annual cash flow and the respective lifetime for all the five projects.

Table 20.14

Project	Initial outlay in Rs.	Annual cash flows in Rs.	Life of the project in years
A	1,00,000	25,000	10
B	70,000	20,000	8
C	30,000	6000	20
D	50,000	15,000	10
E	50,000	12,000	20

Given the discount rate of 15% per annum rank the projects by using net present value.

Solution

In this problem since the yearly cash flows are constants for all the years for all the projects let us calculate present value factor (PVF) for a given constant stream of income of Re. 1 for all the future periods. If Re. 1 is the cash flow for all the n years then we know that the combined present value factor for all the n years is given by $\left[\frac{1-R^{-n}}{R-1}\right]$. Now by using this present value factor formula we first calculate PVF for project A as shown below.

$$\text{PVF for A} = \left[\frac{1-(R)^{-n}}{R-1}\right] = \left[\frac{(1-(1.15)^{-10}}{1.15-1}\right] = 5.0188.$$

Table 20.15

Project	Initial outlay in Rs.	Annual cash flows in Rs.	Life of the project	PV factor at 15%	PV	NPV	Ranking
A	1,00,000	25,000	10	5.0188	125469	25469	1
B	70,000	20,000	8	4.4873	89746	19746	4
C	30,000	6000	20	6.2593	37556	7556	5
D	50,000	15,000	10	5.0188	75282	25282	2
E	50,000	12,000	20	6.2593	75112	25112	3

Similarly, present value factors are obtained for all the projects under consideration and reported in Table 20.15. Once this is done the PV is obtained by multiplying the respective factors by the respective income stream as shown in the table.

The net present value is obtained and reported in column 7. The associated ranking is presented in column 8.

■ Example 23

A company Z wants to buy one among the two machines A and B. The expected cash flows for four years are given in the following table. If initial machine cost is Rs. 10,000 for each type, state the name of the machine to be purchased by using NPV principle, given the market rate of interest at 10% per annum.

Table 20.16

Year	Machine A	Machine B
1	5000	2000
2	4000	3000
3	2000	5000
4	2000	4000

Solution

The following table gives the needed net cash flow for both the machines.

Table 20.17

Year	PV Factor at 10%	Machine A		Machine B	
		Cash flow	PV	Cash flow	PV
0	1.0000	−10,000	−10000.00	−10,000	−10000.00
1	0.9091	5000	4545.45	2000	1818.18
2	0.8264	4000	3305.79	3000	2479.34
3	0.7513	2000	1502.63	5000	3756.57
4	0.6830	2000	1366.03	4000	2732.05
		NPV=	719.90	NPV=	786.15

Since the NPV for machine B is greater we advise the company to go for machine B.

20.16 CONFLICT IN THE NET PRESENT VALUE METHOD

The conclusion obtained in the previous example is true only for 10% discount rate. For example, if we increase the rate to 10.9449% then NPV for both A and B are equal to Rs. 541 as shown in Table 20.18. Any rate more than 10.9449% the preference changes from B to A as illustrated in Table 20.19 for 11%. Thus, our NPV approach fails to give a unique answer.

Table 20.18

Return Rate = 0.109449		Machine A		Machine B	
Year	PV Factor	Cash flow	PV	Cash flow	PV
0	1.000000	−10,000	−10000.00	−10,000	−10000.00
1	0.901348	5000	4506.74	2000	1802.70
2	0.812429	4000	3249.72	3000	2437.29
3	0.732281	2000	1464.56	5000	3661.41
4	0.660041	2000	1320.08	4000	2640.16
		NPV=	541.10	NPV=	541.55

Table 20.19

Return Rate = 0.110000		Machine A		Machine B	
Year	PV Factor	Cash flow	PV	Cash flow	PV
0	1.000000	−10,000	−10000.00	−10,000	−10000.00
1	0.900901	5000	4504.50	2000	1801.80
2	0.811622	4000	3246.49	3000	2434.87
3	0.731191	2000	1462.38	5000	3655.96
4	0.658731	2000	1317.46	4000	2634.92
		NPV=	530.84	NPV=	527.55

20.17 NET PRESENT VALUE PROFILE AND CONFLICT IN NPV

The conflicting situation in the above example can best be illustrated by using the NPV profile graph shown in Figure 20.1. Such a present value profile is obtained by calculating the NPV for varying rates of discounts for the given income stream for both A and B machines. Tables 20.20, 20.21 and the associated Figure 20.1 illustrate the entire summary.

Table 20.20 NPV Profile calculation for machine A

Year	Cash flow	PV at 0%	PV at 5%	PV at 10%	PV at 15%	PV at 20%	PV at 25%
0	−10,000	−10000.00	−10000.00	−10000.00	−10000.00	−10000.00	−10000.00
1	5000	5000.00	4761.90	4545.45	4347.83	4166.67	4000.00
2	4000	4000.00	3628.12	3305.79	3024.57	2777.78	2560.00
3	2000	2000.00	1727.68	1502.63	1315.03	1157.41	1024.00
4	2000	2000.00	1645.40	1366.03	1143.51	964.51	819.20
	NPV	3000.00	1763.10	719.90	−169.06	−933.64	−1596.80

In Figure 20.1 both the profile curves intersect at the rate 10.9449% revealing equal preferences. Any rate below this rate, B is preferable and any rate above this rate, A is preferable as highlighted in the previous example.

Table 20.21 NPV Profile calculation for machine B

Year	Cash flow	PV at 0%	PV at 5%	PV at 10%	PV at 15%	PV at 20%	PV at 25%
0	−10,000	−10000.00	−10000.00	−10000.00	−10000.00	−10000.00	−10000.00
1	5000	5000.00	4761.90	4545.45	4347.83	4166.67	4000.00
2	4000	4000.00	3628.12	3305.79	3024.57	2777.78	2560.00
3	2000	2000.00	1727.68	1502.63	1315.03	1157.41	1024.00
4	2000	2000.00	1645.40	1366.03	1143.51	964.51	819.20
	NPV	3000.00	1763.10	719.90	−169.06	−933.64	−1596.80

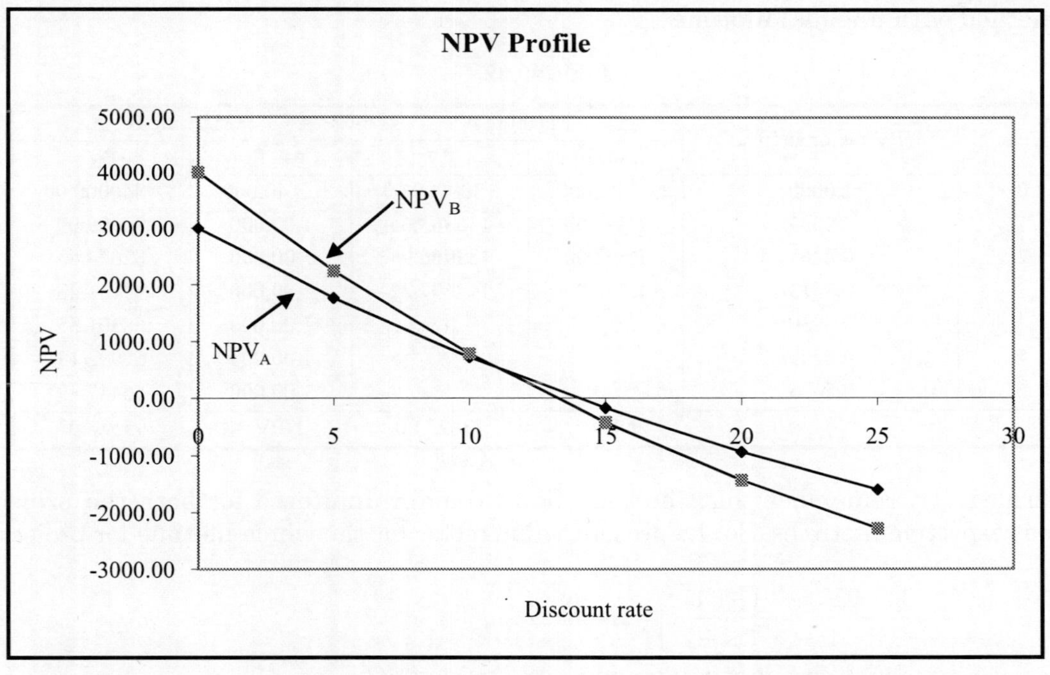

Figure 20.1

20.18 USE OF NET PRESENT VALUE METHOD FOR UNEQUAL LIFETIME PROJECTS

In all the illustrations discussed so far we considered only equivalent lifetime for all the projects under reference. This need not be the case all the time. On several occasions the life time of the projects under reference differs from one another. In such cases, the simple NPV approach leads to misleading conclusions. Such problems are solved by bringing in equalities in lifetime. Alternatively we calculate equivalent annual annuity for all the projects under reference and fix the needed priority. The following example illustrates one such decision making.

Example 24

A company is deciding to choose between two mutually exclusive projects A and B. Project A needs an initial investment of Rs. 3,00,000 and is expected to generate a cash flow of Rs. 1,50,000 per annum for three years of it life span. Project B needs an initial investment of Rs. 3,40,000 with an expected cash flow of Rs. 1,00,000 per annum for it entire life span of 6 years. Given the market rate of interest at 10% per annum list the best viable project.

Solution

NPV method with unequal lifetime

Table 20.22

Year	PV factor at 10%	Machine A		Machine B	
		Cash flow	PV	Cash flow	PV
0	1.0000	−3,00,000	−300000.00	−3,40,000	−340000.00
1	0.9091	1,50,000	136363.64	1,00,000	90909.09
2	0.8264	1,50,000	123966.94	1,00,000	82644.63
3	0.7513	1,50,000	112697.22	1,00,000	75131.48
4	0.6830			1,00,000	68301.35
5	0.6209			1,00,000	62092.13
6	0.5645			1,00,000	56447.39
		NPV=	73027.80	NPV=	95526.07

Alternatively, remember that the cash flows remain unaltered for both the projects over the respective lifetimes. So the present value factor for the whole lifetime for project A

$$= \frac{(1-R^{-n})}{R-1} = \left(\frac{1-1.1^{-3}}{1.1-1}\right) = 2.486852$$

So NPV_A=1,50,000 × 2.486852−3,00,000 = Rs. 73,027.80.
The present value factor for project B for it entire lifetime

$$= \frac{(1-R^{-n})}{R-1} = \left(\frac{1-1.1^{-6}}{1.1-1}\right) = 4.355261$$

NPV_B =1,00,000 × 4.355261 − 3,40,000 = Rs. 95,526.07.
If we simply go by simple NPV, we opt for project B. This is indeed is a wrong procedure because the life of the projects are not equal.

NPV Method after equating the life of A and B

In Table 20.23 a second cycle for A is introduced to make the lifetimes equal. In it during the fourth year the net cash flow is obtained by subtracting the cash flow of Rs. 1,50,000 from the capital invest of Rs. 3,00,000.

Table 20.23

Year	PV factor at 10%	Machine A		Machine B	
		Cash flow	PV	Cash flow	PV
0	1.0000	−3,00,000	−300000.00	−3,40,000	−340000.00
1	0.9091	1,50,000	136363.64	1,00,000	90909.09
2	0.8264	1,50,000	123966.94	1,00,000	82644.63
3	0.7513	1,50,000	112697.22	1,00,000	75131.48
4	0.6830	−1,50,000	−102452.02	1,00,000	68301.35
5	0.6209	1,50,000	93138.20	1,00,000	62092.13
6	0.5645	1,50,000	84671.09	1,00,000	56447.39
		NPV=	148385.07	NPV=	95526.07

In the last column we once again calculate the net present value. Clearly this time the highest present net value project A is preferred.

Merits of this method

1. The time value of money is given due place in this theory.
2. The whole stream of income till the end of the projects lifetime is considered.

Demerits of this method

1. The method is difficult to understand by a common man.
2. The solution depends upon the rate of return. Different rates often give conflicting ranking.

20.19 USE OF EQUIVALENT ANNUITY METHOD FOR DIFFERENT LIFETIME PROJECTS

In this method, though project lifetimes are different, we calculate equivalent annuity for both the projects and take decisions favouring the project having highest equivalent annual annuity.

Equivalent annual annuity = Net present value NPV/PVF

Equivalent annual annuity for A = 73,027.80/2.486852 = Rs. 29,369.5

Equivalent annual annuity for B = 95526.07/2.486852 = Rs. 21,933.49

This time clearly project A is preferable.

20.20 PROFITABILITY INDEX METHOD

In this method we calculate the profitability index by using the formula given below. Once the probability index is obtained, we fix the priority starting from the highest profitability index investment in the usual manner.

$$\text{Profitability index} = \frac{\text{Net present value}}{\text{Initial capital outlay}} \times 100$$

■ Example 25

For the example given in example 20 rank the projects by using profitability index method.

Solution

$$\text{Profitability index for A} = \frac{\text{Net present value}}{\text{Initial capital outlay}} \times 100 = \frac{25,469}{25,000} \times 100 = 25.47$$

Table 20.24

Project	Initial outlay in Rs.	Annual cash flows in Rs	Life of the project	PV Factor at 15%	PV	NPV	Profitability index	Ranking
A	1.00,000	25,000	10	5.0188	125469.22	25469	25.47	4
B	70,000	20,000	8	4.4873	89746.43	19746	28.21	3
C	30,000	6000	20	6.2593	37555.99	7556	25.19	5
D	50,000	15,000	10	5.0188	75281.53	25282	50.56	1
E	50,000	12,000	20	6.2593	75111.98	25112	50.22	2

This way the profitability indices are obtained and reported in Table 20.24 for all the projects under reference. The needed ranking is done in the last column on the basis of the probability index.

Note that though the problem is same our ranking results differ from the previous net present value ranking in the previous method.

20.21 INTERNAL RATE OF RETURN (IRR) METHOD

The internal rate of return is simply the rate of return (interest) at which the present value of all the stream of returns over its entire lifetime is equal to the present cost of the project. The following formula could be used in getting IRR.

$$\sum_{i=1}^{n} \frac{C_i}{(1+\text{IRR})^i} = C_0 \qquad \text{or}$$

$$\sum_{i=0}^{n} \frac{C_i}{(1+\text{IRR})^i} = 0$$

Remember that the IRR equation stated above is same as NPV equation discussed in the previous section. In NPV approach for a given discount rate we calculate the NPV for fixing the priority over investment projects. In IRR approach instead we equate the NPVs to zero and obtain the effective discount rate to fix the priority accordingly.

Merits of this method

1. The time value of money is given due place in this theory.
2. The whole stream of income till the end of the projects lifetime is considered.
3. Unique ranking without any conflict is possible in this method.

Demerits of this method

1. The method is difficult to understand by a common man.
2. IRR and NPV solution often gives conflicting rankings.
3. In the case of multiple investments over time gives rise to multiple solutions as well in some cases.
4. Correct IRR is obtained by trial and error method only.

Example 26

For the project A and B given in example 14 obtain the IRR and state the best investment choice. The needed IRR is obtained by trial and error method for both A and B as highlighted in the following tables.

Solution

$$\sum_{i=0}^{n} \frac{C_i}{(1+IRR)^i} = 0$$

Using this formula, the needed IRR is obtained by trial and error method. In Table 20.25 in the very first column we report the year starting from 0. In the second column we calculate the relevant PV factors for Re.1 cash flow for all the years at 10%. The net present value is found to be positive.

Table 20.25

IRR =	0.1000	Machine A	
Year	PV factor	Cash flow	PV
0	1.0000	−10,000	−10000.00
1	0.9091	5000	4545.45
2	0.8264	4000	3305.79
3	0.7513	2000	1502.63
4	0.6830	2000	1366.03
		NPV =	719.90

So in the second trial shown in Table 20.26 we repeat the calculation for higher discount rate of 13%. In it again the net present value is positive.

Table 20.26

IRR =	0.1300	Machine A	
Year	PV factor	Cash flow	PV
0	1.0000	−10,000	−10000.00
1	0.8850	5000	4424.78
2	0.7831	4000	3132.59
3	0.6931	2000	1386.10
4	0.6133	2000	1226.64
		NPV =	170.10

Table 20.27

IRR =	0.1400000	Machine A	
Year	PV factor	Cash flow	PV
0	1.0000000	−10,000	−10000.00
1	0.8771930	5000	4385.96
2	0.7694675	4000	3077.87
3	0.6749715	2000	1349.94
4	0.5920803	2000	1184.16
		NPV =	−2.06

Accordingly, a further increase is trailed out in Table 20.27 with 14%. At this rate the net present value is negative. So it is clear that our needed IRR will lie in-between 13% to 14%. Once this narrow band is obtained by trial and error method we obtain the needed in-between IRR by linear interpolation method as shown below.

$$13\% + (14\% - 13\%)\frac{170.1}{170.1 + 2.06} = 13.988034\%$$

Corresponding to this IRR the calculations are repeated in Table 20.28. Clearly this time also the needed NPV is not zero as expected. However it is very small. Remember that this is only a linear approximated result.

Table 20.28

IRR = 0.1398034		Machine A	
Year	PV Factor	Cash flow	PV
0	1.0000000	−10,000	−10000.00
1	0.8773443	5000	4386.72
2	0.7697330	4000	3078.93
3	0.6753208	2000	1350.64
4	0.5924889	2000	1184.98
		NPV=	1.27

Alternatively, one can obtain the correct IRR using excel worksheet. To do this type the formula as shown in Table 20.29 and start with an arbitrary IRR in cell B1. Adjust the rate up or down till you get zero NPV in the result row. The final result after such a trial is shown in Table 20.30 for the interest note of 13.98785%.

Table 20.29

	A	B	C	D
1	IRR = 0.1398785		Machine A	
2	Year	PV factor	Cash flow	PV
3	0	=1/(1+$B$1)^A3	−10,000	=C3*B3
4	1	=1/(1+$B$1)^A4	5000	=C4*B4
5	2	=1/(1+$B$1)^A5	4000	=C5*B5
6	3	=1/(1+$B$1)^A6	2000	=C6*B6
7	4	=1/(1+$B$1)^A7	2000	=C7*B7
8			NPV=	=SUM(D3:D7)

Table 20.30

	A	B	C	D
1	IRR = 0.1398785		Machine A	
2	Year	PV factor	Cash flow	PV
3	0	1.0000000	−10,000	−10000.00
4	1	0.8772865	5000	4386.43
5	2	0.7696316	4000	3078.53
6	3	0.6751874	2000	1350.37
7	4	0.5923328	2000	1184.67
8			NPV=	0.00

The needed correct IRR = 13.98785%.

Similarly, in Table 20.31 the final trial results using excel worksheet is reported for project B.

Table 20.31

	A	B	C	D
1	IRR= 0.131609		Machine B	
2	Year	PV factor	Cash flow	PV
3	0	1.000000	−10,000	−10000.00
4	1	0.883697	2000	1767.39
5	2	0.780921	3000	2342.76
6	3	0.690098	5000	3450.49
7	4	0.609838	4000	2439.35
8			NPV=	0.00

The needed IRR = 13.1609%.

According to this criteria investment option A seems better. Further, in this method there is no conflicting situation as in the case of NPV method.

■ Example 27

Using excel worksheet with the help of IRR, name the best yielding project for example 24.

Table 20.32

IRR= 0.23375192		Machine A	
Year	PV Factor	Cash flow	PV
0	1.00000000	–3,00,000	–300000.00
1	0.81053572	1,50,000	121580.36
2	0.65696815	1,50,000	98545.22
3	0.53249615	1,50,000	79874.42
		NPV=	0.00

$RR_A = 23.375192\%$

Table 20.33

IRR = 0.19113938		Machine B	
Year	PV Factor	Cash flow	PV
0	1.00000000	–3,40,000	–340000.00
1	0.83953231	1,00,000	83953.23
2	0.70481451	1,00,000	70481.45
3	0.59171455	1,00,000	59171.46
4	0.49676349	1,00,000	49676.35
5	0.41704900	1,00,000	41704.90
6	0.35012611	1,00,000	35012.61
		NPV=	0.00

$IRR_B = 19.113939\%$

Since 23.375192% > 19.113939% we conclude that A is a better yielding project.

■ Example 28

From the data given below obtain the best investment policy at the discount rate of 10% and 15% respectively using NPV criteria. Also suggest best investment policy using IRR criteria. Also draw NPV profile and explain the conflict if any.

Table 20.34

Initial Investment	Rs. 700	Rs. 800
First Year Cash Flow	Rs. 100	Rs. 600
Second Year Cash Flow	Rs. 400	Rs. 400
Third Year cash Flow	Rs. 600	Rs. 100

Solution

NPV criteria at 10% discount rate

Table 20.35

Interest = 0.1000		Machine A		Machine B	
Year	PV Factor at 10%	Cash flow	PV	Cash flow	PV
0	1.0000	−700	−700.00	−800	−800.00
1	0.9091	100	90.91	600	545.45
2	0.8264	400	330.58	400	330.58
3	0.7513	600	450.79	100	75.13
		NPV = 172.28		NPV = 151.16	

This time the high NPV Project A is preferred.

NPV criteria at 15% discount rate

Table 20.36

Interest = 0.1500		Machine A		Machine B	
Year	PV Factor at 15%	Cash flow	PV	Cash flow	PV
0	1.0000	−700	−700.00	−800	−800.00
1	0.8696	100	86.96	600	521.74
2	0.7561	400	302.46	400	302.46
3	0.6575	600	394.51	100	65.75
		NPV=	83.92	NPV=	89.95

This time the high NPV Project B is preferred.

Net NPV profiles for both A and B

Table 20.37 NPV Profile calculation for Machine A

Year	Cash flow	PV at 0%	PV at 5%	PV at 10%	PV at 15%	PV at 20%	PV at 25%
0	−700	−700.00	−700.00	−700.00	−700.00	−700.00	−700.00
1	100	100.00	95.24	90.91	86.96	83.33	80.00
2	400	400.00	362.81	330.58	302.46	277.78	256.00
3	600	600.00	518.30	450.79	394.51	347.22	307.20
	NPV	400.00	276.35	172.28	83.92	8.33	−56.80

In Figure 20.2 at 10% rate clearly Project A's NPV is more. At 15% rate Project B's NPV is more. Since the IRR = 23.62% is the highest payback rate B is preferred. Thus anything more than 14% both NPV decision and IRR decision coincide.

Table 20.38 NPV Profile calculation for Machine B

Year	Cash flow	PV at 0%	PV at 5%	PV at 10%	PV at 15%	PV at 20%	PV at 25%
0	−800	−800.00	−800.00	−800.00	−800.00	−800.00	−800.00
1	600	600.00	571.43	545.45	521.74	500.00	480.00
2	400	400.00	362.81	330.58	302.46	277.78	256.00
3	100	100.00	86.38	75.13	65.75	57.87	51.20
	NPV	300.00	220.62	151.16	89.95	35.65	−12.80

For any discount rate less than 14% the NPV decision and IRR decision are conflicting to one another.

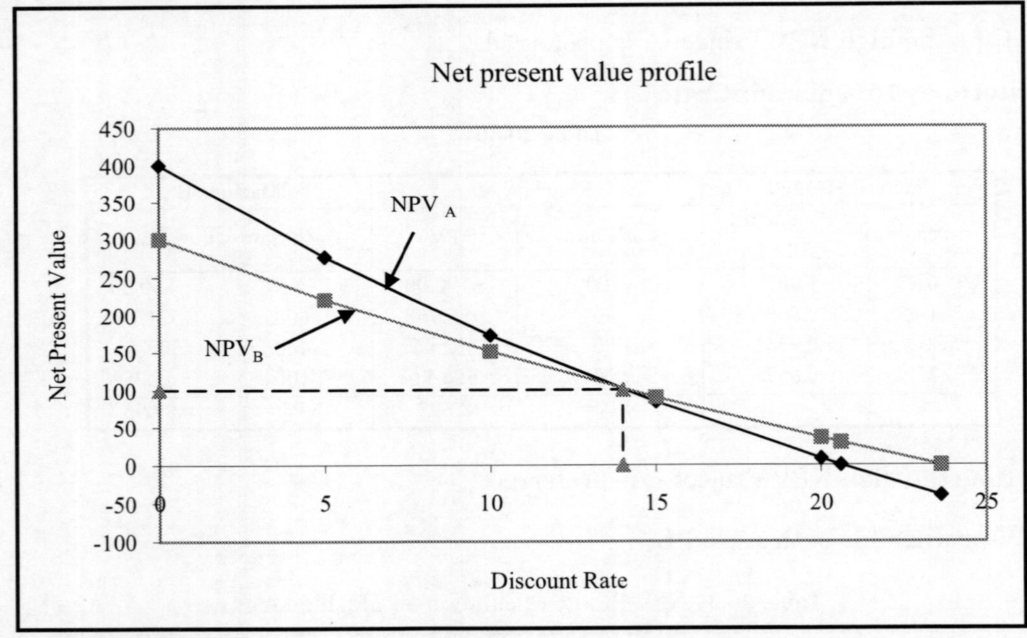

Figure 20.2

IRR Criteria

Table 20.39

IRR = 0.2060		Machine A	
Year	PV factor	Cash flow	PV
0	1.0000	−700	−700.00
1	0.8292	100	82.92
2	0.6876	400	275.02
3	0.5701	600	342.07
		NPV=	0.00

IRR = 20.6%

Table 20.40

IRR = 0.2362		Machine B	
Year	PV factor	Cash flow	PV
0	1.0000	−800	−800.00
1	0.8089	600	485.34
2	0.6543	400	261.73
3	0.5293	100	52.93
		NPV=	0.00

IRR = 23.62%

Thus B is the better project according to this principle. Also note that if the discounting rate is above 14% there is no conflict between NPV criterion and the IRR criterion. Both prefer B over A. But for any discount rate below 14% we notice a conflict between the two principles.

20.22 CAPITAL BUDGETING AND RISK

Till now the stream of returns are assumed to be certain in the respective periods. In other words for every period the probability of receiving the returns was assumed to be unity. In reality in modern business the receivable future returns are only predicted values. Thus, unforeseen events in business make the receivable income risky. In this section we propose to discuss two important types of risk management.

Risk adjusted cash flow (certainty equivalent) approach

In this approach when there is risk in receiving a certain amount of money say Rs. 1000 in the first year. The concerned manager, who does not want to take the risk, will be willing to trade this risky Rs. 1000 return for a risk free certainty amount. Definitely this amount will be less than Rs. 1000. Let this certainty equivalent amount be Rs. 800. In this way the certainty equivalent amount is obtained for the entire income stream including the zero period initial investment. One this is done the suitable investment decision is obtained in the usual manner as discussed earlier. The ratio of the certainty cash flow to the risky cash flow is called the certainty equivalent coefficient.

$$\alpha_i = \frac{\text{Certan cash flow}_i}{\text{Risky cash flow}_i}$$

With this notational modification the relevant NPV is obtained by using the following formula.

$$\text{NPV} = \sum_{i=0}^{n} \frac{\alpha_i C_i}{(1+r)^n}$$

■ Example 29

A company wants to avoid risks of any type. The company wants to choose one among the two proposals for investment. In addition to the respective cash flows the associated risk coefficients are given in the following table. Given the after tax risk free return rate of 8% obtain the best possible proposal.

Table 20.41

Year	Proposal A		Proposal B	
	Cash flow	α_1	Cash flow	α_i
0	–40,000	1.00	–30,000	1.00
1	20,000	0.90	15,000	0.95
2	18,000	0.80	12,500	0.80
3	12,000	0.80	10,000	0.75
4	10,000	0.60	8000	0.70

Solution

Table 20.42

Interest = 0.0800		Proposal A			
Year	PV factor	Cash flow	α_1	$\alpha_i C_i$	PV
0	1.0000	–40,000.00	1.00	–40000.00	–40000.00
1	0.9259	20,000.00	0.90	18000.00	16666.67
2	0.8573	18,000.00	0.80	14400.00	12345.68
3	0.7938	12,000.00	0.80	9600.00	7620.79
4	0.7350	10,000.00	0.60	6000.00	4410.18
				NPV=	1043.31

Table 20.43

Interest = 0.0800		Proposal B			
Year	PV factor	Cash flow	α_1	$\alpha_i C_i$	PV
0	1.0000	–30,000.00	1.00	–30000.00	–30000.00
1	0.9259	15,000.00	0.95	14250.00	13194.44
2	0.8573	12,000.00	0.80	9600.00	8230.45
3	0.7938	10,000.00	0.75	7500.00	5953.74
4	0.7350	8000.00	0.70	5600.00	4116.17
				NPV=	1494.81

Since NPV for proposal B is the better the company can venture for project B.

Risk adjusted discount rate approach

In this method instead of reducing cash flow for risk we propose an increase in return rate by adding a risk premium rate φ to the original rate r. Thus this time the effective risk free return rate will be $k = r + \varphi$. Once the correction is incorporated we obtain the best option in the usual manner by using NPV. So the corrected formula may be written as

$$NPV = \sum_{i=0}^{n} \frac{C_i}{(1+k)^n}$$

■ Example 30

The company in example 29 wants to avoid risks of any type. The company wants to choose one of the two proposals for investment. In addition to the respective cash flows though the market rate of interest is 10% it wants a 15% risk free return. Obtain the best investment option.

Table 20.44

Year	Proposal A	Proposal B
	Cash flow	Cash flow
0	–40,000	–30,000
1	20,000	15,000
2	18,000	12,500
3	12,000	10,000
4	10,000	8000

Solution

This time we calculate the NPV for the risk free discount rate of 15%.

Table 20.45

Interest = 0.1500		Proposal A		Proposal B	
Year	PV factor at 15%	Cash flow	PV	Cash flow	PV
0	1.0000	–40,000	–40000.00	–30,000	–30000.00
1	0.8696	20,000	17391.30	15,000	13043.48
2	0.7561	18,000	13610.59	12,000	9073.72
3	0.6575	12,000	7890.19	10,000	6575.16
4	0.5718	10,000	5717.53	8000	4574.03
		NPV=	4609.62	NPV=	3266.39

Clearly project A is the better proposal.

20.23 STATISTICAL APPROACH FOR RISK

In this statistical approach, the cash flow for a given year is not a fixed amount as in our previous illustrations. In these models the cash flows follow stochastic process based on certain probability distribution. So, ultimately we use the average cash flow obtained as the expected value for further calculations of NPV.

Expected net present value E(NPV) approach

Given the probability distribution, the expected value of the cash flow for each period is obtained by simply multiplying the cash flow by the respective probabilities. Once this is done for all periods we discount them all and obtain the needed present values in the usual manner. The expected net present value for the entire project is obtained by adding all the discounted expected values. While doing so remember that we will not go for discounting the initial capital investment in period 0. Moreover, since this negative cash flow is certain to happen with probability 1 that itself gives the average outflow as highlighted in the following formula.

$$E(NPV) = \bar{C}_0 + \frac{\bar{C}_1}{(1+r)^1} + \frac{\bar{C}_2}{(1+r)^2} + \frac{\bar{C}_3}{(1+r)^3}$$

$$= \sum_{i=0}^{n} \frac{\bar{C}_i}{(1+r)^i}$$

The following example illustrates this method.

Example 31

An IT company is considering a new investment proposal which costs Rs. 1,20,000. The project has three-year lifetime. The relevant cash flow distribution is given in Table 20.44.

Table 20.46

Year 1		Year 2		Year 3	
Cash flow	Probability	Cash flow	Probability	Cash flow	Probability
30,000	0.2	30,000	0.1	40,000	0.3
40,000	0.4	50,000	0.4	60,000	0.3
50,000	0.3	80,000	0.4	80,000	0.2
60,000	0.1	90,000	0.1	1,00,000	0.2

The management expects a minimum return of 10%. Assess the worthiness of this project.

Solution

Given the data, first we calculated the expected (average) cash flow year wise as shown in Table 20.47.

Table 20.47

Year	Cash flow	Probability	Expected value	Expected (average) cash flow
1	30,000	0.2	6000	43,000
	40,000	0.4	16,000	
	50,000	0.3	15,000	
	60,000	0.1	6000	
2	30,000	0.1	3000	64,000
	50,000	0.4	20,000	
	80,000	0.4	32,000	
	90,000	0.1	9000	
3	40,000	0.3	12,000	66,000
	60,000	0.3	18,000	
	80,000	0.2	16,000	
	1,00,000	0.2	20,000	

Once the average or otherwise called the expected cash flows for all the periods are obtained, we calculate the E(NPV) in the usual manner as shown in Table 20.48 for the given discount rate of 10%.

Table 20.48

Rate = 0.1000			
Year	PV factor at 10%	Cash flow	PV
0	1.0000	−1,20,000	−120000.00
1	0.9091	43,000	39090.91
2	0.8264	64,000	52892.56
3	0.7513	66,000	49586.78
		E(NPV)=	21570.25

Since the E(NPV) is greater than zero, we conclude that it is an acceptable proposal.

Importance of variance of E(NPV)

Since the cash flows in the respective years are random variables having their stated chance (probability) of realization, the derived E(NPV) is also a random variable. So before taking a final decision on the investment one must know the variance of the E(NPV) also. Less the variance more will be reliability and vice versa. To obtain the so needed variance of E(NPV) Hiller classifies the relevant data into three categories.

- The cash flows of individual periods are not correlated and hence independent of one another.
- The cash flows between periods are perfectly correlated.
- Part of the cash flows are perfectly correlated and remaining part of the flows are independent.

Variance of NPV for independent cash flows

In the case of independent cash flows between periods we obtained the variance of the E(NPV) by using the following well known statistical formula.

$$V(NPV) = \sigma^2 = \sigma_0^2 + \frac{\sigma_1^2}{(1+r)^2} + \frac{\sigma_2^2}{(1+r)^4} + \ldots\ldots\ldots\ldots + \frac{\sigma_n^2}{(1+r)^{2n}}$$

$$= \sum_{i=0}^{n} \frac{\sigma_i^2}{(1+r)^{2n}}$$

Variance of NPV for perfectly correlated cash flows

If the cash flows are perfectly correlated then the behaviour of cash flows for all the periods will be alike. For example, if the actual cash flow is less by $\alpha\sigma$ times less than its expected value in year one then the actual cash flows for the remaining years also will be $\alpha\sigma$ times less than the corresponding expected values. In other words this is to say that the cash flows for all the years are linearly related. In such cases the expected net present value of the standard deviation is obtained by using the following formula:

$$\sigma(NPV) = \left[\sigma_0 + \frac{\sigma_1}{(1+r)} + \frac{\sigma_2}{(1+r)^2} + \ldots\ldots\ldots\ldots + \frac{\sigma_n}{(1+r)^n} \right]$$

$$= \sum_{i=1}^{n} \frac{\sigma_i}{(1+r)^i}$$

Variance of NPV for partly perfectly correlated cash flows and partly independent cash flows

In the case of partly correlated cash flows and partly independent cash flows the variance of the expected net present value is obtained by using the following formula:

$$V(NPV) = \sigma^2 = \left[V(C_0') + \frac{V(C_2')}{(1+r)^2} + \frac{V(C_3')}{(1+r)^3} + \ldots\ldots + \frac{V(C_n')}{(1+r)^n} \right]$$

$$+ \left[\sqrt{V(C_0^*)} + \frac{\sqrt{V(C_1^*)}}{(1+r)} + \frac{\sqrt{V(C_2^*)}}{(1+r)^2} + \ldots\ldots \frac{\sqrt{V(C_n^*)}}{(1+r)^n} \right]^2$$

$$V(NPV) = \left(\sum_{i=1}^{n} \frac{V(C_i')}{(1+r)^{2i}}\right) + \left(\sum_{i=1}^{n} \frac{\sqrt{V(C_i^*)}}{(1+r)^i}\right)^2$$

■ Example 32

The XYZ company is considering an investment proposal for Rs. 50,000. The project has a lifetime of three years. The respective independent cash flows and their probabilities are given below. Obtain E(NPV) and its variance given the discount rate of 4%.

Table 20.49

Year 1		Year 2		Year 3	
Cash flow	Probability	Cash flow	Probability	Cash flow	Probability
10,000	0.1	10,000	0.2	10,000	0.3
20,000	0.2	20,000	0.3	20,000	0.4
30,000	0.3	30,000	0.4	30,000	0.2
40,000	0.4	40,000	0.1	40,000	0.1

Calculation of E(NPV)

Given the data, first we calculated the expected (average) cash flow year wise as shown the table below.

Table 20.50

Year	Cash flow	Probability	Expected value	Expected (average) cash flow
1	10,000	0.1	1000	30,000
	20,000	0.2	4000	
	30,000	0.3	9000	
	40,000	0.4	16,000	
2	10,000	0.2	2000	24,000
	20,000	0.3	6000	
	30,000	0.4	12,000	
	40,000	0.1	4000	
3	10,000	0.3	3000	21,000
	20,000	0.4	8000	
	30,000	0.2	6000	
	40,000	0.1	4000	

Once the average or otherwise called the expected cash flows for all the periods are obtained we calculate the E(NPV) in the usual manner as shown in Table 20.51 for the given discount rate of 4%.

Table 20.51

Rate = 0.0400			
Year	PV Factor at 4%	Cash flow	PV
0	1.0000	−50,000	−50000.00
1	0.9615	30,000	28846.15
2	0.9246	24,000	22189.35
3	0.8890	21,000	18668.92
		E(NPV)=	19704.43

Since the E(NPV) = Rs. 19,704.43 is greater than zero, we conclude that it is an acceptable proposal.

Calculation of the individual variances

As a next step we calculate the individual variances for all the three years as shown in Table 20.52.

Table 20.52

Year	Cash flow	Probability	Expected value	Expected (average) cash flow	$(C_{ij} - \overline{C}_i)^2 P_{ij}$	σ_i^2
1	10,000	0.1	1000	30,000	40000000	100000000
	20,000	0.2	4000		20000000	
	30,000	0.3	9000		0	
	40,000	0.4	16,000		40000000	
2	10,000	0.2	2000	24,000	39200000	84000000
	20,000	0.3	6000		4800000	
	30,000	0.4	12,000		14400000	
	40,000	0.1	4000		25600000	
3	10,000	0.3	3000	21,000	36300000	89000000
	20,000	0.4	8000		400000	
	30,000	0.2	6000		16200000	
	40,000	0.1	4000		36100000	

Calculation of the variance for E(NPV)

Now after obtaining the individual variances we go for the calculation of the variance of the E(NPV) in the following table by using formula given below.

$$V(NPV) = \sigma^2 = \frac{\sigma_1^2}{(1+r)^2} + \frac{\sigma_2^2}{(1+r)^4} + \ldots\ldots\ldots\ldots + \frac{\sigma_n^2}{(1+r)^{2n}}$$

Table 20.53

Year	Variance σ_i^2	PV Factor $(1+r)^{-2n}$	Present value
1	100000000	0.924556213	92455621.30
2	84000000	0.854804191	71803552.05
3	89000000	0.790314526	70337992.79
SD =	15316.5650894140	$\sigma^2 =$	234597166.14

Thus, $\sigma = \sqrt{234597166.14} = 15316.6.$

Example 33

An electric company is considering to make investment proposal for Rs. 1,20,000. The project has a lifetime of three years over which the following independent cash inflows and their associated probabilities are projected. For a given interest rate of 10% obtain the E(NPV) and its SD.

Table 20.54

Year 1		Year 2		Year 3	
Cash Flow	Probability	Cash Flow	Probability	Cash Flow	Probability
30,000	0.2	30,000	0.1	40,000	0.3
40,000	0.4	50,000	0.4	60,000	0.3
50,000	0.3	80,000	0.4	80,000	0.2
60,000	0.1	90,000	0.1	100,000	0.2

Calculation of E(NPV)

Given the data, first we calculated the expected (average) cash flow year wise as shown the table below.

Table 20.55

Year	Cash flow	Probability	Expected value	Expected (average) cash flow
1	30,000	0.2	6000	43,000
	40,000	0.4	16,000	
	50,000	0.3	15,000	
	60,000	0.1	6000	
2	30,000	0.1	3000	64,000
	50,000	0.4	20,000	
	80,000	0.4	32,000	
	90,000	0.1	9000	
3	40,000	0.3	12,000	66,000
	60,000	0.3	18,000	
	80,000	0.2	16,000	
	1,00,000	0.2	20,000	

Once the average or otherwise called the expected cash flows for all the periods are obtained we calculate the E(NPV) in the usual manner as shown below for the given discount rate of 10%.

Table 20.56

Rate =	0.1000	Cash flow	PV
Year	PV Factor at 4%		
0	1.0000	−1,20,000	−120000.00
1	0.9091	43,000	39090.91
2	0.8264	64,000	52892.56
3	0.7513	66,000	49586.78
		E(NPV)=	21570.25

Since E(NPV) = Rs. 21,570.25 is greater than zero, we conclude that it is an acceptable positive revenue yielding proposal.

Calculation of the individual variance

As a next step we calculate the individual variances for all the three years as shown in Table 20.57.

Table 20.57

Year	Cash flow	Probability	Expected value	Expected (average) cash flow	$(C_{ij}-\overline{C}_i)^2 P_{ij}$	σ_i^2
1	30,000	0.2	6000	43,000	33800000	81000000
	40,000	0.4	16,000		3600000	
	50,000	0.3	15,000		14700000	
	60,000	0.1	6000		28900000	
2	30,000	0.1	3000	64,000	115600000	364000000
	50,000	0.4	20,000		78400000	
	80,000	0.4	32,000		102400000	
	90,000	0.1	9000		67600000	
3	40,000	0.3	12,000	66,000	202800000	484000000
	60,000	0.3	18,000		10800000	
	80,000	0.2	16,000		39200000	
	1,00,000	0.2	20,000		231200000	

Calculation of the standard variance of E(NPV)

Now after obtaining the individual variances we go for the calculation of the variance of the E(NPV) in Table 20.58 by using formula given below.

$$V(NPV) = \sigma^2 = \frac{\sigma_1^2}{(1+r)^2} + \frac{\sigma_2^2}{(1+r)^4} + \ldots\ldots\ldots\ldots + \frac{\sigma_n^2}{(1+r)^{2n}}$$

Table 20.58

Year	Variance σ_i^2	PV Factor $(1+r)^{-2n}$	Present value
1	81000000	0.826446281	66942148.76
2	364000000	0.683013455	248616897.75
3	484000000	0.56447393	273205382.15
SD =	24264.47	σ^2 =	588764428.66

Thus, $\sigma = \sqrt{588764428.66} = 24264.5$.

Example 34

An investment project involves a current outlay of Rs. 10,000. The mean and the standard deviation of cash flows, which are perfectly correlated, are given below.

Table 20.59

Year	Mean	SD
1	5000	1500
2	3000	1000
3	4000	2000
4	3000	1200

Calculate the E(NPV) and its standard deviation given the tax free interest rate of 6% percent.

Solution

Table 20.60

Rate = 0.0600		Cash flow	PV
Year	PV Factor at 6%		
0	1.0000	−10,000	−10000.00
1	0.9434	5000	4716.98
2	0.8900	3000	2669.99
3	0.8396	4000	3358.48
4	0.7921	3000	2376.28
		E(NPV)=	3121.73

Now the needed standard deviation is obtained by using the following formula

$$\sigma(NPV) = \left[\sigma_0 + \frac{\sigma_1}{(1+r)} + \frac{\sigma_2}{(1+r)^2} + \ldots\ldots\ldots\ldots + \frac{\sigma_n}{(1+r)^n} \right]$$

$$= \sum_{i=1}^{n} \frac{\sigma_i}{(1+r)^i}$$

Table 20.61

Rate = 0.0600		Variances	PV
Year	PV Factor at 6%		
1	0.9434	1500	1415.09
2	0.8900	1000	890.00
3	0.8396	2000	1679.24
4	0.7921	1200	950.51
		SD(NPV)=	4934.84

■ Example 35

The management of ABC company wants to invest in a project an initial amount of Rs. 1,20,000. The cash flows in each period consist of two components, one of which varies independently and another one is perfectly correlated. However, each one of them is normally distributed with the mean and standard deviation given below. Given the discount rate of 10% compute the E(NPV) and its standard deviation.

Table 20.62

Year	Component C′		Component C*	
	Mean	SD	Mean	SD
1	24,000	3000	20,000	2800
2	20,000	2800	20,000	2800
3	20,000	3000	20,000	3000
4	20,000	3000	15,000	2000
5	18,000	2000	12,000	2000

Solution

Now the variance of the NPV is given by:

$$V(NPV) = \left(\sum_{i=1}^{n} \frac{V(C'_i)}{(1+r)^{2i}} \right) + \left(\sum_{i=1}^{n} \frac{\sqrt{V(C^*_i)}}{(1+r)^i} \right)^2$$

Now let us calculate the value of the two components separately and later add them together to get the needed value.

Table 20.63

Year	Mean C′	Mean C*	Total Cash Flow	PV Factor at 10%	PV
1	24,000	20,000	44,000	0.909091	40,000.00
2	20,000	20,000	40,000	0.826446	33,057.85
3	20,000	20,000	40,000	0.751315	30,052.59
4	20,000	15,000	35,000	0.683013	23,905.47
5	18,000	12,000	30,000	0.620921	18,627.64
					1,45,643.55
			Less cash outflow		1,20,000.00
				E(NPV)=	25,643.55

Calculation for independent flows

Table 20.64

Year	SD	Variance σ_i^2	PV Factor $(1+r)^{-2n}$	Present value
1	3000	9,000,000	0.826446281	7438016.53
2	2800	7,840,000	0.683013455	5354825.49
3	3000	9,000,000	0.56447393	5080265.37
4	3000	9,000,000	0.46650738	4198566.42
5	2000	4,000,000	0.385543289	1542173.16
	SD=	4859.41	$\sigma^2=$	23613846.97

Calculation for perfectly correlated flows

Table 20.65

Year	SD	PV Factor	Present value
1	2800	0.909090909	2545.45
2	2800	0.826446281	2314.05
3	3000	0.751314801	2253.94
4	2000	0.683013455	1366.03
5	2000	0.620921323	1241.84
		$\sigma=$	9721.32

So V(NPV) = 23613846.97 + 9721.84^2 = 118,113,05

So SD = Rs. 10,868

■ Example 36

A company XYZ wants to choose one of the two investment options A and B. The relevant data for projects A and B are given below. Identify the correct choice.

Table 20.66

Project	E(NPV)	SD
A	15,000	6500
B	15,000	7500

Solution

To identify the correct choice let us calculate the coefficient variation for both A and B.

$$CV_A = \frac{\sigma_A}{\overline{X}_A} \times 100 = \frac{6500}{18000} \times 100 = 36.11\%$$

$$CV_B = \frac{\sigma_B}{\overline{X}_B} \times 100 = \frac{7200}{2200} \times 100 = 32.73\%$$

Thus the project B is the better option.

20.24 NORMAL DISTRIBUTION OF CASH FLOWS

Till now we have not made any specific assumptions of the cash flow distribution. In this section we assume normally distributed cash flows with a given mean and standard deviation. Given the situation we calculate the probability of NPV being less or more than a certain given value of NPV by using the standard normal distribution. The following example illustrates one such case.

■ Example 37

A company X is considering whether to invest Rs. 1,20,000 or not. The management forecast that the new plant will generate a normally distributed cash flow for each period with the mean and standard deviation given below.

Table 20.67

Year	Expected (average) cash flow	Standard Deviation
1	44,000	5800
2	40,000	5600
3	40,000	6000
4	35,000	5000
5	30,000	4000

Given the discount rate of 10% determine:

(i) the mean and variance of E(NPV)

(ii) the probability that the investment has zero E(NPV)

(iii) whether the project should be undertaken provided the management has laid down a target of 70% chance of having NPV at least Rs. 20,000.

Solution (i)

Table 20.68 Computation of E (NPV)

Rate = 0.1000		Cash flow	PV
Year	PV Factor at 10%		
0	1.0000	–120,000	–120000.00
1	0.9091	44,000	40000.00
2	0.8264	40,000	33057.85
3	0.7513	40,000	30052.59
4	0.6830	35,000	23905.47
5	0.6209	30,000	18627.64
		E(NPV)=	25643.55

Table 20.69 Computation of variance for E (NPV)

Year	SD	Variance σ_i^2	PV Factor $(1+r)^{-2n}$	Present value
1	5800	33,640,000	0.8264	27801652.89
2	5600	31,360,000	0.6830	21419301.96
3	6000	36,000,000	0.5645	20321061.48
4	5000	25,000,000	0.4665	11662684.51
5	4000	16,000,000	0.3855	6168692.63
	SD=	9347.3736135271	$\sigma^2 =$	87373393.47

Thus the standard deviation of E(NPV) = 9347.38

Solution (ii)

To compute the needed probability let us calculate the Z value as shown below $Z = \frac{0-25643}{9347.38} = -2.74$. Now from the normal table the probability corresponding to Z = −2.74 is obtained as 0.0031. Thus, there is no chance that the E(NPV) will have zero value in this problem.

Solution (iii)

Now to find P (NPV > 20,000) we calculate the relevant Z first.

$$Z = \frac{20{,}000 - 25{,}643}{9347.38} = \frac{-5643}{9347.38} = -0.07$$

Now from the normal table the probability corresponding to Z = −0.07 is obtained as 0.5300. Thus, there is 53% chance that the E(NPV) will have Rs. 20,000 and above. Thus, the project will be rejected.

Example 38

A company X is considering whether to invest Rs. 1,20,000 or not. The management forecast that the new plant will generate a normally distributed cash flow for each period with the mean and standard deviation given below.

Table 20.70

Year	Expected (average) cash flow	Standard deviation
1	44,000	8800
2	40,000	7600
3	40,000	8000
4	35,000	6000
5	30,000	5000

Given the discount rate of 10% determine:

a) the mean and variance of E(NPV)

b) the probability that the investment has zero E(NPV)

c) the probability that the E(NPV) would be at least Rs. 20,000

Solution (a)

Table 20.71

Rate = 0.1000		Cash flow	PV
Year	PV Factor at 4%		
0	1.0000	−1,20,000	−120000.00
1	0.9091	44,000	40000.00
2	0.8264	40,000	33057.85
3	0.7513	40,000	30052.59
4	0.6830	35,000	23905.47
5	0.6209	30,000	18627.64
		E(NPV)=	25643.55

Table 20.72

Year	SD	Variance σ_i^2	PV Factor $(1+r)^{-2n}$	Present value
1	8800	77,440,000	0.826446281	64000000.00
2	7600	57,760,000	0.683013455	39450857.18
3	8000	64,000,000	0.56447393	36126331.52
4	6000	36,000,000	0.46650738	16794265.69
5	5000	25,000,000	0.385543289	9638582.24
	SD =	12884.4882175668	σ^2 =	166010036.63

Thus the standard deviation of E(NPV) $=\sqrt{166010036.63}=12884.5$

Solution (b)

To compute the needed probability let us calculate the Z value as shown below

$$Z=\frac{0-25643}{12884.5}=-1.9902$$

Now from the normal table the probability corresponding to Z = −1.99 is obtained as 0.0233. Thus there is no chance that the E(NPV) will have zero value in this problem.

Solution (c)

Now to find P(NPV>20,000) we calculate the relevant Z first.

$$Z=\frac{20{,}000-25{,}643}{12884.5}=\frac{-5643}{12884.5}=-0.438$$

Now from the normal table the probability corresponding to Z = −0.438 is obtained as 0.6700. Thus, there is 67% chance that the E(NPV) will have Rs. 20,000 and above.

20.25 DECISION TREE APPROACH TO E(NPV) CALCULATION

The decision tree approach provides an alternative approach to investment analysis involving risk. This method will become handy where ever there is difficulty in classifying the cash flows either as independent or perfectly correlated.

■ Example 39

To illustrate, suppose that a company has two-year project requiring an initial investment of Rs. 1,00,000. The expected year-wise cash flows and their associated probabilities are given in the following tree chart 1. Find the E (NPV) and its SD.

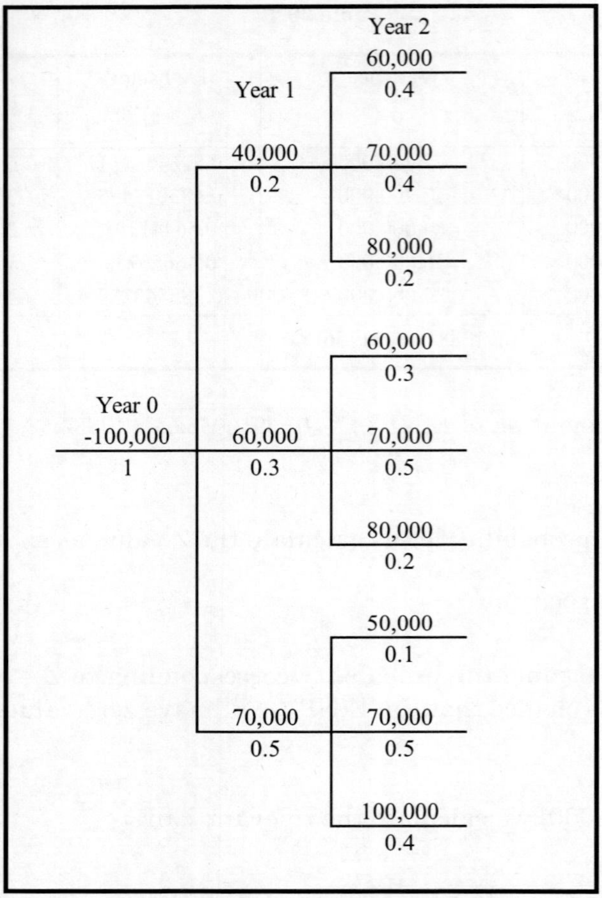

Year 2
60,000
0.4
Year 1
40,000
0.2
70,000
0.4
80,000
0.2
60,000
0.3
Year 0
-100,000
1
60,000
0.3
70,000
0.5
80,000
0.2
50,000
0.1
70,000
0.5
70,000
0.5
100,000
0.4

Figure 20.3

Solution

In the chart altogether we have 9 possible cash flows routes in it, under the head NVP the net present values are obtained in the usual manner by converting all the actual cash flows into the discounted cash flows. For example, for the very first route the NPV at 8% rate is obtained as $-1,00,000\ (1.08)^0 + 40,000\ (1.08)^{-1} + 60,000(1.08)^{-2} = -11,522.63$. Similarly, the calculations are made for other eight possibilities and reported in the chart. Once this is obtained in the last row we calculate the NVP × joint probability as shown in the chart. By adding the last column we get the needed E (NPV). In our problem this result is obtained as Rs. 20,181.76.

Year 0	Year 1	Year 2	NPV	Joint Prob.	NPVx Prob
-100,000 (1)	40,000 (0.2)	60,000 (0.4)	-11522.63	0.08	-921.81
		70,000 (0.4)	-2949.25	0.08	-235.94
		80,000 (0.2)	5624.14	0.04	224.97
	60,000 (0.3)	60,000 (0.3)	6995.88	0.09	629.63
		70,000 (0.5)	15569.27	0.15	2335.39
		80,000 (0.2)	24142.66	0.06	1448.56
	70,000 (0.5)	50,000 (0.1)	7681.76	0.05	384.09
		70,000 (0.5)	24828.53	0.25	6207.13
		100,000 (0.4)	50548.70	0.20	10109.74
				E(NPV) =	20181.76

Figure 20.3 (a)

Calculation of SD

$$\sigma = \sqrt{\sum_{j=1}^{n}\left[NPV_j - E(NPV)\right]^2 P_j}$$

$$\sigma = \{(-11,522.63 - 20,181.76)^2 \times 0.08 + (-2949.25 - 20,181.76)^2 \times 0.08 + ...\} = \text{Rs. } 18,684.$$

Example 40

Suppose that a company has a two-year project requiring an initial investment of Rs. 80,000. The expected year-wise cash flows and their associated probabilities are given in the following tree chart 1. Find the E (NPV).

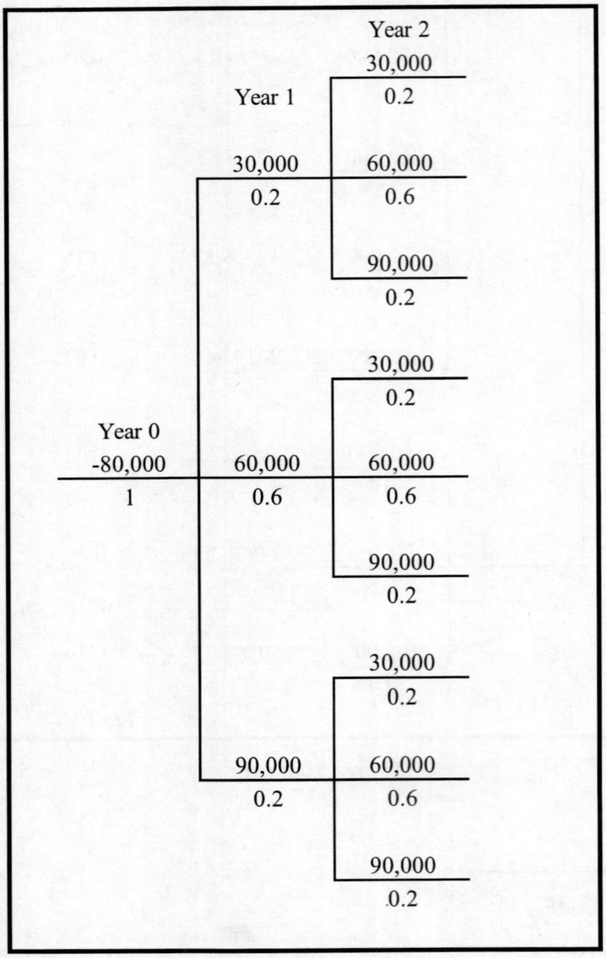

Figure 20.4

Solution

Year 0	Year 1	Year 2	NPV	Joint Prob.	NPVxProb
		30,000 (0.2)	-31228.73	0.04	-1249.15
	30,000 (0.2)	60,000 (0.6)	-8544.42	0.12	-1025.33
		90,000 (0.2)	14139.89	0.04	565.60
		30,000 (0.2)	-5141.78	0.12	-617.01
-80,000 (1)	60,000 (0.6)	60,000 (0.6)	17542.53	0.36	6315.31
		90,000 (0.2)	40226.84	0.12	4827.22
		30,000 (0.2)	20945.18	0.04	837.81
	90,000 (0.2)	60,000 (0.6)	43629.49	0.12	5235.54
		90,000 (0.2)	66313.80	0.04	2652.55
				E(NPV) =	17542.53

Figure 20.4(a)

From the tree chart E(NPV) = Rs. 17,542.53.

20.26 SENSITIVITY ANALYSIS

Sensitivity analysis is a technique designed to measure the response of the NPV/IRR values already obtained by effecting changes in the input variables like the initial capital expenditure, first year cash flow, second year cash flow etc.

Sensitivity of NPV

■ **Example 41**

Given the following NPV table for Machine A conduct sensitivity analysis for

a) a rise in the cost of the project by 5%
b) a decrease in the first year cash flow by 5%
c) a rise in the second year cash flow by 10%
d) a rise in the discount rate by 2%

Original NPV calculations

Table 20.73

Return Rate = 0.100000		Machine A	
Year	PV Factor	Cash flow	PV
0	1.000000	−10,000	−10000.00
1	0.909091	5000	4545.45
2	0.826446	4000	3305.79
3	0.751315	2000	1502.63
4	0.683013	2000	1366.03
		NPV=	719.90

Solution (a)

Effect of a 5% rise in the cost of the project on NPV

Table 20.74

Return Rate = 0.100000		Machine A	
Year	PV Factor	Cash flow	PV
0	1.000000	−10,500	−10500.00
1	0.909091	5000	4545.45
2	0.826446	4000	3305.79
3	0.751315	2000	1502.63
4	0.683013	2000	1366.03
%decrease=	69.454091	NPV=	219.90

Solution (b)

Effect of a 5% decrease in the first year cash flow

Table 20.75

Return Rate = 0.100000		Machine A	
Year	PV Factor	Cash flow	PV
0	1.000000	−10,000	−10000.00
1	0.909091	4750	4318.18
2	0.826446	4000	3305.79
3	0.751315	2000	1502.63
4	0.683013	2000	1366.03
% decrease=	31.570572	NPV=	492.62

Solution (c)

Effect of a 10% rise in the second year cash flow

Table 20.76

Return Rate = 0.100000		Machine A	
Year	PV Factor	Cash flow	PV
0	1.000000	−10,000	−10000.00
1	0.909091	4750	4318.18
2	0.826446	4000	3305.79
3	0.751315	2000	1502.63
4	0.683013	2000	1366.03
% decrease=	31.570572	NPV=	492.62

Solution (d)

Effect of a rise in the discount rate by 2%

Table 20.77

Return Rate = 0.120000		Machine A	
Year	PV Factor	Cash flow	PV
0	1.000000	−10,000	−10000.00
1	0.892857	5000	4464.29
2	0.797194	4000	3188.78
3	0.711780	2000	1423.56
4	0.635518	2000	1271.04
% decrease	51.707476	NPV=	347.66

Sensitivity of IRR

■ Example 42

Given the following IRR table for Machine A conduct sensitivity analysis for

a) a rise in the cost of the project by 5%

b) a rise in the first year cash flow by 10%

Table 20.78 Original data

IRR =	0.1398785	Machine A	
Year	PV Factor	Cash flow	PV
0	1.0000000	−10,000	−10000.00
1	0.8772865	5000	4386.43
2	0.7696316	4000	3078.53
3	0.6751874	2000	1350.37
4	0.5923328	2000	1184.67
		NPV=	0.00

Solution (a)

Effect of a 5% rise in the cost of the project on IRR

Table 20.79

IRR = 0.1116617		Machine A	
Year	PV Factor	Cash flow	PV
0	1.0000000	−10,500	−10500.00
1	0.8995543	5000	4497.77
2	0.8091979	4000	3236.79
3	0.7279174	2000	1455.83
4	0.6548012	2000	1309.60
% Decrease=	20.1723796	NPV=	0.00

IRR comes down to 11.16617%.

Solution (b)

Effect of a 10% rise in the first year cash flow on IRR

Table 20.80

IRR = 0.1116617		Machine A	
Year	PV Factor	Cash flow	PV
0	1.0000000	−10,500	−10500.00
1	0.8995543	5000	4497.77
2	0.8091979	4000	3236.79
3	0.7279174	2000	1455.83
4	0.6548012	2000	1309.60
% Decrease=	20.1723796	NPV=	0.00

IRR goes up to 16.6173%.

20.27 BREAK-EVEN ANALYSIS

Till now we have considered the importance of time value of money in capital budgeting. We were making invest or not to invest decisions on the basis of the arrived NPV. In this section we plan to relate these cash flows otherwise called the revenue to the firm's cost and profit. The relationship between these three variables is often called the cost-volume-profit (CVP).

Definition of Break-Even Point

The breakeven point refers the output level at which the cost of production is equal to the revenue. In other words this is to say that the break-even point refers to the output level at which the form's profit is zero. The following diagram illustrates the break-even output graphically. In it we present both revenue and cost functions. Accordingly, the point of intersection of the revenue and the cost function yields the needed break-even output level as shown in the graph. In the graph the break-even output is 5 units. The corresponding break-even revenue and hence the cost is equal to Rs. 150. More than this break- even output of 5 units the firm will earn positive profit as indicated. Similarly, below this break-even output the firm will incur a loss as indicated.

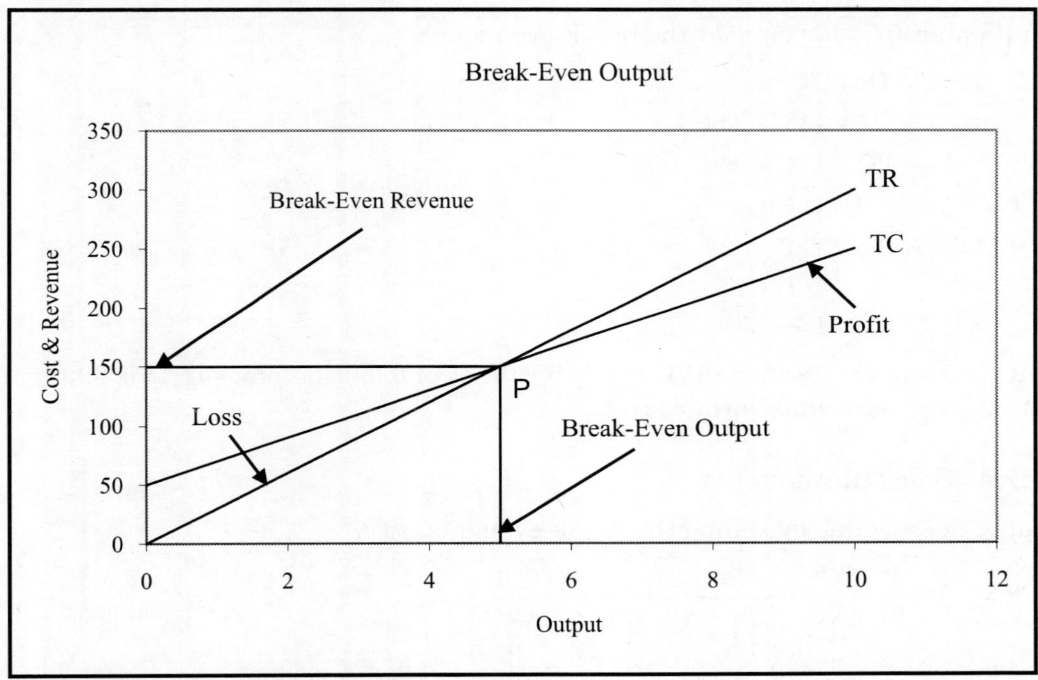

Figure 20.5

■ Example 43

If C = 50 + 20Q and R = 30Q are the cost and the revenue functions obtain the break-even point (break-even output)

Solution

By definition at break-even output

$$\begin{aligned} \text{Revenue} &= \text{Cost} \\ 30Q &= 50 + 20Q \\ 30Q - 20Q &= 50 \\ 10Q &= 50 \\ Q &= 50/10 = 5 \end{aligned}$$

So the needed break-even will occur at 5 units level of production. Clearly below the level of output the firm will incur loss and above this output level the firm will earn a positive profit.

Algebra of break-even point

By definition profit = Revenue at the break-even point.

$$\begin{aligned} TR &= TC \\ &= TVC + TFC \\ PQ &= TFC + AVC \times Q \\ PQ - AVC \times Q &= TFC \\ Q(P - AVC) &= TFC \\ Q &= \frac{TFC}{P - AVC} \end{aligned}$$

In this formula the term P–AVC is often called contribution margin. It is simply the per unit revenue – per unit variable cost.

Break-Even Sales (Revenue)

It is simply the revenue by selling the break-even quantity.

$$R = PQ = \frac{P(TFC)}{P - AVC} = \frac{P(TFC)}{P\left(1 - \frac{AVC}{P}\right)} = \frac{(TFC)}{\left(1 - \frac{AVC}{P}\right)}$$

Note: *Since the cost is equal to the revenue at this point the break-even sale revenue must invariably be equal to the break even total cost.*

■ Example 44

For example 43 obtain the break-even output by using the algebraic formula given above. Also obtain the break-even sales revenue.

Solution

$$\text{Break-even output} = Q = \frac{TFC}{(P - AVC)} = \frac{50}{30 - 20} = \frac{50}{10} = 5$$

$$\text{Break-even sales } = R = \frac{(TFC)}{\left(1 - \frac{AVC}{P}\right)} = \frac{50}{\left(1 - \frac{20}{30}\right)} = \frac{50}{10/30} = \frac{50 \times 30}{10} = \text{Rs. } 150$$

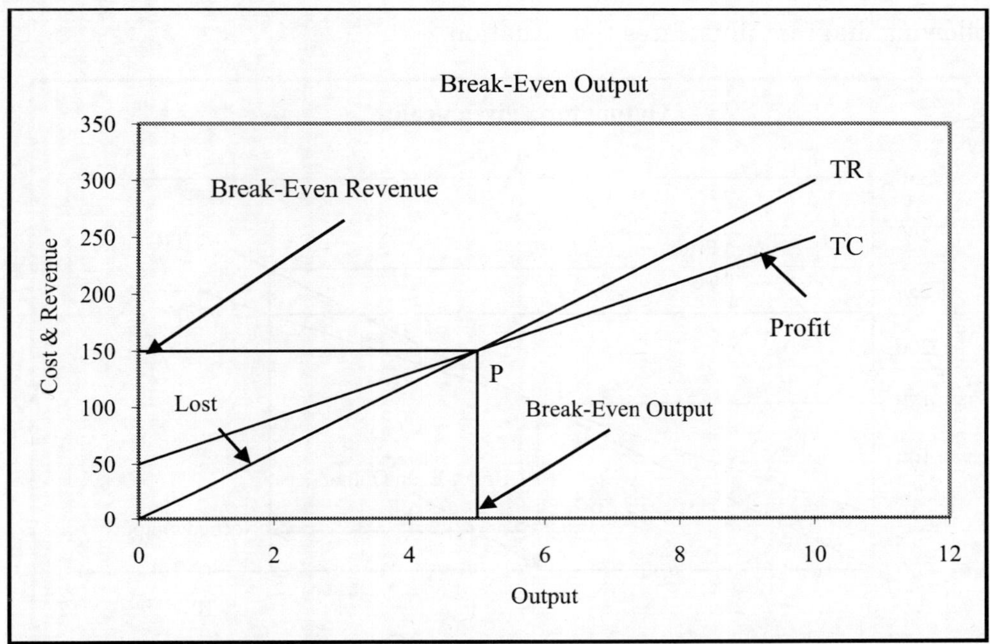

Figure 20.6

Note: The total cost at this point C = 50+20Q = 50+20×5 =150

Output calculation for a given level of profit

By definition Profit = Revenue – Cost.

$$\begin{aligned}\pi &= TR - TC\\ &= PQ - (TFC + TVC)\\ &= PQ - TFC - TVE\\ &= PQ - TFC - AVC \times Q\end{aligned}$$

$$(P - AVC)Q = \pi + TFC$$

$$Q = \frac{\pi + TFC}{(P\text{-}AVC)}$$

Example 45

If C = 50 +20Q and R= 30Q obtain the level of output for a desired level of profit of Rs. 40

$$Q = \frac{\pi + TFC}{(P - AVC)} = \frac{40 + 50}{30.20} = \frac{90}{10} = 9$$

The following diagram illustrates the situation.

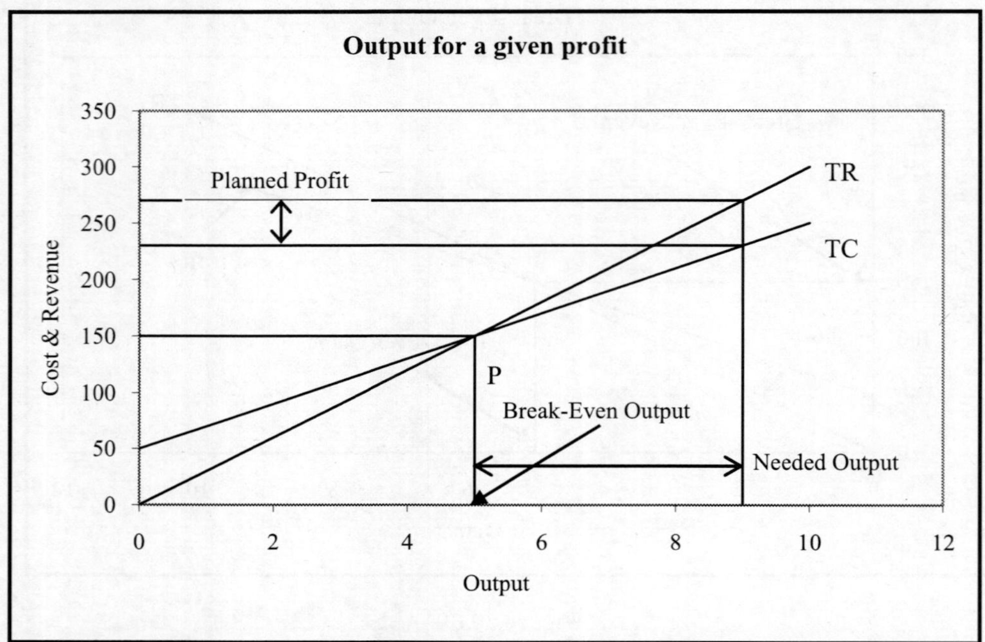

Figure 20.7

Example 46

If C = 50 +20Q and R= 30Q obtain the level of output for a desired level of zero profit. (Break-even point)

Solution

$$Q = \frac{0 + TFC}{(P - AVC)} = \frac{0 + 50}{30.20} = \frac{50}{10} = 5$$

Example 47

ABC Electronics is engaged in producing a certain component which is sold at a uniform price of Rs. 8 per unit. The variable cost of producing the component is Rs. 4.8 per unit, while the total fixed cost is Rs. 24,000.

(i) How many units of the company must be produce so that the company's break-even is achieved?

(ii) How much sales revenue would be made at this level of the activity?

(iii) If the firm wants to have Rs. 16,000 profit obtains the level of output and the sales revenue.

Solution

Data given: TFC = Rs. 24,000 AVC = Rs. 4.80 per unit. Selling price per unit P = Rs. 8

Solution (i)

Break-even Quantity

$$Q = \frac{TFC}{(P - AVC)}$$

$$= \frac{24{,}000}{8 - 4.8} = \frac{24{,}000}{3.2}$$

$$= 7500 \text{ units}$$

Solution (ii)

Break-even sales

$$R = P \times Q = \frac{P(TFC)}{(P - AVC))}$$

$$= \frac{PTFC}{P\left(1 - \frac{AVC}{P}\right)} = \frac{TFC}{\left(1 - \frac{AVC}{P}\right)}$$

$$= \frac{24{,}000}{\left(1 - \frac{4.8}{8}\right)} = \frac{24{,}000}{0.4} = \text{Rs.}60{,}000$$

Alternatively R = P ×Q = 8×7500 = Rs. 60,000

Solution (iii)

$$Q = \frac{\pi + TFC}{(P - AVC)} = \frac{16{,}000 + 24{,}000}{8 - 4.8} = \frac{30{,}000}{3.2} = 12{,}500 \text{ unit}$$

Margin of Safety

The margin of safety is simply the excess of actual output over the break-even output. In rupee terms this is simply the excess of sales revenue over the break-even sales revenue.

$$\text{Marigin of safety ratio} = \frac{\text{Actual Sales} - \text{Break-even Sales}}{\text{Actual sales}} \times 100$$

In the Figure 20.7 the margin of safety in terms of output = 8–5 = 3
In terms of revenue = 240 –150 = Rs. 90

$$\text{Marigin of safety ratio} = \frac{8 - 5}{8} \times 100 = \frac{3}{8} \times 100 = 37.5\%$$

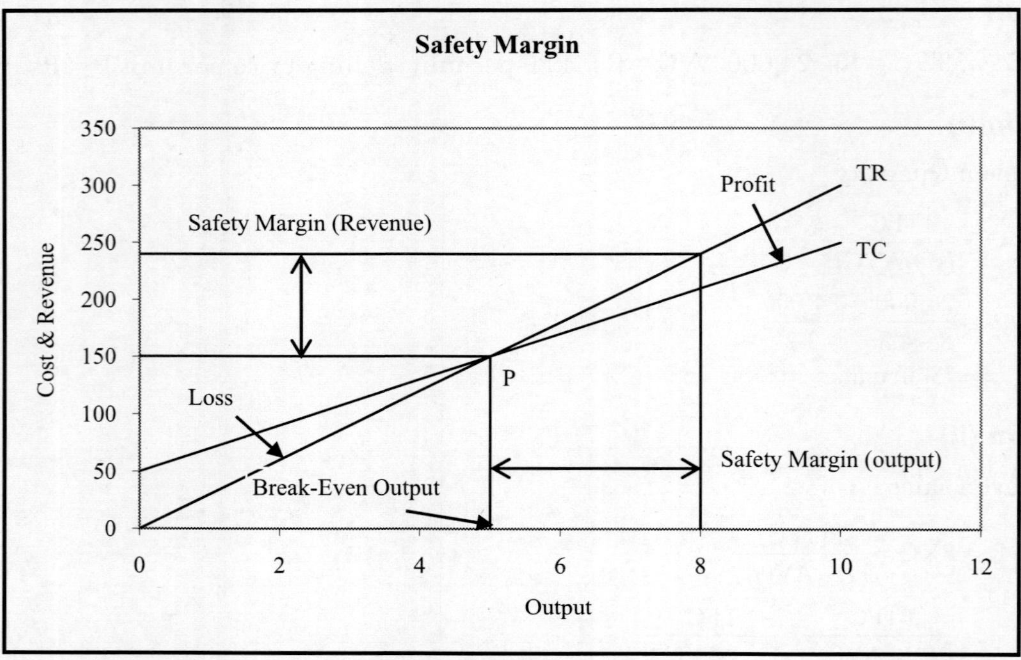

Figure 20.8

Volume Profit Graph

In the previous graph we obtained the required sales for a targeted profit by using the total revenue and cost functions. The same information is easily obtainable from the volume profit graph shown below. In it the profit is obtained as a function of the output as shown below.

$$\pi = R - C = 30Q - (50 + 20Q) = 10Q - 50$$

The break-even output in this graph is simply the intersection of this graph on the X-axis at which the profit is zero.

So from the above equation

$$0 = 10Q - 50$$

$$0 + 50 = 10Q$$

$$Q = 50/10 = 5$$

If Rs. 40 is the desired level of profit then the needed output is obtained from the above equation

$$40 = 10Q - 50$$

$$90 = 10Q$$

$$Q = 90/10 = 9$$

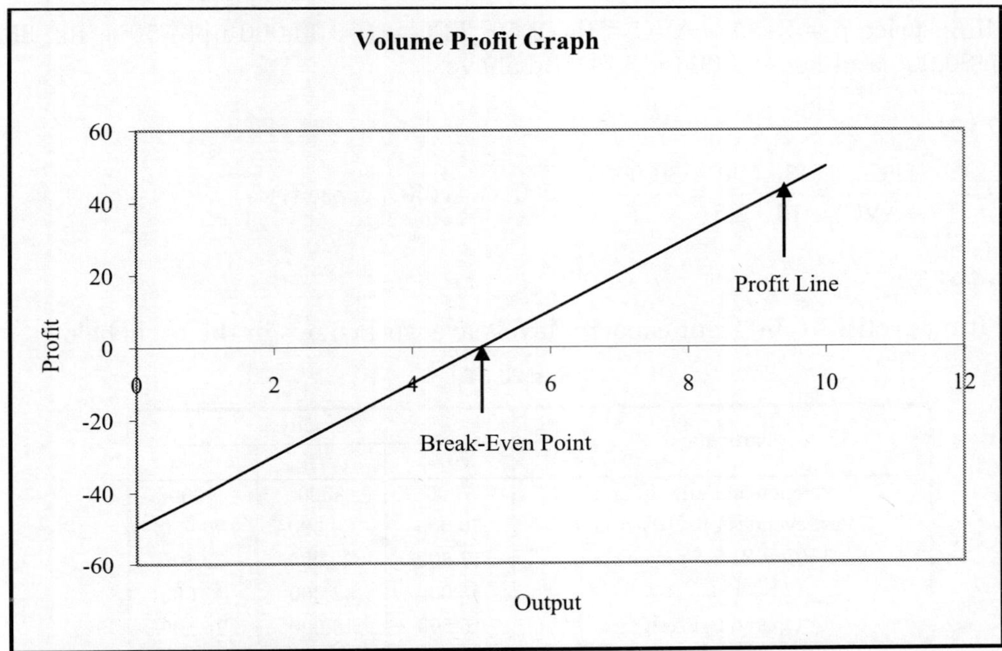

Figure 20.9

Example 48

A firm manufactures a uniform product whose selling price is Rs. 10. It has a capacity of 10,000 units. The variable costs are Rs. 2.50 per unit. The fixed costs are estimated as Rs. 30,000 up to 50% capacity utilization and Rs. 36,000 above that level, and Rs. 42,000 if the utilization is 80% or above.

a) What is the break-even point?

b) What will be the operating profit of the firm at 70%, 80% and 90% utilization?

c) The firm manufactures and sells only 6000 units in a year due to market conditions. What is the margin of safety?

d) What is the lowest level of activity at which the firm can make a profit of Rs. 18,000?

e) Management of the firm is advised that if the selling price is reduced to Rs. 9 per unit, sales will go up to 7500 units, is it worthwhile reducing the price?

f) The management of the firm is also advised that the sales of 7500 units at the present selling price can be achieved if a sum of Rs. 5000 is spent on advertisement.

g) Which alternative is more profitable—reducing the selling price or spending on advertisement?

Data: selling price p = Rs. 10, AVC = Rs. 2.50, TFC = Rs. 30,000 upto 50%, Rs. 36,000 below Rs. 80,000 and Rs. 42,000 for 80% and above.

Solution (a)

$$Q = \frac{TFC}{(P - AVC)} = \frac{30,000}{10 - 2.5} = \frac{30,000}{7.5} = 4000 \text{ units (40\% capacity)}$$

Solution (b)

The operating profits at different capacity levels are worked out in the table below.

Table 20.81

Particulars	Capacity		
	70%	80%	90%
Production and sales in units	7000	8000	9000
Sales revenue at Rs.10 per unit	70,000	80,000	90,000
TVC at Rs.2.5 per unit	17,500	20,000	22,500
TFC	36,000	42,000	42,000
Operating Profit	16,500	18,000	25,500

Solution (c)

Margin of safety = Actual sales revenue – Break-even sales revenue.
= 6,000 × 10 – 4000×10 = Rs. 20,000

Solution (d)

For a desired profit of Rs.18,000 the required sales volume is obtained by using the formula:

$$Q = \frac{\pi + TFC}{(P - AVC)} = \frac{18,000 + 30,000}{10 - 2.5} = \frac{48,000}{7.5} = 6400 \text{ units}$$

Since this is 64% capacity in our calculation the total fixed cost is to be revised from the current level of Rs. 30,000 to the Rs. 36,000 as stated in the problem. Thus the level of needed output is obtained as:

$$Q = \frac{\pi + TFC}{(P - AVC)} = \frac{18,000 + 36,000}{10 - 2.5} = \frac{54,000}{7.5} = 7200 \text{ units}$$

Clearly this time the capacity of 72% is well within total fixed cost limit of Rs. 36,000.

Solution (e)

When the selling price is reduced from Rs.10 to Rs.9 the resulting profit for the given 7500 units of output is obtained as

$$Q = \frac{\pi + TFC}{(P - AVC)} = \frac{\pi + 36,000}{9 - 2.5} = \frac{\pi + 36,000}{6.5} = 75,000 \text{ units}$$

$$\pi = 6.5 \times 75,000 - 36,000 = \text{Rs.}12,750$$

Solution (f)

If the firm opts for incurring an advertisement expenditure of Rs. 5000 then go for price reduction then the profit for 7500 unit of output is obtained as

$$Q = \frac{\pi + TFC + \text{ advertisement cost}}{(P - AVC)} = \frac{\pi + 36,000 + 5000}{10 - 2.5} = \frac{\pi + 41,000}{7.5} = 75,000 \text{ units}$$

$$\pi = 7.5 \times 75,000 - 41,000 = \text{Rs. } 15,250.$$

Solution (g)

Clearly Rs.15,250 > Rs.12,750 the firm can go for advertisement rather than price reduction.

Operating leverage

A firm is said to be highly leveraged provided its fixed costs are large relative to variable cost. A highly leveraged firm will have better profitability maneuver by changing its output than a comparatively lower leveraged firm. The degree of maneuverability otherwise called the degree of operational leverage (DOL) is simply the output elasticity of the profit.

$$DOL = \frac{E\pi}{EQ} = \frac{Q}{\pi}\frac{\Delta\pi}{\Delta Q}$$

Now by definition

$$\pi = \text{R-C} = \text{PQ-TFC-TVC}$$

$$\pi = \text{PQ-TFC-(AVC} \times \text{Q)}$$

$$\Delta\pi = P\Delta Q - AVC \times \Delta Q$$

$$= (\text{P-AVC})\Delta Q$$

$$DOL = \frac{Q}{\pi}\frac{(P - AVC)\Delta Q}{\Delta Q}$$

$$= \frac{Q(P - AVC)\Delta Q}{[PQ - (AVC \times Q) - TFC]\Delta Q}$$

$$= \frac{Q(P - AVC)}{Q(P - AVC) - TFC}$$

To illustrate the importance of operational leverage let us consider two forms with different TFC but same TVC.

Firm A: $C = 50+20Q$; $R = 30Q$

Firm B: C′ = 70 +20Q; R = 30Q

$$DOL_A = \frac{Q(P - AVC)}{Q(P - AVC) - TFC}$$

$$= \frac{8(30 - 20)}{8(30 - 20) - 50} = \frac{80}{30} = 2.667$$

$$DOL_B = \frac{Q(P - AVC)}{Q(P - AVC) - TFC}$$

$$= \frac{8(30 - 20)}{8(30 - 20) - 70} = \frac{80}{10} = 8$$

Clearly $DOL_B > DOL_A$

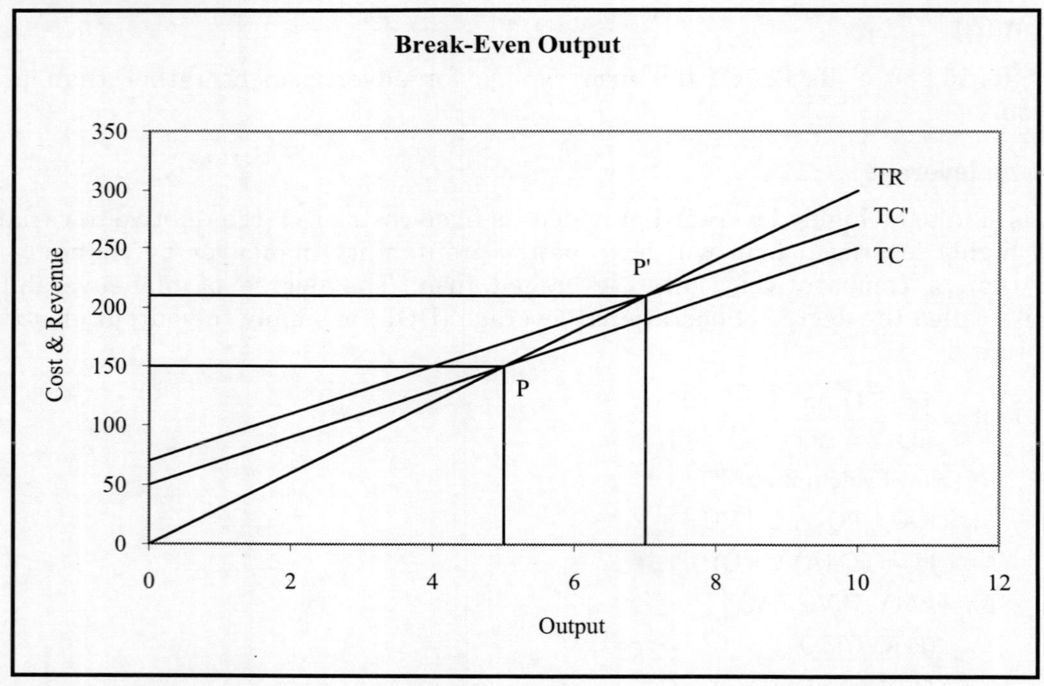

Figure 20.10

Sensitivity analysis

As usual the sensitivity analysis is a technique for examining the changes in the outcome when the data input of the model like the selling price, fixed cost, variable cost etc., change. In general an increase in the selling price will increase the contribution margin and thereby lower the break-even point. An increase in the variable cost will reduce the contribution margin and hence will increase the break-even volume. An increase in the fixed cost will increase the break-even level. If all the three input data levels are changed then the effect on the break-even sales is obtained from the formula:

$$\text{Break-Even Sales (new)} = \frac{TFC \pm \Delta TFC}{1 - \dfrac{AVC \pm \Delta AVC}{P \pm \Delta P}}$$

EXERCISES

1. "In financial decision making, time is of prominent importance". Discuss with suitable illustration.
2. Distinguish between various types of annuity.
3. What is perpetuity? How do you determine the present value of perpetuity?
4. Discuss the importance of discounted cash flow in investment analysis.
5. Distinguish between NPV and IRR methods of investment decision making.
6. "The NPV approach gives different decision rules at different discount rates". Elucidate this statement with a suitable illustration.
7. "NPV and IRR often lead to conflicting decision rules". Explain this statement with suitable illustration.
8. Discuss the utility certainty equivalent and risk adjusted discount rate methods in capital budgeting.
9. Explain with illustration the statistical distribution approach to evaluate risky proposals.
10. What is EVPI of an investment proposal?
11. Discuss the relevance of decision tree in deciding risky investment proposals.
12. What do you understand by sensitivity analysis in the context of investment decision making?
13. What do you mean by break-even analysis?
14. What is margin of safety? How it is measured?
15. Briefly discuss the importance of operation leverage in business decision making.
16. A freehold estate is bought for Rs. 40,000. At what rent should it be let so that the owner receives 5% per annum compounded annually, on his investment?

17. A man borrows Rs. 5000 at 5% rate of interest compounded annually. If the principal and the interest are to be repaid in 10 equal instalments, find the amount of each instalment.

18. An investor has a capital of Rs. 20,000 on which he earns interest at 5% p.a. if he spends Rs. 1800 per year, show that he will be ruined before the end of 17 years.

19. For a project the relevant data on investment and projected cash flows are presented here:

Year	Mean cash flow	Standard deviation of cash flow distribution
0	−40,000	7000
1	30,000	4000
2	30,000	3000

Assume that the cash flows are independently and normally distributed and the discount rate is 0.20. Compute:

a) The probability of the investment having a net present value of zero or less.

b) The expected value of the loss.

Area under the normal curve from Z = 0 to γ

Z:	0.6800	0.6900	0.7000	0.7100	0.7200	0.7300	0.7400	0.7500
Area:	0.2518	0.2549	0.2580	0.2611	0.2642	0.2673	0.2704	0.2734

Unit normal loss integral

Z:	0.6800	0.6900	0.7000	0.7100	0.7200	0.7300	0.7400	0.7500
Loss integral	0.1478	0.1453	9.0000	0.1405	0.1381	0.1358	0.1334	0.1312

(CA, November, 1986)

20. A company has invested Rs. 300 in a new machinery with expected cash flow over two years as follows:

Year 1		Year 2	
Cash flow	Probability	Cash flow	Probability
Rs.200	0.4	Rs.100	0.7
		Rs.200	0.3
Rs.300	0.6	Rs.200	0.4
		Rs.300	0.6

Find the expected NPV and its SD, assuming the cost of capital as 20%

(CA, May, 1987)

21. A housewife is looking at ways of producing domestic hot water and considers two possibilities—an electrical immersion heater having an instillation cost of Rs. 160 and estimated annual electrical charge of Rs. 200, and a gas boiler with an installation cost of Rs. 760 with annual fuel fill of Rs. 80. Assuming yourself as a consultant to this cost conscious housewife, advise her suitably by comparing the two systems, one the basis of (i) total expenditure, and (ii) present value, over five years period. Take interest at 9%. What will be your recommendation if you consider the equipment for a 6 year period?

(CA, May,1993)

22. Anial Corporation provides a multitude of benefits, including a company funded pension plan. Biren is due to retire in five-years and will receive annual pension cheques starting one-year from his date of retirement. Each yearly pension cheque will be equal one-half of his present annual salary of Rs. 12,000. The company will make annual deposits to a Pension Fund Trust during the next five-years. Assume that the pension fund assets are expected to earn 10% annually and that Biren is expected to collect payments for 10 years after retirement. What should be the annual contribution of the corporation to the pension fund?

(CA, November, 1977)

23. A firm is considering the introduction of a new product which will have a life of 5 years. Two alternative methods of promoting the product have been identified.

Alternative I: This will involve employing a large number of agents. An immediate expenditure of Rs. 5,00,000 will be required to advertise the product. This will produce net cash outflows of Rs. 3,00,000 at the end of each of the subsequent five-years. However the agents will have to be paid Rs. 50,000 each year. On termination of the contract, the agent will have to be paid a lump sum Rs. 1,00,000 at end of the fifth year.

Alternative II: Under this alternative, the firm will not employ agents but will sell directly to the consumers. The initial expenditure on advertising will be Rs. 2,50,000. This will bring in cash inflows at the end of each year of Rs. 1,50,000. However, this alternative will involve out-of-pocket costs for sales administration to the extent of Rs. 50,000 the firm also proposes to allocate fixed cost worth Rs. 20,000 per year to this product if this alternative is pursued.

(a) Advise the management as to the method of promotion to be adopted.

(b) Calculate the internal rate of return for alternative 2. You may assume that the firm's cost of capital is 20% p.a

(CA, May, 1986)

CHAPTER

21

Dynamic Programming

- *Introduction*
- *Dynamic programming versus linear programming*
- *Application of dynamic programming in travelling networking*
- *Application of dynamic programming in distribution networking*
- *Exercises*

21.1 INTRODUCTION

Often decision making process involves several decisions to be taken at different points of time. The mathematical technique optimizing a sequence of interrelated decisions over a period of time is called dynamic programming. In such problems the optimum decision is not available in one go as in the case of linear programming. Instead we break the major problem into a series of sub-problem and attempt to solve them backward from the very terminal sub-problem. The following four important steps are to be followed in solving the given dynamic programming problem.

1. Divide the given original problem into sub-problems, normally called stages.
2. Solve the last stage of the problem for all possible states.
3. Moving backward and solve intermediate problems successively. Remember, while doing so we decide the optimum policies up to the stage concerned, starting from the end of the problem.
4. Obtain the complete solution by solving all the stages backward.

21.2 DYNAMIC PROGRAMMING VERSUS LINEAR PROGRAMMING

In linear programming we convert the given problem into a standard mathematical format by incorporating surplus, slack and artificial variables in suitable places before using Simplex algorithm to obtain solution. However, in dynamic programming we do not have any standard mathematical formulation. Also there is no well defined algorithm as well. Instead we divide the given problem into several sub-problems and start solving from the last sub-problem. In simple linear programming we consider the entire problem as single entity and arrive at its solution by one go. In dynamic programming the final optimum decision is achieved by considering all the optimum solutions obtained stage-wise. Thus multistage approach is the key factor in dynamic programming. Richard Bellman developed this technique in early 1950s. The following example illustrates one such dynamic programming.

21.3 APPLICATION OF DYNAMIC PROGRAMMING IN TRAVELLING NETWORKING

■ Example 1

A traveller wants to travel from city 1 to city 11. There is no direct bus route connecting these two cities. He will have to change the bus at three stages in-between. At each break, options are available to the traveller as shown in the network. Obtain the shortest distance route.

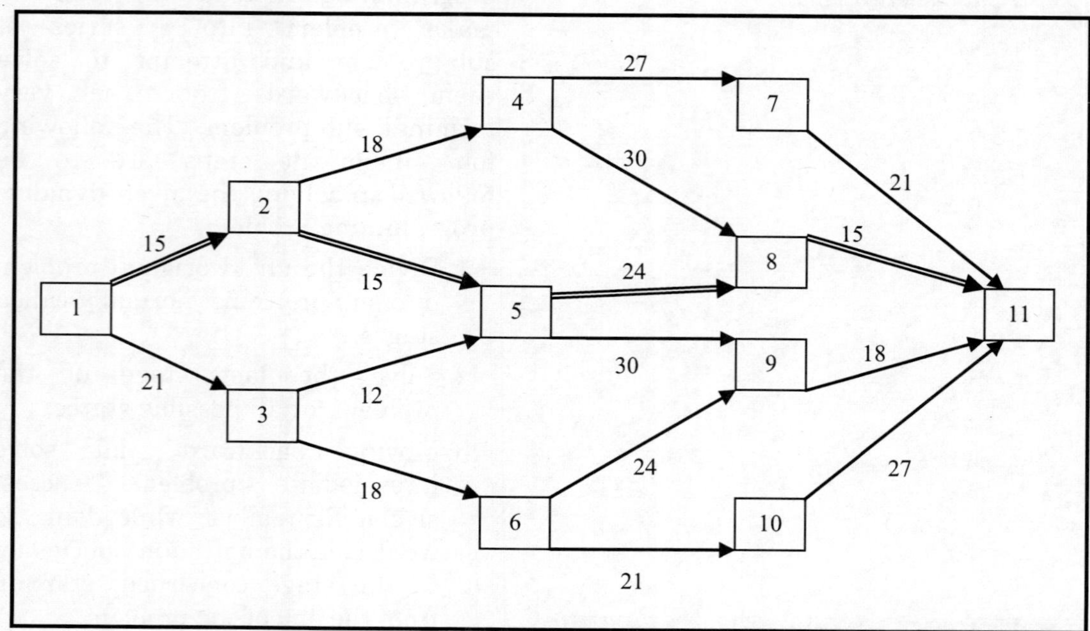

Figure 21.1

1) Solution by complete enumeration

The following table gives the summary of all possible routes and their associated distances. From the Table 21.1, it is clear that the route 1-2-5-8-11 is the least distance yielding route. Note that this type of complete enumeration gives the needed minimum time consuming route provided the problem is simple. In complicated problems with plenty of route possibilities the complete enumeration is not only time consuming but also there is possibility of missing some routes while presenting. In dynamic programming however we approach the problem in a systematic way and obtain the correct result.

Table 21.1

Route	Travel time
1-2-4-7-11	81
1-2-4-8-11	78
1-2-5-8-11	69
1-2-5-9-11	78
1-3-5-8-11	72
1-3-5-9-11	81
1-3-6-9-11	81
1-3-6-11-11	87

2) Solution by dynamic programming tabular method

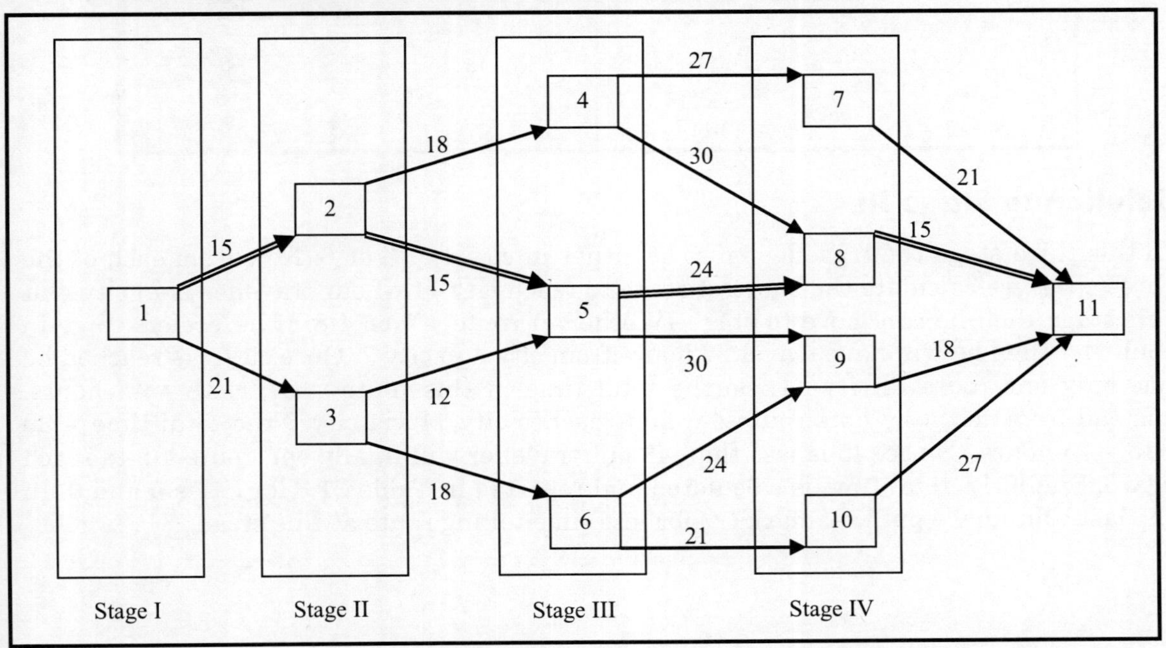

Figure 21.2

As first step we decompose the given problem into smaller sub-problem called stages. A stage is simply the decision making point of our traveller. In stage one our traveller will have to decide whether to go to destination 1 or 2. If, for example he decides for destination 2 then again he will have to choose either destination 5 or 4. This is the second stage of decision making. If chooses for example 4, then at this third stage he will have to take a further decision as whether to go to destination 7 of 8. Remember, once 7 or 8 is reached 11 is reachable only in one route hence no decision is needed at this stage. Thus, altogether our traveller will have to make four decisions to reach the last destination 11. Now let us solve the problem by roll back method from the last stage. To do this task we split the problems into four sub-problems as shown in Figure 21.2.

Solution to Stage IV

In this stage our traveller must by either city 7 or 8 or 9 or 10. Since from these cities only one route is available the traveller will have to travel in the specified route only to city 11 no matter how he reaches the concerned current city. Table 21.2 illustrates the solution values at this stage.

Table 21.2

State	Alternative routes	Travel time	Best route time
7	7-11	21	21
8	8-11	15	15
9	9-11	18	18
10	10-11	27	27

Solution to Stage III

In this third stage the traveller must be either in city 4 or 5 or 6. Now from each of the three cities we calculate the shortest route to reach city 11. If our traveller is in city 4 at this stage then he can move to stage IV ether in route 4-7 or 4-8. If he prefers to go in route 4-7 then he will have travel 27 hours from city 4 to city 7. Once city 7 is reached he has only one route to city 11. So the total time in this route = 27 + 21 = 48 hours. Instead from city 4 he opts route 4-8 and reaches city 11 via city 8 the total time = 30 +15 = 45 hours. Since, 45 is less than 48 our traveller will finally opt, route 4-8 to reach city 8. Similarly, total time is calculated and reported in Table 21.3 for cities 5 and 6. In the last column we put tick mark (✓) for less time taking route at this stage.

Table 21.3

State	Alternative routes	Time in hours			Best route
		To stage IV	From stage IV to city 11	Total	
4	4-7	27	21	48	
	4-8	30	15	45	✓
5	5-8	24	15	39	✓
	5-9	30	18	48	
6	6-9	24	18	42	✓
	6-8	21	27	48	

Solution to Stage II

At this stage clearly our traveller must be either in city 2 or 3. For example, if he is in city 2 then to move to the next third stage he has two routes. If he goes by route 2-4 then he will have to travel for 18 hours to reach city 4. From there onwards we have already worked out the less time consuming route in the previous stage. So the route 2-4-8-11 consumes 45 hours totally. So from second stage the total travel time works out to be 18 + 45 = 63. In Table 21.4 this calculation is reported in the very first row. Similarly, the total time via rout 2-5-8-11 is reported in the second row. The reaming calculations for city 3 are reported in row 3 and 4 in Table 21.4. In the last column, once again we mark the least time consuming route for each branch by tick mark.

Table 21.4

State	Alternative routes	Time in hours			Best route
		To stage 3	From stage 3 to city 11	Total	
2	2-4	18	45	63	
	2-5	15	39	54	✓
3	3-5	12	39	51	✓
	3-6	18	42	60	

Solution to Stage I

At this stage our traveller is in city 1. If he goes by route 1-2 to city 2 then he will have to travel 15 hours. From city 2 we know that least time consuming route is 2-5-8-11. So the total time in this 1-2-5-8-11 is = 15+54 = 69. Similarly, the time economy route is worked out from city 3 and reported in the second row in Table 21.5. Among the two routes clearly the route 1-2-5-8-11 will have the least travel time.

Table 21.5

State	Alternative routes	Time in hours			Best route
		To stage 2	From stage 2 to city 11	Total	
1	1-2	15	54	69	✓
	1-3	21	51	72	

3) Solution for dynamic programming by network diagram method

Alternatively solution to this problem is obtained by using the network diagram itself by applying the back pass method. But this time the back pass calculations are put in the respective nodal box itself as shown in Figure 21.3. In the following network diagram, we obtain the needed minimums at every stage starting from stage VII by back pass method. The final minimum is reported in the stage I node in the left side as 69 hours.

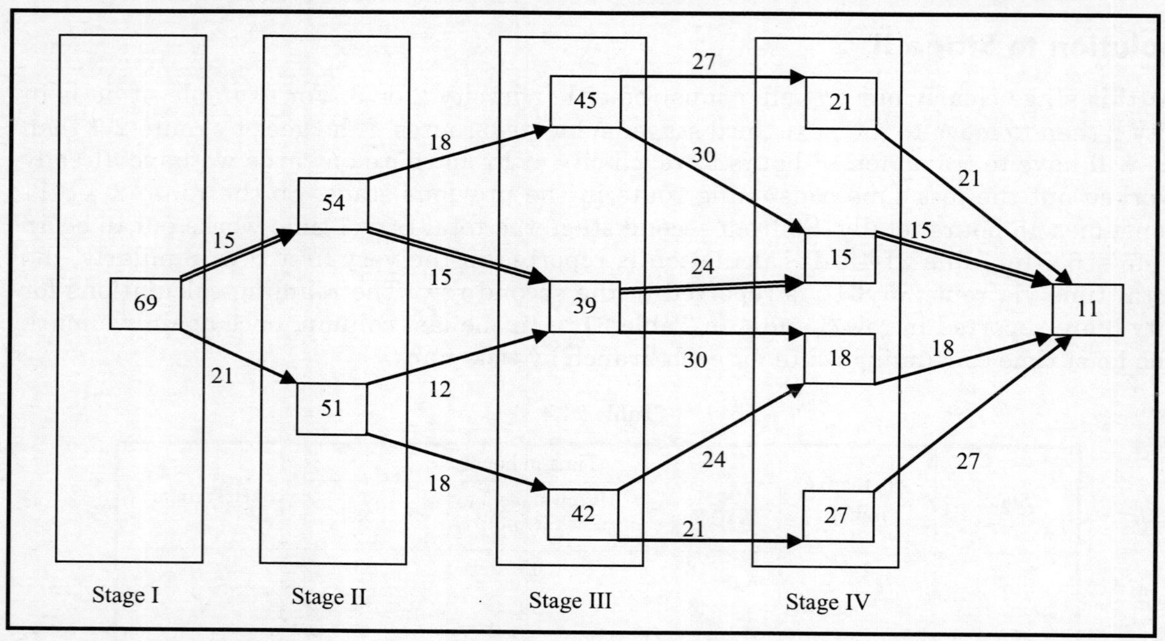

Figure 21.3

In node 7 enter 21 hours, the distance between node 7 and node 11

In node 8 enter 15 hours, the distance between node 8 and node 11

In node 9 enter 18 hours, the distance between node 9 and node 11

In node 10 enter 27 hours, the distance between node 10 and node 11

In node 4 enter 45, {Min (21+27, 15+30)}

In node 5 enter 39, {Min (24+15, 30+18)}

In node 6 enter 42, {Min (24+18, 21+27)}

In node 2 enter 54, {Min (18+45, 15+39)}
In node 3 enter 51, {Min (12+39, 18+42)}
In node 1 enter 69, {Min (15+54, 21+51)}
Thus, the minimum optimum time = 69 hours.

21.4 APPLICATION OF DYNAMIC PROGRAMMING IN DISTRIBUTION NETWORKING

■ Example 2

The Bharat Petroleum has recently found large deposits of oil in Karnataka. To develop this field, the company plans a transportation network from the oil storage in Karnataka to 7 shipping points located on the seashore. In the total transmission line, eight pumping

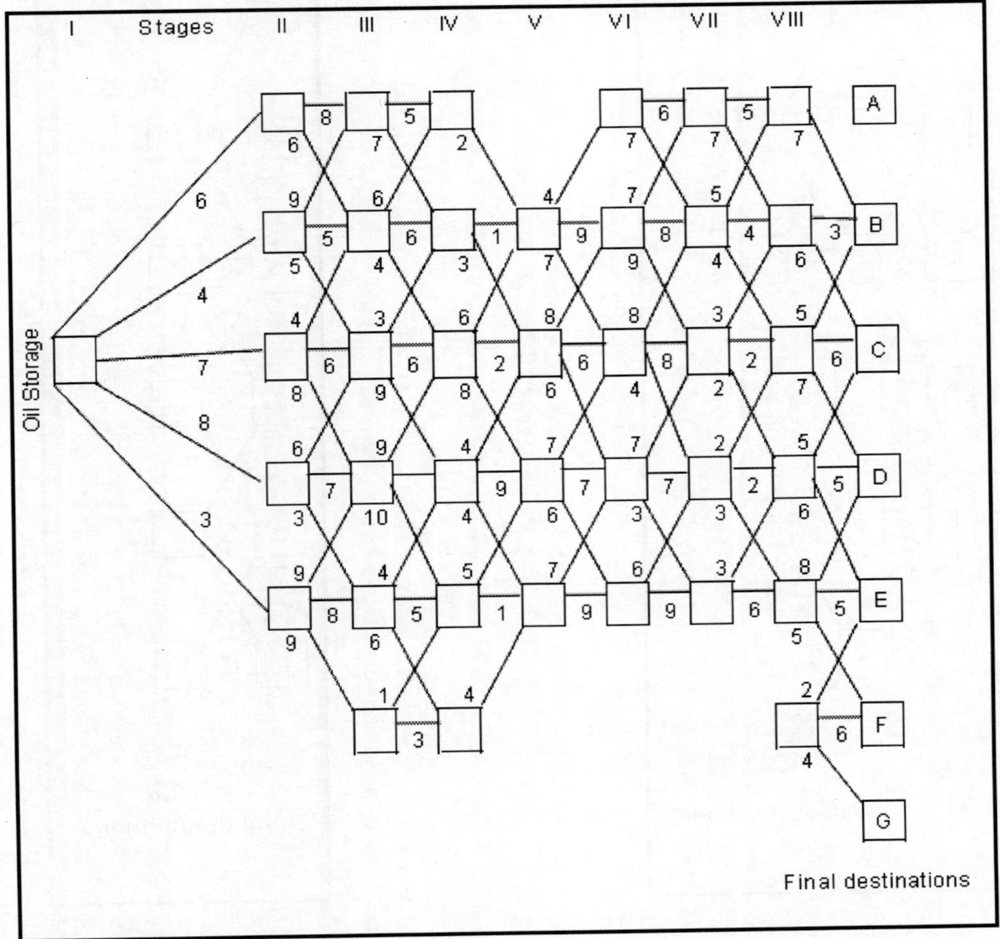

Figure 21.4

stations are to be built to pump the oil from production points to the shipping facilities. A number of sites are available for each sub-pumping stations, but not every site in one area can be reached from every site in the next area for good number of reasons. The following diagram represents the problem with accessible sites connected by arcs, labelled with the associated costs of constructing these arcs.

Solution

This time, workout this problem by using network diagram. The solution to this problem is obtained again by back pass method as described in the previous problem. The very first step is highlighted in the Figure 21.6. Proceed from this step the second backward pass is illustrated in the Figure 21.5.

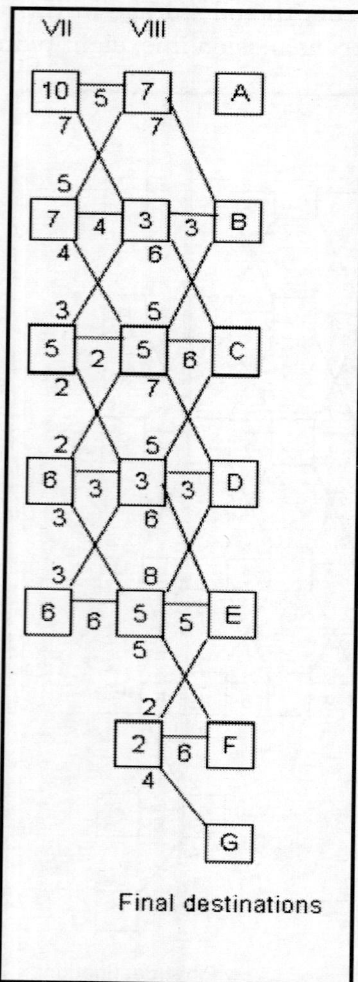

Figure 21.5

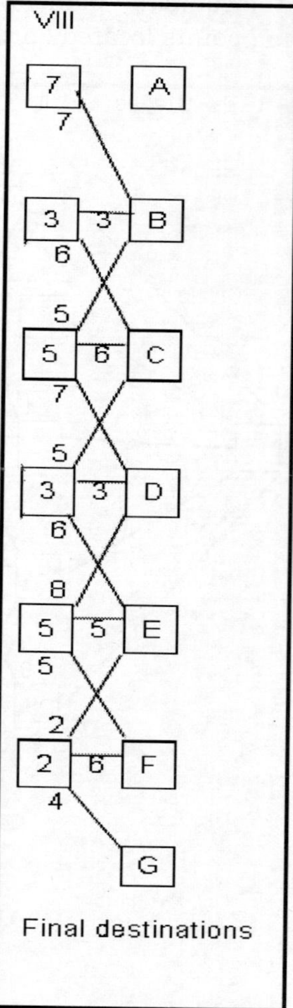

Figure 21.6

Following this procedure of back pass method we obtain the final result as shown in Figure 21.7.

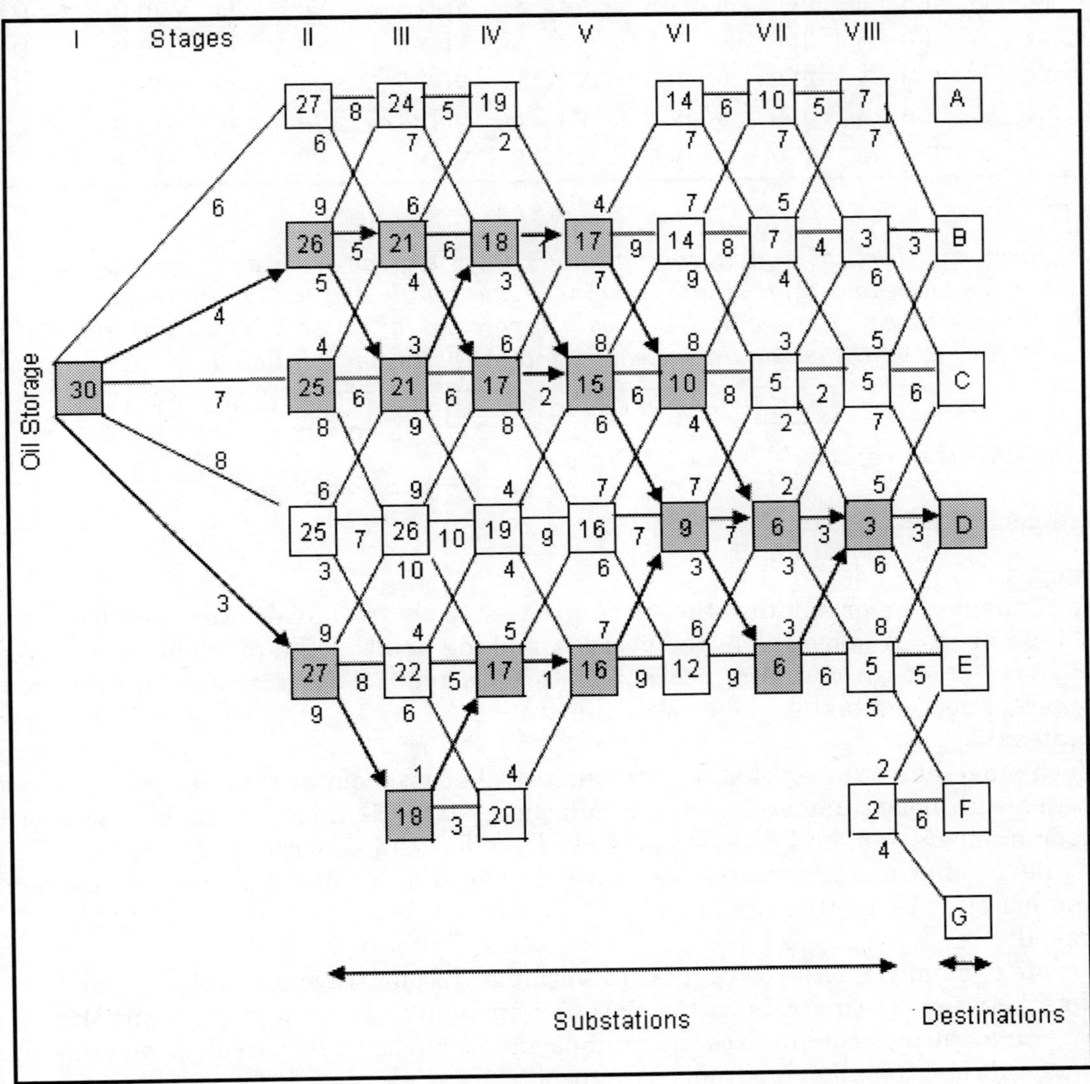

Figure 21.7

In the above figure seven possible solutions are being illustrated and all of them yield a minimum total cost of Rs. 30 Crore.

1)	I-C	to	IIE	to	III-F	to	VI-E	to	V-E	to	VI-D	to	VII-D	to	VIII-D	to	D
2)	I-C	to	IIE	to	III-F	to	VI-E	to	V-E	to	VI-D	to	VII-E	to	VIII-D	to	D
3)	I-C	to	II-B	to	III-C	to	IV-B	to	V-C	to	VI-D	to	VII-D	to	VIII-D	to	D
4)	I-C	to	II-B	to	III-C	to	IV-B	to	V-C	to	VI-D	to	VII-E	to	VIII-D	to	D
5)	I-C	to	II-B	to	III-C	to	IV-B	to	V-B	to	VI-C	to	VII-D	to	VIII-D	to	D
6)	I-C	to	II-B	to	III-B	to	IV-C	to	V-C	to	VI-D	to	VII-D	to	VIII-D	to	D
7)	I-C	to	II-B	to	III-B	to	IV-C	to	V-C	to	VI-D	to	VII-E	to	VIII-D	to	D

Figure 21.8

Note: *From any vertical region, the optimal policy is given from any pumping station with in that region to a final shipping point. For example, suppose the pumping station has been constructed up to node V-C. Now suppose we fail to obtain the building rights from node V-C to VI-D then again one could find the optimum solution as*

V-C to VI-C to VII-D to VIID

However, this time cost could be Rs. 31 Crore.

Mathematical Description:

1. **Stages:**
   In dynamic programming the very first step is to divide the problem into sub-problems. A stage refers the decision making point in the problem. Accordingly at every stage an optimum decision is to be made. In our traveller problem four different decision taking stages were illustrated.
2. **States:**
   Each stage has number of states. In our traveller problem at stage II there are two states with node number 2 and 3. Similarly at stage III there are there states with node numbers 4, 5, and 6. At stage IV we have four states with node 7, 8, 9, and 10. While moving along a route, the traveller moves from one state in one stage to another state in another stage.
3. **Payoff:**
   Payoff is simply refers to the reward variable. In our traveller problem the travel time between destinations is the payoff. Depending upon the type of problem we ether maximizes or minimizes the total payoff. In our traveller problem we minimize the total time from moving from city 1 to city 11.
4. **Principle of optimality:**
   Given the current state, the optimality policy at this stage depends only on the policy decision adopted in the preceding stages. In our traveller problem let the traveller is in state 5 in stage III. Now the decision at state 5 in stage III is based on the decision taken already in stage IV. How he reached state 5 in stage III is not our problem at this juncture.

5. **Recursive Relationship:**
   Formulation of dynamic programming requires creation of recursive relationship. Accordingly, we define optimality function for every stage and state. The defined recursive relationship identifies the optimality policy for stage n by using the optimality policy (n+1) taken already. In general the recursive relationship is formulated as follows.

N = Number of stages

n = index for current stage

n+1 = index for succeeding previous stage

S_n = state of the system in the current stage for which regressive relation holds good

S_{n+1} = state of the system in the previous stage

$f_n(S_n)$ = total payoff for each alternative starting from S_n in stage n to the end of the process

$f_n^*(S_n)$ = optimal total payoff obtained in the previous stage

$C_n(S_n, x_n)$ = immediate payoff in state n when decision x_n is made for the specific value S_n of the stage.

x_n = decision among alternatives made at stage n in the state under consideration

The recursive relation for state S_n at n may be defined for minimization as

$$f_n^*(S_n) = \text{Min}\left[C_n(S_n, x_n) + f_{n+1}^*(S_{n+1})\right]$$

The recursive relation for state S_n at n may be defined for maximization as

$$f_n^*(S_n) = \text{Max}\left[C_n(S_n, x_n) + f_{n+1}^*(S_{n+1})\right]$$

Example 3

A traveller wants to travel from city A to city B. There is no direct bus route connecting these two cities. He will have to catch and change the bus at three stages in-between. At each break many options are available to the traveller as shown in the network. Obtain the shortest time consuming route.

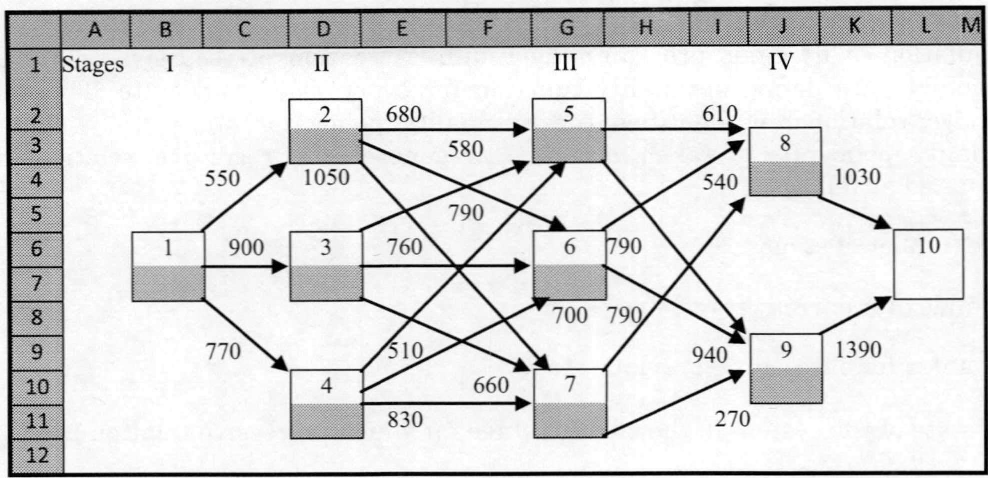

Figure 21.9

Solution by backward working method:

Stage 4 computations

Since there is only one path from each stage 4 city we straightaway write the distances as follows

$$f_4(8) = 1030$$

$$f_4(9) = 1390$$

So the shortest path from city 8 to city 10 =1390 kilometers

Stage 3 computations

By using stage 4 we work backward and find the shortest routes for stage 3 cities as follows. Remember, in stage 3 we have three cities denoted by node 5, 6 and 7.

$$f_3(5) = \min\left[\{c_{58} + f_4(8)\}, \{c_{59} + f_4(9)\}\right]$$

$$= \min\left[\{610 + 1030\}, \{790 + 1390\}\right]$$

$$= \min\left[1640, 2180\right] = 1640 \text{ km}$$

$$f_3(6) = \min\left[\{c_{68} + f_4(8)\}, \{c_{69} + f_4(9)\}\right]$$

$$= \min\left[\{540 + 1030\}, \{940 + 1390\}\right]$$

$$= \min\left[1570, 2330\right] = 1570 \text{ km}$$

$$f_3(7) = f_3(7) = \min\left[\{c_{78} + f_4(8)\}, \{c_{79} + f_4(9)\}\right]$$
$$= \min\left[\{790 + 1030\}, \{270 + 1390\}\right]$$
$$= \min[1829, 1660] = 1660 \text{ km}$$

By using the stage 3 calculations we work out the backward calculations for stage II as follows.

Stage 2 computations

$$f_2(2) = \min\left[\{c_{25} + f_3(5)\}, \{c_{26} + f_3(6)\}, \{c_{27} + f_3(7)\}\right],$$
$$= \min\left[\{680 + 1640\}, \{790 + 1570,\}, \{1050 + 1660\}\right]$$
$$= \min[2320, 2360, 2710] = 2320 \text{ km}$$
$$f_2(3) = \min\left[\{c_{325} + f_3(5)\}, \{c_{36} + f_3(6)\}, \{c_{37} + f_3(7)\}\right],$$
$$= \min\left[\{580 + 1640\}, \{760 + 1570,\}, \{660 + 1660\}\right]$$
$$= \min[2220, 2330, 2320] = 2220 \text{ km}$$
$$f_2(4) = \min\left[\{c_{45} + f_3(5)\}, \{c_{46} + f_3(6)\}, \{c_{47} + f_3(7)\}\right],$$
$$= \min\left[\{510 + 1640\}, \{700 + 1570,\}, \{830 + 1660\}\right]$$
$$= \min[2150, 2270, 2490] = 2150 \text{ km}$$

By using the stage 2 calculations we work out the backward calculations for stage I as follows.

Stage 1 computations

$$f_1(1) = \min\left[\{c_{12} + f_2(2)\}, \{c_{13} + f_2(3)\}, \{c_{14} + f_2(4)\}\right],$$
$$= \min\left[\{550 + 2320\}, \{900 + 2220,\}, \{770 + 2150\}\right]$$
$$= \min[2870, 3120, 2920] = 2870 \text{ km}$$

Hence f1 (1) = 2,870, from the figure the optimum path is obtained as 1-2-5-8-10

Use of excel worksheet in solving dynamic programming problem.

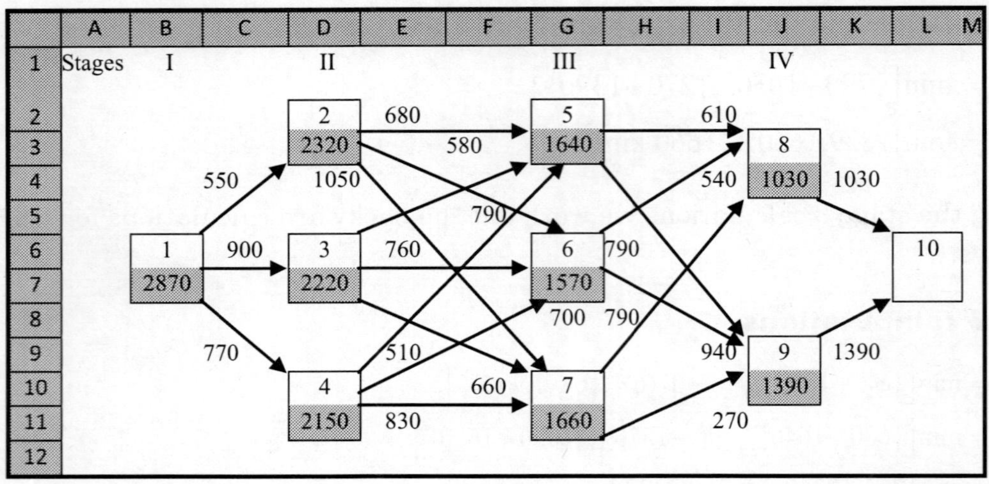

Figure 21.10

Type the formulas given below in the respective cells to get the optimum solution

State 9 = K9

State 8 = K4

State 7 = MIN(J4+H8,J10+I11)

State 6 = MIN(J4+I4,J10+I9)

State 5 = MIN(J4+I2,J10+H6)

State 4 = MIN(E11+G11,G7+G8,G3+E9)

State 3 = MIN(G11+F10,G7+E6,G3+F3)

State 2 = MIN(G11+D4,G7+F5,G3+E2)

State 1 = MIN(D11+C9,D7+C6,D3+C4)

So from state 1, the optimum solution is 2870.

EXERCISES

1. Define the following concepts
   (a) State
   (b) Stage
   (c) Return function
2. In your words explain the 'principle of optimality'.

3. A salesman located in city A decoded to travel to city B. He knew the distances of alternative routes from city A to city B. He draws a highway network map as shown below. In it the city A named as 1 the city B is named as city 10. The other cities on the way are named from 2 to 9. The salesman wants to find the optimum route on which the total distance covered is least. Find the optimum solution and help the salesman.

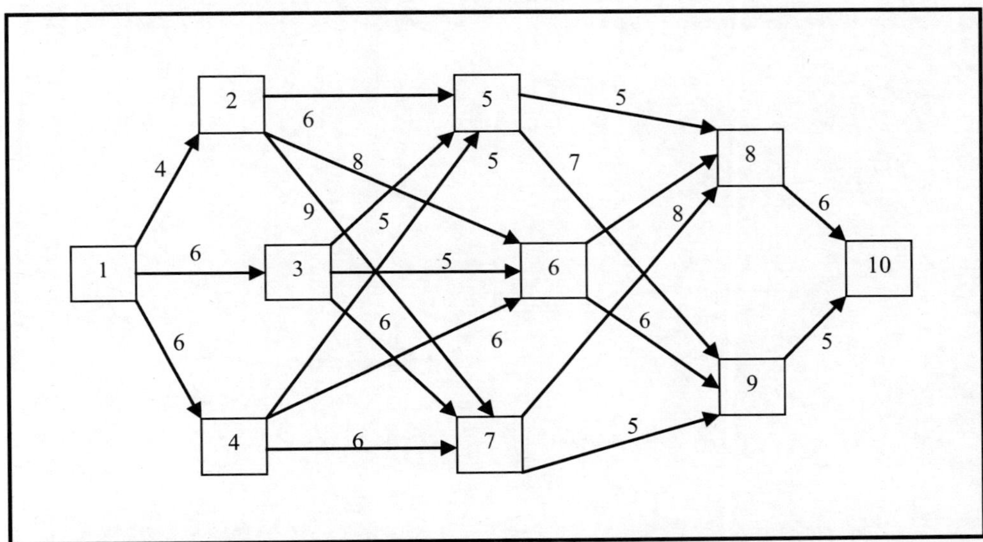

4. Use the dynamic programming to find the shortest path for following network.

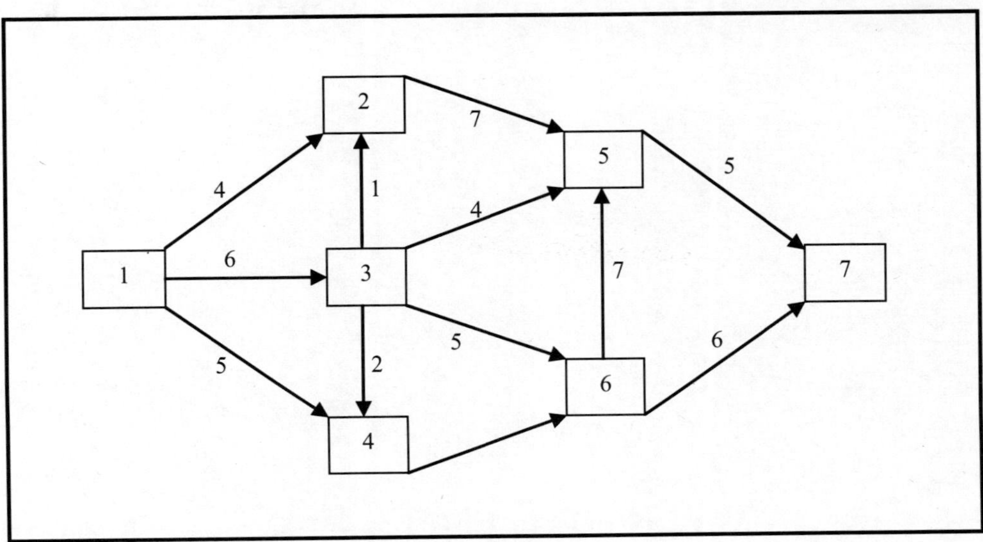

Index